Photophysics and Nanophysics in Therapeutics

Photophysics and Nanophysics in Therapeutics

Edited by

Nilesh M. Mahajan
Dadasaheb Balpande College of Pharmacy, Rashtrasant Tukadoji Maharaj Nagpur University, Nagpur, Maharashtra, India

Avneet Saini
Department of Biophysics, Panjab University, Chandigarh, India

Nishikant A. Raut
Department of Pharmaceutical Sciences, Rashtrasant Tukadoji Maharaj Nagpur University, Nagpur, Maharashtra, India

Sanjay J. Dhoble
Department of Physics, Rashtrasant Tukadoji Maharaj Nagpur University, Nagpur, Maharashtra, India

ELSEVIER

Elsevier
Radarweg 29, PO Box 211, 1000 AE Amsterdam, Netherlands
The Boulevard, Langford Lane, Kidlington, Oxford OX5 1GB, United Kingdom
50 Hampshire Street, 5th Floor, Cambridge, MA 02139, United States

Copyright © 2022 Elsevier Inc. All rights reserved.

No part of this publication may be reproduced or transmitted in any form or by any means, electronic or mechanical, including photocopying, recording, or any information storage and retrieval system, without permission in writing from the publisher. Details on how to seek permission, further information about the Publisher's permissions policies and our arrangements with organizations such as the Copyright Clearance Center and the Copyright Licensing Agency, can be found at our website: www.elsevier.com/permissions.

This book and the individual contributions contained in it are protected under copyright by the Publisher (other than as may be noted herein).

Notices

Knowledge and best practice in this field are constantly changing. As new research and experience broaden our understanding, changes in research methods, professional practices, or medical treatment may become necessary.

Practitioners and researchers must always rely on their own experience and knowledge in evaluating and using any information, methods, compounds, or experiments described herein. In using such information or methods they should be mindful of their own safety and the safety of others, including parties for whom they have a professional responsibility.

To the fullest extent of the law, neither the Publisher nor the authors, contributors, or editors, assume any liability for any injury and/or damage to persons or property as a matter of products liability, negligence or otherwise, or from any use or operation of any methods, products, instructions, or ideas contained in the material herein.

ISBN: 978-0-323-89839-3

For Information on all Elsevier publications visit our website at
https://www.elsevier.com/books-and-journals

Publisher: Andre Gerhard Wolff
Acquisitions Editor: Michelle Fisher
Editorial Project Manager: Susan E. Ikeda
Production Project Manager: Omer Mukthar
Cover Designer: Miles Hitchen

Typeset by Aptara, New Delhi, India

Contents

Contributors   xv

Part I
Phototherapeutics

1. Phototherapy: A critical review
Nilesh Rarokar, Shailendra Gurav, Dadasaheb M. Kokare, Vijay Kale, Nishikant A. Raut

1.1	Introduction	3
1.2	Background	4
	1.2.1 Historical perspective of phototherapy	4
	1.2.2 Overview on various types of phototherapies	5
1.3	Various light sources and methods of phototherapy	7
	1.3.1 Fluorescent tubes	7
	1.3.2 Halogen spotlights	7
	1.3.3 Fiberoptic blankets	7
	1.3.4 Light-emitting diodes	7
	1.3.5 Filtered sunlight	7
1.4	Applications and limitations of phototherapy	8
	1.4.1 Application in neonatal jaundice	8
	1.4.2 Application for morphea, scleroderma, and other sclerosing skin conditions	8
	1.4.3 Application for cancer	8
	1.4.4 Limitations of home phototherapy and sunlight	9
1.5	Recent developments and future scopes	9
	1.5.1 The immunoregulatory effects of phototherapy: Possible pathways	9
	1.5.2 Handheld phototherapy: Targeting difficult-to-treat psoriasis in the office and at home	10
	1.5.3 The excimer laser: A potential new indication and a novel dosimetry protocol	10
	1.5.4 Phototherapy and biologic agents: Combination therapy for recalcitrant psoriasis	11
	1.5.5 Future scope	11
References		11

2. Phototherapy for skin diseases
Renuka K. Mahajan, Dadasaheb M. Kokare, Nishikant A. Raut, Prakash R. Itankar

2.1	Introduction	15
	2.1.1 The epidermis	15
	2.1.2 The hypodermis	17
2.2	Major functions of the skin	17
2.3	Skin diseases and their etiology	18
2.4	Bacterial skin diseases	19
2.5	Fungal skin diseases	20
2.6	Viral skin diseases	20
2.7	Tropical ulcers	20
2.8	HIV related skin diseases	20
2.9	Pigmentation disorders	21
2.10	Parasitic infections	21
2.11	Tumors and cancers	21
2.12	Trauma	21
2.13	Skin tests	21
2.14	Heliotherapy	21
2.15	Naturopathy modalities on inflammation and immunity	22
2.16	Phototherapy for skin diseases	22
2.17	Methods	23
	2.17.1 UVB radiations	23
	2.17.2 UVA radiation	23
	2.17.3 PUVA	23
	2.17.4 Diseases and their treatment using phototherapy	24
	2.17.5 Limitations of phototherapy for skin diseases	25
	2.17.6 Side effects of phototherapy	26
	2.17.7 Recent development and future scope	26
2.18	Concluding remark	26
References		27

3. Phototherapy: The novel emerging treatment for cancer
Sagar Trivedi, Nishant Awandekar, Milind Umekar, Veena Belgamwar, Nishikant A. Raut

3.1	Introduction	31

3.2	Photophysics and photochemistry	32
	3.2.1 Type I mechanism of photodynamic reaction	32
	3.2.2 Type II mechanism of photodynamic reaction	33
3.3	Photodynamic targets at the molecular level	33
	3.3.1 Proteins	33
	3.3.2 Photodynamic therapy-induced lipid peroxidation	34
	3.3.3 Photosensitized modification of nucleic acids	34
3.4	Light source	35
	3.4.1 Near infrared (NIR) light	35
	3.4.2 X-ray	36
	3.4.3 Interstitial light	37
	3.4.4 Internal light	37
3.5	Changes in cell signaling after photodynamic therapy	37
	3.5.1 Calcium	37
	3.5.2 Lipid metabolism	38
	3.5.3 Tyrosine kinases	38
	3.5.4 Transcription factors	39
	3.5.5 Cellular adhesion	39
	3.5.6 Cytokines	39
	3.5.7 Stress response	39
	3.5.8 Hypoxia and angiogenesis	39
3.6	Method of excitation for photosensitizing agents	40
	3.6.1 Intermolecular chemically induced electronic excitation	40
	3.6.2 Resonance energy transfer excitation	40
	3.6.3 Two-stage photosensitizer excitation/excitation by radiation energy transfer intermediary	41
	3.6.4 Cherenkov radiation energy transfer	41
3.7	Photodynamic therapy modifications	42
	3.7.1 Nanotechnology on photodynamic therapy	42
	3.7.2 Application of liposomes and lipoproteins	42
	3.7.3 Photodynamic therapy supported by electroporation	43
3.8	Conclusion	43
	Acknowledgment	43
	Statement of informed consent	43
	Conflict of interest	43
	References	43

4. Fundamentals of photodynamic therapy

Mrunal M. Yawalkar, Samvit Menon, Hendrik C. Swart, Sanjay J. Dhoble

4.1	Introduction	51
4.2	Basic concept of photodynamic therapy	52
	4.2.1 Photosensitizers	52
4.3	Working mechanism	62
	4.3.1 Mechanism of cell death following photodynamic therapy	64
4.4	Advantages and disadvantages of photodynamic therapy	65
	4.4.1 Apoptosis in photodynamic therapy	65
	4.4.2 Immunological effects of photodynamic therapy	67
	4.4.3 Biological effects of photodynamic therapy	69
	4.4.4 Summarizing the advantages and disadvantages of photodynamic therapy	71
4.5	Essential wavelength region in photodynamic therapy	72
4.6	Recent developments in photodynamic therapy	74
	4.6.1 Metal-organic frameworks	74
	4.6.2 Photoactive materials for wavelength response	75
	4.6.3 Photodynamic therapy and hypoxia-controlled nanomedicine	76
4.7	Future scopes and perspectives	77
	References	79

5. Photodynamic therapy for cancer treatment

Sagar Trivedi, Anita Paunikar, Nishikant Raut, Veena Belgamwar

5.1	Introduction	89
5.2	Background of photodynamic therapy	90
	5.2.1 Origin of photodynamic therapy	90
	5.2.2 Mechanism of photodynamic therapy	90
	5.2.3 Working principle of photodynamic therapy	90
	5.2.4 Mechanism of photodynamic therapy in treatment of cancer	92
5.3	Novel strategies in photodynamic therapy	93
	5.3.1 Metronomic photodynamic therapy	93
	5.3.2 Photodynamic therapy molecular beacons	93
	5.3.3 Nanotechnology in photodynamic therapy	93
5.4	Role of photosensitizing agents in photodynamic therapy	94
5.5	Application of photodynamic therapy in treatment of various cancers	102
	5.5.1 Skin tumors	103
	5.5.2 Head and neck tumors	103
	5.5.3 Digestive system tumors	103

5.5.4	Urinary system tumors	103
5.5.5	Brain tumors	104
5.5.6	Nonsmall cell lung cancer and mesothelioma	105
5.6	Recent developments, future scope, and challenges	105
5.7	Conclusion	106
Acknowledgment		106
References		106

6. Photodiagnostic techniques

Anurag Luharia, Gaurav Mishra, Nilesh Haran, Sanjay J. Dhoble

6.1	Introduction	115
	6.1.1 Ionizing radiations	115
6.2	Fundamentals of light used in diagnostic techniques	116
	6.2.1 X-ray production	118
	6.2.2 X-ray beam intensity	119
	6.2.3 Target material	120
	6.2.4 Voltage applied	120
	6.2.5 X-ray tube current	120
6.3	Various photo diagnostic techniques	120
	6.3.1 Plain radiography and digital radiography	120
	6.3.2 Computed tomography	121
	6.3.3 Fluoroscopy	123
	6.3.4 Digital subtraction angiography	123
	6.3.5 Digital radiography and picture archival and communication system	124
	6.3.6 Dual energy X-ray absorptiometry	124
	6.3.7 Dual energy computed tomography	124
	6.3.8 Orthopantomography	124
6.4	Physics of photodiagnostic techniques	125
	6.4.1 Interaction of radiation with matter	125
	6.4.2 Importance of interaction in tissue	127
	6.4.3 Picture archiving and communication system	130
6.5	Opportunities, challenges, and limitations of photodiagnostic techniques	134
References		135

7. The role of physics in modern radiotherapy: Current advances and developments

Anurag Luharia, Gaurav Mishra, D. Saroj, V. Sonwani, Sanjay J. Dhoble

7.1	Introduction	139
7.2	Role of radiotherapy in cancer treatment	140
	7.2.1 What is radiotherapy and how it works?	140
	7.2.2 Types of radiotherapy	141
	7.2.3 Types of external beam radiation therapy	141
	7.2.4 General indications for the radiotherapy	143
	7.2.5 Intent of radiotherapy treatment	143
	7.2.6 Types of cancer treated using radiotherapy	144
	7.2.7 The role of radiotherapy in cancer control	144
7.3	Development of radiation physics	145
	7.3.1 History	145
	7.3.2 External radiotherapy	146
	7.3.3 Clinical radiation generators	147
	7.3.4 Dose planning	148
7.4	Recent advancement in radiotherapy	149
	7.4.1 Instigation	149
	7.4.2 Radiotherapy principle and mechanism	149
	7.4.3 Technology development	150
	7.4.4 Image-guided radiotherapy treatment	151
	7.4.5 Adaptive radiotherapy	152
	7.4.6 Stereotactic radiosurgery and radiotherapy	152
	7.4.7 Particle therapy	152
	7.4.8 Summary	153
7.5	Radiosurgery for noncancerous tumor and diseases	153
	7.5.1 Introduction	153
	7.5.2 History	153
	7.5.3 Treatment	154
	7.5.4 Systems overview	154
7.6	Summary and conclusion	157
References		157

8. Physics in treatment of cancer radiotherapy

Ravindra B. Shende, Sanjay J. Dhoble

8.1	Introduction	163
	8.1.1 Physics of radiotherapy	163
	8.1.2 Structure of matter	163
	8.1.3 Atom	163
	8.1.4 Nucleus	164
	8.1.5 Types of radiation	164
	8.1.6 X-rays	165
	8.1.7 Gamma rays	166
	8.1.8 Particulate radiation	166
	8.1.9 Interaction of radiation with matter	166
	8.1.10 Interaction of photon beam (X-rays or γ rays)	166
	8.1.11 Coherent scattering	169
	8.1.12 Photoelectric effect	169
	8.1.13 Compton effects	170
	8.1.14 Pair production	170
	8.1.15 Photodisintegration	171

8.1.16 Interaction of charged particle	171
8.1.17 Electron and electron interaction	172
8.1.18 Electron and nucleus interaction	172
8.1.19 Interaction of heavy charged particle	172
8.1.20 Biological effect of radiation	173
8.1.21 Linear energy transfer	174
8.1.22 Relative biological effectiveness	174
8.2 Principle of radiotherapy	175
8.2.1 Radiotherapy facility	175
8.3 Traditional facility in treatment of radiotherapy	175
8.3.1 Superficial therapy	175
8.3.2 Orthovoltage therapy or deep therapy	175
8.3.3 Supervoltage therapy machines	176
8.3.4 Cobalt-60 teletherapy unit	176
8.3.5 Betatron and microtron	176
8.3.6 Advance facility in treatment of radiotherapy	177
8.3.7 Linear accelerator (Linac)	177
8.3.8 Tomotherapy	178
8.3.9 CyberKnife	178
8.3.10 Proton and light ion therapy	178
8.3.11 Cyclotron	178
8.3.12 Synchrotron and synchrocyclotron	179
8.3.13 Add-on facility in treatment of radiotherapy	179
8.3.14 Conventional simulator	179
8.3.15 CT simulator	180
8.3.16 Commissioning of radiotherapy facility and quality assurance	180
8.3.17 Technique of radiotherapy	180
8.3.18 External beam radiation therapy	181
8.3.19 Conventional treatment techniques in EBRT	181
8.3.20 Three-dimensional conformal radiation therapy	181
8.3.21 Intensity modulated radiation therapy	182
8.3.22 Rotational therapy or volumetric modulated arc therapy (VMAT)	184
8.3.23 Stereotactic radiosurgery and stereotactic radiotherapy	184
8.3.24 Image-guided radiotherapy	184
8.3.25 Internal beam radiation therapy or brachytherapy	185
8.3.26 Process and treatment of radiotherapy	186
8.4 Patient preparation and simulation	187
8.5 Target delineation and treatment planning	187
8.5.1 Treatment verification and treatment delivery	187
8.5.2 Dosimetry in radiation therapy	188
8.5.3 Activity	188
8.5.4 Particle fluence	188
8.5.5 Energy fluence	188
8.5.6 Exposure	188
8.5.7 Kerma	188
8.5.8 Absorbed dose	189
8.5.9 Methods of radiation dosimetry and dosimeters in radiation therapy	189
8.5.10 Ionization chamber dosimetry	189
8.5.11 Film dosimetry	189
8.5.12 Luminescence dosimetry	190
8.5.13 Thermoluminescence	190
8.5.14 Optically stimulated luminescence	191
8.5.15 Semiconductor dosimetry	191
8.5.16 Physical and clinical dosimetry in radiotherapy	191
8.5.17 Physical dosimetry	191
8.5.18 Clinical dosimetry	192
References	192

9. Role of carbon ion beam radiotherapy for cancer treatment

Vibha Chopra, Nirupama S. Dhoble, Balkrishna Vengadaesvaran, Sanjay J. Dhoble

9.1 Introduction	193
9.2 Radiation therapy for the treatment of cancer	193
9.2.1 Gamma ray therapy	194
9.2.2 Proton therapy	194
9.2.3 Ion beam therapy	194
9.3 Role of carbon ion beam therapy	195
9.4 Development of TLD materials for carbon ion beam therapy	195
9.4.1 Lithium-based phosphors	195
9.4.2 Calcium-based phosphors	198
9.4.3 Some other phosphors	200
9.5 Conclusion	202
References	202

Part II
Nanotherapeutics

10. Nanomaterials physics: A critical review

Khushwant S. Yadav, Sheeba Jacob, Anil M. Pethe

10.1 Introduction	207
10.2 Fundamental concepts of nanomaterial physics	208
10.2.1 Structure sensitive and structure insensitive properties	209
10.2.2 Phases and their distribution	209
10.2.3 Defects in body nanomaterials	209

10.3	Properties of materials	210		
	10.3.1 Factors affecting properties of a material	210		
10.4	Rationale of nanoparticle physics with diverse functions involving nanomaterials	211		
10.5	Self-assembly of nanostructures	212		
10.6	Clinical applications of nanomaterials physics	212		
	10.6.1 Applications of nanomaterials physics in cancer	212		
10.7	Conclusion: Nanotechnology, physics, and clinical outcome	213		
Acknowledgments		214		
References		214		

11. Nanotherapeutic systems for drug delivery to brain tumors

Keshav S. Moharir, Vinita Kale, Mallesh Kurakula

11.1	Introduction	217
11.2	An overview of brain tumors	218
	11.2.1 Malignant brain tumors	218
	11.2.2 Benign brain tumors	218
11.3	Barriers and challenges in the treatment of brain cancer	219
	11.3.1 BBB as a main hurdle	219
	11.3.2 Chemoresistance and efflux	220
	11.3.3 Tumor microenvironment (TME) dynamics and lack of brain tumor classification based on genetics	220
	11.3.4 Resistance due to cancer stem cells (CSCs) of gliomas and GBM	220
	11.3.5 Lack of proper brain cancer mimicking models	221
11.4	Conventional vs nanomedicines in drug delivery for brain cancers	221
11.5	Approaches and mechanisms of nanocarriers for chemotherapeutic drug delivery to brain tumors	222
	11.5.1 Passive targeting	222
	11.5.2 Active targeting	222
	11.5.3 Stimuli responsive nanocarriers systems	224
11.6	Types of nanotherapeutic platforms for drug delivery to treat brain cancer	225
	11.6.1 Inorganic (metallic) nanoparticles	225
	11.6.2 Lipid-based and polymeric nanoparticles	228
11.7	Novel therapies to treat brain cancers	229
	11.7.1 Artificial intelligence (AI)-enabled nanocarriers for oncotherapy	229
	11.7.2 Gene-based nanotherapy	231
	11.7.3 CRISPR/Cas 9-associated brain tumor therapy	232
	11.7.4 Nose to brain drug delivery	232
11.8	Clinical translation of nanotherapeutic systems for brain cancers: From bench to bedside	232
11.9	Conclusion and future prospects	232
References		233

12. Progress in nanotechnology-based targeted cancer treatment

Shagufta Khan, Vaishali Kilor, Dilesh Singhavi, Kundan Patil

12.1	Introduction	239
12.2	Tumor microenvironment: Comparison with normal cells	239
	12.2.1 Angiogenesis and endothelial permeability in cancer	240
	12.2.2 Microenvironment pH	240
	12.2.3 Microenvironment temperature	240
12.3	Nanotechnology-based diagnosis of cancer	240
12.4	Nanotechnology-based drug targeting strategies in cancer	241
	12.4.1 Passive targeting	241
	12.4.2 Active targeting	242
	12.4.3 Physical targeting	245
12.5	Progress in nanotherapeutics for treating breast and lung cancer	245
	12.5.1 Breast cancer	245
	12.5.2 Lung cancer	246
12.6	Future of nanotechnology in cancer treatment	247
12.7	Conclusion	248
References		248

13. Nanotherapeutics for colon cancer

Nilesh M. Mahajan, Alap Chaudhari, Sachin More, Purushottam Gangane

13.1	Introduction	251
	13.1.1 Anatomy	251
	13.1.2 Pathogenesis and molecular pathways for CRC	252
	13.1.3 Risk factors	253
	13.1.4 Stages of CRC	254
	13.1.5 Signs and symptoms	254
13.2	Diagnosis	254
	13.2.1 Endoscopy	255
	13.2.2 Imaging	255
	13.2.3 Laboratory	255
	13.2.4 Pathology	255

13.3 Current therapies 256
    13.3.1 Conventional treatment strategies 256
    13.3.2 Targeted therapy 259
    13.3.3 Targeted therapies using nanocarriers 260
13.4 Nanodrug delivery in cancer therapy 261
    13.4.1 Polymers used in formulations of NPs 261
13.5 Polymeric nanoparticles (PNPs) 262
    13.5.1 Lipid-based nanoparticles 263
    13.5.2 Superparamagnetic iron oxide nanoparticles (SPIONs) 263
    13.5.3 Gold nanoparticles (AuNPs) 263
    13.5.4 Enteric-coated nanoparticles 264
13.6 Conclusion 264
References 265

14. Nanoparticles for the targeted drug delivery in lung cancer

Veena Belgamwar, Vidyadevi Bhoyar, Sagar Trivedi, Miral Patel

14.1 Introduction 269
    14.1.1 Stages of LC 269
    14.1.2 Current treatment strategies on LC 270
    14.1.3 Novel strategies for LC treatment by pulmonary route of administration 272
    14.1.4 Pulmonary physiology and drug absorption 273
    14.1.5 Role of nanoparticulate technology in the diagnosis and treatment of LC 273
    14.1.6 Nanocarriers used for the diagnosis of lung diseases 274
14.2 Nanocarriers in LC treatment 275
    14.2.1 Solid–lipid nanocarriers 275
    14.2.2 Polymeric nanocarriers 276
    14.2.3 Nanoemulsions as potential carrier in LC 276
    14.2.4 Metal-based NPs 277
    14.2.5 Dendrimers-based drug delivery 277
    14.2.6 Target-mediated targeted therapy 279
    14.2.7 Quantum dots (QDs) as a drug delivery system 279
    14.2.8 Bio-NPs for LC 280
    14.2.9 Hydrogel-based drug delivery for pulmonary cancer 281
    14.2.10 Inhalation-based nanomedicine for pulmonary cancer 281
14.3 Marketed formulation 282
14.4 Toxicity issues of inhaled NPS 283
14.5 Conclusion 284
References 285

15. Role of nanocarriers for the effective delivery of anti-HIV drugs

Rohini Kharwade, Nilesh M. Mahajan

15.1 Introduction 291
    15.1.1 HIV life cycle and pathogenesis 291
    15.1.2 Pathophysiology 293
15.2 Conventional antiretroviral therapy 293
15.3 Types of nanocarriers for antiretroviral drugs delivery 295
    15.3.1 Pure drug nanoparticles 296
    15.3.2 Polymeric nanoparticles 297
    15.3.3 Dendrimers 299
    15.3.4 Polymeric micelles 301
    15.3.5 Liposomes 302
    15.3.6 Solid lipid nanoparticles 303
15.4 Nanaotechnological approaches for antiretroviral therapy 304
    15.4.1 Immunotherapy for antiretroviral 304
    15.4.2 Gene therapy 305
    15.4.3 Vaccines 305
15.5 Nanotechnology for improving latency reservoir 306
15.6 Conclusion 307
References 307

16. Drug delivery systems for rheumatoid arthritis treatment

Mangesh Bhalekar, Sachin Dubey

16.1 Introduction 311
    16.1.1 Stages of rheumatoid arthritis 311
    16.1.2 Causes of RA 311
    16.1.3 Symptoms of RA 312
    16.1.4 Pathology of rheumatoid arthritis 312
16.2 Management of rheumatoid arthritis 314
16.3 Targeted delivery strategies to inflamed synovium 314
16.4 Passive targeting 315
    16.4.1 Enhanced permeability and retention (EPR) effect 315
    16.4.2 Hypoxia and acidosis 315
    16.4.3 Stimuli responsive drug delivery 316
    16.4.4 Angiogenesis 316
16.5 Active targeting 316
16.6 Factors for the selection of delivery system 316
    16.6.1 Carrier type 316
    16.6.2 Particle size 316
    16.6.3 Shape 317
    16.6.4 Surface modifications 317
    16.6.5 Prolonged circulation time 317
    16.6.6 Strategies for active targeting 317

16.7 Drug delivery vehicles for rheumatoid arthritis 318
    16.7.1 Liposomes 318
    16.7.2 Dendrimers 319
    16.7.3 Nanoparticles 319
    16.7.4 Polymeric micro- and nanoparticles 320
    16.7.5 Macromolecules and the enhanced permeability and retention effect 320
    16.7.6 Arthritis-specific antigens 321
    16.7.7 The complement system 321
    16.7.8 Specific surface receptors 321
    16.7.9 Monoclonal antibodies 322
    16.7.10 mAbs targeted against B cells 322
    16.7.11 mAbs directed against IL-6function 322
    16.7.12 mAb directed against NFKB ligand 323
16.8 Conclusion 323
References 323

17. Peptide functionalized nanomaterials as microbial sensors

Shubhi Joshi, Sheetal Sharma, Gaurav Verma, Avneet Saini

17.1 Introduction 327
17.2 Conventional techniques for microorganism detection 328
    17.2.1 Pure culture-based protocols 328
    17.2.2 Immunological techniques 328
    17.2.3 Nucleic acid-based assays 329
17.3 Principle behind using biosensors for microorganism detection 330
17.4 Commonly used biosensing recognition elements 331
    17.4.1 Antibodies as biosensing recognition elements 331
    17.4.2 Aptamers as biosensing recognition elements 332
    17.4.3 Bacteriophages as biosensing recognition elements 332
    17.4.4 Carbohydrates as biosensing recognition elements 332
    17.4.5 Peptides as biosensing recognition elements 333
17.5 Advantages and challenges of using peptide-based detection of microorganisms 335
17.6 Properties of nanomaterials making them suitable for construction of microbial sensors 335
    17.6.1 Carbon-based nanoparticles 335
    17.6.2 Metallic nanoparticles 336
    17.6.3 Magnetic nanoparticles 336
    17.6.4 Quantum dots 337
17.7 Techniques enabling microorganism detection 337
    17.7.1 Colorimetric detection 337
    17.7.2 Fluorescence-based detection 338
    17.7.3 Microscopic techniques 338
    17.7.4 Spectroscopic detection 338
17.8 Recent advances in on-site detection of microorganisms using peptide functionalized nanosensors 339
    17.8.1 Bacteria detection 339
    17.8.2 Detection of fungal spores 339
    17.8.3 Virus detection 340
17.9 Conclusion and future perspectives 341
References 341

18. Theranostic nanoagents: Future of personalized nanomedicine

Vidya Sabale, Shraddha Dubey, Prafulla Sabale

18.1 Introduction 349
    18.1.1 Theranostics 349
    18.1.2 Nanoagents 349
    18.1.3 Nanotheranostics 349
18.2 Recent approaches versus theranostic nanoagents 350
    18.2.1 Contemporary treatment methods and their drawbacks 350
18.3 Nanotheranostics and neurological disorders 350
    18.3.1 Blood–brain barrier 350
    18.3.2 Theranostic nanoparticles employed in neurology 351
    18.3.3 Theranostic applications of nanosystems in neurological disorders 355
18.4 Nanotheranostics and rheumatoid arthritis 360
    18.4.1 Rheumatoid arthritis (RA) 360
    18.4.2 Current treatments and their drawbacks 360
    18.4.3 Nanotheranostic approach for rheumatoid arthritis 361
18.5 Nanoparticle-based theranostic agents 363
    18.5.1 Iron oxide nanoparticle-based theranostic agents 363
    18.5.2 Quantum dot-based theranostic agents 365
    18.5.3 Gold nanoparticle-based theranostic agents 366
    18.5.4 Carbon nanotube-based theranostic agents 367
    18.5.5 Silica nanoparticle-based theranostic agents 368

- 18.6 Theranostic nanoagents: future of nanomedicine — 369
- 18.7 Conclusion — 369
- References — 370

19. Improving the functionality of a nanomaterial by biological probes
Panchali Barman, Shweta Sharma, Avneet Saini

- 19.1 Introduction to nanomaterials — 379
- 19.2 Classifications of nanoparticles — 380
  - 19.2.1 Metallic nanoparticles — 380
  - 19.2.2 Semiconductor quantum dots — 383
  - 19.2.3 Metal oxide nanoparticles — 384
  - 19.2.4 Organic nanoparticles — 385
  - 19.2.5 Upconversion nanoparticles — 387
- 19.3 Common conjugation approaches for biomolecule functionalized nanomaterials — 389
  - 19.3.1 Conjugation approaches — 389
  - 19.3.2 Functionalization of nanoparticles — 391
- 19.4 Basic chemistries behind conjugation approaches — 397
  - 19.4.1 Functional groups and conjugation reactions — 397
  - 19.4.2 Polyhistidine–nitrilotriacetic acid chelation — 398
  - 19.4.3 Biotin–avidin chemistry — 399
- 19.5 Applications — 400
  - 19.5.1 Detection of DNA, protein, and metal ions — 400
  - 19.5.2 Detection of human pathogens — 401
  - 19.5.3 Enhancement of antibacterial and anti-inflammatory activity — 402
  - 19.5.4 Theranostics — 403
- 19.6 Conclusion and future perspective — 404
- References — 405

20. Nanostructures for the efficient oral delivery of chemotherapeutic agents
Ravindra Satpute, Nilesh Rarokar, Sunil Menghani, Anjali Ganjare, Vivek S. Dave, Nishikant A. Raut, Pramod B. Khedekar

- 20.1 Introduction — 419
  - 20.1.1 Limitations of conventional chemotherapy — 420
  - 20.1.2 Edges of nanoparticles over the other delivery system — 420
  - 20.1.3 Components of nanoparticles as a targeting system — 420
  - 20.1.4 Characteristics features of ideal targeting moieties — 421
  - 20.1.5 The potential of nanocarriers as drug delivery systems — 421
  - 20.1.6 Nanoparticle properties — 421
  - 20.1.7 Cancer therapy: Selective targeting of tissues by nanotechnology — 421
- 20.2 Nanodrug carriers — 422
  - 20.2.1 Classification of nanoparticles as drug carriers — 422
  - 20.2.2 Micelles — 423
  - 20.2.3 Solid-lipid nanoparticles (SLNs) — 423
  - 20.2.4 Cubosomes — 423
  - 20.2.5 Drug-polymer conjugates — 424
  - 20.2.6 Antibody-drug conjugates — 424
  - 20.2.7 Inorganic nanoparticles — 425
  - 20.2.8 Carbon nanotubes (CNTs) — 425
  - 20.2.9 Gold nanoparticles (GNPs) — 426
  - 20.2.10 Porous silicon particles (PSiPs) — 426
  - 20.2.11 Quantum dots (QDs) — 426
  - 20.2.12 Iron oxide nanoparticles (IONPs) — 427
  - 20.2.13 IONPs — 427
- References — 428

21. Photo-triggered theranostics nanomaterials: Development and challenges in cancer treatment
Neha S. Raut, Divya Zambre, Milind J. Umekar, Sanjay J. Dhoble

- 21.1 Introduction of nanomaterials in phototherapeutics — 431
- 21.2 Types of nanomaterials — 432
  - 21.2.1 Magnetic nanoparticles — 432
  - 21.2.2 Properties and materials for preparation of photo-based nanomaterials — 433
  - 21.2.3 Gold-based nanoparticles — 433
  - 21.2.4 Carbon nanotubes — 433
- 21.3 Polymeric nanocarriers for photosensitizer/dye encapsulation — 434
- 21.4 Nanoconstructs for photodynamic therapy — 434
- 21.5 Photo-triggered theranostic nanocarriers — 435
- 21.6 Approaches to measure drug release through theranostic nanomedicine — 436
  - 21.6.1 Silicon photonic crystals with pores — 436
  - 21.6.2 Fluorescent nanoparticles — 437
  - 21.6.3 Upconversion nanoparticles — 437
  - 21.6.4 Radioluminescent nanoparticles — 437

- 21.7 Magnetic resonance imaging for monitoring release of drug — 437
- 21.8 Photo-triggered theranostics nanomaterials: Principle and applications — 438
  - 21.8.1 Applications of photo-triggered theranostics nanomaterials in cancer treatments — 438
  - 21.8.2 Therapeutic applications of photo-based theranostic nanoparticles — 438
- 21.9 Opportunities and limitations of nanomaterials — 439
- 21.10 Preclinical challenges — 439
- 21.11 Future aspects of nanomaterials in the therapeutics — 439
- References — 440

22. Nanocrystals in the drug delivery system
Raju Ramesh Thenge, Amar Patel, Gautam Mehetre

- 22.1 Introduction to nanocrystals and nanosuspension — 443
  - 22.1.1 Properties of nanocrystals — 443
  - 22.1.2 Nanocrystals and bioavailability — 444
  - 22.1.3 Various methods of characterization of nanocrystals formulations — 444
- 22.2 Production methods and technology of nanocrystals — 445
  - 22.2.1 Top down technology — 445
  - 22.2.2 Bottom up technology — 446
  - 22.2.3 Top down and bottom up technology — 446
  - 22.2.4 Spray drying — 447
- 22.3 Advantages and Disadvantages of nanocrystals — 448
  - 22.3.1 Potential advantages and disadvantages of nanocrystals — 448
  - 22.3.2 Disadvantages of nanocrystals — 448
- 22.4 Pharmaceutical Nanocrystals of API — 448
  - 22.4.1 Case studies of drug loaded in the nanocrystals — 448
  - 22.4.2 Application of nanocrystals-loaded carrier — 449
- 22.5 Conclusion — 452
- References — 452

Index — 455

Contributors

Nishant Awandekar Smt. Kishoritai Bhoyar College of Pharmacy, Kamptee, Nagpur, India

Panchali Barman Institute of Forensic Science and Criminology (UIEAST), Panjab University, Chandigarh, India

Veena Belgamwar Department of Pharmaceutical Sciences, Rashtrasant Tukadoji Maharaj Nagpur University, Nagpur, Maharashtra, India

Mangesh Bhalekar Department of Pharmaceutics, AISSMS College of Pharmacy, Pune, India

Vidyadevi Bhoyar University Department of Pharmaceutical Sciences, Rashtrasant Tukadoji Maharaj Nagpur University, Nagpur, Maharashtra, India

Alap Chaudhari Formulation R&D, Teva Pharmaceuticals, Weston, FL, USA

Vibha Chopra P.G. Department of Physics & Electronics, DAV College, Amritsar, Punjab, India

Vivek S. Dave Department of Pharmaceutical Sciences, St. John Fisher College, Wegmans School of Pharmacy, Rochester, NY, USA

Nirupama S. Dhoble Deaprtment of Chemistry, Sevadal Mahila Mahavidhyalaya, Nagpur, Maharashtra, India

Sanjay J. Dhoble Department of Physics, Rashtrasant Tukadoji Maharaj Nagpur University, Nagpur, Maharashtra, India

Sachin Dubey Drug Product and Analytical Development, Ichnos Sciences SA, La Chaux-de-Fonds, Switzerland

Shraddha Dubey Inselpital University of Bern, Bern, Switzerland

Purushottam Gangane Dept of Pharmaceutics, Dadasaheb Balpande College of Pharmacy, Besa, Nagpur, MS, India

Anjali Ganjare Department of Pharmaceutical Sciences, Rashtrasant Tukadoji Maharaj Nagpur University, Nagpur, Maharashtra, India

Shailendra Gurav Department of Pharmacognosy, Goa College of Pharmacy, Goa University, Panaji, Goa, India

Nilesh Haran Department of Radiology HCG-NCHRI Cancer Centre, Near Automotive Square, Nagpur, Maharashtra, India

Prakash R. Itankar Department of Pharmaceutical Sciences, Rashtrasant Tukadoji Maharaj Nagpur University, Nagpur, Maharashtra, India

Sheeba Jacob Virginia Commonwealth University, Richmond, VA, USA

Shubhi Joshi Energy Research Centre, Panjab University, Chandigarh, India

Vijay Kale College of Pharmacy, Roseman University of Health Sciences, South Jordan, UT, United States

Vinita Kale Pharmaceutics Deptt., Gurunanak College of Pharmacy, Nagpur, Maharashtra, India

Shagufta Khan Department of Pharmaceutics, Institute of Pharmaceutical Education and Research, Borgaon (Meghe) Wardha, Maharashtra, India

Rohini Kharwade Dadasaheb Balpande College of Pharmacy, Rashtrasant Tukadoji Maharaj Nagpur University, Nagpur, Maharashtra, India

Pramod B. Khedekar Department of Pharmaceutical Sciences, Rashtrasant Tukadoji Maharaj Nagpur University, Nagpur, Maharashtra, India

Vaishali Kilor Department of Pharmaceutics, Gurunanak College of Pharmacy, Nagpur, Maharashtra, India

Dadasaheb M. Kokare Department of Pharmaceutical Sciences, Rashtrasant Tukadoji Maharaj Nagpur University, Nagpur, Maharashtra, India

Mallesh Kurakula Product Development, CURE Pharmaceutical, Oxnard, CA, USA

Anurag Luharia Department of Radiology, Datta Meghe Institute of Medical Science (Deemed to be University), Sawangi, Wardha, Maharashtra, India

Renuka K. Mahajan Department of Pharmaceutical Sciences, Rashtrasant Tukadoji Maharaj Nagpur University, Nagpur, Maharashtra, India

Nilesh M. Mahajan Dadasaheb Balpande College of Pharmacy, Rashtrasant Tukadoji Maharaj Nagpur University, Nagpur, Maharashtra, India

Gautam Mehetre Dr. Rajendra Gode College of Pharmacy, Malkapur, Buldana, MS, India

Sunil Menghani Department of Pharmaceutical Sciences, Rashtrasant Tukadoji Maharaj Nagpur University, Nagpur, Maharashtra, India

Samvit Menon Department of Physics, University of the Free State, Bloemfontein, South Africa

Gaurav Mishra Department of Radiology, Datta Meghe Institute of Medical Science (Deemed to be University), Sawangi, Wardha, Maharashtra, India

Keshav S. Moharir Pharmaceutics Deptt., Gurunanak College of Pharmacy, Nagpur, Maharashtra, India

Sachin More Dept of Pharmacology, Dadasaheb Balpande College of Pharmacy, Besa, Nagpur, MS, India

Amar Patel Bristol Myers Squibb, New Jersey, USA

Miral Patel Department of Pharmaceutical Sciences, Arnold and Marie Schwartz College of Pharmacy and Health Science, Long Island University, Brooklyn Campus, NY; Office of Pharmaceutical Quality, Center for Drug Evaluation and Research, Food and Drug Administration, Silver Spring, MD, USA

Kundan Patil Department of Pharmaceutics, Government College of Pharmacy, Amrawati, Maharashtra, India

Anita Paunikar Department of Pharmaceutical Sciences, Rashtrasant Tukadoji Maharaj Nagpur University, Nagpur, Maharashtra, India

Anil M. Pethe Datta Meghe College of Pharmacy, Datta Meghe Institute of Medical sciences, (Deemed-to-be) University, Sawangi (Meghe), Wardha, India

Nilesh Rarokar Department of Pharmaceutical Sciences, Rashtrasant Tukadoji Maharaj Nagpur University, Nagpur, Maharashtra, India

Nishikant A. Raut Department of Pharmaceutical Sciences, Rashtrasant Tukadoji Maharaj Nagpur University, Nagpur, Maharashtra, India

Neha S. Raut Department of Pharmaceutical Chemistry, Smt. Kishoritai Bhoyar College of Pharmacy, Kamptee, India

Vidya Sabale Dadasaheb Balpande College of Pharmacy, Besa, Nagpur, Maharashtra, India

Prafulla Sabale Department of Pharmaceutical Sciences, Rashtrasant Tukadoji Maharaj Nagpur University, Nagpur, Maharashtra, India

Avneet Saini Department of Biophysics, Panjab University, Chandigarh, India

D. Saroj Department of Radiotherapy, Allexis Hospital, Mankapur, Nagpur, India

Ravindra Satpute Toxicology Laboratory, Defense R & D Establishment, Nagpur, Maharashtra, India

Sheetal Sharma Department of Biophysics, Panjab University, Chandigarh, India

Shweta Sharma Institute of Forensic Science and Criminology (UIEAST), Panjab University, Chandigarh, India

Ravindra B. Shende Department of Radiation oncology, Balco Medical Centre, New Raipur, Chhattisgarh, India

Dilesh Singhavi Department of Pharmaceutics, Institute of Pharmaceutical Education and Research, Borgaon (Meghe) Wardha, Maharashtra, India

V. Sonwani HCG NCHRI Cancer Center, Nagpur, India

Hendrik C. Swart Department of Physics, University of the Free State, Bloemfontein, South Africa

Raju Ramesh Thenge Dr. Rajendra Gode College of Pharmacy, Malkapur, Buldana, MS, India

Sagar Trivedi Department of Pharmaceutical Sciences, Rashtrasant Tukadoji Maharaj Nagpur University, Nagpur, Maharashtra, India

Milind J. Umekar Department of Pharmaceutics, Smt. Kishoritai Bhoyar College of Pharmacy, Kamptee, India

Balkrishna Vengadaesvaran Higher Institution Centre of Excellence (HICoE), UM Power Energy Dedicated Advanced Centre (UMPEDAC), Level 4, Wisma R&D University of Malaya, Kuala Lumpur, Malaysia

Gaurav Verma Dr. S.S. Bhatnagar University Institute of Chemical Engineering & Technology (Dr. SSBUICET), Panjab University, Chandigarh, India; Centre for Nanoscience and Nanotechnology (UIEAST), Panjab University, Chandigarh, India

Khushwant S. Yadav Shobhaben Pratapbhai Patel School of Pharmacy & Technology Management, SVKM's NMIMS, Mumbai, India

Mrunal M. Yawalkar Department of Physics, Rashtrasant Tukadoji Maharaj Nagpur University, Nagpur, India

Divya Zambre Department of Pharmaceutics, Smt. Kishoritai Bhoyar College of Pharmacy, Kamptee, India

Section 1

Phototherapeutics

1. Phototherapy: A critical review — 3
2. Phototherapy for skin diseases — 15
3. Phototherapy: The novel emerging treatment for cancer — 31
4. Fundamentals of photodynamic therapy — 51
5. Photodynamic therapy for cancer treatment — 89
6. Photodiagnostic techniques — 115
7. The role of physics in modern radiotherapy: Current advances and developments — 139
8. Physics in treatment of cancer radiotherapy — 163
9. Role of carbon ion beam radiotherapy for cancer treatment — 193

Chapter 1

Phototherapy: A critical review

Nilesh Rarokar[a], Shailendra Gurav[b], Dadasaheb M. Kokare[a], Vijay Kale[c], Nishikant A. Raut[a]
[a]Department of Pharmaceutical Sciences, Rashtrasant Tukadoji Maharaj Nagpur University, Nagpur, Maharashtra, India
[b]Department of Pharmacognosy, Goa College of Pharmacy, Goa University, Panaji, Goa, India
[c]College of Pharmacy, Roseman University of Health Sciences, South Jordan, UT, United States

1.1 Introduction

Phototherapy can be broadly defined as the use of photons for the treatment of diseases without the addition of an exogenous photosensitizer. Ultraviolet B (UVB) radiations (280–320 nm) are the most biologically active radiation in sunlight and are mainly responsible for erythema. The term "phototherapy" was used as a synonym for UVB radiation in the management of psoriasis. Phototherapy has wide clinical applications like pityriasis lichenoides, vitiligo, atopic eczema, polymorphous light eruption, pruritus, etc. (Morison, 1993). Recent work in phototherapy explores the selective use of UV with a narrow wavelength centered on 311 ± 1 nm. It has been proven more effective and less erythemogenic than conventional broadband phototherapy (Larkö, 1989). Treatments are usually given three to five times a week. The most widely used initial dose is 70% of the predetermined minimal erythema dose (MED) (Green et al., 1988). Treatment is continued until the condition is resolved (Wainwright et al., 1998). Selective UV phototherapy is the treatment of choice in children, which was used in the various study. The short-term side effects of phototherapy are usually mild and consist of xerosis and erythema, partly due to occasional overexposure. Another risk can be the photoactivation of herpes-virus. Long-term side effects of UVB phototherapy include premature photoaging and carcinogenesis with an increased incidence of wrinkling, actinic keratoses, lentigines, telangiectasia, and basal, and squamous cell carcinomas (De Gruijl, 1986). Radiation received from UVB phototherapy is cumulative with chronic sunlight exposure. UVB in combination with UVA phototherapy (UVAB) has been shown to produce better therapeutic success than UVB per se in treating mild to moderate atopic dermatitis (Falk, 1985; Hannuksela et al., 1985; Pašić et al., 2003). Phototherapy uses UV radiation or visible light for the treatment of different diseases by the exposure of small, well-defined anatomical areas to non-ionizing radiations using dichromic lamps, fluorescent lamps, light-emitting diodes, lasers, polychromatic polarized light, or very bright, full-spectrum light for therapeutic advantages. The origin of phototherapy can be traced back to 1500 BC, when Hindus treated vitiligo, an autoimmune skin disorder, with photosensitizing plant extracts and subsequent sunlight exposure. For many centuries only natural sunlight (heliotherapy) was used to treat different skin conditions. Interestingly, it is still highly popular for psoriasis and atopic dermatitis in many geographic areas in the world, especially in the Dead Sea region (Matos et al., 2016). As heliotherapy is only feasible in certain periods of the year with additional dosing variables depending on the geographic locations, artificial light sources are developed to emit selective wavelengths of electromagnetic radiation. Furthermore, identifying photosensitizers from plant extracts with unique photochemical properties resulted in the development of the photochemotherapeutic approach. It was found to be most effective in the treatment of inflammatory skin diseases. Lasers and intense pulse light (IPL) are well-established therapeutic tools to treat congenital and acquired vascular lesions. The erythema, skin inflammation can be treated by targeting blood vessels with lasers or IPLs, and it has proven a good alternative for the treatment of various skin diseases. Decades of research have been shown to improve inflammatory skin conditions with variable success by using vascular lasers. In recent years, low-level light/laser treatments (LLLT) emitting low-intensity visible light were tried for psoriasis and atopic dermatitis. Still, their efficacy and the mechanism of action need further clarification. When selecting the appropriate treatment for patients, many different conditions, such as comorbidities, age, and disease severity, must be considered (Kaushik and Lebwohl, 2019a, 2019b). Although there are several traditional and biologically active agents for psoriasis and atopic dermatitis, phototherapy approaches are still widely utilized (Kemény et al., 2019).

1.2 Background

1.2.1 Historical perspective of phototherapy

Modern scientific discoveries and technological inventions created the basis for applying artificial and modified light sources in phototherapy. Undoubtedly, these achievements included Isaac Newton's (1642–1727) splitting of a light beam into seven basic colors, using a prism and his discovery of the color wheel, Friedrich Wilhelm Herschel's (1738–1822) discovery of the infrared spectrum of the sun in 1800, and the discovery of ultraviolet radiation (Roelandts, 2002) in 1801 independently by Johann Wilhelm Ritter (1776–1810) and William Hyde Wollaston (1766–1828). Michel Eugène Chevreul (1786–1889) expanded upon Newton's theory of seven colors by formulating the concept of simultaneous contrast in 1830. He described the phenomenon of an interaction of two colors, side by side, changing human perception. This contrasting effect is more distinct during the interaction between complementary colors (e.g., blue and yellow) (Chevreul, 1839).

The advancement of research on electricity and artificial light sources preceded the application of phototherapy in clinical therapy. Thereafter, Hans Christian Oerstedt (1777–1851) discovered that electric current creates a magnetic field (1777–1851). Michael Faraday (1791–1867) described electromagnetic induction as a source of electric power and built the first electric generator and the motor. Subsequently, Thomas Alva Edison (1847–1931) invented the electric light bulb and the battery as a source of light and electric power, respectively.

In the same historical period, scientific attempts were performed to explain the positive influence of light on humans. The first modern scientific data on the effects of light and colors on human health was published in the early nineteenth century by the German poet and writer Johann Wolfgang von Goethe (1749–1832). In 1810, he published a work on the perception of color vision and the influence of light and colors on the human emotional state (Goethe, 1810). It is considered as the very first report on the psychological effects of colors. However, it is far from perfection as it includes several erroneous assertions, like the thesis on light's homogeneity inherent in the polemics of Newtonian optics. For example, he contradicted Newton with reference to the view that colors arise from the decomposition of light emerging from a prism into tiny particles called corpuscles but suggested that colors derive from the interaction of light and dark, and light is indivisible into any particles.

In the second half of the 19th century, scientific reports pointed to the healing properties of sunlight and reported the bactericidal properties of sunlight along with its therapeutic application in the treatment of rickets (Downes et al., 1877; Palm, 1890).

Activities of sanatoria, using natural solar radiation, were an important element in the historical process of creating contemporary phototherapy. The end of the nineteenth century saw the development of these "sun sanatoria." They became the centers for heliotherapy and hydrotherapy. In addition, attempts were made to combat the tuberculosis epidemic by associating phototherapy with climatic treatment (i.e., therapy by bathing in cold or warm water and walking in the fresh air) (Roelandts, 2002). Pioneers in this therapeutic trend included the "*Sunapostle*" Arnold Rikli (1823–1906) (Levental, 1977), Oskar Bernhard (1861–1939), and August Rollier (1874–1954). Although from 1855, balneotherapy might include light treatment, as found in the Alpine Bed in Slovenia (Zupanic-Slavec and Toplak, 1998), Rikli applied the principle "Water is good, the air is better, and most of all the sunlight." Bernhard promoted heliotherapy at the beginning of 1899 at a private clinic in St. Moritz, Switzerland. Finally, Rollier applied climatic treatment in combination with phototherapy to treat tuberculosis of the bone, beginning in 1903 at a sanatorium in Leysin, Switzerland (Rollier and Rosselet, 1923).

The discoveries (e.g., ultraviolet radiation) and inventions (e.g., the electric generator or the electric-light bulb), as well as balneological experiences of the treatment with the sunlight, contributed to the development of modern phototherapy and the transition from heliotherapy to artificial light phototherapy at the end of the nineteenth century. Nils Ryberg Finsen's (1860–1904) studies on phototherapy led to its rise as a new field in physiotherapy (Grzybowski and Pietrzak, 2012). Finsen is also famous for creating the Medical Light Institute in Copenhagen, Denmark in 1896 (Finsen, 1896) and use of an electric carbon arc torch to treat lupus vulgaris patients with ultraviolet radiation (Finsen and Forchhammer, 1904).

1.2.1.1 Progress in the twentieth century

Phototherapy was developed for neonatal jaundice in the late 1950s. Sister Jean Ward at Rochford General Hospital in Essex, England noted in 1956 that sunshine decreased neonatal jaundice. Concurrently, hospital biochemists observed significantly low bilirubin levels in samples exposed to sunlight before processing (Dobbs and Cremer, 1975). Thus, it was the first evidence for effective light therapy for infantile hyperbilirubinemia (Cremer, Perryman and Richards, 1958). Jerold Lucey, the editor of the journal Pediatrics, published the "1968 landmark randomized controlled trial" results showing the efficacy of phototherapy (Lucey et al., 1968), which continued an important method for treating newborn jaundice (Weiss and Zimmerman, 2013).

Goethe's idea on the impact of light on emotional states was resurrected when phototherapy was employed to treat depression. In 1946, in Scandinavia, the first descriptions of depression treatment using light were published (Marx, 1946); however, the fundamental development of phototherapy in the treatment of depression did not occur for four more decades (Rosenthal et al., 1984).

Phototherapy was also explored in ophthalmology. Gerhard Meyer-Schwickerath (1920–1992) investigated the use of natural sunlight for treating retinal disorders. Gerhard Meyer-Schwickerath performed a successful surgical operation with a photocoagulator in 1949 (Meyer-Schwickerath, 1960). It was a device designed and placed on the clinic's roof to aggregate light onto a mirror in the operating room, paving the way for the application of laser therapy in retinal diseases (Grzybowski et al., 2016).

1.2.2 Overview on various types of phototherapies

The traditional phototherapy units using fluorescent tubes contain standard blue (Westinghouse F20T 12B), daylight (F20 T12D), and cool white (F20 T12CW) lamps. The most effective lights are those with a high energy output near the maximum absorption peak of bilirubin (450 to 460 nm) as shown in Fig. 1.1 (Weiss and Zimmerman, 2013). Special blue lamps (Phillips TL 52/20W, Westinghouse 20 watt F20 T12BB) are the most efficient for neonatal phototherapy because it has more than twice the energy output at 450 nm than the standard blue bulb (Ennever et al., 1983). Investigators reported the significance of blue light in phototherapy with rapid reduction of serum bilirubin than with daylight or standard blue bulbs. On the other hand, these special blue bulbs cause nausea and dizziness among the neonatal care staff. A combination of four special blue lamps placed in the center of the phototherapy unit with two days' light lamps on either side has been found to provide excellent irradiance without producing significant discomfort to staff. Non-fluorescent halogen lamps make a more intense light over a smaller surface. If lamps are placed closer than 50 cm, halogen lamps, unlike fluorescent bulbs, incur a risk of burns to an infant.

Light-Emitting Diode (LED) lights are now commercially available for use in the United States (Maisels and McDonagh, 2008). The Neo Blue LED systems incorporate optimal blue LED technology and are manufactured by Natus Medical Inc., San Carlos, CA, USA. Neo Blue LED's emit blue light in the 450–470 nm spectrum. It is one of the safest phototherapy devices available as they do not emit light in the harmful ultraviolet and infrared radiation range. Further, the absence of heat when delivering overhead neo blue phototherapy is less likely to cause physiological water loss (Seidman et al., 2000). Fiber-optic phototherapy systems first appeared in the market around 1989. They are widely considered to be equally effective and more convenient than overhead lights. Light is delivered from a halogen bulb through a fiber optic cable and is

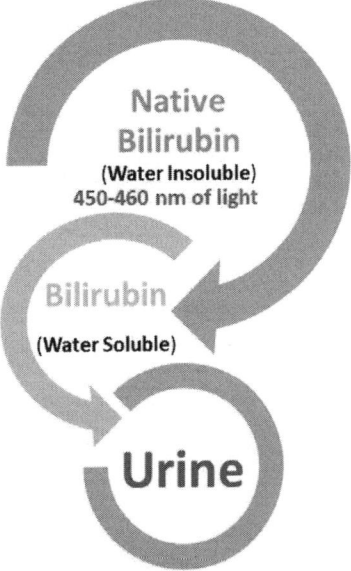

FIG. 1.1 Principle of phototherapy showing conversion of water-insoluble bilirubin to the soluble form upon exposer to a light wavelength of 450–460 nm.

emitted from the sides and ends of fibers inside a plastic blanket protected by a disposable cover. Infants lie on the blanket or are held with the blanket wrapped around them, and the need for eye patches, otherwise required in neonatal phototherapy, is eliminated.

Technology currently provides the clinician with three different modes of phototherapy delivery: fiber optic, low intensity, and high-intensity phototherapy. For low-intensity phototherapy, overhead lamps are typically set at a distance of 50 cm from the patient. The American Academy of Pediatrics has defined High-intensity phototherapy as a spectral irradiance of at least 30 MW per square meter per nanometer. High-intensity phototherapy is achieved by using a unit with eight special blue lamps, or Neo Blue LED systems 25 cm above the unclothed infant on a fiber optic phototherapy blanket in a bassinet while wearing a tie-on surgeon's mask as a diaper. This method allows maximum skin exposure and achieves irradiance as high as 50 uw/cm^2/nm.

However, as overhead lamps to infant distance are decreased, there is an increase in the heterogeneity of irradiation, with a much greater expansion at the center than at the periphery. Lining the bassinet with a white cloth produces greater homogeneity or irradiance and an increase in indirectly reflected irradiance (Maisels et al., 2007; Zauk, 2015).

1.2.2.1 UVB therapy

The UVB therapy includes NB-UVB and BB-UVB. The UVB therapy exerts effect by damaging nuclear DNA within epidermal-dermal junctional cells leading to apoptosis and cellular death of keratinocytes, immune cells, and fibroblasts (Bulat et al., 2011; Bolognia and Schaffer, 2012; Vangipuram and Feldman, 2016). NB-UVB (wavelength 311–312 nm) has largely replaced the use of BB-UVB (290–320 nm) due to its greater efficacy and emission duration at lower cumulative doses resulting in a reduction of associated long-term complications. NB-UVB is indicated as a first-line treatment of moderate-to-severe psoriasis and is also efficacious in treating numerous other dermatologic conditions (Ighani et al., 2018). The excimer laser is a targeted form of UVB treatment. It uses an active medium composed of excited dimers (noble gas argon/krypton/xenon) and reactive halogen gas (fluoride or chloride), delivering 308 nm light and facilitating the delivery of high doses of UVB to localized areas of skin (Mehraban and Feily, 2014). This focal treatment is beneficial in areas, which are difficult to treat such as the scalp, palms, and soles. It allows for a higher initial dose, less treatments, and thus, low long-term side effects. The excimer laser is currently approved in the USA for the treatment of conditions including psoriasis, and vitiligo, in addition to other localized inflammatory dermatoses (Alshiyab et al., 2015).

1.2.2.2 UVA therapy

The UVA is absorbed in the dermis and exerts its apoptotic effect on dermal blood vessel components, dendritic cells, fibroblasts, endothelial cells, and mast cells. UVA is further categorized into UVA1 (340–400 nm) and UVA2 (320–340 nm). UVA1 utilizes non-erythemogenic wavelengths and is indicated in the treatment of atopic dermatitis, localized scleroderma (morphea), systemic sclerosis, urticaria pigmentosa, cutaneous T cell lymphoma, dyshidrotic eczema, as well as other dermatoses (York and Jacobe, 2010).

1.2.2.3 PUVA therapy

PUVA involves administering either oral, topical, or bath psoralen followed by subsequent exposure to UVA, either via phototherapy or direct sunlight (PUVAsol). Psoralens (commonly methoxsalen or 8-methoxypsoralen) intercalate between DNA base pairs and, upon photon absorption, chemically activate to crosslink DNA. It results in several antiproliferative, antiangiogenic, apoptotic, and immunosuppressive effects. Melanogenesis is also stimulated via unknown mechanism. PUVA is effective in the treatment of several dermatologic conditions. PUVA treatments can result in severe blistering photoreactions/burns. PUVA is associated with a higher risk of skin cancer compared to other forms of phototherapy (Youssef et al., 2016).

1.2.2.4 Home phototherapy

Home phototherapy is another type that is limited to otherwise healthy term infants older than 48 hours with bilirubin levels 15-20 mg/dl with no hemolysis. Parents are required to be able to monitor the temperature and hydration status of infant. Home visits by a skilled nurse in evaluating newborns are performed, and bilirubin levels are assessed periodically. UVB therapy is equally effective when performed at home versus in the outpatient office setting and could be cost effective and more convenient to patients. Home phototherapy was associated with higher rates of dose reduction and/or discontinuation of biologics and apremilast, especially in patients with multiple comorbidities. There are several advantages assessing the efficacy of home phototherapy in treating other dermatologic conditions (Click et al., 2017; Howell et al., 2018). Tanning beds are an efficacious method of UV exposure and are recommended for patients who benefit from a more accessible, less costly alternative to office and/or home phototherapy. The effect of tanning beds may be from UVA exposure, though common tanning bulbs emit varying degrees of UVB light. Of concern, use of tanning bed require limited supervision of control settings and duration of treatments, leading to the potential for increased adverse events (AEs) (Radack et al., 2015), however the amount of UV radiation received is likely to be better controlled with tanning beds than with sun exposure (Krenitsky et al., 2020).

1.3 Various light sources and methods of phototherapy

There is a considerable selection of several custom-made and commercial phototherapy devices, which have been produced for investigative and clinical applications. The phototherapy devices are categorized by their light source as follows: (1) fluorescent tube (TL12, 60 cm, 20W) devices with different colors of light (cool white [CW], blue, special blue [BB, 52, and 03], turquoise, or green) of straight or U-shaped (18 cm, 18W tubes), (2) metal halide bulbs used in spotlights and incubator lights, (3) metal halide bulb and fiberoptic light guide combinations as used in pads, blankets or spotlights, and (4) high-intensity LEDs used presently as canopies.

1.3.1 Fluorescent tubes

The most commonly used light source in the U.S. is the special blue tube, such as F20 T12/BB or TL52/20W (Philips, The Netherlands). CW light has also been used together with special blue tubes to ameliorate caregivers' complaints regarding the blue hue of the light (Sisson and Kendall, 1973). Still, this combination of tubes dramatically decreases efficacy by 50% depending on the proportion of CW to special blue tubes. At a standard distance of 40 cm, the devices with a 1:1 ratio of tubes can deliver up to 11 $W/cm^2/nm$, while a unit containing only special blue tubes can deliver up to 24 $W/cm^2/nm$. However, the use of CW light typically provides only homeopathic doses of phototherapy. In addition, it may be inadequate in sufficiently decreasing total bilirubin levels in a jaundiced infant unless the lights are positioned in close proximity, such as directly above the infant (De Carvalho et al., 1999).

1.3.2 Halogen spotlights

Halogen spotlight systems utilize single or multiple metal halide lamps as the light source and provide fairly high irradiance, often exceeding 20 $W/cm^2/nm$. However, these units can generate considerable heat, which can, in turn, cause thermal injury to the infant and staff if applied too closely and can emit ultraviolet (UV) radiation if not appropriately shielded. The use of spotlights is sometimes preferred in the neonatal intensive care unit because with premature or critically ill neonates on radiant warmers, its design allows for ad hoc positioning of these devices for the convenience of caregivers. However, their variable positioning with respect to the distance from the infant and angle of application and their irradiance heterogeneity can lead to unreliable dosing and unpredictable clinical response (Vreman et al., 2004).

1.3.3 Fiberoptic blankets

Fiberoptic devices contain a tungsten-halogen bulb that delivers light via a cable into a plastic pad containing fiberoptic fibers. The pad remains cool and can be placed directly under an infant to increase the skin surface area that is exposed. The pad can also be wrapped around the infant's midsection to provide phototherapy (Vreman et al., 2004). The spectral power of the pad alone is low, therefore it is commonly used in conjunction with overhead lights to provide double phototherapy.

1.3.4 Light-emitting diodes

The gallium nitride LED is one of the most recent innovations in phototherapy. These devices provide high irradiance in the blue to blue-green spectrum without excessive heat generation (Maisels, 2005) Light-emitting-diode units are efficient, long-lasting, and cost-effective. The latest models incorporate amber LEDs to counteract the "blue hue" effect that can irritate caregivers (Stokowski, 2006).

1.3.5 Filtered sunlight

The lack of devices and/or reliable electric power limits the use of phototherapy in underdeveloped countries. Further, modern phototherapy devices are not affordable, often break down because of electrical power surges, and are difficult to maintain due to the unavailability of replacement parts. Even where phototherapy devices are available, most hospitals lack the resources necessary to replace fluorescent lamps. Thus, it is not uncommon, especially in areas without access to phototherapy, for the parents/guardians of jaundiced infants to place their babies in direct sunlight unaware of the potential harm. Using direct sunlight for phototherapy has several clinical and practical drawbacks that could make its use undesirable. Sunlight contains altitude-, seasonal-, and time-of-day-dependent levels of harmful ultraviolet A, B, and C radiation, which can cause serious and permanent damage to human skin. It also contains significant levels of warming infrared radiation, which could raise core body temperature to unsafe levels in the absence of sufficient cooling. It must be underlined that the use of sunlight, when filtered to exclude the harmful spectral radiation, is a novel, practical, and inexpensive method of phototherapy. It potentially offers a safe and efficacious treatment strategy for management of neonatal jaundice in economically disadvantaged countries where conventional phototherapy treatment is not available (Slusher et al., 2014). The most practical and low-cost sunlight filters are the commercially available window-tinting films, widely used in vehicles and residential and commercial structures in sunny climates. Window tinting films can effectively reduce ultraviolet and infrared radiation and offer a range of significant attenuations of therapeutic blue light. Although window-tinting films are traditionally affixed to a glass surface, these films can also be applied over a support frame, under which an infant basket, bassinet, or crib can be placed (Vreman et al., 2013; Yurdakök, 2015).

1.4 Applications and limitations of phototherapy

1.4.1 Application in neonatal jaundice

There is the use of visible light for the phototherapeutic treatment of hyperbilirubinemia in newborns. This relatively common therapy lowers the serum bilirubin level by transforming bilirubin into water-soluble isomers that can be eliminated without conjugation in the liver. The dose of phototherapy largely determines how quickly it works; the dose, in turn, is determined by the light's wavelength, the light's intensity (irradiance), the distance between the light source and the infant, and the body surface area exposed to the light. Commercially available phototherapy systems include delivery of light via fluorescent bulbs, halogen quartz lamps, light-emitting diodes, and fiberoptic mattresses. Proper neonatal and infant care enhances the effectiveness of phototherapy and minimizes complications. Caregiver responsibilities include ensuring effective irradiance delivery, maximizing skin exposure, providing eye protection and carefully monitoring thermoregulation, maintaining adequate hydration, promoting elimination, and supporting parent-infant interaction (Stokowski, 2006).

1.4.2 Application for morphea, scleroderma, and other sclerosing skin conditions

As discussed above, UVA1 phototherapy potentially exerts its therapeutic effect through modulation of the three predominant pathologic mechanisms in sclerosis: immune dysregulation, imbalance of collagen deposition, and endothelial dysfunction. This beneficial effect is predominantly reported in morphea (localized scleroderma), systemic sclerosis (scleroderma), lichen sclerosis, and chronic graft versus host disease (GVHD). Morphea has been the subject of several studies utilizing UVA1 therapy. These studies have shown that patients with morphea respond to low, medium, and high-dose UVA1 therapy. A comparative study of low dose versus high dose therapy revealed that high-dose UVA1 treatment is more effective. In a second comparative study, low and medium doses were compared with narrowband UVB therapy. This study revealed that medium-dose UVA1 therapy was superior to both low-dose UVA1 therapy and narrowband UVB therapy. A study showed no significant difference in improvement with low-dose UVA1 and narrowband UVB (NBUVB). The existing evidence indicates that medium to high dose UVA1 therapy delivered over treatment significantly benefits patients with morphea. These studies specifically included patients with linear and plaque subtypes of morphea. In all these studies, no adverse effects were reported, with the worst adverse effect being transient headaches. UVA1 is unlikely effective in burned-out atrophic lesions, deep morphea Parry Romberg/facial hemiatrophy, and eosinophilic fasciitis. Depth of the pathogenicity decides whether systemic treatment should be considered as first-line therapy in these cases (York and Jacobe, 2010). Hither to, phototherapy has evolved through more complex treatment, and is widely used to treat various diseases, such as atopic dermatitis (Patrizi et al., 2015), psoriasis (Diffey, 1980), vitiligo (Adauwiyah and Suraiya, 2010), acne vulgaris (Hession et al., 2015; Pei et al., 2021), and cancer (Morton et al., 2002).

1.4.3 Application for cancer

Conventional use of chemotherapy and radiation has some drawbacks in clinical practice, which are minimized by re-modulating the current therapy using photosensitizers (PS). Combining PS with light generates reactive oxygen species (ROS) to kill cancerous cells is a novel noninvasive approach known as photodynamic therapy (PDT). It is a selective and targeted approach that enhances specificity towards cancer cells. However, success of PDT depends on the selection of PS, as the administered PS must be deep entered into the tumor cells and then activated by irradiation of light. The cell death that occurs during PDT is the apoptotic response of the sub-cellular localization of PS. There are two approaches generally used in PDT: electron transfer (eT) happens due to the generation of radical species at the excited state. Another is energy transfer (ET) created by electronic excitation and energy transfer from an excited triplet of PS to a triplet oxygen molecule. Hence, the clinical use of PDT in combination with chemotherapy would be the better option as it works at the electronic level inside the cell.

Phototherapy involves the irradiation of tissues with light and is commonly implemented in PDT and photothermal therapy (PTT). Photosensitizers (PSs) are often needed to improve the efficacy and selectivity of phototherapy via enhanced singlet oxygen generation in PDT and photothermal responses in PTT. In both cases, efficient and selective delivery of PSs to the diseased tissue is of paramount importance. PTT monotherapy typically cannot completely eradicate tumors due to non-homogeneous heat distribution in tumor tissues. Several strategies have been adopted to increase the anticancer efficacy of PTT and PTT-based combination therapies. First, better PTT agents with high light absorptivity in the near-infrared spectrum, high photothermal conversion efficiency, long blood circulation times, and enhanced tumor uptake are being sought to improve photothermal therapy. Second, synergistic effects of PTT and other therapeutic modalities are being explored to enhance anticancer efficacy. The combination of PTT with reactive oxygen species, small interference RNAs, or chemotherapeutics can drastically increase treatment efficacy. Third, image-guided PTT with theranostic agents (which are used for treatment and diagnosis both) based on multifunctional nanomaterials can also increase treatment efficacy of PTT via selective delivery of PTT agents to tumors (Morton et al., 2002).

1.4.4 Limitations of home phototherapy and sunlight

The irradiance and surface area exposure produced by the home phototherapy unit is lower than typical units used in the hospital, making them less efficient at lowering the serum bilirubin level. Although phototherapy is indicated as treatment, current guidelines state that clinically high bilirubin warrants treatment should be managed in the hospital.

Lack of control treatment guidelines, duration of therapy, and variable sunlight prevented its use to treat hyperbilirubinemia and/or poor treatment, leading to adverse outcomes. Although in the past, parents had been told to expose jaundiced infants to the sunlight at courtyard of an English hospital, this practice is not considered a safe or reliable way to treat jaundice. There are reports in the literature of infants developing kernicterus after their parents were instructed to treat their jaundice at home by exposing them to sunlight, in some cases for 15 min per day. Sunlight is ineffective and contributes to delays in recognizing the severity of the hyperbilirubinemia (Stokowski, 2006; Lan et al., 2019).

1.5 Recent developments and future scopes

1.5.1 The immunoregulatory effects of phototherapy: Possible pathways

Psoriasis is caused by abnormal interactions among innate immune cells, T cells, and keratinocytes, leading to activation of the T helper cell type 1/T helper cell type 17 (Th1/Th17) immune axes and related cytokines. It contributes to the hyperproliferation, and inflammation seen in psoriasis. There are various mechanisms by which phototherapy may be effective for psoriasis. First, UV light induces apoptosis of keratinocytes and T cells in the epidermis and dermis. Second, UV light promotes immunosuppression by promoting migration of Langerhans cells of the epidermis and decreasing mast cell degranulation and histamine release. Lastly, UV light induces alteration in the cytokine profile of psoriasis (Bulat et al., 2011). The schematic representation for mechanism of phototherapy has been depicted in Fig. 1.2.

Research carried out recently has led to a better understanding of the specific pathways and alteration of cytokines by phototherapy. This therapy shifts the immune response away from the Th1/Th17 pathway toward the counter regulatory Th2 axis. The Th1/Th17 pathway is suppressed by NB-UVB, leading to decreased interleukin-12 (IL-12), IL-17, IL-20, IL-22, and IL-23. These effects on cytokines appear to be systemic and not just localized to psoriatic lesions. PUVA and NB-UVB lower plasma levels of tumor necrosis factor-alpha, IL-17, IL-22, and IL-23 at the end of 6 weeks of treatment. Furthermore, regulatory T (Treg) cells in patients with severe psoriasis display an enhanced propensity to convert into IL-17A-producing cells, linked to the loss of forkhead box P3 (Foxp3). UVB increases Foxp3-positive Treg cells in psoriatic skin lesions. This increase in Foxp3 expression improves Treg cell stability and reduces pro-inflammatory Th1/Th17 cytokines in psoriatic skin lesions (Wong et al., 2013).

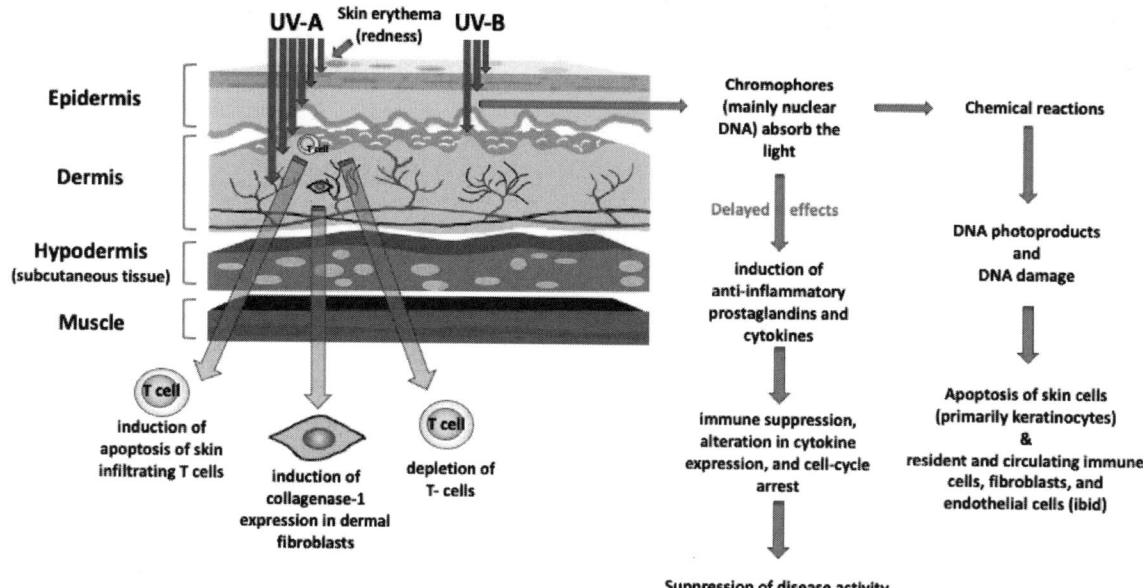

FIG. 1.2 Schematic explanation depicting how phototherapy works.

1.5.2 Handheld phototherapy: Targeting difficult-to-treat psoriasis in the office and at home

Whole-body phototherapy is limited as needless exposure of uninvolved skin and no benefit to unexposed skin (such as the hair-covered scalp). In the last decade, various portable and lightweight handheld phototherapy units have become available to treat localized psoriasis in the clinic and at home. These handheld devices, as compared to full-body irradiance in a booth or by a panel, have the added benefit of limiting skin exposure to UV light. The handheld devices are useful for the treatment of scalp psoriasis as well as recalcitrant localized psoriasis plaques. For example, the Dermalight 90 by National Biological Corporation, USA has a comb attachment that permits the direct application of light to scalp lesions (https://www.natbiocorp.com/our-units/dermalight-90/). Although large-scale clinical trials are lacking, such light combs appear to be productive with longer remission than other phototherapy treatments. The handheld devices typically deliver NB-UVB, but some devices use BB-UVB or UVA. The Dua Light by Theralight, Inc. and Psoria-Light by Psoria-Shield can provide UVB and UVA, but these devices are used only in the clinic setting (National Biological Corporation, 2021).

1.5.3 The excimer laser: A potential new indication and a novel dosimetry protocol

In addition to targeted therapy in the form of handheld devices, the 308 nm excimer laser was developed in 1997 as a targeted NB-UVB source to treat psoriasis. The advantage of using the excimer laser is that because psoriasis plaques can take higher doses of light than normal skin, targeted treatment of psoriasis lesions using higher doses permits quicker time to clearance. In a multicenter open-label trial, 72% of patients with mild to moderate psoriasis achieved at least 75% improvement of the target plaque in an average of 6.2 treatments. In addition, compared with traditional whole-body phototherapy, the excimer laser required fewer visits.

In another study, 13 out of 26 patients with plaque-type psoriasis had continued clearance or long-term improvement after 1 year. These initial studies tested the excimer laser in patients with localized psoriasis. Although the excimer laser is currently indicated by the US Food and Drug Administration for treating mild to moderate psoriasis; however, the excimer laser may also be safe and effective for use in generalized, moderate to severe psoriasis. In a single-center pilot study, patients with greater than 10% but less than 30% body surface area involvement (moderate to severe psoriasis) were treated with excimer laser twice weekly for 12 weeks for a total of 24 treatment sessions. Fifty-four percent achieved PASI-75, and 83% achieved PASI-50, maintained without further treatment for six months. However, the downside to treating moderate to severe psoriasis is the long duration required per session to treat a large body surface area, which may not be feasible in many dermatology office settings (Abrouk et al., 2016).

1.5.4 Phototherapy and biologic agents: Combination therapy for recalcitrant psoriasis

Although several biologic agents showing excellent efficacy in treating moderate to severe psoriasis have been developed in the last decade, phototherapy appears to play an important role in a subset of patients with severe, recalcitrant psoriasis despite treatment with biologic agents. Several studies have demonstrated the efficacy of using etanercept and NB-UVB in combination. These studies evaluated this combination therapy in patients who had not previously received treatment, patients who had an inadequate response with etanercept alone (50 mg once-weekly or 50 mg twice-weekly dosing), or patients who had an inadequate response to NB-UVB alone. Overall, combination therapy was superior, and time to clearance was reduced. A study by Lynde et al. also demonstrated the importance of high adherence to the NB-UVB regimen to achieve significant clinical improvement. High adherence to the NB-UVB regimen was defined as missing two treatments in any four weeks (Nakamura et al., 2016; Choi et al., 2021).

1.5.5 Future scope

Future research areas for the excimer laser include large-scale, long-term studies evaluating its use for the treatment of moderate to severe psoriasis as a potential new indication. In addition, assessing the safety, efficacy, and practicality of the "plaque-based sub-blistering dosimetry" in a large-scale, long-term study is also of interest. Over many decades, phototherapy evolved and will continue to grow further in due course. None the less besides its usefulness in hyperbilirubinemia, it's a valuable treatment option for many other skin disorders.

References

Abrouk, M., et al., 2016. Excimer laser for the treatment of psoriasis: safety, efficacy, and patient acceptability. Psoriasis (Auckl) 6, 165–173. doi:10.2147/PTT.S105047.

Adauwiyah, J., Suraiya, H., 2010 Dec. A retrospective study of narrowband-UVB phototherapy for treatment of vitiligo in Malaysian patients. Med J Malaysia 65 (4), 297–299.

Alshiyab, D., et al., 2015. Targeted ultraviolet B phototherapy: definition, clinical indications and limitations. Clin. Exp. Dermatol. 40 (1), 1–5. doi:10.1111/CED.12441.

Bolognia, J., Schaffer, J., 2012. Ultraviolet Therapy, 3rd ed. Elsevier, New York.

Bulat, V., et al., 2011. The mechanisms of action of phototherapy in the treatment of the most common dermatoses. Coll Antropol. 35 (Suppl 2), 147–151.

De Carvalho, M., et al., 1999. Intensified phototherapy using daylight fluorescent lamps. Acta Paediatr. 88 (7), 768–771. doi:10.1080/08035259950169071.

Chevreul, M., 1839. De la loi du contraste simultané des couleurs (eBook, 1839) [WorldCat.org] [Internet]. [cited 2015 Aug 10]. Pitois-Levrault, Paris. Available from: https://www.worldcat.org/title/de-la-loi-du-contraste-simultane-des-couleurs/oclc/797011105 (Accessed 1 November 2021).

Choi, G., et al., 2021. Inorganic–inorganic nanohybrids for drug delivery, imaging and photo-therapy: recent developments and future scope. Chem. Sci. 12 (14), 5044–5063. doi:10.1039/D0SC06724E.

Click, J., et al., 2017. Effect of availability of at-home phototherapy on the use of systemic medications for psoriasis. Photodermatol. Photoimmunol. Photomed. 33 (6), 345–346. doi:10.1111/PHPP.12349.

Cremer, R., Perryman, P., Richards, D., 1958. Influence of light on the hyperbilirubinaemia of infants. Lancet 1 (7030), 1094–1097. doi:10.1016/S0140-6736(58)91849-X.

Diffey, B.L., 1980. Ultraviolet radiation physics and the skin. Phys. Med. Biol. 25 (3), 405. doi:10.1088/0031-9155/25/3/001.

Dobbs, R.H., Cremer, R.J., 1975. Phototherapy. Arch. Dis. Child. 50, 833–836. doi:10.1136/adc.50.11.833.

Downes, A., et al., 1877. Researches on the effect of light upon bacteria and other organism. RSPS 26, 488–500. Available from: https://ui.adsabs.harvard.edu/abs/1877RSPS...26..488D/abstract (Accessed 1 November 2021).

Ennever, J.F., McDonagh, A.F., Speck, W.T., 1983. Phototherapy for neonatal jaundice: optimal wavelengths of light. J. Pediatr. 103 (2), 295–299. doi:10.1016/S0022-3476(83)80370-9.

Falk, E., 1985. UV-light therapies in atopic dermatitis – PubMed. Photodermatology 2, 241–246. Available from:. https://pubmed.ncbi.nlm.nih.gov/3903678/ (Accessed 1 November 2021).

Finsen, N., 1896. Om Anvendelse I Medicinen af koncentrerede kemiske Lysstraaler. [On the application in medicine of concentrated chemical rays of light]. Gyldendalske Boghandels Forlag, Kjøbenhavn. Available from: https://www.jameslindlibrary.org/finsen-nr-1896/ (Accessed 1 November 2021).

Finsen, N., Forchhammer, H., 1904. Resultate der Lichtbehandlung bei unseren ersten 800 Fällen von Lupus vulgaris [results of light therapy in our first 800 cases of lupus vulgaris]. Mitt Fins Med Lichtinst 5 (6), 1–48.

Goethe, J.W., 1810. In: Cotta, J. (Ed.), Zur Farbenlehre (Goethe's Theory of Colours: Translated from the German; with Notes by Charles Lock Eastlake, R.A., F.R.S. 1840. Archived from the original on 12 December 2016. Retrieved 18 October 2017 – via Internet Archive.). John Murray, London. Available from: https://www.christies.com/en/lot/lot-6246603 (Accessed 1 November 2021).

Green, C., et al., 1988. 311 nm UVB phototherapy–an effective treatment for psoriasis. Br. J. Dermatol. 119 (6), 691 696. doi:10.1111/J.1365-2133.1988.TB03489.X.

De Gruijl, F., 1986. Long-term side effects and carcinogenesis risk in UVB therapy. In: Hönigsmann, H., Jori, G., Young, A. (Eds.), The Fundamental Bases of Phototherapy. OEMF Spa, Milan, Italy, pp. 153–170.

Grzybowski, A., Pietrzak, K., 2012. From patient to discoverer–Niels Ryberg Finsen (1860–1904) – the founder of phototherapy in dermatology. Clin. Dermatol 30 (4), 451–455. doi:10.1016/J.CLINDERMATOL.2011.11.019.

Grzybowski, A., Sak, J., Pawlikowski, J., 2016. A brief report on the history of phototherapy. Clin. Dermatol. 34 (5), 532–537. doi:10.1016/J.CLINDERMATOL.2016.05.002.

Hannuksela, M., et al., 1985. Ultraviolet light therapy in atopic dermatitis. Acta Derm. Venereol. Suppl. (Stockh) 114 (SUPPL. 114), 137–139. doi:10.2340/00015555114137139.

Hession, M., Markova, A., Graber, E., 2015. A review of hand-held, home-use cosmetic laser and light devices. Dermatol Surg 41 (3), 307–320. doi:10.1097/DSS.0000000000000283.

Howell, S.T., Cardwell, L.A., Feldman, S.R., 2018. A review and update of phototherapy treatment options for psoriasis. Curr. Dermatol. Rep 7 (1), 43–51. doi:10.1007/S13671-018-0211-3.

Ighani, A., et al., 2018. Comparison of management guidelines for moderate-to-severe plaque psoriasis: a review of phototherapy, systemic therapies, and biologic agents. J. Cutan. Med. Surg. 23, 204–221. doi:10.1177/1203475418814234.

Kaushik, S., Lebwohl, M., 2019a. Psoriasis: Which therapy for which patient: focus on special populations and chronic infections. J. Am. Acad. Dermatol. 80 (1), 43–53. doi:10.1016/J.JAAD.2018.06.056.

Kaushik, S., Lebwohl, M., 2019b. Psoriasis: Which therapy for which patient: Psoriasis comorbidities and preferred systemic agents. J. Am. Acad. Dermatol. 80 (1), 27–40. doi:10.1016/J.JAAD.2018.06.057.

Kemény, L., Varga, E., Novak, Z., 2019. Advances in phototherapy for psoriasis and atopic dermatitis. Expert. Rev. Clin. Immunol. 15 (11), 1205–1214. doi:10.1080/1744666X.2020.1672537.

Krenitsky, A., Ghamrawi, R.I., Feldman, S.R., 2020. Phototherapy: a review and update of treatment options in dermatology. Cur. Dermatol. Rep. 9 (1), 10–21. doi:10.1007/S13671-020-00290-6.

Lan, G., Ni, K., Lin, W., 2019. Nanoscale metal-organic frameworks for phototherapy of cancer. Coord. Chem. Rev. 379, 65–81. doi:10.1016/J.CCR.2017.09.007.

Larkö, O., 1989. Treatment of psoriasis with a new UVB-lamp – PubMed. Acta Derm. Venereol. 69 (4), 357–359. Available from: https://pubmed.ncbi.nlm.nih.gov/2568064/ (Accessed 1 November 2021).

Levental, Z., 1977. Der 'Sonnendoktor' Arnold Rikli (1823–1906). Gesnerus 34, 394–403.

Lucey, J., Ferreiro, M., Hewitt, J., 1968. Prevention of hyperbilirubinemia of prematurity by phototherapy. Pediatrics 41 (6) 1047–1054.

Maisels, M., 2005. Jaundice. In: MacDonald, M., Mullett, M., Seshia, M. (Eds.), Avery's Neonatology. 6th ed. Lippincott Williams & Wilkins, Philadelphia, PA, pp. 768–846.

Maisels, M., Kring, E., DeRidder, J., 2007. Randomized controlled trial of light-emitting diode phototherapy. J. Perinatol. 27 (9), 565–567. doi:10.1038/SJ.JP.7211789.

Maisels, M., McDonagh, A., 2008. Phototherapy for neonatal jaundice. N. Engl. J. Med. 358 (9), 920–928. doi:10.1056/NEJMCT0708376.

Marx, H., 1946. Zur Klinik des Hypophysenzwischenhirnsystems. Klin. Wochenschr. 24 (1), 18–21. doi:10.1007/BF01635615.

Matos, T., Ling, T., Sheth, V., 2016. Ultraviolet B radiation therapy for psoriasis: pursuing the optimal regime. Clin. Dermatol. 34 (5), 587–593. doi:10.1016/J.CLINDERMATOL.2016.05.008.

Mehraban, S., Feily, A., 2014. 308 nm excimer laser in dermatology. J. Lasers Med. Sci 5 (1), 8. Available from: /pmc/articles/PMC4290518/ (Accessed 2 November 2021).

Meyer-Schwickerath, G., 1960. Light Coagulation. CV Mosby, St. Louis.

Morison, W., 1993. Photochemotherapy. In: Lim, H., Soter, N. (Eds.), Clinical Photomedicine. Marcel Dekker, New York, pp. 327–343.

Morton, C., et al., 2002. Guidelines for topical photodynamic therapy: report of a workshop of the British Photodermatology Group. Br. J. Dermatol. 146 (4), 552–567. doi:10.1046/J.1365-2133.2002.04719.X.

Nakamura, M., Farahnik, B., Bhutani, T., 2016. Recent advances in phototherapy for psoriasis. F1000Res. 13 (5), 1684. doi:10.12688/F1000RESEARCH.8846.1.

National Biological Corporation, 2021. *Dermalight® 90*. https://www.natbiocorp.com/our-units/dermalight-90/ (accessed 07-08-2021).

Palm, T., 1890. The geographical distribution and aetiology of rickets. Practitioner, 10–11.

Pašić, A., et al., 2003. Phototherapy in pediatric patients. Pediatr. Dermatol. 20 (1), 71–77.

Patrizi, A., Raone, B., Ravaioli, G., 2015. Management of atopic dermatitis: safety and efficacy of phototherapy. Clin. Cosmet. Investig. Dermatol. 8, 511–520. doi:10.2147/CCID.S87987.

Pei, 2021. Light-based therapies in acne treatment. Indian Dermatol. Online J 6 (3), 145. doi:10.4103/2229-5178.156379.

Radack, K.P., et al., 2015. A review of the use of tanning beds as a dermatological treatment. Dermatol. Ther. 5 (1), 37–51. doi:10.1007/S13555-015-0071-8.

Roelandts, R., 2002. Bicentenary of the discovery of the ultraviolet rays. Photodermatol. Photoimmunol. Photomed. 18, 208.

Roelandts, R., 2002. The history of phototherapy: something new under the sun? J. Am. Acad. Dermatol. 46 (6), 926–930. doi:10.1067/MJD.2002.121354.

Rollier, A., Rosselet, A., 1923. Heliotherapy. London, H. Frowde; Hodder & Stoughton, London. Available from: https://www.worldcat.org/title/heliotherapy/oclc/14777476 (Accessed 1 November 2021).

Rosenthal, N., et al., 1984. Seasonal affective disorder. A description of the syndrome and preliminary findings with light therapy. Arch. Gen. Psychiatry 41 (1), 72–80. doi:10.1001/ARCHPSYC.1984.01790120076010.

Seidman, D.S., et al., 2000. A new blue light-emitting phototherapy device: a prospective randomized controlled study. J. Pediatr. 136 (6), 771–774. doi:10.1016/S0022-3476(00)75202-4.

Sisson, T., Kendall, N., 1973. Avoidance of undesirable effects of blue light in phototherapy. J. Pediatr. 82 (1), 163–164. doi:10.1016/S0022-3476(73)80040-X.

Slusher, T., et al., 2014. Safety and efficacy of filtered sunlight in treatment of jaundice in African neonates. Pediatrics 133 (6), 1568–1574. doi:10.1542/PEDS.2013-3500.

Stokowski, L., 2006. Fundamentals of phototherapy for neonatal jaundice. Adv. Neonatal Care 6 (6), 303–312. doi:10.1016/J.ADNC.2006.08.004.

Vangipuram, R., Feldman, S., 2016. Ultraviolet phototherapy for cutaneous diseases: a concise review. Oral Dis. 22 (4), 253–259. doi:10.1111/ODI.12366.

Vreman, H., et al., 2013. Evaluation of window-tinting films for sunlight phototherapy. J. Trop. Pediatr. 59 (6), 496–501. doi:10.1093/TROPEJ/FMT062.

Vreman, H., Wong, R., Stevenson, D., 2004. Phototherapy: current methods and future directions. Semin. Perinatol. 28 (5), 326–333. doi:10.1053/J.SEMPERI.2004.09.003.

Wainwright, D., Ferguson, 1998. Narrowband ultraviolet B (TL-01) phototherapy for psoriasis: which incremental regimen? Br. J. Dermatol. 139 (3), 410–414. doi:10.1046/J.1365-2133.1998.02403.X.

Weiss, E.M., Zimmerman, S.S., 2013. A tale of two hospitals: the evolution of phototherapy treatment for neonatal jaundice. Pediatrics 131 (6), 1032–1034. doi:10.1542/PEDS.2012-3651.

Wong, B.T., Hsu, B.L., Liao, M., 2013. Phototherapy in psoriasis: a review of mechanisms of action. J. Cutan. Med. Surg. 17 (1), 6. doi:10.2310/7750.2012.11124.

York, N., Jacobe, H., 2010. UVA1 phototherapy: a review of mechanism and therapeutic application. Int. J. Dermatol. 49 (6), 623–630. doi:10.1111/J.1365-4632.2009.04427.X.

Youssef, R., et al., 2016. Phototherapeutic modalities pose no significantly increased risk of oxidative damage to DNA in dark skinned individuals. Indian J. Dermatol. Venereol. Leprol 82 (6), 666–672. doi:10.4103/0378-6323.186485.

Yurdakök, M., 2015. Phototherapy in the newborn: what's new? J. Pediatric Neonatal Individ. Med. 4 (2), e040255.

Zauk, A.M., 2015. Phototherapy: a simple and safe treatment for neonatal jaundice. J. Pediatr. Neonatal Care 2 (3). doi:10.15406/JPNC.2015.02.00070.

Zupanic-Slavec, Z., Toplak, C., 1998. Water, air and light: Arnold Rikli (1823–1906). Gesnerus 55, 58–69.

Chapter 2

Phototherapy for skin diseases

Renuka K. Mahajan, Dadasaheb M. Kokare, Nishikant A. Raut, Prakash R. Itankar
Department of Pharmaceutical Sciences, Rashtrasant Tukadoji Maharaj Nagpur University, Nagpur, Maharashtra, India

2.1 Introduction

The organs, such as hair, glands, and nails together with the skin make up the integumentary system. Moreover, the skin is the outlying of the integumentary system (Lawton, 2019; Abdo et al., 2020). The skin, is a sizeable organ of the body, covering 16% of total body weight and 1.5–2.0 m^2 of total surface area in an average adult. This multifunctional organ, besides providing protection from pathogens, physical abrasions and radiation from the sun, plays a vital role in synthesis of vitamin D, protection of vitamin B folates, and also makes us aware of external stimuli. Maintaining a constant body temperature with the help of sweating or shivering is another major role performed by skin. Integument is made up of epithelial and connective tissue. Fig. 2.1 illuminates the layers of skin namely, the epidermis, the dermis, and the underlying hypodermis.

2.1.1 The epidermis

The outer most protective wrap of the skin, composed of stratified squamous epithelium and dendritic cells is known as epidermis. It consists specific constellation of four types of cells. **Keratinocytes:** Keratinocytes produces keratin, a protein that hardens the skin. They also contain a special type of fat cells which makes the skin waterproof. The cells from basal layer (column-shaped keratinocytes) emigrate to the surface, resulting in keratinization. In this process keratinocytes pass from synthetic to degenerative phase. The mature keratinocytes at the surface are dead and filled entirely with keratin (Piotrowska et al., 2016). **Melanocytes:** They are dendritic, melanin synthesizing cells. Melanin is a pigment produced in membrane bound organelle called melanosome. This pigment allots color to skin and safeguard one from ultraviolet radiation. Melanocytes are responsible for transfer of melanin from melanosomes to the keratinocytes. This layer exuviates continuously and renews every 15–30 days giving rise to derivative structures, such as sweat glands and nails. **Langerhans cells:** These are phagocytic macrophages, derived from bone marrow. They interact with white blood cells (WBCs) during an immune response and are involved in T cells response. They are distributed over squamous, granular and basal cell layer. **Merkel cells:** These are oval shaped, discs-forming cells appearing deep into the epidermis at the boundary of epidermis and dermis. Merkel cells produce chemical signal in adjoining neuron and perform sensory functions.

The epidermis is further sub divided into five layers as shown in Fig. 2.2. **Stratum corneum** is at the periphery containing innumerable layers of dead, anucleate keratinocytes filled with keratin. This layer has tightly packed cells, which empowers the skin to be tough and impermeable. Its function is prevention from external things like bugs and bacteria. **Stratum lucidum** is generally found on the palms and soles of feet; it contains two to three layers of anucleate cells. The **stratum granulosum**, second layer in epidermis, contains 2–4 sheets of cells held in conjunction by desmosomes (a cell structure specialized for adhesion of cells). These cells hold keratohyaline granules, which give contribution in the development of keratin for skin strengthening. The layer, **stratum spinosum** contains 8–10 sheets of cells, which changes shape and are quite active in mitosis. Stratum basale is the bottomless layer of the epidermis containing one sheet of columnar cells actively splitting up by mitosis to construct cells that trek into the upper epidermal layers and eventually to the exterior of the skin. The stratum basale is segregated from the succeeding layer by a basement membrane, made up of proteins and collagen (Bergfelt, 2009; Abdo et al., 2020; Theoret and Stashak, 2014).

2.1.1.1 Dermis

The dermis is the supporting core layer of the integumentary system, made up of elastin, fibrillar structured protein called collagen, capillaries, and nerves. The collagen is responsible for strengthening, while elastin maintains elasticity of skin.

16 SECTION | 1 Phototherapeutics

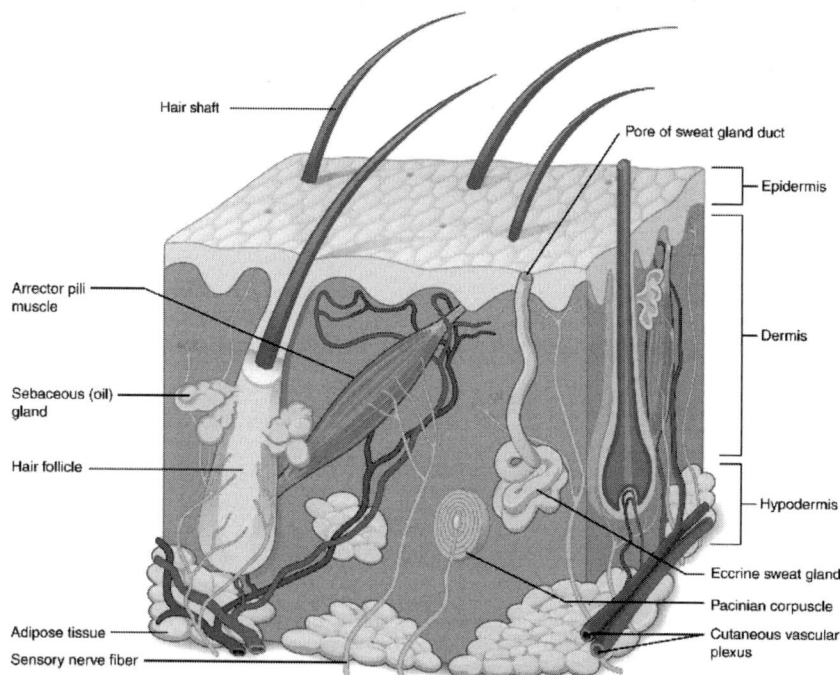

Fig. 2.1 A cross-section of skin with various accessory organs.

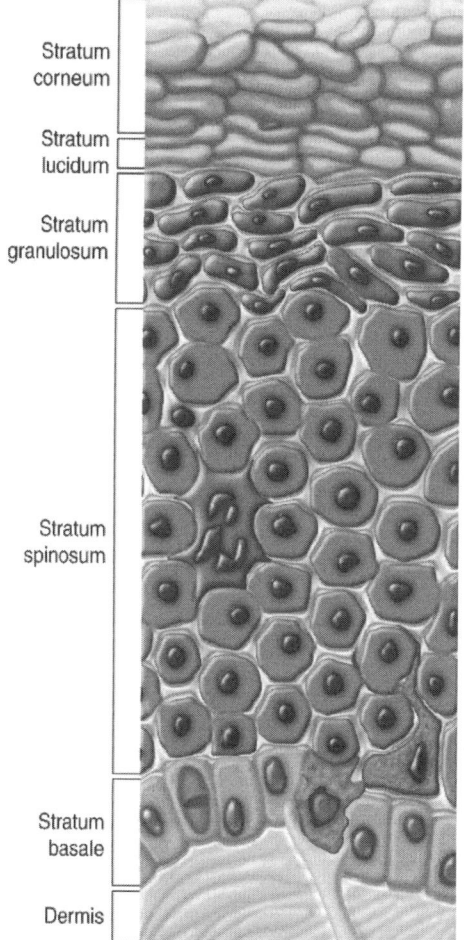

Fig. 2.2 A cross-section of epidermis.

The network of capillaries supplies nutrients to the layers of the skin. This layer lies on panniculus, a subcutaneous tissue, which shows presence of fat cells called lipocytes. The breadth of the layers varies as per the anatomical location, for instance, dermis on back is 30–40 times thicker, as compared to the outermost epidermis. The layers of dermis are divided into two levels, as represented in Fig. 2.3; the upper layer called the papillary region is made of untied connective tissue, and the lower level is made of closely packed tissue called the reticular layer. The papillary layer is an outer thin layer with fingerlike projections called dermal papillae that protrude into the epidermis. The reticular layer is thick, below the papillary layer that makes up most of the dermis. The dermis contains many specialized cells and structures which include: hair follicles, sweat glands, sebaceous glands (produce sebum which helps lubricate skin & hair), and nails. Dermis is accountable for cushioning effect against mechanical injury and also controls skin temperature. When injured, the dermis forms granulation tissue (some tissue rich in new blood vessels and many different cells) to heal itself. This tissue pulls the edges back together in case of wounds or cut. Our body takes three days to three weeks to create these tissues (Bergfelt, 2009; Abdo et al., 2020; Theoret and Stashak, 2014).

2.1.2 The hypodermis

The hypodermis (subcutaneous layer) lies between the dermis, underlying tissues and organs. Mostly, it consists of adipose tissue and is the storage site of body fat. It fastens skin to the basal surface, provides thermal padding, and suck up shocks from impacts to the skin.

2.2 Major functions of the skin

Thermoregulation: It is maintained through sweating and blood flow regulation throughout the skin.

Metabolism: When we are exercising and body is hot, sweat glands in our skin excrete water, wastes (ammonia and urea), salts and proteins.

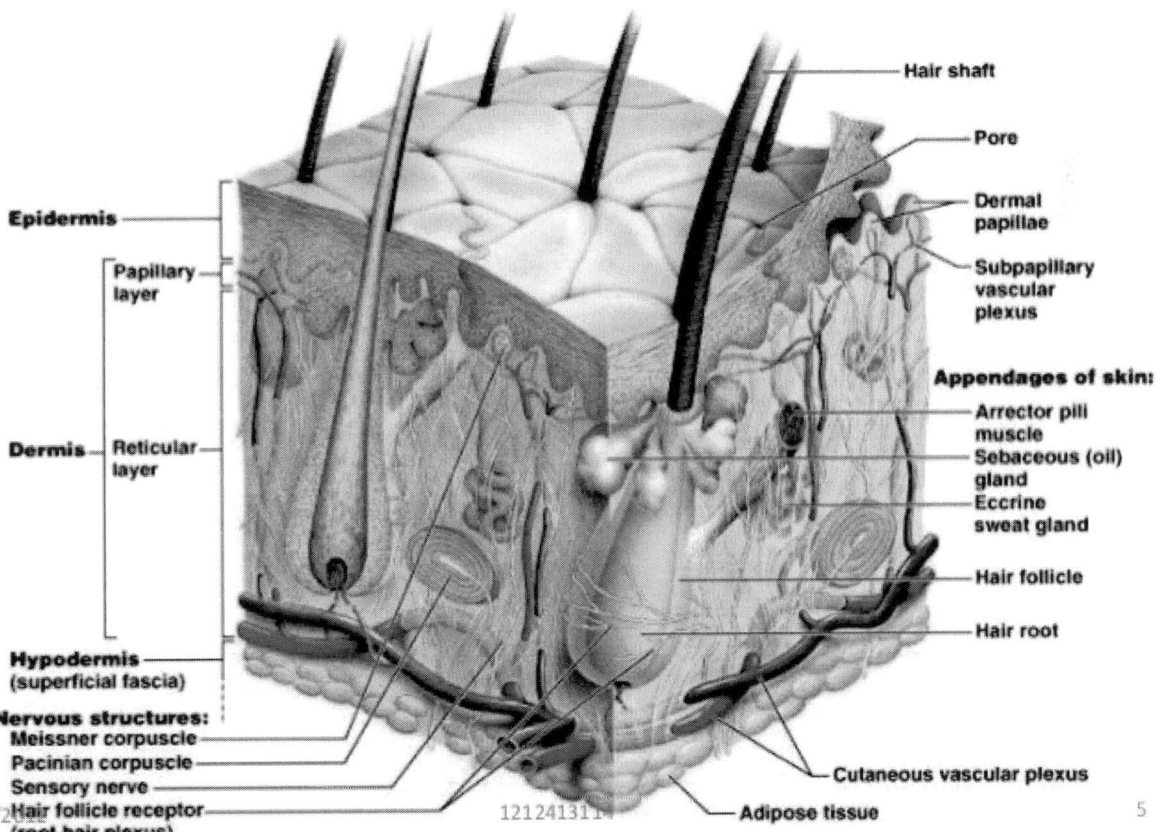

Fig. 2.3 Dermis and its appendages.

Sensation: Sensation for touch, pain and heat is provided by nerve endings (Lawton, 2019). **Protection:** Protection is provided against biological invasion, physical damage and ultraviolet radiation (Piotrowska et al., 2016).

Synthesis of vitamin D: When exposed to the sunrays, the skin produces vitamin D. This is essential for building strong and well-shaped bones (Young Hee Choi, 2019; Piotrowska et al., 2016).

2.3 Skin diseases and their etiology

Skin diseases are contemplated to be the most common cause of human illness. The international classification of disease stated more than 1000 skin or skin related diseases (Hay et al., 2014). Globally, skin disease has shown to be fourth leading nonfatal disease worldwide (Seth et al., 2017). These diseases are mostly caused due to bacteria, fungi, viruses, and parasites (Gelman et al., 2015). Such organisms enter the body through broken skin caused by injury or insect bite. The underlying disease and systemic infections are the other reason for ill health of skin for instance cutaneous lupus erythematosus, HIV, intertrigo, toe web infection, and measles (Tabassum and Hamdani, 2014). The skin diseases are classified as primary and secondary type. The primary diseases are settled by primary care while the secondary may cause disability or disfigurement likewise in leprosy or limb enlargement and immobility due to filariasis (Clebak and Malone, 2018). *Herpes simplex*, *Tricophyton rubrum* are the common virus and fungi causing variety of skin diseases (Karimkhani et al., 2017). Table 2.1 represents various diseases and their treatment.

TABLE 2.1 Skin diseases and treatment.

Sr. no.	Disease	Treatment
1	Nonbullous impetigo	Topical antibiotic creams, mupirocin and fusidic acid and oral antibiotic
2	Bullous impetigo	Dicloxacillin or clindamycin and cephalexin
3	Cellulitis Pneumococcal cellulitis	β-lactam antibiotics are recommended. In MRSA infection drugs used are clindamycin, doxycycline, sulfamethoxazole-trimethoprim, minocycline, along with fluoroquinolones and linezolid (Rodvold and McConeghy, 2014)
4	Orbital Perseptal cellulitis	Second and third generation cephalosporin and clindamycin; metronidazole for anaerobic infection.
5	Folliculitis	Topical application of mupirocin, fusidic acid ointment, or benzoyl peroxide. Followed by 7-day oral antibiotics course of cephalosporin, dicloxacillin or flucloxacillin.
	Pseudomonas folliculitis	MRSA infection if condition gets worse. Ciprofloxacin
	Fungal folliculitis	Itraconazole
6	Skin abscess (boil or furuncle)	MRSA requires doxycycline, clindamycin, trimethoprim, minocycline and/or surgical interventions.
7	Scabies	Application of sulphur ointments for a week or benzyl benzoate emulsion / lindane (Gamma benzyl hexachloride) for 24 h. Single use of malathion 0.5% in aqueous base, crotamiton cream or monosulfiram, permethrin cream or oral and topical administration of ivermectin.
8	Pyoderma	Topical antibacterial like mupirocin or fusidic acid daily for 10 days. Chlorhexidine solution, povidone iodine and potassium permanganate, hexachlorophene scrubbing, and neomycin/ polymyxin B-bacitracin (Neosporin) cream. Gentian violet with 0.5% to 1.0%; Oral administration of antibiotics like cloxacillin or erythromycin
9	Atopic dermatitis	Treatment includes emollients, antipruritic agents, anti-inflammatory therapy with corticosteroids, calcineurin inhibitors and secondary treatment with phototherapy using UVB (290–320 nm), and UVB (311–313 nm)
10	Cutaneous and lymphocutaneous Sporotrichosis	Itraconazole for 2 to 4 weeks; For severe case higher dosage or terbinafine or a saturated solution of potassium iodide.
11	Tinea capitis and Tinea imbricata	Griseofulvin for 6 to 8 weeks, Whitfield's ointment (a benzoic compound), oral antifungals, counting itraconazole, fluconazole, and terbinafine for children older than 4 years.

Sr. no.	Disease	Treatment
12	Onychomicosis	Oral antifungals
13	Herpes simplex, pox virus, shingles (herpes zoster) and warts	Oral or intravenous antiviral agents like acyclovir are effective treatments against viral skin diseases (Clebak and Malone, 2018).
14	Topical ulcer	Cleansing and treatment with penicillin, or grafting in severe cases, metronidazole and topical dressings, penicillin, and split skin grafting.
15	Vitiligo	Treatment includes psoralen UVA and narrow band UVB therapy along with topical agents (afamelanotide) and excimer laser therapy.
16	Melisma	Responds to hydroquinones
17	Psoriasis	Corticosteroids, vitamin D analogues, retinoid, calcineurin inhibitors, salicylic acid, coal tar, goeckerman therapy, and anthralin along with light therapy.
18	Mycosis fungiodes	Topical steroids, phototherapy, topical nitrogen mustard, carmustine, and bexarotene gel.
19	Scleroderma	Immunosuppressants, intralesional, topical, and oral steroids, topical vitamin D and phototherapy.

2.4 Bacterial skin diseases

Staphylococcus aureus, β hemolytic streptococci, and coryneform are the common bacteria causing skin infection. They infect the topmost layer of skin, the follicles and penetrate into the deeper layers of skin causing necrotic ulcers. If left untreated, these infections may spread throughout the body. Examples include folliculitis, trichomycosis, erysipeloid, impetigo, cellulitis, boils, skin, staphylococcal scalded syndrome, and Lyme disease. Some secondary infections are intertrigo and athlete foot. Prolong bacterial diseases may lead to proteinuria. Antibiotics are the effective treatment for bacterial skin infections.

Dermatitis: Also known as eczema, identified by rash, itchy skin, redness and swelling. The commonest type of dermatitis is atopic dermatitis (AD). It is a common inflammatory skin disease, characterized by pruritus in infant and hyperkeratosis and lichenification. Itching predominates in AD and *S. aureus* takes over the lesions formed by scratching (Kapur et al., 2018). Erythroderma is an inflammatory dermatitis caused due to malfunctioning of skin metabolism. This rapidly spreading disease is accompanied with shivering, and pyrexia.

Impetigo: It is the third most common disease caused in children ranging between age 2 to 5 years. The causative agents of the infection are *S. aureus,* β hemolytic streptococci or anaerobic bacteria. A typical form, nonbullous impetigo begins with single red lesion that quickly advances to vesicle, which may further rupture to form dry honey colored crust. Bullous impetigo is *Staphylococcus* induced scaled skin syndrome affecting neonate. This disease affects anogenital areas and buttocks in infants (Cole and Gazewood, 2007).

Pyoderma: These infections are primarily impetigo and secondarily effects of other lesions caused due to invasion of a streptococci or *S. aureus* (Powell et al., 1996).

Scabies: A common eco parasitic, directly transmitted infection caused by mites, *Sarcoptes scabei*, that burrow into the skin causing scabies. It is a bacterial infection caused by streptococci. An intensely itchy papule in the webs of fingers, wrists, elbows, and buttocks is typical of scabies (Hay et al., 2013).

Erysipelas in contrast to cellulitis, is superficial infection caused by Group A beta-hemolytic streptococci affecting dermis while later affects reticular dermis and subcutaneous fats.

Cellulitis: It spreads rapidly, and gives painful and warmth feeling in the indurated area. The conditions like ulcers, trauma, eczema etc. damage the protective covering of epidermis and allow bacterial access of group A beta-hemolytic Streptococcus and *S. aureus* to the interior tissue. General symptoms of infection involve chills, fever, and malaise. People suffering from systemic lupus erythematosus, hematologic cancer, alcohol abuse, diabetes mellitus, or nephrotic syndrome are more prone to Pneumococcal cellulitis (Parada and Maslow, 2000). Orbital cellulitis (Hauser and Fogarasi, 2010) is infection of soft tissue lying anterior of orbital septum. This infection occurs in winter as it is associated with sinus and respiratory tract infection. Perseptal cellulitis is commoner than orbital cellulitis. Tenderness of upper and lower eyelid, unilateral swelling, and erythema are observed in perseptal cellulitis. However, decreased visual accuracy, limited extra ocular movements and chemosis are observed in orbital cellulitis. Perseptal cellulitis may form local abscess while orbital cellulitis if left untreated may cause neurologic deficits, everlasting vision loss, or death (Chaudhry et al., 2007).

Folliculitis: It is the superficial inflammation of hair follicle with papules or postules at the base. It is differentiated as bacterial and fungal. Bacterial folliculitis mostly appears in men. Pseudo folliculitis occurs when hair shaft pierce in the wall of hair follicle causing skin appear pigmented.

Rashes: The red, inflamed, scaly, or itchy spots on skin are called rash. Allergy, irritation, structural defects like malfunctioning oil glands or blocked pores and infection due to any underlying disease are some causes of rash. Examples: dermatitis, eczema, acne, psoriasis, hives, and pityriasis rosea.

Skin abscess (boil or furuncle): The collection of pus under skin due to localized skin infection is known as boil. Sometimes abscesses may need draining by a doctor to get cured. Furuncles is an infection of hair follicle progressing to the adjoining subcutaneous tissue forming abscess generally caused by *S. aureus* or MRSA (Stevens et al., 2014).

2.5 Fungal skin diseases

Group of fungi known as dermatophytes are responsible for fungal skin diseases. These dermatophytes are recognized as lipophilic yeasts, *Malassezia* species, *Epidermophyton, Trichophyton,* and *Microsporum* (rarely). They are commonly known as ring worm fungi or tinea. The infection takes place when the fungi present on skin surface enters into the body. These infections affect nails, skin and hairs; examples are ringworm, athlete's foot, and lock itch. However, fungi may also affect severely on those having underlying diseases, like suppressed immune systems or oropharyngeal candidosis and HIV.

Tinea Capitis: It is a contagious childhood disease causing high inflammatory lesion on the scalp along with scarring and local hair loss. Initial symptoms of infection are local patches of hair loss and scaling. This disease may lead to permanent alopecia.

Tinea Imbricata: Unusual infection occurring in remote areas of India, Brazil, Malaysian pacific, and Indonesia. The disease appears as scaling in the form of concentrated rings. This infection may persist from childhood to old age.

Tinea versicolor: A benign skin infection which creates light colored areas of less pigmentation on the skin.

Sporotrichosis: An infection caused because of fungus *Sporothrix schenckii*, also known as, gardener's disease. This disease is caused due to contact of skin with contaminated plants, soil, or other materials (Kauffman et al., 2007). Immunosuppression, diabetes, alcohol suppression are some risk factors in this infections. Lymphocutaneous sporotrichosis is a form representing a hard nodule at the place of injury. This disease can occur in any age and typically involves hands and arms.

2.6 Viral skin diseases

These include herpes simplex, pox virus, shingles (herpes zoster), and warts. In these diseases virus makes way through the stratum corneum and contaminate the deeper sheets of the skin (Clebak and Malone, 2018). The herpes viruses HSV-1 and HSV-2 may cause blisters or irritation around lips or genitals. Shingles (herpes zoster), a painful rash on a side of the body, are caused by chickenpox virus which can be prevented by an adult vaccine. Wart is a disease in which a virus causes excess skin growth, creating a small hard benign growth on the skin. One can treat this at home using chemicals or duct tape and even can be removed with the consultation of physician. Large areas of the skin can cause a rash due to infection of virus, generally known as viral exantham, which is common in children.

2.7 Tropical ulcers

These diseases are characterized by sudden deep ulceration on lower limbs. Ulcers occur due to infection through combination of different bacteria with a bacterium called *Fusobacterium ulcerans*. People undergoing poor living condition or associated with stagnant water, mud, or flood are affected with these diseases. The lesion progress from mild discomfort, hyperpigmentation to breakdown revealing ulcers, which sometimes becomes chronic (Minatel et al., 2009; Caetano et al., 2009).

2.8 HIV related skin diseases

The life threatening skin diseases caused due to underlying HIV are epidermal necrolysis and Kaposi's sarcoma, which may lead to skin failure. The itchy papular pruritic eruption is the common complications of HIV (Meola et al., 1993).

2.9 Pigmentation disorders

Disorders caused due to variable amount of production of melanin by the body are pigmentation disorders. These disorders range from albinism to hyperpigmentation. Albinism may cause life threatening skin cancers. Malfunctioning of cells, absence of melanocytes, exposure to chemicals or cold, and certain infections cause hypo pigmentation, for example vitiligo, a chronic cutaneous autoimmune disorders characterized by depigmentation of skin (Bae et al., 2017). Hyperpigmentation may be caused due to metabolic disorders, aging, hormonal changes, or skin irritation for instance, melisma, age spots, and freckles. The disorders affecting cheeks and forehead in women are known as melisma. Overuse of corticosteroids ranges from thinning of skin to increased infection rates. Leprosy also leads to skin depigmentation.

2.10 Parasitic infections

Scabies and lice are causative agents for parasitic infections.

2.11 Tumors and cancers

Tumorous and cancerous growths arise when cells multiply faster than normal. Currently, 2–3 million non-melanoma, and 1,32,000 melanoma skin cancers are occurring globally every year (Seth et al., 2017). The major skin cancers are the manifestations of sun exposure. This disease is classified in three distinct forms as, the most curable basal cell cancer, the form which may grow and spread is squamous cell cancer and the deadliest is malignant melanomas. Skin cancers can be intercepted by protecting one from ultraviolet radiations. Early diagnosis increases chances of cure.

Melanoma: The most threatening type of skin cancer is melanoma. A skin biopsy can identify melanoma. Basal cell carcinoma: A common and less dangerous form of skin cancer as it grows and advances slowly.

Squamous cell carcinoma: It begins as an unhealing ulcer, or an uncommon growth. Sun-exposed areas are prone to develop squamous cell carcinoma.

2.12 Trauma

Once the exterior of the skin is damaged, body becomes susceptible to infection, for example burn or cut.

2.13 Skin tests

Skin biopsy: A small segment of skin is drawn out and studied under a microscope to recognize a skin condition.

Skin testing (allergy testing): Extracts (such as pollen) are applied to the skin and observed for any hypersensitive reactions.

Tuberculosis skin test (purified protein derivative or PPD): Proteins from the tuberculosis (TB) bacteria are injected under the skin causing firm skin.

2.14 Heliotherapy

The sunlight has been used to treat dermatological diseases since ancient times. The therapeutic action of sunlight is attributed to the invisible ray spectrum called UV radiations. These rays are absorbed by the melanin present in epidermis, blood plasma, and lipochromes situated in adipose tissue. Use of such natural day light to treat the medical conditions is called as heliotherapy or climate therapy. It was used by ancient Egypt physician Ephesus and Avicenna to treat variety of diseases since 1400 B.C. (Gupta et al., 2016; Vreman et al., 2004). The artificial light used for such treatments is known as actinotherapy. The history highlights Niel Ryberg Finsen (second noble prize winner for medicine in 1903), who in 1890 treated diseases using carbon arc lamp and in 1903, treated lupus vulgaris and tuberculosis of skin with the assistance of light radiations (Nee, 1997; Gupta et al., 2016; Singer and Berneburg, 2018). The use of these modern UV lamps concreted the way from heliotherapy to phototherapy. The ultraviolet part of sunlight spectrum (UVB and UVA) induces production of vitamin D and other chemicals, which are helpful in protection of skin cells. These radiations serve as antiproliferative, anti-inflammatory, and immune suppressant. The heliotherapy has been found beneficial in skin conditions like psoriasis, atopic dermatitis, and vitiligo.

To reduce or avoid disadvantages allied with heliotherapy, the regimens included Dead Sea basin method and minimal erythema dose (MED) method. In the former, the person was exposed for 10–15 minutes two times per day with increment of 10 min every day till reach of 3–6 h per day for at least 3–4 weeks while in latter, MED is determined as dose or time required to the skin to get noticeably pink after 24 h of exposure. During sun bath one has to cover the head with wet towel and body should be enfolded with banana leaves for best results; otherwise, no clothes, dry towel, wet towel, mud pack, or oil massage can be incorporated in treatment. Once body starts sweating, cold water bath, and rest is recommended. The UVB radiations obtained from sunlight provides vitamin D3, which improves muscle strength, fitness, and power in young Olympic athletes is mentioned by Philostratus, an ancient Greek physician, which was not benefited by supplements (Ceglia and Toni, 2017).

This organic compound cannot be synthesized in body but are required for various physiological processes. In presence of UVB radiations, obtained from sunlight, vitamin D3 (1, 25 dihydroxyvitamin) is synthesized from dehydrocholesterol in the skin. This vitamin blocks rise of IFN-γ and promote IL-4 production (van Etten and Mathieu, 2005). In addition, it also decreases IgG secretion, β cell proliferation and plasma cell differentiation, which suggest immunomodulatory effects (Selvaraj, 2011). Its antimycobacterial activity modulates immune system against *M. tuberculosis* (Kaufmann, 2006).

2.15 Naturopathy modalities on inflammation and immunity

Naturopathy treatments are gleaned from fundamental constituents including water, air, fire, earth and ether also known as panchamahabhootas (Yadav et al., 2012). Mud therapy, heliotherapy, hydrotherapy, yoga, diet, fasting, acupressure, acupuncture, etc., are the other naturopathy treatments (Nair and Salwa, 2018). In past few decades, various studies revealed anti-inflammatory effects of naturopathy which occur due to decreased ESR (Erythrocyte sedimentation rate), reduced cytokinins, and improved cell function (Nair, 2015).

2.16 Phototherapy for skin diseases

Phototherapy includes the exposure of different light sources viz. sunlight, LED, halogen or fluorescent light for medical treatment. The effectiveness of phototherapy depends upon interaction between skin and UV radiations. In this method, light open pores, exfoliates skin and allow penetration below surface of skin. The quality of spectra, light intensity, exposed skin surface area, skin pigmentation and thickness, and exposure time are some parameters deciding efficacy of phototherapy (Vreman et al., 2004). UV radiations encompass within 100–380 nm, which are further divided as UVA (320–380 nm), UVB (280–320 nm), and UVC (100–280 nm). UVA is further subdivided as UVA1 (340–400 nm) and UVA2 (320–340 nm). The radiations are absorbed by exterior surface of skin, that is, epidermis and perforate intensely in the dermis. UVA has most effective penetrating ability. Wavelengths determine application of phototherapy in treating dermatological diseases. UVB has higher potency to treat superficial diseases while deep diseases are treated with UVA. The efficacy of radiations is enhanced by use of photosensitizer, psoralen (Singer and Berneburg, 2018). The skin diseases, if not controlled with topical treatment, light therapy proves an appropriate choice especially use of narrow band UVB (NBUVB). Features of phototherapy are highlighted in Table 2.2 (Patrizi et al., 2015).

TABLE 2.2 Features of phototherapy.

Phototherapy	Wavelength	Features	Limitations
Broad band UVB	280–315 nm	Initial radiation method, recently substituted by NBUVB	Sunburn, xerosis, photodamage, erythema
Narrow band UVB	311–313 nm	Less erythemogenic, short acting, more effective	Less side effects as compared to BBUVB
UVAB	280–400 nm	Combination of UVA and UVB	Combined side effect of UVA and UVB
UVA and UVA1	(315–400 nm) (340–400 nm)	Deep dermal action, less sunburn or erythema	Need longer exposure than UVB
Photochemotherapy (PUVA and balenephototherapy)		UVA with psoralens, good outcomes	Systemic toxicity, photoonycholysis, carcinogenic effects. Balenephototherapy more effective than PUVA
Skin Tests			

The phototherapy is also effectual in other mainstay disorders including dermatitis, eczema, atopic dermatitis, cutaneous T-cell lymphoma, urticaria, photo dermatoses, pityriasis rubra, urticaria pigmentosum, pilaris, alopecia areata, aquagenic pruritus, lichen planus, neonatal jaundice, pityriasis lichenoides, graft vs. host disease, and granuloma annulare (Ibbotson, 2018).

2.17 Methods

2.17.1 UVB radiations

UVB radiations emitted from sun are majorly filtered by atmosphere of earth and only 15% of total rays, a person is exposed with, are absorbed by epidermis and outstretched up to superficial part of dermis (Singer and Berneburg, 2018). Conventional broad band UVB, efficient in treatment of lesions caused by psoriasis was reformed by development of Philips TL01 fluorescent lamp emitting narrow UVB spectra (311–312 nm) and was found superior in psoriasis (Petrazzuoli, 2000). Presently, narrow band UVB is been widely used treatment for skin disease (Ortiz-Salvador and Pérez-Ferriols, 2017). The energy from these radiations is absorbed by DNA, where it forms covalent bond between adjacent pyrimidines, resulting in formation of photoproducts (cyclobutane pyrimidine dimersiate). These photoproducts interfere with the physiological functions of enzymes involved in transcription and replication, resulting in mutations. The cells being equipped with repair mechanisms like translesion DNA synthesis, nucleotide excision effectively eliminates UV activated DNA damage before next replication. If the damage is formidable, the cells initiate apoptosis, a chief mechanism for action of UVB radiations. Immunomodulation is an additional effect associated with UVB rays. Narrow band UVB (UVB311) are meant to increase tolerability and are majorly used in the ministration of vitiligo, atopic dermatitis and psoriasis. NBUVB along with highly concentrated salt water act synergistically and facilitates penetration of radiations in dermis. This therapy is known as Bal neo-phototherapy (Klein et al., 2011).

Narrow band UVB reduces inflammation causing cytokinins, that is, IL17, IL22, and IL23 and upregulates IL10, which is accountable for inhibition of macrophage function in psoriasis (Hönigsmann, 2001). Apoptosis of inflammatory cells is another mechanism involved in treatment of various diseases. The rays are absorbed by a chromophore, urocanic acid, present in Stratum corneum causing photoprotective effects. Absence of urocanic acid leads to increase in efficiency of treatment along with increased production of vitamin D, antimicrobial peptides, and cytokinins (Mysore and Shashikumar, 2016). Vitamin D3 and topical calcineurin inhibitors topically enhances response during phototherapy (Singer and Berneburg, 2018).

2.17.2 UVA radiation

Radiations received by the body aforementioned, are absorbed by chromophores in dermis. These chromophores either occur naturally in the body as flavins, melanins, vitamins, porphyrins or introduced from outside, psoralen. The chromophore molecules absorb electromagnetic energy and transfer it to molecular oxygen. The molecular oxygen thus formed, attacks guanine in DNA, resulting in formation of 8 oxoguanine, which act as indicator of UVA treatment. The hydroxyl radicals and superoxide anion, formed along with singlet oxygen are proficient in damaging DNA (Krutmann and Morita, 1999). Chromophores also transfer energy to DNA in the form of single electron oxidation. UVA therapy results in apoptosis of cells capable of causing inflammation due to oxidation and cell damage, which results in anti-inflammatory effects and oxidation of lipids of membrane. UVA upregulated hemeoxygenase 1 has antiapoptotic and cytoprotective effects. It protects against oxidative stress, shows antifibrotic effects and causes debasement of dermal collagen, thus found helpful in treatment of sclerotic disorders. Another advancement of UVA1 (340–400 nm) phototherapy introduced in 1990 was capable to treat inflammatory disorders avoiding DNA damage (Petrazzuoli, 2000). Photobiological effects are based on suppression of cytokinins, induction of apoptosis, and activation of collagenase for abasement of dermal collagen. UVA1 has adapting effects on systemic lupus erythematosus, atopic dermatitis, skin disorders, morphea, and necrobiosis lipoidica (McGrath, 2017).

2.17.3 PUVA

Psoralen, a chromophore (artificial) is used to intensify UVA effects. The furocoumarin 8-methoxypsoralen is inserted in the DNA of keratinocytes, henceforth exposed to UV, resulting in formation of covalent bonds with nucleotides. Further, in next photonic excitation forms another bond on opposite DNA strand creating DNA interstrand cross link. The mechanism implemented in UVA therapy is cell apoptosis and cycle arrest. It has immunomodulatory effect leading to increase in IL-2 and IFN-γ and decrease in IL-4, IL-5, and IL-10. It shows excellent effect in mycosis fungoides, dermatitis, and lymphomatoid papulosis (Singer and Berneburg, 2018). PUVA combined with retinoid is used in graft-versus-host disease, T-cell cutaneous lymphoma, plaque, hand and foot eczema, lichen planus, and severe form of pustular psoriasis. PUVA as compared to other radiations was found better in psoriasis treatment.

2.17.4 Diseases and their treatment using phototherapy

Phototherapy helps skin to make more vitamin D, which eases inflammation, fight tumors and heals wound. Several studies have proved that prolong treatment with narrow and long band UV radiations are safe, well tolerated, and highly effective (Ibbotson, 2018; Weischer et al., 2004). UV light is used to manage vitiligo, psoriasis, dermatitis, lymphoma and acne along with skin wounds in diabetes and jaundice in babies. Substaintially more diseases are presently ministrated using recent techniques of phototerapy. Table 2.3 represents diseases and spectra used for the treatment.

2.17.4.1 Vitiligo

Phototherapy twice a week for year or more is used for disease characterized by pigment disorder, known as vitiligo. This disease, majorly affects the quality of life, self esteem, and social life. Since no melanocytes are present in fully developed vitiligo macules, the autoimmune mechanism cause progressive destruction by autoantibody dependent cellular cytotoxicity or cytotoxic T lymphocytes (Fitzpatrick, 1997), aberrant cytokinin expression, and antimelanocyte antibodies (Grimes et al., 2013). Destruction of T lymphocytes improves vitiligo. PUVA (immunosuppressive) and NBUVB (311 nm) are the principle modalities in the regimen of vitiligo (Bae et al., 2017) which cause rapid reproduction and movement of melanocytes to the epidermis. PUVA stimulates active melanocytes in outer root sheath to divide, reproduce and migrate towards surface here they form a visible pigment island.

2.17.4.2 Atopic dermatitis

Atopic dermatitis (eczema), a red and itchy skin condition, accompanied by hay fever and asthama is common in adolescent. Broad band UVB, NBUVB, and combined UVA and UVB are effective treatments for patients suffering with atopic dermatitis (Rodenbeck et al., 2016). These radiations targets inflammatory cells, inducing positive immunosuppression by altering cytokinin production, inducing apoptosis and inhibiting antigens of langerhans cells causing decrease in cutaneous inflammation (Patrizi et al., 2015). Certain *in vitro* studies indicated dose dependent effect of blue light (405–470 nm) on *P. aeruginosa* and *S. aureus* (Guffey and Wilborn, 2007).

2.17.4.3 Psoriasis

PUVA therapy is used in severe, plaque and pustular psoriasis. It acts on T lymphocytes and also inhibits DNA synthesis in keratinocytes (Fitzpatrick, 1997). Psoriasis a pruritic clinical manifestation characterized by scaling and plaque was

TABLE 2.3 Phototheray for various diseases.

Diseases (Disorders)	Types	Range of Spectra
Vitiligo	NBUVB and PUVA	310–315 nm, 340–400 nm
Atopic dermatitis	Broad band UVB	280–315 nm
	Narrow band UVB	311–313 nm
	UVAB (Combination)	280–400 nm
Psoriasis	Narrow band UVB	311–313 nm
Impetigo	Narrow band UVB, PUVA	310–315 nm, 340–400 nm
Folliculitis	Narrow band UVB	311–313 nm
Acne vulgaris	High intensity blue light	400–450 nm
Fungal skin diseases	UVC	254 nm
Lymphomas	NBUVB, PUVA	310–315 nm, 340–400 nm
Scleroderma	UVA1	340–400 nm

found suppressed with the exposure of UVA or UVB radiations. But the emergence of side effects including erythema evoked the need of modification in treatment, which was further modified with addition of crude coal tar by Dr William Goeckerman and anthralin by Dr John Ingram. Recently, narrow band UVB is considered the most efficient treatment for psoriasis (Yu et al., 2013).

2.17.4.4 Impetigo

A common disease, caused due to β hemolytic streptococci, *S. aureus,* or anaerobic bacteria. It appears as red lesion, which further form vesicle. Impetigo Herpetiformis, also considered as pustular psoriasis, is a rare life threatening dermatosis. Etiology involves mutation of interleukin 36 receptor antagonist (IL36RN) which encodes IL-36 receptor antagonist. This IL-36 is induced by cytokinins IL 17A, IL-22 which take part in pustular dermatoses. NBUVB is considered safe in this condition but associated with folate deficiency. Thus, PUVA is found to be safest for impetigo (Namazi and Dadkhahfar, 2018).

2.17.4.5 Folliculitis

Most common condition in folliculitis includes burning, swelling, itching, pain, inflammation, and tenderness. Light therapy, that is, NBUVB has clinically been proved to reduce symptoms.

2.17.4.6 Acne vulgaris

The disease caused due to invasion of *Propionibacterium acnes* on sebaceous glands was effectively eradicated with the utilization of high intensity blue light (Papageorgiou et al., 2000). Inflammation increased sebum excretion, colonization with *P. acnes* and hypercornification are the causative parameters for acne (Sami et al., 2008). Phototherapy affects postules, infiltrates, and shows slight effect on papules.

2.17.4.7 Fungal skin diseases

The dermatophytes such as *Trichophyton rubrum, Epidermophyton floccosum, Microsporum canis,* and *T. mentagrophytes* were successfully eradicated using UVC (254 nm) radiant exposure. Antifungal blue light (ABL) 415 nm, in vitro on *Candida albicans* and vaginal epithelial cells had effective antifungal effects (Wang et al., 2020). Onychomycosis, a nail fungal infection increases risk of recurrent ulceration and cellulitis (Vila et al., 2015). Oral antifungals may lead to liver toxicity therefore, a nontoxic, clear, noninvasive and reliable method was developed (Dai et al., 2008).

2.17.4.8 Lymphoma

Cutaneous T cell lymphoma (CTCL) constitutes 80% cutaneous lymphomas and 20%–25% B cell lymphomas. Mycosis fungoides (MF) a variant of CTCL includes nodules, tumors and psoriasiform plaques (Ramsay et al., 1992). NBUVB thrice a week or PUVA 2–3 times a week in MF patients is preferred for plaque clearance (Carter and Zug, 2009).

2.17.4.9 Scleroderma

A long-lasting skin disease, affecting internal organs, joints, and cutaneous skin, is scleroderma, which is categorized as local and systemic. In this disease the immune system cause excess production of protein collagen leading thicken and tight skin plus scars on lungs and kidneys. Increased fibroblastic activity causing hypertrophic dermal collagen indicates cutaneous scleroderma while localized form of scleroderma includes morphea, linear scleroderma and does not involve internal organs. UVA1 treatment increases protein expression of fibroblast, mRNA, and collagenase gene expression. The aforementioned therapy 3–4 times per week for 30 treatments is found to be efficacious in scleroderma (Hassani and Feldman, 2016).

2.17.5 Limitations of phototherapy for skin diseases

Even though skin has variety of defense mechanism the UV radiations having mutagenic potential can increase incidence of melanoma and nonmelanoma skin cancers. PUVA has the greatest carcinogenic risk. Also, the cross interaction occurs in simultaneous treatment of cyclosporine and phototherapy, resulting in increased carcinogenic risk (Singer and Berneburg, 2018; Alhasaniah et al., 2018). The other limitations include expense, limited efficacy, logistic and special equipments, trained staff, etc.

TABLE 2.4 Side effects of phototherapy.

Common	Uncommon	Uncommon
Actinic damage	Nonmelanoma skin cancer	Folliculitis
Increased skin aging	Melanoma with PUVA	Folic acid depletion
Local erythema and tenderness	Lentigines	Photo-onycholysis (PUVA)
Pruritus, burning, stinging	Photosensitive eruptions	Herpes simplex virus reactivation
Depigmentation	Polymorphous light eruption	Facial hypertrichosis
Heat-induced flares	Lupus flare	Cataract formation

2.17.6 Side effects of phototherapy

The side effects associated with heliotherapy are photosensitivity, skin ageing, skin cancers and sunburns. Exposure to UVB in addition to immunosuppression, skin cancers and photosensitive diseases is erythemogenic and can lead to DNA damage (Singer and Berneburg, 2018). Long-term treatment leads to photo aging, induction of cutaneous malignancies (Ortiz-Salvador and Pérez-Ferriols, 2017). PUVA leads phototoxic effects, and acute side effects like cataract and may develop non melanoma skin cancer and melanomas. Narrowband UV-B has proved superior although associated with undesirable effects such as itching, erythema, mild burning, and pain, which are tolerable and disappear spontaneously in some times after treatment. Long wave UVA (UVA1) cause photo ageing and carcinogenesis (Baron and Stevens, 2003). Any type of phototherapy may trigger photo induced dermatoses like solar urticarial. The other side effects are mentioned in Table 2.4 (Rodenbeck et al., 2016).

2.17.7 Recent development and future scope

During past few decades, new appliances emitting more specific spectra has increased the ability of therapeutics in treatment of photosensitive dermatoses. Recently, use of UVA, UVB, and PUVA therapies for variety of diseases has increased armamentarium against skin diseases (Petrazzuoli, 2000). The narrow band UVB 311 nm and UVA1 along with tropical medicines have led a revolution in phototherapy. The combination therapies, including vitamin D analogues, glucocorticoids, anthralin, etc., counted great success in the treatment. Currently, not only the methods are being improved but the spectrum of treatment has been broadened. The diseases associated with connective tissue for instance scleroderma can be managed with these new modalities. The therapeutic range, cost effective, quick and simple implication has made phototherapy indispensable in dermatological treatment. In future, the phototherapy will remain mainspring for management of these inexplicable diseases. The developments in this field are incessant. The newer evaluation techniques, modalities, combinations, broad spectrum, and safety profiles are progressing day by day and finding beneficial in wide range of diseases (Esmat and Hegazy, 2017).

2.18 Concluding remark

Heliotherapy, phototherapy plays a vital role in management of various disorders and diseases. In recent years, the efficacy of therapy has increased with the use of novel approaches of combining radiations with targeted molecule based treatment leaving behind its promising effects.

Abbreviations

ABL	Antifungal blue light
AD	Atopic dermatitis
B.C.	Before Christ
CTCL	Cutaneous T cell lymphoma
DNA	Deoxyribonucleic acid
ESR	Erythrocyte sedimentation rate
HIV	Human immunodeficiency virus

HSV-1	Herpes simplex virus 1
HSV-2	Herpes simplex virus 2
IFN-γ	Interferon γ
IgG	Immunoglobulin G
IL	Interleukin
IL36RN	Interlukin 36 receptor antagonist
MED	Minimal erythema dose
MF	Mycosis fungoides
MRSA	Methicillin-resistant Staphylococcus aureus
NBUVB	Narrow band UVB
PUVA	UVA plus 8-methoxypsoralens
UVB	Ultraviolet B
UVA	Ultraviolet A
UVA-1	Long wavelength ultraviolet A1
UVA-2	Long wavelength ultraviolet A2
UVC	Ultraviolet C

References

Abdo, J.M., Sopko, N.A., Milner, S.M., 2020. The applied anatomy of human skin: a model for regeneration. Wound Med. 28, 100179. https://doi.org/10.1016/j.wndm.2020.100179.

Alhasaniah, A., Sherratt, M.J., O'Neill, C.A., 2018. The impact of ultraviolet radiation on barrier function in human skin: molecular mechanisms and topical therapeutics. Curr. Med. Chem. 25 (40), 5503–5511. https://doi.org/10.2174/0929867324666171106164916.

Bae, J.M., Jung, H.M., Hong, B.Y., Lee, J.H., Choi, W.J., Lee, J.H., Kim, G.M., 2017. Phototherapy for vitiligo: a systematic review and meta-analysis. JAMA Dermatol. 153 (7), 666–674. https://doi.org/10.1001/jamadermatol.2017.0002.

Baron, E.D., Stevens, S.R., 2003. Phototherapy for cutaneous T-cell lymphoma. Dermatol. Ther. 16 (4), 303–310. https://doi.org/10.1111/j.1396-0296.2003.01642.x.

Bergfelt, D.R., 2009. Anatomy and physiology of the mare. Equine Breeding Management and Artificial Insemination. W.B. Sanders Company, Philadelphia, USA, 113–131. https://doi.org/10.1016/B978-1-4160-5234-0.00011-8.

Caetano, K.S., Frade, M.A.C., Minatel, D.G., Santana, L.A., Enwemeka, C.S., 2009. Phototherapy improves healing of chronic venous ulcers. Photomed. Laser Surg. 27 (1), 111–118. https://doi.org/10.1089/pho.2008.2398.

Carter, J., Zug, K.A., 2009. Phototherapy for cutaneous T-cell lymphoma: online survey and literature review. J. Am. Acad. Dermatol. 60 (1), 39–50. https://doi.org/10.1016/j.jaad.2008.08.043.

Ceglia, L., Toni, R., 2017. Vitamin D and muscle performance in athletes. Vitamin D (Fourth Edition, Vol. 2). Elsevier, Amsterdam, Netherlands. https://doi.org/10.1016/B978-0-12-809963-6.00113-9.

Chaudhry, I.A., Shamsi, F.A., Elzaridi, E., Al-Rashed, W., Al-Amri, A., Al-Anezi, F., Arat, Y.O., Holck, D.E., 2007. Outcome of treated orbital cellulitis in a tertiary eye care center in the middle East. Ophthalmology 114 (2), 345–354. https://doi.org/10.1016/j.ophtha.2006.07.059.

Clebak, K.T., Malone, M.A., 2018. Skin Infections. Primary Care – Clinics in Office Practice 45 (3), 433–454. https://doi.org/10.1016/j.pop.2018.05.004.

Cole, C., Gazewood, J., 2007. Diagnosis and treatment of impetigo. Am. Fam. Physician 75 (6), 859–864.

Dai, T., Tegos, G.P., Rolz-Cruz, G., Cumbie, W.E., Hamblin, M.R., 2008. Ultraviolet C inactivation of dermatophytes: implications for treatment of onychomycosis. Br. J. Dermatol. 158 (6), 1239–1246. https://doi.org/10.1111/j.1365-2133.2008.08549.x.

Esmat, S., Hegazy, R.A., 2017. Phototherapy and combination therapies for vitiligo. Dermatol. Clin. 35 (2), 171–192. https://doi.org/10.1016/j.det.2016.11.008.

Fitzpatrick, T.B., 1997. Mechanisms of phototherapy of vitiligo. Arch. Dermatol. 133 (12), 1591–1592. https://doi.org/10.1001/archderm.1997.03890480113020.

Gelman, A.B., Norton, S.A., Valdes-Rodriguez, R., Yosipovitch, G., 2015. A review of skin conditions in modern warfare and peacekeeping operations. Mil. Med. 180 (1), 32–37. https://doi.org/10.7205/MILMED-D-14-00240.

Grimes, P.E., Hamzavi, I., Lebwohl, M., Ortonne, J.P., Lim, H.W., 2013. The efficacy of afamelanotide and narrowband UV-B phototherapy for repigmentation of vitiligo. JAMA Dermatol. 149 (1), 68–73. https://doi.org/10.1001/2013.jamadermatol.386.

Guffey, J.S., Wilborn, J., 2007. In Vitro Bactericidal Effects of 405-nm and 470-nm Blue Light. Photomed. Laser Surg. 24, 684–688. https://doi.org/10.1089/pho.2006.24.684.

Gupta, N., Kangri, G., View, W.M., 2016. Ancient light therapies: a boon to medical science 82, 231–236.

Hassani, J., Feldman, S.R., 2016. Phototherapy in scleroderma. Dermatol. Ther. 6 (4), 519–553. https://doi.org/10.1007/s13555-016-0136-3.

Hauser, A., Fogarasi, S., 2010. Periorbital and orbital cellulitis. Pediatr. Rev. 31 (6), 242–249. https://doi.org/10.1542/pir.31-6-242.

Hay, R.J., Johns, N.E., Williams, H.C., Bolliger, I.W., Dellavalle, R.P., Margolis, D.J., Marks, R., Naldi, L., Weinstock, M.A., Wulf, S.K., Michaud, C., J.l. Murray, C., Naghavi, M., 2014. The global burden of skin disease in 2010: an analysis of the prevalence and impact of skin conditions. J. Invest. Dermatol. 134 (6), 1527–1534. https://doi.org/10.1038/jid.2013.446.

Hay, R.J., Steer, A.C., Chosidow, O., Currie, B.J., 2013. Scabies: a suitable case for a global control initiative. Curr. Opin. Infect. Dis. 26 (2). https://journals.lww.com/co-infectiousdiseases/Fulltext/2013/04000/Scabies___a_suitable_case_for_a_global_control.2.aspx.

Hönigsmann, H., 2001. Phototherapy for psoriasis. Clin. Exp. Dermatol. 26 (4), 343–350. https://doi.org/10.1046/j.1365-2230.2001.00828.x.

Ibbotson, S.H., 2018. A perspective on the use of NB-UVB phototherapy vs. PUVA photochemotherapy. Front. Med. 5, 184. https://doi.org/10.3389/fmed.2018.00184.

Kapur, S., Watson, W., Carr, S., 2018. Atopic dermatitis. Allergy Asthma Clin. Immunol 14 (Suppl 2), 52. https://doi.org/10.1186/s13223-018-0281-6.

Karimkhani, C., Dellavalle, R.P., Coffeng, L.E., Flohr, C., Hay, R.J., Langan, S.M., Nsoesie, E.O., Ferrari, A.J., Erskine, H.E., Silverberg, J.I., Vos, T., Naghavi, M., 2017. Global skin disease morbidity and mortality an update from the global burden of disease study 2013. JAMA Dermatol. 153 (5), 406–412. https://doi.org/10.1001/jamadermatol.2016.5538.

Kauffman, C.A., Bustamante, B., Chapman, S.W., Pappas, P.G., 2007. Clinical practice guidelines for the management of sporotrichosis: 2007 update by the Infectious Diseases Society of America. Clin. Infect Dis. 45 (10), 1255–1265. https://doi.org/10.1086/522765.

Kaufmann, S.H.E., 2006. Tuberculosis: back on the immunologists' agenda. Immunity 24 (4), 351–357. https://doi.org/10.1016/j.immuni.2006.04.003.

Klein, A., Schiffner, R., Schiffner-Rohe, J., Einsele-Krämer, B., Heinlin, J., Stolz, W., Landthaler, M., 2011. A randomized clinical trial in psoriasis: synchronous balneophototherapy with bathing in Dead Sea salt solution plus narrowband UVB vs. narrowband UVB alone (TOMESA-study group). J. Eur. Acad. Dermatol. Venereol. 25 (5), 570–578. https://doi.org/10.1111/j.1468-3083.2010.03840.x.

Krutmann, J., Morita, A., 1999. Mechanisms of ultraviolet (UV) B and UVA phototherapy. J. Investig. Dermatol. Symp. Proc. 4 (1), 70–72. https://doi.org/10.1038/sj.jidsp.5640185.

Lawton, S., 2019. Skin 1: the structure and functions of the skin. Nurs. Times 115 (12), 30–33.

McGrath, H., 2017. Ultraviolet-A1 irradiation therapy for systemic lupus erythematosus. Lupus 26 (12), 1239–1251. https://doi.org/10.1177/0961203317707064.

Meola, T., Soter, N.A., Ostreicher, R., Sanchez, M., Moy, J.A., 1993. The safety of UVB phototherapy in patients with HIV infection. J. Am. Acad. Dermatol. 29 (2, Part 1), 216–220. https://doi.org/10.1016/0190-9622(93)70171-O.

Minatel, D.G., Frade, M.A.C., França, S.C., Enwemeka, C.S., 2009. Phototherapy promotes healing of chronic diabetic leg ulcers that failed to respond to other therapies. Lasers Surg. Med. 41 (6), 433–441. https://doi.org/10.1002/lsm.20789.

Mysore, V., Shashikumar, B.M., 2016. Targeted phototherapy. Indian J. Dermatol. Venereol. Leprol. 82 (1), 1–6. https://doi.org/10.4103/0378-6323.172902.

Nair, P.M.K., 2015. Effectiveness of naturopathic interventions on reducing the erythrocyte sedimentation rate in patients with chronic inflammatory disorders. INT J. Innov. Res. Develop. 4 (8), 183–186.

Nair, P.M.K., Salwa, H., 2018. Naturopathy lifestyle interventions in boosting immune responses in HIV-positive population Immunity and Inflammation in Health and Disease. Elsevier Inc. https://doi.org/10.1016/B978-0-12-805417-8.00031-7.

Namazi, N., Dadkhahfar, S., 2018. Impetigo herpetiformis: review of pathogenesis, complication, and treatment. Dermatol. Res. Pract. 2018, 5801280. https://doi.org/10.1155/2018/5801280.

Nee, T.S., 1997. Phototherapy. Clinics in Dermatology 15 (5), 753–767. https://doi.org/10.1016/S0738-081X(97)00016-3.

Ortiz-Salvador, J.M., Pérez-Ferriols, A., 2017. Phototherapy in atopic dermatitis. Adv. Exp. Med. Biol. 996, 279–286. https://doi.org/10.1007/978-3-319-56017-5_23.

Papageorgiou, P., Katsambas, A., Chu, A.C., 2000. Phototherapy with blue (415 nm) and red (660 nm) light in the treatment of acne vulgaris. Br. J. Dermatol. 142 (5), 973–978. https://doi.org/10.1046/j.1365-2133.2000.03481.x.

Parada, J.P., Maslow, J.N., 2000. Clinical syndromes associated with adult pneumococcal cellulitis. Scand. J. Infect. Dis. 32 (2), 133–136. https://doi.org/10.1080/003655400750045213.

Patrizi, A., Raone, B., Ravaioli, G.M., 2015. Management of atopic dermatitis: safety and efficacy of phototherapy. Clin. Cosmet. Investig. Dermatol. 8, 511–520. https://doi.org/10.2147/CCID.S87987.

Petrazzuoli, M., 2000. Advances in phototherapy. Curr. Probl. Dermatol. 12 (6), 282–286. https://doi.org/10.1016/S1040-0486(00)90026-7.

Piotrowska, A., Wierzbicka, J., Zmijewski, M.A., 2016. Vitamin D in the skin physiology and pathology. Acta Biochim. Pol. 63 (1), 17–29. https://doi.org/10.18388/abp.2015_1104.

Powell, F.C., Su, D., W., P., Perry, H.O., 1996. Pyoderma gangrenosum: classification and management. J. Am. Acad. Dermatol. 34 (3), 395–409. https://doi.org/10.1016/S0190-9622(96)90428-4.

Ramsay, D.L., Lish, K.M., Yalowitz, C.B., Soter, N.A., 1992. Ultraviolet-B phototherapy for early-stage cutaneous T-cell lymphoma. Arch. Dermatol. 128 (7), 931–933. https://doi.org/10.1001/archderm.1992.01680170063007.

Rodenbeck, D.L., Silverberg, J.I., Silverberg, N.B., 2016. Phototherapy for atopic dermatitis. Clin. Dermatol. 34 (5), 607–613. https://doi.org/10.1016/j.clindermatol.2016.05.011.

Rodvold, K.A., McConeghy, K.W., 2014. Methicillin-resistant Staphylococcus aureus therapy: past, present, and future. Clin. Infect. Dis. 58 (Suppl 1), S20–S27. https://doi.org/10.1093/cid/cit614.

Sami, N.A., Attia, A.T., Badawi, A.M., 2008. Phototherapy in the treatment of acne vulgaris. J. Drug. Dermatol. 7 (7), 627–632.

Selvaraj, P., 2011. Vitamin D, vitamin D receptor, and cathelicidin in the treatment of tuberculosis. In: Vitamins and the Immune System 86. Elsevier Inc, Amsterdam, Netherlands. https://doi.org/10.1016/B978-0-12-386960-9.00013-7.

Seth, D., Cheldize, K., Brown, D., Freeman, E.F., 2017. Global burden of skin disease: inequities and innovations. Curr. Dermatol. Rep. 6 (3), 204–210. https://doi.org/10.1007/s13671-017-0192-7.

Singer, S., Berneburg, M., 2018. Phototherapy. J. Ger. Soc. Dermatol. 16 (9), 1120–1131. https://doi.org/10.1111/ddg.13646.

Stevens, D.L., Bisno, A.L., Chambers, H.F., Dellinger, E.P., Goldstein, E.J.C., Gorbach, S.L., Hirschmann, J.V, Kaplan, S.L., Montoya, J.G., Wade, J.C., 2014. Practice guidelines for the diagnosis and management of skin and soft tissue infections: 2014 update by the Infectious Diseases Society of America. Clin. Infect. Dis. 59 (2), e10–e52. https://doi.org/10.1093/cid/ciu444.

Tabassum, N., Hamdani, M., 2014. Plants used to treat skin diseases. Pharmacogn. Rev. 8 (15), 52–60. https://doi.org/10.4103/0973-7847.125531.

Theoret, C.L., Stashak, T.S., 2014. Integumentary system. Equine Emergencies. John Wiley & Sons, Hoboken, New Jersey, 238–267. https://doi.org/10.1016/b978-1-4557-0892-5.00019-2.

van Etten, E., Mathieu, C., 2005. Immunoregulation by 1,25-dihydroxyvitamin D3: basic concepts. J. Steroid Biochem. Mol. Biol. 97 (1–2), 93–101. https://doi.org/10.1016/j.jsbmb.2005.06.002.

Vila, T.V.M., Rozental, S., de Sá Guimarães, C.M.D., 2015. A new model of in vitro fungal biofilms formed on human nail fragments allows reliable testing of laser and light therapies against onychomycosis. Lasers Med. Sci. 30 (3), 1031–1039. https://doi.org/10.1007/s10103-014-1689-y.

Vreman, H.J., Wong, R.J., Stevenson, D.K., 2004. Phototherapy: current methods and future directions. Semin. Perinatol. 28 (5), 326–333. https://doi.org/10.1053/j.semperi.2004.09.003.

Wang, T., Dong, J., Yin, H., Zhang, G., 2020. Blue light therapy to treat candida vaginitis with comparisons of three wavelengths: an in vitro study. Lasers Med. Sci. 35 (6), 1329–1339. https://doi.org/10.1007/s10103-019-02928-9.

Weischer, M., Blum, A., Eberhard, F., Röcken, M., Berneburg, M., 2004. No evidence for increased skin cancer risk in psoriasis patients treated with broadband or narrowband UVB phototherapy: a first retrospective study. Acta Derm. Venereol. 84 (5), 370–374. https://doi.org/10.1080/00015550410026948.

Yadav, V., Jayalakshmi, S., Singla, R.K., 2012. Traditional Systems of Medicine-Now & Forever. WebmedCentral 3, WMC003299.

Young, Hee Choi, A.-M., Y., 2019. 乳鼠心肌提取 HHS Public Access. Physiology & Behavior. Lasers Med. Sci. 176 (3), 139–148. https://doi.org/10.1034/j.1600-0625.2002.00112.x.What.

Yu, J.J., Zhang, C.S., Zhang, A.L., May, B., Xue, C.C., Lu, C., 2013. Add-on effect of chinese herbal medicine bath to phototherapy for psoriasis vulgaris: A systematic review. Evid. Based Complement. Alternat. Med. 2013, 1–14. https://doi.org/10.1155/2013/673078.

Chapter 3

Phototherapy: The novel emerging treatment for cancer

Sagar Trivedi[a], Nishant Awandekar[b], Milind Umekar[b], Veena Belgamwar[a], Nishikant A. Raut[a]

[a]Department of Pharmaceutical Sciences, Rashtrasant Tukadoji Maharaj Nagpur University, Nagpur, Maharashtra, India
[b]Smt. Kishoritai Bhoyar College of Pharmacy, Kamptee, Nagpur, India

3.1 Introduction

Even after decades of studies and exploration for prevention, remedy and therapy, cancers nevertheless stay one of the pinnacle reasons of mortality and burden of ailment worldwide (Wang et al., 2016). Every year, thousands of people are diagnosed with different types of cancers around the world and most of them died even after the treatment. In several countries, cancer rank the second cause of death after cardiovascular diseases (Patrice et al., 2003). The extensive development in remedy and prevention for cardiovascular diseases, cancer is emerging as the primary killer in several parts of world. Currently, existing remedies for cancers in clinics are chiefly comprised of chemotherapy, surgical procedures and radiotherapy, or combination of any two or three (Liou and Storz, 2010). A range of newly emerging advanced healing modalities are selectively and specifically used along with gene therapy, targeted drug delivery systems, immunotherapy, phototherapy (photothermal and photodynamic), magnetic hyperthermia, and other nonmainstream treatments (Tapeinos and Pandit, 2016). Most of the modern treatment modalities include destruction of tumor cells through distinctive features of extrinsic stimuli with chemotherapeutic agents, or certain highly energetic radiations, or external high-frequency magnetic field in thermotherapy and external light energy sources in phototherapy (Fan et al., 2016; Zhou et al., 2016). Such external techniques for the management of cancers resulted in enhanced antitumor effects and helped in successful removal of primary solid tumors and possibly can control the recurrence too.

Various cancer therapies usually involve several adverse effects experienced by the individuals however, in case of phototherapy, which uses nonionizing light as driving energy for killing carcinomas offers less side effects and lowest systemic lethal effects (Patrice et al., 2003; Zhao et al., 2016). The two major components of phototherapy include photothermal therapy (PTT) and photodynamic therapy (PDT), which basically destructs the tumor cells by heat or reactive oxygen species when exposed under irradiation of light. The light radiation ranging from 400 nm to 700 nm is generally employed for treatment and inhibition of tumors. PTT and PDT efficiently kills the cancerous cells when light of appropriate frequency and wavelength falls on cell in presence of sufficient photosensitizing agents (PAs) (Ackroyd et al., 2007; Real et al., 2002). A PAs enables the absorption of light and further conversion of light into excessive reactive oxygen species (ROS) or hyperthermia and causes damage to cell, ultimately leading to apoptosis (Jiayuan et al., 2017; Pavani et al., 2012). The basic nature of PAs should give a higher phototherapeutic result and an ideal photosensitizing agent ought to have proficient phototherapeutic result with less or no adverse effects requiring prevalent photosensitization and high explicitness upon cancer cells (Bacellar et al., 2015; Daniell and Hill, 1991; dos Santos et al., 2019). To enhance photosensitization, various nanomaterials have been created by consolidation of photothermal PAs and a photodynamic PS for synchronic PDT and PTT treatment (Agostinis et al., 2004).

The outcomes of PDT can be enhanced by chemically conjugating the PAs with the target ligands, which are capable of attaching to the biomarkers which are overexpressed and found abundantly on the malignant cell's membrane, the dual targeting phototherapy and the newer PAs have enhanced the results of cancer therapy (Banerjee et al., 2017; Delaey et al., 2000). Phototherapy is considered as a harmless and non-invasive technique for management of cancer in which light waves of stipulated wavelengths are used for activation of PAs, these light waves are capable of reaching directly to tumor cells irrespective of their type and location (Tsubone et al., 2017). The contemporary clinical usage of phototherapy in contradiction of cancer in the form of photodynamic therapy (PDT) was initially started by Roswell Park Cancer Institute during the 1970s. The irradiation makes the cancer therapy comparatively less invasive and possibly controls the invasion of cancer cells (Bacellar et al., 2018; Daniell and Hill, 1991; Tsubone et al., 2017). Phototherapy finds its application in management of carcinomas and early diagnosis of tumors demonstrating excellent results in targeting various cancers. This is mainly achieved by releasing the diagnostic and therapeutic agents on the site on action, secondly under the influence of tissue immunological reaction or by supplementing and synergizing immunotherapy (Kwiatkowski et al., 2018; Spring et al., 2015).

PDT induces immunological responses by acting as an immunological agent and causes stimulus of innate immune system. The photochemical and photothermal interactions are responsible for killing of tumor cells by releasing tumor antigens (Anand et al., 2012; Feng et al., 2018). The synergistic effects of Photo-immunotherapy increase the opportunity to expand clinical outcomes. The advantages like lower invasiveness of the applied therapy, precision in cancer targeting can be achieved by dual solicitation of light and photochemotherapeutic agents, the recent developments in the light sources and PAs have proven promising. Similarly, the amalgamation of nanomedical agents and phototherapy modalities has also demonstrated improvement in consequences than discrete therapies (Agostinis et al., 2011). Hence the future research must be focused on various novel combination of PAs and phototherapy for improving the outcome and minimizing the damage to healthy cells and tissues. This chapter deals with latest advances made in methods of treating cancer with phototherapy as a successful anticancer tool.

3.2 Photophysics and photochemistry

The underlying molecular mechanism of PDT relies on interaction of its three basic nontoxic components including the photosensitizers, the light of suitable wavelength and the amount of oxygen dissolved in the cells for generating an effective antitumor effect (Kwiatkowski et al., 2018). The PDT and PTT mainly works on two important mechanism of the photodynamic response and both are diligently reliant on oxygen molecules present inside the cells (Ogilby, 2010; Robertson et al., 2009). The basic mechanism of both the types is parallel in nature, in which a PAs enters inside the cell subsequently activated by light of specific wavelength, and further converted to excited singlet state S1 from basic energy state, that is, S° by overlapping with the absorption spectrum of PAs, during the process of photons absorption. Emission of fragments is seen in form of a quantum of fluorescence and the residual energy (Verma et al., 2007; Yu et al., 2021) causes the PAs molecules to get excited to the triplet state T1 and leading to a suitable, therapeutic understanding of the compound (Shanshan et al., 2021; Xiao et al., 2020).

Many of the PAs in their ground (i.e., singlet) state have two electrons and are arranged in such a manner that they possess opposite spins in an active molecular orbital, the impact of light causes transmission of 1 electron to higher energy orbital (Fig. 3.1) and the excited PAs become unbalanced emitting extra energy in the form of fluorescence and/or heat (Ogilby, 2010). The excited PAs undergo an intersystem overpass to generate a new stable triplet state with reversed spin of one electron (Xiao et al., 2020), the triplet state excited molecule transfers its energy to the ground state or to the molecular oxygen (O_2). This causes generation of $^1O_2$ and commonly termed as type II reaction or mechanism. The underlying mechanism of type I process happens in similar manner in which the PAs responds through the organic molecules present in the microenvironment of cells and further coverts to a radical by attaining a hydrogen atom or electron (Algorri et al., 2021; Oleinick and Evans, 1998).

The superoxide anion radical (O_2) are formed by auto-oxidation of the condensed PAs and mutation or the rejection of one electron of O_2 gives hydrogen peroxide (H_2O_2), which endures one electron and causing reduction to practically unselective oxidant hydroxyl radical (HO). Type II mechanism is more responsible for generation of reactive oxygen species (ROS) when systematically compared with type 1 reactions and the mechanism of working for most of the PAs occurs by type II mechanism than type I mechanism (Dima et al., 2002).

3.2.1 Type I mechanism of photodynamic reaction

The biomolecules gain energy from the excited triplet state T1 PAs with the help of various environmental factors and the substrates of carcinogenic tissue initiates the movement of a single hydrogen molecule which further causes generation of free radicals and anions of photosensitizer and the substrate (Dima et al., 2002; Kwiatkowski et al., 2018). The electrons start forming conjugations with the oxygen particles which assists them to stay in the elementary energetic state and the cycle continues and promotes the formation reactive oxygen species (ROS); initially as superoxide anion radicals ($O_2 \bullet-$) and later on the formation of ROS occurs inside the cells, which causes the oxidative pressure to bring about apoptosis or similar damage to the malignant cells (Dima et al., 1998; Verma et al., 2007).

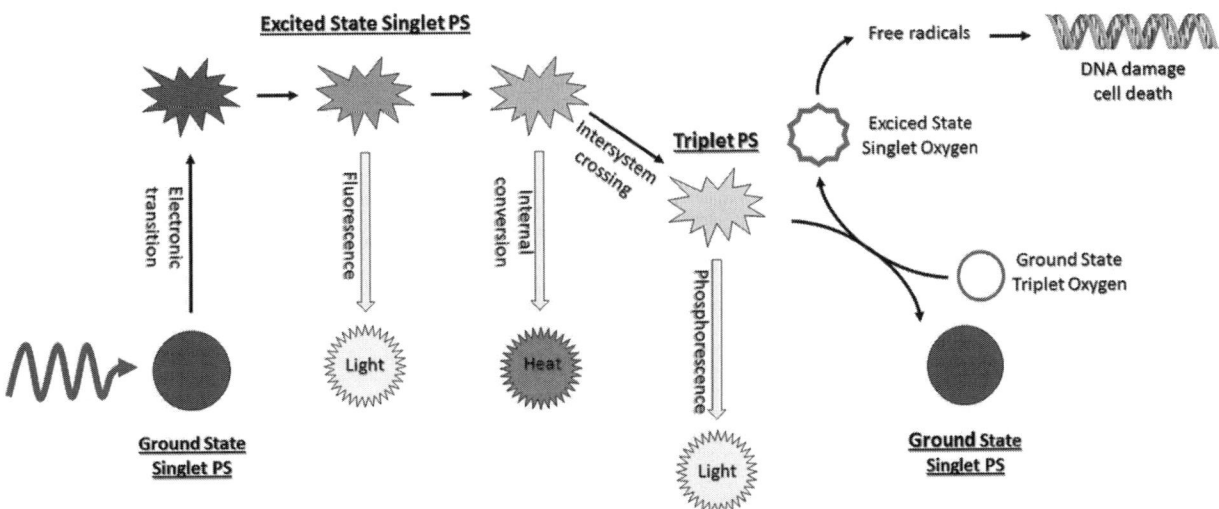

FIG. 3.1 Mechanism of the photodynamic reaction. *From Robertson et al. (2009).*

3.2.2 Type II mechanism of photodynamic reaction

The oxygen molecule which is in basic energy state receives the energy from excited triplet state PAs, this continuous transfer of energy from PAs → O2 confirms the identical rotations of the molecules. The excited oxygen molecules are spawned and mainly classified by remarkably tough oxidizing properties (Algorri et al., 2021; Yu et al., 2021). Among the various biologically active compounds, usually exist in basic singlet state, while the oxygen molecules exhibit triplet state as the basic and the singlet state as the excited state (Allison, 2014). Due to this nature of excited PAs, they possibly do not harm the cell structures and respond solitary with oxygen molecules thawed in the cytoplasm (Koo Lee and Kopelman, 2011; Sundaram and Abrahamse, 2020). It is anticipated that the process of type II reaction plays an important role in acclimatizing the effectiveness of PDT and these reactions are influenced by several factors such as the concentration of oxygen, pH and dielectric constant of tissues and PAs, while the oxygen becomes the first molecule to give positive outcome through the said mechanism (Kato, 1996).

This highly reactive species of oxygen serves as underlying mechanism for substantial damage to the proteins, fats and various additional molecules at photosensitized site and results in death of tumor cells and by the process of apoptosis and or necrosis (Nowis et al., 2005). The intracellular engulfing of PAs facilitates the cell death and the injury to mitochondria resulted in apoptosis, damage to cell membrane, lysosomes or endoplasmic reticulum, disturbance in the integrity of cellular structures and promotion of necrosis that can exaggerate autophagy (Koo Lee and Kopelman, 2011; Oliveira et al., 2020).

3.3 Photodynamic targets at the molecular level

The presence of various biomolecules inside the cell serves as substrate for oxidation of $^1O_2$ and numerous similar reactive species are generated during action of PDT. The factors affecting these reactions are mainly the space between PAs and the site of action, the quantity of targets available for action and finally the rate constants for specific reactions (You et al., 2003). The presence of singlet oxygen species plays a vital role in damaging the tumor cells through PDT and concurrently PAs work by the mechanism of Type I reactions serving as a factor to enhance the efficacy of PDT. The PDT stimulates the production of complex of $^1O_2$ and some radical species rarely involving the type I and type II reactions. Some recent investigations suggest that combination of type I and type II reactions along with the species generated can serve as synergistic effect and possibly enhancing the outcome (Dabrowski et al., 2011; Dąbrowski et al., 2007).

3.3.1 Proteins

The most abundant biomolecule present in cells are proteins having high-rate constants for reaction with $^1O_2$ and other reactive oxygen species (ROS), which qualifies them as an ideal target for PDT. The presence of different types of amino acid possesses different types of rate constants causing selective damage, especially to the amino acid side chains (Dabrowski et al., 2010). These reactions are very complex in nature and various substrates are produced during this process and they also serve as initial sites for oxidative modifications of proteins. The different amino acids including cysteine, methionine, tyrosine, histidine, and tryptophan reacts with $^1O_2$ with a reaction rate of $>10^7$ M^{-1} s^{-1} under physiological conditions. Among these amino acids, cysteine and methionine are primarily oxidized to sulfoxides and further through a thermic reaction histidine produces endoperoxide and by similar mechanism tryptophan yields N-formyl kynurenine and tyrosine undergo phenolic oxidative coupling (Davies, 2004; Dogra and Kim, 2020). It was observed that protein sulfhydryl groups are susceptible to photo oxidation when exposed to pH range of 7.0–8.5 and further followed by histidine, tryptophan, and tyrosine. When the hydrogen atoms are removed from the residual cysteine it leads to generation of a thiyl radical which is capable of cross linking to second thiyl radical and binds via disulfide bonds and at the diffusion-controlled rates the hydroxyl radical reacts with amino acids (Davies, 2005).

The destruction of the tumor cells is governed by the amount of hydroxyl ions (HO) present at the site of action and the reactivity of initially produced conjugates. The hydroxyl radicals are formed by tyrosine by the oxidation process and the rate of reaction is diffusion limited and the tyrosyl radical reacts with O_2 which further causes formation of tyrosine hydroperoxide (Ehrenshaft et al., 2015; Headlam et al., 2004; Höhn et al., 2013). The protein binding of amino acids to PAs, crosslinking of proteins to two tyrosine units by coupling reaction are some alterations occurring to proteins other than photo-oxidation. The coupling reaction is secondary reaction which possibly yields amino acid residues and additional groups in the protein (Benov et al., 2010; Mutairi et al., 2007).

This alternation of proteins is not dependable only on the type but also on the location of susceptible amino acid residues in the creased protein and therefore increases the selectivity of PDT and improves the efficiency. This further facilitates the localization of PAs to the target tumor sites (Atlante et al., 1989) and possibly some photo-induced variations cause the proteins to eliminate the catalytic process (Mutairi et al., 2007) as well as the cell signaling process and ultimately leads to cell death (Boyle and Dolphin, 1996).

3.3.2 Photodynamic therapy-induced lipid peroxidation

When it comes to abundancy the lipids are comparatively low in amount when compared to proteins but simultaneously the biomembranes contain a higher amount of unsaturated fatty acids which tends to make membranous organelles and plasma membrane as a primary target for photo-generated $^1O_2$ and other ROS (Davies, 2004). The rate of reaction of these unsaturated fats with the singlet oxygen occurs in the series of $0.74^{-2.4} \times 10^5$ M^{-1} s^{-1} and mainly depends on degree of unsaturation of fatty acids (Davies, 2005). The enhanced solubility of oxygen in lipids causes the PDT-mediated lipid damage, during the excited state the PAs in the lipid environment has increased chance of reacting with O_2 and forms $^1O_2$ and oxygen-derived radicals. Other than the radicals formed which leads to hydrogen deliberation by initiating lipid peroxidation and $^1O_2$ binds to unsaturated fats and leads to formation of lipid peroxides (Headlam et al., 2004; Höhn et al., 2013).

The presence of various transition metals causes the local deterioration of lipid peroxides and generating alkoxyl and peroxyl radicals (Ehrenshaft et al., 2015), the radicals formed initially induce the free-radical chain responses and possibly destroys the biomembranes and improves the outcome (Benov et al., 2010; Mutairi, 2007). Therefore, the lipid peroxidation is not limited to immediate impairment of lipids and capable of incorporating additional modifications to proteins and polynucleotides (Atlante et al., 1989), as well as influencing membrane capacities (Mutairi et al., 2007). Lipid peroxidation can prompt adjustments in metabolism and cell flagging, coming about eventually in cell apoptosis (Boyle and Dolphin, 1996).

3.3.3 Photosensitized modification of nucleic acids

The damage to DNA produced by oxidative reactions is among the major reason for cell killing via PDT, it can possibly be mediated with the help of 1 electron oxidation $O_2^- \cdot /HO \cdot$ and $^1O_2$ (Cadet et al., 2010). This type of damage to the tumor cells occur when an excited PAs in the T1 state causes alterations to DNA base by eliminating an electron/hydrogen atom (Kryston et al., 2011). The guanine among the various bases is specifically exposed as it possess low ionization potential (Magnander and Elmroth, 2012) and the guanine cation radical (G + ·) acts as quick substrate and is changed to 8-oxo-7,8-dihydro-2′-deoxyguanosine and 2,6-diamino-4-hydroxy-5-formamidopyrimidine by opposing one electron oxidation and decrease responses (Figueroa-González and Pérez-Plasencia, 2017) or is deprotonated to a profoundly receptive guanine radicals, Gua (– H)·. As the addition reactions occurs with $O_2^- \cdot$, Gua (– H) · is modified to 2,2,4-triamino-5(2H)-oxazolone (Pouget et al., 2000). This guanine cation radicals can respond to lysine, arginine, and serine in proteins via nucleophilic addition reaction and causing DNA-protein cross-linking (Kumar et al., 2020).

From the type I photo process the hydroxyl radicals are generally formed via ROS and further reaction with all the constituents of DNA occurs at diffusion-limited rates. The initial way by which the HO· induces base alterations is only possible by addition of double bonds (Rosenthal, 1991). This reaction mainly occurs when an hydrogen atoms via HO from the methyl gathering of thymine and the 2-amino gathering of guanine (Pouget et al., 2000) enters inside the cell. This reaction primarily from the DNA backbone starts initiating responses for damaging the DNA strand (Benov, 2015; Sies and Menck, 1992). When compared to an HO·, singlet oxygen (1 Δ g state) is more specific DNA modifier (Cadet, 2012) and its objective is conversion of guanine to 8-oxo-7,8-dihydro-2′-deoxyguanosine at rate of $k = 5 \times 10^6$ M^{-1} s^{-1} (Cadet et al., 2006) and during the PDT treatment a specific $^1O_2$ oxidation moiety (Dedon, 2008) that can be identified.

The alteration of the DNA backbone occurs as the $^1O_2$ doesn't respond to 2-deoxyribose and supplementary actions can be achieved by oxidation of 8-oxo-7,8-dihydro-2′-deoxyguanosine by $^1O_2$ (Cadet et al., 2012) DNA injury prompted by PDT can end in genotoxicity and mutagenicity (Uchida, 2007). The disablement on DNA incited by other anticancer therapy regimens, in any case, the degree of nucleic corrosive harm is viewed as a lot more modest (Epe, 1991) and there are no reports that PDT causes auxiliary tumors (Cooke et al., 2010).

3.4 Light source

The three basic necessities for management of malignancies through PDT includes the suitable light source of appropriate wavelength, sufficient amount of PAs and abundance of O_2 at the site of tumor growth (Huang et al., 2010). Among this the choice of light used plays a key role and is mainly selected based on homogeneous nature of light reaching at the target region for effective PDT. Several PAs have more retention especially in visible region of 400–700 nm. At present there are four different types of laser light sources available [argon-siphoned lasers, metal-fume siphoned lasers (Au- or Cu-fume lasers), strong state lasers (Nd:YAG lasers, Ho:YAG lasers, KTP:YAG/color lasers), and diode lasers] and three non-laser light sources [light, light-emitting diodes (LEDs), and daylight] have been applied in PDT (Daayana et al., 2011; Gaio et al., 2016).

Amongst these sources, many of them such as for example red light argon color laser, Nd:YAG laser, a red light (635 nm) LED light, a 570–670 nm frequency red light, green light (520 nm), 420-nm blue light-transmitting diode, and sunlight find clinical applications. However, the permeation of this light is very less i.e. near about 1–6 mm due to presence of various endogenous chromophores in natural tissues, like cytochromes and hemoglobin, as these can ingest visible light and other reasons include the light scattering, diffusing, and disorienting, occurring due to the assorted design of biological tissues collectively affecting the productivity of PDT (Yang et al., 2019). Consequently, the intensity of penetration is enhanced with the use of near infrared (NIR) light, X-beam, interstitial light, and interior light (Etcheverry et al., 2016; Sousa et al., 2021). On this premise, ongoing advancements in of PDT in light source will be examined exhaustively in this segment.

3.4.1 Near infrared (NIR) light

It is observed that NIR light of 700–1350 nm, is mainly classified into NIR I (700–1000 nm) and NIR II (1000–1350 nm) regions commonly termed as "optical window" for biological tissues. Long-frequency NIR light restricts the stages of tissue dissipating with the penetrability of 1 cm and this light is capable of entering the tumor cells and decreases the phototoxicity on solid tissues (Hornung et al., 2000; Kercher et al., 2020). A study conducted by Chang et al. (2020), investigated the effect of combination of porphyrinic metal-organic framework (MOFs) using nanosystem mediated PDT for upconversion nanoparticles (UCNPs). The formulation comprising of MOFs on Nd^{3+}-sharpened UCNPs to bring about Janus nanostructures. These nanostructures were proven to carry out conversion of low-energy NIR light to high-energy UV for activation of porphyrinic MOFs to generate ROS and the in-vivo and in-vitro analysis assures the positive impact of NIR-mediated PDT (Chang et al., 2020).

The synthesis of earth doped UCNPs with graphene quantum speck (GQD) to frame GQD-designed (UCNPs-GQD) was proposed by Zhang et al. (2018), in which UCNPs were excited by NIR and they emitted UV-vis light stimulating prominent $^1O_2$ generation of GQD for intensely satisfactory. Hence, this method was found to be effective in enhancing the penetration (Zhang et al., 2018). As the issue concerning about the low energy charges a study conducted by Gao et al. (2018) in NIR-initiated PDT helped in overcoming this drawback by ensuring high effectiveness by NIR-mediated PDT. The work includes complexing the isophthalic acid into the layered double hydroxides (LDHs) interlayer gallery, which framed an IPA/LDH nanohybrid Fig. 3.2B. The microenvironments interface was restricted by the LDH monolayers and concurrently 2 photon mediated formation of triplet excited $^1O_2$ quantum yield up to 0.74. Hence these benefits of IPA/LDH

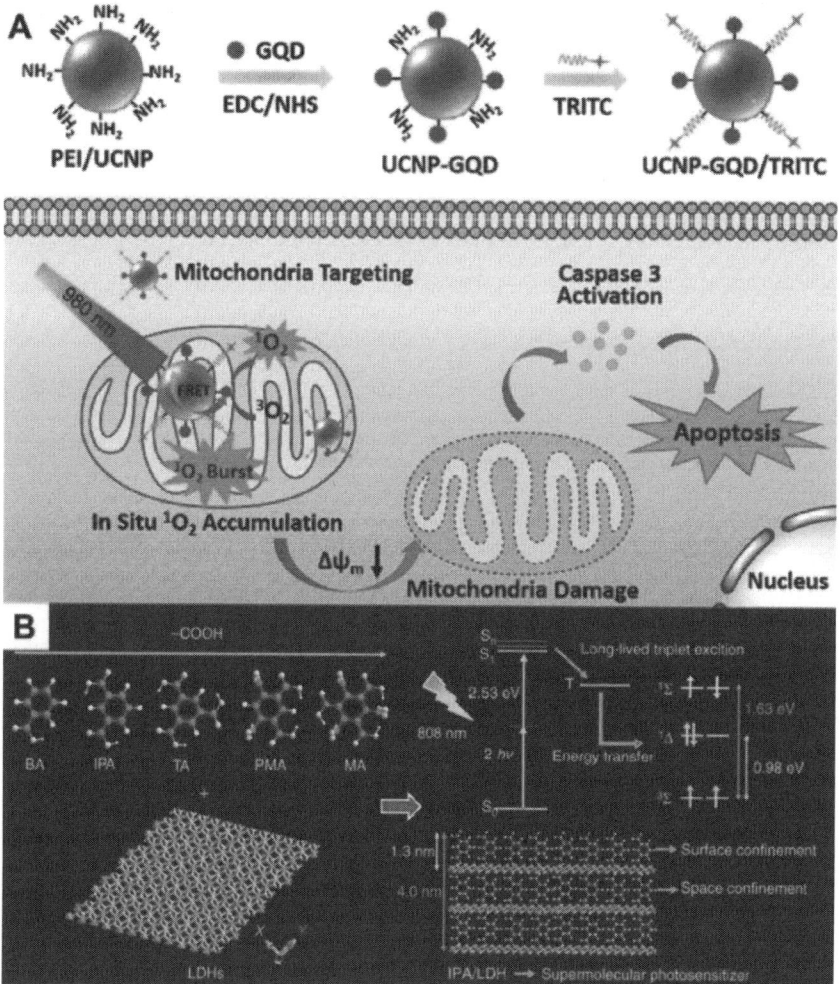

FIG. 3.2 Schematic diagram of UCNPs-GQD with mitochondria-targeting potency for high-efficient cells apoptosis upon laser irradiation. Reproduced from (B) Schematic illustration of IPA/LDH nanohybrids as two-photon PAs for 1O2 generation under 808 nm NIR laser. GQD, graphene quantum dot; IPA, isophthalic acid; LDH: layered double hydroxides; NIR, near-infrared; Pas, photosensitizers; UCNPs, upconversion nanoparticles; 1O2, singlet oxygen. *From Hu et al. (2021).*

nanohybrid like high penetration at 808 nm light causes to exhibit a highly effective anticancer activity with very low side effects. Hence RTP blends can be used as two-photon-mediated PAs for NIR-initiated PDT (Gao et al., 2018).

3.4.2 X-ray

The application of NIR light is profound for the management of carcinomas but the major drawback of penetrability about 10–15 mm, on the other hand deeply penetrating X-rays can be proved to be an excellent alternative for activation of PAs by enhancing the tumor penetration with clinical tumor imaging for treatment of deeply located tumors (Hu et al., 2021; Van Den Bergh, 1998). The major types of X-ray mediated PDT (X-PDT) includes metal-based X-PDT, uncommon earth-component based X-PDT, quantum dab (QD)-based X-PDT, silicon-based X-PDT (Brancaleon and Moseley, 2002; Yang et al., 2019). A work conducted by Sun et al. (2020), comprising of synthesis of PAs complex of (R-AIE-Au) in the presence of light of rose Bengal (RB)-aggregation mediated emission gold clustoluminogens (AIE-Au) for low-portion metal-based X-PDT. The Au molecules gets excited when illuminated with X-ray and further this energy is transferred to RB for PDT. The supporting in vitro and in vivo results showed that R-AIE-Au enables the generation of ROS and serves as a triggering factor with this X-PDT component and proves as powerful therapy of radioresistant malignant growths (Sun et al., 2020).

Other study involving the use of cerium (Ce)-doped intensely fluorescent NaCeF4:Gd,Tb scintillating nanoparticles (ScNPs) for management of carcinomas using X-PDT, among which Ce and Tb are able to retain their energy gained by X-ray and produce ROS for tumor damage (Zhong et al., 2019). QD-Photofrin was used as QD-based X-PDT QD (CdSe center with a ZnS shell) for excitation of PAs through fluorescence resonance energy move (FRET) and the photons reaching to the QD is directly corresponding to the radiation portion. The ability of FRET moving toward 100% when the photofrin/QD proportion came to 291:1. The in-vitro studies demonstrated that the blend of QD-Photofrin and X-ray, comparatively enhanced the killing of H460 cells as compared to X-ray alone. A study comprising of SiC/SiOx center/shell scintillating nanowire (ScNW) for formation of tetra (4-carboxyphenyl) porphyrin (H2TCPP) as PAs (SiC/SiOx-H2TCPP) for silicon-based X-PDT. This SiC/SiOx-H2TCPP could create $^1O_2$ when presented to low-portion X-ray illumination. Additionally, the illumination just took 20 s, which is more limited than the 40 s and 90 s utilized in typical clinical treatment (Yang et al., 2008).

3.4.3 Interstitial light

The tissues and cells which situated deeply are always challenging to treat, hence Interstitial PDT (I-PDT) is found to be most suitable light for treatment by delivering numerous diffusing strands implanted on the site of tumor target (Hu et al., 2021). A study involving use of Photofrin-intervened I-PDT for treatment of locally progressed head and neck malignancies (HNC). The results showed that when the tumor sites were exposed to I-PDT with Photofrin with light of 630 nm, enhanced the treatment outcomes when compared with light alone. The in-vivo study reveals that local stains of VX2 were observed in mouse bearing tumors with I-PDT at 16.5-398 mW/cm^2 and ≥45 J/cm^2 (Shafirstein et al., 2018), and this technique enables the tumor site targeting and spares the healthy tissues. Formulation of round and hollow diffuse optical strands (CDFs) was carried out by Ismael et al. (2020), as the light source (630–760 nm) and the CDFs have capability of radiating the light portion (20–50) J•cm^{-2} at the tumor site, and enabling a powerful therapy of breast malignancy (Ismael et al., 2020).

3.4.4 Internal light

The various drawbacks associated with external light source like the transmission output, hence the exploration moves towards internal light source. As a study done by Jiang et al. (2019), involves the use of hemoglobin (Hb)-linked polymer nanoparticles (CPNs) for activation of PAs. At the site of H_2O_2, Hb causes the catalysis of luminol and produces chemiluminescence and which is further transferred to CPNs via chemiluminescence resonance energy transfer (CRET) to O_2, generating ROS for treatment of malignancy (Jiang et al., 2019). In another study, Sun et al. (2019), fabricated a glow PersLum material hydrogel (PLM-hydrogel) for enabling higher result oriented constant luminescence-sensitized PDT. The PLM-hydrogel was prepared by scattering high-temperature calcined PLM into a biocompatible alginate-Ca^{2+} hydrogel, which was capable of effectively penetrating into the tumor site for triggering the constant production of $^1O_2$ through a powerful light source. In vivo examinations successfully demonstrated the steady radiance, light inexhaustibility, and strong fixing capacity of PLM-hydrogel in tumors (Sun et al., 2019).

3.5 Changes in cell signaling after photodynamic therapy

One of the most important area for enhancing the ultimate results of PDT is exploring the cell biology and signal transduction pathways, transcription factors, cell cycle regulation, inflammation and cell death, and in initiation of this process PDT actuation plays an important role (Fig. 3.3) (Mohanty et al., 2013; Stranadko and Ivanov, 2004). These changes in the cellular metabolism promote the apoptosis and cell death following PDT by the route of mitochondrial pathway comprising caspases and arrival of cytochrome C, or by pathways concerning ceramide or passing receptors. If under specific pathological conditions the cells exposed to PDT, it can cause severe damage to malignant cells. The impact of PDT on signal transduction pathways can create cell stress reactions like hypoxia and angiogenesis and increases the effectiveness of PDT (Benachour et al., 2012). Calcium expression levels, transcription factors, cell adhesion molecules, lipid metabolism effects, tyrosine kinase expression, transcription factors, and cytokines are the major areas of attention regarding signal transduction pathways.

3.5.1 Calcium

Some in-vitro studies indicate that after PDT there is sudden rise in the levels of intracellular calcium in the cells and this mechanism have assisted for improving the management of tumor cells and causing cell deaths under special events and conditions (Sorrin et al., 2020). The influx of Ca^{2+} through ion channels and increase in stimulation of ion exchange mechanisms is mainly considered as basic underlying mechanism behind the rise in intracellular Ca^{2+} after PDT treatment.

The stimulation of ion channels causes the release of Ca^{2+} further starting accumulation in the vittles in the endoplasmic reticulum and mitochondria, and/or stimulation of ion exchange mechanisms (Konopka and Goslinski, 2007). As the Ca^{2+} finds its substantial relation with various pathways relating to PDT and plays a vital role, some important cellular functions like analyzing the effect of calcium levels being expressed inside cells, as it has crucial role in cell death reactions like apoptosis to PDT.

3.5.2 Lipid metabolism

Many of the pathways find a common connection between the Ca^{2+} and lipid metabolism, since the initiation of phospholipases occurs by the normal cell signaling pathway and the stimulation of arachidonic acid metabolites occurs after the exposure to PDT and this results in formation of phospholipase A2 (PLA2), which is among the one of the membrane proteins endorsed by Ca^{2+} and leads to cell apoptosis or survival (Fig. 3.3) (Nowis et al., 2005). The secondary stress-induced messenger known as Ceramide is mainly responsible for breakdown of sphingophospholipids (which are part of the cell membrane) to sphingomyelin, which ultimately leads to formation of ceramide and phosphorylcholine. These are responsible for inducing apoptosis, while cell senescence, the cell cycle, and cell differentiation can be induced by sphingolipid ceramide (Castano et al., 2005). The different compound involved in stress signaling cascades comprises both protein kinases and protein phosphatases and enhanced levels of ceramide are seen and leads to cell apoptosis which influences the complete metabolic pathways of metastases (Fig. 3.3) (Robertson et al., 2009).

3.5.3 Tyrosine kinases

Signal transduction plays a vital role in cells for obtaining an external stimulus and enables to respond this signal in suitable mode. The mitogen-activated protein kinase (MAPK) signaling pathway occurs inside the eukaryotic cells and produces signal transduction, which possibly modify various cellular events including extracellular signal-regulated kinases, such as (ERK-1/2) (Sahu et al., 2009). The role of ERK's in cell survival has been explored which gave positive results, ERK's expression is fundamentally diminished comparable to an increment in cell survival thus it is suggested that two cycles are associated (Fig. 3.3) (Jiang et al., 2014). The proliferation of cells, angiogenesis and the cell death can be initiated by blocking the epidermal growth factor receptor (EGFR) a type of tyrosine kinase (Robertson et al., 2009). Several investigations on cancerous cell lines have discovered that PDT can incite the total loss of EGFR on the cell layer thus initiate against proliferative reactions (Malatesti et al., 2017). Enhanced down regulation of protein expression and phosphorylation of EGFR is generally observed after PDT which induces apoptosis and leading to tumor regression state in metastases (Fig. 3.3) (Davids et al., 2009).

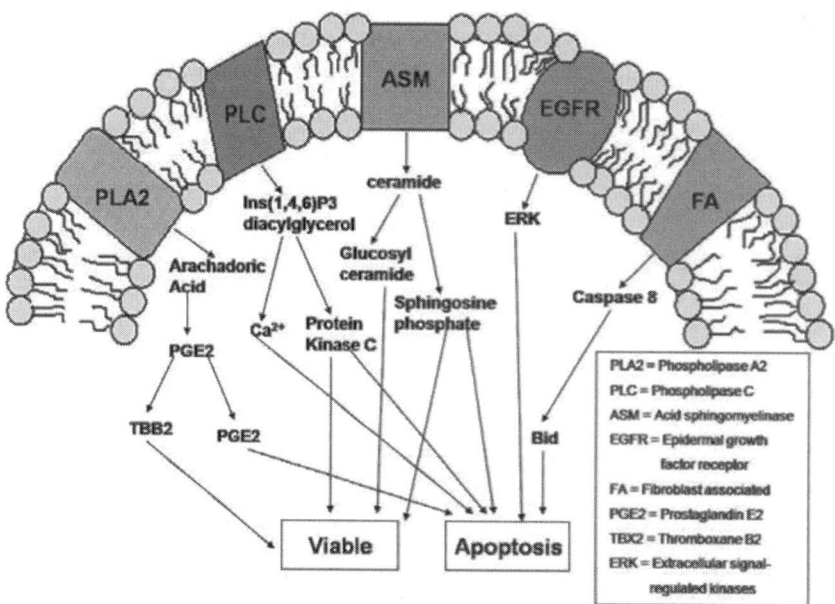

FIG. 3.3 Signal transduction pathways activated after PDT. Events occur at receptors located on the plasma membrane and lead to changes in cellular metabolism. These may tend toward increasing apoptosis or increasing cell survival. *From Robertson et al. (2009).*

3.5.4 Transcription factors

Transcription factors are the type of proteins which links to various genes and interfere in the transcription pathways and promotes the cell death (Theodossiou et al., 2009). These transcription factors act as an intracellular messenger that initiates signals for activation of various genes modifications and the gene transcription occurs through various transcription factors and auxiliary proteins (Davids et al., 2008). Several PDT studies have showed that cell multiplication is associated with transcription regions of many genes and endorsed by mixture of physical or synthetic reactions and causes induction as well as prevention of apoptosis (Li et al., 2020). Hence this factor can contribute for improving PDT outcomes as it is responsible for deactivation or triggering of numerous signaling pathways, causing damage to cell components, and inducing apoptotic state.

3.5.5 Cellular adhesion

The mammalian cell has nature of binding with the several specific membrane protein receptors and to the extracellular networks (Malatesti et al., 2017). The membrane proteins are generally arranged in sequence like selectins and cadherins, as intracellular cell adhesion molecule-1 (ICAM-1), integrins, immunoglobulin G superfamily, and vascular cell adhesion molecule-1 (VCAM-1) (Li et al., 2020) supplements the congregations. The intracellular signaling pathways are responsible for cell behavior, development and tissue inflammation with migration and penetration of underlying tissues through these membrane proteins by enacting as transduction receptors (Buytaert et al., 2007). These progressions are to a great extent brought about by the damage of adhesion molecules situated inside the cell membranes and studies have shown that a reduction in cell attachment straightforwardly relates with abatement in metastatic capability of carcinogenic cells (Dewaele et al., 2010).

3.5.6 Cytokines

In mediating and managing the inflammation, hematopoiesis and immunity, the small secreted proteins called cytokines play an important role (Robertson et al., 2009). They are capable of producing de novo in response to a stimulus and signals the cells with the help of second messengers like tyrosine kinases (Zou et al., 2017) by confining to specific membrane receptors for altering the gene expression. Reactions to cytokines incorporate expanding or diminishing articulation of membrane proteins (counting cytokine receptors), proliferation, and secretion of effector molecules. By and large PDT might instigate over or underarticulation of cytokines, which frequently identifies with tumor relapse and conceivable destruction (Robertson et al., 2009).

3.5.7 Stress response

When the cells are exposed to conditions like heat, oxidizing conditions or to various toxic agents, heat shock proteins (Hsps) and glucose-regulated proteins (Grps) are observed. All the heat shock proteins do not bind to molecular chaperones and mediate stability and protection of cells at intermediate stages of folding, assembly, translocation across membranes and degradation (Castano et al., 2005). The linking mechanism of heat-shock factors (HSFs) to specific heat-shock elements (HSEs) is responsible for transcription of Hsp family genes. Some readings suggest that PDT serves as light activated targeted inducer of specific gene expression (for instance suicide genes), if the gene of concern might be associated to the heat shock or Grp promoters, and so can moreover root cause of cell death or cellular resistance to PDT mediated laser therapy (Buytaert et al., 2007).

3.5.8 Hypoxia and angiogenesis

The malignant growth, resistance developed due to chemotherapy, enhanced metastasis, and poor prognosis are the factors contributing to tumor hypoxia (Castano et al., 2005). The major impacts in hypoxic tumor cells are mediated by oxygen-regulated transcriptional activation of particular genes. Hypoxia-inducible factor (HIF)- 1 is the major transcriptional activator of oxygen-regulated genes, which is involved in metabolism, angiogenesis, cell survival, cell invasion, and drug resistance (Verma et al., 2007). Since PDT is prepared to do quickly devouring critical measures of tissue oxygen and furthermore closing down blood that conveys oxygen supply to the tumor district, the actual treatment might create extreme degrees of hypoxia. Consequently, significant degrees of HIF-1 articulation are frequently connected with a helpless generally speaking cell reaction to PDT treatment, so further examination is justified to comprehend and defeat this marvel (Ell and Gossner, 1994; Stranadko and Ivanov, 2004).

Therefore, the PDT can consume the large amount of tissue oxygen and restricts the supply of blood which enables the oxygen supply to the tumor site and the exposure of PDT may significantly enhance the levels of hypoxia. The higher levels of HIF-1 expression may indicate the lower cellular response to PDT treatment and hence further investigations are needed to understand this phenomenon.

3.6 Method of excitation for photosensitizing agents

The excitation of a PAs plays a vital role in the treatment through PDT, because the compound which possess chemiluminescence (CL) or are capable of emitting beta radiation are very less. Therefore, for enhancing the outcomes of PDT, the PAs are generally excited with various fundamental techniques for improving the treatment regimens.

3.6.1 Intermolecular chemically induced electronic excitation

Molecular chemistry-initiated electron excitation is the most essential, yet may be the least frequently applied, technique for excitation of PAs, the Chemically Initiated Electron Exchange Luminescence (CIEEL) was initially proposed by Schuster (Koo and Schuster, 1977; Schuster, 1979). Although CIEEL has improved efficiency for excitation of PAs but some studies denies the efficiency of this hypothesis, especially in case of low quantum efficiencies for certain CL responses (Blum et al., 2020; Lin et al., 2013).

The chemiexcitation of various compounds like substitute mechanisms (e.g., ICIC) (Blum et al., 2020) has not yet fully understood (Augusto et al., 2013). Chemically initiated electron exchange luminescence (CIEEL) causes the direct chemical excitation of various CL and PAs due to complexity of CIEEL when contrasted with different strategies, that is, resonance energy transfer (RET) and CRET. The natural mechanism behind the excitation of PAs involves oxidative reactions by means of RET and CRET that requires the data regarding the emission and excitation spectra of the molecule (Bartoloni et al., 2015). The emission of chemiluminescence by the bioluminescent compounds mainly occurs because of presence of 1,2-dioxetanone ring. The excited electronic state and existence of an electron donor within the electron donating compound or inside the dioxane containing compound itself, confirms to intermolecular or intramolecular CIEEL-like CL, respectively, after the dioxetane ring decomposes (Blum et al., 2020).

3.6.2 Resonance energy transfer excitation

The most commonly used technique, for the excitation of self-mediated PDT treatments in absence of light involves the use of a chemiluminescent compound, which transfers the energy to PAs via RET. The RET in broad sense is employed for excitation purposes and subtype Forster resonance energy transfer (FRET) excites the PAs by CL in light giving rise to numerous tiny Forster break down in the biological systems at nano scale. The mechanism of energy transfer to RET by bioluminescence resonance energy transfer (BRET) and chemiluminescent resonance is indistinguishable (Hananya and Shabat, 2019). The ideal chemiluminescent compound emits the light in blue wavelengths, as the energy require for excitation is comparatively high via RET and it has been observed that constant loss of energy occurs during this RET energy transfer process (Jares-Erijman and Jovin, 2003). The blue light emission spectra in RET generally do not require donor, as the efficacy of RET depends on the overlay of the emission spectra generated by donor as well as the acceptor (Somssich and Simon, 2017).

Various chemifluorescent compounds transmit to the blue or green wavelengths, which possess higher attenuation in biological tissues (Bunt and Wouters, 2017) assuring the clinical excitation of PAs with precaution that the excitation shall not come from actual emission of the photons from the donor, nor the RET interaction itself. These types of mechanisms are quite different from CIEEL-like compounds as CIEEL excitation occurs through oxidative mechanism. The excitation of RET PAs occurs due to chemiluminescent molecule through a CIEEL-like instrument and therefore at this point the energy is transferred to PAs by means of FRET (Müller et al., 2013).

The most commonly used chemiluminescent compound in various studies is luminol, which is being used since 1928 as it is capable of responding to blood and generating fluoresce, due to presence of iron and some similar compounds which catalyzes the oxidation of luminal (Blum et al., 2020). In one of the studies conducted, PLGA-RD BRET formulation was investigated using in-vivo model and the results indicated the identical tumor growth suppression curves as seen with PLGA-RD nanoparticles when illuminated with light and the endpoint tumor volumes were compared.

The clinical outcomes were promising, as they deduce the force of catalyzed radiance by fLuc and other luciferases, because they possess lower excitation power as compared to external light sources and can be viable for PDT. Several investigations have been reported for assessing the efficiency of PDT when exposed with internal tumor light source. A study indicating the bioluminescence generated by luciferins is comparatively weak when compared to conventional PDT lacking remarkable tumor destruction (Lale et al., 2015; Wachter et al., 2003; Xu et al., 2018).

Hence because of such drawbacks the application of similar techniques for excitations in PDT has been discontinued, however implication of nanoparticles in the said excitation mechanism have shown some positive results in management of cancer cells and tumors. The nano-formulation mediated-RET enables the significant excitation of PAs in various in-vivo studies (Magalhães et al., 2016; Zhang et al., 2020). The plausible reasons include irregularities in such investigations or incorporation of PDT mechanism in a nanoparticle formulation considering enhanced toxicity over free PDT systems.

3.6.3 Two-stage photosensitizer excitation/excitation by radiation energy transfer intermediary

The possible mechanism for enhancing the RET mediated PDT is addition of an intermediary possessing excitation band gap between the CL compound which is among the one of PAs acceptor. The PAs and CL generally do not have an overlapping absorption and emission spectra, and in case if they overlap or cross-over, the intermediary fluorescent compounds are employed. Necessarily the intermediary's absorption spectrum should overlap with the emission of CL compounds and the emission spectrum must cover the PAs absorption subsequently decreeing a second RET step (Algorri et al., 2021; Sai et al., 2021). During the second RET step, extra loss of energy is observed because of the energy emitted by increased overlapping of absorption/emission spectra between the intermediate of PAs and CL compounds. Concurrently, the overlapping between CL and PAs compound promotes the CL and enhances the activation of PAs (Magalhães et al., 2016; Wachter et al., 2003).

This type of conjugation possesses more selectivity and possibly enhances the ultimate outcome in terms of biological systems (Grigalavicius et al., 2019; Tseng et al., 1994). The existence of an intermediate does not eliminate the chances of direct CL/PAs RET, but accounts for higher treatment efficiency (Algorri et al., 2021; Beljonne et al., 2009; Colombo et al., 2005; Nowis et al., 2005). It is anticipated that quantum dots are ideal for enacting the role of FRET intermediate, as their nanosize enables wide range of absorption/emission and PAs can be employed followed by their activation. A study conducted by Kim et al. (2015) in which conjugation of RLuc8 an chemiluminator was done with quantum dot which have a maximum absorption at 655 nm causing the shifting of emission peak from blue light (peak around 500 nm) to approximately around 670 nm and allowing the Luc-QD complex for excitation of Ce6 PAs, and therefore the red shifting of chemifluorescence signal can be seen in the Luc8-QD conjugated system. The in-vivo analysis of nanoconjugation confirmed that, as the dose of RLuc-QD and number of RET PDT (N = 9) is increased, there is significant supersession of tumor as compared to regimen N = 3 of superficially illuminated PDT. The dose related issues can be overcome application of these nanocarriers suggesting at least thrice the dose of Luc-QD is required in comparison to conventional PDT treatment (Kim et al., 2015).

3.6.4 Cherenkov radiation energy transfer

Cherenkov radiation energy transfer (CRET) was first discovered by Cherenkov, based on the phenomenon of beta emission from radioactive particles (Tomasi and Kassal, 2020; Wittmann et al., 2020). Cherenkov radiations arises from a charged particles (usually an electron), which finds their path via dielectric medium like water and further travels at higher speed in the medium (Nelson et al., 2020; Worster et al., 2019). An electromagnetic field constantly produce a charged molecule travels through a medium, and this can induce a related polarization in the electrons of close by atoms. This is neither excitation nor an ionization, in fact the formation of circular wavelets are observed due to the fluctuations of charges among the adjacent electrons and each charged molecule travels through a unique path (Dimitriev et al., 2019; Wilhelm et al., 2019). When the speed of charged molecules surpasses the speed of light in the medium, it causes the interference of wavelets to produce a sonic blast and produces electromagnetic waves of the visible frequencies and this seems to be blue because of the proportionality of the radiations to the frequency (Czar et al., 2019; Jang, 2019).

These radiations at this point possess adequate energy to excite fluorophores, quantum dots, PAs and any other moiety having capacity to excite in the band gap of blue light region (Qi et al., 2019). Therefore, Cherenkov radiation plays an important role in CRET, by employing various radioactive isotopes possessing high beta- emissions (e.g., 18F, 64Cu, or 68Ga, 90Y, and so on) and few investigations proposed that CRET may excite fluorophores and photosensitizers too. The efficacy of FRET, CRET basically relay on the amount of emitting isotope (the donor) and the concentration of the fluorophore (acceptor). Ultimately, the overlapping between Cherenkov luminescence emission peak and the absorbance peak of acceptor (Zhu and Van Voorhis, 2019). The excitation of fluorophores by Cherenkov radiation mechanism is probably not equivalent for the molecules transferring to the fluorophores due to the difference in the electronic energy source. The excitation occurs because of the chemical reactions and similarly due to the electromagnetic waves. Subsequently RET is supposed to have radiation-less energy transfer and the Cherenkov arouses due to the transmission of radiation itself, however the underlying mechanism of both RET as well as CRET is not appropriately stated.

3.7 Photodynamic therapy modifications

3.7.1 Nanotechnology on photodynamic therapy

Since decades remarkable advancements have been witnessed in the arena of nanotechnology, the said technique utilizes the various nanostructured carriers for the diagnostic and therapeutic purposes and enabling the site specific delivery and enhancing the effectiveness of anti-cancer therapy (Ferrari, 2005). The combination of PAs to nanostructures can possibly improve the efficacy of the PDT and would likely eliminate the side effects (Peer, 2007; Shi et al., 2016). The nanocarriers assist to achieve the targeting of PAs to various specific receptors and simultaneously improve the selectivity of photodynamic treatment. The PAs can be encapsulated or immobilized to nanofabricated materials by covalent and noncovalent interactions (Bertrand, 2014).

Most of the PAs are usually hydrophobic in nature which has property of aggregating in the aqueous environment and this interaction lowers the tumor cells killing ability of PDT. The monomeric structure of PAs is mainly responsible for the photoactivity, this integrity can be achieved by conjugation of PAs with nanoparticles (Shi, 2010; Sun et al., 2018). The interlinking of PAs with hydrophilic polymer particles leads to increase the bioavailability of hydrophobic porphyrins (Endo, 2017; Liang et al., 2021). The application of polymeric nanoparticles like micelles in PDT can probably improve the targeted delivery of more photosensitizer particles to the tumor environment and circumvents degradation of the photosensitizer before reaching the target tumor tissue (Matsumoto et al., 2017; Nakamura et al., 2017). The image mediated diagnosis with the help of contrast materials or fluorescent markers can also be done using polymeric nanoparticles conjugated with auxiliary ligands to the PAs molecules (Liu et al., 2019; Mesquita et al., 2018; Wang et al., 2017). The meso-tetra-4-hydroxyphenylporphyrin molecules (mTHPP) combined with polyethylene glycol molecule (PEG) has increased the solubility of porphyrin and reduced the aggregation in the aquatic environment (Hao et al., 2016; Kun et al., 2019).

Exploring the hydrophilic properties of the PAs leads to improve selectivity and higher efficacy of the photodynamic therapy for prevention of aggregation of photosensitizers such as polyacrylamide (PAA). PAA is basically nontoxic in nature and is administered fundamentally due to the exceptionally well solvent properties (Rong et al., 2014). Several studies conclude that, the use of PAA as PAs nanostructure can improve the efficiency of PDT, Polyacrylamide can be used with photosensitizers: methylene blue (MB) (Hao et al., 2016; Ozturk et al., 2014) and porphyrins (Chepurna et al., 2020). A rise in PDT effectiveness with the use of PAA in the treatment of C6 glioma cells was noticed (Wang et al., 2011). Few other polymers applied for enhancing the efficacy of PDT are N-(2-hydroxypropyl) methacrylamide (HPMA) which were employed for successful treatment of neuroblastoma and human ovarian carcinoma in vitro (Prieto-Montero et al., 2021).

3.7.2 Application of liposomes and lipoproteins

The targeting PAs to the tumor site can be convincingly achieved the application of liposomes and lipoproteins, the lipid vesicles comprising of symmetrically arranged phospholipid bilayers and are the topic of research for enhancing the ultimate outcome of PDT (Dellinger et al., 1990; Milanesi et al., 1989). Various experimental and clinical studies revealed that liposomes as a carrier for transport of PAs to the tumor microenvironment for improving the PDT (Ricchelli et al., 1993). The increased vascular penetrability and retention time of active moieties on the site of tumor and relating tissues can happen by application of lipidic nanocarriers, which consecutively improves the target specificity. The leaky nature of blood vessels creates an unusual endothelial obstruction which improves the vascular penetrability inside the tumor, which grounds for accretion of PAs to the target area by mechanism of simple diffusion (Brault et al., 1986; Mojzisova et al., 2007).

Hematoporphin derivative (HpD) was the first PAs encapsulated inside the liposomes and the study conducted by Cozzani et al. (1985), utilized the HpD loaded liposomal mediated PDT and the results showed improved outcomes (Cozzani et al., 1985).The another moiety under study for liposomes facilitated PDT was benzoporphyrin derivative monoacid (BPDMA) which was utilized for treatment of cancers in Switzerland and USA (Jori et al., 1986). The presence of numbers of receptors for low-density lipoproteins (LDL) on the surface of tumor cells influenced the PAs to combine with LDL (Milanesi et al., 1989). Lipoproteins have noteworthy mechanism behind the carriage mechanism and target specificity of PAs toward the malignant cells. Some studies revealed that PAs binds the LDL by non-covalent bonds, even before the administration which leads to enhancement of PDT outcomes when compared to the administration of the photosensitizer itself (Ricchelli et al., 1990).

3.7.3 Photodynamic therapy supported by electroporation

Electroporation is a technique involving opening of cell membranes by mechanism of reversible or irreversible strategy with the help of electrical pulses. The reversible EP assists in enhancing the penetrability of various drugs through cell membranes and simultaneously enabling gene transfection (Jia et al., 2021; Silantyev et al., 2019). The strong electric field causes the reorganization and structural changes of the cell membranes directly affecting the lipids present there in and concurrently the hydrophilic pores serves as an extra route for transport of various moieties across cell membrane (Gehl, 2003; Hanson et al., 2021). The dependency of this technique on utilization of electroporation for enabling cytostatic is termed as electrochemotherapy (ECT) (Rols, 2006; Rols and Pierre, 2008). Hence the temporal destabilization in the cell membranes due to the pulsed electric field causes the improved accumulation of cytostatic agents in the tumor prone area and ultimately leading improvements in chemotherapy regimen and lessens the severe side effects (Becker and Kuznetsov, 2007; Mir, 2014) and EP can succors the higher amassing of PAs inside the cells.

A study conducted by Labanauskiene et al. (2007) for enhancing the cytotoxic action using the combination of EP and PDT, comprising of chlorine e6 and AlPcS4 phthalocyanine as PAs, the protocol focused on Chinese hamster lung fibroblast cells DC-3 F (Labanauskiene et al., 2007). Studies also suggest that combination of EP and PDT affects the substantial decrease in the incubation time of PAs with mammary gland adenocarcinoma cells and results in 10~ fold increase in efficacy of PDT (Kou et al., 2017). Some additional exploration for effective management of carcinomas through PDT-EP combination includes use of clinically approved Photofrin (Cucu et al., 2021; Gupta et al., 2021; Rols, 2006).

3.8 Conclusion

Phototherapies, such as photothermal therapy (PTT) and photodynamic therapy (PDT), have enhanced the target mediated damage to the cancer tissues and cells via accumulation of photosensitizers and causing minimal harm to healthy cells and tissues as compared to conventional carcinoma management strategies. PTT and PDT differs from other oncologic treatment methods by the way of their selectivity accompanied with enhanced therapeutic outcomes. The improved sensitivity and non-invasive nature of these optical therapies makes them choice for treatment of carcinomas and is gaining interest worldwide as investigations are being made for improving the efficiency of PDT. Various photoexcitation techniques and combination of PDT with other techniques have improved the treatment outcome and reduced the side effects. Due to these advantages of the phototherapies, they are gaining more and more eminent positions among clinicians. Speedy expansion and progresses in technology and optoelectronic equipment can favor the implementation of the phototherapy methods in clinical practice.

Acknowledgment

First author (SST) acknowledges Government of India, Ministry of Science and Technology, Department of Science and Technology (DST), India for financial support through the DST-INSPIRE Fellowship (IF 190486).

Statement of informed consent

All authors give consent for publication.

Conflict of interest

The authors declare no conflict of interest, financial or otherwise.

References

Ackroyd, R., Kelty, C., Brown, N., Reed, M., 2007. The history of photodetection and photodynamic therapy. Photochem. Photobiol. 74 (5), 656–669. https://doi.org/10.1562/0031-8655(2001)0740656THOPAP2.0.CO2.

Agostinis, P., Berg, K., Cengel, K.A., Foster, T.H., Girotti, A.W., Gollnick, S.O., Hahn, S.M., Hamblin, M.R., Juzeniene, A., Kessel, D., Korbelik, M., Moan, J., Mroz, P., Nowis, D., Piette, J., Wilson, B.C., Golab, J., 2011. Photodynamic therapy of cancer: an update. CA Cancer J. Clin. 61 (4), 250–281. https://doi.org/10.3322/CAAC.20114.

Agostinis, P., Buytaert, E., Breyssens, H., Hendrickx, N., 2004. Regulatory pathways in photodynamic therapy induced apoptosis. Photochem. Photobiol. Sci. 3 (8), 721–729. https://doi.org/10.1039/B315237E.

Algorri, J.F., Ochoa, M., Roldán-Varona, P., Rodríguez-Cobo, L., López-Higuera, J.M., 2021. Light technology for efficient and effective photodynamic therapy: a critical review. Cancers 13 (14), 3484. https://doi.org/10.3390/CANCERS13143484.

Allison, R.R., 2014. Photodynamic therapy: oncologic horizons. Future Oncol. 10 (1), 123–142. https://doi.org/10.2217/FON.13.176.

Anand, S., Ortel, B.J., Pereira, S.P., Hasan, T., Maytin, E.V., 2012. Biomodulatory approaches to photodynamic therapy for solid tumors. Cancer Lett. 326 (1), 8–16. https://doi.org/10.1016/J.CANLET.2012.07.026.

Atlante, P., Quagliariello, M., Salet, 1989. Haematoporphyrin derivative (Photofrin II) photosensitization of isolated mitochondria: inhibition of ADP/ATP translocator. J. Photochem. Photobiol. B 4 (1), 35–46. https://doi.org/10.1016/1011-1344(89)80100-9.

Augusto, S., Júnior, S., Khalid, Baader, 2013. Efficiency of electron transfer initiated chemiluminescence. Photochem. Photobiol. 89 (6), 1299–1317. https://doi.org/10.1111/PHP.12102.

Bacellar, I.O.L., Oliveira, M.C., Dantas, L.S., Costa, E.B., Junqueira, H.C., Martins, W.K., Durantini, A.M., Cosa, G., Di Mascio, P., Wainwright, M., Miotto, R., Cordeiro, R.M., Miyamoto, S., Baptista, M.S., 2018. Photosensitized membrane permeabilization requires contact-dependent reactions between photosensitizer and lipids. J. Am. Chem. Soc. 140 (30), 9606–9615. https://doi.org/10.1021/JACS.8B05014.

Bacellar, I.O.L., Tsubone, T.M., Pavani, C., Baptista, M.S., 2015. Photodynamic efficiency: from molecular photochemistry to cell death. Int. J. Mol. Sci. 16 (9), 20523–20559. https://doi.org/10.3390/IJMS160920523.

Banerjee, S.M., MacRobert, A.J., Mosse, C.A., Periera, B., Bown, S.G., Keshtgar, M.R.S., 2017. Photodynamic therapy: inception to application in breast cancer. Breast 31, 105–113. https://doi.org/10.1016/J.BREAST.2016.09.016.

Bartoloni, F.H., De Oliveira, M.A., Ciscato, L.F.M.L., Augusto, F.A., Bastos, E.L., Baader, W.J., 2015. Chemiluminescence efficiency of catalyzed 1,2-dioxetanone decomposition determined by steric effects. J. Org. Chem. 80 (8), 3745–3751. https://doi.org/10.1021/ACS.JOC.5B00515.

Becker, S.M., Kuznetsov, A.V., 2007. Numerical assessment of thermal response associated with in vivo skin electroporation: the importance of the composite skin model. J. Biomech. Eng. 129 (3), 330–340. https://doi.org/10.1115/1.2720910.

Beljonne, D., Curutchet, C., Scholes, G.D., Silbey, R.J., 2009. Beyond förster resonance energy transfer in biological and nanoscale systems. J. Phys. Chem. B 113 (19), 6583–6599. https://doi.org/10.1021/JP900708F.

Benachour, H., Bastogne, T., Toussaint, M., Chemli, Y., Sève, A., Frochot, C., Lux, F., Tillement, O., Vanderesse, R., Barberi-Heyob, M., 2012. Real-time monitoring of photocytotoxicity in nanoparticles-based photodynamic therapy: a model-based approach. PLoS One 7 (11). https://doi.org/10.1371/JOURNAL.PONE.0048617.

Benov, L., 2015. Photodynamic therapy: current status and future directions. Med. Princ. Pract. 24 (Suppl. 1), 14–28. https://doi.org/10.1159/000362416.

Benov, L., Craik, J., Batinic-Haberle, I., 2010. Protein damage by photo-activated Zn(II) N-alkylpyridylporphyrins. Amino Acids 42 (1), 117–128. https://doi.org/10.1007/S00726-010-0640-1.

Bertrand, 2014. Cancer nanotechnology: the impact of passive and active targeting in the era of modern cancer biology. Adv. Drug Deliv. Rev. 66, 2–25. https://doi.org/10.1016/j.addr.2013.11.009.

Blum, N.T., Zhang, Y., Qu, J., Lin, J., Huang, P., 2020. Recent advances in self-exciting photodynamic therapy. Front. Bioeng. Biotechnol. 8. https://doi.org/10.3389/FBIOE.2020.594491.

Boyle, R.W., Dolphin, 1996. Structure and biodistribution relationships of photodynamic sensitizers. Photochem. Photobiol. 64 (3), 469–485. https://doi.org/10.1111/j.1751-1097.1996.tb03093.x.

Brancaleon, L., Moseley, H., 2002. Laser and non-laser light sources for photodynamic therapy. Lasers Med. Sci. 17 (3), 173–186. https://doi.org/10.1007/S101030200027.

Brault, D., Vever-Bizet, C., Dellinger, M., 1986. Fundamental aspects in tumor photochemotherapy: interactions of porphyrins with membrane model systems and cells. Biochimie 68 (6), 913–921. https://doi.org/10.1016/S0300-9084(86)80109-2.

Bunt, G., Wouters, F.S., 2017. FRET from single to multiplexed signaling events. Biophys. Rev. 9 (2), 119–129. https://doi.org/10.1007/S12551-017-0252-Z.

Buytaert, Dewaele, Agostinis, 2007. Molecular effectors of multiple cell death pathways initiated by photodynamic therapy. Biochim. Biophys. Acta 1776 (1), 86–107. https://doi.org/10.1016/J.BBCAN.2007.07.001.

Cadet, 2012. Photoinduced damage to cellular DNA: direct and photosensitized reactions. Photochem. Photobiol. 88 (5), 1048–1065. https://doi.org/10.1111/J.1751-1097.2012.01200.X.

Cadet, J., Douki, T., Ravanat, J.L., 2006. One-electron oxidation of DNA and inflammation processes. Nat. Chem. Biol. 2 (7), 348–349. https://doi.org/10.1038/NCHEMBIO0706-348.

Cadet, J., Douki, T., Ravanat, J.L., 2010. Oxidatively generated base damage to cellular DNA. Free Radic. Biol. Med. 49 (1), 9–21. https://doi.org/10.1016/J.FREERADBIOMED.2010.03.025.

Cadet, J., Ravanat, J.L., TavernaPorro, M., Menoni, H., Angelov, D., 2012. Oxidatively generated complex DNA damage: tandem and clustered lesions. Cancer Lett. 327 (1–2), 5–15. https://doi.org/10.1016/J.CANLET.2012.04.005.

Castano, Demidova, Hamblin, 2005. Mechanisms in photodynamic therapy: part two-cellular signaling, cell metabolism and modes of cell death. Photodiagn. Photodyn. Ther. 2 (1), 1–23. https://doi.org/10.1016/S1572-1000(05)00030-X.

Chang, B., Zhao, J., Zhenghan, C., Daquan, Gu, Lele, Zhao, 2020. Nd3+-sensitized upconversion metal–organic frameworks for mitochondria-targeted amplified photodynamic therapy. Angew. Chem. Int. Ed. 59 (7), 2634–2638. https://doi.org/10.1002/ANIE.201911508.

Chepurna, O.M., Yakovliev, A., Ziniuk, R., Nikolaeva, O.A., Levchenko, S.M., Xu, H., Losytskyy, M.Y., Bricks, J.L., Slominskii, Y.L., Vretik, L.O., Qu, J., Ohulchanskyy, T.Y., 2020. Core-shell polymeric nanoparticles co-loaded with photosensitizer and organic dye for photodynamic therapy guided by fluorescence imaging in near and short-wave infrared spectral regions. J. Nanobiotechnol 18 (1). https://doi.org/10.1186/S12951-020-0572-1.

Colombo, L.L., Vanzulli, S.I., Villanueva, A., Cañete, M., Juarranz, A., Stockert, J.C., 2005. Long-term regression of the murine mammary adenocarcinoma, LM3, by repeated photodynamic treatments using meso-tetra (4-N-methylpyridinium) porphine. Int. J. Oncol. 27 (4), 1053–1059.

Cooke, M.S., Loft, S., Olinski, R., Evans, M.D., Bialkowski, K., Wagner, J.R., Dedon, P.C., Møller, P., Greenberg, M.M., Cadet, J., 2010. Recommendations for standardized description of and nomenclature concerning oxidatively damaged nucleobases in DNA. Chem. Res. Toxicol. 23 (4), 705–707. https://doi.org/10.1021/TX1000706.

Cozzani, J., Bertoloni, M., Carlini, S., Ruschi, 1985. Efficient photosensitization of malignant human cells in vitro by liposome-bound porphyrins. Chem. Biol. Interact. 53 (1–2), 131–143. https://doi.org/10.1016/S0009-2797(85)80091-0.

Cucu, C.I., Giurcăneanu, C., Popa, L.G., Orzan, O.A., Beiu, C., Holban, A.M., Grumezescu, A.M., Matei, B.M., Popescu, M.N., Căruntu, C., Mihai, M.M., 2021. Electrochemotherapy and other clinical applications of electroporation for the targeted therapy of metastatic melanoma. Materials 14 (14), 3985. https://doi.org/10.3390/MA14143985.

Czar, M.F., Breitgoff, F.D., Sahoo, D., Sajid, M., Ramezanian, N., Polyhach, Y., Jeschke, G., Godt, A., Zenobi, R., 2019. Linear and kinked oligo(phenyleneethynylene)s as ideal molecular calibrants for förster resonance energy transfer. J. Phys. Chem. Lett. 10 (21), 6942–6947. https://doi.org/10.1021/ACS.JPCLETT.9B02621.

Daayana, S., Winters, U., Stern, P.L., Kitchener, H.C., 2011. Clinical and immunological response to photodynamic therapy in the treatment of vulval intraepithelial neoplasia. Photochem. Photobiol. Sci. 10 (5), 802–809. https://doi.org/10.1039/C0PP00344A.

Dabrowski, Arnaut, Pereira, Monteiro, Urbańska, Simões, Stochel, 2010. New halogenated water-soluble chlorin and bacteriochlorin as photostable PDT sensitizers: synthesis, spectroscopy, photophysics, and in vitro photosensitizing efficacy. ChemMedChem 5 (10), 1770–1780. https://doi.org/10.1002/CMDC.201000223.

Dąbrowski, J.M., Pereira, M.M., Arnaut, L.G., Monteiro, C.J.P., Peixoto, A.F., Karocki, A., Urbańska, K., Stochel, G., 2007. Synthesis, photophysical studies and anticancer activity of a new halogenated water-soluble porphyrin. Photochem. Photobiol. 83 (4), 897–903. https://doi.org/10.1111/J.1751-1097.2007.00073.X.

Dabrowski, J.M., Urbanska, K., Arnaut, L.G., Pereira, M.M., Abreu, A.R., Simões, S., Stochel, G., 2011. Biodistribution and photodynamic efficacy of a water-soluble, stable, halogenated bacteriochlorin against melanoma. ChemMedChem 6 (3), 465–475. https://doi.org/10.1002/CMDC.201000524.

Daniell, Hill, 1991. A history of photodynamic therapy. Aust. N. Z. J. Surg. 61 (5), 340–348. https://doi.org/10.1111/J.1445-2197.1991.TB00230.X.

Davids, Kleemann, Kacerovská, Pizinger, Kidson, 2008. Hypericin phototoxicity induces different modes of cell death in melanoma and human skin cells. J. Photochem. Photobiol. B, Biol. 91 (2–3), 67–76. https://doi.org/10.1016/J.JPHOTOBIOL.2008.01.011.

Davids, L.M., Kleemann, B., Cooper, S., Kidson, S.H., 2009. Melanomas display increased cytoprotection to hypericin-mediated cytotoxicity through the induction of autophagy. Cell Biol. Int. 33 (10), 1065–1072. https://doi.org/10.1016/J.CELLBI.2009.06.026.

Davies, M.J., 2005. The oxidative environment and protein damage. Biochim. Biophys. Acta 1703 (2), 93–109. https://doi.org/10.1016/J.BBAPAP.2004.08.007.

Davies, M.J., 2004. Reactive species formed on proteins exposed to singlet oxygen. Photochem. Photobiol. Sci. 3 (1), 17–25. https://doi.org/10.1039/B307576C.

Dedon, P.C., 2008. The chemical toxicology of 2-deoxyribose oxidation in DNA. Chem. Res. Toxicol. 21 (1), 206–219. https://doi.org/10.1021/TX700283C.

Delaey, E., Van Laar, F., De Vos, D., Kamuhabwa, A., Jacobs, P., De Witte, P., 2000. A comparative study of the photosensitizing characteristics of some cyanine dyes. J. Photochem. Photobiol. B 55 (1), 27–36. https://doi.org/10.1016/S1011-1344(00)00021-X.

Dellinger, M., Vever-Bizet, C., Brault, D., Moreno, G., Salet, C., 1990. Uptake and photodynamic efficiency of hematoporphyrin, hydroxyethylvinyldeuteroporphyrin and hematoporphyrin derivative (Photofrin ii®): a study with isolated mitochondria. Photochem. Photobiol. 51 (2), 185–189. https://doi.org/10.1111/J.1751-1097.1990.TB01701.X.

Dewaele, M., Verfaillie, T., Martinet, W., Agostinis, P., 2010. Death and survival signals in photodynamic therapy. Methods Mol. Biol. 635, 7–33. https://doi.org/10.1007/978-1-60761-697-9_2.

Dima, V.F., Ionescu, M.D., Balotescu, C., Dima, S.F., 2002. Photodynamic therapy and some clinical applications in oncology. Roum. Arch. Microbiol. Immunol. 61 (3), 159–205.

Dima, V.F., Vasiliu, V., Dima, S.V., 1998. Photodynamic therapy: an update. Roum. Arch. Microbiol. Immunol. 57 (3–4), 207–230.

Dimitriev, O.P., Piryatinski, Y.P., Slominskii, Y.L., 2019. Abnormal emission in the heterogeneous J-aggregate system. J. Phys. Chem. C. https://doi.org/10.1021/ACS.JPCC.9B08248.

Dogra, V., Kim, C., 2020. Singlet oxygen metabolism: from genesis to signaling. Front. Plant Sci. 10. https://doi.org/10.3389/FPLS.2019.01640/FULL.

dos Santos, A.F., de Almeida, D.R.Q., Terra, L.F., Baptista, M.S., Labriola, L., 2019. Photodynamic therapy in cancer treatment – an update review. J. Cancer Metastasis Treat. 5. https://doi.org/10.20517/2394-4722.2018.83.

Ehrenshaft, M., Deterding, L.J., Mason, R.P., 2015. Tripping up Trp: Modification of protein tryptophan residues by reactive oxygen species, modes of detection, and biological consequences. Free Radic. Biol. Med. 89, 220–228. https://doi.org/10.1016/J.FREERADBIOMED.2015.08.003.

Ell, C., Gossner, L., 1994. Photodynamic therapy: Its potential for the treatment of gastrointestinal malignancies and precancerous conditions. Endoscopy 26 (2), 262–264.

Endo, K., 2017. Development of neighboring electrophilic activation of active center in catalytic reactions via organometallic intermediates. Bull. Chem. Soc. Jpn. 90 (6), 649–661. https://doi.org/10.1246/BCSJ.20170015.

Epe, B., 1991. Genotoxicity of singlet oxygen. Chem. Biol. Interact. 80 (3), 239–260. https://doi.org/10.1016/0009-2797(91)90086-M.

Etcheverry, Pasquale, Garavaglia, 2016. Photodynamic therapy of HeLa cell cultures by using LED or laser sources. J. Photochem. Photobiol. B, Biol. 160, 271–277. https://doi.org/10.1016/J.JPHOTOBIOL.2016.04.013.

Fan, W., Bu, W., Shi, J., 2016. On the latest three-stage development of nanomedicines based on upconversion nanoparticles. Adv. Mater. 28 (21), 3987–4011. https://doi.org/10.1002/ADMA.201505678.

Feng, X., Shi, Y., Xie, L., Zhang, K., Wang, X., Liu, Q., Wang, P., 2018. Synthesis, characterization, and biological evaluation of a porphyrin-based photosensitizer and its isomer for effective photodynamic therapy against breast cancer. J. Med. Chem. 61 (16), 7189–7201. https://doi.org/10.1021/ACS.JMEDCHEM.8B00547.

Ferrari, M., 2005. Cancer nanotechnology: opportunities and challenges. Nat. Rev. Cancer 5 (3), 161–171. https://doi.org/10.1038/nrc1566.

Figueroa-González, G., Pérez-Plasencia, C., 2017. Strategies for the evaluation of DNA damage and repair mechanisms in cancer. Oncol. Lett. 13 (6), 3982–3988. https://doi.org/10.3892/OL.2017.6002.

Gaio, E., Scheglmann, D., Reddi, E., Moret, F., 2016. Uptake and photo-toxicity of Foscan®, Foslip® and Fospeg® in multicellular tumor spheroids. J. Photochem. Photobiol. B 161, 244–252. https://doi.org/10.1016/J.JPHOTOBIOL.2016.05.011.

Gao, R., Mei, X., Yan, D., Liang, R., Wei, M., 2018. Nano-photosensitizer based on layered double hydroxide and isophthalic acid for singlet oxygenation and photodynamic therapy. Nat. Commun. 9 (1), 1–10. https://doi.org/10.1038/s41467-018-05223-3.

Gehl, 2003. Electroporation: theory and methods, perspectives for drug delivery, gene therapy and research. Acta Physiol. Scand. 177 (4), 437–447. https://doi.org/10.1046/J.1365-201X.2003.01093.X.

Grigalavicius, M., Mastrangelopoulou, M., Berg, K., Arous, D., Ménard, M., Raabe-Henriksen, T., Brondz, E., Siem, S., Görgen, A., Edin, N.F.J., Malinen, E., Theodossiou, T.A., 2019. Proton-dynamic therapy following photosensitiser activation by accelerated protons demonstrated through fluorescence and singlet oxygen production. Nat. Commun. (1), 10. https://doi.org/10.1038/S41467-019-12042-7.

Gupta, P., Kar, S., Kumar, A., Tseng, F.-G., Pradhan, S., Mahapatra, P.S., Santra, T.S., 2021. Pulsed laser assisted high-throughput intracellular delivery in hanging drop based three dimensional cancer spheroids. Analyst 146 (15), 4756–4766. https://doi.org/10.1039/D0AN02432E.

Hananya, N., Shabat, D., 2019. Recent advances and challenges in luminescent imaging: bright outlook for chemiluminescence of dioxetanes in water. ACS Cent. Sci. 5 (6), 949–959. https://doi.org/10.1021/ACSCENTSCI.9B00372.

Hanson, S.M., Forsyth, B., Wang, C., 2021. Combination of irreversible electroporation with sustained release of a synthetic membranolytic polymer for enhanced cancer cell killing. Sci. Rep. (1), 11. https://doi.org/10.1038/S41598-021-89661-Y.

Hao, Zhang, Zheng, Niu, Guo, Zhang, Chang, Zhang, Wang, Zhang, 2016. Multifunctional nanoplatform for enhanced photodynamic cancer therapy and magnetic resonance imaging. Colloids Surf. B Biointerfaces 151, 384–393. https://doi.org/10.1016/J.COLSURFB.2016.10.039.

Headlam, Henrietta, Davies, Michael, 2004. Markers of protein oxidation: Different oxidants give rise to variable yields of bound and released carbonyl products. Free Radic. Biol. Med. 36 (9), 1175–1184. https://doi.org/10.1016/J.FREERADBIOMED.2004.02.017.

Höhn, A., König, J., Grune, T., 2013. Protein oxidation in aging and the removal of oxidized proteins. J. Proteomics 92, 132–159. https://doi.org/10.1016/J.JPROT.2013.01.004.

Hornung, R., Fehr, M.K., Walt, H., Wyss, P., Berns, M.W., Tadir, Y., 2000. PEG-m-THPC–mediated photodynamic effects on normal rat tissues. Photochem. Photobiol. 72 (5), 696.

Hu, T., Wang, Z., Shen, W., Liang, R., Yan, D., Wei, M., 2021. Recent advances in innovative strategies for enhanced cancer photodynamic therapy. Theranostics 11 (7), 3278–3300. https://doi.org/10.7150/THNO.54227.

Huang, R.B., Mocherla, S., Heslinga, M.J., Charoenphol, P., Eniola-Adefeso, O., 2010. Dynamic and cellular interactions of nanoparticles in vascular-targeted drug delivery (review). Mol. Membr. Biol. 27 (7), 312–327. https://doi.org/10.3109/09687688.2010.522117.

Ismael, Amasha, Bachir, 2020. Optimized cylindrical diffuser powers for interstitial PDT breast cancer treatment planning: a simulation study. Biomed. Res. Int. 2020. https://doi.org/10.1155/2020/2061509.

Jang, S.J., 2019. Effects of donor-acceptor quantum coherence and non-markovian bath on the distance dependence of resonance energy transfer. J. Phys. Chem. C 123 (9), 5767–5775. https://doi.org/10.1021/ACS.JPCC.8B12481.

Jares-Erijman, E.A., Jovin, T.M., 2003. FRET imaging. Nat. Biotechnol. 21 (11), 1387–1395. https://doi.org/10.1038/NBT896.

Jia, Yifan, Zhihan, Chenyang, Ahmed, Lukui, 2021. Emerging exosomes and exosomal mirnas in spinal cord injury. Front. Cell Dev. Biol. 9. https://doi.org/10.3389/FCELL.2021.703989.

Jiang, Bai, Liu, Lv, Ren, Wang, 2019. Luminescent, oxygen-supplying, hemoglobin-linked conjugated polymer nanoparticles for photodynamic therapy. Angew. Chem. Int. Ed. 58 (31), 10660–10665. https://doi.org/10.1002/ANIE.201905884.

Jiang, Z., Shao, J., Yang, T., Wang, J., Jia, L., 2014. Pharmaceutical development, composition and quantitative analysis of phthalocyanine as the photosensitizer for cancer photodynamic therapy. J. Pharm. Biomed. Anal. 87, 98–104.

Jiayuan, Dou, Liming, 2017. Porphyrin photosensitizers in photodynamic therapy and its applications. Oncotarget 8 (46), 81591–81603. https://doi.org/10.18632/ONCOTARGET.20189.

Jori, G., Reddi, E., Cozzani, I., Tomio, L., 1986. Controlled targeting of different subcellular sites by porphyrins in tumour-bearing mice. Br. J. Cancer 53 (5), 615–621. https://doi.org/10.1038/BJC.1986.104.

Kato, H., 1996. History of photodynamic therapy past, present and future. Jpn. J. Cancer Chemother. 23 (1), 8–15.

Kercher, E.M., Zhang, K., Waguespack, M., Lang, R.T., Olmos, A., Spring, B.Q., 2020. High-power light-emitting diode array design and assembly for practical photodynamic therapy research. J. Biomed. Opt. 25 (06), 1. https://doi.org/10.1117/1.JBO.25.6.063811.

Kim, Y.R., Kim, S., Choi, J.W., Choi, S.Y., Lee, S.-H., Kim, H., Hahn, S.K., Koh, G.Y., Yun, S.H., 2015. Bioluminescence-activated deep-tissue photodynamic therapy of cancer. Theranostics 5 (8), 805. https://doi.org/10.7150/THNO.11520.

Konopka, K., Goslinski, T., 2007. Photodynamic therapy in dentistry. J. Dent. Res. 86 (8), 694–707. https://doi.org/10.1177/154405910708600803.

Koo, J., Schuster, G.B., 1977. Dioxetane chemiluminescence. The effect of deuterium substitution on the thermal decomposition of trans-3,4-diphenyl-1,2-dioxetane. J. Am. Chem. Soc. 99 (16), 5403–5408. https://doi.org/10.1021/JA00458A029.

Koo Lee, Y.E., Kopelman, R., 2011. Polymeric nanoparticles for photodynamic therapy. Methods Mol. Biol. 726, 151–178. https://doi.org/10.1007/978-1-61779-052-2_11.

Kou, J., Dou, D., Yang, L., Kou, J., Dou, D., Yang, L., 2017. Porphyrin photosensitizers in photodynamic therapy and its applications. Oncotarget 8 (46), 81591–81603. https://doi.org/10.18632/ONCOTARGET.20189.

Kryston, T.B., Georgiev, A.B., Pissis, P., Georgakilas, A.G., 2011. Role of oxidative stress and DNA damage in human carcinogenesis. Mutat. Res. – Fundam. Mol. Mech. Mutagen. 711 (1–2), 193–201. https://doi.org/10.1016/J.MRFMMM.2010.12.016.

Kumar, N., Raja, S., Van Houten, B., 2020. The involvement of nucleotide excision repair proteins in the removal of oxidative DNA damage. Nucleic Acids Res. 48 (20), 11227–11243. https://doi.org/10.1093/NAR/GKAA777.

Kun, H., Zhiquan, B., Xueying, Y., Meixing, Q., Hui, Q., Huang, W., 2019. Endogenous oxygen generating multifunctional theranostic nanoplatform for enhanced photodynamic-photothermal therapy and multimodal imaging. Theranostics 9 (25), 7697–7713.

Kwiatkowski, S., Knap, B., Przystupski, D., Saczko, J., Kędzierska, E., Knap-Czop, K., Kotlińska, J., Michel, O., Kotowski, K., Kulbacka, J., 2018. Photodynamic therapy – mechanisms, photosensitizers and combinations. Biomed. Pharmacother. 106, 1098–1107. https://doi.org/10.1016/J.BIOPHA.2018.07.049.

Labanauskiene, J., Gehl, J., Didziapetriene, J., 2007. Evaluation of cytotoxic effect of photodynamic therapy in combination with electroporation in vitro. Bioelectrochemistry 70 (1), 78–82. https://doi.org/10.1016/J.BIOELECHEM.2006.03.009.

Lale, S.V., Kumar, A., Prasad, S., Bharti, A.C., Koul, V., 2015. Folic acid and trastuzumab functionalized redox responsive polymersomes for intracellular doxorubicin delivery in breast cancer. Biomacromolecules 16 (6), 1736–1752. https://doi.org/10.1021/ACS.BIOMAC.5B00244.

Li, X.Y., Tan, L.C., Dong, L.W., Zhang, W.Q., Shen, X.X., Lu, X., Zheng, H., Lu, Y.G., 2020. Susceptibility and resistance mechanisms during photodynamic therapy of melanoma. Front. Oncol. 10. https://doi.org/10.3389/FONC.2020.00597.

Liang, X., Chen, H., Li, L., An, R., Komiyama, M., 2021. Ring-structured DNA and RNA as key players in vivo and in vitro. Bull. Chem. Soc. Jpn. 94 (1), 141–157. https://doi.org/10.1246/BCSJ.20200235.

Lin, Z., Chen, H., Lin, J.M., 2013. Peroxide induced ultra-weak chemiluminescence and its application in analytical chemistry. Analyst 138 (18), 5182–5193. https://doi.org/10.1039/C3AN00910F.

Liou, G.Y., Storz, P., 2010. Reactive oxygen species in cancer. Free Radic. Res. 44 (5), 479–496. https://doi.org/10.3109/10715761003667554.

Liu, J., Du, P., Liu, T., Córdova Wong, B.J., Wang, W., Ju, H., Lei, J., 2019. A black phosphorus/manganese dioxide nanoplatform: Oxygen self-supply monitoring, photodynamic therapy enhancement and feedback. Biomaterials 192, 179–188.

Magalhães, C.M., Esteves da Silva, J.C.G., Pinto da Silva, L., 2016. Chemiluminescence and bioluminescence as an excitation source in the photodynamic therapy of cancer: a critical review. ChemPhysChem, 2286–2294. https://doi.org/10.1002/CPHC.201600270.

Magnander, K., Elmroth, K., 2012. Biological consequences of formation and repair of complex DNA damage. Cancer Lett. 327 (1–2), 90–96. https://doi.org/10.1016/J.CANLET.2012.02.013.

Malatesti, N., Munitic, I., Jurak, I., 2017. Porphyrin-based cationic amphiphilic photosensitisers as potential anticancer, antimicrobial and immunosuppressive agents. Biophys. Rev. 9 (2), 149–168. https://doi.org/10.1007/S12551-017-0257-7.

Matsumoto, A., Suzuki, M., Hayashi, H., Kuzuhara, D., Yuasa, J., Kawai, T., Aratani, N., Yamada, H., 2017. Studies on pyrene and perylene derivatives upon oxidation and application to a higher analogue. Bull. Chem. Soc. Jpn. 90 (6), 667–677. https://doi.org/10.1246/BCSJ.20160337.

Mesquita, M.Q., Dias, C.J., Gamelas, S., Fardilha, M., Neves, M.G.P.M.S., Faustino, M.A.F., 2018. An insight on the role of photosensitizer nanocarriers for photodynamic therapy. An. Acad. Bras. Cienc. 90 (1), 1101–1130.

Milanesi, C., Sorgato, F., Jori, G., 1989. Photokinetic and ultrastructural studies on porphyrin photosensitization of hela cells. Int. J. Radiat. Biol. 55 (1), 59–69. https://doi.org/10.1080/09553008914550071.

Mir, L.M., 2014. Electroporation-based gene therapy: recent evolution in the mechanism description and technology developments. Methods Mol. Biol. 1121, 3–23. https://doi.org/10.1007/978-1-4614-9632-8_1.

Mohanty, N., Jalaluddin, M., Kotina, S., Routray, S., Ingale, Y., 2013. Photodynamic therapy: the imminent milieu for treating oral lesions. J. Clin. Diagn. Res. 7 (6), 1254–1257. https://doi.org/10.7860/JCDR/2013/5767.3088.

Mojzisova, H., Bonneau, S., Brault, D., 2007. Structural and physico-chemical determinants of the interactions of macrocyclic photosensitizers with cells. Eur. Biophys. J. 36 (8), 943–953. https://doi.org/10.1007/S00249-007-0204-9.

Müller, S.M., Galliardt, H., Schneider, J., George Barisas, B., Seidel, T., 2013. Quantification of Förster resonance energy transfer by monitoring sensitized emission in living plant cells. Front. Plant Sci., 4. https://doi.org/10.3389/FPLS.2013.00413/FULL.

Mutairi, 2007. Inactivation of metabolic enzymes by photo-treatment with zinc meta N-methylpyridylporphyrin. Biochim. Biophys. Acta 1770 (11), 1520–1527. https://doi.org/10.1016/j.bbagen.2007.06.006.

Mutairi, Craik, Batinic, Benov, 2007. Induction of oxidative cell damage by photo-treatment with zinc N-methylpyridylporphyrin. Free Radic. Res. 41 (1), 89–96. https://doi.org/10.1080/10715760600952869.

Nakamura, M., Tahara, Y., Fukata, S., Zhang, M., Yang, M., Iijima, S., Yudasaka, M., 2017. Significance of optimization of phospholipid poly(ethylene glycol) quantity for coating carbon nanohorns to achieve low cytotoxicity. Bull. Chem. Soc. Jpn. 90 (6), 662–666. https://doi.org/10.1246/BCSJ.20170003

Nelson, T.R., White, A.J., Bjorgaard, J.A., Sifain, A.E., Zhang, Y., Nebgen, B., Fernandez-Alberti, S., Mozyrsky, D., Roitberg, A.E., Tretiak, S., 2020. Non-adiabatic excited-state molecular dynamics: theory and applications for modeling photophysics in extended molecular materials. Chem. Rev. 120 (4), 2215–2287. https://doi.org/10.1021/ACS.CHEMREV.9B00447.

Nowis, D., Makowski, M., Stokłosa, T., Legat, M., Issat, T., Gołąb, J., 2005. Direct tumor damage mechanisms of photodynamic therapy. Acta Biochim. Pol. 52 (2), 339–352.

Ogilby, P.R., 2010. Singlet oxygen: there is indeed something new under the sun. Chem. Soc. Rev. 39 (8), 3181–3209. https://doi.org/10.1039/B926014P.

Oleinick, N.L., Evans, H.H., 1998. The photobiology of photodynamic therapy: cellular targets and mechanisms. Radiat. Res. 150 (5 SUPPL), 146–156.

Oliveira, H., Correia, P., Pereira, A.R., Araújo, P., Mateus, N., de Freitas, V., Oliveira, J., Fernandes, I., 2020. Exploring the applications of the photoprotective properties of anthocyanins in biological systems. Int. J. Mol. Sci. 21 (20), 1–33. https://doi.org/10.3390/IJMS21207464.

Ozturk, M.S., Rohrbach, D., Sunar, U., Intes, X., 2014. Mesoscopic fluorescence tomography of a photosensitizer (HPPH) 3D biodistribution in skin cancer. Acad. Radiol. 21 (2), 271–280. https://doi.org/10.1016/J.ACRA.2013.11.009.

Patrice, T., Moan, J., Peng, Q., 2003. An outline of the history of PDT. Photodyn. Ther. 1–18. https://doi.org/10.1039/9781847551658-00001.

Pavani, C., Iamamoto, Y., Baptista, M.S., 2012. Mechanism and efficiency of cell death of type II photosensitizers: effect of zinc chelation. Photochem. Photobiol. 88 (4), 774–781. https://doi.org/10.1111/J.1751-1097.2012.01102.X.

Peer, D., 2007. Nanocarriers as an emerging platform for cancer therapy. Nat. Nanotechnol. 2 (12), 751–760. https://doi.org/10.1038/nnano.2007.387.

Pouget, J.P., Douki, T., Richard, M.J., Cadet, J., 2000. DNA damage induced in cells by γ and UVA radiation as measured by HPLC/GC-MS and HPLC-EC and comet assay. Chem. Res. Toxicol. 13 (7), 541–549. https://doi.org/10.1021/TX000020E.

Prieto-Montero, R., Prieto-Castañeda, A., Katsumiti, A., Cajaraville, M.P., Agarrabeitia, A.R., Ortiz, M.J., Martínez-Martínez, V., 2021. Functionalization of photosensitized silica nanoparticles for advanced photodynamic therapy of cancer. Int. J. Mol. Sci. (12), 22. https://doi.org/10.3390/IJMS22126618.

Qi, Q., Taniguchi, M., Lindsey, J.S., 2019. Heuristics from modeling of spectral overlap in förster resonance energy transfer (FRET). J. Chem. Inf. Model. 59 (2), 652–667. https://doi.org/10.1021/ACS.JCIM.8B00753.

Real, P.J., Sierra, A., De Juan, A., Segovia, J.C., Lopez-Vega, J.M., Fernandez-Luna, J.L., 2002. Resistance to chemotherapy via Stat3-dependent overexpression of Bcl-2 in metastatic breast cancer cells. Oncogene 21 (50), 7611–7618. https://doi.org/10.1038/SJ.ONC.1206004.

Ricchelli, F., Gobbo, S., Jori, G., Moreno, G., Vinzens, F., Salet, C., 1993. Photosensitization of mitochondria by liposome-bound porphyrins. Photochem. Photobiol. 58 (1), 53–58. https://doi.org/10.1111/J.1751-1097.1993.TB04903.X.

Ricchelli, F., Jori, G., Moreno, G., Vinzens, F., Salet, C., 1990. Factors influencing the distribution pattern of porphyrins in cell membranes. J. Photochem. Photobiol. B 6 (1–2), 69–77. https://doi.org/10.1016/1011-1344(90)85075-8.

Robertson, C.A., Evans, D.H., Abrahamse, H., 2009. Photodynamic therapy (PDT): a short review on cellular mechanisms and cancer research applications for PDT. J. Photochem. Photobiol. B 96 (1), 1–8. https://doi.org/10.1016/J.JPHOTOBIOL.2009.04.001.

Rols, M.P., 2006. Electropermeabilization, a physical method for the delivery of therapeutic molecules into cells. Biochim. Biophys. Acta–Biomembr. 1758 (3), 423–428. https://doi.org/10.1016/J.BBAMEM.2006.01.005.

Rols, Pierre, M., 2008. Mechanism by which electroporation mediates DNA migration and entry into cells and targeted tissues. Methods Mol. Biol. 423, 19–33. https://doi.org/10.1007/978-1-59745-194-9_2.

Rong, P., Yang, K., Srivastan, A., Kiesewetter, D.O., Yue, X., Wang, F., Nie, L., Bhirde, A., Wang, Z., Liu, Z., Niu, G., Wang, W., Chen, X., 2014. Photosensitizer loaded nano-graphene for multimodality imaging guided tumor photodynamic therapy. Theranostics 4 (3), 229–239. https://doi.org/10.7150/THNO.8070.

Rosenthal, I., 1991. Phthalocyanines as photodynamic sensitizers. Photochem. Photobiol. 53 (6), 859–870. https://doi.org/10.1111/J.1751-1097.1991.TB09900.X.

Sahu, K., Bansal, H., Mukherjee, C., Sharma, M., Gupta, P.K., 2009. Atomic force microscopic study on morphological alterations induced by photodynamic action of toluidine blue O in Staphylococcus aureus and Escherichia coli. J. Photochem. Photobiol. B 96 (1), 9–16.

Sai, D.L., Lee, J., Nguyen, D.L., Kim, Y.P., 2021. Tailoring photosensitive ROS for advanced photodynamic therapy. Exp. Mol. Med. 53 (4), 495–504. https://doi.org/10.1038/S12276-021-00599-7.

Schuster, G.B., 1979. Chemiluminescence of organic peroxides. conversion of ground-state reactants to excited-state products by the chemically initiated electron-exchange luminescence mechanism. Acc. Chem. Res. 12 (10), 366–373. https://doi.org/10.1021/AR50142A003.

Shafirstein, G., Bellnier, D.A., Oakley, E., Hamilton, S., Habitzruther, M., Tworek, L., Hutson, A., Spernyak, J.A., Sexton, S., Curtin, L., Turowski, S.G., Arshad, H., Henderson, B., 2018. Irradiance controls photodynamic efficacy and tissue heating in experimental tumours: implication for interstitial PDT of locally advanced cancer. Br. J. Cancer 119 (10), 1191–1199. https://doi.org/10.1038/s41416-018-0210-y.

Shanshan, G., Zhong, B., Bin, L., 2021. Recent advances of AIE light-up probes for photodynamic therapy. Chem. Sci. 12 (19), 6488–6506. https://doi.org/10.1039/D1SC00045D.

Shi, 2010. Nanotechnology in drug delivery and tissue engineering: from discovery to applications. Nano Lett. 10 (9), 3223–3230. https://doi.org/10.1021/nl102184c.

Shi, J., Kantoff, P.W., Wooster, R., Farokhzad, O.C., 2016. Cancer nanomedicine: progress, challenges and opportunities. Nat. Rev. Cancer 17 (1), 20–37. https://doi.org/10.1038/nrc.2016.108.

Sies, H., Menck, C.F.M., 1992. Singlet oxygen induced DNA damage. Mutat. Res. DNAging 275 (3–6), 367–375. https://doi.org/10.1016/0921-8734(92)90039-R.

Silantyev, A.S., Falzone, L., Libra, M., Gurina, O.I., Kardashova, K.S., Nikolouzakis, T.K., Nosyrev, A.E., Sutton, C.W., Mitsias, P.D., Tsatsakis, A., 2019. Current and future trends on diagnosis and prognosis of glioblastoma: from molecular biology to proteomics. Cells 8 (8), 863. NLM (Medline). https://doi.org/10.3390/cells8080863.

Somssich, M., Simon, R., 2017. Studying protein–protein interactions in Planta using advanced fluorescence microscopy. Methods Mol. Biol. 1610, 267–285. https://doi.org/10.1007/978-1-4939-7003-2_17.

Sorrin, A.J., Ruhi, K., M., F., N., A., Karimnia, V., Polacheck, W.J., Celli, J.P., Huang, H.C., Rizvi, I., 2020. Photodynamic therapy and the biophysics of the tumor microenvironment. Photochem. Photobiol. 96 (2), 232–259. https://doi.org/10.1111/PHP.13209.

Sousa, R., Leandro, B., Cruz, C., Rosana, U., Paulo, S., THiago, F., Luiz, M., Paula, S., Lívio, A., Luciana, ….M., 2021. In vitro evaluation of physical and chemical parameters involved in aPDT of Aggregatibacter actinomycetemcomitans. Lasers Med. Sci. https://doi.org/10.1007/S10103-021-03267-4.

Spring, B.Q., Rizvi, I., Xu, N., Hasan, T., 2015. The role of photodynamic therapy in overcoming cancer drug resistance. Photochem. Photobiol. Sci. 14 (8), 1476–1491. https://doi.org/10.1039/C4PP00495G.

Stranadko, E.F., Ivanov, A.V., 2004. Current trends in photodynamic therapy of neoplasms and non-neoplastic diseases. Biofizika 49 (2), 380–383.

Sun, Luo, Feng, Cai, Zhuang, Xie, Chen, Chen, 2020. Aggregation-induced emission gold clustoluminogens for enhanced low-dose x-ray-induced photodynamic therapy. Angew. Chem. Int. Ed. Engl. 59 (25), 9914–9921. https://doi.org/10.1002/ANIE.201908712.

Sun, Wu, Wang, Zhou, Zhang, Cheng, Kan, Zhang, Yu, 2019. Turning solid into gel for high-efficient persistent luminescence-sensitized photodynamic therapy. Biomaterials 218. https://doi.org/10.1016/J.BIOMATERIALS.2019.119328.

Sun, Z., Matsuno, T., Isobe, H., 2018. Stereoisomerism and structures of rigid cylindrical cycloarylenes. Bull. Chem. Soc. Jpn. 91 (6), 907–921. https://doi.org/10.1246/BCSJ.20180051.

Sundaram, P., Abrahamse, H., 2020. Phototherapy combined with carbon nanomaterials (1d and 2d) and their applications in cancer therapy. Materials 13 (21), 1–22. https://doi.org/10.3390/MA13214830.

Tapeinos, C., Pandit, A., 2016. Physical, chemical, and biological structures based on ROS-sensitive moieties that are able to respond to oxidative microenvironments. Adv. Mater. 28 (27), 5553–5585. https://doi.org/10.1002/ADMA.201505376.

Theodossiou, T.A., Hothersall, J.S., De Witte, P.A., Pantos, A., Agostinis, P., 2009. The multifaceted photocytotoxic profile of hypericin. Mol. Pharm. 6 (6), 1775–1789. https://doi.org/10.1021/MP900166Q.

Tomasi, S., Kassal, I., 2020. Classification of coherent enhancements of light-harvesting processes. J. Phys. Chem. Lett. 11 (6), 2348–2355. https://doi.org/10.1021/ACS.JPCLETT.9B03490.

Tseng, S.C.G., Feenstra, R.P.G., Watson, B.D., 1994. Characterization of photodynamic actions of rose bengal on cultured cells. Invest. Ophthalmol. Vis. Sci. 35 (8), 3295–3307.

Tsubone, T.M., Martins, W.K., Pavani, C., Junqueira, H.C., Itri, R., Baptista, M.S., 2017. Enhanced efficiency of cell death by lysosome-specific photodamage. Sci. Rep. 7 (1). https://doi.org/10.1038/S41598-017-06788-7.

Uchida, K., 2007. Lipid peroxidation and redox-sensitive signaling pathways. Curr. Atheroscler. Rep. 9 (3), 216–221. https://doi.org/10.1007/S11883-007-0022-7.

Van Den Bergh, H., 1998. On the evolution of some endoscopic light delivery systems for photodynamic therapy. Endoscopy 30 (4), 392–407. https://doi.org/10.1055/S-2007-1001289.

Verma, S., Watt, G.M., Mai, Z., Hasan, T., 2007. Strategies for enhanced photodynamic therapy effects. Photochem. Photobiol. 83 (5), 996–1005. https://doi.org/10.1111/J.1751-1097.2007.00166.X.

Wachter, E., Dees, C., Harkins, J., Scott, T., Petersen, M., Rush, R.E., Cada, A., 2003. Topical rose bengal: pre-clinical evaluation of pharmacokinetics and safety. Lasers Surg. Med. 32 (2), 101–110. https://doi.org/10.1002/LSM.10138.

Wang, S., Fan, W., Kim, G., Hah, H.J., Lee, Y.-E.K., Kopelman, R., Ethirajan, M., Gupta, A., Goswami, L.N., Pera, P., Morgan, J., Pandey, R.K., 2011. Novel methods to incorporate photosensitizers into nanocarriers for cancer treatment by photodynamic therapy. Lasers Surg. Med. 43 (7), 686–695. https://doi.org/10.1002/LSM.21113.

Wang, S., Huang, P., Chen, X., 2016. Hierarchical targeting strategy for enhanced tumor tissue accumulation/retention and cellular internalization. Adv. Mater. 28 (34), 7340–7364. https://doi.org/10.1002/ADMA.201601498.

Wang, L., Li, D., Hao, Y., Niu, M., Hu, Y., Zhao, H., Chang, J., Zhang, Z., Zhang, Y., 2017. Gold nanorod-based poly(lactic-co-glycolic acid) with manganese dioxide core-shell structured multifunctional nanoplatform for cancer theranostic applications. Int. J. Nanomed. 12, 3059–3075.

Wilhelm, P., Vogelsang, J., Höger, S., Lupton, J.M., 2019. Homo-FRET in π-conjugated polygons: intermediate-strength dipole-dipole coupling makes energy transfer reversible. Nano Lett. 19 (8), 5483–5488. https://doi.org/10.1021/ACS.NANOLETT.9B01998.

Wittmann, B., Wenzel, F.A., Wiesneth, S., Haedler, A.T., Drechsler, M., Kreger, K., Köhler, J., Meijer, E.W., Schmidt, H.W., Hildner, R., 2020. Enhancing long-range energy transport in supramolecular architectures by tailoring coherence properties. J. Am. Chem. Soc. 142 (18), 8323–8330. https://doi.org/10.1021/JACS.0C01392.

Worster, S., Stross, C., Vaughan, F.M.W.C., Linden, N., Manby, F.R., 2019. Structure and efficiency in bacterial photosynthetic light harvesting. J. Phys. Chem. Lett. 10 (23), 7383–7390. https://doi.org/10.1021/ACS.JPCLETT.9B02625.

Xiao, Y.F., Chen, J.X., Li, S., Tao, W.W., Tian, S., Wang, K., Cui, X., Huang, Z., Zhang, X.H., Lee, C.S., 2020. Manipulating exciton dynamics of thermally activated delayed fluorescence materials for tuning two-photon nanotheranostics. Chem. Sci. 11 (3), 888–895. https://doi.org/10.1039/C9SC05817F.

Xu, Y., Yang, X., Wang, T., Yang, L., He, Y.Y., Miskimins, K., Qian, S.Y., 2018. Knockdown delta-5-desaturase in breast cancer cells that overexpress COX-2 results in inhibition of growth, migration and invasion via a dihomo-γ-linolenic acid peroxidation dependent mechanism. BMC Cancer 18 (1). https://doi.org/10.1186/S12885-018-4250-8.

Yang, M., Yang, T., Mao, C., 2019. Enhancement of photodynamic cancer therapy by physical and chemical factors. Angew. Chem. Int. Ed. 58 (40), 14066–14080. https://doi.org/10.1002/ANIE.201814098.

Yang, Read, Mi, Baisden, Reardon, J.M., Helmke, L., Sheng, 2008. Semiconductor nanoparticles as energy mediators for photosensitizer-enhanced radiotherapy. Int. J. Radiat. Oncol. Biol. Phys. 72 (3), 633–635. https://doi.org/10.1016/J.IJROBP.2008.06.1916.

You, Y., Gibson, S.L., Hilf, R., Davies, S.R., Oseroff, A.R., Roy, I., Ohulchanskyy, T.Y., Bergey, E.J., Detty, M.R., 2003. Water soluble, core-modified porphyrins. 3. Synthesis, photophysical properties, and in vitro studies of photosensitization, uptake, and localization with carboxylic acid-substituted derivatives. J. Med. Chem. 46 (17), 3734–3747. https://doi.org/10.1021/JM030136I.

Yu, L., Yang, X., Li, X., Qin, L., Xu, W., Cui, H., Jia, Z., He, Q., Wang, Z., 2021. Pink1/PARK2/mROS-dependent mitophagy initiates the sensitization of cancer cells to radiation. Oxid. Med. Cell. Longev. 2021, 5595652. http://www.ncbi.nlm.nih.gov/pubmed/34306311.

Zhang, Wen, Huang, Wang, Hu, Xing, 2018. Mitochondrial specific photodynamic therapy by rare-earth nanoparticles mediated near-infrared graphene quantum dots. Biomaterials 153, 14–26. https://doi.org/10.1016/J.BIOMATERIALS.2017.10.034.

Zhang, Y., Hao, Y., Chen, S., Xu, M., 2020. Photodynamic therapy of cancers with internal light sources: chemiluminescence, bioluminescence, and cerenkov radiation. Front. Chem. 8. https://doi.org/10.3389/FCHEM.2020.00770.

Zhao, P., Zheng, M., Luo, Z., Fan, X., Sheng, Z., Gong, P., Chen, Z., Zhang, B., Ni, D., Ma, Y., Cai, L., 2016. Oxygen nanocarrier for combined cancer therapy: oxygen-boosted ATP-responsive chemotherapy with amplified ROS lethality. Adv. Healthc. Mater. 5 (17), 2161–2167. https://doi.org/10.1002/ADHM.201600121.

Zhong, X., Wang, X., Zhan, G., Tang, Y., Yao, Y., Dong, Z., Hou, L., Zhao, H., Zeng, S., Hu, J., Cheng, L., Yang, X., 2019. NaCeF4:Gd, Tb scintillator as an X-ray responsive photosensitizer for multimodal imaging-guided synchronous radio/radiodynamic therapy. Nano Lett. 19 (11), 8234–8244. https://doi.org/10.1021/ACS.NANOLETT.9B03682.

Zhou, Z., Song, J., Nie, L., Chen, X., 2016. Reactive oxygen species generating systems meeting challenges of photodynamic cancer therapy. Chem. Soc. Rev. 45 (23), 6597. https://doi.org/10.1039/C6CS00271D.

Zhu, T., Van Voorhis, T., 2019. Unraveling the fate of host excitons in host-guest phosphorescent organic light-emitting diodes. J. Phys. Chem. C 123 (16), 10311–10318. https://doi.org/10.1021/ACS.JPCC.9B02820.

Zou, Z., Chang, H., Li, H., Wang, S., 2017. Induction of reactive oxygen species: an emerging approach for cancer therapy. Apoptosis 22 (11), 1321–1335. https://doi.org/10.1007/S10495-017-1424-9.

Chapter 4

Fundamentals of photodynamic therapy

Mrunal M. Yawalkar[a], Samvit Menon[b], Hendrik C. Swart[b], Sanjay J. Dhoble[a]
[a]Department of Physics, Rashtrasant Tukadoji Maharaj Nagpur University, Nagpur, Maharashtra, India
[b]Department of Physics, University of the Free State, Bloemfontein, South Africa

4.1 Introduction

According to the World Health Organization, cancer caused 9.6 million deaths in 2018, which is about 17% of all the deaths worldwide (Wild et al., 2020). This is a 4% increase from what was reported in 2008 where cancer deaths accounted for 13% of all deaths worldwide (Ferlay et al., 2008). Even though regular testing and monitoring help in early detection, diagnosis and treatment of various cancers are the most effective ways to improve survival, there is a need to improve on current therapeutic methods to make them more efficient, effective and affordable to patients. Chemotherapy, radiotherapy, immunotherapy, small molecule-based therapies, and surgery are the conventional therapeutic strategies adopted to treat cancer. Among these, while chemotherapy is the most common mode of treatment, it is also associated with severe systemic side-effects (Olsen et al., 2019). In case of surgery, there is always a high possibility of recurrence associated with it and since radiotherapy involves the use of ionizing radiation, there exists a possibility of inducing widespread damage to normal tissues (Cohen–Jonathan et al., 1999). While improvising on existing cancer treatment modalities is necessary, there is also a need for research on alternate treatment modalities that are safe, effective, and affordable.

Light has been used as a therapeutic agent for several thousands of years to treat several diseases such as rickets, psoriasis, skin cancer, etc. (Daniell and Hill, 1991, Ackroyd et al., 2001, Bensasson, 2013, Finsen, 1901). Niels Rydberg Finsen, regarded as the Father of Phototherapy, advocated use of light of different colors for the treatment of various medical conditions. His most notable finding was that the light from sunlight or carbon arc could be successfully used for the treatment and cure of lupus vulgaris, a tuberculotic condition of the skin, which was then widespread in Nordic countries during winter. He believed implicitly in the healing power of light and worked strenuously for the benefit of his affected and suffering countrymen. A Medical Light Institute named after him was set up at Copenhagen. Finsen was awarded the Noble prize for Physiology Medicine in 1903 (Bonnett, 2000; Christensen, 1923). Over decades, phototherapy extended to diverse areas and is also used for psoriasis treatment, jaundice in newborn and apoptosis of tumours and vitamin D synthesis to cure rickets or osteomalacia.

Oscar Raab, a medical student, in 1897–98 was part of a career building programme at the pharmacology laboratory of Professor Hermann von Tappeiner at the Ludwig Maximillian University in Munich. He studied the effect of various fluorescing dyes like acridine, quinine, methylphosphin, eosin, etc., and discovered that paramecia were killed in presence of low concentrations of acridine in bright light. Conversely, they survived in absence of light. This laid the foundation for use of light in addition to photosensitizers and presence of molecular oxygen for beneficial termination of target cells.

It was in 1904, that von Tappeiner and Jodbauer introduced "photodynamische wirkung," the term, best translated as "photodynamic action" for this process which mainly depends on the absorption spectra of the dyes. Tappeiner proposed that dye-sensitized photodamage possess therapeutic values essential in dermatology. Later, he and Jesionek also published the result of using eosin and light for the treatment of various diseases like psoriasis, herpes, skin cancer which marked the starting point of photodynamic therapy (PDT) (Bonnett, 2000; Krasnovsky, 2007).

However, the use of light in cancer therapeutics began much later in 1960 through studies carried out by by R. L. Lipson and S. Schwartz at the Mayo Clinic in Rochester. During their studies, they observed that when crude hematoporphyrin was injected, neoplastic lesions fluoresced that could be seen during surgery. To optimize tumor localization, hematoporphyrin was treated with various chemicals, less active porphyrins-monomers were removed and a purified version of hematoporphyrin derivative (HpD) was obtained. This derivative named as Photofrin, served as one of the most widely used photosensitizers (PSs) in modern day PDT for cancer treatment (Dougherty, 1996). With the aid of this commercially acceptable photosensitizing drug, suitable light sources and adequate clinical trials T.J. Dougherty working as a research associate, at Roswell Park Cancer Institute in Buffalo, New York successfully demonstrated clinical importance of PDT as cancer therapy in 1970s. For this, he is known as Father of PDT. However, due to the phototoxicity effects of Photofrin and its ability to absorb radiation only up to 640 nm, novel PSs are being developed with lower phototoxicities and stronger absorption between 650 and 900 nm (Gomer, 1991; Pass, 1993).

PDT is an alternative tumor-removing oncologic intervention that has evolved into an effective site-specific therapeutic modality. It involves injecting a tumor localizing PS that is activated upon illumination at a particular wavelength. The activated PS transfers energy to molecular oxygen which then forms a reactive oxygen species (ROS) such as cytotoxic singlet oxygen ($^1O_2$) that oxidizes cellular macromolecules leading to tumour cell death (Oleinick et al., 2002). These ROS have short lifetimes (<0.04 μs) in biological media and consequently, also have a small reactive radius (<0.02 μm). Thus, compared to chemotherapy that possesses systemic toxicity and radiotherapy that uses ionizing radiation that can damage healthy tissues, PDT does not contain any toxic components. Besides, PDT is minimally invasive, repeatable and reduces long term morbidity, all of which improves a patient's quality of life. Over the years, PDT has been effective in the treatment of superficial bladder cancer (Yavari et al., 2011), early and obstructive lung cancer (Allison et al., 2011), skin cancer (Kostovic et al., 2012), oral cancer (Grant et al., 1993), and colorectal cancer (Herrera-Ornelas et al., 1986). The discovery that PDT leads to apoptosis in malignant cells by Agarwal et al. in 1991 (Agarwal et al., 1991) provided a benchmark for the widespread use of this method in cancer therapeutics.

Unfortunately, despite several publications on the successful use of PDT for treating tumours in animal models, it has not yet garnered success as a first-line clinical intervention in humans due to the lack of an effective PS, challenges in designing an ideal PS, determining the correct light dosimetry for efficacious therapeutics, problems in monitoring treatment response, etc. Nanoparticles have been approached as a potential solution to overcome some of the limitations shown by various PSs in PDT......

4.2 Basic concept of photodynamic therapy

Photodynamic therapy also known as photoradiation therapy or photochemotherapy involves administration of a PS, followed by a distribution interval, and subsequent illumination of the tumor area with light of a suitable wavelength. Appropriate dose of PS is administered to the patient, topically, orally or systemically and allowed to accumulate in the malignant cells. The PS confined to specific tumor cells or tissues is inactive in dark and needs to be activated by the low power, tissue penetrating light of specific wavelength that matches the PS absorption spectrum . After photon absorption, PS goes to an excited singlet or triplet state that can react with ambient oxygen to produce free radicals and singlet oxygen that are cytotoxic to targeted tissue (O'Connor et al., 2009; Dai et al., 2012). These cytotoxic reactive oxygen series (ROS) destroy the malignant cells by apoptosis and/or necrosis, cease microvasculature in the tumor and activate the host immune system. ROS are highly reactive but short-lived, so tumor damage occurs only in the proximity of production site. In contrast to immunosuppressive treatments like radiotherapy and chemotherapy, PDT results in acute inflammation, expression of heat-shock proteins, invasion and infiltration of the tumor by leukocytes, and might increase the presentation of tumor-derived antigens to T cells (Castano et al., 2006). Success of PDT depends on administration of photosensitizer, presence of oxygen, and local illumination of tumor tissue by light. PS, oxygen and light are individually non- toxic, but when irradiated, initiate photochemical reactions that generate highly cytotoxic reactive singlet oxygen imperative for damage and death of cancer cells (Muniyandi et al., 2020).

4.2.1 Photosensitizers

PSs absorb the light energy in the phototherapeutic window of 600–900 nm and generate photochemical reactions conducive for the termination of abnormal cell growth. PSs employed in clinical practices have been mostly developed from the families of porphyrins, chlorophylls and dyes with unique properties (Allison and Downie, 2004). PSs used in PDT, porphyrins, and nonporphyins (O'Connor et al., 2009) are nontoxic natural or synthetic dyes or chemical compounds with unique photophysical and photochemical properties. Tetrapyrrole biomolecules of haem, chlorophyll, and bacteriochlorophyll mostly forms the basis of several PSs used for cancer treatment (Abrahamse and Hamblin, 2016).

An ideal PS ought to be a single pure compound with known chemical composition, good quality control, cost-effective and easily reproducible. It should be soluble in body's tissue fluids and exhibit highly selective incorporation in tumor tissue for greater vascular damage. At the same time it must be non toxic for the normal cells and can be quickly expelled from the organisms upon completion of the treatment, to minimize the adverse effects of phototoxicity. It must be activated in the phototherapeutic window (Abrahamse and Hamblin, 2016; Ion, 2000) with a high extinction coefficient. Light absorption within this optical window (600–900 nm) by the common physiological chromophores including melanin, haemoglobin and water is low; which enhances the vascular permeability of this incident light and leads to increased tumoral accumulation of PS (Mallidi et al., 2016). PS must possess low dark toxicity and display toxicity only after excitation by illumination (Sanadi, 2013). Its triplet excited state must be long-lived to raise singlet oxygen quantum yield essential for tumor devastation. Efficient fluorescence of these therapeutic agents helps envisage malignant cells and facilitate their destruction (Ormond and Freeman, 2013).

4.2.1.1 Porphyrin-based photosensitizers

Most PSs used in cancer treatment are porphyrins based, a class of tetrapyrroles with intense colors and marvellous biochemical and photochemical characteristics and are essential to sustain life. They have derived name from Greek word "porphura" meaning purple (O'Connor et al., 2009). These are mostly made up of highly conjugated, heterocyclic macrocycle called porphine made up of four pyrrolic sub-units. These sub-units are joined on opposing sides at α position (1, 4, 6, 9, 11, 14, 16, 19) of the porphine macrocycle in Fig. 4.1 by four methine (CH) bridges (5, 10, 15, 20) known as the *meso*-carbon positions. This conjugated porphine macrocycle may be replaced at (2, 3, 7, 8, 12, 13, 17, 18) β- and/or meso-positions. If meso- or β-hydrogens are replaced with non-hydrogen atoms or groups, the derived compounds are termed as porphyrin (Josefsen and Boyle, 2008). They show characteristic intense energy absorption band in UV-VIS region near 400 nm and weaker absorptions in 450–700 nm range. The strong Soret or B band near 400 nm occurs on account of electronic transition from ground state to second excited singlet state ($S_0 \rightarrow S_2$). The transitions from ground to first excited singlet state ($S_0 \rightarrow S_1$) are the red-edge Q band witnessed in the form of four bands with weak intensity. Fluorescence occurs mainly due to the electron transition from first excited singlet state to the lower energy ground state owing to 22π electron system. In addition, porphyrins are soluble in various polar solutions. They can easily penetrate the membrane cells, incorporate easily in neoplastic tissues but prove non-toxic for the healthy cells and are hence considered promising sensitizing candidates for PDT (Josefsen and Boyle, 2008). Depending on their development PSs are further listed as first, second and third generation photosensitizers.

4.2.1.1.1 First generation photosensitizers

1841 marked the birth of PS when Scherer separated hematoporphyrin from dried blood by the removal of iron (Kou et al., 2017). Hausmann in 1908 used this hematoporphyrin (Fig. 4.2A) as a PS and reported successful destruction of paramecia and red blood cells and sensitization in mice on exposure to light (Lipson and Baldes, 1960). However, the first human study of PS was done by Meyer-Betz in 1913 when he injected himself with hematoporphyrin and experienced sensitivity to light for two months (Valles). Modern era of PDT began in 1960 when S. Schwartz and R.L. Lipson at Mayo Clinic discovered fluorescence of neoplastic lesions on injection of crude hematoporphyrin. Schwartz treated hematoporphyrin with sulfuric and acetic acid to obtain the optimal mixture, which he termed as HpD. In 1976, Kelly and Snell used HpD first time for the detection and PDT of cancer in humans (Dougherty et al., 1998). This water soluble PS proved effective for treating the carcinomas of brain, lungs, larynx, skin, esophagus, and stomach to a certain extent (Kou et al., 2017).

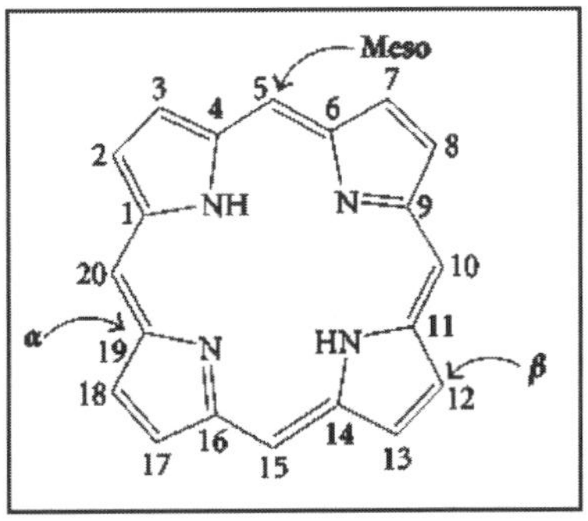

FIG. 4.1 Porphine macrocycle (Josefsen and Boyle, 2008).

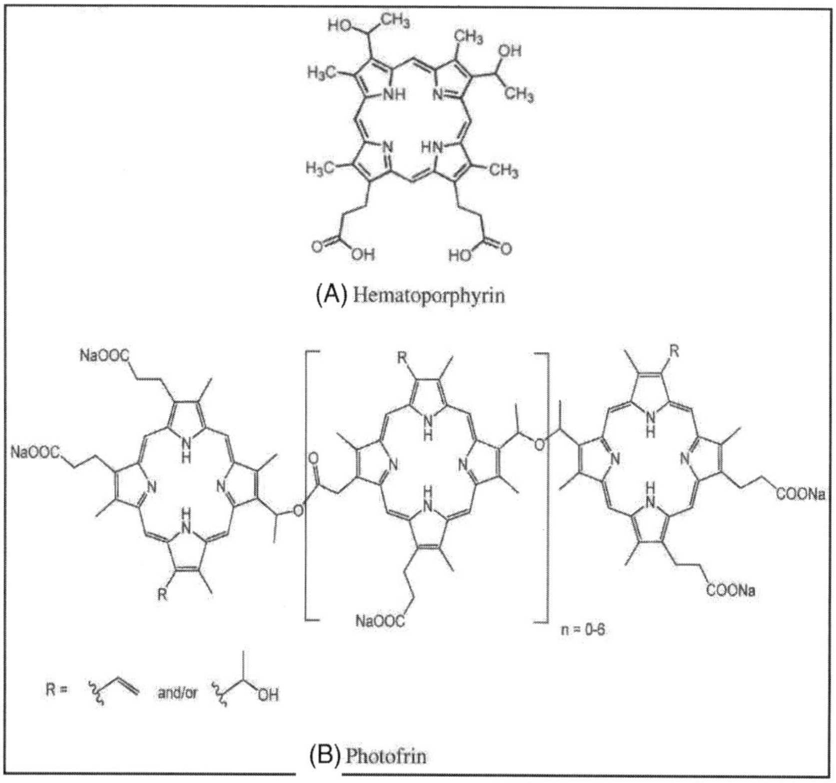

FIG. 4.2 First generation photosensitizer: (A) hematoporphyrin (Gunaydin et al., 2021) and (B) photofrin (Gallardo-Villagrán et al., 2019).

HpD was partially purified by removing less active porphyrins, the monomers to obtain Porfimer sodium. HpD and Porfimer sodium (Photofrin - QLT Phototherapeutics inc., Vancouver, BC) (Zane et al., 2001) are the first most widely used systemic PS in clinical PDT. Photofrin (Fig. 4.2B) also known as dihematoporphyrin ether or Porfimer sodium is a complex mixture of dimeric and oligomeric compounds with two to nine porphyrin unit linked mostly by ether bonds. Photofrin is used successfully in the treatment of early and late stage lung, esophageal and bladder cancers, cancers of head, brain, neck and breast, Kaposi's sarcoma, Barrett's esophagus with high-grade dysplasia, psoriasis and several other malignant and nonmalignant diseases (O'Connor et al., 2009; Tjahjono, 2006). Extensive scientific and clinical efforts put in by Dr T. J. Dogherty et.al. at Roswell Park Cancer Institute in Buffalo, NY gained regulatory approval for Photofrin by the Food and Drug Administration Modernization Act of 1997 (the Modernization Act) (FDA) and was used for cancer treatment across 40 countries worldwide including USA (Abrahamse and Hamblin, 2016) Canada, France, Japan, Netherlands and Italy. Photofrin is usually administered intravenously in either 5% dextrose or 0.9% sodium chloride at a concentration of 2 mg/kg of patient's weight. 24 to 48 h post administration of Photofrin, the tumor is irradiated by a 630 nm laser. The light activates Photofrin and sets necrosis in the tumor 48 hours post-irradiation resulting in partial or total tumor destruction. Normally, it takes 6 weeks for healing and the patient needs to avoid direct contact to sunlight during the period. Quantitative histalogical diagnosis carried out reveals the PDT response. The patient may be treated again if the tumor exists (Caliskan and Bilgin, 2018).

Photofrin is considered as "gold standard" for its frequent use as a PDT agent in oncology. However, Photofrin has low molar extinction coefficient of $1170 \, M^{-1}cm^{-1}$, which necessitates high dosage for therapeutic effect. It causes prolonged skin photosensitivity that lasted for months after PDT. It's sub optimal tumor selectivity, complex nature consisting of more than 60 components, poor tissue penetration on account of wavelength absorption at 630 nm paved way for the improved second generation PS (O'Connor et al., 2009).

4.2.1.1.2 Second generation photosensitizers

To surmount the inadequacies of first generation PSs-HpD and its commercial variants like Photofrin, Photogen, Photosan, and Photocarcinorin (Alexandrova et al., 2004), and enhance the efficacy of PS molecules; other tetrapyrrole structures like chlorins, phthalocyanines, pheophorbides, purpurins, etc., with strong absorbance in deep spectral region were studied. Second generation PS are mainly derived from porphyrin and porphyrin related structures and possess increased chemical purity and reproducibility that ensured better absorbance in the phototherapeutic window with enhanced light penetration, reduced drug dosage, less skin sensitivity and faster clearance from normal tissue with favorable pharmacokinetics (Straten et al., 2017). Some Second generation PSs with potential benefits and tried in clinical settings are listed here.

1. Protoporphyrins
   Protoporphyrin IX, the precursor of heme, brings about metabolism of heme by combination of mitochondrial transport proteins. This potent PS has a longer absorption wavelength in erythroleukemia cells. 5-Aminolaevulinic acid (ALA) has been found to be a biosynthetic precursor of protoporphyrin IX (Kou et al., 2017; Chen et al., 2012). On systemic, topical or oral administration of ALA (Fig. 4.3A), it is absorbed by the epidermal cells and metabolically converted into protoporphyrin IX. Though protoporphyrin IX is a precursor of haem, limited supplies of iron, inhibit its conversion to haem and is endogenously accumulated in target tissue. It shows maximum light absorption at 409 nm with smaller peaks occurring at 509 nm, 584 nm and 625 nm. Post ALA administration, protoporphyrin IX peak accumulation level can be achieved in less than 14-18 hours (Cantisani et al., 2014). It can also be excreted rapidly from the patient's body as compared to other PSs and hence minimizes photosensitivity (Juzeniene et al., 2009). Oral ALA-based PDT has efficiently cured conditions of Barrett's esophagus, pre-cancerous dysplasia and early esophageal carcinomas (Hasan et al., 2002). According to Fehr et al. the uptake of ALA and photodynamic destruction of glandular epithelium causes

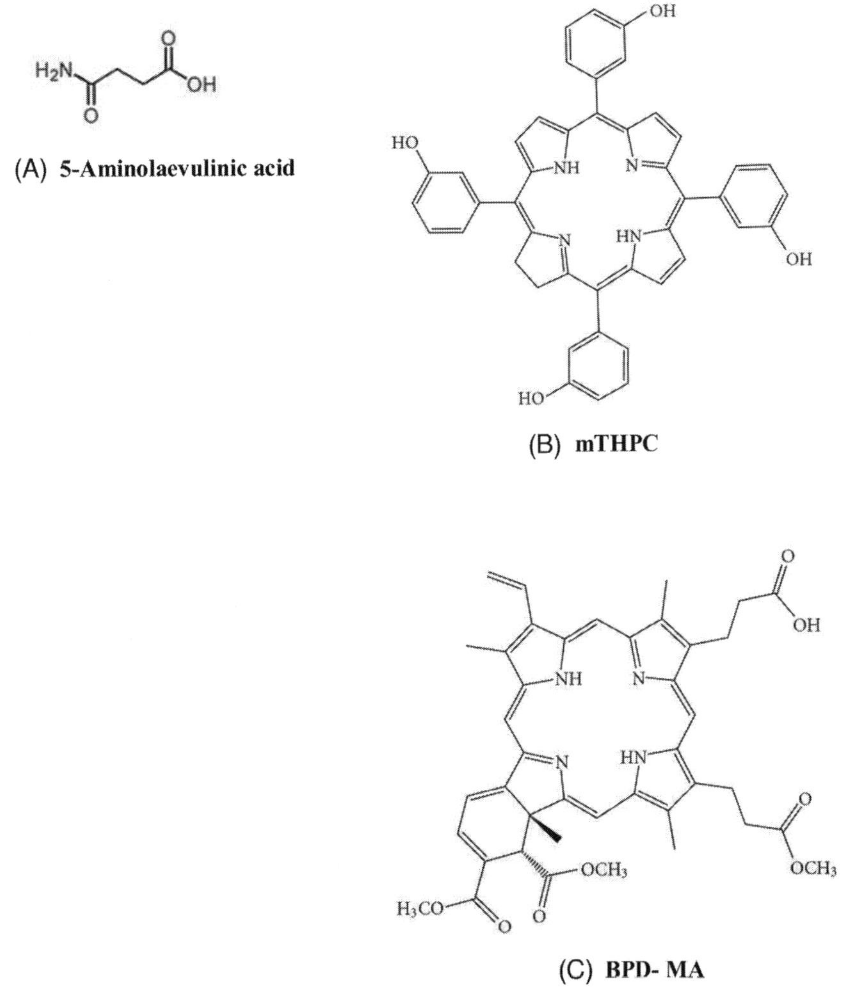

FIG. 4.3 Second generation photosensitizers (A) (Gunaydin et al., 2021), (B–H) (Josefsen and Boyle, 2008).

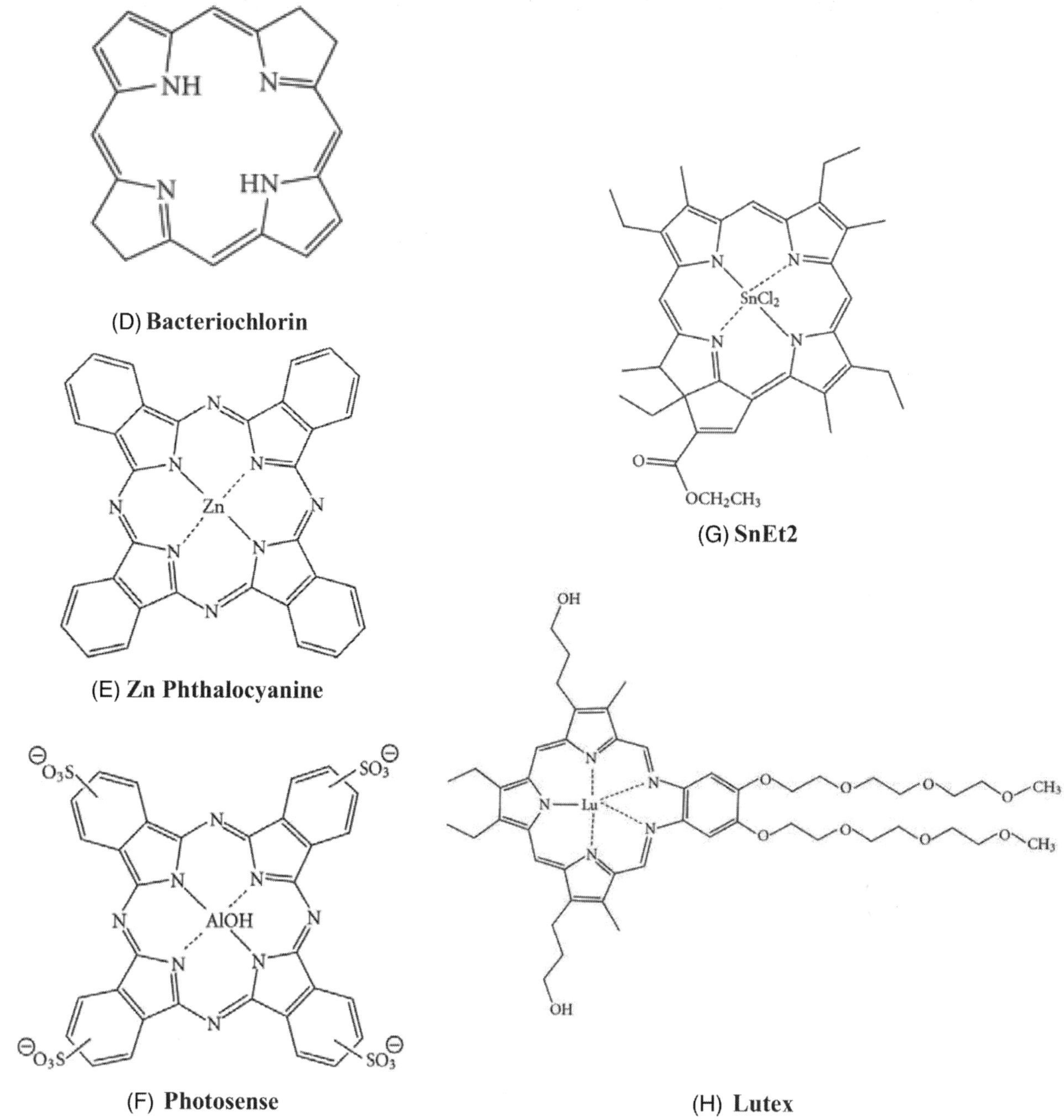
NH
N
HN
(D) Bacteriochlorin
Zn
(E) Zn Phthalocyanine
AlOH
SO3
O3S
(F) Photosense
SnCl2
OCH2CH3
O
(G) SnEt2
OH
Lu
CH3
(H) Lutex

FIG. 4.3 (Cont'd)

irreversible endometrial ablation (Fehr et al., 1996). A phase I trial of ALA-mediated PDT in 11 patients with oral leukoplakia helped to establish its benefits and safe dose of up to 4 J/cm^2. ALA approved by FDA and marketed as Levulan by DUSA Pharmaceutical Incorporation, Toronto USA (Josefsen and Boyle, 2008) is found effective to treat many diseases like cutaneous superficial and nodular basal cell carcinoma, superficial head and neck cancer, gastrointestinal cancers, etc. (Chilakamarthi and Giribabu, 2017). Esterified ALA, methyl aminolevulinate (MAL) with enhanced tissue penetration capability, is marketed as Metvix by Photocure ASA (Oslo, Norway) as a potent PS for the treatment of basal cell carcinoma and other skin lesions and got approval in Europe and Australia. Other PSs like benzyl (Benvix) and hexyl ester (Hexvix) derivatives registered by Photocure ASA are used for gastrointestinal cancer and diagnosis of bladder cancer (Josefsen and Boyle, 2008; Chilakamarthi and Giribabu, 2017).

2. Chlorins

Chlorins are natural photosensitive chlorophyll derivatives with twenty π electrons in the aromatic ring. Owing to their strong absorptions in the red region, they have been probed as PS or PDT of cancerous diseases (Pavlíčková et al., 2019). Chlorin based PSs show moderate Q absorption band in the red region (600–700 nm) due to a reduced double bond in the pyrrole ring that enhances its tissue penetration capability. They can be obtained by isolation of naturally occurring chlorophylls, semi-synthesis using naturally occurring chlorophylls as the starting material, synthetic modulation using porphyrin precursors and de novo synthesis of gem-dialkylchlorins (Dias and Mfouo-Tynga, 2020).

Water solubility of chlorin based PSs was improved by their conjugation with amino acids, peptides, boron, etc. Chlorin derivates with hydrophobic carbon chains displayed good phototoxicity, cell permeability and low toxicity in the dark. A pure chemical, meta-tetrahydroxyphenylchlorin (mTHPC) (Fig. 4.3B) with generic name temoporfin and trade name Foscan (Biolitec Pharma,Germany (Senge and Brandt, 2011)), and with a intense absorption peak at 652 nm was approved in 2001 in Europe to treat head and neck cancers. Another chemically pure hydrophilic chlorin, NPe6 (mono L-aspartyl chlorin) or taloporfin (Laserphyrin for Injection, Meiji Seika Pharma; Meiji Seika Pharma, Tokyo, Japan (Yano et al., 2021)), localize in lysosomes and cause apoptosis by lysosomal damage (Chilakamarthi and Giribabu, 2017). It was approved in Japan to cure early stage centrally located lung cancer. NPe6 proved safe and effective for PDT with its major absorption band at 664 nm, molar extinction coefficient of 4×10^4 $M^{-1}cm^{-1}$ and rapid clearance from the body. Optimal irradiation of tumor areas within 2 to 4 h after NPe6 administration and minimal resultant skin photosensitivity thereafter, vascular stasis resulting from platelet aggression and thrombos formation, and direct tumor cytotoxicity was found exploitable for PDT (O'Connor et al., 2009; Gomer and Ferrario, 1990). Clinical work carried out later at Light Sciences Oncology established that LSII, formerly known as NPe6 could successfully be used even for PDT of large tumors (Gomer, 2010).

The so-called benzoporphyrin derivative mono-carboxylic acid BPD-MA (Fig. 4.3C) is a most promising second generation PS. BPD-MA has a strong absorption peak near 700 nm, which allows deeper tissue penetration of light and improved activation than Photofrin. Previous studies carried out in animal tumor models have proved that BPD-MA serves as an effective PS if the animals are treated 3 h post intravenous injection of the drug. Tumor response was absent if the treatment was provided 24 h after drug administration (Meunier et al., 1994). Studies undertaken by Ritcher et. al on 80 μg of drug distribution in normal and P815 (mastocytoma) or M1 (rhabdomyosarcoma) tumor-bearing DBA/2 J mice further revealed that ^{3}H-BPD-MA localizes well in tumors than in the normal tissues except that in kidney, liver and spleen. ^{3}H-BPD-MA levels were found highest in all tumor tissues at 3 h post injection and receded rapidly during the first 24 h. 60% of the injected dose cleared from the body through bile and feces while only 4% cleared through kidneys and urine. Also, at 3h the drug retained its total photosensitizing traits whereas only 39% was present at 24 h and even less in liver and kidney (Richter et al., 1990).

BPD-MA or Verteporfin of QLT Inc. Vancouver, British Columbia, Canada (Samkoe et al., 2007) with brand name Visudyne got approval in North America and Europe to be used for the treatment of age-related macular degeneration (AMD) (Delmarre et al., 2001). On account of the success of clinical trials it turned out to be the first PS that got approval for the treatment of subfoveal choroidal neovascularisation (CNV), an acceptable treatment that stabilizes or slows visual acuity loss in adult patients with predominantly classic or occult with no classic subfoveal CNV secondary to AMD, or subfoveal CNV secondary to pathological myopia or POHS (Keam et al., 2004). This hydrophobic chlorin- like photsensitizer is highly effective in vivo with greatly reduced skin sensitivity as compared to other PSs. Its hydrophobic nature assists fast accumulation at critical intracellular membranous organelles. However, its self aggregation tendency in aqueous media hampers interaction through cell membrane which has unfavorable effect on drug pharmacokinetics and biodistribution. So BPD-MA in monomeric form is formulated in lipids for intravenous delivery (Chowdhary et al., 2003). Oku et al. liposomalized BPD-MA into long-circulating glucuronate-modified liposomes and injected it intravenously into Balb/c mice having Meth A sarcoma. Tumor regression and complete curing with 6mg/kg drug dose and irradiation by laser light gave four times better result than the antiangiogenic PDT in animals on treatment with BPD-MA solution or BPD-MA entrapped in conventional liposomes (Oku et al., 1997; Ichikawa et al., 2004).

3. Bacteriochlorins

Reduction of two double bonds in the pyrrole ring produced a class of highly active tetrapyrrole macrocycles – bacteriochlorins (Fig. 4.3D). This unsaturated ring structure of bacteriochlorins naturally occur in the photosynthetic pigments bacteriochlorophylls a and b of purple anaerobic phototrophic bacteria of the Rhodospirillales and Rhizobiales orders. Bacteriochlorins show strong absorption in 700–900 nm, near infrared region (NIR), which is crucial for light harvesting and favorable to treat deep tumors located up to 8 mm (Zhu et al., 2018).

However, naturally occurring derivates of bacteriochlorins depict low stability, tendency towards quick photodegradation, conversion to its corresponding chlorin and high production cost. To develop efficient bacteriochlorins, various de novo synthetic procedures viz; presence of electron-withdrawing substituents, presence of exocyclic rings in the macrocycle, insertion of appropriate metal ions into the macrocycle, introduction of halogen atoms in the meso-tetraphenyl bacteriochlorins were followed. These structural modifications helped to create varied compounds like bacteriopyropheophorbides, bacteriopurpurinimides and tetraphenylbacteriochlorin derivatives.

4. Pthalocyanines (Pc)

Phthalocyanines (Pcs) were discovered by Braun and Tcheriac in 1907 during the synthesis of o-cyanobenzamide from acetic anhydride andphthalamide. Sir Reginald Linstead thenceforth determined the structure and coined the term "phthalocyanines" in 1933 for this class of macrocyclic compounds (Chen et al., 2020). Pcs derivates were the most promising PS, with their large extinction coefficients in 670–750 nm regions, deep tissue penetration, and ease of physical and chemical modification, good photostability and high competence to produce singlet oxygen. Pcs are planar aromatic macrocycles made up of four isoindole units linked together by nitrogen atoms. Fused benzene ring present at the internal and external positions are normally termed as α- and β-positions, respectively. The electronic structure of outer benzene rings remains unchanged and electronic delocalization generally occur at the inner ring, formed by 16 atoms and 18 π-electrons (Ion, 2017).

But, π–π stacking and low solubility in these molecules restricted their clinical application. To subdue their disadvantages, incorporation of cationic or anionic groups, peptides, crown ethers, glycerinum, etc., were carried on (Zhang and Jiang et al., 2018). Researchers in the oncological field are working extensively toward developing and optimizing such photosensitive compounds. Several metallo- Pcs complexes were produced in the early 1930s through variations in the central metal ion and the peripheral substituents. Heavy diamagnetic metals as the central ion promoted longer lifetimes and a balance between hydrophobicity and hydrophilicity of the peripheral substituents provoked direct tumor cell killing as in cis-disulfonated phthalocyanine complexes (Chen et al., 2020). Biodistribution and pharmacokinetics of Al and Ga metal complexes with lower degree of sulfonation formed PSs with cytotoxicity as good as Photofrin. Ce displayed higher cytotoxicity but led to cutaneous photosensitivity. Arrangement of sulfonated groups further improved the PDT efficiency. Amphiphilic nature of the adjacent sulfonated groups proved beneficial than the oppositely placed ones. Russian endorsement of Photosense - sulfonated Aluminium phthalocyanine (AlPcS$_{2-4}$) (Fig. 4.3F) had shown appreciable results in the clinical trials for cervical cancer. disulfo-di-phthalimidomethyl ZnPc significantly inhibited tumor weight when given in the form of cremophor based emulsion (Chilakamarthi and Giribabu, 2017). Encapsulation of hydrophobic unsubstituted zinc (II) pthalocyanine ZnPc (Fig. 4.3E) by liposomes or insertion into the nanochannels of mesoporous silica NPs or incorporation into layered double hydroxide enhanced its water solubility and conjugation with a tumor targeting ligand. Such ZnPcs with increased singlet oxygen exhibited high photodynamic activity, low dark toxicity and low drug and light dose (Dennis, 2014).

5. Purpurins

Purpurins are potent hydrophobic second generation photo sensitizers of high purity with substantial absorption between 650–780 nm in the red region and easy synthesis from porphyrins (Garbo, 1996). Miravant Medical Technologies (USA) developed light activated cytotoxic drug Rostaporfin (SnET2, Sn[IV] etiopurpurin, tin ethyl etiopurpurin, PhotoPoint SnET2 (Fig. 4.3G) with Purlytin as trademark) as part of their Photopoint PDT program. On injection into the patient, Rostaporfin disseminates and selectively attaches to plasma lipoproteins, present in high concentration on account of production by the hyper-proliferating cancer cells. After a time interval of 24 h, the targeted cells are stimulated by red light of 664 nm that activates Rostaporfin. This sets in formation of highly reactive radical species that destroy the cancerous cells without harming the surrounding healthy tissue (Hunt, 2002). Tin ethyl etiopurpurin (SnET2), a hydrophobic photoreactive purpurin, is weakly soluble in aqueous media and is hence administered for PDT as an emulsion. SnET2 binds to low density lipoproteins LDL, high density lipoproteins HDL and high density proteins, HDP depending on its concentration in the human plasma. Experiments were carried out to study the mode delivery effect on the binding of SnET2 when Cremophore EL (CRM) or dimethyl sulfoxide (DMSO) was used for administration of SnET2 into human plasma. SnET2 binds as monomeric entities whose environmental-sensitive fluorescent properties depend on the type of protein or solvent like DMSO, CRM or H_2O, with which it interacted. SnET2 concentrations; however does not change with the mode of delivery (Kongshaug et al., 1993). In SnET2, absorption occurs at 660 nm which is almost five times greater than that of HPD at 630 nm. This facilitates deeper penetration of light in tissues leading to increased damage to the malignant tissue. Also the high singlet oxygen yield enhances the cytotoxicity, vital for PDT. Major et al. reported destruction of all endometrial glands of a rat model by 2 mg/kg intravenous administration of SnET2 and subsequent illumination of the uterine horn with a 0.5 W laser light dose of 12.5 J/cm, 1.5 h after drug administration. SnET2 exhibited strong endometrial damages at lower light doses than other photosensitizers (Major and Tromberg, 1999). Studies carried by Kessel et al. on male rats bearing the N-[4-(5-nitro-2-furyl)-2- thiazolyl1bdformamide-induced tumor proved that SnET2 distribution remains unaffected by the substantial reduction in circulating plasma lipoprotein levels (Kessel et al., 1993). Morgan et.al carried out a series of test on metallopurpurins to explore their photodynamic activity against transplantable/V-[4-(5-nimtro-2-furyl)-2-thiazoyl]formarnide-induced urothelial tumors growing in male Fischer CDF (F344/CrlBR) rats. Though mettalopurpurins absorbed at lower wavelengths, they showed increased histological tumor destruction capacity than their corresponding free-base macrocycle as in tin ethyl etiopurpurin (SnET2) and zinc derivative ZnNT2H2. SnET2 has an extinction coefficient of 39,272 (637 nm) comparable to its corresponding free-base macrocycle value of 39,455 (660 nm) whereas ZnNT2H2 showed an increased intensity of 59,444 at 633 nm (Morgan et al., 1988). Kaplan et al. treated 13 lesions of metastatic adenocarcinoma of the skin in three human patients using SnET2 Purlytin, of Miravant Medical Technologies, Santa Barbara, CA. 24 h post drug administration, lesions were irradiated by a laser light of 664 nm, in multiple fields. The patients were safely and effectively cured from the disease with no evidence of reappearance; which had earlier failed other treatment options (Kaplan et al., 1998). Allison et al. successfully destroyed Kaposi's sarcoma lesions of a 36 year old man without severe morbidity infusing SnET2 and its activation by red light 24 hours post infusion (Allison et al., 1998).

6. Pheophorbides

Pheophorbides are usually produced by photosynthetic plants and microalgae as a result of chlorophyll breakdown during leaf senescence and/or fruit ripening. Pheophorbide a (Pa) is obtained in algae and higher plants by dephytylation and demetallation of chlorophyll *a* interceded by chlorophyllase and Mg-dechelatase. It has a tetrapyrrolic microcycle structure with four methyls, one ethyl, one vinyl, one propionyl and one methoxycarbonyl as substituents. UV-visible spectrum of Pa bears resemblance to that of chlorophyll a with a Soret band centered at 390 nm and Q bands situated between 500 and 700 nm. The dominant Q-band at 670 nm makes Pa a potential molecule for PDT. Pa also shows anticancer, antiviral, anti-inflammatory, antioxidant, immunostimulatory and antiparasitic bioactivities (Saide et al., 2020). Pa-mediated PDT exerts significant anticancer activity against various human cancers like Jurkat leukemia, pancreatic carcinoma, colon cancer, pigmented carcinoma, hepatoma, hepatitis B virus induced Hep3B cell line, hepatitis C virus induced hepatoma cell line etc. This takes place through ROS which poses hindrance to phosphorylation of extracellular signal regulated protein kinase (ERK). Pa-mediated PDT of B16F10 skin melanoma shows greater efficiency than commercial Photofrin (Ming- Keun Tang and Bui-Xuan, 2010). The decarboxylated derivative of Pa, Pyropheophorbide a (PyroPa), and pyropheophorbide a methyl ester (PyroPA-ME) are also employed in anticancer PDT (Lee et al., 2021).

Tookad-palladium bacteriopherophorbide is derived from photosynthetic pigment BChl, the bacterial equivalent of plant chlorophyll. Pd-bacteriopheophorbide, padeliporfin and redaporfin (LUZ11) with long absorption and fluorescence emission wavelength in the phototherapeutic window, low dark toxicity and high phototoxicity have entered clinical trials as NIR fluorescence imaging agents. Tookad also known as Palladium bacteriopheoforbide (Pd-BPheid) or WST 09 is bacteriochlorophyll devoid of phytol group with magnesium as the core BChl replaced by palladium. These bacteriochlorophyll (BChl) c/d/e/f molecules exist in chlorosomes, the unique light harvesting antenna of green sulphur bacteria that live phototrophically at extremely low-light conditions. BChl get organized as self-assembling aggregates on account of intermolecular coordination, hydrogen bonding and $\pi-\pi$ interaction between the chlorin rings. This aggregation forms large J-type oligomers based on 3^1-hydroxy, central magnesium, and 13-keto carbonyl moieties. This pigment concentration leads to red shift of the absorption spectra and excitation delocalization (Katayama and Tamiaki, 2020; Malina et al., 2020). Pre clinical trials of WST 09- PDT were carried out on normal canine prostate. On intravenous administration of the drug at 2 mg/ kg dose,the surgically exposed prostate of the animal model was exposed to 763 nm to different fluences. PDT-induced lesions showed uniform hemorrhagic necrosis and atrophy with full depth necrosis at 80 J/cm^2, without any damage to the surrounding healthy tissue and rapid clearance from the body (Chen et al., 2002). Tookad induces greater damage to vascular architecture, stops blood supply to the treatment region and effectively destroys prostate cancer than other PSs (Handoko et al., 2015).

7. Texaphyrins

The texaphyrins are exemplary metal-coordinating expanded synthetic porphyrins that are recently used in various medical therapies (Sessler and Miller, 2000). Lutetium texaphyrin (Lu-tex, Pharmacyclics Inc (Baskaran et al., 2018)) (Fig. 4.3H), a metallotexaphyrin, exhibits rapid tissue clearance and absorption in deep red region of electromagnetic spectrum (Panjehpour and Panella, 1998). This water soluble PS depicting large absorption band at 732 nm and molar extinction coefficient of 4.2×10^4 M^{-1}cm^{-1} was studied by Young et al. for its utility in PDT in a murine mammary cancer model. Drug dose of 20 mumol/kg could efficiently treat neoplasms even of larger size with drug administratin and subsequent photoirradiation after 3 h at 150 J/cm^2 at 150 mW/cm^2 (Young and Woodburn, 1996). This deep tissue penetrating second generation PS with rapid clearance induces tumor involution in the murine EMT6 sarcoma model by apoptosis and necrosis (Renno et al., 2000).

4.2.1.1.3 Third generation photosensitizers

Second generation photosensitizers with promising characteristics of chemical purity, longer absorption wavelength, higher excitation coefficient and higher singlet oxygen quantum yield (You et al., 2009) have been widely used in PDT for the selective destruction of cancerous as well as non-neoplastic tissues and have received regulatory approval worldwide. As the PS has affinity for tumor tissue, light used in PDT is focused specifically at the treatment site and hence a local PDT effect is observed rather than systemic one (Josefsen and Boyle, 2008). However, selective destruction of cancer cells could not be realized without damage to the neighbouring normal tissue. Lipophilicity of some PSs caused skin photosensitization (You et al., 2009). Experimental as well as clinical studies carried out with second-generation PSs have thus indicated the urgent need to develop enhanced PSs for better therapeutic outcomes. This prepared for the development of third generation PSs, mostly by chemical modifications of second generation PSs with emphasis on bioconjugation and encapsulation with targeting moieties (Mfouo-Tynga et al., 2021).

Tumor-targeted PDT focuses on the morphological and physiological differences between normal and tumor tissues. Due to uncontrolled proliferation, the tumor cells become leaky with irregular vessels, basement membrane and endothelial cells and hence accumulate and retain the PS in high ratio than the normal cells, which is a requisite for a successful PDT. The hyperexpressed HER2 receptors on breast cancer cells are targeted by PDT in combination with trastuzumab. Trastuzumab, the first HER2-targeted therapeutic monoclonal antibody, in 1998 received approval by the FDA for the treatment of metastatic HER2+ breast cancer. Conjugation of PSs to specific ligands has also increased PDT efficiency (Gunaydin et al., 2021).

Recently, nanotechnology has received widespread attention for the development of third generation PSs. Synthetically prepared NPs can easily transport hydrophobic drugs in blood. Modification of surface area of the NPs with functional groups cultivates desired change in chemical or biochemical properties, their large distribution volume enables easy absorption by cells and they (Mesquita and Dias, 2018). The photoactive chromophore is combined with targeting moiety and is delivered to the cancer cells for enhancing the tumor selectivity using polymeric NPs, micelles, liposomes, nanostructured lipid carriers, and metal NPs. It has also been widely applied in targeted PS delivery for preparation of dendrimer and conjugation of PSs with biomolecules like sugar, monoclonal antibodies or peptides. Upconversion nanoparticles (UCNPs) based cancer therapies are found successful in vivo and in vitro. On excitation by near-infrared light, these UCNPs emit visible or ultraviolet light that activates the PS in its vicinity and produce singlet oxygen conducive for PDT. Low toxicity, narrow emission peaks, good photostability, large stokes shifts of upconversion emission has enabled UCNPs for loading, targeting, and controlling the release of drugs in PDT, and in bioimaging and detection (Qiu et al., 2010).Third generation PSs with improved pharmacokinetics, appreciable reduction in the required PS dose and decreased side effects are promising photosensitizers for PDT of cancer. Water-soluble silicon-phthalocyanine derivative, IRDye700DX (IR700) conjugated to Cetuximab saratolacan, is the first PS conjugated to an antibody. This PS received approval ofthe Japanese government for the treatment of recurrent or locally advanced head and neck cancer targets the hyperexpressed epidermal growth factor receptor. On intravenous administration, cetuximab saratolacan binds to head and neck cancer cells expressing high levels of epidermal growth factor receptor. Illumination of the tumor with red light (690 nm) initiates immunogenic cell death and a strong anticancer immune response (Alzeibak et al., 2020). Hydrophobic PS molecule with high aggregation tendency encapsulated in nanodrug carrier comes with increased photochemical efficiency and the competence to localize in tumors after injected intravenously (Abrahamse and Hamblin, 2016). The second generation PS mTHPC, Temoporfin solution (Foscan) is now also available as nanoformulations (Fospeg, Foslip) and new chemical derivatives related to the basic hydroxyphenylporphyrin framework (Senge, 2012). Gaio et al. studied PDT effect of Fospeg, Foscan, Foslip and in spheroids of HeLa cells. Foslip and Fospeg caused more severe spheroid disruption than Foscan® with absence of cytotoxicity in dark (Gaio et al., 2016). Kataoka et al. implemented the Warbug effect- where cancer cells absorb higher glucose levels than normal cells; in the synthesis of third generation G-chlorin, a chlorine PS, conjugated to four molecules of glucose. G-chlorin was found to be more than 20–50 times cytotoxic second generation talaporfin PDT and useful against gastric and colon cancer. On irradiation with red light G-chlorin undergoes fluroscence with emission of red light that also helps in tumor imaging by photodynamic diagnosis (Kataoka and Nishie, 2017). Ce6 has a strong absorption between 650–800 nm in the near infra-red region. To further increase the absorption of PS by the tumor and the levels of ROS generation Chlorin E6 (Ce6), was incorporated into NPs by formation of ion complexes. It was encapsulated in the gold vesicles as well for improvement in cancer imaging and treatment (Kou et al., 2017).

4.2.1.2 Nonporphyrin-based photosensitizers

Though not extensively, nonporphyrin photosensitizers are also being developed and employed for cancer treatment. A few predominant such PSs are discussed here.

1. Hypericin
   Hypericin (Fig. 4.4A), the pigment found in flowers, leaves and stems of Saint John's wort, Hypericum perforatum (Guttiferae), is used as an anticancer agent (Mariewskaya and Tyurin, 2021). This hydrophobic molecule needs a drug-delivery vehicle like liposomes, micelles, NPs. Hypericin with an absorption peak near 600 nm efficiently localizes in the endoplasmic reticulum. ER stress produced on exposure to light, establish damage associated molecular patterns that adeptly activate the immune system (Abrahamse and Hamblin, 2016) This naphthoadiantrone derivative was studied as a PS in PDT on human RA FLS (MH7A cells) with exposure to light at 593 nm and a LiD of 1.5 J/cm^2. Studies revealed that hypercin increases the ROS production resulting in apoptotis and death of MH7A cells (Gallardo-Villagrán et al., 2019).
2. Phenothiazinium salts
   i. Methyelene blue
      Methylene blue (tetramethylthionine chloride) (Fig. 4.4B) is a cationic dye that belongs to phenothiazines (neuroleptic drug group) class of compounds. It is soluble in water and organic solvents. It appears dark blue in color, and has maximum absorption between 609–668 nm wavelengths (Kayabasi and Erbas, 2020). It shows toxicity in vitro against human, rat, and mouse bladder carcinoma, Ehrlich ascites, human HeLa cervical adenocarcinoma, and B-cell lymphoma. With an extinction coefficient of 82,000 M^{-1}cm^{-1} in aqueous solutions and 5.25 h of plasma half-life in humans a considerable reduction in tumor size, as well as total tumor ablation was achieved in mice with solid Ehrlich

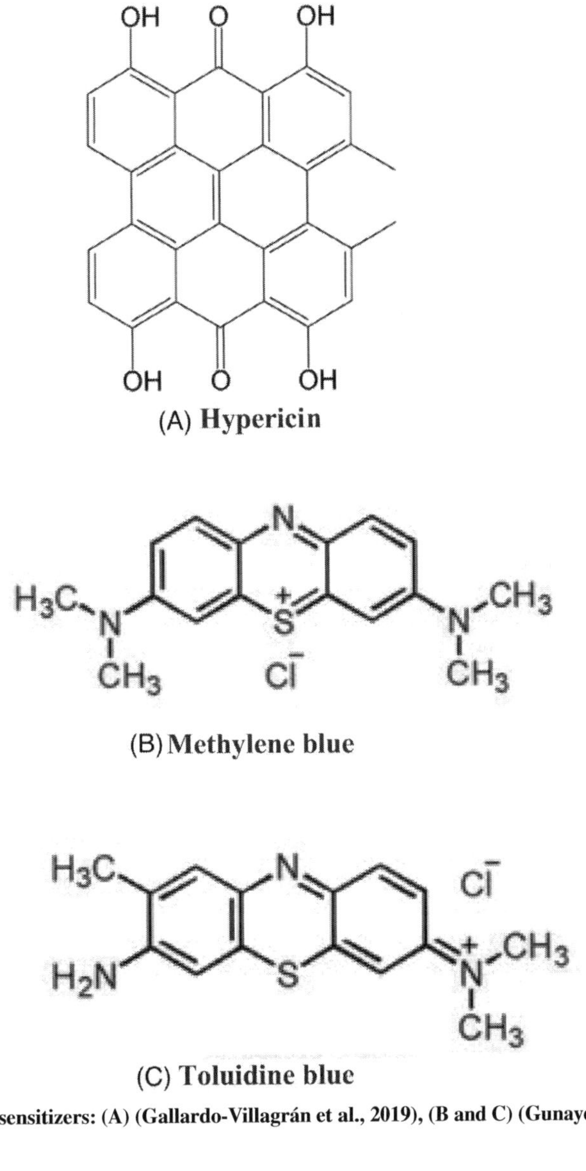

(A) **Hypericin**

(B) **Methylene blue**

(C) **Toluidine blue**

FIG. 4.4 Nonporphyrin-based photosensitizers: (A) (Gallardo-Villagrán et al., 2019), (B and C) (Gunaydin et al., 2021).

carcinomas on intratumoral administration of methylene blue excited by red laser (O'Connor et al., 2009). Santos et al. studied the cytotoxic effect of PDT with methylene blue as PS on common breast cancer types in human breast epithelial cell lines: MCF-7, an ER, PR and HER-2-positive, luminal A cell line; MDAMB-231, a TNBC cell line; and MCF-10A, a normal-like cell line. PDT induces massive tumor cell death with 2 or 20 µM dye succeeded by irradiation with 4.5 J/cm^2 with no harm to the non-malignant cells (dos Santos et al., 2017). Experiments undertaken by Santos et al. proved that oral rinse with methyelene blue promises an efficient, low risk, easy-to-use treatment for refractory pain from oral mucositis related to cancer therapy (Roldan et al., 2020).

  ii. Toluidine blue

   Toluidine blue (TB) (Fig. 4.4C), an acidophilic metachromatic dye and a member of the thiazine group was discovered in 1856 by William Henry Perkin in 1856 and was basically used in the dye industry. Also known as aminotoluene or methylanaline, it mainly has 3 isoforms, namely, ortho-toluidine, meta-toluidine, and para-toluidine. It is partially soluble in water as well as in alcohol. It has great affinity for nucleic acids and so attaches to nuclear material of tissues with a high DNA and RNA content. In medical field, toulidine blue has been widely used as a imperative stain for mucosal lesions as it selectively stains acidic tissue components like phosphate, carboxylates, and sulfates radicals on account of its metachromatic property (Sridharan and Shankar, 2012). It shows absorption in the range of 626 to 632 nm with a molar extinction coefficient of 3×10^4 M^{-1} cm^{-1} (O'Connor et al., 2009). When clinically it is not possible to determine lesions at high risk of progression, touilidine blue helps in early diagnosis of oral cavity and oropharyngeal cancer (Allegra et al., 2009). Eichelbaum et al. used toulidine blue to stain gastric lesions in nine patients. The lesions of adenocarcinoma in six patients stained an intense blue while the benign ones did not stain. The dyes can guide and direct the gastroenterologists, endoscopists, and surgeons to gain information of small lesions, incipient or multiple stomach lesions (Eichelbaum et al., 1977).

3. Merocyanine 540

   Merocyanine 540 is a fluorescent anionic dye that binds itself to cholesterol free domains of the outer leaflets of the cell membranes (Sikurová and Franková, 1993). This 3-sulphopropyl–2–(3H)-benzoxazolilidine-2-butenylidene-1,2-dibutyl-2-thiobarbituric acid, merocyanine 540 is a negatively charged heterocyclic chromophore with excitation maxima at 501 nm and 535 nm and fluroscence emission maxima at 572 nm. Dual excitation peak indicates that the dye exists in both monomeric as well as dimeric form in equilibrium in aqueous environment. This nonmutagenic dye with low systemic toxicity was found efficacious against a wide range of leukemia, lymphoma, and neuroblastoma cells (Smith and Sieber, 1990). Structural modifications of the dye obtained by varying the alkyl substituents on the thiobarbiturate subunit, change its lipophilicity, and hence alter its effect on the leukemic cells (Benniston et al., 1994).

4.3 Working mechanism

PDT, the minimally invasive therapeutic modality depends on the presence and interaction of light, photosensitizer, and oxygen. It includes administration of PS followed by its distribution interval and exposure of tumor area to light of specific wavelength and intensity, usually in the visible or near infra red region (Nowis and Makowski, 2005; Cramers and Ruevekamp, 2003). PS administered intravenously, intraperitoneally, or topically gets selectively incorporated in the rapidly proliferating tumor tissue due to physiological differences between healthy and tumor tissue (Ormond and Freeman, 2013). For this a PS should be lipophilic in its free state so that it easily passes through cellular membranes and gets localized in different organelles such as lysosomes, mitochondria, Golgi apparatus, endoplasmic reticulum, and plasma membranes. PS accumulation in the tumor region may occur via a mechanism analogous to that of low-density lipoprotein receptor. Leaky neovasculature may also aid enhanced permeability and accumulation of PS. Accumulation of PS into malignant cells and realization of a maximal tumor-to-normal cell concentration ratio may take several minutes to hours. Subsequent illumination of this tumor area at a specific wavelength of nonthermal monochromatic light activates the inert PS that interacts with the surrounding oxygen to produce cytotoxic reactive oxygen species (ROS). This leads to a series of molecular and biological reactions conducive for tumor and vascular damage via apoptotic or necrotic mechanisms. Localization of PS at tumor site and subsequent irradiation by non-ionizing light beam brings about selective killing of tumor cells. The destruction of tumor cells is sometimes noticeable in the form of swelling and formation of necrotic tissue. The destroyed tissue is shed or resorbed and the treated site then heals normally (Cabuy, 2012). PDT destroys a tumor through three inter-related mechanisms: direct cytotoxic effects on tumor cells, irreversible damage to the tumor vasculature and stimulation of a strong inflammatory reaction that leads to the development of systemic immunity against the tumor tissue (Cabuy, 2012; Wang et al., 2021).

Most PSs are organic dyes characterized by excited singlet or triplet electronic state as with total electronic spin momentum s as 0 and 1, respectively. Each electronic state is further subdivided in vibrational states. Absorption of photon promotes

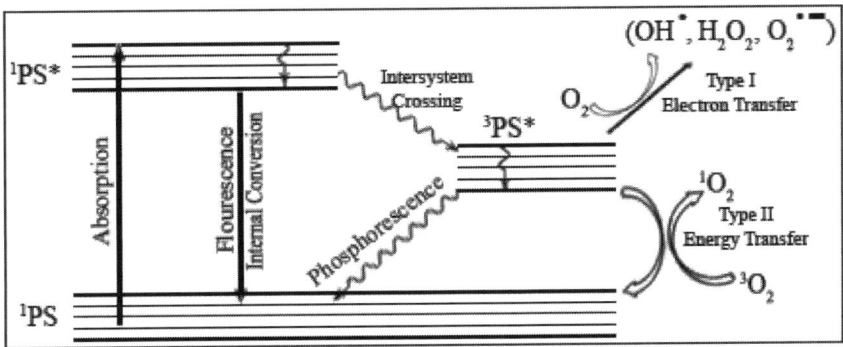

FIG. 4.5 Modified Jablonski diagram depicting photodynamic action.

a PS from stable singlet ground state to an excited singlet or triplet state. The singlet ground state (^{1}PS) and excited singlet state (^{1}PS*) of PS has paired electrons while the excited triplet state (^{3}PS*) has unpaired electrons. PS in the first excited singlet state may undergo photobleaching by transfer of electron to the biological substrate and modify it (Alexandrova et al., 2004). Photoexcitation of PS to higher vibrational levels of singlet excited state results in non-radiative decay from these levels to the lowest vibrational level in the form of heat in 10^{-11}–10^{-14} s. Transition of electron from this lowest vibrational level of highly unstable excited singlet state to singlet ground state results in spontaneous emission. This allowed radiative transition (10^{-8} s) termed as fluorescence (Pucelik et al., 2020) helps in identification and delineation of the tumor target site (Fig. 4.5) (Dias and Mfouo-Tynga, 2020).

Transitions from singlet ground state to triplet excited state are forbidden by spin selection rule; however, spin-orbit coupling occurring due to the presence of heavy metal ions, oxygen, and halogen in the PS structure allows these transitions. As a result, spin gets reoriented and the PS molecule enters the metastable triplet excited state by intersystem crossing (ISC) (Pucelik et al., 2020). Lewis and Kasha reported that the metastable states have 3–6 times longer decay lifetime than fluorescence (Krasnovsky, 2014). Longer decay lifetime and unpaired electrons imparts high reactivity to the triplet state of PS molecule. Decay of ^{3}PS* may take place via two alternate photo-oxidative reaction mechanisms with the surrounding molecules. Type I: an electron transfer process or type II: an energy transfer to ground state molecular oxygen.

Type I pathway: the triplet state PS directly reacts with the target or substrate molecule other than oxygen through electron or hydrogen transfer mechanism and produce free radicals of PS or biological substrate. These free radicals react with ground state oxygen (^{3}O$_2$) to form ROS such as superoxide anions ($O_2^{\bullet-}$), hydrogen peroxide (H_2O_2), and hydroxyl radical ($OH^{\bullet}$) (Josefsen and Boyle, 2008; Krasnovsky, 2007) that causes irreversible oxidation of the biological targets like DNA, proteins, membranes (Papakonstantinou et al., 2018) in close proximity and induce apoptosis or necrosis, causing cell death (Fahmy et al., 2021) and are termed as reactions of primary photodehydrogenation by Schenck and Terenin (Krasnovsky, 2007).

Type II pathway: the triplet state PS may transfer energy directly to molecular oxygen ^{3}O$_2$, owing to its triple multiplicity. According to Mulliken and Kautsky, molecular oxygen has triplet ground state $^3\Sigma_g^-$ and two relatively low-lying singlet levels. On energy gain from PS molecule, ^{3}O$_2$ with excitation energy of 157 kJ mol^{-1} transits from $^3\Sigma_g^-$ to the higher singlet $^1\Sigma_g^+$ level at 762 nm. However, it quickly deactivates to the second singlet level $^1\Delta g$ with lower excitation energy of 94 kJ mol^{-1} and transition at 1268 nm. Foote claimed that it is this electrophilic, $^1\Delta g$ singlet state of oxygen with higher oxidizing properties that brings about the photodynamic reactions and is vital for PDT (Krasnovsky, 2007). During its short lifetime ^{1}O$_2$ oxidizes the target tissues in vicinity causing permanent damage and cell death by apoptosis, necrosis, or autophagy (Mesquita and Dias, 2018). Photodynamic action takes place in PDT as given below:

When a PS absorbs a photon from the incident electromagnetic radiation, an electron is raised to a higher energy molecular orbital thus exciting the PS in the ground singlet state (^{1}PS) to first excited singlet state (^{1}PS*). The PS may relax quickly to ground state or undergo spin inversion and populate the lower energy first excited triplet state (^{3}PS*) by intersystem crossing (Josefsen and Boyle, 2008). ^{3}PS* and ^{1}PS react to form PS anion and cation radicals, PS$^{\bullet-}$ and PS$^{\bullet+}$, respectively. Excited triplet state PS reacts via two pathways type I and type II. Type I process takes place either through electron or hydrogen transfer. In type I process, transfer of electron from the substrate to the PS in close proximity produces a PS radical anion ^{3}PS$^{\bullet-}$ and a substrate radical cation Subs$^{\bullet+}$. The superoxide anion, $O_2^{\bullet-}$ is formed via electron transfer from (1) PS$^{\bullet-}$ to molecular oxygen ^{3}O$_2$ and (2) ^{3}PS* to ^{3}O$_2$. It is this $O_2^{\bullet-}$ generation from ^{3}PS* that competes with the production of singlet oxygen by type II process (Ormond and Freeman, 2013). $O_2^{\bullet-}$ is converted to H_2O_2 and ^{3}O$_2$ by superoxide dismutase. Subsequently the ferric ion is reduced to ferrous ion which reacts later with hydrogen peroxide. This reaction known as Fenton reaction causes hemolytic fission of the oxygen - oxygen bond in H_2O_2 to yield hydroxide ion OH^- and hydroxyl radical $OH^\bullet$. $OH^\bullet$ is a strong electrophile that chemically assail biological molecules and form cytotoxic free radicals for oxidative damage to tumor cell (Dai et al., 2012). Transfer of hydrogen to $^3PS^*$ in type I process, also produces free radicals that readily react with molecular oxygen and generate reactive oxygen species including reactive peroxides conducive for cell damage and death (Josefsen and Boyle, 2008). In the type II reaction, the excited PS in triplet state readily reacts with ground state molecular oxygen. As a result of triplet-triplet annihilation, spin of molecular oxygen is inverted with production of highly reactive oxygen $^1O_2$ that has a major contribution in destruction of cellular mechanism and structure.

Excitation of PS

$^1PS + h\upsilon \rightarrow {}^1PS^*$
$^1PS^* \rightarrow {}^1PS$
$^1PS^* \rightarrow {}^3PS^*$ (Ion, 2000; Ormond and Freeman, 2013)

Type I process by electron transfer

$^3PS^* + {}^1PS \rightarrow PS^{\bullet-} + PS^{\bullet+}$
$^3PS^* + Subs \rightarrow PS^{\bullet-} + Subs^{\bullet+}$
$PS^{\bullet-} + {}^3O_2 \rightarrow {}^1PS + O_2^{\bullet-}$
$^3PS^* + {}^3O_2 \rightarrow PS^{\bullet+} + O_2^{\bullet-}$
$2O_2^{\bullet-} + 2H^+ \rightarrow {}^3O_2 + H_2O_2$
$O_2^{\bullet-} + Fe^{3+} \rightarrow Fe^{2+} + {}^3O_2$
$Fe^{2+} + H_2O_2 \rightarrow Fe^{3+} + OH^\bullet + OH^-$
$O_2^{\bullet-} + OH \rightarrow {}^1O_2 + OH^-$

Type I process by hydrogen transfer

$^3PS^* + R\text{-}H \rightarrow {}^1PS\text{-}H^\bullet + R^\bullet$
$R^\bullet + {}^3O_2 \rightarrow RO_2^\bullet$
$RO_2^\bullet + R\text{-}H \rightarrow RO_2H + R^\bullet$

Type II process

$^3PS^* + {}^3O_2 \rightarrow {}^1PS + {}^1O_2$
$^1O_2 + Subs \rightarrow Subs^+ +$ Oxidative damage (Josefsen and Boyle, 2008)

The two pathways type I and II occur simultaneously and compete with each other. Toxic oxygen species and free radicals produced within the cell membrane during the mechanisms are highly reactive and can cause photo-oxidative damage to nucleic acids, lipids, proteins, and other cellular components within the photosensitive binding site. This probable macroscopic mechanism causes cell damage and death (Caliskan and Bilgin, 2018; Li et al., 2020). Type II mechanism plays a vital role in photodynamic cytotoxicity in presence of adequate oxygen in tumor microenvironment (Cui et al., 2019). Extent of cytotoxicity and damage caused by PDT is decided by the PS type, its location: intracellular or extracellular, total PS dose administered to the patient, light exposure, its fluence and fluence rate, time between PS administration and exposure to light and availability of ambient oxygen. Quantum yield of the singlet oxygen formation and lifetime of its triplet excited state are important features of a PS. Owing to high reactivity but short half-life time of ROS, only the cells which are near to the PS, are greatly affected by PDT (Correia et al., 2021). To obtain a high quantum yield of singlet oxygen generation, a PS molecule should have a triplet energy state higher than the molecular oxygen's lower excited state and long decay lifetime for high energy transfer efficiency (Kashyap et al., 2021).

4.3.1 Mechanism of cell death following photodynamic therapy

The antitumor effect of PDT occurs by the three inter-related processes namely, direct tumor killing by reactive oxygen species through apoptosis and necrosis, damage to the tumor vasculature leading to an disruption of oxygen and nutrients supply causing indirect cell death by hypoxia and activation of antitumor immune response (Mroz et al., 2010). Though it is not possible to define the contribution of each process, however, by changing the fluence rate of light, exposure time or a combination of both, their contribution to tumor damage can be varied. At optimum PDT conditions the tumor cell is destroyed by necrosis while in other cases it undergoes apoptosis (Nowis and Makowski, 2005).

4.3.1.1 Apoptosis

Apoptosis is a genetically programmed process of orderly cell death without inflammatory effects triggered by both extracellular and intracellular signals. It is elicited by PDT via two leading pathways-mitochondria-mediated or intrinsic and death receptor- mediated or extrinsic (Jung et al., 2020).

1. Mitochondria-mediated apoptosis
   It occurs preferentially when PSs accumulate in the mitochondria of tumor cell. Photosensitising agents target the Bcl-2 family of proteins which is classified into two groups depending on their influence on apoptosis. Bcl-2, Bcl-w, Bcl-X_L, Mcl, and A1 are antiapoptotic while Bax, Bfm, Bok, Bcl-Xs and others are pro-apoptotic. Enhanced levels of Bcl-2 genes in tumor cells have a protective effect against PDT induced apoptotis, while the Photodamaged Bcl-2 allows excess of pro-apoptotic protein Bax to interact with the mitochondrial membrane disrupting the trans-membrane potential. This causes opening of Permeability Transition Pore Complex (PTPC), which is a large conductance channel present in the mitochondrial membrane, with subsequent release of cytochrome C. Recent evidences show that cytochrome C release may also occur independently of PTPC due to peroxidation of lipid cardiolipin which is unique to mitochondria. Presence of cytochrome C in the cytoplasm stimulates number of caspases which hydrolyse and cleave many cellular proteins and DNA into fragments, ultimately causing cell death.
2. Death receptor-mediated apoptosis
   It occurs mainly when PS targets the cell membrane. Multimerization of tumor necrosis factor (TNF) receptors primarily of Fas, offsets PDT induced apoptosis. This Fas multimerization may take place in ligand–independent and light-dependent and manner. Fas, Fas-associated protein with death domain (FADD) and procaspase-8 form a death inducing signaling complex. It causes proteolytic self activation of procaspase-8 and activation of other caspases. An activated hydrolytic enzyme caspase-8 initiates intrinsic apoptotic pathway eventually leading to cell lysis. Other pathways which influence PDT-induced apoptotic death are calcium homeostasis, Mitogen-Activated Protein Kinases (MAPK) and ceramide formation (Nowis and Makowski, 2005).

4.3.1.2 Necrosis

Necrosis is an and unregulated irreversible form of cell death (Amaravadi and Thompson, 2007) caused due to cell injury by ischemia, toxins or trauma induced by physical or environment factors, characterized by swelling of cells and inflammation of adjoining cells, resulting in the spillage of cell contents into the extracellular space (Sato et al., 2020). It is markedly differentiated from apoptosis by occurrence of pyknosis, karyolysis, karyorrhexis and the absence of caspases activation. Localization of the PS in the lysosome or cell membrane most likely drives cells to necrosis as it blocks the apoptotic pathway (Kou et al., 2017).

4.3.1.3 Autophagy

The term autophagy is derived from the Greek language, meaning "self-eating." In 1957, de Duve and Wattiaux first described it systemically by, in mammalian cells, as a process of bulk segregation of cellular constituents. Autophagy causes degradation of long-lived proteins and can degrade organelles, such as mitochondria. Autophagy, in mammalian cells, abet cell death in response to certain chemotherapeutic drugs, radiation, hypoxia, ischemia in the brain, cytokines such as INF-g (76), and ligands such as HIV-1-encoded envelope glycoproteins. The deletion or RNAi-mediated knockdown of key autophagy genes in these stressful conditions considerably reduces cell death, while overexpressing these genes can support it. Prolonged ER stress in apoptosis-deficient fibroblasts lacking both Bak andBax, results in autophagy (Gao et al., 2009).

4.4 Advantages and disadvantages of photodynamic therapy

4.4.1 Apoptosis in photodynamic therapy

The two ways through which eukaryotic cells die are through necrosis and apoptosis. Necrosis takes place when cells suffer irreversible injury induced through noxious stimuli that include infections caused by agents such as bacteria, viruses, parasites, etc., hypoxia, heat, or exposure to UV or radioactive radiations (Kim-Campbell et al., 2019). These are non-ATP driven processes where the damage induced to the cell reduces or destroys the cell's ability to produce ATP, resulting in its death. Apoptosis, however, is programmed cell death that is triggered by physiological or pathological processes that occur during early development or ageing and as a homeostatic mechanism to sustain a fixed number of cells in tissues. Apoptosis also works as a defence mechanism to combat cells damaged by noxious agents (Norbury and Hickson, 2001).

PDT follows two kinds of pathways. In the type 1 pathway, photosensitizers (PS) react with triplet oxygen species ($^3O_2$) or ambient water and undergoes electron transfer to generate an oxide anion radical (O_2^-) and a hydroxyl radical (OH^-) (Pucelik et al., 2016, Yang et al., 2019). In the type 2 mechanism, energy transfer takes place between the PSs and the neighbouring $^3O_2$ species, generating cytotoxic singlet oxygen ($^1O_2$) (Zhang et al., 2019) which is the controlling factor in nanomedicine mediated PDT processes (Cai et al., 2018; Tan et al., 2018). PDT processes in hypoxic tumours take place through type 1 pathways when PSs undergo electron transfer to produce cytotoxic reactive oxygen radicals (Lan et al., 2019; Zhang et al., 2019). Because of the highly reactive Type 2 pathway and the hypoxia resistance of type 1 (Castano et al., 2004), researchers focus on the design and synthesis of nanomaterials that integrate types 1 and 2 mechanisms that significantly enhance the biomedical applications of PDT (Kuncewicz et al., 2019).

Apoptotic response in cells is usually a lengthy process that takes from several hours to days to manifest. However, Aggarwal et al (Agarwal et al., 1991) demonstrated how PDT on mice models caused apoptosis in L5178Y lymphoma cells in just 30 min *in-vitro*. The authors suggest that cellular response to PDT-induced damage circumvents the regular apoptotic processes and directly initiates the final stage mechanisms. Another group led by Zaidi et al observed similar apoptotic response *in-vivo* (Zaidi et al., 1993) when they treated RIF-1 tumours with Photofrin-PDT. The authors argue that the apoptotic cells may have originated from the host cells penetrating the tumour instead of the tumour cells. In 1994, He et al. showed that PDT does not result in apoptosis in all *in-vitro* cell lines (He et al., 1994). He demonstrated that an LD_{50} dose of Photofrin-PDT was needed to initiate apoptosis in MTF7 cell lines from mice while a dose of LD_{85} was needed to initiate apoptosis in PC3 human prostatic adenocarcinoma cells, and no apoptosis was observed in H322 carcinoma cells, suggesting that there are multiple pathways for PDT induced apoptosis (He et al., 1994).

Kessel et al. (Kessel et al., 1997) and Luo et al. (Luo and Kessel, 1997) have demonstrated that the extent of apoptosis is a function of the PS used. For instance, cultured cells were treated with tin ethyl etiopurpurin (SnEt2)-PDT, apoptosis was induced under 60 min whereas when using tin octaethylpurpurin amidine (SnOPA)-PDT, apoptosis required over 24 h. This was because SnOPA locally distributed itself uniformly throughout the cells and their membranes while SnEt2 distributed only to lysosomes and mitochondria thereby damaging cell membranes and inhibiting the cell's ability to control apoptosis. Agarwal et al. demonstrated the activation of phospholipases C and A_2 when Photofrin-PDT was used (Agarwal et al., 1993). They showed that when inhibitors were used to block the phospholipases, the increase in intracellular calcium levels and DNA fragmentation, common traits of apoptosis, were not observed.

Kick et al. (Kick et al., 1996) showed that treating cultured cells with Photofrin-PDT activated the proto-oncogenes that are known to regulate apoptosis, *c-jun* and *c-fos*. Activating these proto-oncogenes assisted in transcribing apoptosis mediating factors. When higher levels of the tumour suppressing gene p53 were present in cultured HL-60 cells, the sensitivity to Photofrin-PDT was much lower compared to those with normal levels or no p53 (Fisher et al., 1997). Once it was established that mitochondria can also regulate and induce nuclear apoptosis, it didn't take long before groups such as Kessel et al. hypothesized and demonstrated that PDT can kill cells through mitochondrially regulated apoptosis (Kessel and Luo, 1999; Kessel and Luo, 1998). Consequently, several mechanisms through which mitochondrially regulated apoptosis occur following PDT have been identified. Photosensitizers may distribute locally on mitochondria targets such as the permeability transition (PT) pore membranes or specific proteins within the pore. During PDT, ROS attacks the proteins resulting in an alteration of the amino acid configurations that inhibits the protein's folding conformation, which disrupts their regular functioning.

One of the functionalities of the PT pores is to allow water to flow into ions and antioxidants to flow out of them as the system equilibrates. When the pore membrane potential reduces, the electron transport chain detaches and releases destructive ROS that is scavenged by whatever antioxidants are remaining. However, if the PT is not reversed, damage builds up and breakdowns the electron transport chain (Rück et al., 1998). Interestingly, it was also demonstrated that apoptosis in cells expressed with apoptosis suppressor genes, bcl-2, was a function of the PSs used. For instance, aluminium phthalocyanine-PDT selectively disables the bcl-2 gene, which improves apoptotic response post-PDT (Kim et al., 1999) while cells with an overexpression of bcl-2 were more immune to Photofrin-PDT (He et al., 1996). Both these observations suggest that PT pore formation is one of the pathways for PDT-induced cell death. Below a certain threshold membrane potential, the mitochondria would stop functioning and swell, resulting in the outer membrane to burst due to its lesser surface area compared to the inner membrane. This explosion helps release cytochrome *c* and apoptosis-inducing factor (AIF) that's present between the two membranes. Once these are released, apoptosis is inevitable and irreversible. Various authors have demonstrated that cytochrome *c* is released during PDT treatment that improves the apoptotic response to PDT (Kessel and Luo, 1999; Kessel and Luo, 1998; Kim et al., 1999; Varnes et al., 1999).

In a recent study, Wang et al. (Wang et al., 2020) used a single atom enzyme, OxgenMCC-r supported on single-atom ruthenium (Ru) doped $Mn_3[Co(CN)_6]_2$ metal-organic framework with encapsulated PS chlorin e6 to induce apoptosis even in hypoxia environment. The single-atom enzyme with Ru catalytically degraded intracellular H_2O_2 to O_2 that, in turn, eliminated the hypoxia environment of the tumour. This resulted in an improved production of ROS that causes apoptosis *in-vitro* and in tumour carrying mice *in-vivo*. Nanomaterials, particularly upconversion nanoparticles (UCNPs), is a relatively new class of materials that has found considerable success in various applications of medicine such as PDT PSs, hypoxia sensing, bioimaging, etc. Liang et al. (Liang et al., 2020) designed a multifunctional nanocarrier comprising of a UCNP to target the mitochondria and improve the efficiency of PDT. Biofunctionalized UCNPs targeting mitochondria were excited by a 980 nm laser to activate a Ce6 PS which had an absorption peak that overlapped with the UCNP emission peak at 650 nm. The activated Ce6 generated ROS in addition to the ROS generated by the degradation of H_2O_2 that helped overcome tumour hypoxia and enhance apoptosis. Yan et al. (Yan et al., 2015) demonstrated the use of surface modified reduced graphene oxide (r-GO) helped in the generation of ROS and PDT mediated lysosomal damage that resulted in enhanced tumour cell apoptosis.

Vetha et al. (Vetha et al., 2020) demonstrated the use of a natural plant-based photosensitizer, curcuminoids, sensitized to blue light for the generation of ROS in A549 cancer cell lines *in-vitro*. The curcuminoids were incorporated into liposome NPs which enhanced the effects of Blue-LED initiated PDT and increased the ROS in the A549 cell lines beyond its survivable threshold, promoting apoptosis. Li et al. (Li et al., 2020) synthesized a mitochondria targeting iridium metal heterocyclic carbene complex that reduces mitochondria membrane potential and increases the intracellular ROS levels under 450 nm and 630 nm LED excitation. Studies conducted *in-vitro* on different cancer and HeLa cell lines showed increasing levels of ROS under 450 LED irradiation for 10 min. Similar effects were observed under 630 nm irradiation for 30 min, given the weaker absorption of the complex at this wavelength. *In-vivo* studies on mice under 630 nm over a period of 14 days did not result in mortality and also showed an inhibition in the tumour growth with no change in body weight during this period.

Shi et al. (Shi et al., 2020) integrated UCNPs and meso-tetra(4- carboxyphenyl)porphine metalorganic frameworks that were additionally coated with TiO_2 NPs to trigger Type 1 and Type 2 PDT mechanisms excited by a 980 nm laser. The porphyrin molecules in the metalorganic framework and TiO_2 functioned as photodynamic reagents with various functionalities that were excited by the radiation emitted by UCNPs to produce different kinds of ROS that met the potential requirements of PDT *in-vivo*. Nam et al. (Nam et al., 2020) employed magnetic Fe_3O_4 NPs functionalized with Ce6 and folic acid to show that this system induced apoptosis through cellular and nuclear damage *in-vitro*. Apoptosis in cancer cells was a function of the duration of LED excitation and the nanoparticle loading concentration. The effects of image-based PDT were attributed to the various physical and optical properties of the NPs such as super-paramagnetism, luminescence intensity, and the singlet oxygen quantum yields from Ce6.

4.4.2 Immunological effects of photodynamic therapy

Even though this is an area of research that is constantly evolving, there are reports that suggest that PDT may activate or suppress the immune system, depending on certain conditions. Using PDT to produce tumour antigens that result in specific immunological response to tumor cells is a potential method to treat primary tumours and their metastases. However, there are other reports which show that PDT suppresses the functioning of the immune system which hinders the required immunological responses. PDT appears to vary the balance of factors that control the immune response and can push it either towards activation or suppression through the release of particular cytokines. This mechanism is regulated by several complex components such as the PS employed, the PDT dose and dosimetry, etc.

The development of tumours that can be transplanted, grown in mice or rat strains that have the same major histocompatibility complex haplotype (MHC) possessing uncompromised immune systems has been a major resource to help researchers investigate post PDT tumour immunity. Canti et al. employed both immunosuppressed and regular mice possessing MS-2 fibrosarcomas to evaluate the effects of aluminium disulphonated phthalocyanine PDT on the anti-tumour immune response (Canti et al., 1994). They showed that while all the mice were completely cured, their resistance to MS-2 depended on the initial immune conditions that the mice strains were subject to with the regular non-immunosuppressed, cured mice showing resistance while those that were immunosuppressed and cured by PDT died after tumour reintroduction. Using syngeneic BALB/c mice, Korbelik et al. (Korbelik et al., 1996) demonstrated that while a simple Photofrin-PDT completely and comprehensively cured EMT6 mammary sarcomas, there were no permanent cures for nonobese diabetic, immunodeficient, or nude mice. The transfer of splenic T-cell lymphocytes from naive mice to the immunodeficient mice prior to PDT subjection delayed the recurrence of the tumours while the transfer conducted immediately or one week after PDT showed no benefits.

Adoptive transfer of non-adherent $CD4^+$ and $CD8^+$ T-cells combined with some B-cells, NK-cells, and monocyte spleen cells from regular mice that were cured of EMT6 sarcomas by PDT 5 weeks prior was able to completely reinstate the curing effects of PDT on the EMT6 sarcomas that developing inside immunosuppressed mice. Spleen cells cured by X-ray therapy were much less effective. It was observed that the reduction of particular T-cell clusters from the spleen cells suggested that CD8⁺ cytotoxic T-lymphocytes had the highest curing effect while the CD4⁺ T-cells took on a more supporting role (Korbelik and Dougherty, 1999). Castano et al. (Castano et al., 2006) demonstrated that while benzoporphyrin derivative PDT of weak immunogenic murine sarcoma RIF1 tumours, in C3H/HeN mice, resulted in initial tumor suppression, the local recurrence prevented a permanent cure for the sarcoma. However, in contrast to this result, when genetically modified tumours in jellyfish that expressed the green fluorescent protein were used, PDT not just completely cured the tumour but also long-term resistance was also observed. The authors suggested that PDT resulted in immune recognition of the green fluorescent protein as a model tumour antigen. Several more tumour rejecting antigens were identified in mice models, leading to the possibility of a more logical approach in investigating the various parameters that control the strength of anti-tumour immunity triggered by PDT on various tumours and PDT procedures (Uenaka and Nakayama, 2003).

Korbelik et al. (Korbelik, 1996) were one of the first groups to propose the mechanism through which PDT initiates specific immune response when they demonstrated that tissues treated with PDT release large amounts of cell debris along with inflammatory indicators, cytokines and chemotactic mediators. Within just a few minutes of irradiation, neutrophils invade the treated tissue in large quantities and release lysosomal enzymes and ROS which promptly destroy endothelial and cancerous tissues, which additionally trigger the inflammatory response. Upon neutrophilic death, they release their cellular content that behaves as chemotactic signals for additional groups of inflammatory cells. As the intensity of neutrophilic cells decrease, mast cells flow into the damaged tissues and secrete granules comprising of vasoactive agents and cytokines. The surviving tumor clusters are destroyed when monocytes and macrophages spread across and collect cell debris. On completing the inflammatory response, the monocytes and macrophages release immunosuppressants that down-control the immune response and may affect future specific immunological response.

A study lead by Musser et al. (Musser and Fiel, 1991) showed that immune suppression through PDT depended on the PS used. PDT that used porphyrin $TSSP_4$ and HpD as PSs suppress the immune system almost instantly after treatment while when Photofrin or THPC were used, immune suppression required up to 72 h to develop in animals. The authors suggest that various PSs have varying hydrophobicity and localize to varying subcellular targets and these differences are what control the variation in immune suppression. In another study lead by Lynch et al. (Lynch et al., 1989), the authors suggest macrophages regulate immune suppression since during PDT, they accumulate large quantities of extremely aggregated compounds such as Photofrin and then release immune suppressors. The authors verified this theory by the adoptive transferring of macrophages from mice treated by PDT to dinitrofluorobenzene (DNFB) sensitized mice. The authors observed a 60% reduction in the immune response following a second DNF administration. Additionally, the authors also demonstrated that macrophages treated with PDT generate and release cytokines such as interleukin-1 (IL-1). IL-1 is known to assist in the generation of acute phase proteins such as serum amyloids A and P and C-reactive proteins which regulate the suppression of the immune system and were detected post PDT treatment.

Korbelik et al. showed that PDT can result in an indelible specific immunity using vitamin D_3 binding protein-derived macrophage activating factor (DBPDMAF), that hyperstimulates macrophage multiplication and possibly retrieves antigens, to repress immune suppression (Korbelik and Dougherty, 1999). The use of DBPDMA and PDT assisted in an increase in selective tumour destruction (Korbelik et al., 1997). The cytokines produced by the immune system take on several roles to modulate itself, depending on the situation. The cytokine signals have been investigated during and after PDT to evaluate how the immune system responds at a specific time. Nseyo et al. observed that the urine of patients subjected to bladder PDT contained IL-1β, IL-2, and tumour necrosis factor-α (TNF-α) (Nseyo et al., 1990), suggesting that the patients' immune system remained active in the tissues even after PDT. It is, however, unclear whether the cytokines were generated as an inflammatory or immune response.

Herman et al. demonstrated that the use of protoporphyrin-IX (PP-IX) and Photofrin as PSs in the PDT treatment of human and mice monocytes, initiated the release of IL-2, IL-3, TNF-α, interferon-γ (NF-γ) (Herman et al., 1996). The observed that PP-IX sensitized the immune system without using light and that the immune response to PDT treatment under light was enhanced due to this pre-sensitisation process. It must be noted while interpreting these results that the PP-IX doses employed in these studies were extremely high so that even a small amount of light irradiation could initiate PDT. Bellnier showed that administering a low dose of TNF-α when combined with Photofrin-PDT helped double the phototoxicity of Photofrin and enhanced the site selectivity of treatment (Bellnier, 1991). Since TNF facilitated macrophage growth and differentiation, Bellnier proposed that exogenous TNF at the treatment sites could improve the inflammatory response. Evans et al. (Evans et al., 1990) and Schumaker et al. (Shumaker and Hetzel, 1987) independently detected the presence of TNF in mouse macrophages and *in-vitro* human tissues treated with PDT.

Sun et al. developed a sorafenib and Ce6 nanoparticle system-PDT to enhance anti-tumour response through the intratumour release of sorafenib during PDT treatment (Sun et al., 2020). The authors demonstrated that PDT rapidly breaks down the sorafenib loaded carrier resulting in the increased release of sorafenib. This process enhances the anti-tumour effects of PDT through synergistic tumour tissue apoptosis even under low PDT dosimetry. Additionally, the authors note that the combination of low PDT doses and the rapidly released sorafenib resulted in an increase in the generation of tumor-invading CD8+ T-cells quickly after PDT treatment, suggesting the induction of enhanced anti-tumour immune response. Turubanova et al. employed photosens (PS) and photodithazine (PD)-PDT induced immunogenic cancer cell death and activated anti-tumour immune response in murine glioma GL 261 and fibrosarcoma MCA cell lines *in-vitro* and *in-vivo* in a tumour prophylactic vaccination model (Turubanova et al., 2019). They showed that PS localized in the lysosomes and that while PS-PDT induced apoptosis was inhibited using apoptosis and ferroptosis inhibitors, necroptosis inhibitors fail to stop cell death. PD, however, localized in the endoplasmic reticulum and in the Golgi apparatus and PD-PDT induced cell death was inhibited only by the apoptosis inhibitor. The dying cells released calreticulin, HMGB1 and ATP that were completely consumed by the bone-marrow deprived dendritic cells that released IL-6, suggesting the efficacy of subjecting PDT induced dying cells as a vaccine for immunogenic cancer cell death.

Ni et al. employed a cationic metal-organic framework (MOF), W-TBP, to combine PDT and the delivery of immuno-stimulatory CpG oligodeoxynucleotides to improve antigen presentation in cancer immunotherapy (Mariewskaya et al., 2021). The W-TBP MOF helps in the efficient loading and internalization of CpG by dendritic cells. PDT facilitates immunogenic cell death that induces the release of tumour associated antigens while the CpG augments dendritic cell development, resulting in a new method to help antigen presentation in cancer immunotherapy. Uthaman et al. designed and synthesized light-sensitive nanomicelles through the self-assembly of PEG-stearamine (C18) conjugate (PTS) with an ROS-sensitive thioketal binder (McMahon et al., 1994). This system was also loaded with doxorubicin (DOX) and PS pheophorbide A (PhA) in the hydrophobic core. Upon irradiation, PhA generates ROS that triggers the release of DOX that inhibits tumour growth and improves anti-tumour immunity. The authors *in-vivo* demonstrate that the nanomicelle system comprising of DOX and PhA has a superior ability to accumulate inside the tumour and inhibit its growth.

In a recent study, Udartseva et al. demonstrated the effects of PDT on stromal cells (Mesquita et al., 2012). Stroma is an important constituent of the tumour microenvironment and has a significant influence on the tumour response to PDT and various cancer therapeutics. In an *in-vitro* study conducted on human mesenchymal stromal cells (MSCs) subject to low PDT doses, the authors show the activation of Erk1/2 and the suppression of GSK-3 signals. They also demonstrate that co-cultivation of PDT-treated MSCs with lymphocytes results in an enhanced reduction in MSC viability while increasing its immunogenicity, leading to enhanced anti-tumour immunity. Norum et al explored photochemical internalisation (PCI) as a new method of drug delivery where visible light was used to induce endocytosis and translocation to cytosol (Norum et al., 2017). The authors demonstrate an increase in anti-tumour immunity as a response to $AlPcS_{2a}$-PDT and PCI, used for the delivery of bleomycin. *In-vivo* experiments on thymic mice showed enhanced curative effects in PDT by over 70% and in PCI by over 90% using an irradiation dose half of what was used on athymic mice, confirming the role of T-cells in the curative results through the suppression of tumour cell growth. Antitumor immunity built by PCI equalled PDT, albeit at a lower dose, suggesting the influence of bleomycin on antitumour immunity.

4.4.3 Biological effects of photodynamic therapy

1. Vascular and inflammatory effects
   Even though hematoporphyrin derived (HpD) PDT kills cancer tumours *in-vitro*, Henderson et al. reported severe vascular damage when PDT was performed *in-vivo* (Henderson et al., 1984). They observed that tumor cells were starved off nutrition and oxygen due to haemorrhaging and blood stasis. Fingar et al. confirmed this observation when they treated tumour bearing nice soon after injecting them with Photofrin, so that photodamage could be confined to the vasculature. The authors noticed specific regions of necrotic tumour cells that were far from the damaged vessels (Fingar et al., 1992). Henderson et al. further showed that the destruction or the damage caused to the vasculature is critical to reducing the survivability of EMT-6 and RIF cells *in-vivo*. They measured the oxygen levels in tumour tissues to directly demonstrate that the lack of oxygen kills tumor cells post PDT (Henderson et al., 1985; Henderson and Fingar, 1987).
   Korbelik observed and reported similarities between vascular destruction post PDT and the inflammatory response to tissue trauma or infection (Korbelik, 1996). Hence, several inflammatory agents, such as eicosanoids, have been investigated as PDT response intermediates. Endothelial cells maintain a balance between vasoactive and vasoconstrictive materials. For instance, prostacyclin and endothelium-based growth factor are vasodilators that stop platelet accumulation. They are generated together with endothelin-1, a mediator that triggers endothelial cells to vasoconstricting stimuli. When events, such as PDT, open the basement membrane to the blood serum, waves of eicosanoids push the system towards vasoconstriction (Chaudhuri et al., 1987). Vessel constrictions form at the injury sites. Platelets and neutrophils aggregate on the walls and roll towards the constriction, following which they move into the surrounding tissues through chemokine gradients (Steele et al., 1985).

McMahon et al. demonstrated *in-vivo* that Photofrin-PDT results in the release of large quantities of thromboxane and various other vasoactive leukotrienes. They also showed that the inflammatory response caused by Photofrin cannot be generalized since other PSs do not always result in the same kind of vascular response (McMahon et al., 1994). For instance, NPe6 resulted in blood stasis due to platelet aggregation on artery walls while thromboxane was not detected, and arteriolar constriction did not occur during or after NPe6-PDT. They suggested that the vascular response was due to the lack of thromboxane generation after NPe6-PDT, even though this has not been confirmed directly. The authors propose that NPe6-PDT causes the release of prostacyclin that impairs thromboxane production. SnEt2 causes an inflammatory response but does not cause the constriction of vessels or platelet aggregation, although the mechanism through which this happens is not clearly established.

Several eicosanoids are produced arachidonic acid that is created by phospholipases which hydrolyze phospholipids contained in cells (Fuster et al., 1992). Prostaglandin endoperoxide synthase then converts arachidonic acid to prostaglandins which subsequently catalyzes the inclusion of two oxygen molecules to arachidonic acid (Voet and Voet, 1990). Agarwal et al. *in-vitro* showed the presence of active phospholipases C and A_2 in RIF-1 cells after Photofrin-PDT (Agarwal et al., 1993). Constraining these phospholipases decreased the apoptotic tendency of the cells. Fingar et al. also demonstrated the ability of Indomethacin to inhibit the cyclooxygenase activity of prostaglandin endoperoxide synthase and decrease the quantity of thromboxane produced after Photofrin-PDT, which reduces the PDT response by restricting vessel constriction (Fingar et al., 1992).

In recent years, Xu et al. developed an algorithm to quantify the extent of vasoconstriction in blood vessel images post vascular-PDT (V-PDT) (Xu et al., 2020). The authors develop an automated protocol to quantify vasoconstriction in blood vessels *in-vivo* on nude mice by tracking the pixels of blood vessel images before and after V-PDT. The reduction in the number of pixels post treatment confirmed that the blood vessels were constricted, providing a method to construct a vasoconstriction image for pixel distribution. Kraus et al. used verteporfin-PDT as a vascular damaging modality, in combination with mTORC1 inhibitor rapamycin or mTORC1/C2 or the investigational drug and multiple inhibitor AZD2014 to enhance therapeutic efficiencies in SVEC endothelial cells (Kraus et al., 2019). The authors demonstrate that the enhancement of PDT induced endothelial cell apoptosis through the specific targeting PI3K-mTOR signal pathways enhances the overall treatment efficacy.

Buzzá et al. demonstrated *in-vivo* that the vascular response in the tumour vascular network was a function of the PS used (Buzzá et al., 2014). When porphyrin derived Photogem and chlorin derived Photodithazine were administered topically, considering the total vascular effects, Photodithazine was more efficient since faster effects occurred at low concentrations compared to Photogem. The authors suggest that this is due to the smaller molecular size of Photodithazine that helps in the faster and deeper penetration of the PS, making larger quantities available to the endothelial cells and the blood. The authors also observed that the mode of administration (topical or intravenous) of the PS played a crucial role in determining the efficacy of the method with intravenous injection showing the most efficient method for inducing vascular PDT response (Buzzá et al., 2014). Mesquita et al. studied tumor hemodynamics as a function of murine strains to show how blood flow varies across different mouse strains (Mesquita et al., 2012). Fibrosarcoma was induced on C3H and nude mice following which fluctuations in blood flow were monitored during PDT induced vascular stress. The authors demonstrated that PDT induced vascular stress resulted in a larger decrease in blood flow in the tumours grown on C3H mice compared to nude mice as were the variations in blood flow. Such differences in the hemodynamics were attributed to the variations in the vascular structure such as tumor blood vessel sizes between various murine models.

2. *Cellular effects*

Initially, it was assumed that PDT destroyed tissues through cellular damage caused by the localization of sufficient PS that received enough light to generate lethal levels of ROS (Agarwal et al., 1993; Voet and Voet, 1990). Henderson et al. demonstrated that a larger portion of tumor cells die *in-vivo* after PDT when vascular shutdown deprives them of oxygen and nutrients (Henderson et al., 1985; Henderson and Fingar, 1987; Henderson and Malone, 1984). They did, however, suggest that direct cell cytotoxicity cannot be entirely ruled out either. Fingar et al. confirmed this by injecting two different concentrations of Photofrin at 24 h and 5 min before PDT treatment in SMT-F bearing DBA mice. While both the doses resulted in similar circulating concentrations of the PS, the 24 h gap provided more time for the PS to accumulate in the tumor cells and the 5 min gap resulted in the PS accumulating in the vasculature. The 5 min treatment method induced no tumour response, while the 24 h method resulted in curative responses, showing that cellular damage to a certain extent was necessary to acquire complete tumour destruction (Fingar, 1989, Fingar et al., 1987). Tumour response can be induced in shorter durations with Photofrin doses up to 10 mg/kg, however, with higher doses, selectivity is compromised, resulting in extensive damage to surrounding healthy tissues. Various other authors have employed Photofrin-PDT resistant RIF cells to show the necessity of direct cell cytotoxicity (Sharkey et al., 1993; Varnes et al., 1999; Wilson et al., 1997; Adams et al., 1996). Even though the cells were produced *in-vitro*, PDT resistance was also found *in-vivo* even though the treatment resulted in similar vascular damage.

3. DNA damage
Initially, through *in-vitro* studies conducted by Gutter et al. (Gutter et al., 1977) and Fiel et al. (Fiel et al., 1981) using HpD mediated PDT, it was assumed that DNA was vulnerable to PDT modification. The most common observation was the modification of guanine. Indirect evidence of DNA damage due to PDT was shown *in-vitro* by Gomer et al. who observed a reduction in cell survivability in tissues with slower DNA-damage repair mechanisms (Gomer et al., 1986). Comparing PDT-induced DNA damage to that caused by X-rays after equitoxic doses of PS, less DNA damage was observed from PDT than X-rays. Moan et al. (Moan et al., 1980) and Evensen et al. (Evensen and Moan, 1982) demonstrated that X-ray treatment resulted in 80% higher strand breaks, 5% higher sister-chromatid exchanges, and higher chromatid aberrations compared to PDT. Additionally, strand breaks due PDT repaired more efficiently than those induced by X-ray irradiation while the number of chromatid aberrations did not compare with HpD phototoxicity. Evans et al. did an *in-vitro* comparative study on the carcinogenic potential of Photofrin and aluminum phthalocyanine-PDT to X-ray irradiation in B-lymphoblast cells (Evans et al., 1997). While both the cell lines were p53 deficient, one was transfected with a functioning p53 gene. The line with the functional p53 gene did not amass the number of mutations as the p53 mutants. However, the authors showed that the PDT mutation potential was insignificant compared to that of X-ray irradiation.

Recently, Toupin et al. demonstrated *in-vitro* that DNA intercalation by photosensitizers was not necessary to achieve PDT induced cell death (Toupin et al., 2020). The authors synthesized various Ru(II) complexes as photosensitizers for PDT on breast and prostate cancer cells and showed cytotoxicity was promoted by singlet oxygen species that resulted in cell death through necrosis within 4 h of treatment which also allowed overcoming resistance to apoptotic agents and enhance anti-tumour immunity. However, Li et al. demonstrated that DNA targeted PDT does enhance the overall treatment efficiency by synthesizing dinuclear Ir(III) based multi-stranded metallohelices as photosensitizers that targeted mitochondrial DNA (Li et al., 2020). The length and the odd-even traits of the diamine alkyl binders of the PS resulted in tunable PDT efficiency, suggesting that controlling such features of the PS could improve the DNA binding efficiencies, and thereby enhance PDT treatment efficacies.

Several groups have tried to identify the threshold irradiation dose rate using aminolevulinic acid (ALA) as the PS on melanoma bearing mice (Takahashi et al., 2013; Takahashi et al., 2016; Yamamoto et al., 2012; Kaneko et al., 2018). Reports show that ALA results in the localization of protoporphyrin IX (PpIX) in cancer tissues by preventing the conversion of PpIX to heme (Babilas et al., 2006; Goff et al., 1992), resulting in fluorescence and the subsequent production of ROS upon irradiation by light, making it ideal for PDT. Hasegawa et al. evaluated the effect of X-ray irradiation on PpIX PS in generating ROS and subsequent DNA damage in PDT on B16 melanoma cell lines *in-vitro* (Hasegawa et al., 2020). The authors demonstrated that PpIX significantly enhanced DNA damage in the melanoma nuclei under X-ray irradiation, overall improving the PDT and radiotherapy efficiency.

4.4.4 Summarizing the advantages and disadvantages of photodynamic therapy

Compared to conventional cancer treatment modalities, PDT is a minimally or non-invasive technique that costs a lot less than other known methods. PDT has been employed in the clinical studies of various cancer treatments such as colorectal cancer, obstructive oesophageal cancer, nonmelanoma skin cancer, etc. (Lucena et al., 2015; Yano et al., 2011). Liu et al. (Liu, 2005) showed that patients with obstructive oesophageal carcinoma with severe eating disabilities were able to restore a normal diet post PDT. Similarly, patients with gastric cancer showed no relapse even one year after PDT despite not going through any surgery. PDT helps improve host antitumor immunity in patients with early stage tumour or oesophageal cancer after Photofrin-PDT (Shams et al., 2015; Mroz et al., 2011) and can be applied repeatedly after or before surgery or radiotherapy. Foscan-PDT had an impressive cure rate of 67% in patients with secondary and 85% in primary head and neck cancer (Serra et al., 2008). Apart from being a stand-alone treatment modality, PDT can be combined with several other treatment methods to optimize clinical results. In a study conducted on 140 patients with advanced oesophageal and cardiac cancer (Serra et al., 2008), 42 were subject to PDT while 98 were treated with 5-fluororacil and PDT. The patients treated with 5-fluororacil and PDT showed improved results than those who were subject to PDT alone. In a study conducted by Lucena et al. patients with Bowen's disease treated with ALA-PDT showed no remission or recurrence with all their lesions disappearing in 14 months (Lucena et al., 2015).

Despite its obvious advantages, PDT still has some drawbacks. For instance, PDT is a function of the irradiation wavelength and if light having a wavelength outside the biological transparency window is used, deep tissue penetration cannot be achieved. Further, tumour hypoxia impedes PDT efficiency while the consumption of oxygen during PDT results in additional undesirable effects on local tumour hypoxia, reducing the therapeutic efficiency of PDT. PDT can also cause mild side effects such as urticaria anaphylaxis, itchiness or light sensitivity on the skin, erythema, and mild edema on PDT treated regions, etc. (Wulf et al., 2020; Öfner et al., 1996; Chen et al., 2019). A study reported by Serra et al. showed that

TABLE 4.1 The advantages and disadvantages of PDT in cancer therapy.

Advantages	Disadvantages
Minimally or noninvasive	Moderate skin photosensitization
Improves antitumor immunity	Tissue penetration is a function of irradiation wavelength
Can be combined with other treatment modalities	Low efficiencies in hypoxia environment
Low side effects	Oxygen consumption increases tumor hypoxia
Excellent repeatability	
Inexpensive treatment method	

out of 22 patients with gastric cancer that was subject to PDT, seven patients showed mild skin photosensitivity while 12 patients suffered from local pain 1–10 days after the treatment (Serra et al., 2008). To overcome these drawbacks, researchers focus on developing new light sources such as LEDs, argon-pumped dye laser, or up-conversion NPs that can be excited by longer wavelengths to emit in shorter wavelengths to facilitate deep tissue penetration (Agostinis et al., 2011; Fan et al., 2016; Gai et al., 2018; Yang et al., 2015). Materials that enrich molecular oxygen are being coupled with various photosensitizers to reduce the tumour hypoxia environment to enhance PDT efficiencies (Tang et al., 2017). Certain PSs that can breakdown cellular H_2O_2 to generate oxygen were also developed to reduce tumour hypoxia (Chen et al., 2015). To reduce fluorescence quenching and subsequent phototoxicity during PS transportation to the target *in-vivo*, various nanostructured materials possessing unique photothermal and photoacoustic properties are being designed (Li et al., 2017). The various advantages and disadvantages of PDT are summarized in Table 4.1.

4.5 Essential wavelength region in photodynamic therapy

Any light sources having appropriate spectral characteristics to activate photosensitizers can be used in PDT. Classically, Argon-dye lasers that excite Kiton red or Rhodamine B to generate 5 W red light are used. Like most lasers, these too generate highly coherent monochromatic light that can be coupled with specially designed fibre optics to deliver focused light to the target site. However, these Argon-dye lasers are bulky and expensive to operate, making them useful only in a laboratory environment and not so much in clinical use. These have since been replaced by portable, lightweight, and cheaper diode lasers, making PDT economically more viable.

One of the major issues of PDT is the tissue penetration depth of the irradiation source. The light penetration depth is a function of the absorption and scattering cross sections of the various tissue components such as water and haemoglobin, which in turn depends on the wavelength of light used. Owing to lower absorption cross sections of water and hemoglobin between 650 and 950 nm, irradiation wavelengths in this region penetrate larger depths. At shorter wavelengths, light absorption and scattering by haemoglobin increase, making irradiation wavelengths shorter than 600 nm unsuitable for PDT. Between 630 and 800 nm, the light penetration depth is between 3 and 8 mm.

The wavelengths used in PDT depend on the photosensitizers used. PSs should ideally have high absorption coefficients in the red/NIR regions, especially between 700 and 900 nm. Porphyrin based PSs such as Photofrin® have a "Soret" band, which is a strong absorption band at 400 nm and a weaker satellite band between 600 and 800 nm known as the Q-band, which is the only band that is useful for PDT (Abrahamse and Hamblin, 2016). Most traditionally used PSs such as porphyrins, chlorins, bacteriochlorin, etc. are aromatic molecules with 22, 20, and 18-π electrons, respectively, which results in the specific absorption wavelengths. For instance, for porphyrins, apart from the Soret band, the strongest Q-band absorption is between 600 and 650 nm while chlorins have their Q-band transitions between 630 and 700 nm and bacteriochlorins have their Q-bands between 700 and 800 nm, making them very useful in the PDT treatment of deeply located tumours.

Purpurin PSs have one lesser pyrole group making them strongly absorb between 630 and 710 nm (Spikes, 1990). These are water insoluble compounds and are delivered by emulsions or liposomes. Free-base, tin and zinc metallopurins are reported to cause extensive tumour necrosis while tin based purpurins cause membrane damage and the free-base PSs prevents DNA biosynthesis (Morgan et al., 1989). Benzoporphyrin (BP), synthesized from protoporphyrin, has an absorption peak at 690 nm which is four times more intense than Photofrin®-II at 623 nm. This PS can exist as a monoacid or a diacid, with the monoacid form showing higher photodynamic activity. BP achieves an optimal tissue accumulation within 3 h of administering and reduces to 50%–60% of its initial concentration in 48 h through biotransformation into various organic moieties (Richter et al., 1990). BP sensitizes both *in-vitro* and *in-vivo* and has better tissue penetration than Photofrin. Its lipophilicity facilitates cell membrane association and optimum irradiation and PDT efficiency is obtained 3 h after administering the PS with up to 80% cure rates *in-vivo* (Kessel, 1989; Richter et al., 1991).

Pcs are porphyrin derivatives with absorption coefficients highest in the 675–770 nm range. These PSs can be chelated with several metals such as aluminium and zinc to improve their phototoxicity (Dougherty, 1987). *In-vivo* studies on mice show that tumour necrosis takes place without perforation because of photodegradation and sensitizer loss from submucosal collagen (Dougherty, 1987). However, Kochubeev et al. showed that skin photosensitivity due to Pcs are lower than that caused by Photofrin (Kochubeev et al., 1988) while Anholt et al. showed that because of dye relocalisation upon light irradiation, PDT with fractionated Pcs is more effective than that with continuous light exposure (Anholt and Moan, 1992). Verteporfin is a benzoporphyrin derivative that is used primarily for ocular PDT and for age-related macular degeneration. Verteporfin-PDT involves administering Visudyne® intravenously followed by activation using a 690 nm diode laser while the PS is still in circulation.

Since 2010, ZnO NPs are being modified and explored as suitable PSs in PDT (Hackenberg et al., 2010). Hackenberg et al. demonstrated that ZnO NPs produced significant phototoxicity when they injected these NPs to head and neck squamous cell carcinoma cell lines. A decrease in cells was observed at ZnO concentrations of 2 μg/mL after 15 min of UVA-1 irradiation (Hackenberg et al., 2010). Zhang et al. demonstrated that ZnO generated ROS under UV irradiation, suggesting its use as a PS in PDT (Zhang et al., 2014). Unfortunately, since UV cannot be used as an excitation source in PDT, research is still ongoing to tune this wavelength to the red/NIR regions by doping with suitable rare-earth ions.

Synthetic dyes constitute the third class of PS agents used in PDT. The most commonly used dye in PDT is Rose Bengal that has both anticancer and antimicrobial properties that shows strong fluorescence on excitation at 520 nm. However, the dye is strongly hydrophilic, leading to poor cellular uptake and its excitation wavelength prevents tissue penetration (Gianotti et al., 2014). To overcome delivery restrictions of the dye, various nanocarriers, such as Au NPs (Cheng et al., 2008), PEG-Au NPs (Camerin et al., 2016), Zn-phthalocyanine NPs (Wang et al., 2019), etc., are used to enhance transmembrane transport of the drug. Methylene blue, with two strong absorption peaks at 635 and 670 nm (Giannelli and Bani, 2018) is a commonly employed dye in medicine and PDT due to its favourable excitation wavelengths, low dark toxicity, and the production of ROS along with hydroxyl radicals that facilitate cell apoptosis (Orth et al., 2000).

Recently, there has been an increased interest in the use of upconversion NPs (UCNPs) in the development of enhanced PDT drugs for the light-activated generation of ROS. UCNPs can convert highly penetrating red/NIR light to higher energy wavelengths that match the activation wavelengths of the PSs used. This unique property of UCNPs is exploited to activate PSs at deep tissue layers that are otherwise inaccessible to UV or visible light. Besides, UCNPs can be suitably surface modified to deliver the PSs to tumour sites. By formulating the hydrophobic PSs in the nanoparticle form, their previous drawbacks of reaching a stable dispersion in suitable solvents for administering into the body can be resolved.

The principal component of the UCNP-PDT system is the phosphor core which is tasked with converting lower energy NIR radiation to higher energy UV or visible light. By satisfying the irradiation wavelength component in PDT, careful selection of the host and dopant ions is important so that suitable emission wavelengths that activate PSs can be achieved. The most commonly used UCNP host is $NaYF_4$ because of its low phonon vibrational energy that facilitates effective charge transfer between dopant ions with low quenching effects. The classically used Yb^{3+}-Er^{3+} dopant combination is also the most widely used pair in PDT that produces a range of emission wavelengths from blue to green and red (Krämer et al., 2004). Xia et al. attempted to improve the red upconversion emission by varying to Yb^{3+} doping concentration from 18%–20% to 25% (Xia et al., 2014). To increase the intensity of the red upconversion band, Mn^{2+} ions were doped into the $NaYF_4$:Yb,Er matrix to obtain a spectrally pure, single band red emission peak at 650–670 nm with much higher intensity compared to the material without Mn^{2+}-doping (Tian et al., 2013, Wang et al., 2013).

When UV or blue light between 450 and 475 nm is required to activate lesser used PSs such as TiO_2 and hypocrellin A, Yb^{3+} is paired with Tm^{3+} (Jin et al., 2013). Jin et al. showed that doping Gd^{3+} ions into the host matrix results in a cubic to hexagonal phase transition in the $NaYF_4$ host which enhanced the blue upconversion intensity (Jin et al., 2013), showing that geometry of the host plays an important role in controlling the luminescence intensities. To further improve luminescence intensities and reduce quenching, the UCNPs can be coated with an undoped host shell such as $NaGdF_4$ (Zhao et al., 2012; Park et al., 2012) or CaF_2 (Qiao et al., 2012) to form a core-shell structure. In another interesting UCNP design, Liu et al. synthesized a $NaYF_4$ Yb,Er core on which they deposited a $NaYF_4$Yb,Tm shell that resulted in multicolour features of the material for imaging and PDT (Liu et al., 2013).

In designing a UCNP-PDT system, the selection of the PS depends on the emission wavelengths of the phosphors used or vice-versa. Since the emission spectra of UCNPs lie in the visible region, this may overlap with the PS absorption spectra, thus providing a certain amount of flexibility in picking the PS. The most widely coupled PS with UCNPs is zinc phthalocyanine (ZnPc). This is because ZnPc has (a) a strong red absorption at 670 nm (Whalley, 1961) that matches well with the 660 nm upconversion emission of Yb-Er systems with excess Yb dopant and (b) a relatively high $^1O_2$ ROS quantum

TABLE 4.2 Recent trends in UCNPs used in PDT.

Phosphor	Surface coating/ functionalization	Photosensitizer	Excitation wavelength (nm)	Emission wavelength used for PS activation	In-vivo/in-vitro model used	Ref.
$NaYF_4$:Yb,Er	N-Succinyl-N'-octyl chitosan	ZnPc	980	660 nm	Bel-7402 hepatocellular carcinoma cells. Mice bearing murine S180 sarcoma	(Cui et al., 2013)
$NaYF_4$:Yb,Er @ $NaYF_4$:Yb,Tm	PEG succinimidyl carbonate	Monomalonic fullerene	980	450, 475, 540, and 650 nm	HeLa cells	(Liu et al., 2013)
Fe_3O_4 @ $NaYF_4$:Yb,Er	PEG	Photofrin-derivative aluminum phthalocyanine chloride tetrasulfonic acid ($AlPcS_4$)	980	654–674 nm	MCF-7 breast cancer cells	(Varnes et al., 1999)
$NaYF_4$:Yb,Er, Mn	a-Cyclodextrin	Ce6, ZnPc, MB	980	660 nm	A-549 human epithelial lung cancer cell	(Tian et al., 2013)
$NaYF_4$:Yb,Er, Gd	None	ALA	980	405 and 540 nm	MKN45 human gastric cancer cell line	(Vetha et al., 2020)
$NaGdF_4$:Yb,Er	PAA, BSA	Rose Bengal	980	540 nm	4T1 murine breast cancer cells	(Chen et al., 2014)
$NaYF_4$:Yb,Er @ $NaGdF_4$	PEG-phosphorlipid	Ce6	980	660 nm	U87-MG glioblastoma. Male nude mice bearing U87-MG glioblastoma	(Park et al., 2012)
$NaGdF_4$:Yb,Er @$NaGdF_4$	Silica	AlC4Pc	980	650–670 nm	MEAR murine liver cancer cells	(Zhao et al., 2012)
$NaGdF_4$:Yb,Er @CaF_2	Mesoporous silica	Silicon phthalocyanine dihydroxide	980	660 nm	HeLa cells	(Qiao et al., 2012)
$NaYF_4$:Yb,Tm	None	TiO_2	980	470 nm	A-5RT3 squamous cell carcinoma	(Wang et al., 2013)

yield of 0.67 (Gürol et al., 2007). Even though other clinically approved Pcs such as Photofrin derived tetra-sulfonic phthalocyanine aluminium (AlPcS4), tetrasubstituted carboxy aluminium phthalocyanine (AlC4Pc) and silicon phthalocyanine dihydroxide have been used as PSs in various UCNP-PDT systems, they have lower molar absorption coefficients and ROS quantum yields compared to ZnPc, making them less popular. The only other PS that has photophysical and photobiological properties comparable to that of ZnPc is Chlorin Ce6, because of which it is fairly popular in UCNP-PDT systems. Its ROS quantum yield is 0.64 while it's molar absorption coefficient at 405 nm (blue) is about 2,00,000 $M^{-1}cm^{-1}$ and about 40,000 $M^{-1}cm^{-1}$ at 654 nm. ZnPc, on the other hand, has strong absorption in the red region with a molar absorption coefficient of 2,82,000 $M^{-1}cm^{-1}$ at 672 nm.

Some commonly used UCNPs with their excitation and emission wavelengths are shown in Table 4.2.

4.6 Recent developments in photodynamic therapy

4.6.1 Metal-organic frameworks

Recently, photoactive nanoscale metal-organic frameworks (N-MOFs) are being explored to overcome the oxygen dependence of PDT (Lan et al., 2019). MOFs are porous hybrid materials that have high crystallinity and surface areas and are formed through coordination bonds between metal ions or clusters and organic ligands. Because of their high porosity and surface areas, biocompatibility, and ease in modification, MOFs are widely used in catalysis, energy, fuel cells, medicine, etc. (Cui et al., 2015; García-García et al., 2014; Sava Gallis et al., 2017; Deng et al., 2017). The structural features of NMOFs make it ideal for effective drug delivery with high loading capacity and enhanced permeability and retention (EPR) effect that allows better accumulation of the drug at tumor sites.

Lu et al. pioneered the application of MOFs in PDT when they used benzoporphyrin functionalized MOF, DBP-UiO, with over 50 wt% of PS and by assembling the DBP ligands with Hf atoms, resulting in the additional generation of ROS due to heavy atom effects (Lu et al., 2014). Zhou et al. developed a porphyrin-based NMOF through direct assembly and functionalized with folic acid for PDT (Park et al., 2016). They accurately varied the size of the PCN-224 MOF to a broad range between 30 and 190 nm that enhanced the cellular uptake and PDT effects of the NMOF. Additionally, modifying the NMOF surface with folic acid onto a Zr_6 cluster improved the targeting efficiency. Recent focus has shifted away from self-assembly to alternate methods to develop photoactive NMOFs. For instance, Wang et al. synthesized the fluorescent dye BODIPY containing NMOFs (UiO-PDT) through a ligand exchange method (Wang et al., 2016). While the as-prepared UiO-PDT could generate ROS under specific light irradiation, its PDT effects were low due to its heterogeneous nature. However, the MTT assay against B16F10 cell lines *in-vitro* showed that UiO-PDT possessed significant cytotoxicity under light irradiation. Because of its high surface areas and porosity, NMOFs can be used to synthesize hierarchical photoactive composites. Zheng et al. demonstrated that MOF-porous organic polymer (POP) core-shell composites (MOF@POP) exhibited excellent aqueous dispersibility and had enhanced ROS singlet oxygen generation and cytotoxicity under light irradiation on HepG2 cells and HeLa cells *in-vitro* (Zheng et al., 2017).

Zheng et al. reported the synergistic treatment of PDT using NMOFs by incorporating the PS tetratopic chlorin (TCPC) ligands into a Hf-UiO-66 structure without changing its fundamental topology (Zheng et al., 2018). The authors show that unlike porphyrin incorporated NMOFs that had homogeneous porphyrin arrangements, the heterogenous TCPC component in the UiO-TCPC structure facilitated in efficient PDT effects. In another work reported by Zhang et al., si-RNA loaded zirconium-ferriporphyrin nanostructures were synthesized, through a simple hydrothermal method that produced hydroxyl radicals (·OH) du to catalytic effects on H_2O_2 (Zhang et al., 2018). Wu et al. developed Pt nanoenzyme encapsulated chlorin coated with an Au shell (Pt@UiO-66-NH2@Au-Ce6) that can continuously generate O_2 to significantly enhance the PDT effects under a 670 nm irradiation bot *in-vivo* and *in-vitro* (Liu et al., 2018).

As discussed earlier, ROS generated by PDT not just destroys cancer tumours but also triggers an immune response leading to immunogenic cell death that enhances anti-tumour immunity. The immunogenic cell death releases DAMPs (damaged associated molecular patterns) which enhances tumour immunity in the tumour microenvironment. In 2016, Lin et al. synthesized a photoactive NMOFs called TBC-Hf that had its channels encapsulated with indoleamine 2,3-dioxygenase inhibitor (IDOi) that achieved photodynamic and anti-tumour effects in colorectal cancer models (Lu et al., 2016). Later, a PEG modified benzoporphyrin based NMOF was used for PDT in combination with αPD-1 checkpoint blockade therapy, which successfully inhibited primary and metastasis tumours *in-vivo* (Zeng et al., 2018). NMOFs have since been used for PDT in combination with other therapeutic modalities. For instance, redox homeostasis is essential in a tumor microenvironment to sustain regular cancer tissue activities. Several Cu-based NMOFs have been developed to inhibit the redox environment in tumors through GSH consumption, H_2O_2 dissociation, molecular O_2 loading, etc., to enhance PDT efficacy (Zhang et al., 2018; Cai et al., 2019; Wang et al., 2019). When photoactive NMOFs are used to load therapeutic nucleic acids, synergistic therapeutic PDT and gene therapy can be achieved to improve the overall curative effects of PDT. This was demonstrated by Wang et al. who encapsulated Ce6-DNAzyme into a ZIF-8 MOF to form a nanoplatform to evaluate the combined effects of PDT and gene therapy *in-vivo* and *in-vitro* (Wang et al., 2019).

4.6.2 Photoactive materials for wavelength response

As research on red light activation for PDT and other photobiological therapeutics is intensifying, optogenetic proteins and ruthenium complexes have begun to show promising results. Ruthenium polypyridyl complexes can undergo ligand exchange processes under visible light that, based on some reports, can extend into the NIR region while optogenetic proteins that undergo photodissociation have also been recently used as useful biomaterials in various applications. The performance of these materials is complemented by the development of new photo-responsive organic molecules that are activated in the visible spectrum (Rapp and DeForest, 2020).

In 2017, a group of researchers lead by Campbell developed a monomeric protein, PhoCl, that undergoes irreversible spinal peptide cleavage when irradiated by 400 nm UV light (Zhang et al., 2017). PhoCl is a first of its kind photocleavable monomeric protein reported to date and has been further developed to reduce the distance between the cleavable region and any C-terminal species. Recently, Shadish et al showed the expression of protein fusions where the bioactive components were genetically attached to PhoCl's C-terminal (Shadish et al., 2019). Making use of a chemoenzymatic process to introduce a single azide species at the protein's N-terminal, the photocleavable PhoCl was conjugated to PEG hydrogels formed through a process known as strain promoted azide-alkyne cycloaddition (Agard et al., 2004).

Ruthenium polypyridyl complexes are being used as biomaterials in PDT by restructuring the complex to promote ligand exchange in a local solvent instead of ROS generation (Zayat et al., 2003; Zayat et al., 2006). These complexes have been used previously *in-vivo* and *in-vitro* as caging groups for several small bioactive materials (Li et al., 2018) and have recently been employed in biomaterial settings (Sun et al., 2016; Zeng et al., 2018). Rapp et al. pioneered the reuse of Ru-complexes as crosslinkers when they used two monodentate pyridine ligands functionalized with aldehydes to produce a photocleavable crosslinker that had an absorption maximum at 450 nm with a tail that extended up to 520 nm. The photophysical properties of this aldehyde functionalized Ru-complex facilitated the rapid degradation of hydrogels due to its reaction with hydrazine-modified hyaluronic acid.

Theis et al. synthesized a ruthenium bipyridine pendant amine complex as a crosslinker in hydrogel compounds that had photophysical properties similar to that of aldehyde functionalized Ru-complexes with an additional feature of the 4-(aminomethyl)pyridine ligand exchange upon irradiation in an acetonitrile solvent (Theis et al., 2017). This ligand exchanged Ru-bipyridine 4-(aminomethyl) pyridine complex easily photocleaved when irradiated with UV-visible light of wavelengths >395 nm, yielding a product within 1 min at high dilutions. Comparatively, Ru-polypyridine complexes cleaved under multiphoton irradiation at high frequency pulsed 800 nm NIR wavelengths. Another photoactive material that is currently popular is the light-oxygen-voltage sensing domain-2 (LOV-2) protein that links noncovalently and has a high affinity to the Jα, small α-helical protein (Renicke et al., 2013; Fenno et al., 2011). Under low intensity blue irradiation at 470 nm, LOV-2 goes through a large conformational realignment that displaces Jα by several tens of angstroms. This process is reversible in the dark and thus, under pulsed irradiation, the LOV-2-Jα system can undergo alternate softening and stiffening cycles.

Recently, a group of researchers lead by Gasser, Ciofini, and Spingler combined DFT calculations with experimentally known photophysical and biological properties of Ru-polypyridine complexes to tune its wavelength absorption to the transparency widow, enabling their use as PSs in PDT (Karges et al., 2020). The photophysical properties of Ru complexes such as its absorption, emission, excited state lifetime, are all functions of the ligands bound to the Ru ion. Thus, through a combined computational and experimental approach, the authors developed a Ru-4,7-diphenyl-1,10-phenanthroline-bipyridine complex that had an absorption maximum at 595 nm with an absorption tail that extended into the biological window. This complex was stable in plasma and localized in the cytoplasm of HeLa cells *in-vitro* and showed no structural differences upon irradiation, making it a suitable PS candidate for deep tumour PDT. Such rational designs that combine theoretical models with experimental design can help develop new ligands that tune the optical properties of the metal complexes as PSs in PDT.

4.6.3 Photodynamic therapy and hypoxia-controlled nanomedicine

When PSs absorb light, they go from a ground state to an excited singlet state which then de-excites through two distinct pathways; one is through fluorescence or through a non-radiative transition and the second is by going into an excited triplet state that has a longer luminescence lifetime due to cross relaxations (Abrahamse and Hamblin, 2016; Castano et al., 2004) The PS in the triplet state can transfer energy to molecular oxygen through type-I and type-II processes (discussed in section 4.3) (Baptista et al., 2017) generating ROS that attacks vasculatures and other tissues, resulting in photo-induced cell death (Yanovsky et al., 2019). Thus, being an oxygen consuming process, PDT increases tumor hypoxia that leads to resistance, making this an important feature to discuss.

Tumor hypoxia is the region of solid tumours where the oxygen pressure is lower than 10 mm Hg, compared to healthy tissues where the oxygen pressure is 40–60 mm Hg. While PDT shows resistance when the oxygen pressure goes below 20 mm, the differences in the PSs used, light dosimetry, tumor characteristics play a role in controlling the oxygen demand during PDT (Song et al., 2019). Currently, tumor hypoxia is measured by three strategies; first, an Eppendorf oxygen electrode, considered a gold standard in evaluating the local oxygen concentrations, is used to determine the local oxygen pressure; second, the tumour hypoxia is determined by monitoring specific endogenous markers such as HIF1α, HIF2α, osteopontin, lysyl oxidase, etc.; and third, exogenous probes such as pimonidazole etanidaole pentafluoride can calculate the oxygen levels in a tumour microenvironment. Hypoxia is one of the major contributors for lowering PDT efficiencies due to its inhibition of ROS production (Qian et al., 2016). Besides, it also inhibits immune response and enhances tumour invasion and metastasis, thus conferring the need for research on methods to reduce tumour hypoxia during PDT (Atkuri et al., 2007; Atkuri et al., 2005).

Currently, nanomedicine-based drug delivery systems provide solutions for improving tumour selective delivery in PDT and combinational therapy. Nanomedicine delivery with multifunctional strategies such as oxygen supply, down-controlling oxygen consumption, and oxygen independent PDT are some of the core areas of research (Luo et al., 2016; Cheng et al., 2015; Yang et al., 2018).

Oxygen carriers are primarily haemoglobin and non-haemoglobin carriers which, through a Trojan effect, preserve and release large amounts of oxygen required for use in PDT. The superior oxygen transportability of haemoglobins can be exploited to deliver oxygen during PDT. For instance, Luo et al. loaded haemoglobin and indocyanine green PS on an artificially developed nanoscale red cell to improve the oxygen supply for PDT consumption (Luo et al., 2016). The additional oxygen made available through haemoglobin facilitated the additional generation of ROS, which subsequently oxidized haemoglobin into the more cytotoxic ferryl-hemoglobin which synergized with PDT to enhance anti-tumor immunity. Thus, multifunctional nanocarriers like hemoglobin not just helps in overcoming the hypoxic tumor environment and improve PDT efficiency, but also exerts a strong immune response for anti-tumour immunity.

Among non-haemoglobin nanocarriers, perfluorohexane (PFH) is known to be a remarkable oxygen carrier with excellent biocompatibility. Wang et al. developed multifunctional PFH-MoSx hollow NPs loaded with human serum albumin and chloroaluminium phthalocyanine as modifiers for triple modal imaging and PDT-photothermal combined treatment (Wang et al., 2018). Results obtained *in-vivo* from 4T1 tumour bearing nude mice showed that the oxygen rich PFH NPs provided sufficient oxygen to overcome tumour hypoxia and improve PDT efficiency. Smart oxygen generators are also an interesting class of materials that react with water of hydrogen peroxide through a self-sufficient method to produce oxygen. Enzymes such as catalase and metal oxides such as MnO_2, CaO_2, and CeO_2, metals such as gold nanoclusters and Pt (IV) prodrugs are some of the materials that show self-sufficient smart oxygen generation. For instance, through non-covalent interactions with fluorinated polyethyleneimine, Li et al. developed chlorin-catalase NPs that demonstrated excellent transmembrane, transmucosal and intratumor permeability (Li et al., 2019).

Similarly, water splitting nanomaterials can be used to generate oxygen to improve PDT activity (Zheng et al., 2016). Yang et al. synthesized carbon nanodot-TiO_2 nanotubes with excellent photostability and high catalytic performance that overcome the drawbacks posed by organic-inorganic metal hybrids in PDT. Carbon nanodots helped in tuning the bandgap of TiO_2 and facilitated in effective electron-hole charge separation that enhanced the photosensitization effect that helped decompose water and remove tumour hypoxia and provide oxygen for PDT.

4.7 Future scopes and perspectives

PDT has already been established as a promising therapeutic method in treating cancer. However, it is still not a first line treatment option because of the various drawbacks of conventional PSs. From this review, it is evident that the application of nanotechnology through the development of smart functional materials is the way forward to achieve desired breakthroughs in PDT. Surface functionalized and modified NPs can be loaded with various hydrophobic PSs and not just transport them to tumour sites through the EPR effect but can also improve site selectivity for enhanced loading of the PS at the target. Currently, a lot of effort is being invested in developing multifunctional materials that can perform various diagnostic and therapeutic functions, into nanoplatforms for uses in medicine and healthcare. Such nanocarriers facilitate drug delivery to target sites with improved spatial specificity while simultaneously achieving instantaneous imaging for monitoring diseased/affected areas and possessing therapeutic abilities.

Understanding various facets of nanomaterial chemistry has significantly evolved over the years and so have the various techniques involved in nanomaterial fabrication and synthesis. Currently, it is possible to develop nanomaterials that are loaded with multiple sophisticated functionalities for applications in biology, medicine, physics, and chemistry. However, despite their uses in biology and medicine, there is still a lack of thorough understanding and plenty to be explored on how these nanomaterials perform and behave in complex biological media. Thus, there is an immediate need to identify comprehensive characterisation of such multifunctional nanosystems as therapeutic agents to investigate their crosstalk, quenching, and interference with each other's performance that affect overall PDT activity. Additionally, physicochemical characterisation, authentication, and safety protocols need to be established to synthesize homogeneous NPs that produce accurate and reproducible results.

The fact that none of these nanosystems is FDA approved for clinical trials is concerning. The major holdup in translating these materials into clinical practice is the lack of standardized methods to evaluate its *in-vivo* biodistribution, a comprehensive pharmacokinetic pharmacodynamic understanding, the *in-vivo* effects of the nanomaterials both short and long term, and its excretion from the system. To overcome these issues, the National Characterization Laboratory (NCL) was set up by the National Cancer Institute in the United States in 2004. The NCL developed several standardized protocols for the comprehensive study of the various nanomaterial physicochemical properties, *in-vitro* immunological and cytotoxic properties, administration, distribution, metabolism, and elimination and toxicity profiles in various *in-vivo* animal models to promote the regulated use of nanotechnology-driven cancer therapeutics and expedite clinical trials. Today, over a decade since its inception, several hundreds of nanocancer therapeutic agents have been identified and characterized with several more in the pipeline. More similar regulatory bodies need to be established to provide solutions for the step-by-step evaluation of such nanomedical systems, given the quantity of research being conducted by both academia and the pharmaceutical industry.

PDT using UCNPs, even though an exciting and promising treatment modality, faces several limitations, such as PS loading, that affect the efficiency of the modality. Currently, most PDT *in-vivo* studies that employ UCNPS, administer the material through intratumoural routes partly due to the premature leakage of PS that resulted in lower quantities of PS being delivered to the tumour site and partly due to a lack of effective targeting through proper surface functionalization methods that lead to its accumulation at other organs, resulting in severe toxicity. Although several studies report surface functionalization and modification of NPs with suitable biocompatible layers and including tumour identifying motifs for accurate targeting of the NPs to the tumour, the amount of material that eventually accumulated on the tumor after intravenous injection was relatively low, thus lowering the PDT efficiency and resulting in tumour recurrence in several of these studies.

This indicates the need to further optimize the molecular weight, surface structure and morphology, and surface coverage ratio of the biocompatible layers such as PEG that are used to enhance the biodistribution of NPs *in-vivo*. Additionally, it is well-known that targeting ligands in high densities can result in undesirable interactions with endothelial and non-tumour cells that prompt immune responses resulting in opsonization-mediated elimination of the nanomaterials from the system (Ferrari, 2008, Wang and Thanou, 2010). Designing truly tumour targeting nanomaterials might require strict optimisation of the number of targeting ligands coated on the surface of the nanoparticle, targeting several receptors that are expressed in the tumour and tumour vasculature, and the addition of cleavable antibiofouling agents that block the target ligands until it accumulates in the tumor. Having said that, it must be understood that most *in-vivo* toxicology studies are restricted to small mammals such as mice, which might be difficult to translate to a clinical level because the duration of the therapeutic and toxicology studies in these cases are relatively short at 2 weeks and 3 months, respectively.

Since PDT is an oxygen consuming process, tumor hypoxia severely impairs its performance and thus, it is imperative to investigate hypoxia mechanisms and the challenges it throws at PDT, at the same time, developing new drugs that help in curbing some of these challenges and improve PDT action at tumour sites. This has resulted in the development of nanoparticle-driven drug delivery systems that control hypoxic environments to improve PDT efficiency. NPs functionalized and loaded with tumour hypoxia modulators such as oxygen carriers and oxygen generators or subcellular oxygen independent methods and hypoxia-responsive drugs enhance PDT effects because of their ability to manipulate the hypoxia environment in a tumour. Research should focus on developing nanosystems for effective oxygen delivery to tumour microenvironments and investigate more subcellular organelle-targeted drugs in a clinical setting. Hopefully, such hypoxia reducing PDT enhancing drugs provide more scope and opportunity for the development of PDT at a clinical level.

Despite the advances made in the development of multifunctional NPs for oxygen delivery to tumours, there are still several challenges that need to be addressed in PDT and tumour hypoxia. For instance, it is difficult to accurately control tumour hypoxia because of tumour characteristics, such as the heterogeneity of tumor species and steric heterogeneity within a tumor. Another concern is that a one-shot PDT treatment does not deliver adequate therapeutic effects because of tissue penetration problems light sources face and due to the use of PSs with phototoxicity. Thus, PDT needs to be combined with other therapeutic methods such as photothermal therapy or photochemotherapy or radiotherapy to deliver desired effects in treating tumours. Even though there are several multifunctional nanomaterials that have been designed and developed to tackle the hypoxia concerns in PDT *in-vivo*, as discussed earlier, very few of these are FDA approved for clinical use. How do we bridge the gap between preclinical tests using *in-vivo* animal models and clinical trials? Finding adequate answers to this question is the way forward in not just employing nanodrug delivery systems in PDT, but also for cancer research in general.

Considering the multifunctional materials for light-based co-therapy, the immediate concern is on utilizing suitable synergistic effects by carefully choosing appropriately designed materials. Until now, only thermal effects of carriers upon drug release have been confirmed while the synergistic effects of PDT-chemo and PDT-PTT need to be clearly established before designing appropriate nanosystems. Nevertheless, there is still plenty of scope to improve PDT performance by reducing particle size, improving the stability of PS loading on nanoplatforms, and optimizing PDT parameters. This is truly a diverse, multidisciplinary field and hence, it is important that there is frequent and regular communication between scientists, academicians, clinicians, and investors to share and discuss the various challenges and opportunities in bringing PDT in as a first choice therapeutic method in oncology. There is no doubt that versatile and efficient, multifunctional NPs that take up an array of responsibilities including complex tasks such as bioimaging, are the future of PDT.

References

Abrahamse, H., Hamblin, M.R., 2016. New photosensitizers for photodynamic therapy. Biochem. J. 473 (4), 347–364.
Ackroyd, R., Kelty, C., Brown, N., Reed, M., 2001. The history of photodetection and photodynamic therapy. Photochem. Photobiol. 74, 656–669.
Adams, K., Espiritu, M., Wilson, B., Singh, G., 1996. Photodynamic therapy-resistant cells maintain their resistance in vivo. Photochem. Photobiol 63, 98S.
Agard, N.J., Prescher, J.A., Bertozzi, C.R., 2004. A strain-promoted [3 + 2] azide– alkyne cycloaddition for covalent modification of biomolecules in living systems. J. Am. Chem. Soc. 126, 15046–15047.
Agarwal, M.L., Clay, M.E., Harvey, E.J., Evans, H.H., Antunez, A.R., Oleinick, N.L., 1991. Photodynamic therapy induces rapid cell death by apoptosis in L5178Y mouse lymphoma cells. Cancer Res. 51, 5993–5996.
Agarwal, M.L., Larkin, H.E., Zaidi, S., Mukhtar, H., Oleinick, N.L., 1993. Phospholipase activation triggers apoptosis in photosensitized mouse lymphoma cells. Cancer Res. 53, 5897–5902.
Agostinis, P., Berg, K., Cengel, K.A., Foster, T.H., Girotti, A.W., Gollnick, S.O., Hahn, S.M., Hamblin, M.R., Juzeniene, A., Kessel, D., 2011. Photodynamic therapy of cancer: an update. CA Cancer J. Clin. 61, 250–281.
Alexandrova, R., Stoykova, E., Ion, R.-.M., 2004. Photodynamic therapy of cancer. Exp. Pathol. Parasitol. 7/3, 3–23.
Allegra, E., Lombardo, N., Puzzo, L., Garozzo, A., 2009. The usefulness of toluidine staining as a diagnostic tool for precancerous and cancerous oropharyngeal and oral cavity lesions. Acta Otorhinolaryngol. Ital 29 (4), 187–190.
Allison, R., Moghissi, K., Downie, G., Dixon, K., 2011. Photodynamic therapy (PDT) for lung cancer. Photodiagn. Photodyn. Ther. 8, 231–239.
Allison, R.R., Downie, G.H., et al., 2004. Photosensitizers in clinical PDT. Photodiagn. Photodyn. Ther. 1 (1), 27–42.
Allison, R.R., Mang, T.S., Wilson, B.D., Vongtama, V., 1998. Tin ethyl etiopurpurin-induced photodynamic therapy for the treatment of human immunodeficiency virus-associated Kaposi's sarcoma. Curr. Ther. Res. 59 (1), 23–27.
Alzeibak, R., Mishchenko, T.A., Shilyagina, N.Y., et al., 2020. Targeting immunogenic cancer cell death by photodynamic therapy: past, present and future. J. Immunother. Cancer 9, e001926.
Amaravadi, R.K., Thompson, C.B., 2007. The roles of therapy-induced autophagy and necrosis in cancer treatment. Clin. Cancer Res 13 (24), 7271–7279. doi:10.1158/1078-0432.CCR-07-1595.
Anholt, H., Moan, J., 1992. Fractionated treatment of CaD2 tumors in mice sensitized with aluminium phthlocyanine tetrasulfonate. Cancer Lett. 61, 263–267.
Atkuri, K.R., Herzenberg, L.A., Herzenberg, L.A., 2005. Culturing at atmospheric oxygen levels impacts lymphocyte function. Proc. Natl. Acad. Sci. 102, 3756–3759.
Atkuri, K.R., Herzenberg, L.A., Niemi, A.-K., Cowan, T., Herzenberg, L.A., 2007. Importance of culturing primary lymphocytes at physiological oxygen levels. Proc. Natl. Acad. Sci. 104, 4547–4552.
Babilas, P., Landthaler, M., Szeimies, R.-M., 2006. Photodynamic therapy in dermatology. Eur. J. Dermatol. 16, 340–348.
Baptista, M.S., Cadet, J., Di Mascio, P., Ghogare, A.A., Greer, A., Hamblin, M.R., Lorente, C., Nunez, S.C., Ribeiro, M.S., Thomas, A.H., 2017. Type I and type II photosensitized oxidation reactions: guidelines and mechanistic pathways. Photochem. Photobiol. 93, 912–919.
Baskaran, R., Lee, J., Yang, Su-Geun, 2018. Clinical development of photodynamic agents and therapeutic applications. Biomater Res. 26, 22–25.
Bellnier, D.A., 1991. Potentiation of photodynamic therapy in mice with recombinant human tumor necrosis factors-α. J. Photochem. Photobiol. B 8, 203–210.
Benniston, A.C., Harriman, A., Gulliya, K.S., 1994. Photophysical properties of merocyanine 540 derivatives. J. Chem. Soc., Faraday Trans. 90, 953–961.
Bensasson, R.V., 2013. Primary Photo-Processes in Biology and Medicine. Springer Science & Business Media, Plenum Press, London and New York.
Bonnett, R., 2000. Chemical Aspects of Photodynamic Therapy. CRC Press, London. Advanced chemistry texts, ISSN 1027-3654.
Buzzá, H., Silva, L., Moriyama, L.T., Bagnato, V.S., Kurachi, C., 2014. Evaluation of vascular effect of photodynamic therapy in chorioallantoic membrane using different photosensitizers. J. Photochem. Photobiol. B 138, 1–7.
Cabuy, E., 2012. Photodynamic therapy in cancer treatment. Reliable Cancer Therapies. 3 (2), 1–54.
Cai, X., Luo, Y., Song, Y., Liu, D., Yan, H., Li, H., Du, D., Zhu, C., Lin, Y., 2018. Integrating in situ formation of nanozymes with three-dimensional dendritic mesoporous silica nanospheres for hypoxia-overcoming photodynamic therapy. Nanoscale 10, 22937–22945.
Cai, X., Xie, Z., Ding, B., Shao, S., Liang, S., Pang, M., Lin, J., 2019. Monodispersed copper (I)-based nano metal–organic framework as a biodegradable drug carrier with enhanced photodynamic therapy efficacy. Adv. Sci. 6, 1900848.
Caliskan, S.O., Bilgin, M.D., 2018. Chapter 93: Phtodynamic therapy. Book: The most recent studies in science and art, 2. Gece Publishing, Ankara, Turkey. ISBN • 978-605-288-357-0.
Camerin, M., Moreno, M., Marín, M.J., Schofield, C.L., Chambrier, I., Cook, M.J., Coppellotti, O., Jori, G., Russell, D.A., 2016. Delivery of a hydrophobic phthalocyanine photosensitizer using PEGylated gold nanoparticle conjugates for the in vivo photodynamic therapy of amelanoma. Photochem. Photobiol. Sci. 15, 618–625.
Canti, G., Lattuada, D., Nicolin, A., Taroni, P., Valentini, G., Cubeddu, R., 1994. Antitumor immunity induced by photodynamic therapy with aluminum disulfonated phthalocyanines and laser light. Anticancer Drugs 5, 443–447.
Cantisani, C., Paulino, G., Faina, V., et al., 2014. Overview on topical 5-ALA photodynamic therapy use for non melanoma skin cancers. Int. J. Photoenergy, 7. http://dx.doi.org/10.1155/2014/304862.
Castano, A., Liu, Q., Hamblin, M., 2006. A green fluorescent protein-expressing murine tumour but not its wild-type counterpart is cured by photodynamic therapy. Br. J. Cancer 94, 391–397.
Castano, A.P., Demidova, T.N., Hamblin, M.R., 2004. Mechanisms in photodynamic therapy: part one—photosensitizers, photochemistry and cellular localization. Photodiagn. Photodyn. Ther. 1, 279–293.

Castano, A.P., Mroz, P.l, Hamblin, M.R., 2006. Photodynamic therapy and anti-tumour immunity. Nat. Rev. Cancer 6, 535–545.

Chaudhuri, K., Keck, R.W., Selman, S.H., 1987. Morphological changes of tumor microvasculature following hematoporphyrin derivative sensitized photodynamic therapy. Photochem. Photobiol. 46, 823–827.

Chen, D., Song, M., Huang, J., et al., 2020. Photocyanine: a novel and effective phthalocyanine-based photosensitizer for cancer treatment. J. Innov. Opt. Health Sci. 13 (3), 2030009. https://doi.org/10.1142/S1793545820300098.

Chen, D., Tang, Y., Zhu, J., Zhang, J., Song, X., Wang, W., Shao, J., Huang, W., Chen, P., Dong, X., 2019. Photothermal-pH-hypoxia responsive multifunctional nanoplatform for cancer photo-chemo therapy with negligible skin phototoxicity. Biomaterials 221, 119422.

Chen, H., Tian, J., He, W., Guo, Z., 2015. H2O2-activatable and O2-evolving nanoparticles for highly efficient and selective photodynamic therapy against hypoxic tumor cells. J. Am. Chem. Soc. 137, 1539–1547.

Chen, H.-M., Yu, C.-H., Lin, H.-P., Cheng, S.-J., Chiang, C.-P., 2012. 5-Aminolevulinic acid-mediated photodynamictherapy for oral cancers and precancers. J. Dental Sci. 7, 307–315.

Chen, Q., Huang, Z., Luck, D., Beckers, J., Brun, Pierre-Herve, Wilson, B.C., Scherz, A., Salomon, Y., Hetzel, F.W., 2002. Preclinical studies in normal canine prostate of a novel palladium-bacteriopheophorbide (WST09) photosensitizer for photodynamic therapy of prostate cancer. Photochem. Photobiol. 76 (4), 438–445.

Chen, Q., Wang, C., Cheng, L., He, W., Cheng, Z., Liu, Z., 2014. Protein modified upconversion nanoparticles for imaging-guided combined photothermal and photodynamic therapy. Biomaterials 35, 2915–2923.

Cheng, Y., Cheng, H., Jiang, C., Qiu, X., Wang, K., Huan, W., Yuan, A., Wu, J., Hu, Y., 2015. Perfluorocarbon nanoparticles enhance reactive oxygen levels and tumour growth inhibition in photodynamic therapy. Nat. Commun. 6, 1–8.

Cheng, Y., Samia, A.C., Meyers, J.D., Panagopoulos, I., Fei, B., Burda, C., 2008. Highly efficient drug delivery with gold nanoparticle vectors for in vivo photodynamic therapy of cancer. J. Am. Chem. Soc. 130, 10643–10647.

Chilakamarthi, U., Giribabu, L., 2017. Photodynamic therapy: past, present and future. Chem. Rec. 17, 1–29.

Chowdhary, R.K, Shariff, I., Dolphin, D., 2003. Drug release characteristics of lipid based benzoporphyrin derivative. J. Pharm. Pharmaceut. Sci. 6 (1), 13–19.

Christensen, C.E., 1923. Finsen's Medical light institute 1896 October 23 - 1921. Acta Radiol. 2 (3), 210–242.

Cohen–Jonathan, E., Bernhard, E.J., McKenna, W.G., 1999. How does radiation kill cells? Curr. Opin. Chem. Biol. 3, 77–83.

Correia, J.H., Rodrigues, J.A., Pimenta, S., Dong, T., Yang, Z., 2021. Photodynamic therapy review: principles, photosensitizers, applications, and future directions. Pharmaceutics 13 (9), 1332.

Cramers, P., Ruevekamp, M., et al., 2003. Foscans uptake and tissue distribution in relation to photodynamic efficacy. Br. J. Cancer 88, 283–290.

Cui, S., Yin, D., Chen, Y., Di, Y., Chen, H., Ma, Y., Achilefu, S., Gu, Y., 2013. In vivo targeted deep-tissue photodynamic therapy based on near-infrared light triggered upconversion nanoconstruct. ACS Nano 7, 676–688.

Cui, X., Zhang, J., Wan, Y., et al., 2019. Dual fenton catalytic nanoreactor for integrative type-I and type-II photodynamic therapy against hypoxic cancer cells. ACS Appl. Bio Mater. doi:10.1021/acsabm.9b00456.

Cui, Y., Song, R., Yu, J., Liu, M., Wang, Z., Wu, C., Yang, Y., Wang, Z., Chen, B., Qian, G., 2015. Dual-emitting MOF⊃ dye composite for ratiometric temperature sensing. Adv. Mater. 27, 1420–1425.

Dai, T., Fuchs, B.B., Coleman, J.J., Prates, R.A., et al., 2012. Concepts and principles of photodynamic therapy as an alternative antifungal discovery platform. Front. Microbiol. 3, 1–16.

Daniell, M., Hill, J., 1991. A history of photodynamic therapy. Aust. N. Z. J. Surg. 61, 340–348.

Delmarre, D., Hioka, N., Boch, R., Sternberg, E., Dolphin, D., 2001. Aggregation studies of benzoporphyrin derivative. Can. J. Chem. 79, 1068–1074.

Deng, J., Wang, K., Wang, M., Yu, P., Mao, L., 2017. Mitochondria targeted nanoscale zeolitic imidazole framework-90 for ATP imaging in live cells. J. Am. Chem. Soc. 139, 5877–5882.

Dennis, K.P., 2014. Phthalocyanine-based photosensitizers: more efficient photodynamic therapy? Future Med. Chem. 6, 18. https://doi.org/10.4155/fmc.14.139.

Dias, L.D., Mfouo-Tynga, I.S., 2020. Learning from nature: bioinspired chlorin-based photosensitizers immobilized on carbon materials forcombined photodynamic and photothermal therapy. Biomimetics 5, 53. doi:10.3390/biomimetics5040053.

dos Santos, A.F., Terra, L.F., Wailemann, R.A.M., 2017. Me thylene blue photodynamic therapy induces selective and massive cell death in human breast cancer cells. BMC Cancer 17, 194.

Dougherty, T.J., 1987. Photosensitizers: therapy and detection of malignant tumors. Photochem. Photobiol. 45, 879–889.

Dougherty, T.J., 1996. A brief history of clinical photodynamic therapy development at Roswell Park Cancer Institute. J. Clin. Laser Med. Surg. 14, 219–221.

Dougherty, T.J., Gomer, C.J., Henderson, B.W., et al., 1998. Photodynamic therapy. J. Natl. Cancer Inst. 90 (12), 889–905.

Eichelbaum E., Navas AG., Arreaza G.C., 1977. Toluidine blue and gastric cance. G.E.N.: organo oficial de la Sociedad Venezolana de Gastroenterología, Endocrinología y Nutrición 31 (4), 283–289.

Evans, H.H., Horng, M.F., Ricanati, M., Deahl, J.T., Oleinick, N.L., 1997. Mutagenicity of photodynamic therapy as compared to UVC and ionizing radiation in human and murine lymphoblast cell lines. Photochem. Photobiol. 66, 690–696.

Evans, S., Matthews, W., Perry, R., Fraker, D., Norton, J., Pass, H.I., 1990. Effect of photodynamic therapy on tumor necrosis factor production by murine macrophages. J. Natl. Cancer Inst. 82, 34–39.

Evensen, J.F., Moan, J., 1982. Photodynamic action and chromosomal damage: a comparison of haematoporphyrin derivative (HpD) and light with X-irradiation. Br. J. Cancer 45, 456–465.

Fahmy, S.A., El-Said Azzazy, H.M., Schaefer, J., 2021. Liposome photosensitizer formulations for effective cancer photodynamic therapy. Pharmaceutics 13, 1345.

Fan, W., Bu, W., Shi, J., 2016. On the latest three-stage development of nanomedicines based on upconversion nanoparticles. Adv. Mater. 28, 3987–4011.

Fehr, M.K., Tromberg, B.J., Svaasand, L.O., Ngo, P., Berns, M.W., Tadir, Y., 1996. Structural and functional effects of endometrial photodynamic therapy in a rat model. Am. J. Obstet. Gynecol. 175 (1), 115–121.

Fenno, L., Yizhar, O., Deisseroth, K., 2011. The development and application of optogenetics. Annu. Rev. Neurosci. 34.

Ferlay, J., Shin, H., Bray, F., Forman, D., Mathers, C., Parkin, D., 2008. GLOBOCAN v1. 2, Cancer Incidence and Mortality Worldwide: IARC Cancer Base No. 10 [Internet]. International Agency for Research on Cancer and 2010, Lyon, France.

Ferrari, M., 2008. Nanogeometry: beyond drug delivery. Nat. Nanotechnol. 3, 131.

Fiel, R.J., Datta-Gupta, N., Mark, E.H., Howard, J.C., 1981. Induction of DNA damage by porphyrin photosensitizers. Cancer Res. 41, 3543–3545.

Fingar, V.H., 1989. Drug, Light, and Oxygen Dependence of Photodynamic Therapy of Murine Tumors. Roswell Park Cancer Institute, State University of New York, Buffalo, New York, pp. 1.

Fingar, V.H., Potter, W.R., Henderson, B.W., 1987. Drug and light dose dependence of photodynamic therapy: a study of tumor cell clonogenicity and histologic changes. Photochem. Photobiol. 45, 643–650.

Fingar, V.H., Wieman, T.J., Wiehle, S.A., Cerrito, P.B., 1992. The role of microvascular damage in photodynamic therapy: the effect of treatment on vessel constriction, permeability, and leukocyte adhesion. Cancer Res. 52, 4914–4921.

Finsen, N., 1901. Phototherapy: 1. The Chemical Rays of Light and Smallpox, 2. Light as a Stimulant, 3. The Treatment of Lupus Vulgarisby Concentrated Chemical Rays. Edward Arnold, London.

Fisher, A.M., Danenberg, K., Banerjee, D., Bertino, J.R., Danenberg, P., Gomer, C.J., 1997. Increased photosensitivity in HL60 cells expressing wild-type p53. Photochem. Photobiol. 66, 265–270.

Fuster, V., Badimon, L., Badimon, J.J., Chesebro, J.H., 1992. The pathogenesis of coronary artery disease and the acute coronary syndromes. N. Engl. J. Med. 326, 310–318.

Gai, S., Yang, G., Yang, P., He, F., Lin, J., Jin, D., Xing, B., 2018. Recent advances in functional nanomaterials for light–triggered cancer therapy. Nano Today 19, 146–187.

Gaio, E., Scheglmann, D., Reddi, E., Moret, F., 2016. Uptake and photo-toxicity of Foscan®, Foslip® and Fospeg® in multicellular tumor spheroids. J. Photochem. Photobiol. B 161, 244–252.

Gallardo-Villagrán, M., Leger, D.Y., Liagre, B., Therrien, B., 2019. Photosensitizers used in the photodynamic therapy of rheumatoid arthritis. Int. J. Mol. Sci. 20 (13), 3339.

Gao W., Kang Jeong-Han, Liao Y., Li M., Yin Xiao-Ming. Chapter 30 autophagy and cell death. doi:10.1007/978-1-60327-381-7_30.

Garbo, G.M., 1996. Purpurins and benzochlorins as sensitizers for photodynamic therapy. J. Photochem. Photobiol. B 34 (2-3), 109–116.

García-García, P., Müller, M., Corma, A., 2014. MOF catalysis in relation to their homogeneous counterparts and conventional solid catalysts. Chem. Sci. 5, 2979–3007.

Giannelli, M., Bani, D., 2018. Appropriate laser wavelengths for photodynamic therapy with methylene blue. Lasers Med. Sci. 33, 1837–1838.

Gianotti, E., Martins Estevão, B., Cucinotta, F., Hioka, N., Rizzi, M., Renò, F., Marchese, L., 2014. An efficient rose bengal based nanoplatform for photodynamic therapy. Chem. Eur. J. 20, 10921–10925.

Goff, B.A., Bachor, R., Kollias, N., Hasan, T., 1992. Effects of photodynamic therapy with topical application of 5-aminolevulinic acid on normal skin of hairless guinea pigs. J. Photochem. Photobiol. B 15, 239–251.

Gomer, C.J., 1991. Preclinical examination of first and second generation photosensitizers used in photodynamic therapy. Photochem. Photobiol. 54, 1093–1107.

Gomer, C.J., 2010. Photodynamic Therapy: Methods and Protocols. 635. Humana press, Springer Science + Business Media, Totowa, NJ. e-ISBN 978-1-60761-697-9.

Gomer, C.J., Ferrario, A., 1990. Tissue distribution and photosensitizing properties of mono-l-aspartyl chlorin e6 in a mouse tumor model. Cancer Res. 50, 3985–3990.

Gomer, C.J., Rucker, N., Ferrario, A., Murphree, A.L., 1986. Expression of potentially lethal damage in Chinese hamster cells exposed to hematoporphyrin derivative photodynamic therapy. Cancer Res. 46, 3348–3352.

Grant, W., MacRobert, A., Bown, S., Hopper, C., Speight, P., 1993. Photodynamic therapy of oral cancer: photosensitisation with systemic aminolaevulinic acid. Lancet North Am. Ed. 342, 147–148.

Gunaydin, G., Gedik, M.Emre, Ayan, Seylan, 2021. Photodynamic therapy for the treatment and diagnosis of cancer–a review of the current clinical status. Front. Chem. 9, 1–26. doi:10.3389/fchem.2021.686303.

Gürol, I., Durmuş, M., Ahsen, V., Nyokong, T., 2007. Synthesis, photophysical and photochemical properties of substituted zinc phthalocyanines. Dalton Trans., 3782–3791.

Gutter, B., Speck, W.T., Rosenkranz, H.S., 1977. The photodynamic modification of DNA by hematoporphyrin. Biochim. Biophys. Acta 475, 307–314.

Hackenberg, S., Scherzed, A., Kessler, M., Froelich, K., Ginzkey, C., Koehler, C., Burghartz, M., Hagen, R., Kleinsasser, N., 2010. Zinc oxide nanoparticles induce photocatalytic cell death in human head and neck squamous cell carcinoma cell lines in vitro. Int. J. Oncol. 37, 1450–1457.

Handoko, Y.A., Rondonuwu, F.S., Limantara, L., 2015. The photosensitizer stabilities of Tookad® on aggregation, acidification, and day-light irradiation. Procedia Chem. 14, 474–483.

Hasan, T., Ortel, B., Moor, A.C.E., Pogue, B., 2002. Photodynamic therapy of cancer. Cancer Med, 605–622.

Hasegawa, T., Takahashi, J., Nagasawa, S., Moriyama, A., Iwahashi, H., 2020. DNA strand break properties of protoporphyrin IX by X-ray irradiation against melanoma. Int. J. Mol. Sci. 21, 2302.

He, J., Agarwal, M.L., Larkin, H.E., Friedman, L.R., Xue, L.y., Olelnick, N.L., 1996. The induction of partial resistance to photodynamic therapy by the protooncogene BCL-2. Photochem. Photobiol. 64, 845–852.

He, L., Mao, C., Brasino, M., Harguindey, A., Park, W., Goodwin, A.P., Cha, J.N., 2018. TiO2-capped gold nanorods for plasmon-enhanced production of reactive oxygen species and photothermal delivery of chemotherapeutic agents. ACS Appl. Mater. Interfaces. 10 (33), 27965–27971.

He, X.Y., Sikes, R.A., Thomsen, S., Chung, L.W., Jacques, S.L., 1994. Photodynamic therapy with photofrin II induces programmed cell death in carcinoma cell lines. Photochem. Photobiol. 59, 468–473.

Henderson BW, D.T., Malone, P.B., 1984. Studies on the mechanism of tumor destruction by photoirradiation therapy. In: Doiron, G.C. (Ed.), Porphyrin Localization and Treatment of Tumors. Alan R. Liss, New York, pp. 601–612.

Henderson, B., Dougherty, T., Malone, P., 1984. Studies on the mechanism of tumor destruction by photoradiation therapy. Prog. Clin. Biol. Res. 170, 601.

Henderson, B.W., Fingar, V.H., 1987. Relationship of tumor hypoxia and response to photodynamic treatment in an experimental mouse tumor. Cancer Res. 47, 3110–3114.

Henderson, B.W., Waldow, S.M., Mang, T.S., Potter, W.R., Malone, P.B., Dougherty, T.J., 1985. Tumor destruction and kinetics of tumor cell death in two experimental mouse tumors following photodynamic therapy. Cancer Res. 45, 572–576.

Herman, S., Kalechman, Y., Gafter, U., Sredni, B., Malik, Z., 1996. Photofrin II induces cytokine secretion by mouse spleen cells and human peripheral mononuclear cells. Immunopharmacology 31, 195–204.

Herrera-Ornelas, L., Petrelli, N.J., Mittelman, A., Dougherty, T.J., Boyle, D.G., 1986. Photodynamic therapy in patients with colorectal cancer. Cancer 57, 677–684.

Hunt, D.WC., 2002. Rostaporfin (Miravant Medical Technologies). IDrugs 5 (2), 180–186.

Ichikawa, K., Takeuchi, Y., Yonezawa, S., Hikita, T., Kurohane, K., Namba, Y., Oku, N., 2004. Antiangiogenic photodynamic therapy (PDT) using Visudyne causes effective suppression of tumor growth. Cancer Lett. 205, 39–48.

Ion, R.M., 2000. Porphyrins for tumor destruction in photodynamic therapy. Curr. Top. Biophys. 24 (1), 21–34.

Ion, R.M., 2017. Porphyrins and phthalocyanines: photosensitizers and photocatalysts. http://dx.doi.org/10.5772/intechopen.68654.

Jin, S., Zhou, L., Gu, Z., Tian, G., Yan, L., Ren, W., Yin, W., Liu, X., Zhang, X., Hu, Z., 2013. A new near infrared photosensitizing nanoplatform containing blue-emitting up-conversion nanoparticles and hypocrellin A for photodynamic therapy of cancer cells. Nanoscale 5, 11910–11918.

Josefsen, L.B., Boyle, R.W., 2008. Photodynamic therapy: novel third-generation photosensitizers one step closer? Br. J. Pharmacol. 154, 1–3.

Josefsen, L.B., Boyle, R.W., 2008. Photodynamic therapy and the development of metal-based photosensitisers. Hindawi: Metal Based drugs. https://doi.org/10.1155/2008/276109.

Jung, S., Jeong, H., Yu, Seong-Woon, 2020. Autophagy as a decisive process for cell death. Exp. Mol. Med. 52, 921–930.

Juzeniene, A., Kaliszewski, M., Bugaja, A., Moan, J., 2009. Clearance of protoporphyrin IX induced by 5-aminolevulinic acid from WiDrhuman colon carcinoma cells. Photodynamic Therapy: Back to the Future. In: Kessel, David H. (Ed.), Proc. of SPIE, 7380. SPIE • CCC code: 1605-7422/09/$18 •. doi:10.1117/12.822944.

Kaneko, T., Tominaga, M., Kouzaki, R., Hanyu, A., Ueshima, K., Yamada, H., Suga, M., Yamashita, T., Okimoto, T., Uto, Y., 2018. Radiosensitizing effect of 5-aminolevulinic acid and protoporphyrin IX on carbon-ion beam irradiation. Anticancer Res. 38, 4313–4317.

Kaplan, M.J., Somers, R.G., Greenberg, R.H., Ackler, J., 1998. Photodynamic therapy in the management of metastatic cutaneous adenocarcinomas: case reports from phase 1/2 studies using tin ethyl etiopurpin (SnET2). 1998, 121–125.

Karges, J., Heinemann, F., Jakubaszek, M., Maschietto, F., Subecz, C., Dotou, M., Vinck, R., Blacque, O., Tharaud, M., Goud, B., 2020. Rationally designed long-wavelength absorbing ru (ii) polypyridyl complexes as photosensitizers for photodynamic therapy. J. Am. Chem. Soc. 142 (14), 6578–6587.

Kashyap, A., Ramasamy, E., Ramalingam, V., Pattabiraman, M., 2021. Supramolecular control of singlet oxygen generation. Molecules 26, 2673. https://doi.org/10.3390/molecules26092673.

Kataoka, H., Nishie, H., et al., 2017. New photodynamic therapy with next-generation photosensitizers. Ann. Transl. Med. 5 (8), 183.

Katayama, A., Tamiaki, H., 2020. Synthesis of zinc bacteriochlorophyll-d analogs bearing an alkoxyimino group at the 131-position and their self-aggregation in an aqueous micelle solution. Tetrahedron Lett. 61 (3), 151386.

Kayabasi, Y., Erbas, O., 2020. Methylene blue and its importance in medicine. D J Med. Sci. 6 (3), 136–145.

Keam, S.J, Scott, L.J, Curran, M.P. 2004. Spotlight on verteporfin in subfoveal choroidal neovascularisation. Drugs Aging 21 (3), 203–209.

Kessel, D., 1989. In vitro photosensitization with a benzoporphyrin derivative. Photochem. Photobiol. 49, 579–582.

Kessel, D., Garbo, G.M., Hampton, J., 1993. The role of lipoproteins in the distribution of tin etiopurpurin (SnET2) in the tumor-bearing rat. Photochem. Photobiol. 57 (2), 298–301.

Kessel, D., Luo, Y., 1998. Mitochondrial photodamage and PDT-induced apoptosis. J. Photochem. Photobiol., B: Biol. 42, 89–95.

Kessel, D., Luo, Y., 1999. Photodynamic therapy: a mitochondrial inducer of apoptosis. Cell Death Differ. 6, 28–35.

Kessel, D., Luo, Y., Deng, Y., Chang, C., 1997. The role of subcellular localization in initiation of apoptosis by photodynamic therapy. Photochem. Photobiol. 65, 422–426.

Kick, G., Messer, G., Plewig, G., Kind, P., Goetz, A., 1996. Strong and prolonged induction of c-jun and c-fos proto-oncogenes by photodynamic therapy. Br. J. Cancer 74, 30–36.

Kim, H.-R.C., Luo, Y., Li, G., Kessel, D., 1999. Enhanced apoptotic response to photodynamic therapy after bcl-2 transfection. Cancer Res. 59, 3429–3432.

Kim-Campbell, N., Gomez, H., Bayir, H., 2019. Cell death pathways: apoptosis and regulated necrosis. Critical Care Nephrology, 3rd Elsevier, pp. 113–121.e112.

Kochubeev, G., Frolov, A., Zenkevich, E., Gurinovich, G., 1988. Chlorin e6 complexation with serum human and bovine albumins. Mol. Biol. 22, 968–975.

Kongshaug, M., Moan, J., Cheng, L.S., Garbo, GM., Kolboe, S., Morgan, A.R., Rimington, C., 1993. Binding of drugs to human plasma proteins, exemplified by Sn(IV)-etiopurpurin dichloride delivered in cremophor and DMSO. Int. J. Biochem. 25 (5), 739–760.

Korbelik, M., 1996. Induction of tumor immunity by photodynamic therapy. J. Clin. Laser Med. Surg. 14, 329–334.

Korbelik, M., Dougherty, G.J., 1999. Photodynamic therapy-mediated immune response against subcutaneous mouse tumors. Cancer Res. 59, 1941–1946.

Korbelik, M., Krosl, G., Krosl, J., Dougherty, G.J., 1996. The role of host lymphoid populations in the response of mouse EMT6 tumor to photodynamic therapy. Cancer Res. 56, 5647–5652.

Korbelik, M., Naraparaju, V., Yamamoto, N., 1997. Macrophage-directed immunotherapy as adjuvant to photodynamic therapy of cancer. Br. J. Cancer 75, 202–207.

Kostovic, K., Pastar, Z., Ceovic, R., Bukvic Mokos, Z., Stulhofer Buzina, D., Stanimirovic, A., 2012. Photodynamic therapy in dermatology: current treatments and implications. Coll. Antropol. 36, 1477–1481.

Kou, J., Dou, D., Yang, L., 2017. Porphyrin photosensitizers in photodynamic therapy and its applications. Oncotarget 8 (46), 81591–81603.

Krämer, K.W., Biner, D., Frei, G., Güdel, H.U., Hehlen, M.P., Lüthi, S.R., 2004. Hexagonal sodium yttrium fluoride based green and blue emitting upconversion phosphors. Chem. Mater. 16, 1244–1251.

Krasnovsky Jr, A.A., 2007. Primary mechanisms of photoactivation of molecular oxygen. History of development and the modern status of research. Biochemistry (Moscow) 72 (10), 1065–1080.

Krasnovsky Jr, A.A., 2007. Singlet oxygen and primary mechanisms of photodynamic therapy and photodynamic diseases. Photodynamic therapy at cellular level, 17–62. ISBN 978-81-308-0174-2.

Krasnovsky Jr, AA., 2014. Phosphorescence of triplet chlorophylls. Handbook of porphyrin science. Applications II, 33. World Scientific Publishing Company. ISBN 978-981-4425-09-4.

Kraus, D., Palasuberniam, P., Chen, B., 2019. Therapeutic enhancement of verteporfin-mediated photodynamic therapy by mTOR inhibitors. Photochem. Photobiol. 96 (2), 358–364. doi:10.1111/php.13187.

Kuncewicz, J., Dąbrowski, J.M., Kyzioł, A., Brindell, M., Łabuz, P., Mazuryk, O., Macyk, W., Stochel, G., 2019. Perspectives of molecular and nanostructured systems with d-and f-block metals in photogeneration of reactive oxygen species for medical strategies. Cood. Chem. Rev. 398, 113012.

Lan, G., Ni, K., Lin, W., 2019. Nanoscale metal–organic frameworks for phototherapy of cancer. Coord. Chem. Rev. 379, 65–81.

Lan, G., Ni, K., Veroneau, S.S., Feng, X., Nash, G.T., Luo, T., Xu, Z., Lin, W., 2019. Titanium-based nanoscale metal–organic framework for type I photodynamic therapy. J. Am. Chem. Soc. 141, 4204–4208.

Lee, H., Park, Ho-Yong, Jeong, Tae-Sook, 2021. Pheophorbide a derivatives exert antiwrinkle effects on UVB-induced skin aging in human fibroblasts. Life 11, 147.

Li, A., Turro, C., Kodanko, J.J., 2018. Ru (II) polypyridyl complexes as photocages for bioactive compounds containing nitriles and aromatic heterocycles. Chem. Commun. 54, 1280–1290.

Li, G., Yuan, S., Deng, D., Ou, T., Li, Y., Sun, R., Lei, Q., Wang, X., Shen, W., Cheng, Y., 2019. Fluorinated polyethylenimine to enable transmucosal delivery of photosensitizer-conjugated catalase for photodynamic therapy of orthotopic bladder tumors postintravesical Instillation. Adv. Funct. Mater. 29, 1901932.

Li, X., Kim, C.-y., Lee, S., Lee, D., Chung, H.-M., Kim, G., Heo, S.-H., Kim, C., Hong, K.-S., Yoon, J., 2017. Nanostructured phthalocyanine assemblies with protein-driven switchable photoactivities for biophotonic imaging and therapy. J. Am. Chem. Soc. 139, 10880–10886.

Li, X., Wu, J., Wang, L., He, C., Chen, L., Jiao, Y., Duan, C., 2020. Mitochondrial-DNA-targeted iriii-containing metallohelices with tunable photodynamic therapy efficacy in cancer cells. Angew. Chem. 132, 6482–6489.

Li, X-Y., Tan, L-C., Dong, L.-.W., et al., 2020. Susceptibility and resistance mechanisms during photodynamic therapy of melanoma. Front. Oncol. 10, 597.

Li, Y., Wang, K.-N., He, L., Ji, L.-N., Mao, Z.-W., 2020. Synthesis, photophysical and anticancer properties of mitochondria-targeted phosphorescent cyclometalated iridium (III) N-heterocyclic carbene complexes. J. Inorg. Biochem. 205, 110976.

Liang, S., Sun, C., Yang, P., Huang, S., Cheng, Z., Yu, X., Lin, J., 2020. Core-shell structured upconversion nanocrystal-dendrimer composite as a carrier for mitochondria targeting and catalase enhanced anti-cancer photodynamic therapy. Biomaterials 240, 11985.

Lipson, R.L., Baldes, E.J., 1960. The photodynamic properties of a particular hematoporphyrin derivative. Arch. Dermatol. 82 (4), 508–516. doi:10.1001/archderm.1960.01580040026005.

Liu, C., Luo, L., Zeng, L., Xing, J., Xia, Y., Sun, S., Zhang, L., Yu, Z., Yao, J., Yu, Z., 2018. Porous gold nanoshells on functional NH2-MOFs: facile synthesis and designable platforms for cancer multiple therapy. Small 14, 1801851.

Liu, D.Q.L.H.L., 2005. Clinical application of photodynamic therapy. Chin. J. Biomed. Eng. 2, 72–84.

Liu, X., Zheng, M., Kong, X., Zhang, Y., Zeng, Q., Sun, Z., Buma, W.J., Zhang, H., 2013. Separately doped upconversion-C 60 nanoplatform for NIR imaging-guided photodynamic therapy of cancer cells. Chem. Commun. 49, 3224–3226.

Lu, K., He, C., Guo, N., Chan, C., Ni, K., Weichselbaum, R.R., Lin, W., 2016. Chlorin-based nanoscale metal–organic framework systemically rejects colorectal cancers via synergistic photodynamic therapy and checkpoint blockade immunotherapy. J. Am. Chem. Soc. 138 (38), 12502–12510.

Lu, K., He, C., Lin, W., 2014. Nanoscale metal–organic framework for highly effective photodynamic therapy of resistant head and neck cancer. J. Am. Chem. Soc. 136, 16712–16715.

Lucena, S.R., Salazar, N., Gracia-Cazaña, T., Zamarrón, A., González, S., Juarranz, Á., Gilaberte, Y., 2015. Combined treatments with photodynamic therapy for non-melanoma skin cancer. Int. J. Mol. Sci. 16, 25912–25933.

Luo, Y., Kessel, D., 1997. Initiation of apoptosis versus necrosis by photodynamic therapy with chloroaluminum phthalocyanine. Photochem. Photobiol. 66, 479–483.

Luo, Z., Zheng, M., Zhao, P., Chen, Z., Siu, F., Gong, P., Gao, G., Sheng, Z., Zheng, C., Ma, Y., 2016. Self-monitoring artificial red cells with sufficient oxygen supply for enhanced photodynamic therapy. Sci. Rep. 6, 1–11.

Lynch, D.H., Haddad, S., King, V.J., Ott, M.J., Straight, R.C., Jolles, C.J., 1989. Systemic immunosuppression induced by photodynamic therapy (PDT) is adoptively transferred by macrophages. Photochem. Photobiol. 49, 453–458.

Major, A., Tromberg, B.J., 2021. Photodynamic therapy of the rat endometrium by systemic and topical administration of tin ethyl etiopurpurin. J. Gynecol. Surg. 15 (2), 71–80.

Malina, T., Koehorst, R.B.M., Bína, D., et al., 2020. Superradiance of bacteriochlorophyll c aggregates in chlorosomes of green photosynthetic bacteria. Sci. Rep. 11 (1), 8354.

Mallidi, S., Anbil, S., Bulin, A.L, 2016. Beyond the barriers of light penetration: strategies, perspectives and possibilities for photodynamic therapy. Theranostics 6 (13), 2458–2487.

Mariewskaya, K.A., Tyurin, A.P., et al., 2021. Photosensitizing antivirals. Molecules 26 (13), 3971.

McMahon, K.S., Wieman, T.J., Moore, P.H., Fingar, V.H., 1994. Effects of photodynamic therapy using mono-L-aspartyl chlorin e6 on vessel constriction, vessel leakage, and tumor response. Cancer Res. 54, 5374–5379.

Mesquita, R.C., Han, S.W., Miller, J., Schenkel, S.S., Pole, A., Esipova, T.V., Vinogradov, S.A., Putt, M.E., Yodh, A.G., Busch, T.M., 2012. Tumor blood flow differs between mouse strains: consequences for vasoresponse to photodynamic therapy. PLoS One 7.

Meunier, I., Pandey, R.K., Senge, M.O., Dougherty, T.J., Smith, K.M., 1994. Benzo porphyrin derivatives: synthesis, structure and preliminary biological. J. Chem. Soc. Perkin Trans. 961–969.

Mfouo-Tynga, I.S., Dias, L., Inada, N.M., Kurachi, C., 2021. Biophysical and biological features of third generation photosensitizers used in anticancer photodynamic therapy: review. Photodiagn. Photodyn. Ther. 1 (1), 102091.

Ming-Keun Tang, P., Bui-Xuan, Ngoc-Ha, et al., 2010. Pheophorbide a-mediated photodynamic therapy triggers HLA class I-restricted antigen presentation in hepatocellular carcinoma. Transl. Oncol. 3 (2), 114–122.

Moan, J., Waksvik, H., Christensen, T., 1980. DNA single-strand breaks and sister chromatid exchanges induced by treatment with hematoporphyrin and light or by X-rays in human NHIK 3025 cells. Cancer Res. 40, 2915–2918.

Morgan, A.R., Garbo, G.M., Keck, R.W., 1988. New photosensitizers for photodynamic therapy: combined effect of metallopurpurin derivatives and light on transplantable bladder tumors. Cancer Res. 48, 194–198.

Morgan, A.R., Rampersaud, A., Garbo, G.M., Keck, R.W., Selman, S.H., 1989. New sensitizers for photodynamic therapy. Controlled synthesis of purpurins and their effect on normal tissue. J. Med. Chem. 32, 904–908.

Mroz, P., Hashmi, J.T., Huang, Y.-Y., Lange, N., Hamblin, M.R., 2011. Stimulation of anti-tumor immunity by photodynamic therapy. Exp. Rev. Clin. Immunol. 7, 75–91.

Mroz, P., Huang, Ying-Ying, Szokalska, A., et al., 2010. Stable synthetic bacteriochlorins overcome the resistance of melanoma to photodynamic therapy. FASEB J. 24 (9), 3160–3170.

Muniyandi, K., George, B., Parimelazhagan, T., Abrahamse, H., 2020. Role of photoactive phytocompounds in photodynamic therapy of cancer. Molecules 25, 4102.

Musser, D.A., Fiel, R.J., 1991. Cutaneous photosensitizing and immunosuppressive effects of a series of tumor localizing porphyrins. Photochem. Photobiol. 53, 119–123.

Nam, K.C., Han, Y.S., Lee, J.-M., Kim, S.C., Cho, G., Park, B.J., 2020. Photo-functionalized magnetic nanoparticles as a nanocarrier of photodynamic anticancer agent for biomedical theragnostics. Cancers 12, 571.

Ni K., Luo T., Lan G., Culbert A., Song Y., Wu T., Jiang X., Lin W., 2020. A nanoscale metal–organic framework to mediate photodynamic therapy and deliver cpg oligodeoxynucleotides to enhance antigen presentation and cancer immunotherapy. Angew. Chem. Int. Ed. Engl 59 (3), 1108–1112.

Norbury, C.J., Hickson, I.D., 2001. Cellular responses to DNA damage. Annu. Rev. Pharmacol. Toxicol. 41, 367–401.

Norum, O.-J., Fremstedal, A.S.V., Weyergang, A., Golab, J., Berg, K., 2017. Photochemical delivery of bleomycin induces T-cell activation of importance for curative effect and systemic anti-tumor immunity. J. Control. Release 268, 120–127.

Nowis, D., Makowski, M., et al., 2005. Direct tumor damage mechanisms of photodynamic therapy. Acta Biochim. Pol. 52 (2), 339–352.

Nseyo, U.O., Whalen, R.K., Duncan, M.R., Berman, B., Lundahl, S.L., 1990. Urinary cytokines following photodynamic therapy for bladder cancer a preliminary report. Urology 36, 167–171.

O'Connor, A.E., Gallagher, W.M., Byrne, A.T., 2009. Porphyrin and nonporphyrin photosensitizers in oncology: preclinical and clinical advances in photodynamic therapy. Photochem. Photobiol. 85, 1053–1074.

Öfner, J., Schlögl, H., Kostron, H., 1996. Unusual adverse reaction in a patient sensitized with Photosan 3. J. Photochem. Photobiol. B 36, 183–184.

Oku, N., Saito, N., Namba, Y., Tsukada, H., Dolphin, D., Okada, S., 1997. Application of long-circulating liposomes to cancer photodynamic therapy. Biol. Pharm. Bull. 20 (6), 670–673.

Oleinick, N.L., Morris, R.L., Belichenko, I., 2002. The role of apoptosis in response to photodynamic therapy: what, where, why, and how. Photochem. Photobiol. Sci. 1, 1–21.

Olsen, M.M., LeFebvre, K.B., Brassil, K.J., 2019. Chemotherapy and Immunotherapy Guidelines and Recommendations for Practice. 1st edition, Oncology Nursing Society.

Ormond, A.B., Freeman, H.S., 2013. Dye sensitizers for photodynamic therapy. Materials 6, 817–840.

Orth, K., Beck, G., Genze, F., Rück, A., 2000. Methylene blue mediated photodynamic therapy in experimental colorectal tumors in mice. J. Photochem. Photobiol. B 57, 186–192.

Panjehpour, M., Panella, T.J., et al., 1998. Detection of lutetium texaphyrin (Lutex) using its absorption of tissue autofluorescence, Proc. SPIE 3247, Optical Methods for Tumor Treatment and Detections: Mechanisms and Techniques in Photodynamic Therapy VII.

Papakonstantinou, E., Löhr, F., Raap, U., 2018. Photodynamic therapy and skin cancer. Dermatologic Surgery and Procedures. Published by Intech Open. http://dx.doi.org/10.5772/intechopen.70309.

Park, J., Jiang, Q., Feng, D., Mao, L., Zhou, H.-C., 2016. Size-controlled synthesis of porphyrinic metal–organic framework and functionalization for targeted photodynamic therapy. J. Am. Chem. Soc. 138, 3518–3525.

Park, Y.I., Kim, H.M., Kim, J.H., Moon, K.C., Yoo, B., Lee, K.T., Lee, N., Choi, Y., Park, W., Ling, D., 2012. Theranostic probe based on lanthanide-doped nanoparticles for simultaneous in vivo dual-modal imaging and photodynamic therapy. Adv. Mater. 24, 57.

Pass, H.I., 1993. Photodynamic therapy in oncology: mechanisms and clinical use. J. Natl. Cancer Inst. 85, 443–456.

Pavlíčková, V., et al., 2019. PEGylated purpurin 18 with improved solubility: potent compounds for photodynamic therapy of cancer. Molecules 24 (24), 4477.

Pucelik, B., Arnaut, L.G., Stochel, G.y., Dabrowski, J.M., 2016. Design of Pluronic-based formulation for enhanced redaporfin-photodynamic therapy against pigmented melanoma. ACS Appl. Mater. Interfaces 8, 22039–22055.

Pucelik, B., Sułek, A., Dabrowski, J.M., 2020. Bacteriochlorins and their metal complexes as NIR-absorbing photosensitizers: properties, mechanisms, and applications. Coord. Chem. Rev. 416, 213340.

Q.Mesquita, M., Dias, C.J., et al., 2018. An insight on the role of photosensitizer nanocarriers for photodynamic therapy. Ann. Braz. Acad. Sci. 90 (1Suppl. 2), 1101–1130.

Qian, C., Yu, J., Chen, Y., Hu, Q., Xiao, X., Sun, W., Wang, C., Feng, P., Shen, Q.D., Gu, Z., 2016. Light-activated hypoxia-responsive nanocarriers for enhanced anticancer therapy. Adv. Mater. 28, 3313–3320.

Qiao, X.-F., Zhou, J.-C., Xiao, J.-W., Wang, Y.-F., Sun, L.-D., Yan, C.-H., 2013. Triple-functional core–shell structured upconversion luminescent nanoparticles covalently grafted with photosensitizer for luminescent, magnetic resonance imaging and photodynamic therapy in vitro. Nanoscale 4, 4611–4623.

Qiu H., Tan M., Ohulchanskyy T.Y., Lovell J.F., Chen G., 2018. Recent progress in upconversion photodynamic therapy. Nanomaterials (Basel). 8 (5), 344.

Rapp, T.L., DeForest, C.A., 2020. Visible light-responsive dynamic biomaterials: going deeper and triggering more. Adv. Healthc. Mater. 9, 1901553.

Renicke, C., Schuster, D., Usherenko, S., Essen, L.-O., Taxis, C., 2013. A LOV2 domain-based optogenetic tool to control protein degradation and cellular function. Chem. Biol. 20, 619–626.

Renno, R.Z., Delori, F.C., Holzer, R.A., 2000. Photodynamic therapy using Lu-tex induces apoptosis in vitro, and its effect is potentiated by angiostatin in retinal capillary endothelial cells. nvest. Ophthalmol. Vis. Sci. 41, 3963–3971.

Richter, A., Waterfield, E., Jain, A., Allison, B., Sternberg, E., Dolphin, D., Levy, J., 1991. Photosensitising potency of structural analogues of benzoporphyrin derivative (BPD) in a mouse tumour model. Br. J. Cancer 63, 87–93.

Richter, A.M., Cerruti-Sola, S., Sternberg, E.D., Dolphin, D., Levy, J.G., 1990. Biodistribution of tritiated benzoporphyrin derivative (3H-BPD-MA), a new potent photosensitizer, in normal and tumor-bearing mice. J. Photochem. Photobiol. B 5 (2), 231–244.

Roldan, C.J, Chung, M., Lin, F., Bruera, E., 2020. Methylene blue for the treatment of intractable pain from oral mucositis related to cancer treatment: an uncontrolled cohort. J. Natl. Compr. Canc. Netw. doi:10.6004/jnccn.2020.7651.

Rück, A., Dolder, M., Wallimann, T., Brdiczka, D., 1998. Reconstituted adenine nucleotide translocase forms a channel for small molecules comparable to the mitochondrial permeability transition pore. FEBS Lett. 426, 97–101.

Saide, A., Lauritano, C., Ianora, A., 2020. Pheophorbide a: state of the art. Mar. Drugs 18, 257.

Samkoe K.S., Clancy A.A., Karotki A., Wilson B.C., Cramb D.T., 2007. Complete blood vessel occlusion in the chick chorioallantoic membrane using two-photon excitation photodynamic therapy: implications for treatment of wet age-related macular degeneration. J. Biomed. Opt. 12 (3), 034025.

Sanadi, R.M., 2013. Photodynamic therapy: an overview. IJSR 2 (2), 271–272.

Sato, A., Hiramoto, A., Kim, Hye-Sook, Wataya, Y., 2020. Anticancer strategy targeting cell death regulators:switching the mechanism of anticancer floxuridine-induced cell death from necrosis to apoptosis. Int. J. Mol. Sci. 21, 5876.

Sava Gallis, D.F., Rohwer, L.E., Rodriguez, M.A., Barnhart-Dailey, M.C., Butler, K.S., Luk, T.S., Timlin, J.A., Chapman, K.W., 2017. Multifunctional, tunable metal–organic framework materials platform for bioimaging applications. ACS Appl. Mater. Interfaces 9, 22268–22277.

Senge, M.O., 2012. mTHPC–a drug on its way from second to third generation photosensitizer? Photodiagn. Photodyn. Ther. 9 (2), 170–179.

Senge, M.O., Brandt, J.C., 2011. Temoporfin (foscan®5,10,15,20-tetra(m-hydroxyphenyl)chlorin)—a second-generation photosensitizer. Photochem. Photobiol. 87, 1240–1296.

Serra, A., Pineiro, M., Pereira, N., Gonsalves, A.R., Laranjo, M., Abrantes, M., Botelho, F., 2008. A look at clinical applications and developments of photodynamic therapy. Oncol. Rev. 2, 235–249.

Sessler, J.L., Miller, R.A., 2000. Texaphyrins: New drugs with diverse clinical applications in radiation and photodynamic therapy. Biochem. Pharmacol. 59 (7), 733–739.

Shadish, J.A., Strange, A.C., DeForest, C.A., 2019. Genetically encoded photocleavable linkers for patterned protein release from biomaterials. J. Am. Chem. Soc. 141, 15619–15625.

Shams, M., Owczarczak, B., Manderscheid-Kern, P., Bellnier, D.A., Gollnick, S.O., 2015. Development of photodynamic therapy regimens that control primary tumor growth and inhibit secondary disease. Cancer Immunol. Immunother. 64, 287–297.

Sharkey, S.M., Wilson, B.C., Moorehead, R., Singh, G., 1993. Mitochondrial alterations in photodynamic therapy-resistant cells. Cancer Res. 53, 4994–4999.

Shi, Z., Zhang, K., Zada, S., Zhang, C., Meng, X., Yang, Z., Dong, H., 2020. Upconversion nanoparticles-induced multi-mode photodynamic therapy based on metal-organic framework/titanium dioxide nanocomposite. ACS Appl. Mater. Interfaces 12 (11), 12600–12608. doi:10.1021/acsami.0c01467.

Shimoyama A., Watase H., Liu Y., Ogura S.-I., Hagiya Y., Takahashi K., Inoue K., Tanaka T., Murayama Y., Otsuji E., Access to a novel near-infrared photodynamic therapy through the combined use of 5-aminolevulinic acid and lanthanide nanoparticles, Photod.

Shumaker, B., Hetzel, F., 1987. Clinical laser photodynamic therapy in the treatment of bladder carcinoma. Photochem. Photobiol. 46, 899–901.

Sikurová, L., Franková, R., Distribution of merocyanine 540 in phospholipid membranes. J. Fluoresc. 3 (4), 261–263.

Smith, O.M., Sieber, Fritz, 1990. Antineoplastic and virucidal effects of merocyanine 540. Trends Photobiol. Photochem. 1. https://www.researchgate.net/publication/235979082.

Song, R., Peng, S., Lin, Q., Luo, M., Chung, H.Y., Zhang, Y., Yao, S., 2019. pH-responsive oxygen nanobubbles for spontaneous oxygen delivery in hypoxic tumors. Langmuir 35, 10166–10172.

Spikes, J.D., 1990. New trends in photobiology: chlorins as photosensitizers in biology and medicine. J. Photochem. Photobiol. B 6, 259–274.

Sridharan, G., Shankar, A.A., 2012. Toluidine blue: A review of its chemistry and clinical utility. J. Oral Maxillofac. Pathol. 16 (2), 251–255.

Steele, P.M., Chesebro, J.H., Stanson, A.W., Holmes Jr, D.R., Dewanjee, M.K., Badimon, L., Fuster, V., angioplasty, Balloon, 1985. Natural history of the pathophysiological response to injury in a pig model. Circ. Res. 57, 105–112.

Straten, D.van, Mashayekhi, V., de Bruijn, H.S., Oliveira, S., Robinson, D.J., 2017. Oncologic photodynamic therapy: basic principles, current clinical status and future directions. Cancers (Basel) 9 (2), 19.

Sun, W., Parowatkin, M., Steffen, W., Butt, H.J., Mailänder, V., Wu, S., 2016. Phototherapy: ruthenium-containing block copolymer assemblies: red-light-responsive metallopolymers with tunable nanostructures for enhanced cellular uptake and anticancer phototherapy. Adv. Healthc. Mater. 5 (4), 467–473.

Sun, X., Cao, Z., Mao, K., Wu, C., Chen, H., Wang, J., Wang, X., Cong, X., Li, Y., Meng, X., 2020. Photodynamic therapy produces enhanced efficacy of antitumor immunotherapy by simultaneously inducing intratumoral release of sorafenib. Biomaterials 240, 119845.

Takahashi, J., Misawa, M., Iwahashi, H., 2016. Combined treatment with X-ray irradiation and 5-aminolevulinic acid elicits better transcriptomic response of cell cycle-related factors than X-ray irradiation alone. Int. J. Radiat. Biol. 92 (12), 774–789.

Takahashi, J., Misawa, M., Murakami, M., Mori, T., Nomura, K., Iwahashi, H., 2013. 5-Aminolevulinic acid enhances cancer radiotherapy in a mouse tumor model. Springerplus 2, 602.

Tan, L., Li, J., Liu, X., Cui, Z., Yang, X., Zhu, S., Li, Z., Yuan, X., Zheng, Y., Yeung, K.W., 2018. Rapid biofilm eradication on bone implants using red phosphorus and near-infrared light. Adv. Mater. 30, 1801808.

Tang, X., Cheng, Y., Huang, S., Zhi, F., Yuan, A., Hu, Y., Wu, J., 2017. Overcome the limitation of hypoxia against photodynamic therapy to treat cancer cells by using perfluorocarbon nanodroplet for photosensitizer delivery. Biochem. Biophys. Res. Commun. 487, 483–487.

Theis, S., Iturmendi, A., Gorsche, C., Orthofer, M., Lunzer, M., Baudis, S., Ovsianikov, A., Liska, R., Monkowius, U., Teasdale, I., 2017. Metallo-supramolecular gels that are photocleavable with visible and near-infrared irradiation. Angew. Chem. Int. Edit. 56, 15857–15860.

Tian, G., Ren, W., Yan, L., Jian, S., Gu, Z., Zhou, L., Jin, S., Yin, W., Li, S., Zhao, Y., 2013. Red-Emitting upconverting nanoparticles for photodynamic therapy in cancer cells under near-infrared excitation. Small 9, 1929–1938.

Tjahjono, D.H., 2006. Porphyrin structure-based molecules for photodynamic therapy of cancer. Acta Pharm. XXXI (1), 1–12.

Toupin, N.P., Nadella, S., Steinke, S.J., Turro, C., Kodanko, J.J., 2020. Dual-action Ru (II) complexes with bulky π-expansive ligands: phototoxicity without DNA intercalation. Inorg. Chem. 59, 3919–3933.

Turubanova, V.D., Balalaeva, I.V., Mishchenko, T.A., Catanzaro, E., Alzeibak, R., Peskova, N.N., Efimova, I., Bachert, C., Mitroshina, E.V., Krysko, O., 2019. Immunogenic cell death induced by a new photodynamic therapy based on photosens and photodithazine. J. Immunother. Cancer 7, 350.

Udartseva O.O., Zhidkova O.V., Ezdakova M.I., Ogneva I.V., Andreeva E.R., Buravkova L.B., Gollnick S.O., Low-dose photodynamic therapy promotes angiogenic potential and increases immunogenicity of human mesenchymal stromal cells, J. Photochem.

Uenaka, A., Nakayama, E., 2003. Murine leukemia RL 1 and sarcoma Meth A antigens recognized by cytotoxic T lymphocytes (CTL). Cancer Sci. 94, 931–936.

Uthaman S., Pillarisetti S., Mathew A.P., Kim Y., Bae W.K., Huh K.M., Park I.-K., Long circulating photoactivable nanomicelles with tumor localized activation and ROS triggered self-accelerating drug release for enhanced locoregional chemo-photodynamic th.

Valles M.A.. HpD and second generation photosensitizers for the photodynamic therapy of cancer. Afinidad -Barcelona- 50(448):469–479.

Varnes, M.E., Chiu, S.-M., Xue, L.-Y., Oleinick, N.L., 1999. Photodynamic therapy-induced apoptosis in lymphoma cells: translocation of cytochrome c causes inhibition of respiration as well as caspase activation. Biochem. Biophys. Res. Commun 11, 1–13.

Vetha, B.S.S., Oh, P.-S., Kim, S.H., Jeong, H.-J., 2020. Curcuminoids encapsulated liposome nanoparticles as a blue light emitting diode induced photodynamic therapeutic system for cancer treatment. J. Photochem. Photobiol. B 205.

Voet, D., Voet, J., 1990. Arachidonate metabolism: prostaglandins, prostacyclins, thromboxanes, and leukotrienes. Biochemistry, 658–665.

Wang, C., Cheng, L., Liu, Y., Wang, X., Ma, X., Deng, Z., Li, Y., Liu, Z., 2013. Imaging-guided pH-sensitive photodynamic therapy using charge reversible upconversion nanoparticles under near-infrared light. Adv. Funct. Mater. 23, 3077–3086.

Wang, D., Wu, H., Phua, S.Z.F., Yang, G., Lim, W.Q., Gu, L., Qian, C., Wang, H., Guo, Z., Chen, H., 2020. Self-assembled single-atom nanozyme for enhanced photodynamic therapy treatment of tumor. Nat. Commun. 11, 1–13.

Wang, H., Chen, Y., Wang, H., Liu, X., Zhou, X., Wang, F., 2019. DNAzyme-loaded metal–organic frameworks (MOFs) for self-sufficient gene therapy. Angew. Chem. Int. Ed. 58, 7380–7384.

Wang, J., Liu, L., You, Q., Song, Y., Sun, Q., Wang, Y., Cheng, Y., Tan, F., Li, N., 2018. All-in-one theranostic nanoplatform based on hollow MoSx for photothermally-maneuvered oxygen self-enriched photodynamic therapy. Theranostics 8, 955.

Wang, M., Thanou, M., 2010. Targeting nanoparticles to cancer. Pharmacol. Res. 62, 90–99.

Wang, W., Wang, L., Li, Z., Xie, Z., 2016. BODIPY-containing nanoscale metal–organic frameworks for photodynamic therapy. Chem. Commun. 52, 5402–5405.

Wang, X., Luo, D., Basilion, J.P., 2021. Photodynamic therapy: targeting cancer biomarkers for the treatment of cancers. Cancers 13, 2992.

Wang, Y., Wu, W., Liu, J., Manghnani, P.N., Hu, F., Ma, D., Teh, C., Wang, B., Liu, B., 2019. Cancer-cell-activated photodynamic therapy assisted by Cu (II)-based metal–organic framework. ACS Nano 13, 6879–6890.

Wang, Z., Gai, S., Wang, C., Yang, G., Zhong, C., Dai, Y., He, F., Yang, D., Yang, P., 2019. Self-assembled zinc phthalocyanine nanoparticles as excellent photothermal/photodynamic synergistic agent for antitumor treatment. Chem. Eng. J. 361, 117 1.

Whalley, M., 1961. 182. Conjugated macrocycles. Part XXXII. Absorption spectra of tetrazaporphins and phthalocyanines. Formation of pyridine salts. J. Chem. Soc., 866–869.

Wild, e.C.P., Weiderpass, E., Stewart, B.W., 2020. World Cancer Report: Cancer Research for Cancer Prevention. International Agency for Research on Cancer, Lyon, France.

Wilson, B., Olivo, M., Singh, G., 1997. Subcellular localization of Photofrin and aminolevulinic acid and photodynamic cross-resistance in vitro in radiation-induced fibrosarcoma cells sensitive or resistant to photofrin-mediated photodynamic therapy. Photochemist. Potobio. 65, 166–176.

Wulf, H.C., Nissen, C.V., Philipsen, P.A., 2020. Inactivation of protoporphyrin IX in erythrocytes in patients with erythropoietic protoporphyria: a new treatment modality. Photodiagn. Photodyn. Ther. 65, 101582.

Xia, L., Kong, X., Liu, X., Tu, L., Zhang, Y., Chang, Y., Liu, K., Shen, D., Zhao, H., Zhang, H., 2014. An upconversion nanoparticle–zinc phthalocyanine based nanophotosensitizer for photodynamic therapy. Biomaterials 35, 4146–4156.

Xu Q.C., Zhang Y., Tan M.J., Liu Y., Yuan S., Choong C., Tan N.S., Tan T.T.Y., Anti-cAngptl4 Ab-conjugated N-TiO2/NaYF4: Yb, Tm nanocomposite for near infrared-triggered drug release and enhanced targeted cancer cell ablation, Adv. Healthc. Mater.

Xu, X., Lin, L., Li, B., 2020. Automatic protocol for quantifying the vasoconstriction in blood vessel images. Biomed. Opt. Exp. 11, 2122–2136.

Yamamoto, J., Ogura, S.-I., Tanaka, T., Kitagawa, T., Nakano, Y., Saito, T., Takahashi, M., Akiba, D., Nishizawa, S., 2012. Radiosensitizing effect of 5-aminolevulinic acid-induced protoporphyrin IX in glioma cells in vitro. Oncol. Rep. 27, 1748–1752.

Yan, X., Niu, G., Lin, J., Jin, A.J., Hu, H., Tang, Y., Zhang, Y., Wu, A., Lu, J., Zhang, S., 2015. Enhanced fluorescence imaging guided photodynamic therapy of sinoporphyrin sodium loaded graphene oxide. Biomaterials 42, 94–102.

Yang, B., Chen, Y., Shi, J., 2019. Reactive oxygen species (ROS)-based nanomedicine. Chem. Rev. 119, 4881–4985.

Yang, D., Hou, Z., Cheng, Z., Li, C., Lin, J., 2015. Current advances in lanthanide ion (Ln 3+)-based upconversion nanomaterials for drug delivery. Chem. Soc. Rev. 44, 1416–1448.

Yang, D., Yang, G., Sun, Q., Gai, S., He, F., Dai, Y., Zhong, C., Yang, P., 2018. Carbon-dot-decorated TiO2 nanotubes toward photodynamic therapy based on water-splitting mechanism. Adv. Healthc. Mater. 7, 1800042.

Yano, S., Hirohara, S., Obata, M., Hagiya, Y., Ogura, S.-i., Ikeda, A., Kataoka, H., Tanaka, M., Joh, T., 2011. Current states and future views in photodynamic therapy. J. Photochem. Photobiol. C 12, 46–67.

Yano, T., Minamide, T., Takashima, K., Nakajo, K., Kadota, T., Yoda, Y., 2021. Clinical practice of photodynamic therapy using talaporfin sodium for esophageal cancer. Cancer. J. Clin. Med. 10, 2785.

Yanovsky, R.L., Bartenstein, D.W., Rogers, G.S., Isakoff, S.J., Chen, S.T., 2019. Photodynamic therapy for solid tumors: a review of the literature. Photodermatol. Photoimmunol. Photomed. 35, 295–303.

Yavari, N., Andersson-Engels, S., Segersten, U., Malmstrom, P.-U., 2011. An overview on preclinical and clinical experiences with photodynamic therapy for bladder cancer. Can. J. Urol. 18, 5778.

You, Y., Ngen, E.J., Rajaputra, P., 2009. Conjugate systems using delocalized cationic dyes as a carrier of photosensitizers to mitochondria. photodynamic therapy: back to the future. In: Kessel, David H. (Ed.), Proc. of SPIE, 7380 738064-1.

Young, S.W., Woodburn, K.W., 1996. Lutetium texaphyrin (PCI-0123): a near-infrared, water-soluble photosensitizer. Photochem. Photobiol. 63 (6), 892–897.

Zaidi, S.I., Oleinick, N.L., Zaim, M.T., Mukhtar, H., 1993. Apoptosis during photodynamic therapy-induced ablation of RIF-1 tumors in C3H mice: electron microscopic, histopathologic and biochemical evidence. Photochem. Photobiol. 58, 771–776.

Zane, C., De Panfilis, G., Calzavara-Pinton, P., 2001. Chapter 7 Photosensitizers—systemic sensitization. Compr. Ser. Photosci. 2, 101–114.

Zayat, L., Calero, C., Alborés, P., Baraldo, L., Etchenique, R., 2003. A new strategy for neurochemical photodelivery: metal– ligand heterolytic cleavage. J. Am. Chem. Soc. 125, 882–883.

Zayat, L., Salierno, M., Etchenique, R., 2006. Ruthenium (II) bipyridyl complexes as photolabile caging groups for amines. Inorg. Chem. 45, 1728–1731.

Zeng, J.-Y., Zou, M.-Z., Zhang, M., Wang, X.-S., Zeng, X., Cong, H., Zhang, X.-Z., 2018. π-extended benzoporphyrin-based metal–organic framework for inhibition of tumor metastasis. ACS Nano 12, 4630–4640.

Zeng L., Xiang L., Ren W., Zheng J., Li T., Chen B., Zhang J., Mao C., Li A., Wu A., Multifunctional photosensitizer-conjugated core–shell Fe3O4 @ NaYF 4: Yb/Er nanocomplexes and their applications in T 2-weighted magnetic resonance/upconversion lumines.

Zeng, X., Zhou, X., Wu, S., 2018. Red and near-infrared light-cleavable polymers. Macromol. Rapid Commun. 39, 1800034.

Zhang, H., Shan, Y., Dong, L., 2014. A comparison of TiO2 and ZnO nanoparticles as photosensitizers in photodynamic therapy for cancer. J. Biomed. Nanotechnol. 10, 1450–1457.

Zhang, J., Jiang, C., et al. 2018. An updated overview on the development of new photosensitizers for anticancer photodynamic therapy. Acta Pharm. Sin. B 8 (2), 137–146.

Zhang, K., Meng, X., Cao, Y., Yang, Z., Dong, H., Zhang, Y., Lu, H., Shi, Z., Zhang, X., 2018. Metal–organic framework nanoshuttle for synergistic photodynamic and low-temperature photothermal therapy. Adv. Funct. Mater. 28, 1804634.

Zhang, K., Yu, Z., Meng, X., Zhao, W., Shi, Z., Yang, Z., Dong, H., Zhang, X., 2019. A bacteriochlorin-based metal–organic framework nanosheet superoxide radical generator for photoacoustic imaging-guided highly efficient photodynamic therapy. Adv. Sci. 6.

Zhang, S., Li, Q., Yang, N., Shi, Y., Ge, W., Wang, W., Huang, W., Song, X., Dong, X., 2019. Phase-change materials based nanoparticles for controlled hypoxia modulation and enhanced phototherapy. Adv. Funct. Mater. 29, 1906805.

Zhang, W., Lohman, A.W., Zhuravlova, Y., Lu, X., Wiens, M.D., Hoi, H., Yaganoglu, S., Mohr, M.A., Kitova, E.N., Klassen, J.S., 2017. Optogenetic control with a photocleavable protein, PhoCl. Nat. Methods 14, 391.

Zhang, W., Lu, J., Gao, X., Li, P., Zhang, W., Ma, Y., Wang, H., Tang, B., 2018. Enhanced photodynamic therapy by reduced levels of intracellular glutathione obtained by employing a nano-MOF with CuII as the active center. Angew. Chem. Int. Ed. 57, 4891–4896.

Zhao, Z., Han, Y., Lin, C., Hu, D., Wang, F., Chen, X., Chen, Z., Zheng, N., 2012. Multifunctional core–shell upconverting nanoparticles for imaging and photodynamic therapy of liver cancer cells. Chemistry 7, 830–837.

Zheng, D.-W., Li, B., Li, C.-X., Fan, J.-X., Lei, Q., Li, C., Xu, Z., Zhang, X.-Z., 2016. Carbon-dot-decorated carbon nitride nanoparticles for enhanced photodynamic therapy against hypoxic tumor via water splitting. ACS Nano 10, 8715–8722.

Zheng, X., Wang, L., Liu, M., Lei, P., Liu, F., Xie, Z., 2018. Nanoscale mixed-component metal–organic frameworks with photosensitizer spatial-arrangement-dependent photochemistry for multimodal-imaging-guided photothermal therapy. Chem. Mater. 30, 6.

Zheng, X., Wang, L., Pei, Q., He, S., Liu, S., Xie, Z., 2017. Metal–organic framework@ porous organic polymer nanocomposite for photodynamic therapy. Chem. Mater. 29, 2374–2381.

Zhu, W., Gao, Y.H., Liao, P.Y., 2018. Comparison between porphin, chlorin and bacteriochlorin derivatives for photodynamic therapy: Synthesis, photophysical properties, and biological activity. Eur. J. Med. Chem. 160, 146–156.

Chapter 5

Photodynamic therapy for cancer treatment

Sagar Trivedi*, Anita Paunikar*, Nishikant Raut, Veena Belgamwar
Department of Pharmaceutical Sciences, Rashtrasant Tukadoji Maharaj Nagpur University, Nagpur, Maharashtra, India
Both the authors contributed equally.

5.1 Introduction

Among the various diseases responsible for high mortality rates, cancer remains a matter of concern when it comes to global health and it puts a great impact, even though numerous advancements had been made in the field for its treatment, recurrence, and metastasis, but remains tailback for effective treatment and management of various cancers (Minn et al., 2005). Usually, the cancer metastasis or the abnormal growth of tumor cells involves a sequential and various complex process that possibly invades the specific organs and these cancerous cells start multiplying at the site and start becoming resistant to conventional anticancer therapies (Weigelt et al., 2005).

Studies reveal that single or multiple therapies are required for different types of cancers which includes cesarean techniques, immunotherapy (Immunosuppressant), radiotherapy, chemotherapy and targeted drug delivery or small-molecule targeted therapies in combinations with chemotherapies (Lucky et al., 2015; Zhang et al., 2018).

Along with the need of new drug developments, novel technological advancements are required to improve the conditions in recurrence, multi drug resistance and metastasis of tumor cells, however, due to the limitations in the cancer treatments, there is higher worldwide mortality rate in cancer (Angelis, 2008). Even long-term use of treatments worsens the condition of patients due to adverse effects of drugs such as risk of infection (due to neutropenia), anemia (as radiotherapy, surgery or chemotherapy can suppress bone marrow and reduce red blood cells), bruising and bleeding due to Chemotherapy. Targeted drug therapies leads to decrease in pallets count which is responsible for blood clot, hair loss depending on dose, sore mouth, tiredness (fatigue), feeling sick (nausea), ulcers, diarrhea, hormonal changes, skin rashes, baldness, induced depression, allergies, etc. are some more adverse effects caused due to chemotherapy (Daniell and Hill, 1991).

The researchers are keener and more curious towards new strategies for the treatment of cancers which are appropriate, safe, active at low dose, environmentally acceptable, cost-effective and have higher patients' compliance. Hence, this raffles to develop a therapy regimen that proves to be effective for curbing cancer and its future recurrence. Literature suggests that light possess therapeutic potential for treatment of various diseases and has been used over past 3000 years by the ancient Indian and Chinese civilizations in combination with chemical agents leads to fluorescence phenomenon and shows the cytotoxicity (Ackroyd et al., 2007; Kou et al., 2017; Muniyandi et al., 2020).

One of the most appealing technique involving use of light for management of carcinomas is Photodynamic therapy (PDT), in which photosensitizers (PS) are localized on target and then excited through specific light source to generate variable reactive oxygen species (ROS) for cytotoxicity. Photosensitizers are chemical agents having insignificant toxicity (dark toxicity) in absence of light, are allowed to get accumulated in cellular targets like nuclear envelope, mitochondria, or lysosomes. Further assessment of accumulation of photosensitizers in tissues and its elimination from them is done through fluorescence quantum yield (ΦF) and excited with specific electromagnetic waves between red/near infrared (600–850 nm) spectrums. In excited state its interaction with molecular triplet oxygen forms singlet oxygen, hydroxyl radicals and superoxide ions which ultimately causes oxidative damage to cellular components resulting in apoptosis.

PDT is broad application treatment modality indicated for cancer, dermal, ophthalmic, immune, and vascular pathologies. Thus, PDT is more efficacious since it covers immunogenic form of cell death (ICD). This therapy now stands as more effective after removing the challenges associated with PS of less pathological coverage and solubility at physiological pH with development of second and third generation PS respectively. Advantageously PDT also destroys multidrug resistant tumor and clinically it is indicated for pre- and post-operative localized effects, radical and palliative management. It does not affect normal tissues and is marginally invasive. In cancer patients with other long-term co-morbidities PDT proved effective over intolerable surgery and radiation. Recently the development of nanoparticle-based delivery and linking of PDT with these delivery systems have given enhanced results in the treatment of various cancers and similarly combinations of PDT with immunotherapies (neutrophil-mediated therapeutic delivery) also resulted in enhanced outcome. In the healthcare economy and compliance prospects PDT is simple and convenient to be performed as an outpatient therapy.

Continuous advancements and development of novel treatment regimens are being proposed for treatment of cancer, however there is always scope for improvement. The advancements in this field are based on the shortcomings or limitations of the existing technologies and therapies which can be accessed through reviewing the literature. In this chapter, we have reviewed various aspects of PDT including its applications which may enable identifying the gaps to suggest development of novel photosensitizers for future therapy.

5.2 Background of photodynamic therapy

The roots of light as a therapy in the treatment of cancer are sketched from antiquity to the modern era and it became more important because of promising results for management of carcinomas alone and in combination with laser for photocoagulation. Earlier the basic principle which was considered behind the PDT involvement in treating cancer was, as interaction of light with cells is measured as vital component for its survival and simultaneously under various circumstances can lead to their destruction.

5.2.1 Origin of photodynamic therapy

Phototherapy is also known as heliotherapy, while the PDT was originated by various ancient civilizations in countries like India, Egypt, Rome, Greece, and China. Initially people unaware about PDT and hence the knowledge of PDT was declined (Mahmoud, 2016). In the early era, sunbathing and heliotherapy were considered as a scientific foundation for various diseases after the ending of Ming and initiation of Qing era in 17th and 18th centuries (Downes and Blunt, 1877). The scientists from England known as Downes and Thomas Blunt (1877) established a theory of actinic rays (ultraviolet) of sunlight causing bactericidal and antimicrobial action that proves the strong relationship between sunlight and diseases in human beings. Hence, sunlight was applied for various clinical practices as a therapeutic modality (McDonagh, 2001).

Also in the 1890s, another scientist Oscar Raab serendipitously observed that the previously transferred visible light ray from acridine dye causing extermination of paramecia, and it was the major evidence substantiating cytotoxicity of fluorescence produced by acridine red dye. Later in 1908, neurologist Jean Prime suggested the oral eosin was used to treat epilepsy. Whereas in 1903 von Tappeiner and Jesionek suggested topical application of eosin with light exposure to treat dermal carcinoma (McDonagh, 2001; Zhang et al., 2018).

In 1999 the Arnold Rikli reintroduced PDT drug and the main parts of PDT process which includes photosensitizing agent, a light source and oxygen source. After that sunlight was introduced as a therapy to cure multiple disorders, such as vitiligo (Bowcock and Fernandez-Vina, 2012), psoriasis (Wong et al., 2013), psychosis (Vocks, 2006), rickets and skin cancer (Jarrett and Scragg, 2017). In 1550 BC an ancient Egyptians Ebers Papyrus, discovered phototherapy by using plant extract or powder of plants such as parsnip, Saint-John's-wort, Ammi majus and parsley were applied to the affected skin lesions and then patients allowed to take sunrays on a body part. Also, in Atharva-Veda 1400 BC suggested that, the plants extract like the Bavachee plant and the *Psoralea corylifolai* with sun rays can be used to treat the vitiligo (Ackroyd et al., 2007; Mahmoud, 2014; Siboni et al., 2002).

5.2.2 Mechanism of photodynamic therapy

The better understanding of the underlying mechanism behind the PDT as tool for dealing with cancer, possibly come with defining the components and exploring their role individually. The basic principle involved introducing a PS agent which is further activated by light of set frequencies and in existence of oxygen it gives the therapeutic effect. The process takes place in a very multifarious interplay of time and space; however, it must eventually end in ablation of the lesions or the carcinomas and should spare the healthy cells and tissues (Allison and Moghissi, 2013).

5.2.3 Working principle of photodynamic therapy

The elementary working principle for an effective PDT is based on the combined functional action of three vital components namely, photosensitizer (PS), a light source, and molecular oxygen (Fig. 5.1) (Turksoy et al., 2019). Major light

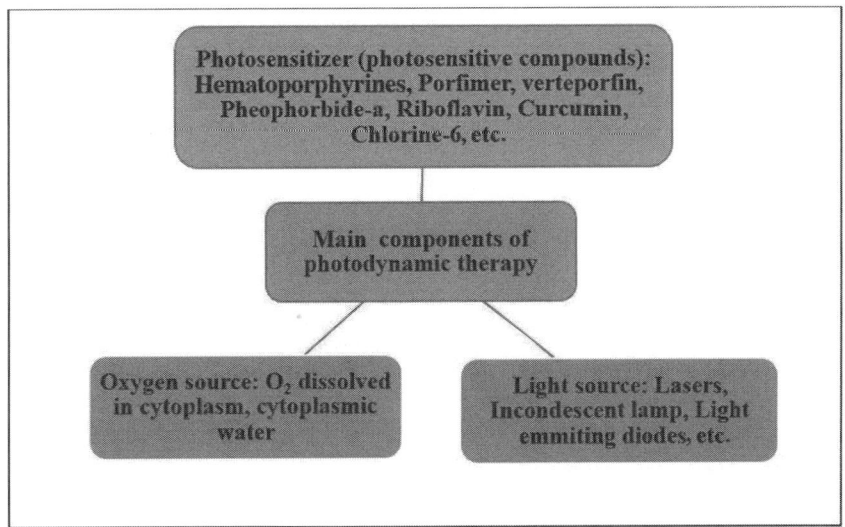

Fig. 5.1 Main components of photodynamic therapy.

sources employed in PDT includes lasers, light-emitting diodes, and lamps, they are mainly selected on the basis of location of cancer, that is, the target location, absorption spectrum of the used photosensitizer, and required light dose (Aksakal et al., 2019). When the nontoxic photosensitizing compounds are placed on the target site, under the influence of suitable light irradiation, it causes activation of PS materials which enables it to absorb and transfer electrons, otherwise in situ form of oxygen acts as an electron receiver (Bapat et al., 2021). The process tends to form cytotoxic reactive oxygen species (ROS), which causes irretrievable injury to the cells and tissues at the target site by rupturing the membrane of the cells or by inducing necrosis or apoptosis causing cell death (Atriwal et al., 2021).

There are mainly two basic types of ROS which corresponds to a distinctive PDT mechanism, firstly by transferring electrons, oxygen radicals are produced e.g. superoxide anion O^{2-}, hydroxyl radical HO, hydroperoxyl radical HOO and secondly by energy transfer which gives singlet oxygen ($^1O_2$) (Cheng et al., 2021). The mechanism initially discussed causes the transitional changes in PS molecules from ground state to the singlet excited state and to the triplet excited state, which further via electron transfer causes these excited PS molecules to react with the substrate to give free radicals (Fig. 5.2). The second photodynamic reaction causes the energy transfer of excited PA moiety to molecular oxygen to give highly active singlet oxygen that further interacts with lipids, proteins, and nucleic acids, causing cell death by necrosis or apoptosis (Dong et al., 2019).

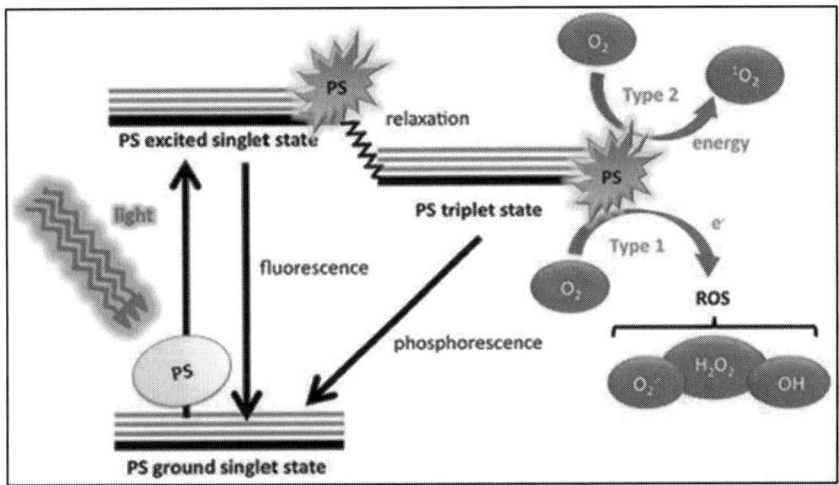

Fig. 5.2 Schematic illustration of photodynamic therapy, including the Jablonski diagram. The PS ground state molecule ready to absorb photon leads to excitation to the first excited singlet state. After relaxation of excited singlet molecule converted to the triplet state at which molecule is stable for longer duration. This triplet PS shows photodynamic activity by two different ways as shown in type I and in type II, results in formation of cytotoxic reactive oxygen species (ROS) and singlet oxygen, respectively (Dai et al., 2012).

5.2.4 Mechanism of photodynamic therapy in treatment of cancer

As highlighted above the general mechanism of PDT comprises the formation of singlet oxygen which can cause damage up to a range of several nanometers by substantial and complex cascade of various events which leads to local, regional, and systemic alteration of both tumor and immune response for the management of cancer. Hence this section deals with the action of PDT on various cellular, vascular, and systemic immune events for improving the anticancer output (Allison and Moghissi, 2013).

5.2.4.1 Tumor cell destruction

The basic underlying mechanism behind the destruction or inhibition of tumor by the PDT is programmed (apoptotic) (Oleinick et al., 2002) pathways and nonprogrammed (necrosis) pathways (Lai et al., 2006). The PAs are brought to the tumor site via various mechanisms like receptor mediated phagocytosis/endocytosis, low density lipoprotein receptor binding, lipid binding, uptake via tyrosine kinase/epidermal growth factor receptor, diffusion, biodistribution at cellular or subcellular level (Piette et al., 2003; Płonka et al., 2015; Wang et al., 2021). Each cell uptakes PAs through different process like Photofrin is penetrated by multiple pathways in cancerous cell but also gets accumulated in other cells too. When the tumor cells are exposed to light of increased intensity the cell in tumor get rapidly ablated by necrosis, resulting in fast cellular and sub cellular membrane destruction because of release of various metabolic byproducts like calcium which are not suitable for appropriate cellular functions leading to cell death (Aniogo et al., 2020; Gunaydin et al., 2021).

Exposure of PDT to the tumor site is also responsible for release of cytokines and toxic chemicals, which leads to bystander effect producing a regional and systemic reaction for cell death (Zhenya et al., 2021). When the site of tumor is exposed to low light doses apoptotic death may be caused by PDT by ceasing the functions of cells and further which undergoes an orderly programmed dissolution but it doesn't cause any bystander effect or immune response because a low doses no toxic chemicals are leaked (Brillo et al., 2021; Fang et al., 2021; Piette et al., 2003). The PDT induced apoptosis affects both tumor and normal cells mechanism, but still a better mechanism for removal of damaged cells (Usuda et al., 2003). The PAs get concentrated in the rapidly multiplying cancerous cells along with normal cells, causing lethal action of PDT to normal cells too. The carcinogenic cells undergo necrosis or apoptosis and if at the same time excess amount of PAs gets accumulated in normal cells or tissues severe morbidity can be observed (Aniogo et al., 2020; Ichinose et al., 2006; Płonka et al., 2015; Wang et al., 2021).

5.2.4.2 Vascular events

The same mechanism for concentration of PAs is followed by the endothelial cells of the vascular systems as followed by the tumor cells including receptors, diffusion, and multiple other pathways (Gunaydin et al., 2021). In addition to this, the neovasculature of malignancy is leaky in nature and the borders of the cell are not properly arranged and organized, which serves as an additional means for accumulation of PAs at the site (Fang et al., 2021). The PAs get activated by the basic mechanism of PDT and results in happening of several destructive events like disruption of vascular walls (Brillo et al., 2021), insufficiency of blood reaching to the tumor site and creating scarcity of oxygen and ultimately leading to necrosis of neovasculature (Allison and Moghissi, 2013).

Another process involved is discharge of various lethal chemicals and supplementary intra cellular debris causing obstruction and failure of the microvasculature nourishing the tumor (Castano et al., 2004; Mroz et al., 2011). All these events create a local platelet aggregation and a reckless loss of blood supply to the tumor leading to a lethal damage (Castano et al., 2005; Robertson et al., 2009). If the PDT is given at low doses, apoptosis may also occur causing tumor hypoxia and obliteration but deprived of cytokine and immune stimulation (Gunaydin et al., 2021; Magalhães et al., 2021; Wang et al., 2021). Hence, the major mechanism behind the apoptosis and necrosis inducing damage to tumor is not only the lowered blood supply or hypoxia (Fang et al., 2021) but also the release of various toxic compounds like thromboxanes, platelet aggregators (Lange et al., 2021), and toxic cytokines, which provokes the immune system (Castano et al., 2004).

5.2.4.3 Immune system mediated photodynamic therapy

PDT improves the immune repose for combating cancer and tumors by enhancing the surveillance, improving the immune system with the help of PDT which can possibly induce long term tumor control by targeting the lesion site and vasculature site (Allison and Moghissi, 2013). During the exposer of PDT to the tumors and their related vasculature, it induces necrosis along with release of inflammatory mediators at the site of treatment like various cytokines, growth factors; proteins and additionally an immune cascade is also instigated (Korbelik and Sun, 2001; Kwitniewski et al., 2008). Thereafter the released immunomodulators stimulate the white blood cells to activate neutrophils and macrophages congregating on the treatment region (Autiero et al., 2010; Raza et al., 2014). The macrophages phagocytes the cancer cells which are damaged by PDT mechanism and proteins from tumor like CD4 helper T lymphocytes activates CD8 cytotoxic T lymphocytes (Krosl and Korbelik, 1994; Sandbo et al., 2013). The produced T cells induce necrosis and apoptosis in tumor cells over a long period of time even after completion of PDT (Hang et al., 2015). Clinical data suggest that patient who are treated with PDT, gives a higher level of numerous cytokines and histology of tumors suggests regular immune cell infiltration (Korbelik et al., 2020).

5.3 Novel strategies in photodynamic therapy

The basic process behind an appropriate use of PDT includes the use of PA which is organic in nature and its activation is done by continuous use of light, moreover it is applied as an acute, high dose single treatment (Agostinis et al., 2011). In addition there are several fundamental approaches for effective use of PDT and some of which are under preclinical investigations, including several photophysics, chemistry, and/or photobiological mechanisms (Montaseri et al., 2021; Zhang et al., 2021). One of the approach is 2-photon PDT, in which short laser pulses having capacity of 100 femtoseconds, possessing high peak power are used and this enables ease in absorption of two light photons by the PA (Starkey et al., 2008).

Generally, a single photon adds up only one-half of the excitation energy and the deep tissues penetration can be achieved by use of near-infrared light and similarly photochemistry and photobiological works on the same mechanism (Karges et al., 2020). The excellent site targeting to individual blood vessels can be achieved by application of strongly focused laser beams which reduces the chances of damage to the healthy tissues which are adjacently placed (Chennoufi et al., 2020; Dib et al., 2019). Hence use of these novel strategies by designing the regimen to employ novel PA to get very high 2-photon cross-sections. It will perhaps help in overcoming light attenuation limitations specifically in pigmented tumors like melanoma.

5.3.1 Metronomic photodynamic therapy

The metronomic type of PDT (mPDT) aims at delivering the light and the pharmacologically active moiety at the same time over a sustained period of time ranging from few hours to some days which leads to tumor cell-specific apoptosis, with a minimum necrosis of tissues (Lilge and Wilson, 1998). Metronomic PDT finds its application in the management of gliomas in curtailing the direct effect of photodynamic light damaging the normal tissues and cells, which further produce harm due to inflammation caused in response to PDT-induced tumor necrosis (Dabrowski et al., 2011; Gunaydin et al., 2021). *In-vitro* studies demonstrated the dose-dependent tumor responses and treatment using intracranial model, aminolevulinic acid (ALA) and also by implanting optical fiber source (Hambsch et al., 2017; Moriyama et al., 2014; Orenstein et al., 1996), this concludes that mPDT molecular pathways may act differently at high-dose PDT (Salman et al., 2015).

5.3.2 Photodynamic therapy molecular beacons

The fundamentals of PDT molecular beacons (MBs) state its application for high target specificity by employing MBs as the fluorescent probes (Zheng et al., 2007). In MBs the PA is attached to the probes and hence it remains inactive till its reaction occurs with the target specific enzyme. The linkage between the two compounds may be served by an antisense oligonucleotide (hairpin) loop which is unlocked by hybridization to corresponding mRNA (Chen et al., 2008). Initially, the PDT MBs showed the potential effects on caspase-3 linker amongst pyropheophorbide and a carotenoid quencher, achieving eightfold and fourfold quenching and unquenching and correspondingly, as validated by the $^1O_2$ yield (Chen et al., 2005; Pinthus et al., 2006).

Similarly, *in-vitro* and in vivo studies confirmed that matrix metalloproteinase (MMP)-based beacons showed high affinity towards MMP-positive and MMP-negative tumors (Sharman et al., 2004; Zhang et al., 2003; Zheng et al., 2002). The Hairpin-type beacons are the another type of MBs targeting raf-1 mRNA and possessing high degree of tumor-to-nontumor affinity enabling nearly complete restoration in human breast cancer cells *in-vitro* upon hybridization (MacPherson et al., 1997; Stefflova et al., 2007). The mechanistic character of MBs showing affinity toward the tumor site not only depends on the delivery of PA but also on unquenching interaction and specificity of beacons toward the interaction (O'Connor et al., 2009).

5.3.3 Nanotechnology in photodynamic therapy

Nanoparticles in PDT possess multiple impending roles as PA delivery, PA site targeting and as energy transducers (Wang et al., 2007). Liposomal nanoformulation aids in delivering the water insoluble PAs compounds for example verteporfin. The nanoparticles enable high loading efficiency of PAs to the target site with modification of surface with targeting moieties such as antibodies or peptides. Similarly the localization of PAs at the tumor site can be achieved by various other approaches including biodegradable polymers and ceramic (silica) and metallic (gold, iron oxide) nanoparticles (Chatterjee et al., 2008) and hybrid nanostructures assist as a theranostic molecule allowing both PDT and therapeutic approaches such as hyperthermia or an imaging technique such as MRI (Sajja et al., 2009). Silicon nanoparticles and quantum dots provides delivery of 2-photon PA because they have very less water solubility and they themselves potentiate to generate $^1O_2$ on photoexcitation, further they may also be linked to various organic PAs, so that they can absorb light energy with high efficiency and relocate it to PAs (Sinha et al., 2006).

Some modification or upgradations of nanoparticles influence them to absorb the light of long wavelength (near infrared) and converting it to shorter wavelength for activation process of PAs. Thus, this advantage of nanoparticle mediated PDT eliminates the photophysical and photochemical attributes of the PS for activation of PAs (Karthikeyan et al., 2011; Kim et al., 2007). A study facilitating the co-delivery of PAs encapsulated inside polymeric nanoparticles accumulating in a liposomal formulation comprising an antiangiogenic agent (or vice versa) has been reported to give the therapeutic synergism in a PDT (Gunaydin et al., 2021).

5.4 Role of photosensitizing agents in photodynamic therapy

The Photosensitizing Agents (PAs) are the vital elements among the three key constituents of PDT including oxygen and light. In the 1970s, Dr. Thomas Dougherty familiarized the world with "hematoporphyrin derivative" (HpD) which is a mixture of deuteroporphyrin, water-soluble protoporphyrin and their derivatives, monomers, hematoporphyrin, dimers, and oligomers and their ester derivatives. Such types of photosensitizing mixture were termed as Photosensitizing Agents (Panzarini et al., 2014). Since ancient era, various photosensitizers have been developed and awarded with two Nobel awards in the arena of PDT, mentioned in Table 5.1 and relatively less amount of work has been carried out in the field of elementary and clinical applications using PDT for the treatment of atherosclerosis and tumor (Kou et al., 2017).

The selection of compounds as Photosensitizing Agents mainly depends on the photochemistry and the uptake efficiency of PAs, hence some of the clinically approved agents for the treatment of cancer are listed in Table 5.2. The PAs like

TABLE 5.1 Brief history about development of PDT.

Sr. No.	Scientist	Discovery
1	Scherer	In 1841, Scherer discover hematoporphyrin by extraction of haem from dried blood.
2	L. Pasteur and P. Bert	Discovery of phototoxicity in 1861–1871
3	J. L. W. Thudichum	In 1867, discovery of fluorescence spectrum from red substance (hematoporphyrin).
4	F. Hoppe-Seyler	In 1871 new terminology "hematoporphyrin" was given to red substance.
5	Schultz	1874 describe the reason behind porphyria in patients (errors in heme biosynthesis).
6	N. R. Finsen	In 1895–1903 discovered the phototherapy and received Nobel prize in 1903.
7	O. Raab and H. von Tappeiner	In 1897–1904 first time reported phototherapy for cancer.
8	H. von Tappeiner	1904 the term "photodynamic action" was introduced first and in1903–1905 discovered topical use of eosin and light.
9	W. Hausmann, F. Meyer-Betz	In 1908–1913 introduction to PDT experiments with hematoporphyrin on paramecia, erythrocytes, mice, guinea pigs, and humans.
10	A. Policard	In1924 A. Policard saw red porphyrin fluorescence in tumors.
11	H. Fischer	In 1925 discovery of porphyrins and received second Nobel prize in 1929.
12	S. Scwarz	In 1945 discovery of radio sensitization with porphyrins.
13	D. Harman	In 1959 proposed the free radical theory of ageing and disease.
14	R. Lipson E.Baldes	In 1960–1967 synthesis of HpD.
15	H. Kautsky G. Herzberg	In 1970 theory of active oxygen.
16	Z. Malik M. Djaldetti	In 1975 discovery of ALA for PpIX induction.
17	T. J. Dougherty, et al.	1983–1993 discovery of Photofrin
18	J. Kennedy R. Pottier	1990 discovery of Clinical application of ALA.

TABLE 5.2 Various generation photosensitizers and characteristics properties (Dai et al., 2012; Kataoka et al., 2017; Maiya, 2000; Muniyandi et al., 2020).

First generation photosensitizers

Photosensitizer	Wavelength (nm)	Manufacturer	Applications
Hematoporphyrin derivative (HpD)	630	Axcan Pharma, Canada	Oesophageal cancer, Lung adenocarcinoma, Endo-bronchial cancer
Photofrin Photoheme	635, 625, 630		

Second generation photosensitizers

Photosensitizer	Wavelength (nm)	Manufacturer	Applications
Meso-tetrakis (4-sulfonatophenyl) porphyrin (TPPS)	535	-	Enhancing tumor affinity, Improving tumor retention for treatment of various tumors.
N-aspartyl chlorin e6, NPe6	660	Japan	Treatment of fibrosarcoma, liver, brain, and oral cancer primary or secondary cancer of the skin and mucosal surfaces
Ameluz®/Levulan®	635	DUSA, USA	Mild to moderate actinic keratosis
Metvix®/Metvixia®	570–670	Galderma, UK	Non-hyperkeratotic actinic keratosis and basal cell carcinoma
Foscan®/ Meta-tetra(hydroxyphenyl) chlorin	652	Biolitec, Germany	Advanced Head and neck cancer, Approved drug for the treatment of bronchial and oesophageal cancer
Laserphyrin®	664	Meiji Seika, Japan	Early centrally located lung cancer
Visudyne®/ Benzoporphyrin/ Benzo-porphyrin derivative monoacid ring A (BPD-MA), vertoporfin	690, 689	Novartis, Switzerland	Age-related macular degeneration, Prostate and skin cancer.
Redaporfin®	749	Luzitin, Portugal	Biliary tract cancer

Under clinical trails

Photosensitizer	Wavelength (nm)	Manufacturer	Applications
Fotolon	665	Apocare Pharma, Germany	Nasopharyngeal, sarcoma
Radachlorin	662	Rada-pharma, Russia	Skin cancer
Photochlor	664	Rosewell Park	Head and neck cancer
TOOKAD	762	Negma-Lerads	Prostate cancer
Texaphyrins =Lutrin, Antrin, Optrin, Xcytrin	732	Pharmacyclics	Coronary artery disease, hepatocellular cancer, nasopharyngeal carcinoma, colon, prostate, Leukaemia
Photrex	664	Miravant, USA	AMD
Talaporfin	664	Meiji Seika, Japan	Colorectal neoplasms, Liver metastasis

Third generation photosensitizers

Photosensitizer	Wavelength (nm)	Manufacturer	Applications
Gold-Nano clustered Hyaluronan Nano-Assemblies	500-550	-	Tumor ablation
Chlorin E6 (Ce6) +Up conversion nanoparticles, MACEDACEPhotoditazin	980, 405	-	Gynecological diseases, prostate cancer, fibrosarcoma, Liver, brain, lung, and oral cancers.
Photofrin+ gap junctional intercellular communication (Connexin 32)	570	-	Inhibit the migration and invasion of cervical cancer cells
Ce6+tumor-targeting nanogel	400	-	Fluorescence imaging of tumor sites and an enhanced PDT to induce complete death of cancer cells.
Ce6+ChitoUDCA nanoparticles	200–400	-	Treatment of primary biliary cirrhosis
ICG-loaded nanospheres coated with chitosan	800–805	-	Breast, Brain and Skin cancer

porphyrin precursors [e.g., aminolevulinic acid (ALA), phenothiazines (e.g., methylene blue), cyanines (e.g., merocyanine, indocyanine green), hypericin, and xanthenes (e.g., Rose Bengal) have been considered as good candidates for PDT. The PAs are also used particularly in different types of cancers for example Porfimer sodium (Photofrin), mTHPC (Foscan), NPe6 (Laserphyrin), SnEt2 (Purlytin), Visudyne (Veteporfin), and motexafin lutetium (LuTex) showed specificity toward breast cancer tumor cells and found ideal for cellular and vascular-targeted PDT.

Chemical structure of photosensitizer plays an important role to differentiate its therapeutic action between anti-cancer and antimicrobial drugs. Anti-cancer drugs have more lipophilic with positive or negative or carry no charge. Antimicrobial structures carry cationic charges and selectively kills gram-negative bacteria (dos Santos et al., 2017), tetrapyrrole structure containing derivatives such as chlorophyll, hem, and bacteriochlorophyll mostly showed anticancer properties by type-II mechanism (with the exception of bacteriochlorin's) singlet oxygen as compared with the type I ROS (such as hydroxyl radicals) (Ferreira et al., 2013).

An ideal photosensitizer should have characteristics property as follows:

- It should not show any allergic reactions or hyposensitivity, lower dark toxicity
- It should be feasible to administer by different routes of administration without any pain
- It should be readily activated in tumor by light, clears normal tissue and should be excreted easily from body of a patient
- It should be easily detected in visible spectrum
- It should have high absorption peak value in the near infrared (NIR) in between 600 and 800 nm, because photosensitizers having longer wavelength than 800 nm does not exert energy to excite oxygen molecule to its singlet state and reactive oxygen species
- It must have better tissue penetrability and should readily emit adequate energy to produce singlet oxygen
- It should readily form ROS and kill selectively carcinogenic or metastasis tumor cells
- It should be easily manufactured, purify, and have longer stability and compatibility on storage
- It should possess enough water solubility to achieve better blood circulation of photosensitizers and imaging disease sites through clinical imaging devices.

Recent studies showed that the PAs are classified based on their efficacy and selectivity. PAs possessing higher absorption wavelengths allows deep penetration of the illuminating sources and ultimately enhance the targeting of tumor, generally termed as second-generation PAs. The next generation PAs are the third-generation compounds developed with the aim of targeting tumor sites with active carriers like antibody-directed PA and PA-loaded nanocarriers. These cause increase in power and efficiency of PDT enabling to explore a wide range of tissues, cells, and diseases. Hence, to achieve all ideal criteria, various photosensitizers were introduced in the ancient era and are classified into three different generations depending on their discovery and special features.

- **First generation:** They were discovered in 19th century and called as hematoporphyrins (Hp), obtained from desiccated blood through concentrated sulfuric acid treatment in 1841 by Schere and were mixed with several porphyrins to enhance special feature of hematoporphyrin. It showed fluorescent phenomenon utilized in diagnostic aid in various cancers. However, due to its complex structure, poor solubility, and penetrability, it requires higher doses of hematoporphyrin. To resolve all these problems, various hematoporphyrin derivatives were introduced having better solubility, penetrability, and tumor selectivity with PDT. The European Medicine Agency (EMA) and U.S. Food and Drug Administration (FDA) approved first hematoporphyrin derivative known as Porfimer sodium, or Photofrin® to cure cancers. Nonetheless, due the complex nature of derivative it also showed relatively poor absorptivity and penetrability, prolonged half-life, short wavelength of light and intense gathering in normal tissues. Thus it required larger doses to get therapeutic activity to result in photo-sensitive toxicity (Bayona et al., 2017; O'Connor et al., 2009; Ormond and Freeman, 2013).
- **Second generation:** It includes water soluble derivatives of porphyrins, such as porphyrins, chlorins, pheophorbides, bacteriopheophorbides, texaphyrins, and phthalocyanines. They possess higher solubility, penetrability, reproducibility, and tumor selectivity requiring lower doses of drug which can easily be utilized to treat large variety of deeper carcinomas with low wavelength of UV rays or light sources. The second-generation photosensitizers include Temoporfin (Foscan; Motexafin lutetium (Lutrin and Lutex; Pharmacyclics Inc), Palladium bacteriopheophorbide (Tookad; Negma-Lerads), Bio- litec AG), Protoporphyrin IX precursors (Hex- vix, purpurins (Purlytin), Metvix and Levulan) and Verteporfin (Visu-dyne; Novartis) (Bayona et al., 2017; O'Connor et al., 2009; Ormond and Freeman, 2013).
- **Third generation:** It includes various porphyrins, either synthetics or derivatives containing nonporphyrin derivatives linked with biological carriers, various antibodies, or liposomes to achieve better stability, compatibility, penetration of light, physicochemical characteristics of drug and their therapeutic efficacy. These modification in compound shows

higher specificity, selectivity toward cancerous tissues and cells with less or no damage to the normal body tissues (Bayona et al., 2017; Bellnier et al., 2006; O'Connor et al., 2009).

To minimize limitations and drawbacks of individual generations, following techniques used to enhance specificity, selectivity following modifications made during photodynamic therapy:

- Second generation of photosensitizers were used in combinations with molecules to target the specific target receptor.
- The LDL lipoprotein biomolecule used in combinations with photosensitizers as the proliferating tumor cells required cholesterol for the synthesis of cell walls.
- The monoclonal antibodies used as a conjugate with photosensitizer to target specific antigen of cancer cell.
- To improve targeted drug delivery of photosensitizer transferrin receptors, growth factor receptors or hormones (e.g., insulin) used as tumor surface markers during therapy.

All these combinations used to improve selectivity, specificity, reduces drug dose and relative accumulation of drug, and thus lowers the toxicity (Kataoka et al., 2017; Yoon et al., 2013).

PAs obtained from various sources, such as natural, synthetic, and semisynthetic sources, and are capable of transferring light energy (Niculescu and Grumezescu, 2021). Wide variety of structural molecules were screened for their photosensitizing activity and among them plant and bacterial chlorophyll derivatives showed acceptable ideal properties needed to be a photosensitizer. Number of plants, bacterial pigments, dyes, and porphyrins were introduced in ancient era however, only about a dozen employed in clinical trials, characterized for activity, and approved by FDA. Some of the PAs have not been approved although some of the countries licensed those to use within the territories of a country. Also, some PAs were approved but not available for purchase or having an extremely high cost so as it has not preferentially being used in therapies (Bellnier et al., 2006; Ormond and Freeman, 2013).

Presently, wide variety of structural molecules are used as photosensitizers in the market. Individual agents have its own characteristics properties, so it is worth to have a brief discussion on each agent as depicted in Table 5.2 (Kataoka et al., 2017; Yoon et al., 2013).

Recent trends showed that, researchers are greatly attracted toward the eco-friendly plant-based sources showing reasonable anticancer activity, and now they are also incorporated in phototherapy as a natural potent drug source. The natural photoactive compounds, activated by light with the applications of photochemistry (Bullous et al., 2011; Mang, 2004) are used as potent third-generation photosensitizers.

Natural photoactive compounds from plants include some plant derived photoactive components, such as alkaloids, thiophenes, curcumins, anthraquinones, and furanocoumarins. The advanced and hyphenated technology helps in isolation, purification, identification, characterization, and derivatization. These techniques are utilized for improvement of extraction with bioactivity, photoactivity and the anticancer action.

Some natural photoactive compounds with sources are depicted in Table 5.3.

As time passes, different types of photosensitizers were developed in order to explore new clinical applications. So, here we shortly discuss about various photosensitizers used in phototherapy.

1. **Hematoporphyrin derivative (HpD):** It is a combination of monomeric, dimeric and oligomeric with ester and ether linkages containing species (Bellnier et al., 2006; O'Connor et al., 2009; Ormond and Freeman, 2013). The first derivative called Photofrin II (Porfimer sodium) was discovered by Rosewell Park Cancer Institute, Buffalo, NY (USA) in 1988 and later bought by QLT Photo Therapeutics, Vancouver, British Columbia (Canada) and approved in Canada (bladder cancer), Japan (early stage lung, oesophageal, gastric, and cervical cancers as well as cervical dysplasia), Germany (early stage lung cancer), Netherlands and United States (advanced oesophageal cancer) and 11 additional countries in Europe as a tumor-photosensitizing agent for the PDT (Dougherty et al., 1998; Lindenmann et al., 2011). Photofrin is injected intravenously and exposed with laser light of 630 nm wavelength. Initially The activation required minimum 20 min or more per lesion. The drug is accumulated in several organs, comprising the skin for 6 to 8 weeks. During and after drug therapy, revelation of skin to daylight (a strong light source) can result in severe burn or cellular damage due to propagation of photodynamic reaction (Breskey et al., 2013; Reynolds, 1997; Schweitzer, 2001). According to literature survey, due to certain restricted features of photofrin such as its complex structure and low absorption rate of light, only a small amount of light able to penetrate and concentrate photofrin into the tumor cells through the skin and get accumulated into the skin surface causing cutaneous photosensitive toxicity (Gomer, 1991; Oleinick et al., 2002; Sun et al., 2016; Toratani et al., 2016).

Second generation:

2. **m-Tetrahydroxophenyl chlorine (mTHPC) (Foscan):** Foscan (Temoporfin; Biolitec) is a plant derived chlorine derivative of second-generation photosensitizer. It is synthetically pure and yields a speedy and substantial photodynamic reaction (PDR). It has higher tumor selectivity at 652 nm wavelength and after taking drug, patients should halt in a dark

TABLE 5.3 Natural photoactive compounds with sources.

Biological source	Chemical constituent	Chemical property and groups	Absorption wavelength (nm)	Possible mode of action	References
Angelicae dahuricae, Tetradium daniellii, Glehnia littoralis, Heracleum persicum, Syzygium Sps, Ruta graveolens, Ficus sps.	Furano-coumarins	furan ring.	333	DNA intercalation under dark type 2 PDT reaction. Cross-linking and adduct formation with DNA and RNA. Cell membrane damage.	(Muniyandi et al., 2020)
Curcuma longa	Curcumin	Dicinnamoylmethane, curcumin, curcuminoids, demethoxycurcumin, bisdemethoxycurcumin	420–480 nm	Cell membrane is the primary target of curcuminoids. Induction of caspase-mediated cell death.	(Muniyandi et al., 2020)
Polyathia suberosa, Dactylopius coccus, Xanthoria parietina, Drechslera avenae, Polygonum cuspidatum, Heterophyllaea pustulata, H. lycioides	Anthraquinones	1,5-dihydroxy przewalsquinone B, ziganein, uredinorubellins, caeruleoramularin, hypericin, cercosporin, elsinochromes A-C pleichrome, hypocrellin.	437 nm	Type 1 and 2 PDT action.	(Comini et al., 2007; Daub et al., 2005; Montoya et al., 2005; Theodossiou et al., 2009)
Aloe vera, Rheum palmatum, Rumex crispus		Hydroxyanthraquinones, rhein, physcion, emodin, rubiadin, damnacanthol, soranjidiol, alizarin, purpurin, rubiadin, aloe-emodin.			
Asteraceae spp, Heliopsisa, Rudbeckia spp,	Polyacetylenes and Thiophenes	Furanoacetylines, thiarubrines, thiophenes, polyacetylene (aliphatic compounds with more than three conjugated triple bonds), thiophenes (aromatic acetylenes, e.g., phenylheptatriyne).	488	Membrane damage or erythrocyte leakage; type 1 and type 2 PDT reaction, as well as type 1 and 2 PDT mixed reaction.	(Ghannoum et al., 2004; Ibrahim et al., 2016, LIMA et al., 2009; Seth et al., 2014)
Arnica, Centaurea scabiosa, Tagetes erecta					
Porophyllum obscurum, Echinops, Bidens					
Ambrosia chamissonis, T. minuta, E. latifolius, E. sgrijissi, Rhaponticum uniflorum.					

room for 24 hours as due to room light the drug may get activated. The light exposure must not be less than 90 h and not more than 110 h after Foscan injection. It is advised to take in a short time interval of 4 weeks between treatments otherwise it may cause dark toxicity along with the painful treatment and hence, illumination also required during procedure of anaesthesia. But still it is recommended for primary and recurrent head and neck cancers (Lou et al., 2004) and for breast and pancreatic cancer (Story et al., 2013), prostate cancer after radiotherapy with prostate specific antigen (PSA) measurements and prostate biopsies (Wyss et al., 2001). The drug shows some side effects such as headache, dysphagia, hemorrhage and oedema (Moore et al., 2006).

3. **Mono-L-aspartyl chlorine e6 (NPe6):** Mono-L-aspartyl chlorine e6 (NPe6) is sold as LS11, MACE, NPe6 etc. as its trade name and the derivative is also known as Fotolon (RUE Belmedpreparaty, Minsk, Republic of Belarus). It is a plant derived chlorine analogue and highly effective for generation of the PDR. It is free of toxicity such as dark toxicity which is showed by Foscan® (Biolitec Pharma Ltd., Dublin, Ireland) and also can be injected as infusion for several hours (Grant et al., 1993). The use of Photofrin or Foscan is permitted simultaneously for therapy and it is convenient for patients and practitioners, economical and have high patient compliance rate for the use in the clinical PDT of tumors (Spikes et al., 1993).

TABLE 5.4 Summary of combined effect of PDT and different therapeutic approaches for management of cancer.

Drug/treatment approaches	Findings	References
Antioxidant agents		
Ascorbate combined with Photofrin/PDT	Antioxidants like ascorbic acid (AA) may prevent cancer cells of cellular free radical damage. AA on markers of oxidative stress and apoptosis in rat DS-sarcoma cells on 5-aminolevulinic acid-based photodynamic therapy (ALA-PDT).	(Frank et al., 2006)
Ascorbate combined verteporfin mediated PDT	Ascorbate lowered the conversion of highly reactive singlet oxygen to less reactive hydrogen peroxide, and simultaneously high accumulation of verteporfin was also seen.	(Kramarenko et al., 2006)
Chemotherapeutic agents		
Porphyrin-platinum complexes combining cisplatin/oxaliplatin	Complex exhibited a synergistic antiproliferative effect compared to cisplatin and hematoporphyrin alone.	(Lottner et al., 2004)
Photoactivated aluminum disulfonated phthalocyanine (AlS2Pc) combined with the Adriamycin (ADR) and cisplatinum (CDDP)	Combination of drugs + PDT enhanced the antitumor effect. The effective doses of drugs was also reduced and lowering their toxic effects on normal host tissues.	(Canti et al., 1998)
Doxorubicin mediated photodynamic therapy with methylene blue	Surfactant-polymer hybrid nanoparticles formulated using an anionic surfactant, which possibly enhanced tumor accumulation of both doxorubicin and methylene blue and has substantial therapeutic potential against drug-resistant tumors.	(Khdair et al., 2010)
Immunotherapy		
Benzoporphyrin derivative (BPD)/PDT	Granulocyte-macrophage colony-stimulating factor (GM-CSF) immunotherapy caused tumor-localized immune stimulation and PDT-induced antitumor immune reaction along with tumor control.	(Shi et al., 1999)
Streptococcal preparation OK-432 mediated PDT	Antitumor effect of PDT was enhanced by administration of OK-432.	(Uehara et al., 2000)
Angiogenesis inhibitors		
Hypericin-mediated photodynamic therapy in combination with Avastin (bevacizumab)	Avastin along with PDT can improve tumor responsiveness and minimal expression of vascular endothelial growth factor (VEGF) in tumors treated with combination therapy of PDT and Avastin.	(Bhuvaneswari et al., 2007)
5-aminolaevulinic acid (ALA)-based photodynamic therapy coupled with nimesulide	In-vitro experiments suggest a combination of COX-2 selective inhibitor (nimesulide) and ALA-based PDT for more effective treatment of various skin and oral diseases.	(Akita et al., 2004)
Receptor ihibition		
Erbitux combined PDT	PDT and Erbitux efficiently hinder tumor growth and useful in treatment of bladder tumors.	(Bhuvaneswari et al., 2009)

4. **Aminolevulinic acid (ALA):** 5-aminolevulinic acid (ALA) is a novel endogenous photosensitizing agent used in PDT (Kennedy and Pottier, 1994; Wolf et al., 1994). The active ALA achieved by the hem biosynthesis to protoporphyrin IX (PpIX). The synthesis of ALA is the primary and rate determining step which includes condensation reaction among glycine and succinyl-CoA. The reaction is catalysed by ALA synthase (ALAS) and requires pyridoxal-5-phosphate (PLP) as a cofactor for the synthesis in mitochondria (Luo et al., 2006). It is used in cancer treatment such as superficial skin cancers cells, oral cavity tumours (Brown and Kesselt, 1990), and anecdotally in gastrointestinal cancer and drug is specifically concentrated in superficial skin (Allison and Sibata, 2010). The ALA acts as a prodrug activated enzymatically and converted to active Proto-Porphyrin IX as a potent photosensitizing agent. The ALA can be administered by topical, oral and intravenous route, while it has been observed that topical application of drug is uncomfortable and leads to patient non-compliance (Ascencio et al., 2008).

5. **Lutex:** The Lutex (Motexafin lutetium, Pharmacyclics Inc.) being a porphyrin derived photosensitizer approved for the treatment of prostate cancer. Motexafin lutetium is activated at specific wavelength between 730 and 770 nm due to

TABLE 5.5 Details of ongoing clinical trials with the usage of PDT for treatment of various cancers.

Sr.no	Indication	Estimated study completion date	Estimated participants	Clincal trials. gov identifier
1.	Bladder cancer	May 2022	125	NCT03945162
2	Skin cancer	December 2025	20	NCT02751151
3	Lung cancer	February 2022	03	NCT04753918
4	Bile duct cancer	August 30, 2024	20	NCT03003065
5	Esophageal cancer	December 30, 2024	1000	NCT01842555
6	Neoplastic diseases	July 2021	400	NCT02159742
7	Pancreatic cancer	December 30, 2021	30	NCT03033225
8	Actinic keratosis and nonmelanoma skin cancer	January 2022	20	NCT03110159
9	Locally advanced lung cancer and advanced cancers obstructing the airway-pilot study	February 4, 2022	18	NCT03735095
10	Recurrent head and neck cancer	January 16, 2023	82	NCT03727061
11	Nonsmall cell lung cancer with pleural disease	May 28, 2023	12	NCT03678350
12	Esophagogastric cancer	April 2022	30	NCT03133650
13	Prostate cancer	September 2025	200	NCT03492424
14	Anal dysplasia and microinvasive anal cancer	January 1, 2023	12	NCT02698293
15	Upper tract urothelial cancer	July 4, 2024	100	NCT04620239

presence of macrocyclic modification. The dose of drug is studied for its potency by varying various parameters such as PAs dose, light dose and PS-light interval with lower (0.5 mg/kg) to higher (2 mg/kg) doses of Motexafin lutetium and it was found that the higher doses of Lutex® is promising than low dose for PDT (Allison and Sibata, 2010). However, it's prolonged use causes PDT-induced photobleaching in prostate cancer (Allison and Sibata, 2010). Its metalloporphyrin derivative, known as Motexafin gadolinium (Xcytrin) is a highly selective agent used for brain metastasis and lung cancer (Rodriguez et al., 2016). It acts by generating ROS via intracellular oxygen and disrupts redox-dependent pathways and enhance tumor cells apoptosis (Henry et al., 2008), while it's use is restricted in the treatment of non-small cell lung cancer patients with brain metastases. As compared to other photosensitizers, Lutex is found in pure form, water soluble compound having higher selectivity towards tumor cells. The drug can cause dermal photosensitivity, only when exposed continuously for 24–48 h. Also, the drug activation selectively at 730 nm helps drug to penetrate deeper into the abdominal tissues and shows normal cell toxicity (Bown, 2013; Denis et al., 2013; Muniyandi et al., 2020).

6. **Visudyne:** The benzoporphyrin derivative of verteporfin known as Visudyne® (Verteporfin, Novartis) used as first-line treatment for age-related macular degeneration, serious ocular diseases and myopic choroidal neovascularization (Raquel et al., 2020). The dose of Verteporfin injected by intravenous routes of administration. The injected drug activated specifically at 690 nm wavelength and get eliminated within 5–6 h. When the penetrability of Visudyne was compared, it was 50 times more than Photofrin® at 690 nm. The treatment with Visudyne shows minor side effects including pain at site of injection, inflammation, oedema, discoloration of skin and hemorrhage, photopsia, blurred and fuzzy vision, visual impairments including scotoma and black spots. Verteporfin is clinically most potent in pancreatic cancer (Park et al., 2018).

7. **Tookad:** Tookad is a palladium bacteriopheophorbide (Negma Ler- ads/Steba Biotech) permitted for the treatment of prostate cancer and vascular targeted PDT which is selectively activated at the specific higher wavelength of light such as 762 nm. It acts by generating super oxide and hydroxyl radicals, and initiates apoptosis of tumor cells (Baskaran et al., 2018) and of prostate cancer cells (Bapat et al., 2021). To treat prostate cancer, the intravenous single dose of Tookad administered for 10 min and activated specifically with light of 762 nm wavelength. Tookad has fast absorption, penetration, and half-life near about 0.02–0.03 h. It gets accumulated in the tumor vascular cell causing tumor vessel

destruction, tissue phototoxicity, and sometimes death of the patient (Spanò et al., 2015). Hence such toxicity of drug can be reduced by reducing the dose and by increasing the rate of elimination of drug (Allison and Sibata, 2010).

Third generation: As we have discussed earlier, the limitations of second-generation such as poor water-solubility, poor elimination rate and PDT induced photobleaching born the third generation of PDT. The novel third-generation includes various drug conjugation with antibodies, gene engineering mediated PDT and use of nanotechnology in PDT (Josefsen and Boyle, 2008) and some potent photosensitizers from natural sources for the treatment of cancer (Allison and Moghissi, 2013).

8. **Alkaloid:** Alkaloids are nitrogen containing heterocyclic structure secondary metabolites of plants having photoactivity such as Cinchona (Quinine and Cinchonamine) (Muniyandi et al., 2020), Camptothecin (Denis et al., n.d.) and Vinblastine (Muniyandi et al., 2020), Berberine and Curcumin (Kou et al., 2017) reported to possess analgesic, antipyretics, antimalarial, and anticancer activity. The alkaloids mainly act as anticancer agents by arresting or disturbing tumor cell cycle at various steps such as G1 or G2/M phases, preventing spindle fibers or tubulin, promoting apoptosis by regulating Bcl-2, Bax, NF-κB, Bcl-xL and various caspase proteins, regulating cyclin-dependent kinase (CDK) or autophagy in tumor cells (Muniyandi et al., 2020). The combination of alkaloids with irradiation of light will synergize the therapeutic effects of drugs (Fuchs et al., 2000). The anticancer action of berberine and other main phytochemicals such as β-carboline and harmine in treatment of brain cancer cell is potentiated by UV and blue light irradiation at about 410 nm (Martins et al., 2021). They have significant photosensitization activity to produce ROS after light irradiation at about 410 nm (Muniyandi et al., 2020).
9. **Curcumin:** Curcumin and other chemical constituents are colored pigment obtained from rhizomes of *Curcuma longa* L. and other species were found to have photosensitizer property. Curcumin absorbs specific wavelength of light about 420 nm and shows photodynamic action (Ormond and Freeman, 2013). It acts as antibacterial, antimicrobial and disinfectant action via photodynamic activity (Atriwal et al., 2021). The curcumin shows its anticancer activity by ROS-generating and arresting or disturbing various transcription factors which includes AP-1, VEGF, NF-κB, iNOS, 5-LOX, COX-2, MMP-2, IL-8, and MMP-9 which are responsible for disturbance in tumor cells angiogenesis and growth (Eldridge et al., 2016; Goldman et al., 1993; Muniyandi et al., 2020).
10. **Furanocoumarins:** The furanocoumarin analogue psoralen and its derivatives such as 8-methoxypsoralen (8-MOP) and 5-methoxypsoralen (5-MOP) acts as photosensitizer and absorb specific wavelength such as 300–400 nm. The psoralen and its derivatives have been approved for the treatment of psoriasis, eczema, dermatitis and other skin problems (McCormick et al., 2016). Apart from this, various furanocoumarin analogues have been reported to treat skin cancer, breast cancer and leukaemia via inhibiting various signal transducer and activating transcription 3 (STAT3), phosphatidylinositol-3-kinase, nuclear factor-κB (NF-κB) and AKT protein expression (Urbach et al., 1976). Various studies reported that, furanocoumarin analogue with PDT have potent action toward STAT3 protein expression, cutaneous T-cell lymphoma and B16F10 murine melanoma (Muniyandi et al., 2020).
11. **Anthraquinones (AQ):** Near about 700 compounds of anthraquinone, naphthoquinones, and benzoquinones derivatives have found to possess photosensitizing activity which includes catenarin, physcion, emodin, and rhein (Muniyandi et al., 2020). The anthraquinone derivatives of *Melaleuca elliptica* such as soranjidiol, morindone and rubiadin were found to act against the lymphocytic leukaemia (P-388) (Eldridge et al., 2016). Whereas the anthraquinone derivatives of *Haliotis rugosa pustulata* extracted from its leaves and stem have found to possess photosensitizing activity and shows cytotoxic behavior by generating superoxide anion radicals or singlet oxygen. The anthraquinone derivatives such as soranjidiol and rubiadin absorb specific wavelength of light in between 380–480 nm (Rodriguez et al., 2016) and shows its antiproliferative effect on MCF-7 breast cancer cells (Montoya et al., 2005).
12. **Polyacetylene and thiophenes:** The polyacetylene and thiophenes with pyrimidine derivatives play an important role of photosensitizers by producing radicals and ROS by type 2 PDT reaction. The polyacetylenes absorb specific wavelength of light in between 314–350 nm and get activated or excited yields $^1O_2$ under irradiation and thiophenes can generate high photo yield to show their photosensitizing, cytotoxic, and potent biological activities, including antitumor, anti-inflammatory, analgesic, and antimicrobial activities (Rosenblum et al., 2018). The polyacetylene and thiophenes and its pyrimidines in combination with PDT found to possess potent anti-inflammatory, antimicrobial and antitumor activities whereas its thiazolopyrimidine and thioxopyrimidine derivatives with PDT have antitumor action against NCI-H460 (non-small cell lung cancer), MCF-7 (breast adenocarcinoma) and SF-268 (CNS cancer) cells (Ibrahim et al., 2016).
13. **Phthalocyanines:** Phthalocyanines (Pc) contains metal complex like aluminum phthalocyanine tetra sulfonate AlPcS4, Photosens-21 absorb specific wavelength of light at about 670–700 nm and exerts its photosensitizing activity by forming an intersystem crossing to meta complex and shows PDT properties (Yaghini et al., 2009). The Phthalocyanines derivative which is known as AlPcS4, as Photosens is approved in Russia for the treatment of skin, stomach, oral, lip

and breast carcinoma (dos Santos et al., 2017). However, it was found to have skin phototoxicity for several weeks after taking drug (Allison and Moghissi, 2013).

14. **Xanthenes and cyanines:** The Xanthenes and Cyanine derivative 4,5-Dibromorhodamine methyl ester (TH 9409, 27) shows its photosensitizing activity due to presence of the halogen atoms and can absorb light at specific wavelength of 514 nm. It exerts photosensitizing and cytotoxic activity by producing cross link with the triplet state of metal ion and yields singlet oxygen. In combination with PDT, it shows potent activity against graft-versus-host disease, lymphocytes carcinoma (Ormond and Freeman, 2013), leukemia, and neuroblastoma treatment (O'Connor et al., 2009).
15. **Pheophorbides:** Pheophorbide or 2-(1-hexyloxyethyl)-2-devinyl pyropheophorbide (HPPH/Photochlor) is undergoing clinical trial for the treatments of esophageal cancer (NCT00060268-Phase trial I/II) and basal cell skin cancer (NCT00017485-Phase I trial), lung cancer (NCT00528775-Phase-II). Furthermore, Phase I trials for treating dysplasia, carcinoma of the oral cavity, carcinoma of the oropharynx (NCT01140178), oesophageal cancer at precancerous or early-stage conditions (NCT00281736) and head and neck cancer (NCT00670397) and in phase I/II trials involving Barrett's oesophagus are also undergoing (NCT01236443) (Bellnier et al., 2006; Hamdous et al., 2017; Kessel, 2019).
16. **Ph II-loaded Pluronic micelles:** The Ph II in micelles based on Pluronics (P123 and F127) has potent activity in treatment of certain resistant types of cancers (Studies showed positive results of nano therapy in resistant and caspase-3 deficient human cancer cell model). The nano therapy used to carry and inject Photofrin II in the form of nanocarrier formulation (Ph II) and it helps to recover multidrug resistance (MDR) in breast MCF-7/WT (Caspase-3 deficient) and ovarian SKOV-3 (resistant to chemotherapy). The nanotherapy of polymeric micelles containing Photofrin II containing less than 20 nm particles of Photofrin II® easy to absorb and concentrate in tumor cells of ovarian SKOV-3 (resistant to chemotherapy) cells and breast MCF-7/WT (caspase-3 deficient). The nano therapy helps to improve photodynamic activity and efficacy of Ph II in the treatment of cancers resistant to several cytotoxic drugs (cisplatin, diphtheria toxin and adriamycin) and ovarian cancer (Lamch et al., 2014).
17. **CD44-targeted hyaluronan-decorated double coated nanoparticles (dcNPs) delivering the lipophilic chemotherapeutic Docetaxel (DTX) and an anionic porphyrine (TPPS4):** The novel type of CD44-targeted hyaluronan-decorated double coated nanoparticles (dcNPs) help in delivery of lipophilic chemotherapeutic nanoparticles of Docetaxel (DTX) and an anionic porphyrine in combination therapy in the treatment of cancers involving overexpression of CD44 receptor. They act by targeting the tumor cells, get absorb and concentrated and dismantled from the dcNPS surface and shows fluorescent and phototoxic activity in tumor cells (Kou et al., 2017).
18. **Photosensitizer-loaded gold vesicles with strong plasmonic coupling effect:** The novel GV-Ce6 improves the targeted drug delivery of Ce6 into the tumor cells and the therapeutic efficacy and selectivity of photosensitizer-Loaded Gold Vesicles with strong plasmonic coupling shows agonistic effect with PTT/PDT in the treatment of several cancers. The biological activity is improved with the use of vesicular nanocarrier, single wavelength continuous wave laser irradiation. These novel photosensitizer-loaded gold vesicles with strong plasmonic coupling drug qualify the certain ideal criteria of ideal photosensitizers, such as selectivity of drugs toward tumor cell. As the surface of GVs is coded with specific biomarkers to achieve targeting drug delivery, high aqueous solubility and stability, nano size of GVs (50–150 nm) improves the penetrability of drug, biocompatibility and use of biodegradable polymers. Hence it is free from skin phototoxicity and improve clinical applications in treatment of cancers, especially in the cancer theranostics (Yin et al., 2014).

Apart from the PAs capacity of photoactive reactions with the activation of light, the autofluorescence of the compound also plays an important role because it is responsible for imaging properties which enables recognition of pre-cancer lesions and early malignancies, tumor margins (Park et al., 2018; Xu et al., 2020; Yang and Wang, 2021). The autofluorescence of the compound assists in determining the residual dysplastic tissue upon surgical tumor resections and enhances the working of PDT, therefore combining the imaging, detection and therapeutic properties in a single compound which makes it a theranostic agent (Park et al., 2016; Son et al., 2019).

5.5 Application of photodynamic therapy in treatment of various cancers

The PDT finds its application for management of cancers since late 1970s, when five patients suffering from bladder cancer were treated with the combined effects of hematoporphyrin derivative (HPD) plus light (Plail et al., 1990). Subsequently, in 1978 a large group of patients were effectively treated with PDT using HPD, in which fractional or whole positive results were detected in 111 and 113 malignant lesions respectively (Berns, 1984). Later PDT was considered a valid option for treatment of malignant and premalignant tumors, hence the effectiveness of PDT in treatment of carcinomas varies from tumors to tumors and even depend on the site and microenvironment of tumors (Dougherty et al., 1982).

5.5.1 Skin tumors

The porfimer sodium and 5-aminolevulinic acid (ALA) and its derivatives are the two main compounds which find their application for treatment of premalignant and malignant skin tumors through PDT (Braathen et al., 2012). Currently PDT is approved by United States, Canada, and the European Union (EU) for treatment of skin tumors like actinic keratosis (AK) and basal cell carcinoma (BCC) (Huang et al., 2019). PDT has showed positive results for treatment of squamous cell carcinoma (SCC) in situ/ Bowen disease and also potentials for management of extra-mammary Paget disease up to an extent (de Haas et al., 2008; Garcia-Zuazaga et al., 2005). Conversely, the studies suggesting the use of PDT for treatments of skin cancers and tumors, highlights that topical PAs have been quite unsatisfactory and demonstrates higher recurrence rates, approximately to 50% (Kormeili et al., 2004).

5.5.2 Head and neck tumors

PDT has been efficaciously used as a tool for treatment of carcinomas in the various parts of neck and head including oral cavity, pharynx, and larynx (Jerjes et al., 2010), by preventing the damage to the healthy tissues with an aim of preserving the vital process of speech and swallowing (D'Cruz et al., 2004; Hashemipour et al., 2020; Shetler, 1996). A study comprising of randomized clinical trial, in which a comparative measurement was done for treatment of nasopharyngeal carcinoma, between PDT with porfimer sodium and chemotherapy (5-FU and cisplatin) and the results showed that there was a significant increment in the clinical output with PDT accounting to P = .001 and simultaneously enhanced the Karnofsky performance score (Luo et al., 2006; Shetler, 1996).

A study conducted with a large sample size of 300 patients over a long period of 15 years for management of Squamous Cell Carcinomas (SCCs) of the oral cavity, pharynx, or larynx, Kaposi sarcoma, melanoma, and SCC in the head and neck region, with porfimer sodium-mediated PDT. In this study 2.0 mg/kg of porfimer sodium was given 48 hours prior to irradiation with 630 nm of light since an neodymium yttrium aluminum garnet (Nd:YAG) pumped dye laser. The light rays were carried to the site between the range of 50 and 75 joules per square centimetre (J/cm^2) for oral cavity, nasopharyngeal, and skin lesions and 80 J/cm^2 for laryngeal tumors (Biel, 1998).

5.5.3 Digestive system tumors

The use of PDT in treatment of digestive system tumors is mainly bifurcated in two groups on the basis of anatomy of gastrointestinal tract (GIT) i.e. PDT of oesophagus and beyond (Siersema, 2005). The PDT in treatment of GIT cancers has shown better results in management of Barrett oesophagus, early oesophageal cancer and in some grades of dysplasia (Kenudson et al., 2007; Overholt et al., 2005). Moreover, PDT gave enhanced effect for premalignant conditions in Barrett oesophagus having high grade dysplasia, because they are superficially placed and have enlarged mucosal area, so light penetrates easily and up to a greater extent (Chennat et al., 2009; Foroulis and Thorpe, 2006).

A study involving total 102 patients suffering from Barrett oesophagus and high-grade dysplasia (69 patients) or mucosal adenocarcinoma (33 patients) were given treatment with porfimer sodium mediated PDT over a media follow up time of 1.6 years. After giving PDT, it was observed that glandular epithelium was completely ablated in 56% of patients with a single course of PDT, among which strictures demanding expansion seen in 20% of patients, which was accounted as the utmost common and severe side effect, with a failure of ablation in 4% patients. Hence PDT is considered as effective, safe and less invasive regimen for treatment of Barrett dysplasia and mucosal adenocarcinoma (Wolfsen, 2005).

The other uses of PDT includes treatment of tumors in digestive systems, unresectable pancreatic cancers, colon polyps, palliate bulky colon and rectal cancers, while the application of PDT for treating these cancers require strong clinical recommendations (Xingshu et al., 2020). Furthermore, PDT may also be used in combatting hepatocellular carcinoma (Shishkova et al., 2013), where early clinical trials have shown the positive ray of hope with high efficacy in treatment with talaporfin-mediated PDT using interstitial LEDs (Qumseya et al., 2013).

5.5.4 Urinary system tumors

5.5.4.1 Prostate cancer

Treatment of prostate cancer mainly relies on radiotherapy as there are less options available for overcoming isolated local failure. The first line treatment includes surgical removal of the tumors or either by ionizing radiations associated with significant morbidities due to damage caused to the normal and healthy tissues of nerves, bladder, and rectum (Muschter, 2003). Hence to overcome this intrinsic limitation, PDT plays an important role by site specific targeting to the carcinoma and causing negligible harm to healthy cells and tissues of organs. The use of techniques like interstitial brachytherapy linked with radioactive seeds which can possibly enhance the penetration and localization of light to complete prostate gland by implementing interstitial, cylindrically diffusing optical fibers (Huang and Zheng, 2005).

The underlying mechanism of killing the cells by PDT does not depend on disturbing the cell cycle, rather by damaging the DNA, unlike chemotherapy or radiotherapy, minimizing the possibility of therapy cross-resistance and reducing the chances of normal tissues becoming malignant in later stages. Hence all these factors contribute to develop an enhanced therapeutic regimen for prostate cancer. Several clinical trials have been performed for assessing the efficiency of PDT against prostate cancer using second generation PAs. Among these studies, one pilot study deals with application of temeporfin-mediated PDT, in which 14 patients were subjected to biopsy study confirming the localized failure after radiotherapy at early stage prostate cancer and further treated with 8 implanted, interstitial, cylindrically diffusing optical fibers (Nathan et al., 2002). Amongst the patients, 13 of them received the light of significantly higher dose of around 50 J/cm^2 and positive results relating to prostate-specific antigen therapy were seen in 9 patients and sufficient pathologic effects in five patients. One patient undergone development of urorectal fistula afterwards of rectal biopsy, four patients were having stress incontinence and remaining 4 patients were observed suffering from erectile dysfunction.

Conclusively, the use of temeporfin-mediated PDT as a first line therapy tool for management, 6 patients were organ confined with Gleason score of 6, treated with 4 to 8 interstitial fibers with implants designed in such a way to treat biopsy confirmed disease. Four patients experienced second sessions of PDT because of prevailing of disease even after first PDT at third month follow up. Hence this regimen was well tolerated by the patients under treatment and significantly induced necrosis however, 6 patients were having the prevalence of disease even after PDT as confirmed by biopsy reports.

5.5.4.2 Bladder cancer

The diagnosis of bladder cancers which are generally superficial and multifocal in nature are mainly done by debulked endoscopy and the symmetry of bladder enables improved and consistent delivery of light, therefore this characteristic of bladder cancer makes it ideal for treatment with PDT (Nseyo et al., 1998). Statistics shows early response rates ranging from 2 to 3 months towards PDT in 50 to 80% of patients and long term 1 to 2 years significant responses were seen in 20 to 60% of patients (Moghissi et al., 2015; Shackley et al., 2001). It was further noted that patients under study were suffering from recurrence of carcinomas developed after treatment with standard therapies like bacillus Calmette-Guérin (BCG) (Agostinis et al., 2011). Some studies reports the use of HPD-mediated PDT for treatment of bladder cancers, among which one investigation was done by means of focal HPD-mediated PDT for treating 50 superficial bladder transitional cell carcinomas (TCCs), among which 37 patients attained a 74% comprehensive response rate (Prout et al., 1987).

5.5.5 Brain tumors

Currently, several rigorous clinical trials are being undertaken for assessing PDT as an adjunctive treatment for brain tumors as the freshly diagnosed and recurrent tumor lesions showed an enhanced uptake of Photosensitizing Agents (PAs). Treatment of brain tumors with PDT finds its root since 1980s, in which around 1000 patients were treated worldwide with PDT. Muller & Wilson showed the early attempts of treating postresection glioma cavity in humans with PDT (Muller and Wilson, 1995) and simultaneously Schmidt et al. (2004), performed a phase 1/2 trial comprising of 23 patients suffering from glioblastoma multiforme (GBM) and anaplastic astrocytoma (AA) (Schmidt et al., 2004).

The other brain tumor lesions treated with PDT includes malignant ependymomas, malignant meningiomas, melanoma and lung cancer brain metastasis, (Muller and Wilson, 1996) and recurrent pituitary adenomas, the initial trials gave promising results for treatment. The PAs used till date are consisting of various formulations of HPDs (porfimer sodium) and ALA as well as temeporfin and use of lamps, dye lasers, gold vapor potassium titanyl phosphate dye lasers, and diode lasers for activation of PAs (Eljamel, 2003).

The PAs serves as intraoperative diagnostic tools by use of photodetection (PD) and fluorescence-guided resection (FGR) and simultaneously as an adjunctive therapeutic compound during PDT (Eljamel, 2003; Ritz et al., 2008). The malignant cells show high affinity towards uptake of PAs and the recent studies showed that PD, FGR, and PDT possess surplus advantage of delaying the tumor progression in initial phase of tumor growth (Eljamel, 2008; Liu et al., 2021). A study conducted by inclusion of 375 patients out of which 138 patients were recently diagnosed with GBM and 140 patients with recurrent GBM, the patients got 5 mg/kg of HPD 24 hours previous to surgery and the light dose was 70 to 260 J/cm^2 (Stylli et al., 2005). After follow-up it was reported that mean survival rate of both GBM was in the middle of 14.3 and 14.9 months and around 28% to 41% of patients endured more than 2 years. A randomized controlled trial using adjuvant porfimer sodium-mediated PDT for treatment of supratentorial gliomas in which 96 patients were divided in two randomized groups and received PDT with porfimer sodium either 40 J/cm^2 or 120 J/cm^2. Around 48 patients received higher doses and showed the average survival of 10 months, similarly 49 patients on low dose survived approximately for 9 months; the variance among both groups was not statistically significant (P > .05) (Muller and Wilson, 2006).

5.5.6 Nonsmall cell lung cancer and mesothelioma

In the year 1982, PDT was first time employed for management of non-small cell lung cancer (NSCLC) with an aim of achieving tumor necrosis and reopening of the airway (Okunaka and Kato, 1995). PDT in combating lung cancer focuses on patients with disease in progressive state where PDT acts only as palliative care option and secondly patients with early central lung cancer where surgery is not possible (Alexiades-Armenakas, 2006; Ninane, 1995; Venuta et al., 1998). PDT in lung cancer is measured to have more specificity and lesion-oriented and comparatively produces lower collateral damage and accompanied with lesser complications. A randomized trial of PDT in contrasted with neodymium yttrium aluminum garnet (Nd:YAG) laser therapy for obstructing NSCLC lesions (Shepherd and Radchenko, 2019; Valvani et al., 2019), gave similar initial efficacy for both of the treatments with an extended period of response eminent for PDT and PDT liked to palliative radiation enhances the time to bronchus reclusion (Ma et al., 2018; Mohan et al., 2018).

The second-generation PAs have also been employed in treatment of early-stage lung cancers, a series of 70 different cancer lesions were evaluated, symmetrically measuring in size around 1.0 cm or less in diameter and 21 lesions measuring greater than 1.0 cm in diameter for treatment with PDT and talaporfin. The outcome was approximately 94.3% (66 of 70 patients) and 90.4% (19 of 21 patients), respectively, mechanistically PDT employed with talaporfin causes destruction of residual cancer lesions after electrocautery (Spikes et al., 1993). A study on 133 patients out of which 89 individuals were suffering from NSCLC, 31 from metastatic airway lesions, 4 from small cell lung cancer, 7 from benign tumors, and 2 patients with unspecified lung conditions having a common lesion position at the main stem bronchi. The treatment regimen included two treatments with PDT during a span of 3-days of hospitalization and further two additional PDTs were received by patients after 2 weeks, improving dyspnea in some patients simultaneously (Huang and Zheng, 2005).

5.6 Recent developments, future scope, and challenges

PDT has made a remarkable progress in last few decades because it enables high efficiency and feasibility for management of cancer. The inclusion of nanotechnology in PDT has broadened the opportunities and enabled multifaceted openings for treatment of cancer and possibly eliminating the drawbacks of conventional PDT paradigms (Ferroni et al., 2019). The research in field on cancer mainly focuses on developing strategies which significantly provides non-invasive tumor treatment (Huang and Zheng, 2005). The newly developed photosensitizers are mainly includes second and third generation photosensitizers, among which some of PAs of third generation are still under clinical investigations (Uramoto et al., 2014).

The recently introduced PAs are bound with sugars which enhance the solubility and increases the site specificity by enabling them to bind to sugar specific receptors like mannose receptors, which are generally overexpressed at tumor sites and this binding possibly reduce the toxicity (Ismael et al., 2020). To improve the penetration of light in the tumors and carcinomas technique called two photon PDT is employed which uses longer wavelength and generate singlet oxygen particularly in incident of C60 (Huang and Zheng, 2005). The efficiently developed PAs are liked with endoscopic devices and similar techniques which assists in combatting tumors also serves as diagnostic tool.

The substantial development in arena of PDT can be tracked from the various modifications made in the vital components of PDT including source of light, photosensitizing agents and enhancing the specificity and site targeting of the PDT (Cramer et al., 2020). Since long time the penetration of light deep in affected tissues and cells remains a matter of concern and numerous studies suggested that visible light is not considered as ideal for cancer therapy. Similarly, near-infrared (NIR) light restricts the penetration of light to tissues up to a depth of 2 cm only and do not adequately activates the PAs (Jing et al., 2020). Secondly, used source of light is X-ray, which is capable of deep spear in tissues but exhibiting low ROS efficiency and possibly cause damage to healthy tissues. Hence the advancements deal with exploration of light capable of producing effect to deeper tissues and higher photoexcitation, therefore recently PA activators comprises microwaves, radio-waves, ultrasound, Electrical Fields, Magnetic Fields, and electromagnetic fields (EMF) to induce PDT (Sharif et al., 2021).

The newly introduced sources produce a local hyperthermia which leads to tumor sensitization and causing higher blood supply to tumor and increasing ROS production and disrupting the cell walls and enhancing the penetration in cancerous cells and minimal damage to healthy cells and tissues. The numerous progressions made in PDT by application nanomaterials enables as both diagnostic and therapeutic use, combination of PAs and different types of nanomaterials like nanoemulsions, metallic nanoparticles enables high specific targeting and enhanced ROS generation efficiency (Pucelik et al., 2020). These combinations are generally made with an aim to increase antitumor effect by sensitizing the tumor cells towards PDT and by interfering cytoprotective molecular responses triggered by PDT in surviving tumor orstromal cells, this interaction between PDT and PAs in confirmed by illuminated area. PDT can also serve as neoadjuvant, adjuvant, or repetitive adjuvant treatment in surgical procedures by fluorescence image-guided to confine illumination to the most suspicious lesions (McFarland et al., 2020; Shah et al., 2020). Various combinations of PDT and some ongoing clincial trials in feild of photodyamic therapy for cancer treatment are well illustrated in Tables 5.4 and 5.5 respectively.

5.7 Conclusion

Photodynamic therapy (PDT) has been proved to be a promising technique for managing various cancers and malignancies with less long-term adverse effects and morbidity when compared to conventional chemotherapy or radiotherapy regimens. PDT has emerged as an encouraging tool for management of various cancers possessing numerous advantages over chemotherapeutics and radiotherapeutic agents which significantly reduce the immunity. However, PDT-induced immunogenic cell death causing an effective local inflammatory reaction and simultaneously producing an excellent anticancer response and ultimately leading higher destruction of metastases. The interdisciplinary exclusivity of PDT encourages experts in physics, chemistry, biology, and medicine with its auxiliary expansion and unique applications to work in collaborations of each other to flourish it more.

Acknowledgment

First author (SST) acknowledges Government of India, Ministry of Science and Technology, Department of Science and Technology (DST), India for financial support through the DST-INSPIRE Fellowship (IF 190486).

Statement of Informed Consent: All authors give consent for publication.

Conflict of Interest: The authors declare no conflict of interest, financial or otherwise.

References

Ackroyd, R., Kelty, C., Brown, N., Reed, M., 2007. The History of photodetection and photodynamic therapy. Photochem. Photobiol. 74, 656–669. https://doi.org/10.1562/0031-8655(2001)0740656THOPAP2.0.CO2.

Agostinis, P., Berg, K., Cengel, K.A., Foster, T.H., Girotti, A.W., Gollnick, S.O., Hahn, S.M., Hamblin, M.R., Juzeniene, A., Kessel, D., Korbelik, M., Moan, J., Mroz, P., Nowis, D., Piette, J., Wilson, B.C., Golab, J., 2011. Photodynamic therapy of cancer: an update.. Cancer J. Clin. 61, 250–281. https://doi.org/10.3322/CAAC.20114.

Aksakal, N.E., Eçik, E.T., Kazan, H.H., Yenilmez Çiftçi, G., Yuksel, F., 2019. Novel ruthenium(ii) and iridium(iii) BODIPY dyes: insights into their application in photodynamic therapy: In vitro. Photochem. Photobiol. Sci. 18, 2012–2022. https://doi.org/10.1039/C9PP00201D.

Akita, Y., Kozaki, K., Nakagawa, A., Saito, T., Ito, S., Tamada, Y., Fujiwara, S., Nishikawa, N., Uchida, K., Yoshikawa, K., Noguchi, T., Miyaishi, O., Shimozato, K., Saga, S., Matsumoto, Y., 2004. Cyclooxygenase-2 is a possible target of treatment approach in conjunction with photodynamic therapy for various disorders in skin and oral cavity. Br. J. Dermatol. 151 (2), 472–480. https://doi.org/10.1111/J.1365-2133.2004.06053.X.

Alexiades-Armenakas, M., 2006. Long-pulsed dye laser-mediated photodynamic therapy combined with topical therapy for mild to severe comedonal, inflammatory, or cystic acne. J. Drugs Dermatol. 5, 45–55.

Allison, R.R., Moghissi, K., 2013. Photodynamic therapy (PDT): PDT mechanisms. Clin. Endosc. 46 (1), 24–29. https://doi.org/10.5946/CE.2013.46.1.24.

Allison, R.R., Sibata, C.H., 2010. Oncologic photodynamic therapy photosensitizers: a clinical review. Photodiagnosis Photodyn. Ther. 7, 61–75. https://doi.org/10.1016/j.pdpdt.2010.02.001.

Angelis, C.De, 2008. Side effects related to systemic cancer treatment: are we changing the promethean experience with molecularly targeted therapies? Curr. Oncol. 15, 198. https://doi.org/10.3747/CO.V15I4.362.

Aniogo, E.C., George, B.P.A., Abrahamse, H., 2020. Role of Bcl-2 family proteins in photodynamic therapy mediated cell survival and regulation. Molecules 25 (22), 5308. https://doi.org/10.3390/MOLECULES25225308.

Ascencio, M., Collinet, P., Farine, M.O., Mordon, S., 2008. Protoporphyrin IX fluorescence photobleaching is a useful tool to predict the response of rat ovarian cancer following hexaminolevulinate photodynamic therapy. Lasers Surg. Med. 40, 332–341. https://doi.org/10.1002/lsm.20629.

Atriwal, T., Azeem, K., Husain, F.M., Hussain, A., Khan, M.N., Alajmi, M.F., Abid, M., 2021. Mechanistic understanding of candida albicans biofilm formation and approaches for its inhibition. Front. Microbiol. 12. https://doi.org/10.3389/FMICB.2021.638609/FULL.

Autiero, M., Cozzolino, R., Laccetti, P., Marotta, M., Quarto, M., Riccio, P., Roberti, G., 2010. In vivo tumor detection in small animals by hematoporphyrin-mediated fluorescence imaging. Photomed. Laser Surg. 28. https://doi.org/10.1089/PHO.2009.2567.

Bapat, P., Singh, G., Nobile, C.J., 2021. Visible lights combined with photosensitizing compounds are effective against *Candida albicans* biofilms. Microorg 9 (3), 500. https://doi.org/10.3390/MICROORGANISMS9030500.

Baskaran, R., Lee, J., Yang, S.G., 2018. Clinical development of photodynamic agents and therapeutic applications. Biomater. Res. 22, 1–8. https://doi.org/10.1186/s40824-018-0140-z.

Bayona, A.M.D.P., Mroz, P., Thunshelle, C., Hamblin, M.R., 2017. Design features for optimization of tetrapyrrole macrocycles as antimicrobial and anticancer photosensitizers. Chem. Biol. Drug Des. 89, 192. https://doi.org/10.1111/CBDD.12792.

Bellnier, D.A., Greco, W.R., Loewen, G.M., Nava, H., Oseroff, A.B., Dougherty, T.J., 2006. Clinical pharmacokinetics of the PDT photosensitizers porfimer sodium (Photofrin), 2-[1-hexyloxyethyl]-2-devinyl pyropheophorbide-a (Photochlor) and 5-ALA-induced protoporphyrin IX. Lasers Surg. Med. 38, 439–444. https://doi.org/10.1002/LSM.20340.

Berns, M.W., 1984. Hematoporphyrin derivative photoradiation therapy. Lasers Surg. Med. 4, 1–4. https://doi.org/10.1002/LSM.1900040102/PDF.

Bhuvaneswari, R., Gan, Y.Y., Soo, K.C., Olivo, M., 2009. Targeting EGFR with photodynamic therapy in combination with Erbitux enhances in vivo bladder tumor response. Mol. Cancer 8, 94. https://doi.org/10.1186/1476-4598-8-94.

Bhuvaneswari, R., Yuen, G.Y., Chee, S.K., Olivo, 2007. Hypericin-mediated photodynamic therapy in combination with Avastin (bevacizumab) improves tumor response by downregulating angiogenic proteins. Photochem. Photobiol. Sci. 6 (12), 1275–1283. https://doi.org/10.1039/B705763F.

Biel, M.A., 1998. Photodynamic therapy and the treatment of head and neck neoplasia. Laryngoscope 108, 1259–1268. https://doi.org/10.1097/00005537-199809000-00001.

Bowcock, A.M., Fernandez-Vina, M., 2012. Targeting skin: vitiligo and autoimmunity. J. Invest. Dermatol. 132, 13–15. https://doi.org/10.1038/JID.2011.353.

Bown, S.G., 2013. Photodynamic therapy for photochemists. Philos. Trans. A Math. Phys. Eng. Sci. 17 (1995), 20120371, 371. doi:10.1098/rsta.2012.0371.

Braathen, L.R., Morton, C.A., Basset-Seguin, N., Bissonnette, R., Gerritsen, M.J.P., Gilaberte, Y., Calzavara-Pinton, P., Sidoroff, A., Wulf, H.C., Szeimies, R.M., 2012. Photodynamic therapy for skin field cancerization: An international consensus. International Society for Photodynamic Therapy in Dermatology. J. Eur. Acad. Dermatology Venereol. 26, 1063–1066. https://doi.org/10.1111/J.1468-3083.2011.04432.X/PDF.

Breskey, J.D., Lacey, S.E., Vesper, B.J., Paradise, W.A., Radosevich, J.A., Colvard, M.D., 2013. Photodynamic therapy: occupational hazards and preventative recommendations for clinical administration by healthcare providers. Photomed. Laser Surg. 31, 398–407. https://doi.org/10.1089/PHO.2013.3496.

Brillo, V., Chieregato, L., Leanza, L., Muccioli, S., Costa, R., 2021a. Mitochondrial dynamics, ros, and cell signaling: a blended overview. Life (Basel, Switzerland) 11. https://doi.org/10.3390/LIFE11040332.

Brown, S.B., Kesselt, D., 1990. The analysis of hematoporphyrin derivative cancer and porphyrinsm photochemotherapy. Molec. Aspects Med. 11, 99–111. http://doi.org/10.1007/978-1-4684-4406-3.

Bullous, A.J., Alonso, C.M.A., Boyle, R.W., 2011. Photosensitiser-antibody conjugates for photodynamic therapy. Photochem. Photobiol. Sci. 10, 721–750. https://doi.org/10.1039/C0PP00266F.

Castano, A.P., Demidova, T.N., Hamblin, M.R., 2005. Mechanisms in photodynamic therapy: part three – photosensitizer pharmacokinetics, biodistribution, tumor localization and modes of tumor destruction. Photodiagnosis Photodyn. Ther. 2, 91–106. https://doi.org/10.1016/S1572-1000(05)00060-8.

Castano, A.P., Demidova, T.N., Hamblin, M.R., 2004. Mechanisms in photodynamic therapy: part one—photosensitizers, photochemistry and cellular localization. Photodiagnosis Photodyn. Ther. 1 (4), 279–293. https://doi.org/10.1016/S1572-1000(05)00007-4.

Canti, Nicolin, Cubeddu, Taroni, Bandieramonte, Valentini, 1998. Antitumor efficacy of the combination of photodynamic therapy and chemotherapy in murine tumors. Cancer Lett. 125 (1–2), 39–44. https://doi.org/10.1016/S0304-3835(97)00502-8.

Chatterjee, D.K., Fong, L.S., Zhang, Y., 2008. Nanoparticles in photodynamic therapy: an emerging paradigm. Adv. Drug Deliv. Rev. 60, 1627–1637. https://doi.org/10.1016/J.ADDR.2008.08.003.

Chen, B., Pogue, B., Hasan, T., 2005. Liposomal delivery of photosensitising agents. Exp. Opin. Drug Del. 2, 477–487. https://doi.org/10.1517/17425247.2.3.477.

Chen, J., Lovell, J.F., Lo, P.C., Stefflova, K., Niedre, M., Wilson, B.C., Zheng, G., 2008. A tumor mRNA-triggered photodynamic molecular beacon based on oligonucleotide hairpin control of singlet oxygen production. Photochem. Photobiol. Sci. 7, 775–781. https://doi.org/10.1039/B800653A.

Cheng, X., Gao, J., Ding, Y., Lu, Y., Wei, Q., Cui, D., Fan, J., Li, X., Zhu, E., Lu, Y., Wu, Q., Li, L., Huang, W., 2021. Multi-functional liposome: a powerful theranostic nano-platform enhancing photodynamic therapy. Adv. Sci (Weinh) 8 (16), e2100876. doi:10.1002/advs.202100876. PMID: 34085415; PMCID: PMC8373168.

Chennat, J., Konda, V.J.A., Ross, A.S., De Tejada, A.H., Noffsinger, A., Hart, J., Lin, S., Ferguson, M.K., Posner, M.C., Waxman, I., 2009. Complete Barrett's eradication endoscopic mucosal resection: an effective treatment modality for high-grade dysplasia and intramucosal carcinoma - an American single-center experience. Am. J. Gastroenterol. 104, 2684–2692. https://doi.org/10.1038/AJG.2009.465.

Chennoufi, R., Trinh, N.D., Simon, F., Bordeau, G., Naud-Martin, D., Moussaron, A., Cinquin, B., Bougherara, H., Rambaud, B., Tauc, P., Frochot, C., Teulade-Fichou, M.P., Mahuteau-Betzer, F., Deprez, E., 2020. Interplay between cellular uptake, intracellular localization and the cell death mechanism in triphenylamine-mediated photoinduced cell death. Sci. Rep. 10, 6881. https://doi.org/10.1038/S41598-020-63991-9.

Comini, L., Núñez, M., Martín, Cabrera, J., Argüello, G., 2007. Characterizing some photophysical, photochemical and photobiological properties of photosensitizing anthraquinones. J. Photochem. Photobiol. A-chem 188, 185–191. doi:10.1016/j.jphotochem.2006.12.011.

Cramer, G.M., Moon, E.K., Cengel, K.A., Busch, T.M., 2020. Photodynamic therapy and immune checkpoint blockade. Photochem. Photobiol. 96, 954–961. https://doi.org/10.1111/PHP.13300.

D'Cruz, A.K., Robinson, M.H., Biel, M.A., 2004. mTHPC-mediated photodynamic therapy in patients with advanced, incurable head and neck cancer: A multicenter study of 128 patients. Head Neck 26, 232–240. https://doi.org/10.1002/HED.10372.

Dabrowski, J.M., Urbanska, K., Arnaut, L.G., Pereira, M.M., Abreu, A.R., Simões, S., Stochel, G., 2011. Biodistribution and photodynamic efficacy of a water-soluble, stable, halogenated bacteriochlorin against melanoma. ChemMedChem 6, 465–475. https://doi.org/10.1002/CMDC.201000524.

Dai, J.M., Fuchs, B.B., Coleman, J.J., Prates, R.A., Astrakas, C., St Denis, T.G., Ribeiro, M.S., Mylonakis, E., Hamblin, M.R., Tegos, G.P., 2012. Concepts and principles of photodynamic therapy as an alternative antifungal discovery platform. Front. Microbiol. 1664-302X 3, 120. doi:10.3389/fmicb.2012.00120. 22514547.

Daniell, Hill, 1991. A history of photodynamic therapy. Aust. N. Z. J. Surg. 61, 340–348. https://doi.org/10.1111/J.1445-2197.1991.TB00230.X.

Daub, M.E., Herrero, S., Chung, K.R., Nov 15 2005. Photoactivated perylenequinone toxins in fungal pathogenesis of plants. FEMS Microbiol. Lett. 252 (2), 197–206. doi:10.1016/j.femsle.2005.08.033. Epub 2005 Sep 6. PMID: 16165316.

de Haas, E.R.M., de Vijlder, H.C., Sterenborg, H.J.C.M., Neumann, H.A.M., Robinson, D.J., 2008. Fractionated aminolevulinic acid-photodynamic therapy provides additional evidence for the use of PDT for non-melanoma skin cancer. J. Eur. Acad. Dermatology Venereol. 22, 426–430. https://doi.org/10.1111/J.1468-3083.2007.02445.X.

Denis, Tyler, Huang, Hamblin, R.M., n.d. Cyclic tetrapyrroles in photodynamic therapy: the chemistry of porphyrins and related compounds in medicine.

Denis, Tyler, Huang, Ying-Ying, Hamblin, Michael, 2013. Cyclic tetrapyrroles in photodynamic therapy: the chemistry of porphyrins and related compounds in medicine. 255–301. https://doi.org/10.1142/9789814407755_0016

Dib, S., Aggad, D., Mauriello Jimenez, C., Lakrafi, A., Hery, G., Nguyen, C., Durand, D., Morère, A., El Cheikh, K., Sol, V., Chaleix, V., Dominguez Gil, S., Bouchmella, K., Raehm, L., Durand, J.O., Boufatit, M., Cattoën, X., Wong Chi Man, M., Bettache, N., Gary-Bobo, M., 2019. Porphyrin-based bridged silsesquioxane nanoparticles for targeted two-photon photodynamic therapy of zebrafish xenografted with human tumor. Cancer Rep 2. https://doi.org/10.1002/CNR2.1186.

Dong, Z., Feng, L., Chao, Y., Hao, Y., Chen, M., Gong, F., Han, X., Zhang, R., Cheng, L., Liu, Z., 2019. Amplification of tumor oxidative stresses with liposomal fenton catalyst and glutathione inhibitor for enhanced cancer chemotherapy and radiotherapy. Nano Lett 19, 805–815. https://doi.org/10.1021/ACS.NANOLETT.8B03905.

dos Santos, A.F., Terra, L.F., Wailemann, R.A.M., Oliveira, T.C., de Morais Gomes, V., Mineiro, M.F., Meotti, F.C., Bruni-Cardoso, A., Baptista, M.S., Labriola, L., 2017. Methylene blue photodynamic therapy induces selective and massive cell death in human breast cancer cells. BMC Cancer 17. https://doi.org/10.1186/S12885-017-3179-7.

Dougherty, T.J., Boyle, D.G., Weishaupt, K.R., 1982. Photoradiation therapy of human tumors. Sci. Photomed. 625–638. https://doi.org/10.1007/978-1-4684-8312-3_23.

Dougherty, T.J., Gomer, C.J., Henderson, B.W., Jori, G., Kessel, D., Korbelik, M., Moan, J., Peng, Q., 1998. Photodynamic therapy. J. Natl. Cancer Inst. 90, 889. https://doi.org/10.1093/JNCI/90.12.889.

Downes, A., Blunt, T.P., 1877. The influence of light upon the development of bacteria. Natur 16, 218. https://doi.org/10.1038/016218A0.

Eldridge, B.N., Bernish, B.W., Fahrenholtz, C.D., Singh, R., 2016. Photothermal therapy of glioblastoma multiforme using multiwalled carbon nanotubes optimized for diffusion in extracellular space. ACS Biomater. Sci. Eng. 2, 963–976. https://doi.org/10.1021/acsbiomaterials.6b00052.

Eljamel, M.S., 2008. Brain photodiagnosis (PD), fluorescence guided resection (FGR) and photodynamic therapy (PDT): past, present and future. Photodiagnosis Photodyn. Ther. 5, 29–35. https://doi.org/10.1016/J.PDPDT.2008.01.006.

Eljamel, M.S., 2003. New light on the brain: The role of photosensitizing agents and laser light in the management of invasive intracranial tumors. Technol. Cancer Res. Treat. 2, 303–309. https://doi.org/10.1177/153303460300200404.

Fang, T., Xiao, J., Zhang, Y., Hu, H., Zhu, Y., Cheng, Y., 2021. Combined with interventional therapy, immunotherapy can create a new outlook for tumor treatment. Quant. Imaging Med. Surg. 11, 2837–2860. https://doi.org/10.21037/QIMS-20-173.

Ferreira, G.C., Kadish, K.M., Smith, K.M., Guilard, R., 2013. (Volume 27). Handbook of Porphyrin Science, 30. https://doi.org/10.1142/8504-VOL27.

Ferroni, C., Del Rio, A., Martini, C., Manoni, E., Varchi, G., 2019. Light-induced therapies for prostate cancer treatment. Front. Chem. 7. https://doi.org/10.3389/FCHEM.2019.00719/FULL.

Foroulis, C.N., Thorpe, J.A.C., 2006. Photodynamic therapy (PDT) in Barrett's esophagus with dysplasia or early cancer. Eur. J. Cardio-thoracic Surg. 29, 30–34. https://doi.org/10.1016/J.EJCTS.2005.10.033.

Frank, Flaccus, Schwarz, Lambert, Biesalski, 2006. Ascorbic acid suppresses cell death in rat DS-sarcoma cancer cells induced by 5-aminolevulinic acid-based photodynamic therapy. Free Radical Biol. Med. 40 (5), 827–836. https://doi.org/10.1016/J.FREERADBIOMED.2005.10.034.

Fuchs, J., Weber, S., Kaufmann, R., 2000. Genotoxic potential of porphyrin type photosensitizers with particular emphasis on 5-aminolevulinic acid: Implications for clinical photodynamic therapy. Free Radic. Biol. Med. 28, 537–548. https://doi.org/10.1016/S0891-5849(99)00255-5.

Garcia-Zuazaga, J., Cooper, K.D., Baron, E.D., 2005. Photodynamic therapy in dermatology: current concepts in the treatment of skin cancer. Expert Rev. Anticancer Ther. 5, 791–800. https://doi.org/10.1586/14737140.5.5.791.

Ghannoum, M.A., Hossain, M.A., Long, L., Mohamed, S., Reyes, G., Mukherjee, P.K., 2004. Evaluation of antifungal efficacy in an optimized animal model of Trichophyton mentagrophytes-dermatophytosis. J. Chemother. 16, 139–144. https://doi.org/10.1179/JOC.2004.16.2.139.

Goldman, C.K., Kim, J., Wong, W.-L., King, V., Brock4, T., Gillespie, G.Y., 1993. Epidermal growth factor stimulates vascular endothelial growth factor production by human malignant glioma cells: a model of glioblastoma multiforme pathophysiology. Mol. Biol. Cell 4 (1), 121–133.

Gomer, C.J., 1991. Preclinical examination of first and second generation photosensitizers used in photodynamic therapy. Photochem. Photobiol. 54, 1093–1107. https://doi.org/10.1111/J.1751-1097.1991.TB02133.X/PDF.

Grant, C., H., A.J., M., P.M., S., S.G., B., 1993. Photodynamic therapy of oral cancer: photosensitisation with systemic aminolaevulinic acid. Lancet 342, 147–148. https://doi.org/10.1016/0140-6736(93)91347-O.

Gunaydin, G., Gedik, M.E., Ayan, S., 2021. Photodynamic therapy—current limitations and novel approaches. Front. Chem. 9, 691697. https://doi.org/10.3389/FCHEM.2021.691697.

Hambsch, P., Istomin, Y.P., Tzerkovsky, D.A., Patties, I., Neuhaus, J., Kortmann, R.-D., Schastak, S., Glasow, A., 2017. Efficient cell death induction in human glioblastoma cells by photodynamic treatment with tetrahydroporphyrin-tetratosylat (THPTS) and ionizing irradiation. Oncotarget 8, 72411–72423. https://doi.org/10.18632/ONCOTARGET.20403.

Hamdous, Y., Chebbi, I., Mandawala, C., Le Fèvre, R., Guyot, F., Seksek, O., Alphandéry, E., 2017. Biocompatible coated magnetosome minerals with various organization and cellular interaction properties induce cytotoxicity towards RG-2 and GL-261 glioma cells in the presence of an alternating magnetic field. J. Nanobiotechnology 15. https://doi.org/10.1186/s12951-017-0293-2.

Hang, Suthakar, Xiaolei, Yuan, Z.M., Ha, C.S., 2015. P53-based strategy for protection of bone marrow from Y-90 ibritumomab tiuxetan. Int. J. Radiat. Oncol. Biol. Phys. 92, 1116–1122.

Hashemipour, M., Pooyafard, A., Navabi, N., Kakoie, S., Rahbanian, N., 2020. Quality of life in Iranian patients with head-and-neck cancer. J. Educ. Health Promot. 9. https://doi.org/10.4103/JEHP.JEHP_508_20.

Henry, D.H., Viswanathan, H.N., Elkin, E.P., Traina, S., Wade, S., Cella, D., 2008. Symptoms and treatment burden associated with cancer treatment: Results from a cross-sectional national survey in the U.S. Support. Care Cancer 16, 791–801. https://doi.org/10.1007/S00520-007-0380-2.

Huang, A., Nguyen, J.K., Austin, E., Mamalis, A., Jagdeo, J., 2019. Updates on treatment approaches for cutaneous field cancerization. Curr. Dermatol. Rep. 8, 122–132. https://doi.org/10.1007/S13671-019-00265-2.

Huang, Zheng, 2005. A review of progress in clinical photodynamic therapy. Technol. Cancer Res. Treat. 4, 283–293. https://doi.org/10.1177/153303460500400308.

Ibrahim, S.R.M., Abdallah, H.M., El-Halawany, A.M., Mohamed, G.A., 2016. Naturally occurring thiophenes: isolation, purification, structural elucidation, and evaluation of bioactivities. Phytochem. Rev. 15, 197–220. https://doi.org/10.1007/S11101-015-9403-7.

Ichinose, S., Usuda, J., Hirata, T., Inoue, T., Ohtani, K., Maehara, S., Kubota, M., Imai, K., Tsunoda, Y., Kuroiwa, Y., Yamada, K., Tsutsui, H., Furukawa, K., Okunaka, T., Oleinick, N.L., Kato, H., 2006. Lysosomal cathepsin initiates apoptosis, which is regulated by photodamage to Bcl-2 at mitochondria in photodynamic therapy using a novel photosensitizer, ATX-s10 (Na). Int. J. Oncol. 29, 349–355.

Ismael, F.S., Amasha, H., Bachir, W., 2020. Optimized cylindrical diffuser powers for interstitial PDT breast cancer treatment planning: a simulation study. Biomed Res. Int. 2020. https://doi.org/10.1155/2020/2061509.

Jarrett, P., Scragg, R., 2017. A short history of phototherapy, vitamin D and skin disease. Photochem. Photobiol. Sci. 16, 283–290. https://doi.org/10.1039/C6PP00406G.

Jerjes, W., Upile, T., Akram, S., Hopper, C., 2010. The surgical palliation of advanced head and neck cancer using photodynamic therapy. Clin. Oncol. (R. Coll. Radiol). 22 (9), 785–791. https://doi.org/10.1016/J.CLON.2010.07.001.

Jing, Jing, Lei, Q., Zhang, X.Z., 2020. Recent advances in photonanomedicines for enhanced cancer photodynamic therapy. Prog. Mater. Sci. 114, 100685. https://doi.org/10.1016/j.pmatsci.2020.100685.

Josefsen, L.B., Boyle, R.W., 2008. Photodynamic therapy: novel third-generation photosensitizers one step closer? Br. J. Pharmacol. 154, 1–3. https://doi.org/10.1038/bjp.2008.98.

Karges, J., Chao, H., Gasser, G., 2020. Critical discussion of the applications of metal complexes for 2-photon photodynamic therapy. J. Biol. Inorg. Chem. 25, 1035–1050. https://doi.org/10.1007/S00775-020-01829-5.

Karthikeyan, K., Babu, A., Kim, S.J., Murugesan, R., Jeyasubramanian, K., 2011. Enhanced photodynamic efficacy and efficient delivery of Rose Bengal using nanostructured poly(amidoamine) dendrimers: potential application in photodynamic therapy of cancer. Cancer Nanotechnol. 2, 95–103. https://doi.org/10.1007/S12645-011-0019-3.

Kataoka, H., Nishie, H., Hayashi, N., Tanaka, M., Nomoto, A., Yano, S., Joh, T., 2017. New photodynamic therapy with next-generation photosensitizers. Ann. Transl. Med. 5, 183. https://doi.org/10.21037/ATM.2017.03.59.

Kennedy, J.C., Pottier, R.H., 1994. Using δ-aminolevulinic acid in cancer therapy 291–302. https://doi.org/10.1021/BK-1994-0559.CH021

Kenudson, M., Ban, S., Ohana, M., Puricelli, W., Deshpande, V., Shimizu, M., Nishioka, N.S., Lauwers, G.Y., 2007. Buried dysplasia and early adenocarcinoma arising in Barrett esophagus after porfimer-photodynamic therapy. Am. J. Surg. Pathol. 31, 403–409. https://doi.org/10.1097/01.PAS.0000213407.03064.37.

Kessel, D., 2019. Photodynamic therapy: a brief history. J. Clin. Med. 8, 1581. https://doi.org/10.3390/jcm8101581.

Khdair, A., Chen, D., Patil, Y., Ma, L., Dou, Q.P., Shekhar, M.P., Panyam, J., 2010. Nanoparticle-mediated combination chemotherapy and photodynamic therapy overcomes tumor drug resistance. J. Control. Release 141 (2), 137–144. https://doi.org/10.1016/j.jconrel.2009.09.004.

Kim, S., Ohulchanskyy, T.Y., Pudavar, H.E., Pandey, R.K., Prasad, P.N., 2007. Organically modified silica nanoparticles co-encapsulating photosensitizing drug and aggregation-enhanced two-photon absorbing fluorescent dye aggregates for two-photon photodynamic therapy. J. Am. Chem. Soc. 129, 2669–2675. https://doi.org/10.1021/JA0680257.

Korbelik, M., Bánáth, J., Zhang, W., Hode, T., Lam, S.S.K., Gallagher, P., Zhao, J., Zeng, H., Chen, W.R., 2020. N-dihydrogalactochitosan-supported tumor control by photothermal therapy and photothermal therapy-generated vaccine. J. Photochem. Photobiol. B Biol. 204.

Korbelik, M., Sun, J., 2001. Cancer treatment by photodynamic therapy combined with adoptive immunotherapy using genetically altered natural killer cell line. Int. J. Cancer 93, 269–274. https://doi.org/10.1002/IJC.1326.

Kormeili, T., Yamauchi, P.S., Lowe, N.J., 2004. Topical photodynamic therapy in clinical dermatology. Br. J. Dermatol. 150, 1061–1069. https://doi.org/10.1111/J.1365-2133.2004.05940.X.

Kou, J., Dou, D., Yang, L., Kou, J., Dou, D., Yang, L., 2017. Porphyrin photosensitizers in photodynamic therapy and its applications. Oncotarget 8, 81591–81603. https://doi.org/10.18632/ONCOTARGET.20189.

Kramarenko, G.G., Wilke, W.W., Dayal, D., Buettner, G.R., Schafer, F.Q., 2006. Ascorbate enhances the toxicity of the photodynamic action of Verteporfin in HL-60 cells. Free Radic. Biol. Med. 40 (9), 1615–1627. https://doi.org/10.1016/J.FREERADBIOMED.2005.12.027.

Krosl, G., Korbelik, M., 1994. Potentiation of photodynamic therapy by immunotherapy: the effect of schizophyllan (SPG). Cancer Lett. 84, 43–49. https://doi.org/10.1016/0304-3835(94)90356-5.

Kwitniewski, M., Juzeniene, A., Glosnicka, R., Moan, J., 2008. Immunotherapy: a way to improve the therapeutic outcome of photodynamic therapy? Photochem. Photobiol. Sci. 7, 1011–1017. https://doi.org/10.1039/B806710D.

Lai, J.C., Lo, P.C., Ng, D.K.P., Ko, W.H., Leung, S.C.H., Fung, K.P., Fong, W.P., 2006. BAM-SiPc, a novel agent for photodynamic therapy, induces apoptosis in human hepatocarcinoma HepG2 cells by a direct mitochondrial action. Cancer Biol. Ther. 5, 413–418. https://doi.org/10.4161/CBT.5.4.2513.

Lamch, Ł., Bazylińska, U., Kulbacka, J., Pietkiewicz, J., Biezuńska-Kusiak, K., Wilk, K.A., 2014. Polymeric micelles for enhanced Photofrin II® delivery, cytotoxicity and pro-apoptotic activity in human breast and ovarian cancer cells. Photodiagnosis Photodyn. Ther. 11, 570–585. https://doi.org/10.1016/j.pdpdt.2014.10.005.

Lange, N., Szlasa, W., Saczko, J., Chwiłkowska, A., 2021. Potential of cyanine derived dyes in photodynamic therapy. Pharmaceutics 13. https://doi.org/10.3390/PHARMACEUTICS13060818.

Lottner, Knuechel, Bernhardt, Brunner, 2004. Combined chemotherapeutic and photodynamic treatment on human bladder cells by hematoporphyrin-platinum(II) conjugates. Cancer Lett. 203 (2), 171–180. https://doi.org/10.1016/J.CANLET.2003.09.001.

Lilge, L., Wilson, B.C., 1998. Photodynamic therapy of intracranial tissues: a preclinical comparative study of four different photosensitizers. J. Clin. Laser Med. Surg. 16, 81–91. https://doi.org/10.1089/CLM.1998.16.81.

Lima, B., Agüero, M.B., Zygadlo, J., Tapia, A., Solís, C., De Arias, A.R., Yaluff, G., Zacchino, S., Feresin, G.E., Schmeda-Hirschmann, G., 2009. Antimicrobial activity of extracts, essential oil and metabolites obtained from tagetes mendocina. J. Chil. Chem. Soc. 54, 68–72.

Lindenmann, J., Maier, A., Matzi, V., Neuboeck, N., Anegg, U., Porubsky, C., Sankin, O., Fell, B., Renner, H., Swatek, P., Smolle-Juettner, F.M., 2011. Photodynamic therapy for esophageal carcinoma. Eur. Surg. - Acta Chir. Austriaca 43, 355–365. https://doi.org/10.1007/S10353-011-0051-X.

Liu, Z., Xie, Z., Li, W., Wu, X., Jiang, X., Li, G., Cao, L., Zhang, D., Wang, Q., Xue, P., Zhang, H., 2021. Photodynamic immunotherapy of cancers based on nanotechnology: recent advances and future challenges. J. Nanobiotechnol. 19. https://doi.org/10.1186/S12951-021-00903-7.

Lou, P.J., Jäger, H.R., Jones, L., Theodossy, T., Bown, S.G., Hopper, C., 2004. Interstitial photodynamic therapy as salvage treatment for recurrent head and neck cancer. Br. J. Cancer 91, 441–446. https://doi.org/10.1038/SJ.BJC.6601993.

Lucky, S.S., Soo, K.C., Zhang, Y., 2015. Nanoparticles in photodynamic therapy. Chem. Rev. 115, 1990–2042. https://doi.org/10.1021/CR5004198.

Li, L.-b., Luo, R.-c., Liao, W.-j., Zhang, M.-j., Luo, Y.-l., Miao, J.-x., 2006. Clinical study of Photofrin photodynamic therapy for the treatment of relapse nasopharyngeal carcinoma. Photodiagnosis Photodyn. Ther. 3, 266–271. https://doi.org/10.1016/J.PDPDT.2006.09.004.

Ma, K., Sun, F., Yang, X., Wang, S., Wang, L., Jin, Y., Shi, Y., Jiang, W., Zhan, C., Wang, Q., 2018. Prognosis of patients with primary malignant main stem bronchial tumors: 7,418 cases based on the SEER database. Onco. Targets. Ther. 11, 83–95. https://doi.org/10.2147/OTT.S142847.

MacPherson, L., Bayburt, E., Capparelli, M., Carroll, B., Goldstein, R., Justice, M., Zhu, L., Hu, S., Melton, R., Fryer, L., 1997. Discovery of CGS 27023A, a non-peptidic, potent, and orally active stromelysin inhibitor that blocks cartilage degradation in rabbits. J. Med. Chem. 40, 2525–2532. https://doi.org/10.1021/jm960871c.

Magalhães, J.A., Arruda, D.C., Baptista, M.S., Tada, D.B., 2021. Co-encapsulation of methylene blue and parp-inhibitor into poly(Lactic-co-glycolic acid) nanoparticles for enhanced pdt of cancer. Nanomaterials 11. https://doi.org/10.3390/NANO11061514.

Mahmoud, 2016. CHAPTER 1. The journey of PDT throughout history: PDT from pharos to present 1–21. https://doi.org/10.1039/9781782626824-00001

Mahmoud, 2014. History of photodynamic therapy. Photodyn. Ther. From Theory to Appl., 3–22. https://doi.org/10.1007/978-3-642-39629-8_1.

Maiya, B.G., 2000. Photodynamic Therapy (PDT). Reson 5, 6–18. https://doi.org/10.1007/BF02837901.

Mang, T.S., 2004. Lasers and light sources for PDT: past, present and future. Photodiagnosis Photodyn. Ther. 1, 43–48. https://doi.org/10.1016/S1572-1000(04)00012-2.

Martins, W.K., Belotto, R., Silva, M.N., Grasso, D., Lavor, S., Itri, R., Baptista, M.S., Suriani, M.D., 2021. Autophagy regulation and photodynamic therapy: insights to improve outcomes of cancer treatment 10, 1–22. https://doi.org/10.3389/fonc.2020.610472.

McCormick, T., Ayala-Fontanez, N., Soler, D., 2016. Current knowledge on psoriasis and autoimmune diseases. Psoriasis Targets Ther. 6, 7. https://doi.org/10.2147/ptt.s64950.

McDonagh, A.F., 2001. Phototherapy: From ancient egypt to the new millennium. J. Perinatol. 21, S7–S12. https://doi.org/10.1038/SJ.JP.7210625.

McFarland, S.A., Mandel, A., Dumoulin-White, R., Gasser, G., 2020. Metal-based photosensitizers for photodynamic therapy: the future of multimodal oncology? Curr. Opin. Chem. Biol. 56, 23–27.

Minn, A.J., Gupta, G.P., Siegel, P.M., Bos, P.D., Shu, W., Giri, D.D., Viale, A., Olshen, A.B., Gerald, W.L., Massagué, J., 2005. Genes that mediate breast cancer metastasis to lung. Nat. 436, 518–524. https://doi.org/10.1038/nature03799.

Moghissi, K., Dixon, K., Gibbins, S., 2015. A surgical view of photodynamic therapy in oncology: a review. Surg. J. 01, e1–e15. https://doi.org/10.1055/S-0035-1565246.

Mohan, A., Harris, K., Bowling, M.R., Brown, C., Hohenforst-Schmidt, W., 2018. Therapeutic bronchoscopy in the era of genotype directed lung cancer management. J. Thorac. Dis. 10, 6298–6309. https://doi.org/10.21037/JTD.2018.08.14.

Montaseri, H., Kruger, C.A., Abrahamse, H., 2021. Inorganic nanoparticles applied for active targeted photodynamic therapy of breast cancer. Pharmaceutics 13, 1–33. https://doi.org/10.3390/PHARMACEUTICS13030296.

Montoya, Comini, Sarmiento, C., B., I., A., G.A., A., J.L., C., 2005. Natural anthraquinones probed as Type I and Type II photosensitizers: singlet oxygen and superoxide anion production. J. Photochem. Photobiol. B. 78, 77–83. https://doi.org/10.1016/J.JPHOTOBIOL.2004.09.009.

Moore, C.M., Nathan, T.R., Lees, W.R., Mosse, C.A., Freeman, A., Emberton, M., Bown, S.G., 2006. Photodynamic therapy using meso tetra hydroxy phenyl chlorin (mTHPC) in early prostate cancer. Lasers Surg. Med. 38, 356–363. https://doi.org/10.1002/LSM.20275.

Moriyama, E.H., Cao, W., Liu, T.W., Wang, H.L., Kim, P.D., Chen, J., Zheng, G., Wilson, B.C., 2014. Optical glucose analogs of aminolevulinic acid for fluorescence-guided tumor resection and photodynamic therapy. Mol. Imaging Biol. 16, 495–503. https://doi.org/10.1007/S11307-013-0687-Y.

Mroz, P., Yaroslavsky, A., Kharkwal, G.B., Hamblin, M.R., 2011. Cell death pathways in photodynamic therapy of cancer. Cancers (Basel) 3, 2516–2539. https://doi.org/10.3390/CANCERS3022516.

Muller, P.J., Wilson, B.C., 1996. Photodynamic therapy for malignant newly diagnosed supratentorial gliomas. J. Clin. Laser Med. Surg. 14, 263–270. https://doi.org/10.1089/CLM.1996.14.263.

Muller, P.J., Wilson, B.C., 1995. Photodynamic therapy for recurrent supratentorial gliomas. Semin. Surg. Oncol. 11, 346–354. https://doi.org/10.1002/SSU.2980110504.

Muller, Wilson, 2006. Photodynamic therapy of brain tumors–a work in progress. Lasers Surg. Med. 38, 384–389. https://doi.org/10.1002/LSM.20338.

Muniyandi, K., George, B., Parimelazhagan, T., Abrahamse, H., 2020. Role of photoactive phytocompounds in photodynamic therapy of cancer. Mol. 25, 4102. 2020. https://doi.org/10.3390/MOLECULES25184102.

Muschter, R., 2003. Photodynamic therapy: a new approach to prostate cancer. Curr. Urol. Rep. 4, 221–228. https://doi.org/10.1007/S11934-003-0073-4.

Nathan, D.E., W., S.C., C., W.R., L., P.M., R., H., P., L., J., M.C., P., M., E., A.R., G., A.R., M., S.G., B., 2002. Photodynamic therapy for prostate cancer recurrence after radiotherapy: a phase I study. J. Urol. 168, 1427–1432. https://doi.org/10.1097/01.JU.0000030000.81684.7E.

Niculescu, A.-G., Grumezescu, A.M., 2021. Photodynamic therapy—an up-to-date review. Appl. Sci. 11, 3626. 2021Page11, 3626. https://doi.org/10.3390/APP11083626.

Ninane, V., 1995. Endoscopic treatment of bronchial cancer. Rev. Med. Brux. 16, 25–29.

Nseyo, U.O., DeHaven, J., Dougherty, T.J., Potter, W.R., Merrill, D.L., Lundahl, S.L., Lamm, D.L., 1998. Photodynamic therapy (PDT) in the treatment of patients with resistant superficial bladder cancer: a long term experience. J. Clin. Laser Med. Surg. 16, 61–68. https://doi.org/10.1089/CLM.1998.16.61.

O'Connor, A.E., Gallagher, W.M., Byrne, A.T., 2009. Porphyrin and nonporphyrin photosensitizers in oncology: preclinical and clinical advances in photodynamic therapy. Photochem. Photobiol. 85, 1053–1074. https://doi.org/10.1111/J.1751-1097.2009.00585.X.

Okunaka, T., Kato, H., 1995. Laser bronchoscopic therapy for lung cancer. Japanese J. Cancer Chemother. 22, 179–184.

Oleinick, N.L., Morris, R.L., Belichenko, I., 2002. The role of apoptosis in response to photodynamic therapy: what, where, why, and how. Photochem. Photobiol. Sci. 1, 1–21. https://doi.org/10.1039/B108586G.

Orenstein, Kostenich, G., Roitman, L., Shechtman, Y., Kopolovic, Y., Ehrenberg, B., Malik, Z., 1996. A comparative study of tissue distribution and photodynamic therapy selectivity of chlorin e6, Photofrin II and ALA-induced protoporphyrin IX in a colon carcinoma model. Br. J. Cancer 73, 937–944. https://doi.org/10.1038/BJC.1996.185.

Ormond, A.B., Freeman, H.S., 2013. Dye sensitizers for photodynamic therapy. Materials (Basel) 6, 817–840. https://doi.org/10.3390/MA6030817.

Overholt, B.F., Lightdale, C.J., Wang, K.K., Canto, M.I., Burdick, S., Haggitt, R.C., Bronner, M.P., Taylor, S.L., Grace, M.G.A., Depot, M., 2005. Photodynamic therapy with porfimer sodium for ablation of high-grade dysplasia in Barrett's esophagus: International, partially blinded, randomized phase III trial. Gastrointest. Endosc. 62, 488–498. https://doi.org/10.1016/J.GIE.2005.06.047.

Panzarini, E., Inguscio, V., Fimia, G.M., Dini, L., 2014. Rose Bengal acetate photodynamic therapy (RBAc-PDT) induces exposure and release of damage-associated molecular patterns (DAMPs) in human HeLa cells. PLoS One 9. https://doi.org/10.1371/JOURNAL.PONE.0105778.

Park, W., Cho, S., Han, J., Shin, H., Na, K., Lee, B., Kim, D.H., 2018. Advanced smart-photosensitizers for more effective cancer treatment. Biomater. Sci. 6, 79–90. https://doi.org/10.1039/c7bm00872d.

Park, W., Park, S.J., Cho, S., Shin, H., Jung, Y.S., Lee, B., Na, K., Kim, D.H., 2016. Intermolecular structural change for thermoswitchable polymeric photosensitizer. J. Am. Chem. Soc. 138, 10734–10737. https://doi.org/10.1021/JACS.6B04875.

Piette, J., Volanti, C., Vantieghem, A., Matroule, J.Y., Habraken, Y., Agostinis, P., 2003. Cell death and growth arrest in response to photodynamic therapy with membrane-bound photosensitizers. Biochem. Pharmacol. 66, 1651–1659. https://doi.org/10.1016/S0006-2952(03)00539-2.

Pinthus, J., Bogaards, A., Weersink, R., Wilson, B., Trachtenberg, J., 2006. Photodynamic therapy for urological malignancies: past to current approaches. J. Urol. 175, 1201–1207. https://doi.org/10.1016/s0022-5347(05)00701-9.

Plail, R.O., Harty, J.I., Lottmann, H.B., 1990. Photodyn. Ther. 119–140. https://doi.org/10.1007/978-1-4471-1783-4_6.

Płonka, J., Latocha, M., Kuśmierz, D., Zielińska, A., 2015. Expression of proapoptotic BAX and TP53 genes and antiapoptotic BCL-2 gene in MCF-7 and T-47D tumour cell cultures of the mammary gland after a photodynamic therapy with photolon. Adv. Clin. Exp. Med. 24, 37–46. https://doi.org/10.17219/ACEM/38152.

Prout, L., R, B., UO, N., JJ, D., PP, G., J, K., ME, T., YH, L., YZ, M., 1987. Photodynamic therapy with hematoporphyrin derivative in the treatment of superficial transitional-cell carcinoma of the bladder. N. Engl. J. Med. 317, 1251–1255. https://doi.org/10.1056/NEJM198711123172003.

Pucelik, B., Sułek, A., Barzowska, A., Dąbrowski, J.M., 2020. Recent advances in strategies for overcoming hypoxia in photodynamic therapy of cancer. Cancer Lett. 492, 116–135. https://doi.org/10.1016/J.CANLET.2020.07.007.

Qumseya, B.J., David, W., Wolfsen, H.C., 2013. Photodynamic therapy for Barrett's esophagus and esophageal carcinoma. Clin. Endosc. 46, 30–37. https://doi.org/10.5946/CE.2013.46.1.30.

Raquel, D., Almeida, Q.De, Baptista, M., Labriola, L., 2020. Photodynamic therapy in cancer treatment - an update review. https://doi.org/10.20517/2394-4722.2018.83

Raza, A., Khan, M.S., Ghanchi, N.K., Raheem, A., Beg, M.A., 2014. Tumour necrosis factor, interleukin-6 and interleukin-10 are possibly involved in Plasmodium vivax-associated thrombocytopaenia in southern Pakistani population. Malar. J. 13. https://doi.org/10.1186/1475-2875-13-323.

Reynolds, T., 1997. Photodynamic therapy expands its horizons. J. Natl. Cancer Inst. 89, 112–114. https://doi.org/10.1093/JNCI/89.2.112.

Ritz, Müller, Dietz, Duffner, Bornemann, Roser, Tatagiba, 2008. Hypericin uptake: a prognostic marker for survival in high-grade glioma. J. Clin. Neurosci. 15, 778–783. https://doi.org/10.1016/J.JOCN.2007.03.022.

Robertson, C.A., Evans, D.H., Abrahamse, H., 2009. Photodynamic therapy (PDT): a short review on cellular mechanisms and cancer research applications for PDT. J. Photochem. Photobiol. B Biol. 96, 1–8. https://doi.org/10.1016/J.JPHOTOBIOL.2009.04.001.

Rodriguez, F.J., Vizcaino, M.A., Lin, M.T., 2016. Recent advances on the molecular pathology of glial neoplasms in children and adults. J. Mol. Diagn. 18 (5), 620–634. doi:10.1016/j.jmoldx.2016.05.005.

Rosenblum, D., Joshi, N., Tao, W., Karp, J.M., Peer, D., 2018. Progress and challenges towards targeted delivery of cancer therapeutics. Nat. Commun.. https://doi.org/10.1038/s41467-018-03705-y.

Sajja, H., East, M., Mao, H., Wang, Y., Nie, S., Yang, L., 2009. Development of multifunctional nanoparticles for targeted drug delivery and noninvasive imaging of therapeutic effect. Curr. Drug Discov. Technol. 6, 43–51. https://doi.org/10.2174/157016309787581066.

Salman, M., Naseem, I., Hassan, I., Khan, A.A., Alhazza, I.M., 2015. Riboflavin arrests cisplatin-induced neurotoxicity by ameliorating cellular damage in dorsal root ganglion cells. Biomed Res. Int. 2015. https://doi.org/10.1155/2015/603543.

Sandbo, N., Ngam, C., Torr, E., Kregel, S., Kach, J., Dulin, N., 2013. Control of myofibroblast differentiation by microtubule dynamics through a regulated localization of mDia2. J. Biol. Chem. 288, 15466–15473.

Schmidt, M.H., Meyer, G.A., Reichert, K.W., Cheng, J., Krouwer, H.G., Ozker, K., Whelan, H.T., 2004. Evaluation of photodynamic therapy near functional brain tissue in patients with recurrent brain tumors. J. Neurooncol. 67, 201–207. https://doi.org/10.1023/B:NEON.0000021804.50002.85.

Schweitzer, V.G., 2001. PHOTOFRIN-mediated photodynamic therapy for treatment of early stage oral cavity and laryngeal malignancies. Lasers Surg. Med. 29, 305–313. https://doi.org/10.1002/LSM.1133.

Seth, K., Sanjeev, K., Raj, K., Priyank, P., Vachan, S., Rohit, G., Uttam, C., Asit, K., 2014. 2-(2-arylphenyl)benzoxazole as a novel anti-inflammatory scaffold: synthesis and biological evaluation. ACS Med. Chem. Lett. 5 (5), 512–516. doi:10.1021/ml400500e.

Shackley, D.C., Briggs, C., Whitehurst, C., Betts, C.D., O'Flynn, K.J., Clarke, N.W., Moore, J.V., 2001. Photodynamic therapy for superficial bladder cancer. Expert Rev. Anticancer Ther. 1, 523–530. https://doi.org/10.1586/14737140.1.4.523.

Shah, V.M., Sheppard, B.C., Sears, R.C., Alani, A.W., 2020. Hypoxia: friend or foe for drug delivery in pancreatic cancer. Cancer Lett. 492, 63–70.

Sharif, S., Nguyen, K.T., Bang, D., Park, J.O., Choi, E., 2021. Optimization of field-free point position, gradient field and ferromagnetic polymer ratio for enhanced navigation of magnetically controlled polymer-based microrobots in blood vessel. Micromachines 12. https://doi.org/10.3390/MI12040424.

Sharman, W., Lier, J., van, Allen, C., 2004. Targeted photodynamic therapy via receptor mediated delivery systems. Adv. Drug Deliv. Rev. 56, 53–76. https://doi.org/10.1016/j.addr.2003.08.015.

Shepherd, R.W., Radchenko, C., 2019. Bronchoscopic ablation techniques in the management of lung cancer. Ann. Transl. Med. 7, 362. https://doi.org/10.21037/ATM.2019.04.47 362.

Shetler, A.C., 1996. Photodynamic therapy may be alternative to head and neck surgery. Todays. Surg. Nurse 18, 19–22.

Shishkova, N., Kuznetsova, O., Berezov, T., 2013. Photodynamic therapy in gastroenterology. J. Gastrointest. Cancer 44, 251–259. https://doi.org/10.1007/S12029-013-9496-4.

Siboni, G., Weitman, H., Freeman, D., Mazur, Y., Malik, Z., Ehrenberg, B., 2002. The correlation between hydrophilicity of hypericins and helianthrone: internalization mechanisms, subcellular distribution and photodynamic action in colon carcinoma cells. Photochem. Photobiol. Sci. 1, 483–491. https://doi.org/10.1039/B202884K.

Shi, F.S., Weber, S., Gan, J., Rakhmilevich, A.L., Mahvi, D.M., 1999. Granulocyte-macrophage colony-stimulating factor (GM-CSF) secreted by cDNA-transfected tumor cells induces a more potent antitumor response than exogenous GM-CSF. Cancer Gene Ther. 6 (1), 81–88. https://doi.org/10.1038/SJ.CGT.7700012.

Siersema, 2005. Photodynamic therapy for Barrett's esophagus: not yet ready for the premier league of endoscopic interventions. Gastrointest. Endosc. 62, 503–507. https://doi.org/10.1016/J.GIE.2005.07.018.

Sinha, R., Kim, G.J., Nie, S., Shin, D.M., 2006. Nanotechnology in cancer therapeutics: bioconjugated nanoparticles for drug delivery. Mol. Cancer Ther. 5, 1909–1917. https://doi.org/10.1158/1535-7163.MCT-06-0141.

Son, J., Yi, G., Kwak, M.H., Yang, S.M., Park, J.M., Lee, B.I., Choi, M.G., Koo, H., 2019. Gelatin-chlorin e6 conjugate for in vivo photodynamic therapy. J. Nanobiotechnology 17. https://doi.org/10.1186/S12951-019-0475-1.

Spanò, V., Parrino, B., Carbone, A., Montalbano, A., Salvador, A., Brun, P., Vedaldi, D., Diana, P., Cirrincione, G., Barraja, P., 2015. Pyrazolo[3,4-h]quinolines promising photosensitizing agents in the treatment of cancer. Eur. J. Med. Chem. 102, 334–351. https://doi.org/10.1016/j.ejmech.2015.08.003.

Spikes, D., Bommer, J., C., J., 1993. Photosensitizing properties of mono-l-aspartyl chlorin e6 (NPe6): a candidate sensitizer for the photodynamic therapy of tumors. J. Photochem. Photobiol. B Biol. 17, 135–143. https://doi.org/10.1016/1011-1344(93)80006-U.

Starkey, Rebane, Drobizhev, F., M., A., G., A., E., K., M., C.W., S., 2008. New two-photon activated photodynamic therapy sensitizers induce xenograft tumor regressions after near-IR laser treatment through the body of the host mouse. Clin. Cancer Res. 14, 6564–6573. https://doi.org/10.1158/1078-0432.CCR-07-4162.

Stefflova, K., Li, H., Chen, J., Zheng, G., 2007. Peptide-based pharmacomodulation of a cancer-targeted optical imaging and photodynamic therapy agent. Bioconjug. Chem. 18, 379–388. https://doi.org/10.1021/bc0602578.

Story, W., Sultan, A.A., Bottini, G., Vaz, F., Lee, G., Hopper, C., 2013. Strategies of airway management for head and neck photo-dynamic therapy. Lasers Surg. Med. 45, 370–376. https://doi.org/10.1002/LSM.22149.

Stylli, S.S., Kaye, A.H., MacGregor, L., Howes, M., Rajendra, P., 2005. Photodynamic therapy of high grade glioma - long term survival. J. Clin. Neurosci. 12, 389–398. https://doi.org/10.1016/J.JOCN.2005.01.006.

Sun, B., Li, W., Liu, N., 2016. Curative effect of the recent photofrin photodynamic adjuvant treatment on young patients with advanced colorectal cancer. Oncol. Lett. 11, 2071–2074. https://doi.org/10.3892/OL.2016.4179.

Theodossiou, T.A., Hothersall, J.S., De Witte, P.A., Pantos, A., Nov-Dec 2009. The multifaceted photocytotoxic profile of hypericin. Mol. Pharm. 6 (6), 1775–1789. doi: 10.1021/mp900166q. PMID: 19739671.

Toratani, Tani, Kanda, Koizumi, Yoshioka, Okamoto, 2016. Photodynamic therapy using Photofrin and excimer dye laser treatment for superficial oral squamous cell carcinomas with long-term follow up. Photodiagnosis Photodyn. Ther. 14, 104–110. https://doi.org/10.1016/J.PDPDT.2015.12.009.

Turksoy, A., Yildiz, D., Akkaya, E.U., 2019. Photosensitization and controlled photosensitization with BODIPY dyes. Coord. Chem. Rev. 379, 47–64. https://doi.org/10.1016/J.CCR.2017.09.029.

Uehara, Sano, Wang, Sekine, Ikeda, Inokuchi, 2000. Enhancement of the photodynamic antitumor effect by streptococcal preparation OK-432 in the mouse carcinoma. Cancer Immunol. Immunother. 49 (8), 401–409. https://doi.org/10.1007/S002620000134.

Uramoto, Hidetaka, Tanaka, Fumihiro, 2014. Recurrence after surgery in patients with NSCLC. Transl. Lung Cancer Res. 3, 242–249. https://doi.org/10.3978/j.issn.2218-6751.2013.12.05.

Urbach, Forbes, Davies, Berger, 1976. Cutaneous photobiology: past, present and future. J. Invest. Dermatol. 67, 209–224. https://doi.org/10.1111/1523-1747.EP12513042.

Usuda, J., Azizuddin, K., Chiu, S., Oleinick, N.L., 2003. Association between the photodynamic loss of Bcl-2 and the sensitivity to apoptosis caused by phthalocyanine photodynamic therapy. Photochem. Photobiol. 78, 1.

Valvani, A., Martin, A., Devarajan, A., Chandy, D., 2019. Postobstructive pneumonia in lung cancer. Ann. Transl. Med. 7, 357. https://doi.org/10.21037/ATM.2019.05.26.

Venuta, F., De Giacomo, T., Rendina, E.A., Della Rocca, G., Flaishman, I., Ciccone, A.M., Pompei, L., Ricci, C., 1998. Endoscopia Operativa delle vie Aeree. Minerva Chir. 53, 483–488.

Vocks, E., 2006. Climatotherapy in atopic eczema. Handb. Atopic Eczema, 507–523. https://doi.org/10.1007/3-540-29856-8_55.

Wang, Shin, D.M., Simon, J.W., Nie, S., 2007. Nanotechnology for targeted cancer therapy. Expert Rev. Anticancer Ther. 7, 833–837. https://doi.org/10.1586/14737140.7.6.833.

Wang, X., Luo, D., Basilion, J.P., 2021a. Photodynamic therapy: targeting cancer biomarkers for the treatment of cancers. Cancers (Basel) 13. https://doi.org/10.3390/CANCERS13122992.

Weigelt, B., Peterse, J.L., Van't Veer, L.J., 2005. Breast cancer metastasis: markers and models. Nat. Rev. Cancer 5, 591–602. https://doi.org/10.1038/NRC1670.

Wolf, P., Fink-Puches, R., Cerroni, L., Kerl, H., 1994. Photodynamic therapy for mycosis fungoides after topical photosensitization with 5-aminolevulinic acid. J. Am. Acad. Dermatol. 31, 678–680. https://doi.org/10.1016/S0190-9622(08)81742-2.

Wolfsen, H.C., 2005. Endoprevention of esophageal cancer: Endoscopic ablation of Barrett's metaplasia and dysplasia. Expert Rev. Med. Devices 2, 713–723. https://doi.org/10.1586/17434440.2.6.713.

Wong, T., Hsu, L., Liao, W., 2013. Phototherapy in psoriasis: a review of mechanisms of action. J. Cutan. Med. Surg. 17, 6–12. https://doi.org/10.2310/7750.2012.11124.

Wyss, P., Schwarz, V., Dobler-Girdziunaite, D., Hornung, R., Walt, H., Degen, A., Fehr, M., 2001. Photodynamic therapy of locoregional breast cancer recurrences using a chlorin-type photosensitizer. Int. J. Cancer 93, 720–724. https://doi.org/10.1002/IJC.1400.

Xingshu, Lovell, J.F., Yoon, J., Chen, X., 2020. Clinical development and potential of photothermal and photodynamic therapies for cancer. Nat. Rev. Clin. Oncol. 17, 657–674. https://doi.org/10.1038/S41571-020-0410-2.

Xu, X., Lu, H., Lee, R., 2020. Near infrared light triggered photo/immuno-therapy toward cancers. Front. Bioeng. Biotechnol. 8. https://doi.org/10.3389/FBIOE.2020.00488/FULL.

Yaghini, E., Seifalian, A.M., MacRobert, A.J., 2009. Quantum dots and their potential biomedical applications in photosensitization for photodynamic therapy. Nanomedicine 4, 353–363. https://doi.org/10.2217/nnm.09.9.

Yang, Y., Wang, H., 2021. Recent progress in nanophotosensitizers for advanced photodynamic therapy of cancer. JPhys Mater. 4. https://doi.org/10.1088/2515-7639/ABC9CE.

Yin, X., Zhang, L., Wang, Y.-H., Zhang, B.-H., Gan, Y.-H., Ge, N.-L., Chen, Y., Li, L.-X., Ren, Z.-G., 2014. Transcatheter arterial chemoembolization combined with radiofrequency ablation delays tumor progression and prolongs overall survival in patients with intermediate (BCLC B) hepatocellular carcinoma. BMC Cancer 14, 849. https://doi.org/10.1186/1471-2407-14-849.

Yoon, I., Li, J.Z., Shim, Y.K., 2013. Advance in photosensitizers and light delivery for photodynamic therapy. Clin. Endosc. 46, 7–23. https://doi.org/10.5946/CE.2013.46.1.7.

Zhang, J., Jiang, C., Longo, J.P.F., Azevedo, R.B., Zhang, H., Muehlmann, L.A., 2018. An updated overview on the development of new photosensitizers for anticancer photodynamic therapy. Acta Pharm. Sin. B 8, 137. https://doi.org/10.1016/J.APSB.2017.09.003.

Zhang, M., Zhang, Z., Blessington, D., Li, H., Busch, T., Madrak, V., Miles, J., Chance, B., Glickson, J., Zheng, G., 2003. Pyropheophorbide 2-deoxyglucosamide: a new photosensitizer targeting glucose transporters. Bioconjug. Chem. 14, 709–714. https://doi.org/10.1021/bc034038n.

Zhang, Z., Zhang, X.-L., Li, B., 2021. Mesoporous silica-coated upconverting nanorods for singlet oxygen generation: synthesis and performance. Materials (Basel). 14, 3660. https://doi.org/10.3390/MA14133660.

Zheng, G., Chen, J., Stefflova, K., Jarvi, M., Li, H., Wilson, B.C., 2007. Photodynamic molecular beacon as an activatable photosensitizer based on protease-controlled singlet oxygen quenching and activation. Proc. Natl. Acad. Sci. 104, 8989–8994. https://doi.org/10.1073/PNAS.0611142104.

Zheng, G., Li, H., Zhang, M., Lund-Katz, S., Chance, B., Glickson, J., 2002. Low-density lipoprotein reconstituted by pyropheophorbide cholesteryl oleate as target-specific photosensitizer. Bioconjug. Chem. 13, 392–396. https://doi.org/10.1021/bc025516h.

Zhenya, Sidan, Yuting, Fanling, Liang, 2021. Luminescent AIE dots for anticancer photodynamic therapy. Front. Chem. 9. https://doi.org/10.3389/FCHEM.2021.672917.

Chapter 6

Photodiagnostic techniques

Anurag Luharia[a], Gaurav Mishra[a], Nilesh Haran[b], Sanjay J. Dhoble[c]

[a]*Department of Radiology, Datta Meghe Institute of Medical Science (Deemed to be University), Sawangi, Wardha, Maharashtra, India*
[b]*Department of Radiology HCG-NCHRI Cancer Centre, Near Automotive Square, Nagpur, Maharashtra, India*
[c]*Department of Physics, Rashtrasant Tukadoji Maharaj Nagpur University, Nagpur, Maharashtra, India*

6.1 Introduction

In 1895, Prof. William Roentgen has come up with the unknown radiation, which was capable of showing the internal anatomy of the human body called as X-rays, shortly in 1896, another form of penetrating rays was discovered by Prof. Henri Becquerel and came up with natural radioactivity. These two initial physics discoveries were utilized for the clinical diagnosis in June 1896 where the battlefield physicians located bullets in wounded soldiers using X-rays (Manes, 1956; Grammaticos, 2004). Similarly, in 1950s physicians with an endocrine emphasis, used iodine-131 (radioactive element) to diagnose and treat thyroid disease.

Ionizing radiation has validated its existence and effectiveness in modern medicine for both diagnostic and therapeutic use. For the last decade rapid growth in medical radiation application has witnessed in India towards the betterment of mankind (Shaffer et al., 2016; Lehtinen and Racoveanu, 1985). The aim of radio diagnostic practice is to obtain best image quality with reduced radiation dose to patient (Okafor et al., 2018) for this the periodic quality check and radiological safety survey and education of all the radiation generating instruments plays a very important role.

Ionizing radiation is a duel-edged sword so X-rays, CT scans, and other procedures should be used cautiously. In India, Atomic Energy Regulatory Board has approved the medical physicist and radiological safety officer to take care of all the safety aspect in radiology along with the quality assurance checks and its clinical relevance (Board, Atomic Energy Regulatory 2016; Dixon et al., 1993; Vorotyntseva et al., 2018). Various site and technique specific photo diagnostic instruments are available for various photo diagnostic procedures like computed tomography where, whole-body three-dimensional scanning is possible, mammography for the breast soft tissue imaging, dental photoimaging instruments dedicated for the dental imaging, similarly using radioisotopes in PET CT, SPECT for the functional imaging is commonly used (Cohen and Liewehr, 1998; Mettler and Fred, 2008). The diagnostic radiology plays a very important role to define the staging, extent or type of the disease for better clinical management with low intensity radiation is used as compared to therapeutic procedure range and provides anatomical as well as functional information of the human body (Whittaker, 1974) depends upon the type of procedures selected.

Radiation has proved its application and utility in various domains like medical, industries and research as shown in tree diagram. It is omnipresent as UV light from the sun, heat energy from a stove burner, visible light from a candle, X-rays from an X-ray machine, alpha particles released by uranium's radioactive decay sound waves from your stereo, microwaves from your microwave oven, and radiation from your cell phone are all examples of electromagnetic radiation (Chmielewski and Haji-Saeid, 2004). Radiation is the energy which can travel from one place to another in the form of heat, sound, waves, light, etc., basically categories into two types ionizing and nonionizing, further defined on the basis of their charge, mass and ability to penetrate the object. Radiation is basically classified into two main categories, that is, ionizing and nonionizing radiation as shown in Fig. 6.1 which ionizing radiations are used for the medical applications mainly and further classified as a particle radiation and electromagnetic radiation (Chmielewski and Haji-Saeid, 2004; Li et al., 2001) in diagnostic radiology variable intensity of X-rays are used.

6.1.1 Ionizing radiations

An atom consists of electron and photon. The photon forms the central part and electrons revolves in specified orbits along the central proton according to its specified energies just like planets revolve around sun in the solar system (Kohout and Savin, 1996). The orbits are designated as K, L, M, N from inwards to outside. The electrons are placed in the orbits in increasing order depending upon the atomic number of the atom.

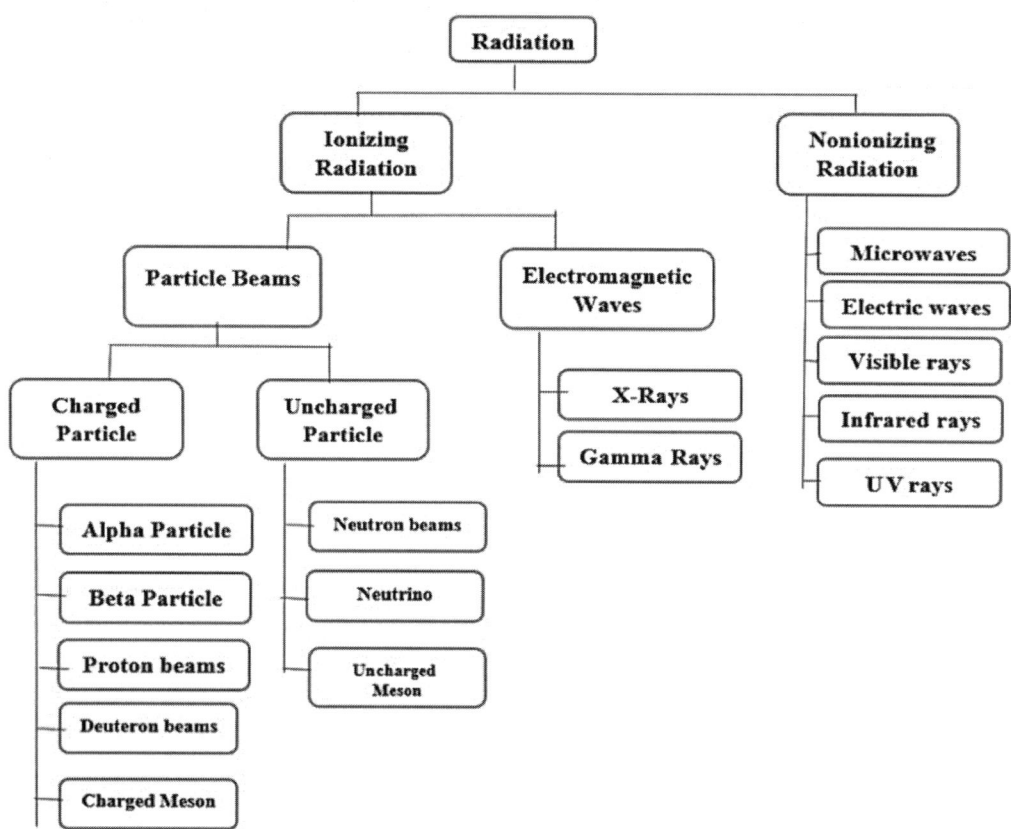

FIG. 6.1 Schematic tree diagram showing basic classification of radiation.

For example: sodium atom has atomic number of 11, so it has 2 K orbital electrons, 8 in M shell and 1 N shell electronwhich is also known as valence electron.

When one of an atom's electrons is entirely gone, the atom is said to be ionised. The detached electron is a negative ion, while the remaining atom is a positive ion, resulting in the formation of an ion pair (Siegfried, 2002). An electron's binding energy in an atom is the amount of energy necessary to entirely remove the electron from the atom against the force of the positive nucleus. This binding energy depends in the atomic energy of the element, as the atomic energy increases so the binding energy.

When an electron is pushed from one shell to another, an atom is said to be excited. This involves the expenditure of energy which causes electrons to raise from their shell to the unoccupied shell, when it falls back to its own shell the energy is re-emitted in the form of single packet or photon of light. In this way the ionizing radiations are produced at the molecular level.

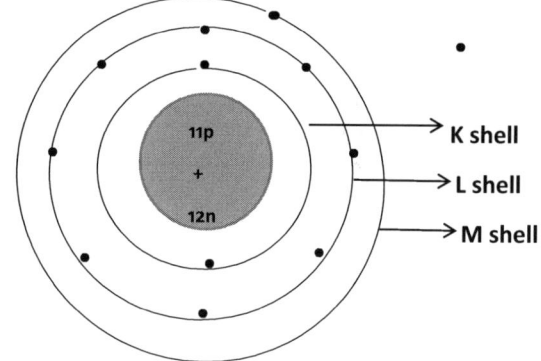

FIG. 6.2 Schematic showing electron configuration in sodium atom.

6.2 Fundamentals of light used in diagnostic techniques

X-ray exams have been an essential element of medical diagnostic radiology practise since its discovery in 1895 (Mould, 1995). The use of X-rays serves society enormously by detecting a wide range of illnesses, bone fractures, and deformations. (Vazquez et al., 2000). Though the usage of X-rays has provided several benefits to civilization, an excessive dosage of X-ray radiation is hazardous to humans. Concerns about radiation safety have developed in tandem with the increased use of X-rays in medical applications (Reed, 2011). Radiation is defined as energy that moves in the form of waves or particles and includes electromagnetic radiation, such as radiowaves, microwaves, visible light, X-rays, and gamma rays (Versoza and Band, 2005; Keiser, 2016). It is present everywhere in the cosmos. X-rays are electromagnetic radiation that contains both particles and waves and is utilized in diagnostic radiology.

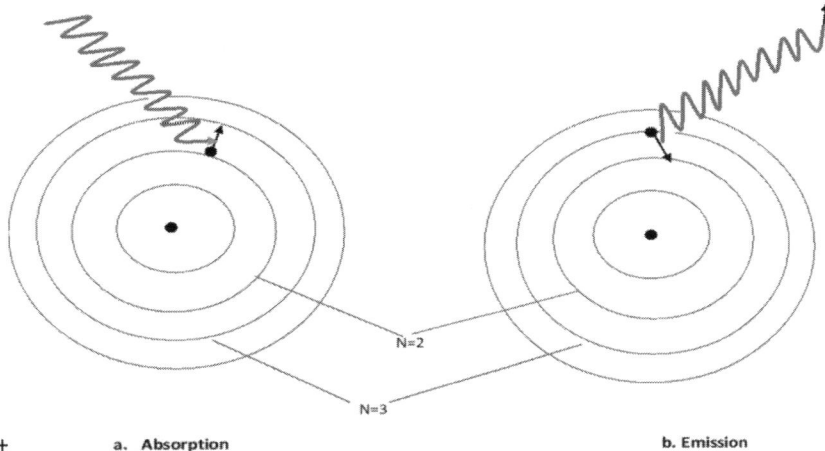
FIG. 6.3 Schematic showing the process of absorption and emission of energy.

1. Wave concept of electromagnetic radiation
   Electromagnetic radiation is propagated through space in the form of waves. They may be compared to waves traveling down a stretched rope when one end is moved up and down in a rhythmic motion. While the waves with which we are familiar must be propagated in a medium; electromagnetic waves need no such medium; that is, they can be propagated through a vacuum. Waves of all types have an associated wavelength and frequency (Keiser, 2016). All types of electromagnetic radiation have the same velocity, and the frequency is inversely proportional to its wavelength.
2. Electromagnetic radiation particle nature
   Short electromagnetic waves like X-rays may interact with materials as if they were particles rather than waves. These particles are unique energy bundles, and each of these energy bundles is referred to as a quantum, or photon. Photons are particles that travel at the speed of light. The quantity of energy carried by each quantum, or photon, is determined by the radiation's frequency (v) (Gustafson et al., 2001; Baird, 2019). The energy of a photon is twice when the frequency (number of vibrations per second) is doubled. The actual amount of energy of the photon may be calculated by multiplying its frequency by a constant.

   The ability to visualize the dual characteristics of electromagnetic radiation presents a true challenge. But we must unavoidably reach the conclusion that EM radiation sometimes behaves as a wave and other times as a particle. The particle concept is used to describe the interactions between radiation and matter. Because we will be concerned principally with interactions, such as the photoelectric effect and Compton scatter, we will use the photon (or quantum) concept in this text (Nikjoo et al., 2012). The electron volt is the unit of measurement for photon energy (eV). The amount of energy acquired by an electron when it is accelerated by a potential difference of one volt is referred to as an electron volt.

   Basically, radiation is of two types:

a. Ionizing radiation
b. Non-ionizing radiation

Ionizing radiation has sufficient energy to eject electrons from atoms of matter or body tissues through which it traverses is called ionizing radiation. Examples: medical and dental radiography, computed tomography (CT), nuclear medicine and fluoroscopy procedures are examples of diagnostic examinations that use ionizing radiation. On the other hand, nonionizing radiation has enough energy to vibrate atoms but does not have enough energy to remove electrons from atoms of matter or body tissues. Examples: Radio waves and microwaves (Holmes-Siedle and Adams, 1993). Ultrasonography and magnetic resonance imaging (MRI) are examples of diagnostic examinations that use nonionizing radiation.

Basically, X-rays are originating from:

a. Natural sources
b. Manmade sources

Ionizing radiation is present everywhere in nature in varying amounts at different locations and it is a part of our daily life, every day, all of us are subjected to natural background radiation. This is emitted by the ground and construction materials around us, as well as the air we breathe, the food we eat, and even space (cosmic rays) (Ryan, 2012). Depending upon the location where one individual lives, each individual is exposed to 1 to 3 mSv every year, with global usual of 2.4 milli Sievert from the natural sources of ionising radiation.

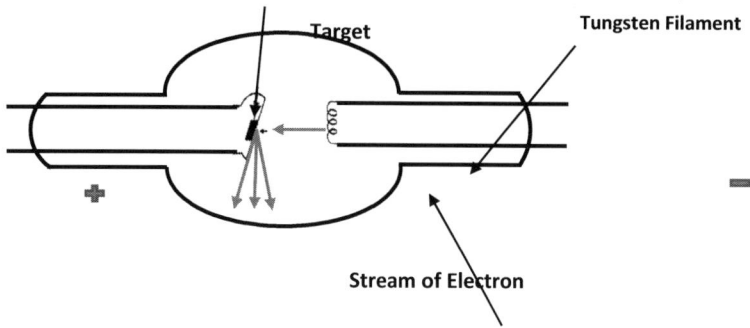

FIG. 6.4 Sketch showing the production of X-ray.

Ionizing radiation is the biggest man-made source of radiation. The most well-known application is diagnostic X-ray equipment, which uses X-rays for examining chest, teeth, broken bones, etc. Another use of man-made ionising radiation is nuclear medicine, in which a slight quantity of radioactive material (isotope) is administrated into a patient's veins and concentrates in a specific organ of interest, such as the skeleton for a bone scan. The radioactive substance generates gamma rays, a kind of radiation that behaves similarly to X-rays. A special camera detects the gamma rays coming out from the body of the patient and builds up an image of what is happening inside the body of the patient. Other manmade source of ionizing radiation includes industrial radiography and nucleonic gauging applications, which are used in numerous industries, such as construction, civil engineering, and oil well-logging applications.

6.2.1 X-ray production

When a fast-moving stream of electrons is suddenly decelerated in an X-ray tube's "target" anode, energy conversion generates X-rays. The X-ray tube is made of pyrex glass and features a vacuum and two electrodes (this is a diode tube) (Merzbacher and Lewis, 1958). Electrons created at the cathode (negative electrode or filament) are pushed toward the anode by a high potential difference (positive or target electrode) (Garcia, 1970). Figure depicts the fundamental elements of an X-ray tube, a schematic of a stationary anode x-ray tube (Akselsson and Johansson, 1974). The heated tungsten filament generates electrons, which are propelled across the tube to reach the tungsten target, where X-rays are generated.

X rays are formed by two distinct processes, which result in (1) the development of a continuous spectrum of X-rays (bremsstrahlung) and (2) the generation of distinctive X-rays. The amount (number) of X-rays produced is proportional to the target material's atomic number (Z), the square of the kilovoltage (kVp), and the milliamperes of X-ray tube current (mA). The quality (energy) of the X-rays produced is nearly completely determined by the X-ray tube potential (kVp).

The amount of (1) nuclear disintegration that occurs per unit time, (2) ionizing photons present in a field, (3) energy transferred from radiation to tissue, (4) energy absorbed in the tissue, and (5) biological effectiveness of energy absorption are all quantities of interest in radiological physics. X-rays are generated by following ways:
 1. General or Bremsstrahlung radiation
 2. Characteristic radiation

1. General radiation
When an electron passes close to the nucleus of a tungsten atom, the positive charge of the nucleus interacts with the electron's negative charge.

This leads to attraction of the electron towards the nucleus and causes deflection of the electron from its original direction (Sommerfeld, 1929).

In this process the electron losses its energy. This kinetic energy lost during this process is emitted directly in the form of a photon or radiation. This radiation is called general radiation. In general, the majority of radiation has little energy and appears as heat. Only 1 percent of the X rays are generated, rest of the energy is converted to heat. This is because an electron will undergo many reactions before coming to rest and energy its losses with reaction is very small.

2. Characteristic radiation
Characteristic radiation is produced when electrons hitting the target expel electrons from the inner orbits of the target atoms. When an electron is taken from a tungsten atom, it gets an additional positive charge and forms a positive ion

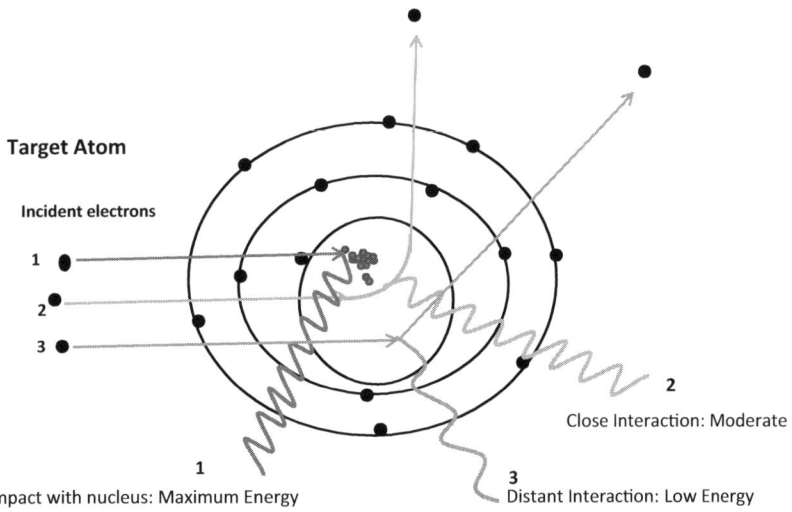

FIG. 6.5 The production of general radiation by deflection of the electron from its path.

(Seibert, 2004). The ionized tungsten atom gets rid of surplus energy in two ways as it returns to its normal condition. Firstly, by removal of the additional electron from the atom and carry off excess energy, however the expelled electron does not produce X-rays and produce only heat. Second, in order to expel surplus energy, the atom releases radiation with x-ray wavelengths. An X-ray is considerably more likely to be produced by a tungsten atom with an inner shell electron than by heat.

The wavelengths of the X-rays generated are distinctive of the atom that has been ionized, therefore they are referred to as characteristic X-rays. The wavelengths of the X-rays generated are distinctive of the atom that has been ionized, therefore they are referred to as characteristic X-rays.

6.2.2 X-ray beam intensity

An X-ray beam's intensity is estimated by multiplying the number of photons in the beam by the energy of each photon. The intensity is expressed in units of Roentgens per minute. The intensity changes depending on the kilovoltage, tube current, target material, and filtering (Mainardi and Barrea, 1989).

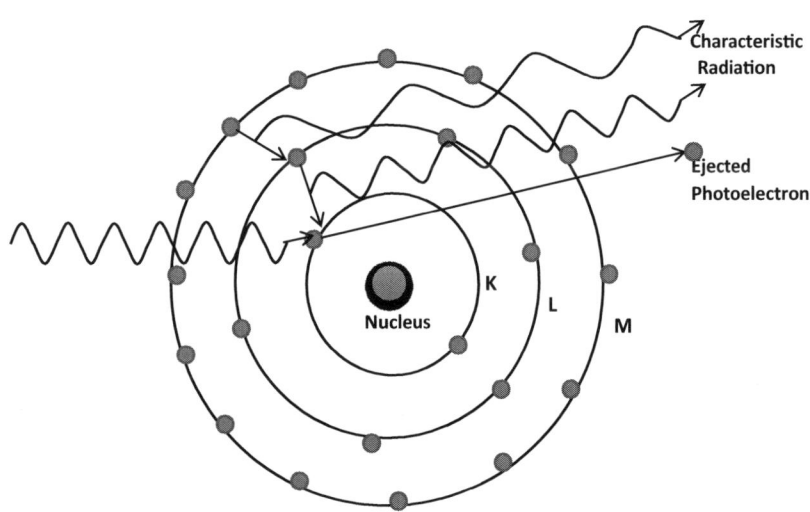

FIG. 6.6 The production of characteristic radiation.

6.2.3 Target material

It decides how much radiation will be produced by a given voltage, as a result, the greater the atomic number of the target substance, the more efficient the generation of x rays. Because of its high atomic number of 74, tungsten is employed as a target material. The vice versa is also true that is the lower the target material atomic number low efficiency electrons will be produced this principle is used in mammography which requires low intensity X-rays for optimum breast tissue visualization, in this case molybdenum with atomic number of 42 is used as target for mammography (Prabhu et al., 2020). For mammography, the maximum tube voltage is 40 kVp. At this voltage, molybdenum's 17.5 keV K alpha and 19.6 keV K-beta characteristic radiation accounts for a sizable percentage of the overall radiation production.

6.2.4 Voltage applied

The maximum energy of the x rays produced determines by the kVp, so as the kVp increases so the amount of X-rays produced. The intensity of X-rays produced is proportional to the square of kilovoltage.

6.2.5 X-ray tube current

The number of electrons that strike the target is proportional to the applied current (mA). As the mA increases, more electrons are created, resulting in more X-rays.

6.3 Various photo diagnostic techniques

The diagnostic radiology plays a very important role to define the staging, extent or type of the disease for better clinical management, today we are using variable intensity X-rays in computed tomography, mammography, dental radiology fluoroscopy units as well as various artificial radioactive element like F-18, I-131, Tc-91, rubidium-82 in PET-CT, SPECT or Gamma Camera for getting the diagnostic information about the patient (Dixon, 1997; Ell, 2006). Basically, for Photo diagnostic procedures low intensity radiation is used as compared to therapeutic procedure range and provides anatomical as well as functional information of the human body depends upon the type of procedures selected.

Taking into account the different perspectives of diagnostic radiology, multiple techniques or methods and means of imaging of the human body have taken birth in the past 5 decades. A detailed historical review of these diagnostic techniques is out of the scope of this draft. The term 'photo diagnostic' literally means deriving diagnostic information from a source by incorporation of light or its various forms in suitable working conditions and appliances and radiological instrumentation, for the means of not only diagnosis of a disease or a disorder, but possibly a way to cure it and have multiple encounters of patient follow up as may be required. The imaging modalities of Ultrasound and Magnetic Resonance Imaging also help and assist in this endeavor. However, ultrasound incorporates the basic principles of sound as form of energy in diagnosis of diseases and the latter makes use of magnetic and electromagnetic properties of hydrogen, one of the most abundantly present elements in the human body for the diagnosis of disease. Here, one is "sonodiagnostic" and the other is "magnetic." So, to stick to the scope of photodiagnostic techniques, the following techniques have to be considered in this text. They are as follows:

1. Plain radiography
2. Computed tomography
3. Fluoroscopy
4. Digital subtraction angiography
5. Digital radiography and PACS
6. Dual energy X-ray absorptiometry
7. Dual energy computed tomography
8. Orthopantomography

6.3.1 Plain radiography and digital radiography

This imaging modality is also known as "projectional radiography" and "conventional radiography" (Shelledy and Peters, 2014). This involves generation of an image which is two dimensions by incorporation of X-ray radiation energy. The use of word 'plain' here indicates that there is no use of advanced imaging modalities capable of generation of three dimensional images of the subject such as computed tomography or any usage of additional chemical agent which provides contrast to

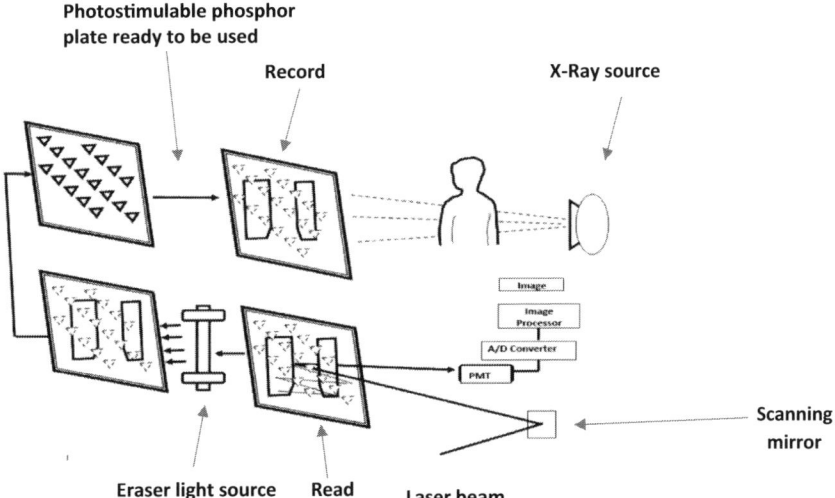

FIG. 6.7 Schematic showing how the image is produced in digital radiography.

the region of interest and has to be intravenously injected into the body of the subject. This is a routinely advised medical imaging investigation and is quite economical in comparison with other imaging modalities which may provide more in detail information regarding the subject. For the same reason, this modality has been included as a "screening" test for many diseases or disorders. Not only in the setting of diseases, but trauma also dictates use of plain radiography in order to check the integrity of the bones which may be subjected to impact or injury. An essential fact to know here is that x-rays follow the same principles of behavior and interaction with matter as light, the only difference there being is the ionizing nature of X-rays unlike light. The term "ionizing" here refers to the property of xrays to be able to remove electrons already revolving around the atomic nucleus in their respective orbits with the energy they possess, resulting into a positive charge developed by the affected atom owing to removal of its electrons by X-ray radiation. The physical details of apparatus responsible for generation of X-ray radiation will be covered in detail in this chapter ahead. The main factors which contribute to the diagnostic value of this imaging modality are its ability to distinguish in between two materials on the basis of their density, which means higher the density of the material of interest, brighter shade of white will be the representation of the material of interest on the resultant film defining the attenuation of the material of interest on the film. The degree of brightness or density is achieved by usage of suitable factors such as suitable kilovoltage in the X-ray machines, which may vary from patient to patient, depending upon their built – less factors for a thin individual to slightly higher factors for an obese one. For certain radiographic procedures usage of iodinated contrast medium is advised depending upon the serum urea and serum creatinine levels of the patient within normal limits, thereby guaranteeing normal renal profile of patient as contrast is excreted mostly through kidneys (Mishra et al., 2016). Any elevation of these above normal prescribed limits, contrast administration is contraindicated. Moreover, plain radiography is known to serve as the initial modality to give initial or basic information regarding most suspected pathologies.

Digital radiography uses detectors which are based on amorphous silicon thin-film transistor (TFT) arrays which amplifies electric signals, and the improved signal is then stored as an electric charge. This charge can then be released by using high potential. It is of two types of direct digital radiography and indirect digital radiography.

Direct radiography uses amorphous selenium as a photoconductor which allows the radiographic images to transfer an electrical charge during exposure and deposit a single layer of material on the amorphous silicon TFT array.

In indirect radiography a phosphor, that is, cesium iodide is used to convert x rays to light which is then detected by the photodiodes incorporated into the TFT arrays.

In this procedure the image can be seen instantly by the viewer without the hassle of removing the cassette, taking it to the reader and then viewing it thus reducing the time and mistakes.

6.3.2 Computed tomography

The use of "ionizing" radiation taken a step ahead can be said to be applied in this perspective. The source of x rays as well as the film responsible for catching the resultant photograph generated of the subject of interest were stationary

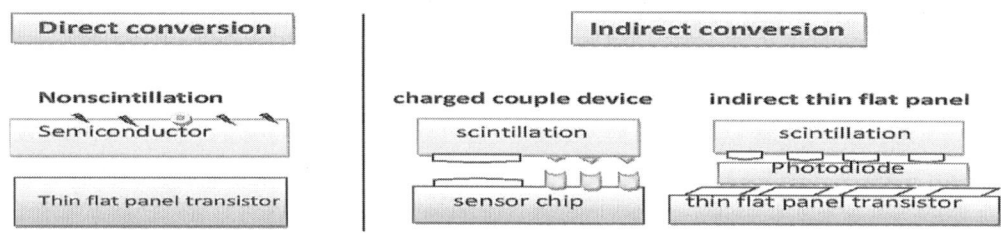

FIG. 6.8 Schematic showing cassette-less digital radiography system.

in plain radiography and could be adjusted according to required or desirable situations and scenarios. Now, these are located in a circular structure which automates their rotatory movement at high speeds called as a "gantry" in computed tomography. The X-ray emitter as well as detectors, multiple in number and situated in the form of 'arrays' within the gantry rotate at high speeds and in between this hollow "gantry," the patient or the subject of interest is required to be situated. The "gantry" is fixed while the table on which patient lies is subjected to translatory motion as desirable to obtain diagnostic images. The images produced in this manner are entered into a computerized software which integrates them into sectional images in the axial plane, and then does the task of 'reconstructing' them in other sectional planes such as sagittal and coronal planes. In this way, the images obtained are in 3 dimensions and better delineate the internal viscera. Therefore, what was obtained as a 2-dimensional image on plain radiography is now a 3-dimensional image on computed tomography. The dose of radiation to which the patient is subjected to here is much higher than that in plain radiography. To reduce the radiation dose, it is highly advisable and should be practised that the computed tomography should be carried out only in the region of interest and unnecessary exposure to radiation be avoided at all costs (37, 38, 39). Some applications of the same are there in the field of neurosciences, dental medicine and surgeries as well as nuclear medicine and are as follows:

6.3.2.1 Computed tomography perfusion imaging

This is a special type of computed tomography imaging which deals with administration of iodinated contrast medium intravenously in the patient's body and taking subsequent images of the required region of interest, for example, brain, and take a note of the blood flow in the area of interest. Through the usage of dedicated sophisticated software, the different parameters related to blood flow are calculated and yielded, such as total volume of blood flow per unit time through a particular brain, transit time taken by blood etc. This application serves highly in diagnosis of brain strokes and gives a map of the area of the brain which is already salvaged and which may be saved by subsequent intervention (Fishman and Jeffrey, 1995; Hsieh, 2003; Stirrup, 2020). Also, computed tomography serves as an excellent tool for diagnosing and follow up of patients of severe diseases such as recent pandemic of CoViD – 19 since 2020 (Mishra et al., 2021).

6.3.2.2 Cone beam computed tomography

Application of computed tomography in the field of dental imaging makes use of cone beam CT. This is also referred to by other names such as C-arm computed tomography, cone beam volume computed tomography, flat panel computed tomography or digital volume tomography (Scarfe et al., 2006). The main principle here is 360-degree motion of the detector and emitter of ionising radiation with the subject of interest situated at the centre. There is collection of imaging data through this motion of apparatus and through dedicated software, there is reconstruction of the acquired images in different anatomical planes. This has useful applications in various dental disciplines such as conservative dentistry and endodontics, orthodontics, oral and maxillofacial surgery in diagnosis as well as planning for oral surgeries (Hatcher, 2010; Orth et al., 2008; Kohout and Savin, 1996). Dedicated software may also aid in reconstruction of the acquired images into a 3-dimensional image which can be rotated, zoomed, aligned, and desirable measurements can be carried out as required. This is highly sought after in dental surgeries where precise planning is required regarding which structures to target and the extent up to which the degree of involvement is going to be there.

6.3.2.3 Positron emission tomography in nuclear medicine

This is special application of computed tomography as an imaging modality which makes usage of special drugs. As the name suggests, the principle here is usage of drugs which have radioactive isotopes attached to them, to be administered intravenously into the patient's body (Bailey et al., 2005). These drugs being radiologically active actually get traced through special equipment which is specifically designed for this purpose – the gamma cameras (Lehtinen and Racoveanu, 1985; Board, Atomic Energy Regulatory, 2006). These special cameras evaluate and display the manner in which these injected radiotracer drugs move through the patient's body and thereby the region of interest on a screen and any irregularities are revealed (Newberg et al., 2002). There are different radioisotopes used for this such as fluorodeoxyglucose, sodium fluoride, and oxygen labeled with radioisotope of fluorine (Som et al., 1980; Kelloff et al., 2005). These are used respectively for the purpose of evaluation of any irregularities or induced responses in the normal metabolic processes taking place in the human body such as cancer detection, new growth of bones or blood flow and monitoring of blood distribution.

6.3.3 Fluoroscopy

This application deals with displaying the interior of the human body in the form of moving pictures. The basic principle lies the same as plain radiography, the only difference being in the usage of a screen to capture the video of the body interior. This special screen is called as "fluorescent screen."

The usage of contrast agents such as barium sulphate for delineation of the gastrointestinal tract and administration of the same by oral or rectal route is made use of for diagnosis of any luminal irregularities, luminal narrowing or stenosis or widening or dilatation, any breach in the mucosa of the gastrointestinal tract such as ulcers which may show pooling or collection of contrast medium in one place or may show delayed passage can be diagnosed. The early systems produced in the 1950's was devoid of image storage but recent technology has made it possible to reproduce the video loops obtained on patient examination which can be replayed as per requirement and proper reporting by the radiologist and subsequent diagnosis may be arrived at (Merriam-Webster, Inc. 1995; Webster, 2016; Manes, 1956; Marchiori, 2014; Menon et al., 2013). This modality also carries a higher radiation dose as compared to plain radiography and therefore should be used minimally, as prescribed by ALARA.

6.3.4 Digital subtraction angiography

Digital subtraction angiography is an imaging modality which can be said to the next level of fluoroscopy. Its dedicated use is for proper delineation of blood vessels in the human body by using radioopaque contrast injected intravenously in the patient's body. Fluoroscopy displays images in the form of a series of images as a video loop on a display. This loop however takes into its aegis the area of interest and all the structures around it, structures which are lying over the area of interest as well as behind it. Therefore, specific display of blood vessel lumen is not proper. What digital subtraction here does here is that by usage of dedicated software, it creates a mask image of the area of interest. This mask image is the image or a scout image – an image of the area of interest taken just before the intravenous contrast makes its way into the blood vessels. This image can be said to the background of the subsequent video loops which are taken simultaneously with injection of contrast medium and recording of the video loops is done. The resultant output is a video loop with a grey background and the contrast medium being displayed as black as the intravenous contrast progresses across the blood vessels in the area of interest. These loops can be viewed and replayed as per requirement to arrive at proper diagnosis and plan a vascular intervention as may be required according to the presenting situation – vascular narrowing or stenosis, vascular dilatation in cases of aneurysms, pseudoaneurysms. This can be carried out in all parts of the human body wherever blood vessels are involved or are suspected to be

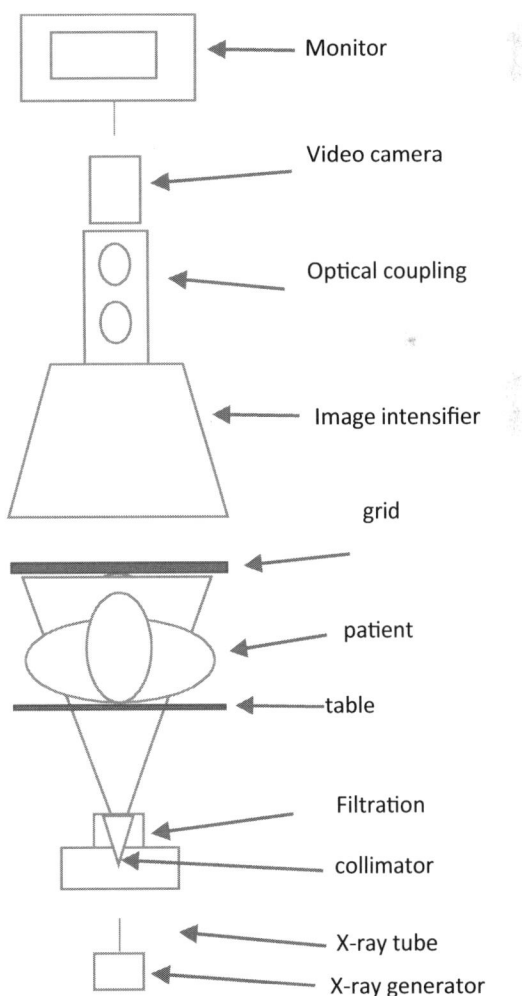

FIG. 6.9 Schematic showing of fluoroscopy system block diagram.

affected (Martin, 2015; Jeans and Stout, 1990; Hanafee and Stout, 1962; Menken et al., 1985). Digital subtraction angiography can be made use of to carry out complex non vascular interventional procedures, such as vertebroplasty, kyphoplasty in the setting of osteoporotic vertebral fractures (Mishra and Banode, 2016).

6.3.5 Digital radiography and picture archival and communication system

Digital radiography is also known as filmless radiography. The crux behind this modality is its ability to digitally acquire a radiograph or a radiological image and transfer it to electronic media to various user interfaces, to various clinicians, radiologists, surgeons as may be needed. This requires usage of specific dedicated software packages. These software packages are available in the market under the category of picture archival and communications system (PACS) and are sold by multiple brands. One of the greatest advantages of digital radiography is its ease of use with regards to the user device – a personal computer, a handheld mobile device, electronic tablets, etc., which is subject to its operating system compatibility, respectively. The software can be continuously made better in operation by proper customer feedback and regular updates by the company/companies providing the said products. Digital radiography can be said to be applicable to all radiological image formats, be it plain radiography, computed tomography, digital subtraction angiography, fluoroscopy, etc., which redefines and cements its utility in the present scenario (Marchiori, 2014). A great advantage again is upgradation of the respective user interface of the software which can be aimed at being more and more user-friendly and task oriented at the same time.

6.3.6 Dual energy X-ray absorptiometry

The term "dual energy" here means usage of two X-ray beams with different energy levels. These X-ray beams are directed toward the area of interest and the X-rays, which interact with adjacent soft tissue component and are absorbed or attenuated are subtracted by using dedicated software packages. Thus, the resultant image invariably renders the density of area of interest post software driven subtraction of soft tissue (Mettler and Fred, 2008; Lorente-Ramos et al., 2011). This imaging modality is incorporated in situations where there are irregularities with reference to density of the human bone, in conditions like osteopenia or osteoporosis where the bone density decreases below the normal limit (El Maghraoui and Roux, 2008; Guglielmi et al., 2011).

6.3.7 Dual energy computed tomography

The same principle applies here as well – application of two X-ray beams at different energy levels a described in the previous imaging modality. The overall functioning here is similar to that computed tomography. There is generation of two data sets of images acquired – one set for each X-ray beam applied at a required energy level unlike an image data set generated by computed tomography. The main advantage here is that application of two X-ray beams with different energy levels makes it possible for the radiologist to allocate substance specific spectra for identification and makes the process of mapping of a substance feasible (Vlahos et al., 2010; Godoy et al., 2011; Lu et al., 2010). This is very useful in the setting of kidney stones or stones in the gall bladder. Also, a musculoskeletal utility of the same is identification of uric acid deposition along the affected joint as in gouty arthritis.

6.3.8 Orthopantomography

This imaging modality is generally used though not entirely limited to dental medicine. It helps to achieve a "panoramic" radiographical image of the required area of interest such as the mandible, the maxilla, teeth, superior alveolar process,

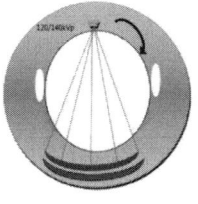
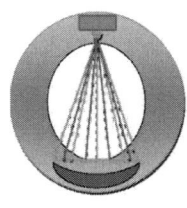
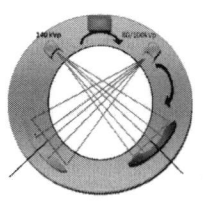

a b c

FIG. 6.10 Schematic showing (A) detector-based spectral CT, (B) single source DECT, and (C) dual source DECT.

inferior alveolar process, oral soft tissue in some cases, etc. (Rushton and Horner, 1996). A "panorama" means displaying a certain part wider than it actually is (Nikjoo et al., 2012). It serves as a highly useful and inexpensive modality in the setting of overall dental evaluation, trauma to the jaw, evaluation of tooth, or oral soft tissue infection or inflammation, dental cyst evaluation, diagnosing salivary gland stones, unerupted teeth, supernumerary teeth, and even possible localization of foreign body within the jaw.

6.4 Physics of photodiagnostic techniques

Most commonly photoelectric and Compton effect are seems to be prominent in diagnostic radiology procedures with X-rays while phenomenon like pair production with radioactive isotopes based procedures in nuclear medicine. Desirable characteristics of radiation for diagnosis should have sufficient penetrating power and can able to produce signals carrying diagnostic information same time it should be safe to patient and controllable with respect to energy, intensity, etc., most important it should produce large output to facilitate short exposure time to avoid movement un-sharpness and economically fit (Johns and Cunningham, 1983; Gambini, 2010). Photography is a transmission imaging (X-rays) procedure as well as emission-based imaging (nuclear medicine) procedure, where primary image is not visible but can be made visualizes on film as static image. Medical radiography is referred to as projection imaging where acquisition of a two-dimensional image of the three-dimensional anatomy is possible using physics and mathematical algorithms (Mahesh, 2002; Penney et al., 2001).

6.4.1 Interaction of radiation with matter

X-rays or photons will interact with an atom and cause electrons to migrate as they pass through a substance. In the above, an electron has been described as travelling through the medium and interacting with other atoms, which causes ionization and excitation. As a consequence, the cells are partly or completely destroyed, depending on the level of the deposited energy. Additionally, an immense quantity of heat is produced (Nikjoo et al., 2012).

Summarizing, X-rays, or photons emit energy, which is transferred to electrons, and the electrons then transmit that energy to the cell system, ultimately producing the biological effect. Thus, indirectly ionizing radiations are referred to as secondary ionization. These wavelike and particle-like properties are referred to as an interaction that is similar to waves and particles. When X and gamma rays interact with structures with wavelengths comparable to their own, the interaction will produce visible output. When photons of different energies engage with atoms, these photons lower the energy level of the atoms, while photons of higher energy interact with electrons, and photons of the highest energy interact with nuclei. Five process types may conduct the aforementioned structural level interactions: coherent scattering, photoelectric absorption, Compton scattering, pair creation, and photodisintegration. Photoelectric absorption and Compton scattering are two of the most important interactions in diagnostic radiology (Klassen, 2011).

6.4.1.1 Attenuation

Attenuation is a result of absorption and scattering. It describes how light may be absorbed and scattered by an object, thereby decreasing the amount of light in a beam. When a beam travels through an absorber of thickness x, absorption, and dispersion take place. Because of this, the transmitted beam will contain fewer photons (McCaffrey et al., 2007).

6.4.1.2 Rayleigh or coherent scattering

The photon interacts with an atom's electron, causing the atom to become excited. Excess energy from the excited atom is released as a dispersed X-ray with a wavelength equal to the incoming photons. The energy of the radiated

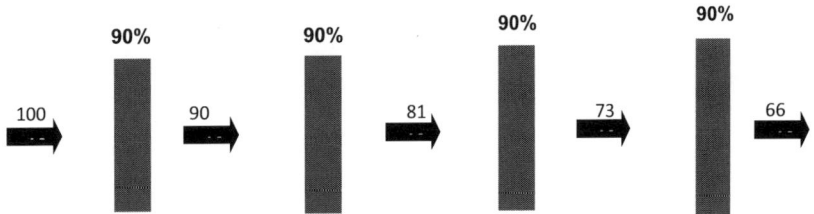

FIG. 6.11 Schematic showing the attenuation of incident by absorber.

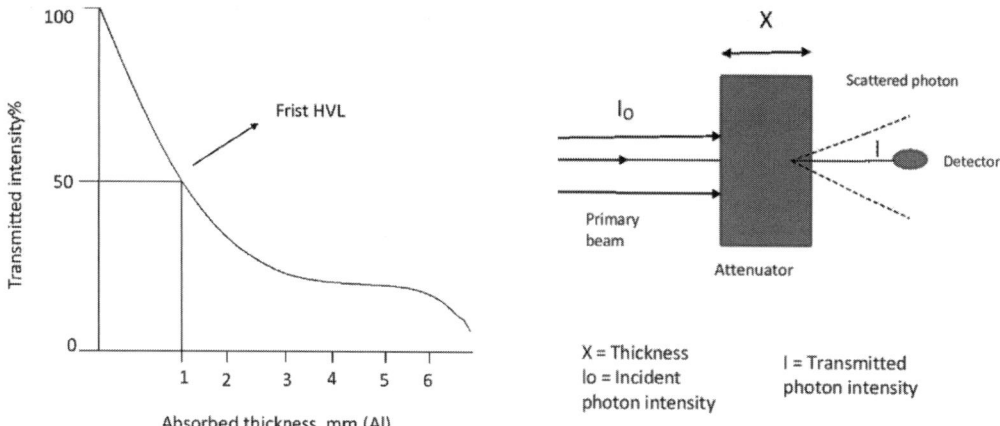

FIG. 6.12 Schematic graph showing decrease in intensity with increase in thickness and the decrease in the intensity due to absorber in the path of the incident beam.

radiation will be the same as that of the incoming photon. However, the scattered photon's direction differs from that of the incoming photon. As a result of coherent scattering, the photon changes direction without changing wavelength. There is no energy transfer or ionization in this process, and the majority of photons are dispersed ahead (Young, 1981). As the X-ray energy drops, so does the scattering angle. This interaction takes place mostly with low-energy photons.

6.4.1.3 Compton scattering

In Compton scattering, a photon interacts with a free electron of an atom and gains some of its energy in the form of an emission. The atom then ejects its electron, resulting in the remainder of the energy being transferred to the valence electron. In addition to damaging tissue, an ejected electron loses energy by ionizing and exciting atoms in the tissue, which raises the patient dose. If Compton scattering or photoelectric absorption are absent, the scattered photon will simply pass through the material without any contact. The photon's wavelength will be longer once it is dispersed. When an incoming photon has the same energy as the scattered photon plus the ejected electron, it is referred to be an incoming photon of E0. Compton scattering is seen in all tissue, from low-energy X-rays to those with very high energy.

An important element of soft tissue contact occurs in the diagnostic energy range (100 keV–10 MeV). Because the scattered X-rays provide little useful information, have decreased picture contrast, and may be hazardous in radiography and fluoroscopy, these images are not very useful (Daniel et al., 2013). Fluoroscopy may emit significant amounts of radiation that adds to occupational radiation exposure.

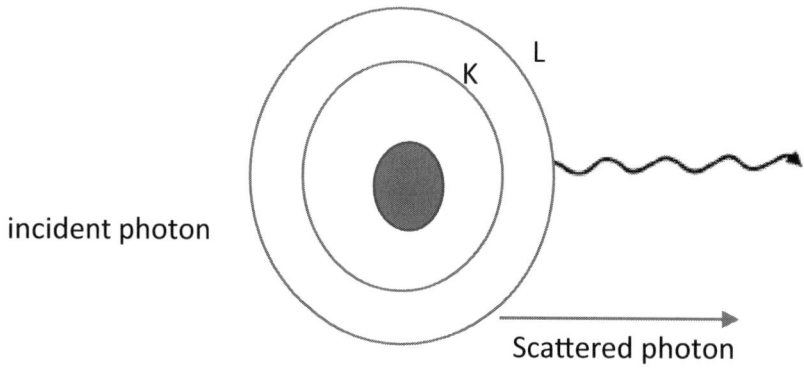

FIG. 6.13 Coherent scattering, the incident photon, and scattered photon have same wavelength.

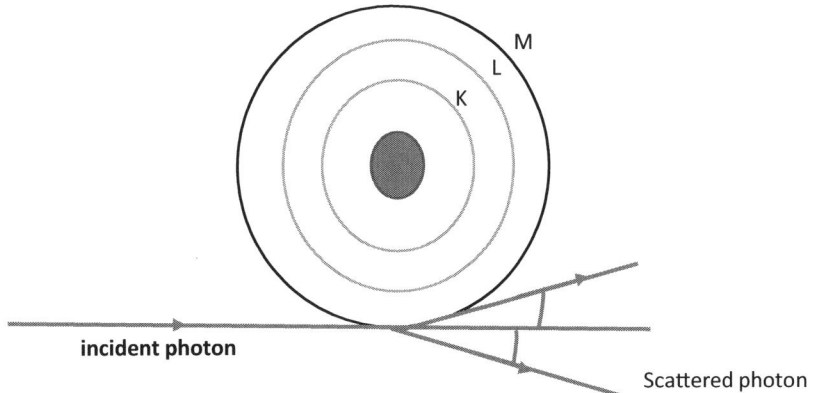

FIG. 6.14 Compton scattering.

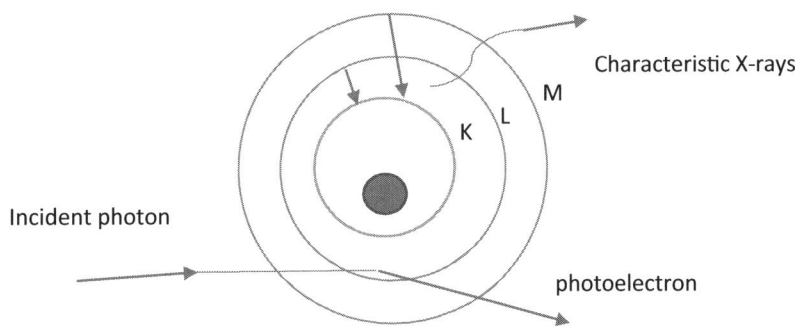

FIG. 6.15 Photoelectric effect.

6.4.1.4 Photoelectric effect

One of the bound electrons is released from the K or L shells through the photoelectric effect (PE) when a photon of energy E collides with an atom. The electron that has been forcefully ejected from an atom is known as a photoelectron, and its kinetic energy equals the sum of the electric binding energy and the kinetic energy of the ejected electron.

During this process, all incoming photon energy is transferred to the electron. Photons with energies equivalent to or larger than the orbital binding energy of electrons will be absorbed and cause the photoelectric effect. After the photoelectric effect, the atom is considered to be ionized and there is a vacancy in the shell. This position has been filled by an electron with lower binding energy from a higher orbit. This will set off a chain reaction of electron transition events, beginning with those in the outer orbit and working inward. The differentiating characteristics The photoelectric effect becomes important for soft tissue imaging when photons have energy of 50 keV (Rosenstein, 1976; Corde et al., 2004). It is used to differentiate between two tissues whose atomic numbers are slightly different.

6.4.1.5 Pair production

A photon with energy higher than 1.02 MeV will be subjected to a strong nuclear field if it gets inside the nucleus of an atom. A rapid transformation may occur, with the photon changing into a positron and an electron pair. Electron and positron both need 0.511 MeV of energy to create. If an additional MeV is supplied, then 0.511 is transferred as kinetic energy to the electron and 0.512 as to the positron. Light interacts with the nuclear field. Einstein's famous energy conversion into mass prediction is on display here. In order to start off, the pair production process requires a minimum energy of 1.02 MeV. When energy is involved, the probability of a pair bond occurring increases. Similarly, when the atomic number increases, the likelihood of pair formation also increases (Johns and Cunningham, 1983). For photons with energy higher than 5 MeV, it is necessary.

6.4.2 Importance of interaction in tissue

6.4.2.1 Differential absorption

The X-ray, as a result of Compton scattering and the photoelectric effect, interacts with a portion of the human body while another portion of the body is exposed to it without touch. Information provided by the Compton scattered X-ray is not

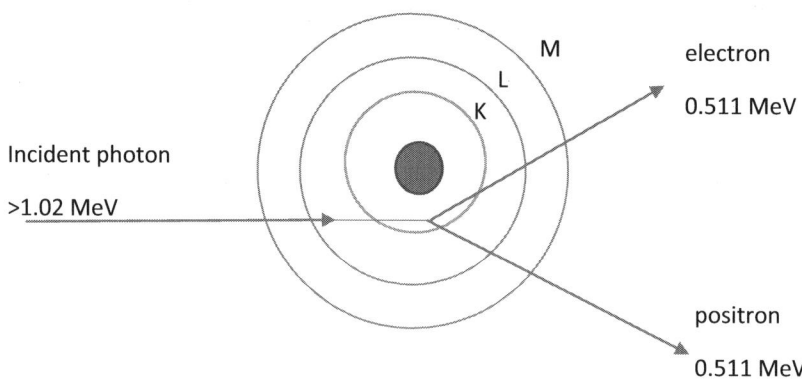

FIG. 6.16 Pair production.

helpful for image production. But it makes it difficult to diagnose the problem because of the added noise. This prevents scatters from reaching the detector, therefore allowing proper treatment procedures to be used. Photodynamic detection uses the photoelectric effect to give information for diagnosis and image generation in the detector. While they have a high absorption coefficient, bone-like anatomical structures are radiopaque, which produces white areas (or bones) on radiographs. Because the X-rays that pass through the body without hindering it arrive at the detector, producing dark areas (black) on the radiograph, the image will appear black on film. X-rays and above show anatomical structures to be radiolucent. The radiographic image is formed by the X-ray absorption differences of the photoelectric process, contrasted with X-rays that aren't exposed to any contact. Differential absorption is a term used to describe this disparity (Tillman et al., 1996). However, reducing kVp results in improved differential absorption and image contrast, albeit at the expense of increasing patient dose.

6.4.2.2 Atomic number

The soft tissue's odds of photoelectric absorption are estimated to be about Z^3. The atomic numbers of bone and soft tissues are 13.8 and 7.4, respectively. The chance of an EPIRB activation is 13.8 times $(7.4/3)3$ higher in bone compared to soft tissue. When the amount of energy rises, the probability of this event diminishes. In other words, at high X-ray energies, there are just a few interactions, and the amount of X-rays transmitted is higher without any interaction. Tissue Z has no effect on Compton scattering. The probability of Compton scatter is the same in bone and soft tissue, which decreases with increasing energy. This degradation is slow compared to photoelectric absorption, which decreases quickly (Phelps et al., 1975). Thus, Compton scattering is the dominant scattering mechanism at high photon energy.

6.4.2.3 Mass density

Whether it is the nature of the contact or the fact that X-rays are being used, the effect of X-rays on tissue is in proportion to the tissue density. A large volume of substance with a mass density of $1\ kg/m^3$. It shows how tightly the atoms are packed relative to their mass. As the mass density increases, the number of electrons, which in turn influences the overall interaction, also increases. The Z-related photoelectric effect, in addition to mass density, contributes to differential absorption. Bone, which is denser than soft tissue, absorbs and disperses X-rays twice as much ($1850/1000 = 18.5$). The mass density of the lungs helps in imaging in radiography. Air-filled soft tissue cavities may be seen as a result of the difference in tissue density. The atomic number and mass density of barium and iodine are two of the highest and greatest contrast agents available. When they are implemented using the low kVp technique, they may be used to inspect inside organs. To make things easier, employing a high kVp technique will let you view the lumen of the organ (Jones et al., 2003). The X-ray helps in furthering the clarification of the organ in this case.

6.4.2.4 Photon energy

The kilovoltage range in diagnostic radiology is around 20–150 kVp, and effective photon energy varies from 15–100 keV. Over this spectrum, the relative relevance of interaction changes. Photoelectric absorption, rather than Compton scatter, is the primary source of attenuation at low energies. As a result, soft tissue and bone look dark and bright on the X-ray film, respectively. The thickness of the tissue also plays a role in the attenuation process. The same amount of attenuation may be produced by a thick layer of soft tissue and a thin layer of bone. However, by combining low energy with photoelectric effect interaction, the thickness effect may be overcome. Compton scatter is dominating at high energies, and the distinction between soft tissue and bone is lost. Despite the fact that Compton scatter is density dependent, tissue distinction is still achievable using high kVp methods. At low kV X-rays, such as mammography, photoelectric absorption accounts for approximately 75% of the attenuation in soft tissues. At this energy, the Compton scatter has just a tiny effect. At higher energies, photoelectric absorption accounts for 15%–20% of attenuation in soft tissues, and Compton is dominating, as shown in chest radiography or gamma imaging. Starting with X-rays how are they produced we have discussed in the separate radiographic section above now how they are producing image we will discuss. The X- rays produced cannot be seen by the human eye, so they must be converted to a visual image. This is done by the property called luminescence. Luminescence refers to emission of light by substance. There are certain substances, which generate light like phosphor. When X-rays fall on phosphor light is generated which then act on the X-ray film to generate image. The whole structure is called intensifying screen. The intensifying screen consists of base or support made of plastic, reflecting layer made of titanium dioxide, a phosphor layer, and a protective coat all glued together. The phosphor used in the intensifying screen is of cesium iodide. It produces light in blue color. The efficiency with which the phosphor converts X-rays into usable light photons is known as intrinsic conversion efficiency of phosphor. The more the thickness of the phosphor layer the maximum amount of light will be generated. The cassette in which the intensifying screens are mounted must be tight contact with the x ray films. This leads to generation of the image. Now a days photostimulable phosphors are used which are made up of europium activated barium fluoride bromide (BaFBr:EU). Europium is used an activator. It makes traps at the activator site in the crystalline structure. To prepare phosphor plate for image formation it is first flooded with light from high intensity sodium discharge lamps, which leads to erasure of any image remaining from prior exposures. The plate is now exposed to X- rays, during the exposure electrons in the phosphor are excited into the higher state, out of theses almost half return to the normal energy levels almost instantly and are not available for image formation. The rest of the excited electrons are trapped and form the latent image. These trapped electrons are removed from the trap using thermal agitation in the X-ray reader. The image is readable for up to 8 hours in phosphor plate at room temperature. The cassette is then removed and then placed in the reader. In the reader it is exposed to a light from helium neon laser (633 nm). The trapped electrons are exposed to this laser which absorbs the laser energy and get freed up from their trapped state they release light photon which has wavelength different from the neon laser light and gets absorbed by the X-ray film and converted to image (Johns and Cunningham, 1983; Tillman et al., 1996; Phelps et al., 1975; Jones et al., 2003; Martin, 2007).

Till now we have seen how X-rays are converted too light photons now we will see how the light photons after falling on X-ray film get converted to image. X-ray film is a kind of photographic film that is coated on both sides with a polyurethane sheet, and contains a radiation-sensitive emulsion that is developed to generate an image. The important ingredients of an emulsion are gelatin and silver halide. Gelatin helps in keeping silver halide grains well dispersed and preventing it from clumping.

The silver halide is the most important component in emulsion. It is made up of silver bromide (90%–99%) and silver iodide (1%–10%). An impurity (allylthiourea) is added to the emulsion, which acts as a sensitivity speck and traps the free electron. Metallic silver is black the silver in the emulsion is the product which causes dark areas in a developed radiograph. The energy absorbed from the light photon gives electron energy to escape. The electron can move in crystal can move for long distances until it gets encountered by an impurity and forms a compound AgS. The electrons give the sensitivity speck a negative charge which attracts the silver ion in the crystal which on adding with the electron forms a neutralized atom. Thus, on subsequent exciting the electrons by the light photons the sensitivity speck grows and forms clumps of silver. Theses clumps of silver acts a latent image centre. These latent image centers are then developed to form a useful image.

Development is a chemical process that amplifies the latent image centers by a factor of millions. Developer is basically a reducing agent which causes reduction of the silver ion by addition of electron to form metallic silver. The developing solutions contain hydroquinone plus phenidone.

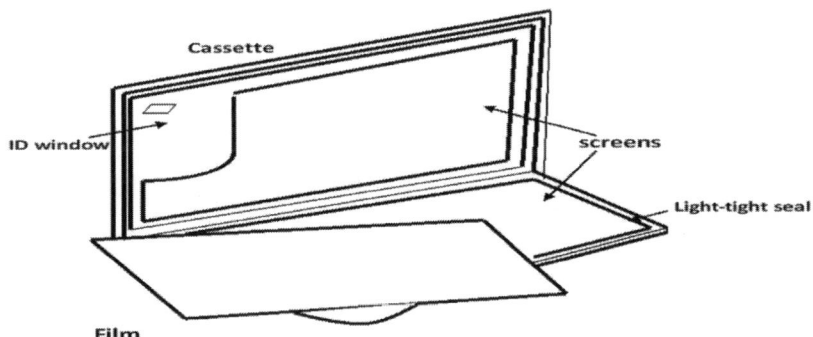

FIG. 6.17 **Schematic showing the luminescent screen.**

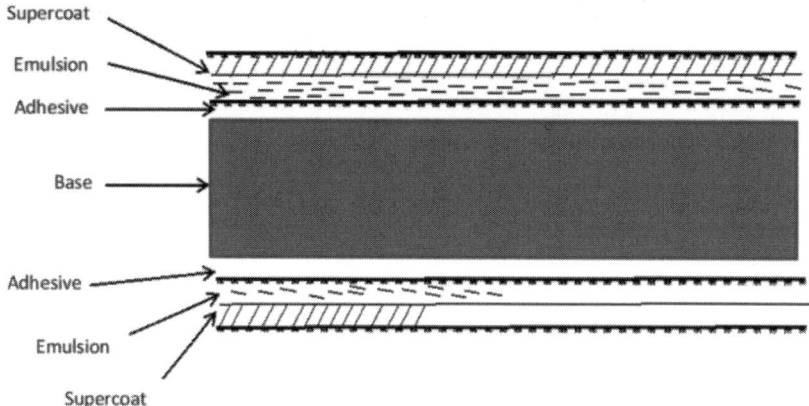

FIG. 6.18 Schematic showing the anatomy of X-ray film.

The areas which get developed appear black and the areas which remain undeveloped remain white thus forming a complete radiographic image to be viewed on a viewer. After developing the film, it is fixed using thiosulfate which helps in immediate usefulness and permanence of the developed radiograph. After fixing the film is washed thoroughly to remove the chemicals to prevent browning of the film on long term storage.

Unlike a traditional x-ray, in which an X-ray tube is stationary and which moves around on a gantry, a CT scanner instead utilizes a motorized X-ray source that revolves around a circular aperture of a gantry. Patients in a CT scan are placed on a table that moves to the gantry, and the X-ray tube revolves, creating a full 360-degree circle with a beam of X-rays while firing much narrower beams of X-rays all around the patient, rather than using film that are used in x-ray detection. X-rays, after they leave the patient, are detected by the metallic detectors and sent to the computers that are specifically programmed to produce a picture.

To generate a 2D image of a patient, each time the X-ray rotates a single time, the Ct computer utilizes specific computational methods. This measurement depends on the CT scanner being used; however, the thickness of the tissue in each picture slice may vary from 1 to 10 mm. The bed in the gantry slides forward after the slice is finished so that another section of the body may be fitted into place. Until the whole field is covered by the scanner, the procedure is resumed. To create a 3D picture of the patient, the image slices are placed on top of each other by a computer to form a stack.

In a fluoroscopy upon hitting the image intensifier tube, the x-ray beam is converted into light photons and then sent to the fluorescent screen. Photo electrons are produced when light photons impact the photocathode. As soon as the electron concentration in the photocathode falls below the cut-off level, the electrons in the photocathode are quickly taken away from the photocathode via the high voltage differential between it and the anode. Electrons travel from the cathode to the anode, where they are concentrated electrostatically and guided to the output fluorescent screen. Electrons which have been emitted from the output screen strike the screen, which transmits the fluoro light that contains the images of the imaging sensors to the eye of the observer.

6.4.3 Picture archiving and communication system

PACS is a facility to view images instantly in any location not only in the hospital but elsewhere too. It eliminates the hassle for storing images and quick retrieval when required (Mezrich, 1988). In PACS the various medical imaging devices and workstations are linked to the PACS workflow manager which controls the flow of images and information (Strickland, 2000). One of the defining characteristics of PACS is that images from different imaging modalities are all formatted in a same universal file format, allowing them to be identified and utilized by the whole system. Digital imaging and communications in medicine (DICOM) format must be utilized for presenting this data.

Images may be imported or exported using the PACS workflow manager. The archive is divided into two parts: current cases are considered a short-term archive while longer-term pictures are defined as a long-term archive.

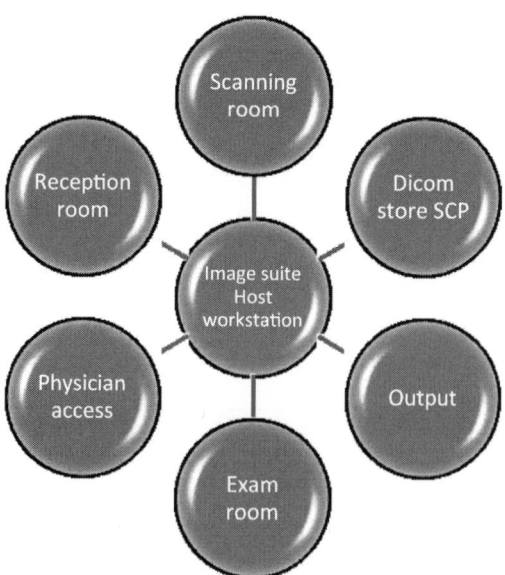

FIG. 6.19 Schematic showing picture archiving and communication system.

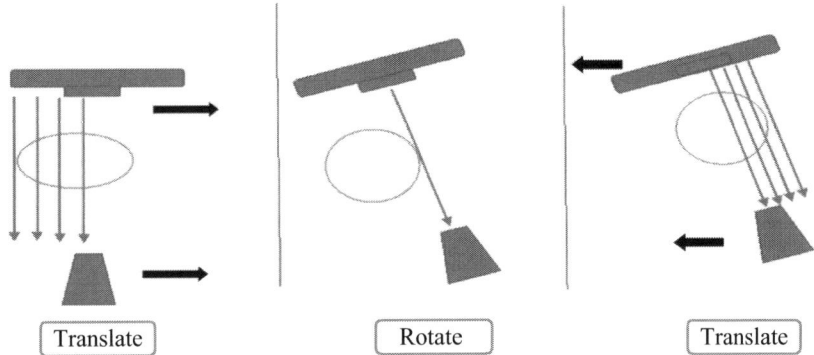

FIG. 6.20 Schematic showing the first generation of CT scanner.

To efficiently use PACS it is to be attached to the hospital information system (HIS) and fully integrated to the radiology information system (RIS). The images are read by workstation for clinical reporting or review (Baker, 1999).

Computed tomography (CT) has a fundamental concept known as the principle of many projections. By obtaining numerous projections of an object, it is possible to build a three-dimensional reconstruction that can be seen in sagittal, axial, and coronal planes (Mezrich, 1988; Ohnesorge et al., 1999). CT scanners are categorized into five different groups. Out-of-the-box thinkers (translate-rotate, one detector). The first-generation CT scanner has a one detector and a pencil-like X-ray beam that it uses to perform tomographic scans of its subjects. Both linear and rotational motions are used in the X-ray tube detector. The time required for a single CT head was around 25–30 minutes.

Second generation (translate-rotate, multiple detectors). The main advantage from second generation was it reduced the scanning time. This was done by replacing the pencil beam with a fan shaped beam and adopting multiple detectors instead of one, however the movements were similar, that is, translate and rotary. In this around, 30 detectors are used. It took around 90 sec for a single CT head.

Third generation (rotate-rotate). In this generation the translate motion was completely removed by rotate and rotate method. Only rotate method was used. Multiple detectors are aligned along the arc of a circle whose center is the x ray tube focal spot and the X-ray beam is collimated into fan shaped.in this around 288 detectors are used. It took around 5 sec for a single CT head.

Fourth generation (rotate-fixed). In this the detectors form a ring that surrounds the patient. The detectors do not move at all; however, the X-ray tube rotates in a circle inside the detector ring. It uses around 1080 detectors in an assembly. Projections are taken at many angles during the rotation of the X-ray tube. Thus, one CT scan will be made up of many projections, each projection taken at a slightly different angle. In today's world, this fourth gen scanners are used.

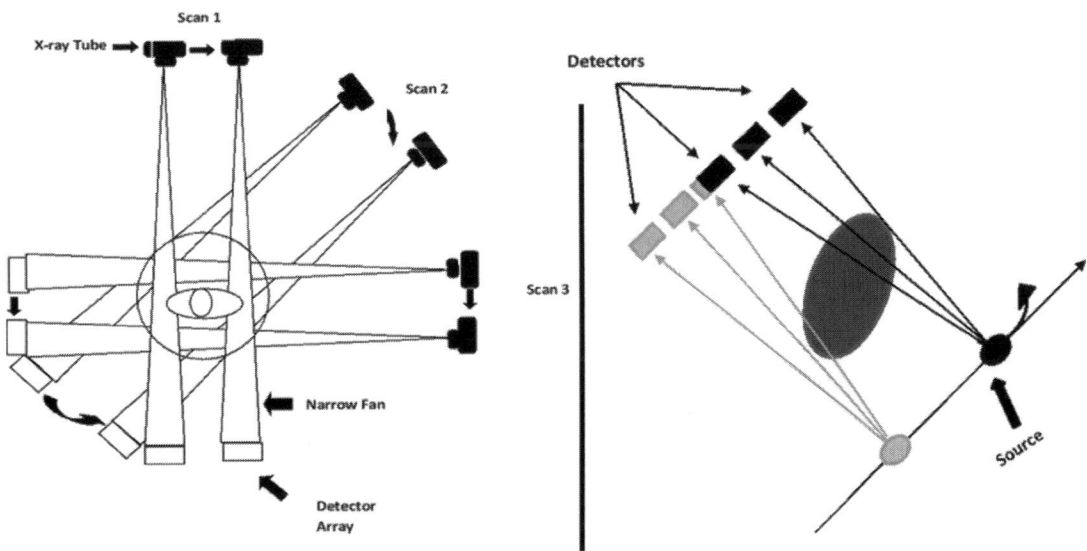

FIG. 6.21 Schematic presentation of second generation of CT scanner.

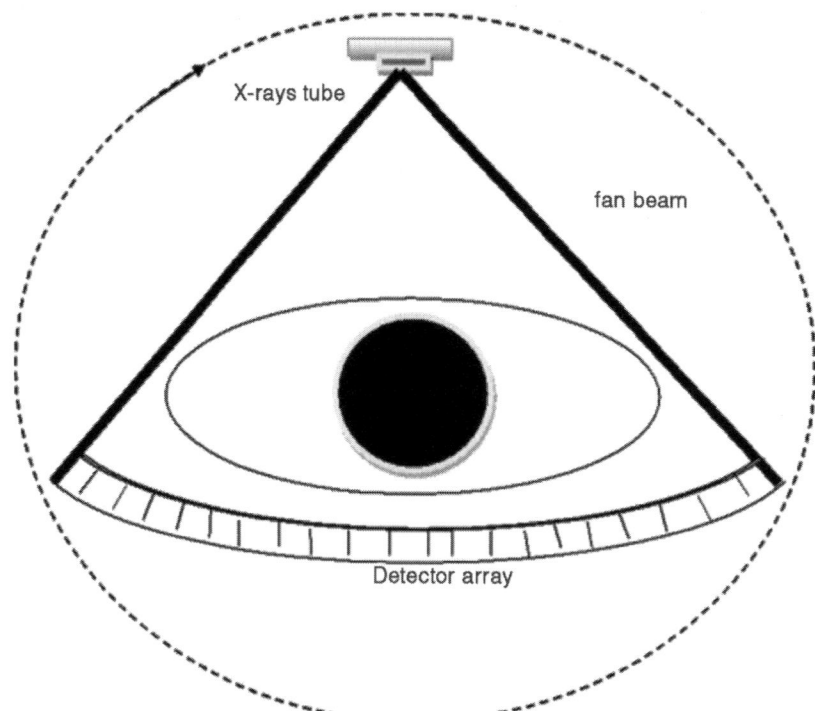

FIG. 6.22 Schematic presentation of third generation of CT scanner.

The detectors are made of scintillation crystals or xenon gas ionization chambers.

Positron emission tomography (PET): This technique uses the energy from the nucleus for production of desired image. The most common positron emitter is fluorine F18. When a F18 emits a positive beta particle (positron) this travels for 2 mm through the patient before annihilation from the negative electron. Their combined mass is converted into energetic photons each of 511keV exactly, which are emitted simultaneously but in opposite directions. PET imaging is based upon identifying these annihilated photons and identifying their true origin from patient to locate the radioactive source (Shreve and Townsend, 2010). A PET scanner consists of a ring surrounding the patient made up of large amount of scintillation detectors around 10,000–20,000 which are of bismuth germanate (BGO). Other substances which can be used for making scintillation detectors are lutetium oxyorthosilicate (LSO) or gadolinium oxyorthosilicate (GSO). These substances have high detection efficiency to absorb and convert 511 keV photons to light have a very short scintillation decay time and good energy resolution.

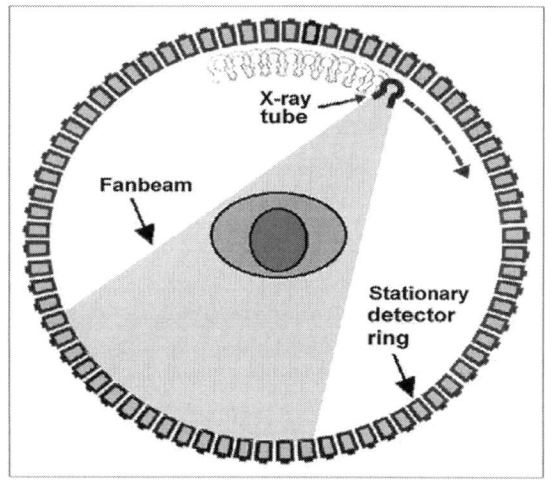

FIG. 6.23 Schematic presentation of fourth generation of CT scanner

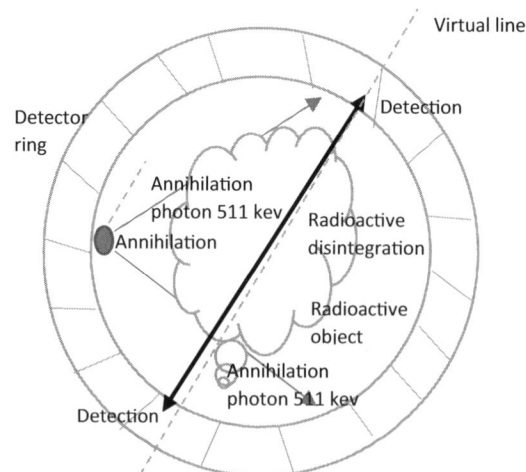

FIG. 6.24 Schematic showing PET CT Scanner.

F18, with a half-life of 110 min, is the primary positron emitter utilised in PET (FDG). Allowing a patient to be administered radioactive material provides them a radiation dosage. The radionuclide will be delivered to the organ of interest, which will then be exposed to the radiation. When the activity in an organ rises, it takes up a greater percentage of the total dosage, absorbs a greater amount of total dose, has a longer effective half-life, and produces a greater amount of beta and gamma radiation. The 3 microgray minimum dosage is administered depends on the clinical condition.

6.4.3.1 Fluoroscopy

For radiology, recording the picture generated by the transmission of X-rays through the body using a digital recording device is referred to as radiography. In contrast, for fluoroscopy, the process of capturing images in real time, rather than producing one image, is called fluoroscopy. Since the introduction of fluoroscopy, both X-rays and fluoroscopy have existed, and fluoroscopy uses zinc sulfide screens. Although the phosphor used in the fluoroscope screen was back by a lead glass, and the viewer looked at the image through the screen, the screen was too dim, requiring a dark room in order for the radiologist to be able to see the contrast of the image (Daly and Templeton, 1999).

The image intensifier consists of an evacuated glass or ceramic envelope that is surrounded by a metal housing which prevents light from getting into the tube and shielding the device from the effects of magnetic fields. Fluoroscopy screening times can be as short as few seconds to minutes. The X-ray generator, which provides power to the tube, is turned on and off at regular intervals to produce short pulses of X-rays. At this rate, the eye is unable to differentiate between a continuous and pulsed image.

Digital subtraction angiography: In order to get a contrast-filled picture of vessels in isolation, digital subtraction angiography is used. This is great for helping provide clarity of a picture, making it easier to identify the target, and using the right amount of contrast medium for better visualization.

Before taking the contrast picture, the noncontrast image is captured. This just depicts the typical anatomy. Once the contrast-filled containers are full, the contrast pictures are taken. It illustrates the full containers seen in the first picture, laid on top of the original image. On a pixel-by-pixel basis, the first picture is subtracted from the second image. This third frame is made by subtracting the original picture from the derivative. In turn, it confirms the existence of vials that are empty yet have the anatomy normally seen in vials. Real-time viewing of the removed pictures is possible.

Images are collected at low and high kV, and dual energy subtraction technique is utilized to combine the results. When running at low kilovolts (KV), high-kV pictures show great contrast between bone and soft tissue, but at high KV, contrast is diminished for bone. If the low kV picture is subtracted from the high kV image, it will lead to a reduction in visibility of bone and only reveal soft tissue, which is the opposite of what happens when high kV image is subtracted from the low kV image (Lu et al., 2012).

6.4.3.2 Orthopantomography

The OPG is one big picture of the jaw, teeth, and a few other things. It is used to evaluate the overall oral health, as well as jaw and tooth fractures, abscesses, tumors, and cysts, as well as joint problems, growth, and development (Nemtoi et al., 2013). The OPG test usually consists of one or both of the x-ray tube and film rotating around the patient while he or she stays steady, from one side of the jaw to the other. The panoramic imaging method developed above, which covers the lower limits of the mandible, upper limits of the maxillary sinuses, and mandibular condyles and TMJ joints (joints in the area where the jaw connects to the cranium), allows for capture of the whole range of the mandible.

Cone beam CT: The diverging cone-shaped source of radiation is guided through the target using computed tomography. X-ray detector with numerous dexdxs (analogous to pixels) is used to detect the attenuated X-rays on the opposite side. With fewer revolutions of the X-ray tube gantry, volume acquisition is achieved. There are numerous benefits to using fan beam CT as opposed to standard fan beam CT. It takes less time to perform, less patient movement is introduced, and the scan tube is more efficient (Nemtoi et al., 2013).

6.4.3.3 Dual energy compute tomography

This imaging technique is also known as spectral CT, which is performed using two distinct X-ray photon energy spectra to create a single picture. A single picture is generated with standard CT; however, dual energy CT may be utilized to produce a multitude of image forms.

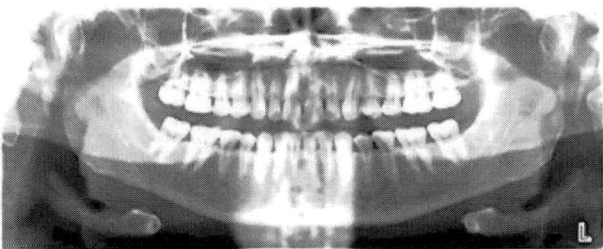

FIG. 6.25 Schematic showing normal orthopantomogram view.

These methods are often divided into those that are used before the patient is scanned (prospective) and those that follow scanning (post scan) (retrospective) (Villari et al., 1999).

6.4.3.4 Prospective techniques

It has two X-ray tubes, one at 90 degrees from the other, that output different voltages. As both the detectors are of the same size, it has a restricted number of files of view. Both datasets are collected at the same time, and therefore temporal resolution is excellent.

Using a single helical scan as well as coregistration, the successive acquisitions occur at two distinct tube potentials and afterwards, for post-processing, the scans are coregistered. The temporal resolution is low since the subject is scanned twice, but it has a complete field of view (field of vision). One source single beam twin-lens: this procedure splits the X-ray beam into high-energy and low-energy spectra, which are sent to the patient along the z-axis. Only one tube at a time. It has a complete FOV, but because of the tube rotation, it has a small decrease in temporal resolution.

6.4.3.5 Retrospective techniques

A dual-layer DECT is one where the lower-energy photons pass through the upper layer of the detector, while the higher-energy photons are absorbed by the lower layer. As both datasets are collected at the same time, it has a complete FOV and exhibits excellent temporal resolution.

The specific binding energy (also known as the K-edge) of any drug is independent of its chemical formula. The DECT uses various photoelectric energies and the KS edge to help it work. Both iodine and calcium have greater K edges, with calcium having a K edge of 33.2 keV and iodine having a K edge of 4 keV, enabling the iodine and calcium to detach themselves from the surrounding structures. DECT (Delay Enhanced Cordless Telecommunications) uses x-ray tubes of 140 kilovolts (kV) and 80 kilovolts (kV) with an angle offset of 90 degrees.

Dual energy X- ray absorptiometry (DEXA) scan: It is a technique used in the diagnosis of osteopenia and osteoporosis. Values are calculated primarily for the lumbar vertebrae and femur bones (Webster, 2016). Bone mineral density (BMD) is calculated specifically in g/cm square and then compared with two reference population giving two score. These are T score and Z score.

T score: It is calculated in comparison by standard deviation to a young population, matched for sex and ethnicity and classified as:

- Score greater than equal to -1 is normal
- Score less than -1.0 to greater than -2.5 is osteopenia.
- Score less than equal to -2.5: osteoporosis
- Score less than 2.5 with fragility fractures is severe osteoporosis.

Z score: It is calculated by comparing standard deviation to an age, sex and ethnicity population. In this <-2.0 is considered low bone density for age.

6.5 Opportunities, challenges, and limitations of photodiagnostic techniques

The responsibility of radiodiagnosis has increased by many folds with advent of computed radiography, digital radiography, MDCT and the nuclear imaging techniques has given a major boost in the timely and definitive management for the severely injured in any setting. Limitation, optimization, and justification of radiation doses are the real concerns, radiological safety of patient and professionals are equally important, real time imaging procedures needs special consideration with increase demand or well-trained radiation professionals to handle the radiation generating equipment's with open research and employment opportunity (Gupta et al., 2017; Villari et al., 1999).

Opportunities: The world of radiology is constantly evolving since time being. The first radiograph was invented by Wilhelm Conrad Roentgen on November 8, 1895, on which day every year international day of radiology is being celebrated worldwide. Since its inception in 1895 it has evolved manifold times providing a boon in diagnostic procedures and have helped medical field unearth various new pathologies. Right from X-rays to MRI the journey has been long and fascinating and have achieved a pivotal role in medical field. Form a subspeciality, it has developed into a new specialty with superspecialty in interventional radiology and that is utterly possible due to photo diagnostic techniques.

In today's world there is high demand of experts in these techniques. As the future is unraveling new techniques are coming up which are safe and less time consuming without compromising on image clarity. These techniques have come up a long way from diagnosing diseases to treating them. The same x rays which are used for diagnosing pathologies are used in the form of DSA in a modified manner to treat them. So, these techniques are playing a pivotal role in patient care.

A whole new branch has come up using these techniques for curative purpose known as Interventional radiology which uses fluoroscopy to treat patients both curative and palliative intent.

The main advantage of photo diagnostic technique is without opening the abdomen surgeon can diagnose the problem and intensivist can look for the tubes, a pulmonologist can look for lung disorders and neurologist can look for brain issues (Bazzocchi et al., 2012; Reyes et al., 2020; Griffith et al., 2019). It not only helps in acute emergencies in saving patients life but also helps in directing appropriate treatment without which devastating effects could have happened.

Limitations: The major limitation associated with photo diagnostic technique is radiation which after a threshold dose can be harmful for the patient. Photodiagnostic techniques cannot be used in pregnant patients as the radiations can cause mutations in the growing embryo leading to catastrophic deformity in new born, hence avoided in them. No one photo diagnostic technique is perfect one, each has to be supplemented with other for able to diagnose a disease, for example, X-rays are useless in diagnosing brain disorders as the X-rays cannot pierce skull bone. Soft tissues cannot be appreciated in x rays. CT does not give real time cine information of the body, for that requires fluoroscopy.

Long-term exposure to these techniques can lead to cancer in the technicians or radiologist in spite of wearing protective equipment's. Protective equipments like gown, shield, glass have to be worn every time while performing these procedures.

These techniques require constant smooth flow of electricity every time they are used without electricity they are of no use. Barring x ray machines these procedures are not portable and cannot be used bedside, the patient has to be shifted to the machine with all its goodies (Kahn and Charles, 2017; Parks et al., 1970; Reiner et al., 2007). These equipment's are extremely costly and wear and tear with maintenance requires lots of money.

Challenges: The major challenge in these techniques is to make them safe for all in terms of minimizing radiations and not compromising on efficiency. Another major challenge is their portability (Kilani et al., 2015). Once installed they are difficult to move to other places. The other challenge is cost effectiveness, these are extremely costly and requires high maintenance. The curative use of these techniques is less as compared to the diagnostic part (Cohen et al., 2005). The component wear and tear is very much and requires regular maintenance. Radiation leakage is a very big challenge of these techniques and proper and efficient shielding is of utmost importance to prevent this.

References

Hudson River Museum. The Panoramic River: The Hudson and the Thames. Hudson River Museum, 2013.

Dual Energy X ray Absorptiometry - Bone Mineral Densitometry". International Atomic Energy Agency. 2017-08-07.

"ALARP Guidance Note" (PDF). Commonwealth Offshore Petroleum and Greenhouse Gas Storage (Safety) Regulations 2009. NOPSEMA. June 2015. Archived from the original (PDF) on 16 June 2016. Retrieved 20 May 2016.

"fluoroscope". Oxford Dictionaries UK Dictionary. Oxford University Press. Retrieved 2016-01-20.

thefreedictionary.com ≥ scintigraphy Citing: Dorland's Medical Dictionary for Health Consumers, 2007 by Saunders; Saunders Comprehensive Veterinary Dictionary, 3 ed. 2007; McGraw-Hill Concise Dictionary of Modern Medicine, 2002 by The McGraw-Hill Companies.

Cone Beam-Computed Tomography in Endodontics" (PDF). www.aae.org. Summer 2011. Retrieved October 21, 2019.

"Individual State Licensure Information". American Society of Radiologic Technologists. Archived from the original on 18 July 2013. Retrieved 19 July 2013.

"Patient Page". ARRT – The American Registry of Radiologic Technologists. Archived from the original on 9 November 2014.

"CT scan – Mayo Clinic". mayoclinic.org. Archived from the original on 15 October 2016. Retrieved 20 October 2016.

Akselsson, R., Johansson, T.B., 1974. X-ray production by 1.5–11 MeV protons. Zeitschrift für Physik 266 (4), 245–255.

Bailey, D.L., Townsend, D.W., Valk, P.E., Maisy, M.N., 2005. Positron Emission Tomography: Basic Sciences. Springer-Verlag, Secaucus, NJ ISBN 978-1-85233-798-8.

Baird, C.S., 2019. Electromagnetic radiation. Access Sci.

Baker, S.R., 1999. PACS and radiology practice: enjoy the benefits but acknowledge the threats. Picture archiving and communications systems. Am. J. Roentgenol. 173 (5), 1173–1174.

Bazzocchi, A., et al., 2012. Incidental findings with dual-energy X-ray absorptiometry: spectrum of possible diagnoses. Calcif. Tissue Int. 91 (2), 149–156.

Board, Atomic Energy Regulatory. "Government of India." 2006. Annual Report. Safety Surveillance of Radiation Facilities. Available at: http://www.aerb.gov.in/AERBPortal/pages/English/t/annrpt/2006/chapter3.pdf (Accessed January 28, 2014) (2016).

Chmielewski, A.G., Haji-Saeid, M., 2004. Radiation technologies: past, present and future. Radiat. Phys. Chem. (1-2), 17–21 71.

Cohen, M.D., et al., 2005. Challenges facing radiology educators. J. Am. Coll. Radiol. 2 (8), 681–687.

Cohen, S., Liewehr, F., 1998. Diagnostic procedures. Pathways of the Pulp 8, 1–30.

Corde, S., et al., 2004. Synchrotron radiation-based experimental determination of the optimal energy for cell radiotoxicity enhancement following photoelectric effect on stable iodinated compounds. Br. J. Cancer 91 (3), 544–551.

Daly, B., Templeton, P.A., 1999. Real-time CT fluoroscopy: evolution of an interventional tool. Radiology 211 (2), 309–315.

Daniel, O., Ogbanje, G., Jonah, S.A., 2013. X-Rays and scattering from filters used in diagnostic radiology. Int. J. Sci. Res. Publ. 3 (7), 1–11.

Dixon, A.K., 1997. Evidence-based diagnostic radiology. Lancet North Am. Ed. 350 (9076), 509–512.

Dixon, A.K., Freer, C.E.L., Appleton, D.S., 1993. Quality assurance in radiology. Proceedings of the Royal Society of Edinburgh, Section B: Biological Sciences 101, 321–342.

El Maghraoui, A., Roux, C., 2008. DXA scanning in clinical practice. QJM 101 (8), 605–617. doi:10.1093/qjmed/hcn022.
Ell, P.J., 2006. The contribution of PET/CT to improved patient management. Br. J. Radiol. 79 (937), 32–36.
Fishman, E.K., Jeffrey, R.B., 1995. Spiral CT: Principles, Techniques, and Clinical Applications. Raven Press (ID). ISBN 978-0-7817-0218-8.
Gambini, D.J., 2010. Basic concepts of radiology physics. J. Radiol. 91 (11 Pt 2), 1186–1188.
Garcia, J.D., 1970. X-ray production cross sections. Phys. Rev. A 1 (5), 1402.
Godoy, M.C., Heller, S.L., Naidich, D.P., 2011. Dual-energy MDCT: comparison of pulmonary artery enhancement on dedicated CT pulmonary angiography, routine and low contrast volume studies. Eur. J. Radiol. 79 (2), e11–e17. doi:10.1016/j.ejrad.2009.12.030.
Grammaticos, P.C., 2004. Pioneers of nuclear medicine, Madame Curie. Hell. J. Nucl. Med. 7 (1), 30–31.
Griffith, B., Kadom, N., Straus, C.M., 2019. Radiology education in the 21st century: threats and opportunities. J. Am. Coll. Radiol. 16 (10), 1482–1487.
Guglielmi, G., Muscarella, S., Bazzocchi, A., 2011. Integrated imaging approach to osteoporosis: state-of-the-art review and update. Radiographics 31 (5), 1343–1364. doi:10.1148/rg.315105712.
Gupta, A.K., Garg, A., Khandelwal, N., 2017. Diagnostic Radiology: Gastrointestinal and Hepatobiliary Imaging. JP Medical Ltd.
Gustafson, B.o.Å.S., et al., 2001. Interactions with electromagnetic radiation: theory and laboratory simulationsInterplanetary Dust. Springer, Berlin, Heidelberg, pp. 509–567.
Hanafee, W., Stout, P., 1962. Subtraction technic. Radiology 79 (4), 658–661. doi:10.1148/79.4.658.
Hatcher, D.C., 2010. Operational principles for cone-beam computed tomography. J. Am. Dent. Assoc. 141 (Suppl 3), 3S–6S. doi:10.14219/jada.archive.2010.0359. PMID 20884933.
Holmes-Siedle, A., Adams L., 1993. "Handbook of Radiation Effects."
Hsieh, J., 2003. Computed Tomography: Principles, Design, Artifacts, and Recent Advances. SPIE Press, pp. 265. ISBN 978-0-8194-4425-7.
Jeans, W.D., Stout, P., 1990. The development and use of digital subtraction angiography. Br. J. Radiol. 63 (747), 161–168. doi:10.1259/0007-1285-63-747-161. PMID 2185864.
Johns, H.E., Cunningham J.R., 1983. "The physics of radiology."
Jones, A.K., Hintenlang, D.E., Bolch, W.E., 2003. Tissue-equivalent materials for construction of tomographic dosimetry phantoms in pediatric radiology. Med. Phys. 30 (8), 2072–2081.
Kahn Jr, Charles E., 2017. "From images to actions: opportunities for artificial intelligence in radiology." 719–720.
Keiser, G., 2016. Fundamentals of light sourcesBiophotonics. Springer, Singapore, pp. 91–118.
Kelloff, G.J., Hoffman, J.M., Johnson, B., Scher, H.I., Siegel, B.A., Cheng, E.Y., et al., April 2005. Progress and promise of FDG-PET imaging for cancer patient management and oncologic drug development. Clin. Cancer Res. 11 (8), 2785–2808. doi:10.1158/1078-0432.CCR-04-2626. PMID 15837727.
Kilani, M.S, et al., 2015. Ethylene vinyl alcohol copolymer (Onyx®) in peripheral interventional radiology: indications, advantages and limitations. Diagn. Interv. Imaging 96 (4), 319–326.
Klassen, S., 2011. The photoelectric effect: reconstructing the story for the physics classroom. Sci. Educ. 20 (7), 719–731.
Kohout, M., Savin, A., 1996. Atomic shell structure and electron numbers. Int. J. Quantum Chem. 60 (4), 875–882.
Lacoste, A.C., Bourguignon M., and Godet J.L.. "Radiation protection of patients in medical radiology: proposal for a worldwide action plan." Session 2: Medical use of radiation: 291.
Lehtinen, E., Racoveanu, N.T., 1985. WHO programme on the rational use of radiodiagnostic investigations, Criteria and methods for quality assurance in medical X-ray diagnosis. Proceedings of the scientific seminar held in Udine. Italy 17-19 April 1984.
Li, L.B., et al., 2001. Medical radiation usage and exposures from medical X ray diagnosis in Shandong province of China. Radiat. Prot. Dosim. 93 (3), 261–266.
Lorente-Ramos, R., Azpeitia-Armán, J., Muñoz-Hernández, A., 2011. Dual-energy x-ray absorptiometry in the diagnosis of osteoporosis: a practical guide. AJR Am. J. Roentgenol. 196 (4), 897–904. doi:10.2214/AJR.10.5416. Pubmed citation.
Lu, G.M., Wu, S.Y., Yeh, B.M., 2010. Dual-energy computed tomography in pulmonary embolism. Br. J. Radiol. 83 (992), 707–718. doi:10.1259/bjr/16337436. Free text at pubmed - Pubmed citation.
Lu, Li, et al., 2012. Digital subtraction CT angiography for detection of intracranial aneurysms: comparison with three-dimensional digital subtraction angiography. Radiology 262 (2), 605–612.
Mahesh, M., 2002. The AAPM/RSNA physics tutorial for residents: search for isotropic resolution in CT from conventional through multiple-row detector. Radiographics 22 (4), 949–962.
Mainardi, R.T., Barrea, R.A., 1989. X-ray spectral determination by successive modifications of the beam intensity. Nucl. Instrum. Methods. Phys. Res. Sect. A. 280 (2-3), 387–391.
Manes, G.I., 1956. The discovery of X-Ray. ISIS 47 (3), 236–238.
Marchiori, D.M., 2014. Clinical Imaging: With Skeletal, Chest, and Abdominal Pattern Differentials. Elsevier Mosby.
Martin, C.J., 2007. The importance of radiation quality for optimisation in radiology. Biomed. Imaging Interv. J. 3 (2).
Martin, E., 2015. Concise Medical Dictionary. Oxford University Press, Oxford. ISBN 9780199687817.
McCaffrey, J.P., et al., 2007. Radiation attenuation by lead and nonlead materials used in radiation shielding garments. Med. Phys. 34 (2), 530–537.
Menken, M., DeFriese, G.H., Oliver, T.R., Litt, I., 1985. Introduction to Digital Subtraction AngiographyThe Cost Effectiveness of Digital Subtraction Angiography in the Diagnosis of Cerebrovascular Disease (PDF). Office of Technology Assessment, U.S. Government Printing Office, Washington D.C., pp. 15. ISBN 9781428923348.
Menon, C., Bloomfield, R.E., Clement, T., 2013. Interpreting ALARP, 8th IET International System Safety Conference incorporating the Cyber Security Conference. IET.
Merriam-Webster, Inc., 1995. Merriam-Webster's Medical Dictionary. Merriam-Webster, Merriam-Webster.

Merzbacher, E., Lewis, H.W., 1958. X-ray production by heavy charged particlesCorpuscles and Radiation in Matter II/Korpuskeln und Strahlung in Materie II. Springer, Berlin, Heidelberg, pp. 166–192.
Mettler, Jr, Fred, A., et al., 2008. Effective doses in radiology and diagnostic nuclear medicine: a catalog. Radiology 248 (1), 254–263.
Mezrich, R.S., 1988. The implication of PACS for radiology practice. Am. J. Roentgenol. 151 (4), 828.
Mishra, G., Banode, P., 2016. Evaluation of outcome of percutaneous vertebral kyphoplasty in vertebral compression fractures. J. Datta Meghe Inst. Med. Sci. Univ 11 (1).
Mishra, G., Dass, A., Mahalaqua, N.K., Quazi, S.Z., 2021. Spectrum of initial computed tomography findings in RT-PCR positive patients with novel coronavirus 2019 disease – a systematic review of 2327 cases. Open Public Health J. 14 (Suppl-1, M2), 118–127. doi:10.2174/1874944502114010118.
Mishra, G., Parihar, P., Banode, P., 2016. Is plain Radiography still essential in diagnosing spinal tuberculosis? J. Datta Meghe Inst. Med. Sci. Univ 11 (1).
Mould, R.F., 1995. The early history of x-ray diagnosis with emphasis on the contributions of physics 1895-1915. Phys Med. Biol. 40 (11), 1741.
Nemtoi, A., et al., 2013. Cone beam CT: a current overview of devices. Dentomaxillofac. Radiol. 42 (8), 20120443.
Newberg, A., Alavi, A., Reivich, M., January 2002. Determination of regional cerebral function with FDG-PET imaging in neuropsychiatric disorders. Semin. Nucl. Med. 32 (1), 13–34. doi:10.1053/snuc.2002.29276. PMID 11839066.
Nikjoo, H., Uehara, S., Emfietzoglou, D., 2012. Interaction of Radiation with Matter. CRC press.
Ohnesorge, B., Flohr, T., Klingenbeck-Regn, K., 1999. Efficient object scatter correction algorithm for third and fourth generation CT scanners. Eur. Radiol. 9 (3), 563–569.
Okafor, C.H., Ugwu, A.C., Okon, I.E., 2018. Effects of patient safety culture on patient satisfaction with radiological services in Nigerian radiodiagnostic practice. J. Patient Exp. 5 (4), 267–271.
Orth, R.C., Wallace, M.J., Kuo, M.D., 2008. C-arm cone-beam CT: general principles and technical considerations for use in interventional radiology. J. Vasc. Interv. Radiol. 19 (6), 814–820. doi:10.1016/j.jvir.2008.02.002.
Parks, T.G., et al., 1970. Limitations of radiology in the differentiation of diverticulitis and diverticulosis of the colon. Br. Med. J. 2 (5702), 136–138.
Penney, G.P., et al., 2001. Validation of a two-to three-dimensional registration algorithm for aligning preoperative CT images and intraoperative fluoroscopy images. Med. Phys. 28 (6), 1024–1032.
Phelps, M.E., Gado, M.H., Hoffman, E.J., 1975. Correlation of effective atomic number and electron density with attenuation coefficients measured with polychromatic x rays. Radiology 117 (3), 585–588.
Prabhu, S., et al., 2020. Production of x-rays using x-ray tube. J. Phys. Conf. Ser. 1712 (1).
Reed, A.B., 2011. "The history of radiation use in medicine." 3S.
Reiner, B.I., Knight, N., Siegel, E.L., 2007. Radiology reporting, past, present, and future: the radiologist's perspective. J. Am. Coll. Radiol. 4 (5), 313–319.
Reyes, M., et al., 2020. On the interpretability of artificial intelligence in radiology: challenges and opportunities. Radiol. Artif. Intell. 2 (3), e190043.
Rosenstein, M., 1976. Organ Doses in Diagnostic Radiology. Bureau of Radiological Health, Rockville, MD (USA) No. PB-255363; FDA/BRH-76/88; DHEW/PUBL/FDA-76/8030.
Rushton, V.E., Horner, K., 1996. The use of panoramic radiology in dental practice. J. Dent. 24 (3), 185–201.
Ryan, J.L., 2012. Ionizing radiation: the good, the bad, and the ugly. J. Invest. Dermatol. 132 (3), 985–993.
Saha, G.B. (Ed.), 2006. Physics and Radiobiology of Nuclear Medicine, 3rd ed. Springer, New York. doi:10.1007/978-0-387-36281-6. ISBN 978-0-387-30754-1.
Scarfe, W.C., Farman, A.G., Sukovic, P., 2006. Clinical applications of cone-beam computed tomography in dental practice. J. Can. Dent. Assoc. 72 (1), 75–80.
Seibert, J.A, 2004. X-ray imaging physics for nuclear medicine technologists. Part 1: Basic principles of x-ray production. J. Nucl. Med. Technol. 32 (3), 139–147.
Shaffer, T.M., Drain, C.M., Grimm, J., 2016. Optical imaging of ionizing radiation from clinical sources. J. Nucl. Med. 57 (11), 1661–1666.
Shelledy, D.C., Peters, J.I., 2014. Respiratory Care: Patient Assessment and Care Plan Development -11-26. Jones & Bartlett Publishers, pp. 430. ISBN 978-1-4496-7206-5.
Shreve, P., Townsend, D.W., 2010. Clinical PET-CT in Radiology: Integrated Imaging in Oncology. Springer Science & Business Media.
Siegfried, R., 2002. "From elements to atoms: a history of chemical composition."
Som, P., Atkins, H.L., Bandoypadhyay, D., Fowler, J.S., MacGregor, R.R., Matsui, K., et al., July 1980. A fluorinated glucose analog, 2-fluoro-2-deoxy-D-glucose (F-18): nontoxic tracer for rapid tumor detection. J. Nucl. Med. 21 (7), 670–675.
Sommerfeld, A., 1929. About the production of the continuous x-ray spectrum. Proc. Natl Acad. Sci. 15 (5), 393.
Stirrup, J., 2020. Cardiovascular Computed Tomography. Oxford University Press. ISBN 978-0-19-880927-2.
Strickland, N.H., 2000. PACS (picture archiving and communication systems): filmless radiology. Arch. Dis. Child. 83 (1), 82–86.
Tillman, C., et al., 1996. Elemental biological imaging by differential absorption with a laser-produced x-ray source. JOSA B 13 (1), 209–215.
Vazquez, H., et al., 2000. Risk of fractures in celiac disease patients: a cross-sectional, case-control study. Am. J. Gastroenterol. 95 (1), 183–189.
Versoza, G., Band, B., 2005. "Electromagnetic radiation."
Villari, N., Stecco, A., Zatelli, G., 1999. Dosimetry in dental radiology: comparison of spiral computerized tomography and orthopantomography. Radiol. Med. (Torino) 97 (5), 378–381.
Vlahos, I., Godoy, M.C., Naidich, D.P., 2010. Dual-energy computed tomography imaging of the aorta. J. Thorac. Imaging 25 (4), 289–300. doi:10.1097/RTI.0b013e3181dc2b4c. Pubmed citation.
Vorotyntseva, N.S., Moshurov, I.P., Ganzya, M.S., 2018. Medical radiology and radiation safety. Med. Radiol. 63 (4), 40–49.
Webster, M., 2016. Merriem Webster' Medical Dictionary. New Edition.
Whittaker, L.R., 1974. Simple principles in radiodiagnosis. Trop. Doct. 4 (2), 87–90.
Young, A.T., 1981. Rayleigh scattering. Appl. Opt. 20 (4), 533–535.

Chapter 7

The role of physics in modern radiotherapy: Current advances and developments

Anurag Luharia[a], Gaurav Mishra[a], D. Saroj[b], V. Sonwani[c], Sanjay J. Dhoble[d]

[a]Department of Radiology, Datta Meghe Institute of Medical Science (Deemed to be University), Sawangi, Wardha, Maharashtra, India.
[b]Department of Radiotherapy, Allexis Hospital, Mankapur, Nagpur, India
[c]HCG NCHRI Cancer Center, Nagpur, India
[d]Department of Physics, Rashtrasant Tukadoji Maharaj Nagpur University, Nagpur, Maharashtra, India

7.1 Introduction

The term "radiotherapy" or "radiation therapy" literally means application of radiation in the field of treating patients and certain diseases and disorders. The ionizing radiations are used for the treatment of diseases such as such as cancer, metastatic deposits and even noncancerous conditions. This modality as mentioned earlier, uses ionizing radiation by properly aiming and bombarding targeted ionizing radiation upon the area of interest which is generally the tumoral tissue. The ionizing nature of radiation is able to remove an orbital electron from the tissue by interacting with it, owing to its high energy content, results in the development of a positive charge at the tissue level. This change of charge is responsible for the cascade of processes which cause pertinent damage to the genetic material of the tissue of interest by ionization. The removal of electrons resulting in the development of positive charge leads to formation of hydroxyl ions which are capable of causing genetic material damage and thereby cease the process of cell division. This principle is of prime importance when ionizing radiation has to deal with cancer and similar metastatic diseases and disorders, the main aim of the therapy being the stoppage or inability of the tumor cells to divide and propagate by damaging genetic material as a result of ionization (Horwich, 1994; Radiosensitivity).

The discipline which deals with use of ionizing radiation for treating cancer in patients is called as 'radiation oncology' and the specialists dealing with planning and execution of treatment carried out in this manner are radiotherapist, medical physicist and radiotherapy technologist (Horwich, 1994). There are various roles of radiation treatment based on the condition of the patient and ongoing treatment for cancer. The radiation treatment may be aimed at curing the tumoral tissue which may be referred to as "curative." The role may be solely to prevent any incidence of the tumor redeveloping or recurring from the site of interest which is known as "adjuvant." There is a third role which may be considered here in cases where cure of the said tumor may not be possible, and usage of ionizing radiation in such scenarios may lead to better quality of life of patient with a sunray of hope of the administered radiation treatment being "curative," and along with this, provision of symptomatic relief from pain or other associated complaints of the patient which may arise as a result of the said tumor(s). This can be called as "palliative" role of radiation therapy (Radiosensitivity). There may be simultaneous administration of chemotherapy and radiotherapy in cases of severe condition of cancer patients, making the overall result more hopeful in order to cease tumor growth.

The administration of ionizing radiation is made possible by the usage of special devices which perform the task of increasing the velocity of subatomic particles and subjecting them to variable electric potentials which move back and forth, around a particular set value of electric potential. This results in acceleration of subatomic particles at high speeds and resultant production of ionizing radiation. These special devices are 'linear particle accelerators' and were first invented in the 1930's by Gustav Ising and Rolf Wideroe respectively (Ising, 1924; Widerøe, 1928). The linear accelerator type to be used takes into account what subatomic particles are required to be subjected to such linear acceleration – it can be electrons, ions or even protons. This is based on the make of the linear accelerators and subatomic particle selection according to requirements therein (CERN Courier, 2002).

The main objective is to treat the patient and for this, it is key to plan the radiation dosage, direction of propagation of ionizing radiation through site of interest at properly planned trajectories, particular angles and should be targeted to involve maximum area of involved tissue. There are however some movements like that of respiration, the heart, filling of urinary bladder, even the superficial markings made on the skin for planning radiation treatment may move depending upon the habitus of the patient. This has led to taking into target a small area of surrounding normal tissue around the tumoral tissue if there is suspected possibility of the tumoral process spreading into the adjacent normal tissues, local lymph nodes etc. (Horwich, 1994; Radiosensitivity). As radiation has potential side effects which can be both stochastic and nonstochastic, it is mandatory to follow the ALARA principle (As Low As Reasonably Achievable) at all times while giving radiation treatment to the patient (ALARP Guidance, ALARP Guidance Note 2015).

This chapter makes an attempt to enlist and describe the various advances and current practices with recent developments in the field of radiation therapy, their methods of planning and execution, targeting tissue of interest, apt dosage with prescribed dosage reduction, if needed and with special reference to recent updates in the field of radiation therapy (Delaney et al., 2005; Chew et al., 2021).

7.2 Role of radiotherapy in cancer treatment

The importance and responsibilities of radiation have grown exponentially in recent years, with almost half of all cancer patients requiring either radical or palliative radiotherapy treatment (Jacob et al., 2010). Treatment with radiation treatment is an essential weapon in the fight against cancer, and it is frequently used in conjunction with other medicines, such as chemotherapy or tumor removal surgery. The primary aims of radiation therapy are to reduce tumors and destroy cancer cells as quickly as possible (Withers et al., 1988; Nonnenmacher et al., 2016). X-rays, charged particles, and gamma rays are just a few of the forms of radiation that are often encountered. Radiation is supplied using a machine, which can come from either outside or within the body to deliver the treatment. External-beam radiation treatment is a term used to describe radiation delivered from the outside, whereas internal radiation therapy, also known as brachytherapy (Chargari et al., 2019; Devlin, 2015), is used to describe radiation delivered from within. The vast majority of cancer patients are required to undergo radiation therapy for a particular period and sometimes throughout their treatment regimen (Perez et al., 1987).

7.2.1 What is radiotherapy and how it works?

By hitting cancer cells with high-energy particles or waves, radiation therapy kills them. Cells develop and multiply in order to replace those that are damaged or old. Cancer cells multiply at a faster rate than normal cells and lack the regulatory mechanisms that normal cells possess. By inflicting genetic damage on cancer cells, the high-energy particles (or waves) destroy them (DNA). DNA is the genetic material that contains the instructions for cell division and growth. Radiation therapy is a precision procedure that directs high-energy beams directly to the tumor. Attempts are undertaken to maintain the greatest number of viable cells possible. Due to the fact that this therapy is limited to a single part of the body, it is ineffective in treating malignancies that have spread to other places of the body. Radiation treatment is used to diagnose and treat cancer in its early stages, to prevent it from returning (recurrence), to treat advanced cancer-related symptoms, and to treat returned cancer (Nonnenmacher et al., 2016; Baskar et al., 2014).

Radiation therapy (or radiotherapy) is a fundamental component of tumor contouring. High-energy waves are used to target cancerous cells. The waves interfere with critical internal cell functions required for cell division, eventually causing the cells to die. As the cells die, the tumor shrinks. The major disadvantage of radiation is that it is not selective for cancer cells and may also destroy healthy cells (Pecorino, 2016). Tumors and normal tissues respond differentially to radiation treatment, depending on their pre- and post-treatment development patterns. By interacting with their DNA and other target molecules, radiation destroys cells. Cell death does not occur quickly; it occurs when cells make an unsuccessful effort at division. Abortive mitosis is the term used to describe this phenomenon. As a result, radiation-induced cell death occurs more quickly in rapidly reproducing tissues (Garau et al., 2011; Wheldon et al., 1982). Normal tissue compensates for the cells lost during radiation therapy by accelerating the division of remaining cells. In comparison, following radiation therapy, tumor cells divide more slowly, and the tumor may diminish in size. The amount to which a tumor shrinks is governed by the ratio of new cells to those that die. Carcinomas are a type of cancer that usually proliferates at a rapid rate. Radiation therapy is frequently successful in the treatment of numerous types of cancer. Depending on the radiation dose used and the type of tumor, the tumor may restart growing following therapy cessation, often at a slower rate than previous. Radiation is frequently used in conjunction with surgery and/or chemotherapy to inhibit regrowth of the tumor (Washington and Leaver, 2015; Rao and Deng, 2010; Luqmani, 2005).

7.2.2 Types of radiotherapy

External beam radiation treatment (EBRT) and internal radiation therapy (IRT) are the two most used kinds of radiation therapy (Martins, 2018). EBRT is administered by the use of a machine that directs radiation at the tumor. The machine is huge and has the potential to be loud. Instead of touching patient, it moves about sending radiation to a specific portion of body from various angles. (Martins, 2018; To, A. Quick Guide 2021).

7.2.2.1 External beam radiation therapy for cancer

EBRT is administered by the use of a machine that directs radiation at the tumor. It is also a localized therapy, affecting a specific portion of patient's body when administered. For example, for lung cancer, patient will only receive radiation to the chest and not the rest of the body (To, A. Quick Guide 2021).

Radiation treatment using external beams is used to treat a wide variety of cancers.

7.2.2.1.1 Types of beams used in radiation therapy

External radiation therapy uses three different types of particles to generate radiation beams (Brahme, 1987):

a. Photons
b. Protons
c. Electrons

a. Photons
Photon beams are used in the majority of radiation treatment devices. X-rays also utilize photons, albeit at lesser levels. Photon beams have the ability to penetrate deep into the body to reach malignancies. Photon beams disperse small amounts of radiation as they move through the body. These beams do not come to a halt when they reach the tumor, but continue on into normal tissue beyond it (Naseri and Mesbahi, 2010).

b. Protons
Protons are positively charged particles. Proton beams, like photon beams, can reach malignancies located deep inside the body. Proton beams, on the other hand, do not disperse radiation as they travel through the body and cease when they reach the tumor. Proton beams have the potential to limit the quantity of normal tissue exposed to radiation. Clinical studies are now being conducted to compare proton-beam radiation treatment to photon-beam radiation therapy. While some cancer institutions employ proton beams in radiation treatment, the equipment' exorbitant cost and size restrict their utilization (Suit and Urie, 1992).

c. Electrons
Electrons are negatively charged particles. Electron rays cannot penetrate far into biological tissues. As a result, their application is confined to malignancies on the skin or near the surface of body (Brahme, 1987).

7.2.3 Types of external beam radiation therapy

EBRT is available in a variety of configurations, all of which have the objective of delivering the maximal recommended dosage of radiation to the tumor while sparing the surrounding normal tissue. Each technique makes use of a computer to evaluate photographs of the tumor in order to determine the most exact dosage and treatment course feasible (Paz-Manrique et al., 2020).

7.2.3.1 Three dimensional conformal radiation therapy

Three dimensional conformal radiation therapy is a frequently utilized type of external beam radiation therapy. It utilizes images from CT, MRI, and PET scans to precisely arrange the treatment zone, a process referred to in medical terms as simulation. The photographs are analyzed using computer software, and the resulting radiation beams are intended to conform to the contour of the tumor. Conformal radiation is a type of radiation that conforms to the shape of a tumor through the use of a range of beam orientations. With the precise structure, it is possible to deliver higher doses of radiation to the tumor while sparing the normal tissue surrounding it. The majority of participants receive therapy once a day from Monday through Friday. The number of treatments necessary vary according to the characteristics of the disease, such as the type and stage of cancer, as well as the size and location of the tumor (Verhey, 1999).

7.2.3.2 Intensity-modulated radiation therapy

IMRT is a kind of conformal radiation treatment that is delivered in three dimensions. radiation beams are directed at the tumor from a variety of orientations, similar to 3-D conformal radiotherapy. Immediate-phase radiation therapy (IMRT) employs much smaller beams than 3D conformal radiation therapy, and the intensity of the beams in particular locations can be altered in order to provide larger doses to specific sections of the tumor. Monday through Friday, the majority of participants receive therapy once a day. There is no set number of therapies for cancer; instead, it is determined by the specificity of illness, such as the kind and stage of the cancer, as well as the size and location of the tumor (Paz-Manrique et al., 2020; Verhey, 1999).

7.2.3.3 Image-guided radiation therapy (IGRT)

Ionizing radiation therapy (IGRT) is a kind of IMRT. Radiation therapy, on the other hand, employs imaging scans not only for treatment planning before to radiation therapy sessions, but also during radiation therapy sessions and subjected to several scans, including CT, MRI, and PET scans. Using computer algorithms, these scans are analyzed to determine whether the tumor's size and position have changed. Because of the frequent imaging, the posture or radiation dose can be modified throughout treatment if necessary. These modifications can assist to increase the precision of therapy while still sparing normal tissue. Monday through Friday, the majority of participants receive therapy once a day. There is no set number of therapies for cancer; instead, it is determined by the specifics of illness, such as the kind and stage of the cancer, as well as the size and location of the tumor (Dawson and Jaffray, 2007; De Los Santos et al., 2013; Xing et al., 2006).

7.2.3.4 Tomotherapy

Tomotherapy is a sort of IMRT that makes use of a machine that is a mix of a CT scanning and an external-beam radiation generator. Tomotherapy devices acquire photographs of the tumor just before treatment sessions begin, allowing for extremely accurate tumor targeting while sparing surrounding normal tissues and organs (Mackie et al., 1999). During treatment, it spins around patient, delivering radiation in a spiral pattern, slice by slice, as it rotates around patient. Tomotherapy may be more effective than 3-D conformal radiation treatment in terms of protecting normal tissue; however, this has not been proven in clinical studies to be the case yet (Jeraj et al., 2004). There is no set number of therapies for cancer; instead, it is determined by the particular type of specific illness, such as the kind and stage of the cancer, as well as the size and location of the tumor (Mackie, 2006).

7.2.3.5 Stereotactic radiosurgery

Stereotactic radiosurgery can be used to treat small tumors with well-defined boundaries in the brain and central nervous system utilizing concentrated, high-energy beams (Leksell, 1983). It may be necessary in some circumstances if surgery is deemed too risky due to patients age or other health problems, or if the tumor cannot be safely reached with surgery alone. Gamma Knife stereotactic radiosurgery is a type of stereotactic radiosurgery that utilizes gamma rays (Chin and Regine, 2010). Patient will be secured in a head frame or some other device to prevent movement during the process. Stereotactic radiosurgery is based on the employment of many small radiation beams targeted at the tumor from a variety of angles. Each individual beam has a negligible effect on the tissue it passes through, but a perfectly targeted dose of radiation is delivered to the point where all of the beams come together (Lutz et al., 1988). Although treatment regimens may vary, medication is normally provided in a single dose. It may be taken up to five times daily in certain circumstances (Podgorsak et al., 1988).

7.2.3.6 Stereotactic body radiation therapy

A comparable technique to radiosurgery, stereotactic body radiation treatment is used to treat tiny, isolated tumors that are located outside the brain and spinal cord, most often in the liver or lung (Lo et al., 2010). When patients are unable to have surgery because of age, health difficulties, or the location of the tumor, this treatment may be a possible cure (Timmerman et al., 2007). Stereotactic body radiation therapy, like stereotactic radiosurgery, involves the use of specific equipment to keep motionless during treatment. It directs an extremely accurate beam of light to a specific region. Tumors that are not located within the brain are more prone to move in conjunction with the regular action of the body, such as while breathing or digesting food or liquids. As a result, the radiation beams cannot be focused as accurately as they can in stereotactic radiosurgery, which is a more precise technique (Chang and Timmerman, 2007). Stereotactic body radiation is therefore typically administered in multiple doses because of this phenomenon.

7.2.3.7 Brachytherapy

A source of radiation is inserted into the patient's body during this sort of treatment. In nature, radiation sources can be solid or liquid (Nag et al., 2000). Internal radiation therapy with a solid source is referred to as brachytherapy. As part of this method of treatment, seeds, ribbons, or capsules containing a radiation source are implanted in patients' body, either inside or around the tumor. Brachytherapy is the localized treatment that targets only a certain area of the body. In contrast to other types of treatments, brachytherapy is limited to one specific area of the body. Breast, cervical, prostate, and ocular malignancies are among the conditions for which it is frequently prescribed (Chargari et al., 2019).

7.2.3.8 Types of brachytherapy

Low-dose rate (LDR) implants: The radiation source is kept in place for 1 to 7 days after the treatment in this kind of brachytherapy and patients are likely to get hospitalized during this time. Medical physicist in association with radiotherapist removes the radiation source and any catheters or applicators used during the RT treatment (Nag et al., 2002).

High-dose rate (HDR) implants: The radiation source is kept in place for just 10 to 20 minutes at a time before being removed. This kind of brachytherapy involves leaving the radiation source in place for just 10 to 20 minutes at a time and then removing it. Two to 5 days of twice-daily treatment or once a week for 2 to 5 weeks may be necessary. The duration of treatment is influenced on the kind of cancer. Catheter or applicator may stay in place during therapy, or it may be withdrawn and reinserted prior to each treatment visit (Kolotas et al., 1999).

Permanent implants: Once the radiation source has been inserted into the patient's body, the catheter is removed. The implants will stay in patients' body, although the radiation will gradually weaken with time, almost all radiation will be eradicated. When the radiation is initially installed, patients may need to limit contact with others and take extra steps to protect personal health and well-being. Avoid spending time with children or pregnant women at all costs for some due course of time suggested (Polgár and Major, 2009).

7.2.4 General indications for the radiotherapy

It is possible to utilize radiotherapy in the early stages of cancer as well as after the disease has progressed (Ang and Garden, 2006).

It can be used to:

- Attempt a full cure for cancer (curative radiotherapy)
- Enhance the efficacy of other medicines – it can be used in conjunction with chemotherapy or prior to surgery, for example (neoadjuvant radiotherapy)
- Decrease the likelihood of the cancer recurring following treatment (adjuvant radiotherapy)
- If a cure is not feasible, to alleviate the symptoms (palliative radiotherapy)

Radiotherapy is widely regarded as the most effective cancer treatment available following surgery; however, the effectiveness of this treatment varies from person to person (Ang and Garden, 2006; Delaney and Barton, 2015).

7.2.5 Intent of radiotherapy treatment

a. Palliative
b. Curative

Radiotherapy has long been recognized as an important component of cancer patients' palliative treatment. Given that it is a local therapy with a limited duration, it is recommended that the number of fractions be maintained to a minimum without compromising efficacy (Milano et al., 2009). Radiation oncology should be well-known to all palliative care programs, and they should have an excellent working connection with the radiation oncology department. Many staff members and patients may be intimidated by the technical components of the issue, and departments should work to strengthen their outreach and education efforts. Among the most common indications are pain reduction (especially for bone pain), management of bleeding, fungus, and ulceration, dyspnea, blocking of hollow viscera, and shrinking of tumors that are producing issues due to their excessive space occupancy on the body (Milano et al., 2009). Furthermore, it plays a crucial role in the treatment of three oncological emergencies: superior vena caval blockage, spinal cord compression, and elevated intracranial pressure owing to brain metastases, to name a few. It is being worked on developing more pragmatic fractionation schedules that are consistent with good results in terms of palliative end objectives, allowing for shorter courses with fewer hospitalizations for the benefit of the patient and their family's comfort and convenience. A greater amount of clinical research and assessment of palliative radiation is needed (Milano et al., 2009; Milano et al., 2008).

Radiation treatment can be used to treat tumor-related bleeding, superior vena cava blockage, spinal cord compression, brain metastases, painful bone metastases, and superior vena cava obstruction.

Radiation therapy is an actually curative treatment that can be administered in standard 2-Gy fractions, hypofractionated fractions, and ablative stereotactic courses, among other configurations. In order to cure cancer and extend patients life expectancy, radical radiation is used to decrease and control the growth of the tumor. Radiotherapy is now a successful treatment for the majority of ICSOL malignancies, including oral cavity tumors, gynecological malignancy, lung cancer, and prostate cancer, among others.

7.2.6 Types of cancer treated using radiotherapy

RT is used to treat a wide array of cancer forms. Radiation therapy is administered to around 50% of cancer patients. Radiation may be used to treat prostate, skin, head and neck, throat, larynx, breast, brain, colorectal, lung, and bone cancers, as well as leukemia, ovarian, and uterine cancers. Radiation therapy is more effective against certain types of cancer than it is against others. The amount and type of radiation required in any specific case is determined by the tumor's size, disease stage, location, the patient's health, the manner of delivery, and the total dose supplied. Certain forms of cancer are thought to respond better to radiation treatment than others. Radiation can be used successfully to retard the growth of malignant tumors on rare occasions without permanently injuring the surrounding normal tissue. All types of cancer include skin and lip cancers, head and neck cancers, breast cancer, cervical cancer, endometrial cancer, prostate cancer, Hodgkin's disease and extranodal lymphoma, testicular seminoma and ovarian dysgerminoma, medulloblastoma, pineal germinoma, ependymoma, retinoblastoma, and choroidal melanoma. Along with Wilms tumor and rhabdomyosarcoma, a number of other malignancies have a poor response to radiation but may be treated with combination therapy, including colorectal cancer, soft tissue carcinoma, and testicular embryonal carcinoma (Jaffray and Gospodarowicz, 2015; Baskar et al., 2012; Hu and Harrison, 2005; Berkey, 2010).

The majority of other malignant tumors are not deemed treatable with radiation because they are difficult to diagnose early enough and/or have a considerably faster development rate than the diseases listed above. Tumors situated in particularly sensitive tissue are unable to be treated with the high doses of radiation required to completely eradicate the tumor. Furthermore, radiation alone is not always effective in the treatment of severely metastatic cancers. In certain cases, only a small percentage of cures are possible as a result of surgery, radiation, or a combination of the two methods of treatment.

7.2.7 The role of radiotherapy in cancer control

Radiotherapy can be used alone or in combination with other treatment methods such as surgery, chemotherapy, and hormone therapy to provide an effective neoadjuvant and adjuvant treatment.

The goal of radiation might be cure, control, or palliation, and it can provide advantages in the following areas (Barton et al., 2006; Jaffray et al., 2015; Delaney et al., 2005):

- Organ preservation
- Quality of life
- Survival outcomes
- Effective palliation of symptoms.

The treating team considers a range of factors when deciding on a course of radiotherapy. Tumor related factors include:

- The site of the cancer
- A histologically proven cell type
- The grade and stage of the tumor
- The radiosensitivity of the tumor.

A variety of individual considerations, such as comorbidities, performance status, and unsuitability for surgical resection or anesthetic, may impact the choice to employ radiation treatment.

Palliative radiation is used to treat a considerable proportion (about 50%) of the population getting radiotherapy for the management of pain and palliation includes:

- Pain at bony metastases and pathological fractures
- Providing relief from symptoms caused by cerebral metastases

- Spinal cord compression
- Superior vena cava obstruction
- Control of bleeding
- Reducing fungating lesions.

7.3 Development of radiation physics

Everyone is exposed to a little quantity of radiation because of their environment. Cosmic rays, radioactive substances in the earth (such as uranium-238), radionuclides in food (such as potassium-40), and radon gas inhalation all contribute to this exposure. It is the movement of energy that makes radiation. EM radiation (EMR) is a type of electromagnetic radiation that flows through the air in the form of waves. Photons are another way to think about electromagnetic radiation (EMR). The photon is the "package" of energy that transmits the energy across the universe (Delaney et al., 2005; Ivanov and Platov, 2004; Cossairt, 1993; Roach et al., 1976).

After Wilhelm Conrad Rontgen defined the radiation that bears his name in 1895, Henri Becquerel discovered natural radioactivity in 1896, and Mane and Pierre Curie isolated radium in 1898. As a result of this research, medical diagnoses and treatments have become more accurate. Clinical experience, physical and technological improvements, and radiation biology research all contributed to a stronger scientific foundation for ionizing radiation treatment approaches in the decades leading up to World War II. Progress has been rapid since World War II, as indicated by the availability of new radiation sources, artificial radionuclides, and physical accelerators for electrons and other charged particles. Radiation therapy has made significant strides in both curative and palliative treatment options with to advances in technology, procedures, and clinical expertise (Khan and Gibbons, 2014; Johns, 1953).

7.3.1 History

Researchers were able to monitor changes in bodily tissues exposed to X-rays or radioactive chemicals over the first 50 years following the 1890s scientific discoveries (Reed, 2011). After World War II, fresh prospects for advancement in radiation treatment were created, but they were also disrupted by the conflict. Using new radionuclides and building physical electron and particle accelerators that were operationally dependable for use in clinical radiation opened new avenues for therapeutic improvement. Even more importantly, the way the public perceives radiation and its effects has evolved significantly. Previously erroneous faith in radiation's miraculous properties has given way to an equally irrational distrust of even its most beneficial applications, such as those in medicine, nuclear weapons, nuclear power plant accidents, and a lack of awareness about our "natural" radiation environment all contribute to this distrust (Giger et al., 2008).

With voltages up to about 200 kV, external radiation therapy has several problems, including a lack of penetration to deep tumors, bone absorption that can lead to an uneven dose distribution and brittle bones, and a high dosage to the outer layer of skin that can cause under-infection. Since the 1930s, people have been aware of the constraints. These limits were attempted to be mitigated by using a variety of methods, such as rotation, pendulum, and convergence irradiation. It is possible to use high-energy X-ray or gamma radiation, such as the facility in Bergen, Norway. This kit, Cobalt 60, a nuclide created by nuclear reactors, was widely available in the 1950s. By using Cobalt 60 devices instead of teleradium guns, new technological advancements were possible (Keevil, 2012).

Gamma radiation can now be used at distances of up to 100 cm from the patient, thanks to new, powerful radiation sources. That means traditional X-ray limitations may be eliminated, and novel uses for radiation therapy could be explored. Devices dubbed "telegamma" are reasonably easy to use but are also quite reliable (Webb, 2009). When it comes to replacing the cobalt 60 radiation source, which has a half-life of around 5 years (for example), the operational cost is relatively modest (Kostylev, 2000). Often, the radiation source is so enormous that the geometric penumbra diminishes the sharpness of the radiation field's edges. In order to lower the radiation source's size, it is possible to use a greater specific activity; however, doing so raises the cost. Most radiation departments have telegamma machines that are adequate for various purposes. Surgery on tiny brain volumes necessitated the use of an advanced form of radiation. By Leksell et al. in Sweden, the technique was first used to treat tiny, stereotactically targeted tumors in the brain (Duck, 2014). It is possible to carry out this process with great accuracy. The gamma knife, a particular instrument created for this purpose, comprises around two hundred cobalt 60 isotopes scattered throughout a vast spherical sector, each one properly collimated toward the sphere's center (Tabakov, 2017). Using a treatment couch that is positioned in the sphere, patients are held in place with geometric precision so that the radiation volume corresponds to the target volume in the patient's skull. A sophisticated computer-based three-dimensional dosage planning approach has attracted substantial international attention and increased Swedish exports of radiation equipment. Instead of using a cobalt 60 pendulum or rotational irradiation, similar approaches have been devised using high energy X-rays from a linear accelerator. It has recently been used to treat bigger tumors as well as in other locations. Accelerators that use electrons Van de Graaff generators, resonance transformers, and betatrons were the first electron accelerators used in radiotherapy (Round, 2013). In the 1930s, several radiation therapy institutions built their own accelerators. Industrial radiography in the 1940s led to the development of electron accelerators for use in radiotherapy. The betatrons and other high-energy electron irradiation devices made it possible to use photon energies in the MeV range for X-ray therapies. In the 1970s, these sorts of accelerators were replaced by linear accelerators because of their low radiation intensity, restricted maneuverability, and sometimes insufficient operational reliability (Thwaites and Tuohy, 2006). There were significant developments in microwave linear accelerators in the 1950s, which have since taken over. Since the 1940s, the Royal Institute of Technology in Stockholm has created the microtron, an electron accelerator based on these principles. Clinical departments around the world have microtrons with a range of 10 MeV to 20 MeV installed. Since 1986, Umei, Sweden has had a prototype of the "race-track" microtron, which produces electrons with energies up to 50 MeV. Treatment options may expand as a result of the use of this device in conjunction with established technology. High-intensity X-rays with energies ranging from 5 MV to 50 MV are produced by electron accelerators with a narrow focus. The use of high-energy photons in the treatment of a wide range of diseases has increased in recent years. To begin with, they provide a more uniform distribution of radiation dose, even in large tumors, and a clearer picture of how much radiation is being delivered to nearby vital organs. Using high-energy photons for whole-body irradiation in conjunction with bone marrow transplant is a particular case. In order to use this procedure, however, the treatment room and equipment must be modified. High-energy electron irradiation is possible with most of the accelerators listed. Apart from the treatment of mycosis fungoides and breast cancer, electron radiation can be used in a variety of other ways. The accelerators are far more complicated and expensive than telegamma devices, making them more difficult to use. Control programs and detectors inside the machine are also necessary to ensure that the radiation field is consistent, and the intensity level is correct (Muller, 1977; Robison, 1995; Lindell and Walstam, 1956; Thwaites and Tuohy, 2006).

Treatment tables are integral to radiotherapy equipment. It must be able to irradiate the patient from nearly every angle, and the radiation must not be affected by supporting structures. Repeated visits for radiation necessitate that the patient's treatment posture be accurately replicated. Manually adjusting the 10 to 15 parameters for each treatment increases the possibility of error. Modern medical/technical gadgets frequently include a "check and confirm" procedure to ensure safety. The first therapy is recorded and documented by a computer. During the treatments, this computer can block radiation if the settings differ from the margins selected. In order to validate and identify any necessary modifications, it is critical to measure radiation exposure in vivo. With integrated TL-detectors or display semiconductor detectors (thermoluminescence).

7.3.2 External radiotherapy

Naturally, skin lesions were the first to be treated with radiotherapy because they are the most visible. An X-ray treatment for basal cell carcinoma of the female nose was pioneered by Dr. Thor Stenbeck in Stockholm in 1899. It was bombarded every day for 3 months, and the tissue responses were closely monitored. At the Second International Radiology Congress in Stockholm in 1928, nearly 30 years after treatment, the patient could be presented as the first cancer patient in the world to be cured by radiotherapy. Radiotherapy's early development was based only on what could be observed in the tissue. Radiotherapy was most commonly used to treat skin, mouth, throat, and other parts of the body that were easily accessible. To power the earliest X-ray tubes, high voltage wires were exposed and exposed. Eventually, the voltage of the tubes could be increased, making the devices more reliable. As recently as the 1950s, X-rays generated at voltages up to 200 kV were used for external radiation. High-energy X-rays (about 1 MeV) were produced in England and the United States, and a few clinics had access to such equipment (2). In 1942, a 1.5 MeV van de Graaff generator was created in Bergen with assistance from the Norwegian Red Cross (3). Radium is suitable for some radiation treatments because of its unique properties. Radium, on the other hand, was prohibitively expensive and scarce. Local applications with modest, enclosed sources were the primary use cases for this technology (e.g., for treating uterine cancer). Teleradium cannons, which used gamma rays to treat cancers in the head and neck area, were also developed. With less energy absorbed by bone and cartilage and a softer effect on the skin than conventional 200 kV X-rays, this procedure is preferable. Short treatment distances, long treatment intervals, and a lack of specific dose distribution customization all contributed to the relatively successful development of teleradium systems in Sweden. As far back as the mid-1920s, a generic biological dosing unit based on radiation-induced skin redness was established. The skin of a doctor or a member of staff was used to calibrate the X-ray equipment to this unit. When radiation was measured by physical means, it became evident that the biologic calibration was extremely unreliable. Rontgen's unit "R" (or "r") is based on ionization in the air, which is the basis for this measurement. At the 1928 Stockholm Congress, this dosing unit was accepted and lasted for several decades as the most important dosing unit. In Sweden, a mobile, dose-monitoring service was launched at that time. An international radiation dosimetry laboratory in Sweden was established, and it served the Nordic countries for decades. Condenser chamber method invented by Rolf Sievert allowed dosage measurements in vivo to be carried out. For patients getting external and intravenous therapy, as well as for radiation-protection evaluations for staff, the approach could be used. The Sievert Chamber is the most common name for this instrument (Thwaites and Tuohy, 2006; Tubiana et al., 1985; Verhaegen and Seuntjens, 2003; Norlund, 2003).

Brachytherapy is the local application of radiation sources. Radium (Ra) gamma radiation therapy was also created through trial and error. Radium salt was injected into tumors through metal tubes or needles put into bodily cavities. Radium capsules filling the entire cavity accomplish two things. The tumor-infiltrated sections of the wall are reduced in thickness as the hollow expands. As a last point, each segment of the cavity receives strong local radiation, but the uterine wall quickly reduces the exposure to neighboring organs. For radium treatments, which have a half-life of around 1600 years, treatment times might range from few hours to days. The doses of radiation received in close proximity were calculated using the geometric shape of the radiation source and container, as well as the square of the distance. In 1921, Sievert presented both a theoretical and numerical examination of these connections. Initially, dosage and treatment durations for local radium applications were determined empirically and defined in mg-Ra hours. Tables indicating the number of mg Ra hours required to treat a particular surface or volume. In the United States, Edith Quimby, Paterson, and Parker developed recommendations and tables for predicting treatment durations and achieving the required dosage distributions in their respective countries. Radium treatment was the most frequently used method for skin, lip, mouth, throat, and uterine malignancies. Gynecologic radiotherapy treatments have been shown to be extremely successful. After Forssell's pioneering work, Heyman created the "Stockholm technique" for cervical cancer and the "packing method" for uterine cancer. Kottmeier refined these methods (Robison, 1995; Kiseleva et al., 1972; Zwahlen et al., 2010; Myint et al., 2018; Skowronek, 2017).

7.3.3 Clinical radiation generators

7.3.3.1 Kilovoltage units

It was common practice to use X-rays rated at up to 300 kVp for external beam radiation until around 1950. Conventional kilovoltage machines were gradually phased out in the 1950s and 1960s due to the rise in popularity of cobalt-60 machines and the development of more powerful machines. In spite of this, these machines have not entirely vanished from the scene. Higher energy beams are still useful, even in today's megavoltage beam era, for treating superficial skin lesions (Hill et al., 2014).

7.3.3.1.1 Grenz-ray therapy

It is called Grenz-ray therapy when a low-energy X-ray beam of less than 20 kV is used to treat a patient. The poor penetration depth of these radiations means that they are no longer used in radiation therapy (Fairris et al., 1985).

7.3.3.1.2 Contact therapy

Irradiation of cancerous lesions at short distances from the source (focal spot) is possible with contact therapy or endocavitary machines that operate at 40 to 50 kV (SSD). A tube current of 2 mA is usual for the machine's operation. The smallest SSD that can be provided by an app on one of these computers is around 2 cm. The very soft component of the energy spectrum is often absorbed by a 0.5 to 1.0 mm thick aluminum filter placed in the beam. This type of radiation is clearly beneficial for malignancies with a depth of 1 to 2 mm. Within 2 cm of soft tissue, the beam is nearly totally absorbed. Slightly deeper forms of rectal cancer have been treated with endocavitary X-ray machines (Hill et al., 2014).

7.3.3.1.3 Superficial therapy

When using X-rays with potentials between 50 and 150 kV, the phrase superficial therapy is used. Filtration (often 1 to 6 mm aluminum) is added in various thicknesses to achieve the necessary level of beam hardness. It is possible to employ a superficial beam of this quality to treat cancers that are less than 5 millimeters deep (90% depth dose). In order to produce enough depth dose; the dose drop-off is so steep that it would require a significant overdose of the skin (Pearse and Chow, 2020).

7.3.3.1.4 Orthovoltage therapy or deep therapy

X-ray treatment with potentials ranging from 150 to 500 kV is known as orthovoltage therapy, sometimes known as deep therapy. At 200 to 300 kV, the majority of orthovoltage equipment operates at a current of 10 to 20 mA. Between 1 and

4 mm Cu, many filters have been created. There are several other features of orthovoltage beams that make them inappropriate for the treatment of cancers behind bone, including higher absorbed dosage in bone and increased scattering, which have been discussed here (Gerig et al., 1994).

7.3.3.1.5 Supervoltage therapy

High-voltage or supervoltage X-ray therapy is defined as a range of 500 to 1000 kV. In the 1950s and 1960s, significant progress was achieved in the development of high-voltage X-ray equipment. The high-voltage transformer insulation was a big issue at the time. Conventional transformer systems were quickly found to be unsuitable for generating potentials greater than 300 kVp. New techniques for designing high-energy devices were devised, as technology advanced swiftly. As an example, there is the resonant transformer, which boosts voltage quite effectively (Earle et al., 1973).

7.3.3.1.6 Megavoltage therapy

Megavoltage X-ray beams have an energy of 1 MV or more. Aside from X-ray radiation, radionuclides can also be included in this group as long as their gamma rays have an energy of at least 1 MeV (Hoppe et al., 1976).

These machines include accelerators such as the Van de Graaff generator and linear accelerators as well as teletherapy units such as cobalt-60, which use megavoltage radiation.

7.3.4 Dose planning

As telegamma and accelerators don't allow direct observation of tissue reaction during treatment, they necessitate novel ways for planning and delivering radiotherapy. There have been methods developed to adjust for uneven body shapes and inhomogeneous tissues while measuring dose distribution in homogeneous mediums and tissue-equivalent phantoms. Devices have been developed to customize the radiation beam based on the specific needs of each patient (Whelan, 2016). The geometry of the radiation field can be easily altered by placing rubberized, lead shields directly on the patient's skin during X-ray irradiation up to 200 kV. An individual lead alloy block, approximately 10 cm thick, must be mounted to the treatment head in order to define the radiation field when employing telegamma devices or accelerators. Radiation therapy for Hodgkin's disease is an example of this method. All main lymph nodes can be treated at the same time using high-energy photon radiation and a tailored field design. This often results in a cure for the condition. There were two-dimensional patient photographs and dosage distribution maps measured in water at the beginning of the process (Pacelli et al., 2019). A mechanical contour device, an anatomic atlas, and perhaps X-ray pictures perpendicular to each other were used to create the patient image. The head and neck region, for example, has a lot of diversity in shape or anatomy, thus a semi-three-dimensional planning process was used to ensure that the treatment area would be able to respond to these differences. Medical dosages require a much lower level of precision than the radiation doses needed to treat cancer and other diseased organs, which are much greater. Even a one-percent difference in dosage can alter tumor response by 2 to 5 percent, according to some recent research. When it comes to radiation treatment, there is frequently a tiny therapeutic range between the dose required to achieve the desired effect on cancer and the level that causes unbearable side effects. Most of the time, the target volume's dose is only varied by 10%. Brachytherapy is the local application of radiation sources. In many nations, the use of radium for interstitial applications has decreased significantly since the 1950s. Radiation protection issues, as well as advances in external radiation technology, play a role in this occurrence. Since the success of intracavitary gynecological radium therapy has been difficult to expand on, there have been advancements in the development of new radiation sources that can be used in its stead. Reviving interest in local applications with dosimetric and radiobiologic benefits has resulted from this. Radionuclides cobalt 60 and cesium 137 are commonly used for interstitial therapy (Meisberger et al., 1968). It can take days for a single treatment at moderate intensity to just a few minutes for a series of treatments at high intensity. After applying gamma-irradiating nuclides to the tumor tissue, another type of interstitial treatment employs biologically inert material that is left inside the tumor for a limited period of time before being removed. Prostate cancer, for example, has recently been studied using this technique. As early as 1903, Strebel proposed surgically inserting a tube into malignant tissue to allow for the rapid insertion and removal of radioactive sources. "After loading" was first developed in the 1950s, but it was not generally adopted until the 1960s when new radionuclides with extremely selective activity were available. After loading can either be done manually or remotely by means of flexible tube attached to a device and other end to the patient. During the 1960s, Radiumhemmet built and tested the first device of this type. There are now numerous commercially available after loading devices. The treatment duration is reduced, and the treatment capacity is increased by utilizing this approach, which handles highly active material. Multiple high-dose fractions are required to replace protracted irradiation that often lasts several days (as is usual with radium therapy). Metabolic ionizing radiation now has more options, thanks to the availability of artificial radionuclides. Ideally, the tumor or organ to be irradiated would be able to absorb a radioactive material from the blood stream sufficiently and selectively. Radiation-induced hyperthyroidism and some thyroid cancers can be treated with radioactive iodine, which accumulates in the thyroid and some malignancies. Radionuclide uptake concentration and release can be measured using a variety of scintigraphic procedures (Cobin et al., 2001). Industrial production of entire body scintigraphic devices has been established at Radiumhemmet. It is possible to treat polycythemia vera with phosphorus 32 and bone metastasis with strontium 89 (van Isselt et al.).

With more revolutionary discoveries in the field of radiation and radioactivity at the end of the nineteenth century, physics-based medical technology developed rapidly, giving rise to the modern medical-physics profession, and revolutionizing the practice of medicine. Radiation-based medical diagnosis and treatment were revolutionized, and the practice of medicine was revolutionized due to the modern medical-physics. We are now on the cusp of a new revolution in postgenomic personalized medicine, with physics-based techniques once again at the forefront of the field of medical research and development.

7.4 Recent advancement in radiotherapy

7.4.1 Instigation

Cancer patients need undergo radiation therapy for successful treatment of a variety of reasons, such as the final treatment of localized tumors and the relief of symptoms produced by widely disseminated disease. It is expected that nearly one-half of all cancer patients will be benefitted from radiation therapy (Barton et al., 2006; Barton et al., 2014; Tyldesley et al., 2011). Radiotherapy is particularly important in the treatment of cancer. In some cases, radiation therapy can achieve disease control rates that are comparable to surgery, but with reduced morbidity and mortality. Consider that many advanced throat tumors are treated with radiation and chemotherapy rather than laryngectomy, allowing most patients to keep their ability to speak after treatment (Lefebvre et al., 1996; Forastiere et al., 2003). It is possible to employ radiation therapy prior to surgery in some situations, allowing for a more limited, safer, and effective surgical procedure. Following local excision (lumpectomy) of early-stage breast cancer, the use of localized radiotherapy has prevented thousands of women from suffering the morbidity and scarring associated with a mastectomy since the 1980s (Veronesi et al., 2002; Fisher et al., 2002; National Comprehensive Cancer Network 2008). In a curative scenario, radiotherapy (RT) may be the only aggressive treatment available. The drug can also be administered during (intraoperative), before (neoadjuvant), or after resection (adjuvant) surgery, as well as in conjunction with systemic therapy for organ preservation (for example, the larynx, breast, and urinary bladder) (Alterio et al., 2016; Gerardi et al., 2016; Mak et al., 2008). Patients with locally advanced or disseminated cancer can benefit from treatment as well, since it has been shown to reduce or eliminate pain associated with bone metastases in 60 percent of instances (Fisher et al., 1995; Salvador-Coloma and Cohen, 2016; Chow et al., 2012).

A therapy procedure that is driven by technological breakthroughs, such as RT, can be described as such. Technology advancements in engineering and computing have resulted in significant advancements in radiation therapy techniques, from simple treatment fields to highly conformal RT techniques like intensity-modulated radiation therapy (IMRT), intensity-modulated arc therapy (IMAT), and stereotactic radiation therapy (SRT), all of which aim to improve outcomes by escalating the dose to tumors. As a result, shorter courses of radiation therapy are being used to treat specific tumors (such as breast and prostate cancer) instead of more extensive conventional therapies. Both patient outcomes and healthcare costs are significantly impacted by this (Budach et al., 2015). The integration of imaging information into all phases of therapy, from simulation to planning to delivery, has resulted in significant technological advancements in recent years. Indeed, treatment planning systems (TPS) are capable of utilizing sophisticated picture registration and fusion algorithms (Brock et al., 2017; Nahum and Uzan, 2012). Furthermore, with the introduction of local radiation damage models into treatment planning optimization, which is becoming more radiobiology-focused (Baskar et al., 2014).

7.4.2 Radiotherapy principle and mechanism

It is possible to kill cancer cells by using radiation as a physical agent. It is employed in medical operations because it generates ions (electrically charged particles) and deposits energy in the cells of tissues it passes through. Some cancer cells can either be destroyed or genetically altered such that they die, depending on how much energy is deposited into the cancer cells. DNA damage from high-energy radiation prevents cells from dividing or reproducing, resulting in cell death. It is important to send as much radiation to the cancer cells as possible while limiting the quantity of radiation that is supplied to healthy cells nearby or in the radiation path. Although radiation can destroy both normal and cancerous cells, it is more effective when it is given to cancerous cells. Malignant cells, on the other hand, have a harder time mending themselves and performing their usual functions. Cancer cells, in general, are less effective at repairing damage induced by radiation treatment than normal cells. Consequently, cancer cells are eliminated in a unique way.

7.4.3 Technology development

It is the purpose of radiotherapy to deliver as much radiation as feasible to the tumor while sparing healthy tissue. New imaging modalities, more powerful computers and software, and new delivery systems like advanced linear accelerators have all played a role in this achievement.

7.4.3.1 Three-dimensional conformal radiotherapy

Three-dimensional conformal radiation (3DCRT) and intensity modulated radiotherapy (IMRT) are examples of more conformal radiotherapy procedures than the typical rectangular treatment fields of the past (IMRT). Increasing the dose supplied to tumor-bearing tissues while reducing the irradiation of organs at risk (OARs) is thought to be the driving force behind these changes in radiotherapy administration. The gross tumor volume (GTV) and microscopic spread areas (clinical target volume, CTV) are included in the planned target volume (PTV), as well as a margin around it. There is a margin of safety around the planning target volume (PTV). Multiple beam arrangements around the PTV are used to generate homogeneous dose distribution inside the PTV. 3DCRT treatment plans are always limited by their inability to treat complex target, dose escalation and low dose gradient outside the PTV which compromises between the PTV coverage and normal organ dose tolerance. Higher CTV to PTV margin is given to accommodate any tumor motion and patient setup error depicted in Fig. 7.1 showing the typical field arrangement in 3DCRT planning technique (Fig. 7.2).

7.4.3.2 Intensity modulated radiotherapy

IMRT is a 3DCRT technique that uses multiple beams and optimized intensity modulated fluency distributions. An inverse planning, dose optimization process determines the optimal modulated intensity for each beam by incorporating dose criteria not only for the target volume but also for the nearby organs at risk. Multileaf collimator (MLC) equipped linacs are

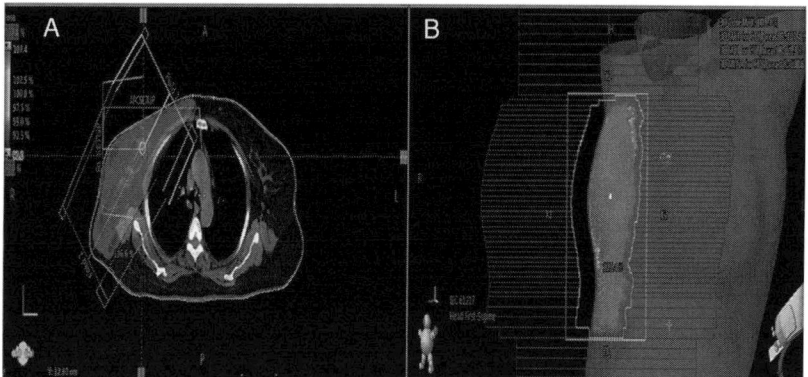

FIG. 7.1 Shows the (A) tangential field arrangement (B) beams eye view for the ca-breast patient.

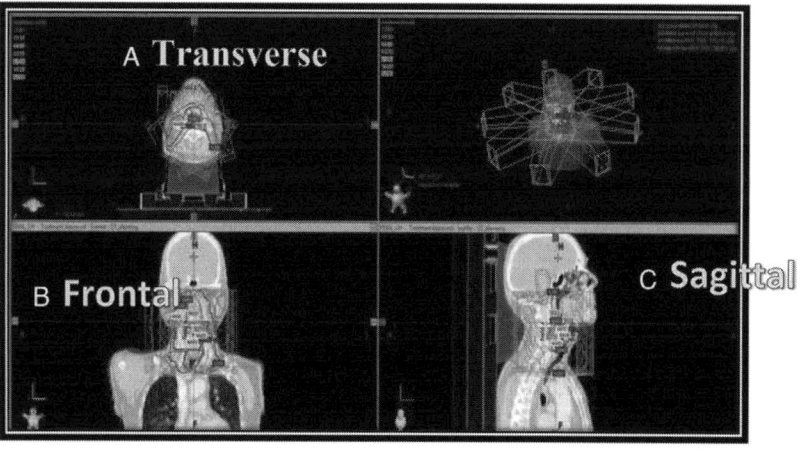

FIG. 7.2 Show the IMRT isodose lines distribution in (A) transverse (B) frontal and (C) sagittal view for head and neck of patient.

typically used to deliver IMRT in either a static or dynamic approach with moving leaves. If gantry rotation is included in dynamic IMRT, it results in intensity modulated arc therapy (IMAT). The use of computerized inverse planning distinguishes IMRT from other radiation treatments. The quantity, shape, and orientation of the beams in conformal radiotherapy are determined by the treatment planner's ability. Inverse planning, on the other hand, determines the plan's outcome in terms of the tumor dose and the normal structure dose limit. To achieve the appropriate configuration, the computer system changes the beams' intensity. An initial dose distribution is established by taking into consideration each beamlet's path through the body.

An initial dose distribution is created by tracing each beamlet through the subject. A slight modification in the weighting of a single beamlet is then made, and this change is approved if it improves the distribution. During a first iteration, this process is repeated for all beamlets, resulting in a better plan. The iterative approach is performed until no further progress can be detected. As a result, the appropriate intensity throughout each beam is achieved, resulting in the desired dose distribution. With a forward-planned manual technique, handling too much iteration would not be possible.

One of the advantages of the IMRT is that it simultaneously allows different doses to be delivered to different target volumes in single phase of treatment plan. Thus, it eliminates the needs for field matching and use of electrons thereby lowers the dosimetric uncertainties.

7.4.3.2.1 Segmental IMRT

This approach is a direct descendant of conformal radiotherapy. Several separate MLC-shaped fields (segments) are formed for each beam orientation. Summing, all of the segments result in modulated field intensity. This procedure is known as step-and-shoot because the radiation is only turned on when the segments are in place. Most linear accelerator manufacturers now offer this option, and it is expected to become the most extensively used technology.

7.4.3.2.2 Dynamic IMRT

On the other hand, MLC-IMRT employs dynamic MLC-IMRT, which keeps the MLC leaves moving during each treatment session. Using computer control, each pair of opposing MLC leaves sweeps across the target volume at a predetermined beam angle to achieve the desired fluence profile. Radiation can be supplied to a specific location by adjusting the pace and distance between leaves. Compared to segmental IMRT, this method yields more uniform dose distributions, making it better suited to difficult-to-treat targets. While the MLC is in continuous motion during beam on, this approach continues to administer radiation, causing greater leakage and increasing the total body dose. Intensity modulated Arc Therapy (IMAT) is a modality in which, in addition to MLC movement, gantry rotation and changes in dose rates occur during the beam on. It is more advanced and sophisticated treatment technique than the IMRT. Fig. 7.3 depicts the two full Arc beam arrangement in IMAT technique.

7.4.4 Image-guided radiotherapy treatment

Tumor and normal tissue movement occurs over time, and this movement can be clinically significant during course of the treatment. Because of the sharp dose gradients in IMRT plans, the tumors may be missed spatially (underdosed) or the OARs may be overdosed. These variations can be classified as intratreatment or inter-treatment in radiotherapy. In practice, each tumor site exhibits both of these effects up to certain extent; some are dosimetrically significant, while others are not. As a result, precise

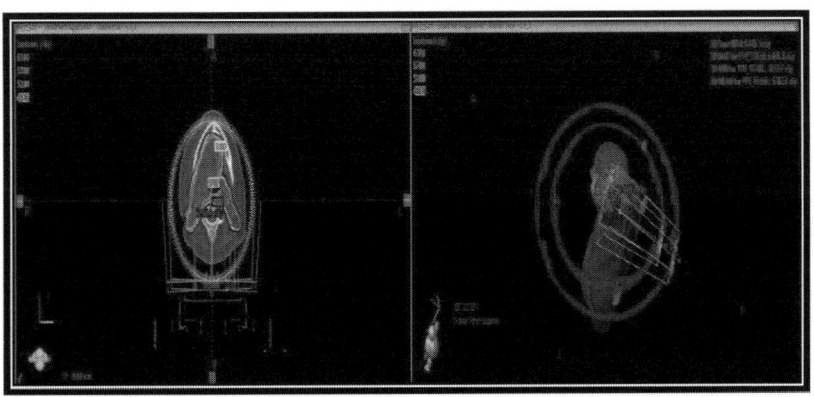

FIG. 7.3 Shows the IMAT technique with two full arc beam arrangement.

image guidance is required for optimal IMRT delivery. The dose to the organs at risk is reduced when this margin is reduced. There are several ways to increase the therapeutic ratio of radiation therapy by altering dosages, the fractions of radiation exposure and dose-escalation. One of the most useful tools for recognizing and correcting treatment problems is IGRT.

7.4.4.1 Intertreatment motion

Radiation oncologists currently have the best chance of improving therapy delivery right now by detecting and correcting changes that occur between therapy fractions. Imaging the patient with megavoltage or kilovoltage cone beam systems referenced to the planning system, or other approaches, can be used to investigate inter-treatment motion. The imaging can be done on a regular, predetermined schedule or at specific points during treatment.

7.4.4.2 Intratreatment motion and their detection

The challenges of imaging, tracking, and managing motion during treatment are difficult, but they may provide the best answers to treatment verification at the end. It is now possible to track passively implanted fiducial seeds in three dimensions using radiographic imaging, as well as radiofrequency tracking of interactively implanted fiducial seeds. Using flat-panel technologies in conjunction with cone beam CT, it is possible to provide fluoroscopic monitoring of anatomy or fiducials at kilovoltage or megavoltage energies. Target or fiducial monitoring algorithms are being developed using cone beam technologies, which are now being applied to construct algorithms for on-line automated monitoring of targets. It is possible that these will include a computer-guided choice regarding when to terminate a therapy and make adjustments in the rare cases where this is required. Instead of implanted markers, technological innovation has enabled the surveillance of local tumor surrogates, such as adjacent bone outlines, rather than the tumor itself.

7.4.5 Adaptive radiotherapy

The goal of modern radiotherapy is to define the target for clinical tumor delineation, treatment planning, and radiation delivery to the target volume. Development in imaging tools technology aids in enhancing tumor delineation and as a result treatment planning process has become more efficient and accurate. Availability of imaging modality inside the treatment room helps in monitoring the target movement and anatomic changes during and between the therapy sessions. Adaptive radiation therapy (ART) is a major goal that involves developing automated mechanisms to detect and correct treatment variations. The concept of ART was first introduced in 1997 (Yan et al., 1997) by the William Beaumont Hospital group. It consists of systematic measurements of treatment variations during radiation treatment process, which finally provides early feedback to modify the treatment plan, and delivers treatment adapted plan to daily patient target volumes.

7.4.6 Stereotactic radiosurgery and radiotherapy

Stereotactic radiosurgery is not a new science, and it was developed on the work of pioneering Swedish researchers, led by L. Leksell (1907–1986), who established the field in the early twentieth century. More than 50 years ago, he began investigating the possibility of combining stereotactic surgery techniques with external radiation therapy. The Swedish researchers Larsson and Leksell pioneered proton radiosurgery and at the same time that analogous research was beginning in the United States. In the succeeding 12 years, these two scientists have developed Gamma knife cranial radiosurgery, and in 1972, Leksell and his son founded Elekta Instruments to manufacture this and other related equipment. In 1983, the first description of a linear accelerator as an alternative to gamma irradiation appeared. Currently, cranial radiosurgery is the gold standard of care for all histologies of benign and malignant brain tumors, as well as a variety of benign conditions such as arteriovenous fistulas, trigeminal neuralgia, and other pain conditions, as well as movement disorders such as epilepsy and Parkinson's disease. Because there is essentially no movement of internal organs, the brain is a perfect location for this type of treatment. For using the Gamma Knife stereotactic radiosurgery system, the patient must first be strapped to the positioning device (also known as a stereotactic frame), and then the treatment unit must be attached. Stereotactic body radiation therapy, which is used outside of the brain, is dependent on patient immobilization equipment to achieve exact tumor targeting. Fig. 7.4 shows the typical Target for SRS treatment.

7.4.7 Particle therapy

The Bragg peak is where most of the energy of charged particles, such as protons, is deposited as they reach the end of their range (depending on their energy). An improved dose uniformity improves normal tissue sparing and reduces radiation dosage to normal tissue. Intensity modulated proton treatment (IMPT) allows for three-dimensional dosage distributions by adjusting fluence and position of the Bragg peak. Proton therapy appears to offer superior dose distributions to IMRT photon-based therapies, particularly in terms of minimizing low and intermediate doses to normal tissue. While proton therapy is novel and theoretically provides dosimetric benefits, independent assessment is necessary to determine its limitations and strengths (Mohan and Grosshans, 2017; Fossati et al., 2016). Indeed, prospective clinical trials comparing proton treatment to photon intensity-modulated radiation therapy are required. Additionally, proton therapy is advancing technologically in the field of motion management, with the move from passively scattered to actively scanned beams (Kubiak, 2016; Bert and Durante, 2011). Proton therapy has been used to treat malignancies of the eye, base of the skull, and spine, with an emphasis on pediatric cancers (Fossati et al., 2016; Ciocca et al., 2016; Matloob et al., 2016; Mishra and Daftari, 2016; Leroy et al., 2016). In comparison to other RT modalities, proton treatment has been demonstrated to have a decreased incidence of vision and hearing impairment, neurocognitive deterioration, and secondary malignancies in children.

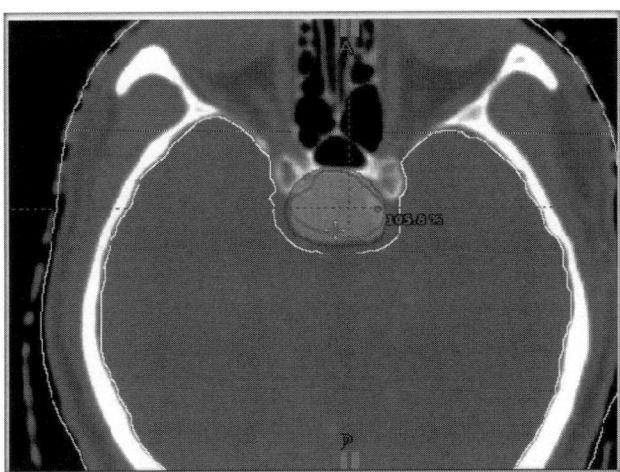

FIG. 7.4 Shows the PTV dose wash in SRS treatment.

7.4.8 Summary

RT has made great development over the years, with technological advances that have changed its clinical usage; yet, we must not neglect the subject's complexity, which serves as a bridge between physics, chemistry, biology, and medicine. We will only be able to produce tailored radiation therapy with superior target delineation, normal tissue avoidance, dose escalation, dose fractionation, and treatment response prediction if we thoroughly investigate all these parameters. Some types of radiation therapy are already in use, while others require more research before becoming widely employed. Despite the development of novel cancer medicines, RT remains an essential approach in the management and control of the majority of cancer types. It also serves as an important organ preservation approach by lowering the necessity for and likelihood of surgery. Over the last decade, rapid advancements in imaging and radiotherapy delivery technology have enabled radiotherapy to be delivered at larger doses and in a more conformal manner. However, 'precision radiation oncology' must embrace more than the ability to give precise technical irradiation in the future. To allow for biological dose optimization, image guidance during therapy, and adaptive radiotherapy, new imaging biomarkers must be included in multiple stages of treatment, including the radiation planning process.

7.5 Radiosurgery for noncancerous tumor and diseases

7.5.1 Introduction

Radiosurgery is the most advanced and precise form of radiotherapy delivery technique. It has a role in management of both malignant and nonmalignant tumors. However, it was initially developed for nonmalignant intracranial tumors (Schültke et al., 2017; Leksell, 1983).

Radiosurgery can be performed on conventional linear accelerators, robotic linear accelerators like Cyberknife or through dedicated radiosurgery systems like Gammaknife (Hoffelt, 2006).

Radiosurgery has changed the way clinicians think and use radiation fractionation. Its principle has been now widely used in management of many nonmalignant tumors, vascular malformations, and functional disorders.

7.5.2 History

Radiosurgery was first described by Horsely and Clark in 1906 (Benedict et al., 2008). They used cranial landmarks to locate intracranial tumors in 3 dimensions. Spiegel and Wycis developed the first frame-based radiosurgery system using a plaster head cap (Kushnirsky et al., 2015).

It was Lars Leksell, a Swedish Neurosurgeon who invented the term "Radiosurgery" in the year 1951. He developed the first metal head frame and used 250 KV X-rays for the treatment of intracranial targets. Subsequently in 1967 used multiple cobalt sources aligned to deliver radiation to the target with sub millimeter accuracy, known as Gamma knife (Ganz, 2014).

In 1980's linear accelerators were adapted to deliver radiosurgery with similar head frames and special dosimetric softwares like X-knife (Martin and Gaya, 2010). Over the years more sophisticated systems including robotics have been developed and with development of on board imaging, precision of such radiosurgery procedures have improved further while it has remained the challenge for the use of rigid radiosurgery frames to frameless radiosurgery.

7.5.3 Treatment

Radiosurgery conventionally delivers high dose radiation in a single fraction however, at times equivalent doses can be delivered over 2 to 5 fractions in fractionated radiosurgery. Such fractionations depend on clinical circumstances. Radiosurgery uses the principle of higher ablative effect with use of higher doses.

Clinicians generally use the LQ model to predict tumor and normal tissue effect with radiation therapy while, with doses more than 10 Gy, the LQ model does not predict such effects precisely. Hence this area is still in progress and varying dose and fraction regimens are used in clinical practice based on institutional experience.

7.5.4 Systems overview

Gammaknife is a specialized intracranial radiosurgery system, comprises up to 201 cobalt-60 sources, each about 30 curies in energy, arranged in a circular array within a strongly protected unit. The machine directs gamma radiation toward a predetermined target. Such brain target locations can be positioned in the center of the radiation focus, allowing for the delivery of a tumoricidal radiation dose in a single treatment session (Li et al., 2016).

X-Knife, Novalis Trilogy, Edge, and Hyperarc are all radiosurgery systems that uses linear accelerators. The Hyperarc technology from Varian enhances the SRS strategy and treatment delivery. Treatment is performed in a single click and includes intrafraction imaging that is predetermined by the system, including for noncoplanar arcs (Wen et al., 2015).

Cyberknife is a radiosurgery system that uses a robotic linear accelerator. The CyberKnife System is a minimally invasive treatment option for malignant and noncancerous tumors, as well as for other illnesses that require radiation therapy. CyberKnife treatments typically last between one and five sessions.

The CyberKnife System is unique in that it is the only one that uses a linear accelerator mounted on a robot to deliver the high-energy X-rays or photons required for radiation therapy. It provides radiation from thousands of beam angles with real-time imaging guidance and the assistance of a robot, setting a new standard for dose delivery precision everywhere in the body and enabling stereotactic radiosurgery. The robot moves and bends around the patient, approaching the tumor from thousands of unique angles, considerably increasing the number of viable sites for concentrating radiation on the tumor while reducing dosage to healthy tissue in the surrounding area (Purdy, 2008).

Indication

1. Vestibular schwannoma.
   - Arises from Schwann cells of nerve sheaths.
   - Incidence of 0.6 to 0.8 per 100,000 person – year.
   - 6 to 8% of all intracranial tumors.
   - Most common at the cerebello pontine angle.
   - Presents with decreased hearing, vertigo, tinnitus, imbalance or facial weakness.
   - Can be sporadic or NF 2 associated.
   - Sporadic has a growth rate of 1 to 2 mm per year, whereas NF 2 associated grows at 2 to 3 mm per year.
   - Staged as per KOOS grading for vestibular schwannoma.
   - High resolution CT and MRI is required for evaluation.
   - Pure tone audiometry to quantify hearing loss.
   - Radiosurgery recommendation is based on age, medical comorbidity, and neurological deficits. Also, for pure intracanalicular lesions or postoperative residual disease.
   - Single fraction treatment if preferred for lesions <3 cm.
   - 12 to 14 Gy single dose is typically prescribed. 25 Gy/5 fractions typically prescribed for fractionated radiosurgery.
   - Post-treatment 3 to 6 monthly follow up is advised.
   - Imaging at 6 to 12 months.
   - PTA if needed.
2. Meningioma
   - Arises from cap cells of arachnoid villi.
   - Incidence of 7.86 per 100,000 person – year.

- 30% of all intracranial tumors.
- More common among females.
- Parasagittal/parafalcine is the most common site.
- Can be NF 2 associated sporadic or ionizing radiation induced.
- CT and MRI is used for evaluation.
- Tumors are graded using WHO classification.
- Radiosurgery recommendation is based on age, medical comorbidity and neurological deficits. Also, for postoperative residual disease.
- Single fraction is preferred for tumors < 3 cm.
- For WHO Grade 1: 12-15 Gy single fraction.
- For WHO Grade 2/3: 16-20 Gy single fraction.
- Fractionated treatment usually with 25 Gy/5 fractions or 37.5 gy/15 fractions.
- Post-treatment 3 to 6 monthly follow up is advised.
- Imaging at 6 to 12 months.

3. Pituitary adenoma
   - Incidence is 2.94 per 100,000 person - year
   - 10 to 15% of all intracranial tumors.
   - Associated with MEN type 1
   - Mostly in the anterior lobe of the pituitary. Can be functional or nonfunctional.
   - Classified as microadenoma of <1 cm, macroadenoma if >1 cm.
   - Presents with pressure symptoms like headache, cranial nerve palsies, visual field deficits or hydrocephalus due to 3rd ventricle obstruction.
   - Secreting tumors may additionally present with endocrine disorders.
   - Classified using Hardy classification of pituitary adenomas.
   - MRI and endocrine workup are a must.
   - Radiosurgery recommendation is based on age, medical comorbidity, and neurological deficits. Also, for postoperative residual disease.
   - Single fraction is preferred for tumors <3 cm.
   - 12 to 20 Gy single fraction for nonsecretory.
   - 15 -30 Gy single fraction for ACTH secreting.
   - 10 to 35 Gy single fraction for GH secreting.
   - Fractionated treatment 21 Gy/3 fractions or 25 Gy/5 fractions.
   - Post-treatment 3 to 6 monthly follow up is advised.
   - Imaging at 6 to 12 months.

4. Paraganglioma
   - Arises from glomus cells.
   - Incidence is 1 per 100,0000 person - year
   - Most commonly seen in Abdomen followed by thorax and head & neck.
   - Classified using Spetzler – Martin grading scale.
   - CT, MRI and MRA required for evaluation.
   - Radiosurgery recommendation is based on age, medical comorbidity, and neurological deficits. Also, for postoperative residual disease.
   - 12 to 30 Gy single fraction.
   - Fractionated treatment 25 Gy/5 fractions.
   - Post-treatment 3 to 6 monthly follow up is advised.
   - Imaging at 6 to 12 months.

5. AV malformation
   - Abnormal communication between arterial and venous vasculature.
   - Point prevalence is 18 per 100,000 person - year
   - Most commonly seen in the supratentorial region.
   - Generally sporadic, it may be associated with hereditary hemorrhagic telangiectasia.
   - Presents with intracranial hemorrhage, seizures, headaches, and focal neurological deficits.
   - Surgical risk associated with AVM is classified using Spetzler – Martin grading scale.
   - Radiosurgery recommendation is based on age, medical comorbidity and neurological deficits, site and size of the AVM.

- Single fraction is preferred for tumors <3 cm.
- 8 to 10 Gy Single fraction for low grade, small volume including those in eloquent areas and not amenable to surgery.
- Fractionated treatment, 12 Gy/ 2 fractions or 28 Gy/4 fractions a week apart.
- Radiosurgery recommendation is based on age, medical comorbidity, and neurological deficits. Also, for postoperative residual disease.
- Post treatment 3 to 6 monthly follow up is advised.
- Imaging at 6 to 12 months.

6. Trigeminal neuralgia
   - Incidence is 4 to 13 per 100,000 person – year
   - Sporadic in nature, rarely familial.
   - Causation is not well understood. Compression of trigeminal nerve root from aberrant vessels is hypothesized. Secondary causes include MS, trauma, tumor and Herpes Zoster.
   - Presents with complaints of episodes of shock like pain in distribution of the trigeminal nerve.
   - Diagnosis of exclusion. Requires MRI
   - International Classification of Headache Disorders describes the diagnostic criteria.
   - Radiosurgery recommendation is based on age, medical comorbidity, and neurological deficits.
   - 70 to 90 Gy Single fraction for medically refractory cases.
   - Post-treatment 3 to 6 monthly follow up is advised.
   - Imaging at 6–12 months.

7. Sphenopalatine neuralgia
   - Etiology is unclear, possibly related to pterygopalatine ganglion.
   - Female predilection
   - Present with unilateral pain in the orbit, nose, and posterior mastoid process. Ipsilateral autonomic stimulation from vasomotor activity.
   - 70 to 90 Gy single fraction for medically refractory cases.
   - Post-treatment 3 to 6 monthly follow up is advised.
   - Imaging at 6 to 12 months.

8. Cluster headache
   - Etiology is unclear, possibly related to pterygopalatine ganglion or trigeminal nerve root.
   - Female predilection
   - Present with unilateral pain along the distribution of cranial nerve V1. Ipsilateral autonomic stimulation from vasomotor activity.
   - Gamma Knife radiosurgery to the ganglion/ nerve root is under investigation.

9. Pineal gland tumor
   0. Residual tumor postoperative or local recurrence after conventional RT.
   1. 15 Gy single fraction SRS
   2. Post treatment 3 to 6 monthly follow up is advised.
   3. Imaging at 6 to 12 months.

10. Others
    - Intractable epilepsy
      Mesial temporal lobe epilepsy is the most common cause for intractable epilepsy. It usually warrants anterior temporal lobectomy. Patients not fit for microsurgery or refusing surgery may be offered Gammaknife based radiosurgery with 24 Gy in single fraction marginal dose.
    - Parkinsons
      Patients identified for unilateral deep brain stimulus but not fit due to age or medical comorbidities may be offered Gammaknife radiosurgery with 110 to 120 Gy in single fraction to subthalamic nucleus.
    - Ventricular tachycardia
      Patients with anatomic defect or failed ablative procedure may be offered LINAC with motion management or Cybeknife radiosurgery with 24 Gy single fraction treatment to the region mapped after detailed electrophysiological mapping.

7.6 Summary and conclusion

Radiation therapy would not exist without physics and this self-evident yet frequently neglected fact forms the basis of this chapter. While radiation therapy "lives" at the nexus of multiple areas, it is almost certainly the most physics-dependent (Beddoe and Burns, 2006; AAPM. ORG 2007–2009; Mould, 2007). This not only involves the reliance on clinical physics to ensure the delivery of radiation as safe and exact, but also on the science and research side of physics in general, and medical physics in particular (Karzmark and Pering, 1973). To emphasize the importance of physics, we frequently refer it to medical physics. Not only the fundamental physics discoveries such as X-rays and protons made substantial contributions to radiation treatment, but also the field of applied physics, though not directly related to medicine, such as linear accelerators, cyclotrons, and synchrotrons made it to breakthrough in the cancer treatment. Without these monumental contributions from fundamental physics, advancements in the radiation therapy would have been substantially slowed, to the point where it may only play a very minor and insignificant role today. It is vital to establish a link between basic scientific study and medicinal application (Webb, 1993; Hounsfield, 1973; Cormack, 1979; Mould, 1993; Webb, 2008).

This is unquestionably advantageous for radiation therapy covering a wide range of fields such as medicine, physics, biology, computer science, and mathematics. Indeed, physicists usually excel at mathematics, which explains their effectiveness in biological and mathematical modeling, such as treatment outcome modeling. The linear–quadratic model and the idea of biologically effective dose (BED) have already been discussed in this series in a review by Jack Fowler and have not been covered in this chapter. Furthermore, physicists have played an important role in optimizing mathematical therapies, particularly IMRT. They have proved especially useful in this area by translating radiation therapy problems into a mathematical language that optimization experts can understand, allowing them to use their specialized tools to optimize radiation treatments. Because of their algorithmic thinking, physicists are often excellent computer programmers, which is an essential talent in an environment where computers are omnipresent. Their ability to see "the forest for the trees" enables them to deal with the huge volumes of data that are typically generated during radiation therapy (Brahme et al., 1982; Rawlinson and Cunningham, 1972; 148, Yan, 2010).

Along with these baffling technological difficulties, there are also deeper physics riddles to solve. These include real four-dimensional (4D) (space-time) imaging with great spatial and temporal resolution, as opposed to existing motion-correlated CT imaging technology, which is based on the assumption of periodic motion and correlation with external surrogates. MRI in conjunction with radiation therapy equipment is a possibility for real 4D imaging. Numerous prototypes of this technology are being developed in academic and industry settings currently (Lagendijk et al., 2008; Fallone et al., 2009). Another difficulty associated with dynamic treatments is developing a biophysical model of the patient with realistic mechanical properties for various tissues and organs in order to simulate and extrapolate realistic motion patterns for individual patients (Eom et al., 2010). Finally, the current strength of radiation therapy is primarily attributable to historical physics research. The scientific method has been fundamental to this advancement during the previous century. Research-oriented physicists face new difficulties and opportunities both within and outside the usual area of radiation therapy (Particle TherapyCo-OperativeGroup 2010); however, the current emphasis on professional rather than scientific aspects of medical physics make resolving such future challenges more difficult (Lomax, 1999). Radiation therapy is biting the hand that feeds it. The development of school programs in medical physics aimed at professional advancement erodes the field's high-end research capacity. There is a paucity of academic programs on medical physics that are comprehensive, rigorous, and focused (Hall, 2006). University research ecosystems are also lacking capability of sustaining a steady flow of new research ideas and overall area progress over an extended period. This imbalance jeopardizes medical physics' long-term sustainability (Enghardt, 2004). Every effort must be taken to reverse this trend and to ensure that, it produces more selective treatments that will specifically target cancer cells while sparing normal cells.

References

AAPM. ORG, 50th anniversary papers. Med. Phys., 2007–2009. A series of articles published in Medical Physics between 2007 and 2009, available from: http://scitation.aip.org/journals/doc/MPHYA6-home/anniversary_papers.jsp.

"ALARP Guidance". UK Health and Safety Executive.

"ALARP Guidance Note" *(PDF)*. Commonwealth Offshore Petroleum and Greenhouse Gas Storage (Safety) Regulations 2009. NOPSEMA. June 2015. Archived from the original *(PDF)* on 16 June 2016. *Retrieved* 20 May *2016*.

Alterio, D., Franco, P., Numico, G., et al., 2016. Non-surgical organ preservation strategies for locally advanced laryngeal tumors: what is the Italian attitude? Results of a national survey on behalf of AIRO and AIOM. Med. Oncol. 33, 76. https://doi.org/10.1007/s12032-016-0781-5. PMID: 27290695.

Ang, K.K, Garden, A.S., 2006. Radiotherapy for Head and Neck Cancers: Indications and Techniques. Lippincott Williams & Wilkins.

Barton, M.B., Frommer, M., Shafiq, J., 2006. Role of radiotherapy in cancer control in low-income and middle-income countries. Lancet Oncol. 7, 584–595. https://doi.org/10.1016/S1470-2045(06)70759-8. PMID: 16814210.

Barton, M.B., Jacob, S., Shafiq, J., et al., 2014. Estimating the demand for radiotherapy from the evidence: a review of changes from 2003 to 2012. Radiother. Oncol. 112, 140–144. https://doi.org/10.1016/j.radonc.2014.03.024. PMID: 24833561.

Baskar, R., et al., 2012. Cancer and radiation therapy: current advances and future directions. Int. J. Med. Sci. 9 (3), 193.

Baskar, R., et al., 2014. Biological response of cancer cells to radiation treatment. Front. Mol. Biosci. 1, 24.

Beddoe, A.H., Burns, J.E., 2006. 50th anniversary issue. Phys. Med. Biol. 51, R1–R504.

Benedict, S.H., et al., 2008. The role of medical physicists in developing stereotactic radiosurgery. Med. Phys. 35 (9), 4262–4277.

Berkey, F.J., 2010. Managing the adverse effects of radiation therapy. Am. Fam. Physician 82 (4), 381–388.

Bert, C., Durante, M., 2011. Motion in radiotherapy: particle therapy. Phys. Med. Biol. 56, R113–R144. https://doi.org/10.1088/0031-9155/56/16/R01. PMID: 21775795.

Brahme, A., 1987. Design principles and clinical possibilities with a new generation of radiation therapy equipment: a review. Acta Oncol. (Madr) 26 (6), 403–412.

Brahme, A., Roos, J.E., Lax, I., 1982. Solution of an integral equation encountered in rotation therapy. Phys. Med. Biol. 27, 1221–1229.

Brock, K.K., Mutic, S., McNutt, T.R., et al., 2017. Use of image registration and fusion algorithms and techniques in radiotherapy: Report of the AAPM Radiation Therapy Committee Task Group No. 132. Med. Phys. https://doi.org/10.1002/mp.12256.

Budach, W., Bölke, E., Matuschek, C., 2015. Hypofractionated radiotherapy as adjuvant treatment in early breast cancer: a review and meta-analysis of randomized controlled trials. Breast Care (Basel) 10, 240–245. https://doi.org/10.1159/000439007.

Chang, B.K., Timmerman, R.D., 2007. Stereotactic body radiation therapy: a comprehensive review. Am. J. Clin. Oncol. 30 (6), 637–644.

Chargari, C., et al., 2019. Brachytherapy: An overview for clinicians. CA Cancer J. Clin. 69 (5), 386–401.

Chew, M.T., et al., 2021. Radiation, a two-edged sword: From untoward effects to fractionated radiotherapy. Radiat. Phys. Chem. 178, 108994.

Chin, L.S., Regine, W.F., 2010. Principles and Practice of Stereotactic Radiosurgery. Springer Science & Business Media.

Chow, E., Zeng, L., Salvo, N., et al., 2012. Update on the systematic review of palliative radiotherapy trials for bone metastases. Clin. Oncol. (R. Coll. Radiol.) 24, 112–124. https://doi.org/10.1016/j.clon.2011.11.004.

Ciocca, M., Mirandola, A., Molinelli, S., et al., 2016. Commissioning of the 4-D treatment delivery system for organ motion management in synchrotron-based scanning ion beams. Phys. Med. 32, 1667–1671. https://doi.org/10.1016/j.ejmp.2016.11.107. PMID:27890567.

Cobin, R.H., et al., 2001. AACE/AAES medical/surgical guidelines for clinical practice: management of thyroid carcinoma. Endocr. Pract. 7 (3), 202–220.

Cormack A.M., 1979. Early two-dimensional reconstruction and recent topics stemming from it.

Cossairt, J.D, 1993. Radiation Physics for Personnel and Environmental Protection. Fermi National Accelerator Lab.(FNAL), Batavia, IL (United States) No. FERMILAB-TM-1834-REV.

Dawson, L.A., Jaffray, D.A., 2007. Advances in image-guided radiation therapy. J. Clin. Oncol. 25 (8), 938–946.

De Los Santos, J., et al., 2013. Image guided radiation therapy (IGRT) technologies for radiation therapy localization and delivery. Int. J. Radiat. Oncol. Biol. Phys. 87 (1), 33–45.

Delaney, G., et al., 2005. The role of radiotherapy in cancer treatment: estimating optimal utilization from a review of evidence-based clinical guidelines. Cancer 104 (6), 1129–1137.

Delaney, G., et al., 2005. The role of radiotherapy in cancer treatment: estimating optimal utilization from a review of evidence-based clinical guidelines. Cancer 104 (6), 1129–1137.

Delaney, G.P., Barton, M.B., 2015. Evidence-based estimates of the demand for radiotherapy. Clin. Oncol. 27 (2), 70–76.

Devlin, P.M., 2015. Brachytherapy: Applications and Techniques. Springer Publishing Company.

Duck, F.A., 2014. The origins of medical physics. Physica Med. 30 (4), 397–402.

Earle, J.D., Bagshaw, M.A., Kaplan, H.S., 1973. Supervoltage radiation therapy of the testicular tumors. Am. J. Roentgenol. 117 (3), 653–661.

Enghardt, W., 2004. Charged hadron tumor therapy monitoring by means of PET. Nucl. Instrum. Methods Phys. Res. Sect. A 525, 284–288.

Eom, J., Xu, X.G., De, S., Shi, C., 2010. Predictive modeling of lung motion over the entire respiratory cycle using measured pressure-volume data, 4DCT images, and finite element analysis. Med. Phys. 37, 4389–4400.

Fairris, G.M., et al., 1985. Conventional superficial X-ray versus Grenz ray therapy in the treatment of constitutional eczema of the hands. Br. J. Dermatol. 112 (3), 339–341.

Fallone, B.G., Murray, B., Rathee, S., Stanescu, T., Steciw, S., Vidakovic, S., et al., 2009. First MR images obtained during megavoltage photon irradiation from a prototype integrated linac-MR system. Med. Phys. 36, 2084–2088.

Fisher, B., Anderson, S., Bryant, J., et al., 2002. Twenty-year follow-up of a randomized trial comparing total mastectomy, lumpectomy, and lumpectomy plus irradiation for the treatment of invasive breast cancer. N. Engl. J. Med. 347 (16), 1233–1241.

Fisher, B., Anderson, S., Redmond, C.K., et al., 1995. Reanalysis and results after 12 years of follow-up in a randomized clinical trial comparing total mastectomy with lumpectomy with or without irradiation in the treatment of breast cancer. N. Engl. J. Med. 3 (33), 1456–1461. https://doi.org/10.1056/NEJM199511303332203. PMID: 7477145.

Forastiere, A.A., Goepfert, H., Maor, M., et al., 2003. Concurrent chemotherapy and radiotherapy for organ preservation in advanced laryngeal cancer. N. Engl. J. Med 349 (22), 2091–2098.

Fossati, P., Vavassori, A., Deantonio, L., et al., 2016. Review of photon and proton radiotherapy for skull base tumors. Rep. Pract. Oncol. Radiother. 21, 336–355. https://doi.org/10.1016/j.rpor.2016.03.007. PMID: 27330419 PMCID: 4899429.

Fowler J.F. 21 years of biologically effective dose. Br. J. Radiol.;83:554–68.

Ganz, J.C., 2014. The History of the Gamma Knife. Elsevier.

Garau, i, Macià, M., Calduch, A.L., López, E.C., 2011. Radiobiology of the acute radiation syndrome. Rep. Pract. Oncol. Radiother. 16 (4), 123–130.

Gerardi, M.A., Jereczek-Fossa, B.A., Zerini, D., et al., 2016. Bladder preservation in non-metastatic muscle-invasive bladder cancer (MIBC): a single-institution experience. Ecancermedicalscience 10, 657. https://doi.org/10.3332/ecancer.2016.657. PMID: 27563352 PMCID: 4970626.

Gerig, L., Soubra, M., Salhani, D., 1994. Beam characteristics of the Therapax DXT300 orthovoltage therapy unit. Phys. Med. Biol. 39 (9), 1377.

Giger, M.L., Chan, H.-P., Boone, J., 2008. Anniversary paper: history and status of CAD and quantitative image analysis: the role of medical physics and AAPM. Med. Phys. 35 (12), 5799–5820.

Hall, E.J., 2006. Intensity-modulated radiation therapy, protons, and the risk of second cancers. Int. J. Radiat. Oncol. Biol. Phys. 65, 1–7.

Heavy ions offer a new approach to fusion". *CERN Courier*. 2002-06-25. Retrieved 2021-01-22.

Hill, R., et al., 2014. Advances in kilovoltage x-ray beam dosimetry. Phys. Med. Biol. 59 (6), R183.

Hoffelt, S.C, 2006. Gamma knife vs. cyberKnife. Oncology Issues 21 (5), 18–20.

Hoppe, R.T., Goffinet, D.R., Bagshaw, M.A., 1976. Carcinoma of the nasopharynx. Eighteen years' experience with megavoltage radiation therapy. Cancer 37 (6), 2605–2612.

Horwich, A., 1994. Walter & Miller's Textbook of Radiotherapy (5th edn). Br. J. Cancer 69,1198.

Hounsfield, G.N., 1973. Computerized transverse axial scanning (tomography). Part 1: Description of system. Br. J. Radiol. 46, 1016–1022.

Hu, K., Harrison, L.B., 2005. Impact of anemia in patients with head and neck cancer treated with radiation therapy. Curr. Treat. Options Oncol. 6 (1), 31–45.

Ising, G., 1924. Prinzip einer Methode zur Herstellung von Kanalstrahlen hoher VoltzahlArkiv för Matematik, Astronomi och Fysik18. Band, pp. 1–4 Nr.S.

Ivanov, L.I., Platov, Yu.M., 2004. Radiation Physics of Metals and Its Applications. Cambridge Int Science Publishing.

Jacob, S., et al., 2010. Estimation of an optimal utilisation rate for palliative radiotherapy in newly diagnosed cancer patients. Clin. Oncol. 22 (1), 56–64.

Jaffray, D.A., et al., 2015. Global task force on radiotherapy for cancer control. Lancet Oncol. 16 (10), 1144–1146.

Jaffray, D.A., Gospodarowicz, M.K., 2015. Radiation therapy for cancer. Disease Control Priorities 3, 239–247.

Jeraj, R., et al., 2004. Radiation characteristics of helical tomotherapy. Med. Phys. 31 (2), 396–404.

Johns, H.E., 1953. The Physics of Radiation Therapy. Charles C Thomas.

Karzmark, C.J., Pering, N.C., 1973. Electron linear accelerators for radiation therapy: history, principles and contemporary developments. Phys. Med. Biol. 18, 321–354 Crossref. PubMed.

Keevil, S.F., 2012. Physics and medicine: a historical perspective. Lancet North Am. Ed. 379 (9825), 1517–1524.

Khan, F.M., Gibbons, J.P., 2014. Khan's the Physics of Radiation Therapy. Lippincott Williams & Wilkins.

Kiseleva, E.S., Falileeva E.P., and Berkgaut V.G. "Role of telegamma therapy and radioactive iodine in treatment of cancer of the thyroid gland." (1972).

Kolotas, C., Baltas, D., Zamboglou, N., 1999. CT-based interstitial HDR brachytherapy. Strahlenther. Onkol. 175 (9), 419–427.

Kostylev, V.A., 2000. Medical physics: yesterday, today, and tomorrow. Biomed. Eng. 34 (2), 106–112.

Kubiak, T., 2016. Particle therapy of moving targets—the strategies for tumor motion monitoring and moving targets irradiation. Br. J. Radiol. 89, 20150275. https://doi.org/10.1259/bjr.20150275.

Kushnirsky, M., Patil, V., Schulder, M., 2015. The history of stereotactic radiosurgeryPrinciples and Practice of Stereotactic Radiosurgery. Springer, New York, NY, pp. 3–10.

Lagendijk, J.J., Raaymakers, B.W., Raaijmakers, A.J., Overweg, J., Brown, K.J., Kerkhof, E.M., et al., 2008. MRI/linac integration. Radiother. Oncol. 86, 25–29.

Lefebvre, J.L., Chevalier, D., Luboinski, B., Kirkpatrick, A., Collette, L., Sahmoud, T., 1996. Larynx preservation in pyriform sinus cancer: preliminary results of a European Organization for Research and Treatment of Cancer phase III trial. J. Natl. Cancer Inst. 88 (13), 890–899.

Leksell, L., 1983. Stereotactic radiosurgery. Journal of Neurology, Neurosurgery & Psychiatry 46 (9), 797–803.

Leroy, R., Benahmed, N., Hulstaert, F., et al., 2016. Proton therapy in children: a systematic review of clinical effectiveness in pediatric cancers. Int. J. Radiat. Oncol. Biol. Phys. 95, 267–278. https://doi.org/10.1016/j.ijrobp.2015.10.025. PMID: 27084646.

Li, W., et al., 2016. The use of cone beam computed tomography for image guided Gamma Knife stereotactic radiosurgery: initial clinical evaluation. Int. J. Radiat. Oncol. Biol. Phys. 96 (1), 214–220.

Lindell, Bo, Walstam, R., 1956. A new telegamma apparatus. Acta Radiol. 45 (3), 236–248.

Lo, S.S., et al., 2010. Stereotactic body radiation therapy: a novel treatment modality. Nat. Rev. Clin. Oncol. 7 (1), 44–54.

Lomax, A., 1999. Intensity modulation methods for proton radiotherapy. Phys. Med. Biol. 44, 185–205.

Luqmani, Y.A, 2005. Mechanisms of drug resistance in cancer chemotherapy. Med. Princ. Pract. 14 (Suppl. 1), 35–48.

Lutz, W., Winston, K.R., Maleki, N., 1988. A system for stereotactic radiosurgery with a linear accelerator. Int. J. Radiat. Oncol. Biol. Phys. 14 (2), 373–381.

Mackie, T.R, et al., 1999. TomotherapySeminars in Radiation Oncology9. WB Saunders.

Mackie, T.R., 2006. History of tomotherapy. Phys. Med. Biol. 51 (13), R427.

Mak, R.H., Zietman, A.L., Heney, N.M., et al., 2008. Bladder preservation: optimizing radiotherapy and integrated treatment strategies. BJU Int. 102, 1345–1353. https://doi.org/10.1111/j.1464-410X.2008.07981.x. PMID: 19035903.

Martin, A., Gaya, A., 2010. Stereotactic body radiotherapy: a review. Clin. Oncol. 22 (3), 157–172.

Martins, P.N., 2018. A brief history about radiotherapy. Int. J. Latest Res. Eng. Technol. 4, 8–11.

Matloob, S.A., Nasir, H.A., Choi, D., 2016. Proton beam therapy in the management of skull base chordomas: systematic review of indications, outcomes, and implications for neurosurgeons. Br. J. Neurosurg. 30, 382–387. https://doi.org/10.1080/02688697.20161181154. PMID: 27173123.

Meisberger, L.L., Keller, R.J., Shalek, R.J., 1968. The effective attenuation in water of the gamma rays of gold 198, iridium 192, cesium 137, radium 226, and cobalt 60. Radiology 90 (5), 953–957.

Milano, M.T., et al., 2008. Descriptive analysis of oligometastatic lesions treated with curative-intent stereotactic body radiotherapy. Int. J. Radiat. Oncol. Biol. Phys. 72 (5), 1516–1522.

Milano, M.T., Philip, A., Okunieff, P., 2009. Analysis of patients with oligometastases undergoing two or more curative-intent stereotactic radiotherapy courses. Int. J. Radiat. Oncol. Biol. Phys. 73 (3), 832–837.

Mishra, K.K., Daftari, I.K., 2016. Proton therapy for the management of uveal melanoma and other ocular tumors. Chin. Clin. Oncol. 5 (4), 50. https://doi.org/10.21037/cco.2016.07.06. PMID: 27558251.

Mohan, R., Grosshans, D., 2017. Proton therapy–present and future. Adv. Drug. Deliv. Rev. 109, 26–44. https://doi.org/10.1016/j.addr.2016.11.006.

Mould, R.F., 1993. A Century of X-Rays and Radioactivity in Medicine: With Emphasis on Photographic Records of the Early Years. Institute of Physics Publishing, Bristol, UK.

Mould, R.F., 2007. Radium history mosaic. Nowotwory 57.

Muller, R.A., 1977. Radioisotope dating with a cyclotron. Science 196 (4289), 489–494.

Myint, A.S., et al., 2018. Dose escalation using contact X-ray brachytherapy after external beam radiotherapy as nonsurgical treatment option for rectal cancer: outcomes from a single-center experience. Int. J. Radiat. Oncol. Biol. Phys. 100 (3), 565–573.

Nag, S., et al., 2000. The American Brachytherapy Society recommendations for high-dose-rate brachytherapy for carcinoma of the cervix. Int. J. Radiat. Oncol. Biol. Phys. 48 (1), 201–211.

Nag, S., et al., 2002. The American Brachytherapy Society recommendations for low-dose-rate brachytherapy for carcinoma of the cervix. Int. J. Radiat. Oncol. Biol. Phys. 52 (1), 33–48.

Nahum, A.E., Uzan, J., 2012. (Radio) biological optimization of external-beam radiotherapy. Comput. Math. Methods Med. 2012. https://doi.org/10.1155/2012/329214.

Naseri, A., Mesbahi, A., 2010. A review on photoneutrons characteristics in radiation therapy with high-energy photon beams. Rep. Pract. Oncol. Radiother. 15 (5), 138–144.

National Comprehensive Cancer Network. NCCN clinical practice guidelines in oncology. Breast Cancer. 2008. http://www.nccn.org/professionals/physician_gls/PDF/breast.pdf. (Accessed October 10, 2008)

Nonnenmacher, L., et al., 2016. Cell death induction in cancer therapy– past, present, and future. Crit. Rev. Oncogen. 21 (3-4).

Norlund, A., 2003. Costs of radiotherapy. Acta Oncol. (Madr) 42 (5-6), 411–415.

Pacelli, R., et al., 2019. Technological evolution of radiation treatment: Implications for clinical applicationsSeminars in Oncology46. WB Saunders.

Particle TherapyCo-OperativeGroup [homepageontheInternet] 2010

Paz-Manrique, R., et al., 2020. Trimodal treatment for localized high risk prostate cancer: a design of a phase II, multi-center, open-label, randomized, parallel-group study to compare the effectiveness and safety of radical prostatectomy, external beam radiotherapy (EBRT) and androgen deprivation therapy (ADT) versus EBRT and ADT for patients with localized high risk prostate cancer. Princ. Pract. Clin. Res. 6 (1), 3–8.

Pearse, J., Chow, J.C.L., 2020. An Internet of Things app for monitor unit calculation in superficial and orthovoltage skin therapy. IOP SciNotes 1 (1), 014002.

Pecorino, L., 2016. Molecular Biology of Cancer: Mechanisms, Targets, and Therapeutics. Oxford University Press, USA.

Perez, C.A., et al., 1987. Long-term observations of the patterns of failure in patients with unresectable non-oat cell carcinoma of the lung treated with definitive radiotherapy report by the radiation therapy oncology group. Cancer 59 (11), 1874–1881.

Podgorsak, E.B., et al., 1988. Dynamic stereotactic radiosurgery. Int. J. Radiat. Oncol. Biol. Phys. 14 (1), 115–126.

Polgár, C., Major, T., 2009. Current status and perspectives of brachytherapy for breast cancer. Int. J. Clin. Oncol. 14 (1), 7–24.

Purdy, J.A., 2008. Dose to normal tissues outside the radiation therapy patient's treated volume: a review of different radiation therapy techniques. Health Phys. 95 (5), 666–676.

Radiosensitivity on GP notebook. http://www.gpnotebook.co.uk/simplepage.cfm?ID=2060451853

Rao, W., Deng, Z.-.S., 2010. A review of hyperthermia combined with radiotherapy/chemotherapy on malignant tumors. Crit. Rev. Biomed. Eng. 38 (1).

Rawlinson, J.A., Cunningham, J.R., 1972. An examination of synchronous shielding in 60 Co rotational therapy. Radiology 102, 667–671.

Reed, A.B. "The history of radiation use in medicine." (2011): 3S-5S.

Roach, W.T., et al., 1976. The physics of radiation fog: I–a field study. Q. J. R. Meteorolog. Soc. 102 (432), 313–333.

Robison, R.F., 1995. The race for megavoltage x-rays versus telegamma. Acta Oncol. (Madr) 34 (8), 1055–1074.

Round, W.H, 2013. The (pre-) history of medical physics. Eng. Phys. Sci. Med. 2013.

Salvador-Coloma, C., Cohen, E., 2016. Multidisciplinary care of laryngeal cancer. J. Oncol. Pract. 12, 717–724. https://doi.org/10.1200/JOP.2016.014225. PMID: 27511718.

Schültke, E., et al., 2017. Microbeam radiation therapy—grid therapy and beyond: a clinical perspective. Br. J. Radiol. 90 (1078), 20170073.

Skowronek, J., 2017. Current status of brachytherapy in cancer treatment–short overview. J. Contemp. Brachyther. 9 (6), 581.

Suit, H., Urie, M., 1992. Proton beams in radiation therapy. J. Natl. Cancer Inst. 84 (3), 155–164.

Tabakov, S., 2017. History of medical physics–a brief project description. Med. Phys. 5 (1).

Thwaites, D.I., Tuohy, J.B., 2006. Back to the future: the history and development of the clinical linear accelerator. Phys. Med. Biol. 51 (13), R343.

Timmerman, R.D., et al., 2007. Stereotactic body radiation therapy in multiple organ sites. J. Clin. Oncol. 25 (8), 947–952.
To, A. Quick Guide. "Radiotherapy." (2021).
Tubiana, M., et al., 1985. External radiotherapy in thyroid cancers. Cancer 55 (S9), 2062–2071.
Tyldesley, S., Delaney, G., Foroudi, F., et al., 2011. Estimating the need for radiotherapy for patients with prostate, breast, andlung cancers: verification of model estimates of need with radiotherapy utilization data from British Columbia. Int. J. Radiat. Oncol. Biol. Phys. 79, 1507–1515. https://doi.org/10.1016/j.ijrobp.2009.12.070.
van Isselt, J.W. "Radioiodine therapy: from medical tradition to clinical science."
Verhaegen, F., Seuntjens, J., 2003. Monte Carlo modelling of external radiotherapy photon beams. Phys. Med. Biol. 48 (21), R107.
Verhey, L.J., 1999. Comparison of three-dimensional conformal radiation therapy and intensity-modulated radiation therapy systemsSeminars in Radiation Oncology, 9. WB Saunders.
Veronesi, U., Cascinelli, N., Mariani, L., et al., 2002. Twenty-year follow-up of randomized study comparing breast-conserving surgery with radical mastectomy for early breast cancer. N. Engl. J. Med. 347 (16), 1227–1232.
Washington, C.M., Leaver, D.T., 2015. Principles and Practice of Radiation Therapy-E-Book. Elsevier Health Sciences.
Webb, S., 1993. The Physics of Three Dimensional Radiation Therapy. IOP Publishing, Bristol, UK.
Webb, S., 2008. Combating cancer in the third millennium—the contribution of medical physics. Phys. Med. 24, 42–48.
Webb, S., 2009. The contribution, history, impact and future of physics in medicine. Acta Oncol. (Madr) 48 (2), 169–177.
Wen, N., et al., 2015. Characteristics of a novel treatment system for linear accelerator–based stereotactic radiosurgery. J. Appl. Clin. Med. Phys. 16 (4), 125–148.
Whelan, B. "Maximising the mutual interoperability of an MRI scanner and a cancer therapy particle accelerator." (2016).
Wheldon, T.E., Michalowski, A.S., Kirk, J., 1982. The effect of irradiation on function in self-renewing normal tissues with differing proliferative organisation. Br. J. Radiol. 55 (658), 759–766.
Widerøe, R., 1928. Über Ein Neues Prinzip Zur Herstellung Hoher Spannungen. Archiv für Elektronik und Übertragungstechnik 21 (4), 387–406.
Withers, H.R., Taylor, J.M.G., Maciejewski, B., 1988. The hazard of accelerated tumor clonogen repopulation during radiotherapy. Acta Oncol. (Madr) 27 (2), 131–146.
Xing, L., et al., 2006. Overview of image-guided radiation therapy. Med. Dosim. 31 (2), 91–112.
Yan, D., 2010. Adaptive radiotherapy: merging principle into clinical practice. Semin. Radiat. Oncol. 20, 79–146.
Yan, D., Vicini, F., Wong, J., Martinez, A., 1997. Adaptive radiation therapy. Phys. Med. Biol. 42, 123–132.
Zwahlen, D.R., et al., 2010. High-dose-rate brachytherapy in combination with conformal external beam radiotherapy in the treatment of prostate cancer. Brachytherapy 9 (1), 27–35.

Chapter 8

Physics in treatment of cancer radiotherapy

Ravindra B. Shende[a], Sanjay J. Dhoble[b]
[a]Department of Radiation oncology, Balco Medical Centre, New Raipur, Chhattisgarh, India
[b]Department of Physics, Rashtrasant Tukadoji Maharaj Nagpur University, Nagpur, Maharashtra, India

8.1 Introduction

Remarkable breakthrough happened soon after discovery of X-ray by a Wilhelm Rontgen dated in 1895 followed by radium discovered by Marie Curie and Pierre Curie in 1898. Radiological physics began with these discoveries. Soon after, these discoveries were found to be miraculous for the medical industry. Effects of interaction of radiation with physical and biological entities are well recognized shortly. Within a short span of time, use of both X-rays and radium become very much popular in the field of medicine. Application of an ionizing radiation for the treatment of cancer initiated dated back late in the 19th century. Rapid and wide uses of ionizing radiation in the treatment of cancer procreate a field of radiotherapy. Radiotherapy mostly utilizes high-energy Mega-voltage (MV) X-ray, gamma rays, electrons, protons, and others to damage cancer cells. In radiotherapy source of radiation either placed externally or internally closed in the vicinity of the tumor. Radiations are used as a curative or palliative care for treatment of cancer patients. Over the period, the side-by-side revolution in engineering technology and information technology (IT) brought change in treatment cancer radiotherapy. In the last few decades, multiple advanced treatment modalities and techniques of radiotherapy have been invented. Several computerized dose calculation algorithms and new radiobiological models were established. Modern radiotherapy involved an automated fully controlled radiation delivery system and advanced treatment planning system (TPS). Every advanced radiotherapy facility undergoes commissioning of radiation generating equipment and corresponding TPS. This goes through several calibrations and quality checks with rigorous procedure (Khan and Gibbons, 2014). In addition, physical and clinical dosimetric protocols with baseline standards are well established. The medical physicist performs the routine and periodic checks before clinical implementation of each radiation treatment plan. Applied medical physics has wide applications and plays a crucial role in delivering the radiation at every stage of radiotherapy.

8.1.1 Physics of radiotherapy

Physics of radiotherapy includes radiological physics particularly deals with ionizing radiations, interaction of ionizing radiations with physical and biological matters, characteristics of radiological quantities, and measurement of an absorbed dose. The measurement of absorbed dose is most essential in radiation therapy. Radiation dosimetry carried the quantification of absorbed dose and energy inside matter of interest.

8.1.2 Structure of matter

Matter is any substance that has mass and occupies physical space composed of individual entities called elements. Every element is made up of a subatomic elementary particle with distinguishable properties in their different physical and chemical properties. Atoms of different kinds of elements with different composition form substances called compounds. Technically atoms are elementary building blocks of each substance in the universe.

8.1.3 Atom

In a Greek lingo the "atomos" mean an indivisible entity that derived the word "atom" from Greek word. Atom is an elementary particle constituting with a positively charged (+) nucleus in the center and negatively charged (−) electron revolving around it. Atoms consist of a central core encompassing primarily large atomic mass called the nucleus and surrounding "cloud" of an electron orbiting around the core (Khan and Gibbons, 2014). The electron is revolving in a specific orbit around the nucleus with electric charge (e = 1.60217662 × 10^{-19} C). Bohr's postulates states, electron is revolving only in the orbits of nucleus for which it satisfies the relationship L = n$\hbar$. L is an electron angular momentum, h ($\hbar$ = h/2Π) is a Planck's constant and n is an orbital quantum integer. Also, electrons do not gain or loss their energy when they remain orbiting in an orbiting state; they only exchange parts of its energy when they transit from one nth orbit to another nth orbit. Electrons orbiting in particular orbit are called shells with a given principal quantum number (n). Number of electrons that can be accommodated in a shell is equal to 2n^2. Shells are designated as K = 1, L = 2, M = 3, N = 4 ....and so on. Electrons in shells are further subdivided into its subshell called orbital and they are labeled as s, p, d, f ...and so on. Each orbital possesses different values of orbital angular momentum associated with l = (n−1) even for the same principal quantum number. The highest number of electrons can hold in the sub-shell is 2(2l + 1), where l is the orbital angular quantum number (l). This distribution of electrons in an atomic sub-shell is form electronic configuration of the atom. Atom of the radius approximately is of order 10^{-10} m. Modern particle physics does not consider atoms as elementary particles.

8.1.4 Nucleus

Advance modern physics claimed and proved that atom can be further subdivided into its subatomic particles. Central core of an atom is called a nucleus. This positively charged core of the nucleus made up of positively charged protons (p$^+$) and neutral neutrons (n^0) called nucleons. The total number of protons in the nucleus representing atomic number (Z) of an atom is equal to the total number of electrons revolving in orbits of an atom. Total number of protons and neutrons in the nucleus (p + n) together form the mass number of an atom, symbolized by A. Atomic number and mass number of an elements usually represented by $_Z X^A$. The classifications of an atom are based upon the proportion of proton and neutron present inside the nucleus (Podgorsak editor, 2005).

a. Two or more atoms having nucleus with the same number of protons but different numbers of neutrons called isotopes.
b. Atoms with the equal number of neutrons but different numbers of protons are called isotones.
c. Atoms with the equal mass number of nuclides but differ in proton number or atomic numbers are called isobars.
d. Atoms having equal atomic number and mass number, while differing their nuclear energy states are called isomers.

This combination of nuclide results in nuclear stability of an atom. The line of stability represents nuclear stability. Elements of lower atomic number (Below Z = 20) are relatively more stable with almost equal number of protons and neutrons. However, as atomic number increases, ratio of neutron to proton ratio increases and goes beyond unity. Neutron to proton ratio increases for stability of elements as Z increases. Nuclear stability is identified in terms of even and odd numbers of neutrons and protons. Maximum number of stable elements is present with an even number of neutrons and protons called elements with even-even nuclides. This clearly states that nuclei gain stability when neutrons and protons are mutually coupled. An element with combination of even-Z and odd-N nuclide is about 20%. About same proportion of elements are present with a combination of odd-N and even-Z nuclei. However, only four stable nuclei exist with odd-Z and odd-N, namely $_1H^2$, $_3Li^6$, $_5B^{10}$, $_7N^{10}$. Nuclear radii is proportional to the number of nuclides present (Mass number A) inside and it is of order 10^{-15} m, R = R$_0$ × A$^{1/3}$. Rest mass of the neutron (m$_n$ = 1.674927471 × 10^{-27} kg) is slightly greater than the proton rest mass (m$_p$ = 1.6726219 × 10^{-27} kg) (Podgorsak editor, 2005). The rest mass of the proton is almost 1836 times higher than mass of electron. Modern science found hadrons (proton and neutron) made up of strongly interacting particles called quark. Modern particle nuclear physics classified fundamental particles in two categories are quark and lepton. Both quark and lepton are considered as fundamental or elementary particles at present. Elementary particles are particles that do not know to have any further substructure.

8.1.5 Types of radiation

Electromagnetic radiation comprises range of radiation frequency from higher to lower frequency. This includes radiation of Gamma rays, X-rays, ultraviolet rays, visible light, infrared waves, microwaves, and radio waves in order of decreasing frequency. These electromagnetic radiations consist of oscillating electric and magnetic field components perpendicular to each other. Both the field act as mode of energy propagation and carries equal energy in the direction of propagation with light speed in vacuum. The speed of light (c) and frequency of oscillation (ν) of waves governed by the relation,

$$c = \nu * \lambda \tag{8.1}$$

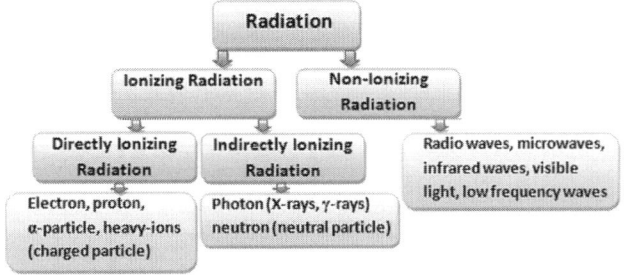

FIG. 8.1 Classification of radiation.

λ is the wavelength of e.m.w. or distance between two consecutive and identical points on the wave. ν is the frequency of the e.m wave. Energy E of particles of light is given by Planck's equation.

$$E = h\nu = hc/\lambda \tag{8.2}$$

where, h = Planck's constant = 6.626×10^{-34} J s.

This spectrum of radiation is categorized based on ability to ionize the matter with which they interact. They are classified in two main categories, ionizing and nonionizing radiations. Ionizing radiation has an ability to remove the electron from an atomic shell and ionize the matter. However, nonionizing radiation belongs to low frequency radiation unable to ionize the matter. Minimum energy required to ionize an atom called ionizing potential. This Ionizing potential of an atom is varying depending on the type of atom. Fig. 8.1 illustrates the classification of radiation.

Radiation therapy utilizes both directly and indirectly ionizing radiation to treat cancer patient (Beyzadeoglu et al., 2010). Directly ionizing radiation like electron, proton and other charged particle deposit their energy through direct coulomb interactions between electronic cloud of an atom in medium and incident charged particles. However, indirectly ionizing radiation follows a different process than directly ionizing radiation. Indirect ionizing radiation like photon or neutron follows the two-step process. In first steps incident uncharged particles released the charged particle in medium through the several interactions (Photon release electron and positron whereas, neutron release proton and heavy ions in the medium). In second steps, these released free charged particles from an atom deposit their energy through coulomb interaction with clouds of electrons of an atom in the medium.

8.1.6 X-rays

Wilhelm Conrad Roentgen discovered the X-rays in 1895. Followed by William Coolidge developed pressurized (10^{-3} mm Hg) glass tube consists of cathode and anode at two opposite ends, between which electric potential of magnitude 10^6 to 10^8 V applied. Electrons produced by the process of thermionic emission by cathode. These electrons with high kinetic energy accelerated toward the anode of high melting point. These high-energy electrons decelerated due to coulomb interaction with nuclei of the target. Electrons lose their part of kinetic energy in the process of deceleration resulting in production of continuous X-rays. Those continuous X-rays are also called Bremsstrahlung (radiative loss). Emission of the bremsstrahlung spectrum may consist of photon energy ranging from zero to maximum kinetic energy of the incident electron. Emission spectrum depends on the kinetic energy of the incident electron, atomic number Z of target material and thickness of a target. On the other hand, spectrum emerging from the target found continuous distribution of energy with discrete energy characteristic peak superimposed on this called characteristic X-rays. These characteristic X-rays are resultant of electronic transition between orbits. Incident electrons with high kinetic energy may eject electrons from the inner orbital of an atom, resulting in vacancy in the inner orbital is filled by another electron transition from an outer atomic orbital of an atom. The energy of each characteristic X-ray emission is equivalent to the difference of binding energy between inner shell and outer shell electron. In the mechanism of characteristic X-ray emission, incoming electrons with energy (E_0) transfer some part of energy to orbital electrons (ΔE). If amount of transferred energy is higher than binding energy of the orbital electron, it ejects an orbital electron with residual energy ($\Delta E - E_K$). Where, E_K is the binding energy of electron in K shell. While the incoming primary electron deflected from its path with remaining energy ($E_0 - \Delta E$). In case, K (n = 1) shell electron knocked out and subsequent transition of L (n = 2) shell electron filled this vacancy then characteristic emission is $K\alpha = (E_L - E_K)$. Fig. 8.2A shows simultaneous emission spectrum of bremsstrahlung and characteristics X-rays. Similarly, Fig. 8.2B illustrates an energy level diagram for characteristic emission due to electronic transition.

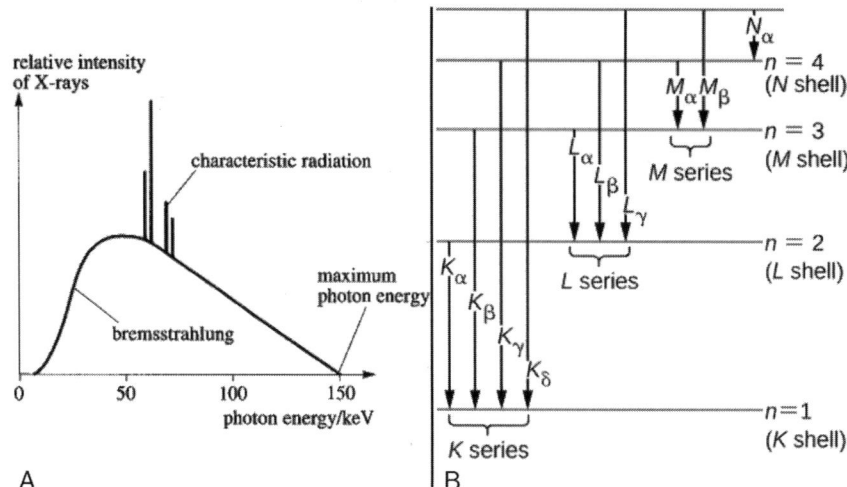

FIG. 8.2 (A) Spectrum of Bremsstrahlung and characteristic x-ray emission, (B) Energy level diagram of characteristic x-ray emission.

Bremsstrahlung or continuous X-rays are heterogeneous in nature having broad energy spectrum, whereas characteristic radiation is monoenergetic in nature due to its discrete electronic transition between different energy levels (Khan and Gibbons, 2014). Usually, low energy X-rays ranging from 20 keV to 150 keV are used in diagnostic radiology. A medical linear accelerator produces a megavoltage beam from an energy range of 4 keV to 25 MeV used in radiation oncology for treatment of cancer (Mayles et al., 2007).

8.1.7 Gamma rays

There is only physical difference between X-rays and gamma rays are their source of origin. Beside this, there no any physical difference these rays. The X-rays are produced due to the deceleration of electron or electronic transition between different energy levels in an atom outside the nucleus. However, gamma rays (γ) are emitted from inside the nucleus of an atom (Intranuclearly). High Z elements are highly unstable due to their electrostatic repulsion between the protons, which eventually leads to emission of surplus energy in the form of alpha and beta particles to achieve nuclear stability. If the nucleus still possesses surplus energy, gamma rays are emitted to attain a state of stability. Emission of gamma rays does not alter the number of nucleons inside the nucleus, but it brings a nucleus to a lower energy state from a higher energy state. Cobalt-60 radionuclide undergoes the decay of the beta particles followed by emission of gamma rays. Unstable $_{27}Co^{60}$ transformed into stable $_{28}Ni^{60}$ with emission of beta rays of energy 0.31MeV and two photons of energy 1.17MeV and 1.33MeV.

8.1.8 Particulate radiation

Particulate radiation is a type of radiation, which exhibits mass and energy but it may or may not have electric charge. Electrons, protons, neutrons, alpha particles, Pi mesons and heavy ions like carbon are all particulate radiation. Most of these particulate radiations are widely being used in treatment of cancer.

8.1.9 Interaction of radiation with matter

Interaction of radiation with matter consists of both direct and indirect ionizing radiation.

8.1.10 Interaction of photon beam (X-rays or γ rays)

As a beam of (X-rays or γ rays) radiation goes through matter (medium), there is interaction between an incoming photon and medium can take place. The resultant interaction may or may not cause transfer of energy to matter. Transfer of energy to the medium could undergo multiple types of interaction. Transfer of energy involves excitation of an atom and ejection of electrons from an atom of the medium. These high-speed electrons further lead to an ionization and excitation of atom in its path. Obviously if the beam does not transfer any energy to medium, it cannot produce any effect in the medium. If the incoming beam transfers a part of its energy, then it produces some physical or chemical changes in the medium, followed by biological changes observed in living organisms.

Before we deal with the types of interactions in detail, we will discuss the attenuation or reduction in primary beam intensity after an interaction with the absorber or medium. Let us consider a narrow beam of monochromatic radiation falls on a uniform absorber of thickness X. N be the number of photons with intensity I incident normally on the front face of the absorber. Let the dN reduction in number of photons suffering collision or interaction proportional to initial number of photon N and thickness of absorber dX.

$$dN \alpha NdX$$
$$dN = -\mu NdX \qquad (8.3)$$

where, μ is constant of proportionality called attenuation coefficient. The reduction of the number of photons with increasing absorber thickness is indicates by negative sign. Eq. (8.3) can be written in term of intensity (I):

$$dI = -\mu IdX \qquad (8.4)$$

$$\mu = -(dI / I\, dX) \qquad (8.5)$$

If thickness x is expressed as a length, then μ called as linear attenuation coefficient. The attenuation coefficient of an absorber depends on the number of atoms, electrons of the absorber as well as on its density. From Eq. (8.5), the unit of μ is cm^{-1}. Where thickness is measured in centimeters. Solving Eq. (8.4) for the condition of reduction in intensity of the primarily incident beams varies from I_0 to I with the change in thickness of 0 to x giving an exponential attenuation of x-rays or gamma rays with thickness of absorber.

$$\int dI / I = -\mu \int dX$$
$$I(x) = I_0 e^{-\mu x} \qquad (8.6)$$

where, I(x) = Intensity of the transmitted beam through thickness of absorber x. I_0 = Intensity of incoming beam incident on absorber.

Eq. (8.6) clears that intensity of radiation constantly decreasing as it propagates inside the medium. Reduction in intensity depends on thickness and type of medium. This is valid only if the attenuation coefficient is actually constant and this is only true if the incident beam is mono-energetic and if the beam is narrow. Intensity I is a function of thickness x. If I(x) is plotted against thickness x for a given narrow monoenergetic beam, the straight line will be obtained on semilogarithmic paper, concluding that the attenuation of a monoenergetic photon beam is described by an exponential function shown in Fig. 8.3A.

Intensity of the incident beam is continuously decreasing as it passes through the absorber, eventually it reaches to half of its original intensity ($I_0/2$). Thickness of the absorber that attenuates intensity of the photon beam become 50% of its original intensity is called the half value layer. From Eq. (8.6), we can derive the relationship between HVL and linear attenuation coefficient as follows,

$$HVL(x_{1/2}) = \ln 2 / \mu \approx 0.693 / \mu \qquad (8.7)$$

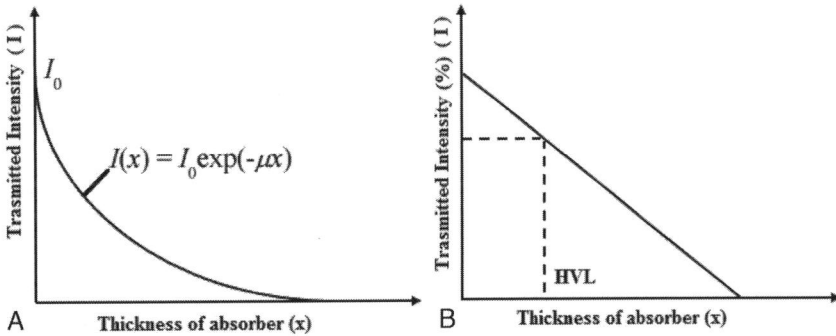

FIG. 8.3 Graph showing (A) Intensity of monoenergetic-transmitted beam follows an exponential function. (B) Percentage transmission of a narrow monoenergetic photon beam function of absorber thickness (x) on semilogrithmic paper indicating HVL.

TABLE 8.1 Attenuation coefficient.

Name of the coefficient	Symbol	Relation between coefficient	Unit of the coefficient
Linear attenuation coefficient	μ	μ	cm^{-1}
Mass attenuation coefficient	μ/ρ	μ/ρ	cm^2/g
Atomic attenuation coefficient	$_a\mu$	$(\mu/\rho)/(N_A/A)$	cm^2/atom
Electronic attenuation coefficient	$_e\mu$	$(\mu/\rho)/(ZN_A/A)$	cm^2/electrons

HVL represents the quality of the beam. For the monoenergetic beam, HVL is constant. However, for heterogeneous beams the first HVL is less than the subsequent due to the beam hardening effect. Fig. 8.3B shows the plot of transmitted intensity on logarithmic scale versus absorber thickness on linear scale. This graphical method helps in determining the value of HVL, even without knowing the value of linear attenuation coefficient.

Some more fundamental attenuation coefficients are defined considering mass and atomic composition of material. If N_A is Avogardo's number of an absorber. Then the number of atoms per unit gram of the absorber is N_A/A, where A is the mass number. Similarly, the number of electrons per unit gram is $N_A Z/A$, where Z is the absorber atomic number. Then, attenuation coefficient can be obtained in the following way shown in Table 8.1.

Photons may undergo various possible types of interaction depending on the incident photon energy and absorber atomic number Z. During this, incoming photons interact with the electrons of an atom inside the material and partial or all of its energy transferred into kinetic energy of the electron. If some part of its energy given to electrons, then the incident photons are scattered with remaining energy. This scattered photon may undergo multiple interactions and lose its energy by converting into electron kinetic energy. The fraction of photon energy transferred into the kinetic energy of the charged particle per unit thickness of the absorber is called the energy transfer coefficient (μ_{tr}). Relationship between μ_{tr} and μ is given as follows:

$$\mu_{tr} = \left(\overline{E}_{tr}/h\nu\right)\mu \tag{8.8}$$

where, $\overline{E}_{tr}$ mean energy transferred into kinetic energy of charged particle in each photon interaction. From this mass energy transferred coefficient is given by μ_{tr}/ρ. The electrons are set into the motion due to an acquired kinetic energy. Electrons lose their kinetic by inelastic collisions such as ionization and excitation with electrons of an atom inside the material. Few of electrons lose their energy by producing bremsstrahlung with nuclear interaction. This bremsstrahlung X-ray energy radiated out from locally and it is not included in calculation for local energy absorption. Fraction of energy lost by secondary charged particles in the process of bremsstrahlung or radiative loss in materials is referred as g. Therefore, another coefficient defined in order to account for energy absorption apart from energy transfer coefficient is energy absorption coefficient (μ_{en}). The energy absorption coefficient is defined as the product of energy transfer coefficient and (1-g). Similarly, the mass-energy absorption coefficient is given by μ_{en}/ρ.

$$\mu_{en} = \mu_{tr}(1-g) \tag{8.9}$$

Interaction with low Z material like soft-tissue, electrons lose their entire energy by excitation and ionization collision. Bremsstrahlung components are very negligible. In such a scenario Eq. (8.9) becomes $\mu_{en} = \mu_{tr}$. These coefficients differ significantly only in case of secondary electrons with higher kinetic energy and for interaction with high Z material. Energy absorption coefficient is very important in radiotherapy, since it allows us for evaluation of energy absorption in human tissue. It also helps in predicting biological effects of radiation.

When a photon beam undergoes attenuation, it mostly caused by five major interactions are coherent scattering phenomenon, photoelectric effects, Compton effects, pair production and photodisintegration. Out of these, photodisintegration is interaction between incoming incident photon and nucleus of an atom occurred at high energy above (>10MeV). Each of these processes possesses their own independent attenuation coefficient depending on photon energy and attenuator atomic number Z. Total mass attenuation coefficient of the photon beam is a sum of all attenuation coefficients due to these processes as follows (Khan and Gibbons, 2014).

$$\mu/\rho = \sigma_{coh}/\rho + t/\rho + \sigma_c/\rho + \pi/\rho \tag{8.10}$$

where, σ_{coh}, t, σ_c, π are attenuation coefficients for coherent scattering phenomenon, photoelectric effect, Compton effect and pair production respectively.

8.1.11 Coherent scattering

The coherent scattering is a Thomson or Rayleigh scattering (Podgorsak editor, 2005). Thomson scattering is elastic scattering that takes place with single unbound electrons of an atom. However, in Rayleigh scattering phenomenon, incoming photons interact with every electrons of an atom. The Rayleigh scattering phenomenon can be explained by looking at wave nature or wavelength of electromagnetic radiation. Here, electromagnetic waves or photons are absorbed by tightly bound electrons of an atom causing it to oscillate with the same frequency of incident wave or photon. Oscillation causes the electrons to rise to the higher energy state. The oscillating electron radiates the photons energy at the same frequency as that of incident EM wave or photon and returns to ground state. These scattered x-rays photons have the same wavelength as the incident beam. There is no transfer of energy taking place through this process. Rayleigh scattering contributes nothing to absorbed dose/kerma, since neither energy transferred to any charged particle nor any ionization/excitation produced. Rayleigh scattering is highly probable at lower energy for high Z material. Rayleigh coherent scattering cross-section is proportional to Z^2 and inversely proportional to square of energy, that is, $s_{coh} \approx (Z^2/(hn)^2)$ cm^2/atom.

8.1.12 Photoelectric effect

Photoelectric effect is process of emission or ejection of inner shell electrons after the absorption of electromagnetic radiation (Mayles et al., 2007). Initially, an atom absorbs the entire incident photon (hν) energy. Then, all of its absorbed energy transferred to electrons of an atom. This transferred energy of the photons overcomes binding energy of an inner shell electron and ultimately rejected from an atomic orbital. In photoelectric effect, incident photon interacts with a bound electron in an atom from K, L, M or N shells. The ejected electrons with kinetic energy equal to (hν-E_B), where E_B is the electron binding energy. The kinetic energy of photoelectrons is related to energy of incident photon by a well-known Einstein photoelectric equation,

$$h\nu = E_B - 1/2 \, mv^2 + \text{KE of residual atom} \tag{8.11}$$

The part of incident photon energy is used up to remove the electron from its orbit, that is, in just setting the electron free and the remaining energy of the photon is split between the photon and the residual atom. Because of the enormous size of an atom, the K.E acquired by an atom is almost zero. Therefore, the energy is conserved practically between 1st two terms on the right side of Eq. (8.11) so we can write,

$$h\nu = E_B - 1/2 \, mv^2 \tag{8.12}$$

Once an electron has been knocked out from an atom, it leaves the atom in a higher excited state. The vacancy created by an ejected electron filled by another outer shell electron with emission of characteristic X-rays. The energy released in emission of characteristic X-rays possibly ejects electrons from an outer shell. These electrons are called Auger electrons and appear with an energy difference between the original atomic excitation energy and BE of the shell from which the electron is ejected. Auger process is favored only in low Z materials for which electron binding energy is small. Photoelectric interaction depends on

1. The position of the electron in an atom: Photoelectric interaction occurs only with bound electrons of an atom. Truly speaking, no electron in an atom is free. However, relative to the incident photon energy some of the electrons may be regarded as free. If, incident photon energy hν much greater than the binding energy of an electron hν >> E_B, then we can say that electron is almost free or can be regarded as free for all practical purposes. However, hν > E_B, the electron is relatively more bound than the previous situation and if h$\nu \approx E_B$ or less, the orbital electron is bound with the atom. Therefore, an incident photon of a particular energy will see all the electrons inside an atom with different degrees of binding or different BE. Thus, the photon will interact only to those electrons, which have BE comparable to the photon energy or just more than that. In other words, K shell electrons will contribute more than any lower shell electrons and similarly L shell electrons will contribute more than those of M shell electrons will and so on.
2. The atomic number of the absorber: If the same photon faces two atoms of low Z and high Z material the BE of high Z atom are more than low Z, one can say qualitatively that high Z materials are more predominantly in producing photoelectric absorption. Since the photoelectric effect strongly depends on Z of the material, the mass attenuation coefficient for photoelectric effect (t/r) is varies atomic number Z of the material as,

$$t/r \alpha Z^3 \tag{8.13}$$

This relation helps in evaluation of photoelectric absorption in Z of various tissues such as bone, muscle, fat in diagnostic radiology.

3. The photon energy: As the photon energy increases, a greater number of electrons can be regarded as free, in other words more shells become ineffective in producing photoelectric effect. As energy of incident photon increases, the probability of photoelectric effect decreases. The photoelectric absorption cross-section differs with photon energy as $1/E^3$. A crude but most commonly used formula is t/r » Z^n/ $(h\nu)^3$ n varies from 3 to 5 depending on energy of photon.

$$\text{For low energy photon, } t/r \approx Z^4 / (h\nu)^3 \text{ cm}^2 / \text{atom} \tag{8.14}$$

$$\text{For high energy photon, } t/r \approx Z^5 / (h\nu) \text{ cm}^2 / \text{atom} \tag{8.15}$$

Directional distribution of photoelectrons mainly depends on energy of photon energy, as energy of photon increases emission photoelectron lies in forward direction. Photoelectric effect is prime interaction at energy less than 35KV predominant in diagnostic radiologyand material with high atomic number.

8.1.13 Compton effects

In Compton effects (Khan and Gibbons, 2014), incident photon energy ($h\nu_0$) is much greater than the binding energy of an electron in an atom. This causes atomic electrons from the outer orbital to act as a free and stationary relative to incident photon energy. In Compton effects, photons interact with these free electrons in an outer orbital. In which incoming photon transferred part of its energy to the outer orbital electron of an atom. Consequently, electrons gain sufficient energy to be knocked out with some kinetic energy and emitted at an angle (θ). The incoming photon scattered at an angle (ϕ) with lesser energy. By using law of conservation of momentum and energy, we can obtained the relationship for kinetic energy of an emitted Compton electron E_K given as,

$$E_K = h\nu_0 \frac{\alpha(1-\cos\varphi)}{1+\alpha(1-\cos\varphi)} \tag{8.16}$$

Scattered photon energy ($h\nu'$) is diminished by amount of energy lost in emission of Compton electron can be derived as below,

$$h\nu' = h\nu_0 - E_K$$
$$h\nu' = h\nu_0 \frac{1}{1+\alpha(1-\cos\varphi)} \tag{8.17}$$

where, $m_0c^2 = 0.511$ MeV is electron rest mass energy and $\alpha = h\nu_0/m_0c^2$. Photon scattering angle (ϕ) and recoiled Compton electron (θ) are related as below,

$$\cot\theta = (1+\alpha)\tan(\phi/2) \tag{8.18}$$

As we discussed the probability of photoelectric interaction decreases with increase in incident photons energy, once energy goes higher than the binding energy of the K shell electron, the photoelectric effect start diminishes rapidly. Then, Compton effects start dominating, at certain energy Compton effects are highly predominant. Compton effects also start decreasing beyond certain energy. Compton effects are independent of Z of absorbing material, since this essentially involves interaction of photons with free electrons only. However, Compton mass attenuation coefficient (σ_c/ρ) is independent of atomic number Z and depends only on number of electrons per gram i.e. density of electron in absorbing material. The density of electron is nearly the same for all elements except hydrogen. Therefore, values of (σ_c/ρ) are almost similar for all material. In radiotherapy the Compton phenomenon is the main source for absorption of ionizing radiation.

8.1.14 Pair production

When the incident photon energy is greater than 1.02 MeV, then other third kinds of interaction becomes greatly important. The photon now prefers to interact with the electric field surrounding a nucleus or an atomic electron. This is known as pair production (Mayles et al., 2007). In this process, incident photon energy ($h\nu$) is completely absorbed by an electromagnetic field of the nucleus producing a pair of positron-electrons whose total energy is just equal to incident photon energy $h\nu$.

$$\begin{aligned} hn &= (TE)_{e-} + (TE)_{e+} \\ &= KE_{e-} + m_0c^2 + KE_{e+} + m_0c^2 \\ &= KE_{e-} + KE_{e+} + 2m_0c^2 \end{aligned} \tag{8.19}$$

Here the energy has been converted into matter, that is, two particles of the same mass have been produced. Thus, the incident photon energy must be at least equal to the energy equivalent of two masses, that is, 2×0.511 MeV = 1.02 MeV which is called the threshold energy for the pair production. In that case, KE's of the electron and positron will be zero. Any surplus incident photon energy over 1.02 MeV will be shared by the electron and positron pair as their KE's. The process cannot take place in vacuum and it occurs only in the field of charged particles. The pair production occurs mainly in the presence of Coulomb field of nucleus and field of an atomic electron for conservation of momentum.

Phenomenon of pair production occurs mainly in the electromagnetic field of nucleus, it is expected that higher the field higher will be the probability of pair production. Probability of this phenomenon increases rapidly with increase in Z of the absorbing material. Attenuation coefficient for process of pair production is proportional to square of atomic number (Z^2) per atom. Also, likelihood of increase in pair production with an incident photon energy above the threshold energy. A curve shows interaction probably almost coincides for all material up to the 20 MeV. At higher energy, the curve for high-Z material falls below low-Z material due to the screening of nuclear charge by the orbital electrons.

8.1.15 Photodisintegration

Photodisintegration reaction is also called a photonuclear reaction, where high-energy photon is absorbed by the nucleus of an atom resulting in an emission of neutrons or protons (Mayles et al., 2007). Photonuclear reactions cannot undergo below some threshold photon energy. The photonuclear cross-section increases with increasing energy of photon (above threshold). It reaches to a maximum value and then decreases for (γ, n) decay. In every case, the maximum value of the total cross-section for all photonuclear reactions is lesser than 5% of combine cross-section of the same atom for Compton and pair production interactions. Hence, it is commonly neglected in dosimetric consideration. The reaction cross-section is proportional to Z. The other processes are (γ, p), (γ, α), etc. The threshold energy for this reaction is order of 10 MeV or more. The (γ, n) interaction has greater practical importance because production of neutrons may cause problems in radiation protection. In case of medical linear accelerators (Linacs, microtrons, betatrons) where, electrons are accelerated to 10 MeV or more to produce x-rays. The interaction of this high energy x-rays with the accelerator body or surrounding matter will produce neutrons. Fig. 8.4 illustrates how the phenomenon of photodisintegration takes place.

$$\gamma +_4 Be^9 \rightarrow_4 Be^8 +_0 n^1 \rightarrow 2(_2He^4)$$

8.1.16 Interaction of charged particle

As we discussed, a photon loses its energy undergoing multiple interactions like Rayleigh coherent scattering, photoelectric phenomenon, Compton and pair production phenomenon represented by their mass attenuation coefficient (Khan and Gibbons, 2014). Similarly, the quantity called stopping power (S) defined to represent the kinetic energy lost per unit path length of an absorber (dE/dx) for a charged particle. Thus, stopping power per unit density of medium ρ is called mass stopping power (S/ρ). It is usually expressed in MeV cm^2/gm.

$$(S/\rho) = \frac{1}{\rho} \text{Mev cm}^2 / \text{gm} \tag{8.20}$$

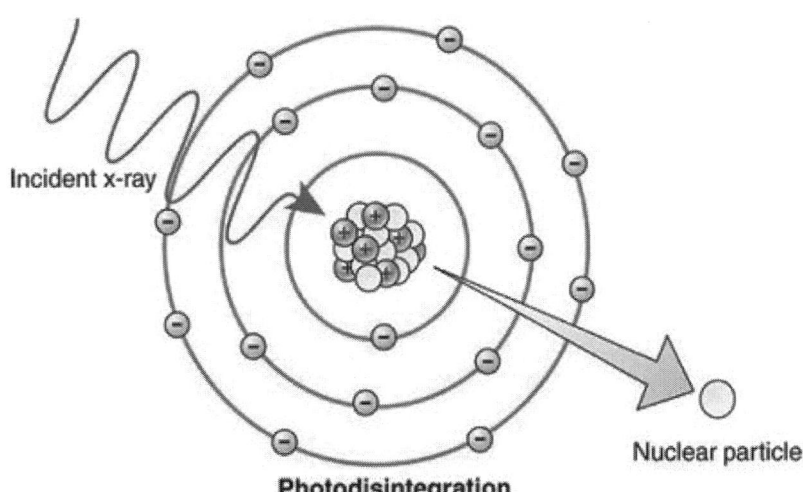

FIG. 8.4 Diagram illustrating the process of photodisintegration.

Mass stopping power S/ρ consists of two components

1. Mass collision stopping power $(S/\rho)_{col}$ due to an interaction between the incident electron and orbital electrons of an atom causes atomic excitation and ionization of an atom in a medium.
2. Mass radiative stopping power $(S/\rho)_{rad}$ resulting from interaction between electronnucleus leads to bremsstrahlung production.

Therefore, the total stopping power of an electron beam passing through an absorber is the addition of both collision and radiative stopping power.

$$(S/\rho)_{Total} = (S/\rho)_{col} + (S/\rho)_{rad} \tag{8.21}$$

Mass stopping power plays asignificant role in radiotherapy in order to find the clinical range of electrons in tissue, dose in a medium and used to determine radiation yield (i.e., Bremsstrahlung efficiency).

Now, let us discuss the interaction of electrons with matter. As a beam of electrons travels through the medium, it undergoes interaction with electrons around the nucleus and nucleus of an atom by electrostatic coulomb interactions. This interaction between the incident electron with orbital electrons and nucleus of an atom can be elastic or inelastic. In case of elastic collision, there is no change in loss of energy of the electron. However, electrons deflected from its original path. While in case of inelastic collision incoming electron loses its energy in interaction with orbital electron leaving an atom ionized or in excitation state and interaction with nucleus producing bremsstrahlung. Electrons experience deflection from its original path after the collision interaction. Incoming electrons undergo a kind of interaction depending on impact parameter. The perpendicular distance between trajectories of an electron before the interaction and nucleus of the host atom is called as impact parameter. Fig. 8.5 illustrates interaction of electrons based on impact parameters.

Case 1: If **b** >> **a** then, the electron undergoes a soft collision with an atom and only a very little amount of energy transferred to the atomic orbital electron.

Case 2: If **b** ≈ **a** then, electrons undergo a hard collision with a significant amount of energy transferred to an atomic orbital electron.

Case 3: If **b** < **a** then, electron undergoes radiative interaction with a nucleus of an atom with emission of bremsstrahlung.

8.1.17 Electron and electron interaction

The incident electrons interact with atomic orbital electron in absorber results in ionization and excitation. Ionization is the ejection of electrons from the orbital of an atom of medium. Due to the absorption energy, the electron of the atom attends the higher electronic orbit called an excitation state. Both atomic excitation and ionization responsible for collision energy losses characterized by collision stopping power $(S/\rho)_{col}$.

8.1.18 Electron and nucleus interaction

The electrostatic coulombic interaction between the incident electron and atomic nucleus of an absorber results in energy losses of an electron. These energy losses caused by declaration of electrons due to a coulomb nuclear attraction leads to production of continuous x-ray or bremsstrahlung. This is also called radiative losses and characterized by radiative stopping power $(S/\rho)_{rad}$.

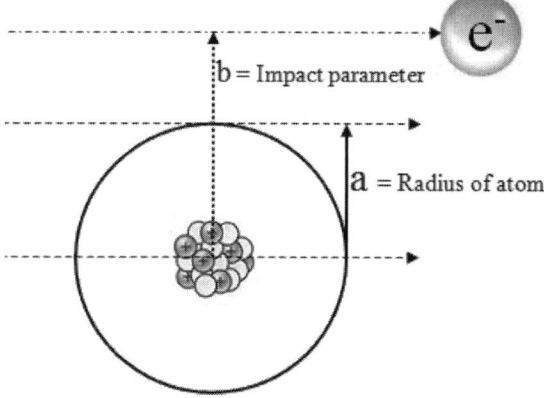

FIG. 8.5 Diagram illustrating interaction of electron with an atom of radius (A) based on impact parameter (B).

8.1.19 Interaction of heavy charged particle

Depending upon mass of the particle, charged particles can be categorized into heavy or light particles. Electrons and positrons are considered as light particles due to their tiny mass. Both electrons and positrons carry rest mass 1840 times lesser than mass of the proton. Particles like Proton, Carbon, α-particle and atomic nuclei are treated as heavy charged particles. When heavy charged particles traversing matter losses their energy through ionization and excitation of an atom. If transferred energy is sufficient to knock out an orbital electron, it leaves the atom ionized or it may leave the atom in an excited state. For heavy charged particle rate of energy loss per unit length is defined as stopping power (dE/dx) is directly proportional to square of particle charge and inversely proportional to square of its velocity. Therefore, as the particle slows down, the rate of energy loss increases, causing absorption of energy in the medium. This rate of energy becomes maximum as article velocity approaches to the zero. The depth dose depositions depend on the rate of energy loss in the medium. Initially dose deposition increases slowly due to the greater particle velocity. At the end of particle range sharp increase in dose deposition, this sharp dose deposition peak observed at end of particle range called Bragg peak. Bragg peaks have greater clinical significance in prescribing dose to tumor radiotherapy.

8.1.20 Biological effect of radiation

Ionizing radiation produces physical, chemical and biological change as they interact with matter of interest. Biological effect of radiation mainly classified into direct effect and indirect effect due to radiation (Beyzadeoglu et al., 2010). Cell is a fundamental functional unit of the living organism consisting of a centrally located critical target for radiation. Deoxyribonucleic acid (DNA) molecules principally considered as a critical target for radiation. It is generally accepted that this critical site within the cells which must be damaged in order to kill the cell. DNA contains genetic information for the development and function of the living organism.

Direct effect: If ionizing radiation of any type x-rays, γ-rays, charged and uncharged particles interacts with biological material, there is probability of radiation to interact directly with critical target DNA in the cell. The atom of DNA molecules itself may undergo excitation or ionization, that initializing to a series of events which may affect the ability of cells to reproduce and survive brings a biological change. This effect of radiation called radiation direct effect. Direct effect of radiation is highly dominant in high linear energy transfer (LET) radiation type, such as proton, carbon, and α-particle.

Indirect effect: Alternatively, the human body contains about 80% water. There is much higher probability of radiation interacting with other than critical targets in the cells, particularly with water molecules. When ionizing radiation interacts with water molecules it results in formation of reactive species called free radical (H^0, OH^0). This free radical is strong enough to interact and damage the critical site DNA of the cell causing biological change over a longer period is called the indirect effect of radiation. The atom or molecule carrying an unpaired electron in its outer most shell called as free radical. Indirect action mainly results from the radiolysis process of water as follows:

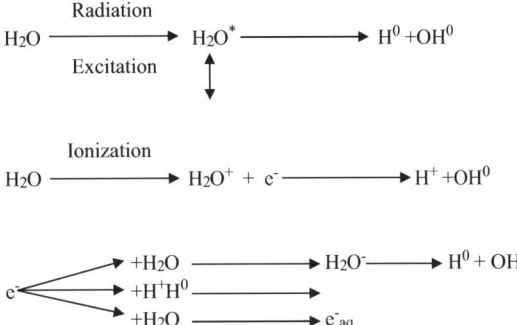

Some of the radicals recombine with each other leading to the formation of H_2 and H_2O_2. Radiolysis of water summarized as follows:

$$H_2O \rightarrow H^0 + OH^0 + e^-_{aq} + H_2 + H_2O_2 + H_3O^+$$

Due to presence of unpaired electrons, the free radicals are highly reactive. Radical OH^0 are responsible for most of the radiobiological effect in microorganisms and mammalian cells.

Direct and indirect action of ionizing radiation causes a variety of several changes in DNA molecules. Broadly, the changes are of two types, namely, cross-linking and degradation. Intermolecular cross-linking occurs between two molecules such as DNA-DNA and DNA-protein cross linking. Following exposure to ionization radiation, DNA degradation appears to be a very important change, even though base damage, base changes etc. also occur. DNA is degraded into smaller fragments because of breaks. In irradiated cells, decreases in molecular weight of DNA resulting from degradation can be detected by gradient centrifugation technique. When the break is located on one of the stands it is referred to as a single strand break (SSB). When two breaks are located on opposite strands, separated by less than 5 bases, the resulting breakage is referred to as double strands break (DSB). Besides this, exposure of radiation leads to the loss of proliferating integrity of cells, delay in mitosis cell division, chromosomal aberration and gene mutation, this eventually can lead to a death of cells or formation of mutate cells. The radiation biological effect varies with the type of radiation evaluated from the amount of energy transferred per unit length referred as LET.

8.1.21 Linear energy transfer

Highly dense ionizing radiation (High LET radiation) deposits large amounts of energy along their path as compared to the sparsely ionizing radiations. Linear energy transfer (LET) is defined as energy transfer per unit length of the track. It is usually expressed in KeV/μm.

Different radiation undergoes the interaction by different mechanisms. Gamma ray or X-ray produces fast electrons, which are sparely ionizing radiation. Neutrons interact with the proton and other heavier nuclei give rise to recoil protons, which brings subsequent ionization. Alpha particles and charged particles are heavy, highly particles and slow moving produces very dense ionizations along their path, so called high LET radiation. While the X-rays and gamma rays are sparely ionizing comes under a low LET radiation category, neutron falls into the category in between are called intermediate LET radiation. Radio sensitivity toward the biological entities increases with high LET radiation. Effect due to different LET radiation evaluated by defining a term called relative biological effectiveness (RBE).

8.1.22 Relative biological effectiveness

Radiation of different quality even at the equal absorbed dose produces different biological and clinical effects. Radiation quality is expressed by type of particle and energy spectrum. So, difference in biological effectiveness is related to energy deposition along the particle track at subcellular level. RBE is a radiobiological concept. RBE is the ratio of dose of 250 KV X-rays required to bring certain biological effect to the dose of any test radiation required to produce the same biological effect (Halperin et al., 2013).

$$\text{RBE} = \frac{250 KV \text{ of } x-\text{ray required to produce specific effect}}{\text{Dose of any test radiation required to produce same spacific effect}}$$

250 KV of X-ray taken as reference radiation for a long period. In modern radiotherapy Co^{60} –γ rays are predominantly referred to as reference radiation quality. Biological effectiveness is related to LET of the radiation. RBE increases gradually with LET, reaches a maximum in the range of 100-200 KeV/μ then decreases beyond about 200 KeV/μ shown in Fig. 8.6. Because if LET is increased to much higher values, a large number of ionization occurs within the cells. However, energy deposition required for killing cells is wasted beyond specific LET, which consequently results in reduction of RBE. It must be remembered that the RBE-LET relationship does not remain the same for different biological systems.

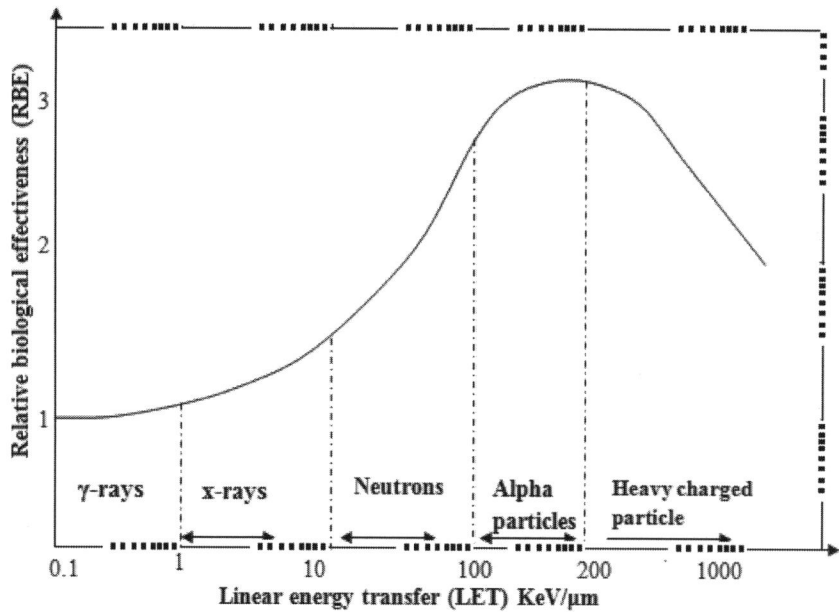

FIG. 8.6 Relationship between RBE and LET for several radiobiological end points.

8.2 Principle of radiotherapy

The goal of radiotherapy is to furnish adequate optimum curative radiation doses to target (tumor) volume and minimum dose to surrounding normal tissue. Principle of radiotherapy based on the term Therapeutic ratio. The therapeutic ratio numerically defined as ratio of tissue tolerance dose (TTD) to tumor lethal dose (TLD) (Halperin et al., 2013).

$$\text{Therapeutic ratio}(TR) = \frac{\text{Tissue tolerance dose}}{\text{Tumor lethal dose}}$$

where, TTD is the dose below which the normal tissue damage is considered as negligible and TLD of radiation, which produces complete and permanent destruction to the tumor in vivo, although it may not be lethal to every individual cell, however radiation dose which destroys 80%–90% tumor cells. Tissue lethal dose depends on several components like type of tumor, type of radiation, rate of dose delivery, fractionation schedule. Principles of radiotherapy usually illustrated Holthusen's Curve by plotting two sigmoid curves representing tumor control and tissue damage shown in Fig. 8.7. Curve for normal tissue damage shifts right of the tumor control curve, which clearly indicates that the percentage of tumor control is higher for a particular dose, for the same instance percentage of incidence of normal tissue damage is much lesser. Larger the difference between both curves, higher the therapeutic ratio. Higher therapeutic ratio generally refers to achievement of radiotherapy therapeutic goal with less complication.

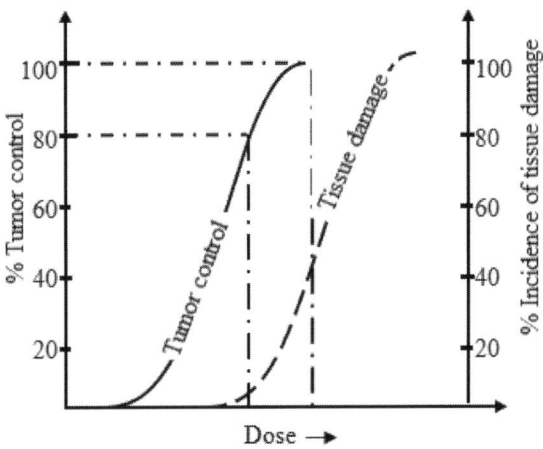

FIG. 8.7 Diagram illustrating tumor control versus normal tissue damage in relation with radiation dose.

8.2.1 Radiotherapy facility

Ionizing radiation like γ-rays (gamma), X-rays, and particulate radiation are used in treatment of cancer over a long period. In earliest days, external beam radiation therapy (EBRT) started with low energy Kilovoltage X-ray radiation for superficial skin cancer treatment. Eventually, engineering and technological revolution brought change in treatment of external beam radiotherapy. Use of Kilovoltage therapy has gradually diminished and it has been replaced by high-energy gamma therapy (Cobalt-60) and megavoltage therapy. Subsequent development in megavoltage machines and gamma therapy rewarded widespread use of high-energy electron and photon beam within a short period.

8.3 Traditional facility in treatment of radiotherapy

Traditionally several x-ray generating equipment are made for an external beam radiotherapy in treatment of cancer. Kilovoltage machines below 500 KV are subcategorized according to their uses and increase in beam penetration.

Grenz-ray therapy: This kind of therapy involved the very low energy X-ray ranging from 10 KV to 20 KV. Due to the very lower penetration, it had discontinued and no longer been used for radiation therapy. This range of X-ray becomes very much popular in diagnostic radiology.

Contact or short distance therapy: Contact therapy machine operated at electric potential order of 40 KV to 50 KV. It was traditionally developed to simulate a condition of radium surface therapy. Typically, 0.5 to 1 mm of aluminum filters are used to eliminate soft components of X-rays. Contact therapy machines are designed to provide Source to skin distance (SSD) below 2cm to produce the rapid dose fall-off inside the tissue. Beam of this type completely absorbed by soft tissue of around 2 cm thickness. 50% of depth dose usually comes around 5 mm. This quality of radiation beam is hardly useful for treatment lesions deeper than 2 mm.

8.3.1 Superficial therapy

Superficial therapy utilizes tube potential ranging from 50 KV to 150 KV with added filtration of 1 mm to 6 mm aluminum. Usual treatment SSD used is in the range of 15 cm to 20 cm. Superficial therapy used to treat superficial lesions at depth around 5 mm which fall around 90% depth dose in tissue. 50% depth dose falls around 2 cm depth in tissue.

8.3.2 Orthovoltage therapy or deep therapy

Deep therapy machines are operated in the range of X-ray generating potential of order 150 KV to 500 KV. However, most of the clinical units are operated at potential of 150 KV to 300 V with added copper filtration of 1 mm to 4 mm. This kind of machine is made with adjustable field size and collimator setting usually set at an SSD of 50 cm. In Orthovoltage machine depth dose distribution depends on multiple factors such as operating potential, field size, treatment SSD. 90% of dose usually lies around 2 cm beneath the skin surface for SSD of 50 cm. 50% percentage depth dose comes at 5 to 7 cm in tissue. Such a machine treating a lesion lies 2 to 3 cm under the skin. One of the major limitations with these machines is to keep skin dose within the tolerance for the relatively deep lesion. In order to treat deep-seated tumors a large amount of dose needs to be delivered, which ultimately leads to increases in skin dose beyond the skin tolerance. Orthovoltage therapy was basic and popular therapy in treatment of cancer before 1950.

8.3.3 Supervoltage therapy machines

X-ray machines are designed with high voltage accelerating potential of 500 KV to 1000 KV called super voltage or high voltage therapy machines. The significant progress was made in order to develop high voltage therapy machines in the 1950s to 1960s. A major issue of that time was designing high voltage transformers. Conventional transformers of that time were unable to produce the voltage beyond 300 KV. However, sooner technological revolution brought the change and high-energy megavoltage machines were developed.

8.3.4 Cobalt-60 teletherapy unit

Cobalt-60 teletherapy machine primarily used for treatment of localized cancer in radiotherapy worldwide for many decades. Teletherapy machines employ an artificially produced cylindrical source (Cobalt- $_{27}Co^{60}$) of 1 to 2 cm diameter. The isotopes $_{27}Co^{60}$ produced by a neutron bombardment on natural stable elements $_{27}Co^{59}$ in a high flux nuclear reactor represented by nuclear reaction $Co^{59}(n, \gamma)\ Co^{60}$. Co^{60} undergoes beta decay followed by two-gamma emission of energy 1.17 and 1.13 MeV with half-life of $T_{1/2} = 5.263$ years. The average energy emission of cobalt-60 sources is 1.25 MeV. The cobalt-60 sources are usually designed in shape of cylinder, disk or pallet contained inside a stainless steel capsule. The source activity used in teletherapy machines is usually between the range of 5000 Ci to 15,000 Ci.

Teletherapy machines are built simple and robust in nature. Teletherapy machines are designed isocentric with SSD of 80 to 100 cm in range. The basic component in design of teletherapy cobalt-60 machine consist of rotating gantry, head of the machine with source drive mechanism, beam collimation system, patient support couch. Source remains in the source drawer other than exposure time. Source capsule comes to the exposure position while treatment for calculated exposure time and sent back to the safe position after completion of treatment with the help of built in drive mechanism. Source drawer and head of the machine surrounded by high Z material lead and alloy of many high-density material such as tungsten. Beam collimation system design to provide rectangular field opening from 1 cm to 35 cm. Teletherapy machines are built to deliver a patient treatment in both SSD and isocentric source to axis distance (SAD) setup. Cobalt-60 theletherapy machines provide greater percentage depth dose compared to other traditional therapy machines as we discussed above. Higher the energy, higher the power penetration corresponds to greater percentage depth dose (PDD). For cobalt-60 teletherapy machines, 50% of depth does fall around a depth of 10 cm. The major disadvantage of theletherapy machine is decrease in its dose rate due to decrease in source activity over a period, which consequently leads to increase in patient treatment time. Even today teletherapy machines are used in many under-developing countries.

8.3.5 Betatron and microtron

Betatron is a particle accelerator that accelerates a lighter particle based on the principle of changing magnetic fields to accelerate an electron in a circular orbit. This consists of accelerating tube shaped like hollow doughnuts, pole of the accelerating tube connected with an alternating current along with changing magnetic field around the tube. Electrons emitted from the filament by the process of thermionic emission are injected into the accelerating tube with the help of an injector. As the magnetic field increases, electrons experience acceleration continuously and spinning with high velocity around the tube. Once desired energy is achieved, a beam of electrons deflected out from the spinning orbit and made it fall on target to produce x-ray. Betatrons were built to produce both high-energy electrons and x-ray was used for treatment of cancer in radiotherapy. Betatrons were first used for radiotherapy in the early 1950s. Although betatron can produce beams over a wide energy range required for radiotherapy, less than 6 MeV to more than 40 MeV. However, soon after its clinical application, limitation of betatrons was recognized. Betatrons were discontinued due to its low dose rate and limited field size.

Microtron is an electron accelerator working contrast to the principle of changing magnetic field used in betatron. In microtron electron moving in uniform magnetic field and electron experience acceleration each time as they passes through radio-frequency (RF) resonant cavity. Once the accelerated electron reached to the required energy level, beam of electron extracted out via exit pipe and made to fall on target to generate x-ray. Circular microtron is bulky because of the large magnetic gap needed to accommodate accelerating RF cavities and the large size of magnetic field needed to encompass electron-spinning orbits. These limitations were overcome soon by using standing waveguides instead of a single accelerating cavity.

8.3.6 Advance facility in treatment of radiotherapy

Advent in technology and computation brought change in radiotherapy. In the last few decades, radiotherapy machines are designed to be extremely sophisticated and advanced. Contemporary machines are equipped with automation. Current radiotherapy machines comprise low energy beams starting from 4 MV to high energy 25 MV with multiple electron beams. In addition to existing electron and photon beam, recent era established clinical benefits of light and heavy ion particle beam therapy.

8.3.7 Linear accelerator (Linac)

The linear accelerator is a machine that produces high-energy electron and x-ray radiation. Linac utilizes high radiofrequency (RF) electromagnetic waves to accelerate a charged particle instead of direct electric potential. Medical linac usually operates in S band (RF level of 2856 MHz or 2998 MHz). Wavelengths corresponding to S band frequencies are 10.5 cm and 10 cm. The charged particles like electrons are accelerated along the tubes called accelerating waveguides. This high-energy electron beam alone can be used to treat a tumor on surface. In order to treat a deep-seated tumor, this high energy-accelerating beam of electrons is made to strike on a thick x-ray target. An incident electron loses its energy undergoing deceleration and produces Bremsstrahlung radiation in forward direction.

Linacs are usually designed isocentrically for source to isocenter distance 100 cm. A basic operational auxiliary consists of major distinct sections as given below (Podgorsak editor, 2005):

- Gantry
- Gantry stand and support system
- Modulator accessories
- Patient support system
- Control console

Modern linac gantry consists of several components that help in producing beam of electron from electron gun and accelerating them through standing or travelling waveguide. External power supply provided to modulator to form a pulsed direct current (DC) power, which is necessary for microwave power production. At the same time, DC power pulsed is supplied to electron gun to produce electrons. Magnetron is a device that serves as a source of microwave as well as amplifying microwave power. However, Klystron is a device used to amplify microwave power that produced by another microwave source RF driver. This microwave power guided the accelerating waveguide to accelerate an electron. Most low single energy machines use a magnetron. Usually a high-energy machine designed with Klystron. Bending magnet type of 90^0, 270^0 (achromatic) or 112.5 (slalom) used to increase a path length of electron which ultimately helps in reducing size of accelerating waveguide. The steering coils and focusing coils, used for steering and focusing the accelerated electron beam. The high-energy electron size of pencil beam of 2 mm to 3 mm in diameter emerging from the exit opening of an accelerator. In addition, Head of the gantry consists of a component for shaping, monitoring and localizing beam of the electron or photon for clinical application. The major basic components consist in a typical head of contemporary linac are:

- Primary collimator
- Retractable X-ray target for photon beam
- Flattening filter for photon and Scattering foil for electron beam
- Dual transmission type ionization chamber
- Field defining and optical distance finding accessories
- Secondary collimator with asymmetric jaw and multileaf collimator

Primary collimator is immobile and situated just beneath the x-ray target. Primary collimator is primarily made to direct the X-ray beam toward the patient and to reduce radiation leakage. Flattening filter is placed on the carousel just below the target assembly to achieve beam uniformity across the field. Similarly, scattering foil is also placed on a carousel for electron beams to produce clinically broadened and flat fields across the treatment field. During the electron beam, scattering foil comes into the central beam line and the flattening filter moves to the side. Output of the beam is continuously monitored by a dual ion chamber placed below the flattening filter, which continually monitors dose, dose rate, beam symmetry etc. Beam passing through the monitor chamber further collimated by a pair of asymmetric jaws designed to provide field size ranging from 0 x 0 cm^2 to 40 x 40 cm^2 defined at isocenter. Jaws are made up of lead or tungsten placed in a secondary collimator. The modern medical linear accelerator comes with a Multileaf collimator (MLC) designed to deliver beamlets based IMRT with conformal dose distribution. These modern Linacs are proficient in delivering computerized treatment plans with techniques such as static and rotational therapy.

8.3.8 Tomotherapy

Mackie et al. first proposed the concept of helical tomotherapy in the early 1990s. Soon after this, many research and publication demonstrated potential benefits of delivering radiotherapy using this innovative approach. Recent commercially available Tomotherapy machines are fully integrated with treatment planning, patient positioning imaging capability and dose delivery. Today's tomotherapy machines are capable of delivering highly conformal dose distribution enabled by a unique binary MLC. These machines are designed with IMRT under an integrated 3D image guidance facility.

Tomotherapy machine is analogy of helical computed tomography (CT) machine. Helical tomotherapy consists of 6 MV linac mounted on CT gantry ring which allows it to rotate around the patient at SAD distance of 85 cm. Similar to the helical CT, radiation source revolves around the patient throughoutits treatment,at a same time patient couch transverse longitudinally across the gantry bore along the y-direction. Tomotherapy machine uses a fan beam to deliver radiation similar to helical CT. Pitch is dimension of the treatment beam.The pitch is defined as quotient of distance travelled by patient couch in each gantry rotation and total field width along y-direction. Modern tomotherapy is designed with fixed field width defined by a pair of jaws in y-direction. Selectable field width for any particular patient can be 1 cm, 2.5 cm or 5 cm. Field of X-dimension is defined by a pair of 64-leaf binary MLC. Each individual leaf projects the width of 6.25 mm at isocenter. Maximum lateral field dimension defined by MLC is 40 cm along x-direction. Intensity modulation accomplished by continuous movement of leaves within the treatment field.

Tomotherapy machine equipped with MVCT scan of patient anatomy in order to verify patient treatment position before, during or after the treatment. One of the major clinical advantages of tomotherapy machines is capable of treating craniospinal irradiation with single isocenter, whereas linac needs multi isocentric technique to cover the entire craniospinal field. However, there are some disadvantages associated with it like these machines do not have electron beam facility, noncoplanar field is not possible.

8.3.9 CyberKnife

Accuray first developed commercial CyberKnife in mid 1990s for treatment of intracranial stereotactic radiosurgery. However, modern CyberKinfe equipped with fully robotic technology allows us to treat lesions almost anywhere in the body using the approach of Stereotactic body radiotherapy (SBRT) or Stereotactic radiotherapy (SRT). CyberKinfe delivers radiation with a miniature small X-band ($\approx$10000 MHz) linac mounted on an industrial robotic arm. CyberKinfe treatments are nonisocentric, where beams are directed from any desired angle. CyberKnife enabled online target imaging followed by automatic adjustment of radiation beam direction to recompense for target motion. CyberKinfe system predetermines treatment position and monitors target motion with respect to reference planning CT by taking a repeated pair of orthogonal imaging throughout the patient treatment. CyberKinfe gives an alternative to invasive stereotactic radiosurgery procedure.

8.3.10 Proton and light ion therapy

The basic fundamental component of proton or light ion therapy facility includes a type of particle accelerator, beam transport system into one or multiple rooms, shielding mechanism, beam-shaping device, patient positioning, and monitoring system. Most proposed proton, light ion therapy facilities uses either cyclotron or synchrotron to accelerate the particle to desire beam energy (150–250 MeV) required for radiotherapy (Khan and Gibbons, 2014). Proton therapy and light ion therapy have their own principle benefits due to the depth dose characteristics of these particles that show advantage over a photon beam. Before discussing interaction of protons with matter let us discuss the fundamental principle behind the type of accelerator being used for proton and light ion therapy. Proton therapy machines are commercially available from vendors like Varian medical system and IBA Ltd.

8.3.11 Cyclotron

Cyclotron accelerates a charged particle such as protons, deuterons and light ion particles using high frequency alternating electric potential applied between two hollow D-shaped conducting dees in evacuated cylinders. The charged particle move in a circular path under an affect of uniform steady magnetic field applied perpendicular to the top of the Ds. Both the Ds are placed face to face with a narrow gap between them and charged particles are accelerated only when they pass through this gap due to an alternating high electric potential. The particles are injected into the center of the gap between the dees.

Applied magnetic field causes particles to bend in a circular way due to the Lorentz force acting perpendicular to direction of propagation. Beam of accelerating particles attending a higher energy at each circular orbit. Once desired energy is achieved, a beam of accelerating particles deflected out of the chamber and guided to the required target.

8.3.12 Synchrotron and synchrocyclotron

Principle differences between synchrotron and synchrocyclotron are based on application of electric and magnetic fields. The both magnetic field and frequency of applied electric field varies in a synchrotron. However, synchrocyclotron varies either magnetic field or frequency of applied electric field only. In synchrocyclotron, preaccelerated proton beam from a linear accelerator is injected into the circulated narrow vacuum tube ring like structure large in diameter. Protons are accelerated rapidly as they pass through the accelerating RF cavity. Accelerating RF cavity is powered by alternating electric fields with frequency that matches with the frequency of circulating protons. Protons are held within the tube ring with the help of a bending magnet mechanism, which allows maintaining trajectories of particles in circulating orbit. Both Magnetic field and RF frequency are increased synchronously to increase beam energy. Once the desired beam energy is achieved the beam is extracted out and guided toward the target.

Synchrotrons have advantages over cyclotrons, synchrotrons can simply produce the beam of variable energy, whereas cyclotron produces continuous beams of fixed energy with higher intensity. Synchrotrons can be operating to produce Spread out Bragg Peak (SOBP) required at any particular depth without use of energy degraders. However, cyclotron operates at its maximum energy and desired energy is achieved with the help of degraders. SOBP is created with the help of degraders to treat deep-seated tumors. Even a single accelerator can be designed to provide a beam to the several gantries in different rooms. Beam transport system selectively guides the beam to a particular room with the help of a bending magnet and selectively energized from the control console. The intensity of beam looses during the transport usually less than 5 %. Particle beams emerging from an accelerator are small in dimension. However, the beam further spread out just before entering the patient treatment room to cover the field dimension of treatment head. Desired beam spreading is achieved either with passive scattering using high Z material or beam scanning with help of magnet over treatment field.

As protons pass through the medium, they interact with the orbital electron and nucleus of an atom through electromagnetic interaction. It undergoes elastic and inelastic collision with atomic electrons. In elastic collision protons scattered without losing energy. However, in case inelastic collision proton reduces its part of kinetic energy in excitation and ionization of atoms of medium resulting in absorption of dose in a medium. Proton also interacts with atomic nuclei producing bremsstrahlung, but negligible. This rate of energy loss in the medium gives rise to depth dose distribution. The monoenergetic proton follows a slow increase in depth dose initially followed by sudden sharp increase in dose as particle velocity loses to its maximum at end of range. This sharp pronounced peak at extreme just before particle range is called Bragg peak. Clinically Bragg peak is too narrow to accommodate the extent of tumor volume. In order to achieve wider depth coverage, Bragg peak is spread out by superimposing beams of different energy called SOBP. Due to the heavy particle nature of protons scattered at a smaller angle than electrons and photons. Hence, protons beam produces a sharp lateral dose distribution than electron and photon beams.

It has been discussed above that greater is LET, greater is RBE. Proton is considered as high LET radiation. Because heavy charged particles proton have greater relative biological effectiveness (RBE) than 1.0, which is simply referred for sparsely ionizing radiation like γ-rays and X-rays. Extensive radiological studies have been carried out to determine values of RBE for protons. RBE found a function of depth inside the tissue. However, the most clinically adopted value of RBE is 1.1 for proton relative to cobalt-60 γ-rays or megavoltage X-radiation.

8.3.13 Add-on facility in treatment of radiotherapy

In addition to treatment delivery systems, radiotherapy involves additional equipment that plays an important role in radiotherapeutic procedures which are not actually related to dose delivery but equally important in radiotherapy. This add-on facility plays a crucial role in target localization, identifying treatment gantry angle, determining extent of disease, defining field bordering, and helps in computerized treatment planning called simulator. At present, radiotherapy simulators consist of conventional simulators and computed tomography (CT) simulators.

8.3.14 Conventional simulator

Conventional simulator is a mimic of our treatment therapy machine that comprises rotating gantry with diagnostic X-ray tube mounted on it, which simulates all geometrical, mechanical properties as that of therapy machine, allows in determining desirable treatment beam angles and clinical treatment properties for patients undergoing radiotherapy as discussed above. Older models of conventional simulators were equipped with conventional film-based imager with image intensifier to define the field bordering with the extent of tumor volume. However, modern simulators have replaced film-based technique with amorphous silicon semiconductor-based detector technology. This allows us to acquire a high-resolution, high-contrast image, and facilitates the idea of filmless technique in radiotherapy.

8.3.15 CT simulator

CT simulators are nothing but CT scanners equipped with special features that allows them to use for computerized radiotherapy treatment planning. Special feature consist in radiotherapy CT simulator are as follows (Halperin et al., 2013):

- A flat tabletop surface couch to lie down patient during simulation that is identical to couch exists in radiotherapy machine.
- Availability of laser marking system for accurate patient orientation and lying fiducial in single plane as a reference that helps in determining treatment isocenter.
- Virtual simulation software package that allows in localizing, contouring, beam aligning, setting up treatment isocenter and simulating treatment using digitally reconstructed radiograph (DRRs).
- Also allows in computerized treatment planning.

Entire process of simulating patient treatment plans carried out using CT data acquisition of patients in virtual simulation software without physical presence of patients is called virtual simulation. One of the major advantages of CT simulator over a conventional simulation is that patient simulation is performed virtually and patients need not to stay after CT acquisition in order to define treatment plan parameters. However, in the case of conventional simulators, patients need to present during the process of simulation, define field geometry and determine treatment plan parameters. The conventional simulator takes much more time than CT simulator for radiotherapy planning.

8.3.16 Commissioning of radiotherapy facility and quality assurance

Actual commissioning of the radiotherapy facility began after the installation followed by customer acceptance tests (CAT) carried by an installation engineer along with medical physicist to ensure that machine meets product specification and agreements. These tests are performed according to the acceptance testing procedure on which both manufacturer's representative and radiotherapy facility physicist agreed. Radiotherapy facility CAT consists of electrical, mechanical, radiation, dosimetric test procedures which insure the entire functional performance of equipment according to IEC standards. Medical physicist is designated an authorized radiation safety officer by a regulatory authority for a particular radiotherapy facility. He ensures the radiation level well within tolerance around facility conduction thorough radiation survey. Medical physicist performs commissioning of radiotherapy facility, which involves independent commissioning of radiotherapy equipment such as linear accelerator, brachytherapy unit, and commissioning of treatment planning system (TPS) related to it. This process undergoes many multiple checks and calibration, beam data measurements, standardizing baseline values, feeding beam data to TPS and its validation, setting up periodic quality assurance programs for future verification, etc.

"Quality assurance in radiotherapy" is all procedures that ensure integrity of machine performance to deliver safe use of radiation electrically, mechanically, and dosimetrically to the target of interest. This involves daily, weekly, monthly, quarterly, and yearly quality assurance test procedures of the machine. In addition to this there are patient's specific quality checks called patient specific QA performed just before patient treatment for each individual patient. This ensures the quality of treatment plan and its intended delivery.

8.3.17 Technique of radiotherapy

Depending on mode of delivery, radiotherapy is classified into two basic categories as external beam radiotherapy (EBRT) and internal radiation therapy known as brachytherapy. In addition to this, radiotherapy is also classified according to time before or after any kinds of treatment given to patient as follows (Halperin et al., 2013):

- Radiotherapy given prior to any sorts of treatment modality called neoadjuvant radiotherapy.
- Radiotherapy prescribed subsequent to any kinds of treatment modality called adjuvant radiotherapy.
- Radiotherapy given before surgery called preoperative radiotherapy.
- Radiotherapy prescribed after surgery called postoperative radiotherapy.
- Radiotherapy prescribed concurrently with chemotherapy called radio chemotherapy.

Patients come for radiation therapy with different stages and grades of cancer. A status of disease decides the aim of radiation therapy. Goal of radiation therapy can be curative, palliative, prophylactic, and total body radiation therapy.

Application of curative therapy made to cure early stage cancer like nasopharyngeal cancer, Hodgkin's lymphoma, early glottic cancers, and some skin cancers. The curative radiotherapy is also called definitive radiation therapy. However, palliative treatment is given to obtain temporary relief and mitigate of cancer symptoms by giving palliative doses of radiation. Prophylactic radiation given for prevention of possible metastases and in case of recurrence occurred through application of preradiation. Total body irradiation (TBI) is given to suppress the immune system for ablation of bone marrow by radiation. This assists in space availability for transplanting cells and eradicates leukemic cells during bone marrow transplantation.

8.3.18 External beam radiation therapy

Radiotherapy is delivered by an external source kept outside the body at a certain distance called teletherapy or EBRT. EBRT begins with very simple convention fields like anterior, posterior, lateral, orthogonal, oblique, and two parallel opposed fields. However, over period lots of technological modification brought change in delivery of EBRT to meet radiotherapy goals. The application of an automated multileaf collimator (MLC) brought one of the major innovative techniques to deliver radiation therapy in late twenties. Modern EBRT can be delivered by static gantry angle or by the rotational motion of gantry around the patients. EBRT techniques from conventional to modern radiotherapy are discussed in the following sections.

8.3.19 Conventional treatment techniques in EBRT

Conventional EBRT mostly accompanied with conventional simulator, where the manual planning carried out with few conventional fields. Conventional fields like interior, posterior, lateral, oblique or combination of parallel opposed fields are chosen to deliver radiation. Conventional radiotherapy also called 2-dimensional radiation therapy (2DRT), since it is carried out with a two-dimensional radiographic image. Treatments are given either in SSD or SAD setup. Energy is determined as per tumor depth. The electron or lower energy photon beams are preferred for superficial tumors. Treatment field dimensions with adequate setup margin are given to accommodate patient setup uncertainties. Field dimension defined with jaws, conventional lead blocks or modern machines equipped with MLC also used to shield critical organs or normal tissue. Treatment time or monitor unit required to deliver prescribed dose is calculated manually. Modern TPS allow us to perform radiographic image-based planning with conventional field approach. Also, it has been utilized to calculate dose for conventional planning. Figs. 8.8 and 8.9 are showing geometry of a field defined for conventional radiotherapy planned for a whole brain and pelvis cases.

8.3.20 Three-dimensional conformal radiation therapy

Three-dimensional conformal radiation therapy (3DCRT) is CT-based treatment planning technique. That uses three-dimensional CT volumetric data for treatment planning that distinguishes 3DCRT from conventional treatment planning. CT data on which target (tumor) volume and closely spaced organ at risk (OAR) is contoured. "3DCRT is a conformal radiation technique, which creates conformal dose distribution around the three-dimensional target volume andminimum

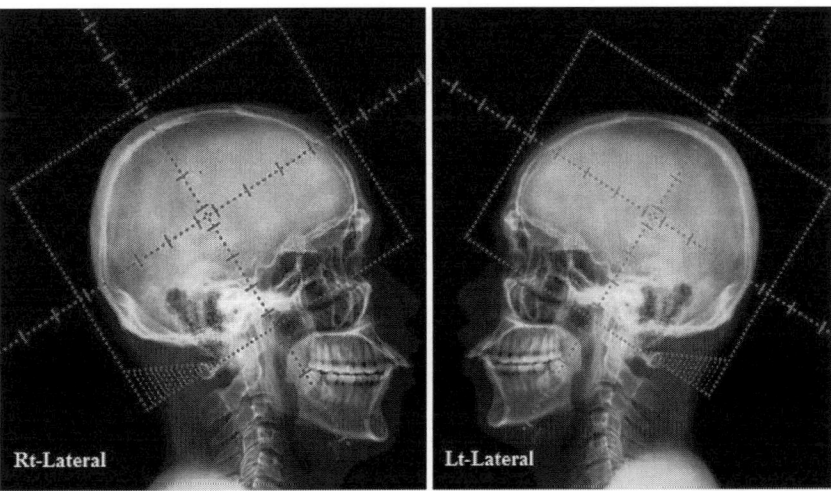

FIG. 8.8 Radiographic image showing conventionally planned whole brain radiotherapy with bilateral field.

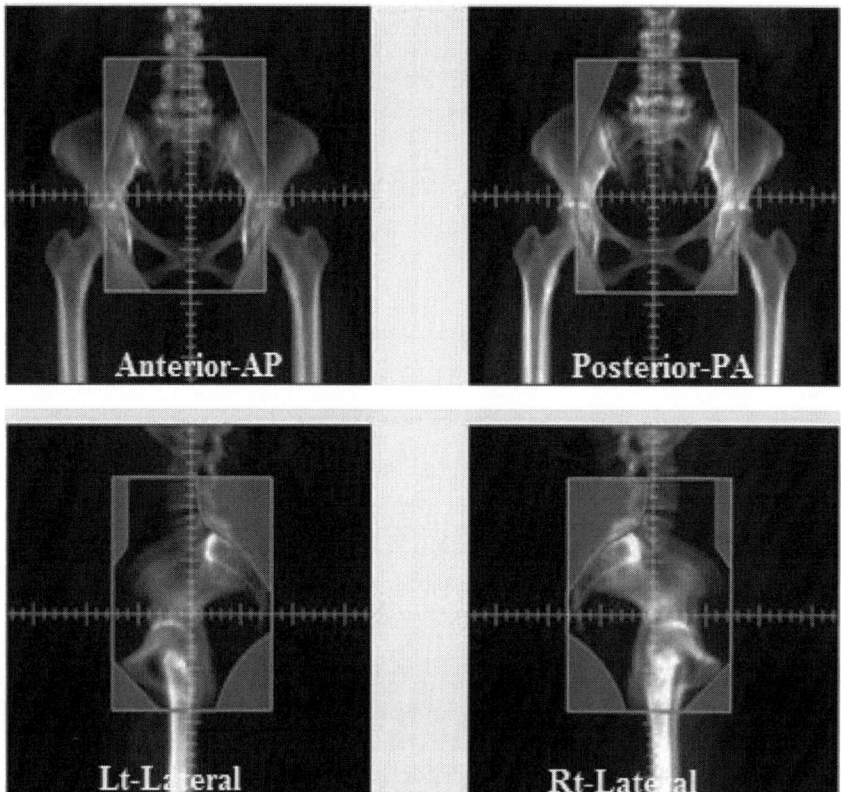

FIG. 8.9 Radiographic image showing conventionally planned pelvis radiotherapy with box field technique.

dose to the surrounding normal tissue by uniform radiation field through conforming 3D anatomical target with the help of MLC." Traditionally lead or customized cerrobend blocks are used for conforming the target and shielding surrounding normal tissue. However, modern machines equipped with computer-controlled MLC automatically conform to the target. 3DCRT carried out through modern TPS allows in dose calculation and plan evaluation. Beam eye view (BEV) and DRR helps the planner to view target and OAR so that intended planning can be performed. DRR from the planning system can also be used as a reference image for treatment verification. Treated volume in case of advanced 3DCRT is lesser compared to conventional therapy, this produces higher conformity plans and allows in more organ sparing which ultimately helps in reducing radiation toxicity with 3DCRT.

Fig. 8.10A illustrates technical comparison between 2D conventional therapy and 3D conformal therapy. Whereas, Fig. 8.10B depicts dose conformity of one of the pelvis cases planned with 3DCRT technique in advance Varian Eclipse treatment-planning system. Even though 3DCRT produces much more conformal dose distribution than conventional therapy, it fails to produce satisfactory dose distribution for complex target shapes like concave or convex. It limits the dose in proximity of OAR.

8.3.21 Intensity modulated radiation therapy

Yu proposed IMRT in 1995, which is an advanced form of 3DCRT (Bortfeld et al., 2006). "IMRT is a radiation therapy technique where multiple beam incidents from different angles to produce the composite uniform dose distribution through beam of nonuniform fluence modulated with the help of MLC." 3DCRT is a kind of forward planning process. However, IMRT is based on an inverse planning optimization approach, in which computerized optimization algorithm optimizes dose based on objective cost function given by the planner. All the beams are optimized to deliver high dose to the target volume and minimal dose to adjacent OAR and surrounding normal tissue.

An intensity modulation in IMRT technique is achieved by either step and shoots or dynamic movements of MLC. In step and shoot IMRT, radiation dose is delivered by small multiple segmented field called subfields of a static gantry field. Where each subfield with uniform beam intensity implying that the radiation is only turned on when the MLC leaves are halted in each of the specific subfield positions. There is no movement of MLCs while the radiation on. However, in case

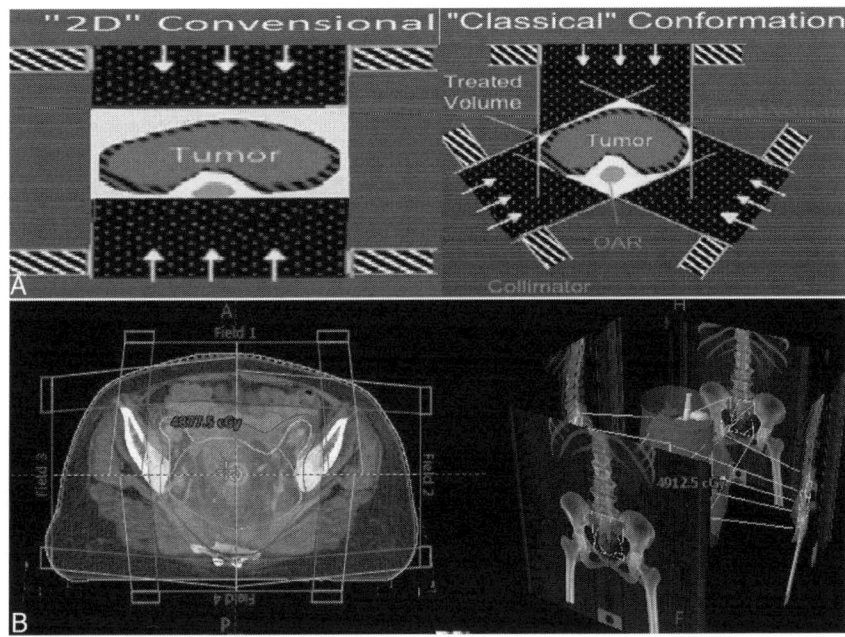

FIG. 8.10 (A) Illustrating technical comparison of 2DRT and 3DCRT. (B) Picture showing conformal dose distribution created with 3DCRT technique planned in Eclipse treatment planning system for pelvis site.

of dynamic IMRT radiation dose is delivered in a dynamic fashion with the leaves of the MLC set in the motion during the patient irradiation. In dynamic mode, gantry is at fixed position and field opening formed by each pair of opposing MLC leaves that are sweeping across the tumor target volume under the computer control. Simultaneously the radiation beam turned on to produce the prescribed calculated fluence map.

IMRT is the most advanced form of radiation delivery technique producing highly conformal dose distribution increasing Tumor control probability (TCP) and lowering treatment morbidity by reducing normal tissue complication probability (NTCP). IMRT has potential advantages due to its ability to manipulate optimal fluence around complex targets and OAR close to it. IMRT has become the most popular and usable technique for delivering external beam radiotherapy. The number of monitor unit (MU) in the IMRT plan is two to three times higher than conventional therapy, which consequently leads to the longer treatment time, higher integral dose, and greater risk due to setup uncertainties are disadvantageous of IMRT. Fig. 8.11A illustrates the technical comparison of 3DCRT and IMRT techniques. Similarly, Fig. 8.11B is showing

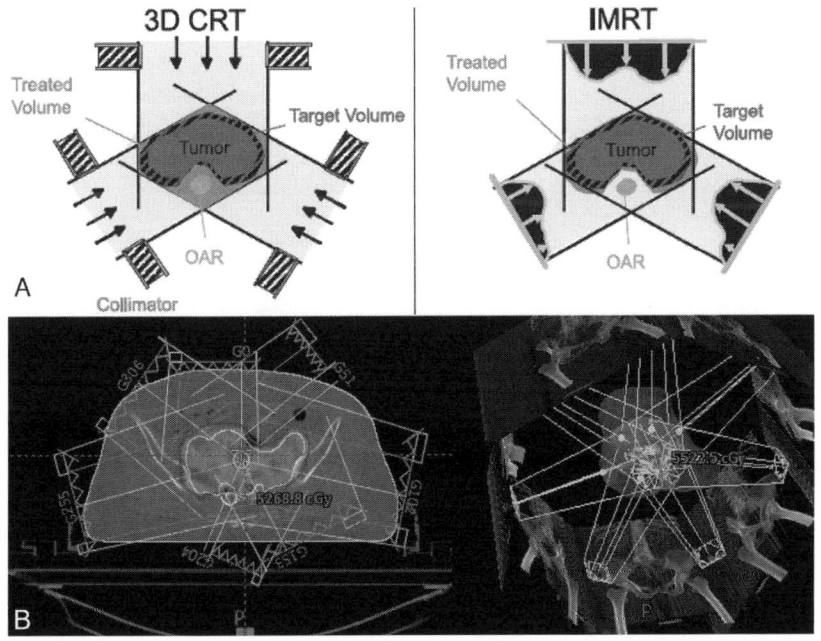

FIG. 8.11 (A) Illustrating technical comparison of 3DCRT and IMRT techniques. (B) Picture showing conformal dose distribution around target and sparing of OAR such as rectum and bowel for pelvis case planned with IMRT technique in Eclipse treatment planning.

conformal dose distribution around target with sparing of OAR such as rectum and bowel in pelvis case planned with IMRT technique in eclipse treatment planning system.

8.3.22 Rotational therapy or volumetric modulated arc therapy (VMAT)

Volumetric modulated arc therapy or rotational therapy is an advanced form of IMRT. There are two main ways of rotational therapy. First one VMAT delivered with a linear accelerator and another with a Tomotherapy machine. "VMAT is a technique where radiation is delivered continuously through the rotation of treatment machine around patient." This delivered by simultaneous dynamic motion of MLC, constant or variable dose rate and rotation of gantry (Bortfeld et al., 2006). VMAT produces a highly conformal dose distribution encompassing the target volume with OAR sparing. Dose distribution produced by VMAT is as good as or better than IMRT depending on location of tumor and sites. One of the major advantages with VMAT compared to IMRT is its lower MU, which consequently helps in decreasing treatment time. VMAT usually takes 1.5 to 3 mit to deliver 200 cGy. However, IMRT does take 4 to 5 minutes to deliver the same treatment. In addition, VMAT has great ability to control high dose spillage near to the target compared to IMRT. However low dose spillage is observed to be more and VMAT does not have much control over it. Several literatures have mentioned potential clinical benefits of VMAT. Fig. 8.12 shows conformal dose distribution falls around the target with organ sparing such as bladder and rectum planned under VMAT techniques in Eclipse treatment planning system.

8.3.23 Stereotactic radiosurgery and stereotactic radiotherapy

Stereotactic radiosurgery (SRS) is single-fraction external beam radiotherapy technique for treatment of intracranial lesion, where a very high dose of radiation given in a single fraction. However, stereotactic radiotherapy (SRT) is fractionated radiotherapy given over a small number of fractions such as 3 to 5. In modern radiotherapy, these techniques are used to treat a variety of malignant and benign lesions as well as functional disorders in the brain such as arteriovenous malformation, acoustic neuroma, solitary primary brain tumor, single metastasis, pituitary adenoma. These treatments can be delivered by several modalities such as linear accelerator with help of micro MLC or cone, CyberKnife, GammaKnife, and proton therapy with multiple beams from different directions. High degree of dose gradient is achieved during planning to spare normal tissue adjacent to the target.

8.3.24 Image-guided radiotherapy

Patient positional uncertainties for advanced techniques like IMRT, VMAT and higher dose per fraction therapies like SRS, SRT are most sensitive, even small errors in patient setup can lead to significant dosimetric variation. Image-guided radiation therapy (IGRT) is form of radiotherapy guided by imaging techniques such as MV-KV, KV-KV or volumetric Cone beam computed tomography (CBCT) prior to the patient treatment. The imaging procedure potentially ensuring relative position of target and reference anatomical landmark with respect to simulated and planned CT. Radiotherapy machines, which facilitate IGRT

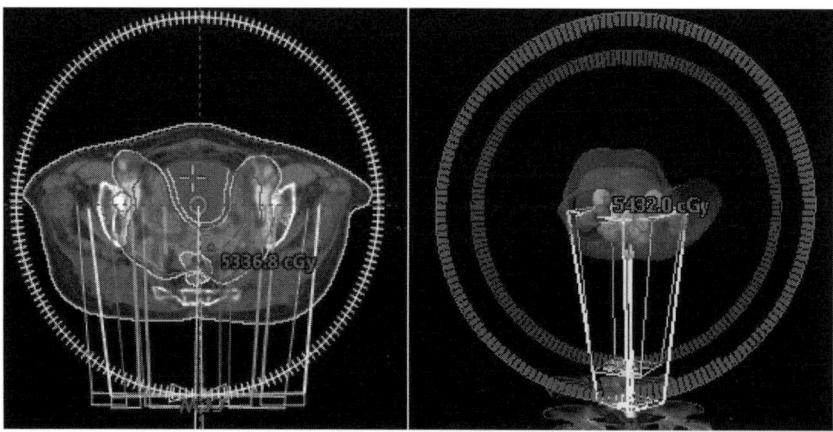

FIG. 8.12 Showing conformal dose distribution around target and sparing of OAR planned with VMAT technique in Eclipse treatment planning system for pelvis site.

techniques, are equipped with imaging acquisition facilities. This consists of imaging source, detector panel, and processing software. Modern radiotherapy machines are integrated with Electronic portal imaging devices (EPID) based on an array of amorphous-silicon (a-Si) detectors that allows imaging with batter contrast and special resolution (Bortfeld et al., 2006).

8.3.25 Internal beam radiation therapy or brachytherapy

Technique in which radiation source planted inside the tumor or vicinity of the tumor for temporarily or permanently is called internal radiation therapy also known as brachytherapy. The term "brachytheray" used to referred for short distance therapy (Joslin et al., 2001). This is used to treat locally accessible tumor, usually used in treatment of cervical, uterine, prostate, rectal, skin and breast cancer. Brachytherapy started soon after the discovery of radium discovered by Marie and Pierre curie in 1898. Clinical application of radium was widely recognized and used extensively all over the world. However, radium has a several disadvantages associated such as radium is α-emitters, radium releases Radon gas that is hazardous in nature and absorbed in tissue, γ-energy emitted from radium is of higher energy beyond brachytherapy usable range and radium has lower specific activity. Very soon radium source was discontinued and substitutes for radium source were produced artificially. Most commonly used brachytherapy source is gamma emitters which have the property of rapid-dose fall-off. Encapsulated sources of Cobalt-60, Gold-198, Strontium-90, Yttrium-90, Iodine-125, Pladdnium-103, Cesium-137, and Irridium-192 are being used for internal therapy. Source of Ce-137 mostly used in LDR brachytherapy. Since most of the LDR facilities are replaced by a HDR facility, Ce-137 becomes less usable. Strontium-90 and Yttrium-90 is beta emitter whereas Iodine-125 low energy gamma emitters are mostly used in treatment of eye plaque therapy. Co-60 and Ir-192 sources become very much popular for treatment of HDR brachytherapy (Halperin et al., 2013).

Treatment of brachytherapy has evolved over the period from manual brachytherapy to remote after-loading brachytherapy. Modern era brought safe use of radionuclide with the remote after-loading brachy and advanced TPS. Remote after loader brachy is a computer operated fully automated machine that drives the encapsulated miniaturized source to a planned treatment position with the help of a guide tube as the Medical physicist defines it in TPS. In recent times, most of the brachytherapy techniques are carried out through remote after loading machines. Brachytherapy procedure involved small catheters like a tube or small fine needle or large specific device called applicators depending on techniques of brachytherapy. Catheters, needles or specified applicators are implanted inside or closed to the tumor during brachytherapy procedure executed by clinical Radiation oncologist. Modern brachytherapy planning based on volumetric CT data involved contouring, catheter reconstruction, defining source position, dose optimization, dose calculation and plan evaluation established higher standard in treatment of modern brachytherapy.

Brachytherapy treatments are mostly classified based on treatment duration and dose rate (Joslin et al., 2001).

Brachytherapy is classified based on treatment duration as follows,

- Temporary brachytherapy: Dose is delivered for short duration and source is removed once desired dose has been delivered.
- Permanent brachytherapy: Source is planted into the tumor site and dose is delivered for entire lifespan of source untilit decayscomplete.

Similarly, brachytherapy is classified based on dose rate as follows,

- Low dose rate (LDR): 0.4 to 2 Gy/hr
- Medium dose rate (MDR): 2 to 12 Gy/hr
- High dose rate (HDR): 12 Gy/hr

HDR has become standard of brachytherapy due to optimized dose distribution, outpatient treatment and elimination of staff radiation exposure. Recent HDR afterloading brachytherapy mostly designed with 10-15 Curie activity for Ir-192 source and 2.2 Curie for C0-60 source.

Usually, techniques of brachytherapy are categorized based on placement of source as follows,

- Intracavitary brachytherapy: Radiation sources placedwithin the body cavities in closed vicinity to tumor volume.
- Interstitial brachytherapy: Radiation sources implanted surgically into the tumor volume.
- Intraluminal brachytherapy: Radiation sources are placed into lumen.
- Intraoperative brachytherapy: Radiation doses are given to tumor bed while surgery performed.
- Surfaced mould therapy: Radiation sources are placed over the surface of the tumor. Surface mould therapy usually used to treat skin cancer.

Different HDR brachytherapy machines from various vendors shown in Fig. 8.13.

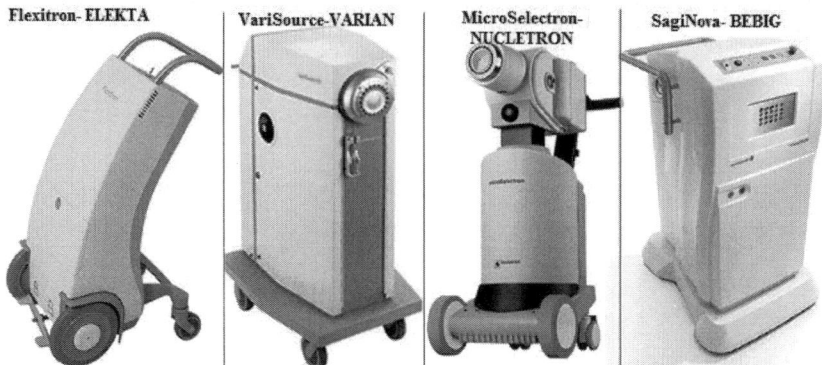

FIG. 8.13 Showing modern HDR brachytherapy machines from various vendors.

8.3.26 Process and treatment of radiotherapy

Radiotherapy carried out by a group of members that involves radiation oncologist, medical physicist, dosimetrist, radiation therapist, and nurses. Every individual has his or her specific role during radiation therapy. The success of radiotherapy depends on availability of technical equipment and works performed by each individual at their level. When the patient is referred to the radiation oncologist for radiotherapy, he investigates the patient through patient history, physical examination, pathology reports, radio diagnostic CT reports, MRI reports, and nuclear imaging reports and creates short case detail. Case report gives important details about the type of cancer, location, size, stage, and grade of disease based on which radiation oncologist would plan radiation therapy and decide dose prescription. Patient should be well intimated about his type of cancer and availability of treatment, course of treatment duration, and side effects of radiation before starting radiation therapy. Patients undergoing treatment for advanced radiation therapy go through several procedures (Halperin et al., 2013) illustrated in Fig. 8.14.

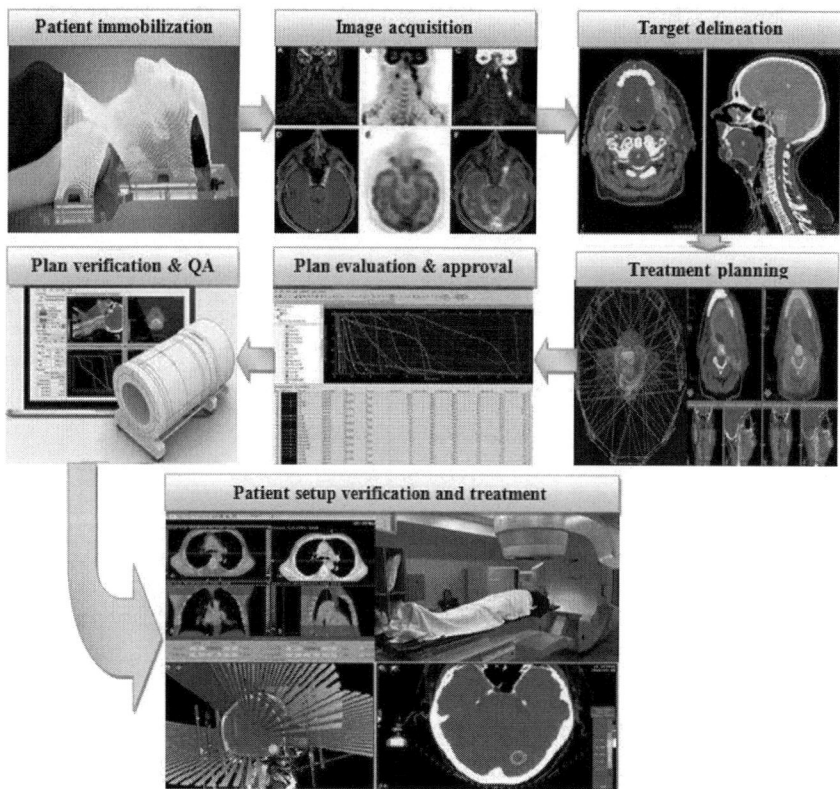

FIG. 8.14 Flow diagram illustrating radiation therapy patient undergoing various procedures.

8.4 Patient preparation and simulation

Patient preparation followed by simulation is the first process of radiotherapy where the patient is immobilized using several immobilization devices. Aim of immobilization is to restrict the movement of patients during radiation therapy to avoid change in the treatment area and minimize corresponding errors due to shift in patient position. Patients should be immobilized in easily reproducible and most comfortable positions. Most usual and frequent way to immobilize the patient is with a thermoplastic mask. There are several devices used to make a comfortable position for a patient during radiotherapy such as headrest, knee rest, velcro belt, elastic belt, etc. Patient is simulated in an ideal comfortable position for the patient, reference tattoos are marked with respect laser followed by image acquisition for planning purposes. Patient CT data are acquired keeping in mind need of treatment planning and treatment protocol going to be implemented during radiation therapy.

8.5 Target delineation and treatment planning

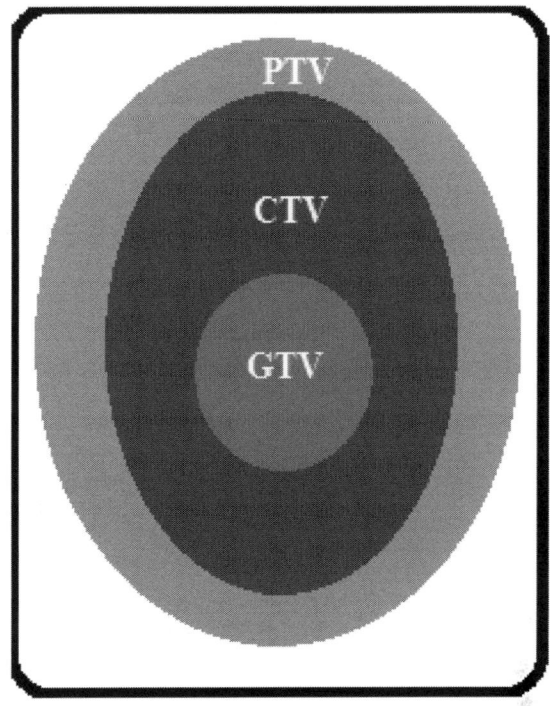

FIG. 8.15 Illustrating the demarcation of GTV, CTV, PTV as per guideline given in ICRU report-50.

Acquired CT data are being transferred to contouring workstation where radiation oncologist does contour as per international clinical practice guideline mentioned in NCCN and other related clinical RTOG guidelines. Contouring includes several targets clinically named as gross tumor volume (GTV), clinical target volume (CTV), planning target volume (PTV), and relevant OARs closed to the target. Fig. 8.15 demonstrates the demarcation of various volumes likePTV, CTV, GTV as per guideline given in the International commission on radiation units and measurements (ICRU) reports-50 (Podgorsak editor, 2005).

- GTV: It is the gross palpable visible or verified extent and local malignant growth.
- CTV: The volume of tissue contains a gross tumor volume and subclinical microscopic malignant disease.
- PTV: Tissue volume contains CTV and additional volume of margin surrounding CTV to accommodate for daily patient setup uncertainties.

There are various computerized TPS available commercially like Eclipse, Monaco, Oncentra, Adac Pinnacle, etc. The Advanced treatment planning system consists of different plan optimization and dose calculation algorithms. Most of the linear accelerators can be integrated with their own TPS or even TPS from another commercial vendor. The dosimetrist or medical physicist makes relevant treatment plans following (ICRU, AAPM TG, and RTOG) guidelines. Treatment plan is created in TPS based on type of treatment techniques; volume contoured and dose prescription. A planner tries to achieve prescribed dose to target volume and minimum dose outside the target to OARs and normal tissues. Manipulation of several optimization parameters could result in a desired plan. Once the desired plan is made, it is evaluated by radiation oncologist and approved for further treatments. Medical physicists performed the quality check for those treatment plans called "patient specific quality assurance." When a quality parameters of the corresponding plan meet the desired goal, Medical physicist approves the patient treatment and the plan is transferred the treatment machine.

8.5.1 Treatment verification and treatment delivery

Treatment verification is a process of verifying treatment position with respect to simulated obtained during simulation. As we have discussed above modern radiation therapy machines are equipped with imaging facilities in order to verify treatment position. Radiation therapist sets the patient on treatment machines similar to a simulated position. One should reproduce patient setup geometry exactly as it is simulated during simulation to avoid errors in treatment delivery. Orthogonal images or volumetric CBCT are acquired before the treatment to verify the treatment position. Difference between planned reference images and acquired image of the day is determined and applied against any patient shift found. Medical physicists and radiation oncologists both verify the patient treatment position before the first day treatment.

8.5.2 Dosimetry in radiation therapy

Medical radiation dosimetry refers to the measurement of absorbed dose at a certaingiven point in a medium of interest produced by an ionizing radiation. Radiation dosimetry deals with quantitative determination of dose and energy. There are different types of dosimetry systems available for measurement and quantification of radiation. At present, ionization chambers, film, luminescence, semiconductor and MOSFET dosimetry are used at various levels in the radiation field. Radiation dosimeters used measure exposure, kerma, absorbed dose, effective dose and related radiation quantities. Ionization chamber-based dosimetry is most often routine and standardized dosimetry in radiation therapy. Radiation dosimetry in radiation therapy particularly deals with calibration of therapeutic machines and measurement of absorbed dose inside the human tissue. Medical radiation dosimetry is usually performed in tissue equivalent material or water. Water is an ideal and recommended medium in radiation therapy dosimetry for megavoltage photon and electron beam due its readily available and to soft tissue equivalent. However, dosimetric measurements are often carried out in water equivalent solid material such as solid water (RW1, RMI-457), polystyrene, lucite, plastic water and virtual water are mimicked in terms of number of electrons per gram, mass density, effective atomic number of water. There are several quantities and their corresponding units are defined in order to quantify the radiation in a medium. Before discussing various types of radiation dosimetry some of radiation quantities used often are described below (Landberg et al., 1993).

8.5.3 Activity

The number of nuclear transformations or disintegration per second called as activity. Activity measures in SI unit Becquerel or old unit Curie. Radioactive decay is a process by which unstable atomic nuclei gain stability.

$$1\text{ Bq} = 1\text{ dist. Per sec} = 1\text{ dps.}$$
$$1\text{ Ci} = 3.7 \times 10^{10}\text{ dist. per sec.}$$
$$= 3.7 \times 10^{10}\text{ Bq}$$

8.5.4 Particle fluence

The total number of particle (dN) incidents on the sphere of cross section area (dA) is called particle fluence. Unit of particle fluence is m^{-1}.

$$\Phi = dN / dA \tag{8.22}$$

8.5.5 Energy fluence

The radiant energy (dE) incident on the sphere of cross section area(dA) is called energy fluence. Unit of energy fluence is J/m^2.

$$\Psi = dE / dA \tag{8.23}$$

8.5.6 Exposure

It is a measure of energy fluence of photons (Gamma rays or X-rays) at any point in air. It is defined as the amount of charge (in coulomb) produced by radiation in 1 kg of air. Unit of exposure is Coulomb per kg (C/kg) or old unit is Roentgen.

$$1R = 1\text{ Electrostatic unit of charge}(esu)/1\text{ cc of air at STP.}$$
$$= 2.58 \times 10^{-4}\text{ C}/\text{kg}$$
$$1\text{ air kerma}(Gy) = 114\text{ R}$$

8.5.7 Kerma

It is kinetic energy release per unit mass. The quantity Kerma is defined for indirectly ionizing radiation like X-rays or gamma radiation only. Kerma is defined as the sum of the initial kinetic energies of all charged particles liberated or transferred ($d\bar{E}_{tr}$) by an indirectly ionizing radiation in material of mass (dm). Unit of kerma is joule per kilogram (J/kg) named as gray (Gy).
where, 1 Gy = 1 J/kg.

$$K = d\bar{E}_{tr} / dm \tag{8.24}$$

8.5.8 Absorbed dose

The mean energy imparted ($d\bar{E}_{ab}$) per unit mass (dm) of matter by an ionizing radiation is called as absorbed dose. It is a quantity applicable for both directly and indirectly ionizing radiation. It is a measure of the amount of energy absorbed per unit mass of the matter at the point of interest. Unit of absorbed dose is gray (Gy) and the old unit is radiation absorbed dose (Rad). Dose of 1 Gy corresponds to energy absorption of 1 Joule per 1 kg of the medium.

$$D = d\bar{E}_{ab} / dm$$
$$1 \, Gy = 100 \, Rad$$
$$1 \, Rad = 100 \, ergs/gm = 10^{-2} \, Gy \quad (8.25)$$
$$1 \, Rad = 1 \, cGy$$

8.5.9 Methods of radiation dosimetry and dosimeters in radiation therapy

Medical radiation dosimetry often performed by professional Medical physicists. A different type of dosimetry provides a different level of accuracy. The measurement of absorbed dose is very difficult directly in medium. However, there are direct methods to determine absorbed dose is carried out in some laboratories by Calorimetry or Fricke dosimetry. Several other types of dosimetry are being used in radiation therapy as follows (Mayles et al., 2007)

8.5.10 Ionization chamber dosimetry

Ionization chamber is a dosimeter that most widely used in diagnostic radiology and radiation therapy (Khan and Gibbons, 2014). These detectors are commercially obtainable in various shapes and sizes depending on requirements for radiation dosimetry. Ionization chambers work in conjunction with electrometers. Operation of ionization chamber is based on collection of pairs of charges produced by ionizing radiation undergoing interaction with medium inside the chamber through application of electric field. Ionization chambers are gas filled detectors consisting of a central collecting electrode surrounded by an outer conducting wall which acts opposite to the central electrode. Usually a chamber wall is made up of the low Z (air or tissue equivalent) material. Basic properties of ionization chamber as given below:

- Ionization chambers are operated in current or charge mode in radiation dosimetry.
- In ionization chamber-based detector pulse size or measured radiation intensity depend on the number of ions produced inside the detector, also it is independent of applied electric potential.
- Ionization chambers are usually made in cylindrical or parallel plates in shape used in radiation therapy. Most popular cylindrical ionization chamber of 0.6 cc farmar type thimble chamber used for calibration of photon beam in radiation therapy. However, parallel plate ionization chambers are endorsedfor electron beam calibration.
- All air vented ionization chambers required to apply temperature pressure correction due to change in mass of air because of change in ambient temperature and pressure, which consequently affect ionization produced in the chamber.

Ionization chambers are established as the golden standard for absolute as well as relative dosimetry in radiation therapy. Similarly, well type ionization chambers specially designed for routine calibration or source strength of brachytherapy source at clinical sites. Because of low air kerma rate of brachtherapy source chamber is designed with 4Π geometry and sufficiently large volume (about 250 cc or more) to attain required sensitivity. The Well type chambers usually used in clinics are calibrated in terms of reference air kerma rate from standard laboratories.

8.5.11 Film dosimetry

There are categories of film radiographic and radiochromic films are available commercially in the market. Radiographic films are radiation sensitive emulsion (AgBr suspended in gelatin) with a base of thin plastic. Development of radiographic films needs to undergo several processes through developer, fixer, washing and drying. However, radiochromic film is a contemporary self-developing film used for radiation dosimetry. It consists of a special dye fixed on a polymer base from single or double sided. When an ionizing radiation exposed to radiographic or radio chromic film it undergoes ionization of crystalline material such as AgBr or polymerization of dye turns the film into blackening. Degree of blacking depends on exposed radiation and is proportional to energy absorbed is measured by determining Optical density (OD) by the densitometer. The optical density is defined as follows:

$$OD = \log_{10}(I_0 / I_T) \quad (8.26)$$

where, I_0 = Initial light intensity (Light detected without film in place)
I_T = Transmitted light intensity (Light detected with film in place)

In radiation dosimetry inherent optical density of film i.e. fog density is subtracted to obtain net optical density. Usually films are exposed to get sufficient optical density between 1.3 to 1.7. A plot of optical density versus radiation exposure or dose is called Hunter–Driffield (H&D) curve. Fig. 8.16 illustrates characteristics of radiographic film. A film H&D characteristic curve has three distinct regions with different contrast. Very low contrast region corresponds almost to the background of the film called toe region. Middle of the curve where the contrast is relatively higher and optical density follows linearly with respect to exposure is called linear region. If the curve is nonlinear then appropriate correction factor should be applied to convert OD to absorbed dose. Shoulder region shows the inability of film to produce further contrast and follows saturation in OD with higher exposure.

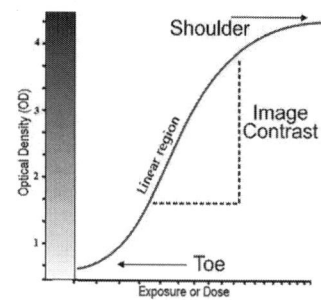

FIG. 8.16 Characteristic H&D curve for radiographic film.

Basic characteristic of film dosimetry as given below:

- Film dosimery offers excellent 2D special resolution that gives special areal distribution of radiation in areas of interest.
- Usable dose range for radiographic film is bounded and strongly energy dependent. However, radiochromic film can be used for higher dose ranges up to 100 Gy.
- Sensitivity of radiographic film is much higher than radiochromic film. Even very short exposure of radiation can be noticed on radiographic film.
- Film dosimetry can be used for quantitative or qualitative measurement of dose and fluence verification.
- Slope of linear portion of H&D curve called gamma of film.
- Films are mostly used for relative dosimetry and quality control of radiation therapy machines such as optical field congruence, isocenter spoke test, MLC leakage.
- Radiochromic film have advantages over radiographic film such ease of use, self-developing, useful for higher dose range, dose rate independent, insensitive to ambient condition.
- Radiation dose from X-rays, gamma and beta can be measured by film dosimeter.

8.5.12 Luminescence dosimetry

Phenomenon of spontaneous emission of ultraviolet, visible or infrared light after the absorption of radiation by substance is called luminescence. Once absorption of radiation is over by a luminescence material, it retains part of energy in its metastable states. The stored energy in metastable state is released once it stimulated by an external agent. The luminescences are of two kinds' fluorescence and phosphorescence. They are classified based on time delay between stimulation by a stimulating agent and emission of light. The phenomenon of fluorescence takes place with a time delay of 10^{-10} to 10^{-8}s. However, phenomenon of phosphorescence occurs with a time delay exceeding 10^{-8}s. An external agent likes heat or light can stimulate the phenomenon of phosphorescence.

- If heat is act as stimulating agent, then phenomenon is known as thermoluminescence.
- If light is act as stimulating agent, then phenomenon is known as optically stimulated luminescence (OSL).
- Luminescence detectors can measure gamma and beta separately.

Most of the personal monitoring devices used in radiation fields are based on thermoluminescence or OSL.

8.5.13 Thermoluminescence

Thermoluminescence is thermally stimulated phosphorescence (Podgorsak editor, 2005). When a crystalline material is exposed to radiation some portion of energy is absorbed by the material and stored. This stored energy is being released once exposed crystalline material is exposed by heat externally. This mechanism of luminescence can be explained by electron band theory. All luminescence material doped externally by an extrinsic impurity creates a difference between valence band and conduction band. Extrinsic impurity causes lattice defect that forms its localized energy level between valence and conduction band. Usually electrons lie in the ground state in a valence band. The absorption of sufficient energy by an exposed radiationleads the electrons to rise in conduction band departing holes in the valence band. Some electrons are trapped in a lattice defect formed between valence and conduction band; this is also called a metastable state where energy is being stored. When the crystal is being heated the electrons are released from their metastable state and come down to ground state with emission of ultraviolet, visible or infrared light. The thermoluminescence reader detects this emitted light and analyzed in order to quantify radiation.

Most thermoluminescence materials used for radiation therapy are LiF:Mg,Ti, LiF:Mg, P, Cu and $Li_2B_4O_7$:Mn, because of their tissue equivalence. Other $CaSO_4$:Dy, Al_2O3:C and CaF_2:Mn. material with high sensitivity often used for making thermoluminescence detector (TLD) for personal monitoring.

8.5.14 Optically stimulated luminescence

OSL carried similar principle to thermoluminescence dosimetry (Mayles et al., 2007). The only the difference is in stimulating agents used in OSL. The heat used as a stimulating agent in the thermoluminescence process. However, light or laser is used to release trapped electrons in OSL. OSL is used for invivo dosimetry in radiation therapy for quantification and verification delivered dose to patients. Optically stimulated luminescence dosimeters (OSLD) have also been used for personal monitoring over the world. Highly sensitive $Al_2 O_3$: C is commonly used material for OSLD.

8.5.15 Semiconductor dosimetry

Semiconductor diode detector offers many advantages in clinical radiation therapy. Silicon diode detectors are a p-n junction diode that is made up of either n or p type silicon material. The diode is called an n or p type semiconductor detector. Both type of detector are available commercially in the market. However, only the p type silicon diode used in radiation therapy. Because p-type diodes produces very less dark current and less affected by radiation damage. Semiconductor diodes have higher sensitivity, real-time readout, robustness, independence of air pressure.

Once the semiconductor detector is exposed to the radiation, ionizing radiation enters the sensitive volume of the detector and interacts with material inside. Electron-hole pairs are produced both inside and outside the depletion layer. These pair of charges swept across the depletion layer and collected rapidly under the action of intrinsic electric potential difference. Current is generated flowing opposite to direction of diode current, which can be measured and related to absorbed dose. Few important characteristics of semiconductor are discussed below,

- Diode detectors usually operated without external bias that reduces leakage current.
- Diodes exhibit linear relation between dose and charge due to its short circuit mode.
- Diodes are usually used for relative and stereotactic radiosurgery or high dose gradient small field dosimetry.
- Diodes show the variation dose with temperature, dependency on dose rate, angular dependency and energy dependency.

Semiconductor diodes detector usually used in measurement of PDD, TPR, beam profiles and output factors for small field dosimetry like SRS, SRT.

8.5.16 Physical and clinical dosimetry in radiotherapy

Radiotherapy demands measurement of absorbed dose in the tissue. ICRU has recommended accuracy in overall radiation treatment delivery to the patient should be within ± 5%. The deliver accurate dose to the patient, radiation therapy machines need to calibrate accurately as well as individual patient treatment delivery need to evacuate before the actual treatment. Usually dosimetry in radiation therapy includes both physical and clinical dosimetry. Physical dosimetry looks into the physical aspect of dose output by machine. However, clinical dosimetry looks after actual dose delivery of the patient.

8.5.17 Physical dosimetry

Medical physicists perform calibration of radiotherapy machines as per codes or guidelines given for reference dosimetry for practice of radiotherapy. Codes and guidelines given are different for different types of beam. Usually code from technical reports by an International atomic energy agency (IAEA) or American association of physicist in medicine (AAPM) task group are used for calibration of photon or electron beam in EBRT (IAEA Technical Reports Series No. 398, 2000). The code for brachytherapy is also available. They mostly focused on use of dosimetric equipment, procedure for measurements and methods of calculation.

Output of the EBRT machine is nothing but the dose rate at the reference point under the reference condition. Usually dose rate at depth of dose maxima in tissue equivalent material is considered as machine output. Reference condition for calibration of photon and electron beam given as per TRS-398, photon beam is calibrated either in SSD or SAD setup at reference depth (5cm or 10 cm) for collimator opening of 10x10 cm^2 in water using cylindrical chamber. However, electron beam is calibrated in SSD setup at depth of ($Z_{ref} = 0.6*R_{50}–0.1$) g/cm^2, where, R_{50} is half value depth referred to as electron beam quality. Plane parallel plate chamber is recommended for electron beam calibration (Podgorsak editor, 2005). Output of the teletherapy machine is given in cGy/mit, whereas output medical linear accelerator is given in cGy/MU.

8.5.18 Clinical dosimetry

This mostly involves the patient specific dosimetry. This is performed before the actual patient treatment for verification of treatment plan and its delivery on machines. Patient specific clinical dosimetry assures the multiple treatment plan parameters while delivered on machine. It verifies dose being delivered throughout the tumor volume of the patient within the tolerance. There are various types of devices available commercially for verification of treatment plan and delivery. Mostly they are based on an array of ionization chambers or diode-based detectors.

References

Beyzadeoglu, M., Ozyigit, G., Ebruli, C., 2010. Basic Radiation Oncology. Springer-Verlag, New York.

Bortfeld, T., Schmidt-Ullrich, R., De Neve, W., Wazer, DE., 2006. Image Guided IMRT. Springer, Berlin & Heidelberg.

Halperin, EC., Wazer, DE., Perez, CA., Brady, LW., 2013. Perez and Brady's Principles and Practice of Radiation Oncology, 5th ed. Wolters Kluwer Health/Lippincott Williams & Wilkins, Philadelphia.

IAEA Technical Reports Series No. 398, 2000. Absorbed Dose Determination in External Beam Radiotherapy. International Atomic Energy Agency, Vienna.

Joslin, C.A.F, Flynn, A., Hall, E.J, 2001. Principles and Practice of Brachytherapy Using After Loading Systems. Arnold, London.

Khan, F.M., Gibbons, J.P., 2014. The Physics of Radiation Therapy, 5th ed. Lippincott Williams & Wilkins, Philadelphia, PA.

Landberg T., Chavaudra J., Dobbs J., Hanks G., Johansson K.-A., Möller T., Purdy J. Journal of the International Commission on Radiation Units and Measurements, Volume os 26, Issue 1, 1 September 1993.

Mayles, P., Nahum, A., Rosenwald, J.C., 2007. Handbook of Radiotherapy Physics: Theory and Practice. Taylor & Francis Group, CRC Press, Boca Raton, FL.

Podgorsak editor, E.B., 2005. Radiation Oncology Physics: A Handbook for Teachers and Students. International Atomic Energy Agency (IAEA), Vienna, Austria.

Chapter 9

Role of carbon ion beam radiotherapy for cancer treatment

Vibha Chopra[a], Nirupama S. Dhoble[b], Balkrishna Vengadaesvaran[c], Sanjay J. Dhoble[d]

[a]P.G. Department of Physics & Electronics, DAV College, Amritsar, Punjab, India
[b]Deaprtment of Chemistry, Sevadal Mahila Mahavidhyalaya, Nagpur, Maharashtra, India
[c]Higher Institution Centre of Excellence (HICoE), UM Power Energy Dedicated Advanced Centre (UMPEDAC), Level 4, Wisma R&D University of Malaya, Kuala Lumpur, Malaysia
[d]Department of Physics, Rashtrasant Tukadoji Maharaj Nagpur University, Nagpur, Maharashtra, India

9.1 Introduction

A variety of materials are known for thermoluminescence dosimetric (TLD) applications with varying efficiencies for different R&D areas viz. medical (Efstathopoulos et al., 2003; Kron, 1999), accidental (Veronese et al., 2010), retrospective (Kazakis et al., 2016; Mesterhzy et al., 2012), personal (Gai et al., 2015), thermal neutron (Bhadane et al., 2016) dosimetry, solid state lighting (Dalal et al., 2016; Naik et al., 2016), and 2D OSL mapping (Oliveira et al., 2016). A large number of standard commercial dosimeters are available at national as well as international level which is also well-known under their commercial names for use in particular observed dose range. The most famous dosimeters developed are LiF: Mg, Cu, P (TLD-700H), Al_2O_3 (TLD- 500), $CaSO_4$:Dy (TLD-900) and CaF_2:Dy (TLD-200) and many more are expected to be available in the near future (Fox et al., 1988; Hasan et al., 1985; Noh et al., 2001; Sahare et al., 2007). Most of the phosphors can be used as TLDs within a specific range of radiation doses and are not applicable for all ranges of doses since their response depends on various factors viz. linearity, precision, dose rate, fading, reproducibility, and others. Thus, there is a need to explore more sensitive materials that show linearity of TL response up to wide range, energy independent, thermally stable and exhibit low fading. Moreover, there is a continuous increasing demand of efficient TL dosimeters to monitor high dose levels of swift heavy ions (SHI) since they are used for treatment of cancer and tumor cells. In this regard especially the doses from carbon ion beam are needed to be monitored as these ions are extensively used in medical applications (Kanagasekaran et al., 2008). The significance of the SHI ions over the older available technology of photon radio therapy lies in the fact that SHI delivers a better mean energy per unit length at a particular depth. SHI irradiation also results in the modification of the luminescence properties of the material by inducing defects because of huge energy deposition through electronic excitation (Wesch et al., 2004). In this view, number of reports are made by different groups on the thermoluminescence response of SHI exposed phosphors for high energy space dosimetry, ion beam dosimetry for personnel applications, radiotherapy, diagnostic purposes, etc. (Bedyal et al., 2016; Dutta et al., 2016; Kore et al., 2015; Kore et al., 2016; Som et al., 2014; Som et al., 2016; Yerpude et al., 2016). Research community is focused on the development of dosimeters for carbon ion beam therapy owing to its large number of benefits over gamma radiation therapy and proton therapy in the field of cancer treatment. Carbon ion beam radiotherapy was first used at the National Institute of Radiological Sciences (NIRS), Japan using world's first heavy ion accelerator complex (Heavy Ion Medical Accelerator in Chiba, HIMAC) in 1994. Currently, the facility is available in 13 centers worldwide and is under process for CIRT in Yamagata, Japan, and Yonsei University, Korea along with several facilities to be planned around the globe.

9.2 Radiation therapy for the treatment of cancer

Radiation therapy is usually a local treatment that affects only the cancerous part of the body to be cured unlike the other cancer-fighting drugs, that expose the whole body to radiations and causes harm to the healthy tissues as well. Commonly, electromagnetic and particulate radiation which includes X-rays, gamma rays, electron beams, proton beams, and ion beams are used in radiation therapy for the treatment of tumours.

9.2.1 Gamma ray therapy

When radiation passes through matter it may interact with the material, transferring some or all of its energy to the atoms of that material. Gamma rays are absorbed or scattered causing number of processes. The prominent effects are Photo electric effect, Compton effect, and pair production (Knoll, 1999; Leo, 1994). Gamma rays interact with matter in different way as compared to charged particles due to greater penetration power and have no definite range. Unlike charged particles, a well collimated beam of gamma rays is exponentially absorbed in matter because photons are absorbed or scattered in a single event. The measurement of radiation dose of such beam is very important and essential as well. For this reason continuous efforts have been done by the research community around the globe for the development of new materials and to improve dosimetric properties of already available TLD materials as an efficient TLD which exhibits tissue equivalency (low Z_{eff}) as well as high Z_{eff} to ensure its use in low and high dosimetry (Emen et al., 2016; Gupta et al., 2016; Nyenge et al., 2016; Shivaramu et al., 2015). Further, nanomaterials are found to be more stable at high doses of gamma radiations, as their glow curve structure remains unchanged with dose and do not saturate even at very high doses where microcrystalline phosphors saturate (Chopra et al., 2013; Lochab et al., 2007; Salah et al., 2004; Salah et al., 2007; S.P. et al., 2007; Singh et al., 2011). So, in the current scenario research is mainly focused on the development of nanomaterials to be used as gamma dosimeters. In the past years, only Co-60 therapy was used for cancer treatment in the hospitals, hence only gamma ray dosimeters were explored, but later the ion beam therapy came into scene due to their advantages over Co-60 therapy. So, the current area of research is devoted to explore the dosimeters for proton and heavy ion beam therapy.

9.2.2 Proton therapy

Proton therapy is an emerging field in the area of research. It is considered as a highly effective treatment for tumors present in lung, prostate, brain, neck and head. Proton is a positively charged subatomic particle that deposits energy differently than X-ray beam and gamma ray. Compared to gamma or X-ray, the proton beam has a high dose 'bragg peak' region and low entrance dose that is designed to cover the entire tumor and not beyond that. So, by proton beam the dose is focused and delivered maximally to the malignant cancers tissue volume while minimizing the exposure to healthy tissues (Devicienti et al., 2010; Lawrence and Feng, 2013; Newhauser and Zhang, 2015; Podgorsak, 2006). Ionization chamber is the basic tool for dosimetry of proton beam therapy. However, various lithium fluoride based detectors, that is, LiF:Ti, Mg, LiF: Cu, Mg, P which are basically TL detectors are being widely used to measure the dose obtained from proton beam. Dosimetric studies of nanocrystalline boron based TL phosphors such as $Li_2B_4O_7$:Cu, MgB_4O_7:Dy have been performed using gamma rays and proton beam of energy 3 MeV, 150 MeV, electron beam of energy 4 MeV and 9 MeV respectively (Bahl et al., 2013; Chopra et al., 2014; Salah et al., 2013). Gamma rays and proton beam irradiated nanocrystalline $K_2Ca_2(SO_4)_3$: Eu phosphor is also reported to exhibit very good TL properties (Pandey et al., 2011). Since the proton therapy possesses very low risk of showing adverse effects, so, it is a clear choice of treatment in near future and there is a very wide scope to find best suitable TL dosimeter for proton beam therapy.

9.2.3 Ion beam therapy

The field of therapy with ions is maturing rapidly. Ion beams are being used as ultimate tool to cure very advanced hypoxic as well as radiation resistant tumors having complex local spread. Ion beams are heavier, their trajectories are stiffer, they have less multiple scattering, less range straggling, more favorable dose deposition profile, thus the normal healthy tissues beside the surrounding of cancerous tissue are exposed to very less energy of ion beam. Ion beam on entering the target material simultaneously interacts with many electrons causing excitation and ionization of its atoms. Charged ion transfers its energy to the electron, as a result it slows down due to decrease in its velocity. The average energy loss per unit path length is known as the stopping power or simply dE/dx. It was first calculated by Bohr using classical mechanics and later corrected by Bethe, Bloch, and others using quantum mechanics. A plot of specific energy loss along the track of charged particles, that is, distance of penetration is known as Bragg curve. The heavy ions shows sharp bragg peak. Different research groups have reported and investigated the TL response of SHI irradiated phosphors for high-energy space dosimetry, radiotherapy, personal dosimetry and diagnostic purposes in clinical use etc. (Dutta et al., 2016; Kore et al., 2015; Kore et al., 2016; Som et al., 2016; Som et al., 2014; Yerpude et al., 2016). C^{5+} ions being heavier charged particles than constituent particles present in conventional radiation sources ensure more focused penetration with minimum scattering to the tumor even located deep inside the body. Thus carbon ion beam therapy plays very important role in the field of radiotherapy. So, the scientists all around the world are in search of dosimeters for carbon ion beam therapy.

9.3 Role of carbon ion beam therapy

Nowadays carbon ion is among the most favorable option available for heavy ion beam therapy. Effectiveness of carbon ion beam is 2 to 3 time more as compared to X rays in biological testing (Heilmann et al., 1996; Kraft, 2000). Depending upon the mass of atomic nuclei, ion beams are categorized into different categories in heavy ion beam therapy. Beam of particles having mass of atomic nuclei greater than protons are referred as heavy particle beams. Heavy ions like neon and argon cause irreparable loss on the entrance itself, so the healthy tissues located in the opening front of the tumor gets damaged. In case of very light ions like protons, scattering effect is greater as compared to the carbon ions so localized treatment cannot be achieved (Kraft, 2000). So, in heavy ion therapy carbon ion beam has found to be a most favorable option. A strong increase in the linear energy transfer (LET) in the affected part makes carbon ion beam a potential tool to treat tumors of all types especially which are surrounded by complex and critical structures (Scholz, 2000; Schulz-Ertner et al., 2006; Suit et al., 2010). In case of carbon ion beam, the energy is delivered to the tumor cells in a very narrow zone which appears as an explosion inside the tumor. Hence, very small damage has been done to the healthy tissues present in the path of the tumor. The secondary electrons induced by carbon ions cause multiple sites splitting of double stranded DNA even at low level of oxygen allowing its access or reach to the hypoxic regions of the tumor (Ahnström and Edvardsson, 1974; Ohno et al., 2011; Ohno, 2013; Scifoni et al., 2010). The DNA damage mechanism using carbon ion was studied in details (Ahnström et al., 2000). The dose deposition in case of photon beam irradiation decreases with the penetration depth however it increases gradually for the ion beams then decreases after obtaining a sharp maximum called Bragg peak. It is the deciding factor that decides the best dose localization of the tumors to be treated (Oza et al., 2018). Further the ion beams have low straggling, stiffer particle trajectory, minimum scattering and sharp field edges etc. (Eickhoff et al., 1999). They have well defined range and their angular scattering is also small as compared to photon radiotherapy (Tsujii et al., 2007). Carbon ion beam possess higher value of relative biological effectiveness (RBE) than a proton beam. RBE corresponds to the rate with which the biological cells are destroyed by ion beam with respect to a photon beam of same dose. Its scattering effect is also very small as compared to that of light particle (e.g., photon) because of its relatively greater atomic nuclei mass. Keeping all these properties in mind it can be concluded that the loss can be much more complex which makes the repair of cancer cells very difficult. So, the ion beam dosimetry requires great accuracy and precision while treating humans. For this purpose ion beam dosimeters are needed to be developed that satisfy all the properties required for dosimetry (Amaldi and Kraft, 2005; Karger et al., 2010; Rivera, 2012; Schulz-Ertner et al., 2004). The commercially used dosimeters such as $CaSO_4$:Dy and LiF:Cu, Mg, P are not yet standardized for carbon beam therapy so there is a lot of scope for the development and tailoring new materials to be suitable for carbon ion beam therapy. In this chain, $CaMg_3(SO_4)_4$:Dy^{3+} phosphor has been developed which is 3-5 times more sensitive than commercially used $CaSO_4$:Dy for carbon beam therapy (Kore et al., 2014). Further, most of the materials in microcrystalline form shows saturation at very low dose for carbon ion beam. So, better nanocrystalline phosphors are needed to be searched which possess wider TL linearity range for carbon ion beams and gamma radiations as well. For this, TL response of 85 MeV carbon beam irradiated nanocrystalline $BaSO_4$: Eu (Sharma et al., 2018) and $CaSO_4$:Dy (Salah, 2008) has been reported. Strontium tetraborate was found to be five times more efficient than TLD-100 (Santiago et al., 1998) for carbon ion beam dosimetry. Many more such works are reported and new phosphors are developed in the search of best suited dosimeter for Carbon ion beam dosimetry.

9.4 Development of TLD materials for carbon ion beam therapy

Planning of treatment using carbon ion beam therapy is a challenging due to complexity and limited knowledge of chemical, physical and biological processes involved in the same. Typically, planning of the treatment involves the process to design maximum yield of beam of radiation to maintain the balance between accurate dose to the target dose and excluding normal tissues. The crucial role is played by the TL materials used for the treatment of tumors. So, the dosimeter material should possess the characteristics so as to deliver the dose to the tumor with precision and accuracy. The commercial dosimeters LiF and $CaSO_4$ are not best suited for carbon ion beam therapy, so the scientists are in search of TLD material best suited for the same. Many phosphors have been tried that yields good results, few of them are discussed below:

9.4.1 Lithium-based phosphors

Lithium fluoride (LiF), a greatly acknowledged highly sensitive TL material is widely used as commercial radiation dosimetric material. It is having effective atomic number 8.1 i.e. closer to biological tissue. For the better sensitivity, the material was activated with different impurities or activators such as P, Mg, Cu (Shinde et al., 2001). Further, LiF:Mg,Cu,P in nanocrystalline form was produced by Salah et al. (2007). Europium activated LiF was found to have the most TL sensitivity towards gamma rays. In this light, the TL studies of LiF:Eu nanostructure with carbon ion irradiation having a wide range of fluence were done (Salah et al., 2015b). LiF:Eu in nanocubes shape were synthesized by the co-precipitation method and evaluation of their TL properties was performed using 85 MeV C^{6+} ion beam irradiation in a wide 10^9–10^{13} ions/cm^2 fluence range. A simpler TL glow curve was obtained than the one created by gamma rays. Its TL linearity lies in 10^9–10^{12} ions/cm^2 range that shows its possibility to be used as mixed field radiation dosimeter.

Solution combustion method was used to synthesize $LiMgBO_3:Dy^{3+}$, a low Z_{eff} material in its nanocrystalline form (Yerpude et al., 2019). TL studies of $LiMgBO_3:Dy^{3+}$ were performed by exposing to γ-rays and C^{5+} ion beam. TL glow curves for C^{5+} ion beam irradiation were observed at different fluence ranges from 2×10^{10} ions/cm^2-1×10^{12} ions/cm^2 and it was found that the TL intensity shows increasing trend with fluence up to 1×10^{11} ions/cm^2 and then decreases after further increment in the fluence. TL linearity was observed from 2×10^{10} to 1×10^{11} ions/cm^2 fluence range and then decreased. Therefore, the phosphor is found to be fit for use in the above observed range for C^{5+} ion beam irradiation.

$LiNaSO_4$ phosphor was reported to exhibit more sensitivity than $CaSO_4$:Dy when doped with Eu (Sahare and Moharil, 1990). TL properties of the same with Ti and Dy ions as impurities were also reported (Kassem et al., 1993; Vidya and Lakshminarasappa, 2015). Its effective atomic number is about 11.59. So, the material is significant for measurement of low dose areas such as environmental dosimetry and radiotherapy. Further 75 MeV C^{6+} ion beam irradiated $LiNaSO_4$ phosphor was also studied to observe its effects on structure and luminescence processes (Gupta et al., 2017). Glow curves after exposing $LiNaSO_4$ to γ-rays obtained from a ^{60}Co source with 10 Gy, from ^{137}Cs source with 2200 mRad, and from C^{6+} ion beam of energy 75 MeV having 2×10^{10} ions per cm^2 fluence are shown in Fig. 9.1. The TL glow curve of commercially available TLD ($CaSO_4:Dy^{3+}$) was also recorded to compare with the above said TL glow curves.

TL glow curve shows a main TL peak observed at 151°C having a small shoulder around 106°C. The structure of TL glow curve of $LiNaSO_4$ was same for different doping concentrations of Eu whereas TL sensitivity was reported to be

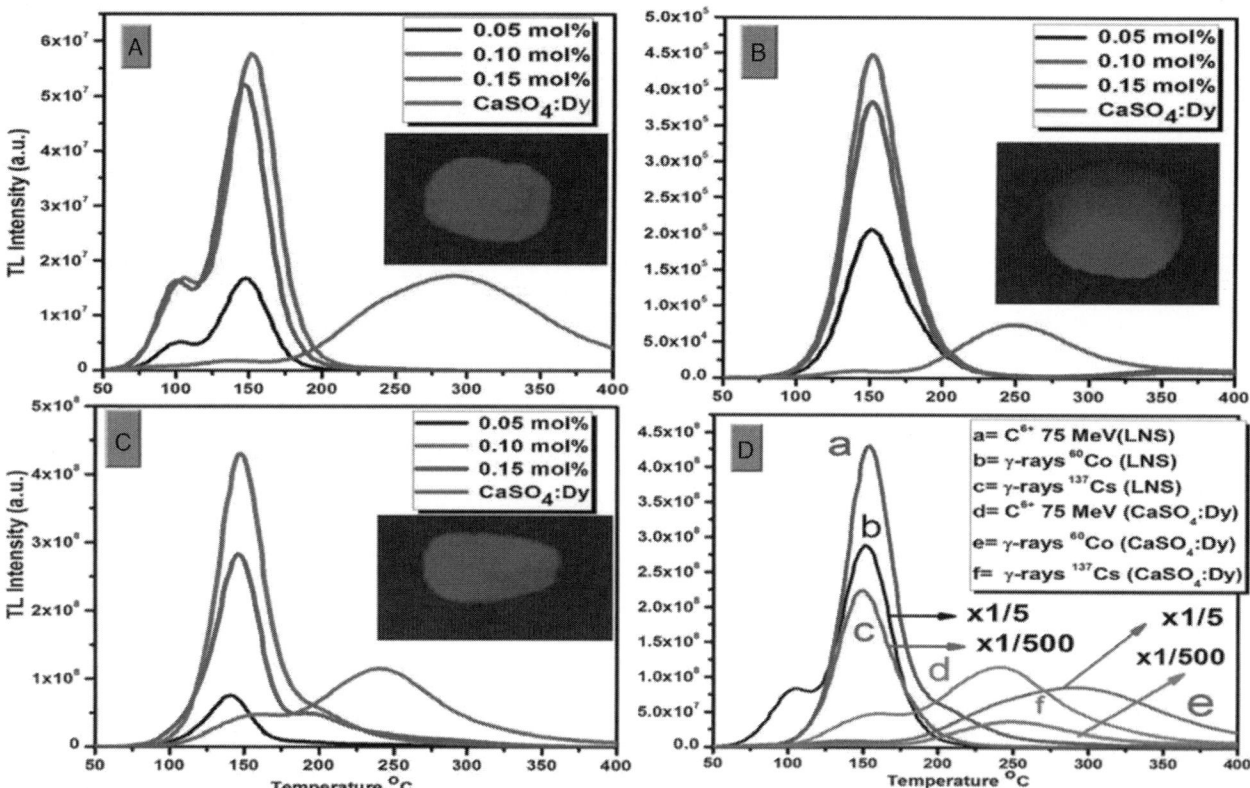

FIG. 9.1 TL glow curves of the LNS phosphor exposed to (A) γ-rays from ^{60}Co source with 10 Gy, (B) gamma rays from a γ-Cs source with 2200 mRad and (C) C^{6+} ion beam of energy 75 MeV having 2×10^{10} ions per cm^2 fluence, and (D) comparison of TL glow curves of LNS phosphor and commercial TLD $CaSO_4$:Dy phosphor irradiated with a C^{6+} ion beam of energy 75 MeV energy, and γ-rays from ^{60}Co source with 10 Gy and a ^{137}Cs source with 2200 mRad. To get value of the relative intensities the ordinate has to be multiplied by the numbers denoted to the curves. The inset represents the blue luminescence of samples obtained during thermal excitation. *(Reproduced with the permission from the ref. Gupta et al. (2017) © 2017 Royal Society of Chemistry).*

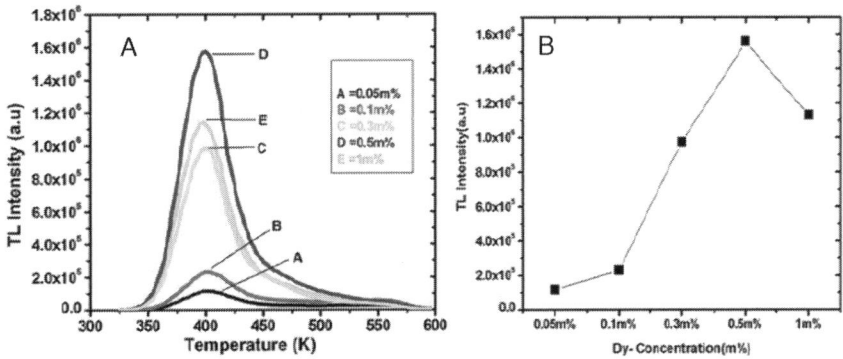

FIG. 9.2 TL glow curves for $Li_2BaP_2O_7$:Dy obtained after irradiating with C^{5+} ion beams at fluence of 15×10^{11} ions/cm^2. *(Reproduced with the permission from the ref. Wani et al. (2015) © 2015 Elsevier Ltd.).*

changed. Eu (0.1 mol%) doped sample shows the highest TL intensity. The γ-ray irradiated phosphor from a ^{60}Co source shows 3.34, 3.32, and 3.30 times more TL sensitivity as compared to commercial $CaSO_4$:Dy^{3+} phosphor at 10 Gy, 20 Gy, and 30 Gy doses, respectively. However, the γ-ray irradiated phosphor from a ^{137}Cs source possesses 3.05, 3.08 and 3.07 times more TL sensitivity than that observed from a commercial $CaSO_4$:Dy^{3+} phosphor at 22 mGy, 44 mGy and 66 mGy doses, respectively. C^{6+} ions (75 MeV) exposed $LiNaSO_4$ and γ-irradiated samples shows identical TL glow curves. Further, the shape of glow curve of standard TLD phosphor $CaSO_4$:Dy^{3+} varies when irradiated with C^{6+} ion beam as compared to prepared phosphor. It has 4.35, 4.30 and 4.33 times more TL sensitivity after irradiation with C^{6+} ion beam than that of the commercial $CaSO_4$:Dy phosphor with 2×10^{10}, 5×10^{10} and 1×10^{11} ions per cm^2 fluences, respectively. Although it shows significant intensity but 35% fading rate within 28 days of storage was reported which is not suitable for TLD materials. So, the material is still open for the further research in future, may be in nanocrystalline form, to study its response for C^{6+} ion beam for applications in radiation dosimetry.

High temperature solid state diffusion method was employed to synthesize $Li_2BaP_2O_7$:Dy phosphor (Wani et al., 2015). Fig. 9.2 depicts TL response of Li_2BaP_2O: Dy at different concentrations of Dy and subjected to C^{5+} ion beams irradiation having fluence of 15×10^{11} ions/cm^2.

The shape of TL glow curve is found to be simple, does not vary with concentration whereas the TL intensity increases upto 0.5 mol % and then starts decreasing. Further $Li_2BaP_2O_7$:Dy phosphor has very less TL intensity (around 30%) as compared to commercial $CaSO_4$:Dy TLD phosphor as shown in Fig. 9.3.

Gamma irradiated $Li_2BaP_2O_7$:Dy shows approximately 28% less TL intensity than $CaSO_4$:Dy TLD phosphor. Irradiation of the phosphor with a wide range of C^{5+} ion beam fluences ranging from 15×10^{10}–30×10^{12} ions cm^{-2} results in the fact that the phosphor was incapable of maintaining TL linearity over the entire studied range. So, this material in the present form cannot be used for radiation dosimetry. The material needs to be explored in different dimensions.

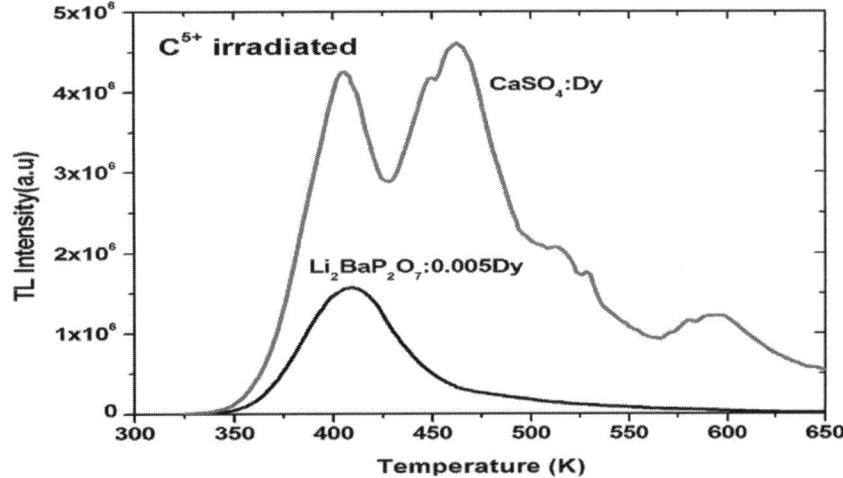

FIG. 9.3 Represents TL glow curves for $Li_2BaP_2O_7$: 0.005 Dy and $CaSO_4$:Dy irradiated with C^{5+} ion beams having fluence of 15×10^{11} ions/cm^2. *(Reproduced with the permission from the ref. Wani et al. (2015) © 2015 Elsevier Ltd.).*

9.4.2 Calcium-based phosphors

Calcium-based TLDs are well known for their higher TL sensitivity towards carbon ion beam radiation. The family includes $CaMg_3(SO_4)_4$ (CMS) phosphor, $CaMg_2(SO_4)_3$:Dy^{3+}, $LiCaAlF_6$:Ce, $LiCaBO_3$.

Acid distillation procedure was employed to synthesize $CaMg_3(SO_4)_4$ (CMS): Dy^{3+} TLD phosphor (Kore et al., 2014) and its TL properties were studied after giving γ-rays and carbon ion beam exposure. It shows stable TL glow curve for both types of radiations. $CaMg_3(SO_4)_4$ (CMS): Dy^{3+}(0.2 mol%) was found to be 3.5 times more TL sensitive than commercially available $CaSO_4$:Dy^{3+} TLD phosphor when given carbon ion beam exposure. TL glow curves of the CMS phosphor, were observed after irradiating the samples with gamma rays obtained from a ^{60}Co source with 15 Gy dose and ^{137}Cs source at a 2300 mRad dose. Further TL glow curves were observed for 75 MeV and 50 MeV C^{5+} ion beam for 216 kGy dose. A comparison of all TL glow curves was done with that of commercially available $CaSO_4$: Dy TLD phosphor and are shown in Fig. 9.4.

For ^{60}Co source irradiated samples, two peaks are shown at 146°C and 260°C and the second peak is found to be more dominating as compare to ^{137}Cs source. The shape of TL glow curve changes on irradiating the samples with 50 and 75 MeV C^{5+} ions. Further TL response curves of the CMS phosphor for ^{60}Co source irradiation shows linearity up to 1000 Gy. TL linearity ranging from 100 mRad to 5 Rad is also shown on irradiating the phosphor with γ-rays obtained from the ^{137}Cs source. But early TL saturation is observed for CMS phosphor when exposed to the carbon ion beam varying a dose range from 22 kGy - 4 MGy. Its TL sensitivity decreases after 23 kGy C^{5+} ion beam dose. The sensitivity of phosphor was observed higher for 50 MeV irradiated sample than 75 MeV carbon irradiated sample. These results makes CMS a suitable candidate to be used potentially for ion beam irradiation dosimetry of higher doses.

Triple sulfate $CaMg_2(SO_4)_3$: Dy is synthesized using solid state diffusion method (Tamboli et al., 2018). Material was irradiated with C^{6+} ion beam and γ-rays to study TL properties. C^{6+} ion beam irradiated $CaMg_2(SO_4)_3$: Dy TL glow curves are shown in Fig. 9.5.

It was observed that at lower doses, three peaks are obtained whereas at higher doses the structure of glow curves changes completely, and only two peaks i.e. at 193°C and 315°C respectively are observed. Increase in TL intensity was observed with increment in dose from C^{6+} ion beam upto 53 kGy and after that it got saturated. On comparing the TL

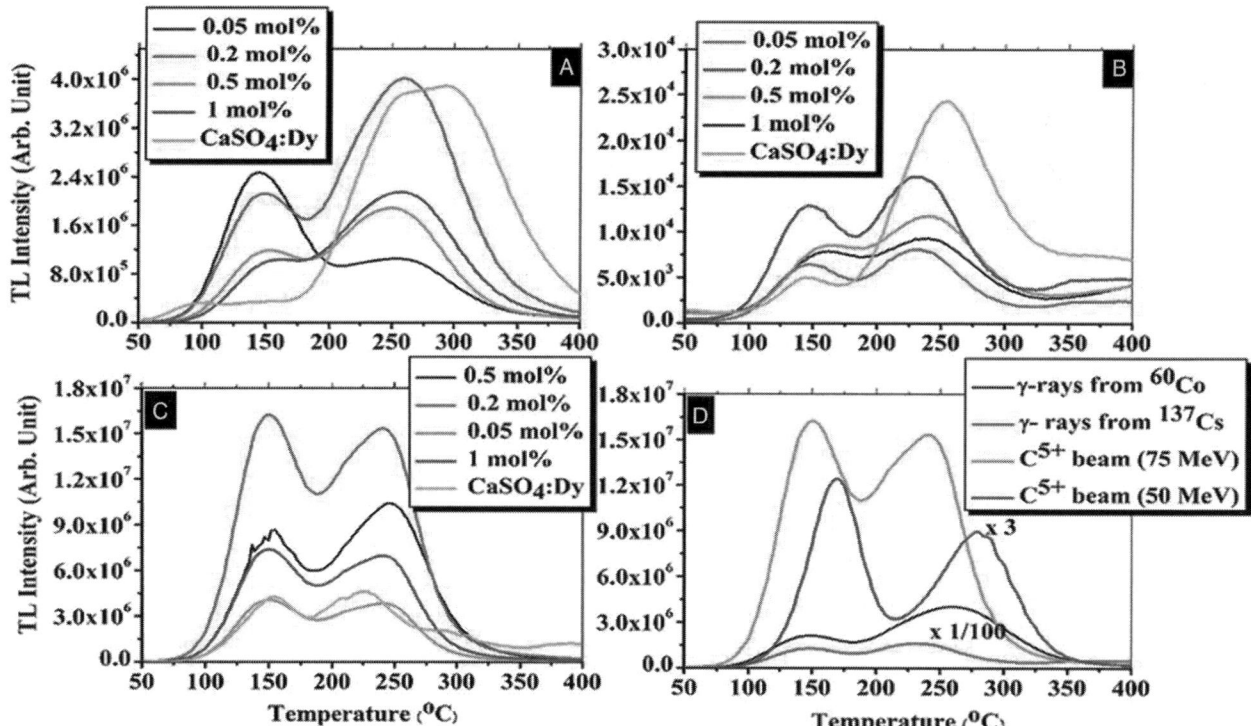

FIG. 9.4 TL glow curves of the CMS TLD phosphor exposed to (A) γ-rays from a ^{60}Co source with 15 Gy, (B) γ-rays from a ^{137}Cs source with 2300 mRad, (C) C^{5+} ion beam of energy 75 MeV having a 216 kGy dose, and (D) comparison between TL glow curves of the CMS irradiated with C^{5+} ion beam with energies 50 and 75 MeV, and γ-rays from a ^{60}Co source with 15 Gy and ^{137}Cs source with 2300 mRad. *(Reproduced with the permission from the ref. Kore et al. (2014) © 2014 Royal Society of Chemistry).*

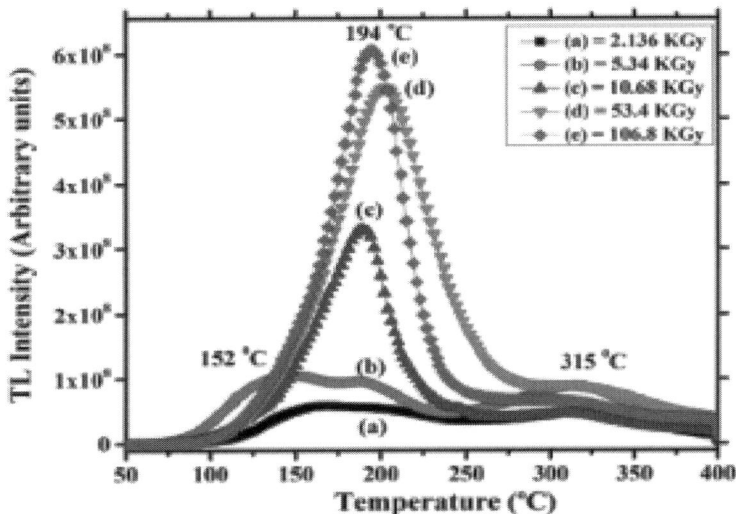

FIG. 9.5 TL glow curve of $CaMg_2(SO_4)_3:Dy^{3+}$ exposed to C^{6+} beam with different doses. *(Reproduced with the permission from the ref. Tamboli et al. (2018) © 2018 Elsevier Ltd.).*

intensities after irradiation with gamma-ray (100 Gy) and C^{6+} beam (5.34 Gy) of $CaMg_2(SO_4)_3:Dy^{3+}$ (Dy^{3+} = 0.2 mol%) and commercial $CaSO_4:Dy$ phosphor are shown in Fig. 9.6.

It can be revealed from the figure that $CaSO_4:Dy$ phosphor exhibits quite higher TL Intensity than $CaMg_2(SO_4)_3:Dy^{3+}$. So, the phosphor still needs to be studied to improve the dosimetric properties for its applications.

$LiCaAlF_6:Ce$ was synthesized by using in-house precipitation method and concentration of dopant was varied (Yerpude et al., 2016). Ce^{3+}-activated $LiCaAlF_6$ is a well-known phosphor which finds diverse applications in scintillators and solid-state tunable lasers (Bayramian et al., 1996; Dubinskii et al., 1993; Marshall et al., 1994; Nikl et al., 2013; Yokota et al., 2011). The effect of C^{5+} ion beam irradiation upon the TL behavior of $LiCaAlF_6:Ce$ is also studied. TL glow curves of this phosphor exposed to 15×10^{11} ions/cm² of fluence and 75 MeV and 50 MeV C ion beam and 10 Gy gamma-rays from ^{60}Co were observed. Glow curve of $CaSO_4:Dy$ was also irradiated with 15×10^{11} ions/cm² fluence having energy 75 MeV carbon to compare the glow curve. The reported phosphor was found to be almost eight times more sensitive towards carbon ion beam irradiation as compared to the standard TLD. So, this shows that the material is very useful for its dosimetric applications.

Solid state reaction method resulted in polycrystalline $LiCaBO_3$ phosphor doped with Dy or Ce (Oza et al., 2015a). Different doses range from 3.75×10^{12} to 7.5×10^{13} ion cm⁻² of carbon beam were used to irradiate $LiCaBO_3:Ce$ and

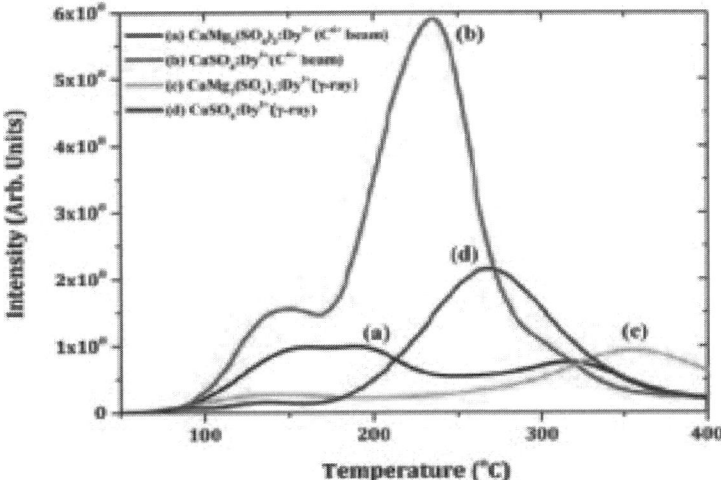

FIG. 9.6 Comparison of TL intensity after irradiation with gamma-ray and C^{6+} beam $CaSO_4:Dy^{3+}$ and $CaMg_2(SO_4)_3:Dy^{3+}$. *(Reproduced with the permission from the ref. Tamboli et al. (2018) © 2018 Elsevier Ltd.).*

LiCaBO$_3$:Dy^{3+} phosphors and their TL properties were studied. Their TL sensitivity showed decreasing trend with increase in ion fluence. Both the phosphors showed 60% fading within 20 days for the lower temperature glow peak. So, the phosphor cannot be used as dosimeter for carbon ion beam therapy in the present form and should be explored further to improve the properties.

9.4.3 Some other phosphors

In the past few years some mixed phosphors based on sulfates, phosphates, borates, aluminates were investigated for TL properties. Few of these materials are K$_2$CaMg(SO$_4$)$_3$:Dy (Kore et al., 2016), Sr$_2$B$_5$O$_9$Cl (Oza et al., 2018), NaCaPO$_4$:Dy^{3+} (Nair et al., 2020), Al$_2$O$_3$:Tb (Salah et al., 2015a), Mg$_2$BO$_3$F:Dy (Kore et al., 2015). The response of Carbon ion beam was studied for all these phosphors seeking the best suitable material for carbon ion beam therapy.

K$_2$CaMg(SO$_4$)$_3$:Dy TLD material was synthesized by SSR method and was studied for its TL properties with C^{5+} ion beams of energy 50 and 75 MeV having fluence range varying from 15×10^{10} ions/cm^2 to 30×10^{12} ions/cm^2. The comparative analysis of TL glow curves obtained after exposure to γ-rays from ^{60}Co and ^{137}Cs sources was also done. Finally, the comparison of all the response curve was performed with commercial CaSO$_4$:Dy TLD phosphor. The shape of TL glow curve of phosphor irradiated with C^{5+} ion beam was different from that of phosphor irradiated by γ-ray. Further, C^{5+} ion beam irradiated phosphor shows 7.8, 5.6 and 4.9 times more TL intensity for 35 kGy, 197 kGy and 2000 kGy doses, respectively as compared to commercial CaSO$_4$:Dy TLD phosphor. On increasing the energy of the C^{5+} ion beam from 50 MeV to 75 MeV, the TL intensity was reported to increase. Such important TL properties make this potential phosphor suitable for dosimetry of gamma-rays and carbon ion beams.

Al$_2$O$_3$ (Aluminum Oxide) is a highly efficient luminescent phosphor widely in use for radiation dosimetry using TL phenomena. Al$_2$O$_3$ doped with Tb was irradiated with 85 MeV C^{6+} ion beam with fluence ranging from 10^9 to 10^{13} ions/cm^2 (Salah et al., 2015a). It showed a simple glow curve exhibiting a peak at 230°C, which is better than those obtained by γ-rays irradiation. TL linearity was observed in dose range from 10^9 to 10^{11} ions/cm^2, which corresponds to the equivalent absorbed doses from 0.285 to 28.5 kGy. Its wide TL linearity along with the low fading effect makes this low cost nanophosphor a potential dosimeter for C ion beam dosimetry.

Mg$_2$BO$_3$F:Dy synthesized using wet chemical method shows marvelous TL properties when irradiated with γ-rays as well as carbon ion beam of energy 50 and 75 MeV at different fluences ranging from 15×10^{10} to 30×10^{12} ions/cm^2 (Kore et al., 2015). The TL intensity of all such glow curves are depicted in Fig. 9.7.

The TL intensity of Mg$_2$BO$_3$F:Dy phosphor irradiated with ion beam was reported to increase from 50 MeV C^{5+} irradiated sample to high energy (75 MeV) C^{5+} irradiated sample. This may be due to increment in penetration depth of carbon ion beam with energy. TL efficiency of 75 MeV and 50 MeV C^{5+} ion beam irradiated Mg$_2$BO$_3$F:Dy phosphor relative to gamma-rays of ^{60}Co was reported to be 0.83 and 0.37, respectively. Its strong TL properties make this material important for carbon ion beam dosimetry. But large fading (about 70 % in 16 days) could be a major drawback regarding phosphor and further study of this phosphor should be done to solve this problem.

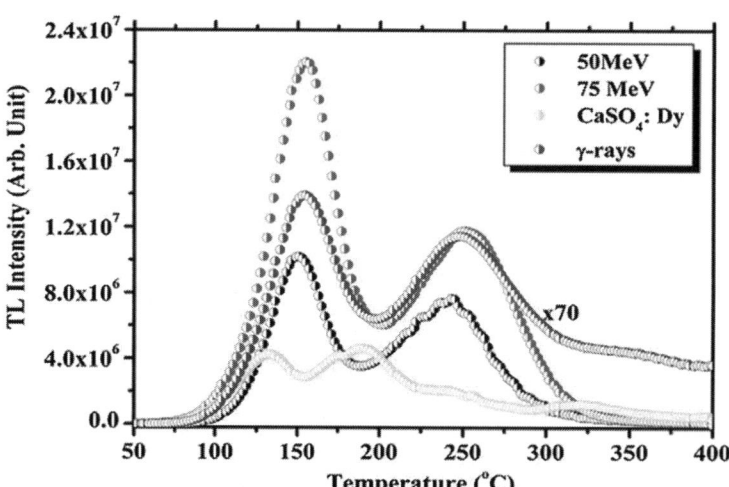

FIG. 9.7 TL glow curves of samples irradiated with C^{5+} ion of energy 50, 75 MeV with CaSO$_4$:Dy TLD phosphor for comparison. *(Reproduced with the permission from the ref. Kore et al. (2015) © 2015 Elsevier Ltd.).*

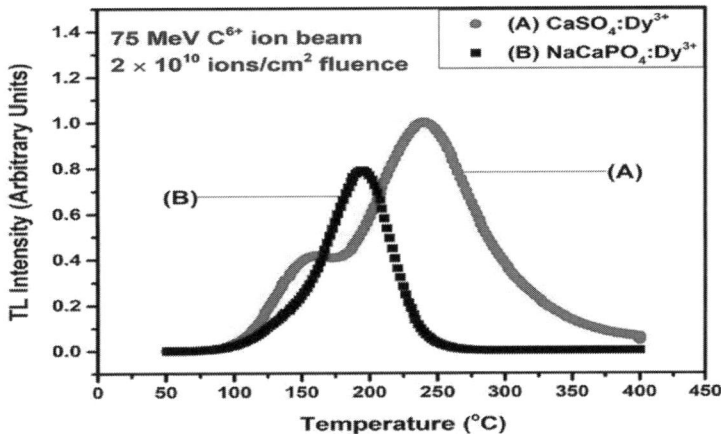

FIG. 9.8 Comparison between the TL intensities of the Carbon ion-beam irradiated NaCaPO$_4$:Dy^{3+} and the commercial CaSO4:Dy^{3+} phosphors. *(Reproduced with the permission from the ref. Nair et al. (2020) © 2020 Elsevier Ltd.).*

Sr$_2$B$_5$O$_9$Cl:Dy phosphor was synthesized using modified SSR method and reported to be a good TLD material applicable in personnel dosimetry and found to be two times less intense as compared to standard TLD phosphor CaSO$_4$:Dy when irradiated with γ-rays (Dhoble and Moharil, 2000; Oza et al., 2015c; Oza et al., 2015b). Sr$_2$B$_5$O$_9$Cl: Dy phosphor shows intense TL signal for C^{5+} ion-beam irradiation, for fluence varying from 1.5×10^{11} to 30×10^{11} ion/cm^2 (Oza et al., 2018). A noticeable change was reported in peak positions and peak intensity in TL glow curve as compare to TL glow curve for γ-ray irradiation. It was further reported that the phosphor is incapable of maintaining linear response over the entire dose range (i.e., 1.5×10^{11}–30×10^{11} ion/cm^2) for C^{5+} ion-beam irradiation and more than 50% TL signal was lost in 15 days. So, the material cannot be used in the present form for carbon beam radiotherapy. Material needs to be explored using different fluences of beam or different dopants before making it suitable for application in carbon ion beam therapy.

Phosphate materials contribute special applications in radiation dosimetry, due to their chemical and thermal stability and easiness of synthesis methods. TL properties of NaCaPO$_4$:Dy^{3+} were studied using γ-rays (Shinde and Dhoble, 2012). NaCaPO$_4$:Dy^{3+} and NaCaPO$_4$:Ce^{3+} also shows excellent properties to be exploited for optically stimulated luminescence (OSL) dosimetry (Palan and Omanwar, 2017; Xu et al., 2020). Dosimetric aspects of 75 Mev C^{6+} ion-beam exposed NaCaPO$_4$:Dy^{3+} phosphor were studied and compared with the behavior of material after γ-irradiations from ^{60}Co and ^{137}Cs sources (Nair et al., 2020). Fig. 9.8 represents the comparison between the TL intensities of the C^{6+}ion-beam irradiated NaCaPO$_4$:Dy^{3+} and commercial available CaSO$_4$:Dy^{3+} TLD phosphors.

It was found that the NaCaPO$_4$:Dy^{3+} TLD phosphor constituted 0.8 times the TL sensitivity than commercial CaSO$_4$:Dy^{3+} phosphor whereas for 150 Gy γ-ray dose from the ^{60}Co, the TL sensitivity was 0.9 times and for 4 mGy dose of the gamma-rays from ^{137}Cs source, is 11 times more TL sensitive than the commercial CaSO$_4$:Dy^{3+} phosphor. Further The TL glow curve for C^{6+} ion-beam irradiated NaCaPO$_4$:Dy^{3+} is shown in Fig. 9.9.

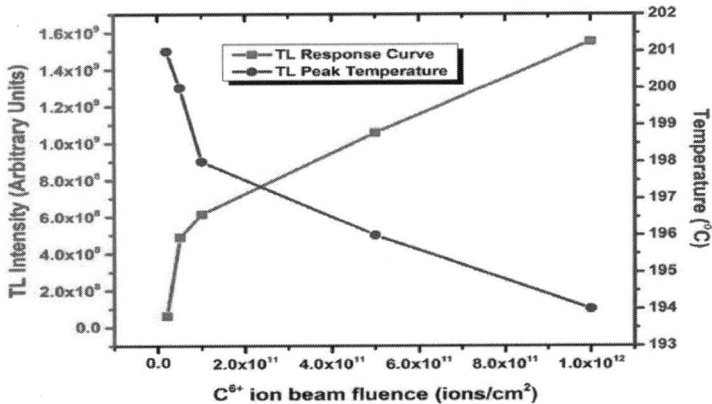

FIG. 9.9 Variation of the TL glow curve and TL glow peak temperature for NaCa$_{1-x}$PO$_4$:xDy^{3+} ($x = 0.01$) phosphor, irradiated with different fluences of the 75 MeV C^{6+} ion-beam. *(Reproduced with the permission from the ref. Nair et al. (2020) © 2020 Elsevier Ltd.).*

Fig. 9.9 shows a fast increase during the lower ion-beam fluence and after 1×10^{11} ions/cm^2, the linearity is observed. Moreover, phosphor does not get saturated in its TL response even when the beam fluence is 1×10^{12} ions/cm^2. The fading is also low for C^{6+} ion-beam irradiated NaCaPO$_4$:Dy^{3+} samples. So, NaCaPO$_4$:Dy^{3+} phosphor can be a potential candidate and a good dosimeter during C^{6+} ion beam therapy of cancer-treatment.

9.5 Conclusion

Carbon ion beam therapy plays a promising role in the treatment of a variety of malignancies and almost all types of tumors, taking into consideration the unique radiobiological and physical properties of carbon ion beam in radiotherapy. Research community is in search of the best suitable TL dosimeter that can work for the wide linear range. All over the globe, there are only few centers that are using carbon ion radiotherapy, but these centers are publishing promising data for the patients treated and recovered. Now a days, carbon ion radiotherapy is being used for almost every type of malignant tissue, including intracranial, neck and head, tumors in lungs, tumors present in the gastrointestinal tract, prostate and genitourinary cancers, breast cancer, gynecologic malignancies, and pediatric cancers. Further, extensive efforts are needed to define the role of carbon ion beam therapy for its clinical practice. Till now there is no suitable TL dosimeter available for carbon beam radiotherapy. So, it is seen that there is a lot of scope of research in the field of carbon ion beam radiotherapy in the present and near future.

References

Ahnström, G., Edvardsson, K.A., 1974. Radiation-induced single-strand breaks in DNA determined by rate of alkaline strand separation and hydroxylapatite chromotography: an alternative to velocity sedimentation. Int. J. Rad. Bio. 26, 493–497.

Ahnström, G., Nygren, J., Eriksson, S., 2000. The effect of dimethyl sulphoxide on the induction and repair of double-strand breaks in human cells after irradiation with gamma-rays and accelerated ions: rapid or slow repair may depend on accessibility of breaks in chromatin of different compactness. Int. J. Rad. Bio. 76, 533–538.

Amaldi, U., Kraft, G., 2005. Recent applications of synchrotrons in cancer therapy with carbon ion. Europhys. News 36, 114–118.

Bahl, S., A., P., S.P., L., Aleynikov, V.E., A.G., M., Kumar, P., 2013. Synthesis and thermoluminescence characteristics of gamma and proton irradiated nanocrystalline MgB$_4$O$_7$:Dy,Na. J. Lumin. 134, 691–698.

Bayramian, A.J, Marshall, C.D, Wu, J.H, Speth, J.A, Payne, S.A, Quarles, G.J, 1996. Ce:LiSrAlF6 laser performance with antisolarant pump beam. J. Lumin. 69, 85–94.

Bedyal, A.K., Kumar, V., Ntwaeaborwa, O.M., Swart, H.C., 2016. Effect of swift heavy ion irradiation on structural, optical and luminescence properties of SrAl$_2$O$_4$:Eu^{2+}, Dy^{3+} nanophosphor. Radiat. Phys. Chem. 122, 48–54.

Bhadane, M.S., Mandlik, N., Patil, B.J., Dahiwale, S.S., Sature, K.R., Bhoraskar, V.N., Dhole, S.D., 2016. CaSO$_4$:Dy microphosphor for thermal neutron dosimetry. J. Lumin. 170, 226–230.

Chopra, V., Singh, L., Lochab, S.P., 2013. Thermoluminescence characteristics of gamma irradiated Li$_2$B$_4$O$_7$:Cu nanophosphor. NIM A 717, 63–68.

Chopra, V., Singh, L., Lochab, S.P., Aleynikov, V.E., Oinam, A.S., 2014. TL dosimetry of nanocrystalline Li$_2$B$_4$O$_7$: Cu exposed to 150 MeV proton, 4 MeV and 9 MeV electron beam. Radiat. Phys. Chemis. 102, 5–10.

Dalal, M., Taxak, V.B., Chahar, S., Khatkar, A., Khatkar, S.P., 2016. A promising novel orange-red emitting SrZnV$_2$O$_7$:Sm^{3+} nanophosphor for phosphor-converted white LEDs with near-ultraviolet excitation. J. Phy. Chem. Solids 89, 45–52.

Devicienti, S., Strigari, L., D'Andrea, M., Benassi, M., Dimiccoli, V., Portaluri, M., 2010. Patient positioning in the proton radiotherapy era. J. Exp. Clin. Cancer Res. 13, 29–47.

Dhoble, S.J., Moharil, S.V., 2000. Preparation and characterisation of Eu2+ activated Sr2B5O9Cl TLD phosphor. Nucl. Instrum. Method. Phy. Res. B 160, 274–279.

Dubinskii, M.A., Semashko, V.V., Naumov, A.K., Abdulsabirov, R.Y., Korableva, S.L., 1993. Ce^{3+} doped colquiriite: a new concept of all-solid-state tunable ultraviolet laser. J. Mod. Optic. 40, 1–5.

Dutta, S., Sharma, S.K., Lochab, S.P., 2016. CaMoO$_4$:Dy phosphor as effective detector for swift heavy ions - depth profile and traps characterization. J. Lumin. 170, 42–49.

Efstathopoulos, E.P., Makrygiannis, S.S., Kottou, S., Karvouni, E., Giazitzoglou, E., Korovesis, S., Tzanalaridou, E., Raptou, P.D., Katritsis, D.G., 2003. Medical personnel and patient dosimetry during coronary angiography and intervention. Phys. Med. Bio 48, 3059–3068.

Eickhoff, H., Haberer, T., Kra, G., Krause, U., Richter, M., Steiner, R., Debus, J., 1999. The GSI cancer therapy project. Strahlenther. Onkol. 175, 21–26.

Emen, F.M., Altinkaya, R., Kafadar, V.E., Avsar, G., Yes¸ilkaynak, T., Kulcu, N., 2016. Luminescence and thermoluminescence properties of a red emitting phosphor, Sr$_4$Al$_{14}$O$_{25}$:Eu^{3+}. J. Alloys Compd. 681, 260–267.

Fox, P.J., Akber, R.A., Prescott, J.R., 1988. Spectral characteristics of six phosphors used in thermoluminescence dosimetry. J. Phys. D. Appl. Phys. 21, 189–193.

Gai, M.Q., Chen, Z.Y., Fan, Y.W., Yan, S.Y., Xie, Y.X., Wang, J.H., Zhang, Y.G., 2015. Synthesis of LiMgPO4:Eu,Sm, B phosphors and investigation of their optically stimulated luminescence properties. Radiat. Meas. 78, 48–52.

Gupta, K.K., Kadam, R.M., Dhoble, N.S., Lochab, S.P., Singh, V., Dhoble, S.J., 2016. Photoluminescence, thermoluminescence and evaluation of some parameters of Dy^{3+} activated Sr$_5$(PO$_4$)$_3$F phosphor synthesized by sol-gel method. J. Alloys Compd. 688, 982–993.

Gupta, K.K., Kadam, R.M., Dhoble, N.S., Lochab, S.P., Dhoble, S.J., 2017. On the study of the C^{6+} ion beam and c-ray induced effect on structural and luminescence properties of Eu doped LiNaSO4: explanation of TSL mechanism using PL, TL and EPR study. Phys.chem.chem.phys. 20, 1540–1559. http://doi.org/10.1039/c7cp05835g.

Hasan, F., Kitis, G., Charalambous, S., 1985. The thermoluminescence behavior of CaSO4:Dy (TLD-900) for doses up to 30 Mrad. Nucl. Instr. Meth. Phy. Res. Sec. B9, 218–222.

Heilmann, J., Taucher-Scholz, G., Haberer, T., Scholz, M., Kraft, G., 1996. Measurement of intracellular dna double-strand break induction and rejoining along the track of carbon and neon particle beams in water. Int. J. Radiat. Oncol. Biol. Phys. 34, 599–608.

Kanagasekaran, T., Mythili, P., Srinivasan, P., Vijayan, N., Kanjilal, D., Gopalakrishnan, R., et al., 2008. On the observation of physical, chemical, optical and thermal changes induced by 50 MeV silicon ion in benzimidazole single crystals. Mater. Res. Bull. 43, 852–863.

Karger, C.P., Jäkel, O., Palmans, H., Kanai, T., 2010. Dosimetry for ion beam radiotherapy. Phys. Med. Biol. 55, R193–R234.

Kassem, M.E., El-Kolaly, M.A., Ismail, K.Z., 1993. Thermal and thermoluminescence properties of γ-irradiated LiNaSO4 crystals doped with Tl. Mater. Lett. 16, 102–107.

Kazakis, N.A., Tsetine, A.T., Kitis, G., Tsirliganis, N.C., 2016. Insect wings as retrospective/accidental/forensic dosimeters: an optically stimulated luminescence investigation. Radiat. Meas. 89, 74–81.

Knoll, G.F., 1999. Radiation Detection and Measurement, 3rd ed. John Wiley & Sons Inc., New York.

Kore, B.P., Dhoble, N.S., Lochab, S.P., Dhoble, S.J., 2014. A new highly sensitive phosphor for carbon ion dosimetry. RSC Adv. 4, 49979–49986.

Kore, B.P., Dhoble, N.S., Kadam, R.M., Lochab, S.P., Dhoble, S.J., 2015. A comparative investigation of g-ray and C^{5+} ion beam impact on thermoluminescence response of Mg2BO3F: Dy phosphor. Mater. Chem. Phys. 161, 96–106.

Kore, B.P., Dhoble, N.S., Kadam, R.M., Lochab, S.P., Singh, M.N., Dhoble, S.J., Swart, H.C., 2016. Thermoluminescence and EPR study of K2CaMg(SO4)3:Dy phosphor: the dosimetric application point of view. J. Phys. D. Appl. Phys. 49, 095102.

Kraft, G., 2000. Tumor therapy with ion beams. Nucl. Instr. Methods Phys. Res. A 454, 1–10.

Kron, T., 1999. Applications of thermoluminescence dosimetry in medicine. Radiat. Prot. Dosim. 85, 333–340.

Lawrence, T.S., Feng, M., 2013. Protons for prostate cancer: the dream versus the reality. J. Natl. Cancer Inst. 105, 7–8.

Leo, W.R., 1994. Techniques for Nuclear and Particle Physics Experiments-A How to Approach, 2nd revised edition. Springer-Verlag, Berlin.

Lochab, S.P., Pandey, A., Sahare, P.D., Chauhan, R.S., Salah, N., Ranjan, R., 2007. Nanocrystalline MgB4O7: Dy for high dose measurement of gamma radiation. Phys. Status Solidi(a) 204, 2416–2425.

Marshall, C.D., Payne, S.A., Speth, J.A., W.FKrupke, G.JQ., Castillo, V., 1994. Ultraviolet laser emission properties of Ce^{3+} doped LiSrAlF6 and LiCaAlF6. J. Opt. Soc. Am. B 11, 2054–2065.

Mesterhzy, D., Osvay, M., Kovcs, A., Kelemen, A., 2012. Accidental and retrospective dosimetry using TL method. Radiat. Phys. Chem. 81, 1525–1527.

Naik, R., Prashantha, S.C., Nagabhushana, H., Sharma, S.C., Nagaswarupa, H.P., Girish, K.M., 2016. Effect of fuel on auto ignition route, photoluminescence and photometric studies of tunable red emitting Mg2SiO4:Cr^{3+} nanophosphors for solid state lighting applications. J. Alloys Compd. 682, 815–824.

Nair, G.B., Tamboli, S., Dhoble, S.J., Swart, H.C., 2020. Comparison of the thermoluminescence properties of NaCaPO4:Dy^{3+} phosphors irradiated by 75 MeV C^{6+} ion and γ-rays. J. Lumin. 224, 117274–117279.

Newhauser, W.D., Zhang, R., 2015. The physics of proton therapy. Phys. Med. Biol. 60, R155–R209.

Nikl, M., Bruza, P., Panek, D., Vrbova, M., Mihokova, E., Mares, J.A., 2013. Scintillation characteristics of LiCaAlF6 based single crystals under X-ray excitation. Appl. Phys. Lett. 102, 161907–161911.

Noh, A.M., Amin, Y.M., Mahat, R.H., Bradley, D.A., 2001. Investigation of some commercial TLD chips/discs as UV dosimeters. Radiat. Phys. Chem. 61, 497–499.

Nyenge, R.L., Swart, H.C., Poelman, D., Smet, P.F., Martin, L.I.D.J., Noto, L.L., Som, S., Ntwaeaborwa, O.M., 2016. Thermal quenching, cathodoluminescence and thermoluminescence study of Eu^{2+} doped CaS powder. J. Alloys Compd. 657, 787–793.

Ohno, T., 2013. Particle radiotherapy with carbon ion beams. EPMA J 4, 4–9.

Ohno, T., Kanai, T., Yamada, S., Yusa, K., Tashiro, M., Shimada, H., Torikai, K., Yoshida, Y., Kitada, Y., Katoh, H., Ishii, T., Nakano, T., 2011. Carbon ion radiotherapy at the Gunma University Heavy Ion Medical Center: new facility set-up. Cancers (Basel) 26, 4046–4060.

Oliveira, L.C., Yukihara, E.G., Baffa, O., 2016. MgO: Li, Ce, Sm as a high-sensitivity material for optically stimulated luminescence dosimetry. Sci. Rep. 6, 24348.

Oza, A.H., Dhoble, N.S., Lochab, S.P., Dhoble, S.J., 2015a. Luminescence study of Dy or Ce activated LiCaBO3 phosphor for γ-ray and C^{5+} ion beam Irradiation. Luminescence 30, 967–977.

Oza, A.H., Dhoble, N.S., Dhoble, S.J., 2015b. Influence of P ion on Sr2B5O9Cl:Eu for TL dosimetry. Nucl. Inst. Methods Phys. Res. B 344, 96–103.

Oza, A.H., Dhoble, N.S., Park, K., Dhoble, S.J., 2015c. Synthesis and thermoluminescence characterizations of Sr2B5O9Cl:Dy^{3+} phosphor for TL dosimetry Lumin. J. Bio. Chem. Lumin. 30, 768–774.

Oza, A.H., Chopra, V., Dhoble, N.S., Dhoble, S.J., 2018. Impact of C^{5+} ion beam on Dy activated Sr2B5O9Cl TL phosphor. J. Mater. Sci: Mater. Electron. 29, 7621–7628.

Palan, C.B., Omanwar, S.K., 2017. Synthesis and preliminary OSL studies of NaCaPO4:Ce phosphor for radiation dosimetry. Res. Chem. Intermed. 43, 4043–4050.

Pandey, A., Bahl, S., Sharma, K., Ranjan, R., Kumar, P., Lochab, S.P., Aleynikov, V.E., Molokanov, A.G., 2011. Thermoluminescence properties of nanocrystalline K2Ca2(SO4)3:Eu irradiated with gamma rays and proton beam. Nucl. Instrum. Meth. Phys. Res. B 269, 216–222.

Podgorsak, E.B., 2006. Radiation Physics for Medical Physicists. Springer-Verlag, Berlin, Heidelberg, pp. 10–14.

Rivera, T., 2012. Thermoluminescence in medical dosimetry. Appl. Radiat. Isotopes 71, 30–34.

Lochab, S.P., Sahare, P.D., Chauhan, R.S., Ranjan, R., Pandey, A., 2007. Thermoluminescence and photoluminescence study of nanocrystalline Ba$_{0.97}$Ca$_{0.03}$SO$_4$: Eu. J. Phys. D: App. Phys. 40, 1343–1350.

Sahare, P.D., Moharil, S.V., 1990. Thermoluminescence in LiNaSO$_4$ Radiat. Eff. Defects Solids 114, 167–172.

Sahare, P.D., Ranjan, R., Salah, N., Lochab, S.P., 2007. K$_3$Na(SO$_4$)$_2$:Eu nanoparticles for high dose of ionizing radiation. J. Phys. D. Appl. Phys. 40, 764–795.

Salah, N., 2008. Carbon ions irradiation on nano- and microcrystalline CaSO$_4$: Dy J. Phys. D: App. Phys. 41, 155302–155306.

Salah, N., Alharbi, N.D., Sami, S., Habiba, S.P., 2015a. Lochab Thermoluminescence properties of Al$_2$O$_3$:Tb nanoparticles irradiated by gamma rays and 85 MeV C^{6+} ion beam. J. Lumin. 167, 59–64.

Salah, N., Alharbi, N.D., Habib, SS., Lochab, S.P., 2015b. TL response of Eu activated LiF nanocubes irradiated by 85 MeV carbon ions. Nuc. Instrum. Meth. Phys. Res. B 358, 201–205.

Salah, N., Habib, S., Babkair, SS., Lochab, S.P., Chopra, V., 2013. TL response of nanocrystalline MgB$_4$O$_7$:Dy irradiated by 3 MeV proton beam, 50 MeV Li^{3+} and 120 MeV Ag^{9+} ion beams. Radiat. Phys. Chem. 86, 52–58.

Salah, N., Sahare, P.D., Nawaz, S., Lochab, S.P., 2004. Luminescence characteristics of K$_2$Ca$_2$(SO$_4$)$_3$:Eu,Tb micro- and nanocrystalline phosphor. Radiat. Eff. Def. solids 159, 321–334.

Salah, N., Sahare, P.D., Rupasov, A.A., 2007. Thermoluminescence of nanocrystalline LiF: Mg, Cu, P.. J. Lumin. 124, 357–364.

Santiago, M., Lavat, A., Caselli, E., Lester, M., Perisinotti, L.J., de Figuereido, A.K., Spano, F., Ortega, F., 1998. Thermoluminescence of strontium tetraborate. Phys. Stat. Sol. A 167, 233–236.

Scholz, M., 2000. Heavy ion tumour therapy. Nucl. Instrum. Methods B 163, 76–82.

Schulz-Ertner, D., Jäkel, O., Schlegel, W., 2006. Radiation therapy with charged particles. Semin. Radiat. Oncol. 16, 249–259.

Schulz-Ertner, D., Nikoghosyan, A., Thilmann, C., Haberer, T., Jäkel, O., Karger, C., 2004. Results of carbon ion radiotherapy in 152 patients. Int. J. Radiat. Oncol. Biol. Phys. 58, 631–640.

Scifoni, E., Surdutovich, E., Solovyov, A.V., 2010. Radial dose distribution from carbon ion incident on liquid water. Eur. Phys. J. D 60, 115–119.

Sharma, K., Bahl, S., Singh, B., Kumar, P., Lochab, S.P., Pandey, A., 2018. BaSO$_4$: Eu as an energy independent thermoluminescent radiation dosimeter for gamma rays and C6+ ion beam. 145, 64–73.

Shinde, K.N., Dhoble, S.J., 2012. Thermoluminescence and photoluminescence in the NaCaPO$_4$:Dy^{3+} phosphor. Radiat. Protect. Dosim. 152, 463–467.

Shinde, S.S., Dhabekar, B.S., Gundu Rao, T.K., Bhatt, B.C., 2001. Preparation, thermoluminescent and electron spin resonance characteristics of LiF:Mg,Cu,P phosphor. J Phys. D 34, 2683–2689.

Shivaramu, N.J., Lakshminarasappa, B.N., Nagabhushana, K.R., Singh, F., 2015. Thermoluminescence of sole gel derived Y$_2$O$_3$:Nd^{3+} nanophosphor exposed to 100 MeV Si^{8+} ions and gamma rays. J. Alloys Compd. 637, 564–573.

Singh, L., Chopra, V., Lochab, S.P., 2011. Synthesis and characterization of thermoluminescent Li$_2$B$_4$O$_7$ nanophosphor. J. Lumin. 131, 1177–1183.

Som, S., Das, S., Dutta, S., Pandey, M.K., Dubey, R.K., Visser, H.G., Sharma, S.K., Lochab, S.P., 2016. A comparative study on the influence of 150 MeV Ni^{7+}, 120 MeV Ag^{9+} and 110 MeV Au^{8+} swift heavy ions on the structural and thermoluminescence properties of Y$_2$O$_3$: Eu^{3+}/Tb^{3+} nanophosphor for dosimetric applications. J. Mater. Sci. 51, 1278–1291.

Som, S., Dutta, S., Chowdhury, M., Kumar, V., Kumar, V., Swart, H.C., Sharma, S.K., 2014. A comparative investigation on ion impact parameters and TL response of Y$_2$O$_3$:Tb^{3+} nanophosphor exposed to swift heavy ions for space dosimetry. J. Alloys Compd. 589, 5–18.

Suit, H., De Laney, T., Goldberg, S., Paganetti, H., Clasie, B., Gerweck, L., 2010. Proton vs carbon ion beams in the definitive radiation treatment of cancer patients. Radiother. Oncol. 95, 3–22.

Tamboli, S., Kadam, R.M., Rajeswari, B., Singh, B., Dhoble, S.J., 2018. Correlated PL, TL and EPR study in γ-rays and C^{6+} ion beam irradiated CaMg$_2$(SO$_4$)$_3$:Dy^{3+} triple sulphate phosphor. J. Lumin. 203, 267–276.

Tsujii, H., Mizoe, J., Kamada, T., Baba, M., Tsuji, H., Kato, H., Kato, S., Yamada, S., Yasuda, S., Ohno, T., Yanagi, T., Imai, R., Kagei, K., Kato, H., Hara, R., Hasegawa, A., Nakajima, M., Sugane, N., Tamaki, N., Takagi, R., Kandatsu, S., Yoshikawa, K., Kishimoto, R., Miyamoto, T., 2007. Clinical results of carbon ion radiotherapy at NIRS. J. Radiat. Res. 48, 1–13.

Veronese, I., Galli, A., Cantone, M.C., Martini, M., Vernizzi, F., Guzzi, G., 2010. Study of TSL and OSL properties of dental ceramics for accidental dosimetry applications. Radiat. Meas. 45, 35–41.

Vidya, Y.S., Lakshminarasappa, B.N., 2015. Influence of Li+ and Dy3+ on structural and thermoluminescence studies of sodium sulfate. Appl. Phys. A: Mater. Sci. Process. 118, 249–260.

Wani, J.A., Dhoble b, N.S., Lochab c, S.P., Dhoble, S.J., 2015. Luminescence characteristics of C^{5+} ions and ^{60}Co irradiated Li$_2$BaP$_2$O$_7$:Dy^{3+} phosphor. Nucl. Instrum. Methods Phys. Res. B 349, 56–63.

Wesch, W., Kamarou, A., Wendler, E., 2004. Effect of high electronic energy deposition in semiconductors. Nucl. Instrum. Methods. B. 225, 111–128.

Xu, J., Chen, Z., Gai, M., Fan, Y., He, C., 2020. Optically stimulated luminescence of Dy^{3+}-doped NaCaPO$_4$ glass-ceramics. J. Rare Earths 38, 927–932.

Yerpude, M.M., Chopra, V., Dhoble, N.S., Kadam, R.M., Aleksander, R.K, Dhoble, S.J., 2019. Luminescence study of LiMgBO$_3$: Dy for gamma ray and carbon ion beam exposure. Rad. Phys. Chem. 34, 933–944.

Yerpude, M.M., Dhoble, N.S., Lochab, S.P., Dhoble, S.J., 2016. Comparison of thermoluminescence characteristics in gamma-ray and C^{5+} ion beam-irradiated LiCaAlF$_6$:Ce phosphor. Luminescence 31, 1115–1124.

Yokota, Y., Fujimoto, Y., Yanagida, T., Takahashi, H., Yonetani, M., Hayashi, K., 2011. Crystal growth of Na co-doped Ce:LiCaAlF$_6$ single crystals and their optical, scintillation, and physical properties. Cryst. Growth Des. 11, 4775–4779.

Section 2

Nanotherapeutics

10. Nanomaterials physics: A critical review — 207
11. Nanotherapeutic systems for drug delivery to brain tumors — 217
12. Progress in nanotechnology based targeted cancer treatment — 239
13. Nanotherapeutics for colon cancer — 251
14. Nanoparticles for the targeted drug delivery in lung cancer — 269
15. Role of nanocarriers for the effective delivery of anti-HIV drugs — 291
16. Drug delivery systems for rheumatoid arthritis treatment — 311
17. Peptide functionalized nanomaterials as microbial sensors — 327
18. Theranostic nanoagents : Future of personalized nanomedicine — 349
19. Improving the functionality of a nanomaterial by biological probes — 379
20. Nanostructures for the efficient oral delivery of chemotherapeutic agents — 419
21. Photo-triggered theranostics nanomaterials: Development and challenges in cancer treatment — 431
22. Nanocrystals in the drug delivery system — 443

Chapter 10

Nanomaterials physics: A critical review

Khushwant S. Yadav[a], Sheeba Jacob[b], Anil M. Pethe[c]

[a]Shobhaben Pratapbhai Patel School of Pharmacy & Technology Management, SVKM's NMIMS, Mumbai, India
[b]Virginia Commonwealth University, Richmond, VA, USA
[c]Datta Meghe College of Pharmacy, Datta Meghe Institute of Medical sciences, (Deemed-to-be) University, Sawangi (Meghe), Wardha, India

10.1 Introduction

The interdisciplinary field of nanophysics has connected physics to nanotechnology in a way that has impacted the diverse aspects of life sciences. The evolution of technology at nanoscale has led to developments in the field of nanomaterial physics that it has become an interface to all these sciences. Studying the concepts of nanomaterials electronic, structural, optical or thermal characteristics behavior becomes not only crucial but also instrumental in connecting them for newer developments (Wu, 2019; Rahman et al., 2020). Nanostructured materials have established wider applications in diverse fields due to their attractive physical properties like ultra-small sizes and surface area along with high specific interface areas (Wang et al., 2020). The physics of ultra-small angle neutron scattering (USANS) techniques to think on the metamorphism aspects of evolution of substances (Anovitz et al., 2009).

Materials to show applications in diverse fields of technology have to be understood for every possible characterization. Nanotechnology applied to materials makes them novel when appropriately designed and engineered at the nanometers scale (Ayres et al., 2010). This crossing of junction of is crucial element in the physics aspects of materials. In this context, materials at the nanometer scale play a fundamental role in connecting the basic research on them to their applied application which can lead to formation of a marketable product. For engineered materials the new properties by the virtue of dimensions replace other materials by newer functions empowering new products.

The fact that the dimensions of nanomaterials are analogous to those of natural biological structures such as proteins and DNA allows for the direct integration of nanomaterials into biological systems. With these biological applications of the nanoscale, it is important to ponder on the fact how do the properties of materials change at nanoscale (Zhang et al., 2012; Nguyen et al., 2018). This brings in the role of nanophysics in biology and eventually in medicine too.

The physics changes in a material at the nanoscale level makes impactful role that material may have. Some of these diverse properties that come into the existence at the nanoscale include size, surface area, electric conductance, magnetic behavior, chemical reactivity and fluorescence to name some (Yao et al., 2014). The newer insights bring in the biosensing, and bioimaging possible due to nanophysics playing an instrumental role. One of the best examples is the unique properties gold material exhibit at the nanoscale (Jain et al., 2008).

Gold at the nanoscale forms clusters due to presence of oxide surfaces and presents itself as having unique catalytic activity responsible for the rearrangement of electronic structure of the gold cluster (Häkkinen et al., 2003).

The simple and precise explanation to this is, an increase in surface area or a bigger surface-to-volume ratio would make the material to perform in a different way (faster action). This fact is because of the number of particles increasing when they become smaller making way for greater percentage of atoms on their surface. Connecting the nanomaterials to the exhibition of properties different from the having small particle sizes exhibit enhanced optical emission as well as nonlinear optical properties due to the quantum confinement effect. The magnetic properties of materials change due to phenomenon of agglomeration at the nano scale (Issa et al., 2013).

Over the last three decades nanomaterial physics has taken a varied leap from just developing newer components to use of modified materials which consider the physics aspects and nanomaterials. The nanomaterials applications reach to medicine due to nanobiointerface. In this context the present chapter discusses the fundamental concepts of nanomaterial physics highlighting the role of physics in formation of the nanoparticles. The chapter also puts forward the rationale of nanoparticle physics with diverse functions in allied areas. Clinical applications are the last thing the chapter ends with to catch up with the challenges associated with them.

10.2 Fundamental concepts of nanomaterial physics

The fundamentals of nanomaterials with respect to physics methodically explain and elaborate the basic principles associated with structures and formation of nanomaterials. The mechanisms associated with physics aspect of these nanomaterials are important to both comprehend and understand the relevant phenomena involved. When there is a combination of nanotechnology with these nanomaterials there are many elements involved and a physical base can best describe these different aspects.

To have a comprehension on the properties of materials, fundamental concepts of nanomaterial physics need to be explained well (Xia et al., 2013). Nanomaterial characterization is complicated as nanomaterials built up with unique groundwork methods. Moreover, multidisciplinary aspect of nanomaterials is involved. There is involvement of wide range of tools needing different aspects of data analysis.

Dominant characteristics of material properties:

Here the basic concepts of nanomaterials are explained.

There are four noteworthy dominant characteristics of material properties which affect their inherent properties.

i. First is the *composition* of the material,
ii. Second are the *Phases* along with their distribution profile.
iii. Third is defect structure and
iv. Residual stress is the last one.

These four characteristics that control the properties of material are illustrated in the Fig. 10.1. The defect structure may be either present in the phases or between the phases dislocations often for the defect structure can strictly weaken a material (especially a crystal type of material) (Hirth and Pond, 1996; Sakai et al., 1997). Whereas, the residual stress may have to travel through the length scales impacting the surface compressive behavior of the material (example the glass material toughens by stress) (Wang et al., 2001).

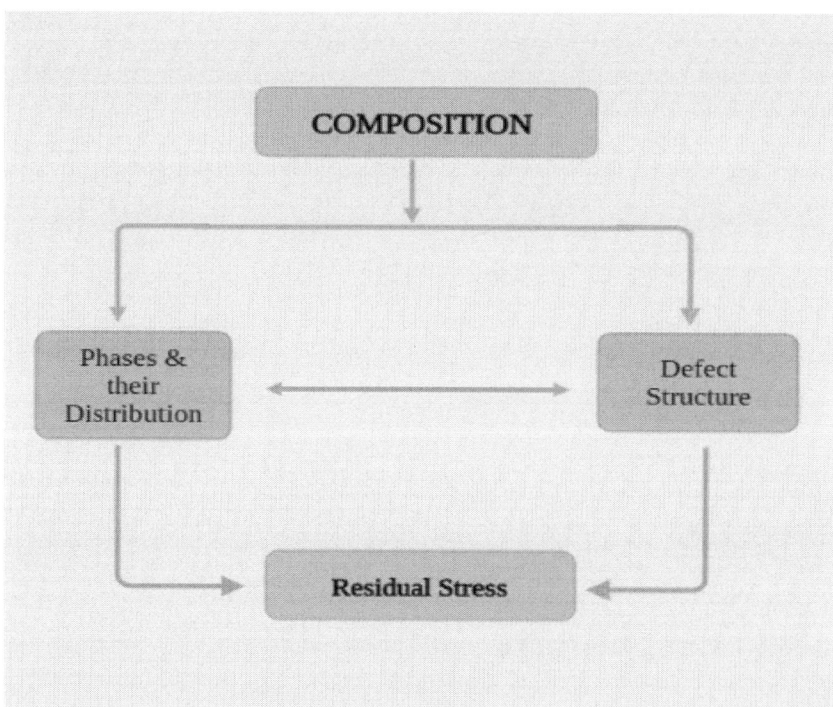

FIG. 10.1 Dominant characteristics that controls the properties of materials.

10.2.1 Structure sensitive and structure insensitive properties

To begin with understanding of sensitive and insensitive properties of structure the relation between density and vacancy can be correlated. It is imperative to link the density associated with a material to the amounts of vacancies present in that material. If the vacancies concentration is small the impact of associated density of that material would not be sensitive to these vacancies (Zhang et al., 2012). For a density function that is dislocated, the elastic modulus may not be sensitive. However, yield stress which is a structure sensitive property will be more dependent on the aspects of dislocations. There will be varied response from a GPa to an MPa for an absence to presence of dislocations respectively for the yield strength, showing the impact on the magnitude. Whenever structure comes in the context, be it the nano or micro level, the structure sensitivity has a role in determining the defect structure of the material in question which eventually affects the overall property of that material.

10.2.2 Phases and their distribution

The definition of a 'phase' is dependent on the atomic entity that material is and structure is the physical property. Both ferromagnetic and ferroelectric terms can be associated with the matter's physics aspect. Material behavior is dependent on the existence of phases and to the extent their distribution profile.

Materials can be monolithic or hybrids (Fig. 10.2). When there are two or more solid components involved, they are termed as the composites. These composites, if they have any one of the components involved in the nano-scale they become nanocomposites. The sandwich structures are known to exist whenever one of the materials involved as a surface which plays the important role of defining its property. In contrast, the lattice structure has one or more sides of a core material that may be combined with space to build a final form. The segmented structures can be consisting of 1 or more materials and be classified in 1D, 2D or 3D.

The *distribution of phases* is a vital feature which governs the properties associated with the material. For a weight or volume fraction in the phase, its distribution will impact not only the basic shape but also its connectivity to other materials.

10.2.3 Defects in body nanomaterials

In nanocrystalline materials, there are majorly four types of defects: point defects, line defect, surface defects and lattice distortions. In the point defects there is presence of voids and vacancy. Whereas, in line defects there are dislocations and in

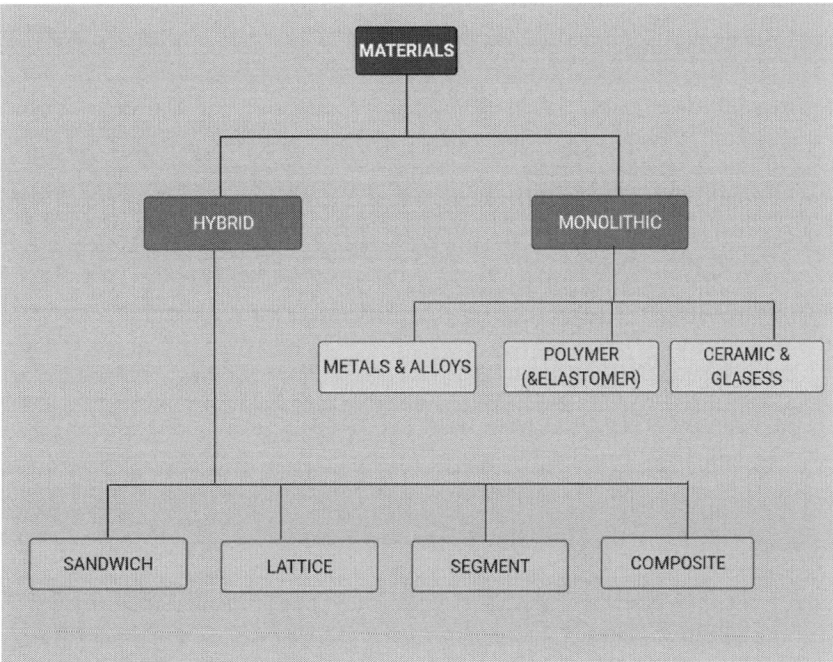

FIG. 10.2 The monolithic and hybrid classification of materials.

surface defects there is presence of grains which could comprise of stacking faults and twin grains. The lattice distortions are unlike the point, line and surface defect.

Nanocrystalline materials can have three types of point defects based on the causing times (Gleiter, 1989; Wang et al., 1995). First type is for a short time, second is medium time and the third is the permanent time. These can lead to development of free volumes, nanovoids, and holes in the nanomaterial. For materials that are compacted the developed nanovoids cannot be removed even by the process of annealing.

10.3 Properties of materials

Nanomaterial properties are affected by the impact of residual stress of surface which may be a part of interface of any matrix or inclusion (Skordaris et al., 2017). There is a size-based effect of these interface properties acting at the atomic scale of the nanoparticulate materials. Depending on the type of material the surface phenomenon can behave in elastic or rigid manner and present a bulk or exclusive surface phenomenon. The particles in the nanosized component or the nanocomposites can have different distribution of particles which can be random or discrete.

The term "stochastic" can be best described in this case to comprehend the properties of materials that are defined by a random probability distribution. In the nanostructure if the property of being stochastic, the local fields can be avoided to explain the beneficial of the ergodicity property that will govern the overall properties of the material (Mook et al., 2010; Liu et al., 2017). There can be dominance of the surface as well as the interface effects. Hence it is important to understand the basics of the factors affecting properties of materials.

10.3.1 Factors affecting properties of a material

The overall property of a nanostructure material is depended on various factors such the surface characteristics, interface energy or the surface/interface to volume ratio. These factors contribute too many of the specific applications of nanomaterials such as biomedical imaging. The optical properties are associated with divergent aspects such as the metals which display the excitation of surface plasmons or the magnetic excitation of nanoparticles when they come in magnetic fields by displaying superparamagnetic characteristics.

10.3.1.1 Thermal properties

Thermal properties of material are related to properties concern with application of heat or temperature. In this process there will be characteristic change in the solid material with a change in response of heat and heat conduction. The main types of thermal properties are: thermal conductivity, thermal expansion, thermal stress, thermal diffusivity, thermal effusivity, and heat capacity.

The first property is the thermal conductivity which explains the ability of a material to conduct by or through itself. It is defined as "the amount of heat conducted in a unit time." For example, thermal conductivity of copper is 100% but lead is only 9%. When material expands on subjecting to increased temperature or heat there is a considerable change in their shape, this phenomenon is termed as material expansion. During the expansion, the atoms of the material accumulate heat energy. Application of heat can either expand or contract the material, this gives the material a stress, known as thermal stress. The stress can be even explosive. In diffusivity there is diffusion of thermal energy which makes the energy to flux through the atoms in the materials due to vibration. In contrast, effusivity is the resistance of material to bring in the change with the change in heat in the surrounding medium. Heat capacity is the amount of heat (in joules or calories) needed to bring in change material temperature by 1 degree Celsius or Kelvin.

10.3.1.2 Mechanical properties

The aptitude and propensity of a material to be shaped into different molds or designs is understood by their mechanical properties. Some of the representative mechanical properties of materials are: strength (resistance to deformation or breakdown), hardness (resist to permanent shape), ductibility (ability to get deformed under stress), brittleness (opposite of ductility), creep (propensity to move slowly and deform permanently), and resilience (tendency to absorb the energy).

10.3.1.3 Optical properties

Optical properties are the result of exposing the material to visible light (electromagnetic radiations). The optical properties of matter include reflection, refraction, absorption, transmittance, polarization, diffraction, or fluorescence. The incident radiation is somewhat either transmitted or absorbed and sometimes partially reflected and partially absorbed.

10.3.1.4 Electrical properties

The materials which would permit electric flow of current through itself will show typical electrical properties. These materials will simply allow flow of electric current through them. These electrical properties of a material are needed to apply to the electrical applications like conductivity of electricity or thermoelectric areas. Other electrical properties are resistivity and coefficient of resistance.

10.3.1.5 Magnetic properties of nanomaterials

The magnetic field decides the way the attraction of the delivery carrier and finally the positioning. The magnetic field should be strong enough at the carrier location, parallel to this the magnet field should be strong enough at the site of action. There are two possible sources of magnetic field, one the external other internal. In external types, the magnets are placed external (outside) of the body. From this place also the external magnet is responsible to give enough opportunity to magnetize the carrier and in turn a magnetic field gradient is created which is used for targeting. The biggest disadvantage associated with this type is the distance of the external magnet. In the internal types, the magnet or the device that can be magnetized is implanted locally to the target. The internal types can be preferentially used to target blood vessels or even deeper tissues.

10.4 Rationale of nanoparticle physics with diverse functions involving nanomaterials

Some of the nanoscale materials need a mention whose structure specifically exist in the nanoscale only. This is possible because they do not have any bulk counterpart. Carbon nanotube (Li et al., 2007) and fullerenes are the best examples of these. If their sizes are increased then they would eventually lose the basic properties which are due to the virtue of their basic nano size. These activities may be termed as their super-activities such as the super-catalytic activity or the superparamagnetic activity. The nano-hairs of the lotus leaf give it the superhydrophobicity which is due to the non-buckling effect only possible at the nano-scale (Cheng and Rodak, 2005; Balani et al., 2009).

While these nanomaterials reach the nanoscale their properties change drastically like the increase in fracture strength of Ni to a high 900 MPa (otherwise is just 100 MPa) at the nanometer-size (Sun et al., 2014). Another interesting example is of some amalgamation of properties such as the role of abalone shell (Espinosa et al., 2011) showing fracture toughness of higher dimensions (>1000 times) than that of calcium carbonate. Entity such as carbon nanotubes show electrical signal in response to a stimulus at the nanoscale only (Kam et al., 2009).

Nano scale surface can enable both the functions of hydrophobicity and roughness to nano-hydrophobicity and nano-roughness that these surfaces would not permit water deposition or water wetting onto their surfaces. Eventually such surfaces will not permit growth of algae or fungi. This phenomenon is also known as super hydrophobicity.

Ceramics generally appear to be opaque because of light scattering. However, the nano scale of the grains involved in the ceramics enable these ceramics to become transparent. Aluminum oxide nano ceramics are an example of the same (Sharifitabar et al., 2011).

Increase in catalytic activity will happen due to the magnitude order becoming multi-fold due to availability of more surface area. This increased surface area also allows functionalization easy as a greater number of functional groups can be attached. The functionalization can be site-selective with desired densities of the functional groups. The functional groups may be selectively placed in internal part of (mesopores) nanoparticle or solely be kept on external surface of nanoparticles (Kecht et al., 2008).

Nanoporous membrane filters will have membranes which are designed to have pores preferably below hundred nanometers thereby suitable for size-based selective separations. Hence, this can be mostly be beneficial in water-filtrations or fluid transport across a nanoporous membrane that removes bacteria and other pollutants from water. Fluid transport through a nanoporous membrane with appropriate molecular dynamic simulations can be used for cation removal from water (Köhler et al., 2018).

As a new technology the applications of nanoporous membranes have led to biotechnological use in extraction of DNA. The nanoporous aluminum oxide membrane correctly patterning can simplify the extraction of DNA from multiple samples concurrently (Kim et al., 2006).

There are *these basic physics principles* which in combination with latest biomaterials enable biological functions that were otherwise not attainable or not likely to occur. Loading of the drug molecule with appropriate sensors is possible due to nano size which eventually can permit the targeted drug delivery by reaching the correct site for showing the effect.

Finally, the biotechnology too has examples which work only at the nanoscale. The aspect of receiving the signals by the devises which need to extract signals locally from one DNA strand works best at the nanoscale only (Benson et al., 2015; Bathe and Rothemund, 2017; Deluca et al., 2020).

10.5 Self-assembly of nanostructures

Nanostructures when spontaneously organize themselves into ordered structures on their own without any external interference. There are two machine components in this phenomenon. The first one is the 'self' characteristic feature which signifies unrestrained moment. In its essence, the second aspect is about 'spontaneous' which comprises of uninhibited and promiscuous effect resulting in formation of a new ordered structure (Grzelczak et al., 2010). The ordered structure so created can be of 2-dimension or 3-dimension. The formation here is occurring from the basic building blocks and resulting in a new form. One important feature of the said system is thermodynamic equilibrium. Distinct and separate components in this process will instinctively establish into a different component due to some particular interactions. The interactions in such procedure are of two types, direct and indirect. The interparticle forces can be responsible for the direct interactions, whereas, external fields are types of indirect method.

Nanotechnology offers intriguing potential for addressing the challenges which the spontaneous organization makes ordered structures without any human involvement. This is interesting promising possible solutions for nano sized fabrication with high-end engineering approach and is thought to be a major breakthrough in nanophysics science as there is minimal human intervention. The concept of self-assembly, like a paradigm shift, has crossed many multitudes of issues and is all set to translate the technology. Some of the major applications of self-assembly include:

i. *The role of anisotropy related to the surface stress* which is responsible for breaking the symmetry of layers and can lead to development of new forms and patterns (Lu and Suo, 2002).
ii. *Self-assembly by bottom-up strategy* accountable for forming superstructures of nano size. These structures are both flexible and transformative to permit more application pathways (Deng et al., 2020).
iii. *Self-assembly by use of thermodynamics at the interface* to be used in making of a nanodevice (Hu et al., 2012). At the oil-water interface there is a feasibility for the low-dimensional structures too right from 1-D to 2-D.
iv. *Exploration of the contact angle at the interface* of the water and oil will provide nanostructures with varied applications of magnetic, electrical and optic properties. The self-assembly of these structures is related to the formation 90° angle at the interface and the available functional groups and terminal groups (Duan et al., 2004). The principle behind these liquid-liquid interfaces is change in the interfacial energy which generally gets reduced so as to allow the process of self-assembly.

10.6 Clinical applications of nanomaterials physics

Application of nanomaterials physics at the nano-bio interface impacts the toxicology or therapeutic outcome of the material in clinical settings (Gagner et al., 2012).

Atomic and molecular spatial arrangements of the nanomaterials affect the interface physics which subsequently changes the biological applications (Mahmoudi, 2018).

Application of nanomaterials physics at the nanobiointerface impacts the toxicology or therapeutic outcome of the material in clinical settings. Some of the clinical applications include medical devices, imaging, cancer therapy, targeted drug delivery, quantum physics in therapy, nanomotors, and nanorobotics.

However, the present section takes the specific example of *cancer* where the role of nanomaterial physics in clinical application is elaborated. Difficulties to detect cancers (such as pancreatic cancer) continue to be associated with poor outcomes. Numerous diagnostic tools now exist for cancer screening. Once a cancer is identified, surgery, radiation, and chemotherapies are staples of care. New treatment modalities such as targeted therapies offer hope of better treatment outcomes. In context to this the subsequent section connects the applications the nanomaterial physics has with cancer in terms of clinical outcomes.

10.6.1 Applications of nanomaterials physics in cancer

Cancer screening is the cornerstone of cancer detection. Adoption of guidelines for breast exams, and colonoscopies, based on age and other risk factors, has led to early identification and reduction of the deadliest breast and colon cancers. For breast cancer, mammograms provide are the critical imaging technology. CT scans provide a mechanism for imaging tumors generally. For certain soft tissue cancer, or neurological tumors, MRI scans are more appropriate. Newer technologies combine traditional CT or MRI scans with radioactive tracers. These PET scans can identify tumors based on antibodies specific to the tumor or based on glucose uptake. Diagnostic protocols are specific to the cancer. Certain cancers can be identified definitively based on radiographic study, while others require biopsy for confirmation. Tumors obtained after biopsy are graded depending on the tumor stage and localization. Tumor are further tested for expression of genetic markers that help in the better treatment of the cancer. For example, depending on the origin of the tumor cell, and expression of ER and HER2 receptors, the breast tumor can be classified as basal type A/B or luminal type A/B (Park et al., 2012). The expression of these receptors helps in determining the aggressiveness and prognosis of the cancer.

Surgery, radiation, and chemotherapies are the mainstays of cancer treatment. Depending on the nature of cancer and the sensitivity to the treatment, surgery, radiation, or chemotherapy can be the primary treatment. While surgery removes the tumor or part of the organ where the tumor is localized, radiation uses high-powered energy beams, such as X-rays to kill the cancer cells. Radiation can be provided outside the body known as external beam radiation or placed inside the body localized to the tumor known as brachytherapy. Local removal via surgery or targeted radiation is always preferable, However, local invasion can complicate surgical removal, and both surgery and radiation carry risk of morbidity. Chemotherapy uses drugs to kill cancer cells, chemotherapy, radiation, and hormone therapy can also be used as adjuvant therapies after primary treatment of cancer. These therapies are given to kill any remaining cancer cell in the body to reduce the chance of reoccurrence of the cancer. Additionally, medications or palliative treatments are provided to relieve patients of the side effects caused by cancer and treatments. Bone marrow transplant, also known as stem cell transplant, is another treatment option for cancer such as blood cancer where the bone marrow inside the bone which makes the blood cells for the body are replaced by cells from a donor.

Some of the cancers such as the breast and prostate are hormone dependent cancers. These cancers require hormones such as estrogen, progesterone or androgen produced by the body for the growth of the cancer cell. Removing these hormones or blocking their effect greatly reduces the progression of the tumor. However, not all tumors of breast and prostate would be hormone responsive. Some tumors are hormone-independent where they do not require hormones for their growth and progression. Such tumors do not respond to hormone therapies and are generally aggressive and can be easily metastasized to other region of the body.

Targeted therapies are currently the focus of much anticancer drug development. The outcome of patients diagnosed and treated for certain cancers such as lung cancer and breast cancer have continued to increase with the development of targeted therapies. Targeted therapies differ from the standard chemotherapies in several ways. Targeted therapies act on specific molecular targets that are associated with cancer, whereas most standard chemotherapies act on all rapidly dividing normal and cancerous cells.

Cancer physics gives information on mechanism involved in spatial growth, proliferation, and other properties of cancer (Korgaonkar and Yadav, 2019). Nanocarriers targeting have shown progress in pharmacokinetic parameters of therapeutics by altering their delivery sites and duration of action at the desired location and can reduce associated systemic toxicities (Bates et al., 2017; Upadhya et al., 2021; Yadav et al., 2021). The nanocarriers do this type of action because of making use of molecular communication with the targets at nano scale level (Soni and Yadav, 2017; Yadav et al., 2018; Yadav and Prabhakar, 2020). Nanocarriers are also useful in overcome resistance in different types of cancer by having better understanding of principle mechanisms (Fulfager and Yadav, 2021). Over the last three decades nanotechnology has emerged in medicine and from first generation drug delivery system to the milestones that the technology has achieved to the second-generation nanosystems (Mohan and Yadav, 2021). Polymers used in engineering of these nanoparticles play an important role; especially the responsiveness makes lot of drastic changes within the body which makes the delivery better (Pethe and Yadav, 2019). Polylactic co-glycolic acid (PLGA)-based nanocarrier (Joshi et al., 2021) or polycaprolactone based have tremendous role in engineering these nanosized carriers (Kurakula et al., 2021).

Immunotherapies are the latest advancement in the treatment of cancer. Immunotherapies uses one's immune system to recognize cancer cell as foreign cell and destroy it. Many immunotherapy studies are in the phase II and phase III stages of clinical trials. These therapies are hopeful of providing successful treatment to cancer. Combining nanotechnology to immunotherapy had led to next-generation drug delivery systems in cancer therapy with a promise to reduce the side effects (Le et al., 2018). In this relationship between the two is thought to be bringing in the change with eventually allows the cargos loaded in these nanoparticles to be delivered. There is also reversion of immunosuppression (Huang et al., 2019). As a frontier in cancer therapy, immunotherapy along with nanotechnology approach is efficacious in syngeneic induction of immunogenic cell death (ICD) which is a step towards the cancer therapy (Lu et al., 2017).

10.7 Conclusion: Nanotechnology, physics, and clinical outcome

Although we have made considerable advancement in the diagnosis and treatment of cancer, we still lack technology to detect and treat early cancer cells. Many clinically approved drugs are unable to target cancer cells located at distant regions due to their large size and various barrier system of the human body. Although nanotechnology has not been deployed clinical in the diagnosis and treatment of cancer, many pharmaceuticals use nanoparticles to develop various diagnosis tests. Nanogels are nanoparticles that are composed of synthetic polymers or biopolymers and are chemically or synthetically crosslinked. The weakly crosslinked polymers help in the increased incorporation of biomacromolecules. Furthermore, the large surface area to volume ratio of the nanoparticles relative to bulk materials can be densely covered with antibodies, small molecules, peptides and other moieties that help in recognizing cancer cells with increased sensitivity and specificity. As nanoparticles are smaller in size ranging from 10 to 150 nm, it helps them to easily extravasate from bloodstream and blood-brain barrier into surrounding tumor tissues. Additionally, surface modification of nanoparticles such as PEGylation helps in the intravenous delivery of the drug, stability of the drug, and longer retention of the drug in circulation. Surface modification of the nanoparticles also help in the encapsulation of hydrophobic drugs and poorly soluble anticancer drugs. Nanoparticles can also be modified to be stimuli responsive where they have ability to swell or shrink in response to tumor microenvironment such as pH and temperature. Thus, the increased sensitivity, specificity and multiplexed measuring capacity of nanotechnology-based cancer diagnosis and treatment can help it move into the clinic soon.

Application of nanomaterials physics at the nano-bio interface impacts the toxicology or therapeutic outcome of the material in clinical settings. Some of the clinical applications include medical devices, imaging, cancer therapy, targeted drug delivery, quantum physics in therapy, nanomotors and nanorobotics.

Acknowledgments

The authors are thankful to Ms. Karishma Shetty and Ms. Soma Yasaswi, M. Pharm. students of SPPSPTM, SVKM's NMIMS, Mumbai, for drawing the figures.

References

Anovitz, L.M., et al., 2009. A new approach to quantification of metamorphism using ultra-small and small angle neutron scattering. Geochim. Cosmochim. Acta 73, 7303–7324. http://doi.org/10.1016/j.gca.2009.07.040.

Ayres, C.E., et al., 2010. Nanotechnology in the design of soft tissue scaffolds: innovations in structure and function. Wiley Interdiscip. Rev. Nanomed. Nanobiotechnol 2, 20–34. http://doi.org/10.1002/wnan.55.

Balani, K., et al., 2009. The hydrophobicity of a lotus leaf: a nanomechanical and computational approach. Nanotechnology 20, 305707. http://doi.org/10.1088/0957-4484/20/30/305707.

Bates, P.J., et al., 2017. G-quadruplex oligonucleotide AS1411 as a cancer-targeting agent: uses and mechanisms. Biochimica et Biophysica Acta - General Subjects 1861, 1414–1428. http://doi.org/10.1016/j.bbagen.2016.12.015.

Bathe, M., Rothemund, P.W.K., 2017. DNA nanotechnology: a foundation for programmable nanoscale materials. MRS Bull 42, 882–888. http://doi.org/10.1557/mrs.2017.279.

Benson, E., et al., 2015. DNA rendering of polyhedral meshes at the nanoscale. Nature 523, 441–444. http://doi.org/10.1038/nature14586.

Cheng, Y.T., Rodak, D.E., 2005. Is the lotus leaf superhydrophobic? Appl. Phys. Lett. 86, 144101. http://doi.org/10.1063/1.1895487.

Deluca, M., et al., 2020. Dynamic DNA nanotechnology: toward functional nanoscale devices. Nanoscale Horizons 5, 182–201. http://doi.org/10.1039/c9nh00529c.

Deng, K., et al., 2020. Self-assembly of anisotropic nanoparticles into functional superstructures. Chem. Soc. Rev. 49, 6002–6038. http://doi.org/10.1039/d0cs00541j.

Duan, H., et al., 2004. Directing self-assembly of nanoparticles at water/oil interfaces. Angewandte Chemie - Int. Ed. 43, 5639–5642. http://doi.org/10.1002/anie.200460920.

Espinosa, H.D., et al., 2011. Tablet-level origin of toughening in abalone shells and translation to synthetic composite materials. Nat. Commun. 2, 173. http://doi.org/10.1038/ncomms1172.

Fulfager, A.D., Yadav, K.S., 2021. Understanding the implications of co-delivering therapeutic agents in a nanocarrier to combat multidrug resistance (MDR) in breast cancer. J. Drug Delivery Sci. Technol. 62, 102405. http://doi.org/10.1016/j.jddst.2021.102405.

Gagner, J.E., et al., 2012. Engineering nanomaterials for biomedical applications requires understanding the nano-bio interface: a perspective. J. Phys. Chem. Lett. 3 (21), 3149–3158. http://doi.org/10.1021/jz301253s.

Gleiter, H., 1989. Nanocrystalline materials. Prog. Mater Sci. 33, 223–315. http://doi.org/10.1016/0079-6425(89)90001-7.

Grzelczak, M., et al., 2010. Directed self-assembly of nanoparticles. ACS Nano 4 (7), 3591–3605. http://doi.org/10.1021/nn100869j.

Häkkinen, H., et al., 2003. Structural, electronic, and impurity-doping effects in nanoscale chemistry: supported gold nanoclusters. Angewandte Chemie - Int. Ed. 42 (11), 1297–1300. http://doi.org/10.1002/anie.200390334.

Hirth, J.P., Pond, R.C., 1996. Steps, dislocations and disconnections as interface defects relating to structure and phase transformations. Acta Mater. 44 (12), 4749–4763. http://doi.org/10.1016/S1359-6454(96)00132-2.

Hu, L., et al., 2012. Oil–water interfacial self-assembly: a novel strategy for nanofilm and nanodevice fabrication. Chem. Soc. Rev. 41, 1350–1362. http://doi.org/10.1039/c1cs15189d.

Huang, P., et al., 2019. Nano-, micro-, and macroscale drug delivery systems for cancer immunotherapy. Acta Biomater. 85, 1–26. http://doi.org/10.1016/j.actbio.2018.12.028.

Issa, B., et al., 2013. Magnetic nanoparticles: surface effects and properties related to biomedicine applications. Int. J. Mol. Sci. 14 (11), 21266–21305. http://doi.org/10.3390/ijms141121266.

Jain, P.K., et al., 2008. Noble metals on the nanoscale: Optical and photothermal properties and some applications in imaging, sensing, biology, and medicine. Acc. Chem. Res. 41, 1578–1586. http://doi.org/10.1021/ar7002804.

Joshi, G., et al., 2021. Polylactic coglycolic acid (PLGA)-based green materials for drug deliveryApplications of Advanced Green Materials. Woodhead Publishing, Elsevier. http://doi.org/10.1016/b978-0-12-820484-9.00017-9.

Kam, N.W.S., Jan, E., Kotov, N.A., 2009. Electrical stimulation of neural stem cells mediated by humanized carbon nanotube composite made with extracellular matrix protein. Nano Lett. 9, 273–278. http://doi.org/10.1021/nl802859a.

Kecht, J., Schlossbauer, A., Bein, T., 2008. Selective functionalization of the outer and inner surfaces in mesoporous silica nanoparticles. Chem. Mater. 20, 7207–7214. http://doi.org/10.1021/cm801484r.

Kim, J., Voelkerding, K.V., Gale, B.K., 2006. Patterning of a nanoporous membrane for multi-sample DNA extraction. J. Micromech. Microeng. 16, 33. http://doi.org/10.1088/0960-1317/16/1/005.

Köhler, M.H., Bordin, J.R., Barbosa, M.C., 2018. 2D nanoporous membrane for cation removal from water: effects of ionic valence, membrane hydrophobicity, and pore size. J. Chem. Phys. 148, 222804. http://doi.org/10.1063/1.5013926.

Korgaonkar, N., Yadav, K.S., 2019. Understanding the biology and advent of physics of cancer with perspicacity in current treatment therapy. Life Sci. 239, 117060. http://doi.org/10.1016/j.lfs.2019.117060.

Kurakula, M., Rao, G.S.N.K., Yadav, K.S., 2021. Fabrication and characterization of polycaprolactone-based green materials for drug delivery in Applications of Advanced Green Materials. Woodhead Publishing, Elsevier. http://doi.org/10.1016/b978-0-12-820484-9.00016-7.

Le, Q.V., Choi, J., Oh, Y.K., 2018. Nano delivery systems and cancer immunotherapy. J. Pharm. Investig. 48, 527–539. http://doi.org/10.1007/s40005-018-0399-z.

Li, C., et al., 2007. A fullerene-single wall carbon nanotube complex for polymer bulk heterojunction photovoltaic cells. J. Mater. Chem. 17, 2406–2411. http://doi.org/10.1039/b618518e.

Liu, H., et al., 2017. Carbon nanostructures in biology and medicine. J. Mater. Chem. B 5, 6437–6450. http://doi.org/10.1039/c7tb00891k.

Lu, J., et al., 2017. Nano-enabled pancreas cancer immunotherapy using immunogenic cell death and reversing immunosuppression. Nat. Commun. 8, 1811. http://doi.org/10.1038/s41467-017-01651-9.

Lu, W., Suo, Z., 2002. Symmetry breaking in self-assembled monolayers on solid surfaces: anisotropic surface stress. Phys. Rev. B – Condens. Matter Mater. Phys. 65, 085401. http://doi.org/10.1103/PhysRevB.65.085401.

Mahmoudi, M., 2018. Debugging nano–bio interfaces: systematic strategies to accelerate clinical translation of nanotechnologies. Trends Biotechnol. 36, 755–769. http://doi.org/10.1016/j.tibtech.2018.02.014.

Mohan, V., Yadav, K.S., 2021. Potentiality of Q3 characterization of nanosystem: surrogate data for obtaining a biowaiver. Drug Dev. Res. 82, 27–37. http://doi.org/10.1002/ddr.21731.

Mook, W.M., et al., 2010. Compression of freestanding gold nanostructures: from stochastic yield to predictable flow. Nanotechnology 21, 055701. http://doi.org/10.1088/0957-4484/21/5/055701.

Nguyen, P.Q., et al., 2018. Engineered living materials: prospects and challenges for using biological systems to direct the assembly of smart materials. Adv. Mater. 30, 1704847. http://doi.org/10.1002/adma.201704847.

Park, S., et al., 2012. Characteristics and outcomes according to molecular subtypes of breast cancer as classified by a panel of four biomarkers using immunohistochemistry. Breast 21, 50–57. http://doi.org/10.1016/j.breast.2011.07.008.

Pethe, A.M., Yadav, K.S., 2019. Polymers, responsiveness and cancer therapy. Artif. Cells, Nanomed., Biotechnol. 47, 395–405. http://doi.org/10.1080/21691401.2018.1559176.

Rahman, A., et al., 2020. Structural, optical and photocatalytic studies of trimetallic oxides nanostructures prepared via wet chemical approach. Synth. Met. 259, 116228. http://doi.org/10.1016/j.synthmet.2019.116228.

Sakai, A., Sunakawa, H., Usui, A., 1997. Defect structure in selectively grown GaN films with low threading dislocation density. Appl. Phys. Lett. 71, 2259. http://doi.org/10.1063/1.120044.

Sharifitabar, M., et al., 2011. Fabrication of 5052Al/Al2O3 nanoceramic particle reinforced composite via friction stir processing route. Mater. Des. 32, 4164–4172. http://doi.org/10.1016/j.matdes.2011.04.048.

Skordaris, G., et al., 2017. Effect of PVD film's residual stresses on their mechanical properties, brittleness, adhesion and cutting performance of coated tools. CIRP J. Manuf. Sci. Technol. 18, 145–151. http://doi.org/10.1016/j.cirpj.2016.11.003.

Soni, G., Yadav, K.S., 2017. Communication of drug loaded nanogels with cancer cell receptors for targeted deliveryModeling and Optimization in Science and Technologies. Springer Nature, Switzerland AG, pp. 503–515. http://doi.org/10.1007/978-3-319-50688-3_21.

Sun, J.Y., et al., 2014. Improvement in ductility of high strength polycrystalline Ni-rich Ni 3Al alloy produced by EB-PVD. J. Alloys Compd. 614, 196–202. http://doi.org/10.1016/j.jallcom.2014.06.071.

Upadhya, A., Yadav, K.S., Misra, A., 2021. Targeted drug therapy in non-small cell lung cancer: clinical significance and possible solutions-Part I. Expert Opin. Drug Deliv. 18, 73–102. http://doi.org/10.1080/17425247.2021.1825377.

Wang, N., et al., 1995. Effect of grain size on mechanical properties of nanocrystalline materials. Acta Metall. Mater. 43, 519–528. http://doi.org/10.1016/0956-7151(94)00253-E.

Wang, Y.U., et al., 2001. Nanoscale phase field microelasticity theory of dislocations: model and 3D simulations. Acta Mater. 49, 1847–1857. http://doi.org/10.1016/S1359-6454(01)00075-1.

Wang, Y., et al., 2020. Effect of surface hydroxyl group of ultra-small silica on the chemical states of copper catalyst for dimethyl oxalate hydrogenation. Catal. Today 350, 127–135. http://doi.org/10.1016/j.cattod.2019.06.031.

Wu, W., 2019. Stretchable electronics: functional materials, fabrication strategies and applications. Sci. Technol. Adv. Mater. 20, 187–224. http://doi.org/10.1080/14686996.2018.1549460.

Xia, T., et al., 2013. Implementation of a multidisciplinary approach to solve complex nano EHS problems by the UC center for the environmental implications of nanotechnology. Small 9, 1428–1443. http://doi.org/10.1002/smll.201201700.

Yadav, K.S., Prabhakar, B., 2020. Nanogels in MedicineHandbook of Materials for Nanomedicine. Jenny Stanford Publishing. http://doi.org/10.1201/9781003045113-10.

Yadav, K.S., Saxena, R., Soni, G., 2018. Chapter 9: Nanogels as targeted drug delivery vehicles. RSC Smart Materials. Royal Society of Chemistry, pp. 143–160. http://doi.org/10.1039/9781788010481-00143.

Yadav, K.S., Upadhya, A., Misra, A., 2021. Targeted drug therapy in nonsmall cell lung cancer: clinical significance and possible solutions-part II (role of nanocarriers). Expert Opin. Drug Deliv. 18, 103–118. http://doi.org/10.1080/17425247.2021.1832989.

Yao, J., Yang, M., Duan, Y., 2014. Chemistry, biology, and medicine of fluorescent nanomaterials and related systems: new insights into biosensing, bioimaging, genomics, diagnostics, and therapy. Chem. Rev. 114, 6130–6178. http://doi.org/10.1021/cr200359p.

Zhang, W., et al., 2012. Role of vacancies in metal-insulator transitions of crystalline phase-change materials. Nat. Mater. 11, 952–956. http://doi.org/10.1038/nmat3456.

Zhang, X.Q., et al., 2012. Interactions of nanomaterials and biological systems: implications to personalized nanomedicine. Adv. Drug. Deliv. Rev. 64, 1363–1384. http://doi.org/10.1016/j.addr.2012.08.005.

Chapter 11

Nanotherapeutic systems for drug delivery to brain tumors

Keshav S. Moharir[a], Vinita Kale[a], Mallesh Kurakula[b]
[a]*Pharmaceutics Deptt., Gurunanak College of Pharmacy, Nagpur, Maharashtra, India*
[b]*Product Development, CURE Pharmaceutical, Oxnard, CA, USA*

11.1 Introduction

Cancer is a leading cause of mortality and key obstacle to improvement in the life expectancy across the globe, indicated by factors like aging, changing lifestyles, socioeconomic development, and changes in risk factor distribution. The year 2020 has witnessed 19.3 million new cases of all cancers combined resulting in 9.9 million deaths. Same report with current prevalence rate projects likelihood of 28.4 million new cancer cases in 2040, a massive 47% rise as compared to 2020 (Sung et al., 2021). The Global Cancer Observatory (GCO) lists 36 different types of cancers based on different tissues/organ systems involved. Among several types of cancers, lung cancer has highest mortality rate (18%) (https://seer.cancer.gov/statfacts/html/lungb.html accessed on 23rd March 2021) while cancers of brain-CNS are much difficult to treat having 5 year average survival rate of only 33% (https://seer.cancer.gov/statfacts/html/brain.html accessed on 23rd March 2021).

Among several other cancers, brain cancers are group of one of the most difficult types of tumors to treat. According to a worldwide database provided by GLOBOCAN and International Agency for Research on Cancer (IARC), an estimated 3,08,102 cases of brain and CNS cancers with 2,51,329 deaths have been reported in the year 2020, spanning all the categories of sex and age groups (Sung et al., 2021). Although there is better understanding of cancer biology, pathogenesis, diagnostic aids and clinical outcome with respect to brain tumors, complex nature of central nervous system as well as lack of specificity and targeted drug delivery to cancerous cells has impeded the progress in designing ideal regimens to treat brain tumors. In previous couple of decades, there is significant development in cancer therapies by exploring genetic make-up of patients, that include novel medicines such as immune check point inhibitors, personalized chemo – radio therapy approach accompanying improvement from complicated macrosurgery to sophisticated micro level precision surgeries. Even though great efforts are being taken, there remain some gaps between existing scientific knowledge and actual development of 'carrier systems' that can practically "deliver" chemotherapeutic agents to the very location of brain tumor crossing the Blood Brain Barrier (BBB) and Blood Brain – Tumor Barrier (BBTB). Several other obstacles such as, but not limited to metastasis, gene mutations, multidrug resistance, active efflux of drugs by tumor cells, and complex tumor microenvironment need to be resolved for better prognosis (Khaitan et al., 2018).

Recent studies in last two decades have unleashed the ability of nanoparticulate drug delivery systems to aid in chemotherapy of brain cancers and promising future owing to their beneficial biological and physicochemical properties. Nanotherapeutic systems not only cross the BBB effectively, but also aim to carry drug molecules to the target cells or tissues. Additionally, as compared to traditional drug delivery systems, they stay in circulation for longer time, can be used as diagnostic tool, biocompatible, stable and exhibit enhanced permeability and retention (EPR) effect (Zottel et al., 2019). Scores of nanotechnology based strategies have been tested for overcoming hurdles in drug delivery to the precise location of tumors inside brain and many are under development in order to gain better clinical outcome. This chapter thus attempts to throw light on advances in nanotherapeutics for brain cancers. This is preceded by brief information on brain tumors, challenges in drug delivery and limitations of conventional therapies. And finally, chapter concludes with role of Artificial Intelligence (AI) and clustered regularly interspaced short palindromic repeats (CRISPR-Cas9) and future course of development in treating brain tumors.

11.2 An overview of brain tumors

It is essential to know the types of tumors, their exact location, pattern of growth (benign or malignant), histopathology, nature of cells involved, and molecular markers implicated to maximize impact of targeted cancer therapies. Grading and classification of the brain tumor stage is mainly based on advanced imaging techniques, surgical pathology, and analysis of anatomical as well as histopathological attributes. Over the recent times, diverse multimodal techniques of diagnosis and therapeutic alternatives for brain tumors have evolved with deeper investigations focused on signaling pathways, tumor markers and tumor microenvironment (Ali et al., 2020).

The literature shows more than 150 diverse brain tumors. Depending on the origin, cancers of brain can be classified in main two groups – primary and metastatic. Primary cancer growth originates within the brain or immediate vicinity. With respect to cells or tissues involved, primary brain cancers are further grouped as glial (involving neuroglial cells of CNS) and nonglial (structural components of brain including nerves, blood vessels and glands). Any of these may develop into benign or malignant cancerous growth. Metastatic brain tumors have their origin outside the brain (Such as lungs, head and neck, bones, breast, etc.) and reach to the brain often through normal blood supply. Earlier, patients with metastatic brain cancers had poor prognosis with mean survival rate of few weeks to few months. Advent of precision diagnostic tools, targeted drug delivery, novel surgical procedures, improved radiation techniques, and progressive palliative care has subdued cancer woes and improved quality of life in brain cancer patients. Brain tumors have gradation from Grade 1 to 4 which is categorized according to their ability to proliferate. Grade 1 and 2 are generally noncancerous since they grow slowly and less likely to invade in surrounding locations (American Association of Neurological Surgeons, 2021).

11.2.1 Malignant brain tumors

Most commonly observed malignant type of brain tumors are gliomas arising from brain supporting cells called the glial cells. Gliomas are further differentiated based on subtypes of glial cells astrocytes, oligodendroglial cells, and ependymal cells involved

- Astrocytomas – common glioma, originate from star-shaped cells called astrocytes. More common as pediatric low-grade and middle-aged high-grade brain cancers.
- Ependymomas – they originate from proliferation of ependymal cells of ventricles. They are generally observed in the spaces of brain and surrounding the spinal cord. Prevalence is in adolescents and children.
- Medulloblastomas – common in children, generally involve cerebellum. Although high grade, response to chemotherapy and radiation is often positive.
- Oligodendrogliomas – originate from myelin covering of neuronal cells. These are rare, slow growing and many times occupy part of cerebrum. Observed in middle aged adults, prognosis is better than other gliomas.
- Glioblastoma multiforme (GBM) – most invasive, aggressive, common grade IV astrocytic brain cancer. GBM is a highly vascular, resistant to treatment and shows poor prognosis. Tumor microenvironment also exhibits heterogeneity and extensive angiogenesis that makes drug penetration and surgical removal difficult. The mean survival time for GBM patients is about 14 months.

11.2.2 Benign brain tumors

Benign brain tumors are noncancerous, do not spread and recurrence is seldom once removed completely and correctly. Several types of benign brain tumors exist corresponding to type of brain cells involved. A brief description of important benign brain tumors is given below (American Cancer Society Medical and Editorial Content Team, no date)

- Chordomas – slow growing, rare, infrequently invasive and commonly observed at the base of the skull and lower part of spine. Affected population is mostly 50 to 60 years age group. Of all types of primary brain tumors, only 0.2% contribution is by chordomas.
- Craniopharyngiomas – although benign, these tumors are complicated and complex to remove because of their deep locations near critical functional areas of the brain. Involvement of pituitary gland portion is not uncommon.
- Gangliocytomas – certain gangliomas like anaplastic gangliomas have low occurrence. These are generally well-differentiated found primarily in young adults.
- Meningiomas – originate from protective layers of the membrane-like formations called meninges that envelop the brain and spinal cord. These are the most commonly occurring benign intracranial tumors, contributing 10 to 15 percent among all types of brain cancers. Their malignant nature is rarely observed.

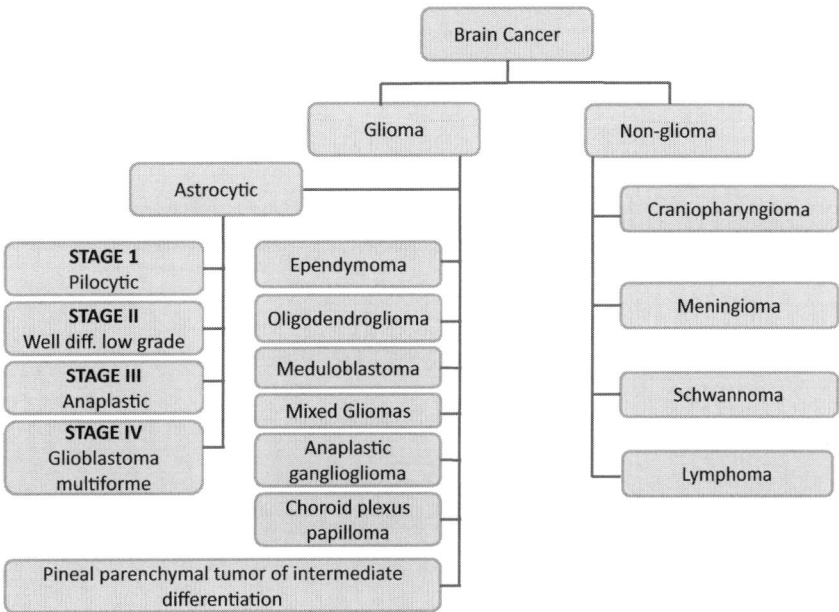

FIG. 11.1 A simple classification of brain tumors based on neural cells involved.

- Pituitary adenomas – benign, noncancerous that mainly grow in anterior lobe and do not spread beyond the skull. Most of them are treated successfully; however some cases have shown secondary effects such as excess pituitary hormone secretions leading to endocrinal complications.
- Schwannomas – originating from Schwann cells, these are fairly common nonmalignant brain tumors in adults. Notably, schwannomas usually take place of normal Schwann cells sparing the invasion. They may also affect the spinal and cranial nerves, for instance acoustic neuromas or vestibular schwannomas are found on vestibulocochlear nerve responsible for the function of hearing and body balance.

Apart from above listed and those mentioned in Fig. 11.1, several rare types of brain cancers have been reported. They include hemangioblastomas (slow-growing, usually located in the cerebellum), rhabdoid tumors are rare, highly aggressive, metastatic, and extend to entire central nervous system. Among other cranial region, tumors of pineal gland in central area of the brain appear as pineoblastoma (fast growing) and pineaocytoma (slow growing), often difficult to remove due to their deep seated positions. A whole different approach is given in pediatric brain tumor classification and treatments. The major types of pediatric brain tumors include ependymomas, medulloblastomas, brainstem gliomas, low-grade astrocytoma (pilocytic), and craniopharyngiomas. Further details about gradation of brain cancers can be obtained from WHO system of classification (Gupta and Dwivedi, 2017).

11.3 Barriers and challenges in the treatment of brain cancer

The main barriers in treating brain cancers include – (1) cancer cell heterogeneity and metastasis, (2) difficulties in marking tumor boundaries, (3) vicinity of highly significant functional areas within brain, (4) lack of precision in selecting drug and tumor targeting strategies, (5) resistance development to chemotherapeutic agents, and (6) unsupportive tumor microenvironment (blood brain tumor barrier) accompanying blood-brain-barrier (BBB). The integrity of tight junctions is further consolidated by astrocytes sitting below them (Li et al., 2020). Fig. 11.2 gives an idea in brief about commonly observed obstacles and challenges in the treatment for brain cancer.

11.3.1 BBB as a main hurdle

The blood brain barrier is the most significant parameter that restricts the entry of foreign substances inside the brain. Astrocytes, endothelial cells, adherence junctions, and tight junctions are complex components that form BBB. Tight junctions of endothelial cells control the influx of polar chemotherapeutic agents past the BBB while molecules which are

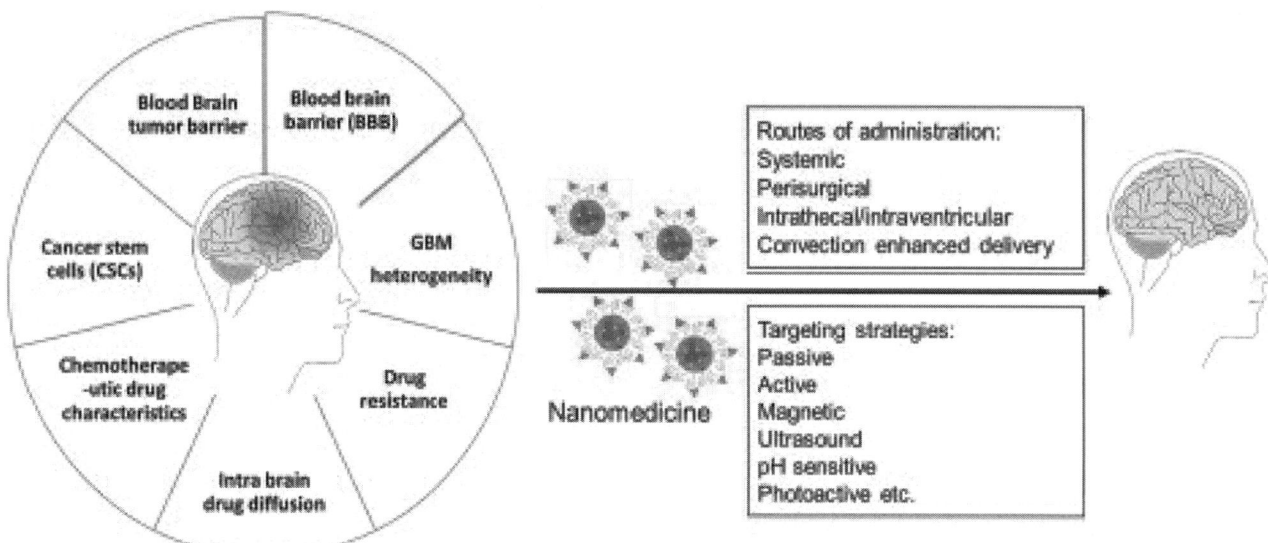

FIG. 11.2 Challenges in drug delivery to GBM. *Adapted with permission from Ganipineni et al. (2018).*

nonpolar and smaller in size can get across BBB plasma membrane easily (Ferraris et al., 2020). Even if the drug molecules cross the BBB, catabolic enzymes of brain tissues and surrounding fluid inactivate them (Foti et al., 2015).

11.3.2 Chemoresistance and efflux

Resistance to chemotherapeutic drugs as well as drug efflux from glioma cells is attributed to the presence of P-glycoprotein (P-gp) associated with ATP binding cassettes (ABC) multidrug-efflux transporters. ABCs are actively involved in efflux of molecules across the cell membrane (Aldape et al., 2019). Thus, dual mode nanotherapeutic systems with P-gp inhibiting drug are under investigation that can push drug across the BBB and prevent chemoresistance by P-gp (Kuntner et al., 2010).

Although it is theoretically said that lipophilic drugs cross BBB effectively, predictions are often unmet in reality, particularly when we talk about anticancer drugs for brain tumors. The reason for this anomaly was found to be high affinity of drugs towards efflux transporters within the BBB (Lingineni et al., 2017).

11.3.3 Tumor microenvironment (TME) dynamics and lack of brain tumor classification based on genetics

Tumor micro-environment (TME) and genetics in brain cancers have unique characteristics which are difficult to predict in individual patient due heterogeneous nature of tumor cells, such as postsurgery recurrence of glioblastoma has completely changed epigenetics, signaling pathways and probably set of biomarkers involved. On the similar lines, lack of precision in selection of drugs due to indecisive or vague sub-classification of brain cancers can deprive the patient of accurate anticancer therapy. This is evident from the fact that genomic sub-typing of brain tumors is yet not available completely that can improve the drug selection criteria for better clinical outcomes (Aldape et al., 2019; Pajtler et al., 2017).

Further, barriers for anticancer drugs to reach the site of metastasized and widely infiltrated malignant GBM are observed. The diffusivity of drug is limited to short distance or within same local area, making it almost difficult to target infiltrated and delocalized GBM cells.

11.3.4 Resistance due to cancer stem cells (CSCs) of gliomas and GBM

Stem cells have role in initiation and proliferation of cells. Similarly, CSCs play same role in cancer development with hallmark stem cell marker development. CSCs pose challenges to the anticancer treatment by (1) creating signaling pathway to resist apoptosis (2) response pathways to repair damaged DNA (3) strengthening drug efflux mechanism (4) overexpressing cancer cells responsible for metastasis and resistance, and (5) overexpression of cells that result in angiogenesis (Bao et al., 2006; Kim et al., 2015).

11.3.5 Lack of proper brain cancer mimicking models

The simulations of all types of temporal brain cancer related changes that actually take place in humans are practically difficult to achieve in animal models. Pathophysiological changes, TME, heterogeneous nature of cancer cells and presence of possible listed markers are difficult to achieve. Further, the clinical trials that take place are not easy to relate with actual patients in future as generalization is complicated owing to patient to patient heterogeneity. Ideal features of GBM preclinical model demand that they should be stable, reproducible, have same characteristics of GBM as in humans, and make fairly accurate predictions about therapeutic strategies and outcome (Pierce and Keating, 2014).

11.4 Conventional vs nanomedicines in drug delivery for brain cancers

Chemotherapy, surgery, radiotherapy and combination of either of these form basis of conventional strategies in treatment of brain cancers, albeit they have their own pros and cons.

Dual mode drug delivery to achieve more than one objective seems difficult in conventional drug delivery systems to treat brain cancers. On the contrary, nanocarriers exhibit multiple advantages within the same formulation such as enhanced capacity to cross BBB, selective targeting, and longer retention time (Woodworth et al., 2014). In addition, nanotherapeutic systems can have "theranostic" approach that incorporates therapy with diagnostic arm (Mallidi et al., 2015). Nanotechnology offers formulation strategies and platforms that can overcome drug deactivating adverse surroundings effects of glioma cells such as acidic conditions, degrading enzymes, efflux, drug leakage from angiogenesis related vasculature, and reactive oxygen species (Brannon-Peppas and Blanchette, 2012). Fig. 11.3 shows a comparative assessment of conventional formulation with nanodevices covering possible applications and disadvantages.

Nanotherapeutic systems have been tested and evaluated in various forms and sizes. They may be made of organic, inorganic, metallic, polymeric, or composite materials having different shapes. Compared with traditional drug delivery systems, nanocarriers have morphological advantages that include shape, size, structure, surface charge, solubility. These structural features influence BBB crossing, mechanism of drug release, perfusion, polydispersity, interaction with immune cells, and drug uptake by cancerous cells (Richards and Endres, 2016). Larger surface area due to smaller particle size allows higher drug loading capacity resulting in better drug utilization, hence reduction in drug toxicity. NPs having particle size below 100 nm undergo enhanced permeation and retention (EPR) effect mainly in solid tumors. EPR effect involves release of drug molecules from NPs and uptake within the tumor cells by gaining entry through leaking endothelium and angiogenic blood vessels. Smaller NPs having particle size (<20 nm) are further filtered through kidneys while those with higher particle size (>150 nm) are phagocytosed by reticuloendothelial system (RES) (Lalatsa et al., 2014; Nance et al., 2012). However, EPR effect for drug uptake from NPs in brain cancer is debatable due to dense tissues obstructing drug diffusion, highly dependent on the nature of delivery system, pH change in TME and elevated fluid pressure in the vicinity of tumor cells (Séhédic et al., 2015).

It is not feasible for conventional drug delivery systems to undergo modification and have longer retention time. Notably, surface modification of NPs is a pragmatic approach for increasing half-life with longer retention time near tumor cells, and they are often coated for example with nonionic surfactant like polysorbate 80, low molecular weight poly (ethylene)

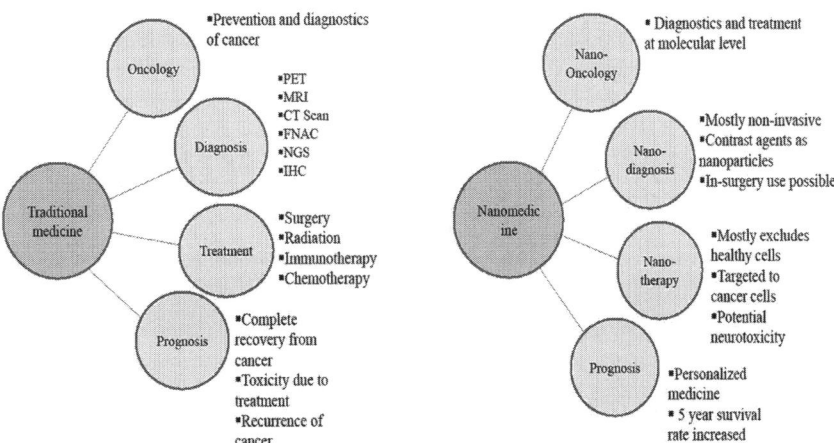

FIG. 11.3 Conventional drug delivery system vs nanomedicine in brain cancer treatment.

glycol (PEG), chitosan, albumin, etc. to achieve the above mentioned objective (Lalatsa et al., 2015; Wang et al., 2009). It is also tried and tested to prove that specifically positive charge on drug nanocarriers show more uptake by tumor cells as compared to conventional systems of drug delivery, probably due to the net negative charge on surface of the cell membrane (Kim et al., 2010). This ongoing discussion thus shows prospects of NPs as beneficial systems that have several advantages over conventional drug delivery systems employed in brain cancer therapy as well as diagnosis.

11.5 Approaches and mechanisms of nanocarriers for chemotherapeutic drug delivery to brain tumors

11.5.1 Passive targeting

Passive targeting of drugs to brain tumor cells is an EPR dependent phenomenon which involves release of drug molecules from NPs and uptake within the tumor cells by gaining entry through leaking endothelium, loss of tight junction, and poorly structured angiogenic blood vessels. Passive targeting does not require external energy for drug propagation across the BBB. However, the closeness of EPR is said to be more prominent in preclinical animal models rather than human patients and has influence of factors such as brain tumor type, vasculature, size, involvement of immune cells, and lymphatic drainage and interstitial fluid pressure. Recently, nanomedicines have been employed to make maximum use of EPR effect. For example, Bort et al. have demonstrated a successful experiment in which chemotherapeutic agent loaded polysiloxane nanoparticles with about 4 nm diameter delivered to brain metastatic cancer (Wu et al., 2019). Duan et al. summarizes different physical methods like acoustic waves, thermal effects, nanobubbles and ultrasound to augment EPR effect of NPs (Wu et al., 2019). Fig. 11.4 helps to understand advantages of active targeting over passive drug delivery, such as (1) crossing BBB and (2) tumor cell selectivity sparing healthy neural cells in brain cancer nanotherapy.

11.5.2 Active targeting

Active targeting of therapeutic and diagnostic agents using nanomedicine aims at selectivity towards brain tumor cells with precision and accuracy. Active targeting can be achieved by employing ligands, aptamers, hormones, peptides, monoclonal antibodies, oligonucleotides or markers in order to have better penetration of brain metastases, enhanced retention, and circulation (Cho et al., 2008; Wu et al., 2019).

The basic mechanism of active targeting involves identification and selection of overexpressed receptors on the brain tumor cell surface. The overexpressed receptors are then targeted with suitable NPs. For this to achieve, surface of NPs is coated with ligands mentioned earlier (Danhier et al., 2010). NPs are internalized after binding with target – the transportation takes place via receptor mediated endocytosis. Active targeting has an added advantage of ability to target metastasized tumor cells in other areas of the brain (Marcucci and Lefoulon, 2004). Active targeting can be grouped into three classes – absorptive-mediated transcytosis, receptor-mediated endocytosis, and transporter mediated transcytosis.

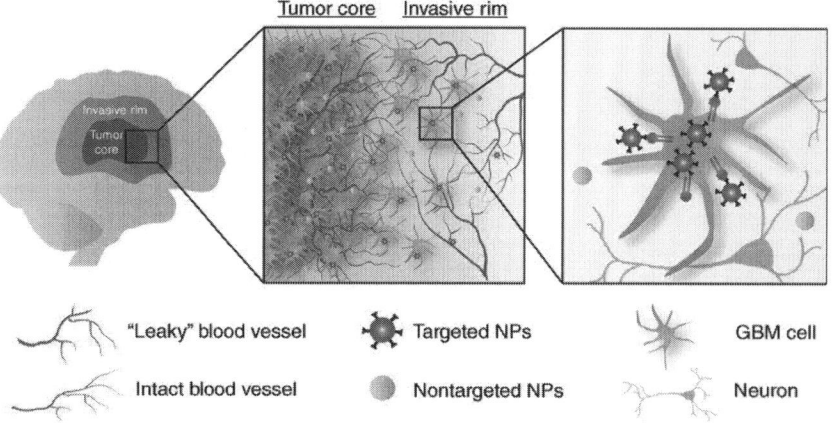

FIG. 11.4 **Active vs passive targeting of glioma and GBM.** Note the efficacy of active targeting NP approaches in reaching tumor cell sparing normal neural cells. *Adapted with permission from Wadajkar et al. (2017).*

11.5.2.1 Absorptive-mediated transcytosis (AMT)

The mechanism of AMT in drug delivery from nanoparticles is electrostatic interactions. For this, surface modification/functionalization and activation of system by cell cationic or proteins penetrating peptides (CPPs) is utilized. CPPs are minimally invasive meaning the transcytosis of drug load across the membrane is without disturbing its integrity is achieved (Xu et al., 2009). However, electrostatic interaction does not discriminate between normal tissues and cancer tissues (Higa et al., 2015).

11.5.2.2 Transporter- or carrier-mediated transcytosis (TMT/CMT)

A common but important pathway that carries hormones and nutrients like glucose across the BBB membrane, works by inducing conformational changes in membrane proteins and follows concentration gradient. Glucose transporter (GLUT1) that transports glucose to the brain has high presence and association with brain capillary endothelial cells. Quin et al. have reported that bioavailability of coumarin-6 increased more than 3 fold in glioma cells when embedded into glycosylated cholesterol nanoliposomes targeted towards GLUT1 transporter and compared with simple liposomes (Qin et al., 2010).

11.5.2.3 Receptor-mediated endocytosis (RME)

The principle mechanism of chemotherapeutic drug transport across the tight junctions of BBB includes transportation with ligands like antibodies via specific and selective receptors. Active targeting by this approach involves selective receptors which are up-regulated and overexpressed in certain specific brain tumors. Transferin receptor (TfR) that helps iron transport into CNS is commonly evaluated receptor as it has shown promising results. Zhang et al. reported TfR modified c[RGDfK]- hybrid micelles loaded with Paclitaxel in conjugated form that enhanced survival time of mice with (orthotropic) U87MG tumor to 42.8 days from 34.8 days (only paclitaxel containing micelles) (Zhang et al., 2012). This effect was attributed to TfR mediated drug endocytosis leading to higher drug accumulation in the brain. Apart from TfR, several other receptors such as LDL receptors, insulin receptors and glutathione receptors have been assessed for nanotherapeutic drug delivery previously (Zhang et al., 2015).

11.5.2.4 Peptide conjugated

Amino acid chains from 2 to 50 in number are termed as polypeptides. These peptides gain entry across the membrane when bound to cell surface receptors. If peptides are used as ligand for drug loaded nanocarriers, receptor mediated endocytosis takes place that results in drug-nanocarrier entry inside the cell through receptors. This principle has been assessed in active transport of anticancer drug molecules into the tumor cells. Certain CPPs, for example rabies derived peptide (RDP) is explored as a ligand for carrying anticancer drugs in nanocarriers. One such study of nanoliposome-RDP complex with curcumin as active agent has supported the positive outcome in brain cancer mouse model on parameters of survival time, drug bioavailability and tumor suppression (Zhao et al., 2018).

11.5.2.5 Small molecule ligand mediated

Small molecules as ligands for delivering antineoplastic drugs to brain cancer tissue have several benefits such as active targeting, easy to decorate, nonimmunogenic, safe and general compatibility. Biotin, folic acid (FA), and bisphosphonate are well studied targeted ligands for active drug delivery to tumors. FA is most promising till date since folate receptor is overexpressed in many brain cancers like glioma, GBM and astrocytoma but remain unchanged in normal cells. An insightful work by Xu et al. confirms clinical utility of folic acid in targeting glioma cells. FA-borneol-PAMAM(G5) dendrimers loaded with doxorubicin were evaluated on C6 glioma induced rats *in vivo* and compared with plain unmodified doxorubicin solution. Dual role of modified dendrimers was observed with drug transport across BBB doubled, survival time improved from 18 to 28 days and *in vivo* tumor growth inhibition increased to 57.4% on comparison with unmodified doxorubicin (Xu et al., 2016). Thus, modification of nanocarriers with small molecules assures higher drug concentration at tumor site with improved tumor suppression.

11.5.2.6 Oligonucleotide (aptamer) mediated

Targeted drug delivery selectively to cancer tissues as prime objective has been studied with oligonucleotides as conjugating ligand molecules. Among various examples of single stranded oligonucleotide sequences, aptamers (also called as "chemical antibodies") are well established that bind to specific target at molecular level with high affinity. They easily enter tumor cells by binding with oncoproteins. They can also deliver antineoplastic drugs or siRNA inside cancer tissues.

Aptamers may be used repeatedly as they have negligible immunogenicity (Fu and Xiang, 2020). Retention of paclitaxel in GBM tissues at higher concentrations, longer retention and improved survival time in mice cancer model was observed with aptamer AS1411 containing NPs (Luo et al., 2017). The exclusive feature of aptamers to identify specific binding targets such as nucleic acids, phospholipids, and proteins plays significant role in penetrating tumor tissues. Few aptamers employed in GBM treatment modalities include AS1411, GBM128, GMT3-9, U2, and aptamer32 (Hays et al., 2017).

11.5.2.7 Cytokine-targeted nanocarriers

Cytokines are released as natural inflammatory and immune response. They pose anticancer activity by helping in killing cancer cells and normal cells survive more. Stimulated T cells secret interleukin-13 (IL-13) which exhibit two receptors, namely glioblastoma associated with IL-13Rα2 but absent in normal tissues and IL-13/4R is present in noncancer cells and binds to IL-4. Receptor IL-13Rα2 is overexpressed in pediatric astrocytomas, thus a viable target for anticancer drugs (Joshi et al., 2008). The research work by A. Madhankumar and group shared the details backing this hypothesis. PEGylated quantum dots (QD) of cadmium selenide conjugated with interleukin-13 (IL-13) were formulated with objective of targeting IL-13Rα2 high affinity receptors which are overexpressed by glioma and GBM stem cells. Competitive binding study and transmission electron microscope confirmed preferential binding of cadmium selenide-PEG-IL13QD with glioma cells (and their stem cells) and tumor related exosomes respectively. Thus, IL13-QD complex can be used as a marker for early diagnosis, cancer metastasis, and prognosis of glioma owing to its selective binding with glioma associated exosomes and IL-13Rα2 (Madhankumar et al., 2017).

11.5.2.8 Cancer stem cells (CSCs) targeted nanoparticles

CSCs initiate, propagate, and aid to proliferate cancerous cells in tumor milieu. They are one of the major causes for brain tumor relapse, metastasis, and chemotherapy resistance. Conventional therapy systems can not eliminate CSCs completely. Thus, novel strategies to destroy stem cells are required, such as PLGA NPs of curcumin exhibited induction of autophagy, improved half-life and halted augmentation of glioma initiating SU-2 and SU-3 CSCs (Zhuang et al., 2012). The resistance of gene O6-methylguanine-DNAmethyltransferase (MGMT) related within brain CSCs to temozolomide can be controlled by incorporation of monoclonal antibodies in carbon nanotubes (Wang et al., 2011). This strategy works with better outcome when dual mechanism likes efflux inhibition or CSCs specific markers are targeted.

11.5.2.9 Dual-targeted/multifunctional nanocarriers

The two aspects associated with drug delivery to brain cancers are (1) delivering maximum drug across the BBB and (2) targeting selectively the cells of glioma, GBM, astrocytomas, and schwannomas while sparing normal brain tissues surrounding the tumor. However, all passively targeted drug delivery systems and many active drug delivery approaches lack these bi-functional features in the same formulations. Thus, an ideal and model system of drug delivery to brain tumors should have dual targeting ligand taking into consideration BBB hurdle and tumor cells selectivity (Gao, 2016). One such dual function, actively targeting drug delivery system is reported by Gao et al. In doxorubicin liposomes, transferrin (Tf) and folate (F) acted as ligands having confirmed role of selectively targeting glioma cells and effectively crossing BBB respectively. Drug accumulation was evaluated by eosin stained radiolabeled assay that indicated accumulation of ligand linked doxorubicin in brain while plain doxorubicin liposomes aggregated in cardiac tissue region (Gao et al., 2013).

11.5.3 Stimuli responsive nanocarriers systems

The longevity and stability of anticancer drugs at the site of tumor cells is unpredictable and often shows uneven distribution. To address this issue, the biochemical and pathological changes in TME can be used to some extent for targeted drug delivery (Roy and Li, 2016). We can call them as endogenous stimuli for drug targeting that point towards changes in pH, temperature, extracellular matrix (ECM), overexpression of certain markers, photosensitivity, and changes in enzyme levels taking place in the vicinity of tumors, that is, TME (Bruschi et al., 2017). As external or endogenous environmental conditions (physical, chemical, or biological) change, then only drug release and delivery takes place in triggered manner near the site of action. This can reduce the distribution of drugs in normal tissue and improve drug availability near tumor locations. These internal and external triggering factors for drug delivery can be explained with following examples.

11.5.3.1 Photosensitive (physical) drug delivery systems

The mechanism of this drug delivery system is based on the principles of photothermal therapy (PTT) and photodynamic therapy (PDT). External Photothermal agent or photosensitive agent present in formulation near tumor location will absorb the specific wavelength light thereby releasing reactive oxygen species (ROS) or increase in temperature surrounding tumor cells. Stable NPs of silk fibroin containing indocyanine green dye were prepared by Xu et al for glioma therapy. Irradiation with near infrared (NIR) waves resulted in hyperthermia in short time to temperature rise by 33.9°C. Controlled release was observed for 72 hours exhibiting prolonged cytotoxic effect while red fluorescence of dye helped in imaging for surgical intervention (Xu et al., 2018).

11.5.3.2 pH sensitive (chemical) drug delivery systems

The result of rapid glycolysis, ineffective lymphatic drainage mechanism, and insufficient perfusion in growing cancer cells create excess lactic acid resulting in an acidic extracellular tumor microenvironment. Although this phenomenon is spatial and temporal (related to tumor space and dynamic related to time), shift in stromal pH of tumor (approximate extracellular pH 5.6) as compared to stable pH of normal tissues (pH 7.4) can be explored to target with anticancer drugs (He et al., 2020; Rao et al., 2017). Several organic, and metal organic frameworks have been reported as linkers for this type of tumor targeted drug delivery systems which are stable at pH 7.4 and face hydrolyzation at acidic conditions, typically below pH7. In one such recent attempt, Gawali et al. have reported intra-tumoral delivery of doxorubicin using pH sensitive magnetic nanocarriers as dual mode drug delivery system. Differential release of doxorubicin was achieved with breakage of carbamate and hydrazone linker bonds at acidic pH of tumor marked by internalization of drug and improved cytotoxicity (Gawali et al., 2019).

11.5.3.3 Redox-sensitive nanocarriers

Nanocarrier moiety and drug molecules when linked by reducible disulfide (S-S) bond, the system is called as reduction sensitive targeted drug delivery system. These types of systems form nano-micelles in formulation and are highly stable. The reducing tumor microenvironment has four times higher glutathione concentration than normal tissues. Disulfide bonds are conveniently broken down by reducing glutathione to sulfhydryl groups, resulting in nanocarrier degradation and release of anticancer drugs. These disulfide bonds are commonly attached to polymer chain or matrix and help in linking nanoparticles with genes, drugs or targeting agents (Sun et al., 2018).

In one study, black phosphorus quantum dots (BPQDs) were embedded with cysteine containing poly-(disulfide amide) (Cys-PDSA) polymeric system to prepare biodegradable and stable nanocarriers. BPQD was rapidly released after glutathione broke disulfide linkage in novel polymer Cys-PDSA. The formed NPs exhibited novel dual features of therapy as well as imaging on irradiation with near infrared (NIR) photothermal ablation (Chen et al., 2021).

11.6 Types of nanotherapeutic platforms for drug delivery to treat brain cancer

In context of brain cancers, several types of nanocarriers systems have been prepared, assessed, characterized and evaluated for their efficacy in preclinical as well as clinical trials. They can take form of but not restricted to quantum dots and magnetic NPs, liposomes (long-circulating, conventional, active-targeting, cationic liposomes, stimuli-sensitive, and surface modified), nanomicelles, nanogels, prodrugs, dendrimers, gold and silver NPs, iron oxide NPs, and polymersomes (Lombardo et al., 2019). Theranostics nowadays is a buzzword that gives unique opportunity of diagnosis as well as therapies for brain tumors in the same nano-formulation. For example, plain gadolinium NPs and metallofullerene conjugated gadolinium NPs provide better imaging by delineating tumor line marking of cancer cells from normal brain tissues (Wu et al., 2019). Figure 11.5 represents important nanotherapeutic agents under investigation for treating brain tumors. Adapted with permission from Tang et al. (Tang et al., 2019).

In order to reach maximum advantageous point of NPs, they should accomplish basic criteria such as (1) biocompatible and stable in the body (2) minimum loss of drug during binding/complexing/conjugating and transporting the drug (3) selectively target tumor cells after effectively crossing BBB (4) physical and chemical characteristics suitable and feasible for brain tumor targeting (5) convenient to scale-up from lab to large scale (6) easy internalization within tumor cells and (7) not being taken up by RES. This section further covers various types of nanotherapeutic systems and representative examples recently employed in treatment of brain cancers.

11.6.1 Inorganic (metallic) nanoparticles

Several inorganic NPs are tried and tested for theranostic of brain tumor, namely carbon nanotubes, quantum dots, calcium phosphates NPs, dendrimers, iron oxide NPs, lipid containing NPs. Although they have established therapy potential in the glioma therapy, primary limitation exists as cytotoxicity of the released metal ions due to their longer stay in tissues (Shubayev et al., 2009). Few representative NPs of metallic origin are described below.

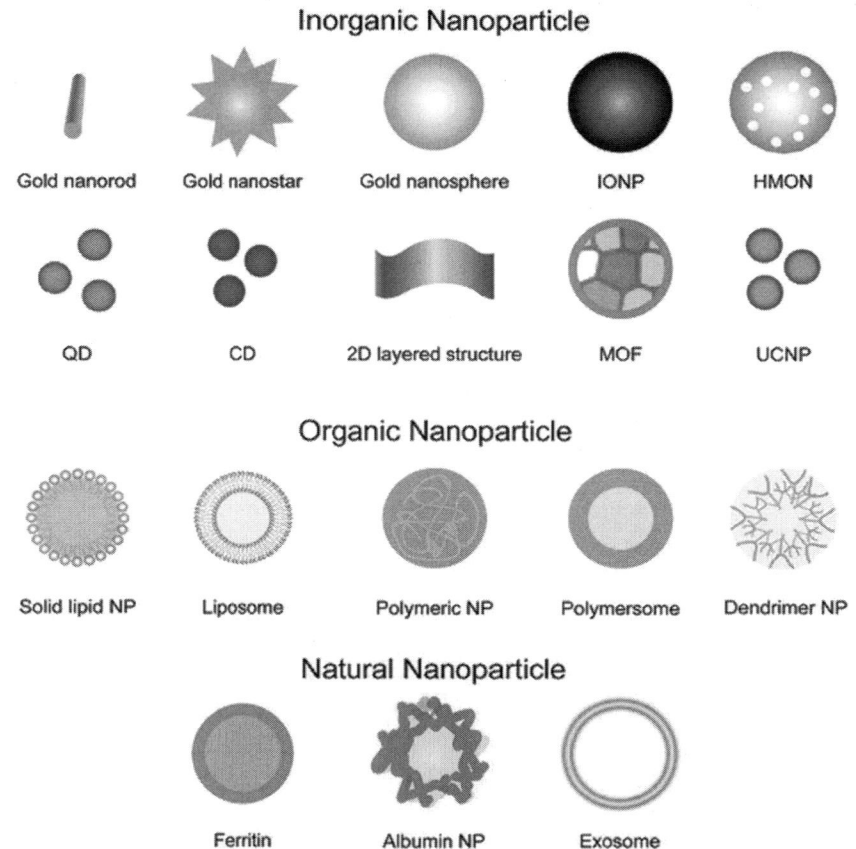

FIG. 11.5 **Nanoparticles employed in brain cancer therapies.** *Adapted with permission from Tang et al. (2019).*
Abbreviations: IONP, iron oxide nanoparticle; HMON, hollow mesoporous organosilica; QD, quantum dot; CD, carbon dot; 2D, two dimensional; MOF, metal–organic framework; UCNP, upconversion nanoparticle; NP, nanoparticle.

11.6.1.1 Gold nanoparticles

The distinctiveness of gold NPs in brain tumor imaging and therapy is marked by easy surface modification, biocompatibility, inertness as compared to other metallic NPs, shapes and size manipulation, amenable to conjugate or complex formation with therapeutic moieties and ease of fabrication. Apart from these, many advanced functionalities are exhibited by gold NPs such as monodispersity, easy core formation, large surface area, high binding capacity and light scattering capacity. The electrostatic interactions and covalent association are popular mechanisms used to bind therapeutic agents to the gold NPs (Sonali et al., 2018). The versatility in management of gold NP size is significant feature that helps to target specific tumor cells of brain escaping normal cells.

Microwave irradiation developed gold NPs coated with modified xanthan gum (carboxymethyl xanthan gum) was reported by Alle et al. (2020). Carboxymethyl xanthan ensured nontoxic, economical and eco-friendly method of fabrication accompanied by ability with efficient reducing agent. Doxorubicin loaded with electrostatic interactions on NPs was released by pH responsive trigger in acidic tumor microenvironment. The doxorubicin release from these coated gold NPs was 4.6 times higher than its free form accompanied by significantly increased uptake in acidic tumor cell conditions and tumor cell-kill capacities. In yet another study, pH controlled release of Curcin (a potent type 1 ribosomal inactivating protein) from gold NPs was reported as successful drug delivery system in acidic glioma microenvironment. Significant tumor cell targeting and killing capability was reported on near infrared (NIR) inflicted photo thermal ablation of gold NPs. Cytotoxic action of curcin was attributed to reactive oxygen species (ROS) formation, destabilization of cytoskeletal structural integrity and aberrations in mitochondrial functions. An enhanced lethality was observed on PEGylation of gold NPs and with potential to deliver transferrin and folate antibody specifically to the glioma cells (Mohamed et al., 2014).

11.6.1.2 Carbon nanotubes and nanodots

The cylindrical nanosized particles called carbon nanotubes (CNTs) have gained immense importance as carriers of therapeutic agents. These are generally rolled, seamless, thin graphene sheets with lowest diameter 1 nm and length several micrometers yielding very high aspect ratio. Single sheet forms single-walled nanotubes (SWNTs) and multiple graphene sheets stacked in a roll form multiwalled nanotubes (MWNTs) with high surface area. The CNTs have unique electronic, mechanical and surface characteristics that can be modified to conjugate or complex with polymers such as peptides, carbohydrates, and organic compounds for targeted cancer therapies (Mahajan et al., 2018). CNTs have shown enhanced immune response involvement in targeting glioma cells. By virtue of the macrophages to preferential target CNTs on comparison with glioma cells, it was demonstrated by VanHandel and team that MWNTs enhanced extent of macrophage infiltration in glioma cells. The GL261 murine glioma model confirmed immuno-modulatory effects of MWNTs by increased IL-10 levels (VanHandel et al., 2009). This also highlighted that once in the vicinity of tumor cells, CNTs enter the tumor cells by endocytosis or direct diffusion sparing normal cells. Further, CNTs are known to have longer residence time in brain as compared to other NPs making them attractive drug molecule carriers for brain tumors (Rastogi et al., 2014).

11.6.1.3 Quantum dots

Quantum dots (QDs) are nanocrystals of small size (2–10 nm) with fluorescent and semiconducting properties. The core of QDs is metalloid or metallic that may be coated to make it biocompatible and to protect from scavenging by immune response. If inside tumor cell, QDs are helpful in cancer diagnostics as changes in tumor environment changes fluorescent light intensity. This can aid in detecting proteins, metabolites and biomarkers formed by tumors (Pietilä et al., 2013). Another example highlights application of QDs in drug delivery to tumors by NPs. An siRNA containing nanoliposomes conjugated with QDs (Lipofectamin 2000) when delivered to tumor region, intensity of fluorescence increases gradually as delivered siRNA shows its activity. However, detailed evaluation of QD toxicity is needed to eliminate any possible side effects (Chen et al., 2005).

11.6.1.4 Mesoporous silica nanoparticles (MSNs)

As the name suggests, MSNs have porous nature with pores that can absorb high quantity of therapeutic agents. These silica particles are presented in spherical shape in varying sizes. MSNs are taken up by tumor cells in the same way as other NPs, that is, endocytosis and internalization. Since MSNs are attracted towards phospholipid head part, they may also attack normal cells. Thus, coating or conjugation of MSNs is sometimes necessary. Europium (Eu) and Gadolinium (Gd) MSNs containing folic acid in coated form prepared by Chan et al were found highly effective in targeting tumor cells that overexpressed folate receptors. In the same NPs, loaded camptothecin exhibited cytotoxicity in cancer cells than healthy cells (Chan and Lin, 2015). Advanced version of MSNs called as Hierarchical Pores Silica Nanoparticles (HPSNs) was used as drug delivery platform by Dai et al. Doxorubicin and folic acid loaded HPSNs delivered and distributed pH dependent and controlled drug release in tumor cells. Thus, actively targeted MSNs show benefit in tumor targeting when used with suitable conjugating agent.

11.6.1.5 Superparamagnetic iron oxide nanoparticles (SPION)

Superparamagnetic iron oxide nanoparticles (SPIONs) contain paramagnetic capacity iron oxide when put in an external magnetic field. This property can be used for drug delivery as well as contrast agent in magnetic resonance imaging (MRI) technique.

SPION most extensively explored and evaluated is Ferumoxtran-10. Features of this compound such as lymph node affinity and powerful paramagnetic signals make it favorable agent in diagnosis as well as nanotherapy of brain cancer (Trivedi and Kompella, 2010). The lymph node involvement of metastatic cancer can be tracked and mapped by images produced with Ferumoxtran-10 (Harisinghani et al., 2003). Apart from this, heat generated by external magnet is found to be helpful in imaging and thermoresponsive drug delivery at cancer locations (Hervault and Thanh, 2014). Iron containing cores of SPION when coated and formulated as liposome or micelles are safe, however high content is fatal for normal cells due to DNA damage and oxidative stress. Hyperthermia due to heat of external magnetic field breaks the bond releasing drug from SPIONs, drug release is further controlled and permeability is increased when core is coated with polymeric NPs. Thus proper combination of magnetic field and electromagnetic waves control rate of drug delivery (Kumar and Mohammad, 2011). Contemporary studies are involved in generating animal models for imaging and targeted drug delivery using SPION.

11.6.1.6 Zinc oxide NPs

Of the several types of NPs used in drug delivery systems to target gliomas, zinc oxide NPs have demonstrated several characteristic advantages such as quick electron exchange kinetics, biocompatibility, ability to absorb drug moieties, and efficient catalytic activity. Ultrasonication, hydrothermal synthesis, and thermal evaporation are some of the methods employed in development of various zinc oxide nanostructures such as nanorods, nanowires, nanoribbons, nanowires, etc. (Wang, 2004). Numerous studies suggest zinc oxide reach the brain disrupting the BBB or by neural transportation. The entry of these NPs across the BBB is mediated by surface protein, mainly apolipoprotein E (Ostrovsky et al., 2009; Shim et al., 2014).

Zinc oxide NPs induce cytotoxicity by oxidative stress and significantly lower energy in the tumor cells. Anticancer studies on cell lines, namely U87 for brain tumor and HeLa cells for cervical cancer have been reported that include crystalline zinc oxide NPs of different nanostructures. These NPs exhibited anticancer activities depending upon dose selected. Profound cytotoxic effect was observed at higher drug concentrations leading to inhibition of DNA strand ligation and nascent micro nuclei formation inside target tumor cells (Wahab et al., 2011).

11.6.2 Lipid-based and polymeric nanoparticles

11.6.2.1 Liposomes

Conventionally classified as small unilamellar, large unilamellar, and multilamellar liposomes have size range around 10 – 100 nm, 100 – 1000 nm, and 500 – 5000 nm respectively incorporating both aqueous and nonaqueous layers in concentric manner (Torchilin, 2005). Liposomes consist of aqueous cores and phospholipid bilayers encompassing them that can be modified in several ways. Liposomes are versatile nanocarriers since amphiphilic phospholipids allow hydrophilic as well as hydrophobic therapeutic agents as loading cargo (Bozzuto and Molinari, 2015).

Several methods are reported for liposome development. Primarily, in the first step organic solvent dissolves lipid. Evaporation of solvent gives dry film to which an aqueous phase is added. Notable methods of preparation include sonication, film hydration, ether injection, membrane extrusion, freeze thawing, French pressure cell, etc. Components, preparation method, surface modifications, drug properties govern the trigger for drug release for example, pH dependent, external electromagnetic field, redox potential, and photothermal activation. However, stability of liposomes remains the issue that can be overcome by incorporating phosphatidylcholine, cholesterol, and PEGylation.

Liposomes either release the drugs in extracellular tumor region or they release the drug within tumor cell after endocytic engulfment by cancerous cells. Conjugation with or applying poly (ethylene) glycol layer on liposomal surface (called as PEGylation) is known to escape reticulo-endothelial system (RES) and host immune system, a phenomenon that helps to target tumor cells selectively. PEGylation also improves half-life, water solubility and reduces toxicity of NPs (AlSawaftah et al., 2021). Further, the EPR effect supports retention and localization of the drug inside tumor cells. To achieve targeting to tumor cells several parameters are taken into consideration, for example, liposomes having particle size above 100 nm are taken up easily by immune cells for phagocytosis while those below this size attain tumor targeting far easily. In addition, active targeting with liposomes can be accomplished by antibody-liposome conjugation that is specific for antigenic tumor markers.

Apart from PEGylation, coating of drug loaded liposomes with D-α-tocopherol polyethylene glycol 1000 succinate (TPGS) has proven enhanced efficacy in targeting glioma cells. In one such study, docetaxel loaded liposomes layered with TPGS were evaluated for brain tumor therapy. The cytotoxic activity of TPGS coated liposomes was compared with unmodified and PEGylated liposomes on C6 glioma tumor cells. The TPGS covered liposomes performed better in comparison with bare and PEG-coated liposomes on parameters of IC50 value and total cellular uptake in vitro (Muthu et al., 2011).

11.6.2.2 Polymeric micelles

These are nanocarriers fabricated in core-shell arrangement. They are formed in aqueous solution from amphiphilic block co-polymers self-assembly or spontaneous arrangement. Just like other micellar structures, they show hydrophilic shell (for loading hydrophilic drugs) and hydrophobic core (for loading lipophilic drugs). Their characteristic properties such as high stability, core-in-shell structure, size distribution, micellar arrangement, biocompatibility, low toxicity, ease of formulation, and ability of surface modification make them one of the preferred drug delivery systems in cancer treatment. Commonly applied fabrication methods that classically produce 10 – 100 nm diameter micelles include direct dissolution, solvent casting, and dialysis (Zhang and Mi, 2019).

Glycyrrhetinic acid and doxorubicin dual delivery for cancer therapy is reported by Yang et al. (2019). Free doxorubicin demonstrated 27.2% against HepG2 cancer cells while significantly higher that is 47.2% apoptosis of cancer cells was observed with doxorubicin-incorporated polymer prodrug micelles. In addition, doxorubicin micellar formulation showed favorable pharmacokinetics in terms of prolonged retention time, higher drug bioavailability, greater drug accumulation at tumor site and lower IC50 value on comparison with free doxorubicin. Thus, dual mode of drug delivery for added antitumor effect can be achieved with polymeric micelles as nanocarriers.

11.6.2.3 Nanoliposomes

Nanoliposomes (NLs) are bilayered, simple, with aqueous core covered phospholipid vesicles. NL structures are able to carry and accommodate both hydrophilic and hydrophobic drugs (Miura et al., 2013).

The nanoliposomal evaluation for therapy of GBM was studied by Belhadj et al. In this study, cyclic RGD c(RGDyK) and p-hydroxybenzoic acid (pHA) were formulated into nanoliposomes (c(RGDyK)/pHA-LS) to target overexpressed glioma protein integrin $\alpha_v\beta_3$ and dopamine receptor respectively. Studies on GBM cell lines U87 revealed efficacy of (c(RGDyK)/pHA-LS) nanoliposomes to target neural endothelial cell capillaries. Here, doxorubicin loaded nanoliposomes exhibited higher accumulation inside glioma tissue crossing BBB followed by BBTB. It was also observed that survival time of c(RGDyK)/pHA-LS/DOX inside intracranial glioma was 35 days, multifold on comparison respectively with free c(RGDyK), pHA-LS and doxorubicin (Belhadj et al., 2017).

11.6.2.4 Dendrimers

Dendrimers, also known as "starburst polymers" are polymeric structures with core that expands with successive branches. They are spherical and appear as layered structures, each layer of branches being called as 'generations'. The branch closest to center G0 is called as G1 while the successive branches are called G2, G3 and so on. They are about 2 to 10 nm in size with three dimensional structures, hyperbranched and can be tailor made to the need. The inter-branch spaces can be used to load the drug easily. Their typical features make them suitable for easy structural modifications and conjugation with nucleic acids, that is, RNA and DNA (Palmerston Mendes et al., 2017). They can be flanked with different functional groups on the periphery thus modifying its entrapment capacity. They have distinctive properties like good polydispersity index, water solubility, space available, and modifiable structure in the internal structure. Most commonly studied dendrimers are polyamidoamine (PAMAM) (Ali et al., 2020). PAMAM dendrimers are nonimmunogenic, biocompatible, water-soluble, with peripheral amine functional groups that can be easily modified for targeted and controlled drug delivery. Controlled drug delivery can be achieved by covalently bonding drug molecules to dendrimer surface that will release drug only on specific trigger in TME such as change in pH, temperature, some tumor characteristic enzyme or redox potential (Kim et al., 2018).

Dendrimers can target specific tissues and cells. Conjugates of dendrimers-antibody have proven better than free antibody in terms of efficiency (Patri et al., 2004). PEGylated dendrimers have shown improved solubility, higher bioavailability and lower toxicity (Kurmi et al., 2011). It has also been observed that surface modification of dendrimers by peptides, folic acid and sugar groups improve efficacy of bound or conjugated drug as these agents accommodate anticancer drugs in better manner (Kesharwani et al., 2014). Table 11.1 shows representative list of nanocarriers for treatment of brain tumors.

11.7 Novel therapies to treat brain cancers

Over the decades, there is continuous improvement in methods to deal with brain tumors. Technological advances have resulted in improved diagnostics, radiotherapy, chemotherapeutic agents, and novel nanocarrier systems to treat brain cancers. This part of the chapter covers newer ways to tackle brain cancers.

11.7.1 Artificial intelligence (AI)-enabled nanocarriers for oncotherapy

The advances in drug delivery technology and diagnostics are of little use and unlikely to get best therapeutic outcome owing to heterogeneous nature of cancer in different patients. Also, generalization of same cancer in different individuals is not possible due to complexity of brain tissues. AI integrated with nanotechnology works by well-developed algorithms and pattern analysis software to suggest customized or tailored therapy. Further, AI when presented with a problem finds solution based on patterns and huge datasets.

Individual patient's biomarker profiles are created by nanosensors. This data when fed to trained datasets, artificial neural network (ANN) algorithms workout to best possible solution that can suggest possible drug therapy, duration and likely outcome for that patient. Not only this, a prediction of nanocarrier and drug crossing BBB, retention time is also possible if previous trained data is present.

For example, Yamankurt et al. have developed high throughput screening algorithm to evaluate the efficacy of nearly 1000 nanocarrier formulations of spherical nucleic acids (SNAs) and the results were used to train an algorithm that can predict the efficacy and activity of SNA nanoformulations (Yamankurt et al., 2019). In another study, an AI model consisting of gene expression data of 82 triple negative breast cancer patients exposed to combinatorial therapies with doxorubicin, fluorouracil, paclitaxel and cyclophosphamide was developed. This pharmacogenetic predictor model was successful in prediction of dose-response correlation in 92% of the total number of patients screened (Hess et al., 2006). Putting into

TABLE 11.1 Representative list of Nanocarriers for treatment of brain cancer.

Type	Nanocarriers	Chemotherapeutic moiety	Application	Reference
Inorganic	Gold nanoparticles	Brain targeted exosomes	Improved transportation through BBB	(Khongkow et al., 2019)
	Iron oxide nanoworms	LinTT1 peptide	Tumors apoptosis	(Säälik et al., 2019)
	Magnetic nanoparticles	External heating source	Killing of glioma cells	(Sanz et al., 2017)
	Gold nanoparticle with PAA	Cisplatin	Enhanced permeability across BBB, slows growth of GBM cells	(Coluccia et al., 2018)
	Mesoporous silica nanoparticles	Asn-Gly-Arg, NGR tripeptide	High drug loading, improved drug stability	(Hu et al., 2016)
Organic	PLGA nanoparticles	Letrozole	Decreased cell proliferation in gliomas	(Tivnan et al., 2017)
	PLGA nanoparticles	Paclitaxel	Enhanced tumor cell selectivity	(Dancy et al., 2020)
	PAMAM dendrimer	Doxorubicin	Dual targeting, high stability	(Liu et al., 2019)
	Liposome	Irinotecan	Higher retention time, better BBB penetration	(Clarke et al., 2017)
	MPEG-PCL liposome	Doxorubicin and Honokiol	Suppression of proliferation, angiogenesis inhibition, cell apoptosis	(Gao et al., 2017)
	PLGA nanoparticles	Methotrexate	Sustained release, pH dependant cytotoxic killing of tumor cells	(Hassan et al., 2017)
	PEG-PLGA nanoparticles	Disulfiram	Reactive oxygen species (ROS) formation, EPR effect, enhanced apoptosis	(Madala et al., 2018)
	NGO/SPION/PLGA	5-iodo-2-deoxyuridine	Improved radiosensitizing effect, tumor volume reduction in presence of external magnetic field	(Shirvalilou et al., 2018)
Carbon nanotubes	Multiwalled, PEGylated with angiopep-2 peptide for dual targeting	Doxorubicin	High drug loading, low toxicity, good biocompatibility	(Ren et al., 2012)
Quantum dots	QD labeled A32 aptamer	Aptamer oligonucleotide	Fluorescence-guide tumor surgery, aptamer bing to epidermal growth factor receptor variant III (EGFR-vIII), on the glioma cell surface	(Tang et al., 2017)
Biomimetic nanoparticles	Albumin nanoparticles	Paclitaxel	Delayed tumor growth, enhanced survival rate	(Lin et al., 2016)
Magnetite	PVA-PEI- fluorecein isothiocyanate complex (Polyplex) coated iron-oxide magnetic nanoparticles	Carmustine (BCNU)	High drug loading, selectivity towards GBM cells	(Akilo et al., 2016)
Thermo-responsive nanoparticles	Liposomes	Paclitaxel	Quick drug release rate at hyperthermic tumor surroundings	(Rehman et al., 2017)
Virus derived platform	Gelatin-siloxane nanoparticles	Tumor-targeting aptamer (TTA) from Human immuno-deficiency virus	Enhanced BBB crossing	(Lam et al., 2018)

MPEG-PCL, methoxy poly(ethylene glycol)-poly(ε-caprolactone); PLGA, poly lactic-co-glycolic acid; PAA, polyacrylic acid; PEG, polyethylene glycol; NGO/SPION/PLGA, nano-graphene oxide-superparamagnetic iron oxide nanoparticles; PVA-PEI, polyvinyl alcohol-polyethyleneimine.

TABLE 11.2 Brief account of nanotherapeutic systems in clinical trials for treatment of brain tumors.

Drug name	Nanoparticle system	Clinical objective	Phase	NCT identifier code
Temozolomide and SGT-53 (p53DNA sequence)	Liposome For glioblastoma	Evaluation of safety, nanoparticle delivery to tumor site, and the induction of apoptosis in the tumor	2	NCT02340156
Rhenium, an isotope produced by reactor	Liposome For glioma	Packaging radioactive isotopes in nanoparticle formulation increases doses of radiation to the brain tumor site with reduced toxicity	1/2	NCT01906385
Nucleic acid component NU-129	Spherical gold nanoparticle for GBM	Nucleic acid targets Bcl2L12 gene responsible for tumor growth and apoptosis inhibition	Early phase 1	NCT03020017
ABI-009 (nab-Rapamycin)	Albumin nanoparticles for recurrent glioma and newly diagnosed GBM	Rapamycin bound to albumin nanoparticle evaluated for antineoplastic and antiangiogenic activity	2	NCT03463265
AGuIX gadolinium-based nanoparticle	Nanoparticles for metastatic brain cancer	Gadolinium-based nanoparticles make radiation work more effectively in the treatment of patients with brain metastases	2	NCT04899908
Nanoliposomal Irinotecan and metronomic temozolomide	Nanoliposome for recurrent glioblastoma	Assessment of the preliminary response rate and progression free survival of nanoliposomal irinotecan with continuous low-dose temozolomide	1/2	NCT03119064
Camptothecin	Nanoliposome for GBM, astrocytoma	Increased survival rate, enhanced drug availability	1/2	NCT00734682
Doxorubicin	Nanocells for recurrent glioblastoma multiforme (GBM).	Safety and tolerability and immunogenicity of EGFR(V)-EDV-Dox	1	NCT02766699
Irinotecan	Nanoliposome for diffuse intrinsic pontine glioma	Early efficacy study of convection enhanced delivery (CED) of irinotecan	1	NCT03086616
Doxorubicin	PEGylated nanoliposome for GBM	To determine the dose limiting toxicity and improved survival rate	2 Completed	NCT00944801

All references obtained from (U.S. National Library of Medicine, https://clinicaltrials.gov/, no date) Accessed on March 03, 2021.

practice, these dosing-treatment efficacy correlation predictors to applications of nanomedicine will improve their performance in therapies of brain cancer.

11.7.2 Gene-based nanotherapy

Gene therapy for brain cancer using nanobiotechnology is suggested in combination with radiotherapy/chemotherapy or for multi-drug resistant tumors. Gene delivery for gliomas and GBM have been explored in many forms such as cell based, using viral vectors, apoptosis inducible gene, immuno-gene therapy, small interfering RNA (siRNA), and antisense therapy.

For instance, genetically modified (GM) adenovirus VB-111 is found to selectively target brain tumor endothelial cell layer in glioma mouse model (Gruslova et al., 2015). Apart from gene delivery as therapy, reducing chemotherapy related adverse effects have been alleviated by gene therapy, for example O6-benzylguanine (O6BG) reduces GBM resistance to temozolomide (Adair et al., 2014). Also, engineered neural stem cells get distributed inside tumor cells and delivery payload in the form of gene or drug (Jeon et al., 2008). Similarly, Chimeric Antigen Receptors (CAR)-T cells are T-cells from patients' blood which are modified in laboratory with particular receptors (CAR) that bind to cancer cell proteins in medulloblastoma and ependymoma.

11.7.3 CRISPR/Cas 9-associated brain tumor therapy

Clustered Regularly Interspaced Short Palindromic Repeats (CRISPR) and CRISPR-associated protein 9 (Cas9) is a unique biotechnology tool that can edit genome sequence by cutting DNA strands at target location so that desired DNA can be removed. This works with a guide RNA (gRNA) to lead Cas 9 nuclease at target location (van der Weyden et al., 2021). Routabi et al. have reported nucleic acid-lipid nanoparticles (SNALPs) for *in vivo* transportation of CRISPR/Cas9 to GBM stem cells by destroying the epidermal growth factor receptor variant III (EGFRvIII). This EGFRvIII causes tumor cell proliferation, invasion and angiogenesis of growing GBM. NPs deliver the cargo inside the tumor cell followed by gRNA guiding Cas9 to nick DNA strands responsible for tumor growth and proliferation (Rouatbi et al., 2019).

11.7.4 Nose to brain drug delivery

Although not new and introduced in early 1980s, drug delivery to brain via nasal route has regained renewed interest mainly due to limitations of drug delivery by other routes. Nasal route of drug administration skips the BBB, non-invasive and diminishes the systemic adverse chemotherapeutic effects. The innermost region of the nasal cavity, that is, olfactory region bears olfactory nerves that directly open in olfactory bulb within CNS. Axons of olfactory dendrites extend through mucosal connective tissue to olfactory bulb in the brain. This mucosa renewed periodically is the basis for nose-to-brain drug delivery. However, drug delivery to brain is heavily dependent on factors such as physico-chemical properties of the drug, mucosa turnover rate, nasal drainage and nature, and location of the brain tumor. Many compounds such as curcumin, rhein, 5-fluorouracil (5-FU), temozolomide, and methotrexate have been tried for nose-to-brain delivery for BGM and glioma (Bruinsmann et al., 2019; Li et al., 2014; Mukherjee et al., 2016).

11.8 Clinical translation of nanotherapeutic systems for brain cancers: From bench to bedside

Many hurdles and difficulties persist in treatment options for brain tumors. All of these have their own advantages and certain limitations. Nanotechnological advances offer promising future to overcome barriers associated with conventional therapies. Several preclinical studies are reported and published some of which have entered clinical trials.

Phase I/II trials of few drugs such as temozolomide (NCT02340156), nab-rapamycin (NCT03463265), irinotecan (NCT03119064), doxorubicin (NCT02766699) are being pursued for determination of safety, efficacy, toxicity and treatment regimens. Table 11.2 shows brief overview of current clinical trials of nanocarriers for brain cancers. The scope of brain cancer nanotherapy has reached another milestone by 'theranostic' approach signifying early diagnosis alongwith possible therapy. Not only this, nanoparticles and nanocarriers are also improving imaging techniques for better understanding of tumor demarcation. Actual use of these novel theranostic systems in patients is possible only when clinical data is validated through clinical trials. Current clinical trials have aim and objective to find and understand whether nanotechnology based therapies have superiority and distinct advantages over conventional drug delivery systems. Unfortunately, conclusive data is not available in most of the studies, while those completed have translated at commercial scale in very low number in case of brain cancers (Ganipineni et al., 2018). Apart from clinical trials, establishing preclinical data to correlate animal model of GBM with that of humans is difficult task due to lack of proper animal models.

On the technical front, for large scale manufacturing of nanomaterials, it is often difficult to ascertain reproducibility in quality, safety and efficacy (Ragelle et al., 2017). Compared to conventional drug delivery systems, NPs have to undergo rigorous QC testing such as release kinetics, particle size distribution, encapsulation efficiency, and many more. All these factors make clinical translation from lab to bedside slow process (Goldberg et al., 2013).

11.9 Conclusion and future prospects

Nanotechnology and nanoengineering associated therapeutics and diagnostics have value added applications over conventional systems. They certainly help in crossing BBB, controlled drug delivery, targeting mostly tumor cells, and incorporating dual mechanism targeted drug delivery systems.

This chapter summarizes various kinds of nanocarriers, such as liposomes, nanosized micelles, carbon nanotubes, dendrimers, carbon dots and NPs (gold, zinc oxide, and silver NPs), for efficient drug delivery in the treatment of brain cancer. Smart nanotherapeutics that works on stimulation or triggering effect for drug release are also available with Emphasis towards nano-platforms responsive to the TME (i.e., reductive, acidic, enzymatic) or external stimulus (i.e., ultrasound, light, magnetic, heat, etc.) However, challenges of reproducible, scalable, and quality controllable NP production as well as commercial scale NP evaluation and screening will make it easy for faster clinical translation to clinics.

Another key challenge in design and development of nanocarriers for cancer therapy is heterogenic nature of brain tumors. There exists intra-patient as well as inter-patient variability that is unpredictable. Further, predetermination of therapeutic and clinical outcomes of onco-nanotherapeutics based on AI may help to customize nanotherapy for individual patients. More the robust dataset, better nanotherapy-outcome correlation from AI can be achieved. New genetic tool like CRISPR-Cas9 show promising future but need technical niche and is not cost effective at this juncture. This technique convincingly works in case when exact set of overexpressed gene is targeted for cleaving.

Finally, safety issues regarding nano sized therapeutic systems cannot be ignored. Toxicity issues of NPs must be given due importance and significant consideration before placing into clinical service. As NPs have their own systemic toxicity, estimation and assessment of NPs affecting nontarget brain tissues must also be looked after as a matter of concern. In conclusion, it will not be wrong to anticipate that progress of nanotherapeutics in the field of oncotherapy is set to improve patient survival time and quality of life in future.

References

Adair, J.E., et al., 2014. Gene therapy enhances chemotherapy tolerance and efficacy in glioblastoma patients. J. Clin. Invest. 124 (9), 4082–4092. The American Society for Clinical Investigation. http://doi.org/10.1172/JCI76739.

Akilo, O.D., et al., 2016. AN in vitro evaluation of a carmustine-loaded Nano-co-Plex for potential magnetic-targeted intranasal delivery to the brain. Int. J. Pharm. 500 (1), 196–209. http://doi.org/10.1016/j.ijpharm.2016.01.043.

Aldape, K., et al., 2019. Challenges to curing primary brain tumours. Nat. Rev. Clin. Oncol. 16 (8), 509–520. Springer US. http://doi.org/10.1038/s41571-019-0177-5.

Ali, E.S., et al., 2020. Targeting cancer cells with nanotherapeutics and nanodiagnostics: current status and future perspectives, Seminars in Cancer Biology. Elsevier Ltd. January. http://doi.org/10.1016/j.semcancer.2020.01.011.

Alle, M., et al., 2020. Doxorubicin-carboxymethyl xanthan gum capped gold nanoparticles: microwave synthesis, characterization, and anti-cancer activity. Carbohydr. Polym. 229, 115511. http://doi.org/10.1016/j.carbpol.2019.115511.

AlSawaftah, N., Pitt, W.G., Husseini, G.A., 2021. Dual-targeting and stimuli-triggered liposomal drug delivery in cancer treatment. ACS Pharmacol. Transl. Sci. 4 (3), 1028–1049. American Chemical Society. http://doi.org/10.1021/acsptsci.1c00066.

American Association of Neurological Surgeons, 2021. *Classification of Brain Tumors*. Available at: https://www.aans.org/en/Media/Classifications-of-Brain-Tumors.

American Cancer Society Medical and Editorial Content Team (no date) 'Brain and Spinal Cord Tumors in Adults What are adult brain and spinal cord tumors ?', American Cancer Society, pp. 1–21. Available at: https://www.cancer.org/cancer/brain-spinal-cord-tumors-adults/about/types-of-brain-tumors.html.

Bao, S., et al., 2006. Glioma stem cells promote radioresistance by preferential activation of the DNA damage response. Nature 444 (7120), 756–760. http://doi.org/10.1038/nature05236.

Belhadj, Z., et al., 2017. Multifunctional targeted liposomal drug delivery for efficient glioblastoma treatment. Oncotarget Vol 8 (No 40). Available at: https://www.oncotarget.com/article/17976/text/.

Bozzuto, G., Molinari, A., 2015. Liposomes as nanomedical devices. Int. J. Nanomed. 10, 975–999. http://doi.org/10.2147/IJN.S68861.

Brannon-Peppas, L., Blanchette, J.O., 2012. Nanoparticle and targeted systems for cancer therapy. Adv. Drug. Deliv. Rev. 64, 206–212. http://doi.org/10.1016/j.addr.2012.09.033.

Bruinsmann, F.A., et al., 2019. Nasal drug delivery of anticancer drugs for the treatment of glioblastoma: preclinical and clinical trials. Molecules 24 (23), 4312. MDPI. http://doi.org/10.3390/molecules24234312.

Bruschi, Marcos L., Borghi-Pangoni, Fernanda B., Junqueira, Mariana V., de Souza Ferreira, Sabrina B., da Silva, Jessica B., 2017. Environmentally Responsive Systems for Drug Delivery. Recent Pat Drug Deliv Formul 2212-4039, 11 (2), 89–100. http://doi.org/10.2174/187221131166617032815 1455.

Chan, M.-H., Lin, H.-M., 2015. Preparation and identification of multifunctional mesoporous silica nanoparticles for in vitro and in vivo dual-mode imaging, theranostics, and targeted tracking. Biomaterials 46, 149–158. Department of Bioscience and Biotechnology, National Taiwan Ocean University, 2 Pei Ning Road, Keelung, Taiwan 20224. http://doi.org/10.1016/j.biomaterials.2014.12.034.

Chen, A.A., et al., 2005. Quantum dots to monitor RNAi delivery and improve gene silencing. Nucleic Acids Res. 33 (22), e190. http://doi.org/10.1093/nar/gni188.

Chen, H., et al., 2021. Redox responsive nanoparticle encapsulating black phosphorus quantum dots for cancer theranostics. Bioactive Mater. 6 (3), 655–665. http://doi.org/10.1016/j.bioactmat.2020.08.034.

Cho, K., et al., 2008. Therapeutic nanoparticles for drug delivery in cancer. Clin. Cancer Res. 14 (5), LP1310–LP1316. http://doi.org/10.1158/1078-0432.CCR-07-1441.

Clarke, J.L., et al., 2017. A phase 1 trial of intravenous liposomal irinotecan in patients with recurrent high-grade glioma. Cancer Chemother. Pharmacol. 79 (3), 603–610. Division of Neuro-oncology, Department of Neurological Surgery, University of California, San Francisco (UCSF), 505 Parnassus Avenue M779, San Francisco, CA, 94143, USA. clarkej@neurosurg.ucsf.edu.. http://doi.org/10.1007/s00280-017-3247-3.

Coluccia, D., et al., 2018. Enhancing glioblastoma treatment using cisplatin-gold-nanoparticle conjugates and targeted delivery with magnetic resonance-guided focused ultrasound. Nanomed. Nanotechnol. Biol. Med. 14 (4), 1137–1148. http://doi.org/10.1016/j.nano.2018.01.021.

Dancy, J.G., et al., 2020. Decreased nonspecific adhesivity, receptor-targeted therapeutic nanoparticles for primary and metastatic breast cancer. Sci. Adv. 6 (3), eaax3931. http://doi.org/10.1126/sciadv.aax3931.

Danhier, F., Feron, O., Préat, V., 2010. To exploit the tumor microenvironment: Passive and active tumor targeting of nanocarriers for anti-cancer drug delivery. J. Control. Release 148 (2), 135–146. http://doi.org/10.1016/j.jconrel.2010.08.027.

Ferraris, C., et al., 2020. Overcoming the blood–brain barrier: Successes and challenges in developing nanoparticle-mediated drug delivery systems for the treatment of brain tumours. Int. J. Nanomed. 15, 2999–3022. http://doi.org/10.2147/IJN.S231479.

Foti, R.S., et al., 2015. "Target-Site" drug metabolism and transport. Drug Metab. Dispos. 43 (8), LP1156–LP1168. http://doi.org/10.1124/dmd.115.064576.

Fu, Z., Xiang, J., 2020. Aptamers, the nucleic acid antibodies, in cancer therapy. Int. J. Mol. Sci. 21 (8). http://doi.org/10.3390/ijms21082793.

Ganipineni, L.P., Danhier, F., Préat, V., 2018. Drug delivery challenges and future of chemotherapeutic nanomedicine for glioblastoma treatment. J. Control. Release 281 (May), 42–57. Elsevier. http://doi.org/10.1016/j.jconrel.2018.05.008.

Gao, H., 2016. Progress and perspectives on targeting nanoparticles for brain drug delivery. Acta Pharmaceutica Sinica B 6 (4), 268–286. http://doi.org/10.1016/j.apsb.2016.05.013.

Gao, J.-Q., et al., 2013. Glioma targeting and blood–brain barrier penetration by dual-targeting doxorubincin liposomes. Biomaterials 34 (22), 5628–5639. http://doi.org/10.1016/j.biomaterials.2013.03.097.

Gao, X., et al., 2017. Enhancing the anti-glioma therapy of doxorubicin by honokiol with biodegradable self-assembling micelles through multiple evaluations. Sci. Rep. 7 (1), 43501. http://doi.org/10.1038/srep43501.

Gawali, S.L., et al., 2019. pH-labile magnetic nanocarriers for intracellular drug delivery to tumor cells. ACS Omega 4 (7), 11728–11736. American Chemical Society. http://doi.org/10.1021/acsomega.9b01062.

Goldberg, M.S., et al., 2013. Biotargeted nanomedicines for cancer: six tenets before you begin. Nanomedicine 8 (2), 299–308. Future Medicine. http://doi.org/10.2217/nnm.13.3.

Gruslova, A., et al., 2015. VB-111: a novel anti-vascular therapeutic for glioblastoma multiforme. J. Neurooncol. 124 (3), 365–372. http://doi.org/10.1007/s11060-015-1853-7.

Gupta, A., Dwivedi, T., 2017. A simplified overview of World Health Organization classification update of central nervous system tumors 2016. J. Neurosci. Rural Pract. 08 (04), 629–641.

Harisinghani, M.G., et al., 2003. Noninvasive detection of clinically occult lymph-node metastases in prostate cancer. N. Engl. J. Med. 348 (25), 2491–2499. Massachusetts Medical Society. http://doi.org/10.1056/NEJMoa022749.

Hassan, M., et al., 2017. Methotrexate-loaded PLGA nanoparticles: preparation, characterization and their cytotoxicity effect on human glioblastoma U87MG cells. Int. J. Med. Nano Res. 4 (1), 1–9. http://doi.org/10.23937/2378-3664/1410020.

Hays, E.M., Duan, W., Shigdar, S., 2017. Aptamers and glioblastoma: their potential use for imaging and therapeutic applications. Int. J. Mol. Sci. 18 (12). http://doi.org/10.3390/ijms18122576.

He, Q., et al., 2020. Tumor microenvironment responsive drug delivery systems. Asian J. Pharm. Sci. 15 (4), 416–448. http://doi.org/10.1016/j.ajps.2019.08.003.

Hervault, A., Thanh, N.T.K., 2014. Magnetic nanoparticle-based therapeutic agents for thermo-chemotherapy treatment of cancer. Nanoscale 6 (20), 11553–11573. The Royal Society of Chemistry. http://doi.org/10.1039/C4NR03482A.

Hess, K.R., et al., 2006. Pharmacogenomic predictor of sensitivity to preoperative chemotherapy with paclitaxel and fluorouracil, doxorubicin, and cyclophosphamide in breast cancer. J. Clin. Oncol. 24 (26), 4236–4244. Wolters Kluwer. http://doi.org/10.1200/JCO.2006.05.6861.

Higa, M., et al., 2015. Identification of a novel cell-penetrating peptide targeting human glioblastoma cell lines as a cancer-homing transporter. Biochem. Biophys. Res. Commun. 457 (2), 206–212. http://doi.org/10.1016/j.bbrc.2014.12.089.

Hu, J., et al., 2016. Asn-Gly-Arg-modified polydopamine-coated nanoparticles for dual-targeting therapy of brain glioma in rats. Oncotarget 7 (45). Available at: https://www.oncotarget.com/article/12047/text/.

Jeon, J.Y., et al., 2008. Migration of human neural stem cells toward an intracranial glioma. Exp. Mol. Med. 40 (1), 84–91. http://doi.org/10.3858/emm.2008.40.1.84.

Joshi, B.H., et al., 2008. Identification of interleukin-13 receptor α2 chain overexpression in situ in high-grade diffusely infiltrative pediatric brainstem glioma. Neuro-oncol. 10 (3), 265–274. http://doi.org/10.1215/15228517-2007-066.

Kesharwani, P., Jain, K., Jain, N.K., 2014. Dendrimer as nanocarrier for drug delivery. Prog. Polym. Sci. 39 (2), 268–307. Elsevier.

Khaitan, D., Reddy, P.L., Ningaraj, N., 2018. Targeting Brain Tumors with Nanomedicines: Overcoming Blood Brain Barrier Challenges. Curr. Clin. Pharmacol. 13 (2), 110–119. http://doi.org/10.2174/1574884713666180412150153.

Khongkow, M., et al., 2019. Surface modification of gold nanoparticles with neuron-targeted exosome for enhanced blood–brain barrier penetration. Sci. Rep. 9 (1), 8278. http://doi.org/10.1038/s41598-019-44569-6.

Kim, B., et al., 2010. Tuning payload delivery in tumour cylindroids using gold nanoparticles. Nat. Nanotechnol. 5 (6), 465–472. http://doi.org/10.1038/nnano.2010.58.

Kim, S.-S., et al., 2015. Effective treatment of glioblastoma requires crossing the blood–brain barrier and targeting tumors including cancer stem cells: The promise of nanomedicine. Biochem. Biophys. Res. Commun. 468 (3), 485–489. http://doi.org/10.1016/j.bbrc.2015.06.137.

Kim, Y., Park, E.J., Na, D.H., 2018. Recent progress in dendrimer-based nanomedicine development. Arch. Pharm. Res. 41 (6), 571–582. http://doi.org/10.1007/s12272-018-1008-4.

Kumar, C.S.S.R., Mohammad, F., 2011. Magnetic nanomaterials for hyperthermia-based therapy and controlled drug delivery. Adv. Drug. Deliv. Rev. 63 (9), 789–808. http://doi.org/10.1016/j.addr.2011.03.008.

Kuntner, C., et al., 2010. Dose-response assessment of tariquidar and elacridar and regional quantification of P-glycoprotein inhibition at the rat blood-brain barrier using (R)-[11C]verapamil PET. Eur. J. Nucl. Med. Mol. Imaging 37 (5), 942–953. http://doi.org/10.1007/s00259-009-1332-5.

Kurmi, B.D., et al., 2011. Lactoferrin-conjugated dendritic nanoconstructs for lung targeting of methotrexate. J. Pharm. Sci. 100 (6), 2311–2320. Elsevier. http://doi.org/10.1002/jps.22469.

Lalatsa, A., et al., 2015. Chitosan amphiphile coating of peptide nanofibres reduces liver uptake and delivers the peptide to the brain on intravenous administration. J. Control. Release 197, 87–96. http://doi.org/10.1016/j.jconrel.2014.10.028.

Lalatsa, A., Schatzlein, A.G., Uchegbu, I.F., 2014. Strategies to deliver peptide drugs to the brain. Mol. Pharm. 11 (4), 1081–1093. American Chemical Society. http://doi.org/10.1021/mp400680d.

Lam, P., Lin, R.D., Steinmetz, N.F., 2018. Delivery of mitoxantrone using a plant virus-based nanoparticle for the treatment of glioblastomas. J. Mater. Chem. B 6 (37), 5888–5895. The Royal Society of Chemistry. http://doi.org/10.1039/C8TB01191E.

Li, J., et al., 2020. Nanoparticle drug delivery system for glioma and its efficacy improvement strategies: a comprehensive review. Int. J. Nanomed. 15, 2563–2582. http://doi.org/10.2147/IJN.S243223.

Li, Y., et al., 2014. Intranasal administration of temozolomide for brain-targeting delivery: therapeutic effect on glioma in rats. Nan fang yi ke da xue xue bao= J. Southern Med. Univ. 34 (5), 631–635.

Lin, T., et al., 2016. Blood–brain-barrier-penetrating albumin nanoparticles for biomimetic drug delivery via albumin-binding protein pathways for anti-glioma therapy. ACS Nano 10 (11), 9999–10012. American Chemical Society. http://doi.org/10.1021/acsnano.6b04268.

Lingineni, K., et al., 2017. The role of multidrug resistance protein (MRP-1) as an active efflux transporter on blood–brain barrier (BBB) permeability. Mol. Divers. 21 (2), 355–365. http://doi.org/10.1007/s11030-016-9715-6.

Liu, C., et al., 2019. Enhanced blood-brain-barrier penetrability and tumor-targeting efficiency by peptide-functionalized poly(amidoamine) dendrimer for the therapy of gliomas. Nanotheranostics 3 (4), 311–330. Ivyspring International Publisher. http://doi.org/10.7150/ntno.38954.

Lombardo, D., Kiselev, M.A., Caccamo, M.T., 2019. Smart nanoparticles for drug delivery application: development of versatile nanocarrier platforms in biotechnology and nanomedicine. J.Nanomater. 2019. http://doi.org/10.1155/2019/3702518.

Luo, Z., et al., 2017. Precise glioblastoma targeting by AS1411 aptamer-functionalized poly (l-γ-glutamylglutamine)–paclitaxel nanoconjugates. J. Colloid Interface Sci. 490, 783–796. http://doi.org/10.1016/j.jcis.2016.12.004.

Madala, H.R., et al., 2018. Brain- and brain tumor-penetrating disulfiram nanoparticles: Sequence of cytotoxic events and efficacy in human glioma cell lines and intracranial xenografts. Oncotarget 9 (3), 3459–3482. http://doi.org/10.18632/oncotarget.23320.

Madhankumar, A.B., et al., 2017. Interleukin-13 conjugated quantum dots for identification of glioma initiating cells and their extracellular vesicles. Acta Biomater. 58, 205–213. http://doi.org/10.1016/j.actbio.2017.06.002.

Mahajan, S., et al., 2018. Functionalized carbon nanotubes as emerging delivery system for the treatment of cancer. Int. J. Pharm. 548 (1), 540–558. http://doi.org/10.1016/j.ijpharm.2018.07.027.

Mallidi, S., et al., 2015. Prediction of tumor recurrence and therapy monitoring using ultrasound-guided photoacoustic imaging. Theranostics 5 (3), 289–301. http://doi.org/10.7150/thno.10155.

Marcucci, F., Lefoulon, F., 2004. Active targeting with particulate drug carriers in tumor therapy: fundamentals and recent progress. Drug Discov. Today 9 (5), 219–228. http://doi.org/10.1016/S1359-6446(03)02988-X.

Miura, Y., et al., 2013. Cyclic RGD-linked polymeric micelles for targeted delivery of platinum anticancer drugs to glioblastoma through the blood–brain tumor barrier. ACS Nano 7 (10), 8583–8592. American Chemical Society. http://doi.org/10.1021/nn402662d.

Mohamed, M.S., et al., 2014. Type 1 ribotoxin-curcin conjugated biogenic gold nanoparticles for a multimodal therapeutic approach towards brain cancer. Biochimica et Biophysica Acta (BBA) – Gen. Subjects 1840 (6), 1657–1669. http://doi.org/10.1016/j.bbagen.2013.12.020.

Mukherjee, S., et al., 2016. Curcumin changes the polarity of tumor-associated microglia and eliminates glioblastoma. Int. J. Cancer 139 (12), 2838–2849. John Wiley & Sons, Ltd. http://doi.org/10.1002/ijc.30398.

Muthu, M.S., et al., 2011. Vitamin E TPGS coated liposomes enhanced cellular uptake and cytotoxicity of docetaxel in brain cancer cells. Int. J. Pharm. 421 (2), 332–340. http://doi.org/10.1016/j.ijpharm.2011.09.045.

Nance, E.A., et al., 2012. A dense poly(ethylene glycol) coating improves penetration of large polymeric nanoparticles within brain tissue. Sci. Transl. Med. 4 (149), 149ra119. LP-149ra119. http://doi.org/10.1126/scitranslmed.3003594.

Ostrovsky, S., et al., 2009. Selective cytotoxic effect of ZnO nanoparticles on glioma cells. Nano Res. 2 (11), 882–890. http://doi.org/10.1007/s12274-009-9089-5.

Pajtler, K.W., et al., 2017. The current consensus on the clinical management of intracranial ependymoma and its distinct molecular variants. Acta Neuropathol. (Berl) 133 (1), 5–12. http://doi.org/10.1007/s00401-016-1643-0.

Palmerston Mendes, L., Pan, J., Torchilin, V.P., 2017. Dendrimers as nanocarriers for nucleic acid and drug delivery in cancer therapy. Molecules 22, 1–21. http://doi.org/10.3390/molecules22091401.

Patri, A.K., et al., 2004. Synthesis and in vitro testing of J591 antibody–dendrimer conjugates for targeted prostate cancer therapy. Bioconjug. Chem. 15 (6), 1174–1181. American Chemical Society. http://doi.org/10.1021/bc0499127.

Pierce, A.M., Keating, A.K., 2014. Creating anatomically accurate and reproducible intracranial xenografts of human brain tumors. JoVE (91), e52017. MyJoVE Corp. http://doi.org/10.3791/52017.

Pietilä, M., et al., 2013. Mortalin antibody-conjugated quantum dot transfer from human mesenchymal stromal cells to breast cancer cells requires cell-cell interaction. Exp. Cell Res. 319 (18), 2770–2780. National Institute of Advanced industrial Sciences and Technology, Tsukuba, Ibaraki 305 8562, Japan.. http://doi.org/10.1016/j.yexcr.2013.07.023.

Qin, Y., et al., 2010. In vitro and in vivo investigation of glucose-mediated brain-targeting liposomes. J. Drug Target. 18 (7), 536–549. Taylor & Francis. http://doi.org/10.3109/10611861003587235.

Ragelle, H., et al., 2017. Nanoparticle-based drug delivery systems: a commercial and regulatory outlook as the field matures. Expert Opin. Drug Deliv. 14 (7), 851–864. Taylor & Francis. http://doi.org/10.1080/17425247.2016.1244187.

Rao, J.U., et al., 2017. Temozolomide arrests glioma growth and normalizes intratumoral extracellular pH. Sci. Rep. 7 (1), 7865. http://doi.org/10.1038/s41598-017-07609-7.

Rastogi, V., et al., 2014. Carbon nanotubes: an emerging drug carrier for targeting cancer cells. J. Drug Deliv. 2014, 1–23. http://doi.org/10.1155/2014/670815.

Rehman, M., et al., 2017. Enhanced blood brain barrier permeability and glioblastoma cell targeting via thermoresponsive lipid nanoparticles. Nanoscale 9 (40), 15434–15440. The Royal Society of Chemistry. http://doi.org/10.1039/C7NR05216B.

Ren, J., et al., 2012. The targeted delivery of anticancer drugs to brain glioma by PEGylated oxidized multi-walled carbon nanotubes modified with angiopep-2. Biomaterials 33 (11), 3324–3333. School of Pharmacy & Key Laboratory of Smart Drug Delivery, Ministry of Education & PLA, Fudan University, Shanghai 200032, China. http://doi.org/10.1016/j.biomaterials.2012.01.025.

Richards, D.M., Endres, R.G., 2016. Target shape dependence in a simple model of receptor-mediated endocytosis and phagocytosis. PNAS 113 (22), 6113–6118. 2016/05/16. National Academy of Sciences. http://doi.org/10.1073/pnas.1521974113.

Rouatbi, N., et al., 2019. CRISPR/Cas9 gene editing of brain cancer stem cells using lipid-based nano-delivery. Neuro-oncol. 21 (Supplement_4), iv7. http://doi.org/10.1093/neuonc/noz167.029.

Roy, A., Li, S.-D., 2016. Modifying the tumor microenvironment using nanoparticle therapeutics. Wiley Interdiscip. Rev. Nanomed. Nanobiotechnol. 8 (6), 891–908. 2016/04/01. http://doi.org/10.1002/wnan.1406.

Säälik, P., et al., 2019. Peptide-guided nanoparticles for glioblastoma targeting. J. Control. Release 308, 109–118. http://doi.org/10.1016/j.jconrel.2019.06.018.

Sanz, B., et al., 2017. Magnetic hyperthermia enhances cell toxicity with respect to exogenous heating. Biomaterials 114, 62–70. http://doi.org/10.1016/j.biomaterials.2016.11.008.

Séhédic, D., et al., 2015. Nanomedicine to overcome radioresistance in glioblastoma stem-like cells and surviving clones. Trends Pharmacol. Sci. 36 (4), 236–252. Elsevier. http://doi.org/10.1016/j.tips.2015.02.002.

Shim, K.H., et al., 2014. Analysis of zinc oxide nanoparticles binding proteins in rat blood and brain homogenate. Int. J. Nanomed. 9, 217–224. http://doi.org/10.2147/IJN.S58204.

Shirvalilou, S., et al., 2018. Development of a magnetic nano-graphene oxide carrier for improved glioma-targeted drug delivery and imaging: in vitro and in vivo evaluations. Chem. Biol. Interact. 295, 97–108. Department of Medical Physics, School of Medicine, Iran University of Medical Sciences, Tehran, Iran.. http://doi.org/10.1016/j.cbi.2018.08.027.

Shubayev, V.I., Pisanic, T.R., Jin, S., 2009. Magnetic nanoparticles for theragnostics. Adv. Drug. Deliv. Rev. 61 (6), 467–477. http://doi.org/10.1016/j.addr.2009.03.007.

Sonali, 2018. Nanotheranostics: emerging strategies for early diagnosis and therapy of brain cancer. Nanotheranostics 2 (1), 70–86. Ivyspring International Publisher. http://doi.org/10.7150/ntno.21638.

Sun, H., Zhang, Y., Zhong, Z., 2018. Reduction-sensitive polymeric nanomedicines: an emerging multifunctional platform for targeted cancer therapy. Adv. Drug. Deliv. Rev. 132, 16–32. http://doi.org/10.1016/j.addr.2018.05.007.

Sung, H., et al., 2021. Global cancer statistics 2020: GLOBOCAN estimates of incidence and mortality worldwide for 36 cancers in 185 countries. CA Cancer J. Clin. n/a (n/a). American Cancer Society. http://doi.org/10.3322/caac.21660.

Tang, J., et al., 2017. Aptamer-conjugated PEGylated quantum dots targeting epidermal growth factor receptor variant III for fluorescence imaging of glioma. Int. J. Nanomed. 12, 3899–3911. http://doi.org/10.2147/IJN.S133166.

Tang, W., Fan, W., Lau, J., Deng, L., Shen, Z., Chen, X., 2019. Emerging blood-brain-barrier-crossing nanotechnology for brain cancer theranostics. Chem. Soc. Rev. 1460-4744, 48 (11), 2967–3014. http://doi.org/10.1039/c8cs00805a.

Tivnan, A., et al., 2017. Anti-GD2-ch14.18/CHO coated nanoparticles mediate glioblastoma (GBM)-specific delivery of the aromatase inhibitor, Letrozole, reducing proliferation, migration and chemoresistance in patient-derived GBM tumor cells. Oncotarget Vol 8 (No 10). Available at: https://www.oncotarget.com/article/15073/text/.

Torchilin, V.P., 2005. Recent advances with liposomes as pharmaceutical carriers. Nat. Rev. Drug Discovery 4 (2), 145–160. http://doi.org/10.1038/nrd1632.

Trivedi, R., Kompella, U.B., 2010. Nanomicellar formulations for sustained drug delivery: strategies and underlying principles. Nanomedicine 5 (3), 485–505. Future Medicine. http://doi.org/10.2217/nnm.10.10.

U.S. National Library of Medicine https://clinicaltrials.gov/ (no date) Clinical Studies Database. Available at: https://clinicaltrials.gov/ (Accessed: 3 March 2021).

van der Weyden, L., Jonkers, J., Adams, D.J., 2021. The use of CRISPR/Cas9-based gene editing strategies to explore cancer gene function in mice. Curr. Opin. Genet. Dev. 66, 57–62. http://doi.org/10.1016/j.gde.2020.12.005.

VanHandel, M., et al., 2009. Selective uptake of multi-walled carbon nanotubes by tumor macrophages in a murine glioma model. J. Neuroimmunol. 208 (1), 3–9. Elsevier. http://doi.org/10.1016/j.jneuroim.2008.12.006.

Wadajkar, A.S., et al., 2017. Tumor-targeted nanotherapeutics: overcoming treatment barriers for glioblastoma. Wiley Interdiscip. Rev. Nanomed. Nanobiotechnol. 9 (4). http://doi.org/10.1002/wnan.1439.

Wahab, R., et al., 2011. Fabrication and growth mechanism of ZnO nanostructures and their cytotoxic effect on human brain tumor U87, cervical cancer HeLa, and normal HEK cells. JBIC J. Biol. Inorganic Chem. 16 (3), 431–442. http://doi.org/10.1007/s00775-010-0740-0.

Wang, C.-H., et al., 2011. Photothermolysis of glioblastoma stem-like cells targeted by carbon nanotubes conjugated with CD133 monoclonal antibody. Nanomed. Nanotechnol. Biol. Med. 7 (1), 69–79. http://doi.org/10.1016/j.nano.2010.06.010.

Wang, C.-X., et al., 2009. Antitumor effects of polysorbate-80 coated gemcitabine polybutylcyanoacrylate nanoparticles in vitro and its pharmacodynamics in vivo on C6 glioma cells of a brain tumor model. Brain Res. 1261, 91–99. http://doi.org/10.1016/j.brainres.2009.01.011.

Wang, Z.L., 2004. Nanostructures of zinc oxide. Mater. Today 7 (6), 26–33. http://doi.org/10.1016/S1369-7021(04)00286-X.

Woodworth, G.F., et al., 2014. Emerging insights into barriers to effective brain tumor therapeutics. Front. Oncol. 4 JUL (July), 1–14. http://doi.org/10.3389/fonc.2014.00126.

Wu, X., et al., 2019. Nanoparticle-based diagnostic and therapeutic systems for brain tumors. J. Mater. Chem. B 7 (31), 4734–4750. The Royal Society of Chemistry. http://doi.org/10.1039/C9TB00860H.

Xu, F., et al., 2009. Brain delivery and systemic effect of cationic albumin conjugated PLGA nanoparticles. J. Drug Target. 17 (6), 423–434. Taylor & Francis. http://doi.org/10.1080/10611860902963013.

Xu, H.-L., et al., 2018. Silk fibroin nanoparticles dyeing indocyanine green for imaging-guided photo-thermal therapy of glioblastoma. Drug Deliv. 25 (1), 364–375. Taylor & Francis. http://doi.org/10.1080/10717544.2018.1428244.

Xu, X., et al., 2016. A novel doxorubicin loaded folic acid conjugated PAMAM modified with borneol, a nature dual-functional product of reducing PAMAM toxicity and boosting BBB penetration. Eur. J. Pharm. Sci. 88, 178–190. http://doi.org/10.1016/j.ejps.2016.02.015.

Yamankurt, G., et al., 2019. Exploration of the nanomedicine-design space with high-throughput screening and machine learning. Nat. Biomed. Eng. 3 (4), 318–327. http://doi.org/10.1038/s41551-019-0351-1.

Yang, T., et al., 2019. Glycyrrhetinic acid-conjugated polymeric prodrug micelles co-delivered with doxorubicin as combination therapy treatment for liver cancer. Colloids Surf. B 175, 106–115. http://doi.org/10.1016/j.colsurfb.2018.11.082.

Zhang, F., Xu, C.L., Liu, C.M., 2015. Drug delivery strategies to enhance the permeability of the blood–Brain barrier for treatment of glioma. Drug Des. Dev. Ther. 9, 2089–2100. http://doi.org/10.2147/DDDT.S79592.

Zhang, H., Mi, P., 2019. 12 - Polymeric micelles for tumor theranostics. In: Cui, W., Zhao, X.B.T.-T.B. (Eds.), Micro and Nano Technologies. Elsevier, pp. 289–302. http://doi.org/10.1016/B978-0-12-815341-3.00012-2.

Zhang, P., et al., 2012. Transferrin-modified c[RGDfK]-paclitaxel loaded hybrid micelle for sequential blood-brain barrier penetration and glioma targeting therapy. Mol. Pharm. 9 (6), 1590–1598. American Chemical Society. http://doi.org/10.1021/mp200600t.

Zhao, M., et al., 2018. Targeted therapy of intracranial glioma model mice with curcumin nanoliposomes. Int. J. Nanomed. 13, 1601–1610. http://doi.org/10.2147/IJN.S157019.

Zhuang, W., et al., 2012. Curcumin promotes differentiation of glioma-initiating cells by inducing autophagy. Cancer Sci. 103 (4), 684–690. John Wiley & Sons, Ltd. http://doi.org/10.1111/j.1349-7006.2011.02198.x.

Zottel, A., Videtič Paska, A., Jovčevska, I., 2019. Nanotechnology Meets Oncology: Nanomaterials in Brain Cancer Research, Diagnosis and Therapy. Materials (Basel) 1996-1944, 12 (10). http://doi.org/10.3390/ma12101588.

Chapter 12

Progress in nanotechnology-based targeted cancer treatment

Shagufta Khan[a], Vaishali Kilor[b], Dilesh Singhavi[a], Kundan Patil[c]

[a]*Department of Pharmaceutics, Institute of Pharmaceutical Education and Research, Borgaon (Meghe) Wardha, Maharashtra, India*
[b]*Department of Pharmaceutics, Gurunanak College of Pharmacy, Nagpur, Maharashtra, India*
[c]*Department of Pharmaceutics, Government College of Pharmacy, Amrawati, Maharashtra, India*

12.1 Introduction

Cancer is a principal cause of death across the globe. The World Health Organization estimates that 13 million people will die by 2030 due to cancer (World Health Organization, 2008).

Despite so much progress in the diagnosis and understanding of cancer, its cure still seems to be out of grip. There are still many types of cancer that are very difficult to treat. Therefore, for effective cancer therapy, it is necessary to understand the cancer pathophysiology, discover new anticancer drugs and develop novel delivery systems which can deliver drug precisely to the cancer cells. It is highly desirable to have innovative treatments that are effective against tumors that have become resistant to conventional therapies.

Until recently, chemotherapy was the only choice to treat most advanced cancers. But these drugs destroy healthy cells in addition to the cancerous ones, causing toxic side effects such as nausea and a debilitated immune system due to a low blood count. Older patients, having anemia or poor kidney function are especially susceptible to these complications.

With the surge in targeted therapies, there has been a drastic reduction in the toxic side effects. Targeted drugs can interact with the specific genes and proteins in cancer cells related to cancer growth. These drugs can identify the differences between the cancer cell and a normal cell, thus target specifically the cancer cell while, healthy cells are protected. A well-known example is Herceptin, a breast cancer drug.

Another possibility gaining massive interest in the recent years, to fight cancer is immunotherapy. Immunotherapy stimulates a person's immune system to fight cancer.

Immunotherapy treatments used to treat cancer include monoclonal antibodies that are artificial versions of immune system proteins, which help the body to recognize and attack cancer cells.

A new area of immunotherapy about which researchers are poised and excited is chimeric antigen receptor (CAR) T-cell therapy, where T cells, a type of body's immune are genetically altered in the lab to better fight cancer cells. Yescarta and Kymriah, belonging to this therapy already exist to treat certain blood cancers such as leukemia and lymphoma, but researchers are now exploring this therapy to treat other common cancers like breast cancer (Dees et al., 2000).

12.2 Tumor microenvironment: Comparison with normal cells

Understanding the differences in the tumor microenvironment and normal tissues is important to design new treatment strategies. Researchers have tried to explore vascular abnormalities, pH and, temperature of tumor cells to target a drug to cancer cells while protecting the normal cells. In this section, the focus will be laid on vascular perfusion, hypoxia, pH, and temperature in the tumor microenvironment.

12.2.1 Angiogenesis and endothelial permeability in cancer

Angiogenesis is the formation of new blood vessels from existing ones. In solid tumors, with volume 1 to 2 mm^3, oxygen, and nutrients can reach the center of the tumor by simple diffusion. Because of the nonfunctional or nonexistent vasculature, in these nonangiogenic tumors, they are highly dependent on their microenvironment for oxygen and the supply of nutrients. However, when tumors grow beyond 2 mm^3, a state of cellular hypoxia is developed, initiating angiogenesis (Bergers and Benjamin, 2003). Local hypoxia at the tumor site causes increased transcription of hypoxia-inducible factor (HIF)-1 which upregulates proangiogenic proteins such as vascular endothelial growth factor (VEGF), platelet-derived growth factor (PDGF) or tumor necrosis factor-α (TNF-α), and initiates mitogenic and migratory activities of endothelial cells (Carmeliet, 2000).

Activated endothelial cells express the dimeric transmembrane integrin αvβ3, which interacts with extracellular matrix proteins (vibronectin, fibronectin) and regulates the migration of the endothelial cell through the extracellular matrix during vessel formation (Avraamides et al., 2008). The activated endothelial cells synthesize proteolytic enzymes, such as matrix metalloproteinases, used to degrade the basement membrane and the extracellular matrix. The inner layer of endothelial cells undergoes apoptosis leading to the formation of the vessel lumen. Immature vasculature undergoes extensive remodeling during which the vessels are stabilized by pericytes and smooth-muscle cells. This step is often incomplete resulting in irregular shaped, dilated, and tortuous tumor blood vessels (Stollman et al., 2009). As a result of this, the endothelial pore size in tumor vessels varies from 10 to 1000 nm which is in the range of 5 to 10 nm in normal vessels (Hirsjärvi et al., 2011).

12.2.2 Microenvironment pH

In normal cells, the intracellular pH is generally considered to be lower than that in the extracellular space. However, cancer cells have a higher intracellular pH and a lower (acidic) extracellular pH. Normal cells under aerobic conditions, convert glucose to pyruvate via glycolysis in the cytoplasm and transport it to the oxygen-consuming mitochondria to produce carbon dioxide and ATP. Tumor cells, because of hypoxia, convert pyruvate to lactic acid in the cytosol. This anaerobic metabolism produces ~18-fold fewer ATP molecules relative to mitochondrial oxidative phosphorylation. The consequence of increased intracellular production of lactic acid is extracellular tumor acidosis. Tumor cells tend to maintain an intracellular pH (pHi) that is slightly alkaline (~pH 7.4) to prevent apoptosis, and for that, they upregulate several proton extrusion mechanisms such as the Na+/H+ exchanger (NHE), HCO_3- transporter, carbonic anhydrase IX, vacuolar-ATPase, and the H+/K+ ATPase. Excess protons are excreted into the extracellular matrix, causing the extracellular pH (pHe) of the tumor microenvironment to become acidic (Heiden et al., 2009).

12.2.3 Microenvironment temperature

Owing to dysfunctional tumor vascular systems and poor heat exchange, there exist mild hyperthermia in the tumor microenvironment (Qin et al., 2017). This characteristic has been explored for drug release in the tumor microenvironment. Nanocarriers functionalized with peptides that show a change in their confirmation when exposed to a slightly higher temperature in the tumor microenvironment resulted in drug release specifically at the tumor site. Al-Ahmady et al. (2012) revealed that leucine zipper peptide that forms coiled self-assembled aggregates composed of α-helix monomers, dissociate into disordered monomers at temperatures higher than 40°C. With this transformation, the peptide loses its original ordered structure, facilitating the release of drugs.

Elastin-like polypeptides are another group of temperature-sensitive peptides that change conformations in response to temperature (Macewan and Chilkoti, 2012).

12.3 Nanotechnology-based diagnosis of cancer

Most of the methods still being used for detecting cancer can detect cancers only when they have caused a visible (significant) change to the tissue and perhaps thousands of cells have proliferated and metastasized.

In recent years, nanotechnology-based diagnostics have drawn great attention to detect various types of cancer. Nanotechnology-based molecular imaging systems have shown promise because of their unique properties like small size, high atomic number, and biocompatibility. Nanotechnology-based imaging contrast agents being developed have the ability to greatly enhance the detection of tumors in vivo. Current nanoscale imaging platforms are enabling novel imaging modalities like photoacoustic tomography (PAT), Raman spectroscopic imaging, and multimodal imaging to detect cancer at an early stage.

Immune superparamagnetic iron oxide nanoparticles (SPIONs) with high specificity for cancer cells and no known side effects are used for lung cancer MRI imaging.

Nanotechnology-based techniques that can help accurately track living cells and monitor dynamic cellular events in tumors include

1. Near infrared (NIR) quantum dots
   Quantum dots emit fluorescence in the near-infrared spectrum (i.e., 700–1000 nm) that can penetrate deep into tissue and can provide higher temporal and spatial resolution. They have been designed for imaging colorectal cancer, liver cancer, pancreatic cancer, and lymphoma (Parungo et al., 2004).
   Researchers at MSKCC and Cornell University have developed silica-hybrid nanoparticles ("C-dots") with cRGDY peptides that target specific tumors.
2. Nanoshells
   Nanoshells are dielectric cores of silicon and coated with a thin metal shell-like gold. They have a size between 10 and 300 nm. These nanoshells convert plasma-mediated electrical energy into light energy and can be flexibly tuned optically through UV-infrared emission or absorption arrays (Loo et al., 2014). Nanoshells are desirable because their imaging is devoid of heavy metal toxicity.
3. Colloidal Gold Nanoparticles
   Gold nanoparticles (AuNPs) are an excellent contrast agent because of their small size, biocompatibility, and high atomic number. Research shows that AuNPs work in both active and passive ways to target cells. When the energy exceeds 80 kev, the mass attenuation rate of gold becomes higher than alternative elements like iodine, indicating a greater prospect for gold nanoparticles (Fu et al., 2018).
4. Nanobubbles
   Researchers at Stanford University have developed nano-based technologies that give insight into both anatomical size and location of prostate cancer cells (nanobubbles for ultrasound imaging) and functional information to avoid misdiagnosis/treatment as well as to monitor self-assembly of nanoparticles for photoacoustic imaging. The nanobubbles developed are coupled directly to the recently approved handheld transrectal ultrasound and photoacoustic (TRUSPA) device. This technique offers a more effective, integrated, and less invasive technique to image and biopsy prostate cancers for diagnosis and prognostication prior to performing interventions like surgery resection, radiotherapy, etc. (National Cancer Institute, 2017).

Nano-based technology is making the visualization of molecular markers possible that identify specific stages of cancer and cancer cell death induced by therapy. Thus, it allows doctors to envision cells and molecules which are undetectable through conventional imaging. Researchers at Standford developed the Target-Enabled in Situ Ligand Assembly (TESLA) nanoparticle system. This is based on nanoparticles that are formed in the body after IV-injection of molecular precursors. The precursors contain specific sequences of atoms that transform into larger nanoparticles after getting cleaved by enzymes produced by cancer cells during apoptosis (i.e., cell death). They carry various image contrast agents to monitor (PET, MRI, etc.) tumor condition in response to therapies (Ye et al., 2014). As this technology can track cancer cell death in vivo and at the molecular level, it is extremely important for delivering effective dosing regimens and novel targeted therapies.

12.4 Nanotechnology-based drug targeting strategies in cancer

12.4.1 Passive targeting

Enhanced permeation and retention effect and long circulating nanocarriers, as explained in section 12.2, due to faulty vessel formation by cancer cells, there are fenestrations in the vascular endothelium. These fenestrations have been used for passively targeting tumor cells. Through the leaky vasculature, nanoparticles can extravasate and accumulate in the tumor cells. Lymphatic vessels are absent in the tumor microenvironment, therefore after passing through the porous endothelial vessels, the nanoparticles cannot return and retain in the extracellular space. This concept is called as enhanced permeation and retention (EPR) effect. The permeation and retention of nanoparticles depend upon their size. Small-sized nanoparticles, 10 to 200 nm can effectively permeate through the fenestrations but for accumulation, size should be a little bigger (around 100 nm) as smaller particles can return back.

It has been recognized that interstitial fluid pressure (IFP) is higher at the center of the solid tumor mass compared to the periphery. It is around 20 mm Hg whereas it is near to 0 mm Hg in the normal cells. Due to the high pressure at the center, there is the flow of liquid from the center to the periphery, therefore very fine particles are flushed out with this fluid (Jain et al., 2007). However, bigger nanocarriers (~100 nm) endure the flowing liquid and are better accumulated. The cut-off size for extravasation of nanocarriers through most of the tumor endothelium is 200 nm but it may increase based on the nature of the tumor and maturity (Lee et al., 2010).

The major limitation of the particulate carriers is their rapid sequestration by the macrophages. Upon administration into the bloodstream, unprotected nanocarriers are recognized by opsonins such as the immunoglobulins, complement proteins, or receptors present on the surface of the macrophage plasma membrane. Consequently, phagocytic cells mark the opsonized nanocarriers for uptake and elimination before reaching the tumor cells. A prerequisite for the nanocarriers to extravasate through the fenestration, is that they must remain in the circulation long enough to reach the tumor vessels. Therefore, in addition to size, long circulation in the vessel is a must for effective passive targeting to the tumor cells.

Surface modifications prevent nanocarriers from being identified and cleared by the mononuclear phagocytic system (MPS) thereupon increase the stability of the nanocarriers (steric stabilization). These surface-modified particles are called the second generation "stealth" nanocarriers. Stealth nanocarriers can have a circulation half-life from a couple of hours to even a couple of days in humans (Moghimi et al., 2001).

Hydrophobic and charged particles are opsonized rapidly, therefore surface modification with hydrophilic polymers like PEG (poly(ethylene glycol)), ethylene oxide/propylene oxide block copolymers (poloxamers and poloxamines), and polysaccharides have increased circulation time of nanocarriers. Several studies have also investigated polymers such as poly(acrylamide), poly(vinyl pyrrolidone), and poly(vinyl alcohol) for the prolongation of the circulation time of nanocarriers but PEG is unchallenged.

Stealth properties depend upon the chemical nature of the surface modifying molecule, its chain length, density, conformation and flexibility on the surface affect the stealth properties. In the case of PEG, the ideal chain length is expected to be between 1500 and 5000 Da. To provide a proper steric repulsion, the density of PEG on the surface should be high enough (>8 mol% with 2000 Da PEG). Then, the chains arrange themselves to a so-called brush conformation.

Doxorubicin encapsulated in PEGylated liposomes, Caelyx (Europe)/Doxil (USA), was the first product to be marketed in the mid-nineties. It increased the elimination half-life of free doxorubicin from 0.2 hours to 55 hours; however, non-PEGylated doxorubicin liposomes (Myocet) could increase the half-life to barely 2.5 hours (Hofheniz et al., 2005). The first nonliposomal nanoparticle cancer therapeutic was Abraxane, that was approved in the year 2005. The formulation consists of paclitaxel bound with albumin and it is used in the treatment of metastatic breast cancer. It has demonstrated better efficacy and lower toxicity compared to a conventional paclitaxel formulation. In addition to the long half-life of albumin, it is assumed to accumulate in some tumors because it binds to a receptor gp60, thus promotes drug targeting (Gradishar et al., 2005). PEGylated liposome containing irinotecan (Onivyde MM-398 by (Merrimack) has been approved in 2015 for treating secondary metastatic pancreatic cancer. In addition to these examples, several passively tumor targeted nanoparticulate formulations – liposomes, micelles, polymer nanoparticles – are currently in clinical trials and some of those are already approved. The recently approved nanocarriers for targeted cancer treatment are depicted in the Table 12.1.

12.4.2 Active targeting

With an active targeting approach, the specificity of targeting can be improved and, perhaps, systemic exposure and toxicity toward healthy cells caused by the long-circulating nanocarriers could be decreased at the same time.

In active targeting, a targeting moiety is attached to the nanocarrier surface to bind the particle to the receptors expressed on the tumor/endothelial cell surface. After binding to the receptors on the cell surface, there could be two possible downstream processes; 1. internalization of the whole nanocarrier by receptor-mediated endocytosis, and 2. Interference with the cell function.

In receptor-mediated endocytosis, the plasma membrane forms an endocytotic vesicle that detaches from the membrane and moves inside the cell. This internalized endosome is delivered to lysosomes. The lysosomes have acidic pH (approximately 5) which is lethal for acid-labile drugs. The receptors which promote endocytosis include transferrin receptor (fc), folic acid receptors, and epidermal growth factor receptors (EGFR) (Kaasgaard et al., 2010) (Fig. 12.1).

Binding to receptors, such as the vascular endothelial growth factor receptors (VEGFR1/2), $\alpha_v\beta_3$ integrin, and matrix metalloproteinases (MMPs) results in the second effect. As, these receptors are involved in angiogenesis, so when the ligands bind with these receptors, they are blocked which essentially prevent the new blood vessel formation and inhibit tumor growth (Ruoslahti et al., 2010).

Success of targeting depends on (1) receptor density on the cell surface, (2) ligand density on nanocarrier, higher the density greater will be the interaction, (3) enhanced opsonization of nanocarriers.

TABLE 12.1 Nanotherapeutics recently approved for treatment of various cancer.

Product	Type of formulation	Status	Type of targeting	Indication	Comment	Ref
Lipoplatin	PEGylated liposome-loaded with cisplastin	European Medicine Agency gave orphan drug status in 2007	Passive	Nonsmall lung cancer and pancreatic, gastric, breast, head and neck cancers	Improved drug targeting, efficacy and safety specially against nephrotoxicity	EU/3/07/451.
Marqibo	Liposomal Vincristine	Approved by FDA (2012)	Passive	Acute lymphoblastic leukemia	Capable to pass through the fenestrations of tumor noevasculature	NCT00495079
Onivyde	Liposomal irinotecan (PEGylated)	FDA (2015)	passive	Metastatic adenocarcinoma of the pancreas	Improved overall survival	NCT01494506
NBTXR3 Hensify (Nanobiotix)	Hafnium oxide nanoparticles stimulated with external radiation to enhance tumor cell death via electron production	Approved 2019	Physical	Locally advanced squamous cell carcinoma	Application expanded to include prostate and lung cancer with combined immunotherapy	NCT02805894
VYXEOS CPX-351	Liposomal formulation of cytarabine:daunorubicin (5:1M ratio)	Approved 2017	Passive	Acute myeloid leukemia	Provided improved efficacy at a lower cumulative daunorubicin and cytarabine dose as compared to free drug	NCT03575325
ThermoDox	Heat-activated liposomal encapsulation of doxorubicin	Phase III	Physical (temperature triggered)	Hepatocellular carcinoma	Showed improvement in survival patients.	NCT00617981
Genexol-PM	Polymeric micelle conjugated with paclitaxel	Approved	Passive targeting	Breast cancer	Reduced toxicity and side effects	NCT00876486

12.4.2.1 Tumor cell targeting

Tumor cells upregulate the expression of various markers on the cancer cells. The most widely explored for targeting is the folic acid receptor as folic acid utilization is increased for the proliferation of cells.

Patil et al., formulated an ornamented liposome with folic acid hooked on the surface for a prodrug of mitomycin C (Patil et al., 2016). Folate was conjugated to poly(ethylene glycol) (PEG)-1,2-distearoyl-sn-glycerol-3-phosphoethanolamine (DSPE) lipid. These liposomes depicted 9-fold greater membrane binding and uptake in KB HiFR epidermal carcinoma cells in vitro, and showed distinct prodrug delivery to J6456 lymphoma cells in vivo compared with nontargeted liposomes.

The major obstacle, in targeting tumor cells, is essentially multidrug resistance due to overexpression of P-glycoprotein (P-gp). P-gp, is an efflux pump located in the cell membrane that eliminates anticancer drugs from the cells.

In a study, hydroxystearate of PEG (Solutol HS 15) was attached to the surface of a lipid nanoparticles loaded with paclitaxel. When this formulation was intratumorally injected into the subcutaneous F98 glioma model, there was a significant reduction in tumor mass and volume as compared to the conventional formulation Taxol. PEG hydroxystearate blocks the action of PG and prevent the efflux of drug (Lacoeuille et al., 2007).

Similarly, a folate-targeted liposome loaded with nitrooxy-doxorubicin (N-DOX) was fabricated to overcome P-glycoprotein (P-gp)-mediated efflux of DOX from multidrug-resistant (MDR) cells. A nitric oxide (NO)-releasing group on DOX prevents MDR by inducing NO-mediated inhibition of P-gp. Upon administration in a mouse model,

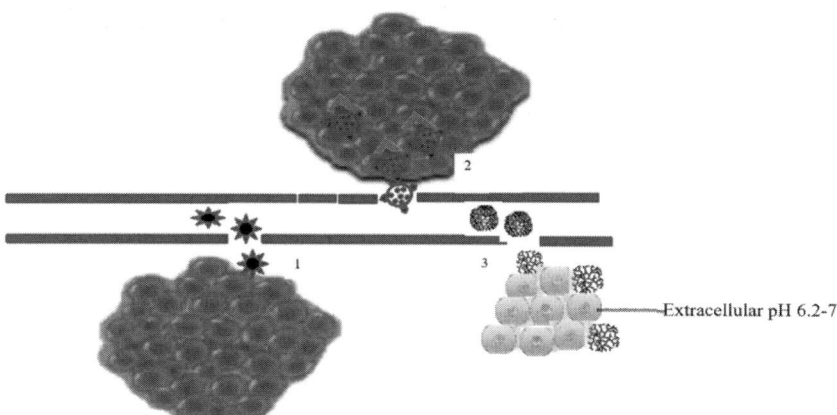

FIG. 12.1 Permeation of different types of nanocarriers (PEGylated liposome, Drug loaded nanoparticle functionalized with ligands and pH-sensitive nanocarrier) through the leaky capillary endothelium of cancer cells and targeting by 1. Passive targeting, 2. Active targeting, and 3. Physical targeting.

superior cellular uptake of N-DOX to DOX-resistant MCF7 breast cancer cells coupled with enhanced antitumor efficacy was observed compared with Caelyx (a liposomal formulation of DOX) which is not engineered to overcome MDR (Pedrini et al., 2014).

Transferrin (Tf) is an iron-binding glycoprotein, responsible for iron transport in the body. Transferrin receptors, participate in the regulation of cell growth, therefore, they are also highly expressed in cancer cells. Nanocarriers decorated with transferrin ligand have shown greater localization in tumor tissue as compared to nonfunctionalized nanocarriers (Choi et al., 2010).

Lectin proteins are found in large quantities on the cancer cell surfaces and they are capable of internalizing certain carbohydrates that can be grafted on the nanocarrier surface.

Epidermal growth factor receptor (EGFR), vascular endothelial growth factor receptors (VEGFR1/2), $\alpha_v\beta_3$ integrin, and matrix metalloproteinases (MMPs) involved in the tumor progression, is overexpressed by several cancer types. These receptors are targeted by monoclonal antibodies, antibody fragments, peptides, and aptamers (Allen, 2002).

12.4.2.2 Tumoral endothelium targeting

By targeting tumoral endothelium, blood vessel formation is restricted and consequently the tumor growth. Because of the enhanced angiogenesis by cancer cells, there is upregulation of several receptors that are involved in the angiogenic process.

The major benefit of this approach is that the nanocarriers do not have to extravasate to reach the tumor cells.

The integrins ($\alpha 2b\beta 3$, $\alpha v\beta 3$, and $\alpha 5\beta 1$) and aminopeptidase N (CD13) are the most common targets for tumor vasculature. They are recognized by cyclic and linear derivatives of the peptide RGD (arginyl-glycyl-aspartic acid), and NGR (asparaginyl-glycyl-argininic acid), respectively.

By amending PEGylated doxorubicin liposomes by cyclic RGD-peptides, tumor growth was prevented in vivo. The effect was due to the inhibition of angiogenesis rather than cytotoxicity. In a study, PEGylated doxorubicin RGD-liposomes were found to accumulate in the tumor neovasculature in an in vivo model but, in this case, with the antiangiogenic effect, there was also cytotoxic effect due to doxorubicin (Murphy et al., 2008).

Nucleolin, is a protein overexpressed in the plasma membrane of cancer cells. Guo et al. (2011) designed PEG–PLGA nanoparticles linked to AS1411, a DNA aptamer that specifically binds nucleolin, and used this combination to deliver paclitaxel to 1347 C6 glioma cells. The nanoparticles provided improved antiglioma efficacy. The authors observed that longer circulation times and increased cytotoxicity were achieved due to the specific binding of the A1411 aptamer to nucleolin and the internalization of the particles in the cells. In vivo, this formulation produced an enhanced inhibition of tumor growth and an increased drug accumulation at the tumor on mice bearing C6 glioma xenografts and intracranial rat C6 gliomas, compared to nontargeted nanoparticles and Taxol.

12.4.3 Physical targeting

In physical targeting stealth characteristic of nanocarrieris should be maintained so that they remain in circulation until they arrive at the tumor site. Upon reaching the tumor site, they are activated by the tumor microenvironment like pH.

Stimuli-sensitive nanocarriers have been prepared using smart polymers or systems that show phase transition under certain stimuli like pH or temperature leading to drug release. In this framework, nanoparticles were prepared using a copolymer based on methoxypoly(ethylene glycol)-poly(ethylenimine)-poly(L-glutamate) using electrostatic interaction and chelate effect to encapsulate simultaneously doxorubicin and cisplatin. In vitro assays demonstrated the increased release of doxorubicin at acidic pH, showing the capacity to release the drug in a cancer setting (Xu et al., 2019).

pH-sensitive nanocarriers have been designed to utilize the slightly acidic extracellular pH of tumor cells. Ligands such as poly(l-histidine) or polysulfonamide are the pH-sensitive polymeric carriers that can cause drug release in the tumor microenvironment. The unsaturated nitrogen in the imidazole ring of histidine has lone pairs of electrons that provide it the pH-dependent amphoteric properties. Poly (histidine) acts as a weak base and acquires a cationic charge when the pH of the environment drops below 6.5. Incorporation of poly(histidine) ligand to polymeric micelles thus leads to physical destabilization in the extracellular, acidic tumor medium. Lee and colleagues documented that the blending of poly (L-lactic acid)/PEG block copolymer with polyhistidine/PEG in polymeric micelles allows destabilization and subsequent drug release in the tumor microenvironment (Lee et al., 2003). ThermoDox is a thermosensitive liposome loaded with doxorubicin. At high temperature, the phase transition starts, and the lipid bilayer of the liposome changes its shape, with doxorubicin being subsequently released. ThermoDox has been tested in phase III clinical trials involving patients with hepatocellular carcinomas and is licensed to Celsion Corporation (Columbia, MD).

12.5 Progress in nanotherapeutics for treating breast and lung cancer

12.5.1 Breast cancer

The most widely used therapeutic strategy to treat breast cancer as of now is chemotherapy. But the major problem with the conventional delivery of chemotherapeutic agents is the toxic effect on normal cells and more importantly multi-drug resistance (MDR). MDR is caused when tumor cells become resistant to structurally and functionally different chemotherapeutic agents. MDR is caused by several processes like efflux by membrane protein, detoxification by reducing the drug activation and potentiating drug metabolism, by enhancing the DNA repair mechanism, blocking apoptosis, and alteration of cell cycle regulation.

Some known substrates of the P-gp transporter are epipodophyllotoxins, vinca alkaloids, taxanes, and anthracyclines that are currently involved in chemotherapeutic regimens.

There are several approaches to inhibit the action of P-gp but the NPs can protect drugs from the action of P-gp by enclosing them. NPs can prevent the degradation of drugs, improve absorption through the epithelial membrane, alter the pharmacokinetics and distribution profile of the drug in the tissue and increase the intracellular concentration in cancer cells.

NPs are designed for both active and passive targeting of anticancer drugs to elevate the intracellular anticancer concentration. Active targeting is achieved by functionalized NPs with ligand targeted toward the receptors preferentially expressed by tumor cells. The cellular uptake, biodistribution, and clearance of NPs depend on particle size, shape, and surface chemistry.

Based on the molecular subtypes, breast cancer is categorized as luminal A, luminal B, human epidermal growth factor receptor 2 (HER2) type, and estrogen (ER)/progesterone receptor (PR) positive. Among all known subtypes ER/PR positive subtypes contribute approximately 70% of all reported cases. While, around 20% of the breast cancers lack HER-2, ER, and PR expression and are called triple-negative breast cancer (TNBC) (Sorlie et al., 2001).

Tamoxifen, doxorubicin, and paclitaxel nanoparticles are currently in clinical use for breast cancer and they have shown better accumulation in the cancer cells.

1. NPs Conjugated with Topoisomerase Inhibitor Drugs. Topoisomerase inhibitors molecularly target topoisomerase enzymes that are essential for unwinding DNA during replication which leads to cell cycle arrest at G1 and G2 and ultimately apoptosis after DNA repair fails (Buchholz et al., 2002).
2. NPs Conjugated with Tubulin Assembly Inhibitor Drugs. Tubulin assembly targeting drugs interfere with mitotic assembly of tubulin, induce chromosome segregations and finally cell death. Paclitaxel is an example tubulin assembly inhibitor.

    Conjugation of paclitaxel with NPs improved treatment effectiveness. Investigations on breast cancers with a metastatic stage show the dramatic eradication of metastasis after administration of paclitaxel in complex with NPs.

Abraxane is paclitaxel bound to albumin approved by FDA for the treatment of breast, lung, and pancreatic cancers. Genexol-PM, a polymeric micelle conjugated with paclitaxel, is approved in Korea for breast cancer treatment, NK-105 conjugated with polymeric micelle is in phase III for metastatic breast cancer treatment (Majumder et al., 2020), EndoTAG-1 is liposomal paclitaxel for pancreatic, hepatic, and breast cancers.

3. NPs Conjugated with Drugs That Target DNA Replication. Cisplastin is a drug which target DNA replication. It interferes with DNA replication primarily via its interaction with purine bases in DNA to form a DNA protein crosslink. Cisplatin-induced DNA damage activates many cellular signaling. It activates cell cycle checkpoints that transitory induces the S phase and subsequently inhibits Cdc2-cyclin A to induce G2 arrest. It also induces cyclin-dependent kinase 4 (Cdk4) inhibitor p16INK4A, which results in cell cycle arrest at G1 (Shapiro et al., 1998).

4. Liposomes
Several liposomal formulations are successful on market including AmBisome (amphotericin B), Doxil (doxorubicin hydrochloride), and Visudyne (verteporfin). These formulations inhibit direct P-gp efflux or bypassing P-gp through an endocytosis pathway. These liposomes were coated with PEG to overcome opsonization and remain in circulation for long so that they can extravasate through leaky vasculature to the cancer site. These PEGylated liposomes have been shown to improve biodistribution of drugs, that is, improved tumor uptake. Therefore, better efficacy and safety as compared to their conventional counterparts are observed (Wang et al., 2012).

5. Polymeric NPs
Polymeric NPs exhibit some unique characteristics like the ease of manufacturing, biocompatibility, and flexibility toward functionalization. They can be made multifunctional to achieve active targeting and diagnosis as well. There are several studies involving functionalized nanoparticles for active targeting of breast cancer yet none has been approved for therapeutic use.

Anhorn et al. developed doxorubicin-doped human serum albumin nanoparticles surface-functionalized with trastuzumab via covalent binding. DOX–NP–trastuzumab bound to 73% of HER2 overexpressing SK-Br-3 breast cancer cells, while the DOX–NP bioconjugated with nonspecific IgG bound to less than 6% of SK-Br-3 cells. The increase in cellular binding was dependent on the concentration of trastuzumab. The binding was not observed with MCF7 breast cancer cells, a cell line that only weakly expresses HER2 (Anhorn et al., 2008). There was enhanced cellular uptake and intracellular distribution of DOX–NP–trastuzumab compared with the control DOX–NP PEGylated with Methoxy-PEG were:

12.5.2 Lung cancer

Lung cancer alone causes more deaths than the deaths caused collectively by colon, breast, and prostate cancer across the globe, and more than 2 million new cases of lung cancer are diagnosed every year (World Health Organization, 2018). Despite advancements in treatment strategies for lung cancer, many patients with NSCLC still have a poor prognosis and limited treatment options (Cappuzzo et al., 2009). This includes the nearly 70% of NSCLC patients who have a genomic mutation (Hirsch et al., 2016). To determine the most appropriate treatment, medical organizations recommend comprehensive genomic testing for patients with lung cancer as part of their upfront diagnosis (Sadiq and Salgia, 2013).

Lung cancer is broadly characterized into small cell lung cancer (SCLC) and nonsmall cell lung cancer (NSCLC). NSCLC is categorized into squamous cell carcinoma (SCC), adenocarcinoma (ADC), and large cell carcinoma (LCC). SCLC is usually found on the main bronchus while SCC initiates on the main stem lobar or segmental bronchi. ADC and LCC develop on peripheral lung tissues.

Generally, all the stages of SCLC are treated with chemotherapy or chemoradiotherapy since surgery is insufficient to control the disease due to the rapid doubling and metastatic characteristics of the malignant cells. Surgery is the treatment of choice for stage 1 NSCLC, during this stage, the tumor is localized to one area of the lung and has not spread to the draining lymph nodes (Cheng et al., 2016).

Current lung cancer treatments often cause adverse short-term and long-term side effects, including nausea, ulcers, cardiotoxicity, increased risk of developing other cancers, and cytotoxicity to healthy cells. Researchers are working hard to develop targeted therapies based on molecular pathways, receptors, and microenvironment of lung cancer.

Polymeric nanoparticles and liposomes for lung cancer:

The encapsulation of chemotherapeutics in nanocarriers, such as liposomes, dendrimers, micelles, reduces cytotoxicity, enables targeted drug delivery, sustained drug release, and avoids drug resistance. In most studies, surface modification of the nanocarriers is done to exploit the tumor microenvironment.

Lactoferrin-chondroitin sulfate nanocomplexes (~190 nm) were prepared to codeliver doxorubicin (Dox) and elagic acid. The nanocomplexes were made by electrostatic interaction between lactoferrin and chondroitin sulfate. Due to the overexpression of CD44 and lactoferrin receptors on the surface of lung cancer cells, these nanocomplexes were able to recognize cancer cells with the help of chondroitin sulfate and lactoferrin content, respectively. The authors suggested that clathrin-mediated endocytosis could have been responsible for the internalization of nanocomplexes, as their size is within the range of the pore size of the clathrin receptor (up to 200 nm) (Rejman et al., 2004).

Chitosan-coated PLGA nanoplexes were proposed to carry an antisense oligonucleotide against the human telomerase RNA component, as telomerase activity is detected in most NSCLC. The potential of the oligonucleotide as a telomerase inhibitor has been described (Dong et al., 2012) although its poor cellular uptake hinders its use in cancer therapy.

Codelivery of gefitinib and vorinostat in hyaluronan-based copolymer nanoparticles reduces the hepatic toxicity, enhances sustained drug release, and inhibits orthotopic lung tumors in mice compared to the free drugs (Jeannot et al., 2018). Conjugating the growth factor, folic acid, on dendrimers improves targeted co-delivery of siRNA and cisplatin resulting in apoptosis of the folate receptor overexpressing H1299 lung cancer cells with negligible toxicity toward MRC9 lung fibroblasts (Amreddy et al., 2018).

Liu et al. have demonstrated the potential of using liposomal nanoparticles in effectively delivering 1,25-dihydroxy vitamin D3, which promotes epithelial differentiation and inhibits NSCLC, into erlotinib resistant HCC827 lung cancer cells (Liu et al., 2018).

To improve the targeted drug release, a combined strategy using physical and active targeting is also attempted wherein a pH and temperature-sensitive nanoparticle was conjugated with folic acid. The shell of the nanoparticle comprised of a copolymer of poly(N-isopropylacrylamide) and carboxymethylchitosan while the core was composed of PLGA and an image contrast agent (superparamagnetic iron oxide, SPIO). PLGA allows the controlled release of the encapsulated drug (gemcitabine in this case), while SPIO performs the dual role of contrasting agent and temperature inducing agent by external application of an alternating magnetic field. SPIO-induced temperature changes lead to the conformational change of polymeric shell, allowing drug release. The pH-sensitive shell gave drug release at the acidic pH of the cancer microenvironment. In addition to physical targeting, active targeting could be achieved owing to the surface conjugation with folic acid, which could bind with the folate receptor in cancer cells (Menon et al., 2017).

A remedy toward a resistant form of cancer was proposed as inhalable self-assembled nanoparticles comprised of human serum albumin (HSA), tumor necrosis factor (TNF)-related apoptosis-inducing ligand (TRAIL), and Dox. Dox was conjugated to HSA and formed nanoparticles were then coated with TRAIL. Tests in H226 cells, which are representative of NSCLC, have shown that the simultaneous presence of Dox and TRAIL enabled increased cytotoxic potential, as cell viability after 3 days of exposure decreased from approximately 60% when only one of the molecules was present in HSA nanoparticles, to 20% to 30% after a dual association of Dox and TRAIL (Choi et al., 2015).

The U.S. Food and Drug Administration has recently approved Tabrecta (capmatinib) for the treatment of adult patients with nonsmall cell lung cancer (NSCLC) that has spread to other parts of the body. This is the first therapy approved by FDA to target metastatic NSCLC.

Tabrecta is a kinase inhibitor and functions by blocking a key enzyme that stops the tumor cells from growing. The FDA approved Tabrecta, based on the results of a clinical trial involving patients with NSCLC having mutations that lead to MET exon 14 skipping, epidermal growth factor receptor (EGFR) wild-type and anaplastic lymphoma kinase (ALK) negative status, and at least one measurable lesion.

Although several strategies have been attempted to improve lung cancer treatment, there is a need to cover a lot of ground. The use of the inhalation route is an attractive possibility for better targeting efficiency to the lungs than the i.v. route, yet it is not quite explored in lung cancer targeting. The success depends upon the deposition of nanocarriers at the right location in the lungs. However, the origin of lung cancer varies upon the type, like small cell lung cancer originates centrally on bronchus epithelium, while nonsmall cell lung cancer, such as adenocarcinoma and large cell carcinoma, stems from peripheral bronchioles and alveolar epithelium, therefore it is challenging to deliver the nanocarriers at these varied locations. Also, when lung cancer is detected at the later stage, cancer has already been spread to the distal lymph nodes, therefore local drug delivery, as well as systemic absorption of the drug is desired. Patients with the advanced stage of lung cancer experience difficulty in breathing, therefore, the inhalable formulation is not appropriate for them.

12.6 Future of nanotechnology in cancer treatment

Nanotechnology-based targeted delivery systems have manifested improvement in therapeutic efficacy and reduction in side effects as evidenced by the results of clinical trials that generated several FDA approvals in recent years. Nonetheless, a lot of nanocarrier-based targeted delivery systems are still at the preclinical stage so there is a lack of clinical data on them. Perhaps, many investigations are not taken forward for clinical trials. Although there is a rising interest in nanotechnology based diagnosis and treatment of cancer among researchers as evidenced by the trends in research publications, there is a serious need of enhanced collaboration between academia and pharmaceutical industries so that translation of these therapeutics from labs to clinics would be accelerated. Additionally, almost all the nano-based targeted delivery systems that are approved for clinical use are based on the principle of passive targeting. Nanotherapeutics capable of actively targeting cancer cells are still not successful to be approved for clinical use. Given the better targeting efficiency of nanocarriers decorated with ligands, the further success of targeted cancer treatment will be implied by the success of active targeting principles in clinics in the future.

12.7 Conclusion

Many investigations in animal models, as well as humans, revealed that nanocarrier-based systems were successful in overcoming limitations of conventional therapy in treating various types of cancers. In recent years, several nanotherapeutics were approved by FDA for clinical application. The recent approval of VYXEOS and Hensify has paved the way to remodel the existing treatment and diagnosis strategies for cancer treatment and nanocarrier-based targeted delivery systems will continue to make breakthroughs to improve cancer therapy and management.

References

Al-Ahmady, Z.S., Al-Jamal, W.T., Bossche, J.V., Bui, T.T., Drake, A.F., Mason, A.J., Kostarelos, K., 2012. Lipid-peptide vesicle nanoscale hybrids for triggered drug release by mild hyperthermia in vitro and in vivo. ACS Nano 6, 9335–9346.

Allen, T.M., 2002. Ligand-targeted therapeutics in anticancer therapy. Nat. Rev. Cancer 2 (10), 750–763.

Amreddy, N., Babu, A., Panneerselvam, J., Srivastava, A., Muralidharan, R., Chen, A., Zhao, Y.D., Munshi, A., Ramesh, R., 2018. Chemo-biologic combinational drug delivery using folate receptor-targeted dendrimer nanoparticles for lung cancer treatment. Nanomedicine 14 (2), 373–384.

Anhorn, M.G., Wanger, S., Kreuter, J., Langer, K., von Briesen, H., 2008. Specific targeting of HER2 overexpressing breast cancer cells with doxorubicin-loaded trastuzumab-modified human serum albumin nanoparticles. Bioconjug. Chem. 19 (12), 2321–2331.

Avraamides, C.J., Garmy-Susini, B., Varner, J.A., 2008. Integrins in angiogenesis and lymphangiogenesis. Nat. Rev. Cancer 8 (8), 604–617.

Bergers, G., Benjamin, L.E., 2003. Tumorigenesis and the angiogenic switch. Nat. Rev. Cancer 3 (6), 401–410.

Buchholz, T.A., Stivers, D.N., Stec, J., et al., 2002. Global gene expression changes during neoadjuvant chemotherapy for human breast cancer. Cancer J. 8 (6), 461–468.

Cappuzzo, F., Marchetti, A., et al., 2009. Increased MET gene copy number negatively affects survival of surgically resected non-small-cell lung cancer patients. J. Clinoncol. 27 (10), 1667–1674.

Carmeliet, P., 2000. Mechanisums of angiogenesis and arteriogenesis. Nat. Med. 6 (4), 389–395.

Cheng, T.Y., Cramb, S.M., Baade, P.D., Youlden, D.R., Nwogu, C., Reid, M.E., 2016. The international epidemiology of lung cancer: latest trends, disparities, and tumor characteristics. J. Thorac. Oncol. 11 (10), 1653–1671.

Choi, C.H., Alabi, C.A., Webster, P., Davis, M.E., 2010. Mechanism of active targeting in solid tumors with transferrin-containing gold nanoparticles. Proc. Natl. Acad. Sci. U S A 107 (3), 1235–1240.

Choi, S.H., Byeon, H.J., Choi, J.S., Tho, L., Kim, I., Lee, E.S., Lee, K.C., Youn, Y.S., 2015. Inhalable self-assembled albumin nanoparticles for treating drug-resistant lung cancer. J. Control Release 197, 199–207.

Dees, S., Ganesan, R., Singh, S., Grewal, S.I., 2020. Emerging car-t cell therapy for the treatment of triple-negative breast cancer. Mol. Cancer Ther. 19 (12), 2409–2421.

Dong, M., Mürdter, T.E., Philippi, C., Loretz, B., Schaefer, U.F., Leher, C.M., Schwab, M., Ammon-Teriber, S., 2012. Pulmonary delivery and tissue distribution of aerosolized antisense 2'-O-Methyl RNA containing nanoplexes in the isolated perfused and ventilated rat lung. Eur. J. Pharm. Biopharm. 81 (3), 478–485.

Fu, F., Li, L., Luo, Q., Guo, T., Yu, M., Song, Y., Song, E., 2018. Selective and sensitive detection of lysozyme based on plasma resonance light-scattering of hydrolyzed peptidoglycan stabilized-gold nanoparticles. Analyst 143 (5), 1133–1140.

Gradishar, W.J., Tjulandin, S., Davidson, N., Shaw, H., Desai, N., Bhar, P., Hawkins, M., O'Shaughnessy, J., 2005. Phase III trial of nanoparticle albumin-bound paclitaxel compared with polyethylated castor oil-based paclitaxel in women with breast cancer. J. Clin. Oncol. 23 (31), 7794–7803.

Guo, J., Gao, X., Su, L., Xia, H., Gu, G., Pang, Z., Jiang, X., Yao, L., Chen, J., Chen, H., 2011. Aptamer-functionalized PEG-PLGA nanoparticles for enhanced anti-glioma drug delivery. Biomaterials 32 (31), 8010–8020.

Heiden, M.G.V., Cantley, L.C., Thompson, C.B., 2009. Understanding the Warburg effect: the metabolic requirements of cell proliferation. Science 324, 1029–1033.

Hirsch, F.R., Suda, K., Wiens, J., Jr Bunn, P.A., 2016. New and emerging targeted treatments in advanced non-small-cell lung cancer. Lancet 388 (10048), 1012–1024.

Hirsjärvi, S., Passirani, C., Benoit, J.P., 2011. Passive and active tumor targeting with nanocarrier. Curr. Drug Discov. Technol. 8 (3), 188–196.

Hofheinz, R.D., Gand-Vogt, S.U., Beyer, U., Hochhaus, A., 2005. Liopsomal encapsulated anti-cancer drugs. Anticancer Drugs 16 (7), 691–707.

Jain, R.K., Tong, R.T., Munn, L.L., 2007. Effect of vascular normalization by antiangiogenic therapy on interstitial hypertension, peritumor edema, and lymphatic metastasis: insights from a mathematical model. Cancer Res. 67 (6), 2729–2735.

Jeannot, V., Gauche, C., Mazzaferro, S., et al., 2018. Anti-tumor efficacy of hyaluronan-based nanoparticles for the co-delivery of drugs in lung cancer. J. Control Release 275, 117–128.

Kaasgaard, T., Andresen, T.L., 2010. Liposomal cancer therapy: exploiting tumor characteristics. Expert Opin. Drug Deliv. 7 (2), 225–243.

Lacoeuille, F., Hindre, F., Moal, F., Roux, J., Passirani, C., Couturier, O., Cales, P., Le Jeune, J.J., Lamprecht, A., benoit, J.P., 2007. In vivo evaluation of lipid nanocapsules as a promising colloidal carrier for paclitaxel. Int. J. Pharm. 344 (1-2), 143–149.

Lee, E.S., Shin, H.J., Na, K., Bae, Y.H., 2003. Poly(L-histidine)-PEG block copolymer micelles and pH-induced destabilization. J. Control Release 90 (3), 363–374.

Lee, H., Fonge, H., Hoang, B., Reilly, R.M., Allen, C., 2010. The effect of particle size and molecular targeting on the intratumoral and subcellular distribution of polymeric nanoparticles. Mol. Pharm. 7 (4), 1195–1208.

Liu, C., Shaurova, T., Shoemaker, S., et al., 2018. Tumor-targeted nanoparticles delivery a vitamin D-based drug payload for the treatment of EGFR tyrosine kinase inhibitor-resistant lung cancer. Mol. Pharamceutics 15 (8), 3216–3226.

Loo, C., Lin, A., Hirsch, L., Lee, M.H., Borton, J., Halas, N., West, J., Drezek, R., 2004. Nanoshell-enabled photonic-based imaging and therapy of cancer. Technol. Cancer Res. Treat. 3 (1), 33–40.

Macewan, S.R., Chilkoti, A., 2012. Digital switching of local arginine density in a genetically encoded self-assembled polypeptide nanoparticle controls cellular uptake. Nano Lett. 12, 3322–3328.

Majumder, N., G Das, N.G., Das, S.K., 2020. Polymeric micelles for anticancer drug delivery. Ther. Deliv. 11 (10), 613–635.

Menon, J.U., Kuriakose, A., Iyer, R., Hernandez, E., Gandee, L., Zhang, S., Takahashi, M., Zhang, Z., Saha, D., Nguyen, K.T., 2017. Dual-drug containing core-shell nanoparticles for lung cancer therapy. Sci. Rep. 7, 13249.

Moghimi, S.M., Hunter, A.C., Murray, J.C., 2001. Long-circulating and target-specific nanoparticles: theory to practice. Pharmacol. Rev. 53 (2), 283–318.

Murphy, E.A., Majeti, B.K., Barnes, L.A., Makale, M., Weis, S.M., Lutu-Fuga, K., Wrasidlo, W., Cheresh, D.A., 2008. Nanaoparticle-mediated drug delivery to tumor vasculature suppresses metastasis. Proc. Natl. Acad. Sci. U S A 105 (27), 9343–9348.

National Cancer Institute, 2017, Nanotechnology and early cancer detection and diagnosis. Available at: https://www.cancer.gov/nano/cancer-nanotechnology/detection-diagnosis. (Accessed 6 June 2021).

Parungo, C.P., Ohnishi, S., De Grand, A.M., Laurence, R.G., Soltesz, E.G., Colson, Y.L., Knag, P.M., Mihaljevic, T., Cohn, L.H., Frangionic, J.V., 2004. In vivo optical imaging of pleural space drainage to lymph nodes of prognostic significance. Ann. Surgoncol. 11 (12), 1085–1092.

Patil, Y., Amitay, Y., Ohana, P., Shmeeda, H., Gabizon, A., 2016. Targeting of pegylated liposomal mitomycin-C prodrug to the folate receptor of cancer cells: intracellular activation and enhanced cytotoxixity. J. Control Release 225, 87–95.

Pedrini, I., Gazzano, E., Chegaev, K., Rolando, B., Marengo, A., Kopecka, J., Fruttero, R., Ghigo, D., Arpicco, S., Riganti, C., 2014. Liposonal nitrooxy-doxorubicin: one step over caelyx in drug- resistant human cancer cells. Mol. Pharmaceutics 11 (9), 3068–3079.

Qin, H., Ding, Y., Mujeeb, A., Zhao, Y., Nie, G., 2017. Tumor microenvironment targeting and responsive peptide-based nanoformulations for improved tumor therapy. Mol. Pharmacol. 92, 219–231.

Rejman, J., Oberle, V., Zuhorn, I.S., Hoekstra, D., 2004. Size-dependent internalization of particle via the pathways of clathrin- and caveolae-mediated endocytosis. Biochem. J. 377 (Pt 1), 159–169.

Ruoslahti, E., Bhatia, S.N., Sailor, M.J., 2010. Targeting of drugs and nanoparticles to tumors. J. Cell Biol. 188 (6), 759–768.

Sadiq, A.A., Salgia, R., 2013. MET as a possible target for non-small-cell lung cancer. J. Clin. Oncol. 31 (8), 1089–1096.

Shapiro, G.I., Edwards, C.D., Ewen, M.E., Rollins, B.J., 1998. P16*INK4A* participates in a G_1 arrest checkpoint in response to DNA damage. Mol. Cell Biol. 18 (1), 378–387.

Sorlie, T., Perou, C.M., Tibshirani, R., Aas, T., Geisler, S., Johnsen, H., Hastie, T., Eisen, M.B., van, P.E., Borresen-Dale, A.L., 2001. Gene expression patterns of breast carcinomas distinguish tumor subclasses with clinical implications. Proc. Natl. Acad. Sci. U S A 98 (19), 10869–10874.

Stollman, T.H., Ruers, T.J., Oyen, W.J., Boerman, O.C., 2009. New targeted probes for radioimaging of angiogenesis. Methods 48 (2), 188–192.

Wang, A.Z., Langer, R., Farokhzad, O.C., 2012. Nanoparticle delivery of cancer drugs. Annu. Rev. Med. 63, 185–198.

World Health Organization, 2008, Key statistics. Available at: www.who.int›cancer›resources›keyfact (Accessed 6 June 2021).

World Health Organization, 2018, Cancer Fact Sheet. Available at: https://www.who.int/news-room/fact-sheets/detail/cancer (Accessed 5 June 2021).

Xu, C., Wang, Y., Guo, Z., Chen, J., Lin, L., Wu, J., Tian, H., Chen, X., 2019. Pulmonary delivery by exploiting doxorubicin and cisplatin co-loaded nanoparticles for metastatic lung cancer therapy. J Control Release, 153–163.

Ye, D., Shuhendler, A.J., Cui, L., Tong, L., Tee, S.S., Tikhomirov, G., FelsherDw, R.J., 2014. Bio-orthogonal cyclization-medicated in situ self-assembly of small-molecule probes for imaging caspase activity in vivo. Nat. Chem. 6 (6), 519–526.

Chapter 13

Nanotherapeutics for colon cancer

Nilesh M. Mahajan[a], Alap Chaudhari[b], Sachin More[c], Purushottam Gangane[a]
[a]*Dadasaheb Balpande College of Pharmacy, Rashtrasant Tukadoji Maharaj Nagpur University, Nagpur, Maharashtra, India*
[b]*Formulation R&D, Teva Pharmaceuticals, Weston, FL, USA*
[c]*Dept of Pharmacology, Dadasaheb Balpande College of Pharmacy, Besa, Nagpur, MS, India*

13.1 Introduction

The term colorectal express the physiology of both the colon and rectal. Cancer of the colon and rectum are together called colorectal cancers (CRC) occurs when abnormal cells grow in this region. This broad term is used due to the fact of the difficulties in distinguishing cancers emerging in the region of the colon ends or the rectum begins. CRC is the third-highest rigorous health problem among the tumor causing conditions in developed and developing countries. It is the third most commonly found cancer and the fourth most cause of death related to cancer. Colon cancer is the prime reason for morbidity and death in the population of western countries (Ryan-Harshman and Aldoori, 2007; Favoriti et al., 2016). Lifestyle and diet being the most apparent risk factors, it is most prevalent in developed countries than the developing countries. It is predicted that by the year 2035 the overall cases of CRC may be increased to 2 to 3 million as a result of changing lifestyles in developing countries (Bray et al., 2018).

After assessing the anatomy and pathophysiology of CRC, we will look forward to the various perspective of CRC such as origination, genesis, initiation and progression. Several factors are responsible for the initiation and progression of CRC including genes, lifestyle, age, personal and family history, smoking being, etc. Environmental factors like chemicals, infectious agents, radiations and genetic factors like mutations, immune system, and hormonal dysfunction can interact in various ways to exaggerate the carcinogenesis (Giovannucci, 2001; Nakaji et al., 2003).

CRC mostly starts in the bowel lining and can grow into muscle layers and then eventually through the bowel wall. The reported etiology is hyperplasia due to activated oncogenes that can save the cancerous cell against apoptosis. Moreover, dysfunction of cellular processes owing to the inactivation of tumor suppressor genes in cancer cells has also been reported. The major intention of cancer therapy is to destroy the cancer cells without making any harm to the normal cells and hence it is very much empirical to search for the selective or targeted drug delivery system as a cancer therapy (Dujovny et al., 2004; Standring, 2019).

13.1.1 Anatomy

The colon and the rectum form the large intestine, which is a part of the gastrointestinal tract (GIT). Anatomically, the large intestine is divided into seven parts (Fig. 13.1) such as the cecum; the ascending, transverse, descending, and sigmoid colon; the rectum; and the anus. The rectum is an eight-inch section of the large intestine at the end of the colon. The anus is the final three inches part of the colon through which fecal matter is expelled. The ring of muscle located at the anus is called the sphincter which helps for controlled removal of feces from GIT (Granados-Romero et al., 2017; Kim, 2020). The wall of the large intestine comprises four structural layers:

A. The innermost layer is the mucosa, which is composed of three separate sublayers: Epithelium, connective tissue, and muscle. The epithelium, which contains crypts (pits or depressions) and encompasses immediate contact with the contents of the colon. The cells deep inside the crypts have a high proliferative index and are the originating site of most CRC.
B. The submucosa is the second layer of the large intestine and contains more connective tissue, blood vessels, lymphatic glands, and nerves.

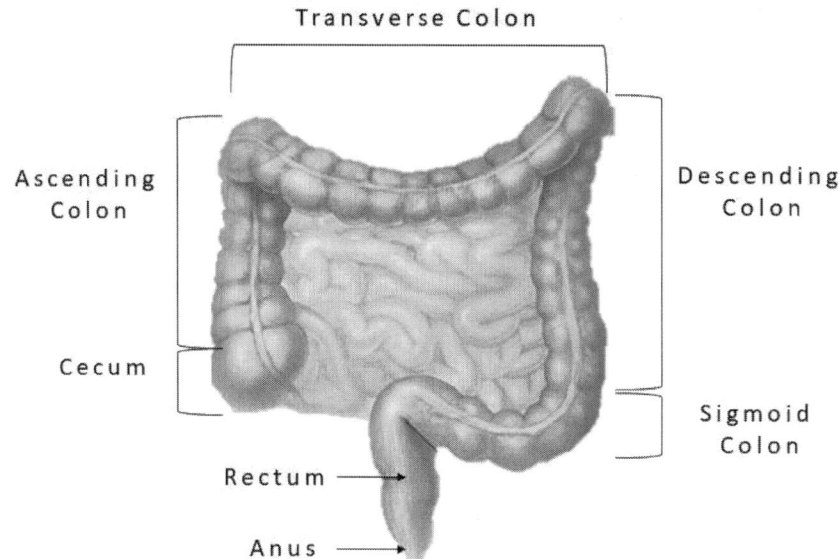

FIG. 13.1 Divisions of the large intestine.

C. The third layer comprises muscle that controls the contraction of the colon and facilitates the movement of digested food through the tract.
D. The fourth and outermost layer of the large intestine is the serosa which is constituted of loose connective tissue. The function of the serosa is to prevent any friction damage from the intestine rubbing against other tissue.

Lymph nodes and lymph vessels form an essential barrier to protect from many bacteria and other pathogens to which the GI tract is exposed. However, like other cancer. The lymphatic system is also responsible for the spreading of CRC to lymph nodes and ultimately to distant organs (Dujovny et al., 2004).

13.1.2 Pathogenesis and molecular pathways for CRC

Approximately 70% of all CRCs occur in the sigmoid colon and rectum while, rarely it occurs in the ascending colon, the transverse colon and splenic flexure, and the descending colon. As happens in other types of cancer, mutations in specific genes are responsible to initiate colorectal cancer. Depending on the origin of the mutation, colorectal carcinomas can be classified as sporadic, inherited and familial (Stewart et al., 2006; Buie and MacLean, 2011).

Point mutations are not associated with inherited syndromes and only affect individual cells and their ancestors. Sporadic cancer is derived from point mutations which are mainly responsible for all CRCs. Adenomatous polyposis coli (APC) is a tumor suppressor gene, in which the first mutation occurs and accelerate the development of nonmalignant adenomas called adenomatous polyps. Because adenomatous polyps and early-stage CRC are usually asymptomatic and therefore difficult to detect. This APC mutation is followed by mutations in KRAS, TP53 and, finally, DCC (Kang et al., 2005; Pickhardt et al., 2013).

Inherited cancers account for only 5% of all CRC cases. These types are caused by inherited mutations that affect only one allele of the mutated gene and other alleles will activate the tumor cell which develops the carcinoma (Romagnoli et al., 1984). Inherited cancers are more classified into two groups, Viz. polyposis and nonpolyposis forms. Polyposis cancer mainly involves familial adenomatous polyposis (FAP), which is indicated by the development of multiple potentially malignant polyps in the colon. In contrast, hereditary nonpolyposis colorectal cancer (HNPCC) is related to mutations in DNA repair mechanisms. The main reason for HNPCC is Lynch syndrome, which is found in 2 to 3% of all CRC cases. This syndrome is caused by inherited mutations in one of the alleles codings for DNA repair proteins such asMSH2, MLH1, MLH6, PMS1, and PMS2. Familial CRC is caused by inherited mutations and approximately accounts for 25% of all CRC cases (Mecklin, 1987; Rabelo et al., 2001).

Researchers have identified three distinct genetic mechanisms that lead to developing the CRC: chromosomal instability (CIN) also known as Loss of heterozygosity (LOH), microsatellite instability (MSI), and CpG island methylator phenotype (CIMP) or serrated neoplasia pathway (Compton et al., 2000; Kanemitsu et al., 2003).

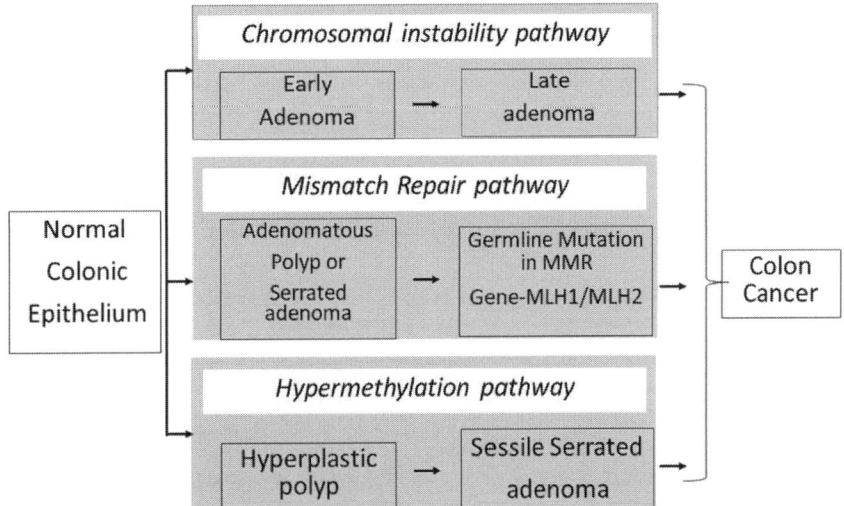

FIG. 13.2 Developmental pathways of colorectal cancer.

Fig. 13.2 showed the chromosomal instability which is characterized by allelic losses of tumor suppressor genes. The mechanism comprises a modification in chromosome segregation, telomere dysfunction, and DNA damage response. These genetic mutations affect critical genes involved in the maintenance of correct cell function, such as APC, KRAS, PI3K, and TP53 among others. An APC mutation also causes the translocation of catenin to the nucleus. It also makes the transcription of genes concerned in tumorigenesis and invasion, whereas mutations in KRAS and PI3K lead to constant activation of MAP kinase. Thus overall effects of these mutations showed increased cell proliferation and cause an uncontrolled entry in the cell cycle (Gryfe et al., 2000).

MSI is illustrated by genetic instability shown by repetitive DNA sequence tracts scattered throughout the genome. As a result of these mutations noncoding as well as codifying microsatellites affect, and develop tumors. Loss of expression of mismatch repair genes (MMR) can be caused by spontaneous events (promoter hypermethylation) or germinal mutations such as those found in Lynch syndrome. Fifteen per cent of CRCs, including nearly all cases of HNPCC display MSI phenotype; the remaining 85% display LOH. Tumors originating from the proximal colon represent a high % of MSI-positive lesion (Peltomaki et al., 1997).

The serrated neoplasia pathway is associated with RAS and RAF mutations, and epigenetic instability. Epigenetic instability, which is responsible for the CpG island methylator phenotype, is another common feature in CRC. The main characteristic of CIMP tumors is the hypermethylation of oncogene promoters, which leads to gene silencing and a loss of protein expression. Genetics and epigenetic are not exclusive in colorectal cancer, and both cooperate in its development, with more methylation events than point mutations (KW and Kinzler B, 1996).

13.1.3 Risk factors

The worldwide CRC occurrence rate is about 4–5% and is mostly related to environmental and genetic factors, in addition to many personal traits or habits are observed to be risk factors as they increase the chances of developing polyps or CRC. The main risk factor for CRC is age; the risk of developing CRC is markedly increased, after the age of fifty while it is rare below the age of fifty. In addition to age, other risk factors include genomic instability, inflammatory bowel disease; positive familial history of CRC in relatives, behaviors strictly related to a sedentary lifestyle, Lynch syndrome, Crohn's disease, alcohol addiction and long-term smoking also play a major role in the development of CRC. Seven-tenths of all CRC are sporadic and usually derive from somatic mutations and dysfunctional Wnt/β-catenin signaling pathways (Mecklin, 1987).

A sitting lifestyle is also related to obesity, in this context, diet is strongly linked to the risk of CRC such that unhealthy nutritional habits increase the chances of developing CRC by up to 70%. It was found that after ingestion, heme group is released from the red meat, which results in increased production of N-nitroso compounds, carcinogenic in nature and formation of cytotoxic and genotoxic aldehydes by lipoperoxidation, Also heterocyclic amines and polycyclic hydrocarbons are released from the meat cooked at high temperature after digestion, both of which are considered to be potential carcinogens (Haiman et al., 2007; Mármol et al., 2017).

TABLE 13.1 Stages of colorectal cancer.

Stage of cancer	Description
0	Early stage: Carcinoma in situ or intramucosal carcinoma
I	Cancer grows through the muscularis mucosa into the submucosa and musclaris propria
IIA	Cancer has spread through the muscular layer of the colon wall to the outermost serosa layer
II B	Cancer has spread through the serosa of the colon wall but has not spread to adjoining organs
II C	Cancer has spread through the serosa of the colon wall and spread to adjoining organs
III	Cancer has spread through the mucosa of the colon wall and spread to at least one nearby lymph node or cancer cells that have formed in tissues adjoining lymph nodes.
IV	Cancer may spread through the colon wall and spread to at least one organ or lymph node which is not near the colon such as lung, liver, etc.

13.1.4 Stages of CRC

Although researchers have developed many staging systems for CRC, only two are commonly used: the modified (combined) Astler-Coller Duke's system also called tumor node metastasis (TNM) system. In this system, patients are placed into one of three categories (stages A, B, and C), and another system is known as the Astler-Coller modified system in which patients are divided into five subdivisions. In the Astler-Coller system, A is limited to the mucosa, B1 involves the muscularis propria but does not penetrate it, B2 penetrates the muscularis propria, and C1 and C2 are counterparts of B1 and B2 with nodal metastases. The combined system has added three more stages: B3 represents the involvement of adjacent structures, C3 is a subpart of B3 with nodal metastasis, and D signifies the presence of distant metastasis (Besson et al., 2011).

More recently, the American Joint Committee on Cancer (AJCC) and the Union International Centre le Cancer (UICC) introduced the TNM staging system. The TNM system was updated in 2003 and further subdivided stage III according to the number of positive regional lymph nodes which is represented in Table 13.1. In stage 0, abnormal cells originate from the colonic wall mucosa. These unusual cells may become cancerous and spread. In the I stage, cancer has initiated from the mucosa of the colon wall and reached the submucosa. IIA, IIB, and IIC are the stages of Colon II cancer although stage III colon cancer subdivided into IIIA, IIIB, and IIIC stages. At last, stage IV colon cancer is divided into stages IVA and IVB. According to various study, researchers conclude that stages 0, I, II, and III are often curable with surgery. After surgery to increase the chance of eradicating the disease, many of the patients with stage III CRC and a few with stage II receive chemotherapy. CRC patients with stages II and III also receive radiation therapy with chemotherapy either before or after surgery. Stage IV CRC is not often curable, but it is treatable and the symptoms of the disease and the growth of cancer can be managed. Clinical trials are also a treatment selection for each stage (Bleiberg et al., 1985; Kilic et al., 2011).

13.1.5 Signs and symptoms

Symptoms begin to appear once the tumor is large enough to cause bowel obstruction. The symptoms vary according to the region of the colon or rectum that is affected (Table 13.2). Pelvic pain occurs at the later stage of the disease and usually indicates the local extension of the tumor to the pelvic nerves (Mandelblatt et al., 1996).

13.2 Diagnosis

Patients can observe a wide range of signs and symptoms such as occult or overt rectal bleeding, change in bowel habits, anemia, or abdominal pain. However, until CRC reaches an advanced stage it remains asymptomatic. In contrast, both benign and malignant tumors show the symptom of rectal bleeding and for further investigation of additional symptoms, colonoscopy is used (Matos et al., 2019). In individuals aged 45 years or older, the new onset of rectal bleeding should be identified by generally prompt colonoscopy. In younger patients to identify the symptoms and risk for colorectal cancer some additional factors are used e.g., change in bowel habits, inexplicable weight loss, having a family history of colorectal cancer, and blood mixed with the stool in contrast with blood on the surface of the stool (Zhang et al., 2002; Hamilton and Sharp, 2004).

Table 13.2 Variation in colorectal cancer symptoms according to tumor location.

Region	Symptoms
Right colon (cecum, ascending colon, left side of the transverse colon)	Abdominal pain, anemia (resulting from chronic blood loss), weakness, weight loss.
Left colon (sigmoid colon, descending colon, right side of transverse colon)	Constipation alternating with diarrhea. Abdominal pain. Nausea and vomiting. Rectum Change in bowel movements. Rectal fullness. Urgency. Bleeding. Tenesmus (straining at stool)

13.2.1 Endoscopy

Colonoscopy is the method of choice for diagnosing CRC. Colonoscopic identification of advanced lesions is relatively easy, but in the initial stage, CRC might appear as very minute mucosal lesions e.g., an innocuous flat laterally spreading polyp. These lesions require complete and careful mucosal inspection and optimal bowel preparation to ensure proper detection. Other factors including adenoma detection by an endoscopist, have been associated with the risk of developing CRC after colonoscopy (postcolonoscopy colorectal cancers) and are used as quality indicators for colonoscopy (Røseth et al., 1993).

13.2.2 Imaging

For the diagnosis of polyps and CRC computed tomography (CT) colonography is used as a complementary imaging method e.g., after incomplete or inadequate colonoscopy. For accurate locoregional and distant staging, however, imaging methods are mostly used. In rectal cancer, MRI is routinely is used for loco-regional staging and helps in further treatment decisions. As that of neoadjuvant systemic therapy used for locally advanced tumors, locoregional staging for colon cancer has become more important.

Although with restrictive accuracy, CT scans are routinely used for this purpose. Distant staging of liver and lungs is routinely done with CT. Magnetic resonance imaging (MRI) is mostly used for further determination of liver lesions. Positron emission tomography (PET) CT imaging is increasingly being used but its exact role for staging and assessment of disease in the higher stage is still debated (Young et al., 2000).

13.2.3 Laboratory

In addition to complete blood count, the diagnosis of carcinoembryonic antigen concentrations is recommended by all guidelines. An increased baseline carcinoembryonic antigen concentration is associated with a worse prognosis, and concentrations that remain elevated in the postoperative phase might indicate residual disease (Barrett et al., 2006; Labianca and Merelli, 2010).

13.2.4 Pathology

Histology is still the basis for pathological staging and subsequent management. Besides the classic TNM staging, histological sub-typing, grading, and histological assessment of lymphatic, perineural, venous invasion, and the value of a multitude of tumor-based markers including mismatch–repair testing and Immunocore are increasingly being recognized. Universal mismatch repair testing is being adopted not only for the identification of Lynch syndrome but also because of implications for adjuvant fluoropyrimidine-based therapy (Jass, 2000; Harada and Morlote, 2020).

13.3 Current therapies

There are many strategies reported to be used for the treatment of CRC. But, the success of any therapy is mostly dependent on the stage at which CRC has been detected and the age of the patient. Intersubject variability has been observed with patients of different age groups and also with the line of treatment. To be precise one line of treatment may prove effective for the particular class of patients whereas, the same may not be beneficial for the other class of patients suffering from CRC. Selection of method of treatment is very tricky in the case of CRC as each tumor responds differently to the different line of treatment. Type of tumor, its stage, age of the patient, the health status of a patient and more importantly lifestyle management are the major contributing factors for the selection of a proper method of treatment. One cannot rely heavily on the single strategy of treatment rather it should be carefully monitored considering the risk factors associated with the patient suffering from the CRC.

Treatment of CRC is closely linked to the disease stage. Treatment modalities include surgery, radiotherapy, and chemotherapy. Surgical resection of the primary tumor and regional lymph nodes is the only curative treatment for CRC and may cure up to 50% of patients. Pharmaceutical therapies play an important adjunct and soothing role in the treatment of most cases in stages III and IV CRC. These pharmacological agents help reduce the incidence of recurrence, prolong survival, and improve the quality of life for CRC patients.

13.3.1 Conventional treatment strategies

The most commonly used conventional methods for treating the CRC are

1. Polypectomy and surgery
2. Radiation therapy
3. Chemotherapy

13.3.1.1 Polypectomy and surgery

Colonoscopy is the most widely used method of detecting the exact location and severity of colon cancer. Removal of polyps during the colonoscopy is referred to as polypectomy. This is the potential way for the removal of precancerous/cancerous tumor. This is an effective option for small and localized tumors wherein, there are maximum chances for the full recovery of the patient. The major problem associated with CRC is recurrence and surgery is mostly the popular way for the treatment of tumors that are resistant to radiation and chemotherapy. Careful and comprehensive risk assessment of postsurgical complications has to be done particularly in the case of old age patients (Kristjansson et al., 2010).

13.3.1.2 Radiation therapy

Radiation is the most common method of using ionizing radiations to suppress the proliferation of malignant cells. The major limitation of this therapy is the risk of exposure to normal tissues that may lead to the formation of radiation-induced cancers. However, techniques like shaped-radiation beams can be effectively used for exposing the cancer site with the maximal optimum dose of radiation. Commonly used radiation therapies include selective internal radiotherapy (SIRT), transarterial chemoembolization (TACE), and radiofrequency ablation (RA). Although these methods are good at targeting radiation exposure to the tumor-specific region, a high recurrence rate has been documented (Marshall, 2008). Radiation therapy may include some side effects including fatigue, mild skin reactions, stomach upset, and so on. It may also cause bloody stools from bleeding through the rectum or blockage of the bowel. Also, sexual problems, as well as infertility in both men and women, may occur after radiation therapy.

13.3.1.3 Chemotherapy

The most commonly heard method of therapy in the review of cancer treatment is chemotherapy. In chemotherapy, the drugs are used to destroy cancer cells, normally by stopping the ability of cancer cells to grow and divide. Different anticancer drugs like alkylating agents, antimetabolites, plant alkaloids, antitumor antibiotics, enzymes, and hormones have been administered in this therapy. These anticancer drugs are known modifiers of biological response that destroy malignant cells, suppress tumor growth or its cell division.

Being cytotoxic they not only destroy the malignant cells but also damaging to the normal tissues and hence their side effects are dangerous. Interference with cell division pathways that includes DNA replication and chromosomal separation is the mainstay mechanism of these kinds of drugs. Unfortunately, they are lacking in specificity and hence pose a significant threat to the normal cells. In this type of therapy, rapidly dividing cells have been targeted and hence it is hazardous to the intestinal lining and immune cells being rapidly dividing cells. Although these cells usually repair them after discontinuation of drug therapy. The use of adjuvant chemotherapy in patients with stage II CRC also supports after curative resection but has a high risk of recurrence. Chemoradiation therapy helps to attempt downstage tumor and reduce the need for radical surgery (Viswanath et al., 2016; Matos et al., 2019). Investigations of direct intrahepatic administration of chemotherapeutic agents for treating CRC liver metastases (via either the hepatic artery or the portal vein) increase the local concentration of a drug with no clinical benefit. Therefore, this approach has not been used for experimental studies. For more than four decades, 5-Fluorouracil (5-FU) plays a major role in chemotherapy for CRC as a single agent and combined with Irinotecan or Oxaliplatin. 5-FU is also commonly used with radiotherapy in the neoadjuvant setting for rectal cancer (Midgley and Kerr, 2000). It is combined with Leucovorin (LV) in adjuvant and metastatic settings with different doses and frequency. Conventionally, 5-FU is commonly administered either as an IV bolus regimen or continually infused over 24 hours. Capecitabine (Roche's Xeloda)—an oral prodrug of 5-FU—is used increasingly in CRC owing to its advantage over intravenous 5-FU. A prospective, integrated analysis of two large randomized Phase III trials in metastatic CRC (mCRC) demonstrated that single-agent Capecitabine has significantly higher tumor response rates compared with 5-FU/LV (Kelly and Goldberg, 2005). Table 13.3 outlines the chemotherapy regimens most commonly used to treat CRC.

Table 13.3 Chemotherapy regimens most commonly used to treat colorectal cancer.

Regimen/Class	Component(s)	Mechanism of action	Adverse effects
Roswell park	5-Flurouracil (600 mg/m^2 bolus of LV at mid infusion) + Leucovorin (500 mg/m^2 given over 2 h repeat every 7 days for 48 weeks)	Antifolate (thymidylate synthase inhibitors) + Lecovorin enhances the activity of 5-FU by facilitating the binding of 5-FU to the enzyme thymidylate synthase	Leukopenia GIT disturbance Nausea vomiting Alopecia
Mayo clinic	5-Flurouracil (425 mg/m^2 IV bolus) + Leucovorin (20 mg/m^2 per days for 5 days, repeat every 4 weeks for 6 months)	Antifolate (thymidylate synthase inhibitors) + Lecovorin enhances the activity of 5-FU by facilitating the binding of 5-FU to the enzyme thymidylate synthase	Leukopenia GIT disturbance Nausea, vomiting Alopecia
De Gramont	5-Flurouracil (400 mg/m^2 IV bolus followed by 600 mg/m^2 IV infusion over 22 h. Repeat on second day; repeat cycle after 2 weeks for 6–12 months) + Leucovorin (200 - 400 mg/m^2 IV over 2 h prior to 5-FU)	Antifolate (thymidylate synthase inhibitors) + Lecovorin enhances the activity of 5-FU by facilitating the binding of 5-FU to the enzyme thymidylate synthase	GIT disturbances Nausea
IFL	5-Flurouracil (500 mg/m^2) + Leucovorin (20 mg/m^2) + Irinotecan (125 mg/m^2 weekly or 350 mg/m^2 IV over 90 min.) repeat every 06 weeks.	Antifolate (thymidylate synthase inhibitors) + Lecovorin enhances the activity of 5-FU by facilitating the binding of 5-FU to the enzyme thymidylate synthase + Irinotecan is a topoisomerase I (a nuclear enzyme that causes reversible, single-strand breaks in DNA during mitosis) Inhibitor.	Neutropenia GIT disturbance Vomiting
FOLFIRI	de Gramont regimen + irinotecan (180 mg/m^2 IV over 90 min on day 1, repeated every 2 weeks.)	Antifolate (thymidylate synthase inhibitors) + Lecovorin enhances the activity of 5-FU by facilitating the binding of 5-FU to the enzyme thymidylate synthase + Irinotecan is a topoisomerase I (a nuclear enzyme that causes reversible, single-strand breaks in DNA during mitosis) Inhibitor.	Myelosuppression GIT disturbance Neurotoxicity Hypersensitivity Alopecia Rash Pain

(Continued)

Table 13.3 (Cont'd)

Regimen/Class	Component(s)	Mechanism of action	Adverse effects
FOLFOX4	de Gramont + oxaliplatin (85 mg/m^2 on day 1 of the infusion repeated every 2 weeks)	Antifolate (thymidylate synthase inhibitors) + Lecovorin enhances the activity of 5-FU by facilitating the binding of 5-FU to the enzyme thymidylate synthase + Oxaliplatin inhibit tumor-cell replication by creating intra- and interstrand DNA cross-links.	Granulocytopenia, Neutropenia, Neurotoxicity, Fever, GIT Disturbance, Nausea and Vomiting
Cetuximab / Irinotecan	Cetuximab 400 mg/m^2 initial dose, followed by 250 mg/m^2 weekly	Cetuximab is a recombinant, chimeric MAb that binds to the extracellular the domain of the human EGFR, thereby blocking EGFR activation+ Irinotecan is a topoisomerase I (a nuclear enzyme that causes reversible, single-strand breaks in DNA during mitosis) Inhibitor.	Acne, Diarrhea, Neutropenia, Nausea and Vomiting
Bevacizumab/ 5-Flurouracil/ Leucovorin	Bevacizumab (5 mg/kg every 2 weeks) + 5-Flurouracil (500 mg/m^2 IV bolus) + Leucovorin (20 mg/m^2 IV)	Bevacizumab is an inhibitor of VEGF. Bevacizumab binds to VEGF and prevents the interaction of VEGF with its receptors on the surface of endothelial cells. VEGF, a multifunctional cytokine and potent permeability factor secreted in response to hypoxia has a major angiogenesis-promoting effect.+ 5- Flurouracil is antifolate (thymidylate synthase inhibitors) + Lecovorin enhances the activity of 5-FU by facilitating the binding of 5-FU to the enzyme thymidylate synthase	Leukopenia, Headache, Rash, Thrombosis
Cetuximab	Cetuximab (400 mg/m^2 initial dose, followed by 250 mg/m^2 weekly until disease progression or unacceptable toxicity	Cetuximab is a recombinant, chimeric MAb that binds to the extracellular the domain of the human EGFR, thereby blocking EGFR activation.	Acne
Irinotecan	Irinotecan (300–350 mg/m^2 IV 90 min infusion on the day, repeat every 3 weeks)	Irinotecan is a topoisomerase I (a nuclear enzyme that causes reversible, single-strand breaks in DNA during mitosis) Inhibitor.	Leukopenia, Neutropenia, GIT Disturbance, Nausea and Vomiting, Alopecia
Capecitabine	Capecitabine (1.25 g/m^2 twice daily for 14 days; subsequent courses repeat every 3 weeks)	Capecitabine is a ßuoropyrimidine. Fluoropyrimidines, which are structural analogues of naturally occurring metabolic products (generally folates, purines, and pyrimidines) needed for the synthesis of nucleic acids, act by inhibiting thymidylate synthase (TS), a key enzyme in DNA synthesis.	Leukopenia, Neutropenia, Thrombocytopenia, GIT disturbance
Tegafur-Uracil	Tegafur (300 mg/m^2)with uracil (672 mg/m^2 daily in 3 divided doses for 28 days; subsequent courses repeated after 7 days interval)	Tegafur-uracil is fluoropyrimidine. (structural analogues of naturally occurring metabolic products (generally folates, purines, and pyrimidines) needed for the synthesis of nucleic acids, act by inhibiting TS, a key enzyme in DNA synthesis.	Myelosuppression, Alopecia, GIT disturbance, Cardiac toxicity, Renal toxicity

Currently, several drugs are approved to treat CRC (Tournigand et al., 2004; Ortega et al., 2010). Chemotherapy may cause nausea, vomiting, diarrhea, neuropathy, or mouth sores. Targeted therapy is a treatment that targets the tissue environment or the specific genes of cancer. This type of treatment-limiting damage to healthy cells while blocks the growth and spread of cancer cells. The most effective treatment for CRC can be developed by carrying out tests to identify the proteins, genes and other factors in a patient's tumor. Further options for CRC are antiangiogenesis therapy, epidermal growth factor receptor (EGFR) inhibitor therapy. The adverse effects of the targeted treatments include rashes on the face and upper body (Ross et al., 2010).

13.3.2 Targeted therapy

Conventional strategies like chemotherapy and radiation therapy are unable to distinguish between malignant and normal tissues and are hence reported to have major side effects. To achieve better control and site-specificity, targeted therapies have been used particularly for the advanced stages of colon cancer. Targeted therapies are known to deliver the drug at tumor vicinity thereby reducing the unintentional exposure to normal cells, ensuring very few side effects with greater safety. Vascular endothelial growth factor (VEGF) and epidermal growth factor receptor (EGFR) are two important molecular markers responsible for the growth and proliferation of tumors in CRC (Marshall, 2008; Krasinskas, 2011).

Both VEGFR and EGFR play important roles in tumor progression, invasion, and metastasis. The therapies mainly focus on the intervention of their signaling pathways which are critical for the generation of malignancies. These can be potentially explored for the targeted drug delivery systems against CRC (Tabernero, 2007).

Carrier systems like nanoparticles including nanoconjugates of monoclonal antibodies (mAbs) can be effectively used to overcome the limitations of cancer chemotherapy. CRC cells express various molecular markers against which mAbs can be designed and used as targeting/therapeutic agents. mAbs have been developed to target specific antigens which are dysregulated in cancer cells and involved in the development and progression of cancer. mAbs can bind and inhibit the target proteins which are overexpressed in tumors. mAbs are reported to target the EGFR signaling pathway and angiogenesis which are supposed to be the critical pathways for carcinogenesis. Some of the mAbs were reported to be effective in patients who are nonresponsive to traditional chemotherapy. Some of the FDA-approved mAbs used for CRCs are cetuximab, panitumumab, and bevacizumab.

Among the mentioned mAbs, cetuximab, and pantitumumab are known to bind to EGFR, later has been reported to be used as the best supportive care for metastatic CRC treatment (Gravalos et al., 2010). Bevacizumab antibody is known to bind with the ligand vascular endothelial growth factor – A (VEGF-A) which increases survival with tolerable toxicity.

In the case of advanced-stage cancer, mAbs are generally used for the targeted therapy. Its success hugely depends on its ability to stimulate the immune system without the side effects of conventional chemotherapy or radiotherapy. Rituximab is the FDA approved mAb in 1997, has been used in the treatment of hematologic malignancies and solid tumors (Weiner et al., 2010).

Antigens or receptor sites on cancer cells are nothing but enzymes or proteins which can be used as a target for targeted immunotherapy. Immunotherapy can be used in combination with antitumor mAbs reported to be effective for targeting immune effector cells that might enhance the efficacy of antitumor mAbs. Various immune cells that have been used for the targeted immunotherapy with mAbs are natural killer (NK) cells, macrophages, T-cells and dendritic cells (DCs) (Houot et al., 2011).

mAbs recognize the cancer cell surface antigens and binds specifically to it for which it has been directed. This eventually leads to inhibit several biochemical events leading to tumor cell death. Other examples of mAbs are Avastin, Erbitux, Rituxan, Herceptin, Mylotarg, Campath, Zevalin, Bexxar, and Vectibix.

There are certain limitations with the use of mAbs like toxicity observed when given in high dose systemically. These are generally used as a second or third line strategy if the case immune system is already weakened by surgery, chemotherapy and radiation. Moreover, chemotherapy and radiation may cause tumor cell mutation leading to a change in the cell surface antigen for which mAbs have been targeted. Because of this change in the target antigen mAbs attempting to target those antigens will remain ineffective. To overcome this problem, targeted therapy should be optimized by identifying a specific category of patients with specific cancer and decoding the mutant antigens.

13.3.2.1 Immunotherapy

Immunotherapy is the most used treatment strategy in oncology, as it is not associated with the limitations as in the case of other conventional therapies. It is based on the body's defense mechanism i.e. the immune system that can recognize the cancer cell and fight to avoid apoptosis. Immune compromised peoples are susceptible to develop certain cancers. It can be developed in peoples with the normal immune system in some cases when the immune system is unable to recognize the cancer cells or cells with underexpressed antigens. This can be best possibly avoided by immunotherapy. The basic principle involved in strengthening the immune system to identify and kill the cancer cells and help to reduce side effects associated with the anticancer drug treatment. It works on a three-tier basis, that is, inhibition of biochemical processes of apoptosis, identification of cancer cells and replacement, or damage repair of normal cells. Inflammatory responsive cells like macrophages, suppressor cells, and T cells in tumors created a microenvironment that resists the activation of the immune system (Morse et al., 2009).

In such cases, T-cell enhancement can be the handy immunotherapy approach as it acts against tumor associated antigens. One more type is adoptive T-cell transfer (ACT) that utilizes autologous tumor-infiltrating lymphocytes proves effective for patients with metastatic melanoma (Halama et al., 2008).

To control or retard the tumor growth, immunotherapy can utilize biological molecules like cytokines and antibodies. One more US FDA approved immunotherapeutic strategy is the use of a patient's dendritic cells for the activation of cytotoxic response toward patient-specific cancer antigens. These dendritic cells are antigen-activated or transfected by using viral vectors. When they are inserted into the patient's body antigens will be presented to the lymphocytes hen present the antigens to effecter lymphocytes which initiate a cytotoxic response to the cancerous cells bearing these antigens.

Broadly there are three main categories of immunotherapy as passive, active, and combination type.

i. Passive immunotherapy: This includes the development of antibodies or immune cell components in-vitro in the laboratory. It can be given to patients for increasing the immunity of patients to fight against infections thereby providing immunity against cancer. But, it is unable to actively stimulate a patient's immune system to respond to the disease as that of vaccines.

ii. Active immunotherapy: This is known to elicit the body's immune response to combat the tumor cells. Cancer vaccines, cellular therapies are examples of active immunity carriers. In the case of adjuvant immunotherapy, adjuvant has been injected with antigenic protein or mAbs. This leads to enhancing the immune response against a particular cancer antigen. Examples of adjuvants are BCG, KLH, IFA, QS21, Detox, DNP, and GM-CSF.

iii. Limitation of adjuvant therapy includes associated toxicities, it can only be administered once or twice to patients, and can only be administered subcutaneously, and cannot be infused.

iv. Combination immunotherapy: This immunotherapy made up of drugs that acquire both active and passive immunotherapy activity. This immunotherapy may also produce adverse effects with other drugs. The side effects of naked mAbs, generally administered intravenously, are mostly mild as compared to the side effects of chemotherapeutic drugs, and are often more like an allergic reaction. Primarily it occurs during the very first administration to patients. The possible side effects may include fever, chills, weakness, headache, nausea, vomiting, diarrhea, low blood pressure, rashes, etc. These side effects can interfere with one's ability to participate in daily activities. Therefore, to be effective immunotherapy is often provided in combination with other treatment modalities, such as surgery, radiation therapy and chemotherapy.

13.3.2.2 Limitations of immunotherapy

Immune evasion is one of the major problems in the development of cancer immunotherapy. It is very difficult to get reproducible results as different targets have been reported for different patients also it can be changed with the mutation of cancer. Immunotherapy can only show good results only when it has started earlier than the development of tolerance in cancer. It should be started before other treatments like radiation, chemotherapy and surgery as these can weaken the immune system of the patient. Also, in the case of CRC tumor is mostly associated with inflammation which causes suppressed immune functions thereby making immunotherapy ineffective (Soudja et al., 2010).

13.3.3 Targeted therapies using nanocarriers

Over the past decade, nanotechnology has been significantly used for the detection of cancer, imaging and therapy (Rampado et al., 2019). Several reports have been published regarding the study of nanovectors for the theranostic purpose and the FDA has also approved some nanoparticles based formulation for the treatment of CRC. Nanovectors are nanoparticles with a 1 to 100 nm used for spatial delivery of bioactive for either therapeutic or diagnostic purposes (Ferrari, 2005).

Nanoparticles of chemotherapeutic agents were found to increase solubility and improve biodistribution. This eventually helps to avoid the toxic effects due to undue exposure to normal cells. Thus efficiency can be improved due to the enhanced therapeutic index and half-life of the drug. Nanovectors can be efficiently engulfed by tumor cells by endocytosis thus help to evade the multidrug resistance (MDR) pumps to avoid the expulsion of free drugs. Nanovectors are well known to improve the pharmacokinetic properties of drug molecules as well as biotechnologically derived drugs like proteins and peptides. This property can be very handy when it comes to the diagnosis where it is expected to deliver the labeled drug molecules at higher concentration in CRC. It improves sensitivity and reduces the toxic effects associated with unintentional exposure to normal cells (Maeda et al., 2000).

Overall *in vivo* performance of nanovectors is hugely dependent on their intrinsic chemical and physical properties, biodistribution and excretion and most importantly their targeting mechanism. There are two well-known targeting mechanisms of nanovectors Viz. passive and active (Sánchez-Moreno et al., 2018).

Passive targeting: It depends on the tumor microenvironment like disorganized and leaky vasculature. This condition is favorable for the enhanced permeability and retention (EPR) effect of nanovectors of particle size of <200 nm diameter. These are reported to extravasate from the systemic circulation through the vascular pores present in the defective solid tumor neovasculature and due to the dysfunction of lymphatic drainage are retained by the tissue (Ferrari, 2005).

Other options for passive targeting include pH-sensitive polymeric nanoparticles or liposomes. Tumor cells use glycolysis to obtain extra energy, resulting in an acidic environment. The pH-sensitive polymeric nanoparticles are designed to be stable at a physiologic pH of 7.4 but degraded to release active drugs in target tissues in which the pH is less than physiologic values, such as in the acidic environment of tumor cells (Lim et al., 2016). Apart from this, thermosensitive nanovectors (Sánchez-Moreno et al., 2018), redox responsive drug delivery system (Guo et al., 2018), have been reported for the successful delivery of drug-using nanovectors for passive targeting of tumors.

Active targeting: This approach is based on the targeting of membrane receptors which are often over expressed on tumor cells. One approach suggested is the inclusion of a targeting ligand or antibody in polymer-drug conjugates. Nanoparticles coated with tumor-specific ligands or antibodies bind to cell-surface receptors, which initiates the entry of nanoparticles into the cell via endosome. In the case of CRC, the reported receptors which are over expressed are the folate receptor alpha (FRα), and the epidermal growth factor receptor (EGFR and HER2) present onto the tumor cells. Some other targets studied are carcinoembryonic antigen (CEA) (Pereira et al., 2018), vascular endothelium growth factor receptor (VEGFR), and the integrin receptors (Cisterna et al., 2016). mAbs like Cetuximab and Bevacizumab have been reported to be used against EGFR and VEGFR respectively. Peptides like arginylglycylaspartic acid (RGD) can be used as a ligand against the integrin receptor and folic acid shows significant activity against FRα.

Colorectal cancer (CRC) is the third most common cancer affecting equally men and women worldwide. In advanced stages it poorly responses to current therapy options and efficient diagnosis is essential in the initial stages of development (Mohd-Zahid et al., 2019; Cabeza et al., 2020). For the treatment, chemotherapy and radiotherapy are available but there are limitations of these therapies due to the development of resistance of cancer cells to chemotherapy drugs and dose-limiting toxicities. Chemotherapy usually damages the healthy cells of our body because there is no site-specific delivery of drugs after administration. To overcome the constraints of present therapy cancer nanomedicine is a rapidly developing interdisciplinary research field. In the last few days, nanoparticles and nanotechnology is showing a new ray of hope in the treatment of CRC therapy (Cabeza et al., 2020).

13.4 Nanodrug delivery in cancer therapy

The application of nanodrug delivery in cancer therapy enhances the conventional way of treatment by reducing the adverse side effects, increasing the bioavailability and specific targeting. Cancer tissues are permeable to nanoparticles this is due to their impaired lymphatic drainage and highly permeable vasculature. Nano drug delivery to the target cell is occurred either actively or passively (Matsumura and Maeda, 1986). Active targeting depends on an interaction between nanoparticles and cell receptors, while passive targeting depends on the long biological half-life, long-circulating time of nanoparticles at the tumor region and several other factors. It is found that the vascular pore cut-off size exhibited by most solid tumors is between 380 nm and 780 nm. Thus, nanosized particles within this size range can access through the cancer vascular pores. This phenomenon is known as the enhanced permeability and retention (EPR) effect. The main obstacle in the delivery of nanoparticles to a tumor cell is to affect the reticuloendothelial system (RES), it uptakes the circulating nanoparticles via macrophages and cleared them in vivo. Nanoparticles with hydrophobic and highly charged surface are prominently recognized by RES, to avoid this, surface modification can be carried out on nanoparticles by coatings with a polymer such as poly (N-vinyl-2-pyrrolidone) (PVP), polyethylene glycol (PEG) and dextran to avoid agglomeration and systemic clearance by macrophages which enhance the bioavailability of chemotherapeutic agents (Matsumura and Maeda, 1986; Maeda et al., 2000).

13.4.1 Polymers used in formulations of NPs

Several polymers like poly(lactide-co-glycolide) (PLGA), polylactide (PLA), polycaprolactone (PCL), poly(D, L-lactide), chitosan, and PLGA–polyethylene glycol (PEG) have been developed for passive and active targeted delivery of therapeutic agents. PLGA is an ideal biodegradable polymer hydrolyzed in the body to produced metabolite monomers of lactic acid and glycolic acid and mostly used for drug delivery or biomaterial applications. PLGA in medical applications shows very little systemic toxicity because its metabolites, lactic acid and glycolic acid are further metabolized by the body in inactive compounds.

Wu et al. designed PLGA nanoparticles modified with Epidermal growth factor (EGF) containing 5Fu and oxygen-transport perfluorocarbon (PFC) to improve therapeutic efficacy against colon cancer (Chandran et al., 2017; Wu et al., 2020).

PLA is also widely used in formulations due to its biocompatible and biodegradable properties. Being able to undergo degradation to monomeric units of lactic acid as a safe material, PLA can also be applied to produce medical implants, biologically active controlled release devices, biocompatible sutures and so many. Polyethylene glycol is a hydrophilic polymer mainly used to modify the surface of NPs from hydrophobic to hydrophilic, which in turn increases the residence time of NPs in blood circulation as it decreases the RES clearance rate.

Chitosan is one of the popular natural polymers obtained from the deacetylation of chitin. Chitosan is insoluble in a neutral environment and high solubility in the acidic microenvironment of tumor tissue. Chitosan is a cationic polymer that shows different rheological characteristics depends on the pH of the surrounding fluid, it may be converted into hydrogels, this property makes it suitable for drug delivery of antioxidants, for example, some researchers formulated chitosan nanocapsules with encapsulated naringenin (polyphenol). From this study, it was proved that naringenin in encapsulated form shows a better anticancer effect than free naringenin (Jain et al., 2019; Detsi et al., 2020).

Alginates are salts derived from alginic acid, the polymer is negatively charged as the carbonyl lateral group liked to a glucose unit. Some researchers formulated curcumin encapsulated novel alginate beads, for colon target therapy; the formulation prevents the release of drug in the upper gastrointestinal tract and the drug immediately releases from the beads after reaching the colon (Patra et al., 2018; Osorio et al., 2020).

Polyacrylates are polymers derived from acrylic acid through free radical polymerization. Polyacrylates are anionic polymers that show maximum swelling from neutral to alkaline pH and minimum swelling under acidic conditions. This property makes it ideal for colon-specific drug delivery systems (Osorio et al., 2020).

Polyols are polymers with hydroxyl groups. A representative polymer of this family is polyvinyl alcohol (PVA), which is synthesized by the hydrolysis of polyvinyl acetate. PVA is a hydrophilic polymer that can generate hydrogels by chemical or physical crosslinking. Some researchers have developed a core-shell electrospun nanofiber, core (PVA and phycocyanin), and shell (Polyoxyethylene) for the targeted therapy of CRC (Osorio et al., 2020).

13.5 Polymeric nanoparticles (PNPs)

The treatment of CRC mainly depends on the tumor size, the local microenvironment and the range of cancer metastasis. Despite this, the tumor tissue has poor lymphatic drainage with leaky vasculature and an acidic microenvironment. The results have shown that appropriate particle size (in the range of 10–200 nm) and stable properties make them able to penetrate the leaky blood vessels and accumulate in tumor cells via EPR effect (Hamzehzadeh et al., 2016).

PNPs are consisting of a solid sphere structure that is manufactured via a self-assembly process. They are called either nanocapsules or nanospheres depending on their method of manufacturing. In nanospheres, the drug is dissolved or dispersed throughout the polymeric matrix while in nanocapsules drug is encapsulated in or confined in a shell-like structure made by a single polymer membrane (Osorio et al., 2020). To decrease immunological interactions the PNPs are mostly covered with nonionic surfactants, for example, opsonization. Polymers used in the formulation of NPs are of either synthetic or natural origin. These PNPs can improve drug bioavailability, control the drug release property, increase the circulation time and decrease nonspecific toxicity in the field of medicine. Moreover, the active and passive targeting functionality of the PNPs is capable of targeting specific tissue sites and enhancing the intracellular penetration of the drugs into the tumor instead of affecting normal tissues. These excellent properties displayed by the biodegradable polymer nanoparticles make them one of the ideal platforms for controlled and targeted drug delivery for colon cancer therapy. PNPs are used as drug carriers for hydrophobic drugs and are widely used for drug discovery. The PNPs constructed from amphiphilic polymers with a hydrophilic and hydrophobic block can perform rapid self-assembly because of the hydrophobic interactions in an aqueous solution. The PNPs can entrap the hydrophobic drugs because of a covalent bond or the interaction via a hydrophobic core (Hamzehzadeh et al., 2016; Elgqvist, 2017).

Due to the least volume and higher thermodynamic stability, the PNPs become suitable drug carriers with good endothelial cell permeability with no rejection from the kidney. Targeted delivery of chemotherapeutic drugs by using NPs can be improved by specific modification of its large surface area results in increased blood circulation time, which may improve outcomes. The attachment of ligands such as antibodies, fragments of antibodies, peptides, aptamers and other small molecules on the surface of nanoparticles for cell recognition has produced a new generation of nanoparticles for cancer therapy with enhanced *in vivo* specificity. The incorporation of these ligands is usually achieved by chemical modification during nanoparticle synthesis or through chemical bonding between ligands and polymers before synthesis.

Targeted nanoparticles are those that contain ligands on their surface and are capable of specifically recognizing cells. In treatment against cancer, the function of targeted nanoparticles is based on the fact that tumors express and/or overexpressed some biomarkers, which can be used as targets for drug delivery. Some researchers described cisplatin prodrug-loaded poly(d,l-lactic-co-glycolic acid)-block-polyethylene glycol nanoparticles targeted with a cyclic pentapeptide c(RGDfK) that bind to the integrin receptor, which is highly upregulated in tumor-associated endothelial cells during angiogenesis (Cisterna et al., 2016; Jin et al., 2020).

13.5.1 Lipid-based nanoparticles

Nanoliposomes are nontoxic, spherical nanocarriers containing an aqueous core with phospholipids bilayers mainly made up of natural lipids like cholesterol, sphingolipids, glycolipids, long-chain fatty acids, and nontoxic surfactants. The outer surface of liposomes is like a biological membrane and so promotes the easy entrapment of them in the cell through the fusion of phagocytises. To deliver the nucleic acids, small peptides and proteins, liposome-based nanoparticles are one of the commonly used nanoparticles in nanoplatform drug delivery (Yang and Merlin, 2020).

Due to the EPR effect, the vesicle of size around 4000 kDa or 500 nm can be entered into the tumor via the gaps in vessels. In tumors they can fuse with cells, are engulfed by endocytosis, and release drugs in the intracellular space.

The liposome up to 100 nm easily penetrate the tumor and stay longer, as the size of liposomes increases they are easily recognized and cleared by the mononuclear phagocyte system results in a shorter half-life. To ensure active targeting, liposome-bound antibodies target tumor-specific antigens and then transport drugs to the tumor.

Nanoliposomes are considered as one of the most effective drug delivery systems at a cellular level due to their nano size, ability to incorporate various substances, and slow-releasing and targeting characteristics, which result in decreasing side effects. Nanoliposomes are of mainly three types (1) long-circulating liposomes or stealth liposomes, these are specially designed phospholipid bilayer structure liposomes coated with gangliosides or polyethylene glycol (PEG) to avoid binding of blood plasma proteins (opsonins) to the liposomal surface which minimizes the RES effect; (2) active nanoliposomes: The target sites of this type of nanoparticles are receptors, peptides hormones and antibodies; and (3) sensitive nanoliposomes: This type of nanoliposomes required some physical or environmental stimuli to get activated, for example, thermosensitive, magnetic or pH-sensitive (Elgqvist, 2017; Ho et al., 2017).

Doxorubicin (Doxil)-liposome is FDA approved nanoliposome for chemotherapy for CRC. Doxil is approximately 100 nm and shows less gastrointestinal and cardiac toxicity. The side effects of Doxil include redness and peeling of the skin. Another recent nanoliposomal drug approved by FDA is Marqibo. It is a cell cycle-dependent anticancer drug and approximately 100 nm in size. Another promising nanoliposomal drug for colorectal liver metastases is (Thermodox) which is a thermosensitive liposome containing doxorubicin used in combination with radiofrequency ablation. Thermodox is a nanoliposomal drug delivery system that releases the drug upon a mild hyperthermic trigger. Thermodox is 25 fold more able to deliver doxorubicin into tumors compared to intravenous administration of doxorubicin (Wathoni et al., 2020).

13.5.2 Superparamagnetic iron oxide nanoparticles (SPIONs)

SPIONs are the core-shell nanoparticles that are most commonly used in medical imaging and therapy. The core-shell nanoparticles are consisting of two or more substances that can be synthesized with different combinations of organic and inorganic materials. The core nanoparticles are coated to enable efficient surface modification, increasing functionality and stability. The core-shell has many applications in the medical field like controlled drug delivery, multimodal imaging, cell labeling, and nuclear medicine therapy.

The SPIONs can load drugs as well as medical radioisotopes owing to their highly active surface and exhibits magnetization only in an applied magnetic field. From the last few years, owing to developments in radiochemistry and radiation sciences, the field of nanomedicine is now applied in nuclear medicine to enable radionuclide therapy (therapeutic isotopes loaded radiolabelled nanoparticles) and multimodal medical imaging (imaging isotopes loaded radiolabelled nanoparticles) of various types of cancer. This has a great help in the diagnosis and therapy of CRC (Ho et al., 2017).

For targeted photothermal therapy in CRC gold SPIONs are developed. These gold SPIONs are coated with single-chain antibodies to target the overexpressed antigen A33 in the cancer cell. It was found that gold SPIONs have five times more affinity toward cells expressing A33 antigen than normal cells. SPION-based magnetic resonance imaging (MRI) is useful in imaging lymph nodes in solid cancer including CRC. This is due to preferential uptake of SPIONs in lymph nodes as well as the ability of SPIONs to differentiate between healthy and cancerous tissues.

13.5.3 Gold nanoparticles (AuNPs)

Since the beginning of the 20th century, gold has been in use in the treatment of various diseases like arthritis/inflammation. The major advantage of AuNPs as compared to other metal NPs is that the gold core is relatively inert, nontoxic and considered to be biocompatible but toxicity may be produced due to surface modification. These AuNPs deliver the drug to the cancer cell via passive targeting, active targeting or both. The reason for passive targeting is the neovasculature produced by the tumor cell. For active targeting, the surface modification of AuNPs is done by attaching ligands like proteins, antibodies or small molecules for specific cell targeting on the targeted surface of the membrane. It was found that the cancer cell overexpressed specific antigens due to this overexpressed surface-specific antibodies binds to it facilitating the entry of NPs in tumor cells. The application of biocompatible coating was found to enhance the efficacy of AuNPs in drug delivery systems. The gold nanoparticles loaded with recombinant human tumor necrosis factor (rhTNF) via PEG linker were found to be three times more effective in the treatment of advanced stages of cancer without any additional side effect (Mohd-Zahid et al., 2019).

The size of AuNPs should be within the specified size range to avoid the interaction with biological macromolecules, which may lead to cytotoxicity and not too large to reduce the chance of macrophage clearance.

The particle size of AuNPs in between 40 and 60 nm was found to have optimal values of membrane bending rigidity and ligand-receptor binding interaction for endocytosis. The hydrodynamic diameter of AuNPs will increase upon coating or surface modification and hence affect the rate of cellular uptake. Thus, it is important to consistently monitor the increase in the size of AuNP before and after modifications to avoid clearance due to larger size (Jin et al., 2020).

13.5.4 Enteric-coated nanoparticles

Enteric-coated nanoparticles are one of the drug delivery technologies that can improve the bioavailability of drugs via oral administration, enhance intracellular penetration and retention time, control the release of encapsulated drugs and targeted delivery in specific parts of the gastrointestinal tract as required in the treatment of colorectal cancer. Cellular uptake and efficacy of nanoparticle drug delivery systems for CRC therapy are influenced by several factors, namely, size, shape, and surface chemistry. The size of the drug delivery carrier plays a crucial role in CRC therapy. Nanoparticles sizes around 100 to 200 nm show better cancer-targeting properties than larger particles. This can improve the selective accumulation of NPs in the colon tissue due to the epithelial-enhanced permeability and retention effect, which may increase selective accumulation in the colon tissue. To be absorbed into the cells, the particle size of the drug should not exceed 104 nm. With a particle size of 102 nm, nanoparticle drugs are generally absorbed via the clathrin pathway.

Enteric coating is usually done by using water-resistant or pH-sensitive polymers like a combination of Chitosan and Eudragit L-100 or Guar gum and Eudragit L30D. Polymer coating uses in enteric-coated systems do not dissolve in gastric fluid and prevents the release of drug compounds in the stomach. Eudragit has the highest entrapment ability compared to other polymers and it is a pH-dependent enteric-coated polymer, dissolves in a pH >5.5 medium. It protects the active ingredient from release and degradation in gastric fluid, improves drug effectiveness, and enables targeting specific areas of the intestine. Polyacrylate polymers like Eudragit are versatile with pH-dependent/independent solubility, which make them suitable for sustained release formulation. Eudragit S100 has solubility characteristics above pH 7, making it suitable for use in colonic release targeting. Therefore, it is mostly used as an enteric coating in the drug delivery system for CRC (Wathoni et al., 2020).

13.6 Conclusion

CRC is the leading cause of death all over the world and conventional treatment of CRC includes polypectomy and surgery, radiation therapy and chemotherapy. The conventional chemotherapy produces multidrug resistance (MDR) of cancer cells. Chemotherapeutic agents are cytotoxic so it damages the normal tissues while destroying the malignant cells, which may lead to severe side effects. Site-specific delivery of the chemotherapeutic agents to the affected area of the colon in a predictable and reproducible manner is the major challenge.

Nanno drug delivery will be definitely helpful to overcome the side effects of conventional chemotherapy. Cancer tissues have impaired lymphatic drainage and highly permeable vasculature, so it helps in permeation of nanoparticles in cancer cell, the phenomenon known as the enhanced permeability and retention (EPR) effect. Nano drug delivery to the target cell is occurred either actively or passively. Active targeting is also possible through interaction between nanoparticles and cell receptors. The major concern in the delivery of nanoparticles to a tumor cell is the uptake of circulating nanoparticles via macrophages and cleared them in vivo. The ligands such as antibodies, fragments of antibodies, peptides, aptamers and other small molecules can be attached on the surface of nanoparticles, which recognize the cancer cell easily and enhanced *in vivo* specificity and reduce the cytotoxic effects on healthy tissue. Hence the nanodrug delivery is found to be very promising way in the treatment of CRC.

References

Barrett, J., et al., 2006. Pathways to the diagnosis of colorectal cancer: an observational study in three UK cities. Fam. Pract. 23 (1), 15–19. http://doi.org/10.1093/fampra/cmi093.

Besson, D., et al., 2011. A quantitative proteomic approach of the different stages of colorectal cancer establishes OLFM4 as a new nonmetastatic tumor marker. Mol. Cell. Proteomics 10 (12), 1–14. http://doi.org/10.1074/mcp.M111.009712.

Bleiberg, H., Buyse, M., Galand, P., 1985. Cell kinetic indicators of premalignant stages of colorectal cancer. Cancer 56 (1), 124–129. http://doi.org/10.1002/1097-0142(19850701)56:1<124::AID-CNCR2820560119>3.0.CO;2-Y.

Bray, F., et al., 2018. Global cancer statistics 2018: GLOBOCAN estimates of incidence and mortality worldwide for 36 cancers in 185 countries. CA Cancer J. Clin. 68 (6), 394–424. http://doi.org/10.3322/caac.21492.

Buie, W.D., MacLean, A.R., 2011. Hereditary nonpolyposis colorectal cancerInherited Cancer Syndromes: Current Clinical Management, 161–179. http://doi.org/10.1007/978-1-4419-6821-0_9.

Cabeza, L., et al., 2020. Nanoparticles in colorectal cancer therapy: latest in vivo assays, clinical trials, and patents. AAPS PharmSciTech 21 (5). http://doi.org/10.1208/s12249-020-01731-y.

Chandran, S.P., et al., 2017. Nano drug delivery strategy of 5-fluorouracil for the treatment of colorectal cancer. J. Cancer Res. Pract. 4 (2), 45–48. Elsevier B.V. http://doi.org/10.1016/j.jcrpr.2017.02.002.

Cisterna, B.A., et al., 2016. Targeted nanoparticles for colorectal cancer. Nanomedicine 11 (18), 2443–2456. http://doi.org/10.2217/nnm-2016-0194.

Compton, C.C., et al., 2000. Prognostic factors in colorectal cancer: College of American Pathologists consensus statement 1999. Arch. Pathol. Lab. Med. 124 (7), 979–994.

Detsi, A., et al., 2020. Nanosystems for the encapsulation of natural products: the case of chitosan biopolymer as a matrix. Pharmaceutics 12 (7), 1–68. http://doi.org/10.3390/pharmaceutics12070669.

Dujovny, N., Quiros, R.M., Saclarides, T.J., 2004. Anorectal anatomy and embryology. Surg. Oncol. Clin. N. Am. 13 (2), 277–293. http://doi.org/10.1016/j.soc.2004.01.002.

Elgqvist, J., 2017. Nanoparticles as theranostic vehicles in experimental and clinical applications-focus on prostate and breast cancer. Int. J. Mol. Sci. 18 (5), 1–53. http://doi.org/10.3390/ijms18051102.

Favoriti, P., et al., 2016. Worldwide burden of colorectal cancer: a review. Updates Surg. 68 (1), 7–11. Springer Milan. http://doi.org/10.1007/s13304-016-0359-y.

Ferrari, M., 2005. Cancer nanotechnology: opportunities and challenges. Nat. Rev. Cancer 5 (3), 161–171. http://doi.org/10.1038/nrc1566.

Giovannucci, E., 2001. Research conference on diet, nutrition and cancer insulin, insulin-like growth factors and colon cancer: a review of the evidence 1. J. Nutr. 131 (11), 3109S–3120S.

Granados-Romero, J.J., et al., 2017. Colorectal cancer: a review. Int. J. Res. Med. Sci. 5 (11), 4667. http://doi.org/10.18203/2320-6012.ijrms20174914.

Gravalos, C., et al., 2010. Integration of panitumumab into the treatment of colorectal cancer. Crit. Rev. Oncol. Hematol. 74 (1), 16–26. http://doi.org/10.1016/j.critrevonc.2009.06.005.

Gryfe, R., et al. 2000. Tumor microsatellite instability and clinical outcome in young patients with colorectal cancer. N. Engl. J. Med. 342, 69–77.

Guo, X., et al., 2018. Advances in redox-responsive drug delivery systems of tumor microenvironment. J. Nanobiotechnol. 16 (1), 1–10. BioMed Central. http://doi.org/10.1186/s12951-018-0398-2.

Haiman, C.A., et al., 2007. A common genetic risk factor for colorectal and prostate cancer. Nat. Genet. 39 (8), 954–956. http://doi.org/10.1038/ng2098.

Halama, N., Zoernig, I., Jäger, D., 2008. Immuntherapie von tumoren – Moderne immunologische strategien in der onkologie. Dtsch. Med. Wochenschr. 133 (41), 2105–2108. http://doi.org/10.1055/s-0028-1091251.

Hamilton, W., Sharp, D., 2004. Diagnosis of colorectal cancer in primary care: the evidence base for guidelines. Fam. Pract. 21 (1), 99–106. http://doi.org/10.1093/fampra/cmh121.

Hamzehzadeh, L., et al., 2016. New approaches to use nanoparticles for treatment of colorectal cancer; a brief review. Nanomed. Res. J. 1 (2), 59–68. http://doi.org/10.7508/NMRJ.2016.02.001.

Harada, S., Morlote, D., 2020. Molecular pathology of colorectal cancer. Adv. Anat. Pathol. 27 (1), 20–26. http://doi.org/10.1097/PAP.0000000000000247.

Ho, B.N., Pfeffer, C.M., Singh, A.T.K., 2017. Update on nanotechnology-based drug delivery systems in cancer treatment. Anticancer Res. 37 (11), 5975–5981. http://doi.org/10.21873/anticanres.12044.

Houot, R., et al., 2011. Targeting immune effector cells to promote antibody-induced cytotoxicity in cancer immunotherapy. Trends Immunol. 32 (11), 510–516. http://doi.org/10.1016/j.it.2011.07.003.

Jain, A., et al., 2019. Minicapsules encapsulating nanoparticles for targeting, apoptosis induction and treatment of colon cancer. Artif. Cells, Nanomed., Biotechnol. 47 (1), 1085–1093. Taylor & Francis. http://doi.org/10.1080/21691401.2019.1593848.

Jass, J.R., 2000. Pathology of hereditary nonpolyposis colorectal cancer. Ann. N. Y. Acad. Sci. 910, 62–74. http://doi.org/10.1111/j.1749-6632.2000.tb06701.x.

Jin, C., et al., 2020. Application of nanotechnology in cancer diagnosis and therapy – a mini-review. Int. J. Med. Sci. 17 (18), 2964–2973. http://doi.org/10.7150/ijms.49801.

Kanemitsu, Y., et al., 2003. Survival after curative resection for mucinous adenocarcinoma of the colorectum. Dis. Colon Rectum 46 (2), 160–167. http://doi.org/10.1007/s10350-004-6518-0.

Kang, H., et al., 2005. A 10-year outcomes evaluation of mucinous and signet-ring cell carcinoma of the colon and rectum. Dis. Colon Rectum 48 (6), 1161–1168. http://doi.org/10.1007/s10350-004-0932-1.

Kelly, H., Goldberg, R.M., 2005. Systemic therapy for metastatic colorectal cancer: current options, current evidence. J. Clin. Oncol. 23 (20), 4553–4560. http://doi.org/10.1200/JCO.2005.17.749.

Kilic, N., et al., 2011. Brachyury expression predicts poor prognosis at early stages of colorectal cancer. Eur. J. Cancer 47 (7), 1080–1085. Elsevier Ltd. http://doi.org/10.1016/j.ejca.2010.11.015.

Kim, I., 2020. Colorectal cancer | 대장암. Dbpia.Co.Kr 66 (11), 1–9.

Krasinskas, A.M., 2011. EGFR signaling in colorectal carcinoma. Pathol. Res. Int. 2011, 1–6. http://doi.org/10.4061/2011/932932.

Kristjansson, S.R., et al., 2010. Comprehensive geriatric assessment can predict complications in elderly patients after elective surgery for colorectal cancer: a prospective observational cohort study. Crit. Rev. Oncol. Hematol. 76 (3), 208–217. Elsevier Ireland Ltd. http://doi.org/10.1016/j.critrevonc.2009.11.002.

Kinzler, K.W., Vogelstein, B., 1996. Lessons from hereditary colorectal cancer. Cell 87 (2), 159–170.

Labianca, R., Merelli, B., 2010. Screening and diagnosis for colorectal cancer: present and future. Tumori 96 (6), 889–901. http://doi.org/10.1177/548.6506.

Lim, E.-K., Chung, B.H., Chung, S.J., 2016. Recent advances in pH-sensitive polymeric nanoparticles for smart drug delivery in cancer therapy. Curr. Drug Targets 19 (4), 300–317. http://doi.org/10.2174/1389450117666160602202339.

Maeda, H., et al., 2000. Tumor vascular permeability and the EPR effect in macromolecular therapeutics: a review. J. Control. Release 65 (1–2), 271–284. http://doi.org/10.1016/S0168-3659(99)00248-5.

Mandelblatt, J., et al., 1996. The late-stage diagnosis of colorectal cancer: demographic and socioeconomic factors. Am. J. Public Health 86 (12), 1794–1797. http://doi.org/10.2105/AJPH.86.12.1794.

Mármol, I., et al., 2017. Colorectal carcinoma: a general overview and future perspectives in colorectal cancer. Int. J. Mol. Sci. 18 (1). http://doi.org/10.3390/ijms18010197.

Marshall, J.L., 2008. Managing potentially resectable metastatic colon cancer. Gastrointest. Cancer Res. : GCR 2 (4 Suppl), S23–S26. Available at: http://www.ncbi.nlm.nih.gov/pubmed/19343144%0Ahttp://www.pubmedcentral.nih.gov/articlerender.fcgi?artid=PMC2661554.

Matos, A.I., et al., 2019. Nanotechnology is an important strategy for combinational innovative chemo-immunotherapies against colorectal cancer. J. Control. Release 307 (April), 108–138. Elsevier. http://doi.org/10.1016/j.jconrel.2019.06.017.

Matsumura, Y., Maeda, H., 1986. A new concept for macromolecular therapeutics in cancer chemotherapy: mechanism of tumoritropic accumulation of proteins and the antitumor agent smancs. Cancer Res. 46 (8), 6387–6392.

Mecklin, J.P., 1987. Frequency of hereditary colorectal carcinoma. Gastroenterology 93 (5), 1021–1025. http://doi.org/10.1016/0016-5085(87)90565-8.

Midgley, R.S., Kerr, D.J., 2000. Adjuvant therapy of colorectal cancer. Hosp. Pract. 35 (5), 55–62. http://doi.org/10.3810/hp.2000.05.198.

Mohd-Zahid, M.H., et al., 2019. Colorectal cancer stem cells: a review of targeted drug delivery by gold nanoparticles. RSC Adv. 10 (2), 973–985. http://doi.org/10.1039/c9ra08192e.

Morse, M.A., Hall, J.R., Plate, J.M.D., 2009. Countering tumor-induced immunosuppression during immunotherapy for pancreatic cancer. Expert Opin. Biol. Ther. 9 (3), 331–339. http://doi.org/10.1517/14712590802715756.

Nakaji, S., et al., 2003. Environmental factors affect colon carcinoma and rectal carcinoma in men and women differently. Int. J. Colorectal Dis. 18 (6), 481–486. http://doi.org/10.1007/s00384-003-0485-0.

Ortega, J., Vigil, C.E., Chodkiewicz, C., 2010. Current progress in targeted therapy for colorectal cancer. Cancer Control 17 (1), 7–15. http://doi.org/10.1177/107327481001700102.

Osorio, M., et al., 2020. Recent advances in polymer nanomaterials for drug delivery of adjuvants in colorectal cancer treatment: a scientific-technological analysis and review. Molecules 25 (10). http://doi.org/10.3390/molecules25102270.

Patra, J.K., et al., 2018. Nano based drug delivery systems: recent developments and future prospects. J. Nanobiotechnology 16 (1), 1–33. http://doi.org/10.1186/s12951-018-0392-8.

Peltomaki, P., et al., 1997. Mutations predisposing to hereditary nonpolyposis colorectal cancer: database and results of a collaborative study. Gastroenterology 113 (4), 1146–1158. http://doi.org/10.1053/gast.1997.v113.pm9322509.

Pereira, I., et al., 2018. Carcinoembryonic antigen-targeted nanoparticles potentiate the delivery of anticancer drugs to colorectal cancer cells. Int. J. Pharm. 549 (1–2), 397–403. Elsevier. http://doi.org/10.1016/j.ijpharm.2018.08.016.

Pickhardt, P.J., et al., 2013. Assessment of volumetric growth rates of small colorectal polyps with CT colonography: a longitudinal study of natural history. Lancet Oncol. 14 (8), 711–720. Elsevier Ltd. http://doi.org/10.1016/S1470-2045(13)70216-X.

Rabelo, R., et al., 2001. Role of molecular diagnostic testing in familial adenomatous polyposis and hereditary nonpolyposis colorectal cancer families. Dis. Colon Rectum 44 (3), 437–446. http://doi.org/10.1007/BF02234746.

Rampado, R., et al., 2019. Nanovectors design for theranostic applications in colorectal cancer. J. Oncol. 2019. http://doi.org/10.1155/2019/2740923.

Romagnoli, P., et al., 1984. Increase of mitotic activity in the colonic mucosa of patients with colorectal cancer. Dis. Colon Rectum 27 (5), 305–308. http://doi.org/10.1007/BF02555636.

Røseth, A.G., et al., 1993. Faecal calprotectin: a novel test for the diagnosis of colorectal cancer? Scand. J. Gastroenterol. 28 (12), 1073–1076. http://doi.org/10.3109/00365529309098312.

Ross, J.S., et al., 2010. Biomarker-based prediction of response to therapy for colorectal cancer current perspective. Am. J. Clin. Pathol. 134 (3), 478–490. http://doi.org/10.1309/AJCP2Y8KTDPOAORH.

Ryan-Harshman, M., Aldoori, W., 2007. Diet and colorectal cancer: review of the evidence. Can. Fam. Physician 53 (11), 1913–1920.

Sánchez-Moreno, P., et al., 2018. Thermo-sensitive nanomaterials: recent advance in synthesis and biomedical applications. Nanomaterials 8 (11), 1–32. http://doi.org/10.3390/nano8110935.

Standring, S., 2019. *The Anatomy of the Large Intestine*. doi: http://doi.org/10.1007/978-3-030-05240-9_2.

Stewart, S.L., et al., 2006. A population-based study of colorectal cancer histology in the United States, 1998-2001. Cancer 107 (SUPPL.), 1128–1141. http://doi.org/10.1002/cncr.22010.

Soudja, S.M., et al., 2010. Tumor-initiated inflammation overrides protective adaptive immunity in an induced melanoma model in mice. Cancer Res. 70 (9), 3515–3525. http://doi.org/10.1158/0008-5472.CAN-09-4354.

Tabernero, J., 2007. The role of VEGF and EGFR inhibition: implications for combining anti-VEGF and anti-EGFR Agents. Mol. Cancer Res. 5 (3), 203–220. http://doi.org/10.1158/1541-7786.MCR-06-0404.

Tournigand, C., et al., 2004. FOLFIRI followed by FOLFOX6 or the reverse sequence in advanced colorectal cancer: a randomized GERCOR study. J. Clin. Oncol. 22 (2), 229–237. http://doi.org/10.1200/JCO.2004.05.113.

Viswanath, B., Kim, S., Lee, K., 2016. Recent insights into nanotechnology development for detection and treatment of colorectal cancer. Int. J. Nanomed. 11, 2491–2504. http://doi.org/10.2147/IJN.S108715.

Wathoni, N., et al., 2020. Enteric-coated strategies in colorectal cancer nanoparticle drug delivery system. Drug Des. Dev. Ther. 14, 4387–4405. http://doi.org/10.2147/DDDT.S273612.

Weiner, L.M., Surana, R., Wang, S., 2010. Monoclonal antibodies: versatile platforms for cancer immunotherapy. Nat. Rev. Immunol. 10 (5), 317–327. Nature Publishing Group. http://doi.org/10.1038/nri2744.

Wu, P., et al., 2020. Enhanced antitumor efficacy in colon cancer using EGF functionalized PLGA nanoparticles loaded with 5-fluorouracil and perfluorocarbon. BMC Cancer 20 (1), 1–10. BMC Cancer. http://doi.org/10.1186/s12885-020-06803-7.

Yang, C., Merlin, D., 2020. Lipid-based drug delivery nanoplatforms for colorectal cancer therapy. Nanomaterials 10 (7), 1–32. http://doi.org/10.3390/nano10071424.

Young, C.J., Sweeney, J.L., Hunter, A., 2000. Implications of delayed diagnosis in colorectal cancer. Aust. N. Z. J. Surg. 70 (9), 635–638. http://doi.org/10.1046/j.1440-1622.2000.01916.x.

Zhang, Y.L., et al., 2002. Early diagnosis for colorectal cancer in China. World J. Gastroenterol. 8 (1), 21–25. http://doi.org/10.3748/wjg.v8.i1.21.

Chapter 14

Nanoparticles for the targeted drug delivery in lung cancer

Veena Belgamwar[a], Vidyadevi Bhoyar[a], Sagar Trivedi[a], Miral Patel[b,c]

[a]University Department of Pharmaceutical Sciences, Rashtrasant Tukadoji Maharaj Nagpur University, Nagpur, Maharashtra, India
[b]Department of Pharmaceutical Sciences, Arnold and Marie Schwartz College of Pharmacy and Health Science, Long Island University, Brooklyn Campus, NY
[c]Office of Pharmaceutical Quality, Center for Drug Evaluation and Research, Food and Drug Administration, Silver Spring, MD, USA

14.1 Introduction

Lung cancer (LC) is the third most typical disease around the world, and common cause of death from cancer, estimated to be responsible for nearly one in five cancer deaths. LC survival is mostly determined by the stage at which it is diagnosed, with later-stage diagnosis having poorer survival (*Lung cancer | World Cancer Research Fund International*). LC, a highly fatal disease (Abdelaziz et al., 2018) and according to the Global cancer statistics GLOBOCAN database, there will be an estimated 18.1 million new cancer cases and 9.6 million cancer deaths reported in 2018 (Bray et al., 2018). The LC is broadly categorized into primary and secondary cancers (Alhajj et al., 2018). Histologically, two types of LC could be distinguished; small cell lung carcinoma (SCLC) and nonsmall cell lung carcinoma (NSCLC) with the latter accounting for 85% of LC cases. Three subtypes of NSCLC were further identified; large cell LC, squamous-cell carcinoma and adenocarcinoma (*Lung Cancer : Risk Factors, Symptoms | Prevention Lung Cancer | India Against Cancer*; Abdelaziz et al., 2018). Despite the advances in the field of oncology, NSCLC is usually diagnosed at a late stage and shows poor prognosis with a 15% overall 5-year survival. Treatment modalities for LC include surgery, chemotherapy, radiotherapy, and/or targeted therapies depending on the cancer stage. However, surgery is usually deemed ineligible due to diagnosis of patients at an advanced stage. Despite the enormous efforts for developing new biomarkers that aid early diagnosis, this goal has not yet been achieved. Whether alone or combined with other therapeutic strategies, chemotherapy is the major key player in the treatment of LC. Despite the several theoretical advantages offered by inhalation chemotherapy of LC, intravenous (i.v.) administration (systemic chemotherapy) is still the mainstay. Many lung barriers need to be overcome first before inhalation treatment gets a foothold in the LC therapy avenue. The upper airways are made of columnar epithelial cells that are ciliated and mucus-producing which both collectively make the mucociliary escalator system. This is the main cleaning system in the upper airways which efficiently sweeps any insoluble particle that gets deposited on it. Deeper in the alveolar sacs, macrophages take the responsibility of getting rid of any insoluble particles that deposit on the alveoli. Chemotherapeutics first need to be inhaled efficiently and deposit deeply in lungs to reach their target regions. The drugs then have to bypass the barriers of mucociliary escalation and macrophage clearance, and lastly to get taken up efficiently by the cancerous cells for them to be effective in treating LC (Abdelaziz et al., 2018).

14.1.1 Stages of LC

There are two types of staging system used to stage LC. They are number system and TNM staging system.

I. Number system of staging LC (*Lung Cancer : Risk Factors, Symptoms | Prevention Lung Cancer | India Against Cancer*; Larbi, 2011):
   - **Stage (I)**: The tumor size is less than 5 cm and it is limited to lungs only. There is no involvement of lymph nodes.
   - **Stage (II)**: Tumor size is more than 5 cm, lymph nodes are involved, larger than 7 cm with no involvement of lymph nodes, spread to the following areas – the chest wall, the muscle under the lung (diaphragm), the phrenic nerve, or

the layers that cover the heart (mediastinal pleura and parietal pericardium). In the main airway (bronchus) close to where it divides to go into each lung, making part of the lung collapse, any size but there is more than one tumor in the same lobe of the lung.
- **Stage (III)**: There is presence of any one of these conditions such as complete lung collapse, has spread into the chest wall, the muscle under the lung (diaphragm), or the layers that cover the heart (mediastinal pleura and parietal pericardium), spread into lymph nodes on the opposite side of the chest, involvement of major structures in the chest include the heart, the wind pipe (trachea), the food pipe (esophagus) or a main blood vessel.
- **Stage (IV)**: The cancer has spread to a distant part of the body such as the liver, bones or the brain.

II. Tumor, Node and Metastases (TNM) staging of LC: This staging system (*Lung Cancer : Risk Factors, Symptoms | Prevention Lung Cancer | India Against Cancer*; Woodard et al., 2016) describes as follows:
- **Tumor (T)**: The T stages of LC are
  - T1 – the tumor is contained within the lung and is smaller than 3 cm across,
  - T2 – the tumor is between 3 and 7 cm across
  - T3 – the tumor is larger than 7 cm,
  - T4 – the tumor has grown into one of the following structures: mediastinum, the heart, a major blood vessel, the wind pipe, food pipe, a spinal bone, the nerve that controls the voice box
- Nodes (N): The N stages for LC (Woodard et al., 2016) are
  - N0 – there is no cancer in any lymph nodes,
  - N1 – there is cancer in the lymph nodes nearest the affected lung,
  - N2 – there is cancer in lymph nodes in the mediastinum but on the same side as the affected lung or there is cancer in lymph nodes just under where the windpipe branches off to each lung,
  - N3 – there is cancer in lymph nodes on the opposite side of the chest from the affected lung or in the lymph nodes above either collar bone or in the lymph nodes at the top of the lung.

Fig. 14.1 depicts the lymph node stations and radiographic borders defined by International Association for the Study of LC (IASLC)

- **Metastases (M)**: The M stages for LC are 1) M0 – there are no signs that the cancer has spread to another lobe of the lung or any other part of the body, 2) M1 – there are signs that the cancer has spread to another lobe of the lung or any other part of the body.

14.1.2 Current treatment strategies on LC

The LC treatment depends on the following parameters (*Lung Cancer : Risk Factors, Symptoms | Prevention Lung Cancer | India Against Cancer*):

- Type of LC (whether its small cell type or nonsmall cell type),
- Size of the tumor,
- Whether the cancer is local or metastatic, and
- General condition of the patient

Treatment modalities: Surgery, chemotherapy, radiotherapy, and targeted therapy may be used alone or in combination depending on the above mentioned factors.

- **Surgery:** Surgery is mostly used treatment strategies for stage I-II nonsmall cell LC and rarely for small cell LC when the disease is in very early stage (Lemjabbar-Alaoui et al., 2015). The surgeon might remove a small section of lung (wedge resection), a larger portion of lung (segmental resection), entire lobe of one lung (lobectomy) or an entire lung (pneumonectomy) (*Lung Cancer : Risk Factors, Symptoms | Prevention Lung Cancer | India Against Cancer*).
- **Chemotherapy**: Chemotherapy is usually well tolerated by patients with fully active, able to carry on all predisease performance without restriction and restricted in physically strenuous activity but ambulatory and able to carry out work of a light or sedentary nature, for example, light house work, office work but rarely effective in patients with capable of only limited self-care, confined to bed or chair more than 50% of waking hours and completely disabled, cannot carry on any self-care, totally confined to bed or chair where palliative care is preferred. Use of chemotherapy is controversial in ambulatory and capable of all self-care but unable to carry out any work activities, up and about more than 50% of waking hours NSCLC patients, which represent nearly 40% of advanced stage NSCLC patients. Chemotherapy is recommended only for patients who are reasonably fit and awake for more than 50 % of the day (Lemjabbar-Alaoui et al., 2015). A combination of chemotherapy drugs are usually given in sessions over a period of weeks or months, with breaks

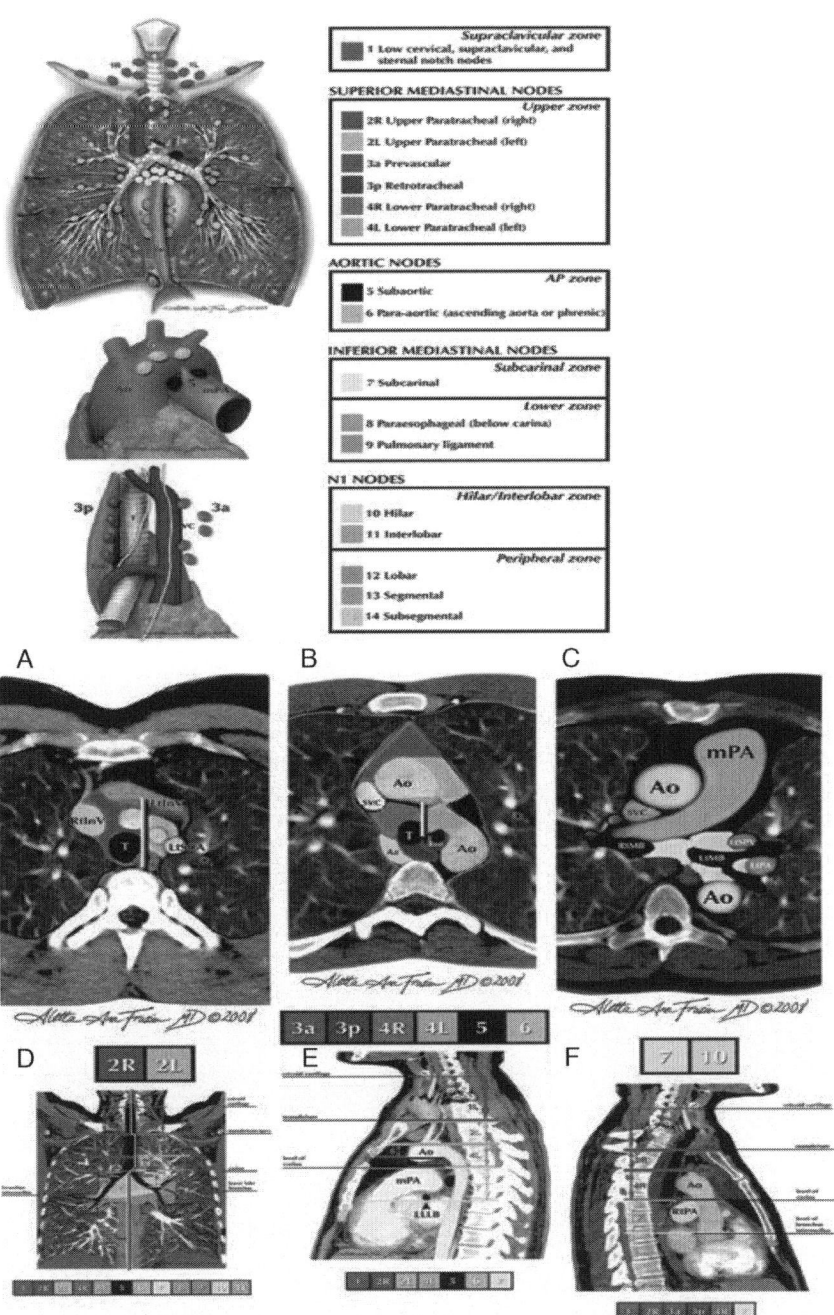

FIG. 14.1 Nonsmall cell LC lymph node stations. (A) International Association for the Study of LC (IASLC) lymph node station map, stations, and CT scan. (B) Application of the IASLC lymph node stations and borders to CT scans. *Adapted from source Woodard et al. (2016).*

in between. It can either be used before (to shrink cancers) or after surgery (to kill any cancer cells that may remain). In some cases, chemotherapy can be used as palliation to relieve pain and other symptoms of advanced cancer. Some of the chemotherapy drugs used in the treatment of LC are Paclitaxel, Carboplatin, Cisplatin, Docetaxel, Etoposide, Gemcitabine, Pemetrexed (*Lung Cancer : Risk Factors, Symptoms | Prevention Lung Cancer | India Against Cancer*).

- **Radiation therapy** (*Lung Cancer : Risk Factors, Symptoms | Prevention Lung Cancer | India Against Cancer*)
Radiation therapy can be directed at LC from outside the body (external beam radiation) or it can be administered through needles, seeds or catheters placed inside the body near the cancerous area (brachytherapy). Radiation therapy can be either used after surgery to kill any cancer cells that may remain or can be used as the palliative therapy to relieve pain and other symptoms in advanced stage LC.

With the current treatment options, the survival rate of LC patients is still very low (Alhajj et al., 2018).
Disadvantages of current treatment strategies (Rocha et al., 2017; Pontes and Grenha, 2020)

- Chemotherapeutic drug moieties are hydrophobic; they show less solubility in water that is why low bioavailability, nonspecificity of drugs.
- Resistance gets developed by the various chemotherapeutic agents.
- Drug gets quickly eliminated from the body.
- Large-spanned distribution into the cancerous tissues required more amount of drug which increases the cost of medication and side effect of medication.
- Difficulties to reach the affected tissues.
- Most of the cytotoxic drugs do not hugely distinguish between tumorous and normal cells, which cause death of normal cells and many side effects, means severity of systemic adverse effects. This immensely poses limitations to the maximum effective dose.

Pontes Jorge and Grenha Ana reviewed and explained the improvement imparted by pulmonary delivery of the drugs over to current strategies that is vectorization to cancer cells reduction, reduction in systemic adverse effects (Pontes and Grenha, 2020).

14.1.3 Novel strategies for LC treatment by pulmonary route of administration

Inhalational route for drug delivery and desired effects has been known since centuries (Rytting et al., 2008; Kaur, 2017). The lung comprises an interesting and attracting route of administration on account of the large surface area of the alveolar region (Rytting et al., 2008; Fernandes and Vanbever, 2009; Kurmi et al., 2010), thin epithelial barrier, high blood flow and the avoidance of first-pass metabolism (Rytting et al., 2008). The pulmonary route is noninvasive, alternative to the intravenous route and can be used to target drugs directly to the lungs, limiting the exposure of secondary organs (Respaud et al., 2015) that is reduced systemic side effects with the administration of minute drug dosages (Shah et al., 2012).The understanding of pulmonary drug delivery and thus its utilization for medical purposes has remarkably advanced over the last decades. It has been recognized that this route of administration offers many advantages and several drug delivery systems have been developed accordingly (Friebel and Steckel, 2010).

Potential advantages of the pulmonary route include (Blank et al., 2011; Shah et al., 2012)

- Provides local action within the respiratory tract
- Accessibility,
- Noninvasiveness,
- Ease of administration,
- Possibility to reach to the target organ in case of LC with an elaborate mucosal network of antigen-presenting cells
- Provides rapid drug action
- Provides reduced dose
- Allows for a reduction in systemic side-effects, can be employed as an alternative route to drug interaction when two or more medications are used concurrently
- Reduce extracellular enzymes levels compared to GI tract due to the large alveolar surface area
- Reduce evasion of first pass hepatic metabolism by absorbed drug
- Offers the potential for pulmonary administration of systemically active materials.

Pulmonary delivery of drugs has become an attractive target in the health care industry as the lung is capable of absorbing pharmaceuticals either for local deposition or for systemic delivery (Chandira and Jayakar; Fernandes and Vanbever, 2009; Kurmi et al., 2010; Shah et al., 2012). Half of all pharmaceuticals are not soluble in water, but are soluble in lipid. As the lung is able to absorb both water and oil into the tissue, this is not a limitation of pulmonary delivery (Chandira and Jayakar). This chapter summarizes and discusses the investigated formulation approaches for the pulmonary delivery. Some of the investigated formulation approaches appear to be promising in overcoming these challenges. However, in order to create products that reach patients, more therapeutically oriented studies are still needed to ensure formulation stability, in-vivo sustained release behavior, pulmonary retention, and bypassing lung clearance mechanisms (Ali et al., 2015). The slow progress in the efficacy of the treatment of severe diseases, has suggested a growing need for a multidisciplinary approach to the delivery of therapeutic agents to targets in tissues. The efficacy of the drug and its treatment can be achieved from the new ideas on controlling the pharmacokinetics, pharmacodynamics, immunogenicity, and biorecognition (Patil and Sarasija, 2012). There is progressive evolution in the use of inhalable drug delivery systems (DDSs) for LC therapy because of its advantages.

This chapter highlights various inhalable colloidal systems studied for tumor-targeted drug delivery including polymeric, lipid, hybrid and inorganic nanocarriers. The active targeting approaches for enhanced delivery of nanocarriers to LC cells were illustrated (Abdelaziz et al., 2018). Nanoparticles (NPs) provide new formulation options for both dispersed liquid droplet dosage forms such as metered dose inhalers and nebulizers, and dry powder formulations. Specifically, pure drug NPs, polymeric NPs, polyelectrolyte complexes, and drug-loaded liposome offer some encouraging results for delivering drugs to and through the lungs (Bailey and Berkland, 2009).

14.1.4 Pulmonary physiology and drug absorption

The chief physiological role of lung is to exchange of oxygen and carbon dioxide between blood and inspired air. The trachea leads from the throat to the lungs, branches into two main bronchi, and one for each lung, which branches into each lobe of the lungs. The right lung is divided into three lobes. Each lobe is like a balloon filled with sponge-like tissue. The left lung is divided into two lobes (Porra, 2006) and they are covered by a thin covering called "pleura" which protects and helps lungs move back and forth as they expand and contract during breathing. A thin, dome-shaped muscle below the lungs called "diaphragm" separates the chest from abdomen. The diaphragm moves up and down during breathing forcing air in and out of the lungs (*Lung Cancer : Risk Factors, Symptoms | Prevention Lung Cancer | India Against Cancer*). The lung is composed of more than 40 different cells (Patil and Sarasija, 2012). Physiology of human and animal body system suggests that lungs are the complex organs, with near about 300 million of alveoli (Patil et al., 2019) having surface area of 70 to 160 m^2, but instead of that pulmonary delivery has more advantage such as noninvasive, avoids first-pass metabolism in the liver and enables targeting of therapeutic agents to the infection site. Inhaled delivery also potentially reduces the dose requirement and the accompanying side effects (Parumasivam et al., 2016). Pulmonary route is extensively studied for the diagnosis and treatment of pulmonary and extra pulmonary disease conditions such as asthma, tuberculosis, emphysema, and bronchitis (Chandel et al., 2019).

Fig. 14.2 illustrates the deposition of drug particles deep into the lungs depends on size of particle, like particle with 1 to 5 um size get deposited deep into lungs.

14.1.5 Role of nanoparticulate technology in the diagnosis and treatment of LC

Nanotechnology can be defined as the science and engineering involved in the design, characterization, and application of materials and devices whose smallest functional organization in at least one dimension is on the nanometers (nm) scale, that is, one billionth of a meter. It is an exciting multidisciplinary field that involves the design and engineering of nano-objects or nanotools with diameters less than 500 nm, and it is one of the most interesting fields of the 21st century. Nanotechnology also offers the ability to detect diseases, such as tumors, much earlier than ever imaginable (Badrzadeh et al., 2016).

Advantages of NPs (Swaroop, no date; Mohanraj and Chen, 2007; Tiruwa, 2016)

- Particle size can be easily altered resulting in attaining both active and passive drug targeting, becomes the most advantageous in the treatment of chronic disease

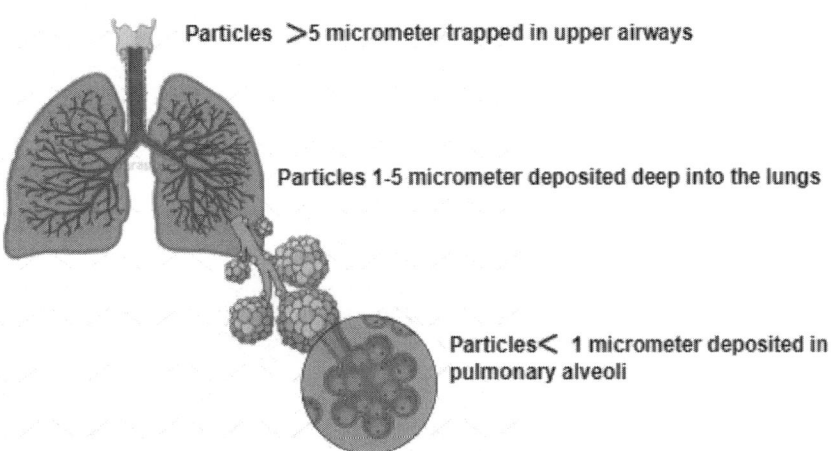

FIG. 14.2 Deposition of drug particle deep into the lung.

- Have ability to control and sustain the drug before reaching the specific site of action and protects drug from rapid degradation
- Drug loading is relatively high and drug can be incorporated into the systems without any chemical reaction, an important factor for preserving the drug activity
- Maintain the drug at specific sites or in target tissue hence with lower doses of drugs shows high therapeutic efficacy
- Site specific targeting can be achieved by attaching targeting ligands to surface of particles or use of magnetic guidance
- As targeted drug carrier NPs reduced drug toxicity
- Enhances aqueous solubility of poorly water soluble drug, which improves bioavailability of drug
- By using polymer, drug release from NPs can be modified, makes polymeric NPs an ideal drug delivery systems for cancer therapy

14.1.5.1 Conventional method of LC diagnosis

LC should be suspected in chronic smokers with protracted or new onset respiratory symptoms. Diagnosis usually starts with a chest X-Ray, if abnormal should lead to a computer tomographic scan. Further diagnosis depends upon the location of the tumor–central tumors are best approached with a bronchoscopy, which allows for direct visualization of the airways as well as tumor, and permits biopsy material to be obtained for histologic diagnosis at the same time. Peripheral tumors can be biopsied via a transthoracic approach under CT or ultrasound guidance. Further invasive techniques including thoracoscopy, mediastinoscopy, and thoracotomy are reserved for those in whom the initial diagnostic techniques mentioned above fail to yield a diagnosis. Even though sputum cytology is a noninvasive method of establishing a diagnosis, negative sputum cytology does not rule out LC (Larbi, 2011).

14.1.6 Nanocarriers used for the diagnosis of lung diseases

Theranostics or personalized cancer treatment is proof based, an individualized prescription that warrants the right treatment at the correct time, leading to significant efficacy and improvements of patient's condition and a decrease in medicinal and services costs (Fathi Karkan et al., 2017). Conjugation of nanocarriers with the conventional imaging techniques helps to amplify the imaging phenomenon several-fold through targeted diagnosis (Hussain, 2019).

Masaoutis et al. reviewed the role of exosomes in intercellular communication during carcinogenesis and treatment, where they outline the function of tumor-derived exosomes as possible biomarkers in LC, regarding which they explained exosomal miRNA and lncRNA in NSCLC diagnosis, exosomal vs. exosome-free circulating miRNA, exosomal proteins, nucleic acids, and exosomal contents as predictive diagnostic biomarkers in NSCLC (de Vries et al., 2010). Diagnosis and treatment of LC have been characterized with a variety of challenges. However, with the advancement in magnetic nanoparticle (MNP) technology, many challenges in the diagnosis and treatment of LC are on the decline (Fathi Karkan et al., 2017).

In radiotherapy, the precise tumor location and extent of spread must be known in order to ensure the therapy is as effective as possible. Within the remit of cancer diagnostics there are two main branches, imaging of the tumor body itself and detection of moieties associated with the tumor. Working with materials at the nanoscale provides a means by which to improve the conventional diagnostic methods within each of these categories. These enhancements stem from the unique material properties that are inherent at the nanoscale. For the patient, this translates to higher image resolution and signal strength as well as less toxicity from administered agents due to more specific delivery (Cryer and Thorley, 2019). De Vries et al. explored highly iodinated nanoemulsions as a possible class of blood pool CT contrast agent, because imaging agents used for CT scans in current clinical practice are based on iodine molecules however due to their small size they are rapidly cleared from the circulation by the kidneys (de Vries et al., 2010).

Hussain (2019) reviewed in his book chapter about the development of novel diagnostic nanoimaging platform for LC, where they have enlisted various examples for diagnosis LC with these novel tools. In his book chapter, they reported that targeted NP-CT agents commonly employ the use of metal-based nanocarriers because of their higher molecular mass and consequently enhanced X-ray contrast, in comparison to iodine. Several experiments have been carried out using bismuth, ytterbium, gold, and tantalum as promising elements for CT imaging probe development. Considering the receptor-based endocytosis (RBE) of diagnostic compounds, significant payloads can be delivered inside the cancerous cells via nanotargeting agents. HER2-specific Au NPs showed enhanced tumor detection compared to passive targeting, and targeted Au NP-CT probes have been investigated for the imaging of PSA (prostate specific antigen) in the case of prostate cancer, CD4-receptors in the case of lymphatic cancer and folate receptor-a associated with LC. Additionally, up-conversion NPs (UCNP) have gained interest for use in X-ray-based diagnosis. The UCNPs are lanthanide-based rare earth NPs (including Lu, Yb, and Gd) with up-conversion luminescence characteristics. PEG based Ba-Gd-F5:Er/Yb NPs are a promising example of UCNPs employed in cancer optical imaging, CT, and MRI [34]. Further they concluded by giving the advantages of the nano-imaging for diagnosis of LC which are as follows (Hussain, 2019):

- Versatility in the design of NPs which enables easy manipulation of composition and morphology for optimum translocation into the tumorous tissues whilst restricting nonspecific targeting,
- Simultaneous loading of multifarious imaging payload without hindering the target-specific biodistribution, that is, functional diversity of NPs, and
- Ultra small particle size (nanorange), surface characteristics (zeta potential), and target-specific delivery. These advantages have slanted the interest of scientists and medical professionals alike for utilizing nanodiagnostic systems in preclinical and clinical setups, especially for the early diagnosis of LC and pulmonary metastasis

14.2 Nanocarriers in LC treatment

Various nanoparticulate systems nowadays are used for the treatment of LC, among which some are discuses below:

14.2.1 Solid–lipid nanocarriers

Solid–lipid NPs (SLN's) are the major class of drug carriers which plays a crucial role in improving the localization of drug at the site of action via active and passive targeting techniques. The important members of the Solid–lipid NPs are Liposomes, oil dispersions (micelles), and lipid NPs which facilitates the upgraded delivery of drugs and genes. The various underlying properties of this nanocarriers like Nano size which aids enhanced permeation and retention (EPR) based on tumor cell abnormalities including hyper-vascularization and poor lymphatic drainage (El-Far et al., 2018).

Solid–lipid Nano carriers are generally lipidic vesicles or bilayered phospholipid vesicles that comprise an aqueous core surrounded by one or more phospholipid bilayers (El-Sherbiny et al., 2015). They are commonly used to deliver hydrophobic and hydrophilic drugs, proteins and nucleic acids in which entrapment efficiency is based on several parameters including drug solubility in lipid, miscibility between drug and lipid, physicochemical structure of lipid matrix, as well as lipid polymorphous state through either incorporation in the lipid bilayer itself or encapsulation in the inner aqueous core, respectively (Gaber et al., 2017).

Drug resistance is one of the major problems associated while targeting lungs because of increased drug efflux mediated by transporters (ABC proteins, etc.), drug inactivation, sequestration by enzymes, DNA repair, target modifications, and apoptosis defects (Shanker et al., 2010). Solid–lipid NPs implies to overcome this barrier by combining anticancer drug with molecules which decreases drug resistance mechanisms within one multifunctional SLN's-based delivery system. Garbuzenko et al. demonstrated Dual-loaded liposomes, containing the chemotherapeutic drug and an agent, either reversing hypoxia-induced drug resistance to KI or suppressing (small molecules and siRNA) cellular resistance mechanisms, have been developed for resistant LC. Codelivery of DOX with suppressors of the pump and nonpump resistance (antisense oligonucleotides [ASO] targeted toMRP1andBCL2) was shown to enhance pro-apoptotic activity and decrease tumor volume in human A549 LC cells in nude mice (Garbuzenko et al., 2010).

Stimuli responsive SLN's are the modified type of nanocarriers which responds to various stimuli like temperature, pH depending on the vacuolization and microenvironment of the tissues and tumor site. Doxorubicin (DOX) is an amphiphilic chemotherapeutic agent this particular drug suffers from poor penetration power, restricted distribution in solid tumors, and ease of protonation at low pH of tumor which subsequently hinder its cellular uptake. Mussi et al. improved the localization of DOX and reported that using anionic hexadecyl phosphate with cationic DOX significantly upgraded its encapsulation efficiency in lipid (Mussi et al., 2013).

Pulmonary administration of SLN's have been proved boon to treatment of various types of LC which facilitates the advantages of unique features of lungs provides large surface area, thin alveolar epithelium, high vascularization, high bioavailability, great capacity for solute exchange, improved pharmacokinetic profile, in addition to great tendency to evade the first-pass effect. CPPD-loaded liposomes for inhalation, named Sustained Release Lipid Inhalation Targeting (SLIT Cisplatin), was developed by Transave, Inc. (Monmouth Junction, NJ) are most commonly used (Agu et al., 2001; Paranjpe and Müller-Goymann, 2014).

Receptor mediated endocytosis mechanism shows a significant in localizing i.e. actively targeting the drug and genes using another approach of surface modifications via attaching ligands to the surface of SLN's. Leiva et al. modified the potent anticancer agent paclitaxel (PTX) which was having poor aqueous solubility thereby paclitaxel was successfully incorporated into glyceryl tripalmitate (tripalmitin) SLNs forming Tripalm-SLNs-PTX, which was used as a solvent emulsion method which managed to control size and monodispersity of particles relative to hot and cold homogenization methods (Leiva et al., 2017).

Gene delivery is a novel approach in which exogenous genetic material is inserted into somatic cells of patients to modulate gene expression of desired proteins (Tian et al., 2012). These strategies have included immune stimulation, transfer of suicide genes, inhibition of driver oncogenes, replacement of tumor-suppressor genes that could mediate apoptosis or antiangiogenesis, and transfer of genes that enhance conventional treatments such as radiotherapy and chemotherapy. Hence gene therapy helps in overcoming multiple genetic defects that present and the molecular profile changes during the disease (Zeng et al., 2012).

14.2.2 Polymeric nanocarriers

Polymeric nanoparticles are generally composed of biodegradable polymers such as poly (lactic acid) (PLA), poly(lactic-co-glycolic) acid (PLGA), gelatin, albumin, chitosan, polycaprolactone, and poly-alkyl-cyanoacrylates. The polymeric material present in the system offers the due advantages of polymers like controlled and sustained release properties and biocompatible nature (Farokhzad et al., 2006). On the other hand di-block and tri-block polymers are gaining popularity, they encapsulate the drug via self-assembly by varying hydrophobicity between blocks and are suitable for systemic administration. The core-shell structure of polymeric NPs facilitates encapsulation of hydrophobic drugs, extension of circulation time, and sustained drug release (Torchilin, 2007).

Targeted delivery of drug can be achieved by modifying the surface of the NPs by conjugating them with various targeting moieties like proteins, transferrin's, nucleotides etc. For example, Abraxane, an FDA-approved albumin-based NP carrying paclitaxel, is indicated for first-line treatment of locally advanced or metastatic NSCLC in combination with carboplatin in patients who are not candidates for curative surgery or radiation therapy (Hyun et al., 2004).

Pulmonary gene delivery through polymeric nanomaterials proves to be an upgraded localization of drug as compared to viral vectors. The most commonly used polymeric materials are polyplexes because they are suitable for large-scale production, nonimmunogenic, less toxic, and highly stable during storage (Hyun et al., 2004).

Among all polymers used for gene delivery, polyethyleneimine (PEI) has been widely used for gene delivery as it binds easily and strongly with nucleic acids and releases during endolysosomal escape of polyplexes. Specifically, positive charges of PEI electrostatically interact with negative charges of phospholipids present in nucleic acids and transfect desired nucleic acids. However, PEI also shows cytotoxicity, but that can be manipulated by varying its configuration and molecular weight (Kafil and Omidi, 2011; Hong et al., 2015; Kim et al., 2015).

Recently hybrid nanomaterials are proving to be windfall for administration of chemotherapeutic agents, usually polymer-lipid hybrid NPs are used for localization of drug into the tumor sites on lungs (Garbuzenko et al., 2014). The core of the NP provides physical stability and structural integrity, whereas the shell assists cellular affinity and uptake. Both hydrophilic and hydrophobic drugs can be entrapped inside polymer-lipid hybrid NPs. The core accommodates hydrophilic drugs, while hydrophobic drugs could be incorporated in the shell (D'Addio et al., 2013; Hadinoto et al., 2013). Transferrin receptor (TfR; CD71) regulates intercellular iron transport and cell growth and is a type II transmembrane glycoprotein. The TfR has three domains: extracellular C-terminal, intracellular N-terminal, and transmembrane domain (Daniels et al., 2006). Like all other cancer cells, LC cells also overexpress several receptors on their cell membrane, such as luteinizing hormone releasing hormone (LHRH), folate receptors alpha (FRA) and glycoproteins, Epidermal Growth Factor Receptors EGFR is a tyrosine kinase protein-based receptor, CD44 is a cell-surface-based glycoprotein receptors which can be used as targets to deliver chemotherapeutic agents into LC without disturbing normal cells. According to Taratula et al., LHRH-targeted PTX and multiple siRNAs (MRP1 and BCL2)-loaded nanostructured lipid nanocarrier-based (NLC) NPs displayed 120- and 16-fold enhanced anticancer effect in comparison to free drug and LHRH-NLC-PTX, respectively, by inhalation. Live imaging demonstrated that LHRH targeted NLC NPs were accumulated only in cancer-affected areas, while nontargeted NLC NPs were distributed all over lung tissues (Taratula et al., 2013).

14.2.3 Nanoemulsions as potential carrier in LC

Nanoemulsions are transparent/translucent, thermodynamically stable, heterogeneous dispersions of two immiscible phases, aqueous and oil nanocarriers which are stabilized by the interfacial layer of suitable surfactant(s) and co-surfactants. Setbacks like conventional dosage form associated with nontargeted delivery, systemic toxicity, and poor bioavailability can be overcome by using nanoemulsions as drug/gene carrier. Nanoemulsion facilitates greater bioavailability and stability, specificity in target site delivery of the drugs, and possibility to decorate the formulation with tumor-specific ligands as needed (Choudhury et al., 2014). Nanoemulsions provide greater stability by protection of system for UV light and oxidative degradation which enhance stability of drug in the formulation. Nanoemulsions can be prepared spontaneously through mixing of two phases with the selected surfactant, which helps in maintenance of low interfacial tension between the two immiscible phases. Such a procedure helps in the formation of nanosized dispersed thermodynamically stable droplets without input of any external energy (Choudhury et al., 2017).

14.2.4 Metal-based NPs

Noble metals such as gold and silver have been extensively investigated for clinical applications, including their use in sensitive diagnostic imaging, detecting, and classifying of LC (Conde et al., 2012). Gold NPs were extensively used for cancer theranostics applications due to easy synthesize and functionalize high biocompatibility, and multifunctional theranostics properties. Several research groups utilized gold NPs for LC theranostics. For example, Nanospectra have developed a silica-gold nanoshells stabilized by (poly)ethylene glycol (PEG) for the photothermal therapy to the solid tumors using an NIR light source. More importantly, in a recent clinical trial, AuroLase was used for the photothermal therapy of primary or metastatic lung tumors (NCT01679470) (Singh et al., 2018).

Peng et al. developed gold NP-based biosensor system with the capacity to detect LC by analyzing an individual's exhaled breath. The sensorusesa combination of an array of chemiresistors based on gold NPs and pattern recognition methods (Peng et al., 2009). Recently, gold NPs have also successfully been tested as sensors for discriminating and classifying different LC histologies. The sensor was able to distinguish between normal and cancerous cells, SCLC and NSCLC, and between two subtypes of NSCLCs. Gold NP conjugates of methotrexate (MTX), a drug with high water solubility and low tumor retention, have shown high tumor retention and enhanced therapeutic efficacy in a Lewis lung carcinoma mouse model (Chen et al., 2007). Supermagnetic iron oxide NPs is extensively used as a MRI contrast agent, and also can be utilized as a delivery carrier in cancer theranostics applications. Iron oxides NPs were long used for various biomedical applications including MRI imaging, drug delivery, magnetic hyperthermia, and cancer theranostics in LC (Wang et al., 2017). Mesoporous silica NPs (MSNs) have been increasingly used in anticancer drug delivery research due to their dynamic capacity for drug loading, controlled drug release property, and multifunctional ability. Human LC cells primarily take up MSNs by endocytosis. MSNs have also been developed as a carrier for radionuclide isotope holmium-165 (Ho165) and tested in a xenograft tumor model (Di Pasqua et al., 2012).

Silver NPs were long used for various biomedical applications including antibacterial applications, anticancer applications, fluorescence imaging, and biosensors silver NPs have demonstrated antiproliferative effects in cancer cells. However, in vitro exposure of human LC cells to silver NPs resulted in reactive oxygen species-induced genotoxicity raising concerns of an unfavorable risk to benefit ratio (Foldbjerg et al., 2011).

In cellular microenvironments, silver NPs (AgNPs) are biodistributed to the major sites of ROS production in the cell, ROS are generated in various tumor cells due to the changes in pH which are linked to cellular stiffness. This stiffness determines cell differentiation that can induce cells to differentiate into human skin cells following AgNP exposure (Asharani et al., 2008). A possible transcription profiling in AgNPs- and Ag+-associated toxicity has been underlined in *in vivo* conditions such as zebra fish embryos. Hydrogen ions (H^+) ions are abundant within mitochondria, where H^+ efflux is the main event (i.e., proton motive force) in ATP synthesis (i.e., energy production) (Van Aerle et al., 2013). Fayez et al, combined action of shikonin and AgNPs for apoptosis in human cancer cells they investigated the synergistic combinatorial effect of shikonin and AgNPs in human LC cells. Shikonin was used as a reducing and capping agent for AgNPs synthesis as a green method avoiding the hazards of chemical methods. Radiolabeling of shikonin AgNPs with radioactive iodine forming [^{131}I] I Shikonin AgNPs was carried out to enable the intracellular tracking of NPs. this study revealed that the combination of shikonin and AgNPs treatment significantly inhibited cell viability and proliferation of A549 cells (human lung carcinoma cell line) with a great potential than the monotherapy (Fayez et al., 2020).

14.2.5 Dendrimers-based drug delivery

Dendrimers and dendrimer-based delivery systems are potential biomedicines in the rapidly growing field of nanomedicine. At present, dendrimers are drawing attention in the scientific domain due to their unique structural features, physiological properties and biological attributes (Sherje et al., 2018). These multifunctional nanomaterials provide unique platforms for targeting, imaging, diagnostics and theranostics (Mintzer and Grinstaff, 2011). Targeted biomedicine using dendrimers as flexible carriers can enhance the therapeutic effects of traditional treatments, circumvent side effects by combination therapy and develop simpler drug administration schemes with better safety. Moreover, dendrimers can readily modulate the physicochemical and biopharmaceutical properties of given therapeutics (Mehta et al., 2019). A nanosized macromolecular dendritic drug delivery system has been established as a potential tool for anticancer drug delivery and for performing as a diagnostic imaging tool because of the prospective research out comes as the newest class of nanocarrier. Dendrimers are biocompatible, highly branched, structurally uniform, and multivalent tree-like synthetic macromolecules. Different dendritic molecules have been explored in the pharmaceutical research for potential drug-delivery tools, such as polyamidoamine (PAMAM), poly-L-lysine (PLL), poly(propylene imine) (PPI), triagine, melamine, polyethylene glycol (PEG), and carbohydrate based citric acid, poly[2,2-bis(hydroxymethyl) propionic acid], poly(glycerol), and poly(glycerol-co-succinic acid) dendrimers (Luong et al., 2016). Among these, two commercial varieties of dendrimers include PAMAM and PPI.

These dendrimers have different generations, where the size of the dendritic molecule largely depends on the number of generations (Kesharwani et al., 2014).

The process of dendrimer synthesis depends on the branching requirements in the final product which are essential to fulfill the desired characteristics of nanocarriers, such as globular structure, nanometric size, etc. (Gorain et al., 2017). This highly branched macromolecule consists of three distinctive areas, which include a central core, generations forming branches originating from the central core, and the terminal functional groups of the branches (Luong et al., 2016).

PAMAM dendrimers have been found to be the most commonly employed carriers for efficient and effective site-specific delivery of chemotherapeutics and other chemicals, including peptides. Their availability in the market, highly controlled size, low systemic toxicity, multiple functions-attached surfaces has made PAMAM dendrimers a promising nanocarrier in chemotherapy against various cancers. These PAMAM dendrimers are hydrophilic in nature with biocompatible and nonimmunogenic characteristics. Their ability to enhance the bioavailability and reduce the frequency of dosing of the chemical entities favors their use as a potential drug delivery tool (Mendes et al., 2017).

PPI dendrimers are also cascades of molecules implementing a divergent method of fabrication. PPI dendrimers are also widely studied for application in pulmonary cancer. These PPI dendrimers are found to contain a greater hydrophobic core than the previously described PAMAM dendrimers due to the presence of amides to the alkyl chains. PPI dendrimers are found to be an effective carrier for chemotherapeutics; however, several architectural modifications are still needed to overcome its drawbacks over other available nanocarriers (Nguyen et al., 2015). PLL dendrimers differ from PAMAM and PPI dendrimers because of their asymmetrical structures, s to reduce the toxicity of dendrimers have brought about newer dendritic tools using amino acids (e.g., lysine). Application of PLL dendrimers on the tumor microenvironment is advantageous because these dendrimers possess antiangiogenic properties, stimulating apoptosis and destroying necrosis (Al-Hamra and Ghaddar, 2005). Small molecules absorb rapidly through pulmonary delivery due to the high surface area and thin epithelial layer of the lung, however multiple dosing at shorter intervals is required to achieve prolonged activity. Therefore, macromolecular drug carriers, such as dendrimers, are a great choice for pulmonary delivery of small-molecule drugs and peptides to prolong activity (Al-Jamal et al., 2010).

Nguyen, H et, al. Improved Method for Preparing Cisplatin-Dendrimer Nanocomplex and its behavior against NCI-H460 LC Cell, preparation of the nanocomplex of a species of aquated cisplatin and carboxylated PAMAM dendrimer G3.5 to evaluate loading capacity as well as plantinum release behavior. In vitro study showed that this drug-nanocarrier complex also help reduce cisplatin's cytotoxicity but can still keep sufficient antiproliferative activity against LC cell, NCI-H460, with help of incorporation of dendrimer G3 (Nguyen et al., 2015). Table 14.1 depicts the dendrimer-based Approaches as improved delivery of chemotherapeutics to LC.

TABLE 14.1 Dendrimer-based approaches in the improved delivery of chemotherapeutics to LC.

Drug/Gene incorporated	Type of dendrimer	Objective (theranostic/diagnostic)	Outcome of the research
Gold NPs	PAMAM dendrimer conjugated with folic acid	Development of folic acid-decorated PAMAM dendrimer containing gold NPs (Au DENPs-FA) for CT imaging tool for LC cells	Brighter image of SPC-A1 cells after incubating with Au DENPs-FA compared to control
Paclitaxel	Peptide-conjugated dendrimer	Preparation and evaluation of PTX loaded dendrimer for treatment of LC	PTX-loaded dendrimers have higher entrapment efficiency (95%) with 25% loading efficiency
Doxorubin (DOX)	PAMAM	For the treatment of lung metastasis through pulmonary delivery	Promising chemotherapeutic efficacy through lung delivery compared to intravenous administration
Gemcitabine	Mannose conjugated PPI dendrimer	Investigation of the targeting potential of mannose conjugated PPI dendrimers containing gemcitabine in order to achieve enhanced efficacy and reduced side effects	Mannose conjugation resulted in higher receptor-mediated binding and high concentration of drug in lung
MDM2siRNA	Triazine-modified dendrimer	Preparation of triazine-modified dendrimers for effective delivery of MDM2 siRNA for treatment of LC	DAT/siMDM2 polyplexes more efficiently silenced the expression of MDM2 compared to commercial transfection reagent Lipo 2000

Adapted from source Gorain et al. (2019b).

14.2.6 Target-mediated targeted therapy

The oligonucleotides aptamers (as DNA-/RNA-based nucleic acids) have also been exploited as both molecular therapy and targeting agents. There exist peptide aptamers that are also considered as attractive biologics, which are able to target the CMMs and thereby inhibit the corresponding biofunctions (Zhu et al., 2012). Despite the biological functions of aptamers being impeded by possible biodegradation, the clinical usefulness of such moieties has been demonstrated by the inhibition of VEGF in age-related macular degeneration using the PEGylated aptamer, pegaptanib (Macugen) (Pednekar et al., 2012). In a study, capitalizing on an RNA aptamer (A-p50) specific to NF-κB, the DOX-induced NF-κB activation that elicits chemoresistance to DOX was successfully suppressed both in vitro and in vivo using a lung tumor xenograft model. In fact, targeted inhibition of NF-κB by A-p50 can affect the regulation of genes involved in proliferation (Ki-67), DNA damage (GADD153), apoptosis (Bcl-XL), and pH regulation (CA9). While NF-κB is able to promote angiogenesis in the presence of DOX via the HIF1α/VEGF pathway, its suppression may inhibit the NSCLC resistance to DOX, which is a clinically important issue (Mi et al., 2008).

Folate acts as an essential agent in proliferation of LC cell. It is deemed that folate transport pathways and folate-dependent metabolic components may be considered as CMMs for targeted therapy of LC. The functional expression of folate receptor alpha (FRα), a folate transporter, has been shown to be a prognostic factor in NSCLC. Using immunehistochemistry and tissue micro arrays, the association of FRα and clinical outcome has been reported, wherein over 70% of adeno carcinomas appear to be positive for this CMM (O'Shannessy et al., 2012). Transferrin (Tf) receptor (TfR), in LC cases and many other types of malignancies, its expression is of prognostic value since this appears to correlate with tumor development and progression. Ina study, analysis of the sTfR in the serum and broncho alveolar lavage (BAL) fluid of patients with LC has been shown to correlate significantly with a high cell association of TfR in BAL patients suffering from NSCLC. Based upon such high expression of TfR in LC, many investigators have focused on developing a targeted therapy of LC by targeting TfR (Gaspar et al., 2012).

Manipulation and analysis of genomic information via genomic technology elucidate associated abnormalities of the genetic material of the LC cell. As a result of such genomic modulation, exploration in this field has implicated several over expressed receptors on the LC cell. These specific receptors have been identified in clinical specimens from malignant patients, including cultured LC cells in laboratories (Gorain, et al., 2019b). Table 14.2 depicts some examples where nanoparticulate systems used for target these over expressed receptors.

14.2.7 Quantum dots (QDs) as a drug delivery system

QDs are nanocrystals consisting of a center of a semiconductor material, sheathed inside the shell of another semiconductor having a larger spectral band gap. QD centers are normally made up of group II and group VI elements (West and Halas, 2003). A normal QD has a breadth of around 2e10 nm, which makes it of a size area that permits one-on-one communication with biomolecules, for example, proteins where the normal size extends from 1 to 20 nm. QDs are considered as one of the novel strategies employed in the treatment of LC (Azzazy et al., 2007). Surface modification has improved aqueous solubility and biocompatibility, which made it superior for use as fluorescent probes compared to organic fluorophores (Cai et al., 2006). QDs have attracted the interest and consideration of researchers because of their concurrent focusing on and diagnostic ability in drug delivery, in biomedical and pharmaceutical applications. The inevitable utilization of QDs is to drastically enhance clinical symptomatic tests for early recognition of cancers. QDs can be synthesized chemically via two main strategies: organic-phase method and water-phase method (aqueous synthesis) (Mishra et al., 2017). The water-phase method is advantageous compared to the organic-phase method in that it is efficacious, has minimal toxicity, and imparts biological applications to QDs by making them soluble in aqueous solution (Bajwa et al., 2016).

Characterization of QDs is done by utilizing several techniques such as spectroscopic techniques, scattering techniques, microscopic techniques, electrical techniques, physical property estimation, and rheological behavior determination. Spectroscopic techniques include: ultraviolet (UV) spectroscopy, infra-red (IR) spectroscopy, X-ray diffraction (XRD) analysis, nuclear magnetic resonance (NMR), Raman spectroscopy, and mass spectrometry (Mishra et al., 2017). Generally, QDs could be embellished with an approach known as PEGylation, which makes use of hydrophilic polymers, for example, polyethyleneglycol (PEG) polymer, for their effective solubilization. Apart from this, QDs can be made more hydrophilic by the conjugation of multidentate phosphine polymers, dihydrolipoic acid (DHLA), and dendrimers onto their surface. Other amphiphilic polymers, triblock copolymer and acrylic acid, could likewise be utilized for solubilization or stabilization of QDs. While amphiphilic phospholipids, calixarenes, and cyclodextrins have been utilized for sheathing of QDs, the microemulsion methodology with silica-coating technique was accounted for to give uniform sizes of QDs. Such

TABLE 14.2 Nanoparticulate systems used for target these over expressed receptors.

Target	Trigger	Route	Model	Type	References
Sigma receptor	pH	iv.	sc. human LC cells iv. murine melanoma cells sc., sur. human LC cells	Lipid/calcium/phosphate Lipid/calcium/phosphate Lipid/calcium/phosphate	(Li et al., 2010; Li et al., 2012; Yang et al., 2012)
CD44	–	N/A	sc. human LC cells sc. human LC cells, iv. murine melanoma cells	Conjugated	(Ganesh et al., 2013b; Ganesh et al., 2013a)
DR 4/5	–	iv. inh.	sc. human LC cells iv. human LC cells	Conjugated Polymeric Protein-based	(Kim et al., 2013; Choi et al., 2015)
Folate receptor	–	iv.	sc. human LC cells	Liposomal	(Morton et al., 2014)
Laminin receptor	–	iv.	sr. murine melanoma cells	Polymeric	(Sarfati et al., 2011)
Phosphatidyl serine	pH	iv.	sc. human LC cells, iv. murine LC cells	Protein/lipid-based	(Zhao et al., 2015)
PSMA	–	iv.	sc. human LC cells	Polymeric	(Hrkach et al., 2012)
HER2	Irradiation	itu.	sc. human LC cells	Au/Ag-based	(Shi et al., 2014)
IGF-1R	Magnetic field	iv.	sc. human LC cells	Magnetic lipoplexes	(Wang et al., 2011)
GC4	–	iv.	iv. murine melanoma cells	Liposomal/protein-based	(Chen et al., 2010)
CD47	–	iv.	iv. murine melanoma cells	Liposomal/protein-based	(Wang et al., 2013)
NSCLC	–	iv.	sc. human LC cells	Peptide based/ dendrimeric	(Liu et al., 2011)
Clotted plasma protein	–	iv.	iv., itr. human LC cells	Lipid-based	(Patel et al., 2014)
Neoplasms	Irradiation	iv.	iv. murine colon carcinoma cells	Protein-based	(Yang et al., 2010)
avb6	–	iv.	sc. human LC cells	Liposomal	(Gray et al., 2013)
avb3, neuropilin-1	–	iv.	sc. human LC cells	Peptide-based	(Shen et al., 2014)
DR 4/5, ES ligand	–	ro.	iv. human colon cancer cells	Liposomal	(Mitchell et al., 2014)
Transferrin receptor	–	iv.	sc. human LC cells	Lipid-based	(Guo et al., 2015; Han et al., 2016)
EphA2	–	iv.	iv., sur. human LC cells	Lipid-based	(Patel et al., 2014)
LDLR	Irradiation	iv.	iv. murine breast cancer cells, transgenic (EML4-ALK)	B/Gd-based	(Alberti et al., 2015)

procedures can be additionally utilized for functionalization of QDs through conjugation with different biofunctional entities (Wu et al., 2010).

14.2.8 Bio-NPs for LC

High biocompatibility, better stability and biodegradability of bio-NPs like protein NPs, solid lipid NPs, viral NPs, aptamers, and apoferritin, where in a bio-mimicking component have proven to be boon for cancer theranostics applications in pulmonary cancer (Perepelyuk et al., 2018). One of the class of bio-NPs is viral NPs (VNPs) obtained from viruses and bacteriophages have gained immense interest for various biomedical applications including drug delivery, biosensing, bioimaging, and vaccine development due to their biocompatibility, flexibility in sizes and shapes, and easy surface modification (Li et al., 2010). The most promising bio nanoparticulate system is Protein NPs prepared from a naturally occurring protein, such as gelatin, gliadin, albumin, and legumin have been recently used for the drug and gene delivery purposes either alone or in a mixture with biodegradable polymers in LC therapy due to their excellent biocompatibility, and lack of inflammation in human bronchial cells and high cellular uptake (MaHam et al., 2009; Wiley et al., 2009). Apoferritin is the hollow protein nanocage without their on core that is composed of self-assembling 24 polypeptide subunits and has internal and external diameters of 8 nm and 12 nm, respectively. Upon removal of the iron core the apoferritin undergoes a process of assembly and disassembly with the change in pH that is extensively utilized for the synthesis of various NPs for LC theranostics (Dostalova et al., 2017).

14.2.9 Hydrogel-based drug delivery for pulmonary cancer

Hydrogels are a binary system of polymer and liquid with a crosslinking agents and area three-dimensional polymer mesh that holds a large amount of water inside. Hydrogels provide Controlled drug-delivery systems that are meant to deliver drugs at a predetermined rate for a preprogrammed period are a good alternative to accomplish and overcome the inadequacy of low bioavailability of conventional dosage forms (Sudhakar et al., 2015). One of the favorable properties of hydrogels is their ability to swell, when put in contact with a thermodynamically compatible solvent. The equilibrium swelling degree and the elastic modulus of hydrogels depend on the crosslink and charge densities of the polymer network as well as on the cross linked polymer concentration after the gel preparation (Okay, 2009). The surface of the hydrogel can be decorated with different materials to give different properties, such as enzyme-linked hydrogels, stimuli-responsive hydrogels, and temperature sensitive hydrogels. Enzymatically degradable hydrogels represent intelligent drug-delivery systems capable of facilitating localized release in response to a cellular event. To enhance the mechanical properties of hydrogels, various crosslinking agents have been studied. The amount of crosslinking agents and types of crosslinking agent are important factors in the extent of the mechanical strength of hydrogel. An increase in crosslinking agent in hydrogel to enhance the mechanical strength can reduce the permeability of the hydrogel; decreasing the hydration of hydrogel during swelling (Vishal Gupta and Shivakumar, 2010). Hydrogel (sol-gel-sol) form concepts were fabricated bearing the polymers (PEG PCL G/DDP, PECE/DDP) containing cisplatin and paclitaxel polymeric micelles. The formulation initially behaves as a solution at room temperature and converts to gel state at the body temperature. The sol gel sol acts as a drug depot for the treatment of LC (Wu et al., 2014). Stimuli-responsive nanomaterial offers an alternative to design a controllable drug delivery system because of its spatiotemporally controllable properties. The intracellular activity and drug depot characteristics of micrometer-sized hydrogels are formed after cellular uptake of a solid polymeric NP that swells in response to mildly acidic conditions as it transforms from a hydrophobic to a hydrophilic structure (Yao et al., 2016).

Lee et al., studied interpenetrating network system (IPN), consisting of polyethylene glycol (PEG) –diacrylate (PEGdA) and modified gelatin, is a biocompatible and biodegradable hydrogel for the local delivery of bioactive molecules and drugs. Gold (III) porphyrin(AuP) stable metal compound in the development for anticancer application when administered systemically. The AuP loaded IPN showed higher cytotoxicity against human LC cell lines compared to IPN only. In mice bearing human LC xenograft, AuP loaded IPN inhibited tumor growth and reduced angiogenesis. AuP loaded IPN provides an improved formulation over systemic delivery for tumor inhibition to complement surgical intervention (Lee et al., 2019).

14.2.10 Inhalation-based nanomedicine for pulmonary cancer

Aerosol therapy has also been evaluated, and is being used for several other conditions and purposes, such as diabetes mellitus, gene therapy, and vaccination (*Effect of nebulizer type and antibiotic concentration on device performance – PubMed*). The areas of this treatment modality that need to be properly addressed are summarized under the headings: prompt inhalation device; lung airway microenvironment; appropriate molecule-chemotherapy selection; deposition evaluation; protection measures; and disease evaluation. Nanotechnology applications in medicine, defined as nanomedicine, have led to a number of applications for imaging or treatment of cancers. Physical properties that have a significant role on the particle size of the inhaled suspension are viscosity ionic strength, osmolarity, and pH. If the values for pH and osmolarity in particular are not in the normal range, broncho-constriction, coughing, and irritation of the lung mucosa is induced (*Alteration in osmolarity of inhaled aerosols cause bronchoconstriction and cough, but absence of a permeant anion causes cough alone – PubMed*). There are various Inhaled particle carriers such as Polymeric nanocarriers, Polymeric NPs, Polymer–drug conjugates, Lipidic nanocarriers, dendrimers and targeted inhalation delivery. Until now, inhaled chemotherapy has been administered for advanced LC and, therefore, although some chemotherapeutic agents provided immediate evidence of dose/time-related toxicity (Zaroglulidis et al., 2012). Table 14.3 depicts Examples of different NPs investigated as targeted anticancer therapeutic carrier for inhalation treatment in LC.

TABLE 14.3 Examples of different NPs investigated as targeted anticancer therapeutic carrier for inhalation treatment in LC.

Nanomedicines	Chemotherapeutics	In vitro/In vivo model	Findings
Polymeric micelles	Paclitaxel	Male Sprague Dawley rats	Prolonged activity in the lungs and reduced systemic toxicity. AUC0–12 in lungs was 45 times greater than iv-administered formulation. The targeting efficiency to lungs through the pulmonary route was 132-fold higher than the iv route
Dendrimers	Doxorubicin	Male Sprague Dawley rats, female F344 rats, MAT 13762 IIIB rat model	Pulmonary administration of drug reduced the lung tumor burden by 95% as compared to control. The number of metastatic foci in the lungs was also reduced significantly
Polymer–drug conjugate	Cisplatin		
	Female Sprague Dawley rats, A549 cell line	Compared to conventional iv infusion, inhalational administration of the hyaluronan–cisplatin conjugate showed higher drug accumulations in the lung tissues	
Magnetic NPs	Quercetin	Female BALB/c mice, A549 cell line	Actively targeted to specific locus cells, the drug-loaded PLGA-MNPs considerably decreased the number of viable A549 cells
Liposomes	Paclitaxel and cyclosporine A	Female BALB/c mice; Renca lung metastases mouse model	Targeted combinational drug delivery through liposomes showed excellent antitumor effect

Adapted from source Lee et al. (2015).

14.3 Marketed formulation

The first chemotherapeutic drug evaluated in clinical trials as an inhaled Nanomedicine for LC treatment was the topoisomerase I inhibitor 9-nitro-20(S)-comptothecin (9-NC), in 2004 (Rosière et al., 2019). Currently various company have formulated NPs in various forms for targeting lungs cancerous cell and Table 14.4 depicts some examples of marketed formulation for LC treatment.

Limitation and challenges with pulmonary route for treatment of LC using inhalational nanoparticulate system (Blank et al., 2011; Rosière et al., 2019)

- Poor payload, drug loading of the nanocarrier is usually 1%-10% (w/w). Consequently, depending on the drug candidate, the aim of delivering sufficient anticancer drug doses to patients by means of these nanocarriers might be unrealistic (Rosière et al., 2019).
- Compared to i.v. injection, pulmonary delivery has led to lower systemic exposure of the encapsulated anticancer drug and possibly by extension, related systemic toxicities. Moreover, the use of drug-loaded nanocarriers locally led to higher and longer drug retention in the lung (up to 7 days) than the free drug. Nanocarriers have also been able to reach the lymphatic system after inhalation (Rosière et al., 2019).
- Fabrication of polymeric nanomaterial (simply NPs) with uniform size requires critical control over each and every step in the synthesis procedure (Babu et al., 2013).

TABLE 14.4 Marketed formulation available for lungs cancer treatment.

Product	Formulation	Company	Indication
Abraxane	Albumin-bound paclitaxel	Celgene Co.	NSCLC
Genexol-PM	Paclitaxel-loaded micelle	Samyang Co.	NSCLC
Paclitaxel poliglumex	Polyglutamate paclitaxel	CTI BioPharma	NSCLC
MPDL3280A	Anti-PD L1 antibody	Genentech	NSCLC
Tecemotide	Liposomal vaccine	Oncothyreon	NSCLC
Doxil	Liposomal doxorubicin	Johnson & Johnson	SCLC
BIND-014	Targeted docetaxel	Bind Therapeutics	NSCLC
CRLX101	Polycyclodextrin camptothecin	Cerulean Pharma	SCLC
NKTR 102	PEGylated irinotecan	Nektar Therapeutics	Lung metastases
Kadcyla	Ab-emtansine conjugate	Genentech	NSCLC
IMMU-132	Ab-SN-38 conjugate	Immunomedics Inc.	NSCLC
IMGN901	Ab-mertansine conjugate	ImmunoGen	SCLC
NC-6004	Micellar Cisplatin	NanoCarrier Co.	NSCLC
MM-398	Liposomal irinotecan	Merrimack Pharmaceuticals	NSCLC
DNIB0600A	Ab-MMAE conjugate	Genentech	NSCLC
AuroShell	Gold-silica nanoshells	Nanospectra biosciences	LC

NSCLC, nonsmall cell LC; SCLC, small cell LC.
Adapted from source Home - ClinicalTrials.gov; Stathopoulos and Boulikas (2012).

- Surface charge determines the fate of NPs *in vivo*. Particle-particle interaction and aggregation tendencies are largely dependent on the zeta potential of NPs. Positively charged NPs have an affinity for the negatively charged cellular membranes of all cells in the body (Babu et al., 2013).
- A NP with large number of surface functional groups provides an avenue for the attachment of multiple kinds of biomolecules for the targeted drug delivery and diagnostic applications for LC (Babu et al., 2013).

14.4 Toxicity issues of inhaled NPS

To use the potential of nanotechnology in nanomedicines, full attention is needed to safety and toxicological issues (De Jong and Borm, 2008). Physicochemical properties like size (aerodynamic, hydrodynamic), size distribution, shape, agglomeration/aggregation, density (material, bulk), surface properties, (area [porosity], charge, chemistry [coating, contaminants, defects]), crystallinity, and bio-contaminants (e.g., endotoxin) and functional properties like solubility/dissolution rate (physiological fluid, in vivo), and surface reactivity (ROS inducing capacity) determine the biological/toxicological properties of NPs (Zoroddu et al., 2014).

The largest database on the toxicity of NPs has originated from inhalation toxicology including the PM_{10} (particulate matter with a size below 10mm). Jong and Borm reviewed about it and described that much of the mass of PM_{10} considered to be nontoxic and so there has arisen the idea that there is a component of PM_{10} that actually drives the proinflammatory effects and combustion-derived NPs seem a likely candidate (De Jong and Borm, 2008).

An understanding of dosimetry and extrapolation modeling is essential for translating results of inhalation tests with NPs to be applied for development aspect. Fig. 14.3 shows the complexity of respiratory tract dosimetry to emphasize the importance of expressing and analyzing data in the form of Exposure-Dose-Response relationships (Oberdörster and Kuhlbusch, 2018).

Bailey and Berkland have reported carbon NPs administered intratracheally in rodents have shown vascular thrombosis, further demonstrated that inhaled iridium particles may migrate from the lung to the systemic circulation may have

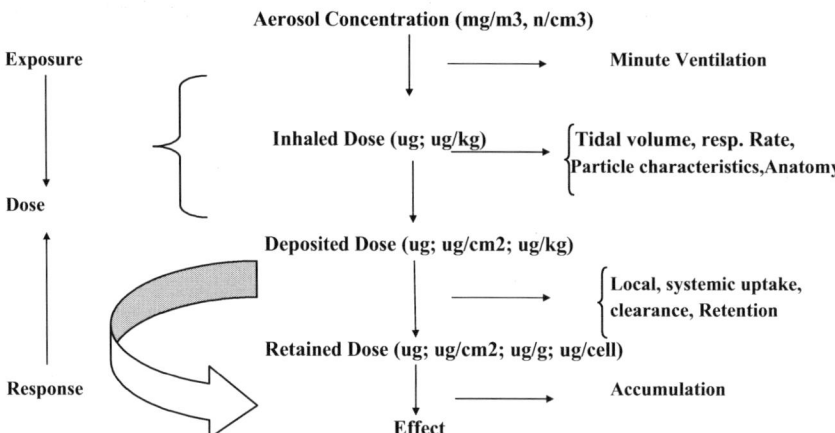

FIG. 14.3 Exposure-Dose-Response relationships. *Adapted from source Oberdörster and Kuhlbusch (2018).*

detrimental vascular effects. Inhaled carbon NPs have been shown to migrate to the brain, although their CNS toxicity remains to be determined. According to them based on their observations, NP toxicity in the lungs may be more dependent on material choice than particle size. Therefore, there may be alternative polymers that can be investigated for use in pulmonary NP drug formulations that could mitigate toxicity (Bailey and Berkland, 2009).

Numerous studies addressed adverse effect of airborn NPs using *in vitro* and *in vivo* systems to compare biological effects micro- and NPs and identify potential NP-specific action. Froehlich E and Salar-Behzadi S reviewed toxicological assessment of inhaled NPs, emphasizing on in vivo, ex vivo, in vitro, and in silico modeling, where they came out on conclusion that with the development of advanced in vitro models, not only in vivo, but also cellular studies can be used for toxicological testing. Where advanced in vitro studies use combinations of cells cultured in the air-liquid interface, useful for particle uptake and mechanistic studies. Whole-body, nose-only, and lung-only exposures of animals could help to determine retention of NPs in the body. Further they explained the simulation programs for lung deposition in humans could help to determine the relevance of the biological findings. Combination of biological data generated in different biological models and in silico modeling appears suitable for a realistic estimation of potential risks by inhalation exposure to NPs (Fröhlich and Salar-Behzadi, 2014).

Bakand et al. critically reviewed on particle toxicity following inhalation exposure, have described, the phagocytosis function of alveolar macrophages to remove inhaled NPs appears to be considerably smaller than for larger particles. Further explained that ultrafine particles can also impair the ability of macrophages to phagocytose and eliminate other particles that may be proinflammogenic. Insoluble particles may remain into lungs indefinitely. Prolong residence of particles in the lung may lead to injury and biological response (Bakand et al., 2012).

The health effects associates with the NPs depends on residence time in the respiratory tract and lungs burden, generate higher inflammatory reaction, impairs macrophages, clearances, inflammation, accumulation of particles, and epithelial cell proliferation, followed by fibrosis, emphysema, and tumors, also may have extra potential of affecting cardiovascular disease directly (Crisponi et al., 2017).

14.5 Conclusion

NPs mediated drug delivery has vast potential with various unique utilizations in cancer diagnosis, detection, imaging, and treatment. Conventional utilization of drug is characterized by poor biodistribution, limited effectiveness, undesirable side effects, and lack of selectivity. Strategies like controlling drug delivery can potentially overcome these limitations. NPs had helped in combating traditional LC therapies such as nonspecific targeting, low therapeutic efficiencies, untoward side effects, and drug resistance as well as surpassing their predecessors with the ability to detect early metastasis. The ability of NPs to be tailored for a personalized medicine strategy makes them ideal vehicles for the treatment of LC. Numerous NPs based experimental therapeutics for LC utilizes a combinatorial approach balancing the design with targeting and tracking moieties and anticancer agents. With the potential advantages of pulmonary route and recent novel drug delivery strategies such as nanocarriers with anticancer drugs may be the promising alternative drug treatment over conventional drug delivery approaches, and for focusing on the same this chapter summarizes and discusses the investigated formulation approaches for the pulmonary delivery. In general, NPs with multi-component structures allow design flexibility in drug delivery of poorly water soluble molecules as well as imparting the ability to overcome biological barriers and selectively target desired sites within the body. The current chapter discusses the various types of NPs which have been extensively studied in treating LC through inhalational route. However, many of these results are still at preclinical stages and these NPs have to show their safety and efficacy in clinical trials and further their use in clinics. Authors tried to compile some marketed formulations and patents over the nanoparticulate system, which put forward a novel treatment strategy "a targeted drug delivery strategy to lung" for LC treatment.

Disclaimer: The views and opinions expressed in this chapter are only those of the authors and do not necessarily reflect the views or policies of any agency, employer or organization.

References

Abdelaziz, H.M., et al., 2018. Inhalable particulate drug delivery systems for lung cancer therapy: nanoparticles, microparticles, nanocomposites and nanoaggregates. J. Control. Release 269 (November 2017), 374–392. Elsevier. http://doi.org/10.1016/j.jconrel.2017.11.036.

Van Aerle, R., et al., 2013. Molecular mechanisms of toxicity of silver nanoparticles in zebrafish embryos. Environ. Sci. Technol. 47 (14), 8005–8014. http://doi.org/10.1021/es401758d.

Agu, R.U., et al., 2001. The lung as a route for systemic delivery of therapeutic proteins and peptides. Respir. Res. 2 (4), 198–209. http://doi.org/10.1186/rr58.

Al-Hamra, M., Ghaddar, T.H., 2005. Facile synthesis of poly-(L-lysine) dendrimers with a pentaaminecobalt(III) complex at the core. Tetrahedron Lett. 46 (34), 5711–5714. Pergamon. http://doi.org/10.1016/j.tetlet.2005.06.078.

Al-Jamal, K.T, et al., 2010. Systemic antiangiogenic activity of cationic poly-L-lysine dendrimer delays tumor growth. PNAS 107 (9), 3966–3971. http://doi.org/10.1073/pnas.0908401107.

Alberti, D., et al., 2015. A theranostic approach based on the use of a dual boron/Gd agent to improve the efficacy of boron neutron capture therapy in the lung cancer treatment. Nanomedicine 11 (3), 741–750. Elsevier Inc.. http://doi.org/10.1016/j.nano.2014.12.004.

Alhajj, N., et al., 2018. Lung cancer: active therapeutic targeting and inhalational nanoproduct design. Expert Opin. Drug Deliv. 1223–1247. Taylor and Francis Ltd. http://doi.org/10.1080/17425247.2018.1547280.

Ali, M.E., McConville, J.T., Lamprecht, A., 2015. Pulmonary delivery of anti-inflammatory agents. Expert Opin. Drug Deliv. 929–945. Informa Healthcare. http://doi.org/10.1517/17425247.2015.993968.

Alteration in osmolarity of inhaled aerosols cause bronchoconstriction and cough, but absence of a permeant anion causes cough alone - PubMed. Available at: https://pubmed.ncbi.nlm.nih.gov/6696320/ (Accessed: 2 June 2021).

Asharani, P.V., et al., 2008. Toxicity of silver nanoparticles in zebrafish models. Nanotechnology 19 (25). http://doi.org/10.1088/0957-4484/19/25/255102.

Azzzay, H.M.E., Mansour, M.M.H., Kazmierczak, S.C., 2007. From diagnostics to therapy: prospects of quantum dots. Clin. Biochem. 917–927. Clin Biochem. http://doi.org/10.1016/j.clinbiochem.2007.05.018.

Babu, A., et al., 2013. Nanoparticle-based drug delivery for therapy of lung cancer: progress and challenges. J. Nanomater. 2013 (November). http://doi.org/10.1155/2013/863951.

Badrzadeh, F., et al., 2016. Drug delivery and nanodetection in lung cancer. Artif. Cells Nanomed. Biotechnol. 44 (2), 618–634. Taylor and Francis Ltd.. http://doi.org/10.3109/21691401.2014.975237.

Bailey, M.M., Berkland, C.J., 2009. Nanoparticle formulations in pulmonary drug delivery. Med. Res. Rev. 196–212. http://doi.org/10.1002/med.20140.

Bajwa, N., et al., 2016. Pharmaceutical and biomedical applications of quantum dots. Artif. Cells Nanomed. Biotechnol. 44 (3), 758–768. Taylor and Francis Ltd.. http://doi.org/10.3109/21691401.2015.1052468.

Bakand, S., Hayes, A., Dechsakulthorn, F., 2012. Nanoparticles: a review of particle toxicology following inhalation exposure. Inhal. Toxicol. 24 (2), 125–135. http://doi.org/10.3109/08958378.2010.642021.

Blank, F., Stumbles, P., Von Garnier, C., 2011. Opportunities and challenges of the pulmonary route for vaccination. Expert Opin. Drug Deliv. 8 (5), 547–563. http://doi.org/10.1517/17425247.2011.565326.

Bray, F., et al., 2018. Global cancer statistics 2018: GLOBOCAN estimates of incidence and mortality worldwide for 36 cancers in 185 countries. CA Cancer J. Clin. 68 (6), 394–424. Wiley. http://doi.org/10.3322/caac.21492.

Cai, W., et al., 2006. Peptide-labeled near-infrared quantum dots for imaging tumor vasculature in living subjects. Nano Lett. 6 (4), 669–676. http://doi.org/10.1021/nl052405t.

Chandel, A., et al., 2019. Recent advances in aerosolised drug delivery. Biomed. Pharmacother. 112, 108601. Elsevier Masson SAS. http://doi.org/10.1016/j.biopha.2019.108601.

Chandira, R.M., Jayakar, B., chiranjib, D., (no date). Recent aspects of pulmonary drug delivery system- an overview. Available at: https://farmavita.net/documents/PULMONARYDRUGDELIVERYSYSTEM.pdf.

Chen, Y., et al., 2010. Nanoparticles modified with tumor-targeting scFv deliver siRNA and miRNA for cancer therapy. Mol. Ther. 18 (9), 1650–1656. Nature Publishing Group. http://doi.org/10.1038/mt.2010.136.

Chen, Y.H., et al., 2007. Methotrexate conjugated to gold nanoparticles inhibits tumor growth in a syngeneic lung tumor model. Mol. Pharm. 4 (5), 713–722. http://doi.org/10.1021/mp060132k.

Choi, S.H., et al., 2015. Inhalable self-assembled albumin nanoparticles for treating drug-resistant lung cancer. J. Control. Release 197, 199–207. Elsevier B.V.. http://doi.org/10.1016/j.jconrel.2014.11.008.

Choudhury, H., et al., 2014. Improvement of cellular uptake, in vitro antitumor activity and sustained release profile with increased bioavailability from a nanoemulsion platform. Int. J. Pharm. 460 (1–2), 131–143. http://doi.org/10.1016/j.ijpharm.2013.10.055.

Choudhury, H., et al., 2017. Pharmacokinetic and pharmacodynamic features of nanoemulsion following oral, intravenous, topical and nasal route. Curr. Pharm. Des. 23 (17). Bentham Science Publishers Ltd.. http://doi.org/10.2174/1381612822666161201143600.

Conde, J., Doria, G., Baptista, P., 2012. Noble metal nanoparticles applications in cancer. J. Drug Deliv. 2012, 1–12. Hindawi Limited. http://doi.org/10.1155/2012/751075.

Crisponi, G., et al., 2017. Toxicity of nanoparticles: etiology and mechanisms, antimicrobial nanoarchitectonics: from synthesis to applications. doi: http://doi.org/10.1016/B978-0-323-52733-0.00018-5.

Cryer, A.M., Thorley, A.J., 2019. Nanotechnology in the diagnosis and treatment of lung cancer. Pharmacol. Ther 198, 189–205. Elsevier Inc.. http://doi.org/10.1016/j.pharmthera.2019.02.010.

D'Addio, S.M., et al., 2013. Aerosol delivery of nanoparticles in uniform mannitol carriers formulated by ultrasonic spray freeze drying. Pharm. Res. 30 (11), 2891–2901. Springer. http://doi.org/10.1007/s11095-013-1120-6.

Daniels, T.R., et al., 2006. The transferrin receptor part I: biology and targeting with cytotoxic antibodies for the treatment of cancer. Clin. Immunol. 121 (2), 144–158. http://doi.org/10.1016/j.clim.2006.06.010.

Dostalova, S., et al., 2017. Apoferritin as an ubiquitous nanocarrier with excellent shelf life. Int. J. Nanomed. 12, 2265–2278. Dove Medical Press Ltd.,. http://doi.org/10.2147/IJN.S130267.

Effect of nebulizer type and antibiotic concentration on device performance – PubMed. Available at: https://pubmed.ncbi.nlm.nih.gov/9141110/ (Accessed: 2 June 2021).

El-Far, S.W., et al., 2018. Phytosomal bilayer-enveloped casein micelles for codelivery of monascus yellow pigments and resveratrol to breast cancer. Nanomedicine 13 (5), 481–499. Future Medicine Ltd.. http://doi.org/10.2217/nnm-2017-0301.

El-Sherbiny, I.M., El-Baz, N.M., Yacoub, M.H., 2015. Inhaled nano- and microparticles for drug delivery. Global Cardiol. Sci. Pract. 2015 (1), 2. http://doi.org/10.5339/gcsp.2015.2.

Farokhzad, O.C., et al., 2006. Targeted nanoparticle-aptamer bioconjugates for cancer chemotherapy in vivo. PNAS 103 (16), 6315–6320. Proc Natl Acad Sci U S A. http://doi.org/10.1073/pnas.0601755103.

Fathi Karkan, S., et al., 2017. Magnetic nanoparticles in cancer diagnosis and treatment: a review. Artif. Cells Nanomed. Biotechnol. 1–5. Taylor and Francis Ltd.. http://doi.org/10.3109/21691401.2016.1153483.

Fayez, H., El-Motaleb, M.A., Selim, A.A., 2020. Synergistic cytotoxicity of shikonin-silver nanoparticles as an opportunity for lung cancer. J. Labelled Compd. Radiopharm. 63 (1), 25–32. John Wiley and Sons Ltd. http://doi.org/10.1002/jlcr.3818.

Fernandes, C.A., Vanbever, R., 2009. Preclinical models for pulmonary drug delivery. Expert Opin. Drug Deliv. 6 (11), 1231–1245. http://doi.org/10.1517/17425240903241788.

Foldbjerg, R., Dang, D.A., Autrup, H., 2011. Cytotoxicity and genotoxicity of silver nanoparticles in the human lung cancer cell line, A549. Arch. Toxicol. 85 (7), 743–750. http://doi.org/10.1007/s00204-010-0545-5.

Friebel, C., Steckel, H., 2010. Single-use disposable dry powder inhalers for pulmonary drug delivery. Expert Opin. Drug Deliv. 7 (12), 1359–1372. http://doi.org/10.1517/17425247.2010.538379.

Fröhlich, E., Salar-Behzadi, S., 2014. Toxicological assessment of inhaled nanoparticles: role of in vivo, ex vivo, in vitro, and in silico studies. Int. J. Mol. Sci. 4795–4822. MDPI AG. http://doi.org/10.3390/ijms15034795.

Gaber, M., et al., 2017. Protein-lipid nanohybrids as emerging platforms for drug and gene delivery: challenges and outcomes. J. Control. Release 269, 75–91. Elsevier B.V.. http://doi.org/10.1016/j.jconrel.2017.03.392.

Ganesh, S., Iyer, A.K., Weiler, J., et al., 2013a. Combination of siRNA-directed gene silencing with cisplatin reverses drug resistance in human non-small cell lung cancer. Mol. Ther. – Nucleic Acids 2 (7), e110. Nature Publishing Group. http://doi.org/10.1038/mtna.2013.29.

Ganesh, S., Iyer, A.K., Morrissey, D.V., et al., 2013b. Hyaluronic acid based self-assembling nanosystems for CD44 target mediated siRNA delivery to solid tumors. Biomaterials 34 (13), 3489–3502. http://doi.org/10.1016/j.biomaterials.2013.01.077.

Garbuzenko, O.B., et al., 2010. Inhibition of lung tumor growth by complex pulmonary delivery of drugs with oligonucleotides as suppressors of cellular resistance. PNAS 107 (23), 10737–10742. National Academy of Sciences. http://doi.org/10.1073/pnas.1004604107.

Garbuzenko, O.B., et al., 2014. Inhalation treatment of lung cancer: the influence of composition, size and shape of nanocarriers on their lung accumulation and retention. Cancer Biol. Med. 11 (1), 44–55. http://doi.org/10.7497/j.issn.2095-3941.2014.01.004.

Gaspar, M.M., et al., 2012. Targeted delivery of transferrin-conjugated liposomes to an orthotopic model of lung cancer in nude rats. J. Aerosol Med. Pulm. Drug Deliv. 25 (6), 310–318. http://doi.org/10.1089/jamp.2011.0928.

Gorain, B., et al., 2017. The use of nanoscaffolds and dendrimers in tissue engineering. Drug Discov. Today, 652–664. Elsevier Ltd. http://doi.org/10.1016/j.drudis.2016.12.007.

Gorain, B., Choudhury, H., et al., 2019a. Dendrimer-Based Nanocarriers in Lung Cancer Therapy in Nanotechnology-Based Targeted Drug Delivery Systems for Lung Cancer. Academic Press, Elsevier, pp. 161–192. http://doi.org/10.1016/b978-0-12-815720-6.00007-1.

Gorain, B., Bhattamishra, S.K., et al., 2019b. Overexpressed receptors and proteins in lung cancer in Nanotechnology-Based Targeted Drug Delivery Systems for Lung Cancer. Academic Press, Elsevier, pp. 39–75. http://doi.org/10.1016/b978-0-12-815720-6.00003-4.

Gray, B.P., McGuire, M.J., Brown, K.C., 2013. A liposomal drug platform overrides peptide ligand targeting to a cancer biomarker, irrespective of ligand affinity or density. PLoS ONE. Public Library of Science 8 (8), e72938. http://doi.org/10.1371/journal.pone.0072938.

Guo, Y., et al., 2015. Transferrin-conjugated doxorubicin-loaded lipid-coated nanoparticles for the targeting and therapy of lung cancer. Oncol. Lett. 9 (3), 1065–1072. Spandidos Publications. http://doi.org/10.3892/ol.2014.2840.

Hadinoto, K., Sundaresan, A., Cheow, W.S., 2013. Lipid-polymer hybrid nanoparticles as a new generation therapeutic delivery platform: a review. Eur. J. Pharm. Biopharm. 85 (3), 427–443. Elsevier B.V.. http://doi.org/10.1016/j.ejpb.2013.07.002.

Han, Y., et al., 2016. Nanostructured lipid carriers as novel drug delivery system for lung cancer gene therapy. Pharm. Dev. Technol. 21 (3), 277–281. Taylor and Francis Ltd. http://doi.org/10.3109/10837450.2014.996900.

Home - ClinicalTrials.gov (no date). Available at: https://clinicaltrials.gov/ (Accessed: 2 June 2021).

Hong, S.H., et al., 2015. Aerosol gene delivery using viral vectors and cationic carriers for in vivo lung cancer therapy. Expert Opin. Drug Deliv. 12 (6), 977–991. Informa Healthcare. http://doi.org/10.1517/17425247.2015.986454.

Hrkach, J., et al., 2012. Preclinical development and clinical translation of a PSMA-targeted docetaxel nanoparticle with a differentiated pharmacological profile. Sci. Transl. Med. 4 (128). http://doi.org/10.1126/scitranslmed.3003651.

Hussain, Z., 2019a. Toward the development of a novel diagnostic nano-imaging platform for lung cancer in Nanotechnology-Based Targeted Drug Delivery Systems for Lung Cancer. Elsevier, pp. 269–292. http://doi.org/10.1016/b978-0-12-815720-6.00011-3.

Hyun, W.K., et al., 2004. Aerosol delivery of glucosylated polyethylenimine/phosphatase and tensin homologue deleted on chromosome 10 complex suppresses Akt downstream pathways in the lung of K-ras null mice. Cancer Res. 64 (21), 7971–7976. http://doi.org/10.1158/0008-5472.CAN-04-1231.

De Jong, W.H., Borm, P.J.A., 2008. Drug delivery and nanoparticles: applications and hazards. Int. J. Nanomed., 133–149. Dove Press. http://doi.org/10.2147/ijn.s596.

Kafil, V., Omidi, Y., 2011. Cytotoxic impacts of linear and branched polyethylenimine nanostructures in A431 cells. BioImpacts 1 (1), 23–30. Tabriz University of Medical Sciences. http://doi.org/10.5681/bi.2011.004.

Kaur, S.S., 2017. Pulmonary drug delivery system: newer patents. Pharm. Pat. Anal. 6 (5), 225–244.

Kesharwani, P., Jain, K., Jain, N.K., 2014. Dendrimer as nanocarrier for drug delivery. Prog. Polym. Sci. 39 (2), 268–307. Pergamon. http://doi.org/10.1016/j.progpolymsci.2013.07.005.

Kim, I., et al., 2013. Doxorubicin-loaded porous PLGA microparticles with surface attached TRAIL for the inhalation treatment of metastatic lung cancer. Biomaterials 34 (27), 6444–6453. Biomaterials. http://doi.org/10.1016/j.biomaterials.2013.05.018.

Kim, Y.D., et al., 2015. Nanoparticle-mediated delivery of siRNA for effective lung cancer therapy. Nanomedicine 10 (7), 1165–1188. Future Medicine Ltd.. http://doi.org/10.2217/nnm.14.214.

Kurmi, B.D., et al., 2010. Micro- and nanocarrier-mediated lung targeting. Expert Opin. Drug Deliv. 7 (7), 781–794. Informa Healthcare. http://doi.org/10.1517/17425247.2010.492212.

Larbi, A.B., 2011. Lung cancer around the world and arab countries. AMAAC Workshop. Algiers

Lee, P., et al., 2019. A multifunctional hydrogel delivers gold compound and inhibits human lung cancer xenograft. Pharm. Res. 36 (4), 61. Springer New York LLC. http://doi.org/10.1007/s11095-019-2581-z.

Lee, W.H., et al., 2015. Inhalation of nanoparticle-based drug for lung cancer treatment: advantages and challenges. Asian J. Pharm. Sci. 10 (6), 481–489. Shenyang Pharmaceutical University. http://doi.org/10.1016/j.ajps.2015.08.009.

Leiva, M.C., et al., 2017. Tripalmitin nanoparticle formulations significantly enhance paclitaxel antitumor activity against breast and lung cancer cells in vitro. Sci. Rep. 7 (1), 1–15. Nature Publishing Group. http://doi.org/10.1038/s41598-017-13816-z.

Lemjabbar-Alaoui, H., et al., 2015. Lung cancer: Biology and treatment options. Biochimica et Biophysica Acta – Rev. Cancer 1856 (2), 189–210. Elsevier B.V.. http://doi.org/10.1016/j.bbcan.2015.08.002.

Li, J., et al., 2010. Biodegradable calcium phosphate nanoparticle with lipid coating for systemic siRNA delivery. J. Control. Release 142 (3), 416–421. J Control Release. http://doi.org/10.1016/j.jconrel.2009.11.008.

Li, J., Yang, Y., Huang, L., 2012. Calcium phosphate nanoparticles with an asymmetric lipid bilayer coating for siRNA delivery to the tumor. J. Control. Release 158 (1), 108–114. J Control Release. http://doi.org/10.1016/j.jconrel.2011.10.020.

Li, K., et al., 2010. Viruses and their potential in bioimaging and biosensing applications. Analyst 135 (1), 21–27. Royal Society of Chemistry. http://doi.org/10.1039/b911883g.

Liu, J., et al., 2011. Novel peptide-dendrimer conjugates as drug carriers for targeting nonsmall cell lung cancer. Int. J. Nanomed. 6 (1), 59–69. http://doi.org/10.2147/IJN.S14601.

Lung Cancer : Risk Factors, Symptoms I Prevention Lung Cancer I India Against Cancer (no date). Available at: http://cancerindia.org.in/lung-cancer/#synved-tabs-1-6 (Accessed: 30 May 2021).

Lung cancer I World Cancer Research Fund International (no date). Available at: https://www.wcrf.org/dietandcancer/lung-cancer/ (Accessed: 30 May 2021).

Luong, D., et al., 2016. Solubility enhancement and targeted delivery of a potent anticancer flavonoid analogue to cancer cells using ligand decorated dendrimer nano-architectures. J. Colloid Interface Sci. 484, 33–43. Academic Press Inc.. http://doi.org/10.1016/j.jcis.2016.08.061.

MaHam, A., et al., 2009. Protein-based nanomedicine platforms for drug delivery. Small 5 (15), 1706–1721. http://doi.org/10.1002/smll.200801602.

Mehta, P., et al., 2019. Dendrimers for pulmonary delivery: Current perspectives and future challenges. New J. Chem. 43 (22), 8396–8409. Royal Society of Chemistry. http://doi.org/10.1039/c9nj01591d.

Mendes, L.P., Pan, J., Torchilin, V.P., 2017. Dendrimers as nanocarriers for nucleic acid and drug delivery in cancer therapy. Molecules 22, 1401. MDPI AG. http://doi.org/10.3390/molecules22091401.

Mi, J., et al., 2008. RNA aptamer-targeted inhibition of NF-κB suppresses non-small cell lung cancer resistance to doxorubicin. Mol. Ther. 16 (1), 66–73. Nature Publishing Group. http://doi.org/10.1038/sj.mt.6300320.

Mintzer, M.A., Grinstaff, M.W., 2011. Biomedical applications of dendrimers: a tutorial. Chem. Soc. Rev. 40 (1), 173–190. http://doi.org/10.1039/b901839p.

Mishra, V., Gurnany, E., Mansoori, M.H., 2017. Quantum Dots in Targeted Delivery of Bioactives and Imaging in Nanotechnology-Based Approaches for Targeting and Delivery of Drugs and Genes. Elsevier Inc., Academic Press, pp. 427–450. http://doi.org/10.1016/B978-0-12-809717-5.00015-4.

Mitchell, M.J., et al., 2014. Trail-coated leukocytes that kill cancer cells in the circulation. PNAS 111 (3), 930–935. National Academy of Sciences. http://doi.org/10.1073/pnas.1316312111.

Mohanraj, V.J., Chen, Y., 2007. Nanoparticles - a review. Trop. J. Pharmaceut. Res. 5 (1), 561–573. African Journals Online (AJOL). http://doi.org/10.4314/tjpr.v5i1.14634.

Morton, S.W., et al., 2014. A nanoparticle-based combination chemotherapy delivery system for enhanced tumor killing by dynamic rewiring of signaling pathways. Sci. Signal 7 (325), 1–27. American Association for the Advancement of Science. http://doi.org/10.1126/scisignal.2005261.

Mussi, S.V., et al., 2013. New approach to improve encapsulation and antitumor activity of doxorubicin loaded in solid lipid nanoparticles. Eur. J. Pharm. Sci. 48 (1–2), 282–290. http://doi.org/10.1016/j.ejps.2012.10.025.

Nguyen, H., et al., 2015. Improved method for preparing cisplatin-dendrimer nanocomplex and its behavior against NCI-H460 lung cancer cell. J. Nanosci. Nanotechnol. 15 (6), 4106–4110. American Scientific Publishers. http://doi.org/10.1166/jnn.2015.9808.

O'Shannessy, D.J., et al., 2012. Folate receptor alpha expression in lung cancer: diagnostic and prognostic significance. Oncotarget 3 (4), 414–425. Impact Journals LLC. http://doi.org/10.18632/oncotarget.519.

Oberdörster, G., Kuhlbusch, T.A.J., 2018. In vivo effects: Methodologies and biokinetics of inhaled nanomaterials. NanoImpact 10, 38–60. Elsevier B.V.. http://doi.org/10.1016/j.impact.2017.10.007.

Okay, O., 2009. General Properties of Hydrogels. Springer, Berlin, Heidelberg, pp. 1–14. http://doi.org/10.1007/978-3-540-75645-3_1.

Paranjpe, M., Müller-Goymann, C.C., 2014. Nanoparticle-mediated pulmonary drug delivery: a review. Int. J. Mol. Sci. 15 (4), 5852–5873. http://doi.org/10.3390/ijms15045852.

Parumasivam, T., et al., 2016. Dry powder inhalable formulations for anti-tubercular therapy. Adv. Drug. Deliv. Rev. 102, 83–101. Elsevier B.V.. http://doi.org/10.1016/j.addr.2016.05.011.

Di Pasqua, A.J., et al., 2012. Tumor accumulation of neutron-activatable holmium-containing mesoporous silica nanoparticles in an orthotopic non-small cell lung cancer mouse model. Inorg. Chim. Acta 393, 334–336. Elsevier S.A.. http://doi.org/10.1016/j.ica.2012.06.016.

Patel, A.R., et al., 2014. Theranostic tumor homing nanocarriers for the treatment of lung cancer. Nanomedicine 10 (5), e1053–e1063. Elsevier Inc.. http://doi.org/10.1016/j.nano.2013.12.002.

Patel, A.R., Chougule, M., Singh, M., 2014. EphA2 targeting pegylated Nanocarrier drug delivery system for treatment of lung cancer. Pharm. Res. 31 (10), 2796–2809. Springer New York LLC. http://doi.org/10.1007/s11095-014-1377-4.

Patil, J.S., Sarasija, S., 2012. Pulmonary drug delivery strategies: a concise, systematic review. Lung India 29 (1), 44–49. Medknow Publications & Media Pvt Ltd. http://doi.org/10.4103/0970-2113.92361.

Patil, T.S., et al., 2019. Evaluation of nanocarrier-based dry powder formulations for inhalation with special reference to anti-tuberculosis drugs. Crit. Rev. Ther. Drug Carrier Syst. 36 (3), 239–276. http://doi.org/10.1615/CritRevTherDrugCarrierSyst.2018024397.

Pednekar, P.P., Jadhav, K.R., Kadam, V.J., 2012. Aptamer-dendrimer bioconjugate: a nanotool for therapeutics, diagnosis, and imaging. Expert Opin. Drug Deliv. 9 (10), 1273–1288. http://doi.org/10.1517/17425247.2012.716421.

Peng, G., et al., 2009. Diagnosing lung cancer in exhaled breath using gold nanoparticles. Nat. Nanotechnol. 4 (10), 669–673. Nature Publishing Group. http://doi.org/10.1038/nnano.2009.235.

Perepelyuk, M., et al., 2018. Evaluation of MUC1-aptamer functionalized hybrid nanoparticles for targeted delivery of miRNA-29b to nonsmall cell lung cancer. Mol. Pharm. 15 (3), 985–993. American Chemical Society. http://doi.org/10.1021/acs.molpharmaceut.7b00900.

Pontes, J.F., Grenha, A., 2020. Multifunctional nanocarriers for lung drug delivery. Nanomaterials 10 (2), 1–24. http://doi.org/10.3390/nano10020183.

Porra, L., 2006. University of Helsinki report series in physics lung structure and function studied by synchrotron radiation. Available at: http://www.ethesis.helsinki.fi (Accessed: 31 May 2021).

Respaud, R., et al., 2015. Nebulization as a delivery method for mAbs in respiratory diseases. Expert Opin. Drug Deliv. 12 (6), 1027–1039. Informa Healthcare. http://doi.org/10.1517/17425247.2015.999039.

Rocha, M., Chaves, N., Bao, S., 2017. Nanobiotechnology for breast cancer treatment in Breast Cancer - From Biology to Medicine. InTech, pp. 411–432. http://doi.org/10.5772/66989.

Rosière, R., Amighi, K., Wauthoz, N., 2019. Nanomedicine-Based Inhalation Treatments for Lung Cancer in Nanotechnology-Based Targeted Drug Delivery Systems for Lung Cancer. Elsevier, pp. 249–268. http://doi.org/10.1016/b978-0-12-815720-6.00010-1.

Rytting, E., et al., 2008. Biodegradable polymeric nanocarriers for pulmonary drug delivery. Expert Opin. Drug Deliv. 5 (6), 629–639. http://doi.org/10.1517/17425247.5.6.629.

Sarfati, G., et al., 2011. Targeting of polymeric nanoparticles to lung metastases by surface-attachment of YIGSR peptide from laminin. Biomaterials 32 (1), 152–161. http://doi.org/10.1016/j.biomaterials.2010.09.014.

Shah, N.D., Shah, V.V., Chivate, N.D., 2012. Pulmonary drug delivery: a promising approach. J. Appl. Pharmaceut. Sci. 2 (6), 33–37. http://doi.org/10.7324/JAPS.2012.2632.

Shanker, M., et al., 2010. Drug resistance in lung cancer Lung Cancer: Targets and Therapy. Dove Medical Press Ltd., pp. 23–36. http://doi.org/10.2147/lctt.s6861.

Shen, J., et al., 2014. IRGD conjugated TPGS mediates codelivery of paclitaxel and survivin shRNA for the reversal of lung cancer resistance. Mol. Pharm. 11 (8), 2579–2591. American Chemical Society. http://doi.org/10.1021/mp400576f.

Sherje, A.P., et al., 2018. Dendrimers: A versatile nanocarrier for drug delivery and targeting. Int. J. Pharm. 548 (1), 707–720. Elsevier B.V.. http://doi.org/10.1016/j.ijpharm.2018.07.030.

Shi, H., et al., 2014. Au@Ag/Au nanoparticles assembled with activatable aptamer probes as smart "nano-doctors" for image-guided cancer thermotherapy. Nanoscale 6 (15), 8754–8761. Royal Society of Chemistry. http://doi.org/10.1039/c4nr01927j.

Singh, P., et al., 2018. Gold nanoparticles in diagnostics and therapeutics for human cancer. Int. J. Mol. Sci 19 (7), 1–16. MDPI AG. http://doi.org/10.3390/ijms19071979.

Stathopoulos, G.P., Boulikas, T., 2012. Lipoplatin formulation review article. J. Drug Deliv. 2012, 1–10. Hindawi Limited. http://doi.org/10.1155/2012/581363.

Sudhakar, C.K., et al., 2015. Hydrogels-promising candidates for tissue engineering in Nanotechnology Applications for Tissue Engineering. Elsevier Inc., William Andrew Publishing, pp. 77–94. http://doi.org/10.1016/B978-0-323-32889-0.00005-4.

Swaroop, H. (no date) 'A Review on Nanoparticles in Targeted Drug Delivery System'. doi: http://doi.org/10.4172/2321-6212.1000r003.

Taratula, O., et al., 2013. Nanostructured lipid carriers as multifunctional nanomedicine platform for pulmonary co-delivery of anticancer drugs and siRNA. J. Control. Release 171 (3), 349–357. Elsevier B.V.. http://doi.org/10.1016/j.jconrel.2013.04.018.

Tian, H., et al., 2012. Enhancement of cisplatin sensitivity in lung cancer xenografts by liposome-mediated delivery of the plasmid expressing small hairpin RNA targeting survivin. J. Biomed. Nanotechnol. 8 (4), 633–641. http://doi.org/10.1166/jbn.2012.1419.

TIRUWA, R., 2016. A review on nanoparticles – preparation and evaluation parameters. Indian J. Pharmaceut. Biol. Res. 4 (2), 27–31. http://doi.org/10.30750/ijpbr.4.2.4.

Torchilin, V.P., 2007. Micellar nanocarriers: pharmaceutical perspectives. Pharm. Res. 24 (1), 1–16. http://doi.org/10.1007/s11095-006-9132-0.

Vishal Gupta, N., Shivakumar, H.G., 2010. Preparation and characterization of superporous hydrogels as gastroretentive drug delivery system for rosiglitazone maleate. DARU, J. Pharmaceut. Sci. 18 (3), 200–210. Available at: /pmc/articles/PMC3304361/ (Accessed: 2 June 2021).

de Vries, A., et al., 2010. Block-copolymer-stabilized iodinated emulsions for use as CT contrast agents. Biomaterials 31 (25), 6537–6544. http://doi.org/10.1016/j.biomaterials.2010.04.056.

Wang, C., et al., 2011. Tumor-targeting magnetic lipoplex delivery of short hairpin RNA suppresses IGF-1R overexpression of lung adenocarcinoma A549 cells in vitro and in vivo. Biochem. Biophys. Res. Commun. 410 (3), 537–542. http://doi.org/10.1016/j.bbrc.2011.06.019.

Wang, Y., et al., 2013. Intravenous delivery of siRNA targeting CD47 effectively inhibits melanoma tumor growth and lung metastasis. Mol. Ther. 21 (10), 1919–1929. Nature Publishing Group. http://doi.org/10.1038/mt.2013.135.

Wang, Z., et al., 2017. Active targeting theranostic iron oxide nanoparticles for MRI and magnetic resonance-guided focused ultrasound ablation of lung cancer. Biomaterials 127, 25–35. Elsevier Ltd. http://doi.org/10.1016/j.biomaterials.2017.02.037.

West, J.L., Halas, N.J., 2003. Engineered nanomaterials for biophotonics applications: improving sensing, imaging, and therapeutics. Annu. Rev. Biomed. Eng. 5 (1), 285–292. Annu Rev Biomed Eng. http://doi.org/10.1146/annurev.bioeng.5.011303.120723.

Wiley, J.A., et al., 2009. Inducible bronchus-associated lymphoid tissue elicited by a protein cage nanoparticle enhances protection in mice against diverse respiratory viruses. PLoS One 4 (9), e7142. Public Library of Science. http://doi.org/10.1371/journal.pone.0007142.

Woodard, G.A., Jones, K.D., Jablons, D.M., 2016. Lung cancer staging and prognosis in Cancer Treatment and Research. Kluwer Academic Publishers, Springer, Cham, pp. 47–75. http://doi.org/10.1007/978-3-319-40389-2_3.

Wu, Y., et al., 2010. PH-Responsive quantum dots via an albumin polymer surface coating. J. Am. Chem. Soc. 132 (14), 5012–5014. American Chemical Society. http://doi.org/10.1021/ja909570v.

Wu, Z.X., et al., 2014. Thermosensitive hydrogel used in dual drug delivery system with paclitaxel-loaded micelles for in situ treatment of lung cancer. Colloids Surf. B 122, 90–98. Elsevier. http://doi.org/10.1016/j.colsurfb.2014.06.052.

Yang, S.G., et al., 2010. 99mTc-hematoporphyrin linked albumin nanoparticles for lung cancer targeted photodynamic therapy and imaging. J. Mater. Chem. 20 (41), 9042–9046. The Royal Society of Chemistry. http://doi.org/10.1039/c0jm01544j.

Yang, Y., et al., 2012. Nanoparticle delivery of pooled siRNA for effective treatment of non-small cell lung cancer. Mol. Pharm. 9 (8), 2280–2289. http://doi.org/10.1021/mp300152v.

Yao, J., Feng, J., Chen, J., 2016. External-stimuli responsive systems for cancer theranostic. Asian J. Pharm. Sci. 11 (5), 585–595. Shenyang Pharmaceutical University. http://doi.org/10.1016/j.ajps.2016.06.001.

Zarogoulidis, P., et al., 2012. Inhaled chemotherapy in lung cancer: future concept of nanomedicine. Int. J. Nanomed. 7, 1551–1572. http://doi.org/10.2147/IJN.S29997.

Zeng, H., et al., 2012. Expression of hPNAS-4 radiosensitizes Lewis lung cancer. Int. J. Radiat. Oncol. Biol. Phys. 84 (4), 533–540. Elsevier Inc. http://doi.org/10.1016/j.ijrobp.2012.06.028.

Zhao, S., et al., 2015. SapC-DOPS nanovesicles as targeted therapy for lung cancer. Mol. Cancer Ther. 14 (2), 491–498. American Association for Cancer Research Inc.. http://doi.org/10.1158/1535-7163.MCT-14-0661.

Zhu, G., et al., 2012. Nucleic acid aptamers: an emerging frontier in cancer therapy. Chem. Commun. 48 (85), 10472–10480. The Royal Society of Chemistry. http://doi.org/10.1039/c2cc35042d.

Zoroddu, M., et al., 2014. Toxicity of nanoparticles. Curr. Med. Chem. 21 (33), 3837–3853. http://doi.org/10.2174/0929867321666140601162314.

Chapter 15

Role of nanocarriers for the effective delivery of anti-HIV drugs

Rohini Kharwade, Nilesh M. Mahajan
Dadasaheb Balpande College of Pharmacy, Rashtrasant Tukadoji Maharaj Nagpur University, Nagpur, Maharashtra, India

15.1 Introduction

The human immunodeficiency virus (HIV) is an enveloped retrovirus belonging to family *lentivirus* with single-stranded RNA molecules (approximately 9 kb). Since its isolation and identification are divided into major two types, human T cell leukemia viruses (HTLV), and human immunodeficiency viruses (HIV). HTLV is subdivided into HTLV-1 and HTLV-2 belongs to the subclass *oncovirinae*, causes adult T cell leukemia and spastic paraparesis (Saura, 1998; Shuh and Beilke, 2005). HIV is subdivided into HIV-1 and HIV-2 from the subclass *lentivirinae* which are responsible for acquired immunodeficiency syndrome (AIDS). From the vast majority of viruses in the developed world, HIV-1 is mainly responsible for global AIDS pandemic and generally called HIV (Wilen et al., 2012). HIV infects CD4+ receptor-bearing helper T cells and ultimately loss of CD4 cells leads to immune deficiency and developing an opportunistic infection. AIDS transmitted by body fluid transfer including blood transfusion, organ transplant, sexual contact, and perinatally from mother to offspring. HIV-1 exist in two predominant forms such as syncytium inducing strain (SI or T-cell tropic) and nonsyncytium inducing (NSI or macrophage-tropic, M-tropic) strain. In SI pathology, HIV infects T-cell by using CXCR4 co-receptor, whereas NSI pathology is associated with CCR5 co-receptor with slower disease progression. Infection with NSI-tropic strain of HIV takes 7.12 years for progression of AIDS however T-tropic strain of the virus may take only 2–3 years (Fauci, 1993; Gardner et al., 2004).

HIV-1 stores its genetic information in RNA instead of DNA; therefore they require DNA when entering into the human cell to make replicate themself. The outer shell of the virus is called envelope which is covered by spike glycoprotein including gp 120 and gp 41which help to enter and lock the virus into the CD4+ receptor. The core of the virus is held a cone-shaped structure called capsid which contains two enzymes, the reverse transcriptase and integrase which is essential for replication of HIV (German Advisory Committee Blood, 2016). Capsid also contains two strands of RNA containing nine genes which hold genetic material and provide instruction to make new viruses. Out of these nine genes, three genes known as *gag*, *pol*, and *env* provide the instructions to make structural and nonstructural proteins for new virus particles (Fig. 15.1). The other six genes such as *tat, rev, nef, vif, vpr,* and *vpu* provide code for proteins which able to make HIV, infects the host cell, produce new virus and release them from infected cells (Margolis, 2010; Palmer et al., 2011).

15.1.1 HIV life cycle and pathogenesis

The life cycle of HIV is complex and roughly divided into the early and late phase of replication (Fig. 15.2). The early phase starts with the attachment of the virion at the host cell surface and ends with the integration of proviral DNA into the host cell genome. However late phase of replication starts with the transcription of proviral DNA and ends with the release of fully infectious progeny virions (Zinkernagel, 1996; Panova, 2020).

15.1.1.1 Viral attachment and binding

In highly activated CD4+ T cells, the life cycle of HIV last in only 1–2 days in association with the planned death of both virally infected and uninfected bystander CD4+ T cell.

Host cell-free HIV virions have 20–30 min half-life thus, the virus must find and infect a new target host cell within a short time. As described above, CD4+ is the primary receptor, and chemokine CCR5 and CXCR4 are important co-receptor for HIV entry into the host cell. Several other receptors, such as poly-glycans, lectins and others can also bind HIV virions

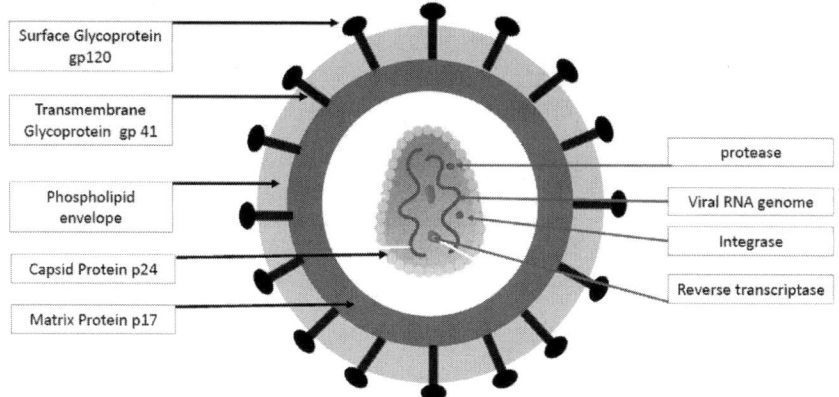

FIG. 15.1 Structure of HIV with all functional proteins.

in an unspecific manner and greatly increase viral infection rate. These all receptors may trap the viral particles at the cell surface and provoke T cell infection (Wilen et al. 2012; Wilen, 2014). HIV infection starts with attachment of envelope protein gp120 to CD4+ receptor on the host cell surface, followed by engagement of chemokine receptors (CXCR4 and CCR5) to form a CD4-gp120-chemokine receptor complex. Subsequently, envelope protein gp41 of the virus mediates the membrane fusion (Margolis and Shattock, 2006).

15.1.1.2 Reverse transcription

After fusion, the genetic information including viral RNA can enter into the host cell. The enzyme reverse transcriptase, in viral RNA, triggers the synthesis of linear double-stranded cDNA (proviral) from viral RNA. During infection, this cDNA inserted into the host cell's genomic DNA by viral integrase. After successful entry of linear proviral cDNA into host cell chromosomes, it replicates along with the host DNA and begins the translation process. Consequently, the infection can be spread either by infection of new cells or multiplication of cell containing proviral cDNA (Alkhatib et al., 1997). However, in some long-living cell such as memory CD4+ T lymphocytes and macrophages, the proviral cDNA may remain silent for many years and acts as a reservoir (Huck et al., 1998). Theses persistence reservoirs of HIV are the unique feature of the viral life cycle which cannot be recognized by the host defense mechanism and will remain as long as the cell is quiescent. Viral reservoirs are established before acute and symptomatic infection is evident. These viral reservoirs cannot be eliminated by conventional antiretroviral therapy (ART) but sometimes completion of ART regimen may become activated and produce infectious HIV (Margolis, 2010; Parrish, 2012).

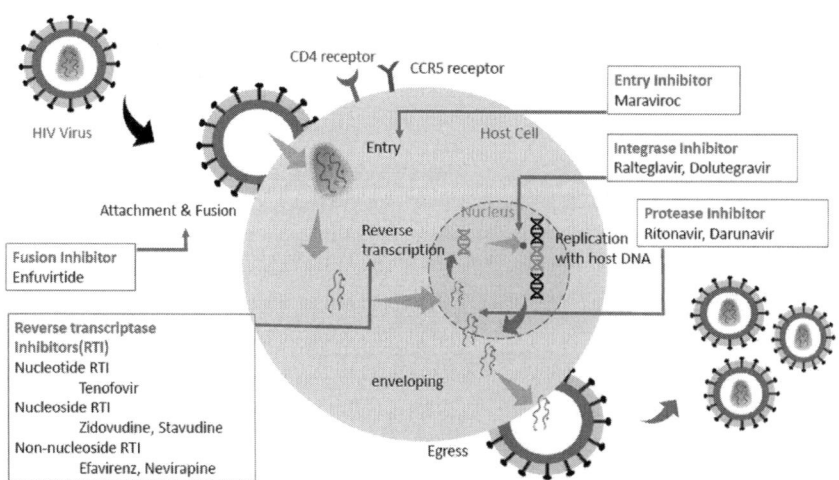

FIG. 15.2 Viral life cycle with add the following drugs in different steps.

15.1.1.3 Transcription

In productively infected cells, cDNA of HIV serves as a template for the transcription of both viral messengers and genomic RNA by the cellular Pol II polymerase. Initially, the transcriptional output is low because elongation of viral transcript is dependent on viral transactivator protein (Tat) which is present in sufficient amount (Swanstrom, 2014). Tat protein is also required for efficient synthesis of full-length HIV transcript, effective viral gene expression and generation of more than 25 different mRNA in three classes (i) unspliced RNA act as a genomic RNA and to produce the Gag and Gag-Pol precursors; (ii) single spliced RNA encoding Vif, Vpr, Vpu and Env; (iii) fully spliced RNA expressing Tat, Rev, and Nef (Mcgettigan, 2003).

15.1.1.4 Translation and assembly

As mentioned above, from the fully spliced mRNA expression Tat enhances viral transcription and RNA elongation. Rev mediates the transport of unspliced mRNA to the cytoplasm. Nef protein makes the infected cell invisible by downregulating several surface receptors so that the infected cell cannot be identified by the host immune system (Swanstrom, 2014). Unspliced mRNA that expresses the Gag and Gag-Pol precursors are processed for major structural and enzymatic proteins. Also, these both precursors are N-terminally myristoylated in the matrix domain and concentrated in lipid rafts at the inner leaflet of the plasma membrane. The viral Env glycoprotein is signed up for the building platforms through interactions with the matrix domain. Finally, two copies of genomic viral RNA, as well as some cellular factors, are recruited to this complex. The accumulation of viral proteins and RNA at the plasma membrane-coated spherical particle (Zhang et al., 2008).

15.1.1.5 Budding

The above membrane-coated spherical particle release from the infected cells by surface budding, or infected cells can undergo lysis with a burst release of new HIV virions, that can infect additional cells. The late domain of the p6 part of Gag and the cellular Tsg101 protein involved in allowing newly formed HIV to pinch off and enter into the circulation (Kafaie et al., 2008; Jiang et al., 2011). During the budding process, HIV protease cleaves viral proteins into their functional forms and cleaves the Gag and Gag-Pol precursors into their mature final components. As a consequence, the configuration of the proteins is reorganized to generate the characteristic electron-dense conical inner core and to render the virus infectious. The infected cells can then release virions (Paul et al., 1996).

15.1.2 Pathophysiology

HIV infects different kinds of immune cells including CD4+ bearing macrophages, T cells, immature dendritic cells, microglia cells, monocytes and stem cells. HIV infection causes destruction of the above mentioned immune cells, which results in immune suppression and leads to several opportunistic infection, signaling of AIDS (Weston and Marett, 2009).

The initial infection is followed by the dominance of M-tropic strain and afterward re-emergence due to T-tropic strain. The switching from M-tropic to T-tropic strain is known as phenotopic switch phenomenon that induces rapid viral replication and increases viral mutation rates. This accelerated rate of disease progression is the effect of high immune system activity, which drops the CD4+ T cell counts and the emergence of T-tropic strain. Studies showed that patients harboring T-tropic virus had faster clinical progression rates, less responsive to treatment and rapid development of ART resistance (Gill et al., 2010).

15.2 Conventional antiretroviral therapy

Since 1984, various strategies have been acquired without much success for the treatment of AIDS. While the researcher's focus shifted to the development of effective and safe antiretroviral drugs intending to block HIV replication. In the early 1990s, the first HIV-1-specific antiviral drugs were given as monotherapy (Wong and Yukl, 2016). Up till now, 28 FDA-approved drugs acting on distinct HIV target sites are available for the treatment of HIV-1 infection. Based on the molecular mechanism these all drugs are distributed into distinct classes (Fig. 15.2). Most of the ARTs using the single drug are unable to penetrate these reservoirs and maintain therapeutic concentrations for prolonged periods, and this is considered as a major factor in the emergence of drug resistance and consequently treatment failure. Therefore with a better understanding of pathogenesis and replication processes of HIV, single-agent antiretroviral therapy has been largely substituted by a cocktail or combination therapy known as highly active antiretroviral therapy (HAART) (Desai et al., 2012). Combination therapy using at least three antiretroviral agents (Table 15.1) directed against different HIV targets and reduces the evolutionary

TABLE 15.1 Marketed available brand under HAART therapy with its characteristics ("ANNEX 3 . WEIGHT-BASED DOSING FOR ARV," no date; "Drugs That FIGHT HIV-1 A reference guide for prescription HIV-1 medications," no date; *FDA-Approved HIV Medicines* | *Understanding HIV/AIDS* | *AIDSinfo*, no date).

Drug category	Generic names	Brand name	Manufacturer	Adult dose	Bioavailability	FDA-approved date
Fusion inhibitors	Enfuvirtide, T-20	Fuzeon P	Hoffmann-La Roche and Trimeris	90 mg, b.i.d. (Subcutaneous)	84 %	13 March 2003
Entry inhibitors—CCR5 coreceptor or antagonist	Maraviroc	Selzentry	Pfizer	150 mg, b.i.d. (Oral) or 300 mg, b.i.d.(Oral) or 600 mg, b.i.d.(Oral)	23–33%	6 August 2007
Protease inhibitors (PIs)	Indinavir	Crixivan P	Merk	800 mg, b.i.d. (Oral) or 800 mg, t.i.d.(Oral)	65 %	13 March 1996
	Nelfinavir	Viracept P	Agouron Pharmaceuticals	1250 mg, b.i.d. (Oral) or 750 mg, t.i.d. (Oral)	20–80%	4 March 2004
	Saquinavir	Invirase P	Roche	1000 mg, b.i.d. (Oral)	Erratic 4%	6 December 1995
	Azantavir	Reyataz	Bristol-Myers Squibb	300 mg, daily (Oral) or 400 mg, daily (Oral)	No data %	20 June 2003
	Darunavir	Prezista P	Tibotec	600 mg, b.i.d. (Oral) or 800 mg, daily (Oral)	37%	23 June 2006
Non-nucleoside reverse transcriptase inhibitors	Etravirine	Intelence P	Tibotec	200 mg, b.i.d. (Oral)	No data	18 January 2008
	Nevirapine	Viramune P	Boehringer Ingelheim	200 mg, b.i.d. (Oral)	20–80	21 June 1996
	Efavirenz	Sustiva P	Bristol-Myers Squibb	600 mg, daily (Oral)	45–80	17 September 1998
Nucleoside reverse transcriptase inhibitors (NRTIs)	Lamivudine	Epivir P	GlaxoSmithKline	150 mg, daily (Oral) or 300 mg, daily (Oral)	86	17 November 1995
	Zidovudine	Retrovir P	GlaxoSmithKline	200 mg, t.i.d (Oral) or 300 mg, b.i.d. (Oral)	60	19 March 1987
	Didanosine	Videx P	Bristol-Myers-Squibb	400 mg, daily (Oral; Delayed-release Capsule)	30–40	31 July 2012
	Stavudine	Zerit P	Bristol-Myers-Squibb	400 mg, b.i.d. (Oral)	78	6 October 2011
Integrase inhibitors	Raltegravir	Insentress P	Merk	400 mg, b.i.d. (Oral)		12 October 2007
HAART combination product	Efavirenz, Emtricitabine, and Tenofovir disoproxil fumarate	Atripla	Bristol-Myers-Squibb and Gilead Sciences	600 mg+ 200 mg+ 245 mg (Film-Coated Tablet, Oral q.d.)		12 July 2006
	Emtricitabine, Rilpivirine, and Tenofovir disoproxil fumarate	Complera	Gilead Sciences	200 mg + 25 mg + 300 mg (Oral, q.d.)		10 August 2011
	Elvitegravir, Cobicistat, Emtricitabine, Tenofovir disoproxil fumarate	Stribild	Gilead Sciences	150 mg + 150 mg + 200 mg + 300 mg (Oral, q.d.)		27 August 2012
	Lopinavir, Ritonavir	Kaletra	AbbVie	200 mg + 50 mg (Oral, q.d.)		15 September 2000

Note: b.i.d. (bis in die): two times a day, q.d. (quaque die): once time a day, t.i.d. (ter in die): three times a day.

drug resistance. HAART dramatically suppresses viral replication and reduces the plasma viral load, below the limit of detection (Piketty et al., 1999; Desai et al., 2012).

As we discussed earlier in the life cycle of HIV, reverse transcriptase plays an important role in the replication of HIV therefore represents one of the targets in the HAART. Both nucleoside reverse transcriptase inhibitors (NRTI) and non-nucleoside reverse transcriptase inhibitors (NNRTI) restrict the reverse transcriptase associated polymerase activity. After the discovery of zidovudine (AZT), NRTIs were the only class of drugs to be approved by the FDA and still it plays a central role in HAART in the struggle of HIV-1 infection. NNRTIs are structurally different molecule but still, it locks the polymerase activity by inducing rotameter conformational changes in some residues of reverse transcriptase (Barton et al., 2016).

Entry inhibitors interact with gp-41 on the viral envelope and blocking its attachment with the cell membrane, while co-receptor inhibitors prevent interaction of the CCR5 receptor with gp120 of the virus. Protease inhibitors interfere in the proteolysis of viral proteins through interaction with the active site of the HIV protease (Qian et al., 2008; Lobritz et al., 2010).

Recently, Archin et al. suggested an extensively debated "shock and kill" approach. They use histone deacetylase inhibitor, Vorinostat 60 to increase the synthesis of HIV RNA. They assume that these infected cells would be killed either by the virus or by the host's immune system (Archin, 2017). The above approach has been proposed with intensified HAART to protect the uninfected cells from infection (Kim et al., 2017).

Combined antiretroviral therapy significantly reduced AIDS-related morbidity and mortality. However, there are several drawbacks in the conventional ART and HAART.

A. *Drugs resistance*: One of the major problems associated with ART is drug resistance. The replication process of HIV is rapid and perfect with generating at least one mutation per genome. This genetic mutation facilitates to develop resistance against ART, specifically when adopted monotherapy as a treatment.
B. *Long-term drug therapy*: Viral reservoirs required long-term continuous combined antiretroviral therapy. Patient poor treatment compliance often plays a major factor in treatment failure and viral rebound (Qian et al., 2008; Lobritz et al., 2010).
C. *Toxicity and drug–drug interaction*: Prolong and continuous treatment with HAART have several side effect including constipation, fever, liver disorder, muscular dystrophy, metabolic disorders, reversible central nervous system adverse effects, and peripheral neuropathy. AIDS developed a number of opportunistic infection, therefore the use of a combination of drugs in conventional therapy can lead to undesirable drug–drug interaction and reducing the efficacy of drugs. For example when a combination of reverse transcriptase inhibitor Nevirapine given with protease inhibitor Saquinavir, then plasma concentration of saquinavir were dropped rapidly. Because Nevirapine induces the liver cytochrome p450 which metabolize saquinavir and reducing its bioavailability and efficacy ("ANNEX 3. WEIGHT-BASED DOSING FOR ARV," no date).
D. *Poor bioavailability*: Most of the antiretroviral drugs are troubled with poor pharmacokinetic when formulated as a conventional solid oral dosage form. However oral route suffers from significant first-pass effect, variation in absorption, expression of multidrug-resistant efflux proteins (P-gp), metabolism, transport barriers and gastrointestinal degradation leads to low and erratic bioavailability. The short half-life of ART drugs required frequent administration of booster dosages which contributes to the development of drug resistance.
E. *Lack of access to target sites*: Most of the ART drugs cannot cross blood–brain barriers (BBB) as a result of which they are unable to enter in microglial cells which act as an HIV reservoir (Barton, Winckelmann and Palmer, 2016).
F. It has been observed that HAART decrease viral load below undetectable levels (HIV RNA detecting blood tests) but impossible to eliminate from the body. The patient experienced a rebound viremia immediately after cessation of ART even though, they continue the therapy for as long as 3 years. In that case, viremia generated from lymphocytes (93.99%), long-lived cells (1.7%), and less than 1% from latently infected CD4+ T cells such as lymphatic system, macrophages, lymphocyte, and microglial cell. Current ART has little impact on latently infected cell due to inaccessibly and short residence time resulting in low concentration at target sites (Oguntibeju, 2012; Kirk and Jacobson, 2014).

15.3 Types of nanocarriers for antiretroviral drugs delivery

To overcome these problems several new drug moiety or improved targeted action, drug delivery system have been proposed concerning the use of nanotechnology. In the context of oral drug delivery, the important characteristics features which are considered for significant antiretroviral effect include solubility and ionization, lipophilicity, permeability, stability in gastrointestinal fluids, metabolism, and viral reservoir targeting strategies of the drug (Lembo et al., 2018). If these properties of drugs are unfavorable then viral loads cannot decrease. Thus nanocarriers play a vital role in sustain release, increased half-life, targeted action with less side effect of the drug. In addition, nanocarriers able to deliver the anti-HIV drugs in a controlled and targeted manner in the latently infected cells (viral reservoir) and improve the life quality of HIV patients (Gupta and Jain, 2010; Lembo et al., 2018).

Nanocarriers include pure drug nanoparticles/nanocrystal, polymeric nanoparticles, polymeric micelles, solid lipid nanoparticles, liposomes, and dendrimers. The purpose of this chapter is to present recent developments in the application of nanotechnology for the delivery of antiretroviral drugs with their mechanism for HIV treatment (Kulkosky and Bray, 2006). Wide varieties of different synthetic, semisynthetic, and natural materials are used to prepare these nanocarriers from nanometers to submicron dimensions. They all are considerably different in physical, chemical, biological properties including their method of preparation and characterization (Iannazzo, 2015).

15.3.1 Pure drug nanoparticles

The pure drug nanoparticle is a promising area for novel anti-HIV therapy for drugs with properties such as insolubility in water and oil, high melting point, low partition coefficient, and high dose. They are characterized by their small size, which gives it unique chemical, physical and biological properties (Ogunwuyi et al., 2016).

According to Noyes–Whitney equation, progressive size reduction of drug particles leads to an increase in the surface area with an increased dissolution rate. Nanometer size range particles facilitate the dissolution rate of the drug by decreasing diffusion layer thickness and increasing concentration gradient between the surface of the particle and solution (Fig. 15.3). Therefore, nanosizing is a suitable approach for increasing bioavailability and systemic absorption rate of those antiretroviral drugs where dissolution is the rate-limiting step. This approach might be possible to overcome many problems of conventional ART such as protect the sensitive drug from degradation, reduce side effects, and improve tissue drug tolerance with targeting specific biological sites either passively or actively (Kumar et al., 2015; Ogunwuyi et al., 2016).

Over the past two decades, drug nanocrystal technology has been highlighted in ART with special benefits to formulate poorly soluble drugs. The nanoparticles can be prepared by various methods, such as self-assembly, vapor, and electrostatic deposition, solvent diffusion, solvent evaporation, and coacervation technique. These nanocrystal particles of drugs are dispersed and stabilized in aqueous media with a minimum amount of surface-active agent, which is known as nanosuspension (Patel et al., 2016). Nanosuspension formulated by bottom-up (precipitation) and top-down (media milling, high-pressure homogenization) technique. In the bottom-up technique, the drug dissolved in the solvent and then precipitates it by adding the dug solution in nonsolvent. The top-down technique can be employed for all insoluble drugs including "brick dust drugs." Drug nanocrystals reveal several advantages including efficient drug loading capacity, easy manufacture, low-cost preparation, and easily administered by various routes such as oral, parenteral, topical, pulmonary delivery. All these advantages of drug nanocrystals promote their impacts on ART (Tuomela et al., 2016).

Van Eerdenbrugh et al. investigated the dissolution rate and in vitro absorption of poorly water-soluble non-nucleoside reverse transcriptase inhibitor (NNRTI) Loviride by nanonization. Author observed the physicochemical properties of Loviride such as high melting point, poor solubility of the drug in water as well as oil illustrate its low bioavailability which makes it brick–dust molecule. So that nanosuspension formulation of Loviride by milling with medium containing 3% sodium lauryl sulfate helps to increase bioavailability by increasing its solubility. It is evaluated by using CaCo-2 cell line for in-vivo drug transportation. In nanosization process, drug molecules experienced stress conditions and changes its physical and chemical properties, polymorphic transition, chemical degradation, and formation of the amorphous phase by process-induced transformation (Van Eerdenbrugh et al., 2010).

Currently, the formulation of Rilpivirine nanosuspension for oral drug delivery in the treatment of HIV showed an enhanced resistance profile against NNRTI-resistant HIV-1. Nanosization of Rilpivirine helps to sustained release action of the drug in the treatment and prophylaxis purpose of HIV-1 infection. Nanosization of Rilpivirine resolves its limited permeability absorption and hepatic first-pass metabolism issue which unfavorably affects bioavailability. However Rilpivirine

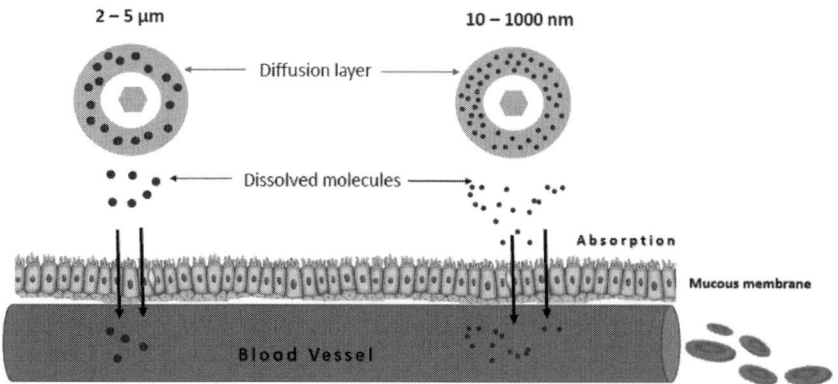

FIG. 15.3 Nanocrystal helps to increase the dissolution and absorption rate.

nanosuspension with particle size 457 nm showed fast release action. However, its macrophage uptake study and gamma scintigraphy characterization confirmed its cellular uptake, targeted action toward spleen, thymus, and lungs, that is, viral reservoir with no cytotoxicity (Mc Crudden et al., 2018).

Administration of water-insoluble drugs in the form of pure drug nanoparticles reduces its bioavailability in the presence or absence of food. Therefore it is necessary to re-disperse the drug nanoparticles in the gastrointestinal fluid after administration. Liversidge et al. developed media milling specifically wet milling technology (Nanocrystal) to avoid these bioavailability variations. Thus, nanocrystallization is a validated technology with well-established process controls and scale-up feasibility which increases bioavailability with a reduced dose of the drug (Merisko-Liversidge et al., 2003).

In media milling technology, the final particle size of the drug depends on the choice of stabilizer. In the process of milling the number of nanosize particles generated with a high surface area which increases surface free energy and interfacial tension between solid and liquid phase. To diminish the surface free energy, particle tends to aggregate which increased particle size and reduced surface area. Therefore it is necessary to add a surface stabilizer to decrease the surface free energy with respect to decrease interfacial tension and particle growth (Wewers et al., 2020). Further, spray drying, lyophilization, and wet granulation technology developed for the conversion of nanosuspension into a tablet compression or powders for stability enhancement and protection against aggregation. These solid drug nanoparticles can be easily reconstituted and ultimately improves patient compliance.

15.3.2 Polymeric nanoparticles

Sizes of polymeric nanoparticles are less than 1000 nm in size which is composed of synthetic and semisynthetic polymers with many desirable features such as manufacturing reproducibility, stability, and sustained drug release. The active drug molecule can be entrapped or encapsulated within the particle, physically adsorbed on the surface, or chemically linked to the surface of the particle (Fig. 15.4). Therefore nanoparticles have been explored for the delivery of anti-HIV drug molecules to modify its release characteristics and dosing frequency with targeted action to viral reservoir sites (Parboosing et al., 2012).

The selection of polymer in the nanoparticle formulation depends on molecular weight, copolymer composition which influences its degradation rate, thermal sensitivity, and pH sensitivity. The synthetic polymers (Table 15.2) most widely used and approved by the FDA for anti-HIV drugs include polyester with copolymers poly (lactic-co-glycolic acid) (PLGA). Natural polymers, such as alginate and chitosan have also been used with lower immunogenicity than synthetic polymers (Mamo et al., 2006).

Nanoparticulate drug delivery mostly uses for targeting the macrophages by the process of opsonization. The mechanism includes in target drug delivery is binding opsonins to the surface of nanoparticles by any interaction such as Van der Waals, electrostatic, or ionic force of attraction. Subsequently, the above-formed complex becomes recognizable by macrophages and phagocytosis takes place. Size, surface charges, and characteristics of nanoparticles have played important role in its clearance and tissue distribution (Auría-Soro et al., 2019).

HIV infection of brain macrophages directly linked to HIV associated dementia complex. Zensi et al. synthesized the apolipoprotein E-coated human serum albumin nanoparticle. They reported brain macrophages uptake of these coated nanoparticles after i.v. administration in mice. They observed that apolipoprotein-coated nanoparticles have more endothelial cells of brain uptake as compared to nonapolipoprotein-coated nanoparticles (Zensi et al., 2009). Additionally, apolipoprotein-coated nanoparticles cannot transport through tight junctions of endothelial cells and luminal membrane therefore nanoparticles follows the transcytosis mechanism of transport for internalization. Biodegradable polymeric nanoparticles have been also used in vaginal microbicide to prevent transmission of HIV during intercourse. In this nanoparticle

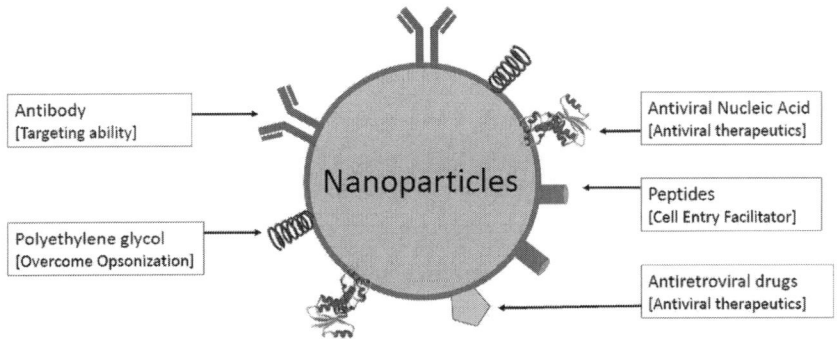

FIG. 15.4 Interaction of active drug molecule and polymeric nanoparticles.

TABLE 15.2 List of synthetic polymers approved by the FDA for antiretroviral treatment (Makadia and Siegel, 2011; Marin et al., 2013; Li et al., 2015).

Type of polymer	Name of polymer	Antiretroviral drug	Testing way	Cell lines/tissue types
Synthetic polymer	(PLGA) Poly (lactic-co-glycolic acid)	Stavudine Nevirapine	In vitro	Macrophages HBMECs
	Poly(n-butyl cyanoacrylate) PBCA	Stavudine	In vitro	Murine BMVECs
	Poly(n-butyl cyanoacrylate) PBCA	Zidovudine Lamivudine	In vitro	Bovine brain endothelial cell line (BBMECs)
	Methyl methacrylate Spm	Zidovudine Lamivudine	In vitro	Bovine brain endothelial cell line (BBMECs)
	Poly(n-butyl cyanoacrylate) PBCA methyl methacrylate Spm	Stavudine	In vitro	HBMEcs cell line producing for blood–brain barrier
	Eudragit RL100	Stavudine	In vitro	Wistar rat skin
	Poly (n-hexyl isocryanate)	Zidovudine	In vitro	Lymphoid tissue
	Polaxomer 388	Rilpivirine	In vivo	Mice and dog
	Polysaccharide, poly ethylene glycol (PEG)	Nevirapine	In vivo	Mice
	Poly (ethylene oxide)	Divalproex sodium	Ex vivo	Pig mucosa
Natural polymer	Chitosan	Lamivudine	In vitro	Drug release study
	Squale	Didanosine	In vivo	PBMCs
	Mannosylated gelatin	Nevirapine	In vivo	Pharmacokinetic study on mice

formulation researcher used chemokines and small interfering ribose nucleic acid (siRNA) as anti-HIV agents. These nanoparticles provide stability to drugs against degradation in the vaginal environment with targeted action at submucosal epithelium. Furthermore, Meng et al. prepared Tenofovir loaded chitosan-based nanoparticles to optimize its mucoadhesion. Its size and encapsulation efficiency were analyzed by dynamic light scattering and UV spectroscopy and its variables were assessed by using Box–Behnken design. When the diameter of nanoparticles decreased from 900 nm to 188 nm, mucoadhesion was doubled from 6% to 12%. The author observed that prepared chitosan nanoparticles were not cytotoxic to vaginal epithelial cell line for 48 h (Magalhães et al., 2020).

Destache et al. studied the release rate of the antiretroviral drug from poly (DL-lactide-co-glycolide) nanoparticles. They compared the release rate of Ritonavir, Lopinavir, and Efavirenz with its nanoparticles after intraperitoneal injection to BALB/mice. In this study, all mice are grouped and received free drugs or nanoparticles of the same drugs. The author observed that free drug concentration peaked within 4 h after injection and was eliminated by 72 h. However, from poly (DL-lactide-co-glycolide) nanoparticles concentration of drugs were detected in all tissues for 28 days. In HIV reservoir sites mostly in macrophages, these nanoparticles showed persistent inhibition of HIV-1 replication (Destache et al., 2010).

Stavudine is known as a first-line antiretroviral drug but the drug had been out of use because of its serious toxicity. Therefore to solve this problem Basu and co-workers prepared colloidal gold-loaded, poly (DL-lactide-co-glycolide) nanoparticles containing antiretroviral drug stavudine. They evaluate and compare the uptake of these nanoparticles by macrophages and its toxicity. These Stavudine nanoparticles emerge a hope that drug may back to use in ART due to advent of nanoparticle formulation technology. The author observed that Stavudine was released from nanoparticle for prolong period (over 63 days) and taken up by macrophages. As a result, these nanoparticles have a promising opportunity for targeted Stavudine delivery with minimum systemic toxicity (Basu et al., 2012). However nanoparticles assist the sustained release of the drug, therefore the above formulation of nanoparticles utilizes for controlled release of drug in HAART with improved regimen adherence (Edagwa et al., 2017).

In the pediatric population, due to poor palatability, some protease inhibitors (Ritonavir, Indinavir) are not used in a liquid formulation. To mask the bitterness of Indinavir sulfate, Chiappetta et al. prepared pH-sensitive microparticles of Eudragit E 100. They prepared different microparticle formulation (MP20, MP 40, MP 60) according to different Eudragit E 100 concentration. At acidic pH 1.5, all formulation showed 100% drug release in 5 min. however, slow release of drug was observed at pH 6.8. From the taste masking study, MP20 scored within the limits set of palatability. These results indicated that microparticles of Eudragit E100 polymer could efficiently entrap 20% Indinavir with acceptable palatability. Therefore it was concluded that polymer-based microparticle technology improved the palatability of indinavir formulations (Chiappetta et al., 2009).

15.3.3 Dendrimers

Among employed nanocarriers, dendrimers are novel monodispersed polymeric molecules with a large number of peripheral functional groups and interior cavities. These functional characteristics make it potential vectors for chemical drugs, peptides, and genes for HIV inhibition (Fig. 15.5). Dendrimers would increase the solubility and stability of chemical drugs. It also promotes cellular uptake via functional end groups and showing targeted action where the virus harbors is a potential way to completely cure the HIV-infected patients. They also act as a potential vector for antiretroviral drugs with low-level cytotoxicity (Hsu et al., 2017).

Various functionalized dendrimers act as an anti-HIV agent at different stages of HIV inhibition by breaking key contact of the virus to target cell action site (Table 15.3). It includes three potential binding sites, that is, CD4 receptor, glycosphingolipids (GSLs) and dendritic cell-specific ICAM-3-grabbing on integrin (DC-SIGN). Hence dendrimers act at several stages of viral infection for prevention of viral genome replication in the host cells and act as an antiretroviral (Peng et al., 2013).

The potential of oral delivery of dendrimer was explored by cationic and anionic PAMAM dendrimers. Researchers pointed out that peripheral positive charge of PAMAM is responsible for PAMAM cytotoxicity, which limits their biomedical application. So that modification of these surface groups is one of the ways which reduces the toxicity, improve biocompatibility and circulation time of PAMAM dendrimers in the body due to bypass the efflux action of P-gp. However, this surface modification has done by neutralizing charge through PEGylation (polyethene glycol conjugation), acetylation, folate, and peptide conjugation or by neutralizing the negative charge (Kesharwani et al., 2014). Permeability of PAMAM dendrimer through Caco-2 monolayer and everted rat intestinal sac also depends on to surface charge as well as their size, charge concentration and surface modification. Anionic PAMAM dendrimers have shown to be more permeable in Caco-2 monolayers as compared to cationic. The mechanisms of dendrimers for transportation across the GI epithelium are depending on energy adsorptive mediated endocytosis and paracellular pathway (Jevprasesphant et al., 2003).

Dutta et al. studied targeting potential of efavirenz (EFV) by encapsulating it in Tuftsin (Tu) conjugated poly (propyleneimine) (PPI) dendrimer. They observed that Tuftsin not only specifically binds to the mononuclear phagocytic cells but also enhances their natural killer activity. The objective of this study is to investigate the in vitro targeting potential and anti-HIV activity of fifth-generation (G5) Tu conjugated PPI. From this study, the entrapment efficiency of PPI and Tu-PPI was found to be 37.43 ± 0.3% and 49.31 ± 0.33%, respectively. TU-PPI was also found to sustain in vitro release of EFV up to 6 days with negligible cytotoxicity and 34.5 times higher cellular uptake in HIV infected macrophage as compared to PPI (Dutta et al., 2008).

Dendrimers have also been shown to have a self-therapeutic effect. However, researchers conclude that HIV capsid could be the target for structure-based design of nanoformulation for discontinuing the life cycle of HIV. As a result, dendrimers have shown to stop in vitro replication of the virus. The researcher observed three to five generation of PAMAM dendrimers easily prevented the binding of TAT protein to TAR RNA. He showed that PAMAM dendrimers could form

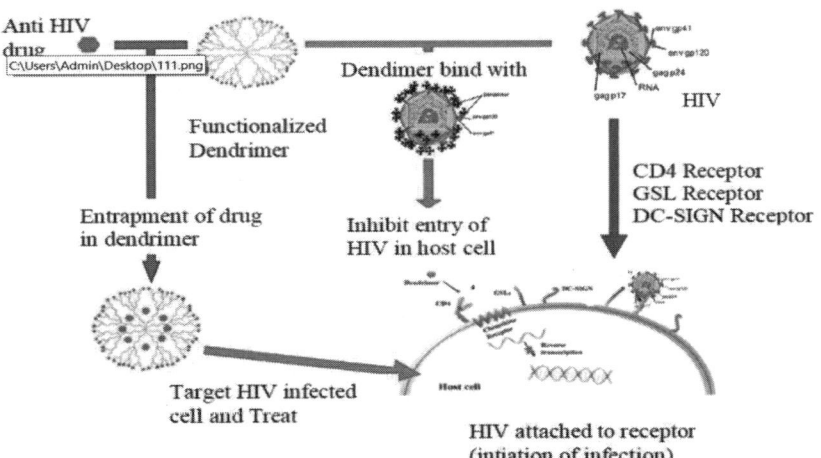

FIG. 15.5 Mechanism action of the dendrimer in antiretroviral therapy (Kharwade et al., 2020).

TABLE 15.3 Various functional dendrimers for HIV inhibition (Tomalia and Fréchet, 2002; Jimenez et al., 2012; Peng et al., 2013).

Code	Dendrimers	Functional surface groups	Interacting site	Inhibition stages
BRI2932 (SPL2932)	PAMAM G4		gp120	Attachment and replication
BRI6195 (SPL6195)	PAMAM G4		gp120	Attachment
PAMAM	PAMAM	$-NH_2$	Tran-acting responsive (TAR)	Replication
SPL7013	Polylysine dendrimer		gp120	Attachment and replication
2G-S16	Carbosilane dendrimer		gp120/CD4	Attachment
PS-GAL 64mer	Polypropylene-imine tetrahexa contamine dendrimers	R = H or SO_3^-	gp120	Attachment
MVC-GBT	Polypropylene-imine dendrimer		gp120	Attachment

complexes with TAR RNA and disrupt the interaction of TAT protein with TAR RNA (Madaan et al., 2014). Furthermore, Telwatte and Moore studied sulfonated poly-L-lysine, SPL7115, and SPL7013 comprised of the divalent benzhydryl-amide core. However, SPL7013 showed the most effective against a broad spectrum of HIV isolates by blocking HIV-1 envelope-mediated cell to cell fusion (Telwatte et al., 2011). Therefore, SPL7013 [3% (w/w)] was formulated as a Vivagel by Starpharma Pvt. Ltd. and submitted to the United States Food and Drug Administration as a first dendrimers-based drug application which preventing the sexual transmission of HIV-1. This formulation showed potent inhibitors against X4 and R5 × 4 HIV-1 in lower epithelial and submucosa layer by disrupting viral particle or loss of gp120 from the viral envelope. Tyssen formulated 3% SPL7013 in carbopol vehicle that has mucoadhesive properties, which may extend its retention in the female genital tract. The clinical trial studies assessed retention and duration of antiviral activity after vaginal administration of Vivagel in Healthy women. They observed potent antiviral activity against HIV-1 immediately after vaginal administration of Vivagel, with activity maintained for at least 3 h postdose. This data provide evidence of antiviral activity in a clinical setting and suggest Vivagel could be administered up to 3 h before coitus (Tyssen et al., 2010).

15.3.4 Polymeric micelles

Polymeric micelles have a size less than 100 nm in diameter. It improves the aqueous solubility, intestinal permeability, targeted action, controlled release, and chemical stability of several drug molecules. The thermodynamic and in vivo stability of polymeric micelles are relatively high as compared to other nanocarriers. Polymeric micelles are composed by block copolymer by self-association at concentration 10^{-6} to 10^{-7} same as that of surfactant-based micelles. Polymeric micelles consist of an internal core (hydrophobic functional groups) and outer shell (hydrophilic functional groups) arrangement which facilitates it for nanocarriers drug delivery. Drug molecule can be encapsulated within the core (nonpolar molecule), conjugate to shell (polar molecule) or positioned in between the core and shell (intermediate polarity) (Xu et al., 2013).

Modification on the surface properties of micelles is necessary for target release (receptor-specific) by attaching hydrophilic blocks to specific antibody or ligands. Lectins receptors are present on HIV viral reservoirs such as T lymphocytes, dendritic cells and macrophages. Therefore in case of viral reservoir targeting drug delivery, it is necessary to target the lectins receptors. For that purpose micelles of polyethylene (PEG)-polylactide copolymer modified with galactose or lactose units are prepared with the entrapment of anti-HIV drug. The above micelles specifically interact with lectins and showed targeted anti-HIV action (Wang, 2018). Structure of polymeric micelle and targeting role in ART are shown in Fig. 15.6.

Li et al. synthesized Lamivudine stearate via easter linkage between Lamivudine and stearic acid. Due to esterification partition coefficient of Lamivudine increased from 0.95 to 1.82. Subsequently above synthesized Lamivudine–stearic acid complex encapsulated in chitosan oligosaccharide polymeric micelles. Lamivudine release from these micelles showed pH-dependent with low cytotoxicity with a higher percentage of cellular uptake (Li et al., no date).

Bomberg et al. recently used polymeric micelles for oral delivery. They studied different polymers for hydrophilic shell-forming blocks in polymeric micelles such as polyethene glycol (PEG), poly-N-vinyl-2-pyrrolidone (PVP), polyvinyl alcohol (PVA), polyethyleneimine, and polyethene oxide (PEO). For hydrophobic core-forming blocks, they used L-lysine, aspartic acid, caprolactone, lactic acid, propylene oxide, and others. From the above study, they concluded that block copolymers of poly (ethylene oxide)-poly (propylene oxide) (PEO-PPO-PEO), known as Pluronics have shown great potential in intestinal permeability for antiretroviral oral drug delivery (Bromberg, 2008). However, Pluronics P-85 modulates the ATP binding cassette transporters activity which reduces the oral bioavailability and CNS permeability of various PIs and NRTIs.

Saquinavir reported low bioavailability due to P-gp mediated efflux transport. Similarly, other reasons also play a key factor for low bioavailability such as poor water solubility, CYP3A4 metabolism and systemic clearance. Simultaneous administration of low dose P-gp and CYP3A4 inhibitor Ritonavir results enhanced bioavailability of Saquinavir. However, in multidrug resistance, inhibition of P-gp by Pluronics P85 with concentration 0.01–1% gives revelation to enhance the bioavailability of Saquinavir. At concentration 0.01% P 85 was present as a unimers, therefore it was concluded that the above-observed effect on P-gp was not due to micelles (Ding et al., 2002). Shaik et al. observed a similar effect of P 85 on Caco-2 monolayers overexpressing with P-gp. Below CMC, P85 unimers showed significant uptake and efflux inhibition of P-gp dependent probe rhodamine in Caco-2 cells as compared to other nonionic detergents, Cremophor EL and Tween 60 (Shaik et al., 2009).

Efavirenz (EFV) is a first-line ART agent in a pediatric therapeutic cocktail with low aqueous solubility and bioavailability. Therefore Chiappetta et al. formulated EFV in polymeric micelles to improve its aqueous solubility and oral bioavailability. In this formulation, EFV was incorporated into the core of linear and branched PEO-PPO-PEO block copolymer micelles. Morphologically it was characterized by dynamic light scattering and transmission electron microscopy

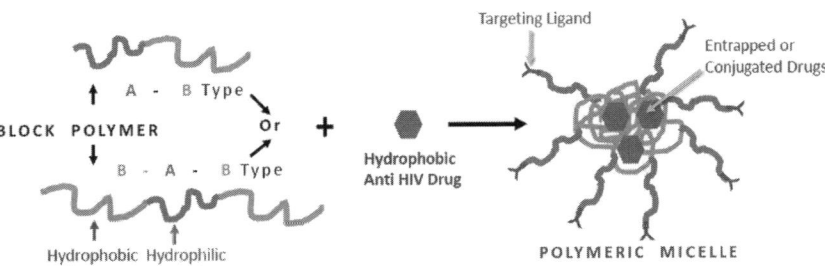

FIG. 15.6 Structure of polymeric micelle its targeting role in antiretroviral therapy.

and bioavailability was assessed by using male Wistar rats. From the result, they conclude that block copolymer significantly improves the oral bioavailability of the EFV which reduces inter-individual variability (Chiappetta, 2009).

15.3.5 Liposomes

Liposomes are vesicular carriers consisting of aqueous core surrounding with phospholipid bilayers. Hydrophilic drugs can be encapsulated in aqueous core whereas hydrophobic and amphiphilic drugs can be solubilized within the phospholipid bilayers. These spherical lipid vesicles improve the solubility and delivering capacity of various drugs by partitioning it into lipid bilayers and aqueous core. In addition to these, liposomes are easily taken up by the reticuloendothelial system and show fast clearance. These distinctive abilities with enhancing biocompatibility, low toxicity, and biodegradability make it very smart carrier in macrophage drug delivery of antiretrovirals. Surface and membrane characteristics of liposomes are mainly depended on the surface charge, steric interactions, and membrane rigidity. Liposomes interact with the cell membrane either by lipid exchange, adsorption, or fusion. In addition to these cells engulf the lipid vesicles by endocytosis process (Yang et al., 2016; Behzadi et al., 2017). There are various modified liposomes developed to improve drug delivery of antiretrovirals with targeted activities such as immunoliposomes, long-circulating liposomes, long-circulating immunoliposomes, pH-sensitive liposomes, surface modified liposomes. Table 15.4 showed some patents of research groups which

TABLE 15.4 Liposomes patents on the treatment or prevention of HIV infection (*Liposomes Patents and Patent Applications (Class 424/450) - Justia Patents Search, no date*).

US patent number	Type	Application
8173767	Liposomes with synthetic peptide	Vaccine peptides, which are immunogenic and elicit a protective immune response against HIV infection in vitro, were reported. The claim included pharmaceutical compositions which may comprise of at least one peptide or salt and a pharmaceutically acceptable vehicle. The composition could be encapsulated with a polymer, liposome or micelle.
8084593	Liposomes–peptide.	A method for treating or preventing a HIV-1 infection of a cell by administration of an isolated nucleic acid encoding a TRIM-cyclophilin polypeptide was reported. Possibility of Liposome-mediated transfer or viral-mediated transfer of the gene transfer vector was reported.
8084250	Liposomes in vaccine	To increase immunogenicity of a tumor antigen thus allowing treatment of cancer and to increase the immunogenicity of a viral antigen, allowing treatment of viral infection, including HIV infection. The composition comprised of a fusion polypeptide comprising an antigen, and an adjuvant. The delivery was through a liposome.
7906641	Liposomes in gene therapy	Liposomes were used to coat the RNA, DNA sequences as well as the expression vectors.
7868162	Cationic or anionic liposomes	Liposomal compositions of protonated compounds with antiviral and antimicrobial activity were reported.
7696179	Liposomes in gene therapy	Treatment of viral infection, e.g., HIV by, e.g., modulating gene expression or protein activity was reported. The formulation comprised of one or more siRNAs directed against CCR5. The carrier was reported to have one or more of the following component selected from a basic peptide, nonoxynol-9-containing spermicide, vaginal lubricant, liposomes, and combinations.
7312305	Liposomes–peptide	Liposomes and microspheres were reported as carriers for envelope protein (Gp 120) and Nef proteins of HIV-1, HIV-2, Simian Immunodeficiency Virus (SIV), Feline Immunodeficiency Virus (FIV), and peptides of these proteins (modulators) for epithelial carcinomas.
6875435	Liposome as carriers	Heat shock protein (hsp) or a portion of an ATP binding domain of an hsp associated with HIV was entrapped into or co-administered with liposomes to induce CD8+CTL response.
5773027	Liposomes encapsulating antiviral drug	This formulation was prepared for the treatment of viral diseases such as HIV and cytomegalovirus infection. by encapsulating the antiviral agents in liposomes.

explore the use of liposomes for ART. However, certain disadvantages limit their use in ART including limited hydrophilic drug-loading capacity, physical and biological instability, poor scale-up, high cost, and toxicity.

Phillips et al. compared the antiretroviral and bone marrow toxicity of Zidovudine (AZT) in the free form and AZT loaded liposomes in C57BL/6 mice. They Observed that AZT loaded liposomes specifically localized in liver, spleen and lung with improved antiretroviral activity as compared to the free drug. Jain et al. formulated AZT encapsulated elastic liposomes for transdermal administration. They observed that transdermal flux of the liposomal AZT approximately 20 times higher than free AZT through rat skin. Furthermore, PEGylated AZT encapsulated elastic liposomes were investigated for in vitro and in vivo lymphatic targetings. They also observed increased transdermal flux of AZT with less opsonization. Therefore PEGylated AZT encapsulated elastic liposomes have long plasma half-life, improved bioavailability and greater localization in lymphatic tissue as compared to non-PEGylated liposomes. Jain et al. also synthesized AZT-myristate, a prodrug of AZT and then it was loaded into the liposomes for targeted delivery (Phillips et al., 1991).

In recent years, researchers investigated surface-modified liposomes by mannose and galactose for their stability and toxicity. They prepared Stavudine loaded mannosylated and galactosylated liposomes and observed enhanced half-life with no hematological or hepatic toxicity (Mitra et al., 2005). In addition to the stability and toxicity of Stavudine researchers further, focus on its biodistribution. For that purpose, they labeled Stavudine loaded galactosylated liposomes with 99mTc and evaluated its biodistribution for an antiretroviral activity using HIV-1-infected MT2 cell line. The scintigraphy imaging and quantitative biodistribution results showed that 99mTc labeled liposomal formulation showed better and prolonged (24 h) uptake by liver and spleen. From the above result, it was concluded that 99mTc labeled liposomal formulation have reduced toxicity and targetability of Stavudine toward the mononucleated phagocytic system (Kawakami et al., 2000).

Malavia et al. formulated liposomes from a combination of naturally occurring and synthetic lipids with different physicochemical properties. They evaluated their ability to inhibit the infection of transformed cells that express virus-specific receptors by using in vivo female mouse model. They identified liposomes with naturally occurring lipid cardiolipin showed favorable antiretroviral activity with an improved therapeutic index. Therefore from the above results author concluded that cardiolipin-based liposomes with synthetic lipids have potential microbicide effect for the prevention of HIV infection in women (Malavia et al., 2011).

Liposomes have been mostly used for the delivery of hydrophobic anti-HIV drugs such as Zidovudine, Didanosine, and zalcitabine. Mostly four types of the liposomes were studies for the anti-HIV drug delivery such as cationic, anionic, sterically stabilized, and immunoliposomes. The above all liposomal formulations have efficient targeted delivery of the antiretrovirals with reduced toxicity and side effect (Nair et al., 2016).

15.3.6 Solid lipid nanoparticles

Solid lipid nanoparticles (SLNs) are at the front position in the rapidly developing field of nanotechnology for solubilization of lipophilic drug molecule, targeted drug delivery, clinical medicine, and research. Due to their unique size, which is dependent on lipids and surfactants offer to develop new therapeutics, long-term stability, drug loading, and release characteristics. SLNs are less toxic and easier scale-up than synthetic polymeric nanoparticles. However poor drug loading capacity, drug expulsion after polymeric transition during storage, unpredictable gelation tendency and inherent low incorporation rates resulting from the crystalline structure of the solid lipid are certain disadvantages that limit their use. Therefore to overcome these problems different types of nanocarriers are synthesized based on SLNs including nanostructured lipid carriers and lipid drug conjugates.

Cells of the reticuloendothelial system (macrophages) play an important role in the pathogenesis of HIV. So that ability of anti-HIV drugs to target these cells are essential for the success of ART. Dandagi et al. investigated the targeting efficiency of Stavudine to RES and brain by using SLNs. For the preparation of SLNs they use various lipids in combination like steric acid, cetyl palmitate, glycerol behenate, and phospholipid. They observed SLNs with cetyl palmitate lipid have been shown 11 fold greater uptake by brain as compared to pure stavudine. Form the above result they demonstrated that Stavudine loaded SLNs provided a mechanism for improving therapy for brain targeting with reducing the dose (Dandagi et al., 2012).

Heiati et al studied SLNs of AZT- palmitate with surface modification by PEG and evaluated its biodistribution. They observed that AZT-palmitate SLNs modified with PEG created a steric barrier which reduces the particle uptake and extending circulation time as compared to the free drug. The AZT loaded SLNs-PEG also increased the targeting efficiency of drug toward the reticuloendothelial system (Chattopadhyay et al., 2008). Bhalekar et al. prepared Darunavir SLNs to overcome solubility and bioavailability associated problems of the drug. Based on the drug solubility and stable dispersion findings, lipid and surfactant were chosen and SLNs were prepared using high-pressure homogenization (HPH) technique. They optimized the variables such as lipid concentration, oil, surfactant, homogenization cycle and evaluate their effect on particle size and entrapment efficiency. They concluded that HPH method can be successfully employed to prepare Darunavir SLN which is believed to have the potential to increase the systemic bioavailability by endocytic uptake and lymphatic transport (Bhalekar et al., 2017).

Makwana et al. developed lymph targeted SLNs formulation of Efavirenz (EFV) and to observe its mechanism of uptake by lymphatics which act as a viral reservoir of HIV. They prepared EFV SLNs using Gelucire 44/14, compritol 888 ATO, Lipoid S 75 and poloxamer 188 by hot homogenization technique followed by ultrasonication method. They observed from DSC and XRPD studies, change in the crystalline index of EFV due to entrapment of drug into SLN. They also observed prolong drug release with particle size 168 nm and well stable after storage at 30 ± 2°C/60 ± 5%RH for 2 months. The lymphatic transport and tissue distribution study indicate that a major percentage of release EFV bypass the liver and therefore enhance the oral bioavailability. From the above observation, it was concluded that EFV SLNs have the potential to target the EFV to the lymphatic system in HIV therapy (Makwana et al., 2015).

15.4 Nanaotechnological approaches for antiretroviral therapy

CD4+ T cells infection is the major effect of HIV infection. These helper T cells are CD4+ T cells are important for immune response in various infections and their loss leads to immune deficiency and developing an opportunistic infection. HIV infection manifested by these helper T cells infection with chronically defunctionalization of B-cells, natural killer cells, and macrophages. Therefore in recent years a lot of interest moves toward the therapeutic use of immune response with nanotechnology to repair the regular function of the immune system to treat HIV infection. Therefore immunotherapies, gene therapy, a vaccine with nanocarriers are the treatment approaches involving the use of immunomodulatory agents or gene vectors to modulate the immune response against HIV infection (Fig. 15.7) (Date and Destache, 2013; Iannazzo, 2015).

15.4.1 Immunotherapy for antiretroviral

Immunotherapy is a treatment approach with the use of immunomodulatory agents to revise the immune response against AIDS. Similar to vaccines, immunotherapy is also based on immunization of patients with nanocarriers-based immunologic formulation however, immunotherapy is only used to treat HIV-infected patients not to protect healthy individuals. In this approach delivering cytokines (IL-2, IL-7, and IL-15) or antigens are used besides a drug that induces target-specific long-lasting HIV-specific T-cell response in the treatment of HIV (Brown and Angel, 2005). Since 1983 immunotherapeutic clinical trial for HIV/AIDS has been carried out and are still in progress. Hence new approaches for targeting DC-SIGN receptors with various nanocarriers for the delivery of immunomodulatory factors, proteins or DNAs were developed to provide immense opportunities for potential immunotherapy. Researchers observed PEG stabilized poly(propylene sulfide) polymeric nanoparticles having an affinity toward DC-SIGN receptors and accumulated in lymph nodes (Bahr et al., 2003).

Recently cross-linked polymeric nanoparticles such as 2-diethylamino ethyl methacrylate, PEG dimethacrylate, and 2-aminoethyl methacrylate are used in immunotherapy. These cross-linked polymeric nanoparticles with pH-responsive core and hydrophilic charged shell were mostly used for the delivery of antigens and proteins toward DC-SIGN receptors. The most recently clinically potential application of nanocarriers in immunotherapy of AIDS is the development of Derma-Vir patch. Derma-Vir patch consists of polymeric nanoparticles with HIV-1 antigen coding plasmid DNA which support the delivery to Langerhans cells. Epidermal Langerhans cells entrap these nanoparticles and transport to lymph nodes. While in transit time, Langerhans cells mature to dendritic cells, and present HIV-1 antigen coding plasmid DNA induce cellular

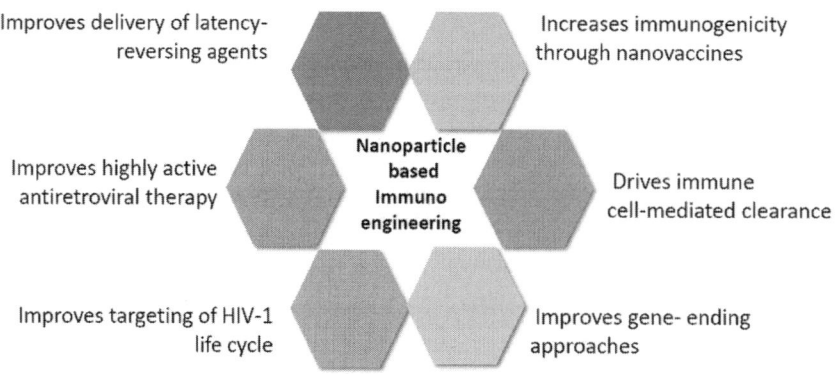

FIG. 15.7 Nanoparticles-based immunoengineering approach.

immunity. DermaVir boost T cell responses which are specific to all 15 HIV antigens of single DNA. For long-lasting immune response repeated DermaVir immunization is required because the DermaVir boosted HIV-specific memory T cells decreased during the 48-week (Lisziewicz and Lorincz, 2012).

15.4.2 Gene therapy

Gene therapy proves immense hope for the treatment and eradication of HIV, in which a gene is inserted into a cell to interfere with viral infection and replication. However, poor cellular uptake and instability of DNA in the physiological environment restricts its therapeutic potential (*Immunotherapy in Patients With HIV Infection and Advanced Cancer—The ASCO Post*, no date). Nucleic acid-based compounds including DNA, siRNA, RNA decoys, ribozymes, protein-based agents, zinc finger nuclease, and aptamers were used to interfere in viral replication for the treatment of acquired as well as a genetic disease. In starting of gene therapy scientist focused on viral vectors as the delivery agents. However, due to the lack of specificity, a low transfection rate, toxicity, immunogenicity, insertion, mutagenesis, and susceptibility to nuclease degradation with limitation in scale-up limits the use of viruses in gene therapy. Therefore the nanotechnology-based nonviral vectors have been recommended as potential candidates (Bowen et al., 2020).

Due to stability, reduced toxicity, reduced risk of immunological rejection and controlled release property liposomes and nanoparticles are act as great candidates for nonviral gene delivery. However cationic liposomes are more commonly used as compared to anionic one. As the positive charge which is present on cationic liposomes binds easily to negatively charged nucleic acids. Properties of dendrimers, such as monodispersity and well-defined structure with surface functional groups also made it potential vectors for gene delivery. The sixth-generation PAMAM dendrimers are mostly used in gene delivery due to their high solubility and reactivity with empty internal cavities and numerous functional groups. A transfection reagent "Superfect TM" is a trademark name for activated dendrimers which carry a high quantity of genetic material than viruses. The transfection efficiency of PAMAM may not only due to the well-defined structure but depend on low pKa of amines (3.9 and 6.9). The low pKa value of the PAMAM dendrimer produces it to buffer for pH change in the endosomal compartment (Kharwade et al., 2020).

These nanocarriers avoid the toxicities and immunological rejections associated with allogeneic transplantation; also maximize the ability to destroy the latently infected cells. Since HIV mutates with multiple strains and different genetic sequences, nanotechnological-based siRNA enhance capability for co-delivery and targeting specificity for HIV infected cell. From that point of view nonviral delivery of siRNA is gaining interest in the treatment of HIV infection. In this approach, a fusion protein, with a peptide transduction domain and double-stranded RNA domain was used for the in vivo delivery of siRNA to T cells with no adverse effect. Similarly, protamine- antibody fusion protein and single-walled nanotubes were also used to deliver siRNA, CXCR4, and CD4-specific siRNA. These siRNAs downregulated the *gag* genes which inhibit HIV replication in T cells. Recently the siRNA carbosilane dendrites was used to target CNS HIV-1 infection and neuro-AIDS without cytotoxicity. Navid et al. investigated the ability of second-generation carbosilane dendrimer to transfect CD4 T lymphocytes with siRNA against Nef HIV-1 gene. He observed as compared to other nonviral vectors carbosilane dendriplex with siRNA potentially transfect CD4 T lymphocytes and inhibit HIV-1 replication (Akhtar and Benter, 2007). These above revolutionary studies demonstrate that nanotechnological-based nonviral siRNA delivery showed more effective for HIV treatment. In a recent development, liposome-based siRNA delivery also optimized the targeting efficiency of HIV infected cells, such as T cell and macrophages. Zou et al. have revealed the great potential of PAMAM dendrimer to deliver siRNAs and suppress HIV infection. These siRNAs target viral as well as cellular transcripts and resulted in prolong HIV Suppression (Zhu et al., 2019).

15.4.3 Vaccines

The continuous search and clinical trials for effective and safe HIV vaccine have been challenging over the three decades since the infection was discovered. Therefore recently researchers encouraged great debate on vaccine research with a major focus on fundamental research as compared to the clinical trial. At present, there are two major hurdles in the field of HIV vaccine development such as design and production of immunogens that can authentically regenerate with preservation of epitope. However successful immunogen development has been challenging due to various dodging pathways of HIV replication with lack of methods which is necessary to neutralize antibodies and cytotoxic T cell (Peek et al., 2008; Victor, 2019).

Initially, scientist focused on viral vector vaccines which consist of a nonreplicating virus that encloses distinct immunogens from the pathogen such as Env-based epitope scaffolds, HIV-1 glycans, BG 505, SOSIP, and polymeric HIV-1 gp120 to which immunity is desired. Such vaccines are commonly known as a live recombinant vaccine which can induce robust T-cell and B-cell immune responses. However, in general population, inherent antivector immunity observed which one is a major obstacle to overcome (Anis, 2019).

The most promising HIV vaccine RV 144 was the combination of two different vaccine types that are prime and booster. The prime recombinant is a canarypox vector vaccine consist of viral vector and booster component (AIDSVAX) is a recombinant glycoprotein 120 subunit vaccine. This prime and boost combination was tested in a clinical trial in Thailand that failed on their own. The initial report shows that the rate of HIV infection among volunteers who received the experimental vaccine was 31% lower than the rate of HIV infection who received the placebo (Gómez et al., 2012). As a result, scientists focus on the development of an effective vaccine that can deliver authenticated immunogens to antigen-presenting cells (lymphoid cell) and produce robust adaptive immune responses.

Therefore in contrast to viral vectors, synthetic polymeric nanoparticles and liposomal nanovesicles offer a potential platform to induce adaptive immune responses while avoiding antivector immunity. Depending on cellular response or humoral response was required, antigens are either encapsulating in their core or adsorb on the surface of nanoparticles (Adhikary, 2015).

Makadia et al. extensively examine the biodegradable and biocompatible poly (lactic-co-glycolic) acid (PLGA) copolymer-based nanoparticles for the controlled release vaccine delivery in different disease. Controlled release nanoparticles showed great advantages for vaccine delivery because of the release of antigen in a controlled manner which gives prolonged and stronger initiation of the immune response (Makadia and Siegel, 2011). An additional major advantage of nanoparticles vaccines is that they can be administered via various routes and easily induce mucosal immunity. PLGA nanoparticles encapsulated with HIV-1 peptide antigens administered via intranasal route elicited Th1/Th2-balanced cellular responses in mucosal surface. However oral administration of PLGA nanoparticles with HIV Env peptides showed a potential response in T-cell mediated protection against viral infection at the site of rectal as well as vaginal mucosa. Surface functionalized nanoparticles provide a major opportunity for the targeted delivery of antigen to dendritic cells. Additionally, surface functionalized nanoparticles provide the ability to put antigens on the surface of the particles for the targeted antibody immune response to B cells (Elmowafy et al., 2019).

Recently MF59 is the only nanoparticles vaccine adjuvant approved for human use although it is not licensed in the United State. MF59 is oil in water emulsion composed by squalene (4.3% v/v) with two surfactants, polysorbate 80 (0.5% v/v) and sorbitan trioleate (0.5% v/v) showed droplet size less than 250 nm. In immunogenicity study of MF59, Ott et al. reported that guinea pigs showed a 34 times increase in antibody titers due to direct stimulation of cytokine production while goat and baboon showed nine and fivefold increase, respectively (Kommareddy et al., 2016).

15.5 Nanotechnology for improving latency reservoir

We have already discussed that HIV proviral DNA is primarily responsible to maintain latency and persistence through clonal expansion. Latently infected viral reservoirs of HIV are the major obstacle in HIV cures which can be found in memory CD4+ T lymphocytes such as central memory (TCM), transitional memory (TTM) and effector memory T cells (TEM) and macrophages. Anatomical reservoirs are also found additional barriers in curative strategies which include lymph nodes, spleen, gut-associated lymphoid tissue and central nervous system. Out of these, a subset of TCM also called helper cells and CD4+ T memory stem cells are highly susceptible to HIV infection and HIV latency because these cells are long-lived and differentiate into mature memory T cells (Malefane, no date; Margolis, 2010).

Several strategies are used to reduce the viral reservoir load by targeting latently infected viral reservoirs cells. This point of view, scientist investigates a new approach called "shock and kill" which reactivates HIV-1 proviruses with maintaining ART to prevent the spread of infection and new cellular reservoir (Cao and Woodrow, 2019). However "shock and kill" is still a controversial strategy due to safety issue and indifferent outcomes from clinical studies. Another strategy is known as "block and lock" was predicted to inhibit the transcription rate and prevent viral latency but cannot produce a functional cure from HIV infection. Prodrug Mycophenolate mofetil and its active metabolite mycophenolic acid have been proposed to inhibit in vitro HIV replication by reducing the substrate (guanosine nucleotides) reverse transcriptase. Further investigations for evaluating its effect on reducing reservoir size are currently under controlled clinical trial (Vansant et al., 2020).

As a result, new approaches are required to tackle these complicated molecular mechanism associated with the latent reservoir in cells and tissue compartments. Nanotechnology has emerged as a promising approach which addressed for HIV cure due to several key points which we have already discussed. Various types of nanocarriers such as liposomes, polymeric nanoparticles, solid lipid nanoparticles, micelles, dendrimers are especially important and beneficial to target viral reservoir (Gunaseelan et al., 2010). Targeted nanocarriers have been investigated for cell-specific accumulation which can be achieved by the use of targeting moieties. It has been suggested that surface-modified negatively charged nanoparticles with anti–CD4 monoclonal antibodies, or its fragments are a good candidate to target and deliver the specific agent to CD4+ T cells. Particularly nanoparticle-based immunoengineered approaches comprise special attention because they offer improved delivery and functionality of classical HIV drugs and latency-reversing agents (LRA) (Bowen et al., 2020).

Euler et al. observed that lipid nanoparticles modified with anti-CD4 monoclonal antibody targeting LRA and antiretroviral to primary CD4+ T cells by activating latent virus and inhibit viral load (Euler and Alter, 2015). Several studies also focused on developing nanocarriers that preferentially distribute antiretroviral to lymph node and the brain to address the insufficient drug concentration with limited accessibility of many drugs. Most investigators focus on nanocarriers for crossing the encapsulated drug through blood–brain barriers by increasing permeability, uptake or transcytosis through microvascular endothelial cells. In addition to those, ligands such as transferring and trans-acting transcriptional activator (TAT) peptide are also used for targeting nanocarriers to the brain (Bagashev and Sawaya, 2013).

15.6 Conclusion

The impact of HIV infection has been demoralizing with leads to immune deficiency developing opportunistic infection every day. The advent of HAART, which involves combination of at least three ARV drugs, has made a tremendous contribution to extend the life span of the HIV-infected patients. However, HAART were unable to eradicate the latent HIV-1 cellular reservoirs, even if administered for decades. For the improvement of this situation several new drug moiety or improved targeted action drug delivery system have been proposed with respect to use of nanotechnology. The studies discussed in this chapter provides the proof that nanotechnology-based approaches offer unique opportunities for the improvement of water solubility, stability, bioavailability and targeting of antiretroviral drugs. Nanocarrier-based immunotherapy, gene therapy, vaccines almost solve the problem associated with biodistribution in HIV reservoirs and emergence of drug resistance. However, the research in this field is still in the early stage due to cost of drug delivery system is high as compared to conventional system and unavailability of references on the efficacy of these nanocarriers in higher animal models and humans. In this regard, development of cost effective nanotechnology-based dosage form with clinical trial is ever growing need to improve the quality of life of HIV infected patients.

References

Adhikary, R.R., More, P., Banerjee, R., 2015. Smart nanoparticles as targeting platforms for HIV infections. Nanoscale 3, 1–14. http://doi.org/10.1039/b000000x.

Akhtar, S., Benter, I.F., 2007. Nonviral delivery of synthetic siRNAs in vivo. J. Clin. Invest. 117 (12), 3623–3632. http://doi.org/10.1172/JCI33494.

Alkhatib, G., et al., 1997. HIV-1 coreceptor activity of CCR5 and its inhibition by chemokines: independence from G protein signaling and importance of coreceptor downmodulation, 348 (234), pp. 340–348.

Anis, H.A., 2019. Gene therapy in the era of nanotechnology/a review of current data. J. Cancer Prevent. Curr. Res. 10 (1), 1–2. http://doi.org/10.15406/jcpcr.2019.10.00380.

ANNEX 3. NACO, 2013. Department of AIDS control, National AIDS control organisation, Ministry of Health and family welfare, Govt of India. WEIGHT-BASED DOSING FOR ARV, pp. 61–75.

Archin, N.M., et al., 2017. Interval dosing with the HDAC inhibitor vorinostat effectively reverses HIV latency Find the latest version: interval dosing with the HDAC inhibitor vorinostat effectively reverses HIV latency. J Clin Invest 127 (8), 3126–3135.

Auría-Soro, C., et al., 2019. Interactions of nanoparticles and biosystems: microenvironment of nanoparticles and biomolecules in nanomedicine. Nanomaterials 9 (10), 1365–1385. http://doi.org/10.3390/nano9101365.

Bagashev, A., Sawaya, B.E., 2013. Roles and functions of HIV-1 Tat protein in the CNS: an overview. Virol. J. 10. http://doi.org/10.1186/1743-422X-10-358.

Bahr, G.M., et al., 2003. Clinical and immunological effects of a 6 week immunotherapy cycle with murabutide in HIV-1 patients with unsuccessful long-term antiretroviral treatment. J. Antimicrob. Chemother. 51 (6), 1377–1388. http://doi.org/10.1093/jac/dkg244.

Barton, K., Winckelmann, A., Palmer, S., 2016. HIV-1 reservoirs during suppressive therapy. Trends Microbiol. xx, 1–11. http://doi.org/10.1016/j.tim.2016.01.006.

Basu, S., et al., 2012. Colloidal gold-loaded, biodegradable, polymer-based stavudine nanoparticle uptake by macrophages: an in vitro study. Int. J. Nanomed. 7, 6049–6061. http://doi.org/10.2147/IJN.S38013.

Behzadi, S., et al., 2017. Cellular uptake of nanoparticles: journey inside the cell. Chem. Soc. Rev. 46 (14), 4218–4244. http://doi.org/10.1039/c6cs00636a.

Bhalekar, M., Upadhaya, P., Madgulkar, A., 2017. Formulation and characterization of solid lipid nanoparticles for an anti-retroviral drug darunavir. Appl. Nanosci. (Switzerland) 7 (1–2), 47–57. http://doi.org/10.1007/s13204-017-0547-1.

Bowen, A., Sweeney, E.E., Fernandes, R., 2020. Nanoparticle-based immunoengineered approaches for combating HIV. Front. Immunol. 11 (April), 1–9. http://doi.org/10.3389/fimmu.2020.00789.

Bromberg, L., 2008. Polymeric micelles in oral chemotherapy. J. Cont. Rel. 128, 99–112. http://doi.org/10.1016/j.jconrel.2008.01.018.

Brown, P.A., Angel, J.B., 2005. Mycobacterial immune reconstitution inflammatory syndrome in HIV-1 infection after antiretroviral therapy is associated with deregulated specific T-cell responses: beneficial effect of IL-2 and GM-CSF immunotherapy. J. Immune Based Therapies Vacc. 7, 1–7. http://doi.org/10.1186/1476-Received.

Cao, S., Woodrow, K.A., 2019. Nanotechnology approaches to eradicating HIV reservoirs. Eur. J. Pharm. Biopharm. 138 (June), 48–63. http://doi.org/10.1016/j.ejpb.2018.06.002.

Chattopadhyay, N., et al., 2008. Solid lipid nanoparticles enhance the delivery of the HIV protease inhibitor, atazanavir, by a human brain endothelial cell line. Pharm. Res. 25 (10), 2262–2271. http://doi.org/10.1007/s11095-008-9615-2.

Chiappetta, D.A., et al., 2009. Indinavir-loaded pH-sensitive microparticles for taste masking: toward extemporaneous pediatric anti-HIV/AIDS liquid formulations with improved patient compliance. AAPS PharmSciTech. 10 (1), 1–6. http://doi.org/10.1208/s12249-008-9168-z.

Chiappetta, D.A., et al., 2009. Efavirenz-loaded polymeric micelles for pediatric anti-HIV pharmacotherapy with significantly higher oral bioavailability. Nanomed 5 (1), 11–23. https://doi.org/10.2217/nnm.09.90.

Dandagi, P.M., et al., 2012. RES and brain targeting stavudine-loaded solid lipid nanoparticles for AIDS therapy. Asian J. Pharm. 6 (2), 116–123. http://doi.org/10.4103/0973-8398.102934.

Date, A.A., Destache, C.J., 2013. A review of nanotechnological approaches for the prophylaxis of HIV/AIDS. Biomaterials 34 (26), 6202–6228. http://doi.org/10.1016/j.biomaterials.2013.05.012.

Desai, M., Iyer, G., Dikshit, R.K., May, 2012. Antiretroviral drugs: critical issues and recent advances. Indian J. Pharmacol. 44 (3), 288–98. doi: 10.4103/0253-7613.96296.

Destache, C.J., et al., 2010. Antiretroviral release from poly(DL-lactide-co-glycolide) nanoparticles in mice. J. Antimicrob. Chemother. 65 (10), 2183–2187. http://doi.org/10.1093/jac/dkq318.

Ding, R., et al., 2002. Dose-dependent increase of saquinavir bioavailability by the pharmaceutic aid cremophor EL. Br. J. Clin. Pharmacol. 53 (6), 576–581.

Drugs That FIGHT HIV-1 A reference guide for prescription HIV-1 medications (no date) U.S department of Health and Human Services.

Dutta, T., Garg, M., Jain, N.K., 2008. Targeting of efavirenz loaded tuftsin conjugated poly(propyleneimine) dendrimers to HIV infected macrophages in vitro. Eur. J. Pharm. Sci. 34 (2–3), 181–189. http://doi.org/10.1016/j.ejps.2008.04.002.

Edagwa, B., et al., 2017. Long-acting slow effective release antiretroviral therapy. Expert Opin. Drug Deliv. 14 (11), 1281–1291. http://doi.org/10.1080/17425247.2017.1288212.

Elmowafy, E.M., Tiboni, M., Soliman, M.E., 2019. Biocompatibility, biodegradation and biomedical applications of poly(lactic acid)/poly(lactic-co-glycolic acid) micro and nanoparticles. J. Pharm. Invest. 49, 347–380. http://doi.org/10.1007/s40005-019-00439-x.

Euler, Z., Alter, G., 2015. Exploring the potential of monoclonal antibody therapeutics for HIV-1 eradication. AIDS Res. Hum. Retroviruses 31 (1), 13–24. http://doi.org/10.1089/aid.2014.0235.

Fauci, R.A.N.D.A.S., 1993. The immunopathogenesis of HIV infection. J. AIDS 41, 377–431.

FDA-Approved HIV Medicines| *UnderstandingHIV/AIDS | AIDSinfo* (no date). Available at: https://aidsinfo.nih.gov/understanding-hiv-aids/fact-sheets/21/58/fda-approved-hiv-medicines (Accessed July 27, 2020).

Gardner, M.B., Carlos, M.P., Luciw, P.A., 2004. Simian Retroviruses. Path. Lab. Med. 195–262.

German Advisory Committee Blood, 2016. Assessment of Pathogens Transmissible by Blood. Human Immunodeficiency Virus (HIV), pp. 203–222. doi: http://doi.org/10.1159/000445852.

Gill, J., Lewden, C., Saag, M., P., R., 2010. Causes of Death in HIV-1 – Infected Patients Treated with Antiretroviral Therapy 1996–2006: Collaborative Analysis of 13 HIV Cohort Studies. Clin. Infect. Dis. 50 (10), 1387–1396. doi: http://doi.org/10.1086/652283.

Gómez, C.E., et al., 2012. Poxvirus vectors as HIV/AIDS vaccines in humans. Hum. Vacc. Immunother. 8 (9), 1192–1207. http://doi.org/10.4161/hv.20778.

Gunaseelan, S., et al., 2010. Surface modifications of nanocarriers for effective intracellular delivery of anti-HIV drugs. Adv. Drug. Deliv. Rev. 62 (4–5), 518–531. http://doi.org/10.1016/j.addr.2009.11.021.

Gupta, U., Jain, N.K., 2010. Non-polymeric nano-carriers in HIV/AIDS drug delivery and targeting. Adv. Drug. Deliv. Rev. 62 (4–5), 478–490. http://doi.org/10.1016/j.addr.2009.11.018.

Hsu, H.J., et al., 2017. Dendrimer-based nanocarriers: a versatile platform for drug delivery. Wiley Interdisc. Rev. Nanomed. Nanobiotechnol. 9 (1), 1–21. http://doi.org/10.1002/wnan.1409.

Huck, W.T.S., et al., 1998. Convergent and divergent noncovalent synthesis of metallodendrimers. J. Am. Chem. Soc. 120 (25), 6240–6246. http://doi.org/10.1021/ja974031e.

Iannazzo, D., et al., 2015. Nanotechnology approaches for antiretroviral drugs delivery. J. AIDS HIV Inf. 1 (2), 1–13. http://doi.org/10.15744/2454-499X.1.201.

Immunotherapy in Patients With HIV Infection and Advanced Cancer - The ASCO Post (no date). Available at: https://www.ascopost.com/News/59734 (Accessed July 29, 2020).

Jevprasesphant, R., et al., 2003. Engineering of dendrimer surfaces to enhance transepithelial transport and reduce cytotoxicity. Pharm. Res. 20 (10), 1543–1550. http://doi.org/10.1023/A:1026166729873.

Jiang, Y., Liu, X., De Clercq, E., 2011. New Therapeutic Approaches Targeted at the Late Stages of the HIV-1 Replication Cycle. Curr. Med. Chem. 18 (1), 16–28. doi: 10.2174/092986711793979751.

Jimenez, L. et al., 2012. Dendrimers as topical microbicides with activity against HIV, pp. 299–309. doi: http://doi.org/10.1039/c1nj20396g.

Kafaie, J., Song, R., Abrahamyan, L., 2008. Mapping of nucleocapsid residues important for HIV-1 genomic RNA dimerization and packaging. Virology 375, 592–610. http://doi.org/10.1016/j.virol.2008.02.001.

Kawakami, S., et al., 2000. Mannose receptor-mediated gene transfer into macrophages using novel mannosylated cationic liposomes. Gene Ther. 7 (4), 292–299. http://doi.org/10.1038/sj.gt.3301089.

Kesharwani, P., Jain, K., Jain, N.K., 2014. Dendrimer as nanocarrier for drug delivery. Prog. Polym. Sci. 39 (2), 268–307. http://doi.org/10.1016/j.progpolymsci.2013.07.005.

Kharwade, R., More, S., Mahajan, N., et al., 2020. Functionalised dendrimers: potential tool for antiretroviral therapy. Curr. Nanosci. 16 (5), 708–722. http://doi.org/10.2174/1573413716666200213114836.

Kharwade, R., More, S., Warokar, A., et al., 2020. Starburst pamam dendrimers: synthetic approaches, surface modifications, and biomedical applications. Arab. J. Chem. 13 (7), 6009–6039. http://doi.org/10.1016/j.arabjc.2020.05.002.

Kim, Y., Anderson, J.L., Lewin, S.R., 2017. Review getting the "Kill" into "Shock and Kill": strategies to eliminate latent HIV. Cell Host Microbe 23 (1), 14–26. http://doi.org/10.1016/j.chom.2017.12.004.

Kirk, G.D., Jacobson, L.P., 2014. Level of adherence and HIV RNA suppression in the current era of highly active antiretroviral therapy (HAART). AIDS Behav. 19 (4), 601–611. http://doi.org/10.1007/s10461-014-0927-4.

Kommareddy, S., Singh, M., O'Hagan, D.T., 2016. MF59: a safe and potent adjuvant for human useImmunopotentiators in Modern Vaccines, second ed. Elsevier, pp. 249–263. http://doi.org/10.1016/B978-0-12-804019-5.00013-X.

Kulkosky, J., Bray, S., 2006. HAART-persistent HIV-1 latent reservoirs: their origin, mechanisms of stability and potential strategies for eradication, (215), pp. 199–208.

Kumar, L., et al., 2015. Nanotechnology: a magic bullet for HIV AIDS treatment. Artif. Cells Nanomed. Biotechnol. 43 (2), 71–86. http://doi.org/10.3109/21691401.2014.883400.

Lembo, D., et al., 2018. Expert opinion on drug delivery nanomedicine formulations for the delivery of antiviral drugs: a promising solution for the treatment of viral infections. Expert Opin. Drug Deliv. 15 (1), 93–114. http://doi.org/10.1080/17425247.2017.1360863.

Li, J., et al., 2015. Polymeric drugs: advances in the development of pharmacologically active polymers. J. Control. Release 219, 369–382. http://doi.org/10.1016/j.jconrel.2015.09.043.

Li, Q. et al. (no date) Synthesis of Lamivudine stearate and antiviral activity of stearic acid-g-chitosan oligosaccharide polymeric micelles delivery system'. doi: http://doi.org/10.1016/j.ejps.2010.08.004.

Liposomes Patents and Patent Applications (Class 424/450) - Justia Patents Search (no date). Available at: https://patents.justia.com/patents-by-us-classification/424/450 (Accessed July 29, 2020).

Lisziewicz, J., Lorincz, O., 2012. HIV-specific immunotherapy with DermaVir, the first pDNA/PEIm pathogen-like nanomedicine. Eur. J. Nanomed. 4 (2–4), 81–87. http://doi.org/10.1515/ejnm-2012-0011.

Lobritz, M.A., Ratcliff, A.N., Arts, E.J., 2010. HIV-1 Entry, Inhibitors, and Resistance. HIV-1 Entry, Inhibitors. and Resistanc. viruses. 2 (5), 1069–1105. http://doi.org/10.3390/v2051069.

Madaan, K., et al., 2014. Dendrimers in drug delivery and targeting: drug-dendrimer interactions and toxicity issues. J. Pharm. Bioallied Sci. 6 (3), 139–150. http://doi.org/10.4103/0975-7406.130965.

Magalhães, J., et al., 2020. Optimization of rifapentine-loaded lipid nanoparticles using a quality-by-design strategy. Pharmaceutics 12 (1), 1–13. http://doi.org/10.3390/pharmaceutics12010075.

Makadia, H.K., Siegel, S.J., 2011. Poly lactic-co-glycolic acid (PLGA) as biodegradable controlled drug delivery carrier. Polymers 3 (3), 1377–1397. http://doi.org/10.3390/polym3031377.

Makwana, V., et al., 2015. Solid lipid nanoparticles (SLN) of Efavirenz as lymph targeting drug delivery system: elucidation of mechanism of uptake using chylomicron flow blocking approach. Int. J. Pharm. 495 (1), 439–446. http://doi.org/10.1016/j.ijpharm.2015.09.014.

Malavia, N.K., et al., 2011. Liposomes for HIV prophylaxis. Biomaterials 32 (33), 8663–8668. http://doi.org/10.1016/j.biomaterials.2011.07.068.

Malefane, M.E. (no date) Applications of Nanotechnology Towards Detection and Treatment of HIV /AIDs : A Review Article, pp. 1315–1321. doi: http://doi.org/10.31031/RDMS.2020.12.000796.

Mamo, T., et al., 2006. Emerging nanotechnology approaches for HIV/AIDS treatment and prevention review. Nanomedicine 5 (2), 269–285.

Margolis, D.M., 2010. Mechanisms of HIV Latency: An Emerging Picture of Complexity. Curr. HIV/AIDS Rep. 37–43. http://doi.org/10.1007/s11904-009-0033-9.

Margolis, L., Shattock, R., 2006. Selective transmission of CCR5-utilizing HIV-1: the "gatekeeper" problem resolved? Current HIV/AIDS Reports . 4(April), pp. 312–317.

Marin, E., Briceño, M.I., Caballero-George, C., 2013. Critical evaluation of biodegradable polymers used in nanodrugs. Int. J. Nanomed. 8, 3071–3091. http://doi.org/10.2147/IJN.S47186.

Mc Crudden, M.T.C., et al., 2018. Design, formulation and evaluation of novel dissolving microarray patches containing a long-acting rilpivirine nanosuspension. J. Control. Release 292, 119–129. http://doi.org/10.1016/j.jconrel.2018.11.002.

Mcgettigan, J.P., et al., 2003. Functional human immunodeficiency virus type 1 (HIV-1) Gag-Pol or HIV-1 Gag-Pol and Env expressed from a single rhabdovirus-based vaccine vector genome. J. Virol 77 (20), 10889–10899. http://doi.org/10.1128/JVI.77.20.10889.

Merisko-Liversidge, E., Liversidge, G.G., Cooper, E.R., 2003. Nanosizing: a formulation approach for poorly-water-soluble compounds. Eur. J. Pharm. Sci. 18 (2), 113–120. http://doi.org/10.1016/S0928-0987(02)00251-8.

Mitra, M., et al., 2005. Targeting of mannosylated liposome incorporated benzyl derivative of *Penicillium nigricans* derived compound MT81 to reticuloendothelial systems for the treatment of visceral leishmaniasis. J. Drug Target. 13 (5), 285–293. http://doi.org/10.1080/10611860500233306.

Nair, M., et al., 2016. Getting into the brain: potential of nanotechnology in the management of neuroAIDS. Adv. Drug. Deliv. Rev. 103, 202–217. http://doi.org/10.1016/j.addr.2016.02.008.

Oguntibeju, O.O., 2012. Quality of life of people living with HIV and AIDS and antiretroviral therapy. HIV/AIDS Res. Palliat. Care 4, 117–124.

Ogunwuyi, O., et al., 2016. Antiretroviral drugs-loaded nanoparticles fabricated by dispersion polymerization with potential for HIV/AIDS treatment. Infect. Dis.: Res. Treat. 9. IDRT.S38108. http://doi.org/10.4137/idrt.s38108.

Palmer, S., Josefsson, L., Coffin, J.M., 2011. HIV reservoirs and the possibility of a cure for HIV infection, pp. 7–9. doi: http://doi.org/10.1111/j.1365-2796.2011.02457.x.

Panova, V., et al., 2020. Antibody-induced internalisation of retroviral envelope glycoproteins is a signal initiation event. PLOS Pathog 120018, 1–25. http://doi.org/10.1371/journal.ppat.1008605.

Parboosing, R., et al., 2012. Nanotechnology and the treatment of HIV infection. Viruses 4 (4), 488–520. http://doi.org/10.3390/v4040488.

Parrish, N.F., et al., 2012. Transmitted/founder and chronic subtype C HIV-1 use CD4 and CCR5 receptors with equal efficiency and are not inhibited by blocking the integrin a 4 b 7. PLOS Pathog 8 (5). http://doi.org/10.1371/journal.ppat.1002686.

Patel, H.M., et al., 2016. Nanosuspension technologies for delivery of poorly soluble drugs - a review. Res. J. Pharm. Technol. 9 (5), 625–632. http://doi.org/10.5958/0974-360X.2016.00120.7.

Paul, W.E. et al., 1996. Structure of the amino-terminal core domain of the HIV-1 capsid protein, 2.

Peek, L.J., Middaugh, C.R., Berkland, C., 2008. Nanotechnology in vaccine delivery. Adv. Drug. Deliv. Rev. 60 (8), 915–928. http://doi.org/10.1016/j.addr.2007.05.017.

Peng, J., et al., 2013. Dendrimers as potential therapeutic tools in HIV inhibition. Molecules 18 (7), 7912–7929. http://doi.org/10.3390/molecules18077912.

Phillips, N.C., Skamene, E., Tsoukas, C., 1991. Liposomal encapsulation of 3′-Azido-3′-deoxythymidine (AZT) results in decreased bone marrow toxicity and enhanced activity against murine AIDS-induced immunosuppression. J. Acquir. Immune Defic. Syndr. 4 (10), 959–966.

Piketty, C. et al., 1999. Efficacy of a five-drug combination including ritonavir, saquinavir and efavirenz in patients who failed on a conventional triple-drug regimen : phenotypic resistance to protease inhibitors predicts outcome of therapy, (March), pp. 71–77.

Qian, K., Morris-natschke, S.L., Lee, K., 2008. HIV entry inhibitors and their potential in HIV therapy. PLOS Pathog 29 (2), 369–393. http://doi.org/10.1002/med.

Saura, C., 1998. HTLV Testing in Blood Transfusion. Vox Sang. 74 (2), 165–169. doi: 10.1111/j.1423-0410.1998.tb05416.x.

Shaik, N., Giri, N., Elmquist, W.F., 2009. Investigation of the micellar effect of pluronic P85 on P-glycoprotein inhibition: cell accumulation and equilibrium dialysis studies. J. Pharm. Sci. 98 (11), 4170–4190. http://doi.org/10.1002/jps.

Shuh, M., Beilke, M., 2005. The human T-cell leukemia virus type 1 (HTLV-1): new insights into the clinical aspects and molecular pathogenesis of adult T-cell leukemia/lymphoma (ATLL) and tropical spastic paraparesis/HTLV-associated myelopathy (TSP/HAM). Microsc. Res. Tech. 196(March), pp. 176–196. doi: http://doi.org/10.1002/jemt.20231.

Swanstrom, R., et al., 2014. HIV-1 Assembly, Budding, and Maturation. Cold Spring Harb Perspect Med. 2 (7), pp. a006924. doi: http://doi.org/10.1101/cshperspect.a006924.

Telwatte, S., et al., 2011. Virucidal activity of the dendrimer microbicide SPL7013 against HIV-1. Antiviral Res. 90 (3), 195–199. http://doi.org/10.1016/j.antiviral.2011.03.186.

Tomalia, D.A., Fréchet, J.M.J., 2002. Discovery of dendrimers and dendritic polymers: a brief historical perspective. J. Polym. Sci. Part A Polym. Chem. 40 (16), 2719–2728. http://doi.org/10.1002/pola.10301.

Tuomela, A., Hirvonen, J., Peltonen, L., 2016. Stabilizing agents for drug nanocrystals: effect on bioavailability. Pharmaceutics 8 (2). http://doi.org/10.3390/pharmaceutics8020016.

Tyssen, D., et al., 2010. Structure activity relationship of dendrimer microbicides with dual action antiviral activity. PLoS One 5 (8). http://doi.org/10.1371/journal.pone.0012309.

Van Eerdenbrugh, B., et al., 2010. Solubility increases associated with crystalline drug nanoparticles: methodologies and significance. Mol. Pharm. 7 (5), 1858–1870. http://doi.org/10.1021/mp100209b.

Vansant, G., et al., 2020. Block-and-lock strategies to cure HIV infection. Viruses 12 (1), 1–17. http://doi.org/10.3390/v12010084.

Victor, O.B., 2019. Nanoparticles and its implications in HIV/AIDS therapy. Curr. Drug Discov. Technol. 16 (August). http://doi.org/10.2174/15701638 16666190620111652.

Wang, J., et al., 2018. Poly (ethylene glycol)—polylactide micelles for cancer therapy. Front. Pharmacol. 9(March), pp. 1–15. doi: http://doi.org/10.3389/fphar.2018.00202.

Weston, R., Marett, B., 2009. HIV infection disease progression, 1(October).

Wewers, M., et al., 2020. Influence of formulation parameters on redispersibility of naproxen nanoparticles from granules produced in a fluidized bed process. Pharmaceutics 12 (4). http://doi.org/10.3390/pharmaceutics12040363.

Wilen, C.B., et al., 2014. HIV: Cell Binding and Entry. Cold Spring Harb Perspect Med. 2 (8), pp. a006866. doi: http://doi.org/10.1101/cshperspect.a006866.

Wilen, C.B., Tilton, J.C., Doms, R.W., 2012. Molecular Mechanisms of HIV Entry. Adv. Exp. Med. Biol. 726, 223–242. http://doi.org/10.1007/978-1-4614-0980-9.

Wong, J.K., Yukl, S.A., 2016. Tissue reservoirs of HIV. Curr. Opin. HIV. AIDS 11 (4), 362–370. http://doi.org/10.1097/COH.0000000000000293.

Xu, W., Ling, P., Zhang, T., 2013. Polymeric micelles, a promising drug delivery system to enhance bioavailability of poorly water-soluble drugs, 2013(1).

Yang, J., et al., 2016. Drug delivery via cell membrane fusion using lipopeptide modified liposomes. ACS Central Sci. 2 (9), 621–630. http://doi.org/10.1021/acscentsci.6b00172.

Zensi, A., et al., 2009. Albumin nanoparticles targeted with Apo E enter the CNS by transcytosis and are delivered to neurones. J. Control. Release 137 (1), 78–86. http://doi.org/10.1016/j.jconrel.2009.03.002.

Zhang, S., Kaplan, A.H., Tropsha, A., 2008. Analysis of protein packing method, (March), pp. 742–753. doi: http://doi.org/10.1002/prot.22094.

Zhu, Y., Liu, C., Pang, Z., 2019. Dendrimer-based drug delivery systems for brain targeting. Biomolecules 9 (12), 1–29. http://doi.org/10.3390/biom9120790.

Zinkernagel, R.M., 1996. Immunology taught by viruses. Sci. 271(January), pp. 173–178.

Chapter 16

Drug delivery systems for rheumatoid arthritis treatment

Mangesh Bhalekar[a], Sachin Dubey[b]
[a]Department of Pharmaceutics, AISSMS College of Pharmacy, Pune, India
[b]Drug Product and Analytical Development, Ichnos Sciences SA, La Chaux-de-Fonds, Switzerland

16.1 Introduction

Rheumatoid arthritis (RA) is an autoimmune disease of systemic origin which leads to chronic inflammation in multiple joints like hands, wrists, and feet. In RA the immune system attacks your own body tissues by mistake (Wasserman, 2011). The resulting inflammation can lead to joint deformity and disability to significant extent. The etiology of RA is not well defined but it is believed genetic as well as environmental factors are involved and it is typically presented between 30 and 50 years of age. About 1% of the population is affected by RA and there is higher incidence in women than in men with a ratio of approximately 3:1 (Scherer et al., 2020). The synovial membrane in joints contains, an acellular structure with an intimallining which is made of macrophages and fibroblast-like cells known as synoviocytes. The pathogenesis of RA is initiated by an immune response that causes inflamed synovium.

16.1.1 Stages of rheumatoid arthritis

- Stage 1: This stage is accompanied by joint pain, stiffness, swelling and inflammation is caused inside the synovium but no damage to bone.
- Stage 2: The second stage of RA is moderate. Synovial inflammation causes damage to the cartilage, causing pain and loss of mobility. Rheumatoid arthritis can be diagnosed in stage 2.
- Stage 3: This is a severe stage of rheumatoid arthritis. In this cartilage and bone both are damaged. The pain and more swelling are enhanced. The cushion between the bones is worn, due to this muscle weakness and more mobility loss than stage 2 occurs. Joints become bent and deformed and symptoms are more visible.
- Stage 4: If rheumatoid arthritis is not treated, joints are destroyed and cannot work. The fusion of joints and bones occur which is known as ankolysis (McInnes and Schett, 2011).

16.1.2 Causes of RA

Exact cause of RA is not clearly understood but it is supposed to be attack by immune system on the synovium. Rheumatoid arthritis may also be triggered due to some infections and bacteria also (Bowen et al., 2020). Some more causes are due to genes, hormones and environmental factors.

Rheumatoid arthritis progression depends upon the factors such as:

- Age, gender
- Family history
- Specific antibodies present in your body
- Environmental factors

The genetic marker HLA is usually found in patients with RA, is considered to control the immune response. Another gene named STAT4 is involved in regulation and activation of the immune system. Genes TRAF1 and C5 both have role in development of chronic inflammation. PTPN22 gene may cause the progression and development of RA. Occurrence of stressful events like physical and emotional trauma can also lead to rheumatoid arthritis. Some factors like obesity, smoking, air pollution, insecticides and occupational exposure to mineral oil and silica may also be involved.

16.1.3 Symptoms of RA

- Swelling.
- Redness and warmth on the affected joint.
- Inflammation inside a joint make it whether you are moving it or not. After disease progression, it may cause damage to the joint.
- Stiffness: in the morning, joint feel harder. It is difficult to move. Some patient takes an hour or several hours to feel joint loose.
- Sweating.
- Dryness in the eyes.

16.1.4 Pathology of rheumatoid arthritis

The immune activation manifests itself in the form of synovial membrane inflammation which reflects externally as swelling of joints. As the result of inflammation leukocytes invade the normal synovial compartment and proinflammatory mediators in synovial fluid interact which leads to inflammatory cascade. The interactions between fibroblast-like synoviocytes (FLSs) and the cells of the innate immune system, including monocytes, macrophages and mast cells is the characteristic feature of this inflammatory cascade.

The T lymphocytes (cells of adaptive immune system) and B-cells (humoral immunity) are also involved (Fig. 16.1).

The hyperplastic synovium in RA contains a mixture of bone marrow-derived macrophages and dysfunction of FLSs. These cells invade and adhere cartilage causing major damage. The synovial cells cause production of cytokines which are central to the pathogenesis of rheumatoid arthritis. Synovitis is caused by the release of cytokines from macrophages (e.g., TNF-α and interleukin-1, 6, 12, 15, 18, and 23), reactive oxygen intermediates, nitrogen intermediates. These cytokines induce macrophage activation, cognate interactions with T cells, immune complexes, and lipoprotein particles (Fig. 16.2). Synthesis of prostaglandins, proteases, and reactive oxygen intermediates by neutrophils contribute to synovitis. Mast cells also play a role through synthesis of high levels of vasoactive amines, cytokines, chemokines, and proteases. Due to release of various cytokines and mediators of inflammation, synovium starts proliferating and spreading, it is known as pannus. The formation of pannus is followed by fibrosis which leads to loss of mobility and it is known as ankolysis. Chemo-attractant cytokine molecules that cause the accumulation of inflammatory cells at the site of inflammation are called chemokines. These chemokines and chemokine receptors are considered to be therapeutic targets in several chronic inflammatory disorders such as RA (Iwamoto et al., 2007). The chemokine receptors are expressed by inflammatory cells that infiltrate into synovial tissue, these including CXCR3, CCR5, CCR3, CCR2, and CXCR2 (Szekanecz and Koch, 2007). Chemokine receptors and their ligands play a role in the migration of leukocytes and their retention in the RA joint (Kehlen et al., 2002; Koch, 2003). Macrophages, chondrocytes, and FLS are considered to be the most potent producers of chemokines in the synovial compartment (Patel et al., 2001) (Fig. 16.3).

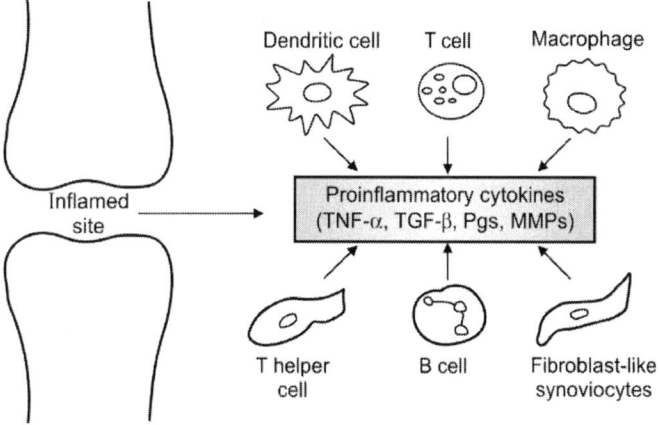

FIG. 16.1 The cells and proinflammatory cytokines in inflammatory cascade.

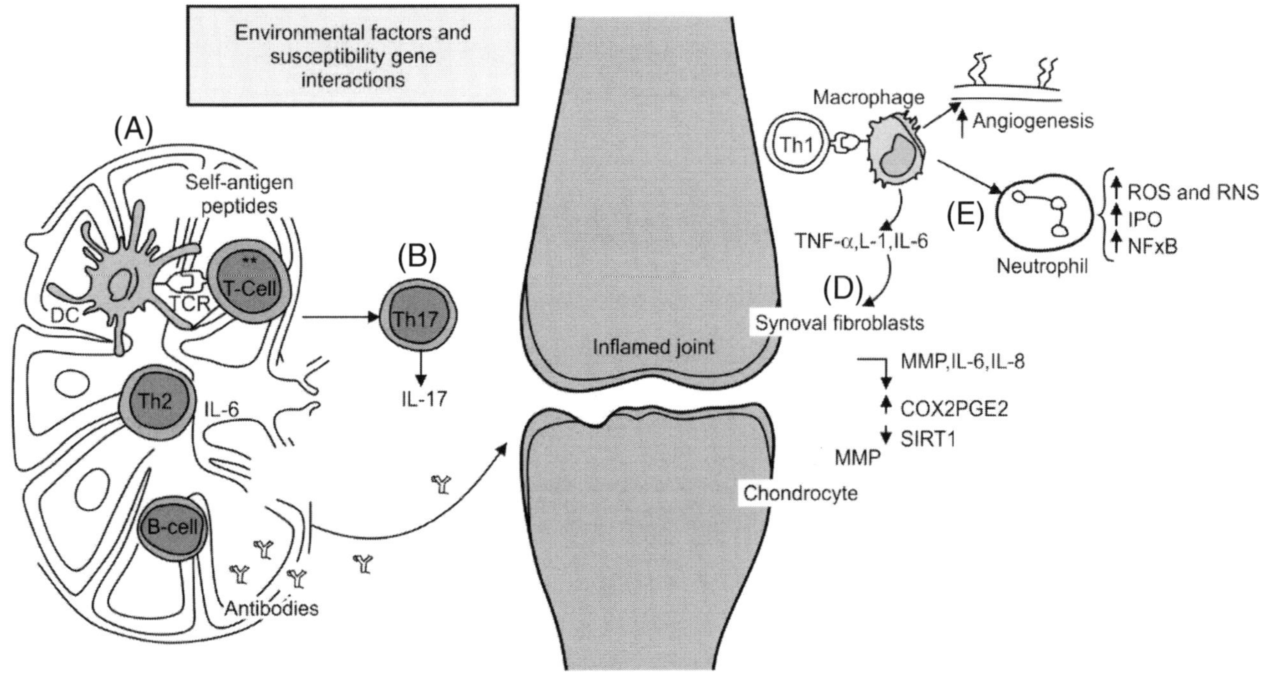

FIG. 16.2 Monocytes and macrophages are important players in RA pathogenesis and secrete proinflammatory cytokines such as TNFa, IL-1, and IL-6. In RA.

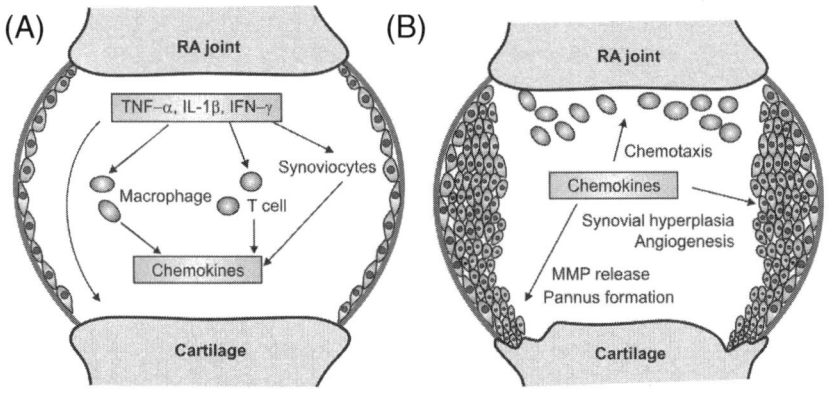

FIG. 16.3 Representation of involvement of chemokines in the joint of RA patients.

Tests for RA (Koch et al., 1992):

- Erythrocyte sedimentation rate
- Magnetic resonance imaging
- Radiology/biopsy
- Blood test
- Serology
- Erythema
- Latex fixation test
- Physical examination
- ELISA
- Ultrasonography
- Immunoassay
- Antinuclear antibody

- Rheumatoid factor test
- Immunoglobulin A
- Anti-CCP antibody test (cyclic citrullinated peptide)
- Complete blood count

16.2 Management of rheumatoid arthritis

Rheumatoid arthritis treatment aims to achieve the following:

- Reduce joint pain
- Prevent damage of bone and cartilage
- Stop inflammation
- Improve the physical function of organs
- Early, aggressive treatment

  The first strategy is to reduce inflammation as immediate as possible.

- Targeting remission: The inflammation in RA is considered as disease activity and immediate concern is to stop it and achieve remission, which means abatement of symptoms of active inflammation.

To achieve this goal a tight control over progression of disease has to be achieved by bringing disease activity to a low level and maintaining it there to minimize further joint and cartilage damage. The treatment of RA includes nonsteroidal anti-inflammatory drugs (NSAIDs), corticosteroids, disease-modifying antirheumatic drugs (DMARDs), biologics, surgery. The progression of RA is halted by blocking TNF-α, IL-1, IL-6. The role of pathways like cyclooxygenase and NF-kβ is also identified. Nonspecific broad range immune suppressants are used as conventional DMARDs, which include methotrexate, sulfasalazine, cyclosporine A, azathioprine, (hydroxy-) chloroquine, cyclophosphamide, and leflunomide.

The precise mechanism of action for these agents is poorly understood. The drawback of using these agents is significant exposure of nontarget organs and tissues leading to adverse effects. The nonsteroidal anti-inflammatory drugs (NSAIDs), particularly the nonselective cyclooxygenase 1 and 2 inhibitors, also can cause gastrointestinal toxicity (Iwamoto et al., 2007). The novel biologic agents designed to target defined molecules are more specific and they act by inhibition of targets such as TNF α and biologic processes such as cellular costimulation. The inhibition of these key immunological processes, however leads to high risk of infectious complications (Stevens et al., 1991).

16.3 Targeted delivery strategies to inflamed synovium

The conventional therapy of RA suffers from various shortcomings. The corticosteroids which otherwise have favorable effect on joint erosion if used in high cumulative doses lead to adverse effects like osteoporosis and diabetes.

The oral route of administration suffers from drawbacks such as poor absorption, first pass metabolism drug degradation. Use of intra-articular route can achieve higher local levels of corticosteroids compared to systemic route, but suffers from rapid clearance from the joint cavity, moreover it is not possible to inject locally to all affected joints. The antiarthritic drugs given in high and repeated dosage lead to high treatment costs. The cost of treatment with biological agents is even more expensive due to their high developmental costs and an elaborate manufacturing process (Bakthavatsalam et al., 2021).

An ideal treatment for RA should have following attributes (Table 16.1)

- Effective at minimal dosage,
- Prolonged half-life and molecular stability in vivo,
- Preferentially accumulate at the sites of inflammation,
- Specifically target the key mechanisms of arthritis
- Not cause nonselective immunosuppression,
- Low systemic toxicity and be affordable.

These objectives are achievable with targeted drug delivery systems. The targeting of the drug is possible with two approaches.

Targets and approaches for drug delivery to RA (Furst et al., 2013; Rhim et al., 2013).

Several unique characteristics and pathophysiological phenomena of this disease can be harnessed as tools or targets to effect improved delivery of therapeutic agents to inflamed joints.

TABLE 16.1 Agents used in treatment of rheumatoid arthritis.

Sr. no.	Category	Examples
1	Nonsteroidal anti-inflammatory drugs (NSAIDs)	Ibuprofen, Naproxen, Meloxicam, Etodolac, Sulindac, Diclofenac
2.	Tumor necrosis factor inhibitors	Etarnercrpt, Infliximab, Adalimumab
3.	T-cell blocking agents	Abatacept
4.	B cell depletion	Rituximab, Anakinra, Interleukin
5.	Corticosteroids	Prednisone, Methylprednisolone
6.	Immunomodulatory and cytotoxic agents	Azathioprine, Cyclosporine, Penicillamine

The phenomenon such as leaky vasculature in inflamed tissues, accumulation of inflammatory cells like macrophages, and T lymphocytes in synovium increased levels of inflammatory mediators such as cytokines and chemokines, tissue hypoxia and acidosis, and angiogenesis. Various tissue-specific antigens that could facilitate the direction of delivery vehicles to the inflamed joints in synovium are identified.

16.4 Passive targeting

16.4.1 Enhanced permeability and retention (EPR) effect

In RA the vasculature becomes hyperpermeable due to inflammatory mediators and causes 6–40 folds increase in blood-joint barrier permeability (Levick, 1981; Bonferoni, 2021). The extravasation of small (of specific size, i.e., 20–200 nm), long-circulating drug carrier systems in the inflamed tissue can take place through this leaky vasculature caused by agents of inflammation. The leaked particles are transported through endothelial pores which are 10 and 1000 nm in diameter and accumulate in the interstitial space (Nichols and Bae, 2014).

The hydrophobic particles in systemic circulation are recognized by the reticuloendothelial system (RES), of liver and spleen, and eliminated by the macrophages present in inflamed environment in the bloodstream. The presence of macrophages may enable selective accumulation of the nanocarrier based drug delivery system in the desired region (Claesson-Welsh, 2015).

The colloidal drug delivery systems such as liposomes, nanoparticles, micelles, and macromolecule–drug conjugates can be used to effect passive accumulation of drug within the pannus, where enhanced permeability is observed (Fig. 16.4).

16.4.2 Hypoxia and acidosis

Cellular infiltration and proliferation of local cells can cause thickening of the inflamed synovium, effusion of synovial fluid in the joint cavity causes the tissue compression leading to tissue hypoxia and acidosis (Stevens et al., 1991). These pathological changes can be used to target the drug delivery passively, in hypoxic and pH sensitive microenvironment of the inflamed tissue. Drug delivery systems that involve bio-reductive mechanism to release the drug under hypoxic conditions have been described for tumor therapy (Naughton, 2001).

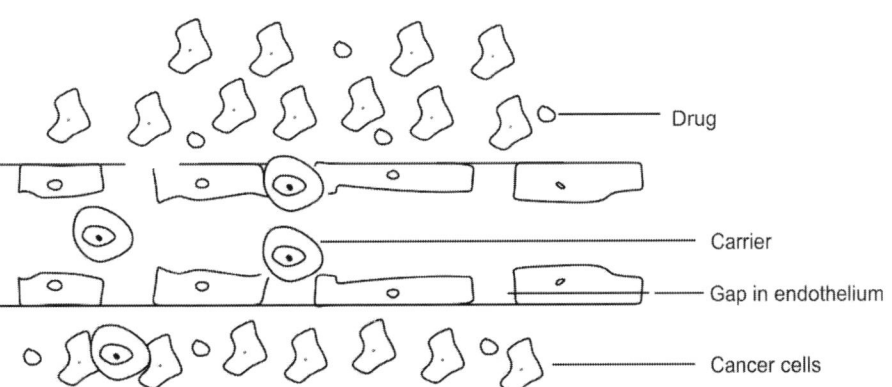

FIG. 16.4 Schematic representation of enhanced permeation and retention for passive targeting.

16.4.3 Stimuli responsive drug delivery

The acidic environment prevailing at the site of inflammation was used to target delivery of dexamethasone. Dexamethasone was conjugated to a pH-sensitive vehicle based on an *N*-(2-hydroxypropyl) methacrylamide (HPMA) co-polymer by a pH-sensitive hydrazone bond. The drug was found to have released at acidic pH of 5.0 in in vitro analysis at body temperature (37°C). The onset of the effect of HPMA conjugated dexamethasone was quick (Wang et al., 2004).

16.4.4 Angiogenesis

Angiogenesis is the process of formation of new blood vessels, as a response to hypoxic conditions. These (neo-)vascular endothelial cells form prime targets for delivery of antiarthritic drugs to get access to the synovium as well as for the inhibition of angiogenesis (Koch and Distler, 2007; Szekanecz and Koch, 2007).

16.5 Active targeting

Two primary cell types present in the pannus tissue are

1. Rheumatoid arthritis synovial fibroblast (RASF's)
2. Rheumatoid arthritis synovial macrophages (RASM's)

These cells selectively express surface receptors such as CD44, folate receptor, and integrin which offer good sites for active targeting. The angiogenic vascular endothelial cells (VECs) which present during neovascularization and the E-selectin adhesion molecule are identified as viable drug delivery target (Iib, 2002; Naor and Nedvetzki, 2003; Nagayoshi et al., 2005).

16.6 Factors for the selection of delivery system

16.6.1 Carrier type

The choice of carrier would depend on physicochemical properties, doses, and administration frequencies, consideration must be given to carrier materials based on drug stability, drug loading capacity and release profiles (van Vollenhoven, 2009). Various types of drug carriers including liposomes and nanoparticulates, micelles, etc. have been used for drug delivery to RA.

Liposomes are lipid vesicles derived from self-assembled enclosed lipid bilayers. Number of studies investigating use of liposomes for drug delivery to RA have been reported. The presence of both a hydrophilic core and a lipid shell enables liposomes to encapsulate both hydrophilic and hydrophobic drugs. The liposomal surface has been modified with polyethylene glycol (PEGylation), targeting ligands and stimuli-sensitive moieties to make it more effective (Gregoriadis and Neeranjun, 1975). However liposomes have a shortcoming of relatively quick release of drugs and poor storage stability (Zhigaltsev et al., 2016).

Compared to liposomes polymeric micro- and nanoparticles made from poly (lactic-co-glycolic acid) (PLGA) and poly(L-lactic acid) (PLA) exhibit high loading capacity and prolonged drug release. Their surface too can be modified by PEGylation and attachment of targeting ligands. Drug targeting using of iron oxide nanoparticles is also reported (Butoescu et al., 2008), the magnetic property prevented rapid clearance of the microparticles by macrophage uptake and lymph drainage by increasing the residence time of the microparticles in the joint (Butoescu et al., 2008). Several other types of carriers such as chitosan nanoparticles, dendrimers and solid lipid nanoparticles (SLNs) have been exploited for formulating suitable drug delivery system for treating RA (Kaur and Harikumar, 2012).

16.6.2 Particle size

Particle size has important role in various intravenous drug delivery processes such as blood circulation, targeting, cellular uptake, and elimination (Yoo et al., 2010). For IV administration particles less than 2–3 μm size are used to avoid capillary clogging. The macrophage uptake of micro/nanoparticles is influenced by size. Larger size particles are more likely to be taken up by macrophages than those of smaller sizes (Fang et al., 2006). Size also plays important role in passive targeting to inflamed synovium. The macromolecules such as serum protein can infiltrate synovium fluids by extravasation, resulting in high concentration of proteins (Wallis et al., 1987). The cut-off particle size for extravasation via leaky vasculature varies from 200 to 800 nm (Torchilin, 2011).

Particles with smaller size range (1–20 μm) have been reported to exhibit more severe toxicity (i.e., inflammation) after intra-articular injection than larger size particles (35–105 μm) (Liggins et al., 2004). The larger particles generally had higher drug-loading capacity and slower release profiles (over several weeks).

16.6.3 Shape

The physical shape of microparticles and nanoparticles has been reported to impact phagocytosis of drug carriers by macrophages. Phagocytosis by macrophages is governed by local geometry and orientation of particles, implying that whether or not particles are phagocytosed can modulated by particle shape (Champion and Mitragotri, 2006; Sharma et al., 2010), targeting ability (Decuzzi et al., 2009) and cellular uptakes (Chithrani et al., 2006). Yoo and Mitragotri have considered design parameters for microparticles used for delivery to RA. All the drug delivery systems such as micro or nanoparticles designed for the target specific drug delivery are spherical in shape. Most shape –related studies have aimed to increase the efficiency of drug delivery to solid tumors and vascular endothelium. The RA greatly resembles solid tumors hence physical geometry of drug carriers can also be important consideration. It has been suggested higher EPR effect may be possible for the nonspherical shape of nanoparticles leading to the higher accumulation (Yoo et al., 2011). It is suggested that nanospherical particles compared to spherical counterparts are likely to tumble and flow more dynamically in the blood due to their anisotropic geometry (Theunis et al., 1994) and due to long axis particles may remain trapped after extravasation, enhancing their retention.

16.6.4 Surface modifications

Another important factor in drug delivery is surface of drug carriers, the surface chemistry can be modified to render beneficial properties, such as prolonged blood circulation, active targeting ability and stimuli-responsiveness, to particles. This can benefit the drug delivery to inflamed synovium.

16.6.5 Prolonged circulation time

A majority of drug carrier particles administered intravenously adsorb plasma proteins, these adsorbed proteins are called opsonins. Opsonization causes the recognition of these circulating particles by reticuloendothelial systems (RES), macrophages followed by phagocytosis and elimination (Paulos, 2004). Due to this circulation time for particles is reduced, this can be prevented by increasing surface hydrophilicity and decreasing surface charge thus preventing binding of opsonin to the particle surface. The most common surface modification is PEGylation which involves attachment of polyethylene glycol (PEG) on the surface of the particles. The PEGylation is also known to interfere with interaction between particles and activated macrophages or other cells in the inflamed synovium.

16.6.6 Strategies for active targeting

Numerous studies have identified activated macrophages to be the main contributors to induction and maintenance of RA by secreting various proinflammatory cytokines, thus serving as a primary target for many RA therapeutics (Kinne et al., 2000).

16.6.6.1 Folate receptor (FR)

Macrophages are present in numerous numbers in inflamed synovium and at the cartilage–pannus junction their number is commensurate to the extent of joint inflammation and tissue degradation (Nakashima-Matsushita et al., 1999). Folate receptor beta (FR-β) is selectively expressed by activated macrophages within the pannus tissue in RA [Nagayoshi et al, 2005]. This fact can be utilized for actively targeting drug delivery to inflamed synovium. The FR-β has a high affinity for folic acid (FA), hence it is used as ligand for targeting the antirheumatic drug to the inflamed sites (Naor and Nedvetzki, 2003; Koning, 2006). Folate is widely used because of favorable properties like low molecular weight, easy availability, ease of conjugation, absence of immunogenicity, water solubility, pH and temperature (Nagayoshi et al., 2005).

16.6.6.2 CD44

It is an omni present multistructural and multifunctional agent which mediates adhesion interaction of cell surface and matrix. In RA the activated macrophages present in the inflamed synovium over express CD 44 a cell surface glycoprotein. CD44 levels detected by Western blot analysis, have been found to be 3.5-fold and 10.7-fold higher in the synovial tissue of RA patients than those of osteoarthritis (OA) patients and joint trauma patients, respectively. The expression of CD44 and degree of progression of the disease are directly related. Thus CD44 can be used as a potential target for targeting drug delivery in rheumatoid arthritis (Koch, 2003). Out of wide range of ligands for CD44, hyaluronic acid (HA) linear polymer consisting of repeating disaccharide units (D-glucuronic acid-[1-β-3]-N acetyl-D-glucosamine-[1-β-4]) is the main one.

16.6.6.3 Antiangiogenesis

The inflamed synovium is hypoxic and this leads angiogenesis process of formation of new blood vessels from existing vessels (Clavel, 2008). The proliferating inflamed tissue receives nutrition and oxygen through these newly formed blood vessels. These vessels being in chronic inflammatory state, attract the inflammatory cells. Their endothelium produces cytokines, adhesion molecules, and other inflammatory stimuli. The synovial membrane infiltrates into cartilage, causing its erosion (Ferrara, 1999; Lowin and Straub, 2011).

The vascular endothelial growth factor (VEGF) is specific to angiogenesis in RA, it is induced by inflammatory markers like TNF-α and IL-1. These markers lead to proliferation and migration of endothelial cells, enhancement in permeability of blood vessels and maturation and stabilization of newly formed blood vessels (Sreedharan et al., 1991; Afuwape et al., 2002).

16.6.6.4 Integrins

They have an important role in adhesion of cells to the extracellular matrix signaling is initiated upon ligand binding, and several intracellular pathways are activated. This leads to a wide variety of effects, depending on cell type. The integrins are produced by synovial fibroblast in the pannus and hence are specific to endothelial cell [Delgado et al, 2002]. The integrin cell-adhesion molecule signaling causes leukocyte inflammatory responses. Molecular modeling studies have found inhibition of integrin with β-lactam antibiotics cefsulodin and ceftazidime, integrin β-subunit binding was inhibited by these antibiotics. This can help in treatment of RA by reduction of inflammation and angiogenesis (Villanueva-romero, 2018). Integrin has been successfully exploited for a targeted delivery strategy. Cyclic peptides containing an Arg-Gly-Asp (RGD) sequence can bind this integrin specifically

16.6.6.5 Vasoactive intestinal peptide (VIP)

VIP is a "neuro immune peptide" present in central and peripheral nervous system. VIP has wide number of actions such as anti-inflammatory, antiproliferative, vasodilatory. The VIP receptors, which are expressed on T-lymphocytes and various inflammatory cells are responsible for peptide action. Activated macrophages and proliferating synoviocytes in RA have overexpressed VIP receptors and ligands specific to VIP receptor can be conjugated to carrier system to achieve active targeting of drug to inflamed joint (Donald, 1996).

16.6.6.6 E-selectin

Inflamed vascular endothelium has a glycol protein called E-selectin expressed on it. E-selectin is involved in leukocyte rolling and adhesion, hence promoting angiogenesis (Veale and Maple, 1996). E selectin is found to be up regulated in the inflammatory condition like RA and its expression is controlled by inflammatory cytokines (Chapman et al., 1996). The selectins are important in the process of rolling of leucocytes along the endothelial cell surface and for the binding of leucocytes prior to trans-migration hence it is important molecular target for therapeutic intervention (Vanniasinghe et al., 2009; Jubeli et al., 2012).

The fucosylated glycoproteins (ESL1, for E-selectin ligand) that have sialylated carbohydrate moieties (sialyl-Lewis x, sLex) epitopes are natural ligands of E-selectin (Ratcliffe et al., 2000; Kapoor et al., 2014).

16.7 Drug delivery vehicles for rheumatoid arthritis

16.7.1 Liposomes

Liposomes have been widely used as drug delivery vehicles and are approved already for clinical use in humans since 1970s. These are lipid vesicles derived from self-assembled enclosed lipid bilayers, the bilayer is formed by phospholipids and cholesterol which resembles cell membrane (Butoescu et al., 2009).

Because of bilayer structure the liposomes can encapsulate hydrophilic drugs in their aqueous core or accommodate lipophilic agents by integration (Fuchs et al., 2004) into the bilayer. Higher accumulation of the drugs entrapped in liposomes has been reported in inflamed joints compared to noninflamed joints (Metselaar et al., 2003). This happens due to the enhanced permeability and retention (EPR) effect prevailing in inflamed tissues due to vasculature being leaky. The liposomes are endocytosed by systemic macrophages and local macrophages in joints (Khoury et al., 2006).

The liposomal surface can be conjugated with other molecules for the purpose of improving pharmacokinetics viz. PEG, poly(amino acid)s (PAA) which avoids their interaction with plasma proteins and phagocytosis by reticuloendothelial system, also referred to as stealth liposomes (Shehata et al., 2008). The surface can also be attached with ligands to achieve targeted drug delivery to inflamed synovium. The performance of liposomes depends on various parameter which influence drug loading capacity, biodistribution and tolerability. These factors are vesicle size, lipid composition and membrane charge which (Williams et al., 2000).

A study by Metselaar and co-workers, claims larger liposomes which are about 450–500 nm diameter, to show higher drug accumulation in the liver and spleen. Whereas smaller liposomes with particle size 90–100 nm accumulated in inflamed paws in the adjuvant arthritis model in the rat (Campo et al., 2012). The PEGylation of liposomes causes lowest splenic uptake and highest targeting to joint tissue.

It was observed that instead of liposomes modification by PEGylation alone, modification by PEG/PVA reduced opsonin binding and receptor-mediated endocytosis in the liver (Moein Moghimi et al., 2006).

Administration of methotrexate with liposomes was found to be faster with non-PEGylated liposomes compared to PEGylated ones but in the long run more accumulation was seen with the former (Horisawa, 2002). Liposomes with 100 nm diameter when conjugated with tetrasaccharidesialyl-Lewis X (SLX), for targeting to E selectin in arthritic mice. Compared to liposomes lacking the ligand the accumulation of SLX-liposomes was higher as was visualized using scanning fluorescent microscopy (Chandrasekar et al., 2007b). siRNA designed to silence TNF α was targeted to the macrophages in the inflamed joints, by encapsulating in a liposomal formulation. Arthritic mice were completely cured after IV treatment with 10 μg of siRNA encapsulated in cationic liposomes (Hood, 2014).

Liposomes with a high cholesterol content >30 mol% can also activate complement via the classical pathway after binding of anticholesterol antibodies, which are abundant in most human sera (Ehrhardt et al., 2004).

Long-term usages of corticosteroids in RA are associated with severe adverse effects and are correlated with the cumulative dose. The majority of studies have in corporated corticosteroids within liposomes. After intra-articular injection of triamcinolone acetonide 21-palmitate encapsulated in liposome in arthritic rabbit, the liposomal formulation showed increased retention in the articular cavity over that of free triamcinolone acetonide. This correlated well with the increased reduction in paw diameter (Lopez-Garcia, 1993).

Cyclic peptides containing an Arg-Gly-Asp (RGD) sequence can bind this integrin specifically. Koning and co-workers used a high-affinity RGD peptide to improve the targeting of dexamethasone phosphate carrying PEGylated liposomes. This caused an increase in of RGD-conjugated PEG-liposomes over unmodified PEG-liposomes into inflamed tissues. The delivery of dexamethasone by these vehicles effectively reduced arthritis in the rat model (Ferrara, 1999).

16.7.2 Dendrimers

Dendrimers are repeatedly branched polymers forming a fractal-like structure of 10–100 nm in size. Due to this structure PEG molecule can be frequently attached with. The branching offers large surface to attach PEG and also allows binding of a large payload of drug to dendrimers. Solid polymer particles are composed of a blend of hydrophobic and hydrophilic biodegradable polymers that can encapsulate therapeutic molecule (Fahmy, 2007).

Dendrimer-associated indomethacin was found to accumulate in inflamed joints in adjuvant induced arthritis at a significantly higher concentration than free indomethacin; this can be attributed to EPR effect due to arthritis. The system was further improvised by conjugation with folate, due to folate moiety binding and internalization of dendrimers took place on macrophage cell line (Qi et al., 2009).

The development of PAMAM based dendrimers and targeting by folate conjugation to inflamed synovium and the folate conjugation through polyethylene glycol conjugates of polyamidoamine dendrimers are also reported.

16.7.3 Nanoparticles

Lipid-based nanoparticles, solid lipid nanoparticles (SLN) and nanostructured lipid carriers (NLC) are composed of physiological lipids; due to which, they have good tolerability, they are usually nontoxic, and are degraded to a nontoxic residue. SLN and NLC are the lipid based carriers that have been investigated for various type of targeted delivery (Yadav, 2016).

SLN depending on their composition can show slow, extended release of the encapsulated drug. Since SLNs have beneficial characteristics of polymeric particles and liposomes, they represent alternative carriers for RA therapeutics.

Actarit is a poorly water-soluble antirheumatic drug, when actarit loaded SLN were administered by intravenous route, the area under the curve (AUC) of the plasma concentration-time was 1.88 times greater than that of the actarit in a 50% propylene glycol solution (Ye, 2008).

Zhoua et al. evaluated the targeting capability of glucocorticoid prednisolone (PD) incorporated in solid lipid nanoparticles (SLNs) and coated with hyaluronic acid (HA). The results of the study indicate that such particles accumulated in inflamed joint, had longer circulation time. The joint swelling, bone erosion and levels of inflammatory cytokines in serum significantly reduced compared to free drug or drug encapsulated in SLNs without HA (Zhou et al., 2018).

Formulation of chloroquine nanoparticles and ex vivo endocytic uptake studies revealed involvement of endocytic pathways in the uptake of SLN from intestine. Plasma drug profile upon pharmacokinetic evaluation demonstrated increased AUC, half-life and decreased elimination rate of the drug (Bhalekar et al., 2016).

16.7.4 Polymeric micro- and nanoparticles

Nanoparticle systems for delivery of rheumatoid arthritis therapeutics have primarily been based upon polymers. Numerous researchers have explored the use of poly (D, L-lactic/glycolic acid) (PLGA) nanoparticles based upon their capacity to extend the circulation time and control the release of encapsulated drugs. In the realm of rheumatoid arthritis drug delivery, a glucocorticoid, betamethasone, was incorporated into PLGA nanoparticles with a size of 100–200 nm. Intravenous administration to arthritic rats and mice showed that the PLGA-betamethasone system was more effective at reducing the inflammatory response than the free glucocorticoid (Higaki et al., 2005). Targeting ability and, consequently, efficacy of the betamethasone was further improved by modifying the PLGA nanoparticles with PEG, forming so-called "stealth nanosteroids" (Ishihara, 2009).

Other polymeric nanoparticle systems involve covalently linking the drug molecule to one of the components so as to slow down therapeutic release, as occurs for polymer-drug conjugates, while still protecting the drug via encapsulation. For example, when methylprednisolone was conjugated to cyclodextrin, the resultant compound self-assembled to yield nanoparticles. This when administered intravenously at a frequency of one dose per week to arthritic mice, a significantly greater reduction in synovitis and pannus formation was achieved than that obtained for free methylprednisolone administered daily at an equivalent cumulative dose (Hwang, 2008).

Another example is the ionic complexation of tumor necrosis factor (TNF)-related apoptosis inducing ligand (TRAIL) conjugated to PEG (PEG-TRAIL), which bears a positive charge, with negatively charged hyaluronic acid (HA) with sizes that range from 100 to 270nm, dependent upon the relative concentration of the two components. One formulation of the HA-PEG-TRAIL complex was capable of significantly reducing the secretion of proinflammatory mediators relative to PEG-TRAIL alone when administered subcutaneously to arthritic mice (Szekanecz et al., 2003) thereby emphasizing the importance of nanoparticulate carrier systems.

Higaki et al. (2005) used poly(D, L-lactic/glycolic acid) (PLGA) nanoparticles to deliver the hydrophilic corticosteroid betamethasone sodium phosphate (BSP) in rat AIA and murine antitype II collagen antibody induced arthritis (AbIA). The nanoparticles encapsulated BSP at a single dose of 100 µg was superior to a threefold dose of free BSP with regard to paw swelling and synovial infiltration by inflammatory cells.

In another study methotrexate (MTX)-loaded poly(ethylene glycol)-poly(lactic-co-glycolic acid) (PLGA) nanoparticles, were prepared with NIR radiation resulting in MTX-loaded MNPs. The synergistic effects of MTX-loaded MNPs with NIR irradiation was found in g RA fibroblast-like synoviocytes (FLSs) and collagen-induced arthritis (CIA) mice (Ha et al., 2020). In vivo tolerability and plasma concentrations of methotrexate following intra-articular injection of methotrexate loaded poly (l-lactic acid) (PLLA) microspheres into healthy rabbits were evaluated. MTX loaded and control microspheres were biocompatible and plasma concentrations of MTX were tenfold higher in rabbits compared to free MTX (Liang, 2004; Liang et al., 2005).

16.7.5 Macromolecules and the enhanced permeability and retention effect

The polymer–drug conjugates have been found to improve the therapeutic efficacy of both conventional DMARDs and biologics (Matsumura and Maeda, 1986). These compounds were originally being developed for cancer and recently tested in rheumatoid arthritis. The EPR effect as mentioned earlier allows macromolecules and artificial particles such as liposomes, dendrimers, and solid polymers to extravasate easily into areas of inflammation in a nonspecific manner (Teng et al., 2007).

The serum proteins including albumin extravasate into inflamed joints in increased amounts (Kraan, 2001). Compared with normal joints, the permeability of inflamed joints for albumin is increased up to sixfold (Hoven et al., 2011). Additionally albumin offers a source of nutrition for metabolically active cells in the inflamed synovium such as macrophages and synovial fibroblasts. Human serum albumin (HSA)-conjugated methotrexate (MTX) conjugate was found to be significantly more effective than free MTX in prevention of onset of arthritis in the collagen induced arthritis model (Fiehn, 2008). Due to this there could be fivefold reduction of dose and reduction of adverse effects which are dose dependent leading to MTX getting tolerated better. A smarter strategy was adopted by Fiehn and colleagues to overcome high cost and low carrier capacity of HAS, they used a prodrug form of MT that can bind to endogenous albumin after intravenous transfer. MTX was converted into a prodrug by conjugation to a polypeptide that binds selectively to the cysteine residue at position 34 of endogenous albumin. The release of MTX in inflamed joints took place by enzymatic cleavage. This approach was equally effective as HSA MTX conjugates (Fiehn, 2008). Methotrexate conjugated to human serum albumin (MTX-HSA) was shown to passively accumulate within the inflamed paws of arthritic mice, a reduction in cellular invasion, reduction of proinflammatory cytokine levels, and a decrease in cartilage damage was reported in arthritic mice.

16.7.6 Arthritis-specific antigens

During the process of angiogenesis, the extracellular matrix is remodeled by proteolysis of existing structures and neosynthesis of its components, paving the path for endothelial cells to proliferate, migrate and align in tube-like structures thus forming the new vessels. During this process of remodeling, certain extracellular matrix components that are usually not found in mature vascular structures are expressed due to alternative splicing. These components are therapeutic targets which are specific to tissues with active angiogenesis (Streit and Detmar, 2003).

Interesting examples include the C domain of tenascin-C and the extra domain B of fibronectin. Specific antibodies have been developed against these domains, termed G11 and L19, respectively (Clavel, 2008). The L19 antibody has been successfully employed for in vivo imaging of angiogenesis and deliver in conjugated therapeutic molecules in tumor models (Kaspar et al., 2007; Rossin et al., 2007).

A recent study investigated the homing capability of both antibodies in the collagen induced arthritis model (Natarajan, 2011). This approach provides drug targets for the inhibition of angiogenesis, by identification of the arthritis-specific endothelial cells. In order to facilitate the physiological extravasation of inflammatory cells, vascular endothelial cells express a variety of different adhesion molecules. Two examples are α v β 3-integrin and E-selectin, both of which are markedly up regulated in inflammation.

E-selectin has served as a target for synovitis-specific scintigraphic imaging, using 111 Indium or 99m technetium as tracers conjugated to an anti-E-selectin monoclonal antibody antigen-binding fragment (Fab) in pigs and in humans (Matsumura and Maeda, 1986; Rhim et al., 2013).

Interesting examples include the C domain of tenascin-C and the extra domain B of fibronectin. Specific antibodies have been developed against these domains, termed G11 and L19, respectively. TheL19 antibody has already been used with great success for in vivo imaging of angiogenesis and for delivery of conjugated therapeutic molecules in tumor models.

A recent study investigated the homing capability of both antibodies in the collagen induced arthritis model. Immunohistochemistry confirmed the presence of the antigens in inflamed synovial tissue, and fluorescence imaging in vivo as well as auto radiography post mortem revealed selective accumulation in inflamed versus noninflamed paws. The enrichment ofL19-conjugated tracers in inflamed paws was up to 8.5-foldhigher than in noninflamed paws. For the G11 conjugates the ratio was up to fivefold (Kaspar et al., 2007; Trachsel, 2007).

16.7.7 The complement system

The synovial inflammation in RA also causes other pathophysiological phenomena that can be used for treatment delivery strategies, including local complement activation and enhanced infiltration by inflammatory cells. Complement activation supports joint inflammation and, among other phenomena, leads to deposition of long-lived complement factor C3 cleavage products such as iC3b, C3dg, and C3d. As complement receptor 2 (CR2) can bind to these cleavage products, Song et al. have constructed a fusion molecule consisting of CR2 as the targeting device and Carry, the murine analog of human soluble CR1, as the therapeutic molecule (Demartis, 2001).

16.7.8 Specific surface receptors

The inflammatory cells such as macrophages represent attractive targets as they play a pivotal role in arthritis pathology and can be identified by several surface molecules. The high-affinity receptor for immunoglobulin G (Fc γ RI, CD64) is highly expressed on activated macrophages in synovial tissue and fluid, whereas CD64 expression is low on quiescent macrophages (Van Roon et al., 2005; Song, 2015; Van Vuuren, 2006). The, cytotoxic drugs such as ricin A or calicheamicin have been used to deplete activated CD64 high macrophages in vitro and in vivo by conjugating to an anti-CD64 antibody. Severity of arthritis as well as bone erosion was drastically reduced (Paulos, 2004; Van Vuuren, 2021).

16.7.9 Monoclonal antibodies

Monoclonal antibodies (mAbs) are man-made proteins that replicate antibodies normally produced by the immune system. These belong to class of therapeutic drugs known as biologics. mAbs are mono-specific antibodies that are produced by immune cells that are all clones of a unique parent cell (Little, 2000; Ha et al., 2020). mAbs are used in immunotherapy to treat autoimmune (AI) diseases like rheumatoid arthritis (RA), as well as certain cancers, viral infections, gastrointestinal diseases, and more. In a healthy immune system, antibodies are produced by white blood cells in response to a foreign invader. Antibodies fight off infection by attaching to foreign proteins called antigens to destroy or neutralize them. mAbs are laboratory-produced antibodies designed to locate and bind to certain molecules, rendering them neutral (Immordino and Cattel, 2006; Koenders and Berg, 2015). The cell base used for cloning the antibodies can be made from mouse or human proteins, or a combination of the two, and is denoted by the final letters in the drug's generic name (Lu et al., 2020).

These include:

- Murine: mAbs are made from mouse proteins. Their name ends in mab.
- Chimeric mAbs are combination of part mouse and part human proteins. Their name ends in -ximab.
- Humanized mAbs are made from small parts of mouse proteins attached to human proteins. Their name ends in -zumab.
- Human mAbs are fully human proteins. Their name ends in -umab.

Since arthritis is autoimmune in nature, that is, antibodies attack joint tissue causing pain, inflammation, and stiffness. mAbs treat RA by targeting specific proteins involved in this destructive inflammatory process. Several cytokine proteins have been identified as contributing to the inflammatory process involved in arthritis. mAbs have been developed to target these proteins (Isomäki et al., 1997; Paulos, 2006). Use of mAbs in RA initially utilized anti-CD4, anti-CD7 as targets, with varying degrees of efficacy and with significant safety concerns. Later the mAbs were directed against proinflammatory cytokines such as TNF-a, CD20-positive B cells, IL-1 and IL-6. TNF-α has central role in the pathogenesis of RA (Saxne et al., 1988; Delgado and Abad, 2002). mAbs to TNF bind soluble and transmembrane TNF, thereby downregulating TNF-induced immune responses including adhesion molecule expression, cytokine production, matrix metalloproteinase production, neutrophil activities, dendritic cell function, and osteoclast differentiation (Demartis, 2001; Scheinecker et al., 2012). It is proved beyond doubt that reduction in TNF-a levels improves the signs and symptoms of RA and reduces radiographic progression. Currently there are four mAbs approved for the treatment of RA (Buchan, 1988; Cohen et al., 2013).

- Infliximab: It is a chimeric IgG1 mAb that consists of human constant regions and murine variable regions.
- Adalimumab: It is a human recombinant IgG1mAb that has no murine component and is produced by phage display technology.
- Golimumab: It is a fully human IgG1 anti-TNF-antibody. It is very similar in structure to infliximab without the mouse protein.
- Certolizumab: It is a humanized Fab fragment (Fc free) fused to a 40-kDa polyethylene glycol (PEG) moiety. The PEGylation improves pharmacodynamics, bioavailability and localization to inflamed tissues (Van Roon, 2003; Goel and Stephens, 2016).

The lack of an Fc region minimizes Fc-mediated effects such as complement-dependent cytotoxicity and antibody-dependent cytotoxicity.

16.7.10 mAbs targeted against B cells

B cells which are critical in the pathogenesis of RA may evolve start producing antibody. B cell and plasma cell infiltration into synovium has consistently detected (Wikaningrum, 1998). B cells may additionally function as antigen-presenting cells and produce inflammatory cytokines (Iwamoto et al., 2007; Becciolini and Favalli, 2018). Rituximab is a B-cell-depleting agent, chimeric/IgG1 mAb which binds to the CD20 cell surface marker found on several maturation stages of B lymphocytes.

16.7.11 mAbs directed against IL-6 function

IL-6 is a pleotopic cytokine produced by cells that has an important role in T-cell activation and immunoglobulin secretion. It also stimulates synovial fibroblast differentiation and osteoclast activation (Harigai et al., 1993; Smits et al., 2019). Tocilizumab is a humanized/ IgG1 mAb directed against IL-6 receptor in its soluble and transmembrane form.

16.7.12 mAb directed against NFKB ligand

Denosumab inhibits bone resorption by binding and inhibiting receptor activator of nuclear factor kappa B ligand (RANKL). It was developed for treatment of osteoporosis and has been shown to slow articular bone loss associated with rheumatoid arthritis. It is given as a subcutaneous injection (Ye, 2008; Yoo et al., 2010).

16.8 Conclusion

The RA being an auto immune disease has unique features in terms of leaky vasculature, neovascularization, expression of certain receptors or presence of N cytokines and chemokines which provide numerous opportunities to formulator to target the drug directly to synovium.

In the view of adverse effects of drugs that are effective in rheumatoid arthritis it should be our effort to keep searching for a robust drug delivery system which will deliver the drug at synovium leading to better bioavailability and reduction of systemic exposure.

References

Afuwape, A.O., Kiriakidis, S., Paleol, E.M., 2002. The role of the angiogenic molecule VEGF in the pathogenesis of rheumatoid arthritis. Histol. Histopathol. 17 (3), 961–972.

Bakthavatsalam, D., et al., 2021. Identification of inhibitors of integrin cytoplasmic domain interactions with Syk. Front. Immunol. 11 (January), 1–15. doi:10.3389/fimmu.2020.575085.

Becciolini, A., Favalli, E.G., 2018. Tocilizumab in the treatment of rheumatoid arthritis: an evidence-based review and patient selection 13, 57–70.

Bhalekar, M.R., Upadhaya, P.G., Madgulkar, A.R., 2016. Fabrication and efficacy evaluation of chloroquine nanoparticles in CFA induced arthritic rats using TNF-α ELISA. PHASCI 84, 1–8. doi:10.1016/j.ejps.2016.01.009.26776969.

Bonferoni, M.C., et al., 2021. Electrochemotherapy of Deep-Seated Tumors: State of Art and Perspectives as Possible "EPR Effect Enhancer" to Improve Cancer Nanomedicine Efficacy. Cancers (Basel) 13 (17), 4437.

Bowen, A., Sweeney, E.E., Fernandes, R., 2020. Nanoparticle-based immunoengineered approaches for combating HIV. Front. Immunol. 11 (April), 1–9. doi:10.3389/fimmu.2020.00789.

Buchan, G., et al., 1988. Interleukin-1 and tumour necrosis factor mRNA expression in rheumatoid arthritis: prolonged production of IL-la. Clin. Exp. Immunol. 73 (3), 449–455.

Butoescu, N., et al., 2008. Co-encapsulation of dexamethasone 21-acetate and SPIONs into biodegradable polymeric microparticles designed for intra-articular delivery. J. Microencaps. 25 (5), 339–350. doi:10.1080/02652040801999551.

Butoescu, N., et al., 2009. Dexamethasone-containing PLGA superparamagnetic microparticles as carriers for the local treatment of arthritis. Biomaterials 30 (9), 1772–1780. doi:10.1016/j.biomaterials.2008.12.017.

Campo, G.M., et al., 2012. Hyaluronan differently modulates TLR-4 and the inflammatory response in mouse chondrocytes. BioFactors 38 (1), 69–76. doi:10.1002/biof.202.

Champion, J.A., Mitragotri, S., 2006. Role of target geometry in phagocytosis. Proc. Natl. Acad. Sci. U S A 103 (13), 4930–4934. doi:10.1073/pnas.0600997103.

Chandrasekar, D. et al., 2007a. Folate coupled poly(ethyleneglycol) conjugates of anionic poly(amidoamine) dendrimer for inflammatory tissue specific drug delivery. J. Biomed. Mater. Res. A 82 (1), 92–103. doi:10.1002/jbm.a.31122.

Chandrasekar, D. et al., 2007b. The development of folate-PAMAM dendrimer conjugates for targeted delivery of anti-arthritic drugs and their pharmacokinetics and biodistribution in arthritic rats. J. Biomed. Mater. Res. 28, 504–512. doi:10.1016/j.biomaterials.2006.07.046.

Chapman, P.T., et al., 1996. Use of a radiolabeled monoclonal antibody against E-selectin for imaging of endothelial activation in rheumatoid arthritis. Arthritis Rheumat. 39 (8), 1371–1375. doi:10.1002/art.1780390815.

Chithrani, B.D., Ghazani, A.A., Chan, W.C.W., 2006. Determining the size and shape dependence of gold nanoparticle uptake into mammalian cells. Nano Lett. 6 (4), 662–668. doi:10.1021/nl052396o.

Claesson-Welsh, L., 2015. Vascular permeability - the essentials. Upsala J. Med. Sci. 120 (3), 135–143. doi:10.3109/03009734.2015.1064501.

Clavel, G., 2008. Angiogenesis markers in rheumatoid arthritis, Angiogenesis markers in rheumatoid arthritis 3 (2), 153–159.

Cohen, M., Omair, M.A., Keystone, E.C., 2013. Review Monoclonal antibodies in rheumatoid arthritis review. Int. J. Clin. Rheumatol. 8 (5), 541–556.

Decuzzi, P., et al., 2009. Intravascular delivery of particulate systems: does geometry really matter? Pharm. Res. 26 (1), 235–243. doi:10.1007/s11095-008-9697-x.

Delgado, M., Abad, C., 2002. Vasoactive intestinal peptide in the immune system: potential therapeutic role in inflammatory and autoimmune diseases. J. Mol. Med. (Berl) 80 (1), 16–24. doi:10.1007/s00109-001-0291-5.

Demartis, S. et al., 2001. Short communication selective targeting of tumour neovasculature by a radiohalogenated human antibody fragment specific for the ED-B domain of fibronectin. Eur. J. Nucl. Med. 28 (4), 534–539. doi:10.1007/s002590100480.

Ehrhardt, C., Kneuer, C., Bakowsky, U., 2004. Selectins—an emerging target for drug delivery. Adv. Drug Deliv. Rev. 56 (4) 527–549. doi:10.1016/j.addr.2003.10.029.

Fahmy, T.M., et al., 2007. Nanosystems for simultaneous imaging and drug delivery to T cells AAPS J. 9 (2), 171–180.

Fang, C., et al., 2006. In vivo tumor targeting of tumor necrosis factor-α-loaded stealth nanoparticles: effect of MePEG molecular weight and particle size. Eur. J. Pharm. Sci. 27 (1), 27–36. doi:10.1016/j.ejps.2005.08.002.

Ferrara, N., 1999. Role of vascular endothelial growth factor in the regulation of angiogenesis. Kidney 56 (3), 794–814.

Fiehn, C., et al., 2008. Targeted drug delivery by in vivo coupling to endogenous albumin: an albumin-binding prodrug of methotrexate (MTX) is better than MTX in the treatment of murine collagen-induced arthritis. Ann. Rheum. Dis. 67 (8), 1188–1191. doi:10.1136/ard.2007.086843.

Fuchs, S., et al., 2004. A surface-modified dendrimer set for potential application as drug delivery vehicles: synthesis, in vitro toxicity, and intracellular localization. Chemistry 10 (5), 1167–1192. doi:10.1002/chem.200305386.

Furst, D.E., et al., 2013. Updated consensus statement on biological agents for the treatment of rheumatic diseases, 2012. Ann. Rheumatic Dis. 72 (Suppl. 2), 2–25. doi:10.1136/annrheumdis-2013-203348.

Goel, N., Stephens, S., 2016. Certolizumab pegol certolizumab pegol, MAbs. 0862(March), 137–147. doi:10.4161/mabs.2.2.11271.

Gregoriadis, G., Neeranjun, E., 1975. Homing of liposomes to target cells. Biochem. Biophys. Res. Commun. 65 (2), 537–544.

Ha, Y., et al., 2020. Methotrexate-loaded multifunctional nanoparticles with near-infrared irradiation for the treatment of rheumatoid arthritis. Arthritis Res. Therapy 22, 1–13. doi:https://doi.org/10.1186/s13075-020-02230-y.

Harigai, M., Hara, M., Inoue, K., 1993. Monocyte Chemoattractant Protein-1(MCP-1) in Inflammatory Joint Diseases and its Involvement in the Cytokine Network of Rheumatoid Synovium. Clin. Immunol. Immunopathol. 69 (1), 83–91.

Higaki, M., Ishihara, T., Izumo, N., Mizushima, Y., 2005. Treatment of experimental arthritis with poly. Ann. Rheum Dis. 64 (8) 25, 1132–1137. doi:10.1136/ard.2004.030759.

Hood, J.D., et al., 2014. Tumor regression by targeted gene delivery to the neovasculature. Science, 2404–2407. doi:10.1126/science.1070200.

Horisawa, E., et al., 2002. Prolonged anti-inflammatory action of DL-lactide /glycolide copolymer nanospheres containing betamethasone sodium phosphate for an intra-articular delivery system in antigen-induced arthritic rabbit, Pharm. Res. 19 (4), 403–410.

Hoven, J.M., Van Den, et al., 2011. Liposomal Drug Formulations in the Treatment of Rheumatoid Arthritis. Mol. Pharm. 8 (4), 1002–1015.

Hwang, J., et al., 2008. α-Methylprednisolone conjugated cyclodextrin polymer-based nanoparticles for rheumatoid arthritis therapy, Int. J. Nanomedicine 3 (3), 359–371.

Iib, A., 2002. Integrin alpha V beta 3 as a target for treatment of rheumatoid arthritis and related rheumatic diseases, pp. 96–99.

Immordino, M.L., Cattel, L., 2006. Stealth liposomes : review of the basic science, rationale, and clinical applications, existing and potential, pp. 297–315.

Ishihara, T., et al., 2009. Treatment of experimental arthritis with stealth-type polymeric nanoparticles encapsulating betamethasone phosphate. J. Pharmacol. Exper. 329 (2), 412–417. doi:10.1124/jpet.108.150276.drug.

Isomäki, P., et al., 1997. Pro-and anti-inflammatory cytokines in rheumatoid arthritis. Ann. Med. 29 (6), 499–507. doi:10.3109/07853899709007474.

Iwamoto, T., et al., 2007. A role of monocyte chemoattractant protein-4 (MCP-4)/CCL13 from chondrocytes in rheumatoid arthritis. FEBS J. 274 (18), 4904–4912. doi:10.1111/j.1742-4658.2007.06013.x.

Jubeli, E., et al., 2012. E-selectin as a target for drug delivery and molecular imaging. J. Controlled Release 158 (2), 194–206. doi:10.1016/j.jconrel.2011.09.084.

Kapoor, B., et al., 2014. Application of liposomes in treatment of rheumatoid arthritis: Quo vadis. Sci. World J. 2014, 1–17. doi:10.1155/2014/978351.

Kaspar, M., Trachsel, E. and Neri, D.., 2007. The antibody-mediated targeted delivery of interleukin-15 and GM-CSF to the tumor neovasculature inhibits tumor growth and metastasis, (10), pp. 4940–4949. doi:10.1158/0008-5472.CAN-07-0283.

Kaur, A., Harikumar, S.L., 2012. Controlled drug delivery approaches for rheumatoid arthritis. J. Appl. Pharm. Sci. 2 (8), 21–32. doi:10.7324/JAPS.2012.2803.

Kehlen, A., et al., 2002. Expression, modulation and signalling of IL-17 receptor in fibroblast-like synoviocytes of patients with rheumatoid arthritis. Clin. Exp. Immunol. 127 (3), 539–546. doi:10.1046/j.1365-2249.2002.01782.x.

Khoury, M., et al., 2006. Efficient new cationic liposome formulation for systemic delivery of small interfering RNA silencing tumor necrosis factor α in experimental arthritis. Arthritis Rheumat. 54 (6), 1867–1877. doi:10.1002/art.21876.

Kinne, R.W., et al., 2000. Macrophages in rheumatoid arthritis. Arthritis Res. 2 (3), 189–202. doi:10.1186/ar86.

Koch, A.E., et al., 1992. Enhanced production of monocyte chemoattractant protein-1 in rheumatoid arthritis. J. Clin. Invest. 90 (3), 772–779. doi:10.1172/JCI115950.

Koch, A.E., 2003. Angiogenesis as a target in rheumatoid arthritis. Ann. Rheum. Dis. 62 (2), 60–67.

Koch, A.E., Distler, O., 2007. Vasculopathy and disordered angiogenesis in selected rheumatic diseases: rheumatoid arthritis and systemic sclerosis. Arthritis Res. Ther. 9 (Suppl. 2), 1–9. doi:10.1186/ar2187.

Koenders, M.I., Van Den Berg, W.B., 2015. Novel therapeutic targets in rheumatoid arthritis. Trends Pharmacol. Sci. 36 (4), 189–195. doi:10.1016/j.tips.2015.02.001.

Koning, G.A., et al., 2006. Targeting of angiogenic endothelial cells at sites of inflammation by dexamethasone phosphate-containing RGD peptide liposomes inhibits experimental arthritis. Arthr. Rheumatol. 54 (4), 1198–1208. doi:10.1002/art.21719.

Kraan, M.C., et al., 2001. Measurement of cytokine and adhesion molecule expression in synovial tissue by digital image analysis 60 (3), 296–298.

Levick, J.R., 1981. Permeability of rheumatoid and normal human synovium to specific plasma proteins. Arthritis Rheumat. 24 (12), 1550–1560. doi:10.1002/art.1780241215.

Liang, L.S., et al., 2004. Methotrexate loaded poly (L-lactic acid) microspheres for intra-articular delivery of methotrexate to the joint, J. Pharm. Sci. 93 (4), 943–956.

Liang, L.S., Wong, W., Burt, H.M., 2005. Pharmacokinetic study of methotrexate following intra-articular injection of methotrexate loaded poly (L-lactic acid) microspheres in rabbits, J. Pharm. Sci. 94 (6), 1204–1215. doi:10.1002/jps.20341.

Liggins, R.T., et al., 2004. Intra-articular treatment of arthritis with microsphere formulations of paclitaxel: biocompatibility and efficacy determinations in rabbits. Inflamm. Res. 53 (8), 363–372. doi:10.1007/s00011-004-1273-1.

Little, M., et al., 2000. Of mice and men: hybridoma and recombinant antibodies, Immunol Today. 21 (8): 364–370 5699 (September). doi:10.1016/S0167-5699(00)01668-6.

Lopez-Garcia, F., et al., 1993. Intra-articular therapy of experimental arthritis with a derivative of triamcinolone acetonide incorporated in liposomes. J. Pharm. Pharmacol. 45 (6), 576–578.

Lowin, T., Straub, R.H., 2011. Integrins and their ligands in rheumatoid arthritis. Integrins and their ligands in rheumatoid arthritis. Mol. Med. (Berl) 80 (1), 16–24. doi:10.1007/s00109-001-0291-5.

Lu, R., et al., 2020. Development of therapeutic antibodies for the treatment of diseases. J. Biomed. Sci. 27, 1–30.

Matsumura, Y., Maeda, H., 1986. A new concept for macromolecular therapeutics in cancer chemotherapy: mechanism of tumoritropic accumulation of proteins and the antitumor agent Smancs. Cancer Res. 46 (8), 6387–6392.

McInnes, I.B., Schett, G., 2011. Mechanism of disease the pathogenesis of rheumatoid arthritis. New Engl. J. Med. 365 (23), 2205–2219.

Metselaar, J.M., et al., 2003. Complete remission of experimental arthritis by joint targeting of glucocorticoids with long-circulating liposomes. Arthritis Rheumat. 48 (7), 2059–2066. doi:10.1002/art.11140.

Moein Moghimi, S., et al., 2006. Methylation of the phosphate oxygen moiety of phospholipid-methoxy(polyethylene glycol) conjugate prevents PEGylated liposome-mediated complement activation and anaphylatoxin production. FASEB J. 20 (14), 2591–2593. doi:10.1096/fj.06-6186fje.

Nagayoshi, R., et al., 2005. Effectiveness of anti-folate receptor β antibody conjugated with truncated *Pseudomonas* exotoxin in the targeting of rheumatoid arthritis synovial macrophages. Arthritis Rheumat. 52 (9), 2666–2675. doi:10.1002/art.21228.

Nakashima-Matsushita, N., et al., 1999. Selective expression of folate receptor β and its possible role in methotrexate transport in synovial macrophages from patients with rheumatoid arthritis. Arthritis Rheumat. 42 (8), 1609–1616. doi:10.1002/1529-0131(199908)42:8<1609::AID-ANR7>3.0.CO;2-L.

Naor, D., Nedvetzki, S., 2003. CD4 in rheumatoid arthritis. Arthritis Res. Ther. 5 (3), 105–115. doi:10.1186/ar746.

Natarajan, V., et al., 2011. Formulation and evaluation of quercetin polycaprolactone microspheres for the treatment of rheumatoid arthritis, J. Pharm. Sci. 100 (1), 195–205. doi:10.1002/jps.

Naughton, D.P., 2001. Drug targeting to hypoxic tissue using self-inactivating bioreductive delivery systems. Adv. Drug Deliv. Rev. 53 (2), 229–233. doi:10.1016/S0169-409X(01)00229-0.

Nichols, J.W., Bae, Y.H., 2014. EPR: evidence and fallacy. J. Control. Release 190, 451–464. doi:10.1016/j.jconrel.2014.03.057.

Patel, D.D., Zachariah, J.P., Whichard, L.P., 2001. CXCR3 and CCR5 ligands in rheumatoid arthritis synovium. Clin. Immunol. 98 (1), 39–45. doi:10.1006/clim.2000.4957.

Paulos, C.M., et al., 2004. Folate receptor-mediated targeting of therapeutic and imaging agents to activated macrophages in rheumatoid arthritis. Adv. Drug. Deliv. Rev. 56 (8) 1205–1217. doi:10.1016/j.addr.2004.01.012.

Paulos, C.M., et al., 2006. Research article Folate-targeted immunotherapy effectively treats established adjuvant and collagen-induced arthritis. Arthritis. Res. 8 (3), 1–10. doi:10.1186/ar1944.

Qi, R., et al., 2009. PEG-conjugated PAMAM dendrimers mediate efficient intramuscular gene expression. AAPS J. 11 (3), 395–405. doi:10.1208/s12248-009-9116-1.

Ratcliffe, J.H., Hunneyball, I.M., Wilson, C.G., Smith, A., Davis, S.S., 2000. Albumin microspheres for intra-articular drug delivery: investigation of their retention in normal and arthritic knee joints of rabbits, 1, 290–295.

Rhim, T., Lee, D.Y., Lee, M., 2013. Drug delivery systems for the treatment of ischemic stroke. Pharm. Res. 30 (10), 2429–2444. doi:10.1007/s11095-012-0959-2.

Rossin, R., et al., 2007. Small-animal PET of tumor angiogenesis using a 76 Br-labeled human recombinant antibody fragment to the ED-B domain of fibronectin. J. Nucl. Med. 48, 1172–1179. doi:10.2967/jnumed.107.040477.

Saxne, T., et al., 1988. Detection of tumor necrosis factor CY but not tumor necrosis factor P in rheumatoid arthritis synovial fluid and serum. Arthritis Rheumat. 31 (8), 1041–1045. doi:10.1002/art.1780310816.

Scheinecker, C., Smolen, J.S., Redlich, K., 2012. Targeting TNF receptors in rheumatoid arthritis. Int. Immunol. 24 (5), 275–281. doi:10.1093/intimm/dxs047.

Scherer, H.U., Häupl, T., Burmester, G.R., 2020. The etiology of rheumatoid arthritis. J. Autoimmun. 110 (December 2019), 102400. doi:10.1016/j.jaut.2019.102400.

Sharma, G., et al., 2010. Polymer particle shape independently influences binding and internalization by macrophages. J. Control. Release 147 (3), 408–412. doi:10.1016/j.jconrel.2010.07.116.

Shehata, T., et al., 2008. Prolongation of residence time of liposome by surface-modification with mixture of hydrophilic polymers. Int. J. Pharm. 359 (1–2), 272–279. doi:10.1016/j.ijpharm.2008.04.004.

Smits, E.A.W., et al., 2019. The availability of drug by liposomal drug delivery. Invest. New Drugs 37 (5), 890–901. doi:10.1007/s10637-018-0708-4.

Song, H., et al., 2015. A complement C3 inhibitor specifically targeted to sites of complement activation effectively ameliorates collagen-induced arthritis in DBA/1J mice. J. Immunol. 179 (1), 7860–7867. doi:10.4049/jimmunol.179.11.7860.

Sreedharan, S.P., et al., 1993. Cloning and expression of the human vasoactive intestinal peptide receptor. Proc. Natl. Acad. Sci. U.S.A. 90 (19), 9233.

Stevens, C.R., et al., 1991. Hypoxia and inflammatory synovitis: observations and speculation. Ann. Rheum. Dis. 50 (2), 124–132. doi:10.1136/ard.50.2.124.

Streit, M., Detmar, M., 2003. Angiogenesis, lymphangiogenesis, and melanoma metastasis. Oncogene 22 (20), 3172–3179. doi:10.1038/sj.onc.1206457.

Su, C., et al., 2019. Enhancing microcirculation on multitriggering. ACS Nano. 13, 4290–4301. doi:10.1021/acsnano.8b09417.

Sudoł-szopińska, I., Jans, L., Teh, J., 2017. Rheumatoid arthritis: what do MRI and ultrasound show. J. Ultrason. 17 (68), 5–16. doi:10.15557/JoU.2017.0001.

Szekanecz, Z., Kim, J., Koch, A.E., 2003. Chemokines and chemokine receptors in rheumatoid arthritis. Semin. Immunol. 15 (1), 15–21.
Szekanecz, Z., Koch, A.E., 2007. Mechanisms of disease: angiogenesis in inflammatory diseases. Nat. Clin. Pract. Rheumatol. 3 (11), 635–643. doi:10.1038/ncprheum0647.
Teng, Y., van Laar, J.M., Huizinga, T.W.J., 2007. Targeted therapies in rheumatoid arthritis: focus on rituximab. Biologics: Targets Ther. 1 (4), 325–333.
Theunis, C.H., et al., 1994. Biodegradable long-circulating polymeric nanospheres. Science 263 (March), 1600. Available at www.sciencemag.org.
Torchilin, V., 2011. Tumor delivery of macromolecular drugs based on the EPR effect. Adv. Drug Deliv. Rev. 63 (3), 131–135. doi:10.1016/j.addr.2010.03.011.
Trachsel, E., et al., 2007. Research article antibody-mediated delivery of IL-10 inhibits the progression of established collagen-induced arthritis. Arthritis Res. 9 (1), 1–9. doi:10.1186/ar2115.
Vanniasinghe, A.S., Bender, V., Manolios, N., 2009. The potential of liposomal drug delivery for the treatment of inflammatory arthritis. Semin. Arthritis Rheumat. 39 (3), 182–196. doi:10.1016/j.semarthrit.2008.08.004.
Van Roon, J.A.G., et al., 2003. Selective elimination of synovial inflammatory macrophages in rheumatoid arthritis by an Fc c receptor I – directed immunotoxin, Arthritis. Rheum. 48 (5), 1229–1238. doi:10.1002/art.10940.
Van Roon, J.A., Bijlsma, J.W., Lafeber, F.P., 2005. Depletion of synovial macrophages in rheumatoid arthritis by an anti-Fc c RI-calicheamicin immunoconjugate. Ann. Rheum. 64 (6), 865–870. doi:10.1136/ard.2004.028845
van Vollenhoven, R.F., 2009. Treatment of rheumatoid arthritis: state of the art 2009. Nat. Rev. Rheumatol. 5 (10), 531–541. doi:10.1038/nrrheum.2009.182.
Van Vuuren, A.J., et al., 2006. CD64-directed immunotoxin inhibits arthritis in a novel CD64 transgenic rat model. 176 (10), 5833-5838. doi:10.4049/jimmunol.
Veale, D.J., Maple, C., 1996. Cell adhesion molecules in rheumatoid arthritis: implications for therapy. Drugs Aging 9 (2), 87–92. doi:10.2165/00002512-199609020-00003.
Villanueva-romero, R. et al., 2018. Review article. The anti-inflammatory mediator, vasoactive intestinal peptide, modulates the differentiation and function of the subsets in rheumatoid arthritis. J. Immunol. Res., 2018, 6043710. doi:10.1155/2018/6043710.
Wallis, W.J., Simkin, P.A., Nelp, W.B., 1987. Protein traffic in human synovial effusions. Arthritis Rheumat. 30 (1), 57–63. doi:10.1002/art.1780300108.
Wang, D., et al., 2004. The arthrotropism of macromolecules in adjuvant-induced arthritis rat model: a preliminary study. Pharm. Res. 21 (10), 1741–1749. doi:10.1023/B:PHAM.0000045232.18134.e9.
Wasserman, A.M., 2011. Diagnosis and Management of Rheumatoid Arthritis. Am. Fam. Physician. 84 (11), 1245–1252.
Wikaningrum, R., et al., 1998. Pathogenic mechanisms in the rheumatoid nodule comparison of proinflammatory cytokine production and cell adhesion molecule expression in rheumatoid nodules and synovial membranes from the same patient. Arthritis. Rheum. 41 (10), 1783–1797.
Williams, A., et al., 2000. The suppression of rat collagen-induced arthritis and inhibition of macrophage derived mediator release by liposomal methotrexate formulations. Inflamm. Res. 49 (4), 155–161. doi:10.1007/s000110050575.
Wong Donald, D.-Z.K., 1996. Regulation by cytokines and lipopolysaccharide of E-selectin expression by human brain microvessel endothelial cells in primary culture, J. Neuropathol. Exp. Neurol. 55 (2), 148–162.
Yadav, N.,et al, 2016. Solid lipid nanoparticles - a review. Int. J. App. Pharmaceutics. 5, 8–18.
Ye, J., et al., 2008. Injectable actarit-loaded solid lipid nanoparticles as passive targeting therapeutic agents for rheumatoid arthritis, Int. J. Pharm. 352 (1–2), 273–279. doi:10.1016/j.ijpharm.2007.10.014.
Yoo, J.-W., Chambers, E., Mitragotri, S., 2010. Factors that control the circulation time of nanoparticles in blood: challenges, solutions and future prospects. Curr. Pharm. Des. 16 (21), 2298–2307. doi:10.2174/138161210791920496.
Yoo, J.W., Doshi, N., Mitragotri, S., 2011. Adaptive micro and nanoparticles: temporal control over carrier properties to facilitate drug delivery. Adv. Drug Deliv. Rev. 63 (14–15), 1247–1256. doi:10.1016/j.addr.2011.05.004.
Zhigaltsev, I.V., et al., 2016. Production of limit size nanoliposomal systems with potential utility as ultra-small drug delivery agents. J. Liposome Res. 26 (2), 96–102. doi:10.3109/08982104.2015.1025411.
Zhou, M., et al., 2018. Targeted delivery of hyaluronic acid-coated solid lipid nanoparticles for rheumatoid arthritis therapy. Drug Deliv. 0 (0), 716–722. doi:10.1080/10717544.2018.1447050.

Chapter 17

Peptide functionalized nanomaterials as microbial sensors

Shubhi Joshi[a], Sheetal Sharma[b], Gaurav Verma[c,d], Avneet Saini[b]

[a]*Energy Research Centre, Panjab University, Chandigarh, India*
[b]*Department of Biophysics, Panjab University, Chandigarh, India*
[c]*Dr. S.S. Bhatnagar University Institute of Chemical Engineering & Technology (Dr. SSBUICET), Panjab University, Chandigarh, India*
[d]*Centre for Nanoscience and Nanotechnology (UIEAST), Panjab University, Chandigarh, India*

17.1 Introduction

Pathogenic microorganisms induce infectious diseases that have the ability to be communicated among individuals and communities at an exponential rate in a short period of time. Most of the communicable ailments are triggered by microbes such as virus, bacteria, parasites, and fungi (Etukudoh et al., 2020). It has been assessed that approximately half of the global population is at the risk of contracting contagious diseases, making microbial diseases one of the most dangerous threats to mankind (Hui and Zumla, 2019; Hwang et al., 2018).

Detection of disease-causing microorganisms is of foremost relevance, particularly for reasons concerning human wellbeing (Chen et al., 2019). In the food processing industry, environment and defense, the identification, monitoring, and control of pathogens is of utmost importance as these have a direct and indirect effect on health care services (Cho and Ku, 2017; Jin, 2018; El-Din, 2015; Charlet, 2018). Although great strides have been achieved in the field of medicine by developing drugs that are capable of curing infectious diseases but the pandemics caused by viruses such as severe acute respiratory syndrome Coronavirus (SARS-CoV), Middle Eastern respiratory syndrome Coronavirus (MERS-CoV), SARS-CoV-2 has explained the world that microorganisms induced diseases have the potential of affecting humans and disrupting the healthcare system globally (Dhama et al., 2020). The impact of such diseases can be limited by formulation of early warning systems that have the ability to indicate early existence of microorganisms. Therefore, construction of efficient sensing methods for detection of microbial toxins is essential.

Well-established chemical and biological analytical methods currently being used rely on inoculating a culture medium with the sample and breeding the microorganisms followed by counting the colonies after a definite amount of time or on polymerase chain reactions (PCR) synthesizing parts of the contaminant genotype followed by a suitable chemical analysis, for example, electrophoresis, which can also be used for separating entire biological species (Kenndler and Blaas, 2001).

All these methods are highly suitable for the task as they yield desired detection limits. However, the analysis time is usually too long as preliminary step involving cell cultivation takes several days. Timely detection and classification of pathogens remains a critical point since biofilms develop and evolve rapidly thereby causing nosocomial infections in patients. To prevent this, specificity regarding microorganism strain is required to select an appropriate clinical treatment (Ruoss and Wynn, 2019).

Hence, there is an unmet need for development of sensitive approaches, which can sense the pathogens rapidly and are cost-effective as well. In this regard, biosensors are considered as rational substitutes to the existing methods to detect pathogens. Even though limited sample collection and preparation steps are involved while using sensors, still they tend to exhibit outstanding characteristics such as enhanced sensitivity, high specificity along with reproducibility. In addition, they are affordable, rapid and reliable, enabling the medical practitioner to promptly diagnose the symptoms of an illness and prescribe a suitable medication (Ahmed et al., 2014; Burlage and Tillmann, 2017; Caygill et al., 2010; Bahadir and Sezgintürk, 2015; Mokhtarzadeh, et al., 2017). This chapter aims to highlight advances made in the field of biosensors with special emphasis on the formulations and current applications of peptide-functionalized nanomaterials for detection of pathogenic microorganisms.

17.2 Conventional techniques for microorganism detection

17.2.1 Pure culture-based protocols

17.2.1.1 Selective approach

There are incidences where in contrast to the overall population of a mixed community, the microbe of interest is available in relatively small number. Additionally, this organism can be the one that tends to grow slowly on the regular cultural media in comparison to other species (Leclerc and Moreau, 2002). It is desirable and important to initially accomplish an increase in the comparative number of organisms so that they become predominant in the population of microbes and can be isolated in pure culture by application of selective approaches (Gracias and McKillip, 2004). These strategies encourage the growth of the target microbe by suppressing or eliminating the other organisms present in the mixed culture. This can be accomplished by application of different physical, chemical and biological strategies.

17.2.1.2 Enrichment media

Enrichment media works by regulating the nutrients present in the growth media along with controlling culture conditions such as temperature, air supply, light, pH so that it exclusively benefits growth of a particular microorganism (Madhuri et al., 2019). Blood agar is the example of enrichment media where the essential nutrients are supplemented with blood supplements. Similarly, Chocolate agar is an enrichment media supplemented with blood that has been treated to a temperature of 40–45°C due to which the agar imparts brown color.

17.2.1.3 Selective media

Selective media comprise of substances that prevent certain microbes from growing but encourage others to grow. This media is selectively used for growth of particular microbes. Therefore, it is important to select a media for different applications, for instance if a microbe is immune to a particular antibiotic, the antibiotic may be introduced in the medium to suppress the growth of other microbes that do not have resistance toward the particular antibiotic (Bonnet et al., 2020). Selective isolation of gram-negative microorganisms can be achieved incubating a mixed culture on MacConkey agar. It has been reported that microbial species capable of fermenting lactose turn pink or red on the agar whereas, microbes unable to ferment lactose do not exhibit transformation of color. Mannitol salt agar (MSA) is a commonly used medium for detection of gram-positive strains of *Staphylococcus* and *Enterococcus* (Kateete et al., 2010).

17.2.1.4 Differential media

Indicator media utilizes biochemical features of microbes to differentiate the target microbe from other microbes present in the same medium. For instance, Eosin-methylene blue agar (EMB) comprises of lactose and sucrose sugar which causes lactose fermenting microbes to produce acid after fermentation reaction leading to a change in color. This change in color enables differentiation between gram-negative bacterial strains (e.g., *E. coli* and coliforms) that have the ability to ferment lactose from those that cannot (Rajapaksha et al., 2019). Similarly, presence of enterobacteria in a mixed culture can be determined by inoculating the mixed culture on triple-sugar iron (TSI) agar slants. The concept behind differentiation is based on enterobacteria's ability of reducing sulfur and fermenting carbohydrates. It should be noted that several selection media such as MacConkey Agar and MSA also function as differential media (Ifeanyi et al., 2014).

17.2.2 Immunological techniques

Immunoassays are quantitative methods that rely on the precise binding of an antigen which may be a bacterial metabolite (Darwish, 2006). Sensitivity and specificity of these methods is largely dependent on the type of antiserum used due to which careful selection of monoclonal antibodies is recommended. In comparison to polyclonal antibodies, monoclonal immunoglobulins are more suitable as they deliver an abundant amount of a specific antibody (Yan et al., 2015). The immunological approaches facilitate quick, precise and on the spot investigation of numerous pathogenic microorganisms and their toxins. Numerous antibodies are currently being used in different analysis for detection of pathogens. It is a well noted fact that physico-chemical processes of a cell are disrupted in response to external stress due to which the effect of physical strain on antibody-antigen reaction should be thoroughly examined and studied.

17.2.2.1 Enzyme-linked immunosorbent assay

The most frequently utilized immunological technique for pathogen identification is enzyme-linked immunosorbent assay (ELISA). It provides specific and reproducible results by identifying the presence of haptens in a specimen (Vernozy-Rozand

TABLE 17.1 Different types of ELISA-based assays.

Types of ELISA	Principle	Advantages	Limitations
Direct ELISA	Works by detecting and measuring a particular antigen in a sample using a specific antibody conjugated to the enzyme.	• Rapid • Reduces incidences of cross-reactivity of secondary antibody	• Limited sensitivity. • Expensive process. • Labeling of every primary antibody is time-consuming.
Indirect ELISA	Two-step method using labeled secondary antibody for detection.	• Enables investigation of wide range of antisera	• High chances of cross-reactivity. • Nonspecific signal issues. • Additional incubation step is mandatory.
Sandwich ELISA	Measures the quantity of antigen sandwiched among two layers of antibodies.	• Rapid • Precise detection • Reproducible results • High sensitivity • Purification of antigen is not required	• Requires use of selective antibodies. • At least two antigenic sites must be present in the antigen to be tested.
Competitive ELISA	Plate-based assay where surface is coated with antigen–antibody complex reactive to target molecule.	• Requires application of nonpurified antibodies	• Higher antigen concentration leads to weak signal production.
Multiplex ELISA	Provides multiple readings using a single sample. It is accomplished through antibody array.	• Quick • Cost efficient • Low sample volume required • Reproducible results	• Requires trained staff. • Cross-reactions need to be minimized.

et al., 2004). Different types of ELISA-based assays along with their advantages and limitations have been summarized in Table 17.1. ELISA-based assays are frequently being used for rapid detection of members belonging to Coronavirus (CoV) family. For instance, Fukushi and his lab partners developed a monoclonal IgG antibody which targeted S protein of MERS virus. They further applied this labeled antibody to formulate a competitive ELISA which could produce results in two and a half hours. The assay exhibited sensitivity of 100% for 1:512-diluted camel serum specimen (Fukushi et al., 2018). Sandwich-based ELISA assay for detection of *Vibrio parahaemolyticus* was reported by Kumar et al. (2011). Monoclonal antibodies were produced against TDH-related hemolysin protein of the pathogenic microbe. The reported assay exhibited decent sensitivity in detecting presence of *Vibrio parahaemolyticus* in sea food specimen and can be applied as a quick method to detect gastrointestinal illness causing microbes in food samples.

In order to prevent fetal impairment caused by Zika virus induced infection, Ehmen and co-workers assessed use of ELISA-based tests for precise ZIKV IgG identification. In this study they formulated two ZIKV IgG ELISAs by merging dissimilar marked ZIKV antigens. High test performance and enhanced detection specificity was observed. It is believed that the developed sensor shall be useful in detecting travelers infected with Zika virus along with their source of origin (Ehmen et al., 2021).

17.2.3 Nucleic acid-based assays

Detection approaches based on nucleic-acid sequencing are considered to be the most efficient techniques in microbiology. Invention of polymerase chain reaction (PCR) proved to be a scientific revolution in molecular biology as the technique rapidly identified disease-causing microorganisms at extremely low detection limits (Mackay, 2004). The technique remains exceptional as it enabled detection of clinically relevant microorganisms without the need of culturing, which is possible as it enables production of numerous copies of the target DNA (Albinana-Gimenez et al., 2009). PCR works by intensifying the nucleic acid target present in a biological specimen by application of a thermostable polymerase enzyme followed by synchronized recognition and quantification. There are three stages in a PCR procedure. Initially, temperature ranging above 90°C is applied to divide a double-stranded DNA by disintegrating the hydrogen bonds present between the complementary base. Further, oligonucleotide primers are annealed at temperature ranging between 50–60°C, followed by optimal primer extension which occurs at 70–78°C. Majority of the oligonucleotide target duplexes have developed at this stage.

Although PCR provides specific and reliable results, it cannot be ignored that the it is time-intensive technique and requires complex processing. Presence of an established sequence data of the target gene is essential to conduct the analysis. Also, because of its high sensitivity, PCR is prone to false positive outcomes. These are usually triggered by contamination caused by extraneous naked nucleic acid, which may come from different sample or contaminated laboratory surroundings. In comparison to PCR, real-time PCR is faster, provides enhanced sensitivity, reproducibility and minimizes the hazard of contamination (Toze, 1999). Nowadays, there are numerous biochips formulated with PCR like characteristics are being formulated for the detection of detection of microbial cells and their toxins. These significant improvements have facilitated a widespread acceptance of PCR in the field of Diagnostic microbiology. But limitations such as high cost of equipment and requirement of trained personnel still hinders its application on a large scale (Botes et al., 2013).

17.3 Principle behind using biosensors for microorganism detection

A biosensor can be described as a compact device that uses a biological substance to identify a target of interest and generate a measurable signal. They are formulated by combining the outstanding selectivity of biological interactions found in nature with the processing power of modern microelectronics (Hoyos-Nogués et al., 2016; Wolfbeis, 2008; Zhang, 2016b). A sensor for detection of pathogenic microorganisms has the ability to detect a microbe using a transducer and produce signals that is quantitative in nature. These signals are proportional to analyte concentration in the specimen (Dai and Choi, 2013; Su et al., 2011). Usually, sensors comprise of two main components, that is, biorecognition element that associates with the target which can be a microbial cell and a processor that translates the interaction into an observable signal (Lazcka et al., 2007). Antibodies, enzymes, peptides, aptamers are frequently used as biorecognition elements as they have the ability to identify the target with enhanced specificity.

Ever since the invention of first sensing device in 1962, formulation and application of sensing devices were intensively studied with an objective to control medical centers and encourage at-home diagnostics, but it did not materialize (Clark Jr and Lyons, 1962). However, in the past decade sensors having the ability to detect and monitor microorganisms from water, food and clinical samples have been reported in significant numbers (Alafeef et al., 2020; Adak et al., 2013; Byeon et al., 2015; He et al., 2014; Hesari et al., 2016; Li et al., 2010).

As mentioned in the above section, earlier, incidences of microorganism induced infections were evaluated by growing the test sample in growth medias followed by their identification by evaluation of their growth along with changes induced by them in the media or surrounding atmosphere during incubation period (Turner et al., 1990).

For instance, in 2012 Filipiak and his team reported detection of Lung infection caused by *Staphylococcus aureus* and *Pseudomonas aeruginosa* by identifying typical metabolic by-products (Filipiak et al., 2012). Guerin provided a method to evaluate efficiency of resazurin in detecting growth of microorganisms in comparison of turbidity analysis. The technique involved incubating the dye with different organic compounds mostly used as growth enhancers (Guerin et al., 2001). This procedure can be completed before determining most probable number but its limitation is that it takes 48 h to give a colorimetry-based shift. U.S. Pat. No. 4,152,213, reported a technique in which occurrence of bacterial species consuming oxygen was estimated. The procedure involved evaluation of oxygen reduction over a period of time in an impenetrable container.

Firstly, the mentioned approaches involve application of radiometric-based assays or destructive techniques which comprise opening of sealed vials with an objective to measure release of gases, evaluation of metabolic residues, fluctuations in pH, color, and loss of transparency. These protocols do not permit taking continuous and multiple readings as there are high chances of contamination during such experiments as every time the container has to be opened exposing its contents to the atmosphere, thereby altering sensitivity. Second, conventional detection procedures require support of other spectroscopic techniques as by performing individual culture-based assay it is difficult to conduct qualitative as well as quantitative detection of microbial communities. Although culture-based strategies are popular among researchers owing to their familiarity with the protocols, user-friendliness and reliability on the obtained results but these approaches are not cost-effective for analyzing large numbers of samples, thereby restricting applicability. Table 17.2 summarizes the desired properties for construction of an ideal pathogen biosensor.

While there has been a great deal of research activity involved in the production of novel sensors specifically for sensing presence of pathogens, the moment has arrived to make them commercially accessible (Wolfbeis, 2008). Advances in field of nanotechnology has provided the world with novel nanostructured substances that have the potential to provide improved biostability, electron transfer characteristic, and exorbitant detection area.

TABLE 17.2 Desirable characteristics for construction of an ideal pathogen biosensor.

Parameters	Quality	Refs.
Sensitivity	10^3–10^1 CFU/mL	Bharrdwaj et al, 2017; Habimana et al., 2018
Specificity	• Capability to distinguish between different serotypes of different pathogenic strains (e.g., can differentiate E. coli Nissle 1917 from E. coli O157:H7). • Negligible background disturbance. • Ability to operate in complex matrices.	Almeida et al., 2018; Piednoir et al., 2015
Speed of detection	• Provide immediate readout.	Cosio et al., 2015; Patel et al., 2016
Reproducibility	• Deliver identical reading for same sample analyzed at different intervals.	Eggins, 2013; Arora, 2013
Size	• Compact, portable device that can operate at the site of interest.	Bhalla et al., 2016; Srinivasan and Tung, 2015
Sample processing	• Label free with minimal sample processing.	Bhardwaj et al., 2017; Matthew, 2006
Stability	• Biorecognition element must be stable at different environmental conditions such as temperature and pH.	Andrews, 1996; Schulz-Schönhagen et al., 2019
Required skills	• Should be easy to use by individuals without any special training.	McMeekin, 2003; Baeumner, 2002

17.4 Commonly used biosensing recognition elements

Recognition components are crucial elements in a biosensor. They assist in the identification of the different substances and impart proficiency during sensing by providing both sensitivity as well as specificity (Zourob et al., 2010; Bhalla et al., 2016). It has been documented that the selection of a recognition component significantly affects the output displayed by the sensing device. Fig. 17.1 illustrates the most commonly used biological substances that are capable of recognizing a specific analyte and are therefore being applied in the construction of biosensors (Matthew, 2006). Besides exhibiting a high binding affinity for the target analyte, these elements should also have a good stability.

Currently, microorganisms are detected using different biorecognition elements like antibodies, aptamers, bacteriophages, carbohydrates and peptides. It is worth noting that any receptor that has the ability to identify the presence of a bacterial cell and attach itself to it can be utilized for the construction of a bacterial sensor (Justino et al., 2015). Each recognition element exhibits unique characteristics due to which understanding the influence of biorecognition element on overall biosensor performance is crucial in the planning stages to promote the success of novel biosensor development (Hatchett and Josowicz, 2008; Morales and Halpern, 2018).

17.4.1 Antibodies as biosensing recognition elements

Naturally occurring proteinaceous constructs having a three-dimensional structural conformation are described as antibodies. Usually, they are ~150 kDa in size. The 3D protein structure of antibodies creates a unique recognition pattern with high specificity and accuracy for the bio analyte detection. Antibodies can be recognized in biological pathways and can be extracted and purified for application as recognition elements in sensors

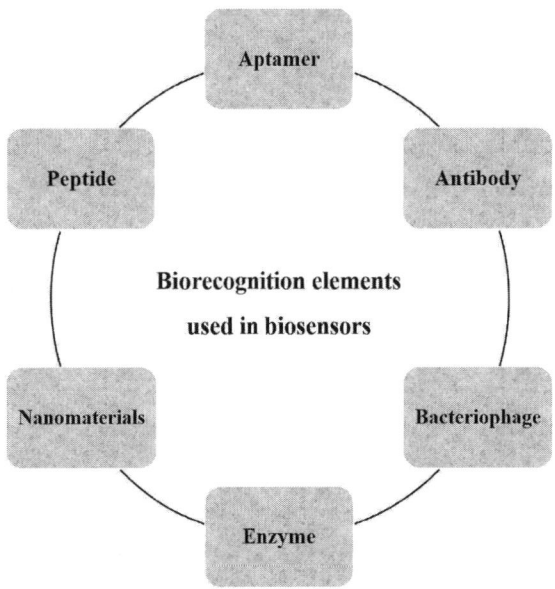

FIG. 17.1 Commonly used biorecognition elements for bacterial detection using a sensor.

(Crivianu-Gaita and Thompson, 2016). Antibody-based receptors are categorized on the basis of affinity. In such sensors, the signal relies on the binding incidence which leads to the formation of an immunocomplex between antibody and antigen. This binding activity can be observed using colorimetric as well as piezometric approaches (Choi et al., 2018; Scarano et al., 2010; Skottrup et al., 2008; Chiu et al., 2017; Su and Li, 2004). The most common method to bind antibodies onto sensor surface is by forming a brush-like array by application of covalent modifications (Caelen et al., 2002; Veiseh et al., 2002). Antibody-based detection techniques are powerful and act as versatile tools for various molecular and cellular analyses, environmental monitoring, and clinical diagnostics due to the specificity and sensitivity imparted by the biological elements.

In spite of the various advantages, there are a few aspects that limits their applicability, for instance, antibody synthesis requires in-depth testing on animal models which is an extremely laborious procedure. There are inconsistencies in quality and purity during batch operations (Miller and Sikes, 2015). Also, they exhibit low stability under harsh environmental conditions and are susceptible to proteolytic degradation (Justino et al., 2015). Often, upon conjugation they display nonspecific binding. There is a need of one binding pair for each sensed target (Sharma et al., 2016; Shen et al., 2017). In spite of the commonly recognized limitations of antibodies as receptors, they are still regarded as an indispensable component in the sensing community.

17.4.2 Aptamers as biosensing recognition elements

Aptamers are single-stranded oligonucleotides designed through a combinatorial selection process called Systemic Evolution of Ligands by Exponential Enrichment (SELEX) (Zhou et al., 2018; Radom et al., 2013). Aptamers are attractive alternatives to antibodies in some cases as it offers numerous advantages such as high affinity toward targets, stability at wide range of parameters such as pH, temperature and ionic strength (So et al., 2008). Aptamers are easy to produce, are nontoxic in nature and exhibit resistance toward denaturation (Zhou and Rossi, 2017; Zelada-Guillén et al., 2009; Lim et al., 2010). SELEX is an iterative process to search a library of randomly generated oligonucleotide sequences for strong binding affinities between the target analyte and oligonucleotide sequences, ensuring a selective and strong interaction pair. The SELEX cycle starts with incubation of the target bioanalyte with an oligonucleotide library containing all potential aptamer sequences. Unbound aptamer sequences are then removed only retaining bound aptamer sequences for polymerase chain reaction (PCR) amplification to regenerate the oligonucleotide library for the next SELEX round. Aptamers are typically 100 base pairs in length compromised of a 20–70 randomized base pair binding region in the center with constant primer binding regions at both ends. SELEX is a beneficial biorecognition element discovery tool providing researchers with the ability to tailor a sequence for a target bioanalyte. A major drawback is that the SELEX method is costly, requiring multiple iterations using a large library of oligonucleotides each time. However, cost is an obstacle that could be mitigated with further research and development (Crivianu-Gaita and Thompson, 2016).

17.4.3 Bacteriophages as biosensing recognition elements

Bacteriophages are viruses that infect and replicate within bacteria. Reports suggest that, they possess specific mechanisms to recognize bacteria and can be used for bacterial detection (Templier et al., 2016). Interestingly, a key feature of phage-based sensors is their ability to distinguish between viable and inactive bacteria, as they can only replicate within viable bacterial cells. Despite the advantages offered by these two approaches, several limitations are associated to their use, including their particularly difficult handling and immobilization on the biosensor, together with their low stability over time (Singh et al., 2014; Richter et al., 2018).

17.4.4 Carbohydrates as biosensing recognition elements

Cell surface carbohydrates (glycans) and adhesin molecules are major components of the outer surface of cells and are major characteristics of the cell types. Many adhesin molecules are lectins that have carbohydrate binding activities. Glycans and adhesins are the first interface to the biotic and abiotic environment of the cell. The interactions of glycans with carbohydrate binding proteins (lectins) are perhaps the most significant and fundamental molecular recognition events in biological systems including bacterial pathogenesis.

Lectin/carbohydrate recognition is the most commonly used approach to detect bacteria. Lectins, which are proteins that recognize carbohydrates, can be used to specifically react with bacterial carbohydrates. Carbohydrate-based sensors are an attractive alternative due to the chemical stability of carbohydrates and good grafting properties. Nonetheless, the detection of bacterial species using these molecules can produce false positives in complex samples because several lectins can bind different carbohydrates, as well as different carbohydrates can bind the same lectin, thus causing a significant reduction in specificity (Brosel-Oliu et al., 2015).

The substrate has binding sites for the tags and activated reactive groups capable of forming covalent bonds with the biomolecules. Zhang and co-workers reported the formulation of a microorganism detection sensor chip. The study reported application of carbohydrate-based sensor working in a synchronized manner with surface plasmon resonance or quartz-crystal microbalance techniques (Zhang et al., 2013). The membrane-based immobilization method presented a higher capture density than the surface construction based on gold-thiolate linkages. This higher capture density allowed the observation of protein biomarkers from both bacteria and virus samples. It can be stated that both lectin- and carbohydrate-based biocapture surfaces are useful analytical tools for the isolation and concentration of microorganisms from complex sample matrices. Carbohydrate complexed surfaces maintain their affinity from microorganisms and are less readily blocked by contaminants in the samples.

17.4.5 Peptides as biosensing recognition elements

In an effort to overcome the limitations exhibited by the above mentioned recognition elements, researchers have started studies using peptides as alternative recognition molecules (Yoo et al., 2014). Peptides compose of amino acids connected by amide linkages leading to formation of short chain like structures (Zhang, 2016a). Peptides are naturally occurring molecules present in all forms of life ranging from humans to microscopic organisms (Zasloff, 2002, 2019). They can be chemically synthesized in large quantities by employing solid-phase peptide synthesis techniques. Peptides comprise of a broad class of biological molecules that have the tendency to exhibit a wide range of properties (Kulagina et al., 2006).

Antimicrobial peptides (AMPs) are a class of peptides that are present as first line of defense against microbial invasion (Porter et al., 2002; Casteels et al., 1989). These peptides are short in length, mostly exhibit a cationic charge and are amphipathic in structure (Ageitos et al., 2017; Nguyen et al., 2011). Besides displaying broad spectrum antimicrobial activity, AMPs exert tumoricidal activity at higher concentration. Immunomodulatory activity exhibited by AMPs has a significant role in suppressing the inflammatory response by stimulating the immune system (Haney and Hancock, 2013). Also, AMPs with membrane translocation activity have the ability to pass through cell membrane barriers without disrupting the membranes are being used as drug delivery vehicles (Brogden, 2005; Henriques et al., 2006). AMPs have come a long way since their discovery, earlier they were considered as agents with the ability to permeabilize microbial membranes followed by growth inhibition. Whereas, nowadays AMPs are regarded as molecules that have the ability to exhibit multidimensional therapeutic as well as sensing related functions (Zhou et al., 2018; Hoyos-Nogués et al., 2018). Application of peptides as biorecognition components is a comparatively recent concept. But due to the potency of peptide molecules to detect phosphate groups present on lipopolysaccharides (LPS) and their ability to retain their specificity and stability under adverse environmental conditions, they are being widely studied to formulate sensors that detect microorganisms (Lim et al., 2015; Etayash et al., 2014; Pavan et al., 2012). Fig. 17.2 illustrates different parts of a peptide-based biosensor.

Novel peptide molecules with desirable characteristics can be designed using *in silico* approaches involving modification of peptide sequence (Floris and Moro, 2012). According to the desired application spacers and anchoring groups can also be incorporated in the peptide sequence to enable their efficient immobilization on different surfaces (Zasloff, 2002).

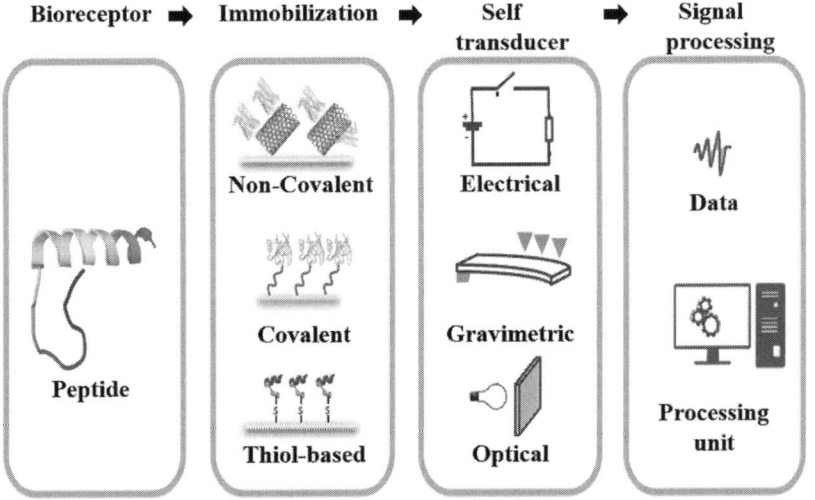

FIG. 17.2 Different parts of a peptide-based biosensor *(reprinted with permission from Barbosa et al., 2018)*.

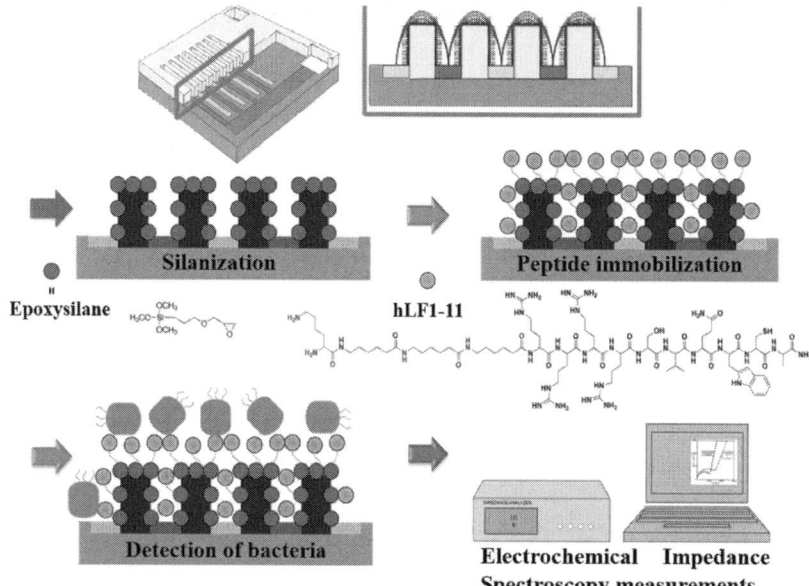

FIG. 17.3 Graphical illustration describing steps involved in construction of peptide immobilized sensor for detection of Streptococcus sanguinis from saliva *(reprinted with permission from Hoyos-Nogués et al., 2016)*.

Majority of the reported sensors which use peptides as recognition elements use Electrochemical impedance spectroscopy (EIS) for evaluation of sensing changes (Fig. 17.3). AMP magainin I (GIGKFLHSAGKFGKAFVGEIMKS) functionalized label-free sensor for detection of *E. coli* was reported by Mannoor et al. (2010). The peptides were conjugated onto gold surface by application of covalent interaction mechanism using C-terminal cysteine residues. The formulated sensor exhibited a sensitivity of 1 bacterium/μL, which was investigated by using impedance spectroscopy. Difference between gram-positive and gram-negative microorganisms could also be determined thereby exhibiting selectivity which is considered as an important characteristic of a nanosensor (Fig. 17.4).

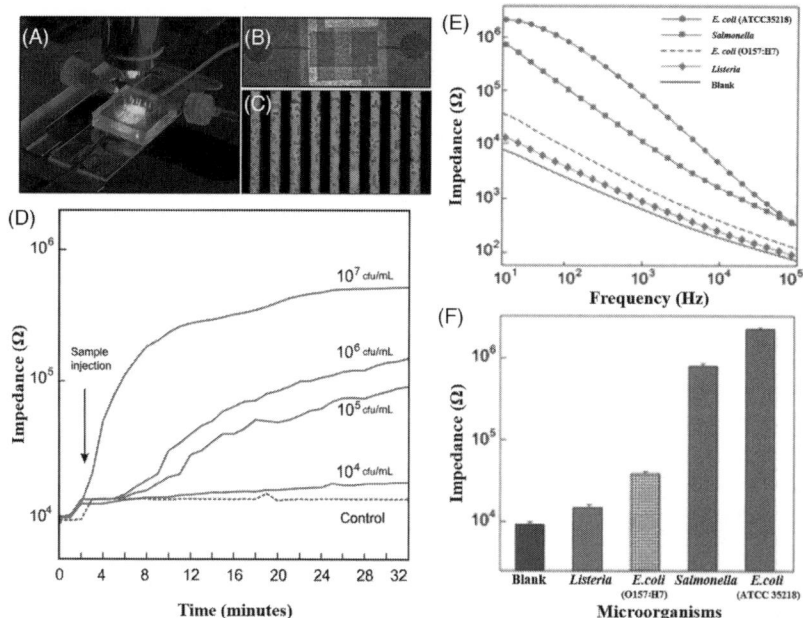

FIG. 17.4 (A) Schematic illustration of a microfluidic flow cell. (B) Optical micrograph of the microfluidic channel with an embedded interdigitated microelectrode array chip. (C) Optical image of the embedded microelectrode array after exposure to 10^7 CFU/mL bacterial cells for 30 min. (D) Real-time monitoring of the interaction of the AMP-functionalized sensor (and an unlabeled control chip) with various concentrations of *E. coli* cells. (E) Impedance spectra of the AMP-functionalized microelectrode array after interaction with various bacterial samples (10^7 CFU/mL). (F) Impedance changes associated with various bacterial species at 10 Hz *(reprinted with permission from Mannoor et al., 2010)*.

17.5 Advantages and challenges of using peptide-based detection of microorganisms

Advances in technology have made it possible for peptides to be used for formulation of detectors. Although application of peptides as recognition components is a new approach, but peptide functionalized biosensors are being used to supplement conventional techniques for microorganism detection since the past decade (Zhou et al., 2018). Among the wide category of peptides, antimicrobial peptides are considered as lucrative substances for designing of sensing devices as they have the ability of identifying a wide spectrum of pathogenic microorganisms (Karimzadeh et al., 2018). Their tendency to exhibit selectivity while interacting with the outer membrane of microbes is the basic principle behind their application as microbial sensors (Sivakumar et al., 2005). Due to the various hydrophobic and electrostatic interactions, relationship between AMP and cell membrane of microbes is formed. Physiochemical properties of peptides such as affinity toward water, charge, presence of hydrophobic and hydrophilic groups plays important role in recognition of microorganism (Shi et al., 2017). Peptides exhibit specificity which can be attributed to distinct chemical affinities of the molecules. Peptides display outstanding properties enabling their applicability as robust, versatile and fast sensors as they are extremely stable under undesirable circumstances. They can come in contact with the natural environment and interact with invariant substances present on the surface of target microorganism without experiencing alteration in its activity. Numerous advantages of peptide-based microbial sensors have been discussed, there are few limitations that need to be addressed while constructing the sensing devices. Microorganism detection using peptide is not always possible in complex samples as there are few reports that state long response time due to existence of diffusion barriers (Liu et al., 2015). Since it is a cumbersome task to identify a single microbial strain from a mixture of nontarget microorganisms. The real-world applicability of peptide-based sensors must be enhanced as majority of the reported studies have been conducted for sensing common microorganisms and under controlled laboratory environment (Qiao et al., 2020). Also, entrapment of antigen also leads to decreased sensitivity of the sensor. By conjugating peptides with nanostructured materials that have vast surface area and favorable properties of forming sensors, peptides applicability can be enhanced. There are limited studies that discuss the analytical problems faced during sensor synthesis, by conducting and providing in-depth validation reports a better insight can be obtained regarding precision, sensitivity and robustness of peptide functionalized sensors.

17.6 Properties of nanomaterials making them suitable for construction of microbial sensors

17.6.1 Carbon-based nanoparticles

Andrade and his colleagues established an electrochemical impedance-based sensing mechanism to detect different gram positive (*Bacillus subtilis*, *Enterococcus faecalis*) and gram negative (*Escherichia coli*, *Klebsiella pneumoniae*) microbes, using peptide fabricated CNT nanostructures. Antimicrobial peptide Clavanin A was covalently linked with CNT using 1-Ethyl-3-(3-dimethylaminopropyl)carbodiimide (EDC) and N-hydroxysuccinimide (NHS) cross linkers (Andrade et al., 2015). The sensor was capable of detecting pathogens in a range of 10^2–10^6 CFU/mL. The formulated sensor efficiently differentiated the bacterial population according to their gram strain. CNTs are usually incorporated in sensing formulations to enhance detection sensitivity and also to reduce overpotential (Ahuja and Kumar, 2009; Jacobs et al., 2010). The sensitivity exhibited by CNTs toward adsorption of molecules along with variations in the surrounding environment makes them ideal for sensor construction (Das et al., 2011; Li et al., 2019). Incorporation of CNTs enables formation of lightweight sensing devices (Tasis et al., 2006; Kuang et al., 2010). Since single walled CNTs display low selectivity against defined targets, they are often functionalized with biomolecules such a enzymes, proteins, nucleic acid, aptamers and peptides in order to enhance their selectivity (Pan et al., 2019). Naik group used molecular dynamics-based simulations to illustrate peptide–CNT interaction *in silico*. The studies shown in Fig. 17.5 demonstrate that the folding conformation of peptide (P1ASP1C) retained its structure after interaction with CNTs but changes in wrapping conformation were detected. The interaction behavior was further confirmed using atomic force microscopy and circular dichroism studies. Although, interactions provide significant information regarding adsorption behavior of a conjugate, selectivity, and adaptability toward different environmental conditions, yet detailed evaluation of the interaction behavior between nanoparticles and peptide molecules is an exciting yet unexplored arena.

Graphitic derivative, graphene oxide (GO) offers significant physio-chemical properties that are favorable for the synthesis of microbial sensors. Since pristine graphene and reduced graphene oxide (rGO) lack presence of oxygen functionalities and is hydrophobic in behavior it is difficult to functionalize them with biomolecules due to which researchers prefer using GO for sensing-based applications (Wisitsoraat and Tuantranont, 2013; Joshi et al., 2020). Due to presence of sp^3 carbon functional groups, high volume-to-surface ratio and structural defects GO forms conjugates via covalent as well as noncovalent interaction mechanisms (Geldert et al., 2017; Joshi et al., 2020). It has been reported that presence of epoxides, hydroxyl along with carboxyl moieties is essential for bonding between peptide molecules and GO to take place.

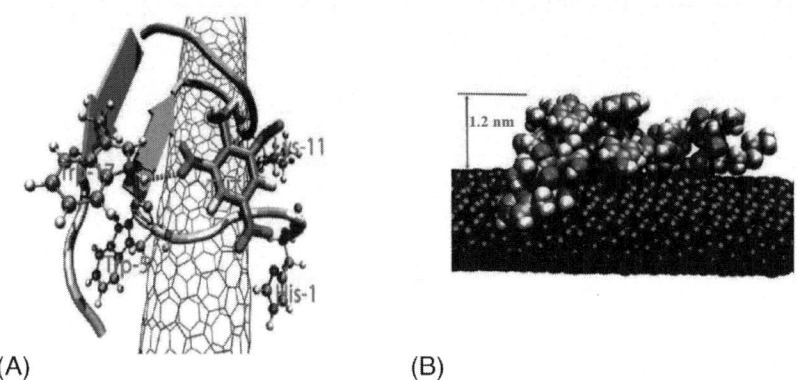

FIG. 17.5 (A) Computational modeling depicting interaction of 2,4,6-trinitrotoluene (TNT) with peptide-CNT hybrid by formation of a hydrogen bond along with π–π interaction with the single-walled nanotube (SWCNT) surface. (B) Illustration depicting adsorption of peptide on (SWCNT) using MD stimulation *(reprinted with permission from Kuang et al. (2010).*

There are numerous reports describing application of peptide functionalize GO as pathogen sensors. But, recently Fan et al. (2020), reported construction of a turn-off sensor with the ability to differentiate between bacterial strains. The sensor was formulated using GO and Ib-AMP4 peptide. Polyhistidine-based adaptor was attached with the conjugate which aided in bacterial species differentiation. The adapter worked by interacting with GO which further facilitated GO in quenching the fluorophore labeled peptidic molecules. The novel sensor exhibited a quenching proficiency of 99.97% and reliable reproducibility. It is believed that with few modifications the sensor applicability can be extended toward detection and differentiation of other microbial organisms as well.

17.6.2 Metallic nanoparticles

Gold nanoparticles have numerous desirable properties such as ease of modification, biocompatibility with biomolecules, robust local surface plasmon resonance (LSPR), and exceptional optical properties which makes them useful for sensing-based applications (Roh et al., 2011; Homola et al., 1999; Xiaohua and El-Sayed, 2011; Mocan et al., 2017; Zheng et al., 2019; Drake et al., 2007). Bacterial sensor for detection of *E. coli* and *S. aureus* was formulated by conjugating gold nanorods with P937 peptide. Upon incubation of bacterial culture with the formulated Au-peptide conjugate the absorbance peak using LSPR displayed improved sensitivity toward variations in microbial concentrations. The sensor enabled quick analysis and can be used for real-time detection of microorganisms (Chen et al., 2018). Gold nanoclusters immobilized with self-assembled Leucocin A peptide was used to design a sensor for detecting presence of *Listeria monocytogenes* in milk samples (Hossein-Nejad-Ariani et al., 2018). The C-terminal region of Leucocin A exhibits enhanced specific binding toward mannose phosphotransferase structure existing on surface of the pathogen. The sensor was constructed by immobilizing the peptide on glass substrate followed by labeling by fluorescent gold nanoparticles having size less than 3 nm. It was observed that the formulated sensor was able to detect the bacteria rapidly with an exposure limit of 2×10^5 CFU/mL (Hossein-Nejad-Ariani et al., 2018).

Similarly, peptide immobilized magnetic sensor was formulated for detection of *S. aureus*. The peptide was inserted among electrode made of gold and magnetic nanoparticles. *S. aureus* protease cleaved the peptide causing displacement of the magnetic particles thereby diminishing appearance of black color from the sensor surface and revealing a golden color. Besides being rapid, the reported sensor provided a reliable and sensitive detection of other microorganisms as well (Eissa and Zourob, 2020).

17.6.3 Magnetic nanoparticles

Magnetic nanoparticles facilitate rapid capture of bacterial cells by separation of the microbe from biological specimen by application of magnetic materials externally (Rios and Zougagh, 2016; Wang and Hu, 2009). Another advantage of using magnetic nanoparticles is the high capture efficacy which can be attributed to the presence of abundant binding sites (Sobczak-Kupiec et al., 2016; Ahmadi et al., 2020). Occurrence of binding events between the external magnetic field and microbe can be monitored and controlled by conjugating biomolecules such as antibodies and peptides with the magnetic nanoparticles (Ravindranath et al., 2009; Chu et al., 2013; Rocha-Santos, 2014; Cardoso et al., 2018). Lipopolysaccharides are extremely toxic components; it is crucial to detect their presence in therapeutic products followed by immediate removal. Currently, the techniques used to detect endotoxins are time consuming.

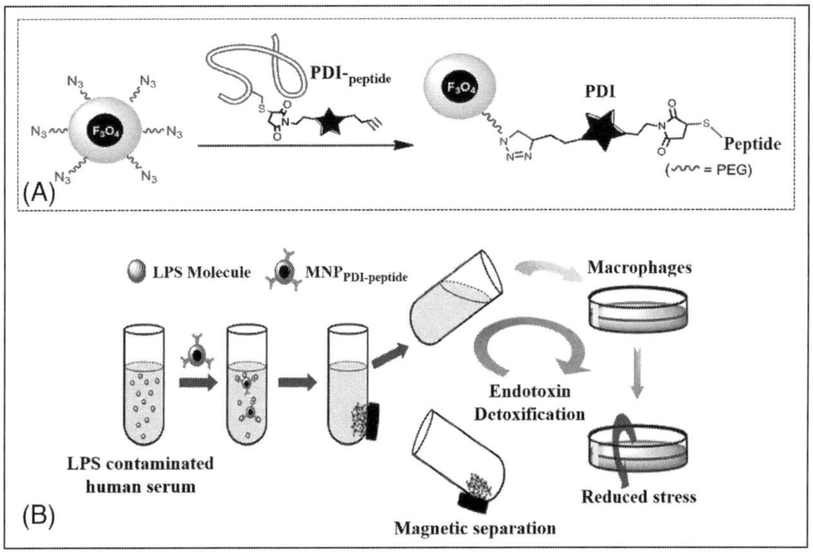

FIG. 17.6 (A) Synthesis of peptide functionalized magnetic nanoconstructs. (B) Schematic illustration representing removal of lipopolysaccharides using peptide functionalized magnetic conjugate *(reprinted with permission from Liu et al., 2014)*.

In order to provide rapid, hassle free identification Liu and co-workers reported a novel approach for detection and elimination of lipopolysaccharides specifically present on cell membrane of gram-negative bacteria. The technique involved coupling of perylene diimide with peptide (YVLWKRKRKFCFI–NH$_2$) and magnetic Fe$_3$O$_4$@SiO$_2$ nanoconstructs using covalent interaction. It was observed that absence of lipopolysaccharide led to self-quenching, whereas its existence would induce conformational variations in the peptide sequence causing disturbance of hydrophobic assembly thereby causing fluorescence retrieval (Fig. 17.6).

Further, the captured lipopolysaccharides could be removed from the sample using an external magnetic source. The technique being simple and cost effective should be used for identification and removal of different pathogens.

17.6.4 Quantum dots

Fluorescent synthetic atoms that display different energy levels and can be modified by adjusting the size of their bandgap are known as Quantum dots (Ghasemi et al., 2009). Their size ranges from 2 to 10 nm and their physical diameter is lesser in comparison to the radius of Bohr (Ghasemi et al., 2009; Valizadeh et al., 2012). These nanosized crystals are semiconducting in nature and exhibit outstanding resistance against photobleaching (Yoon, 2013; Shao et al., 2011). Owing to their surface chemistry and unique optical properties quantum dots are popularly used for formulation of optical sensors (Orellana, 2006; Frasce and Chaniotakis, 2009). Although there is a plethora of studies reporting formulation of quantum dots-based sensors, there are limited studies describing peptide conjugated quantum dots for microorganism sensing (Valizadeh et al., 2012). Chen and his lab mates reported the formation of zinc oxide quantum dots coupled radiolabeled UBI$_{29-41}$ nanocomposite for detection of bacterial infection and discrimination from general swelling. Since, the reported technique is flexible and noninvasive in nature, its applicability can be expanded for image-based detection of other disease-causing agents as well.

17.7 Techniques enabling microorganism detection

17.7.1 Colorimetric detection

Colorimetric approaches have received considerable amount of interest as these techniques are user-friendly and enable on-site recognition without use of highly sophisticated instruments (Le et al., 2020; Alamer et al., 2018; Paniel et al., 2013). Colorimetry is a quantitative technique used to estimate concentration of a specimen by estimating adsorption at a particular wavelength and intensity (Nath and Chilkoti, 2004; Vs et al., 2017). It is used frequently for the detection of microorganisms and it also enables rapid quantification of their growth (Ebralidze et al., 2019; Gilchrist and Nobbs, 1999). Using the principle of colorimetry, an instrument free novel strategy for detection of *E. coli* and *S. aureus* was reported. The strategy used cationic iron oxide-chitosan nanoconjugate which upon interaction with anionic bacterial membrane exhibited a calorimetric response which was observed by naked eye. The reported strategy was facile and rapid as it enabled recognition of 10^2 CFU/mL bacterial cells within 10 min (Le et al., 2020). Using similar principle, a colorimetry-based disposable device to detect Middle East respiratory syndrome coronavirus (MERS-CoV) along with other viral strains was reported by Teengam and her colleagues. The device was formulated using silver nanoparticles as a DNA recognition reagent and was fabricated on a wax paper. The device was simple to fabricate, exhibited enhanced sensitivity and a detection limit of 1.53 nM for MERS-CoV. By extending the application of the reported colorimetry-based strategy, on-site identification of pathogenic microorganisms should be achieved at locations where there is no access to sophisticated equipment (Teengam et al., 2017).

17.7.2 Fluorescence-based detection

Fluorescence-based detection systems have dominated the area of sensing engineering due to the accuracy, sensitivity, and ease of converting optical signals into machine-readable form exhibited by them (Lübbers, 1992). Their applicability is often hindered due to the inability to obtain a low detection limit. This can be attributed to numerous factors such as restricted extinction coefficients; low quantum yields of organic dyes and low dye-to particle labeling concentration (Rajendran et al., 2014; Zhong, 2009). Usually dyes like SYBR green, resazurin, ethidium bromide and propidium iodide (PI) are used but surge in nanotechnology-based research has provided nanomaterials such as quantum dots, metallic materials and silica nanostructures as alternate detection labels (Sun et al., 2015; Tuantranont, 2012). Although in comparison to basic fluorescence-based detections, fluorescence resonance energy transfer (FRET) is considered a superior alternative for detection of microorganisms and evaluation of biomolecule interactions.

17.7.3 Microscopic techniques

Before the discovery of advanced analytical techniques and sophisticated equipment microorganisms were inoculated and grown in suitable culture mediums and studied directly using microscopic techniques. In the year 1676, Anthony Leeuwenhoek was the first to observe the morphology of a bacterial cell by means of a basic microscope. These days the structure, shape, and morphology of the microorganisms can be analyzed using a wide variety of microscopic techniques such as light microscope, phase and differential contrast, fluorescence, field emission scanning electron microscope (FESEM), and transmission electron microscope (TEM). Microscopic techniques are dependable and cost-effective approaches (Mobed et al., 2019). But their usage is limited as it is difficult to observe small sized, transparent microorganisms. In case of nonpathogenic bacterial strains gram staining can be employed to study morphology. But evaluation of infectious microbes using microscope is not a reasonable option (Sobsey, 1999).

17.7.4 Spectroscopic detection

17.7.4.1 Fourier transform infrared spectroscopy

Fourier transform infrared spectroscopy (FTIR) is an analytical technique that enables rapid detection of microbial presence in different specimens by providing a fingerprint spectrum of the isolates. Most of the mentioned spectroscopy-based techniques are nondestructive, label-free and enable on-line identification of samples (Schmitt and Flemming, 1998). Microbes have a unique cellular structure. Their cell wall and outer membrane are composed of different biochemical components such as nucleotides, proteins, carbohydrates and lipids. These components are capable of providing a definite infrared absorption spectrum upon analysis (Seelig, 2004). It has been observed that most of the microbes exhibit similar spectra owing to negligible modifications in their biochemical confirmation (Davis and Mauer, 2010). In the past decade, FTIR has been used to detect and isolate microbes at species level (Wenning and Scherer, 2013). Among all techniques available for species identification, it is the only one that can be used also for typing purposes without carrying out additional experiments, which makes it a valuable tool in epidemiology.

17.7.4.2 Surface-enhanced Raman spectroscopy

Surface-enhanced Raman spectroscopy (SERS) is based on scattering effects which are produced when a specimen is irradiated with a monochromatic light source. It enables structural as well as biochemical evaluation of samples (Craig et al., 2013). Detection of pathogens using Raman spectroscopy primarily involves application of two mechanisms. First involves straight forward recognition of the intrinsic vibrational fingerprint of the microbe whereas the second mechanism involves indirect recognition by application of a SERS nanotag. Xie and co-workers reported separation of different strains of *Enterobacteriaceae* by application of gold-colloid-based Raman spectroscopy coupled with principal component analysis (PCA) and hierarchical cluster analysis (HCA)-based processes.

17.8 Recent advances in on-site detection of microorganisms using peptide functionalized nanosensors

17.8.1 Bacteria detection

Shi and co-workers recently reported the construction of a label free sensor by using electrochemical impedance spectroscopy (EIS) for detection of gram-negative bacterium *E. coli* O157:H7. The sensor was formulated by conjugating affinity peptides having the ability to identify the selected microbe. In the study, Ph.D.™ - 12 Phage Display Peptide Library Kit was used to select an AMP having affinity specifically toward surface of the disease-causing microbe. Further, biopanning technique was employed to recognize peptide contenders that could identify *E. coli* cell membrane. Biosensor was formulated by using self-assembly mechanism during which peptide (GLHTSATNLYLHGGGC) was fabricated on to gold electrode. Upon peptide conjugation the electron transfer rate got stagnant and current peak value decreased exhibiting detection of *E. coli*. In addition to detection studies, sensitivity of the sensor was evaluated by varying *E. coli* concentration from 20 to 2×10^6 CFU/mL (Shi et al., 2020). It was inferred that hydrophobic - hydrophilic interactions along with electrostatic forces played a significant role in cell binding (Zita and Hermansson, 1997). The reported sensor is rapid and provided an accurate and quantifiable detection of the desired microbe.

An impedimetric-based label-free sensor for selective detection of *Salmonella* was formulated by immobilizing nisin on gold electrode (Malvano et al., 2020). Nisin is a bacteriocin composed of 34 amino acids. Bacteriocins are defined as proteinaceous complexes having low molecular weight. They are usually produced by lactic acid bacteria and have a tendency to exhibit inhibitory activity against the same or closely related bacterial strain from which they were produced (Nebbia et al., 2021; Abee and Broughton, 2003; Karpiński and Szkaradkiewicz, 2013). There are numerous reasons for use of nisin as a recognition molecule in this sensor. First, nisin is tasteless, emits no odor and is permitted by Food and Drug Administration for food preservation-based applications, second it exhibits broad spectrum antimicrobial activities. In this study, the peptide functionalized sensor was formulated by covalently conjugating nisin onto gold electrodes using EDC and NHS. After formulation of the sensor milk samples spiked with different microorganisms were evaluated for change in impedance after incubation for a period of 30 min. Since the sensor displayed higher affinity toward gram-negative microbes specifically *Salmonella* species, it can be used as a selective peptide-based sensor in future.

Similarly, Albanese et al. (2019) reported the synthesis of a nisin-based impedimetric sensor for detection of bacterial contamination in water. Nisin was conjugated with gold electrode via its C-terminus and the concentration was standardized (5–10 µg/mL) to obtain enhanced detection of microbes at lower detection parameters. It was observed that even by using peptide at low concentration the formulated sensor detected *E. coli* along with *Listeria innocua* at a range of 1.5 CFU/mL.

Peptide functionalized wireless sensor for detection of *H. pylori* was formulated by Mannoor et al. (2012). The battery free sensor was formulated by obtaining conjugation between a graphitic derivative and graphene binding peptide (HSSYWYAFNNKT-GGG-GLLRASSVWGRKYYVDLAGCAKA). The peptide was bifunctional in nature comprising of a graphene attaching motif. Using self-assembly approach, the peptides were incubated with the graphene coated on silk film, unbound peptides were removed by washing using deionized water. Presence of bacterial cells was determined by conducting electrical measurements followed by monitoring fluorescence activity using microscopy. Optical experiments revealed recognition of *H. pylori* cells by the peptide molecules assembled on graphene sheets. It can be inferred that inherent specificity of peptide molecules enable them to detect disease causing microorganism even in complex environment such as saliva. Since the formulated sensor exhibited enhanced selectivity and sensitivity toward bacterial detection along with biocompatibility and mechanical strength it should be used for detection of other microorganisms as well.

An easy yet efficient approach to formulate an electrochemical nanosensor for identification of gram-positive bacterial species was reported (de Miranda et al., 2017). The sensor was formulated by using cystine doped gold nanoparticles which were further incubated with antimicrobial peptide Clavanin A (ClavA). Detailed sensor surface modification was determined using Atomic Force Microscopy. The motive to conduct a microscopic study was to assess any changes caused by the binding of microbes, thereby affecting specificity of the sensor. After formulation of the sensor impedimetric as well as voltametric measurements were determined by analyzing microbe solutions present in varying concentrations. It was observed that the sensor was successful in distinguishing gram-positive and gram-negative microorganisms by exhibiting a discerning response. Since the formulation displayed promising detection capability even by using low volume of biomaterial with further experimentation it should be used as a low cost, easy to formulate diagnostic alternative.

17.8.2 Detection of fungal spores

Globally, researchers are significantly interested in the formulation of sensors with improved and precise detection of fungal spores due to the detrimental risk they pose to public well-being (Almeida et al., 2019). Infectious diseases caused by

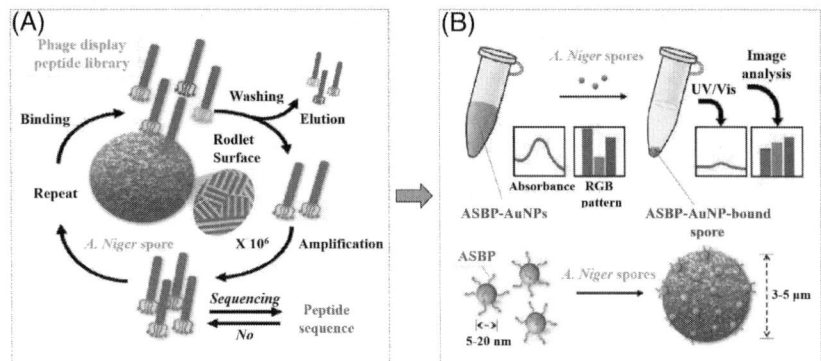

FIG. 17.7 (A) Illustration describing the process involved in identification of a peptide capable of binding to *Aspergillus niger* via screening of phage display library. (B) Colorimetry-based detection of fungal spores by using peptide functionalized gold nanoconjugate *(reprinted with permission from Lee et al., 2020).*

fungi are capable of causing a tremendous amount of damage to plants and vegetable crops, resulting in substantial losses in the production of food and other agriculture related activities (Lee et al., 2021; Casadevall, 2017; Arturo, 2017; Savary et al., 2012). Timely detection and diagnosis of fungal diseases is the major factor that can enable appropriate fungicide treatment (West et al., 2010; Mahlein, 2016).

Based on interactions among fungal spores and peptide functionalized gold nanoparticles Lee and his colleagues reported the formulation of a colorimetry-based sensor with an ability to detect *Aspergillus niger* (*A. niger*). Although United States Department of Agriculture (USDA) has recognized many strains of *A. niger* safe to use in fermentation procedures it is known to cause numerous diseases in fruits and vegetable crops (Krishnan et al., 2009). Even though rare instances of Aspergillosis are observed in humans (Sharma, 2012). Fig. 17.7 displays different steps involved in formulation of the sensor. Using Phage Display Library, a peptide having affinity toward *A. niger* binding peptide ligand was recognized. The selected peptide exhibited enhanced specificity toward the fungi and overlooked the existence of other airborne microorganisms.

Further, peptide-gold nanocomplex was formed by means of covalent interaction mechanism using poly(ethylene glycol) as a cross-linker. It was observed that the formulated calorimetric method identified 50 fungal spores within a period of 10 min. It has been observed that the reported approach is easy and can be replicated for on site, rapid detection of *A. niger*. In future, this strategy can be used to identify a wide range of pathogenic microbes by altering the adhesive properties of peptides and also by using different nanomaterials.

17.8.3 Virus detection

Researchers from Republic of Korea have developed a peptide functionalized polydiacetylene (PEP-PDA) nanosensor for detection of influenza A virus (H1N1) (Von Itzstein, 2007). The virus is infamous for causing Swine flu pandemic in the year 2009 (Webster and Govorkova, 2014; Sakai-Tagawa et al., 2010). A simple colorimetry-based detection system was formulated by functionalizing polydiacetylene (PDA) with peptide molecules by application of noncovalent interaction approach. Peptide molecules were used to efficiently and selectively detect H1N1 virus because of their high inclination toward hemagglutinin (HA) proteins present in the H1 strain. Upon incubation with influenza virus, the sensor produced a colorimetric conversion from blue to red color. This transformation in color was visible through naked eyes and can be attributed to the generation of steric repulsion at the sensor surface (Song et al., 2016).

Yet another peptide functionalized sensor to detect avian influenza virus was reported (Matsubara et al., 2020). The technology used a boron-doped diamond electrode as a substrate. The concentration of peptide (ARLPR) long with dendrimer growth transformed the effectiveness of viral binding on to the electrode surface thereby amplifying the charge-transfer resistance. It was observed that the peptide-immobilized electrodes displayed an outstanding recognition limit of less than one plaque-forming unit of H1N1 and H3N2 viruses. The sensor also exhibited capability to detect avian viruses isolated from H5N3, H7N1, and H9N2. It can be said that in future, the designed detector can act as a credible tool for detection of other influenza A virus subtypes. The ongoing Covid-19 pandemic has taught us that in order to efficiently manage future disease outbreaks formation of such robust, rapid and dependable peptide-based detection methods are need of the hour.

Islam and his fellow researchers reported an approach to construct peptide functionalized graphitic sensor to detect human immunodeficiency virus (HIV). Cardiovascular disorders along with Rheumatoid arthritis is also caused by the virus leading to morbidity and mortality globally (Barnes et al., 2017; Esser et al., 2013). The study described covalent conjugation of antibodies [anti-p24 for HIV, anticardiac troponin 1 (anticTn1) with peptide functionalized graphene complex using EDC and NHS as cross-linkers. Further, electrochemical measurements were accomplished by passing a 100 nA constant current. It was observed that the formulated nanosensor was sensitive in response and depicted a linear relationship under standard conditions (Islam et al., 2019).

17.9 Conclusion and future perspectives

Early detection of pathogenic microorganisms is crucial aspect that needs to be addressed to prevent incidences of outbreaks affecting human population and health care system. Many promising approaches have been formulated to enable microbial detection. Despite all the novel technologies being reported there are still numerous challenges that need to be addressed such as development of stable, rapid, user friendly, cost-effective sensing devices that can be used on-field and still provide enhanced sensitivity and selectivity in detection of microorganisms. Peptides have gained enormous attention in the field of sensors because of the specificity they display toward an analyte, high affinity toward microorganisms and ease of conjugation. There are few drawbacks that limit the applicability of peptides in sensing applications, but they can be overcome by introduction of emerging nanotechnology-based approaches such as formulation of peptide functionalized nanostructures. It is believed that by overcoming the discussed limitations peptide functionalized nanosensors shall be used as reliable new age diagnostic technologies.

References

Abee, T., Delves-Broughton, J., 2003. Bacteriocins—nisin. In: Russell, N.J., Gould, G.W. (Eds.), Food Preservatives. Springer, Boston, MA, pp. 146–178.

Adak, A.K., Boley, J.W., Lyvers, D.P., Chiu, G.T., Low, P.S., Reifenberger, R., Wei, A., 2013. Label-free detection of Staphylococcus aureus captured on immutable ligand arrays. ACS Appl. Mater. Interfaces 5 (13), 6404–6411. doi:10.1021/am4016236.

Ageitos, J.M., Sánchez-Pérez, A., Calo-Mata, P., Villa, T.G., 2017. Antimicrobial peptides (AMPs): ancient compounds that represent novel weapons in the fight against bacteria. Biochem. Pharmacol. 133, 117–138. doi:10.1016/j.bcp.2016.09.018.

Ahmadi, M., Ghoorchian, A., Dashtian, K., Kamalabadi, M., Madrakian, T., Afkhami, A., 2020. Application of magnetic nanomaterials in electroanalytical methods: a review. Talanta 225, 121974. doi:10.1016/j.talanta.2020.121974.

Ahmed, M.U., Saaem, I., Wu, P.C., Brown, A.S., 2014. Personalized diagnostics and biosensors: a review of the biology and technology needed for personalized medicine. Crit. Rev. Biotechnol. 34 (2), 180–196. doi:10.3109/07388551.2013.778228.

Ahuja, T., Kumar, D., 2009. Recent progress in the development of nano-structured conducting polymers/nanocomposites for sensor applications. Sens. Actuators B 136 (1), 275–286. doi:10.1016/j.snb.2008.09.014.

Alafeef, M., Moitra, P., Pan, D., 2020. Nano-enabled sensing approaches for pathogenic bacterial detection. Biosens. Bioelectron. 165, 112276. doi:10.1016/j.bios.2020.112276.

Alamer, S., Eissa, S., Chinnappan, R., Herron, P., Zourob, M., 2018. Rapid colorimetric lactoferrin-based sandwich immunoassay on cotton swabs for the detection of foodborne pathogenic bacteria. Talanta 185, 275–280. doi:10.1016/j.talanta.2018.03.072.

Albanese, D., Garofalo, F., Pilloton, R., Capo, S., Malvano, F., 2019. Development of an antimicrobial peptide-based biosensor for the monitoring of bacterial contaminations. Chem. Eng. Trans. 75, 61–66. doi:10.3303/CET1975011.

Albinana-Gimenez, N., Clemente-Casares, P., Calgua, B., Huguet, J.M., Courtois, S., Girones, R., 2009. Comparison of methods for concentrating human adenoviruses, polyomavirus JC and noroviruses in source waters and drinking water using quantitative PCR. J. Virol. Methods 158 (1-2), 104–109. doi:10.1016/j.jviromet.2009.02.004.

Almeida, F., Rodrigues, M.L., Coelho, C., 2019. The still underestimated problem of fungal diseases worldwide. Front. Microbiol. 10, 214. doi:10.3389%2Ffmicb.2019.00214.

Almeida, M.I.G., Jayawardane, B.M., Kolev, S.D., McKelvie, I.D., 2018. Developments of microfluidic paper-based analytical devices (µPADs) for water analysis: a review. Talanta 177, 176–190. doi:10.1016/j.talanta.2017.08.072.

Amiri, M., Bezaatpour, A., Jafari, H., Boukherroub, R., Szunerits, S., 2018. Electrochemical methodologies for the detection of pathogens. ACS Sens. 3 (6), 1069–1086. doi:10.1021/acssensors.8b00239.

Andrade, C.A., Nascimento, J.M., Oliveira, I.S., de Oliveira, C.V., de Melo, C.P., Franco, O.L., Oliveira, M.D., 2015. Nanostructured sensor based on carbon nanotubes and clavanin A for bacterial detection. Colloids Surf. B 135, 833–839. doi:10.1016/j.colsurfb.2015.03.037.

Andrews, W.H., 1996. Validation of modern methods in food microbiology by AOAC International collaborative study. Food Control 7 (1), 19–29. doi:10.1016/0956-7135(96)00007-2.

Arora, N., 2013. Recent advances in biosensors technology: a review. Octa J. Biosci. 1 (2), 147–150. doi:10.3389/fbioe.2016.00011.

Baeumner, A.J., Schlesinger, N.A., Slutzki, N.S., Romano, J., Lee, E.M., Montagna, R.A., 2002. Biosensor for dengue virus detection: sensitive, rapid, and serotype specific. Anal. Chem. 74 (6), 1442–1448. doi:10.1021/ac015675e.

Bahadır, E.B., Sezgintürk, M.K., 2015. Applications of commercial biosensors in clinical, food, environmental, and biothreat/biowarfare analyses. Anal. Biochem. 478, 107–120. doi:10.1016/j.ab.2015.03.011.

Barbau-Piednoir, E., Botteldoorn, N., Mahillon, J., Dierick, K., Roosens, N.H., 2015. Fast and discriminative CoSYPS detection system of viable *Salmonella* spp. and *Listeria* spp. in carcass swab samples. Int. J. Food Microbiol. 192, 103–110. doi:10.1016/j.ijfoodmicro.2014.09.018.

Barbosa, A.J., Oliveira, A.R., Roque, A.C., 2018. Protein-and peptide-based biosensors in artificial olfaction. Trends Biotechnol. 36 (12), 1244–1258. doi:10.1016/j.tibtech.2018.07.004.

Barnes, R.P., Lacson, J.C.A., Bahrami, H., 2017. HIV infection and risk of cardiovascular diseases beyond coronary artery disease. Curr. Atheroscler. Rep. 19 (5), 20. doi:10.1007/s11883-017-0652-3.

Bhalla, N., Jolly, P., Nello, F., Pedro, E., 2016. Introduction to biosensors. Essays Biochem. 60 (1), 1–8. doi:10.1042/EBC20150001.

Bhardwaj, N., Bhardwaj, S.K., Nayak, M.K., Mehta, J., Kim, K.H., Deep, A., 2017. Fluorescent nanobiosensors for the targeted detection of foodborne bacteria. TrAC Trends Anal. Chem. 97, 120–135. doi:10.1016/j.trac.2017.09.010.

Bonnet, M., Lagier, J.C., Raoult, D., Khelaifia, S., 2020. Bacterial culture through selective and non-selective conditions: the evolution of culture media in clinical microbiology. New Microb. New Infect. 34, 100622. doi:10.1016/j.nmni.2019.100622.

Botes, M., de Kwaadsteniet, M., Cloete, T.E., 2013. Application of quantitative PCR for the detection of microorganisms in water. Anal. Bioanal. Chem. 405 (1), 91–108. doi:10.1007/s00216-012-6399-3.

Brogden, K.A., 2005. Antimicrobial peptides: pore formers or metabolic inhibitors in bacteria? Nat. Rev. Microbiol. 3 (3), 238–250. doi:10.1038/nrmicro1098.

Brosel-Oliu, S., Abramova, N., Bratov, A., Vigués, N., Mas, J., Muñoz, F.X., 2015. Sensitivity and response time of polyethyleneimine modified impedimetric transducer for bacteria detection. Electroanalysis 27 (3), 656–662. doi:10.1002/elan.201400575.

Burlage, R.S., Tillmann, J., 2013. Biosensors of bacterial cells. J. Microbiol. Methods 138, 2–11. doi:10.1016/j.mimet.2016.12.023.

Byeon, H.M., Vodyanoy, V.J., Oh, J.H., Kwon, J.H., Park, M.K., 2015. Lytic phage-based magnetoelastic biosensors for on-site detection of methicillin-resistant *Staphylococcus aureus* on spinach leaves. J. Electrochem. Soc. 162 (8), B230. doi:10.1149/2.0681508jes.

Caelen, I., Gao, H., Sigrist, H., 2002. Protein density gradients on surfaces. Langmuir 18 (7), 2463–2467. doi:10.1021/la0113217.

Cardoso, V.F., Francesko, A., Ribeiro, C., Bañobre-López, M., Martins, P., Lanceros-Mendez, S., 2018. Advances in magnetic nanoparticles for biomedical applications. Adv. Healthcare Mater. 7 (5), 1700845. doi:10.1002/adhm.201700845.

Casadevall, A., 2017. Don't forget the fungi when considering global catastrophic biorisks. Health Security 15 (4), 341–342. doi:10.1089/hs.2017.0048.

Casteels, P., Ampe, C., Jacobs, F., Vaeck, M., Tempst, P.J.T.E.J., 1989. Apidaecins: antibacterial peptides from honeybees. EMBO J. 8 (8), 2387–2391. doi:10.1002/j.1460-2075.1989.tb08368.x.

Caygill, R.L., Blair, G.E., Millner, P.A., 2010. A review on viral biosensors to detect human pathogens. Anal. Chim. Acta. 681 (1-2), 8–15. doi:10.1016/j.aca.2010.09.038.

Charlet, K., 2018. The new killer pathogens: countering the coming bioweapons threat. Foreign Aff. 97, 178.

Chen, H., Liu, K., Li, Z., Wang, P., 2019. Point of care testing for infectious diseases. Clin. Chim. Acta 493, 138–147. doi:10.1016/j.cca.2019.03.008.

Chen, Q., Zhang, L., Feng, Y., Shi, F., Wang, Y., Wang, P., Liu, L., 2018. Dual-functional peptide conjugated gold nanorods for the detection and photothermal ablation of pathogenic bacteria. J. Mater. Chem. B 6 (46), 7643–7651. doi:10.1039/C8TB01835A.

Chiu, N.F., Fan, S.Y., Yang, C.D., Huang, T.Y., 2017. Carboxyl-functionalized graphene oxide composites as SPR biosensors with enhanced sensitivity for immunoaffinity detection. Biosens. Bioelectron. 89, 370–376. doi:10.1016/j.bios.2016.06.073.

Cho, I.H., Ku, S., 2017. Current technical approaches for the early detection of foodborne pathogens: challenges and opportunities. Int. J. Mol. Sci. 18 (10), 2078. doi:10.3390/ijms18102078.

Choi, Y., Hwang, J.H., Lee, S.Y., 2018. Recent trends in nanomaterials-based colorimetric detection of pathogenic bacteria and viruses. Small Methods 2 (4), 1700351. doi:10.1002/smtd.201700351.

Chu, Y.W., Engebretson, D.A., Carey, J.R., 2013. Bioconjugated magnetic nanoparticles for the detection of bacteria. J. Biomed. Nanotechnol. 9 (12), 1951–1961. doi:10.1166/jbn.2013.1701.

Clark Jr, L.C., Lyons, C., 1962. Electrode systems for continuous monitoring in cardiovascular surgery. Ann. N.Y. Acad. Sci. 102 (1), 29–45. doi:10.1111/j.1749-6632.1962.tb13623.x.

Corry, B., Uilk, J., Crawley, C., 2003. Probing direct binding affinity in electrochemical antibody-based sensors. Anal. Chim. Acta 496 (1-2), 103–116. doi:10.1016/j.aca.2003.01.001.

Cosio, M.S., Benedetti, S., Scampicchio, M., Mannino, S., 2015. Electroanalysis in food process control. In: Escarpa, A., Gonzalez, M.C., Lopez, M.A. (Eds.), Agricultural and Food Electroanalysis. Wiley, Chichester, UK, pp. 421–441.

Craig, A.P., Franca, A.S., Irudayaraj, J., 2013. Surface-enhanced Raman spectroscopy applied to food safety. Annu. Rev. Food Sci. Technol. 4, 369–380. doi:10.1146/annurev-food-022811-101227.

Crivianu-Gaita, V., Thompson, M., 2016. Aptamers, antibody scFv, and antibody Fab' fragments: an overview and comparison of three of the most versatile biosensor biorecognition elements. Biosens. Bioelectron. 85, 32–45. doi:10.1016/j.bios.2016.04.091.

Dai, C., Choi, S., 2013. Technology and applications of microbial biosensor. Open J. Appl. Biosens. 2, 83–93. doi:10.4236/ojab.2013.23011.

Darwish, I.A., 2006. Immunoassay Methods and their Applications in Pharmaceutical Analysis: Basic Methodology and Recent Advances. Int. J. Biomed. Sci. 2 (3), 217–235.

Das, M., Dhand, C., Sumana, G., Srivastava, A.K., Vijayan, N., Nagarajan, R., Malhotra, B.D., 2011. Zirconia grafted carbon nanotubes based biosensor for M. Tuberculosis detection. Appl. Phys. Lett. 99 (14), 143702. doi:10.1063/1.3645618.

Davis, R., Mauer, L.J., 2010. Fourier transform infrared (FT-IR) spectroscopy: a rapid tool for detection and analysis of foodborne pathogenic bacteria. Curr. Res. Technol. Educ. Top. Appl. Microbiol. Microbial Biotechnol. 2, 1582–1594.

de Miranda, J.L., Oliveira, M.D., Oliveira, I.S., Frias, I.A., Franco, O.L., Andrade, C.A., 2017. A simple nanostructured biosensor based on clavanin A antimicrobial peptide for gram-negative bacteria detection. Biochem. Eng. J. 124, 108–114. doi:10.1016/j.bej.2017.04.013.

Dhama, K., Patel, S.K., Sharun, K., Pathak, M., Tiwari, R., Yatoo, M.I., Rodriguez-Morales, A.J., 2020. SARS-CoV-2 jumping the species barrier: zoonotic lessons from SARS, MERS and recent advances to combat this pandemic virus. Travel Med. Infect. Dis. 37, 101830. doi:10.1016/j.tmaid.2020.101830.

Drake, C., Deshpande, S., Bera, D., Seal, S., 2007. Metallic nanostructured materials based sensors. Int. Mater. Rev. 52 (5), 289–317. doi:10.1179/174328007X212481.

Ebralidze, I.I., Laschuk, N.O., Poisson, J., Zenkina, O.V., 2019. Colorimetric sensors and sensor arrays. In: Zenkina, O.V. (Ed.), Nanomaterials Design for Sensing Applications. Elsevier, Amsterdam, Netherlands, pp. 1–39.

Eggins, B.R., 2013. Biosensors: An Introduction. Springer-Verlag, Berlin/Heidelberg, Germany.

Ehmen, C., Medialdea-Carrera, R., Brown, D., de Filippis, A.B., de Sequeira, P.C., Nogueira, R.R., Mika, A., 2021. Accurate detection of Zika virus IgG using a novel immune complex binding ELISA. Trop. Med. Int. Health 26 (1), 89–101. doi:10.1111/tmi.13505.

Eissa, S., Zourob, M., 2020. A dual electrochemical/colorimetric magnetic nanoparticle/peptide-based platform for the detection of *Staphylococcus aureus*. Analyst 145 (13), 4606–4614. doi:10.1039/D0AN00673D.

El-Din, S., Rabie, E.M., El-Sokkary, M.M.A., Bassiouny, M.R., Hassan, R., 2015. Epidemiology of neonatal sepsis and implicated pathogens: a study from Egypt. BioMed. Res. Int. 2015, 2015. doi:10.1155/2015/509484.

Esser, S., Gelbrich, G., Brockmeyer, N., Goehler, A., Schadendorf, D., Erbel, R., Reinsch, N., 2013. Prevalence of cardiovascular diseases in HIV-infected outpatients: results from a prospective, multicenter cohort study. Clin. Res. Cardiol. 102 (3), 203–213. doi:10.1007/s00392-012-0519-0.

Etayash, H., Jiang, K., Thundat, T., Kaur, K., 2014. Impedimetric detection of pathogenic gram-positive bacteria using an antimicrobial peptide from class IIa bacteriocins. Anal. Chem. 86 (3), 1693–1700. doi:10.1021/ac4034938.

Etukudoh, N.S., Ejinaka, O., Obeta, U., Utibe, E., Lote-Nwaru, I., Agbalaka, P., Shaahia, D., 2020. Zoonotic and parasitic agents in bioterrorism. J. Infect. Dis. Travel Med. 4 (2), 000139. doi:10.23880/jidtm-16000139.

Fan, X., Xu, W., Gao, W., Jiang, X., Wu, G., 2020. A facile method to classify clinic isolates with a turn-off sensor array based on graphene oxide and antimicrobial peptides. Sens. Actuators B 307, 127607. doi:10.1016/j.snb.2019.127607.

Filipiak, W., Sponring, A., Baur, M.M., Filipiak, A., Ager, C., Wiesenhofer, H., Amann, A., 2012. Molecular analysis of volatile metabolites released specifically by *Staphylococcus aureus* and *Pseudomonas aeruginosa*. BMC Microbiol. 12 (1), 1–16. doi:10.1186/1471-2180-12-113.

Floris, M., Moro, S., 2012. Mimicking peptides... in silico. Mol. Inf. 31 (1), 12–20. doi:10.1002/minf.201100093.

Frasco, M.F., Chaniotakis, N., 2009. Semiconductor quantum dots in chemical sensors and biosensors. Sensors 9 (9), 7266–7286. doi:10.3390/s90907266.

Fukushi, S., Fukuma, A., Kurosu, T., Watanabe, S., Shimojima, M., Shirato, K., Saijo, M., 2018. Characterization of novel monoclonal antibodies against the MERS-coronavirus spike protein and their application in species-independent antibody detection by competitive ELISA. J. Virol. Methods 251, 22–29. doi:10.1016/j.jviromet.2017.10.008.

Geldert, A., Liu, Y., Loh, K.P., Lim, C.T., 2017. Nano-bio interactions between carbon nanomaterials and blood plasma proteins: why oxygen functionality matters. NPG Asia Mater. 9 (8), e422. doi:10.1038/am.2017.129.

Ghasemi, Y., Peymani, P., Afifi, S., 2009. Quantum dot: magic nanoparticle for imaging, detection and targeting. Acta Biomed. 80 (2), 156–165.

Gilchrist, A., Nobbs, J., 1999. Colorimetry, theory. Encyclop. Spectrosc. Spectrom. 337–343.

Gracias, K.S., McKillip, J.L., 2004. A review of conventional detection and enumeration methods for pathogenic bacteria in food. Can. J. Microbiol. 50 (11), 883–890. doi:10.1139/w04-080.

Guerin, P., El Mouatassim, S., Menezo, Y., 2001. Oxidative stress and protection against reactive oxygen species in the pre-implantation embryo and its surroundings. Hum. Reprod. Update. 7 (2), 175–189. doi:10.1093/humupd/7.2.175.

Habimana, J.D.D., Ji, J., Sun, X., 2018. Minireview: trends in optical-based biosensors for point-of-care bacterial pathogen detection for food safety and clinical diagnostics. Anal. Lett. 51 (18), 2933–2966. doi:10.1080/00032719.2018.1458104.

Haney, E.F., Hancock, R.E., 2013. Peptide design for antimicrobial and immunomodulatory applications. Pept. Sci. 100 (6), 572–583. doi:10.1002/bip.22250.

Hatchett, D.W., Josowicz, M., 2008. Composites of intrinsically conducting polymers as sensing nanomaterials. Chem. Rev. 108 (2), 746–769. doi:10.1021/cr068112h.

He, X., Li, Y., He, D., Wang, K., Shangguan, J., Shi, H., 2014. Aptamer-fluorescent silica nanoparticles bioconjugates based dual-color flow cytometry for specific detection of *Staphylococcus aureus*. J. Biomed. Nanotechnol. 10 (7), 1359–1368. doi:10.1166/jbn.2014.1828.

Henriques, S.T., Melo, M.N., Castanho, M.A., 2006. Cell-penetrating peptides and antimicrobial peptides: how different are they? Biochem. J. 399 (1), 1–7. doi:10.1042/BJ20061100.

Hesari, N., Alum, A., Elzein, M., Abbaszadegan, M., 2016. A biosensor platform for rapid detection of *E. coli* in drinking water. Enzyme Microb. Technol. 83, 22–28. doi:10.1016/j.enzmictec.2015.11.007.

Homola, J., Yee, S.S., Gauglitz, G., 1999. Surface plasmon resonance sensors: review. Sens. Actuators B 54 (1-2), 3–15. doi:10.1016/S0925-4005(98)00321-9.

Hossein-Nejad-Ariani, H., Kim, T., Kaur, K., 2018. Peptide-based biosensor utilizing fluorescent gold nanoclusters for detection of *Listeria monocytogenes*. ACS Appl. Nano Mater. 1 (7), 3389–3397. doi:10.1021/acsanm.8b00600.

Hoyos-Nogués, M., Brosel-Oliu, S., Abramova, N., Muñoz, F.X., Bratov, A., Mas-Moruno, C., Gil, F.J., 2016. Impedimetric antimicrobial peptide-based sensor for the early detection of periodontopathogenic bacteria. Biosens. Bioelectron. 86, 377–385. doi:10.1016/j.bios.2016.06.066.

Hoyos-Nogués, M., Gil, F.J., Mas-Moruno, C., 2018. Antimicrobial peptides: powerful biorecognition elements to detect bacteria in biosensing technologies. Molecules 23 (7), 1683. doi:10.3390/molecules23071683.

Hui, D.S., Zumla, A., 2019. Severe Acute Respiratory Syndrome: Historical, Epidemiologic, and Clinical Features. Infect. Dis. Clin. North Am. 33 (4), 869–889. doi:10.1016/j.idc.2019.07.001.

Hwang, H., Hwang, B.Y., Bueno, J., 2018. Biomarkers in Infectious Diseases. Hindawi, London. doi:10.1155/2018/8509127.

Ifeanyi, V.O., Nwosu, S.C., Okafor, J.O., Onnegbu, C.P., Nwabunnia, E., 2014. Comparative studies on five culture media for bacterial isolation. Afr. J. Microbiol. Res. 8 (36), 3330–3334. doi:10.5897/ajmr12.1604.

Islam, S., Shukla, S., Bajpai, V.K., Han, Y.K., Huh, Y.S., Kumar, A., Gandhi, S., 2019. A smart nanosensor for the detection of human immunodeficiency virus and associated cardiovascular and arthritis diseases using functionalized graphene-based transistors. Biosens. Bioelectron. 126, 792–799. doi:10.1016/j.bios.2018.11.041.

Jacobs, C.B., Peairs, M.J., Venton, B.J., 2010. Carbon nanotube based electrochemical sensors for biomolecules. Anal. Chim. Acta 662 (2), 105–127. doi:10.1016/j.aca.2010.01.009.

Jin, D., Kong, X., Cui, B., Jin, S., Xie, Y., Wang, X., Deng, Y., 2018. Bacterial communities and potential waterborne pathogens within the typical urban surface waters. Sci. Rep. 8 (1), 1–9. doi:10.1038/s41598-018-31706-w.

Joshi, S., Sharma, P., Siddiqui, R., Kaushal, K., Sharma, S., Verma, G., Saini, A., 2020. A review on peptide functionalized graphene derivatives as nanotools for biosensing. Microchim. Acta 187 (1), 1–15. doi:10.1007/s00604-019-3989-1.

Joshi, S., Siddiqui, R., Sharma, P., Kumar, R., Verma, G., Saini, A., 2020. Green synthesis of peptide functionalized reduced graphene oxide (rGO) nano bioconjugate with enhanced antibacterial activity. Sci. Rep. 10 (1), 1–11. doi:10.1038/s41598-020-66230-3.

Justino, C.I., Freitas, A.C., Pereira, R., Duarte, A.C., Santos, T.A.R., 2015. Recent developments in recognition elements for chemical sensors and biosensors. TrAC Trends Anal. Chem. 68, 2–17. doi:10.1016/j.trac.2015.03.006.

Karimzadeh, A., Hasanzadeh, M., Shadjou, N., de la Guardia, M., 2018. Peptide based biosensors. TrAC Trends Anal. Chem. 107, 1–20. doi:10.1016/j.trac.2018.07.018.

Karpiński, T.M., Szkaradkiewicz, A.K., 2013. Characteristic of bacteriocines and their application. Pol. J. Microbiol. 62 (3), 223–235. doi:10.33073/pjm-2013-030.

Kateete, D.P., Kimani, C.N., Katabazi, F.A., Okeng, A., Okee, M.S., Nanteza, A., Najjuka, F.C., 2010. Identification of *Staphylococcus aureus*: DNase and mannitol salt agar improve the efficiency of the tube coagulase test. Ann. Clin. Microbiol. Antimicrob. 9 (1), 1–7. doi:10.1186/1476-0711-9-23.

Kenndler, E., Blaas, D., 2001. Capillary electrophoresis of macromolecular biological assemblies: bacteria and viruses. Trends. Analyt. Chem. 20 (10), 543–551. doi:10.1016/S0165-9936(01)00112-1.

Krishnan, S., Manavathu, E.K., Chandrasekar, P.H., 2009. *Aspergillus flavus*: an emerging non-fumigatus *Aspergillus* species of significance. Mycoses 52 (3), 206–222. doi:10.1111/j.1439-0507.2008.01642.x.

Kuang, Z., Kim, S.N., Crookes-Goodson, W.J., Farmer, B.L., Naik, R.R., 2010. Biomimetic chemosensor: designing peptide recognition elements for surface functionalization of carbon nanotube field effect transistors. ACS Nano 4 (1), 452–458. doi:10.1021/nn901365g.

Kulagina, N.V., Shaffer, K.M., Anderson, G.P., Ligler, F.S., Taitt, C.R., 2006. Antimicrobial peptide-based array for *Escherichia coli* and *Salmonella* screening. Anal. Chim. Acta 575 (1), 9–15. doi:10.1016/j.aca.2006.05.082.

Kumar, B.K., Raghunath, P., Devegowda, D., Deekshit, V.K., Venugopal, M.N., Karunasagar, I., Karunasagar, I., 2011. Development of monoclonal antibody based sandwich ELISA for the rapid detection of pathogenic *Vibrio parahaemolyticus* in seafood. Int. J. Food Microbiol. 145 (1), 244–249. doi:10.1016/j.ijfoodmicro.2010.12.030.

Lazcka, O., Del Campo, F.J., Munoz, F.X., 2007. Pathogen detection: a perspective of traditional methods and biosensors. Biosens. Bioelectron. 22 (7), 1205–1217. doi:10.1016/j.bios.2006.06.036.

Le, T.N., Tran, T.D., Kim, M.I., 2020. A convenient colorimetric bacteria detection method utilizing chitosan-coated magnetic nanoparticles. Nanomaterials 10 (1), 92. doi:10.3390/nano10010092.

Leclerc, H., Moreau, A., 2002. Microbiological safety of natural mineral water. FEMS Microbiol. Rev. 26 (2), 207–222. doi:10.1111/j.1574-6976.2002.tb00611.x.

Lee, J.I., Jang, S.C., Chung, J., Choi, W.K., Hong, C., Ahn, G.R., Chung, W.J., 2021. Colorimetric allergenic fungal spore detection using peptide-modified gold nanoparticles. Sens. Actuators B 327, 128894. doi:10.1016/j.snb.2020.128894.

Li, M., Chen, T., Gooding, J.J., Liu, J., 2019. Review of carbon and graphene quantum dots for sensing. ACS Sens. 4 (7), 1732–1748. doi:10.1021/acssensors.9b00514.

Li, S., Li, Y., Chen, H., Horikawa, S., Shen, W., Simonian, A., Chin, B.A., 2010. Direct detection of Salmonella typhimurium on fresh produce using phage-based magnetoelastic biosensors. Biosens. Bioelectron. 26 (4), 1313–1319. doi:10.1016/j.bios.2010.07.029.

Li, Z., Wang, L., Li, Y., Feng, Y., Feng, W., 2019. Carbon-based functional nanomaterials: preparation, properties and applications. Compos. Sci. Technol. 179, 10–40. doi:10.1016/j.compscitech.2019.04.028.

Lim, S.K., Chen, P., Lee, F.L., Moochhala, S., Liedberg, B., 2015. Peptide-assembled graphene oxide as a fluorescent turn-on sensor for lipopolysaccharide (endotoxin) detection. Anal. Chem. 87 (18), 9408–9412. doi:10.1021/acs.analchem.5b02270.

Lim, Y.C., Kouzani, A.Z., Duan, W., 2010. Aptasensors: a review. J. Biomed. Nanotechnol. 6 (2), 93–105. doi:10.1166/jbn.2010.1103.

Liu, F., Mu, J., Wu, X., Bhattacharjya, S., Yeow, E.K.L., Xing, B., 2014. Peptide–perylene diimide functionalized magnetic nano-platforms for fluorescence turn-on detection and clearance of bacterial lipopolysaccharides. Chem. Commun. 50 (47), 6200–6203. doi:10.1039/C4CC01266F.

Liu, Q., Wang, J., Boyd, B.J., 2015. Peptide-based biosensors. Talanta 136, 114–127. doi:10.1016/j.talanta.2014.12.020.

Lübbers, D.W., 1992. Fluorescence based chemical sensors. Adv Bios. 2, 215–260.

Mackay, I.M., 2004. Real-time PCR in the microbiology laboratory. Clin. Microbiol. Infect. 10 (3), 190–212. doi:10.1111/j.1198-743x.2004.00722.x.

Madhuri, R.J., Saraswathi, M., Gowthami, K., Bhargavi, M., Divya, Y., Deepika, V., Buddolla, V. (Eds.), 2019. Recent approaches in the production of novel enzymes from environmental samples by enrichment culture and metagenomic approach. Recent Developments in Applied Microbiology and Biochemistry. Elsevier, Amsterdam, Netherlands, pp. 251–262.

Mahlein, A.K., 2016. Plant disease detection by imaging sensors–parallels and specific demands for precision agriculture and plant phenotyping. Plant Dis. 100 (2), 241–251. doi:10.1094/PDIS-03-15-0340-FE.

Malvano, F., Pilloton, R., Albanese, D., 2020. A novel impedimetric biosensor based on the antimicrobial activity of the peptide nisin for the detection of *Salmonella* spp. Food Chem. 325, 126868. doi:10.1016/j.foodchem.2020.126868.

Mannoor, M.S., Tao, H., Clayton, J.D., Sengupta, A., Kaplan, D.L., Naik, R.R., McAlpine, M.C., 2012. Graphene-based wireless bacteria detection on tooth enamel. Nat. Commun. 3 (1), 1–9. doi:10.1038/ncomms1767.

Mannoor, M.S., Zhang, S., Link, A.J., McAlpine, M.C., 2010. Electrical detection of pathogenic bacteria via immobilized antimicrobial peptides. Proc. Natl. Acad. Sci. 107 (45), 19207–19212. doi:10.1073/pnas.1008768107.

Matsubara, T., Ujie, M., Yamamoto, T., Einaga, Y., Daidoji, T., Nakaya, T., Sato, T., 2020. Avian influenza virus detection by optimized peptide termination on a boron-doped diamond electrode. ACS Sens. 5 (2), 431–439. doi:10.1021/acssensors.9b02126.

Matthew, A., 2006. Current biosensor technologies in drug discovery. Drug Discov. 69, 68–82.

McMeekin, T.A., 2003. Detecting Pathogens in Food. Elsevier, Sawston, Cambridge.

Miller, E., Sikes, H.D., 2015. Addressing barriers to the development and adoption of rapid diagnostic tests in global health. Nanobiomedicine 2, 2–6. doi:10.5772/61114.

Mobed, A., Baradaran, B., de la Guardia, M., Agazadeh, M., Hasanzadeh, M., Rezaee, M.A., Hamblin, M.R., 2019. Advances in detection of fastidious bacteria: From microscopic observation to molecular biosensors. TrAC Trends Anal. Chem. 113, 157–171. doi:10.1016/j.trac.2019.02.012.

Mocan, T., Matea, C.T., Pop, T., Mosteanu, O., Buzoianu, A.D., Puia, C., Mocan, L., 2017. Development of nanoparticle-based optical sensors for pathogenic bacterial detection. J. Nanobiotechnol. 15 (1), 1–14. doi:10.1186/s12951-017-0260-y.

Mokhtarzadeh, A., Eivazzadeh-Keihan, R., Pashazadeh, P., Hejazi, M., Gharaatifar, N., Hasanzadeh, M., Baradaran, B., de la Guardia, M., 2017. Nanomaterial-based biosensors for detection of pathogenic virus. Trac. Trend. Anal. Chem. 97, 445–457. doi:10.1016/j.trac.2017.10.005.

Morales, M.A., Halpern, J.M., 2018. Guide to selecting a biorecognition element for biosensors. Bioconjug. Chem. 29 (10), 3231–3239. doi:10.1021/acs.bioconjchem.8b00592.

Nath, N., Chilkoti, A., 2004. Label free colorimetric biosensing using nanoparticles. J. Fluoresc. 14 (4), 377–389. doi:10.1023/B:JOFL.0000031819.45448.dc.

Nayak, M., Kotian, A., Marathe, S., Chakravortty, D., 2009. Detection of microorganisms using biosensors—a smarter way towards detection techniques. Biosens. Bioelectron. 25 (4), 661–667. doi:10.1016/j.bios.2009.08.037.

Nebbia, S., Lamberti, C., Lo Bianco, G., Cirrincione, S., Laroute, V., Cocaign-Bousquet, M., Pessione, E., 2021. Antimicrobial potential of food lactic acid bacteria: bioactive peptide decrypting from caseins and bacteriocin production. Microorganisms 9 (1), 65. doi:10.3390/microorganisms9010065.

Nguyen, L.T., Haney, E.F., Vogel, H.J., 2011. The expanding scope of antimicrobial peptide structures and their modes of action. Trends Biotechnol. 29 (9), 464–472. doi:10.1016/j.tibtech.2011.05.001.

Orellana, G., 2006. Fluorescence-based sensors. In: Baldini, F., Chester, A.N., Homola, J., Martellucci, S. (Eds.), Optical Chemical Sensors. Springer, Dordrecht, pp. 99–116.

Pan, M., Yin, Z., Liu, K., Du, X., Liu, H., Wang, S., 2019. Carbon-based nanomaterials in sensors for food safety. Nanomaterials 9 (9), 1330. doi:10.3390/nano9091330.

Paniel, N., Baudart, J., 2013. Colorimetric and electrochemical genosensors for the detection of *Escherichia coli* DNA without amplification in seawater. Talanta 115, 133–142. doi:10.1016/j.talanta.2013.04.050.

Patel, S., Nanda, R., Sahoo, S., Mohapatra, E., 2016. Biosensors in health care: the milestones achieved in their development towards lab-on-chip-analysis. Biochem. Res. Int. 2016, 1–12. doi:10.1155/2016/3130469.

Pavan, S., Berti, F., 2012. Short peptides as biosensor transducers. Anal. Bioanal. Chem. 402 (10), 3055–3070. doi:10.1007/s00216-011-5589-8.

Porter, E.A., Weisblum, B., Gellman, S.H., 2002. Mimicry of host-defense peptides by unnatural oligomers: antimicrobial β-peptides. J. Am. Chem. Soc. 124 (25), 7324–7330. doi:10.1021/ja0260871.

Qiao, Z., Fu, Y., Lei, C., Li, Y., 2020. Advances in antimicrobial peptides-based biosensing methods for detection of foodborne pathogens: a review. Food Control 112, 107116. doi:10.1016/j.foodcont.2020.107116.

Radom, F., Jurek, P.M., Mazurek, M.P., Otlewski, J., Jeleń, F., 2013. Aptamers: molecules of great potential. Biotechnol. Adv. 31 (8), 1260–1274. doi:10.1016/j.biotechadv.2013.04.007.

Rajapaksha, P., Elbourne, A., Gangadoo, S., Brown, R., Cozzolino, D., Chapman, J., 2019. A review of methods for the detection of pathogenic microorganisms. Analyst 144 (2), 396–411. doi:10.1039/c8an01488d.

Rajendran, V.K., Bakthavathsalam, P., Ali, B.M.J., 2014. Smartphone based bacterial detection using biofunctionalized fluorescent nanoparticles. Microchim. Acta 181 (15-16), 1815–1821. doi:10.1007/s00604-014-1242-5.

Ravindranath, S.P., Mauer, L.J., Deb-Roy, C., Irudayaraj, J., 2009. Biofunctionalized magnetic nanoparticle integrated mid-infrared pathogen sensor for food matrixes. Anal. Chem. 81 (8), 2840–2846. doi:10.1021/ac802158y.

Richter, Ł., Janczuk-Richter, M., Niedziółka-Jönsson, J., Paczesny, J., Hołyst, R., 2018. Recent advances in bacteriophage-based methods for bacteria detection. Drug Discov. Today 23 (2), 448–455. doi:10.1016/j.drudis.2017.11.007.

Rios, A., Zougagh, M., 2016. Recent advances in magnetic nanomaterials for improving analytical processes. TrAC Trends Anal. Chem. 84, 72–83. doi:10.1016/j.trac.2016.03.001.

Rocha-Santos, T.A., 2014. Sensors and biosensors based on magnetic nanoparticles. TrAC Trends Anal. Chem. 62, 28–36. doi:10.1016/j.trac.2014.06.016.

Roh, S., Chung, T., Lee, B., 2011. Overview of the characteristics of micro-and nano-structured surface plasmon resonance sensors. Sensors 11 (2), 1565–1588. doi:10.3390/s110201565.

Ruoss, L.J., Wynn, J.L., Benitz, W.E., Smith, P.B. (Eds.), 2019. Biomarkers in the Diagnosis of Neonatal Sepsis. Infectious Disease and Pharmacology. Elsevier, Amsterdam, Netherlands.

Sakai-Tagawa, Y., Ozawa, M., Tamura, D., Nidom, C.A., Sugaya, N., Kawaoka, Y., 2010. Sensitivity of influenza rapid diagnostic tests to H5N1 and 2009 pandemic H1N1 viruses. J. Clin. Microbiol. 48 (8), 2872–2877. doi:10.1128/JCM.00439-10.

Savary, S., Ficke, A., Aubertot, J.N., Hollier, C., 2012. Crop losses due to diseases and their implications for global food production losses and food security. Food Security 4, 519–537. doi:10.1007/s12571-012-0200-5.

Scarano, S., Mascini, M., Turner, A.P., Minunni, M., 2010. Surface plasmon resonance imaging for affinity-based biosensors. Biosens. Bioelectron. 25 (5), 957–966. doi:10.1016/j.bios.2009.08.039.

Schmitt, J., Flemming, H.C., 1998. FTIR-spectroscopy in microbial and material analysis. Int. Biodeterior. Biodegrad. 41 (1), 1–11. doi:10.1016/S0964-8305(98)80002-4.

Schulz-Schönhagen, K., Lobsiger, N., Stark, W.J., 2019. Continuous production of a shelf-stable living material as a biosensor platform. Adv. Mater. Technol. 4 (8), 1900266. doi:10.1002/admt.201970045.

Seelig, J., 2004. Thermodynamics of lipid–peptide interactions. Biochim. Biophys. Acta (BBA)-Biomembr. 1666 (1-2), 40–50. doi:10.1016/j.bbamem.2004.08.004.

Seng, J.J.B., Yeam, C.T., Huang, W.C., Tan, N.C., & Low, L.L. (2020). Pandemic related health literacy—a systematic review of literature in COVID-19, SARS and MERS pandemics. Medrxiv. doi:10.1101/2020.05.07.20094227.

Shao, L., Gao, Y., Yan, F., 2011. Semiconductor quantum dots for biomedicial applications. Sensors 11 (12), 11736–11751. doi:10.3390/s111211736.

Sharma, R., 2012. Pathogenecity of *Aspergillus niger* in plants. Cibtech J. Microbiol. 1 (1), 47–51.

Sharma, S., Byrne, H., O'Kennedy, R.J, 2016. Antibodies and antibody-derived analytical biosensors. Essays Biochem. 60 (1), 9–18. doi:10.1042/EBC20150002.

Shen, M., Rusling, J.F., Dixit, C.K., 2017. Site-selective orientated immobilization of antibodies and conjugates for immunodiagnostics development. Methods 116, 95–111. doi:10.1016/j.ymeth.2016.11.010.

Shi, F., Gan, L., Wang, Y., Wang, P., 2020. Impedimetric biosensor fabricated with affinity peptides for sensitive detection of *Escherichia coli* O157: H7. Biotechnol. Lett. 42 (5), 825–832. doi:10.1007/s10529-020-02817-0.

Shi, X., Zhang, X., Yao, Q., He, F., 2017. A novel method for the rapid detection of microbes in blood using pleurocidin antimicrobial peptide functionalized piezoelectric sensor. J. Microbiol. Methods 133, 69–75. doi:10.1016/j.mimet.2016.12.005.

Singh, R., Mukherjee, M.D., Sumana, G., Gupta, R.K., Sood, S., Malhotra, B.D., 2014. Biosensors for pathogen detection: a smart approach towards clinical diagnosis. Sens. Actuators B 197, 385–404. doi:10.1016/j.snb.2014.03.005.

Sivakumar, M., Tominaga, R., Koga, T., Kinoshita, T., Sugiyama, M., Yamaguchi, K., 2005. Studies on visual sensor from self-assembled polypeptides. Sci. Technol. Adv. Mater. 6 (1), 91–96. doi:10.1016/j.stam.2004.06.006.

Skottrup, P.D., Nicolaisen, M., Justesen, A.F., 2008. Towards on-site pathogen detection using antibody-based sensors. Biosens. Bioelectron. 24 (3), 339–348. doi:10.1016/j.bios.2008.06.045.

So, H.M., Park, D.W., Jeon, E.K., Kim, Y.H., Kim, B.S., Lee, C.K., Lee, J.O., 2008. Detection and titer estimation of *Escherichia coli* using aptamer-functionalized single-walled carbon-nanotube field-effect transistors. Small 4 (2), 197–201. https://doi.org/10.1002/smll.200700664.

Sobczak-Kupiec, A., Venkatesan, J., AlAnezi, A.A., Walczyk, D., Farooqi, A., Malina, D., Tyliszczak, B., 2016. Magnetic nanomaterials and sensors for biological detection. Nanomed. Nanotechnol. Biol. Med. 12 (8), 2459–2473. doi:10.1016/j.nano.2016.07.003.

Sobsey, M.D., 1999. Methods to identify and detect microbial contaminants in drinking water Identifying Future Drinking Water Contaminants. The National Academies Press, Washington, DC, pp. 177–203.

Song, S., Ha, K., Guk, K., Hwang, S.G., Choi, J.M., Kang, T., Lim, E.K., 2016. Colorimetric detection of influenza A (H1N1) virus by a peptide-functionalized polydiacetylene (PEP-PDA) nanosensor. RSC Adv. 6 (54), 48566–48570. doi:10.1039/C6RA06689E.

Srinivasan, B., Tung, S., 2015. Development and applications of portable biosensors. J. Lab. Autom. 20 (4), 365–389. doi:10.1177/2211068215581349 10.1177/2211068215581349.

Su, L., Jia, W., Hou, C., Lei, Y., 2011. Microbial biosensors: a review. Biosens. Bioelectron. 26 (5), 1788–1799. doi:10.1016/j.bios.2010.09.005.

Su, X.L., Li, Y., 2004. A self-assembled monolayer-based piezoelectric immunosensor for rapid detection of *Escherichia coli* O157: H7. Biosens. Bioelectron. 19 (6), 563–574. doi:10.1016/S0956-5663(03)00254-9.

Sun, Q., Zhao, G., Dou, W., 2015. Blue silica nanoparticle-based colorimetric immunoassay for detection of *Salmonella pullorum*. Anal. Methods 7 (20), 8647–8654. doi:10.1039/C5AY02073E.

Tasis, D., Tagmatarchis, N., Bianco, A., Prato, M., 2006. Chemistry of carbon nanotubes. Chem. Rev. 106 (3), 1105–1136. doi:10.1021/cr050569o.

Teengam, P., Siangproh, W., Tuantranont, A., Vilaivan, T., Chailapakul, O., Henry, C.S., 2017. Multiplex paper-based colorimetric DNA sensor using pyrrolidinyl peptide nucleic acid-induced AgNPs aggregation for detecting MERS-CoV, MTB, and HPV oligonucleotides. Anal. Chem. 89 (10), 5428–5435. doi:10.1021/acs.analchem.7b00255.

Templier, V., Roux, A., Roupioz, Y., Livache, T., 2016. Ligands for label-free detection of whole bacteria on biosensors: a review. TrAC Trends Anal. Chem. 79, 71–79. doi:10.1016/j.trac.2015.10.015.

Toze, S., 1999. PCR and the detection of microbial pathogens in water and wastewater. Water Res. 33 (17), 3545–3556. doi:10.1016/S0043-1354(99)00071-8.

Tuantranont, A., 2012. Nanomaterials for sensing applications: introduction and perspective. In: Tuantranont, A. (Ed.), Applications of Nanomaterials in Sensors and Diagnostics. Springer, Berlin, Heidelberg, pp. 1–16.

Turner, J.E., Thorpe, T.C., Di Guiseppi, J.L., & Driscoll, R.C. (1990). U.S. Patent No. 4,945,060. Washington, DC: U.S. Patent and Trademark Office.

Valizadeh, A., Mikaeili, H., Samiei, M., Farkhani, S.M., Zarghami, N., Akbarzadeh, A., Davaran, S., 2012. Quantum dots: synthesis, bioapplications, and toxicity. Nanoscale Res. Lett. 7 (1), 1–14. doi:10.1186/1556-276X-7-480.

Veiseh, M., Zareie, M.H., Zhang, M., 2002. Highly selective protein patterning on gold–silicon substrates for biosensor applications. Langmuir 18 (17), 6671–6678. doi:10.1021/la025529j.

Vernozy-Rozand, C., Mazuy-Cruchaudet, C., Bavai, C., Richard, Y., 2004. Comparison of three immunological methods for detecting staphylococcal enterotoxins from food. Lett. Appl. Microbiol. 39 (6), 490–494. doi:10.1111/j.1472-765X.2004.01602.x.

Von Itzstein, M., 2007. The war against influenza: discovery and development of sialidase inhibitors. Nat. Rev. Drug Discov. 6 (12), 967–974. doi:10.1038/nrd2400.

Vs, A.P., Joseph, P., SCG, K.D., Lakshmanan, S., Kinoshita, T., Muthusamy, S., 2017. Colorimetric sensors for rapid detection of various analytes. Mater. Sci. Eng.: C 78, 1231–1245. doi:10.1016/j.msec.2017.05.018.

Wang, F., Hu, S., 2009. Electrochemical sensors based on metal and semiconductor nanoparticles. Microchim. Acta 165 (1-2), 1–22. doi:10.1007/s00604-009-0136-4.

Wang, K., Pu, H., Sun, D.W., 2018. Emerging spectroscopic and spectral imaging techniques for the rapid detection of microorganisms: An overview. Comprehens. Rev. Food Sci. Food Safety 17 (2), 256–273. doi:10.1111/1541-4337.12323.

Webster, R.G., Govorkova, E.A., 2014. Continuing challenges in influenza. Ann. N. Y. Acad. Sci. 1323 (1), 115. doi:10.1111/nyas.12462.

Wenning, M., Scherer, S., 2013. Identification of microorganisms by FTIR spectroscopy: perspectives and limitations of the method. Appl. Microbiol. Biotechnol. 97 (16), 7111–7120. doi:10.1007/s00253-013-5087-3.

West, J.S., Bravo, C., Oberti, R., Moshou, D., Ramon, H., McCartney, H.A., 2010. Detection of fungal diseases optically and pathogen inoculum by air sampling. In: Oerke, E.C., Gerhards, R., Menz, G., Sikora, R.A. (Eds.), Precision Crop Protection—The Challenge and Use of Heterogeneity. Springer, Dordrecht, pp. 135–149. doi:10.1007/978-90-481-9277-9_9.

Wisitsoraat, A., Tuantranont, A., 2013. Graphene-based chemical and biosensors. In: Tuantranont, A. (Ed.), Applications of Nanomaterials in Sensors and Diagnostics. Springer, Berlin, Heidelberg, pp. 103–141.

Wolfbeis, O.S., 2008. Fiber-optic chemical sensors and biosensors. Anal. Chem. 80 (12), 4269–4283. doi:10.1021/ac060490z.

Xiaohua, H., El-Sayed, M.A., 2011. Plasmonic photo-thermal therapy (PPTT). Alexandria J. Med. 47 (1), 1–9. doi:10.1016/j.ajme.2011.01.001.

Yan, H., Liu, L., Xu, N., Kuang, H., Xu, C., 2015. Development of an immunoassay for carbendazim based on a class-selective monoclonal antibody. Food and Agricultural Immunology 26 (5), 659–670. doi:10.1080/09540105.2015.1007446.

Yoo, J.H., Woo, D.H., Chang, M.S., Chun, M.S., 2014. Microfluidic based biosensing for *Escherichia coli* detection by embedding antimicrobial peptide-labeled beads. Sens. Actuators B 191, 211–218. doi:10.1016/j.snb.2013.09.105.

Yoon, H., 2013. Current trends in sensors based on conducting polymer nanomaterials. Nanomaterials 3 (3), 524–549. doi:10.3390/nano3030524.

Zasloff, M., 2002. Antimicrobial peptides of multicellular organisms. Nature 415 (6870), 389–395. doi:10.1038/415389a.

Zasloff, M., 2019. Antimicrobial peptides of multicellular organisms: my perspective. Antimicrob. Pept. 1117, 3–6. doi:10.1007/978-981-13-3588-4_1.

Zelada-Guillén, G.A., Riu, J., Düzgün, A., Rius, F.X., 2009. Immediate detection of living bacteria at ultralow concentrations using a carbon nanotube based potentiometric aptasensor. Angew. Chem. Int. Ed. 48 (40), 7334–7337. doi:10.1002/anie.200902090.

Zhang, L., 2016a. a Gallo, RL: antimicrobial peptides. Curr. Biol. 26, R14–R19. doi:10.1016/j.cub.2015.11.017.

Zhang, S., Zeng, M., Xu, W., Li, J., Li, J., Xu, J., Wang, X., 2013. Polyaniline nanorods dotted on graphene oxide nanosheets as a novel super adsorbent for Cr (VI). Dalton Trans. 42 (22), 7854–7858. doi:10.1039/C3DT50149C.

Zhang, X.C., 2016. Science and Principles of Biodegradable and Bioresorbable Medical Polymers: Materials and Properties. Woodhead Publishing, Sawston, Cambridge.

Zheng, F., Wang, P., Du, Q., Chen, Y., Liu, N., 2019. Simultaneous and ultrasensitive detection of foodborne bacteria by gold nanoparticles-amplified microcantilever array biosensor. Front. Chem. 7, 232. doi:10.3389/fchem.2019.00232.

Zhong, W., 2009. Nanomaterials in fluorescence-based biosensing. Anal. Bioanal. Chem. 394 (1), 47–59. doi:10.1007/s00216-009-2643-x.

Zhou, J., Rossi, J., 2017. Aptamers as targeted therapeutics: current potential and challenges. Nat. Rev. Drug Discov. 16, 181–202. doi:10.1038/nrd.2016.199.

Zhou, C., Zou, H., Li, M., Sun, C., Ren, D., Li, Y., 2018. Fiber optic surface plasmon resonance sensor for detection of *E. coli* O157: H7 based on antimicrobial peptides and AgNPs-rGO. Biosens. Bioelectron. 117, 347–353. doi:10.1016/j.bios.2018.06.005.

Zita, A., Hermansson, M., 1997. Determination of bacterial cell surface hydrophobicity of single cells in cultures and in wastewater in situ. FEMS Microbiol. Lett. 152 (2), 299–306. doi:10.1111/j.1574-6968.1997.tb10443.x.

Zourob, M., Elwary, S., Khademhosseini, A., 2010. Recognition Receptors in Biosensors. Springer, New York, pp. 415–448.

Chapter 18

Theranostic nanoagents: Future of personalized nanomedicine

Vidya Sabale[a], Shraddha Dubey[b], Prafulla Sabale[c]
[a]Dadasaheb Balpande College of Pharmacy, Besa, Nagpur, Maharashtra, India
[b]Inselpital University of Bern, Bern, Switzerland
[c]Department of Pharmaceutical Sciences, Rashtrasant Tukadoji Maharaj Nagpur University, Nagpur, Maharashtra, India

18.1   Introduction

18.1.1   Theranostics

The term "theranostics" is defined as the combination of diagnosis and therapy, an emerging field of medicine supposed to embrace huge prospective to cure many complicated diseases (Sumer and Gao, 2008). The basis is due to the fact that for dreadful diseases, like cancers, at the selective stage of disease development the existing therapies and treatments have proved to be effective on a few patient subpopulations. Therefore, merging diagnosis and therapeutics together could provide more individual specific therapeutic protocols which result in enhanced prognosis (Xie et al., 2010a).

With the advent of science and technology; theranostics is fast-growing field that is a combination of nanotechnology with personalized medicines that significantly enhance the efficacy of treatment through target specific delivery of therapy to any targeted tissues. With the enhanced intelligent signals and sensors corresponding to a slight change of any disease at the molecular level can be detected at ease. Theranostic can provide early detection of cancers in patients through a new process of image guided therapy. The most meticulous imaging methods include magnetic resonance imaging (MRI), computed tomography (CT), positron emission tomography (PET), and optical imaging. From these, optical imaging is more appealing as it can give images in real-time with high spatial resolution and the use of nonionizing irradiation (Wang et al., 2014; Believing in Seeing, 2014; Lim et al., 2015; Fu et al., 2017).

Theranostics can be amalgamated to have the best possible delivery properties, low renal clearance, reduced immunogenicity and antigenicity (for example by PEGylating the surface of theranostic nanoparticles), and high capacity for therapeutic agents, which is required to give the restricted concentrations of specific molecular markers expressed on cancer cells.

18.1.2   Nanoagents

Theranostic nanoagent is a carrier with a combination of therapeutic and diagnostic applications. These nanoagents provide a platform for clinicians to monitor the treatment effect of the suffered areas by merging diagnostic and therapeutic approaches in one system (Bardhan et al., 2011).

18.1.3   Nanotheranostics

Nanotheranostics is a superior type of theranostic that involves "nanotechnology" for the diagnosis and therapy of different diseases with poor diagnosis. This comprises a new generation of different kinds of nanocarriers such as polymer conjugates, dendrimers, micelles, liposomes, metal and inorganic nanoparticles (NPs), carbon nanotubes (CNTs) to develop appealing nanomedicine. The idea of nanotheranostics is the prognosis and alleviation of diseases at their probable initial stages of development. To summarize, nanotheranostic is a recent and superior group of nanomedicine that can detect, treat and check diseases at the cellular and molecular level with the use of nanotechnology (Wang et al., 2014).

18.2 Recent approaches versus theranostic nanoagents

18.2.1 Contemporary treatment methods and their drawbacks

A contemporary treatment technique to treat various disorders consists of surgery, radiotherapy, chemotherapy, and immunotherapy. In recent years, several other recent treatment methods such as gene therapy, hyperthermia therapy, and stem cell therapy (Delhaye et al., 2012; Simonato et al., 2013; Thomas and Omuro, 2014; Yoo et al., 2013) have also been discovered as the treatment of various disorders. However, early diagnosis is a primary requisite for accomplishing efficient therapy. Therefore, strengthening the surgeon with fascinating imaging techniques like positron emission tomography (PET), single-photon emission computed tomography (SPECT), magnetic resonance imaging (MRI), and X-ray computed tomography (CT) will possibly assist their timely decisions, which in turn can reduce the risk of disease recurrence and repeated therapies (Pereira et al., 2014; Small et al., 2006; Perrin et al., 2009; Nasrallah and Wolk, 2014). PET and SPECT both employ imaging techniques based on nuclear medicine using radio probes as the contrast agents. These radio probes produce gamma rays which are gathered to confine or compute specific receptors in the living brain that facilitate to the discovery of the neurological changes occurring during the development of disease (Zhu et al., 2014). Due to its higher sensitivity and better resolution, PET is a more proficient imaging tool than SPECT (Rahmim and Zaidi, 2008). MRI and CT, both are noninvasive imaging tools that can detect morphological changes in diseased tissues. CT uses specialized X-ray equipment to examine different parts of the human body such as brain, sinus, facial bones, dental, spines, cervical, hands, wrist, elbow, shoulder, hip, knee, ankle foot, renal tract. CT helps in the prognosis of the disease, detection of anomaly and treatment is planned accordingly. Nevertheless, CT has certain disadvantages such as little tissue contrast, inability to visualize small groups of tumor cells which are separated from the gross tumor, and lack of providing functional information. MRI is a novel and powerful diagnostic tool, widely used in examining different abnormalities of various parts of the body, such as brain tumors, inflammation in the spine, injuries of joints, assessing blood flow and imaging of cardiac function, etc. As compared to CT, MRI uses radiofrequency waves for signal generation and proposes high-quality images with greater intrinsic soft-tissue contrast and resolution (Pereira et al., 2014; Kasban et al., 2015). In MRI, spin possessing nuclei present within the static magnetic field is excited by applying a second radiofrequency magnetic field at an appropriate resonant frequency and perpendicular to the static magnetic field. The absorption of energy by the nuclei causes a transition from higher to lower energy levels. The absorption and emission of energy by different nuclei (e.g., free water and water bound to tissue) creates a voltage, which is detected, amplified and transformed into an image. Currently, a variety of paramagnetic and superparamagnetic metal nanoparticles are commonly used contrast agents for MRI (Grover et al., 2015; Jain et al., 2008). Likewise, fluorescence imaging is an enormously advantageous tool for real-time assessment of diseased tissues with complementary advantages such as low-cost, high contrast/signal generation, high sensitivity, safety, and can be designed to be activated at the target site through certain specific markers which are over expressed in the affected area while otherwise, they remain inactive. Furthermore, the use of fluorescent probes in the infrared region leads to an increased penetration depth and reduced signal scattering, thus allowing visualization of diseased tissues up to greater depth (Blau et al., 2018). Despite much progress in current diagnostic treatments, successful treatment of numerous disorders is limited owing to many drawbacks of the currently existing methods, such as time consumption and low sensitivity in MRI (Ladd et al., 2018), poor infiltration and more spreading in the tissues in fluorescence-based imaging (Louie, 2010), inability to monitor disease progression along with the administered drug and exertion of toxic effects on the surrounding healthy tissues due to lack of specific/targeted function in case of radioactive tracer-based imaging in PET and SPECT (Lu and Yuan, 2015). Similarly, conventional therapies such as chemotherapy, photodynamic therapy, surgery, radiotherapy, immunotherapy those are generally being followed for managing disorders also experience various challenges such as inability to cross the barriers, inadequate drug uptake in the tissues, inferior biocompatibility, nontargeted delivery and distribution, poor solubility, lower half-lives and retention time and undesirable effects on rapidly dividing healthy cells.

To surmount all these issues and to attain successful therapy, there is an urgent need for the development of highly developed methods that possess high sensitivity, higher resolution, deep penetration power and the ability to facilitate real-time monitoring of disease development along with biosafety. In view of this, ultrasmall multimodal nanotheranostics can prove their potential for biomedical imaging and disease treatment (Zhang et al., 2016a; Miao et al., 2017).

18.3 Nanotheranostics and neurological disorders

18.3.1 Blood–brain barrier

The most important barrier in delivering drugs to the brain is blood–brain barrier (BBB). Due to the presence of tight endothelial junctions between the microvasculature and brain parenchyma, BBB is impermeable to a large number of drugs such as antibiotics, cytostatics, and CNS-active drugs, etc. BBB is composed of an endothelial monolayer, pericytes enclosed within the endothelial basal lamina, and astrocytes touching their end-feet to the abluminal side of the brain vessels. Endothelial cells of the CNS are holding together by tight junctions giving rise to a highly resistant paracellular seal, which hinders the transport of various molecules and ions into the neural tissues (Huber et al., 2001). In addition to endothelial cells, the astrocytic end-feet surrounding BBB regulates barrier permeability and dictates the localization of transporters. There are various transport mechanisms like passive diffusion, endocytosis, receptor-mediated transcytosis, active transport, and carrier-mediated transport, which facilitate the transfer of various molecules, nutrients, and drugs across the BBB. These transport mechanisms implicate various transport proteins which mediate the uptake and expulsion of different metabolites and compounds across the BBB (Bhowmik et al., 2015). Amusingly, in the true sense, BBB is not a barrier; it is dynamic in nature and allows the entry of various essential nutrients, such as amino acids, hexoses, proteins, peptides, and ions to assist the normal functioning of the brain and concurrently restrict the entry of noxious substances. Moreover, the presence of efflux pumps such as P-glycoprotein quickly removes any foreign substance that bypasses the BBB. It has been observed that more than 99.9% of macromolecules and almost more than 98% of small molecules having a size greater than 500 Da cannot penetrate the BBB (Pardridge, 2007). Therefore, these unique features of the BBB microvasculature significantly affect drug delivery to the brain (Hajal et al., 2018), presenting a challenge to the efficacy of neural therapeutics.

18.3.2 Theranostic nanoparticles employed in neurology

Recently, theranostic NPs are coming up as very promising and competent tools for accomplishing targeted drug delivery due to their distinctive physicochemical characteristics such as nanoscale size, encapsulation of drug and adsorption or conjunction of contrast agents on their surface. Moreover, to boost the specificity of theranostic NPs at their acting/target site of action, NPs surface can be altered by various active and passive targeting agents. Active targeting of these NPs can be accomplished through various ligands such as peptides, antibodies, aptamers, and others, which exclusively bind to the expressed receptors on the brain endothelial cell surface enabling these NPs to cross the BBB (Xu et al., 2013). Similarly, for passive targeting, the surface of NPs can be coated with hydrophilic polymers such as polyethylene glycol (PEG), poly (acryloyl morpholine), poly-N-vinylpyrrolidones, polyvinyl alcohol and poly[N-(2 hydroxypropyl) methacrylamide]. These polymers coating not only enhances blood circulation time but concomitantly avoid recognition by macrophages and monocytes. Thus, the risk of opsonization and rapid clearance of NPs from the blood is reduced (Schottler et al., 2016). These multifunctional nanosystems coated with targeting agents, grasp the potential of prognosis, imaging and treatment of disease enacted by a single formulation of biocompatible and biodegradable NPs.

Presently, there are various types of theranostic nanoplatforms such as lipid NPs, polymeric NPs (PNPs), inorganic NPs, and some others being investigated for diagnosis and treatment of neurological disorders (Peng et al., 2015). Table 18.1 (Wen et al., 2012; Sonali et al., 2016a; Guo et al., 2015; Shazeeb et al., 2014; Agulla et al., 2013; Sonali et al., 2016b; Rajora et al., 2017; Saesoo et al., 2018; Bernal et al., 2014; Stephen et al., 2014; Chowdhury et al., 2018; Zhang et al., 2014; Su et al., 2016; Bhojani et al., 2010; Zhang et al., 2016b; Li et al., 2018a; Dehvari et al., 2012; Skaat et al., 2013; Hu et al., 2015; Gao et al., 2018a; Tang et al., 2018) shows applications of these theranostic NPs for the treatment of neurological disorders. Lipid NPs are the most accepted and extensively studied NPs for nanomedicine delivery having clinical acceptance. Various types of lipid NPs such as liposome, solid-lipid NPs, polysomes, lipid coated calcium phosphate, etc. are available derived from the types of lipids and their physiochemical properties (Tang et al., 2018). Both hydrophobic and hydrophilic substances can be delivered using these NPs. Apart from this, their surface can be functionalized or modified with multiple molecules for numerous applications, such as an imaging agent for diagnosis or noninvasive biodistribution monitoring, ligands for specific targeting and PEGylation for prolonged circulation in blood. PNPs are NPs of size less than 1000 nm (colloidal solid), derived from both synthetic polymers, such as poly (lactic acid) (PLA) and poly (lactide-coglycolide) (PLGA), and natural polymers, such as chitosan and collagen which are mostly explored for biomedical applications (Bolhassani et al., 2014). PNPs are widely accepted versatile nanocarrier systems for delivering drugs to CNS across the BBB due to their drug entrapment capability and functionalization with suitable cell-penetrating peptides and/or targeting ligands (Patel et al., 2012). Due to their unique capability of conjugation, entrapment or encapsulation of both drugs and diagnostic agents on their matrix, PNPs are used as strong nanotheranostic agents for treating and diagnosing neurological disorders (Li et al., 2018b). Chitosan has been found to be the most common natural polymer used as theranostic agents for neurodegenerative diseases due to its nontoxicity, biocompatibility, and biodegradability (Duceppe and Tabrizian, 2010). Like natural polymers, synthetic polymers such as PLA and PLGA are also commonly employed to make theranostic agents for neurological disorders owing to their biocompatibility, biodegradability, and low immunogenicity (Danhier et al., 2012).

Moreover, due to their controllable size, great biocompatibility, easy surface modification, and multifunctional nature, inorganic NPs such as iron oxide, ceria, gold, and quantum dots are used in theranostic nanostructures (Vio et al., 2017). Despite gaining remarkable attention inorganic NPs as theranostic systems for neurological disorders, their instability, and potential toxicity to human systems have confined their clinical applications.

TABLE 18.1 Nanoparticles explored as theranostic nanoagents to treat several neurological maladies.

S. no.	Theranostic NPs	Ligand/transport method	Animal model/cell lines	Imaging/diagnostic agent	Therapeutic agent	Neurological disorders	Results	References
1.	Liposome	Endocytosis	Female nude mice	QDs*	Apomorphine	Parkinson's disease	-First liposome drug -QD hybrid -Enhances brain targeting -Greater fluorescence intensity -2.4 fold increases in drug accumulation	(Wen C-J. et al., 2012)
2.	TPGS** Liposomes	Transferrin	Charles Foster (CF) rats	QDs	Docetaxal	Brain tumor	-Targeted Therapy -Higher drug delivery -Twofold more efficient than nontargeted liposome	(Sonali et al., 2016a)
3.	CMD# Magnetoliposome	Passive transport	Human neutroblastoma SH SY5Y cells	SPION##	Doxorubicin	Brain neoplasm	-Greater stability and drug loading ability -Proficient T2-weighted contrast agent -Reduced toxicity as compared to free DOX	(Guo H. et al., 2015)
4.	Liposomes encapsulate with M40401	Mn porphyrins	Mice	Mn(11) cation	Superoxide dismutase	Cerebral ischemia	-Efficient Mn SOD mimetic -T_1 contrast agent for MRI imaging -Neuroprotective effect of mimetic SOD	(Shazeeb et al., 2014)
5.	Stealth immunoliposome	AntiHSP72	Ischemic rats	Rhodamine/gadolinium	Citicoline	Cerebral ischemia	-A novel theranostic nanoplatform -Targeted treatment	(Agulla et al., 2014)
6.	RGD-TPGS embellished liposome	RGD/receptor-mediated endocytosis	Rats	QDs	DTX and QDs	Brain carcinoma	-Biocompatible liposome for targeted codelivery of DTX and QDs -Enhances 70% effectiveness of drug encapsulation -Sustain the release of drug 6.47 fold greater than Docel	(Sonali et al., 2016b)
7.	py-E lipid nanoparticle	Apo-E3/transcytosis	Orthotopic U87 GFP tumor-bearing mice	Porphyrine	Porphyrine	Glioblastoma tumor	-Intrinsic theranostic properties -Act as both Glioma drug and contrast agent -In vitro uptake by LDLR*** mediated endocytosis	(Rajora et al., 2017)
8.	Liposome	AntiCD 20	Athymic nude mice	SPIONs Rituximab	Primary central nervous system lymphoma	Primary central nervous system lymphoma (PCNSL)	-Novel theranostic NP developed for PCNSL -Strong antilymphoma activity -Longer storage ability of NPs -Targeted drug delivery	(Saesoo et al., 2018)

S. no.	Theranostic NPs	Ligand/transport method	Animal model/cell lines	Imaging/diagnostic agent	Therapeutic agent	Neurological disorders	Results	References
9.	Polymeric nanoparticle	Endocytosis	Rat	Supermagnetic iron oxide	Temozolomide (TMZ)	Malignant glioma	-Convection enhanced drug delivery (CED) -Tracked by iron oxide used as MRI contrast agent -Effectively reduced the growth of glioma -Enhances the survival rate	(Bernal et al., 2014)
10.	Chitosan coated iron oxide cores	Chlorotoxin	GBM6-bearing mice	SPION	O6-benzylguanine (+temozolamide)	Brain tumor	-Significant biodistribution in tumor region -Real time monitoring of drug by MRI###	(Stephen et al., 2014)
11.	Polyfluorene-chitosan	Absorptive transcytosis	Endothelial cells (EAhy926.1)	Polyfluorene	Polyfluorene	Alzheimer's and presenile dementia	-Novel polymer conjugate -Hinders or alter amyloid aggregates -Sensed by their distinctive optical properties	(Roy et al., 2018)
12.	Polyethylene glycol Poly (actic acid) NPs	TGN and QSH peptides	AD model mice	DIR****	Coumarin-6	Alzheimer's disease	-Higher uptake and distribution in brain -Improve targeted delivery to amyloid plaque in mice brain	(Zhang et al., 2014)
13.	Algenate coated iron oxide core	G23 peptide/receptor-mediated endocytosis	U87-luc2-bearing mice	Iron oxide	Doxorubicin	Brain tumor	-Safe nanocarrier for brain tumor therapy -MRI contrast agent for in vitro in vivo diagnosis -Significant shrinkage of tumor in treated mice after 7 days	(Su et al., 2016)
14.	Polymeric nanoparticles	F3 peptides	Glioma-bearing rats	Iron oxide	Photofrin	Brain tumor	-Targeted drug delivery -Improved therapeutic efficiency -Significant increase in survival rate	(Bhojani et al., 2010)
15.	PHEMARA-PCBCPP### polymeric nanoparticle	CPP/endocytosis	2 × Transgenic AD mice	SPIONs	siSOX9	Alzheimer's disease	-Efficiently controlled the differentiation of normal stem cells to neuron -Enhances cellular uptake -Ameliorates neurological changes -Real time monitoring and migration of neural stem cells by SPIONs	(Zhang et al., 2016b)

(Continued)

TABLE 18.1 (Cont'd)

S. no.	Theranostic NPs	Ligand/transport method	Animal model/cell lines	Imaging/diagnostic agent	Therapeutic agent	Neurological disorders	Results	References
16.	PCB polymer	Endocytosis	2 × Tg-AD mice	SPIONs	let-7b antisense oligonucleotide and simvastatin	Alzheimer's disease	–Self-assembled Traceable NPs – Traceable NPs with higher drug loading capacity – Effective Control release of drug –Rescued memory deficits	(Li et al., 2018)
17.	SPIONs	Anti-Aβ monoclonal antibodies (aAβmAbs)	Not informed	SPIONs	Not informed	Alzheimer's disease	–Temperature sensitive magnetic drug delivery system –Smart drug delivery system	(Dehvari et al., 2012)
18.	SPIONs	Anti-Aβ monoclonal antibodies (aAβmAbs)	Transgenic mice model of AD*****	Fluorescent SPIONs	BAM10	Alzheimer's disease	–Inhibition of amyloid β40 fibrillation by fivefold –Detected Aβ40 fibrils by MRI and fluorescence imaging	(Skaat et al., 2013)
19.	Iron oxide magnetic nanoparticles	Penetration	APPswe/PS1 dE9 transgenic mouse	Iron oxide	MNPs	Rutin Alzheimer's disease	–Congo red labeled Magnetic NPs –Aβ aggregates detected by MRI –Controlled release of drug in a H2O2-responsive manner –Prevent oxidative stress for AD therapy	(Hu et al., 2015)
20.	Den-RGDReg peptide nanoparticle	c(RGDyK)	Mice having orthotopic U87-GFP Neoplasm	Near Infra-Red fluorophore	IR640B Paclitaxel	Brain neoplasm	–Glioma targeted selectivity –Improved survival rate –Image-guided chemotherapy	(Gao et al., 2018a)

*QDs, quantum dots; **TPGS, D-alpha tocopheryl polyethylene glycol 1000 succinate monoester; #CMD, carboxymehtyl dextran; ## SPIONs, superparamgnetic iron oxide nanoparticles; ***LDLR, low-density lipoprotein receptor; ###MRI, magnetic resonance imaging; **** DIR, 1,1'-dioctadecyl-3,3,3',3'-tetramethylindotricarbocyanine iodide; #### PHEMARA-PCBCPP, poly(2-hydroxyethyl methacrylate)-RA-polycarboxybetaine) cell penetrating peptide; *****AD, Alzheimer's disease.

Furthermore, in recent years nanosystems like upconversion NPs (UCNPs), dendrimers, and CNTs have evoked remarkable interest among various researchers. UCNPs have a unique property of generating high energy visible radiations from low energy near-infrared (NIR) radiations due to the nonlinear optical process. UCNPs get excited within a narrow absorption band at around 975 nm, which falls within the NIR range considered to be the "optical transparency window" (700–1100 nm) of tissues. It has been observed that biological tissues have relatively small scattering and absorption at 975 nm and also less autofluorescence while collecting the NIR UC emission. Therefore excitation of UCNPs at this wavelength does not cause any photodamage to the surrounding biological tissues (Auzel, 2004; Wu et al., 2015; Hilderbrand and Weissleder, 2010). Thus, UCNPs may serve as prevailing smart nanoprobes for prospective nanotheranostic applications (Fang and Wei, 2016) owing to their nontoxicity and stability. Besides the UCNPs, dendrimers are also explored as theranostic nanocarrier systems. A well-organized 3D structure and easy surface modification capability of dendrimers with different functional groups bestow them with high drug loading capacities and make them suitable diagnostic and imaging agents. Furthermore, altered surface dendrimers exhibit less cytotoxicity and more biocompatibility for in vivo applications (Gamage et al., 2016). Similarly, CNTs reveal special electronic and mechanical qualities suitable for use in biomedical applications. Surface functionalization and modification of CNTs with hydrophilic or biocompatible molecules improve their target specific cell/tissue delivery, water dispersibility and minimize toxicity. As compared to other theranostic NPs, intrinsic NIR photoluminescence optical and thermal properties of CNTs have revealed a high potential for brain targeted theranostic applications (Vardharajula et al., 2012).

18.3.3 Theranostic applications of nanosystems in neurological disorders

Various features of nanotheranostic systems are often tremendously helpful in formulating the subsequent generation nano-neuro-therapeutics that can function as both diagnostic and therapeutic modalities to treat dreaded neurological disorders as exemplified in several diseases mentioned below.

18.3.3.1 Glioma (brain tumors)

In recent years, the number of deaths due to cancer has been increasing remarkably. Currently, radiotherapy and chemotherapy are being used in various cancer treatments that are nonspecific and render extinction of normal tissues along with the cancer tissues. Most of the anticancer drugs are unsuccessful to reach the brain because of BBB. Thus, theranostic nanomedicines are being developed to enhance current cancer therapeutics to surmount the existing limitations (Kievit and Zhang, 2011). Ruan et al. developed a tumor microenvironment sensitive size shrinkable theranostic system for improving targeted delivery to glioma tissues as well as for monitoring drug delivery and treatment outcomes. This system incorporates small-sized gold NPs (AuNPs) onto gelatin NPs, which is a substrate for enzyme matrix metalloproteinase-2 (MMP-2), doxorubicin (DOX) (therapeutic moiety) and Cyanine-5.5 (Cy5.5, a fluorescent probe) decorated onto AuNPs through a hydrazone bond, making the system pH-sensitive and arginine–arginine–glycine aspartic acid (RRGD, a tandem peptide of RGD) and oct arginine showing both tumor targeting and tumor penetration ability. The system was surface-modified with a glioma targeting sequence (G-AuNPs-DC-RRGD) to enable active targeting. After 24 h incubation with MMP-2, the size of G-AuNPs-DC-RRGD was observed to shrivel from 188.2 nm to 55.9 nm, whereas pH-dependent release was observed in DOX and Cy5. G-AuNPs-DC-RRGD could enter inside cells and displayed high glioma targeting and accumulation efficiency, capable of delivering DOX to tumors. By virtue of fluorescence imaging, these systems further facilitated simultaneous tracking of drug delivery potential and therapeutic outcomes in vivo (Ruan et al., 2015). In another study, Fan et al. used superparamagnetic iron oxide (SPIO)-conjugated with doxorubicin-loaded microbubbles (DOXSPIO-MBs) for the diagnosis and treatment of brain tumor. For opening the BBB noninvasively and locally, the authors employed a focused ultrasound technique. The microbubble stable nanosystems were fabricated to induce BBB-opening and successful delivery of therapeutic drugs on exposure to focused ultrasound (FUS) in rat C6 glioma model can be a wonderful tool for future image-guided diagnosis and treatment of brain tumors (Fan et al., 2013). Sun et al. fabricated a novel formulation of mixed gold and SPIO-loaded micelles (GSMs) coated with polyethylene glycol-polycaprolactone (PEG-PCL) copolymer (hydrodynamic diameter approximately 100 nm) as both therapeutic and diagnostic tool in the management of glioblastoma multiforme (GBM). To determine the potential theranostic applications of novel GSMs, they investigated the radiosensitizing efficacy of GSMs by quantifying gH2AX (gamma H2AX, a marker of early DNA breaks) DNA damage foci in glioblastoma cell lines (U251 and U373), and found that GSMs administration in conjunction with radiation therapy led to approximately twofold increase in density of double stranded DNA breaks. As a diagnostic tool for GBM, GSMs were used as contrast agents for both CT and MRI of GBM tumors implanted in a mouse model (Sun et al., 2016). Zhou et al. have prepared iNGR-modified doxorubicin-loaded sterically stabilized liposomes (iNGR-SSL/DOX) to overcome the tumor vascular barrier and tumor-stroma barrier by surface modification of liposome with iNGR, a tumor penetrating

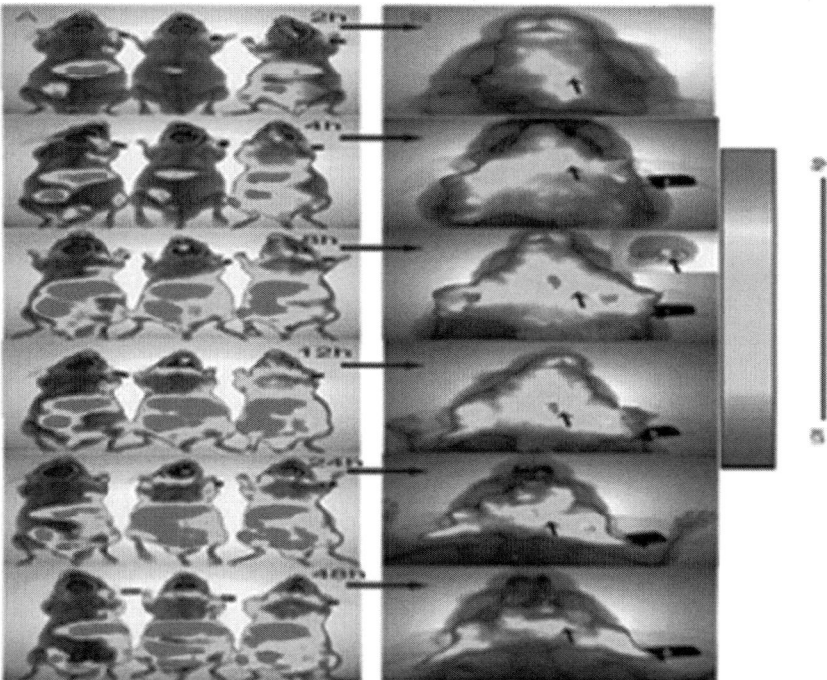

FIG. 18.1 Buildup of liposome quantity in orthotopic tumor tissues of animal model owing to iNGR modification.

peptide. As compared to the unmodified liposome, doxorubicin-loaded iNGR-SSL showed higher uptake of malignant glioma (U87MG) cells and human umbilical vein endothelial cells (HUVEC) with greater cytotoxicity in vitro. In addition, they encapsulated fluorescein amidite (FAM) or DiR as a fluorescent tracer in the nanostructures for aiding in vivo imaging and demonstrated a significant increase in liposome amount in orthotopic tumor tissues of the animal model due to iNGR modification (Fig. 18.1) (Zhou et al., 2017). Ramachandran et al. reported a simple and innovative method for making custom designed theranostic implants by creating a library of drug-loaded polyester nanofibers of PLGA-PLA-PCL blends. For forming a single three-dimensional (3D) composite nanofiber implant containing a separate set of nanofibers able to release the antiglioma drug temozolomide for an explicit period, Nanofibers having different release kinetics (from hours to months) were suitably mixed at suitable weight fractions and electrospun together and then shifted to next set for another period, thus constantly releasing the drug into the orthotopic brain tumor up to 1 month at a steady rate. Furthermore, they loaded nanofiber implant with Fe^{2+} doped calcium phosphate NPs (nCP: Fe) as contrast agents for MRI-guided noninvasive imaging of the nanodevice in vivo (Fig. 18.2) (Ramahandran et al., 2017). Gamage et al. have reported a novel curcumin-conjugated generation-3 (G3-Cur) dendrimer to ameliorate in vivo systemic bioavailability and drug delivery into brain tumors. These conjugates showed improved diagnostic imaging potential with reduced drug-related systemic toxicity in an orthotopic preclinical glioma model. G3-Curc can be a promising new formulation for clinical translation of Curcumin to cancer patients due to superior bioavailability and protracted therapeutic efficacy (Gamage et al., 2016). Jing and the group demonstrated efficient NIR-mediated photo immunotherapy (PIT) in human tumor xenografts developed from cancer stem cells (CSCs), using target specific AC133 monoclonal antibody (mAb) and NIR photosensitizer dye phthalocyanine IR700 conjugates (AC133-IR700 conjugate). Theranostic AC133-IR700 conjugates permitted precise image-guided and spatiotemporally controlled eradication of tumor cells. This study highlighted noninvasive NIR fluorescence imaging of orthotopic gliomas and PIT in glioblastoma stem cells (GBM-SCs). Results indicated rapid shrinkage of both invasively and subcutaneously growing brain tumors in a single treatment. Target binding efficiency and uptake of the AC133-IR700 conjugate was evaluated in nude mice-bearing U251glioma cells over-expressing CD133 by harnessing NIR-fluorescence molecular tomography (FMT). Such designed theranostic systems might be used well for intraoperative imaging and histopathological evaluation of tumor borders in patients toward fluorescence-guided tumor resection (Jing et al., 2016). Sun and coauthors developed highly specific and biocompatible nanoprobes produced from PEG coated iron oxide nanoparticles via functionalizing the surface of the nanoparticles with glioma targeting peptide, chlorotoxin (NP-PEG-CTX nanoprobe). The authors highlight preferential accumulation of NP-PEG-CTX nanoprobes inside glioma tissues, followed by their in vitro demonstration of contrast enhancement in MRI both in 9Lcells (rat malignant glioma) and in xenografts. Further, TEM imaging discovered the internalization of NP-PEG-CTX nanoprobes into the cytoplasm of 9L cells.

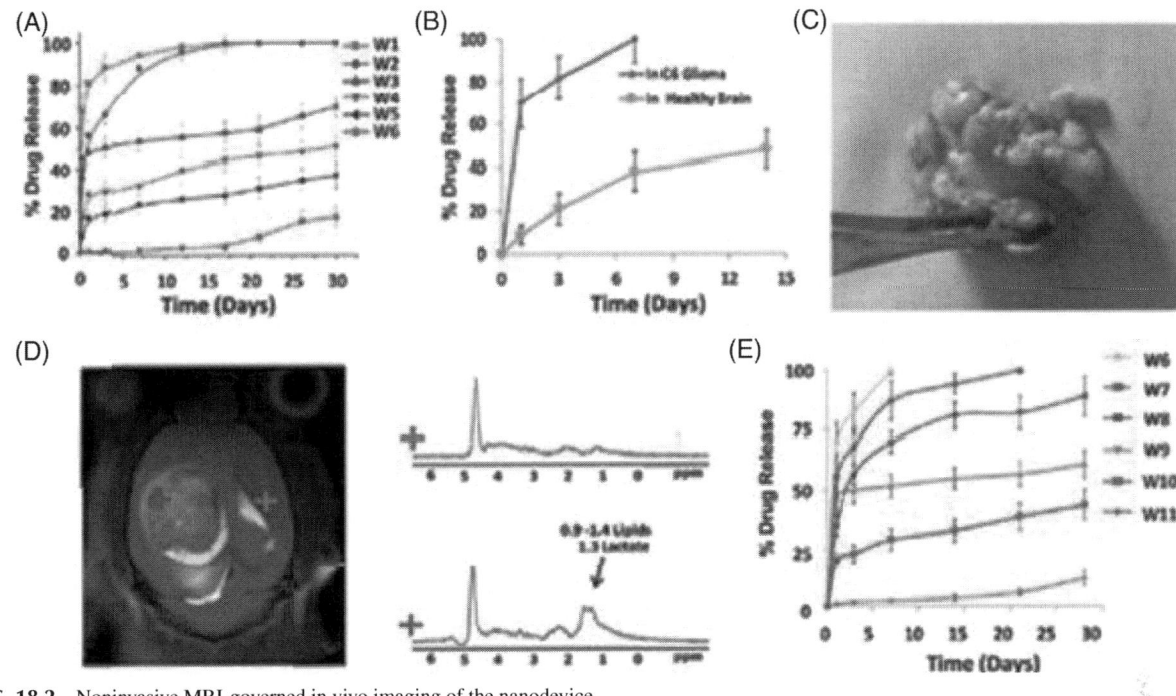

FIG. 18.2 Noninvasive MRI-governed in vivo imaging of the nanodevice.

Concurrent diagnosis and treatment of deadly gliomas and other neurological disorders is possible due to nanoprobe with specific targeting and biocompatibility (Sun et al., 2008).

18.3.3.2 Alzheimer's disease (AD)

AD is one of the most common neurodegenerative diseases that slowly destroy memory and thinking skills. The main feature of AD is accumulated plaque formation due to the aggregation of β-amyloid and intracellular neurofibrillary tangle formation due to phosphorylated tau protein (Dubois et al., 2014). Till date there is no effective treatment for this disease, in spite of having a team of researchers involved. Anti-AD drugs in clinical studies mainly target the CNS as a site of action. Conversely, due to the presence of BBB as a protective as well as a very selective barrier in CNS, the drugs cannot traverse through it in significant concentrations (Potschka, 2010). Gao et al. developed a "sense and treat system" to target amyloid aggregates related to AD, which is one of the best examples to illustrate "image-guided therapy" of Alzheimer's by simultaneous diagnosis and treatment. They used magnetic NPs (MNPs) whose surface was modified with two moieties, first, by napthalimide-based fluorescent probe (NFP) which was an oligomer-specific fluorescent probe designed against the exposed hydrophobic regions on Aβ oligomer surface and second, by Aβ-target peptide, KLVFF. The complex (MNP@NFP-pep) observed the changes in Aβ morphology as a result of the disaggregation of Aβ aggregates due to the local heat production by MNPs that could be sensed by the fluorescence probe. Therefore, it can be used as a real-time, "sense and treat" system for the treatment of AD (Gao et al., 2018b). For AD theranostics, another group of researchers developed a novel tau targeted multifunctional nanocomposite, ceria nanocrystals (CeNC) and iron oxide nanocrystals (IONCs) onto the surface of mesoporous silica nanoparticles (MSNs) functionalized with amino-T807 (PET tau tracer), an amino substituent of T807 and methylene blue (CeNC/IONC/MSN-T807- MB). They focused on an alternative tau targeting approach for treating AD instead of amyloid-β (Aβ) targeted therapy. The CeNC/IONC/MSN-T807-MB nanocomposite was formed by the controlled assembly of ultrasmall ceria nanocrystals (CeNCs), as tau hyperphosphorylation inhibitor and iron oxide nanocrystals (IONCs), as MRI agent, on the surface of uniform mesoporous silica NPs (MSNPs). In addition to this, for active targeting of tau protein, the surface of MSN was functionalized with amino-T807, via1, 4,7-triazacyclononane-1,4,7-triacetic acid (NOTA). Methylene blue (MB), a small inhibitor for tau aggregation was further loaded into the pores of MSNs. In vitro and in vivo studies of CeNC/IONC/MSN-T807-MB nanocomposite formulations using SH-SY5Y cells and AD rat models demonstrated significant suppression of tau hyperphosphorylation and protection of neural death (Chen et al., 2018). Hu et al. reported a nanotheranostics system (Congo red/Rutin-MNPs) based on magnetic iron oxide NPs (MNPs) with ultrasmall size to realize in vivo imaging of amyloid plaques along with targeted delivery and H_2O_2 controlled release of therapeutic agent rutin. Congo red/Rutin-MNPs theranostic system when co-administered with mannitol was able to break and cross BBB of the APPswe/PS1dE9 transgenic mouse and particularly bind to amyloid plaques, allowing amyloid plaques recognition by MRI along with targeted, and stimuli responsive delivery of Rutin via the Aβ-induced production of H_2O_2. Rutin is potent antioxidants which reduce amyloid plaques and neuronal loss, prevents oxidative stress and hinders Aβ aggregation (Hu et al., 2015). Similarly, Li et al.; have reported novel multifunctional peptide-conjugated gold (Au) nanorods for diagnosis and treatment of AD. They combined unique high NIR absorption property of AuNRs with two Aβ inhibitors, Aβ15-20 (Aβ-targeted peptide inhibitor) and polyoxometalates (POMs) for effective inhibition of Aβ aggregation and also dissociation of amyloid deposits by NIR irradiation. Additionally, shape- and size-dependent optical properties of gold nanorods were used as effective diagnostic probes for the detection of Aβ aggregates. Thus, this study exemplifies a multifunctional theranostic nanosystem containing various components such as a targeting ligand, a reporter, and inhibitors in one system for treating AD (Li et al., 2017). Cui et al. have reported a novel smart UCNP-based nanoprobe that can simultaneously serve the purpose for accurate diagnosis and effective therapy of AD. UCNPs nanoprobe is comprised of two major components: UCNPs for detection and imaging, the chelator 8-hydroxyquinoline-2-carboxylic acid (HQC) for capturing of Cu^{2+}, and therapy of AD. These multifunctional UCNPs were capable of detecting and capturing Cu^{2+} both in vitro and in vivo. Fascinatingly, this system has also shown the unique ability to inhibit as well as transforming toxic Aβ intermediates into nontoxic fibers (Cui et al., 2016). Recently, Costa et al. reported the synthesis of functionalized CNTs (f-CNTs) and used them as theranostic carriers for brain delivery of amyloid-targeting drugs/compounds that efficiently cross the BBB. Importantly, attractive intrinsic optical, thermal properties and the ability to cross biological barriers by both energy-dependent and independent mechanisms of functionalized CNTs (f-CNTs), bestows them as useful tools toward brain targeting theranostic applications (Costa et al., 2016). Functionalized multiwalled carbon nanotubes (f MWNTs) capable of crossing an intact BBB have already been demonstrated by Kafa and group (Kafa et al., 2015, 2016; Wang et al., 2016). They functionalized MWNTs with two PiB derivative Gd3+ complexes-Gd (L2) and Gd(L3). PiB is an Aβ-binding molecule and used as a PET imaging agent. In vivo biodistribution results demonstrated considerable uptake and accumulation off-MWNTs in the brain as compared to free metal complexes. Thus, this report claims that f-MWNTs could be used as carriers in theranostic applications involving brain delivery of BBB impermeable compounds (Costa et al., 2018). Li and group showed the theranostic effect of (E)-4-(4-(dibutyl amino)styryl)-1-(2-hydroxyethyl) quinoline-1-ium chloride (DBA-SLOH) NIR dye which accomplished diagnostics via NIR-based in vivo imaging of Aβ plaques in APP/PS1 transgenic(Tg) mouse over-expressing Aβ. Concomitantly therapeutics was achieved by inhibiting Aβ monomers to self-aggregate. This group developed three cationic NIR dyes and confirmed their activity to inhibit Aβ peptide aggregation by the fluorescence assays, CD spectroscopy, in vitro fluorescence staining of Aβ plaques in brain segments and NIR imaging in an animal model. Among them, DBA-SLOH showed the highest BBB permeability, excellent selectivity and binding affinity to Aβ peptides, and intense enhancement of fluorescence signal in the NIR window upon binding with Aβ aggregates. The developed theranostic design with tremendous BBB permeability incorporating both NIR imaging of Aβ species and inhibition of Aβ aggregation in vivo for the diagnosis and therapy of AD may open up new possibilities in AD treatment (Li et al., 2016). In another study, Zhang and the group developed dual functional pegylated poly (lactic acid) NPs (PEG-PLA NPs)-based targeted delivery system for the early diagnosis and treatment of AD. To achieve dual targeting, a 12-amino acid peptide, TGN (targeting ligand for BBB), D-enantiomeric peptide, QSH (targeting ligand for Aβ42 deposits in the brain), were surface functionalized over the NPs. Dual-targeting of the designed system was confirmed via brain distribution studies of NPs using a near infrared dye, DiR, as a probe and ex vivo imaging studies. T3Q3-NP showed precisely improved targeted delivery toward amyloid plaque in the brains of AD mice model pointing toward a valuable theranostics system for AD diagnosis and therapy (Zhang et al., 2014).

18.3.3.3 Parkinson's disease (PD)

PD is the most common age-related neurodegenerative disease, following AD (Ozansoy and Basak, 2013). Neuropathologically, PD is the slow loss of dopaminergic neurons in the substantia nigra pars compacta (SNC) resulting in depletion of dopamine content which causes severe cognitive and motor insufficiency (Kubrusly et al., 2003; Reis et al., 2007). At present, the L-3, 4-dihydroxyphenylalanine (levodopa) precursor of dopamine is a substitute for dopamine which can easily traverse the BBB via large neutral amino acid transporter (LAT1) (Pardridge, 2005). Currently, effective medicine for PD patients is L-DOPA (Connolly and Lang, 2014). In view of this information, Birgitte et al .designed a novel nanosystem of manganese oxide NPs (MONPs) functionalized with L-DOPA capable of releasing Mn2+ ions and L-DOPA concurrently in water after the disintegration of MONPs. They have used MONPs as MRI agents for imaging and diagnosis via time-dependent switch in MR contrast and L-DOPA as stabilizers as well as an active drug for PD (McDonagh et al., 2016). Niu et al. developed an anano-gene delivery technique using magnetic Fe_3O_4 NPs coated with oleic acid and decorated with short hairpin RNA (shRNA) plasmid to interfere with α-synuclein expression in neurons. To render the system to be temperature and pH-responsive for effective release of the gene, N-isopropyl acrylamide derivative (NIPAm-AA) was photoimmobilized onto the oleic acid molecules. Similarly, to promote enhanced neuronal uptake through receptor-mediated endocytosis, nerve growth factor (NGF) was absorbed onto NIPAm AA. In vitro and in vivo results demonstrated that this system is effective in inhibiting apoptosis and repairing a PD model (Niu et al., 2017). Wen et al. have developed a novel OL-PEG-PLGA (odorranalectin-conjugated poly (ethylene glycol)-poly(lactic-co-glycolic acid) nanoconjugate system which was used to increase the delivery of drugs from the nose to the brain for effective treatment of PD. Odorranalectin (OL) is low immunogenic and is the smallest member of the lectin family.

Further, they have functionalized the OL-NP nanoconjugates with a neuroprotective drug, urocortin peptide and a fluorescent tracer, DiR, for in vivo fluorescence imaging. A significant increase in the uptake of modified NPs was observed in vitro and in vivo studies established a good distribution of OL-NP/UCN nanosystems in the brain region along with improved therapeutic effects in hemiparkinsonian rats. Therefore, OL-conjugated nanosystems could be used as potential noninvasive brain drug delivery systems for the treatment of CNS disorders (Wen et al., 2011). From lactoferrin (Lf)-conjugated PEG-PLGA nanoparticles (Lf-NPs), Hu and the group fabricated a novel biodegradable system for efficient brain drug delivery. The group focused on utilizing the iron-binding cationic glycoprotein, Lf as a brain targeting ligand. Lf-NPs were loaded with the cytoprotectant peptide urocortin (UCN), a corticotrophin releasing hormone related peptide to arrest the development of Parkinsonian like features. In vitro and in vivo theranostic and targeting properties of Lf-NPs were evaluated by incorporating coumarin-6 as a fluorescent probe. Qualitative and quantitative uptake studies of Lf-NPs showed a prominent accumulation of Lf-NPs via clathrin-mediated endocytosis in bEnd 3 cells. Lf-NPs loaded with UCN considerably alleviated the striatum lesion caused due to 6-OHDA. All these results proved UCN-loaded Lf-NPs as promising brain drug delivery systems in treating PD (Hu et al., 2011).

18.3.3.4 Neurovascular diseases

Neurovascular diseases are one of the main leading causes of death and disability globally. As per the obtained statistics from American Heart Association (Mozaffarian et al., 2016), major vascular diseases include atherosclerosis, stroke, thrombosis, coronary artery and peripheral artery diseases. For overcoming the drawbacks associated with current interventions/treatments for vascular diseases such as trivial accumulation of drugs at the target site, quick clearance of drugs, and toxicity to normal cells, etc., nanotheransotic systems are designed to target specific vascular regions by surface modification of NPs with specific ligands or peptides (Gupta, 2011). McCarthy et al., developed macrophage targeting fluorescence NPs for the treatment of atherosclerosis. They prepared magnetic fluorescent NPs (MFNPs) by conjugating dextran-coated magnetic NPs with NIR dye, which was then analyzed toward photodynamic therapy. It was observed that upon excitation (at 646 nm) the MFNPs generated singlet oxygen species that killed the macrophages without affecting neighboring normal cells (McCarthy et al., 2010). Another study carried out by Myerson et al. harnessed the concept of nanotheranostics to treat thrombosis. In this study, they developed bivalirudin functionalized perfluorocarbon NPs to target thrombin as a theranostic platform. The NPs bound at the site of active clotting and then quenched local thrombin activity by inhibiting platelet deposition (Myerson et al., 2014). Sun et al. carried out a study in the mice hind limb model by developing VEGF-loaded and IR800-conjugated, graphene oxide NPs as a theranostic platform to image therapeutic angiogenesis (Sun et al., 2013).

In another report, Agyare et al. designed theranostic nanovehicles (TNVs) for targeting, diagnosis, and treatment of cerebrovascular amyloid deposits which cause recurrent hemorrhagic strokes due to the deposition of these amyloid proteins within the walls of the cerebral vasculature. TNVs were composed of apolymeric nanocore formed of chitosan conjugated with gadopentetate dimeglumine (Magnevist), an MRI contrast agent and cyclophosphamide, an immunosuppressant entrapped inside for the treatment of cerebrovascular inflammation. Furthermore, they are also functionalized TNVs, nanocore surface by the F(ab′)2 fragment (F(ab′)24.1) of a novel antiamyloid antibody, IgG 4.1 to target cerebrovascular amyloid. In vitro studies in polarized human microvascular endothelial cell monolayers (hCMEC/D3) and in vivo studies in mice clearly established the capability of TNVs to target as well as image cerebrovascular amyloid by MRI and single photon emission computed tomography techniques (Agyare et al., 2014).

Therefore, TNVs served as new generation advanced theranostic nanocarriers with the capability to simultaneously detect amyloid accumulation and inhibit cytokines produced by Aβ40 in the bovine brain micro vascular endothelial (BBMVE) cells. Similarly, Liu et al. have developed a label-free, CDPC-lipo nanoformulation by the encapsulation of citicoline in liposome for the treatment of ischemic stroke. Citicoline is a well-known neutral neuroprotective used to treat neurodegenerative diseases. This group has discovered its inherent chemical exchange saturation transfer (CEST) MRI signal, which is an interesting property of citicoline for its application as a theranostic agent. In vitro results demonstrated clearly two inherent CEST signals of citicoline at +1 and +2 ppm. Moreover, in vivo results of CDPC-lipo in rat brain model demonstrated enhanced accumulation of the nanoformulation in an ischemic area that could be also monitored by CEST MRI. Therefore, liposomal citicoline may be a potential label-free theranostic system for the treatment of ischemic stroke (Liu et al., 2016a). Recently, Wang et al. have also developed an anti-PirB immune liposome nanoprobe labeled with NIR probe, as a novel theranostic system to target ischemic stroke. PirB is basically a paired immunoglobulin like receptor B which is expressed by neurons and can inhibit neurite outgrowth. They have used soluble PirBectodomain (sPirB) protein as a therapeutic reagent for ischemic stroke. Their results demonstrated that sPirB immunoliposome could significantly accumulate in the ischemic region and improved motor abilities of the cerebral ischemic model of mice (Wang et al., 2018).

18.4 Nanotheranostics and rheumatoid arthritis

18.4.1 Rheumatoid arthritis (RA)

RA is a familiar polyarticular autoimmune inflammatory disease that causes cartilage and bone devastation in the synovial joints resulting in joint impairment and a decrease in life expectancy (Umar et al., 2012; Pham, 2011; Rahman et al., 2015). Though the specific mechanism of RA is not clear, the environmental and genetic factors are known to enhance disease susceptibility and contribute to a diseased state (Umar et al., 2012). The immune system releases certain specific enzymes and chemicals in the synovial tissues leading to erosion of bones and cartilage, which is one of the important factors contributing to inflammation resulting in arthritis. It affects various joints of the body which include the joints of the hands, wrists, elbows, shoulders, knees, hip, cervical spine, ankles, and feet (Kapoor et al., 2014; Chandrasekar and Chandrasekar, 2017). The worldwide prevalence of RA is about 0.51% (Dolati et al., 2016; Alam et al., 2011; Guo et al., 2018; Firestein, 2003); within the developing countries its prevalence varies and various studies indicate that the Western countries have a larger prevalence of RA than others (Alam et al., 2011; Nogueira et al., 2016). As compared with males, the disease is three times more common among females (Dolati et al., 2016; Doan and Massarotti, 2005) and is mostly diagnosed in obese women especially women who are above 65 years of age, which can be attributed to hormonal changes as well (Symmons et al., 2002).

18.4.2 Current treatments and their drawbacks

The present treatments of RA, such as nonsteroidal anti-inflammatory drugs (NSAIDs), disease-modifying antirheumatic drugs (DMARDs), glucocorticoids, and biologic drugs (Umar et al., 2012; Dolati et al., 2016), either offer symptomatic relief or modify the disease process. Though these are successful treatments, their use is limited because of their harmful side effects such as cardiac complications, gastrointestinal damage, and ulcers, along with immunosuppression resulting in the development of opportunistic infections (Umar et al., 2012). Etanercept (Enbrel), infliximab (Remicade), and adalimumab (Humira) are some of the biologics used to treat RA (Dolati et al., 2016; Prasad et al., 2015) and many other inflammatory conditions. In spite of the availability of recent medical advances, there is still an unmet need to develop treatments for RA because of safety and efficacy concerns associated with the currently available drugs (Prasad et al., 2015).

Undesirable consequences owing to nonselective action of the currently available RA therapeutics can be decreased by encapsulating these bioactive molecules in nanocarriers for an extra targeted approach to deliver the medicaments at the preferred location (i.e., the joints) by avoiding frequent or high dosing, to attain an effective drug concentration locally (Pham, 2011; Ngobili and Daniele, 2016). Additionally, nanocarriers can be engineered to protect these bioactive molecules from degradation, thus increasing their bioavailability, while decreasing off-target side effects. There can be targeting of certain specific receptors or of macrophage uptake to invade diseased tissues such as inflamed joints (Prasad et al., 2015; Jiang et al., 2018); and the uptake of nanocarriers by spleen and liver can be prevented by modifying their physiochemical properties such as enhanced penetration through biological barriers and encapsulation (Rahman et al., 2017). Once the nanocarriers reach the target tissue or organ, the drug can be released in a controlled fashion depending on the pH, temperature, solubility, redox conditions, etc. Scientists have been successful in making nanomaterials that respond to these physical or biological stimuli, termed "intelligent" or "smart" delivery systems' (Liu et al., 2014). This allows investigators to examine or re-examine bioactive molecules that were previously thought to be too toxic to deliver through a systemic route (Ngobili and Daniele, 2016). Nanomedicine is the use of nanotechnology in medicine to ease the diagnosis, treatment and monitoring of treatment (Kunjachan et al., 2015) by using nanoparticles, dendrimers, polymeric micelles, drug-loaded liposomes (Oliveira et al., 2018), nanocapsules and nanogels. In addition, polymer protein conjugates, polymer drug conjugates and antibodies also come under the umbrella of nanomedicine (Prasad et al., 2015).

The most imperative fact for the treatment of RA and other chronic inflammatory disorders is better permeability and retention because of the widespread systemic nature of inflammation. Synthesized and surface modified nanoparticles with appropriate functional groups ranging in size between 10 and 1000 nm can help enhance their circulation time and therefore offer maximum benefit in performing their desired task (Katsuki et al., 2017).The nanomedicine actively or passively gathers in the inflamed joints through an enhanced permeation and retention effect. At the site of inflammation, the permeation and accumulation of nanomaterials through passive targeting is assisted by disturbed vasculature and reduced lymphatic drainage underneath inflammatory conditions and at solid tumors. A suitable receptor-specific ligand is attached to a nanocarrier for specific binding in the case of active targeting, which helps in better efficacy and minimum systemic side effects (Prasad et al., 2015; Chen et al., 2017).

In one approach, theranostics, nanomedicine may open up a new way to associate diagnosis and treatment (Pedrosa et al., 2015; Sarmento and Sarmento, 2017). Nanomedicines which are biodegradable, biocompatible, and disease modifying represent a likely therapeutic approach for osteoarthritis and RA. At the inflammation site due to the combination of unique physicochemical properties of nanocarriers and the pathophysiological properties of inflamed joints, selective targeting and minimization of off-target adverse effects have enhanced the bioavailability and bioactivity of DMARNs. Therefore, nanomedicine not only reduces the drug dosage but also decreases the treatment duration (Rubinstein and Weinberg, 2012). The surface alteration of nanomaterials such as nanoparticles (metallic, polymeric, solid-lipid nanoparticles), nanocomposites, nanoconjugates, nanocarriers, nanowhiskers, micelles, specifically coated nanoparticles and receptor-targeted nanoparticles permit the selective targeting of inflamed joints through the linked drugs, ligands, and prognostic or diagnostic markers, as shown in Fig. 18.3 (Qamar et al., 2019). Some of the nanotheranostic and nanotherapeutic approaches utilizing these nanomaterials that have been proved to be effective in RA are summarized in Fig. 18.4 (Qamar et al., 2019).

18.4.3 Nanotheranostic approach for rheumatoid arthritis

The theranostic approach of nanomedicine for RA theranostics will enhance the precision and efficacy of treatment by shifting the generalized approach of existing clinical standards to personalized procedures. Hence, theranostics play the most significant role in diseases, such as RA, cancer, infection, and cardiac disorders which require personalized methods for treatment and monitoring. Depending on the patient's needs and responses, this personalized approach helps the physician to alter the treatment, thus avoiding the untoward effects coupled with the drugs, such as those at a high dosage which

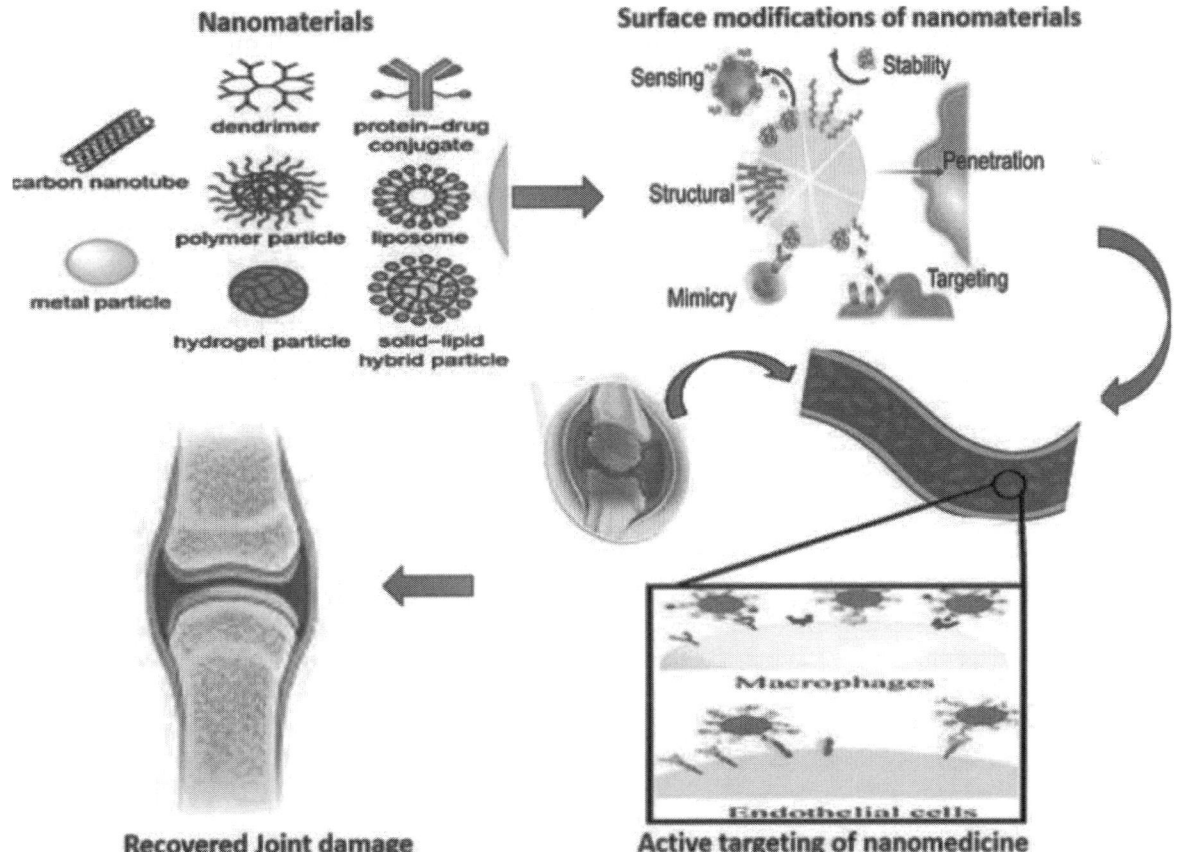

FIG. 18.3 Function of nanomaterials in improvement of joint impairment.

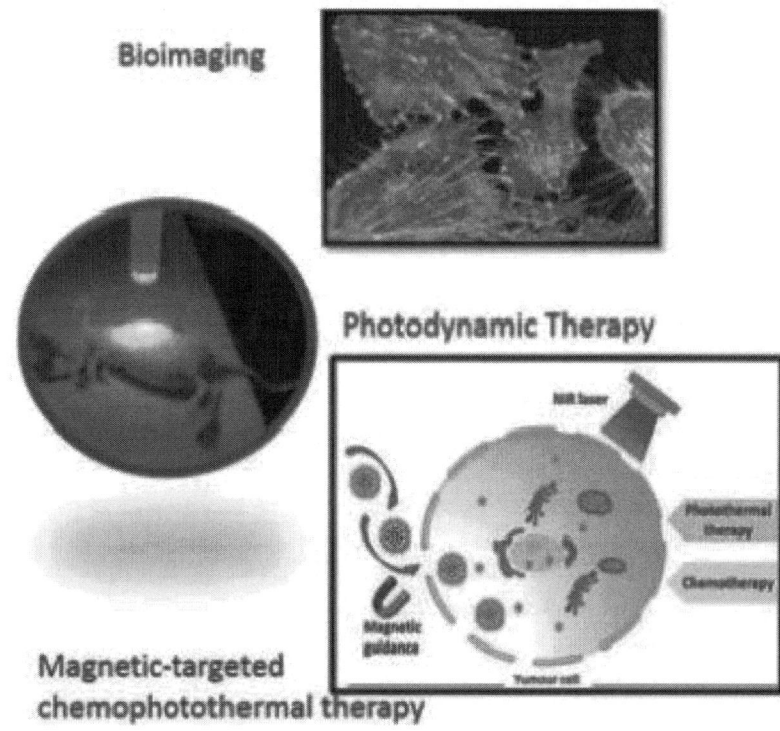

FIG. 18.4 Various theranostic uses of nanomaterials.

may result in drug resistance, relapse and incomplete remission (Pedrosa et al., 2015). Nanotechnology is being considered for the development of new ways to diagnose, fight and monitor the progress of treatment, hence providing a noninvasive and definite imaging tool for RA (Mura and Couvreur, 2012).

18.4.3.1 Bioimaging and photodynamic therapy through nanocomposites in RA

Nanocomposites are thought of as materials of the 21st century due to their unique design and combination of properties. Having at least one phase in the nanometer range, they appear to be appropriate alternatives to coping with the challenges associated with monolithic and microcomposites (Camargo et al., 2009; Eichhorn, 2011). Although the well-timed prognosis and treatment of RA is a challenge, Zhao et al. have revealed the possibility of employing tetra sulfonaphenyl porphyrin with titanium dioxide (TiO_2) nanowhiskers (rod like nanostructures), a novel nanocomposite, as an efficient bioimaging and photodynamic therapeutic agent. The study validated that this nanocomposite solution has an improving effect on RA by decreasing the level of TNF-a and IL-17 in serum, assisted by the use of fluorescence imaging to diagnose disease at early stages and to identify biomarkers in the inflamed joints of RA (Zhao et al., 2015).

18.4.3.2 Magnetic-targeted chemo-photothermal nanotherapy in RA

Kim et al., developed methotrexate-loaded plasmonic nanoparticles consisting of gold in the inner- and outermost layers with iron in the middle layer. Furthermore, certain biomolecules were also attached to target the specific site, provide chemo-photothermal therapy and retain the potential to be imaged in vivo. Infra-red irradiation was employed to generate heat at the inflamed site and release the methotrexate due to resonance. Magnetic resonance imaging along with near-infrared absorbance imaging is permitted by the iron layer in the middle. Additionally, if an external magnetic field was applied, the delivery and retention time of nanoparticles in the inflamed regions may be improved. The study in question was important not only because of the multifunctional nature of nanoparticles but also due to the small dosage of drug required for therapy with the injected nanoparticles (Kim et al., 2015).

18.4.3.3 Combined photodynamic and photothermal therapy in RA

Due to efficient penetration of near-infrared light in inflamed joints, phototherapy (photodynamic therapy and photothermal therapy) offers new treatment modalities for RA. Unique Cu7.2S4 nanoparticles were prepared by Lu et al. (2018) for the treatment of RA, showing that copper-based nanomaterials can serve as photothermal agents as well as photosensitizers, and in the meantime, copper may stimulate chondrogenesis and osteogenesis (Prasad et al., 2015). The Cu7.2S4 nanoparticles in combination with 808 nm near-infrared light irradiation not only inhibited invasion of inflamed synovium, in vivo release of proinflammatory cytokines and erosion of cartilage but also conserved bone by accomplishing higher bone to total volume and superior BMD. Hence phototherapy using multipurpose Cu7.2S4 nanoparticles could be a new treatment approach for RA (Lu et al., 2018).

18.4.3.4 Nanotheranostic approach for macrophage detection and therapy in RA

Pathogenicity, phagocytic nature and abundance are the characteristic features of macrophages that make them potential theranostic targets. Though nanoparticles are phagocytosed by macrophages, in vivo targeting efficacy is influenced by their physicochemical properties such as size, surface charge, functionalization, and ligands. Certain receptors expressed on the surface of macrophages are used for active targeting by ligands, namely extra, mannose, tuftsin, and hyaluronate (Patel and Janjic, 2015). Heo et al. developed nanoparticles through conjugation of 5bcholanicacid to a dextran sulfate backbone for selective delivery of methotrexate to the affected joints in RA. By using a simple dialysis method, methotrexate was loaded into the dextran sulfate nanoparticles with an efficiency of 73% and readily endocytosed by macrophages. In collagen induced arthritic experimental mice, these nanoparticles when administered systemically, efficiently caused about a 12-fold increase in inflamed joints representing their targeted approach. Besides, methotrexate-loaded nanoparticles demonstrated significant therapeutic efficacy in the arthritic mouse model, thus highlighting dextransulfate nanoparticles as a possible nanomedicine for imaging and therapy of RA (Heo et al., 2017).

The inflamed vasculature of the joints offers an opportunity for passive-targeting of nanotheranostic systems owing to the intrinsic disease-causing ability of macrophages and monocytes. Due to this property, macrophages can be used as Trojan horses for the delivery of drugs and imaging agents at the site of disease (Patel and Janjic, 2015).

18.5 Nanoparticle-based theranostic agents

Nanoparticle-based theranostics evolved since nanoparticle (NP)-based imaging and therapy have been investigated separately, to develop a better understanding of them. It has now evolved to a point that can be defined as nanoplatforms that can deliver together the therapeutic and imaging functions. Currently, many nanomaterials are already imaging agents and can be simply employed as theranostic agents by imparting therapeutic functions to them.

18.5.1 Iron oxide nanoparticle-based theranostic agents

Nanocrystals made from magnetite or hematites are called Iron oxide nanoparticles (IONPs) (Morales et al., 1999). With appropriate coatings, IONPs can be easily coupled with drug molecules. For instance, the Zhang group coupled methotrexate (MTX), an anticancer drug, onto an aminated IONP surface (Kohler et al., 2004, 2005, 2006). In vitro studies confirmed the release of drug molecules owing to low pH and occurrence of proteases, where the particles accumulated in lysosomes after internalizing into cells. Hwu et al. (2009) reported on coupling paclitaxel (PTX) to IONP surfaces through a phosphodiester moiety at the (C-2′)-OH position. Per nanoparticles, the average number of PTX molecules was assessed as 83, and the release of the PTX was found to be successful upon exposure to phosphodiesterase. The Cheon group used meso-2,3-dimercaptosuccinic acid (DMSA) to change IONPs, and used SMCC as the crosslinker to couple Herceptin antibody molecules onto the particle surface (Huh et al., 2005; Lee et al., 2007), although Herceptin was employed as a targeting agent rather than a therapeutic agent in the study.

Drug molecules can also be co-capsulated with IONPs into polymeric matrices aside from covalent coupling. Jain et al. (2008) loaded doxorubicin (DOX) and PTX, along with oleic acid coated IONPs, into pluronic-stabilized nanoparticles. Similarly, Yu et al. (2008) loaded DOX into antibiofouling polymer coated IONPs. Application of DOX-loaded nanoconjugates in a Lewis lung carcinoma xenograft model showed better pharmacokinetics and therapeutic effects than DOX alone, presumably due to the antibiofouling feature of the particles. Likewise, protein molecules have also been investigated as drug carriers. For instance, a two-step coating strategy to yield human serum albumin (HSA) coated IONPs have been developed (Xie et al., 2010b). By means of the excellent binding capacity of HSA, a variety of lipophilic drugs can be loaded into such nanoplatforms to gain theranostic agents.

It is well-known that small molecules can be loaded into porous nanostructures via physical absorption and in this direction; efforts have been put to obtain hollow iron oxide nanostructures. Likewise, an effort was reported by the Piao group (Piao et al., 2008). Commencing from spindle shaped β-FeOOH NPs made from FeCl3 hydrolysis, the authors performed a three step, so-called "wrap-bake-peel" treatment to get hollow IONPs. The authors found that DOX could be loaded into such hollow nanoparticles via simple physical absorption and subsequent sustained release from the nanostructures under physiologic conditions (Piao et al., 2008). Recently Cheng group gave a more detailed report on a similar approach used (Cheng et al., 2009). Applying controlled oxidation and acid etching of Fe particles, porous IONPs with a sizable cavity were achieved. Afterward to confer targeting specificity the loaded cisplatin into the cavities of the particles, and coupled Herceptin onto the particle surfaces. As a result of which the conjugates showed selective affinity to ErbB2/Neu positive breast cancer cells and sustained cytotoxicity attributable to the controlled release of cisplatin from the particle carriers (Cheng et al., 2009).

Gene therapy came out as an important therapeutic avenue, where DNAs/RNAs are employed as therapeutics to antagonize abnormal gene regulation. Distinctly small-molecule-based therapeutics, DNAs/RNAs are negatively charged and, by themselves, have trouble in passing through the negatively charged cell membrane. Apart from this, the nucleases that are ubiquitous in living subjects may identify and degrade DNAs/RNAs before they reach their targets and accomplish their task. It is in such cases that nanoparticle-based delivery plays a significant role. Ideally, the nanoparticle carriers can load the therapeutic genes, escort them to the diseased areas, facilitate their shuttling across the target cell membranes and, finally, release them intracellularly and execute their functions.

Medarova and his colleagues established a breakthrough work along this line (Medarova et al., 2007) by coupling thiolated siRNA onto aminated dextran particles using N-succinimidyl-3-(2-pyridyldithio) propionate (SPDP) as a bridging compound. Furthermore, the near-infrared dye Cy5.5 and myristoylated polyarginine peptide (MPAP), a membrane translocation peptide, were attached to the particle surface. First, the probes equipped with siRNA that targets green fluorescence protein (GFP), and the conjugates were tested in a mouse model bearing bilateral 9L-GFP and 9L-RFP tumors. Performing MRI and near-infrared fluorescence (NIRF) imaging appreciable probe accumulation was found in both tumors. On the contrary, the optical signal drop was only noted in the 9L-GFP tumor, and not in the 9L-RFP tumor. siRNA-GFP was then changed to a therapeutic siRNA sequence which targeted the antiapoptotic gene Birc5 (which encodes the protein survivin), and studied the therapeutic potential of the nanoconjugate in an LS174T human colorectal carcinoma xenograft model. Such therapeutic particles induced an amazing drop in surviving transcript level of 97 ± 2% has been confirmed by RT-PCR results which further resulted in increased levels of tumor associated apoptosis and necrosis (Medarova et al., 2007).

In the above investigation, EPR effect was found responsible for the mediation of the particle accumulation in tumor. In order to attain site-specific delivery, biovectors may be added to nanoplatforms. For instance, Lee group coupled siRNA and a PEGylated cyclic Arg-Gly-Asp (RGD) peptide onto magnetic nanoparticles (Lee et al., 2009). In this study, MDA-MB-435 and A549 cells having high and negative integrin $\alpha\nu\beta3$ expression were firmly transfected with GFP and were incubated with nanoparticles loaded with GFP siRNA. The two cell lines demonstrated distinct particle internalization rates, validating RGD-integrin interaction-mediated particle uptake. Such variation in uptake resulted in a spectacular inconsistency in gene regulation efficiency. A significant concentration-dependent decrease in GFP expression has been observed with MDA-MB- 435 cells. Although, the particles alone, with or without RGD coupling, demonstrated no effect on knocking down GFP expression in A 549 cells.

The magnetic properties of IONPs permit them to accumulate due to the presence of an external magnetic field. This feature has been utilized as a targeting mechanism to improve drug delivery efficiency, and related studies in animals (Mornet et al., 2004; Ito et al., 2005) and in humans have been reported. In one such study, seven patients with metastatic breast cancer were infused with epirubicin-loaded IONPs (100 nm in diameter, at 0.5% of the estimated blood volume), and after that, a magnetic field was established around the tumor (Neuberger et al., 2005; Lubbe, 1997). In 50% of the patients, to induce tumor regression, the magnetic field was found useful in directing the ferrofluid to the tumor. Namiki et al. (2009) monitored cationic lipid coated IONPs, and found one formula, designated as LipoMag, that outperformed commercial PolyMag in both transfection and gene knockdown in all 13 tested cell lines. The authors found the one sequence, siRNAEGFR#4, with the greatest percentage knockdown of EGFR mRNA. They then performed 2′-OMe modification on the uridine residues of the sense strand and yielded a modified sequence with a similar knock-down effect but reduced cytokine induction compared with the parent sequence. Then, onto nanoparticles, they loaded this modified siRNA and assessed their therapeutic influence in two gastric cancer models. A 50% reduction in tumor volume was observed in the therapeutic group after a 28-day treatment, along with other desired events, such as inhibition of angiogenesis and induction of apoptosis. Remarkably, during the application of magnetic fields at the tumor sites, such gene knockdown was only noteworthy as the lipid particles alone were not an effective delivery vehicle.

Due to the potential in hyperthermia IONP can itself play a dual role in imaging/therapy. The underlying mechanism behind this is that IONPs can act as antennae in an external alternating magnetic field (AMF) converting electromagnetic energy into heat (Mornet et al., 2004; Ito et al., 2005). This is a promising feature in tumor therapy for tumor cells are more susceptible to elevated temperature than normal cells (Mornet et al., 2004; Ito et al., 2005). In a study, phospholipid coated IONPs were injected into a subcutaneous tumor model in F344 rats and were exposed to an AMF (Ito et al., 2005). The AMF in conjugation with IONPs elevated the temperature of the tumor above 43°C resulting in tumor regression but had no effect on the control group where no IONPs were given. Also, Fab fragment of antihuman MN antigen-specific antibody was chemically anchored onto IONP surfaces and the INOPs were administrated systemically into tumor-bearing mice. The particles showed high tumor uptake, presumably due to an antibody-antigen interaction, and induced efficient tumor hyperthermia when exposed to an AMF (Ito et al., 2005; Shinkai et al., 2001). In one more study, Zn-Pc, a photodynamic therapeutic (PDT) agent-loaded $CoFe_2O_4$ nanoparticles showed superior hyperthermia effects as compared to IONPs (Primo et al., 2008). Combined toxicity from both PDT and magneto hyperthermia was observed after in vitro evaluation with J774-A1 macrophage cells; however, to elucidate the synergy of such combinational in vivo therapy, additional work is needed.

18.5.2 Quantum dot-based theranostic agents

Quantum dots (QDs) are nanocrystals made from semiconductor materials emitting light. QDs are coming up as an important class of biomaterials due to their unique optical properties which are absent in organic dyes or fluorescent proteins, such as being brighter, more photochemical stability and having a narrow emission spectrum.

Mostly due to the innate toxicity of QDs the QD-based drug delivery approach is relatively less investigated. With the first generation QDs this problem is more prominent, where toxic Cd and Pb are frequently used in QD preparation. Current research in QD synthesis has given Cd-free QDs, such as InAs/ZnSe (Zimmer et al., 2006) and InAs/InP/ ZnSe (Xie et al., 2008; Kim et al., 2005). They are potentially more capable carriers, but pertinent applications are so far limited. However, Nurunnabi et al. (2010), reported on making QDs-Herceptin conjugates. The CdTe/CdSe QDs were made water soluble through the addition of PEG-10, 12-pentacosadiynoic acid (PEG-PCDA) and the nanostructure was further stabilized employing UV irradiation. The risk of Cd^{2+} leaking out from the core materials was reduced by crosslinking the coating shell. The nanoparticle depicts an efficient tumor targeting rate and impressive therapeutic effects when tested on an MDA-MB-231 tumor model. Hydrophobic QDs and IONPs, along with DOX were co-encapsulated into micelles formed with PEGylated phospholipid (Park et al., 2008). These conjugates were further attached with a tumor homing peptide F3 and were introduced into an MDA-MB-435 xenograft model. Attributing mediation of the tumor homing peptide tumor targeting was effectively observed by both optical and MR imaging treatments.

Nevertheless, as a proof-of-concept study, no therapeutic studies were pursued and the investigation stopped at the imaging level only.

In another interesting study, Bagalkot et al. (2007) produced a QD aptamer (Apt)-DOX conjugate [QD-Apt(Dox)] and investigated its use for simultaneous cancer imaging, therapy and therapy monitoring. A10 RNA aptamer was linked to the QD surface and was employed as a biovector to target PSMA. While the DOX loading was achieved by its intercalation within the aptamer sequence. The fluorescence activities from QD and DOX were minimized with the help of interaction between DOX and RNA, respectively. On delivering particles into the targeted tumor cells, DOX would be gradually released from the system, which, aside from initiating therapeutic functions, also led to the recovery of QD fluorescence. Likewise, Yuan et al. (2009) loaded MTX onto QD surfaces to induce photoluminescence quenching. The loading was accomplished by means of simple reversible physical adsorption, which can be reversed on exposure to species with greater affinity, such as DNA. This change in coating lead to restoration of the photoluminescence, which can be potentially used to monitor the delivery of drug molecules.

When modified with liopofectamine (Chen et al., 2005) or other positively charged polymers, QDs may act as gene delivery vehicles. For instance, QDs have been encapsulated in poly(maleic-anhydride-alt-1-decene), and further surface-modified with dimethylamino propylamine to be positively charged (Qi and Gao, 2008). As a result, the QDs outperformed polyethylenimine (PEI) by depicting more efficient delivery and significantly reduced toxicity. Also with poly (maleic-anhydride-alt-1-decene) coated QDs, Yezhelyev et al. (2008), converted some carboxyls to tertiary amines with N,N-dimethylethylenediamine resulting in particles with two functional groups on the particle surface, gave both steric and electrostatic interactions which were highly responsive to acidic endosome/ lysosome organelles. On assessing as siRNA delivery vehicles, such QDs have shown a 10- to 20-fold increase in silencing effect and a five to sixfold decrease in toxicity when compared with other common delivery agents, such as liopofectamine, JetPEI, and TransIT. Apart from loading siRNA via electrostatic force, covalent coupling of siRNA to QDs has been explored additionally. Covalent conjugation of siRNA, along with a tumor-homing peptide F3 onto QDs, was explored by Derfus et al. (2007). The feasibility of siRNA release was studied and the coupling was achieved by means of disulfide cross-linker and by forming a nonreducible, thioether linkage. Using eGFP as the target, it was shown that the former formula had greater gene silencing efficiency, which the authors attributed to the susceptibility of disulfide linkage in endosome/lysosome. To modulate the activity of MMP-9, the main component of the BBB, in brain microvascular endothelial cells (BMVEC), recently, QDs modified with amine functionalized polymer polydiallyldimethylammonium chloride (PDDAC) and complexed with MMP-9-siRNA are utilized (Bonoiu et al., 2009). The treatment caused an increase in the expression level of tissue inhibitor of metalloproteinase-1 (TIMP-1), a natural inhibitor of MMP-9 that functions to maintain the basement membrane integrity. Increase in collagen I, IV, V expression and a decrease in endothelial permeability were also observed analogously.

QDs play important role in photodynamic therapy acting as either photosensitizer or as carriers (Samia et al., 2003; Willard and Van Orden, 2003; Bakalova et al., 2004). QDs cause cell damage due to their activation by light and transfer of the triplet state energy to nearby oxygen molecules (Samia et al., 2003). Tsay et al. (2007) studied the modification of QDs with streptavidin and conjugation with biotinylated pDNA. They observed that, when the QDs were photoactivated, they produce reactive oxygen intermediates (ROI) through nitroblue tetrazolium (NBT) assay. In a further study, it was explored that, ROI can cause damage to purine and pyrimidine bases, as established by assays with base excision repair enzymes like amidopyrimidine glycosylase (Fpg) and endonuclease III (Endo III). As compared to small-molecule-based photosensitizers, QD-based PDT offers certain benefits, for example better chemical stability, water solubility, and (for NIRF QDs) less optical interference with biological tissues (Bakalova et al., 2004). However, the disadvantage of QD-based PDT is that the quantum yield is typically less than 5% which is much lower than the yield (40–60%) of classic photosensitizers (Samia et al., 2003, 2006).

QDs have also been explored as carriers for PDT agents. Hsieh et al. (2006) conjugated QDs with Ir-complex. Shi et al. (2006) used meso-tetra (4-sulfonatophenyl) porphine dihydrochloride (TSPP) as a photosensitizer and coupled it onto the QD surface. It was observed that the QDs worked as drug carriers not involving directly in the photodynamic therapy and also acted as an energy hub transferring energy to phthalocyanine for activating its PDT function.

The biomedical applications of QDs include broad absorption spectra, size-dependent narrow and stable emissions spectra, photostability, and the ability to serve as a platform for targeting ligands and therapeutic drugs. Further, based on the multiplexing and FRET capabilities of QDs, there are several unexplored potentials to broaden the use of QD-based theranostic agents. Nevertheless, there is an imperative necessity to identify the metabolism of QDs in the body and to deal with the toxicity issues related to heavy metals.

18.5.3 Gold nanoparticle-based theranostic agents

Gold nanoparticles (Au NPs) gain several distinctive features and have been explored in a variety of imaging related arenas including computed tomography (CT), photoacoustics, and surface-enhanced Raman spectroscopy (SERS).

Au–thiol chemistry is selected to load functional entities onto Au NPs and much more therapeutics have been loaded similarly. For example, PTX was tailored at its C-7 position and covalently attached to 4-mercaptophenol altered Au nanoparticles (Gibson et al., 2007). The conjugates obtained established enhanced therapeutic effects than MTX alone both in vitro and in vivo, which was probably owing to the "concentrated effect" and better pharmacokinetics of the conjugates (Chen et al., 2007). Tumor necrosis factor (TNF), for example, was linked to PEGylated Au nanoparticles, and the resulting conjugates represented improved therapeutic effectiveness and lesser toxicity than native TNF (Paciotti et al., 2004; Goel et al., 2009; Powell et al., 2010).

To prepare Au nanoparticles, Bhumkar et al. (2007) employed chitosan as a reducing agent and coating material. The obtained chitosan–Au nanoparticles were highly positively charged and were found more proficient in loading insulin via electrostatic interaction (53%). For controlling postprandial hyperglycemia, these conjugates were studied in a diabetic model. A drop in the blood glucose level was noted as 30.41% and 20.27% for oral (50 IU/kg) and nasal (10 IU/kg) administration, respectively, after 2 h of administration of these insulin-loaded Au NPs to diabetic rats. It has been reported that a PDT agent and Pc4 can be directly adsorbed onto PEGylated Au nanoparticles with greater effectiveness (Cheng et al., 2008). The Au nanoparticles acted as a drug carrier, reducing the time for sufficient Pc4 delivery (less than 2 h) than that of free drug (2 days). Analogous research using Zn-Pc was stated by Hone et al. (2002). An amphiphilic-block-copolymer-coated Au NP formula for tumor targeting and drug delivery was reported by Prabaharan et al. (2009). The resulted nanostructure comprised of an Au NP core, a hydrophobic PASP inner shell, and a hydrophilic, folate-conjugated PEG outer shell (PEG-OH/ FA). Onto the hydrophobic inner shell, DOX was covalently conjugated with a loading level of 17 wt% using acid-cleavable hydrazone linkage. The obtained nanosystem has both a tumor targeting mechanism (folate on the outer layer) and an intracellular drug release mechanism (hydrazone linkage of DOX on the inner layer).

Au NPs have also been transformed to polyelectrolytes and studied as carriers for gene delivery. For loading plasmid DNA, earlier, Au NPs were functionalized with alkylated quaternary ammonium (Mclntosh et al., 2001; Han et al., 2005, 2006). This nanocarrier when studied in-vitro showed protection of DNA from enzymatic digestion (Han et al., 2006) and GSH triggered release (Han et al., 2005). Afterward, branched PEI was used by Thomas and Klibanov (2003) to give gene loading capacity to Au NPs. As compared to the parent polymer, the transfection strength of Au NPs may be improved by 12 times under optimized conditions. Apart from the electrostatic forces, therapeutic genes can also be loaded onto Au nanoparticles by means of covalent linking. In context to the above studies, thiolated antisense DNA oligos can be directly loaded onto Au nanoparticles with greater effectiveness. In a cellular study with eGFP-expressing C166 cells, Au NPs loaded with antisense DNA demonstrated a greater rate of translocation and a notable gene knockdown effectiveness.

In photothermal therapy, Au NPs are candidate materials owing to their distinctive surface plasmon resonance characteristic. Au NPs used as energy transducers, which upon laser irradiation when concentrated in tumor areas destroy neighboring cancerous cells by converting light into heat. In contrast with conventional drug delivery, such a treatment pattern is active only within the limited illumination area thereby minimizing the normal tissue damage. Taking into consideration above mentioned studies, spherical Au NPs, with a characteristic absorption at 500–600 nm, are not suitable materials for such an application. On the other hand, change in the configuration to a nanorod (Huang et al., 2007), nanocage (Skrabalak et al., 2008), or nanoshell (Ji et al., 2007) can shift the absorption to the NIR region, and reports on gold-nanostructure-based hyperthermia are mounting up. For instance, Chen et.al. validated recently that PEG coated Au nanocages can accumulate in a U87MG xenograft model, and when exposed to NIR light, can elevate the tumor surface temperature to 54°C within 2 min (Chen et al., 2010). Lu and group attached α-melanocyte stimulating hormone (MSH) analog, [Nle4, D-Phe7] α-MSH (NDPMSH), onto Au nanoshells and administered such conjugates to a B16/F10 melanoma model (Lu et al., 2009). Careful histological examination found a high buildup of particles in the tumor and established the homing intervened by NDP-MSH. In the contralateral tumor where no illumination was produced, efficient ablation of B16/F10 melanoma was not observed but it is observed in the tumor exposed to laser illumination. Successful photothermal therapy was validated histologically and, more interestingly, [18F] fluorodeoxyglucose (18F-FDG) PET found a remarkable decrease in tumor uptake, reflecting a metabolic activity drop upon photothermal therapy. Currently, the use of Au nanoshells as light controllable siRNA carriers was stated by the same group (Lu et al., 2010). The siRNA, by means of a sequence targeting NF-κB P65, was prethiolated at the 5′ end of the sense chain and was introduced to the particles surface through thiol–Au interaction. The study concludes that the stable thiol–Au association could carry the siRNA payload on the particle surface, even after cell uptake, but on exposure to NIR light irradiation it could be damaged, which destroy the endolysosomal membrane resulting release of siRNA into the cytoplasm. The observation of light inducible siRNA release and subsequent NF-κB P65 downregulation was established by both in vitro and in vivo studies. Further, studies showed that the downregulation of NF-κB P65 resulted in superior sensitivity to chemotherapy, as confirmed by an improved therapeutic index when such photothermal therapy was combined with irinotecan treatment.

AuNPs have long been employed as a candidate material for building up functional agents for both imaging and therapy applications owing to their distinctive features including strong surface plasmon absorption, stability, biosafety and ease of modification. The high cost of production is its disadvantage which may obstruct bright clinical perspectives of AuNPs. Noticeably though the thiol–Au chemistry is suitable and sufficient for producing stable AuNP conjugates, high concentration of glutathione (GSH) in a living subject, on introduction to a reducing environment may be an issue. Simultaneously, numerous studies are to be carried out at in vivo level, which may result in more stable conjugates that could be highly favorable.

18.5.4 Carbon nanotube-based theranostic agents

Several research groups (Welsher et al., 2008; Liu et al., 2008a; Kam et al., 2005a) studied carbon nanotubes (CNTs) as drug carriers and explored their potential use in Raman and photoacoustic imaging. Carbon nanotubes may be engulfed by the cells and this characteristic has encouraged the researchers to discover their potential role in drug delivery. The complete mechanisms underlying such efficient cell penetration is unknown. As both endocytosis (Kam et al., 2004, 2006; Jin et al., 2008) and passive diffusion (Bianco et al., 2005; Kostarelos et al., 2007) have been reported to be responsible for the ingestion of carbon nanotube, CNTs may be internalized by cells by means of different routes if surface coatings are different.

In 2005, Binaco et al. coupled MTX onto 1,3-dipolar cycloaddition functionalized CNTs. For loading and delivering DNA plasmid, identical, CNT formulas with multiple amine termini were used (Pantarotto et al., 2004; Singh et al., 2005). Nonetheless, it was found that the covalent amide bond was unfavorable to eliciting intracellular drug release in the case of MTX case (Pastorin et al., 2006). In both imaging and therapy, the Dai group used phospholipid- CNT conjugates. For example, they attached siRNA to CNTs by means of a disulfide bond, which was susceptible to enzymatic rupture in the endolysosome (Kam et al., 2005a). This CNT transporter demonstrated greater transfection efficiency, outperforming lipofectamine in generating RNAi. Afterward, by using the same nanostructure they successfully transported siRNA into human T cells and primary cells, which are believed difficult to transfect with conventional cationic liposome-based transfection agents (Liu et al., 2007). A similar group also investigated the coupling of either Pt (IV) prodrug (Dhar et al., 2008) or PTX (Liu et al., 2008b) onto PEGylated CNTs to enhance the pharmacokinetics and therapeutic effects. A branched PEG in the case of PTX was used for phospholipid PEGylation, which was observed to be beneficial against single-chain PEG in giving extra stability to CNTs (Liu et al., 2008b). PTX was linked through a cleavable ester bond to the surface of the nanotube and the construct was then tested in a murine 4T1 breast cancer model. The conjugates established a 10-fold enhancement in tumor homing than PTX alone, which was ascribed to the prolonged circulation half-life of the nanoformula. As a result, this formula demonstrated enhanced tumor suppression outcomes compared to clinically used Taxol.

Aromatic stacking may also serve as a drug loading mechanism. DOX, for instance, was loaded onto SWNT via this route at a remarkably high efficiency of 4 g DOX/g nanotube (Liu et al., 2009a). The interaction suggested a mode of unloading the payloads in acidic endolysosome and tumor micro-environments and is pH-dependent. Earlier, these DOX-loaded nanotubes were assessed in SCID mice bearing Raji lymphoma xenografts, which demonstrated increased therapeutic effectiveness and fewer side effects in contrast to an equimolar amount of free DOX (Liu et al., 2009b).

In photothermal therapy, CNTs in the NIR region are a promising tool due to their strong optical absorbance. Previous studies have shown that, when irradiated by NIR light, CNTs that were internalized in cells were capable of triggering endosomal rupture and cell death (Kam et al., 2005b). More recently, Moon et al. demonstrated in a human epidermoid mouth carcinoma model that the combined treatments of PEGylated SWNT and NIR irradiation led to the eradication of tumors with no observation of recurrence over 6 months (Moon et al., 2009). Also, Ghosh et al. used DNA to encapsulate CNT, which, according to the authors, can lead to improved heat emission efficacy. The conjugates were found to induce complete tumor eradication when injected in a PC3 xenograft model and irradiated (Ghosh et al., 2009). Nonetheless, it is valuable to notice that in both cases the nanotubes were injected intratumorally and not systemically.

By and large, CNTs play a major role in the field of nanomedicine and have distinctive physical and surface characters. Besides, loading drug molecules, to enrich the nanoplatform other nanoparticles such as IONPs or AuNPs may be bound to CNTs. Nonetheless, the nonbiodegradability of CNTs remains a big problem and also, standard protocol to prepare CNTs of high purity and at large scale is lacking.

18.5.5 Silica nanoparticle-based theranostic agents

In general, silica is considered as biosafe material, and earlier it was employed as a surgical implant. It is also well established that, in the synthesis of silica nanoparticles, accurate size and morphology control may be achieved (Jana et al., 2007).

During particle formation, drug molecules can be easily incorporated into silica nanoparticles. Roy et al. incorporated 2-devinyl-2-(1-hexyloxyethyl) pyropheophorbide (HPPH), a hydrophobic photosensitizing anticancer drug, into silica matrices (Roy et al., 2003). It was established that the fluorescent effect of HPPH in the silica matrices is greater as compared to the free form and cancer cells can be efficiently killed with laser irradiation. Recently, the same group co-encapsulated HPPH and a two-photon absorbing dye, 9, 10-bis [4′-(4″-aminostyryl) styryl] anthracene (BDSA), into silica nanoparticles (Kim et al., 2007). It was shown that BDSA can efficiently up-convert the NIR light and intra-partially transfer the energy to HPPH to activate the latter's PDT function.

A discovery in the preparation of silica nanoparticles are that they can be prepared in a mesoporous form and their pore size can be controlled accurately (Slowing et al., 2008). These mesoporous nanostructures embrace immense promise in drug delivery which consists of hundreds of empty channels and a large surface area (N900 m^2/g) acting as excellent reservoirs for small molecules. It is reported that such nanostructures can be achieved using various techniques chemically and physically. Typically in a chemical preparation, n alkyl trialkoxysilane or other surfactants are mixed with other precursors and are introduced into the matrices during the formation of a particle. These surfactants will be discarded later from the nanostructure via postsynthesis solvent extraction or calculations to produce the mesoporous structure (Slowing et al., 2008; Vallet-Regi et al., 2007; Manzano et al., 2009). These mesoporous silica nanoparticles can load many small molecule pharmaceuticals via simple physical interaction with these molecular-sieve structure characteristics (Vallet-Regi et al., 2007). Further, in order to inhibit premature drug release, technologies have been developed to cap the mesopores after drug loading. For example, mesoporous silica nanoparticles were loaded with PTX and the mesopores were then capped with Au NPs (Vivero-Escoto et al., 2009). Such Au NP capping was designed to be photolabile and can be uncapped to liberate guest molecules upon photoirradiation. Similar reports are available for activatable gatekeepers based on QDs (Lai et al., 2003), IONPs (Giri et al., 2005), coumarin (Mal et al., 2003), and diethylenetriamine (Casasus et al., 2004).

Currently, the preparation and application of luminescent porous silicon nanoparticles (LPSiNPs) are explored by Park et al. in 2009. The nanoparticles were prepared in physical ways. A preparation typically comprised of HF etching of single-crystal silicon wafers, lifting-off of the porous silicon film, ultrasonication, filtration of the formed particles through a 0.22-μm filtration membrane and lastly activation of luminescence in an aqueous solution. With the help of quantum confinement effects, the luminescence was produced and the defects were confined to the $Si-SiO_2$ interface. Drug release and cytotoxicity of the DOX-loaded luminescent porous nanoparticles were studied in vitro. A distinctive feature of such silica nanoparticles is that they may be self-destructive in vivo and they can be cleared by kidneys in a relatively short time period, thus minimizing the risk of their getting trapped in and causing harm to normal organs (Park et al., 2009).

18.6 Theranostic nanoagents: future of nanomedicine

In multiplexed molecular imaging several types of exogenous contrast agents can be used, such as conventional fluorescent dyes and quantum dots (QDs). The Conventional fluorescent dyes are well established and advertised; some of them have been approved for clinical use as well. The studies inferred that for in vitro tumor diagnostics the fluorescent dye molecules with emission wavelength ranging from visible to near-infrared could be used as a chromophore even though the most favorable and at present commercially available fluorescent dyes whose emission lies in the near-infrared region (NIR). Prominent background autofluorescence and severe tissue absorption and scattering greatly limit the application of dyes with emission in the visible region, whereas their interferences with long wavelength light are negligible (Xing et al., 2012a).

A combination of nanomaterials and theranostics can be said as nanotheranostics it includes both the drug nanocarriers and efficient imaging agents, which are interesting candidates for a diverse range of imaging-guided therapy applications. Likewise, polymeric nanoparticles responsive to external pH stimuli are also interesting and have taken so much attention for a wide range of applications, especially in biology, such as imaging-guided drug delivery, molecular sensing, tissue engineering, etc. (Xing et al., 2012b).

In context to biomedical applications, polypeptide materials are some sort of biodegradable material that can be potentially applied in biomedical areas. These can be easily acquired by the ring opening polymerization of amino acid N-carboxyanhydrides (NCA). Possibility for the preparation of polypeptide materials with accomplished structures and properties is offered by natural amino acids with various side groups (Fu et al., 2015).

Nowadays, photodynamic therapy (PDT) is of great interest and has been gaining much attention in cancer therapy when the light was turned on in the presence of a photosensitizer, singlet oxygen reactive oxygen species (ROS) can be produced in the presence of oxygen. For example, the preparation of pH-responsive multifunctional polypeptide nanoparticles represented efficiently ROS generation and large drug-carrying ability. It can be readily employed for imaging-guided photodynamic therapy (PDT). The cancer cells can be detected and treated synchronously by near-infrared fluorescence (NIRF) imaging-guided PDT by using the nanoparticles. Under the NIR light irradiation, the BODIPYBr 2 photosensitizer with efficient singlet oxygen yield has been utilized which possesses both fluorescence emission and the ability to produce ROS. It has been inferred by in vitro experiments on HepG2 cancer cells revealed that the efficient cell suppression rate can be increased by more than 40% in the presence of NIR light at an extremely low energy density (12 J/cm^2) and low concentration of photosensitizer (5.4 μM). Also, the uptaking of the nanoparticles by cancer cells can be detected by NIR imaging, indicating the nanoparticles are a potential imaging-guided PDT nanotheranostics (Liu et al., 2016b; Liu et al., 2016a).

Additionally, a pH-Sensitive doxorubicin-conjugated polypeptide has been synthesized which works as both imaging agent and photosensitized and it can be assembled into micelles with entrapping of near-infrared (NIR) photosensitizer BODIPY. The polymer was synthesized by a combination of ring opening polymerization of N-carboxy anhydride with $mPEG-NH_2$ and postchemical modification; for pH-responsive drug delivery, the chemical drug doxorubicin was conjugated to the polymer using hydrazone bond. For imaging-guided PDT, further a NIR dye BODIPY was encapsulated in the micelles. The nanoparticles represented low cytotoxicity and high ROS yield under the NIR light irradiation, which provides a combination of imaging-guided PDT (NIR) and chemotherapy in a nanoparticle system, revealing that the novel kind of polymeric nanoparticle is a potential nanotheranostic for cancers. Also, the energy density needed of the laser for the PDT is also extremely low (Ruan et al., 2017).

18.7 Conclusion

Despite continuous advancement in the scientific field, brain-related maladies remain a serious problem having a high risk of mortality. Nanotheranostics is a promising area that has integrated real-time diagnostic, imaging, targeting and drug delivery features on a single nanoplatform. Copious therapeutic drugs and contrast agents are actually in existence for achieving therapy and diagnosis of neurological disorders. Nonetheless, many elements restrict their applications. Significantly, a major challenge in the pathway of progression of an efficacious and safe theranostic system is BBB which restricts the entry of contrast agents and drugs to the brain. NPs which show low toxicity profile embrace notable opportunities to be developed as nanotheranostic systems. These NPs include lipid NPs, PNPs, inorganic NPs and some others like dendrimers, UCNPs, and CNTs. The authors have also analyzed various diagnosis techniques for neurological disorders that include optical imaging, MRI and SPECT and simultaneously different therapeutic modalities for these ailments are being presented. But, regardless of these promising results, the field of nanothernostics is still in its genesis and so far there are no nanotheranostic systems that can meet clinical standards for successful clinical translation. Therefore, certainly, there is a need for more efforts to surmount multiple difficulties and then only the technology can be transferred from laboratory to patient's bedside for effective and personalized therapy.

In recent trends, the available treatments of RA, by controlling various symptoms, only delay disease progression and have their own side effects. Hence, for presenting a better sustainable and biocompatible solution there is the necessity of a novel therapeutic strategy. Nanomedicine is a contemporary branch of nanobiotechnology offering targeted therapy to inflamed rheumatic joints and thus avoiding untoward off-target side effects.

The benefit of building such function-integrated agents is being imaging agents of many nanoplatforms. Nanoplatforms can be easily loaded with pharmaceuticals owing to their well-developed surface chemistry and helping them to be theranostic nanosystems. In the past, various nanoplatforms such as iron oxide nanoparticles, quantum dots, carbon nanotubes, gold nanoparticles and silica nanoparticles, were explored in the imaging setting and are candidates for establishing nanoparticle-based theranostics.

The dawn of a new era of personalized medicine is emerging and introducing more understanding to the scientific community dealing with the nanosystems as they are capable of diagnosis, drug delivery and monitoring of therapeutic response. It is anticipated with the progression of technology and development of theranostics it will take no time to achieve the goal and offering early detection of various kinds of cancers and their treatments.

References

Agulla, J., Brea, D., Campos, F., Sobrino, T., Argibay, B., Al-Soufi, W., et al., 2013. In vivo theranostics at the peri-infarct region in cerebral ischemia. Theranostics. 4, 90–105.

Agyare, E.K., Jaruszewski, K.M., Curran, G.L., Rosenberg, J.T., Grant, S.C., Lowe, V.J., et al., 2014. Engineering theranostic nanovehicles capable of targeting cerebrovascular amyloid deposits. J. Control. Release. 185, 121–129.

Alam, S.M., Kidwai, A.A., Jafri, S.R., Qureshi, B.M., Sami, A., Qureshi, H.H., et al., 2011. Epidemiology of rheumatoid arthritis in a tertiary care unit, Karachi, Pakistan. J. Pak. Med. Assoc. 61, 123–126.

Auzel, F., 2004. Upconversion and anti-stokes processes with f and d ions in solids. Chem. Rev. 104, 139–174.

Bagalkot, V., Zhang, L., Levy-Nissenbaum, E., Jon, S., Kantoff, P.W., Langer, R., et al., 2007. Quantum dot-aptamer conjugates for synchronous cancer imaging, therapy, and sensing of drug delivery based on bi-fluorescence resonance energy transfer. Nano. Lett. 7, 3065–3070.

Bakalova, R., Ohba, H., Zhelev, Z., Ishikawa, M., Baba, Y., 2004. Quantum dots as photosensitizers? Nat. Biotechnol. 22, 1360–1361.

Bardhan, R., Lal, S., Joshi, A., Halas, N.J., 2011. Theranostic nanoshells: from probe design to imaging and treatment of cancer. Acc. Res. 44, 936–946.

Believing in Seeing, 2014. Nat. Mater. 13, 99. Available at: https://doi.org/10.1038/nmat3879.

Bernal, G.M., LaRiviere, M.J., Mansour, N., Pytel, P., Cahill, K.E., Voce, D.J., et al., 2014. Convection-enhanced delivery and in vivo imaging of polymeric nanoparticles for the treatment of malignant glioma. Nanomedicine 10, 149–157.

Bhojani, M.S., Dort, M.V., Rehemtulla, A., Ross, B.D., 2010. Targeted imaging and therapy of brain cancer using theranostic nanoparticles. Mol. Pharm. 7 (6), 1921–1929.

Bhowmik, A., Khan, R., Ghosh, M.K., 2015. Blood brain barrier: a challenge for effectual therapy of brain tumors. Biomed. Res. Int. 2015, 1–20.

Bhumkar, D.R., Joshi, H.M., Sastry, M., Pokharkar, V.B., 2007. Chitosan reduced gold nanoparticles as novel carriers for transmucosal delivery of insulin. Pharm. Res. 24, 1415–1426.

Bianco, A., Kostarelos, K., Partidos, C.D., Prato, M., 2005. Biomedical applications of functionalised carbon nanotubes. Chem. Commun. 5, 571–577.

Blau, R., Epshtein, Y., Pisarevsky, E., Tiram, G., Dangoor, S.I., Yeini, E., et al., 2018. Image-guided surgery using nearinfrared Turn-ON fluorescent nanoprobes for precise detection of tumor margins. Theranostics 8, 3437–3460.

Bolhassani, A., Javanzad, S., Saleh, T., Hashemi, M., Aghasadeghi, M.R., Sadat, S.M., 2014. Polymeric nanoparticles: potent vectors for vaccine delivery targeting cancer and infectious diseases. Hum. Vaccines Immunother. 10, 321–332.

Bonoiu, A., Mahajan, S.D., Ye, L., Kumar, R., Ding, H., Yong, K.T., et al., 2009. MMP-9 gene silencing by a quantum dot-siRNA nanoplex delivery tomaintain the integrity of the blood brain barrier. Brain Res. 1282, 142–155.

Camargo, P.H.C., Satyanarayana, K.G., Wypych, F., 2009. Nanocomposites: synthesis, structure, properties and new application opportunities. Mater. Res. 12, 1–39.

Casasus, R., Marcos, M.D., Martinez-Manez, R., Ros-Lis, J.V., Soto, J., Villaescusa, L.A., et al., 2004. Toward the development of ionically controlled nanoscopic molecular gates. J. Am. Chem. Soc. 126, 8612–8613.

Chandrasekar, R., Chandrasekar, S., 2017. Natural herbal treatment for rheumatoid arthritis—a review. Int. J. Pharm. Sci. Res. 8, 368–384.

Chen, A.A., Derfus, A.M., Khetani, S.R., Bhatia, S.N., 2005. Quantum dots to monitor RNAi delivery and improve gene silencing. Nucleic Acids. Res. 33, e190.

Chen, J., Glaus, C., Laforest, R., Zhang, Q., Yang, M., Gidding, M., et al., 2010. Gold nanocages as photothermal transducers for cancer treatment. Small. 6, 811–817.

Chen, M., Daddy, J.C.K.A., Xiao, Y., Ping, Q., Zong, L., 2017. Advanced nanomedicine for rheumatoid arthritis treatment: focus on active targeting. Expert Opin. Drug Deliv. 14, 1141–1144.

Chen, Q., Du, Y., Zhang, K., Lian, G.Z., Li, J., Yu, H., et al., 2018. Tau-targeted multifunctional nanocomposite for combinational therapy of Alzheimer's disease. ACS Nano 12, 1321–1338.

Chen, Y.H., Tsai, C.Y., Huang, P.Y., Chang, M.Y., Cheng, P.C., Chou, C.H., et al., 2007. Methotrexate conjugated to gold nanoparticlesinhibits tumor growth in a syngeneic lung tumor model. Mol. Pharm. 4, 713–722.

Cheng, K., Peng, S., Xu, C., Sun, S., 2009. Porous hollow Fe(3)O(4) nanoparticles for targeted delivery and controlled release of cisplatin. J. Am. Chem. Soc. 131, 10637–10644.

Cheng, Y., Samia, A.C., Meyers, J.D., Panagopoulos, I., Fei, B., Burda, C., 2008. Highly efficient drug delivery with gold nanoparticle vectors for in vivo photodynamic therapy of cancer. J. Am. Chem. Soc. 130, 10643–10647.

Chowdhury, S.R., Mondal, S., Muthuraj, B., Balaji, S.N., Trivedi, V., Iyer, P.K., 2018. Remarkably efficient blood–brain barrier crossing polyfluorene–chitosan nanoparticle selectively tweaks amyloid oligomer in cerebrospinal fluid and Aβ1–40. ACS Omega 3, 8059–8066.

Connolly, B.S., Lang, A.E., 2014. Pharmacological treatment of Parkinson disease: a review. JAMA 311, 1670–1683.

Costa, P.M., Bourgognon, M., Wang, J.T., Al-Jamal, K.T., 2016. Functionalised carbon nanotubes: from intracellular uptake and cell-related toxicity to systemic brain delivery. J. Control. Release 241, 200–219.

Costa, P.M., Wang, J.T.-W., Morfin, J-F., Khanum, T., To, W., Sosabowski, J., et al., 2018. Functionalised carbon nanotubes enhance brain delivery of amyloid-targeting Pittsburgh compound B (PiB)-derived ligands. Nanotheranostics 2, 168–183.

Cui, Z., Bu, W., Fan, W., Zhang, J., Ni, D., Liu, Y., et al., 2016. Sensitive imaging and effective capture of Cu^{2+}: towards highly efficient theranostics of Alzheimer's disease. Biomaterials 104, 158–167.

Danhier, F., Ansorena, E., Silva, J.M., Coco, R., Breton, Al., Preat, V., 2012. PLGA-based nanoparticles: an overview of biomedical applications. J. Control. Release 161, 505–522.

Dehvari, K., Lin, K.S., 2012. Synthesis, characterization and potential applications of multifunctional PEO-PPOPEO-magnetic drug delivery system. Curr. Med. Chem. 19, 5199–5204.

Delhaye, C., Mahmoudi, M., Waksman, R., 2012. Hypothermia therapy: neurological benefits. J. Am. Coll. Cardiol. 59, 197–210.

Derfus, A.M., Chen, A.A., Min, D.H., Ruoslahti, E., Bhatia, S.N., 2007. Targeted quantum dotconjugates for siRNA delivery. Bioconjug. Chem. 18, 1391–1396.

Dhar, S., Liu, Z., Thomale, J., Dai, H., Lippard, S.J., 2008. Targeted single-wall carbon nanotube mediated Pt (IV) prodrug delivery using folate as a homing device. J. Am. Chem. Soc. 130, 11467–11476.

Doan, T., Massarotti, E., 2005. Rheumatoid arthritis: an overview of new and emerging therapies. J. Clin. Pharmacol. 45, 751–762.

Dolati, S., Sadreddini, S., Rostamzadeh, D., Ahmadi, M., Jadidi-Niaragh, F., Yousefi, M., 2016. Utilization of nanoparticle technology in rheumatoid arthritis treatment. Biomed. Pharmacother. 80, 30–41.

Dubois, B., Feldman, H.H., Jacova, C., Hampel, H., Molinuevo, J.L., Blennow, K., et al., 2014. Advancing research diagnostic criteria for Alzheimer's disease: the IWG-2 criteria. Lancet Neurol. 13, 614–629.

Duceppe, N., Tabrizian, M., 2010. Advances in using chitosan-based nanoparticles for in vitro and in vivo drug and gene delivery. Expert Opin. Drug Deliv. 7, 1191–1207.

Eichhorn, S.J., 2011. Cellulose nanowhiskers: promising materials for advanced applications. Soft. Matter. 7, 303–315.

Fan, C.H., Ting, C.Y., Lin, H.J., Wang, C.H., Liu, H.L., Yen, T.C., et al., 2013. SPIO-conjugated, doxorubicin-loadedmicrobubbles for concurrent MRI and focused-ultrasound enhanced brain-tumor drug delivery. Biomaterials 34, 3706–3715.

Fang, W., Wei, Y., 2016. Upconversion nanoparticle as a theranostic agent for tumor imaging and therapy. J. Innov. Opt. Health Sci. 9, 1630006. Available at: doi:10.1142/S1793545816300068.

Firestein, G.S.J.N., 2003. Evolving concepts of rheumatoid arthritis. Nature 423, 356–361.

Fu, L., Sun, C., Yan, L., 2015. Galactose targeted pH-responsive copolymer conjugated with near infrared fluorescence probe for imaging of intelligent drug delivery. ACS. Appl. Mater. Interfaces 7, 2104–2115.

Fu, L., Yuan, P., Ruan, Z., Liu, L., Li, T., Yan, L., 2017. Ultra-pH-sensitive polypeptide micelles with large fluorescence off/on ratio in near infrared range. Polym. Chem. 8, 1028–1038.

Gamage, N., Jing, L., Worsham, M.J., Ali, M.M., 2016. Targeted theranostic approach for glioma using dendrimer-based curcumin nanoparticle. J. Nanomed. Nanotechnol. 7, 393.

Gao, X., Yue, Q., Liu, Y., Fan, D., Fan, K., Li, S., et al., 2018a. Image-guided chemotherapy with specifically tuned blood brain barrier permeability in glioma margins. Theranostics. 8, 3126–3137.

Gao, Z.D.N., Guan, Y., Ding, C., Sun, Y., R., J., Qu, X., 2018b. Rational design of a "sense and treat" system to target amyloid aggregates related to Alzheimer's disease. Nano Res. 11, 1987–1997.

Ghosh, S., Dutta, S., Gomes, E., Carroll, D., D'Agostino, R., Olson, J., et al., 2009. Increased heating efficiency and selective thermal ablation of malignant tissue with DNA-encased multiwalled carbon nanotubes. ACS Nano. 3, 2667–2673.

Gibson, J.D., Khanal, B.P., Zubarev, E.R., 2007. Paclitaxel-functionalized gold nanoparticles. J. Am. Chem. Soc. 129, 11653–11661.

Giri, S., Trewyn, B.G., Stellmaker, M.P., Lin, V.S., 2005. Stimuli-responsive controlled release delivery system based on mesoporous silica nanorods capped with magnetic nanoparticles. Angew. Chem. Int. Ed. Engl. 44, 5038–5044.

Goel, R., Shah, N., Visaria, R., Paciotti, G.F., Bischof, J.C., 2009. Biodistribution of TNF-alpha coated gold nanoparticles in an in vivo model system. Nanomedicine. 4, 401–410.

Grover, V.P., Tognarelli, J.M., Crossey, M.M., Cox, I.J., 2015. Magnetic resonance imaging: principles and techniques: lessons for clinicians. J. Clin. Exp. Hepatol. 5, 246–255.

Guo, H., Chen, W., Sun, X., Liu, Y.N., Li, J., Wang, J., 2015. Theranostic magnetoliposomes coated by carboxymethyl dextran with controlled release by low-frequency alternating magnetic field. Carbohydr. Polym. 118, 209–217.

Guo, Q., Wang, Y., Xu, D., Nossent, J., Pavlos, N.J., Xu, J., 2018. Rheumatoid arthritis: pathological mechanisms and modern pharmacologic therapies. Bone Res. 6, 15.

Gupta, A.S., 2011. Nanomedicine approaches in vascular disease: a review. Nanomedicine. 7, 763–779.

Hajal, C., Campisi, M., Mattu, C., Chiono, V., Kamm, R.D., 2018. In vitro models of molecular and nano-particle transport across the blood-brain barrier. Biomicrofluidics 12, 042213. Available at: doi:10.1063/1.5027118.eCollection 2018Jul.

Han, G., Chari, N.S., Verma, A., Hong, R., Martin, C.T., Rotello, V.M., 2005. Controlled recovery of the transcription of nanoparticle-bound DNA by intracellular concentrations of glutathione. Bioconjug. Chem. 16, 1356–1359.

Han, G., Martin, C.T., Rotello, V.M., 2006. Stability of gold nanoparticle-bound DNAtoward biological, physical, and chemical agents. Chem. Biol. Drug Des. 67, 78–82.

Heo, R., You, D.G., Um, W., Choi, K.Y., Jeon, S., Park, J.S., et al., 2017. Dextran sulfate nanoparticles as a theranostic nanomedicine for rheumatoid arthritis. Biomaterials. 131, 15–26.

Hilderbrand, S.A., Weissleder, R., 2010. Near-infrared fluorescence: application to in vivo molecular imaging. Curr. Opin. Chem. Biol. 14, 71–79.

Hone, D.C., Walker, P.I., Evans-Gowing, R., FitzGerald, S., Beeby, A., Chambrier, I., et al., 2002. Generation of cytotoxic singlet oxygen via phthalocyaninestabilizedgold nanoparticles: a potential delivery vehicle for photodynamic therapy. Langmuir. 18, 2985–2987.

Hsieh, J.M., Ho, M.L., Wu, P.W., Chou, P.T., Tsai, T.T., Chi, Y., 2006. Iridium-complex modified CdSe/ZnS quantum dots; a conceptual design for bi-functionality toward imaging and photosensitization. Chem. Commun. 6, 615–617.

Hu, B., Dai, F., Fan, Z., Ma, G., Tang, Q., Zhang, X., 2015. Nanotheranostics: Congo red/Rutin-MNPs with enhanced magnetic resonance imaging and H_2O_2-responsive therapy of Alzheimer's disease in APPswe/PS1dE9 transgenic mice. Adv. Mater. 27, 5499–5505.

Hu, K., Shi, Y., Jiang, W., Han, J., Huang, S., Jiang, X., 2011. Lactoferrin conjugated PEG-PLGA nanoparticles for brain delivery: preparation, characterization and efficacy in Parkinson's disease. Int. J. Pharm. 415, 273–283.

Huang, X., El-Sayed, I.H., Qian, W., El-Sayed, M.A., 2007. Cancer cells assemble and align gold nanorods conjugated to antibodies to produce highly enhanced, sharp, and polarized surface Raman spectra: a potential cancer diagnostic marker. Nano. Lett. 7, 1591–1597.

Huber, J.D., Egleton, R.D., Davis, T.P., 2001. Molecular physiology and pathophysiology of tight junctions in the blood–brain barrier. Trends Neurosci. 24, 719–725.

Huh, Y.M., Jun, Y.W., Song, H.T., Kim, S., Choi, J.S., Lee, J.H., et al., 2005. In vivo magnetic resonance detection of cancer by using multifunctional magnetic nanocrystals. J. Am. Chem. Soc. 127, 12387–12391.

Hwu, J.R., Lin, Y.S., Josephrajan, T., Hsu, M.H., Cheng, F.Y., Yeh, C.S., et al., 2009. Targeted paclitaxel by conjugation to iron oxide and gold nanoparticles. J. Am. Chem. Soc. 131, 66–68.

Ito, A., Shinkai, M., Honda, H., Kobayashi, T., 2005. Medical application of functionalized magnetic nanoparticles. J. Biosci. Bioeng. 100, 1–11.

Jain, T.K., Richey, J., Strand, M., Leslie-Pelecky, D.L., Flask, C., Labhasetwar, V., 2008. Magnetic nanoparticles with dual functional properties: drug delivery and magnetic resonance imaging. Biomaterials. 29, 4012–4021.

Jana, N.R., Earhart, C., Ying, J.Y., 2007. Synthesis of water-soluble and functionalized nanoparticles by silica coating. Chem. Mater. 19, 5074–5082.

Ji, X., Shao, R., Elliott, A.M., Stafford, R.J., Esparza-Coss, E., Liang, G., et al., 2007. Bifunctional gold nanoshells with a superparamagnetic ironoxide-silica core suitable for both MR imaging and photothermal therapy. J. Phys. Chem. C Nanomater. Interfaces. 111, 6245.

Jiang, Y., Fang, R.H., Zhang, L., 2018. Biomimetic nanosponges for treating antibody-mediated autoimmune diseases. Bioconjug. Chem. 29, 870–877.

Jin, H., Heller, D.A., Strano, M.S., 2008. Single-particle tracking of endocytosis andexocytosis of single-walled carbon nanotubes in NIH-3T3 cells. Nano Lett. 8, 1577–1585.

Jing, H., Weidensteiner, C., Reichardt, W., Gaedicke, S., Zhu, X., Grosu, A.I., et al., 2016. Imaging and selective elimination of glioblastoma stem cells with theranostic near-infrared labeled CD133-specific antibodies. Theranostics. 6, 862–874.

Kafa, H., Wang, JT-W., Rubio, N., Klippstein, R., Costa, P.M., Hassan, H.A., et al., 2016. Translocation of LRP1 targeted carbon nanotubes of different diameters across the blood–brain barrier in vitro and in vivo. J. Control. Release. 225, 217–229.

Kafa, H., Wang, JT-W., Rubio, N., Venner, K., Anderson, G., Pach, E., et al., 2015. The interaction of carbon nanotubes with an in vitro blood-brain barrier model and mouse brain in vivo. Biomaterials. 53, 437–452.

Kam, N.W., Jessop, T.C., Wender, P.A., Dai, H., 2004. Nanotube molecular transporters: internalization of carbon nanotube-protein conjugates into Mammalian cells. J. Am. Chem. Soc. 126, 6850–6851.

Kam, N.W., Liu, Z., Dai, H., 2005a. Functionalization of carbon nanotubes via cleavabledisulfide bonds for efficient intracellular delivery of siRNA and potent genesilencing. J. Am. Chem. Soc. 127, 12492–12493.

Kam, N.W., Liu, Z., Dai, H., 2006. Carbon nanotubes as intracellular transporters forproteins and DNA: an investigation of the uptake mechanism and pathway. Angew. Chem. Int. Ed. Engl. 45, 577–581.

Kam, N.W., O'Connell, M., Wisdom, J.A., Dai, H., 2005b. Carbon nanotubes as multifunctional biological transporters and near-infrared agents for selective cancer cell destruction. Proc. Natl. Acad. Sci. USA 102, 11600–11605.

Kapoor, B., Singh, S.K., Gulati, M., Gupta, R., Vaidya, Y., 2014. Application of liposomes in treatment of rheumatoid arthritis: quo vadis. Sci. World J. 2014, 978351.

Kasban, H., El-Bendary, M., Salama, D., 2015. A comparative study of medical imaging techniques. Int. J. Information Sci. Intell. Syst. 4, 37–58.

Katsuki, S., Matoba, T., Koga, J-I., Nakano, K., Egashira, K., 2017. Anti-inflammatory nanomedicine for cardiovascular disease. Front. Cardiovasc. Med. 4, 87.

Kievit, F.M., Zhang, M., 2011. Cancer nanotheranostics: improving imaging and therapy by targeted delivery across biological barriers. Adv. Mater. 23, H217–H247.

Kim, H.J., Lee, S.M., Park, K.H., Mun, C.H., Park, Y.B., Yoo, K.H., 2015. Drug-loaded gold/iron/gold plasmonic nanoparticles for magnetic targeted chemo-photothermal treatment of rheumatoid arthritis. Biomaterials. 61, 95–102.

Kim, S., Ohulchanskyy, T.Y., Pudavar, H.E., Pandey, R.K., Prasad, P.N., 2007. Organically modified silica nanoparticles co-encapsulating photosensitizing drug and aggregation-enhanced two-photon absorbing fluorescent dye aggregates for two-photon photodynamic therapy. J. Am. Chem. Soc. 129, 2669–2675.

Kim, S.W., Zimmer, J.P., Ohnishi, S., Tracy, J.B., Frangioni, J.V., Bawendi, M.G., 2005. Engineering InAs(x)P(1-x)/InP/ZnSe III-V alloyed core/shell quantum dots for the near-infrared. J. Am. Chem. Soc. 127, 10526–10532.

Kohler, N., Fryxell, G.E., Zhang, M.Q., 2004. A bifunctional poly(ethylene glycol) silane immobilized on metallic oxide-based nanoparticles for conjugation with cell targeting agents. J. Am. Chem. Soc. 126, 7206–7211.

Kohler, N., Sun, C., Fichtenholtz, A., Gunn, J., Fang, C., Zhang, M.Q., 2006. Methotrexate immobilized poly (ethylene glycol) magnetic nanoparticles for MR imaging and drug delivery. Small. 2, 785–792.

Kohler, N., Sun, C., Wang, J., Zhang, M.Q., 2005. Methotrexate-modified superparamagnetic nanoparticles and their intracellular uptake into human cancer cells. Langmuir. 21, 8858–8864.

Kostarelos, K., Lacerda, L., Pastorin, G., Wu, W., Wieckowski, S., Luangsivilay, J., et al., 2007. Cellular uptake of functionalized carbon nanotubes is independent of functional group and cell type. Nat. Nanotechnol. 2, 108–113.

Kubrusly, R.C., Guimaraes, M.Z.P., Vieira, A.P.B., Hokoc, J.N., Casarini, D.L., de Mello, M.C.F., et al., 2003. L-DOPA supply to the neuro retina activates dopaminergic communication at the early stages of embryonic development. J. Neurochem. 86, 45–54.

Kunjachan, S., Ehling, J., Storm, G., Kiessling, F., Lammers, T., 2015. Noninvasive imaging of nanomedicines and nanotheranostics: principles, progress, and prospects. Chem. Rev. 115, 10907–10937.

Ladd, M.E., Bachert, P., Meyerspeer, M., Moser, E., Nagel, A.M., Norris, D.G., et al., 2018. Pros and cons of ultra-high-field MRI/MRS for human application. Prog. Nucl. Magn. Reson. Spectrosc. 109, 1–50.

Lai, C.Y., Trewyn, B.G., Jeftinija, D.M., Jeftinija, K., Xu, S., Jeftinija, S., et al., 2003. A mesoporous silica nanosphere-based carrier system with chemically removable CdS nanoparticle caps for stimuli-responsive controlled release of neurotransmitters and drug molecules. J. Am. Chem. Soc. 125, 4451–4459.

Lee, J.H., Huh, Y.M., Jun, Y.W., Seo, J.W., Jang, J.T., Song, H.T., et al., 2007. Artificially engineered magnetic nanoparticles for ultrasensitive molecular imaging. Nat. Med. 13, 95–99.

Lee, J.H., Lee, K., Moon, S.H., Lee, Y., Park, T.G., Cheon, J., 2009. All-in-one target-cell specific magnetic nanoparticles for simultaneous molecular imaging and siRNA delivery. Angew. Chem. Int. Ed. Engl. 48, 4174–4179.

Li, M., Guan, Y., Zhao, A., Ren, J., Qu, X., 2017. Using multifunctional peptide conjugated Au nanorods for monitoring β-amyloid aggregation and chemo-photothermal treatment of Alzheimer's disease. Theranostics. 7, 2996–3006.

Li, Y., Li, Y., Ji, W., Lu, Z., Liu, L., Shi, Y., et al., 2018a. Positively charged polyprodrug amphiphiles with enhanced drug loading and reactive oxygen species-responsive release ability for traceable synergistic therapy. J. Am. Chem. Soc. 140, 4164–4171.

Li, Y., Liu, R., Ji, W., Li, Y., Liu, L., Zhang, X., 2018b. Delivery systems for theranostics in neurodegenerative diseases. Nano Res. 11, 5535–5555.

Li, Y., Xu, D., Ho, S-L., Li, H.W., Yang, R., Wong, M.S., 2016. A theranostic agent for in vivo near-infrared imaging of β-amyloid species and inhibition of β-amyloid aggregation. Biomaterials. 94, 84–92.

Lim, E.K., Kim, T., Paik, S., Haam, S., Huh, Y.M., Lee, K., 2015. Nanomaterials for theranostics: recent advances and future challenges. Chem. Rev. 115, 327–394.

Liu, H., Jablonska, A., Li, Y., Cao, S., Liu, D., Chen, H., et al., 2016a. Label-free CEST MRI detection of citicoline liposome drug delivery in ischemic stroke. Theranostics. 6, 1588–1600.

Liu, J., Huang, Y., Kumar, A., Tan, A., Jin, S., Mozhi, A., et al., 2014. pH-sensitive nano-systems for drug delivery in cancer therapy. Biotechnol. Adv. 32, 693–710.

Liu, L., Fu, L., Jing, T., Ruan, Z., Yan, L., 2016b. pH-triggered polypeptides nanoparticles for efficient BODIPY imaging-guided near infrared photodynamic therapy. ACS Appl. Mater. Interfaces. 8, 8980–8990.

Liu, Z., Chen, K., Davis, C., Sherlock, S., Cao, Q., Chen, X., et al., 2008a. Drug delivery with carbon nanotubes for in vivo cancer treatment. Cancer Res. 68, 6652–6660.

Liu, Z., Fan, A.C., Rakhra, K., Sherlock, S., Goodwin, A., Chen, X., et al., 2009a. Supramolecular stacking of doxorubicin on carbon nanotubes for in vivo cancer therapy. Angew. Chem. Int. Ed. Engl. 48, 7668–7672.

Liu, Z., Li, X., Tabakman, S.M., Jiang, K., Fan, S., Dai, H., 2008b. Multiplexed multi-color Raman imaging of live cells with isotopically modified single walled carbon nanotubes. J. Am. Chem. Soc. 130, 13540–13541.

Liu, Z., Tabakman, S., Welsher, K., Dai, H., 2009b. Carbon nanotubes in biology and medicine: in vitro and in vivo detection, imaging and drug delivery. Nano Res. 2, 85–120.

Liu, Z., Winters, M., Holodniy, M., Dai, H., 2007. siRNA delivery into human T cells and primary cells with carbon-nanotube transporters. Angew. Chem. Int. Ed. Engl. 46, 2023–2027.

Louie, A., 2010. Multimodality imaging probes: design and challenges. Chem. Rev. 110, 3146–3195.

Lu, F.M., Yuan, Z., 2015. PET/SPECT molecular imaging in clinical neuroscience: recent advances in the investigation of CNS diseases. Quant. Imaging Med. Surg. 5, 433–447.

Lu, W., Xiong, C., Zhang, G., Huang, Q., Zhang, R., Zhang, J.Z., et al., 2009. Targeted photothermal ablation of murine melanomas with melanocyte-stimulatinghormone analog-conjugated hollow gold nanospheres. Clin. Cancer Res. 15, 876–886.

Lu, W., Zhang, G., Zhang, R., Flores, L.G., Huang, Q., Gelovani, J.G., et al., 2010. Tumorsite- specific silencing of NF-kappaB p65 by targeted hollow gold nanosphere mediated photothermal transfection. Cancer Res. 70, 3177–3188.

Lu, Y., Li, L., Lin, Z., Wang, L., Lin, L., Li, M., et al., 2018. A new treatment modality for rheumatoid arthritis: combined photothermal and photodynamic therapy using Cu7.2S4 nanoparticles. Adv. Healthcare Mater. 14, 1800013.

Lubbe, A.S., 1997. Preclinical experiences with magnetic drug targeting: tolerance andefficacy and clinical experiences with magnetic drug targeting: a phase I study with 4′ epidoxorubicin in 14 patients with advanced solid tumors-reply. Cancer Res. 57, 3064–3065.

Mal, N.K., Fujiwara, M., Tanaka. Y., 2003. Photocontrolled reversible release of guest molecules from coumarin-modified mesoporous silica. Nature. 421, 350–353.

Manzano, M., Colilla, M., Vallet-Regi, M., 2009. Drug delivery from ordered mesoporous matrices. Expert. Opin. Drug. Deliv. 6, 1383–1400.

McCarthy, J.R., Korngold, E., Weissleder, R., Jaffer, F.A., 2010. A light-activated theranostic nanoagent for targeted macrophage ablation in inflammatory atherosclerosis. Small. 6, 2041–2049.

McDonagh, B.H., Singh, G., Hak, S., Bandyopadhyay, S., Augestad, I.L., Peddis, D., et al., 2016. L-DOPA-coated manganese oxide nanoparticles as dual MRI contrast agents and drug-delivery vehicles. Small. 12, 301–306.

McIntosh, C.M., Esposito, E.A., Boal, A.K., Simard, J.M., Martin, C.T., Rotello, V.M., 2001. Inhibition of DNA transcription using cationic mixed monolayer protected goldclusters. J. Am. Chem. Soc. 123, 7626–7629.

Medarova, Z., Pham, W., Farrar, C., Petkova, V., Moore, A., 2007. In vivo imaging of siRNA delivery and silencing in tumors. Nat. Med. 13, 372–377.

Miao, Q., Xie, C., Zhen, X., Lyu, Y., Duan, H., Liu, X., et al., 2017. Molecular afterglow imaging with bright, biodegradable polymer nanoparticles. Nat. Biotechnol. 35, 1102–1110.

Moon, H.K., Lee, S.H., Choi, H.C., 2009. In vivo near-infrared mediated tumor destructionby photothermal effect of carbon nanotubes. ACS Nano. 3, 3707–3713.

Morales, M.P., Veintemillas-Verdaguer, S., Montero, M.I., Serna, C.J., Roig, A., Casas, L., et al., 1999. Surface and internal spin canting in gamma-Fe_2O_3 nanoparticles. Chem. Mater. 11, 3058–3064.

Mornet, S., Vasseur, S., Grasset, F., Duguet, E., 2004. Magnetic nanoparticle design for medical diagnosis and therapy. J. Mater. Chem. 14, 2161–2175.

Mozaffarian, D., Benjamin, E.J., Go, A.S., Arnett, D.K., Blaha, M.J., Cushman, M., et al., 2016. Heart disease and stroke statistics-2016 Update; A report from the american heart association. Circulation. 133(4), e38–e360.

Mura, S., Couvreur, P., 2012. Nanotheranostics for personalized medicine. Adv. Drug Deliv. Rev. 64, 1394–1416.

Myerson, J.W., He, L., Allen, J.S., Williams, T., Lanza, G., Tollefsen, D., et al., 2014. Thrombin-inhibiting nanoparticles rapidly constitute versatile and detectable anticlotting surfaces. Nanotechnology. 25, 395101.

Namiki, Y., Namiki, T., Yoshida, H., Ishii, Y., Tsubota, A., Koido, S., et al., 2009. A novel magnetic crystal-lipid nanostructure for magnetically guided in vivo gene delivery. Nat. Nanotechnol. 4, 598–606.

Nasrallah, I.M., Wolk, D.A., 2014. Multimodality imaging of Alzheimer disease and other neurodegenerative dementias. J. Nucl. Med. 55, 2003–2011.

Neuberger, T., Schopf, B., Hofmann, H., Hofmann, M., Rechenberg, B.V., 2005. Superparamagnetic nanoparticles for biomedical applications: possibilities and limitations of a new drug delivery system. J. Magn. Magn. Mater. 293, 483–496.

Ngobili, T.A., Daniele, M.A., 2016. Nanoparticles and direct immunosuppression. Exp. Biol. Med. 241, 1064–1073.

Niu, S., Zhang, L-K., Zhang, L., Zhuang, S., Zhan, X., Chen, W.Y., et al., 2017. Inhibition by multifunctional magnetic nanoparticles loaded with alpha-synuclein RNAi plasmid in a Parkinson's disease model. Theranostics. 7, 344–356.

Nogueira, E., Gomes, A.C., Preto, A., Cavaco-Paulo, A., 2016. Folate-targeted nanoparticles for rheumatoid arthritis therapy. Nanomedicine. 12, 1113–1126.

Nurunnabi, M., Cho, K.J., Choi, J.S., Huh, K.M., Lee, Y.K., 2010. Targeted near-IR QDs- loaded micelles for cancer therapy and imaging. Biomaterials. 31, 5436–5444.

Oliveira, I.M., Gonc alves, C., Reis, R.L., Oliveira, J.M., 2018. Engineering nanoparticles for targeting rheumatoid arthritis: past, present, and future trends. Nano Res. 11, 4489–4506.

Ozansoy, M., Basak, A.N., 2013. The central theme of Parkinson's disease: α-synuclein. Mol. Neurobiol. 47, 460–465.

Paciotti, G.F., Myer, L., Weinreich, D., Goia, D., Pavel, N., McLaughlin, R.E., et al., 2004. Colloidal gold: a novel nanoparticle vector for tumor directed drug delivery. Drug Deliv. 11, 169–183.

Pantarotto, D., Singh, R., McCarth, yD., Erhardt, M., Briand, J.P., Prato, M., 2004. Functionalized carbon nanotubes for plasmid DNA gene delivery. Angew. Chem. Int. Ed. Engl. 43, 5242–5246.

Pardridge, W.M., 2005. The blood-brain barrier: bottleneck in brain drug development. NeuroRx. 2, 3–14.

Pardridge, W.M., 2007. Drug targeting to the brain. Pharm. Res. 24, 1733–1744.

Park, J.H., Gu, L., Maltzahn, G.V., Ruoslahti, E., Bhatia, S.N., Sailor, M.J., 2009. Biodegradable luminescent porous silicon nanoparticles for in vivo applications. Nat. Mater. 8, 331–336.

Park, J.H., Maltzahn, G.V., Ruoslahti, E., Bhatia, S.N., Sailor, M.J., 2008. Micellar hybrid nanoparticles for simultaneous magnetofluorescent imaging and drug delivery. Angew. Chem. Int. Ed. Engl. 47, 7284–7288.

Pastorin, G., Wu, W., Wieckowski, S., Briand, J.P., Kostarelos, K., Prato, M., et al., 2006. Double functionalization of carbon nanotubes for multimodal drug delivery. Chem. Commun. 11, 1182–1184.

Patel, S.K., Janjic, J.M., 2015. Macrophage targeted theranostics as personalized nanomedicine strategies for inflammatory diseases. Theranostics. 5, 150–172.

Patel, T., Zhou, J., Piepmeier, J.M., Saltzma, W.M., 2012. Polymeric nanoparticles for drug delivery to the central nervous system. Adv. Drug. Deliv. Rev. 64, 701–705.

Pedrosa, P., Vinhas, R., Fernandes, A., Baptista, P.V., 2015. Gold nanotheranostics: proof-of-concept or clinical tool? Nanomaterials. 5, 1853–1879.

Peng, H., Liu, X., Wang, G., Li, M., Bratlie, K.M., Cochran, E., 2015. Polymeric multifunctional nanomaterials for theranostics. J. Mater. Chem. B 3, 6856–6870.

Pereira, G.C., Traughber, M., Muzic, R.F., 2014. The role of imaging in radiation therapy planning: past, present, and future. Biomed. Res. Int. 2014, 231090. doi:10.1155/2014/231090 Available at:Epub 2014 Apr 10.

Perrin, R.J., Fagan, A.M., Holtzman, D.M., 2009. Multimodal techniques for diagnosis and prognosis of Alzheimer's disease. Nature. 461, 916–922.

Pham, C.T., 2011. Nanotherapeutic approaches for the treatment of rheumatoid arthritis. Wiley Interdiscip. Rev. Nanomed. Nanobiotechnol. 3, 607–619.

Piao, Y., Kim, J., Bin, H., Kim, D., Baek, J.S., Ko, M.K., et al., 2008. Wrap-bake-peel process for nanostructural transformation from beta-FeOOH nanorods to biocompatible iron oxide nanocapsules. Nat. Mater. 7, 242–247.

Potschka, H., 2010. Targeting the brain–surmounting or by passing the blood–brain barrier. Drug Deliv. 2010, 411–431.

Powell, A.C., Paciotti, G.F., Libutti, S.K., 2010. Colloidal gold: a novel nanoparticle for targeted cancer therapeutics. Methods Mol. Biol. 624, 375–384.

Prabaharan, M., Grailer, J.J., Pilla, S., Steeber, D.A., Gong, S., 2009. Gold nanoparticles witha monolayer of doxorubicin-conjugated amphiphilic block copolymer for tumor targeted drug delivery. Biomaterials. 30, 6065–6075.

Prasad, L.K., O'Mary, H., Cui, Z.J.N., 2015. Nanomedicine delivers promising treatments for rheumatoid arthritis. Nanomedicine. 10, 2063–2074.

Primo, F.L., Rodrigues, M.M., Simioni, A.R., Lacava, Z.G., Morais, P.C., Tedesco, A.C., 2008. Photosensitizer-loaded magnetic nanoemulsion for use in synergic photodynamic and magnetohyperthermia therapies of neoplastic cells. J. Nanosci. Nanotechnol. 8, 5873–5877.

Qamar, N., Arif, A., Bhatti, A., John., P., 2019. Nanomedicine: an emerging era of theranostics and therapeutics for rheumatoid arthritis. Rheumatology. 58, 1715–1721.

Qi, L., Gao, X., 2008. Quantum dot-amphipol nanocomplex for intracellular delivery andreal-time imaging of siRNA. ACS Nano. 2, 1403–1410.

Rahman, M., Beg, S., Sharma, G., Anwar, F., Kumar, V., 2015. Emergence of lipid based vesicular carriers as nanoscale pharmacotherapy in rheumatoid arthritis. Recent Patents Nanomed. 5, 11.

Rahman, M., Sharma, G., Thakur, K., Anwar, F., Katare, O.P., Goni, V.G., et al., 2017. Emerging advances in nanomedicine as a nanoscale pharmacotherapy in rheumatoid arthritis: state of the art. Curr. Top. Med. Chem. 17, 162–173.

Rahmim, A., Zaidi, H., 2008. PET versus SPECT: strengths, limitations and challenges. Nucl. Med. Commun. 29, 193–207.

Rajora, M., Ding, L., Valic, M., Jiang, W., Overchuk, M., Chen, J., et al., 2017. Correction: tailored theranostic apolipoprotein E3 porphyrin-lipid nanoparticles target glioblastoma. Chem. Sci. 8, 5371–5384.

Ramachandran, R., Junnuthula, V.R., Gowd, G.S., Ashokan, A., Thomas, J., Peethambaran, R., et al., 2017. Theranostic 3-dimensional nano brain-implant for prolonged and localized treatment of recurrent glioma. Sci. Rep. 7, 43271.

Reis, R.A., Ventura, A.L.M., Kubrusly, R.C., de Mello, M.C.F., de Mello, F.G., 2007. Dopaminergic signaling in the developing retina. Brain Res. Rev. 54, 181–188.

Roy, I., Ohulchanskyy, T.Y., Pudavar, H.E., Bergey, E.J., Oseroff, A.R., Morgan, J., 2003. Ceramic-based nanoparticles entrapping water-insolublephotosensitizing anticancer drugs: a novel drug-carrier system forphotodynamic therapy. J. Am. Chem. Soc. 125, 7860–7865.

Ruan, S., He, Q., Gao, H., 2015. Matrix metalloproteinase triggered size-shrinkable gelatingold fabricated nanoparticles for tumor microenvironment sensitive penetration and diagnosis of glioma. Nanoscale. 7, 9487–9496.

Ruan, Z., Liu, L., Jiang, W., Li, S., Wang, Y., Yan, L., 2017. NIR imaging-guided combined photodynamic therapy and chemotherapy by a pH-responsive amphiphilic polypeptide prodrug. Biomater. Sci. 5, 313–321.

Rubinstein, I., Weinberg, G.L., 2012. Nanomedicines for chronic non-infectious arthritis: the clinician's perspective. Nanomedicine. 8, 577–582.

Saesoo, S., Sathornsumetee, S., Anekwiang, P., Treetidnipa, C., Thuwajit, P., Bunthot, S., et al., 2018. Characterization of liposomecontaining SPIONs conjugated with anti-CD20 developed as a novel theranostic agent for central nervous system lymphoma. Colloids Surf. B Biointerfaces. 161, 97–507.

Samia, A.C., Chen, X., Burda, C., 2003. Semiconductor quantum dots for photodynamic therapy. J. Am. Chem. Soc. 125, 15736–15737.

Samia, A.C., Dayal, S., Burda, C., 2006. Quantum dot-based energy transfer: perspectivesand potential for applications in photodynamic therapy. Photochem. Photobiol. 82, 617–625.

Sarmento, B., Sarmento, M., 2017. Nanomedicines for increased specificity and therapeutic efficacy of rheumatoid arthritis. Eur. Med. J. Rheumatol. 4, 98–102.

Schottler, S., Becker, G., Winzen, S., Steinbach, T., Mohr, K., Landfester, K., et al., 2016. Protein adsorption is required for stealth effect of poly (ethylene glycol)-and poly (phosphoester)-coated nanocarriers. Nat. Nanotechnol. 11, 372–377.

Shazeeb, M.S., Feula, G., Bogdanov, Jr. A.B., 2014. Liposome encapsulated superoxide dismutase mimetic: theranostic potential of an MR detectable and neuroprotective agent. Contrast Media Mol. Imaging. 9, 221–228.

Shi, L., Hernandez, B., Selke, M., 2006. Singlet oxygen generation from water-soluble quantum dot-organic dye nanocomposites. J. Am. Chem. Soc. 128, 6278–6279.

Shinkai, M., Le, B., Honda, H., Yoshikawa, K., Shimizu, K., Saga, S., et al., 2001. Targeting hyperthermia for renal cell carcinoma using human MN antigen-specific magnetoliposomes. Jpn. J. Cancer Res. 92, 1138–1145.

Simonato, M., Bennett, J., Boulis, N.M., Castro, M.G., Fink, D.J., Goins, W.F., et al., 2013. Progress in gene therapy for neurological disorders. Nat. Rev. Neurol. 9, 277–291.

Singh, R., Pantarotto, D., McCarthy, D., Chaloin, O., Hoebeke, J., Partidos, C.D., et al., 2005. Binding and condensation of plasmid DNA onto functionalized carbon nanotubes: toward the construction of nanotube-based gene delivery vectors. J. Am. Chem. Soc. 127, 4388–4396.

Skaat, H., Corem-Slakmon, E., Grinberg, I., Last, D., Goez, D., Mardor, Y., et al., 2013. Antibody-conjugated, dual-modal,near-infrared fluorescent iron oxide nanoparticles for antiamyloidgenic activity and specific detection of amyloid-β fibrils. Int. J. Nanomed. 8, 4063–4076.

Skrabalak, S.E., Chen, J., Sun, Y., Lu, X., Au, L., Cobley, C.M., et al., 2008. Gold nanocages:synthesis, properties, and applications. Acc. Chem. Res. 41, 1587–1595.

Slowing, I.I., Vivero-Escoto, J.L., Wu, C.W., Lin, V.S., 2008. Mesoporous silica nanoparticlesas controlled release drug delivery and gene transfection carriers. Adv. Drug. Deliv. Rev. 60, 1278–1288.

Small, G.W., Kepe, V., Ercoli, L.M., Siddarth, P., Bookheimer, S.Y., Miller, K.J., et al., 2006. PET of brain amyloid and tau in mild cognitive impairment. N Engl. J. Med. 35, 2652–2663.

Sonali, Singh, R.P., Sharma, G., Kumari, L., Koch, B., Singh, S., et al., 2016a. RGD-TPGS decorated theranostic liposomes for brain targeted delivery. Colloids Surf. B Biointerfaces. 147, 129–141.

Sonali, Singh, R.P., Singh, N., Sharma, G., Vijayakumar, M.R., Koch, B., et al., 2016b. Transferrin liposomes of docetaxel for brain-targeted cancer applications: formulation and brain theranostics. Drug Deliv. 23, 1261–1271.

Stephen, Z.R., Kievit, F.M., Veiseh, O., Chiarelli, P.A., Fang, C., Wang, K., et al., 2014. Redox-responsive magnetic nanoparticle for targeted convection-enhanced delivery of O6-benzylguanine to brain tumors. ACS Nano. 8, 10383–10395.

Su, C.H., Tsai, C.Y., Toman, B., Chen, W.Y., Cheng, F.Y., 2016. Evaluation of blood–brain barrier-stealth nanocomposites for in situ glioblastoma theranostics applications. Nanoscale. 8, 7866–7870.

Sumer, B., Gao, J., 2008. Theranostic nanomedicine for cancer. Nanomedicine. 2, 137–140.

Sun, C., Veiseh, O., Gunn, J., Fang, C., Hansen, S., Lee, D., et al., 2008. In vivo MRI detection of gliomas by chlorotoxinconjugated superparamagnetic nanoprobes. Small. 4, 372–379.

Sun, L., Joh, D.Y., Al-Zaki, A., Stangl, M., Murty, S., Davis, J.J., et al., 2016. Theranostic application of mixed gold and superparamagnetic iron oxide nanoparticle micelles in glioblastoma multiforme. J. Biomed. Nanotechnol. 12, 347–356.

Sun, Z., Huang, P., Tong, G., Lin, J., Jin, A., Rong, P., et al., 2013. VEGF-loaded graphene oxide as theranostics for multi-modality imaging-monitored targeting therapeutic angiogenesis of ischemic muscle. Nanoscale. 5, 6857–6866.

Symmons, D., Turner, G., Webb, R., Asten, P., Barrett, E., Lunt, M., et al., 2002. The prevalence of rheumatoid arthritis in the United Kingdom: new estimates for a new century. Rheumatology. 41, 793–800.

Tang, W.L., Tang, W.H., Li, S.D., 2018. Cancer theranostic applications of lipid-based nanoparticles. Drug Discov. Today. 23, 1159–1566.

Thomas, A.A., Omuro, A., 2014. Current role of anti-angiogenic strategies forglioblastoma. Curr. Treat. Options. Oncol. 15, 551–566.

Thomas, M., Klibanov, A.M., 2003. Conjugation to gold nanoparticles enhances polyethylenimine's transfer of plasmid DNA into mammalian cells. Proc. Natl. Acad. Sci. USA 100, 9138–9143.

Tsay, J.M., Trzoss, M., Shi, L., Kong, X., Selke, M., Jung, ME., et al., 2007. Singlet oxygenproduction by peptide-coated quantum dot-photosensitizer conjugates. J. Am. Chem. Soc. 129, 6865–6871.

Umar, S., Asif, M., Sajad, M., Ansari, M., Hussain, U., Ahmad, W., et al., 2012. Anti-inflammatory and antioxidant activity of *Trachyspermum ammi* seeds in collagen induced arthritis in rats. J. Drug. Dev. Res. 4, 210–219.

Vallet-Regi, M., Balas, F., Arcos, D., 2007. Mesoporous materials for drug delivery. Angew. Chem. Int. Ed. Engl. 46, 7548–7558.

Vardharajula, S., Ali, S.Z., Tiwari, P.M., Eroglu, E., Vig, K., Dennis, V.A., et al., 2012. Functionalized carbon nanotubes:biomedical applications. Int. J. Nanomed. 7, 5361–5374.

Vio, V., Jose, Marchant, M.J., Araya, E., Kogan, M.J, 2017. Metal nanoparticles for the treatment and diagnosis of neurodegenerative brain diseases. Curr. Pharm. Des. 23, 1916–1926.

Vivero-Escoto, J.L., Slowing, I.I., Wu, C.W., Lin, V.S., 2009. Photoinduced intracellular controlled release drug delivery in human cells by gold-capped mesoporoussilica nanosphere. J. Am. Chem. Soc. 131, 3462–3463.

Wang, J., Zhang, Y., Xia, J., Cai, T., Du, J., Chen, J., Li, P., et al., 2018. Neuronal PirB upregulated in cerebral ischemia acts as an attractive theranostic target for ischemic stroke. J. Am. Heart Assoc. 7, e007197.

Wang, J.T.-W., Rubio, N., Kafa, H., Venturelli, E., Fabbro, C., Ménard-Moyon, C., et al., 2016. Kinetics of functionalised carbon nanotube distribution in mouse brain after systemic injection: spatial to ultra-structural analyses. J. Control. Release. 224, 22–32.

Wang, Y., Zhou, K., Huang, G., Hensley, C., Huang, X., Zhao, T., et al., 2014. A nanoparticle-based strategy for the imaging of a broad range of tumours by non-linear amplification of microenvironment signals. Nat. Mater. 13, 204–212.

Welsher, K., Liu, Z., Dai, H., 2008. Selective probing and imaging of cells with single walled carbon nanotubes as near-infrared fluorescent molecules. Nano Lett. 8, 586–590.

Wen, C.J., Zhang, L.W., Al-Suwayeh, S.A., Yen, T.C., Fang, J.Y., 2012. Theranostic liposomes loaded withquantum dots and apomorphine for brain targeting and bioimaging. Int. J. Nanomed. 7, 1599–1611.

Wen, Z., Yan, Z., Hu, K., Pang, Z., Cheng, X., Guo, L., et al., 2011. Odorranalectin-conjugated nanoparticles: preparation, brain delivery and pharmacodynamic study on Parkinson's disease following intranasal administration. J. Control. Release. 151, 131–138.

Willard, D.M., Van Orden, A., 2003. Quantum dots: resonant energy-transfer sensor. Nat. Mater. 2, 575–576.

Wu, X., Chen, G., Shen, J., Li, Z., Zhang, Y., Han, G., 2015. Upconversion nanoparticles: a versatile solution to multiscale biological imaging. Bioconjug. Chem. 26, 166–175.

Xie, J., Chen, K., Huang, J., Lee, S., Wang, J., Gao, J., et al., 2010a. PET/NIRF/MRI triple functional iron oxide nanoparticles. Biomaterials. 31, 3016–3022.

Xie, J., Lee, S., Chen, X., 2010b. Nanoparticle-based theranostic agents. Adv. Drug Deliv. Rev. 62, 1064–1079.

Xie, R., Chen, K., Chen, X., Peng, X., 2008. InAs/InP/ZnSe core/shell/shell quantum dots as near-infrared emitters: bright, narrow-band, non-cadmium containing, and biocompatible. Nano Res. 1, 457–464.

Xing, T., Mao, C., Lai, B., Yan, L., 2012a. Synthesis of disulfide-crosslinked polypeptide nanogel conjugated with a near-infrared fluorescence probe for direct imaging of reduction-induced drug release. ACS Appl. Mater. Interfaces. 4, 5662–5672.

Xing, T., Yang, X., Wang, F., Lai, B., Yan, L., 2012b. Synthesis of polypeptide conjugated with near infrared fluorescence probe and doxorubicin for pH-responsive and image-guided drug delivery. J. Mater. Chem. 22, 22290.

Xu, S., Olenyuk, B.Z., Okamoto, C.T., Hamm-Alvarez, S.F., 2013. Targeting receptor mediated endocytotic pathways with nanoparticles: rationale and advances. Adv. Drug Deliv. Rev. 65, 121–138.

Yezhelyev, M.V., Qi, L., O'Regan, R.M., Nie, S., Gao, X., 2008. Proton-sponge coated quantum dots for siRNA delivery and intracellular imaging. J. Am. Chem. Soc. 130, 9006–9012.

Yoo, J., Kim, H.S., Hwang, D.Y., 2013. Stem cells as promising therapeutic options for neurological disorders. J. Cell Biochem. 114, 743–753.

Yu, M.K., Jeong, Y.Y., Park, J., Park, S., Kim, J.W., Min, J.J., et al., 2008. Drug-loaded superparamagnetic iron oxide nanoparticles for combined cancer imaging and therapy in vivo. Angew. Chem. Int. Ed. Engl. 47, 5362–5365.

Yuan, J., Guo, W., Yang, X., Wang, E., 2009. Anticancer drug-DNA interactions measure dusing a photoinduced electron-transfer mechanism based on luminescentquantum dots. Anal. Chem. 81, 362–368.

Zhang, C., Wan, X., Zheng, X., Shao, X., Liu, Q., Zhang, Q., et al., 2014. Dual-functional nanoparticles targeting amyloid plaques in the brains of Alzheimer's disease mice. Biomaterials. 35, 456–465.

Zhang, R., Li, Y., Hu, B., Lu, Z., Zhang, J., Zhang, X., 2016a. Traceable nanoparticle delivery of small interfering RNA and retinoic acid with temporally release ability to control neural stem cell differentiation for Alzheimer's disease therapy. Adv. Mater. 28, 6345–6352.

Zhang, S., Sun, C., Zeng, J., Sun, Q., Wang, G., Wang, Y., et al., 2016b. Ambient aqueous synthesis of ultrasmall PEGylated Cu^{2-}xSe nanoparticles as a multifunctional theranostic agent for multimodal imaging guided photothermal therapy of cancer. Adv. Mater. 28, 8927–8936.

Zhao, C., Rehman, F.U., Yang, Y., Li, X., Zhang, D., Jiang, H., et al., 2015. Bio-imaging and photodynamic therapy with tetra sulphonatophenyl porphyrin (TSPP)-TiO_2 nanowhiskers: new approaches in rheumatoid arthritis. Theranostics Sci. Rep. 5, 11518.

Zhou, J.E., Yu, J., Gao, L., Sun, L., Peng, T., Wang, J., et al., 2017. iNGR-modified liposomes for tumor vascular targeting and tumor tissue penetrating delivery in the treatment of glioblastoma. Mol. Pharm. 14, 1811–1820.

Zhu, L., Ploessl, K., Kung, H.F., 2014. PET/SPECT imaging agents for neurodegenerative diseases. Chem. Soc. Rev. 43, 6683–6691.

Zimmer, J.P., Kim, S.W., Ohnishi, S., Tanaka, E., Frangioni, J.V., Bawendi, M.G., 2006. Size series of small indium arsenide-zinc selenide core-shell nanocrystals and their application to in vivo imaging. J. Am. Chem. Soc. 128, 2526–2527.

Chapter 19

Improving the functionality of a nanomaterial by biological probes

Panchali Barman[a], Shweta Sharma[a], Avneet Saini[b]

[a]Institute of Forensic Science and Criminology (UIEAST), Panjab University, Chandigarh, India
[b]Department of Biophysics, Panjab University, Chandigarh, India

19.1 Introduction to nanomaterials

Nanomaterials are classified as materials ranging from 1 to 100 nm in size having unique optical, magnetic, or electronic properties compared to their bulk counterparts. In recent years, because of these unique properties nanoparticles have attracted significant attention in various fields such as energy, health care, sensing, environment, agriculture, etc. Nanoparticles with a size close to that of biomolecules are efficiently penetrated and delivered to the cells or tissues. A high surface area-to-volume ratio of these particles easily enables their surface loading and modification with drugs, biomolecules and other molecules (Jiang et al., 2013). In various growing areas, nanomaterials have proven their suitability as biosensors and their use has led to improved performance with enhanced sensitivities and lower detection limits of several orders of magnitudes. High specific surface of nanoparticles is a general advantage that enables an increased number of bioreceptor units to be immobilized with them (Holzinger et al., 2014). They have a wide range of distinctive physical properties in accordance with the size, shape, composition, crystal structure (cubic crystal system), surface ligands or capping agents, etc. (Daniel and Astruc, 2004; Redl et al., 2004; Mout et al., 2012). These features make them attractive materials for therapeutic and diagnostic applications. However, application of these particles in biomedicine needs controlled interactions with biomacromolecules (Chou et al., 2011). Nanoparticles have been functionalized using various methods with a number of ligands, viz. small molecules, dendrimers, surfactants, biomolecules, and polymers. Various multivalent surface structures comprised of small molecules and polymeric ligands are incorporated with multiple therapeutics or biomacromolecules by covalent or noncovalent interactions (Mayano and Rotello, 2011). Biomolecule-conjugated nanoparticles also exhibit desirable properties like precise recognition or biocompatibility (Rana et al., 2012). The simplicity of such functionalities enables researchers and chemists to build the necessary features to be used in clinics (Mout et al., 2012). However, there are a range of parameters that must be met for clinical use. First, adequate dispersibility and stability in both in vitro and in vivo environment is required, which is not influenced by ionic strength, pH change, polarity, or temperature changes. Second, they require high tissue selectivity, a prolonged circulation period in the bloodstream and high specificity toward target binding. Third, they require efficient renal clearance to have lower accumulated toxicity. In order to fulfill these parameters, a wide range of coating and conjugation approaches have been established and adapted to functionalize different nanomaterials such as metallic nanoparticles, quantum dots, magnetic nanoparticles, and up-conversion nanoparticles (Jiang et al., 2013). Besides their use in therapeutics, they have also gained great interest in bioimaging. For bioimaging applications, most of the nanoparticle surface modifications are based on chemisorption due to their stronger and stable binding ability compared to physisorption. In the past few years, synthesis and surface functionalization of nanoparticles have been extensively explored in various fields (Erathodiyil and Ying, 2011; Selvan et al., 2010). Biomedical applications of nanoparticles are determined by numerous factors such as size, structure, and charge along with physic-chemical properties. Table 19.1 summarizes the different applications of biomolecules conjugated with the most commonly used nanoparticles.

Nanoparticles can be categorized into various types as per, morphology, physical properties, and chemical properties such as metallic nanoparticles, semiconductors, carbonaceous nanoparticles, metal oxides, etc. which will be discussed further in this chapter. This chapter also discusses different nanotechnology systems that are used in various areas with special focus on biological probes functionalized nanoparticles as novel theranostics, antimicrobial and anti-inflammatory agents and detection of various biological species, metal ions etc. (Fig. 19.1).

TABLE 19.1 Summary of most commonly used nanoparticles, their interaction with biomolecules, and applications.

Type of nanomaterials	Biomolecules	Interaction approaches	Applications	References
Gold nanoparticles (AuNPs)	Deoxyribonucleic acid (DNA), ribonucleic acid (RNA), peptide, antibody, lipids, polymeric shell, ligands	Electrostatic interaction, covalent and noncovalent interaction, encapsulation, entrapment	Drug and gene delivery, diagnostics, bioelectronics, cellular labeling, sensors, photothermal therapy (PTT)	Tiwari et al., 2011; Heddle, 2013; Jazayeri, 2016
Quantum dots	Peptides, biotin, streptavidin, albumin, antibodies	Adsorption, covalent linkage, electrostatic interaction	Targeted therapeutic delivery, biosensors, probes, bioimaging, bioelectronics, multimodal Magnetic resonance imaging (MRI)/fluorescent imaging, photodynamic therapy (PDT)	Xing et al., 2010; Banerjee et al., 2016
Carbon nanotubes (CNTs)	Oligonucleotide, protein, lipids, polymer, ligands	Adsorption/wrapping (noncovalent) or surface functionalization (covalent), ionic, electrostatic interaction, encapsulation	MRI, bioelectronics, bioanalysis, radionuclide imaging drug delivery, PTT	Mallakpour and Soltanian, 2016
Iron oxide	Lipids, protein, polymer, targeting ligand, nucleic acid	Electrostatic interaction, covalent, noncovalent interaction, encapsulation, entrapment	Small interfering RNA (siRNA) and gene delivery, PTT, fluorescent imaging, MRI	Saallah and Lenggoro, 2018; Fang et al., 2010
Silica nanoparticle	Protein, DNA, lipid, saccharide	Entrapment, adsorption, covalent binding, encapsulation, and cross linking	Colorimetric diagnostics, drug delivery, surface enhanced Raman scattering (SERS) detection, PTT	Xing et al., 2010; Wang et al., 2008
Micelle	Biotin, antibody, targeting ligands, DNA, RNA	Encapsulation, cross linking, covalent, and noncovalent interaction	MRI, radionuclide imaging subcellular drug and gene delivery, PDT	O'Reilly et al., 2005; Doerflinger, 2018
Dendrimers	Amino acids, polymer, ligands, folic acid, RNA	Electrostatic interactions, covalent, and noncovalent interactions	Drug and gene delivery, water purifiers, sensors, catalysts, MRI contrast imaging, immunoassay, PDT	Chanphai et al., 2020; Szwed et al., 2016
Liposomes	DNA, protein, lipids, antibody	Covalent interaction, biotin–avidin, NHS-ester	Drug, gene, and protein delivery, immunology, radiology, antifungal, antiviral, and antitumor therapies, dermatology, ultrasound and nuclear imaging	Lopez and Liu, 2018; Riaz et al., 2018

19.2 Classifications of nanoparticles

19.2.1 Metallic nanoparticles

Metallic nanoparticles have been fascinating researchers for more than a century and are now used widely in engineering and biomedical sciences because of their huge potential in nanotechnology. Currently, synthesis and modifications of these materials with different functional groups enable them to couple with various ligands, drugs, and antibodies which widen their potential applications in several fields like biotechnology, diagnostics, therapeutics, and magnetic separation and preconcentration of target analytes. Moreover, various imaging techniques such as computed tomography (CT), MRI, ultrasound, positron emission tomography (PET), SERS, and optical imaging have been developed over time as an aid for imaging various states of diseases (Murphy et al., 2008).

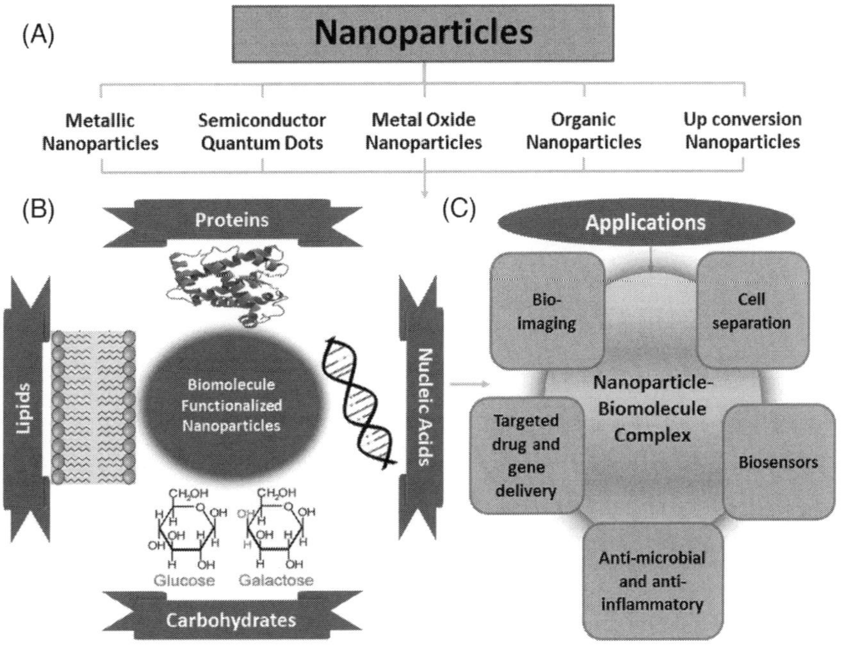

FIG. 19.1 Schematic representation of various nanoparticles (A), types of biomolecules functionalized on nanoparticles (B), and its applications (C).

19.2.1.1 Gold and silver nanoparticles

Gold nanoparticles (AuNPs), also known as colloidal gold, are nanosized suspensions of Au particles. The history of such particles goes back to Roman times, when they were used to stain glasses for decorative purposes. AuNPs are either intense red in color (particles less than 100 nm) or yellow color (for larger particles) (Murphy et al., 2008; Tong et al., 2009) and these interesting optical properties of these AuNPs are due to their unique interaction with light (Jain et al., 2008). The free electrons of the metal nanoparticles undergo oscillation with respect to the metal lattice in the presence of the oscillating electromagnetic field of the light (Jain et al., 2006; Kelly at al., 2003; Link and El-Sayed, 2003). This system resonates at a particular light frequency and is called localized surface plasmon resonance (LSPR).

In the present time, AuNPs are one of most explored nanoparticles because of their unique properties and bring out a wide range of applications in various fields (Fig. 19.2) depending upon its shape. For example, the rod-shaped nanoparticles have two resonances strongly depending upon the nanorod aspect ratio, that is, length-to-width ratio: first, due to plasmon oscillation along the short nanorod axis and second, due to plasmon oscillation along the long axis (Murphy et al., 2005; Link et al., 1999). When length-to-width ratio of the nanorod increases, the long-axis LSPR wavelength shifts from the visible to the near infrared (NIR) range and also increasingly elevates the oscillator strength (Link et al., 1999). For example, longitudinal and transverse absorption peak and anisotropy of the rodlike particles affect their self-assembly (Sharma et al., 2009). Such remarkable optical properties of AuNPs enable them to be the focus of extensive research, with wide range of applications like biomedical imaging, electronics, and science of materials (Rao et al., 2000).

Silver nanoparticles or AgNPs (1–100 nm) make up a significant percentage of silver oxide because of their large surface to bulk silver atoms. They are currently being incorporated into a broad range of clinical equipment, including surgical instruments, surgical masks, bone cement, etc. Moreover, it has also been observed that AgNPs replaces silver sulfadiazine as an effective agent for treating wounds (Qin, 2005; Atiyeh et al., 2007; Lansdown, 2006). Due to their attractive physiochemical properties these nanomaterials have received considerable attention in biomedical imaging using SERS. In fact, the surface plasmon resonance of AgNP and its large effective scattering cross section make them suitable candidates for molecular labeling (Schultz et al., 2000). Thus, many targeted silver oxide nanoprobes are currently being developed and their diverse applications are represented in Fig. 19.3.

19.2.1.2 Palladium and platinum nanoparticles

Alongside catalysis being the primary property of palladium (Pd-) and platinum nanoparticles (PtNPs) (Guo and Wang, 2011), they also have remarkable optical properties of great interest (Xiong et al., 2005). Pt and Pd have analogous catalytic

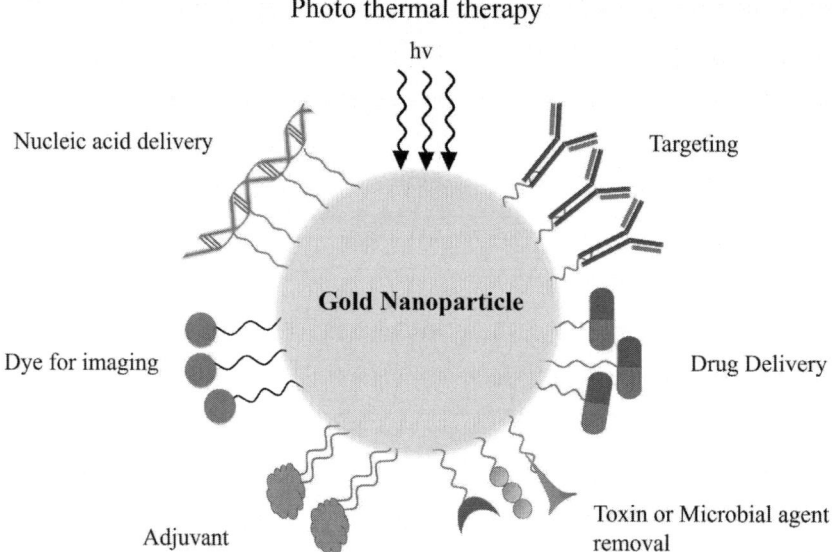

FIG. 19.2 **Different applications of gold nanoparticles in diagnosis and therapy.** *Reprinted with permission from Bagheri et al. (2018).*

properties (Cheong et al., 2010) and are significantly used in hydrogenation and dehydrogenation, crushing of petroleum and catalytic converters. Pt also plays a major role in fuel cells, catalyzing reduction in oxygen and fuel oxidation (e.g., ethanol, methanol, hydrogen, formic acid) (Chen and Holt-Hindle, 2010; Chen et al., 2009). However, in many C–C coupling reactions, such as, Heck, Negishi, Sonogashira, Stille, and Suzuki coupling, Pd materials are the catalyst of choice (Balanta et al., 2011). It is highly beneficial to use these nanoparticles as they give surface-to-volume ratios, high surface-free energies and catalytic activity that can be modified by manipulating their size or shape and stabilizing ligands (Chen et al., 2009; Aiken and Finke, 1999). Synthesis of Pd- and PtNPs are almost similar to that of Ag- and AuNPs, (Jana and Peng, 2003) provided, the precursor salts ($PdCl_4^{2-}$, $PtCl_4^{2-}$) are reduced in Pd- and PtNPs using common reductants (e.g., glycols, sodium borohydrate) in the presence of suitable stabilizers.

In context of biology, Pt shows electrocatalytic behavior against the nonenzymatic oxidation of glucose (Chen et al., 2009; Toghill and Compton, 2010) and the chemical reactions between hydrogen peroxide (H_2O_2) and luminol or lucigenin can be effectively catalyzed by PtNPs (Jana and Peng, 2003; Xu and Cui, 2007). They also potentially mimic the enzyme while catalytically decomposing peroxide and superoxide (Kajita et al., 2007) and have also been observed to be potent antioxidants (Kim et al., 2008).

19.2.1.3 Noble metal nanoclusters

Although nanoclusters have a particle size smaller than 2 nm, they are very different from the counterparts of other nanoparticles. Nanoclusters are usually comprised of around 100 atoms and have intermediate properties to atoms and larger nanoparticles (Shang et al., 2011), which result from the size less than a Fermi unit of the metal's electrons, ensuing in distinct electronic states rather than a continuum. $Au_{25}L_{18}$ (where L is a ligand) (Parker et al., 2010) and Schmid's cluster

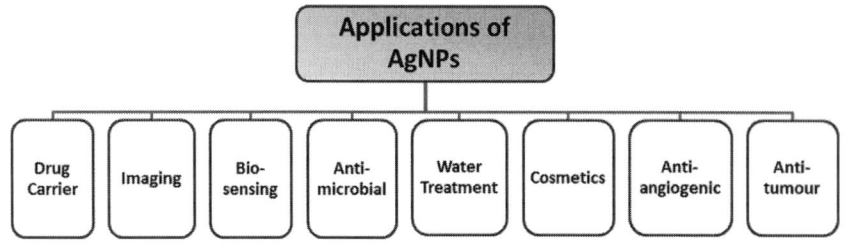

FIG. 19.3 Various applications of silver nanoparticles.

(Jana and Peng, 2003) $Au_{55}[P(C_6H_5)_3]_{12}Cl_6$ are the two structurally best-known nanoclusters. The most important properties of metal nanoclusters are their size-dependent luminescence, good photostability, large shifts in stocks, and high emission levels (Shang et al., 2011; Guo and Wang, 2011). Another promising aspect of these clusters is the broad two-photon absorption cross-sections which is reported to be 2700 GM (Goeppert–Mayer units) and 189740 GM at an excitation of 1290 nm for Au_{25} and Au-nanoclusters, respectively (Ramakrishna et al., 2008; Polavarapu et al., 2011) and 34000 GM and 5000 GM at an excitation of 890 nm and 830 nm for Ag nanoclusters with emissions of 680 and 700 nm, respectively (Patel et al., 2008).

Synthesis of these clusters when compared to that of the other nanoparticles is much simple as they require very less precautions to prevent contamination. Ligands and stabilizers with low affinity toward Ag or Au do not successfully prevent nanoclusters from aggregating and growing into nanoparticles (Xu and Suslick, 2010). Hence, hydroxyterminated second or fourth generation polyamidoamine (PAMAM) dendrimers are an effective template for the synthesis of nanoclusters (Zheng and Dickson, 2002; Zheng et al., 2003).

19.2.2 Semiconductor quantum dots

Semiconductor quantum dots (QDs) are the commonly synthesized nanoparticles using semiconductor materials such as ZnSe, ZnS, CdSe, CdS, CdTe, InP, and others showing a range of surface chemistries as depicted in Fig. 19.4 (de Mello Donega et al., 2011; Murray et al., 2000). Their nanoscale size is correlated with the excitation energy of the bulk constituents producing certain electronic and photonic properties confined to the quantity (Algar et al., 2011; Alivisatos et al., 2005). QDs are significantly larger in size than organic dyes and fluorescent proteins and were primarily designed for electronic applicability only.

However, their composite optical features provided much more versatility and potential for biological applications than the previously available fluorophores (Medintz et al., 2009; Algar et al., 2011; Michalet et al., 2005; Resch-Genger, 2008; Rosenthal et al., 2011). Efficient photophysical characteristics of biological significance include a broad spectrum of absorption that constantly increases toward ultraviolet (UV), narrow, photoluminescent spectra based on size and symmetry ranging from UV to IR (Medintz et al., 2005; Algar et al., 2010, 2011; Michalet et al., 2005; Resch-Genger et al., 2008; Algar and Krull, 2010; Tavares et al., 2011). The specific properties of QDs, such as in vitro and in vivo fluorescent probes, theranostics, biosensors, and PDT agents have been continuously adapted, developed and exploited over the past few years in various biological applications (Resch-Genger et al., 2008; Algar and Krull, 2010; Baker, 2010; Biju et al., 2008; Biju et al., 2010; Zrazhevskiy et al., 2010; Ho and Leong, 2010). QDs are widely used as core/shell structures in biological applications, where the overcoating shell comprises of a material with wider band gap which protects the core and prevents leaching resulting in higher quantum yield (Algar et al., 2011; Dabbousi et al., 1997).

There are different strategies designed for the bioconjugation of QDs which are subclassified into six functional subtypes based on the mechanism of interaction or chemistry used. The subtypes are: (1) covalent conjugation between the ligands on QD surface and biomolecular functional groups, (2) electrostatic interactions between the oppositely charged QDs and biomolecules, (3) dative interactions between the inorganic QD surface and biomolecular thiol surface, (4) high-affinity secondary binding, (5) enzyme-catalyzed bioconjugation, and (6) biological templating (Sapsford et al., 2013).

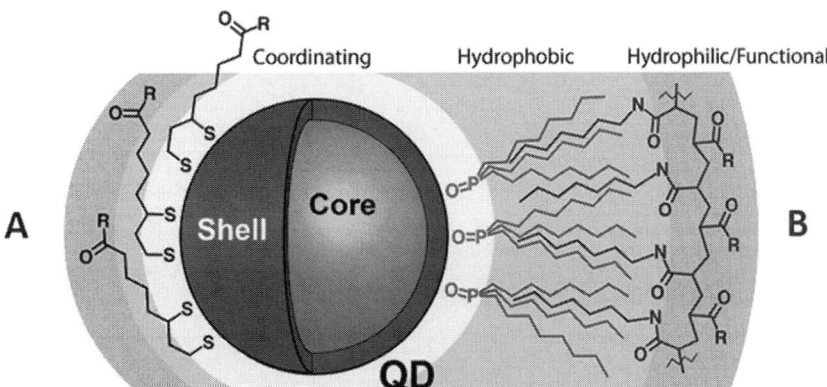

FIG. 19.4 **Surface chemistries of semiconductor quantum dots.** (A) Ligand binding at a quantum dot surface. (B) Amphipol (blue) combines with the native quantum dot ligands (red). Figure is reprinted with/permission from Sapsford et al. (2013). *Copyright 2013 American Chemical Society.*

19.2.3 Metal oxide nanoparticles

Semiconducting metal oxide gas sensors have remarkable advantages over the other conventional techniques such as high sensitivity, rapid detection and are relatively inexpensive. Owing to their unique characteristics, metal oxides (MONPs) are one of the most extensively used engineered nanoparticles. In modern nanotechnology, the characteristics which make the nanophase structures indispensable are their numerous nonlinear optical properties, cold welding properties, increased ductility at high temperatures, super paramagnetic behavior, higher sensitivity and selectivity and their unique catalytic activity (Chavali and Nikolova, 2019). The metallic, semiconductor, or insulating characteristics of the nanoparticles significantly depends upon its electronic structure or configuration. The remarkable physiochemical properties of MONPs are attributed to the high density and the small proportion of edges on their surfaces. Significant analytical uses of MONPs play a major role in attracting the researchers with substantial interest in the areas of materials chemistry, theranostics, agriculture, optics, electronics, catalysis, environment, sensors, etc. (Fig. 19.5) (Chavali and Nikolova, 2019). MONPs are of great benefit as their properties could be modified according to their size and shape. Nanoparticles may have magnetic, electronic and chemical properties depending on a particular size of the nanoparticle compositions (Poole and Owens, 2003).

19.2.3.1 Iron oxide

Iron oxide nanoparticles (IONPs) exist in the form of a crystalline core made up of either Fe(II) or Fe(III) oxide along with a crystal structure of either magnetite (Fe_3O_4) or maghemite (γFe_2O_3). The initial synthesis of IONP often produces Fe_3O_4, which gradually oxidizes to γFe_2O_3 (Sapsford et al., 2013). The increased interest of researchers in IONPs is justified by their diverse applications in materials science and biomedical engineering (McBain et al., 2008; Jeong et al., 2007; Pankhurst et al., 2003; Pankhurst et al., 2009). These nanoparticles have simple and low-cost manufacturing which results in biocompatible colloidal suspensions with remarkable size-dependent magnetic characteristics that are commonly known as super paramagnetism. This phenomenon occurs at nanoscale, the ferromagnetic material dimensions are smaller than the standard size of magnetic domain (Knobel et al., 2008). Suspension of super paramagnetic materials with IONPs referred to as *ferrofluids* show a magnetic behavior that is suitable for numerous biomedical practices (Pankhurst et al., 2003, 2009; Gao et al., 2009; Gupta and Gupta, 2005; Thanh and Green, 2010). Ferrofluids have been clinically employed as MRI contrast agents, where the IONP size and coating defines the biodistribution, and consequently, the organs that become magnetically highlighted (Qiao et al., 2009). IONPs have recently been explored as energy transducers for hyperthermic treatment of cancer (Laurent et al., 2011) and as drug carriers (Kumar and Mohammad, 2011). Besides, IONPs are biocompatible, are readily metabolized by the liver, and are entirely cleared in less than 48 h (Arbab et al., 2005).

However, IONPs can be biofunctionalized either directly by conjugating to the native oxide surface or by conjugating to the surrounding stabilization layer. A wide array of stabilizers, ranging from carboxylates, phosphates, polymers to inorganic materials, such as silica and gold, etc. have been incorporated into the surface of IONPs (Gupta and Gupta, 2005; Laurent et al., 2008; Nakanishi et al., 2003; Faure et al., 2009). Such molecular layers on these particles provide colloidal stabilization that can be electrostatic (Walker et al., 2011) or steric in nature (Ghosh et al., 2011).

19.2.3.2 Silicon dioxide

Silicon dioxide (SiO_2) nanoparticles in the recent time have gained more interest in the fields of theranostics (Barbe et al., 2004; Trewyn et al., 2007; Wang et al., 2008) as they are biocompatible and surface functionalization is much easier as compared to other nanomaterials. This interest has led considerable efforts toward preparation of fluorescent SiO_2 nanoparticles either by functionalizing a fluorophore on the surfaces or encapsulating it in the nanoparticles (Song et al., 2011).

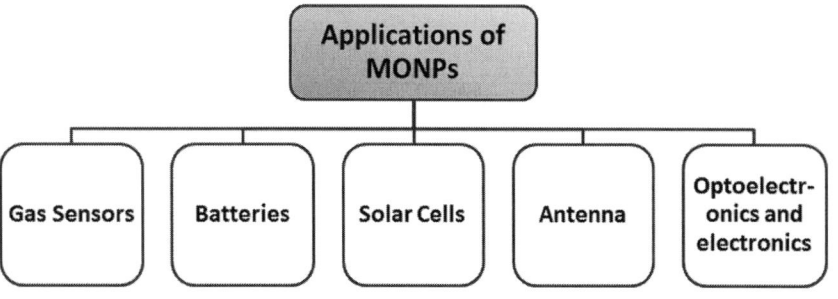

FIG. 19.5 Diverse applications of MONPs (Chavali and Nikolova, 2019).

Wide applications of these nanoparticles have expanded bioconjugation of SiO_2 nanoparticles quite rapidly yet holding on to the silane chemistry as the base for producing such hybrids (Simovie et al., 2011). Stöber method is the typical synthetic approach to access these nanoparticles where tetralkyl silyl esters are hydrolyzed in the presence of ammonia in an alcoholic solution (Park et al., 2002). In an attempt to stabilize and define the size and morphology of subsequent nanoaggregates, the silanes act as oxidized silicon source whereas, ammonia (NH_3) acts as a polymerizing agent for –O–Si–O– silanes (Sapsford, 2013).

19.2.3.3 Titanium dioxide

Titanium dioxide (TiO_2) nanoparticles are extensively used in various common products like pigments (Sapsford, 2013), sunscreens (Popov et al., 2005), and other cosmetic products (Albrecht et al., 2006). In solar cell production (Tan and Wu, 2006; Wei et al., 2006; Wang et al., 2001) and UV-activated photocatalysts development (Kandiel, 2010), TiO_2 nanoparticles are often used as photo-active elements. While exploring the photoactivated cancer treatment (Lagopati et al., 2010) and developing bactericidal composites (Tai et al., 2007), TiO_2 nanoparticles as photocatalysts have been utilized noticeably. Besides, Tai et al. and Zhu et al. have reported a few examples of TiO_2 nanoparticle-based photoelectrochemical gas sensors for gases like NH_3, CO and organic vapors present in air (Tai et al., 2007; Zhu et al., 2002). However, unlike other nanomaterials, TiO_2 nanoparticles show lack of physicochemical properties (optical and magnetic properties) and hence are not widely exploited for sensing applications.

19.2.4 Organic nanoparticles

These nanoparticles are natural molecules and nature has provided a wide range of organic nanoparticles such as carbon allotropes, lipids, milk emulsions, protein aggregates, etc. (Romero and Moya, 2012). Organic nanoparticles are often used in food products (Reza et al., 2008) cosmetics (Discher and Ahmed, 2006), pharmaceutical formulations (Zhu et al., 2011) such as liposome vectors, polymer–protein or polymer–drug complexes. In the present time, ease of encapsulation of materials on organic nanoparticles and their biodegradability make them the most appealing nanomaterials for drug delivery and biomedical applications (Christian et al., 2007).

19.2.4.1 Carbon allotropes

Carbon nanomaterials have a wide array of diversity in materials which includes spherical fullerenes, carbon nanotubes (CNTs), graphene nanomaterials, nanodiamonds (NDs), etc. (Chun et al., 2009; Zhang et al., 2010). Each nanomaterial possesses different chemical and physical properties. Carbon nanomaterials are significant in various biomedical applications including therapeutic agents, drug delivery, diagnostics, and biosensors, etc. (Cai and Chen, 2007).

Carbon nanotubes (CNTs): Among all carbon-based nanomaterials, CNTs are the most explored ones till date in the area of bionanotechnology which includes imaging, therapeutics, sensing, etc. (Tavares et al., 2011). Carbon nanotubes are allotropes of carbon having sp^2-hybridized carbon atoms (Pérez-López and Merkoçi, 2012). These are 1D in structure and have a cylindrical nanosized shape formed by rolling up graphene sheets. CNTs are of two types: single-walled CNTs (SWCNTs) made up of one layer of graphene and multiwalled CNTs (MWCNTs) (Fig. 19.6) made up to multiple layers of graphene. CNTs are suitable for lower detection limit and higher sensitivity values since they show high charge transfer capacity. These nanosized materials are also suitable to be used as genosensors (DNA sensors), enzymatic sensors and immunosensors due to the certain properties such as specific surface area, high adsorption capacity and good biocompatibility (Pérez-López and Merkoçi, 2012). Since last decade, CNTs have made several contributions specifically in the development of electrochemical and optical sensing systems leading to notable applications of CNTs. Several examples of CNT-based biosensors along with the analytes, techniques of detection, limit of detection and their certain applications are cited in Table 19.2 (Pérez-López and Merkoçi, 2012).

Spherical fullerenes: Fullerenes (C_{60}), also known as *Buckminsterfullerene*, is an allotrope of carbon consisting of a carbon-based cage with a diameter of 1 nm. C_{60} has a wide array of usages in the field of biomedicines as represented in Fig. 19.7 (Sapsford et al., 2013). Although C_{60} have been found occurring in nature, they are typically synthesized in laboratories using arc discharge method (Hudhomme et al., 2011). C_{60} conjugated with biological moiety has led to a wide array of applications such as targeted drug delivery, gene delivery (Montellano et al., 2011; Sitharaman et al., 2008; Klumpp et al., 2007; Maeda-Mamiya et al., 2010), and cancer targeting probes (Sapsford et al., 2013). Bioconjugation of fullerenes are usually carried out covalently (e.g., drugs) or through electrostatic interactions (e.g., DNA) (Sitharaman et al., 2008; Klumpp et al., 2007; Maeda-Mamiya et al., 2010; Tagmatarchis and Shinohara, 2001).

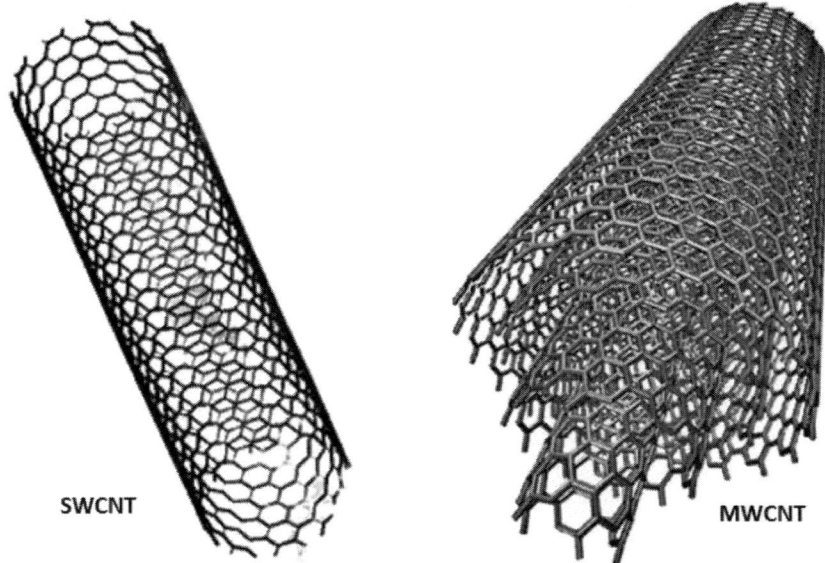

FIG. 19.6 Pictorial representation of SWCNT and MWCNT.

TABLE 19.2 Some reported examples for carbon nanotubes-based biosensing systems.

Materials	Analytes	Detection technique	LOD	Applications
GO-MWCNT	H_2O_2	Electrolytic reduction of H_2O_2	1.17 μM (S/N = 3)	Electrochemical sensing platform
MWCNT	Catechol	Magneto-switchable bio electrocatalytic	7.61 μM (S/N03)	Electrocatalytic magneto-switchable biosensor
SWCNTs	Glycosylatyed protein	NIR fluorescence	2 μg	NIR fluorescent sensor
MWCNTs	Phenol	Amperometric reduction of o-quinone	1.35 μM (S/N03)	Amperometric biosensor
MWCNTs	Dopamine	Amperometric	37 μM (S/N03)	Amperometric biosensor
SWCNTs	Salmonella typhi	Potentiometric	0.2 CFU/mL (S/N03)	Amperometric biosensor
MWCNTs	Dichloroaniline, nitrophenol, and naphthalene	High performance liquid chromatography (HPLC)	0.1–3 ng/mL (S/N03)	Sorbent for microsolid phase extraction (μ-SPE)

Reprinted with permission from Pérez-López and Merkoçi (2012). Copyright © 2012, Springer Nature.

Other carbonaceous nanomaterials: In addition to the above-mentioned carbon-based nanomaterials, there appears to be certain other carbonaceous nanomaterials with different shapes and properties including carbon nanodots (C-dots) (Ray et al., 2009; Baker and Baker, 2010), nanoclusters (Berlin et al., 2011), nanohorns (Zhang et al., 2010), nanorings, nanoribbons (Chun et al., 2009), etc. C-dots having particle size of 10 nm exhibit intense fluorescence/photoluminescence upon surface passivation and are witnessed to have significant potential in bioimaging and biosensing applications (Baker and Baker, 2010; Sun et al., 2006; Cao et al., 2007). Most commonly preferred agents for surface passivation are polymers such as polyethylene glycol (PEG) or poly(propionylethyleneimine-co-ethyleneimine) (Sun et al., 2006; Cao et al., 2007; Yang et al., 2009). Encapsulation of metallofullerenes (trimetallic nitride endohedral), comprising of either Gadolinium or Lutetium, and further conjugating quantum dots to the nanohorn's surface are also used as multimodal imaging agents (Zhang et al., 2010). These allow both phase contrast and fluorescent imaging of various in vitro and in vivo samples.

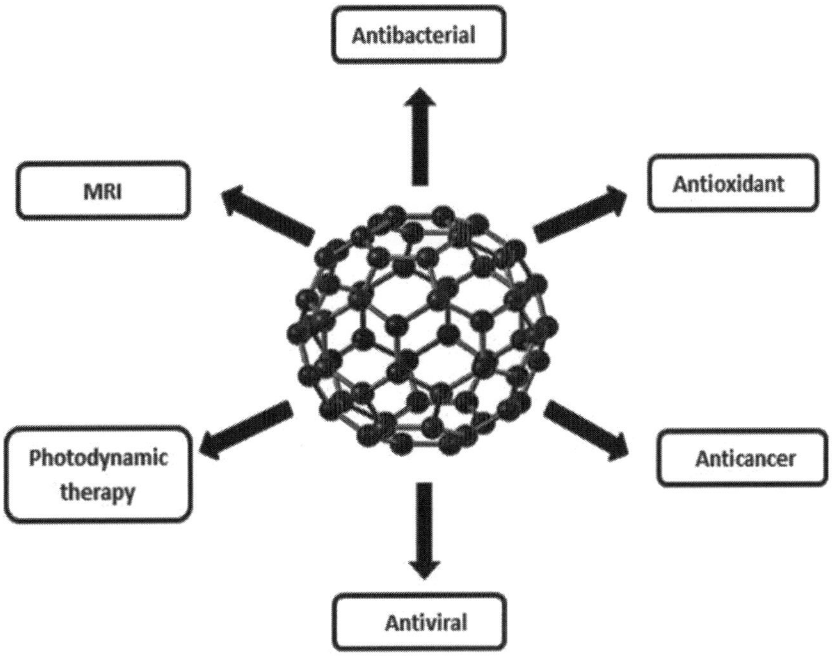

FIG. 19.7 Biomedical applications of spherical fullerenes.

19.2.4.2 Biopolymeric nanomaterials

Biopolymers are long chains of carbohydrates having repeated monosaccharide units which glycosidically linked together (Nitta and Numata, 2013). These nanomaterials are typically attributed to biomolecules such as nucleic acids, polysaccharides, polypeptides, proteins and are formulated as potential alternatives approaches to lipid-based synthetic materials (Dang and Leong, 2006; Nair and Laurencin, 2007; Sundar et al., 2010; De Souza et al., 2010). These nanomaterials are highly charged in nature and have amphiphilic properties which make their formation easier by simple self-assembly, ionic gelation and electrostatic interactions (Sundar, 2010). They are biocompatible, biodegradable, abundant, stable, inexpensive, and are less toxic in nature as compared to their synthetic counterparts (Manivasagan et al., 2017). Such properties lead them to a wide range of biomedical applications as represented in Fig. 19.8 (Nair and Laurencin, 2007; Cormode et al., 2010; Manivasagan et al., 2017). They could be used as drug carriers in gene as well as the scaffolds for bone, cartilage, cardiac, skin and tissue engineering. Besides, these Biopolymeric nanomaterials can also be used for cancer, diabetes, allergy, infection and inflammation (Nitta and Numata, 2013).

19.2.5 Upconversion nanoparticles

Upconversion nanoparticles (UCNPs) are a special and distinctive type of lanthanide ion-doped optical nanomaterials with a range of electronic transitions occurring within the 4f electron shells. They lead to upconversion of two or more photons of lower energy into one photon of high energy (Wen et al., 2018; Liu et al., 2017; Zhou et al., 2015; Wang et al., 2010). On NIR excitation, UCNPs exhibit remarkable upconversion properties, like wide anti-Stokes shift, sharp spectrum of emissions, long lifespan, strong photostability and minimal cytotoxicity (Li and Lin, 2010; Wang and Liu, 2009). UCNPs utilize NIR as it penetrates to a remarkable depth in samples, also reduces photodamage and photobleaching along with minimizing autofluorescence in biological samples. Moreover, in order to induce upconversion luminescence for fluorescence imaging, a continuous laser with low power supply ($1-10^3$ W/cm) is used (Jiang et al., 2013). Hydrophobic UCNPs with uniform shape and size are synthesized with significantly whereas; they are converted into aqueous solutions by using hydrophilic ligands for surface capping in order to make them suitable for biological luminescent labels (Wang et al., 2010, 2011; Kar and Patra, 2012; Haase and Schäfer, 2011). Mostly commonly exploited approaches for coating UCNPs are layer by layer assembly, surface salinization, ligand oxidation, and ligand attraction (Wang and Liu, 2009; Mader et al., 2010; Zhou et al., 2012). With the growing interest and rapid development of these nanoparticles, they have been engineered for a

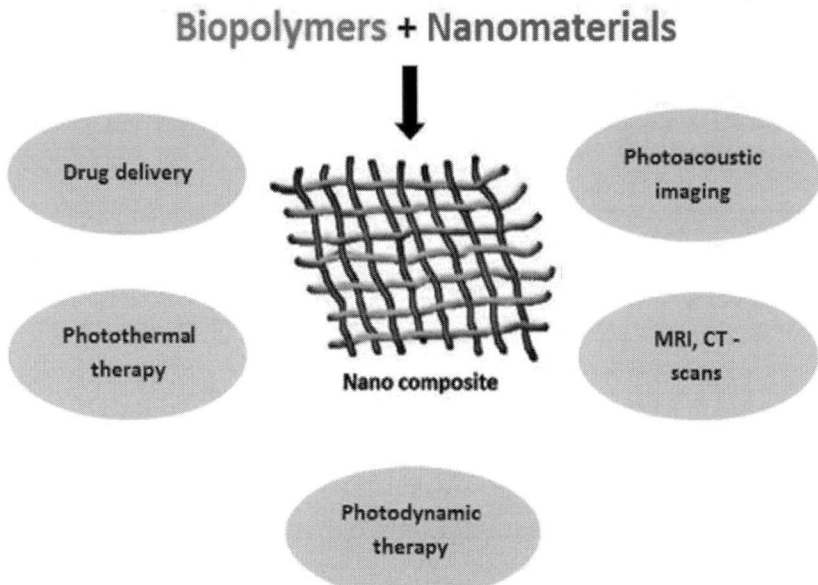

FIG. 19.8 Applications of biopolymeric nanomaterials.

wide range of uses in biomedical field (Fig. 19.9) (Zhou et al., 2012; Wang et al., 2010; Chatterjee et al., 2010; Yang et al., 2012; Jayakumar et al., 2012; Ang et al., 2011).

However, enhancement of properties and range of applications of these various nanoparticles can be obtained by adequate surface chemical modifications which make them more suitable for medical usage without affecting their basic properties. These modifications are necessary as certain "naked" nanoparticles are extremely cytotoxic and cannot be applied directly to humans (Liu et al., 2010). These functionalizations overcome the limitation of cytotoxicity by leading to reproducible immobilization of bioreceptor units along with enhancement in biocompatibility which further highlights the significance of chemistry in development of nanoparticle interfaces and their biomedical applications (Holzinger et al., 2014).

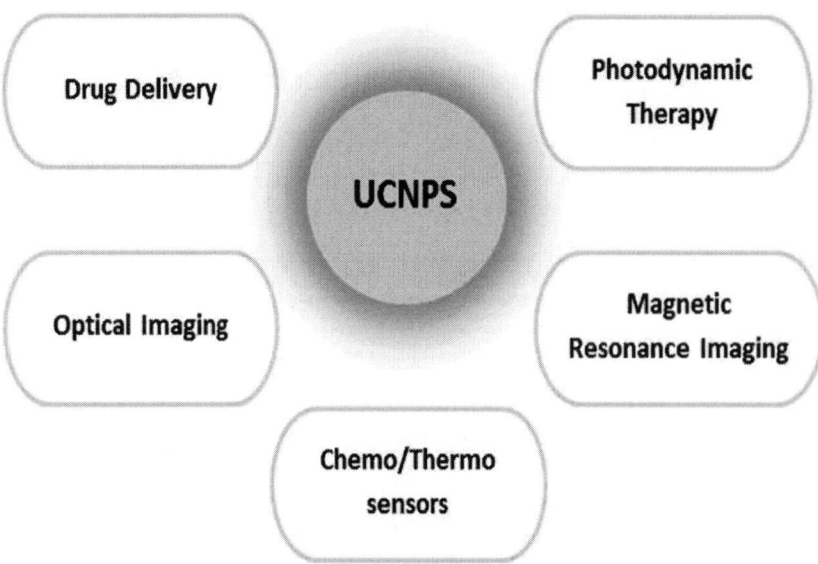

FIG. 19.9 Wide range of applications of upconversion nanoparticles in biomedical field.

19.3 Common conjugation approaches for biomolecule functionalized nanomaterials

19.3.1 Conjugation approaches

While choosing the conjugation strategy of the nanoparticles, various factors play a major role which includes the size, shape, structure and surface chemistry of nanoparticles. Moreover, the nature, size and chemical composition of biomolecule also play a key role. Conjugation can be covalent or noncovalent, with electrostatic binding, certain types of adsorption, and encapsulation (Fig. 19.10) (Sapsford et al., 2013).

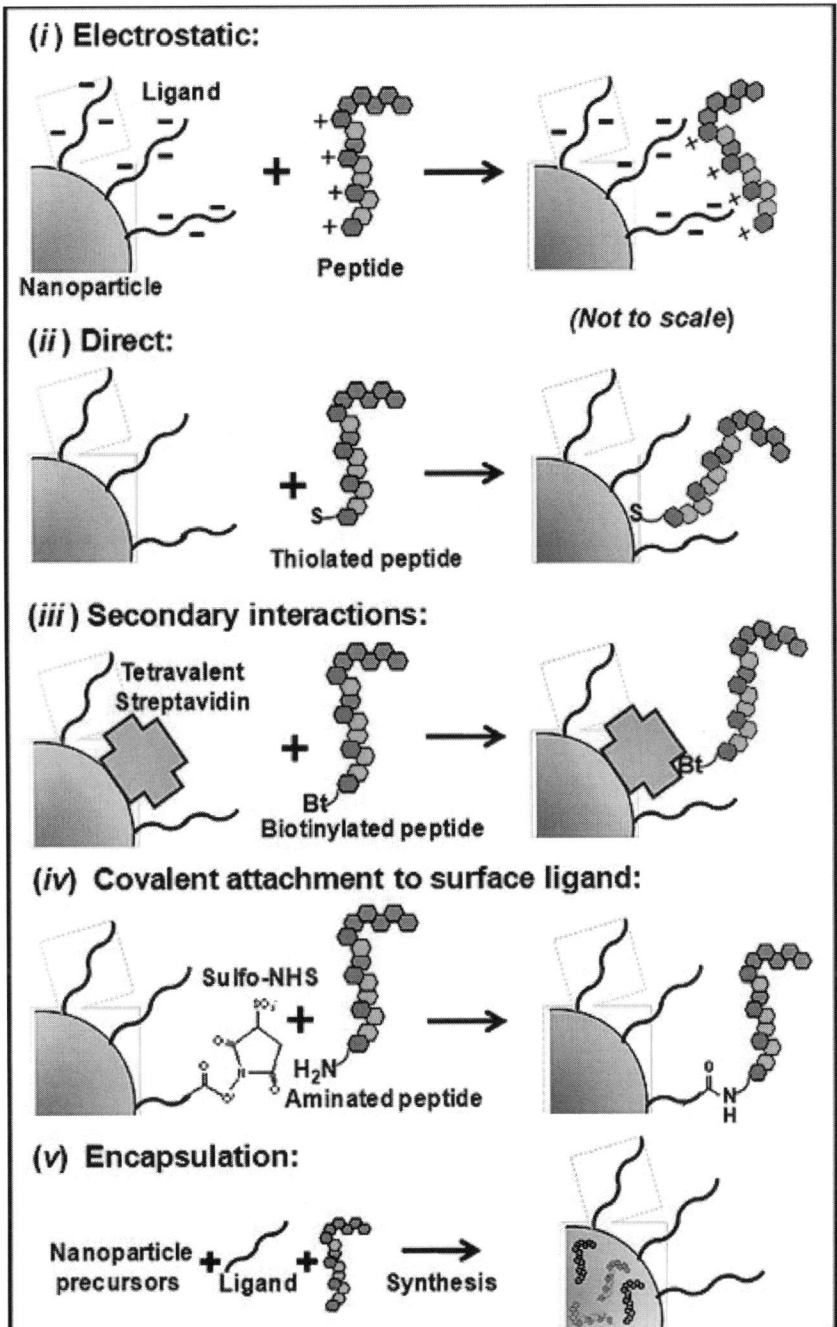

FIG. 19.10 **General strategies usually used for bioconjugation to nanomaterials.** *Reprinted with permission from Sapsford et al. (2013). Copyright 2013 American Chemical Society.*

19.3.1.1 Encapsulation

Encapsulation of therapeutic substances in a nanomaterial is among the most successful chemistries currently being developed for drug delivery (Anton et al., 2008). This encapsulating strategy has the obvious benefits of protecting the active pharmaceutical ingredients from degradation (Delehanty et al., 2009; Emerich and Thanos, 2007; Liong et al., 2008; Harush-Frenkel et al., 2008) due to adverse environmental factors, such as unwanted light, moisture and oxygen, thus leading to an increase in the product's shelf life and facilitating a sustained release of the encapsulate (Shahidi and Han, 1993). In addition, nanoencapsulated drugs can be tagged with fluorescent probes for bioimaging purposes, which is especially useful for assessing drug efficacy during preclinical and clinical trials.

Such nanomaterials can be natural as well as artificial polymers, oxides, metals, liposomes, dendrimers, and micelles. The formulation of these nanomaterials with suitable size, chemical functionality, biocompatibility and, above all, efficient drug releasing properties is an extremely complex process (Fang and Bhandari, 2010).

Various morphologies can be designed for encapsulation, but the most common are: first, mononuclear capsules having a single core enclosed in a shell, second, aggregates, which have multiple cores embedded in a matrix (Schrooyen et al., 2001). There are several techniques used for encapsulating different agents, however, some of the most recent technologies are electrospray, inclusion complexation, liposome trapping, lyophilization, centrifugal suspension separation, cocrystallization, and emulsion, etc. (Augustin and Hemar, 2009). Encapsulation of biological agents needs special care and follows three crucial steps: (i) Wall formation around the material to be encapsulated; (ii) Ensuring that there is no undesired leakage; (iii) Confirming that there are no unnecessary materials. Encapsulation is a promising method for biomolecule-nanoparticle complexes with certain widespread applications in the area of biomedicines and other fields where chemical interactions must be avoided (Fang and Bhandari, 2010).

19.3.1.2 Noncovalent attachment

In several cases, a relatively weak coordination binding appears to show noncovalent interaction and therefore stability is determined by equilibrium dissociation constants. Therefore, during preparation and subsequent application these self-assembled bioconjugates can be very sensitive to nanomaterials and biomolecules concentrations. Again, the stability of the nanomaterial bioconjugates can be enhanced by exploiting multiple contact points, such as through multidentate interactions, and allows use at low concentrations (<10−6 M) which are more appropriate for biological and environmental applications (Algar et al., 2011). Considering that these chemistries usually involve only stoichiometric combinations of the two components, they are especially referred to as self-assembly. These approaches provide quick and simple bioconjugation without any complementary additional reagents required to provide better yield control (Sapsford et al., 2013; Algar et al., 2011).

Electrostatic attachment is the one of the simplest and most commonly explored approaches for bioconjugation of nanomaterials, since it doesn't involve much complex chemical reactions *per se*. The fundamental principle behind the electrostatic immobilization is the attraction between oppositely charged species, that is, charged nanomaterials electrostatically attract an oppositely charged biomolecule, and vice versa. In the nanoscale range, electrostatic interactions are amplified and certain factors viz. charge, reagent concentrations, and ionic strength play a major role to achieve the preferred conjugate (Schneider and Decher, 2004; Schoeler et al., 2003; Liu et al., 2008; Khopade and Carusa, 2002).

Other noncovalent formulations can be accomplished by π interactions, host–guest hydrophobic–hydrophilic interactions etc. along with interaction of hydrogen molecules (Chen and Jiang, 2011; Sakuma et al., 2001; Joshi et al., 2020). However, another class of noncovalent interaction is the secondary interactions between functional groups present on nanomaterial surface and other moieties. These are often characterized by biotin–avidin interactions or biomolecule affinity toward a ligand attached to nanoparticle surface where nanomaterial-biomolecular conjugates are functionalized to show one of the pairs electrostatically attracting an oppositely charged species, and vice versa (Sapsford et al., 2013). Whereas, biomolecules upon immobilization with the nanoparticles also noncovalently interact resulting into a self-assembled complex. The functional groups present on the biomolecule form π interactions with the nanoparticle (Yang et al., 2008, 2012; Li et al., 2016). Noncovalent interactions are also dependent on the geometry and electron density of the biomolecule and offer additional surface adhesion when interacted with the side groups present on the nanoparticle (Joshi et al., 2020; Hassan et al., 2014).

19.3.1.3 Covalent and dative chemistry

Where a biomolecule is coupled to a nanoparticle, it is either coupled to the nanoparticle surface itself or to the ligands present on the particle surface. The conjugation between a biomolecule and a nanomaterial surface is typically triggered by dative bonds, whereas, interactions with ligands is successfully carried out by covalent bonds. Dative bonds arise when two electrons in a bond come from one atom and have greater polarity, lower energies, and longer bond lengths than covalent bonds. A few well-known examples of dative bonding are the chelation of metal ions and Au–thiol chemisorption. Dative bonds as compared to covalent bonds are weaker and can be influenced by oxidation, pH changes and displacement by other similar molecules (Daniel and Astruc, 2004).

Nevertheless, increase number of interactions significantly strengthens these types of bonds. For example, multidentate thiols instead of monothiols strengthen interactions between biomolecule/ligand to Au and semiconductor nanomaterials (Susumu et al., 2007; Willey et al., 2004; Mattoussi et al., 2000). Biological probes attached to the nanomaterial surfaces simultaneously functions as ligands that facilitate improved nanomaterial solubility. Pinaud et al. directly conjugated polycysteine peptides derived from phytochelatin to ZnS capped QDs to uphold solubility (Pinaud et al., 2004). Cross-linkers between nanomaterials and biomolecules as molecular bridges are also required in covalent chemistries.

However, this chapter especially focuses on recent developments in the functionalization of nanoparticles with biological probes based on various chemical processes on the surfaces of nanoparticles along with their underlying chemistries. Biomolecules/biological probes are among the numerous molecules or ions that are produced by cells and living organisms. They have a wide array of dimensions and structures and a broad range of roles. Proteins, nucleic acids, carbohydrates and lipids are the four primary groups of biomolecules. Nucleic acids have the specific purpose of storing the genetic code of an organism—the nucleotide sequence that ascertains amino acid sequence of proteins. The order of amino acids in a protein plays an important role in ascertaining structure and function the protein. Proteins transport various nutrients within and out of the cells, and have a major role in enzymatic and catalytic chemical reactions in the body. Carbohydrates, made up of monosaccharides, disaccharides, oligosaccharides, and polysaccharides, are essential energy sources and structural components of all life and are the most abundantly found on Earth. Lipids, on the other hand, serve as a stored energy source and act as chemical messengers. These often create membranes that isolate cells from their surroundings and compartmentalize the interior of the cell into organelles in complex organisms.

19.3.2 Functionalization of nanoparticles

Functionalization of nanoparticles is essential considering their applications in the biomedical, technical, and electronics fields, as well as for commercialization of nanoproducts. Development of nanoparticles in biological field has given rise to many multifunctional functionalized nanoparticles. By simply modifying the surface by functionalization can bring about nanoparticles with various properties like hydrophobicity, hydrophilicity, corrosion resistance, conductivity, etc. (Kumar et al., 2018). In an attempt to reduce toxicity and enhance the activity of certain chemically synthesized nanoparticles like TiO_2, ZnO, Fe_2O_3, etc. are incorporated with organic/epoxy coating (Fernando, 2009; Domun et al., 2015). These coatings further resist the nanoparticles from adhering to metal substrates. These multifunctional nanoparticles give an enhanced efficacy in clinical studies like targeted drug delivery, cellular drug uptake and make them biocompatible by conjugating with chemical species like small ligands, polymers, or biomolecules (Mout et al., 2012). However, uniform dispersion of nanoparticles in various matrices and their interactions is indispensable to explore their potential in various areas like sensing, imaging, therapeutics, electronics, and other applications. But homogeneous distribution of nanoparticles by surface engineering with this diverse range of chemical species is indeed a tough task and hence, considerable attention and research efforts have been made to design various synthetic strategies for functionalizing them (Fig. 19.11) (Kumar et al., 2018).

19.3.2.1 Small ligands

Nanoparticles are incorporated with a wide range of ligands, which makes them suitable for various applications like sensing of cells (Bajaj et al., 2009, 2010) and biomolecules (Bajaj et al., 2010; Jiang et al., 2010) diagnostics (El-Boubbou et al. 2010), and intracellular delivery (Dreaden et al., 2011). Bajaj and his group in 2010 reported a study in which nanoparticles were functionalized electrostatically with green fluorescent protein (GFP) and the complex were utilized to classify healthy, carcinogenic and metastatic cells derived from cell lines (Bajaj et al., 2010). Here, the fluorescence of the protein was quenched on being functionalized to the surface of nanoparticle. Upon addition of cells, the higher affinity of nanoparticle toward the cell displaces the GFP from the nanoparticle surface and restores the fluorescence (Fig. 19.12) (Moyano et al., 2011). However, surface charge in concurrence with hydrophobicity and aromatic stacking of nanoparticles has an important role to play in these sensing applications. Likewise, surface charge of small molecule functionalized nanoparticles determines their interactions with the molecules at the cell surface (Nel et al., 2009). Moreover, those with positive charged ligands exhibit greater internalization in cells compared with neutral and negatively charged nanoparticles (Cho et al., 2009). Positively charged particles covalently complexed with drug molecules also show higher efficacy while delivering drugs into tumor tissues due to their enhanced uptake (Kim et al., 2010).

In clinical studies, therapeutic delivery comes across certain hurdles like penetrating the cells. Cell embedded nanomaterials are often entrapped in endosomes that prevents their exposure to cytosol and confines their utility. In order to

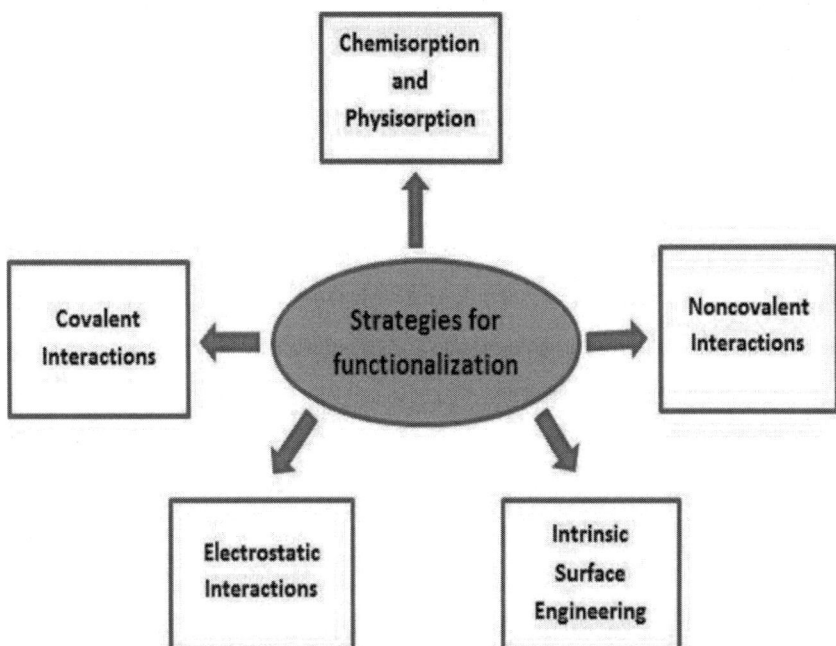

FIG. 19.11 Various synthetic strategies to functionalize nanoparticles.

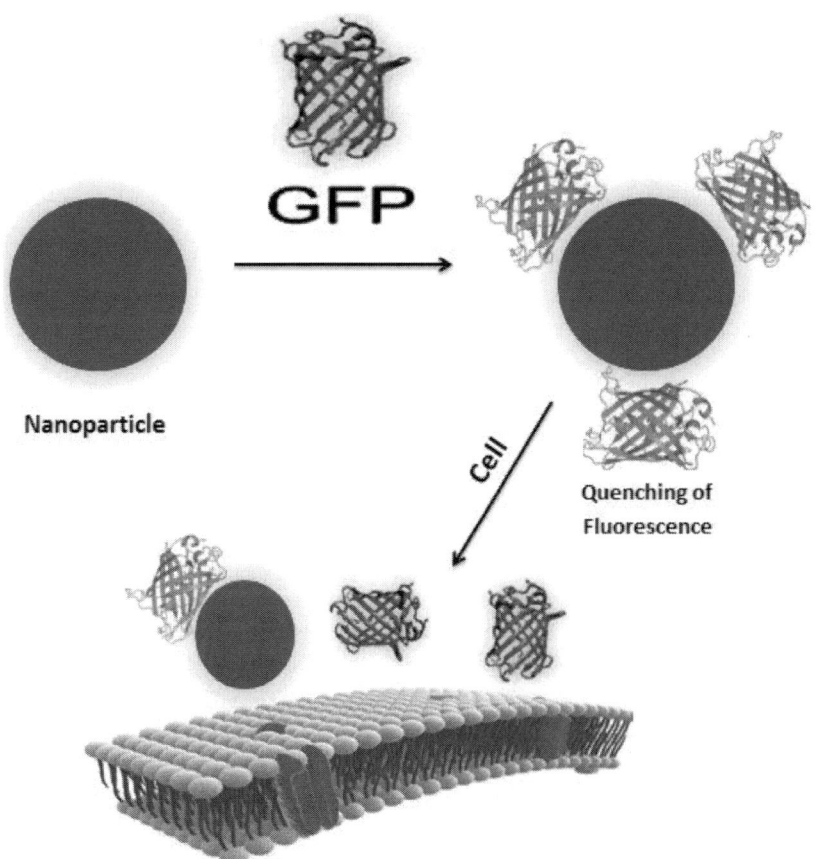

FIG. 19.12 Binding representation of GFP-nanoparticle and cell surface (Mout et al., 2012).

overcome this, nanoparticles are surface functionalized with signaling peptides as they are able to escape endosome and penetrate the nucleus (Kang et al., 2010).

Although majority of the nanoparticle-based therapeutic delivery system explores surface functionalization, an alternate mechanism is followed by the mesoporous silica nanoparticles (MSNs). Ligands molecules such as drugs are loaded into the pores of MSNs by efficiently capping them with the ligands which prevents leakage and leads to an efficient time dependent and targeted delivery system (Roy et al., 2005).

19.3.2.2 Polymers

Polymer encapsulation leads to the coating of amphiphilic polymers on the hydrophobic nanoparticles where the hydrophilic portion of the polymer is exposed to make them water soluble. Water solubility is imparted by the charged species within the hydrophilic portion such as carboxyl or amine groups. Several hydrophobic chains present on the polymer surface leads to multiple hydrophobic interactions with the nanoparticle making the encapsulation highly favorable (Zhang et al., 2011).

Polymers exhibit the characteristics of macromolecular systems to nanoparticles as an alternative to small molecular ligands. Polymer functionalized nanoparticles passively target cells through enhanced permeability and retention (EPR) effect (Petros and DeSimone, 2010) and also inhibits adsorption of blood serum protein (Peer et al., 2007; Niidome et al., 2006). Polymers like PEG coatings enhance the stability of nanoparticles in saline media as compared to unmodified particles. This approach is a versatile process which brings on the requirement of various secondary particles and their core properties. For example, a therapeutic nanosystem comprising of two different nanomaterials presented by Park and his team is observed as a potential therapeutic system (Park et al., 2010). PEG coated gold nanorod was predelivered targeting the tumor cells passively and worked as photothermal antennas irradiating NIR to heat the cells (Zhao and Karp, 2009). Heating the cells accelerated the requirement of the second material; either magnetic nanoworms or liposomes loaded with doxorubicin. Here, a 9-amino acid cyclic peptide was targeted to bind to the protein, p32, which upregulated on the surface of tumor cells due to heating. This therapeutic system comprising of two nanoparticles has shown higher efficacy in reducing the tumor volume as compared to individual nanoparticles.

Polymer functionalized nanomaterials also enables sustained release of drugs (Mout et al., 2012). For example, polymers that are coated on drug loaded nanoparticles change its confirmation and collapse as the temperature increases exposing the nanoparticles and so releasing the drugs. Then soon the temperature drops, the polymers retain its original confirmation and stops the drug release (Fig. 19.13).

19.3.2.3 Biological probes/biomolecules

In the recent time, biomolecule/biological probes functionalized nanomaterials have gained great attention in biomedical field. These nanoparticles have unique properties which are nearly impossible to achieve with synthetic materials such as, gene regulation, intracellular imaging, and higher efficacy in delivering biomacromolecules with minimal cytotoxicity (Rosi and Mirkin, 2005; Lytton-Jean et al., 2011; Massich et al., 2009). In an early study, Mirkin et al. reported a new class

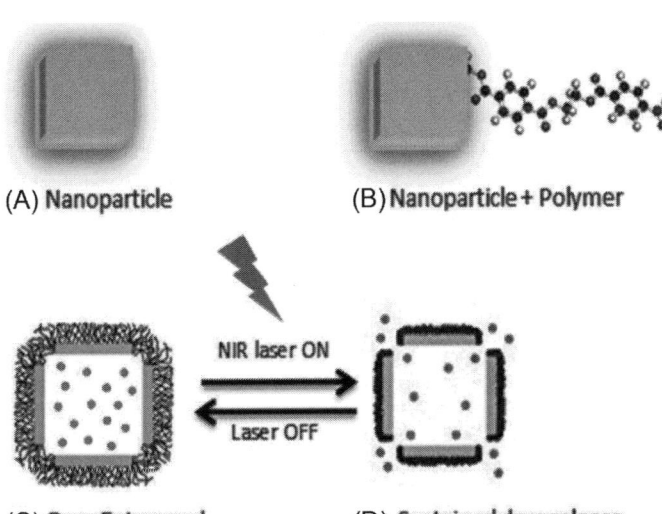

FIG. 19.13 Schematic representation of (A) nanoparticle (B) functionalized with polymer (C) carrying drug molecules. When exposed to NIR, temperature elevates and polymer collapse resulting into (D) release of drugs. On turning off the laser, polymer retains its original confirmation and stop releasing drugs (Mout et al., 2012).

of oligonucleotide-AuNP assembly conjugated through gold–thiol bond formation. This biomolecule functionalized nanoparticle has certain useful optical, electrical, and structural properties depending upon the size and chemical composition of nanoparticle along with length and sequence of the oligonucleotide (Mirkin et al., 1996). Later, in another study Rosi et al. reported gene silencing induced using DNA functionalized AuNP utilizing antisense mechanism (Rosi et al., 2006). siRNA functionalized AuNPs are also reported in several studies where siRNA was covalently linked to AuNP using an alkylthiol and ethylene glycol (Giljohann et al., 2009). This conjugate was found to be stable in serum which makes it suitable for in vivo gene regulation. Nucleotide functionalized nanoparticles also have the advantage of reduced innate immune system, because the enzymes involved to identify foreign nucleotides and activating the immune response are prevented by the particle's local surface environment (Massich et al., 2009).

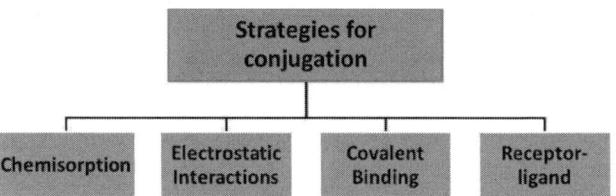

FIG. 19.14 Approaches for biomolecule–nanomaterial conjugation.

Another study by Seferos reports a novel oligonucleotide-AuNP assembly which was developed by combining with fluorophore-labeled complementary sequence. This assembly was used to visualize and quantify messenger-RNA (mRNA) in living cells as a "nanoflare" (Seferos et al., 2007). A nanoflare is a composite formed by coupling AuNP to nucleic acid that releases fluorescence in the presence of the target mRNA (Halo et al., 2014). These probes are nontoxic, highly resistant to enzymatic degradation and can penetrate cells without any additional transfecting reagents.

However, nanoparticles functionalized with biomolecules like protein, peptide, and antibody are able to bind to the receptor cell surface imparting targeted therapeutic delivery. Jiang et al., reported an antibody functionalized multivalent nanoparticle against ovarian and breast tumor cells. The functional affinity among the antibody coated nanoparticle and its receptor plays the role in targeting these particles into the specific tumor cells (Jiang et al., 2008).

Various target-specific biomolecules when functionalized to nanoparticles such as anti-EGFR (antiepidermal growth factor receptor) (El-Sayed et al., 2005; Link and El-Sayed, 2003; Larson et al., 2007; Huh et al., 2005; Hao et al., 2010; Morrison et al., 2018; Breaker and Joyce, 1994), Herceptin antibody (Jiang et al., 2008; Huh et al., 2005), etc. leads to specific detection of wide variety of cancers. Magnetic nanoparticles when coated with protein are also witnessed to be efficient contrast agents for imaging (Hao et al., 2010). The advancement of these ultrasensitive nanoprobes improves the capacity to visualize biological activities that are essential for theranostic purposes. However, conjugation strategies of biomolecules to the nanoparticles usually fall into four classes as depicted in Fig. 19.14.

19.3.2.3.1 DNAzyme and DNA molecule functionalized nanoparticles

DNAzymes, also known as deoxyribozymes, are single stranded DNA-based enzyme mimics which are catalytically active. These are not usually found in nature but are isolated by in vitro selection from random pool of DNA sequence (Morrison et al., 2018). Breaker and Joyce in 1994 had reported the first ever catalytically active DNA molecule developed using in vitro selection (Morrison et al., 2018; Breaker and Joyce, 1994). DNAzymes catalyze RNA-cleavage and other several other chemical reactions such as DNA phosphorylation, RNA, and DNA ligation (Flynn-Charlebois et al., 2003), DNA capping (Li et al., 2000), porphyrin metallation (Li and Sen, 1996), and nucleopeptides formation (Pradeepkumar et al., 2008). DNAzymes/DNA molecules, over the past decade have shown tremendous growth as molecular recognition elements for biosensing applications (Schlosser and Li, 2009; Tram et al., 2012). However, in most instances the reported DNAzyme-based biosensors utilized RNA-cleaving DNAzymes due to their efficient catalytic activity which eventually reduces the response time of these sensors (Schlosser and Li, 2009; Silverman, 2005; Schlosser and Li, 2010). Besides sensing, they also have attracted great attention in the area of therapeutics and diagnostics due to their higher stability and activity (Zhou et al., 2017). They are used to treat various diseases like viral infection, inflammation, cancer etc. Alongside, they have shown significant efficacy in detection of small molecules, nucleic acids or metal ions related to diagnosis of diseases. Over past few decades, it has been clearly known that there are certain benefits of using enzyme mimics instead of natural enzymes: (a) foreign protein-based enzymes generate immunological response in the host body, (b) protein-based enzymes are expensive and prone to contamination, and (c) labeling natural enzymes at specific sites is a difficult task (Zhou et al., 2017). Moreover, DNAzymes are easy to be modified as per requirements and labeled with lesser immunogenicity. Hence, they are conjugated to different nanoparticles for various applications such as signaling and delivery. In theranostics field, nanoparticles have been playing a major role and DNAzymes/DNA molecules can be attached to them by covalent bonding or physisorption. DNA-nanoparticle complexes easily penetrate the skin barrier and internalized by cells (Rosi et al., 2006; Giljohann et al., 2007). Nanomaterials exploited to functionalize and deliver DNAzymes are mostly AuNPs (Wu et al., 2013; Rouge et al., 2015; Yehl et al., 2012), MnO_2 nanosheets, and graphene oxide (GO).

Although therapeutic applications were attempted initially, its analytical applications have been studied more. These molecules are able to detect certain analytes, especially metal ions due to their high metal selectivity (Zhou et al., 2016). Biosensors are designed based on different mechanism like colorimetric, fluorescent, electrochemical, SERS, surface plasmon resonance (SPR), etc. Although metal ions are indispensable elements to human life, but their concentration variations may lead to physiological disorder such as increase in level of Na^+ in blood indicates high blood pressure. Thus, metal detection sensors are always one of the most explored areas to monitor metal concentration in the body. Among all the sensors explored (Grynkiewicz et al., 1985), DNA-based sensors show most significant efficacy in detecting heavy metal ions due to its high stability in blood and serum (Zhou et al., 2015, 2016). First ever DNA-based metal detecting sensor was given by Li and Lu in 2000 for rapid, sensitive and selective detection of Pb^+ (Li and Lu, 2000) followed by a large number of works on DNA-based biosensing system (Gong et al., 2015; Zhang et al., 2011). Another fluorescent DNAzyme-based sensor was designed by Lu and co-workers where DNAzyme was functionalized onto AuNPs by thiol linkage and used as a selective probe for UO_2^{2+} (Wu et al., 2013). This probe was the first intracellular metal ion sensor which could easily enter the cells and can detect metal ions within cellular environment. Fluorescence in general, is monitored as a function of time; increase in metal concentration increases the intensity of florescence, thus quantifying the metal ions (Zhou et al., 2016).

19.3.2.3.2 Protein and antibody functionalized nanoparticles

The functionalization of nanoparticles with amino acids is another approach of enhancing the precision and efficacy of nanoparticles-based delivery systems. Proteins comprising of amino acids have primary amine and carboxylic group at its terminals along with side chains attaching additional functional groups (Ravindran et al., 2013; Di Marco et al., 2010). Enzymes and antibodies are the classes of proteins utilized to functionalize on nanomaterials for molecular recognition. Antibodies (Ab) or immunoglobulins are Y-shaped protective proteins produced by the immune system in response to any foreign antigen in the host body. Functionalization of proteins on nanoparticles has wide applications in biomedical field such as imaging, therapeutics, sensing, catalysis, etc. Several conjugation approaches have been reported as depicted in Fig. 19.14. When charged proteins are incubated with the oppositely charged nanoparticles, they are adsorbed to the particle surface via electrostatic interactions (electric forces attracting one another) and held together by thiol bonds, hydrogen bridges, van der Waal forces, or by hydrophobic interactions. A study was reported by Sastry et al. where water dispersible particles were developed by functionalizing amino acids on the nanoparticles by hydrogen bonding between carboxylic and amino functional groups (Mandal et al., 2001; Selvakannan et al., 2003). Another noncovalent approach of steric stabilization has been utilized for functionalization of polymers or surfactant on the nanoparticle surfaces. Shenton and co-worker have reported a study where colloidal nanoparticles that were synthesized by citrate method are functionalized with Ab at pH slightly above the pI (isoelectric point) of the citrate ligand which actively binds the positively charged amino to the oppositely charged colloidal nanoparticles (Shenton et al., 1999). However, nanoparticles functionalized with biomolecules like Ab or serum albumins having amines, carboxylic acid or thiol groups on the surface are covalently bonded via disulfide, amide or ester bonds (Grynkiewixz et al., 1985).

Surface modifications of nanoparticles with certain biomolecules have enhanced its biocompatibility and intracellular uptake for drug delivery. Metallic nanoparticles have been witnessed to be very useful for imaging and targeted therapies due to its plasmonic nature (Kumar et al., 2007). While treating oncogenic cells, photodynamic properties of nanoparticles play a major role. Hussain et al. reported in a study that AgNPs functionalized with biomolecules easily bind to living cells which allows monitoring the morphological features of normal cells (Hussai et al., 2005). Surface modification of nanoparticles with biomolecules also prevents direct contact between the particle surface and cells reducing its toxic effects on the host body. Several other studies have also reported that functionalization of biomolecules on nanomaterials makes them more stable for applications like imaging probes or deposition on thin films (Balogh et al., 2001).

19.3.2.3.3 Glucosamine functionalized nanoparticles

Glucosamine is one of the most abundant monosaccharides bearing amino group acting as a preferred substrate for biosynthesis of glycosylated proteins and lipids (Veerapandian et al., 2010; Kim et al., 2009; Reginster et al., 2007). It shows potential biological applications such as a cure for osteoarthritis by inhibiting the gene expression of osteoarthritis cartilages in vitro (Reginster et al., 2007) and promotes oncogenic cell deaths by inhibiting transglutaminase 2 (TGase 2) (Kim et al., 2009). This leads to activation of kappa-light-chain-enhancer (NF-αB) in breast cancer cells that are resistant to drugs.

Glucosamine functionalized on nanoparticles are called as glyconanoparticles. Several types of glyconanoparticles such as metallic glyconanoparticles (de Paz et al., 2005; de la Fuente et al., 2001; Barrientos et al., 2003), magnetic glyconanoparticles, and glycol-QDs (Jesus and Penadés, 2005; Penadés et al., 2004) have been reported. Various uses of these glyconanoparticles take in probes for carbohydrate-protein interactions, antiadhesive therapies, biolabels, bioamplification approaches, etc. (Jesus and Penadés, 2006). Besides, Veerapandian and co-workers in their studies have reported copper glyconanoparticles and silver glyconanoparticles which outstanding antibacterial activities (Veerapandian et al., 2010, 2012). AgNP functionalized with glucosamine show distinct antibacterial activity against 8 gram-positive and gram-negative bacteria in a concentration dependent manner. Glucosamine functionalization eases distribution and penetration of the particles into bacterial cell surface leading to enhancement of antibacterial activity (Veerapandian et al., 2010). However, copper glyconanoparticles and their enhanced antibacterial activity were studied by exposing them to UV germicidal irradiation and minimum inhibitory concentration (MIC) against the bacteria. These particles are highly stable, water-dispersible and induce elevation in production of ROS (reactive oxygen species) (Veerapandian et al., 2012). Glyconanoparticles are further exploited as biomedical sensors because of their remarkable properties such as biocompatibility, hydrophilicity and stability.

19.3.2.3.4 Peptide/peptidomimetics functionalized nanoparticles

Peptides as a novel material have gained considerable interest in biomedical fields which exhibits protein functionality and have a high degree of molecular design modularity. Current approaches for designing synthetic bioactive peptides can be widely divided into two categories: (a) The design and screening of peptide libraries in a macromolecular topology from random amino acid compositions (bottom-up approach) and (b) bioactive sequences extraction from natural proteins (top-down approach) (Fig. 19.15) (Jeong et al., 2018; Vanhee et al., 2011; Wang and Yu, 2018; Marasco et al., 2008; Ryvkin et al., 2018). In the last few decades, a substantial number of studies have shown the efficiency of such artificial bioactive peptides.

Several studies have reported that functionalizing peptides with nonbiological materials such as, metal chelates, small molecular compounds, hydrogels, polymers, etc. is likely to overcome the underlying limitations of peptides (Shu et al., 2013; Xiao et al., 2016). Nanoparticles conjugation demonstrated their ability to enhance peptide functionality as well as to incorporate biogenic characteristics that often lead to synergistic effects in a variety of biomedical applications such as diagnosis, imaging, and therapy. Functionalized nanomaterials due to their high surface area to volume ratio and ultrasmall size can interact with variety of biomaterials at ease (Rizvi and Saleh et al., 2018). The simplest approach of engineering peptide-based nanomaterials is self-assembly (Habibi et al., 2016; Jeong et al., 2018). Peptide–nanoparticle conjugation enhances the structural properties of nanostructures, enabling simple and efficient modification of the overall performance, shape, size, and thickness of the conjugates for intended applications.

Peptide functionalized nanoparticles also offer multivalency which enhances their binding kinetics (Fig. 19.16A) (Kitov and Bundle et al, 2003; Gargano et al., 2001; Hong et al., 2007). Most biological system interactions are noncovalent interactions such as, van der Waals forces, ionic bonds, hydrogen bonds, π–π stacking, and hydrophobic interactions. Besides binding strength, these interactions also enhance the selectivity to recognize the target surfaces (Fig. 19.16B) (Martinez-Varacoechea and Frenkel, 2011). Multivalency plays an important role in enabling strong multivalent bindings

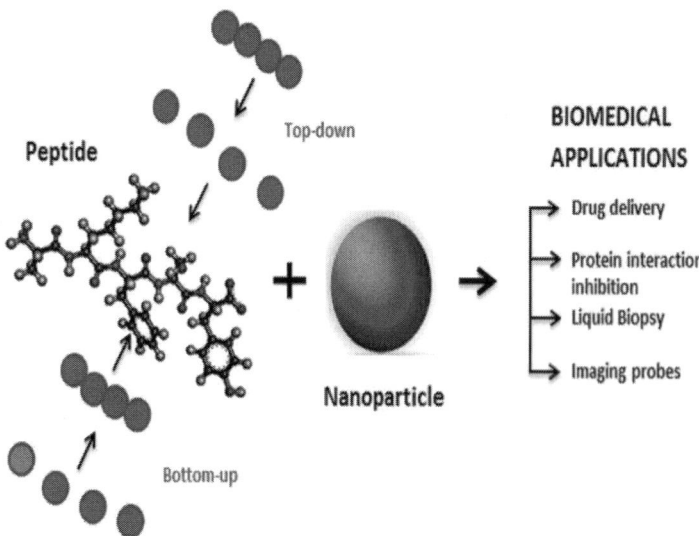

FIG. 19.15 Design of synthetic bioactive peptides and peptide–nanoparticle conjugation for biomedical applications (Jeong et al., 2018).

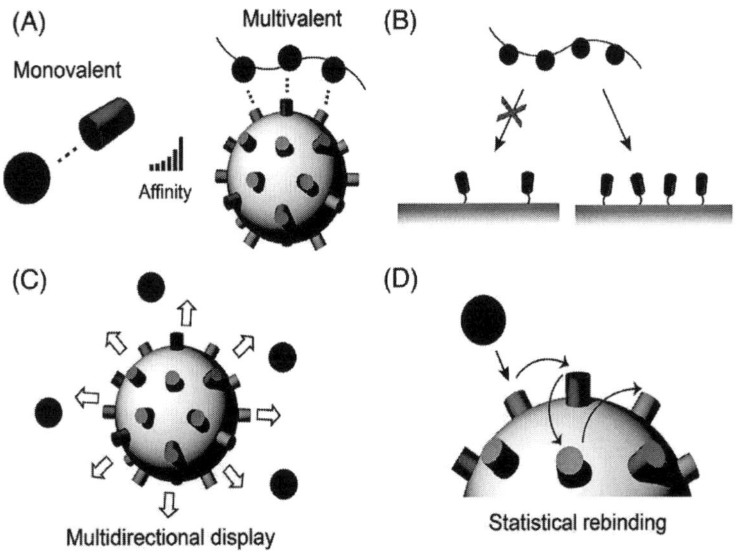

FIG. 19.16 (A) Monovalent vs. multivalent interactions. (B) Multivalent interactions selectivity. (C) Multidirectional ligand display. (D) Statistical rebinding on a multivalent object. *Reprinted with permission from Jeong et al. (2018). Copyright Jeong et al. (2018). Creative Commons Attribution 4.0 International License (http://creativecommons.org/licenses/by/4.0/).*

to multiple targets exposing the peptide in multiple directions and increasing statistical prospects for various monovalent binding events to happen (Fig. 19.16C). Postbinding event, the conjugates express several re-binding sites that may upsurge the target's retention time on the surface, known as the statistical re-binding phenomenon (Fig. 19.16D) (Vauquelin and Charlton, 2010). Moreover, conjugation with various types of peptides or other biological materials provides additional versatility for these synthetic materials such as (Osman et al. 2018), theranostics (Fang and Zhang, 2010), stimulus-responsive property (Borglin et al., 2017), and multitarget, single-material treatment. Recently, peptides have been identified as a strong strategy that can provide versatile selectivity for drug delivery systems, ensuring improved efficiency in the treatment of numerous serious health problems like cancer and neurological disorders (Modi et al., 2014; Jiang et al., 2016). Several studies have been reported conjugating peptides with nanoparticles with enhanced efficacy in therapeutic delivery system (Jeong et al., 2018).

The peptide–nanoparticle conjugates have presented several positive examples of potential targeting of unhealthy cells and permeation through physiological barriers like stratum corneum, external cutaneous layers, blood brain barrier around the blood vessels with access to brain etc. (Niu et al., 2017; Patlolla et al., 2010; Zou et al., 2018; Poellmann et al., 2018; Georgieva et al., 2012; Yao et al., 2015). However, for the ultimate application of this strategy, there are many obstacles that are needed to be addressed, such as long-term toxicity, immunogenicity, and potential for targeting. To overcome these underlying drawbacks of peptides, the approach to conjugate them with nanoparticles has provided a strong platform for potential drug delivery with a high therapeutic index (Jeong et al., 2018)

19.4 Basic chemistries behind conjugation approaches

19.4.1 Functional groups and conjugation reactions

The nanomaterials and the functional groups present on biomolecules such as proteins, DNA, carbohydrates, lipids, etc. that are suitable for conjugations are limited in their diversity. The amino acid residues that are targeted for conjugation in case of proteins and peptides are the N-terminal primary amine, the ε-amino group on lysine side chains, the guanidinium group on arginine side chains, the thiol/sulfanyl group of cysteine, the aspartic acid, glutamic acid, or C-terminus carboxyl groups, and tyrosine phenol (Sapsford et al., 2013). Whereas, hydroxyl groups or aldehydes are the alternate reaction sites for carbohydrates and glycosylated proteins. Commonly preferred chemistries targeting these groups include N-hydroxysuccinimidyl (NHS) amine ester modification along with ethyl(dimethylaminopropyl) carbodiimide-mediated (EDC) condensation of carboxyls with amines, maleimide conjugation to thiols, and diazonium modification of phenolic-side on tyrosine.

Recombinant modification of proteins displays distinctive cysteine residues (i.e., thiol handles) for site specific labeling and conjugation which is very common nowadays (De Lorimier et al., 2002; Medintz et al., 2005). N-termini of proteins can be modified precisely, using the N-terminal transamination chemistry, while polyhistidine (His_n) motif in the protein can serve as site for region-specific NTA modified substrate interactions (Sapsford et al., 2013; Peneva et al., 2008). Peptides and nucleic acid biomolecules show much more flexibility as any functional group can be added to them in the course of initial synthesis or preceding modification. The genomic advancement resulted in the expansion of a range of chemistries for the synthesis and modification of biomolecules such as site-specific functional groups incorporated into their structures, including azides, amines, biotin, thiols, carboxyls, alkynes, and a multitude of organic fluorescent dyes (Wilhelmsson, 2010; Cobb, 2007). The wide availability of thousands of chemically modified precursors provides even more flexibility to synthetic peptides for site-specific incorporation of functional groups (Lu et al., 2010). Further, addition of specific groups like alkynes or azides lets certain functionalities to be integrated in a bioorthogonal manner, that is, without any other alterations to rest of the peptide.

19.4.2 Polyhistidine–nitrilotriacetic acid chelation

The His_n sequences are typically converted into proteins as an affinity tag for their respective purification, by means of immobilized metal-ion affinity chromatography. His_n binds to Co(II), Cr(II), Cu(II), Ni(II), Zn(II), and other divalent metal cations chelated by NTA or functionally similar structural groups at four of the six sites around the available ion (Fig. 19.17) (Sapsford et al., 2013).

The significance for using Hisn Tags and NTA in order to display biomolecular nanomaterials lies in the chromatographic affinity of immobilized metal and surface modification chemistry (Samanta and Sarkar, 2011) and is guided by a wide range of inter-related factors, including: (1) The effectiveness of His_n–NTA interactions is characterized by dissociation constants, $K_d = 10^{-13}$ M. (2) Histidine motifs can be easily synthesized into nascent peptides or genetically engineered into expressed proteins; however, a number of commercial kits are now available and commonly used for these (Sapsford et al., 2013; Durland and Eastman, 1998). (3) There are certain reactive NTA precursors and analogs that can be linked directly to other chemical handles. (4) The His_n tags allow exposure to that which can essentially be regarded as single attachment point. Two-adjacent residues of histidine most often interact with each individual metal–NTA complex. However, the His_n series does not extend a great distance. It helps to prevent undesired cross - linkage and strengthens regulation of biomolecular orientation (Algar et al., 2011). (5) The system has applications in various other areas beyond proteins and peptides. A number of biomolecules have been chemically developed to produce a His_n tag for both purification and immobilization purposes. These typically contain nucleic acids and other materials like PEG (Mailander and Landfester, 2009; Putnam et al., 2003). Ni(II)–NTA modified coatings is the most frequently explored strategy to assist His_n mediated biomolecular assembly to nanoparticles (Algar et al., 2011; Li et al., 2007; Graff et al., 2008). It is mainly accomplished by chemical reaction of the nanomaterial coatings into an NTA nucleophilic derivative. NTA and other structurally analogous chelating agents are subjected to the potential of various closely articulated carboxyl groups to coordinate the metal ion concurrently. In order to have a nanomaterial biomolecular display, use of His_n–Ni(II)–NTA interactions has gained considerable attention and are growing exponentially. Excess amount of NTA can be easily separated from nanoparticles making it less susceptible to unnecessary cross-linking, and more favorable to control biomolecular coordination compared to the other common approaches (Sapsford et al., 2013).

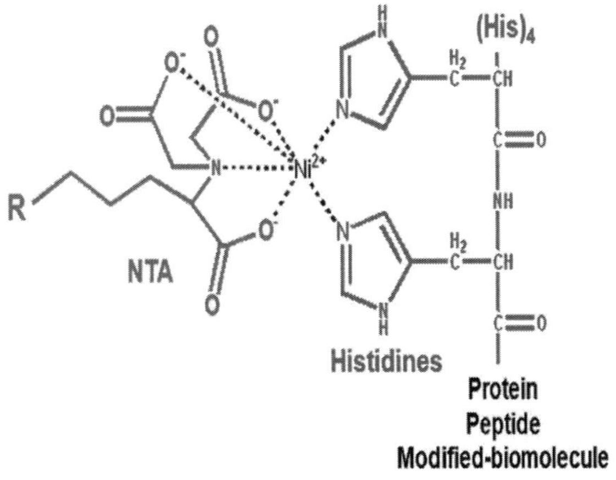

FIG. 19.17 Metal affinity coordination, that is, Ni(II)-chelated nitrilotriacetic acid (NTA)–polyhistidine residue coordination. *Reprinted with permission from Sapsford et al. (2013). Copyright 2013 American Chemical Society.*

19.4.3 Biotin–avidin chemistry

After a new nanomaterial type is synthesized, this chemistry appears to be the first choice for bioconjugation prototyping. Discovery of avidin by E.E. Snell in the 1940s along with the structural identification of biotin by du Vigneaud et al. placed the foundations for developing the biotin–avidin chemistry as a potent tool used in various biomedical and biotechnological applications (Lesch et al., 2010). The interaction between biotin and avidin is one of the best-known noncovalent interactions with an association constant (Ka) of 10^{15} M^{-1} (Sapsford et al., 2013). Besides, it is also a commercially available reagent at a wide range which has resulted in its regular application in nanomaterial research. Biotin–avidin chemistry incorporated studies for conjugation of various nanoparticles are not remarkably monitored and an overview this chemistry is depicted in Fig. 19.18.

Avidin, a glycoprotein, was initially found in egg whites but now identified in many tissues having fairly accurate molecular weight of 67 kDa and consists of a homotetrameric subunit structure and every subunit is proficient enough of binding to a single biotin (Fig. 19.2). It has high stability under a range of extreme conditions, such as change in temperature, denaturants and pH and has a high carbohydrate content of approximately 10 pI (Wilchek et al., 2006). Whereas, biotin, also known as vitamin B7 or vitamin H is fairly a simple organic water-soluble compound that functions as an essential coenzyme for numerous carboxylase enzymes in various different species, including humans. Such enzymes regulate amino acids, fatty acids, and glucose metabolism and also contribute to transcriptional gene expression and stability (Bao et al., 2010; Zempleni et al., 2009).

Avidin–biotin interaction has become a standard system for biorecognition applications due to its high affinity. Moreover, many researchers working on nanomaterials regularly use this system to illustrate proof of concept at the initial phase (Wilchek et al., 2006; Zempleni et al., 2009; Li and Kohli, 2010). Roll et al. reported an aggregation assay using bovine serum albumin (BSA)–biotin-modified AuNPs by observing the spectral shift of UV–visible absorbance of biotin-labeled gold colloids when avidin was added (Roll et al., 2003). Oh et al. reported streptavidin-conjugated quantum dots and biotin-conjugated AuNPs which were used to determine avidin by exploring an inhibition study established on energy transfer quenching signal transduction (Oh et al., 2005). This affinity of the complex is a common way of attaching biomolecules, such as antibodies or DNA, to the surface of a nanomaterial. Bioconjugation using biotin–avidin chemistry is based on two principal methods. First, it involves combination of avidin functional nanoparticle with a biotinylated biomolecule.

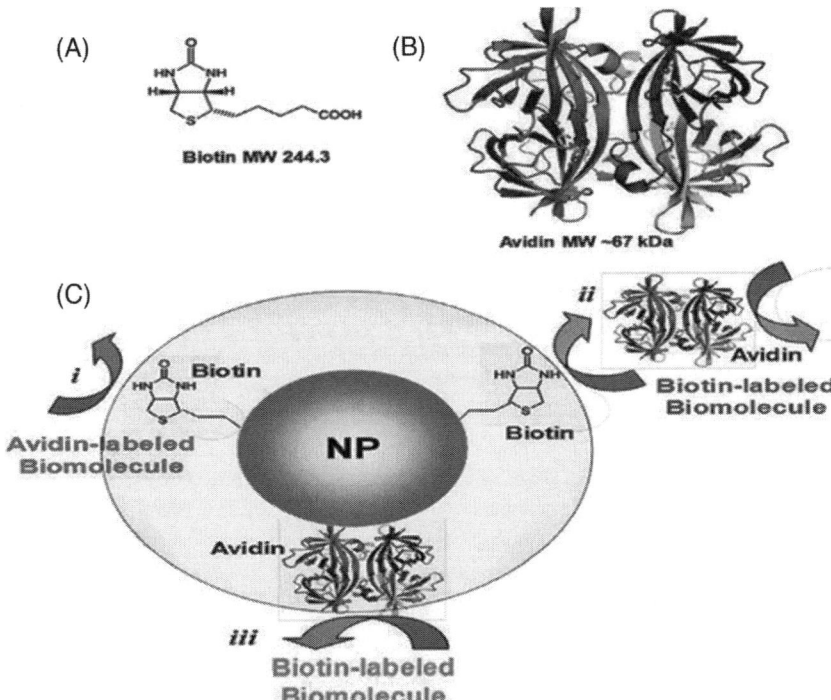

FIG. 19.18 Biotin–avidin chemistry. (A) Chemical structure of biotin (B) a ribbon model of tetrameric avidin, showing the monomers in magenta, blue, cyan, and red and the four biotin molecules in yellow. (C) Three commonly used biotin–avidin approaches for biomolecule–nanoparticle conjugation. *Reprinted with permission from Sapsford et al. (2013). Copyright 2013 American Chemical Society.*

However, avidin attachment to the nanoparticles during the preliminary modification is likely to distort one or more possible binding sites on biotin. This distortion results into overall heterogeneity and the final avidin-functionalized nanoparticles are likely to show a number of diverse orientations and an acceptable range of binding sites. It has also been noted that this particular issue is liable for the heterogeneous orientation of biotinylated DNAs incorporated with streptavidin-functional quantum dots (Boeneman et al., 2010). Second, it is based on polymer functionalized nanoparticles, where, biotin labeling is favored over avidin due to its smaller size. Its smaller size has attracted the researchers due to its minimal interference in the process of self-assembly. Usually the biomolecule is often biotinylated, and the two are interlinked with avidin intermediary. However, the use of streptavidin to attach the 240 kDa β-phycoerythrin light harvesting complex to biotinylated quantum dots is cited as one of the examples of such architecture (Susumu et al., 2007; Medintz et al., 2008). Overall, profound knowledge of avidin–biotin chemistry, its broad utility and efficacy, and on-going evolution for nanoparticle bioconjugation ensure its continued use in the near future.

19.5 Applications

19.5.1 Detection of DNA, protein, and metal ions

Nanoparticles play a major role in creating possibilities and opportunities to develop new generations of sensing tools. Due to their higher sensitivity and selectivity toward biological environment, nanoparticles are widely explored for sensing biological molecules and cells. Different formulations of nanoparticles have been designed to detect biological moieties such as cells, proteins, nucleic acids viz. DNA, RNA, metal ions, small organic compounds, and many more biological specimens as they have major applications in the area of biomedical science, food and drug industries. Several studies have reported schemes involving chemically functionalized AuNP (Zeng et al., 2011; Aldaye and Sleiman, 2006; Huang et al., 2009; Jayasena, 1999; Mei et al., 2003; Gearheart et al., 2001). These methods are simple and use easy surface functionalization strategies that do not typically need expensive instruments.

Many *DNA sensing systems* incorporated with AuNPs reported an increase in sensitivity and limit of detection. Li et al. reported that both single-stranded and double-stranded oligonucleotides on AuNPs show different adsorption tendency (Li and Rothberg, 2004). The single-stranded DNA adsorption stabilized the AuNPs preventing aggregation and the method can be used to detect sequence specific nontagged oligonucleotides. Another study on a fluorescent-based sensing system for hybridization of DNA based on electrostatic properties of DNA was reported by Li et al. (Li and Rothberg, 2004). The sequences tagged with dyed-probes quench their fluorescence when combined with AuNPs which soon regains its fluorescence when exposed to the target. Liu et al. reported an aptazyme functionalized AuNP-based colorimetric adenosine biosensor (Liu and Lu, 2004, 2006). Aptazyme is an adenosine aptamer DNAzyme which can amplify DNAzyme activity via allosteric interactions with adenosine. Adenosine activates the aptazyme that cleaves the substrate strand and impedes the nanoparticle aggregation. Wark et al. attempted to detect unmodified DNA by developing a sensor based on the adsorption of AuNPs to the gold diffraction gratings (Wark et al., 2007). Elghanian et al. reported a colorimetric detection system to detect polynucleotides using AuNPs modified with mercaptoalkyloligonucleotide (Elghanian et al., 1997). This approach focuses on the change in interparticle distances because of the hybridization between the AuNP–mercaptoalkyloligonucleotide complex and the target sequence resulting in a red to-pink/violet color. Huang et al. developed an aptamer modified AuNP sensor unique to PDGFs (platelet-derived growth factors) detection while following changes in color (red to purple) and extinction of bioconjugates resulting from aggregation (Huang et al., 2005).

AuNPs are the most explored particles for detection of target DNA which has led to exclusion of other noble nanoparticles. However, Thompson et al. reported an oligonucleotide-AgNP complex in a format of sandwich assay which offers increase in sensitivity for DNA detection (Thompson et al., 2008). Oligonucleotide probes were attached to AgNPs by using an alkyl thiol group at either the 3′ or 5′ terminus. In this sandwich assay while hybridization of oligonucleotide–AgNP complex was performed, the target DNA sequence was added to a complex solution. Once the complex was hybridized to the target, the nanoparticles in the complex solution aggregated accompanied by a change in surface plasmon of nanoparticles which was observed either visually or by UV–vis spectroscopy. This approach enables detection of a specific DNA sequence at a considerably lower concentration due to their larger extinction coefficient. Another strategy for multiplexed DNA detection was reported by Han et al. (Agasti et al., 2010; Han et al., 2001). Here, semiconductor QDs were incorporated with oligonucleotide functionalized small polymer beads at different ratios which showed different emissions at different wavelengths. Once bound to target DNA, the presence and identity of the target could be read at single-bead level. The core concept was to build advanced microstructures having incorporated codes for rapid identification of the target along with molecular recognition capabilities. Encoded-bead technology is anticipated to be a versatile approach in target selecting target, rapid binding kinetics, and inexpensive to manufacture.

Proteins have corresponding complementary sections identical to strands of oligonucleotides. Biomolecules targeting protein can be attached to the AuNP surface along with other sensing agents for their precise detection. Thanh et al. reported an AuNP-based immunoassay which was intended for the detection of antibodies (Polavarapu et al., 2011). In this study, nanoparticles were modified with protein antigens and this conjugation resulted in changed absorption. This change in absorption was monitored using absorption plate reader. Xiang et al. reported an efficient composite film based on AuNP/ionic liquid/MWNTs modified glass carbon (GC) electrode which was developed using a self-assembly layer-by-layer technique (Thanh and Rosenweig, 2002). Cytochrome C was immobilized with the electrode that was significantly adsorbed by it resulting into electron transfer between cytochrome C and the electrode. This novel sensor further exhibited electrocatalytic activity leading to reduction of H_2O_2. Besides, QDs have also been proven to have efficiency in detecting proteins. A biotin-PEG/polyamine Cadmium sulfide (CdS) QD has been reported for protein detection by Nagasaki et al. (2004). Interaction between this QD and streptavidin leads to an efficient fluorescent energy transfer based on the principle of Förster resonance energy transfer (FRET) which is proportional to the dye-labeled protein concentration. And it can be considered highly sensitive for the target detection. Another luminescent QD bioconjugates were reported by Medintz et al. which detected the proteolytic activity of various enzymes (Medintz et al., 2006). Here, dye-labeled modular peptides having multiple functions were designed having a substrate sequence. The peptides were then self-assembled on a QD surface coated with dihydrolipoic acid in order to have efficient FRET. The substrate strand was cleaved by the proteases present in the system which resulted into alteration in the FRET signature.

Furthermore, real time and on-site sensors to monitor *toxic metal ions* like Pb^{2+}, Cd^{2+}, Hg^{2+}, Se^{2-}, etc. has been a key research interest in academics and health care throughout. These metal ion sensors have wide range of applications like environmental and biological studies, clinical toxicology, and industry waste monitoring. Several approaches in order to widen the applications and range of these metal sensors have been reported until now (Darbha et al., 2008; Huang et al., 2010; Liu and Lu, 2003, 2004; Slocik et al., 2008). DNAzyme-based sensors are the most commonly known sensors for colorimetric detection of metal ions. Strong optical properties of metallic nanoparticles and high selectivity of DNAzymes toward metals make them ideal for the purpose (Liu and Lu, 2003, 2004; Wang et al., 2008). Whereas, Chai et al. reported a glutathione modified AuNP-based colorimetric sensor for Pb^{2+} ions. This sensor detected the colorimetric response of AuNP which aggregated on being exposed to Pb^{2+} (Chai et al., 2010). Another study was reported by Liu et al. where a colorimetric sensor based on DNA–AuNP was designed against Hg^{2+} ions. The basic theory of this sensor has been drawn from the aggregation of AuNP formed by DNA–Hg^{2+}. As DNA–Hg^{2+} complex forms, the ss-DNA conformation changes into folded structures which results into decrease in zeta potential of AuNPs leading to poor electrostatic repulsion between AuNPs. This forms the AuNP aggregation and leads the solution to change it color from red to purple (Liu et al., 2008). Lee et al. have also reported a sensor intended for the detection of Hg^{2+} ions. This is a chip-based colorimetric approach which exploits cooperative binding and catalytic activity of DNA-AuNP complexes with thymine-thymine mismatches in the presence of Hg^{2+} through thymine-Hg^{2+}- thymine complex formation. This approach is suitable for the waste samples management due to its high performance, easy readability and portability (Lee and Mirkin, 2008). However, QDs in FRET process have remarkable significance as donors which has been explored for sensing metal ions and small molecules. Goldman et al. reported a detection system against the explosive 2,4,6-trinitotoluene (TNT) (Goldman et al., 2005). A single chain antibody fragment specific to TNT was functionalized on the QD. A TNT analog comprised of a quenching dye was already pre - fabricated with the binding site of antibody which significantly quenches QD emission. When TNT was introduced in the system, the quencher was displaced which interrupted the energy transfer from the QD to the quencher and the QD emission was then restored.

19.5.2 Detection of human pathogens

When a foreign bacterium enters the bloodstream of a host body, a life-threatening condition named bacteremia occurs. Rapid identification of causative infectious agents is the key requirement when it comes to adequate and effective treatments for these infections. Molecular diagnosis, based on nucleic acid sequences and proteins specific to organisms, has made rapid progress in both the sensitivity and precision of diagnostic procedures. Labeled probes or polymerase chain reaction (PCR) amplification is used in these methods to make the detection more effective (Jamdagni et al., 2017).

The probe functionalized nanoparticles complexes are used to identify complementary DNA sequences and which allows results to be visually detected (Liandris et al., 2009; Bakthavathsalam et al., 2012; Andreaou et al., 2014). Nanoparticles in this approach are hybridized with the genomic DNA sequences that are isolated from the test samples. When exposed to analytical conditions, hybridized and un-hybridized nanoparticles behave differently. Under analytical conditions the oligonucleotide probe hybridized to its complementary sequences has a stabilizing effect and prevents the nanoparticles from aggregating. Whereas, nanoparticles with un-hybridized probes accumulates rapidly and changes the suspension color from red to purple. As a result of this aggregation, a shift in absorption spectrum to longer wavelength is also observed. This strategy leads to sensitive and effective sensing of microorganisms like *Staphylococcus aureus, Mycobacterium tuberculosis, Mycobacterium avium subspecies paratuberculosis, Escherichia coli, Salmonella enteritidis, Salmonella typhimurium, Leishmania* species, and *Plasmodium falciparum* (Jamdagni et al., 2016).

However, several studies have reported using nanoparticle-antibody-based immunosensors (Huang, 2007; Wang et al., 2010a; Verdoodt et al., 2017) for diagnostic applications. As compared to nucleotide-based probe sensor, antibody-based systems are simpler as they do not require genomic DNA sequence isolation from the sample. They are directed toward the whole cell sample. In a study, Huang reported an immunochromatographic assay using AuNPs against *Staphylococcus aureus* (Huang, 2007). The assay was designed using IgG antibodies specific to protein surface of *Staphylococcus aureus*-based on sandwich assay principle. One of the antibodies was conjugated with AuNP whereas, another was immobilized on a porous nitrocellulose membrane. The sample flows through capillary action across the porous membrane and binds to the particles to which protein was already attached. It leads to a red color formation on the strip which enables visualization of the results with 100% sensitivity and 96–100% specificity (Huang, 2007). Wang et al. have outlined an aggregation mediated detection of *Salmonella typhimurium* based on a similar theory that was followed for nanoparticle-probe conjugate-based detection. Oval-shaped AuNPs were mixed with anti-*Salmonella* antibody *typhimurium* and in the presence of target pathogen a prominent change in color was observed from red to bluish purple which was further confirmed by shift in absorption maxima in UV–visible spectrum. This system also carried out photothermal lysis of the targeted bacteria when exposed to NIR (Wang et al., 2010). In addition to antibodies, a cysteine capped AuNP was reported by Raj et al. for visual detection of *Escherichia coli* in urinary tract infection (UTI) patients' clinical samples (Raj et al., 2015). In this method, AuNPs are functionalized with cysteine and this conjugate binds to *Escherichia coli* by electrostatic attraction between positive charge of cysteine and negative charge of *Escherichia coli*. The basic principle behind the detection is the red shift of absorption spectra of cysteine-AuNP change to blue shift with varied concentrations of *Escherichia coli* followed by the color change of the solution from red to blue.

19.5.3 Enhancement of antibacterial and anti-inflammatory activity

Bacterial infections are one of the major sources of chronic infections and mortality. Antibiotics have been the most ideally preferred treatment for these infections throughout whereas, several studies have reported that antibiotics have led to rise of multidrug resistant strains carrying several resistant genes such as NDM-1 (Wang et al., 2017; Hsueh, 2010), *erm* gene, *aadA1* (Nikaido, 2009), *sul1*, *sul2*, *sul3*, *tet* genes (Amador et al., 2019), etc. resistant to carbapenems, macrolides, aminoglycoside, sulfonamide, tetracycline, respectively. Thus, nanoparticles with antimicrobial activity are being widely used as an alternative to antibiotics for targeting drug resistant bacteria. The key processes beneath the antibacterial effects of nanoparticles are: (1) breakdown of the bacterial cell membrane, (2) production of ROS, (3) bacterial cell membrane penetration, and (4) activation of intracellular antibacterial activity (Wang et al., 2017).

Each form of nanoparticle has its own advantages, specificity, and functionality leading to a wide range of applications (Fig. 19.19). The antibacterial activity and mechanism are influenced by parameters including the particle size, shape, surface area, and surface curvature. Nanoparticles have been always an area of interest in therapeutic delivery system to achieve targeted delivery. Antibiotics-loaded nanoparticles compared to other existing drug carriers have significant advantages pertaining to eliminating the side effects of antibiotics. For example, gentamicin loaded on chitosan/fucoidan nanoparticles undergoes sustained release, along with significant antioxidant and antibacterial activities which can be used for the treatment of pneumonia (Reyes et al., 2015). Controlled transport, targeted delivery, biodegradability, and biocompatibility are key features of a drug delivery system (Xia et al., 2012).

Implants such as dental implants, catheters, heart valves, etc. coated with antimicrobial nanoparticle layers leads to inhibition of growth and adhesion of bacteria (*Streptococcus mutans, Staphylococcus epidermis*, and *Escherichia coli*) and inflammation around the implants (Xia et al., 2012). Catheters are often prone to bacterial colonization and nanopolymer coatings show efficient antimicrobial activity to inhibit growth of bacterial biofilms (Samuel and Guggenbichler, 2004; Galiano et al., 2008).

Skin acts as a natural protective shield of the body, although various types of wound infection on skin can be harmful to it. Many microorganisms like *Escherichia coli, Streptococcus, Staphylococcus*, and *Klebsiella* species can cause wound infection. Several infections caused by bacterial species are often chronic that are resistant to antibiotics, but antimicrobial nanoparticles can potentially inhibit bacterial growth and reproduction. Li et al. reported that combination of nanosilver and a mixture of poly(vinyl alcohol) and chitosan show significant efficacy in wound healing (Li et al., 2013).

Besides, polymethyl methacrylate (PMMA) and methyl methacrylate (MMA)-based bone cements, which are widely used to fix joint prostheses in bone replacement procedures, are found to be susceptible to bacterial infections.

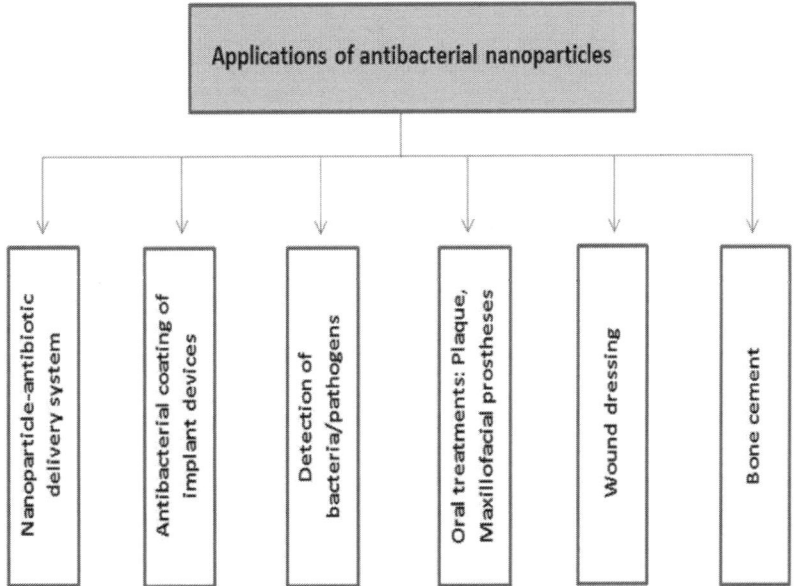

FIG. 19.19 Various applications of biomolecule–nanoparticle conjugates having antibacterial and anti-inflammatory properties.

AgNPs-loaded bone cement significantly reduces surface colonization of *Staphylococcus aureus*, *Staphylococcus epidermidis*, and *Acinetobacter baumannii* and inhibits their growth (Miola et al., 2015; Kose et al., 2016).

The oral cavity consists of several surfaces, each covered with a plethora of bacterial biofilm. Most of these bacteria are involved in oral diseases such as tooth decay and periodontitis (Aas et al., 2005). Plaque is one of common initiating factors for infections in mouth as it allows the microbes to settle on teeth. Lee et al. reported that amoxicillin functionalized with nanodiamonds when combined with gutta-percha significantly reduces bacteria after root canal treatment (Lee et al., 2015). Maxillofacial prostheses also have been shown to grow bacterial biofilms resulting into tissue inflammation around the prostheses. Aboelzahab et al. reported that prostheses coated with nanotitanium dioxide show efficient antibacterial activity following light exposure (Aboelzahab et al., 2012).

Inflammation is an initial homeostatic immune response that defends against tissue trauma. This is characterized by proinflammatory cytokine production and cell activation within the immune system (Wong et al., 2009). The conventional treatments for inflammatory diseases provide nontargeted treatment methods with significant adverse effects. Throughout the years, various research studies have presented nanodrug delivery as a promising approach to the targeted delivery of therapeutic agents, most of which have been shown to minimize side effects and toxicity, enhance the bioavailability and efficacy of a drug at the site. Different types of nanoparticles are studied and found to be significantly effective in treating numerous inflammatory diseases such as rheumatoid arthritis, dermatitis, asthma, inflammatory bowel disease, and Alzheimer's disease (Elsayed et al., 2019).

19.5.4 Theranostics

Theranostics is a growing arena merging therapeutics and diagnostics into single agent. This discipline is powered by advancements in nanoparticle systems that can significantly provide appropriate functionalities and comprehensive results (Cole and Holland, 2015). Theranostics have been widely used in imaging and treatment as well as in designing targeted drug delivery systems for cancer, diabetes, central nervous system (CNS) diseases, immune and genetic disorders (Ramanathan et al., 2018).

The selection of the therapeutic agent has an extremely imperative role in the production of multifunctional nanoparticles. A key point in encapsulation design depends on the therapeutics' solubility and indication capability. Doxorubicin and paclitaxel are the most popular cancer medications that have been encapsulated into nanoparticles for theranostic applications. Barenholz reported a drug formulation "Doxil" including PEGylated liposome drug carrier loaded with doxorubicin for the treatment of sarcoma and ovarian cancer (Barenholz, 2012). Another example of a drug formulation is Abraxane, an albumin-bound paclitaxel used to treat metastatic breast cancer, pancreatic cancer, and lung cancer (Green et al., 2006).

Currently, researchers are focusing on developing an effective nanoplatform for imaging and efficient drug delivery systems for CNS disorders (Gabathuler, 2010; Volicer, 2001). Several studies reported that sufficient surface modifications of a biodegradable polymeric nanoparticles show significant efficacy in delivering drugs to treat neurodegenerative diseases across the blood brain barrier (BBB) (Ramanathan et al., 2018; Roney et al., 2005). Vinogradov et al. reported synthesis of a nanogel drug delivery system which were efficiently carried and delivered phosphorothioate oligonucleotides (SODN) across BBB (Vinogradov et al., 1999). The nanogel was formulated by emulsification solvent evaporation process by cross-linking polyethylene (PEI) and PEG (Agnihotri et al., 2004; Soppimath et al., 2001). In order to load the drug, nanogel was swollen in water and the positively charged PEG and PEI encapsulated the negatively charged SODN. Particle size of this drug loaded nanogel was found to be 80 nm which facilitated greater uptake into cells and deeper penetration in the tissues. Tanifum et al. reported a liposomal nanoparticle incorporated with synthetic amyloid-β-targeted lipid conjugate for amyloid plaque deposition in Alzheimer's disease (AD) (Tanifum et al., 2012). This nanoparticle selectively binds to the parenchymal plaques and deposits of amyloid throughout the brain cells of AD transgenic mice. The potential of the particle to penetrate the BBB and bind to amyloid-β plaque deposits was confirmed by performing in vitro immunohistochemistry and validated co-localization of both the liposome encapsulate and bilayer membrane components on brain cells taken from the treated animals. Fu et al. reported a theranostic application of nanoparticles for acute temporal lobe epilepsy (TLE) using superparamagnetic iron oxide (Fu et al., 2016). Anti-IL-1β monoclonal antibody (mAb) was functionalized on superparamagnetic iron oxide and used for MRI diagnosis and targeted therapy. A number of findings have been considered about the mechanisms underlying anti-IL-1β mAb-superparamagnetic iron oxide crossing the BBB. First, the epileptic seizures induce inflammation of the brain tissues which leads to dysfunction of BBB along with increased cerebral capillary permeability (Sheen et al., 2011). Second, smaller size of nanoparticles lets them to stick to the cell surface and also penetrate through the cells (Vidu et al., 2014). It is reported that antibody functionalized iron oxide nanoparticles can penetrate BBB by using receptor-mediated transcytosis method (Ndong et al., 2015). Therefore, in this study, when mAb was functionalized on nanoparticle, it penetrated and distributed anti-IL-1β mAb-superparamagnetic iron oxide showing selectively binding to overexpressed antigens (IL-1β) on epileptogenic tissues.

For effective diagnostic purposes, molecular imaging techniques play a major role in characterizing biological processes at cellular and subcellular levels (Fig. 19.20) (Tanifum et al., 2012).

The nanoparticles platform provides enormous potential to the theranostic systems along with the ability to quantify the disease condition while simultaneously delivering therapies. This approach contributes in a great extent to the improvement in treatment alternatives, resulting in highly personalized medicine.

19.6 Conclusion and future perspective

Biological probes functionalized nanoparticles have shown a great significance in bioanalytical devices as they evidently enhanced their performances by increasing their sensitivity, selectivity, and bringing down detection limit to minimal concentrations. The nanoparticle surface and core can be functionalized for certain uses, particularly biosensing, bioimaging, and therapeutics. This chapter mainly focused on different types of available nanoparticles, their conjugation chemistry, enhanced performances on being functionalized with various polymers and biological probes and their applications in biomedical field. To name a few, nanomaterials that are presently being exploited are metal nanoparticles and clusters, quantum dots, carbon allotropes, metal oxide nanoparticles, self-assembled DNA structures, protein assemblies, etc. In terms of bioconjugates, these nanomaterials are exploited from molecular electronics to drug delivery systems. The incorporation of biomolecules into nanomaterials leads to a major performance improvement in sensing besides cost efficiency, thermal stability, and high catalytic efficiency. Biological probes functionalized nanostructures have emerged as significant sensors for biomolecules, metal ions, pathogens, etc. and have shown remarkable diagnostic and therapeutic uses likely targeted drug delivery, in vivo imaging, various essential treatments for neurodegenerative diseases and chemotherapies, etc.

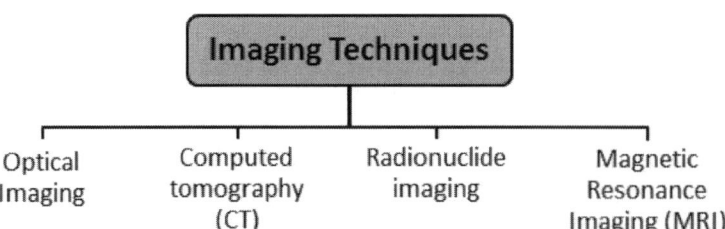

FIG. 19.20 Different types of molecular imaging techniques used in diagnosis (Galiano et al., 2008).

However, a variety of challenges tend to circumvent widespread biomedical applications of nanotechnology and their further integration into therapeutic trials. There is no systematic knowledge of pharmacokinetics, including absorption, distribution, metabolism, and excretion of nanoparticles in the body. There is also a need to enhance their sensitivity toward precise detection of pathogens present in even traces. With these advancements, diagnosis of slow growing microorganisms can be easily done. However, the environmental and health risks corresponding to their relatively smaller size and degradation in the surrounding environment are of great concern. Existing detection technologies are not currently capable of detecting traces of fatal viruses, such as swine flues, or anthrax, due to exposure of biochemicals in the environment which may cause catastrophic effects to humans. However, in the intermediate, future potential research outcomes are expected to resolve these challenges, but to overcome the challenges of ultrasensitive biosensing, greater refinement is required. Alongside, assessment of safety and biocompatibility of nanomaterials in vivo in terms of long time toxicities is also essential, so that substantial biological observations can be made before eventually flowing into therapies. Considering their vast potential in biomedicine and sensing, biological probe functionalized nanoparticles are presumed to show tremendous progress in diagnostic and therapeutic applications.

References

Aas, J.A., Paster, B.J., Stokes, L.N., Olsen, I., Dewhirst, F.E., 2005. Defining the normal bacterial flora of the oral cavity. J. Clin. Microbiol. 43 (11), 5721–5732. https://doi.org/10.1128/JCM.43.11.5721-5732.2005.

Aboelzahab, A., Azad, A.M., Dolan, S., Goel, V., 2012. Mitigation of *Staphylococcus aureus*-mediated surgical site infections with IR photoactivated TiO_2 coatings on Ti implants. Adv. Healthcare Mater. 1 (3), 285–291. https://doi.org/10.1002/adhm.201100032.

Agasti, S.S., Rana, S., Park, M.H., Kim, C.K., You, C.C., Rotello, V.M., 2010. Nanoparticles for detection and diagnosis. Adv. Drug Deliv. Rev. 62 (3), 316–328. https://doi.org/10.1016/j.addr.2009.11.004.

Agnihotri, S.A., Mallikarjuna, N.N., Aminabhavi, T.M., 2004. Recent advances on chitosan-based micro-and nanoparticles in drug delivery. J. Control. Release 100 (1), 5–28. https://doi.org/10.1016/j.jconrel.2004.08.010.

Aiken III, J.D., Finke, R.G, 1999. A review of modern transition-metal nanoclusters: their synthesis, characterization, and applications in catalysis. J. Mol. Catal. A Chem. 145 (1-2), 1–44. https://doi.org/10.1016/S1381-1169(99)00098-9.

Albrecht, M.A., Evans, C.W., Raston, C.L., 2006. Green chemistry and the health implications of nanoparticles. Green Chem. 8 (5), 417–432. https://doi.org/10.1039/B517131H.

Aldaye, F.A., Sleiman, H.F., 2006. Sequential self-assembly of a DNA hexagon as a template for the organization of gold nanoparticles. Angew. Chem. Int. Ed. 45 (14), 2204–2209. https://doi.org/10.1002/anie.200502481.

Algar, W.R., Krull, U.J., 2010. New opportunities in multiplexed optical bioanalyses using quantum dots and donor–acceptor interactions. Anal. Bioanal. Chem. 398 (6), 2439–2449. https://doi.org/10.1007/s00216-010-3837-y.

Algar, W.R., Susumu, K., Delehanty, J.B., Medintz, I.L., 2011. Semiconductor quantum dots in bioanalysis: crossing the valley of death. Anal. Chem. 83 (23), 8826–8837. https://doi.org/10.1021/ac201331r.

Algar, W.R., Tavares, A.J., Krull, U.J., 2010. Beyond labels: a review of the application of quantum dots as integrated components of assays, bioprobes, and biosensors utilizing optical transduction. Anal. Chim. Acta 673 (1), 1–25. https://doi.org/10.1016/j.aca.2010.05.026.

Alivisatos, A.P., Gu, W., Larabell, C., 2005. Quantum dots as cellular probes. Ann. Rev. Biomed. Eng. 7 (1), 55–76. https://doi.org/10.1146/annurev.bioeng.7.060804.100432.

Amador, P., Fernandes, R., Prudêncio, C., Duarte, I., 2019. Prevalence of antibiotic resistance genes in multidrug-resistant Enterobacteriaceae on Portuguese livestock manure. Antibiotics 8 (1), 23. https://doi.org/10.3390/antibiotics8010023.

Andreadou, M., Liandris, E., Gazouli, M., Taka, S., Antoniou, M., Theodoropoulos, G., Tachtsidis, I., Goutas, N., Vlachodimitropoulos, D., Kasampalidis, I., Ikonomopoulos, J., 2014. A novel non-amplification assay for the detection of *Leishmania* spp. in clinical samples using gold nanoparticles. J. Microbiol. Methods 96, 56–61. https://doi.org/10.1016/j.mimet.2013.10.011.

Ang, L.Y., Lim, M.E., Ong, L.C., Zhang, Y., 2011. Applications of upconversion nanoparticles in imaging, detection and therapy. Nanomedicine 6 (7), 1273–1288. https://doi.org/10.2217/nnm.11.108.

Anton, N., Benoit, J.P., Saulnier, P., 2008. Design and production of nanoparticles formulated from nano-emulsion templates—a review. J. Control. Release 128 (3), 185–199. https://doi.org/10.1016/j.jconrel.2008.02.007.

Arbab, A.S., Wilson, L.B., Ashari, P., Jordan, E.K., Lewis, B.K., Frank, J.A., 2005. A model of lysosomal metabolism of dextran coated superparamagnetic iron oxide (SPIO) nanoparticles: implications for cellular magnetic resonance imaging. NMR Biomed. 18 (6), 383–389. https://doi.org/10.1002/nbm.970.

Atiyeh, B.S., Costagliola, M., Hayek, S.N., Dibo, S.A., 2007. Effect of silver on burn wound infection control and healing: review of the literature. Burns 33 (2), 139–148. https://doi.org/10.1016/j.burns.2006.06.010.

Augustin, M.A., Hemar, Y., 2009. Nano- and micro-structured assemblies for encapsulation of food ingredients. Chem. Soc. Rev. 38 (4), 902–912. https://doi.org/10.1039/B801739P.

Bagheri, S., Yasemi, M., Safaie-Qamsari, E., Rashidiani, J., Abkar, M., Hassani, M., Mirhosseini, S.A., Kooshki, H., 2018. Using gold nanoparticles in diagnosis and treatment of melanoma cancer. Artif. Cells Nanomed. Biotechnol. 46 (Suppl. 1), 462–471. https://doi.org/10.1080/21691401.2018.1430585.

Bajaj, A., Miranda, O.R., Kim, I.B., Phillips, R.L., Jerry, D.J., Bunz, U.H., Rotello, V.M., 2009. Detection and differentiation of normal, cancerous, and metastatic cells using nanoparticle-polymer sensor arrays. Proc. Natl. Acad. Sci. 106 (27), 10912–10916. https://doi.org/10.1073/pnas.0900975106.

Bajaj, A., Rana, S., Miranda, O.R., Yawe, J.C., Jerry, D.J., Bunz, U.H., Rotello, V.M., 2010. Cell surface-based differentiation of cell types and cancer states using a gold nanoparticle-GFP based sensing array. Chem. Sci. 1 (1), 134–138. https://doi.org/10.1039/C0SC00165A.

Baker, M., 2010. Nanotechnology imaging probes: smaller and more stable. Nat. Methods 7 (12), 957–962. https://doi.org/10.1038/nmeth1210-957.

Baker, S.N., Baker, G.A., 2010. Luminescent carbon nanodots: emergent nanolights. Angew. Chem. Int. Ed. 49 (38), 6726–6744. https://doi.org/10.1002/anie.200906623.

Bakthavathsalam, P., Rajendran, V.K., Mohammed, J.A.B., 2012. A direct detection of *Escherichia coli* genomic DNA using gold nanoprobes. J. Nanobiotechnol. 10 (1), 8. https://doi.org/10.1186/1477-3155-10-8.

Balanta, A., Godard, C., Claver, C., 2011. Pd nanoparticles for C–C coupling reactions. Chem. Soc. Rev. 40 (10), 4973–4985. https://doi.org/10.1039/C1CS15195A.

Balogh, L., Swanson, D.R., Tomalia, D.A., Hagnauer, G.L., McManus, A.T., 2001. Dendrimer–silver complexes and nanocomposites as antimicrobial agents. Nano Lett. 1 (1), 18–21. https://doi.org/10.1021/nl005502p.

Banerjee, A., Pons, T., Lequeux, N., Dubertret, B., 2016. Quantum dots–DNA bioconjugates: synthesis to applications. Interface Focus 6 (6), 20160064. https://doi.org/10.1098/rsfs.2016.0064.

Bao, B., Rodriguez-Melendez, R., Wijeratne, S.S., Zempleni, J., 2010. Biotin regulates the expression of holocarboxylase synthetase in the miR-539 pathway in HEK-293 cells. J. Nutr. 140 (9), 1546–1551. https://doi.org/10.3945/jn.110.126359.

Barbe, C., Bartlett, J., Kong, L., Finnie, K., Lin, H.Q., Larkin, M., Calleja, G., 2004. Silica particles: a novel drug-delivery system. Adv. Mater. 16 (21), 1959–1966. https://doi.org/10.1002/adma.200400771.

Barenholz, Y.C., 2012. Doxil®—the first FDA-approved nano-drug: lessons learned. J. Control. Release 160 (2), 117–134. https://doi.org/10.1016/j.jconrel.2012.03.020.

Barrientos, Á.G., de la Fuente, J.M., Rojas, T.C., Fernández, A., Penadés, S., 2003. Gold glyconanoparticles: synthetic polyvalent ligands mimicking glycocalyx-like surfaces as tools for glycobiological studies. Chemistry 9 (9), 1909–1921. https://doi.org/10.1002/chem.200204544.

Berlin, J.M., Pham, T.T., Sano, D., Mohamedali, K.A., Marcano, D.C., Myers, J.N., Tour, J.M., 2011. Noncovalent functionalization of carbon nanovectors with an antibody enables targeted drug delivery. ACS Nano 5 (8), 6643–6650. https://doi.org/10.1021/nn2021293.

Biju, V., Itoh, T., Anas, A., Sujith, A., Ishikawa, M., 2008. Semiconductor quantum dots and metal nanoparticles: syntheses, optical properties, and biological applications. Anal. Bioanal. Chem. 391 (7), 2469–2495. https://doi.org/10.1007/s00216-008-2185-7.

Biju, V., Itoh, T., Ishikawa, M., 2010. Delivering quantum dots to cells: bioconjugated quantum dots for targeted and nonspecific extracellular and intracellular imaging. Chem. Soc. Rev. 39 (8), 3031–3056. https://doi.org/10.1039/B926512K.

Boeneman, K., Deschamps, J.R., Buckhout-White, S., Prasuhn, D.E., Blanco-Canosa, J.B., Dawson, P.E., Stewart, M.H., Susumu, K., Goldman, E.R., Ancona, M., Medintz, I.L., 2010. Quantum dot DNA bioconjugates: attachment chemistry strongly influences the resulting composite architecture. Acs Nano 4 (12), 7253–7266. https://doi.org/10.1021/nn1021346.

Borglin, J., Selegård, R., Aili, D., Ericson, M.B., 2017. Peptide functionalized gold nanoparticles as a stimuli responsive contrast medium in multiphoton microscopy. Nano Lett. 17 (3), 2102–2108. https://doi.org/10.1021/acs.nanolett.7b00611.

Breaker, R.R., Joyce, G.F., 1994. A DNA enzyme that cleaves RNA. Chem. Biol. 1 (4), 223–229. https://doi.org/10.1016/1074-5521(94)90014-0.

Cai, W., Chen, X., 2007. Nanoplatforms for targeted molecular imaging in living subjects. Small 3 (11), 1840–1854. https://doi.org/10.1002/smll.200700351.

Cao, L., Wang, X., Meziani, M.J., Lu, F., Wang, H., Luo, P.G., Lin, Y., Harruff, B.A., Veca, L.M., Murray, D., Xie, S.Y., 2007. Carbon dots for multiphoton bioimaging. J. Am. Chem. Soc. 129 (37), 11318–11319. https://doi.org/10.1021/ja073527l.

Chai, F., Wang, C., Wang, T., Li, L., Su, Z., 2010. Colorimetric detection of Pb^{2+} using glutathione functionalized gold nanoparticles. ACS Appl. Mater. Interfaces 2 (5), 1466–1470. https://doi.org/10.1021/am100107k.

Chanphai, P., Thomas, T.J., Tajmir-Riahi, H.A., 2020. Application and biomolecular study of functionalized folic acid-dendrimer nanoparticles in drug delivery. J. Biomol. Struct. Dyn. 39 (3), 1–19. https://doi.org/10.1080/07391102.2020.1717994.

Chatterjee, D.K., Gnanasammandhan, M.K., Zhang, Y., 2010. Small upconverting fluorescent nanoparticles for biomedical applications. Small 6 (24), 2781–2795. https://doi.org/10.1002/smll.201000418.

Chavali, M.S., Nikolova, M.P., 2019. Metal oxide nanoparticles and their applications in nanotechnology. SN Appl. Sci. 1 (6), 607. https://doi.org/10.1007/s42452-019-0592-3.

Chen, A., Holt-Hindle, P., 2010. Platinum-based nanostructured materials: synthesis, properties, and applications. Chem. Rev. 110 (6), 3767–3804. https://doi.org/10.1021/cr9003902.

Chen, G., Jiang, M., 2011. Cyclodextrin-based inclusion complexation bridging supramolecular chemistry and macromolecular self-assembly. Chem. Soc. Rev. 40 (5), 2254–2266. https://doi.org/10.1039/C0CS00153H.

Chen, J., Lim, B., Lee, E.P., Xia, Y., 2009. Shape-controlled synthesis of platinum nanocrystals for catalytic and electrocatalytic applications. Nano Today 4 (1), 81–95. https://doi.org/10.1016/j.nantod.2008.09.002.

Cheong, S., Watt, J.D., Tilley, R.D., 2010. Shape control of platinum and palladium nanoparticles for catalysis. Nanoscale 2 (10), 2045–2053. https://doi.org/10.1039/C0NR00276C.

Cho, E.C., Xie, J., Wurm, P.A., Xia, Y., 2009. Understanding the role of surface charges in cellular adsorption versus internalization by selectively removing gold nanoparticles on the cell surface with a I2/KI etchant. Nano Lett. 9 (3), 1080–1084. https://doi.org/10.1021/nl803487r.

Chou, L.Y., Ming, K., Chan, W.C., 2011. Strategies for the intracellular delivery of nanoparticles. Chem. Soc. Rev. 40 (1), 233–245. https://doi.org/10.1039/C0CS00003E.

Christian, N.A., Milone, M.C., Ranka, S.S., Li, G., Frail, P.R., Davis, K.P., Bates, F.S., Therien, M.J., Ghoroghchian, P.P., June, C.H., Hammer, D.A., 2007. Tat-functionalized near-infrared emissive polymersomes for dendritic cell labeling. Bioconjug. Chem. 18 (1), 31–40. https://doi.org/10.1021/bc0601267.

Chun, H., Hahm, M.G., Homma, Y., Meritz, R., Kuramochi, K., Menon, L., Ci, L., Ajayan, P.M., Jung, Y.J., 2009. Engineering low-aspect ratio carbon nanostructures: nanocups, nanorings, and nanocontainers. ACS Nano 3 (5), 1274–1278. https://doi.org/10.1021/nn9001903.

Cobb, A.J., 2007. Recent highlights in modified oligonucleotide chemistry. Org. Biomol. Chem. 5 (20), 3260–3275. https://doi.org/10.1039/B709797M.

Cole, J.T., Holland, N.B., 2015. Multifunctional nanoparticles for use in theranostic applications. Drug Deliv. Transl. Res. 5 (3), 295–309. https://doi.org/10.1007/s13346-015-0218-2.

Cormode, D.P., Jarzyna, P.A., Mulder, W.J., Fayad, Z.A., 2010. Modified natural nanoparticles as contrast agents for medical imaging. Adv. Drug Deliv. Rev. 62 (3), 329–338. https://doi.org/10.1016/j.addr.2009.11.005.

Dabbousi, B.O., Rodriguez-Viejo, J., Mikulec, F.V., Heine, J.R., Mattoussi, H., Ober, R., Jensen, K.F., Bawendi, M.G., 1997. CdSe) ZnS core–shell quantum dots: synthesis and characterization of a size series of highly luminescent nanocrystallites. J. Phys. Chem. B 101 (46), 9463–9475. https://doi.org/10.1021/jp971091y.

Dang, JM., Leong, KW., 2006. Natural polymers for gene delivery and tissue engineering. Adv. Drug Deliv. Rev. 58 (4), 487–499. https://doi.org/10.1016/j.addr.2006.03.001.

Daniel, M.C., Astruc, D., 2004. Gold nanoparticles: assembly, supramolecular chemistry, quantum-size-related properties, and applications toward biology, catalysis, and nanotechnology. Chem. Rev. 104 (1), 293–346. https://doi.org/10.1021/cr030698 ±.

Darbha, G.K., Singh, A.K., Rai, U.S., Yu, E., Yu, H., Chandra Ray, P., 2008. Selective detection of mercury (II) ion using nonlinear optical properties of gold nanoparticles. J. Am. Chem. Soc. 130 (25), 8038–8043. https://doi.org/10.1021/ja801412b.

de la Fuente, J.M., Barrientos, A.G., Rojas, T.C., Rojo, J., Cañada, J., Fernández, A., Penadés, S., 2001. Gold glyconanoparticles as water-soluble polyvalent models to study carbohydrate interactions. Angew. Chem. Int. Ed. 40 (12), 2257–2261. https://doi.org/10.1002/1521-3773(20010618)40 12≤2257::AID-ANIE2257≥3.0.CO;2-S.

De Lorimier, R.M., Smith, J.J., Dwyer, M.A., Looger, L.L., Sali, K.M., Paavola, C.D., Rizk, S.S., Sadigov, S., Conrad, D.W., Loew, L., Hellinga, H.W., 2002. Construction of a fluorescent biosensor family. Protein Sci. 11 (11), 2655–2675. https://doi.org/10.1110/ps.021860.

de Mello Donega, C., 2011. Synthesis and properties of colloidal heterononocrystals. Chem. Soc. Rev. 40 (3), 1512–1546. https://doi.org/10.1039/C0CS00055H.

de Paz, J.L., Ojeda, R., Barrientos, Á.G., Penadés, S., Martín-Lomas, M., 2005. Synthesis of a Ley neoglycoconjugate and Ley-functionalized gold glyconanoparticles. Tetrahedr. Asymm. 16 (1), 149–158. https://doi.org/10.1016/j.tetasy.2004.11.066.

De Souza, R., Zahedi, P., Allen, C.J., Piquette-Miller, M., 2010. Polymeric drug delivery systems for localized cancer chemotherapy. Drug Deliv. 17 (6), 365–375. https://doi.org/10.3109/10717541003762854.

Delehanty, J.B., Boeneman, K., Bradburne, C.E., Robertson, K., Medintz, I.L., 2009. Quantum dots: a powerful tool for understanding the intricacies of nanoparticle-mediated drug delivery. Expert Opin. Drug Deliv. 6 (10), 1091–1112. https://doi.org/10.1517/17425240903167934.

Di Marco, M., Shamsuddin, S., Razak, K.A., Aziz, A.A., Devaux, C., Borghi, E., Sadun, C., 2010. Overview of the main methods used to combine proteins with nanosystems: absorption, bioconjugation, and encapsulation. Int. J. Nanomed. 5, 37–49.

Discher, D.E., Ahmed, F., 2006. Polymersomes. Annu. Rev. Biomed. Eng. 8, 323–341. https://doi.org/10.1146/annurev.bioeng.8.061505.095838.

Doerflinger, A., Quang, N.N., Gravel, E., Pinna, G., Vandamme, M., Ducongé, F., Doris, E., 2018. Biotin-functionalized targeted polydiacetylene micelles. Chem. Commun. 54 (29), 3613–3616. https://doi.org/10.1039/C8CC00553B.

Domun, N., Hadavinia, H., Zhang, T., Sainsbury, T., Liaghat, G.H., Vahid, S., 2015. Improving the fracture toughness and the strength of epoxy using nanomaterials–a review of the current status. Nanoscale 7 (23), 10294–10329. https://doi.org/10.1039/C5NR01354B.

Dreaden, E.C., Mackey, M.A., Huang, X., Kang, B., El-Sayed, M.A., 2011. Beating cancer in multiple ways using nanogold. Chem. Soc. Rev. 40 (7), 3391–3404. https://doi.org/10.1039/C0CS00180E.

Durland, R.H., Eastman, E.M., 1998. Manufacturing and quality control of plasmid-based gene expression systems. Adv. Drug Deliv. Rev. 30 (1-3), 33–48. https://doi.org/10.1016/s0169-409x(97)00105-1.

El-Boubbou, K., Zhu, D.C., Vasileiou, C., Borhan, B., Prosperi, D., Li, W., Huang, X., 2010. Magnetic glyco-nanoparticles: a tool to detect, differentiate, and unlock the glyco-codes of cancer via magnetic resonance imaging. J. Am. Chem. Soc. 132 (12), 4490–4499. https://doi.org/10.1021/ja100455c.

Elghanian, R., Storhoff, J.J., Mucic, R.C., Letsinger, R.L., Mirkin, C.A., 1997. Selective colorimetric detection of polynucleotides based on the distance-dependent optical properties of gold nanoparticles. Science 277 (5329), 1078–1081. https://doi.org/10.1126/science.277.5329.1078.

El-Sayed, I.H., Huang, X., El-Sayed, M.A., 2005. Surface plasmon resonance scattering and absorption of anti-EGFR antibody conjugated gold nanoparticles in cancer diagnostics: applications in oral cancer. Nano Lett. 5 (5), 829–834. https://doi.org/10.1021/nl050074e.

Elsayed, M.A., Norredin, A., 2019. The potential contribution of nanoparticles in the treatment of inflammatory diseases. In: Ane C.F, Nunes (Ed.), Translational Studies on Inflammation. IntechOpen, London, pp. 171–185. https://doi.org/10.4018/978-1-7998-3594-3.ch002.

Emerich, D.F., Thanos, C.G., 2007. Targeted nanoparticle-based drug delivery and diagnosis. J. Drug Target. 15 (3), 163–183. https://doi.org/10.1080/10611860701231810.

Erathodiyil, N., Ying, J.Y., 2011. Functionalization of inorganic nanoparticles for bioimaging applications. Acc. Chem. Res. 44 (10), 925–935. https://doi.org/10.1021/ar2000327.

Fang, C., Veiseh, O., Kievit, F., Bhattarai, N., Wang, F., Stephen, Z., Li, C., Lee, D., Ellenbogen, R.G., Zhang, M., 2010. Functionalization of iron oxide magnetic nanoparticles with targeting ligands: their physicochemical properties and in vivo behavior. Nanomedicine 5 (9), 1357–1369. https://doi.org/10.2217/nnm.10.55.

Fang, C., Zhang, M., 2010. Nanoparticle-based theragnostics: integrating diagnostic and therapeutic potentials in nanomedicine. J. Control. Release 146 (1), 2. https://doi.org/10.1016/j.jconrel.2010.05.013.

Fang, Z., Bhandari, B., 2010. Encapsulation of polyphenols – a review. Trends Food Sci. Technol. 21 (10), 510–523. https://doi.org/10.1016/j.tifs.2010.08.003.

Faure, C., Meyre, M.E., Trépout, S., Lambert, O., Lebraud, E., 2009. Magnetic multilamellar liposomes produced by in situ synthesis of iron oxide nano particles:"magnetonions. J. Phys. Chem. B 113 (25), 8552–8559. https://doi.org/10.1021/jp901105c.

Fernando, R.H., 2009. Nanocomposite and Nanostructured Coatings: Recent Advancements. ACS Publications, Washington, pp. 2–21. https://doi.org/10.1021/bk-2009-1008.ch001.

Flynn-Charlebois, A., Wang, Y., Prior, T.K., Rashid, I., Hoadley, K.A., Coppins, R.L., Wolf, A.C., Silverman, S.K., 2003. Deoxyribozymes with 2′-5′ RNA ligase activity. J. Am. Chem. Soc. 125 (9), 2444–2454. https://doi.org/10.1021/ja028774y.

Fu, T., Kong, Q., Sheng, H., Gao, L., 2016. Value of functionalized superparamagnetic iron oxide nanoparticles in the diagnosis and treatment of acute temporal lobe epilepsy on MRI. Neural Plast. 2016, 1–12. https://doi.org/10.1155/2016/2412958.

Gabathuler, R., 2010. Approaches to transport therapeutic drugs across the blood–brain barrier to treat brain diseases. Neurobiol. Dis. 37 (1), 48–57. https://doi.org/10.1016/j.nbd.2009.07.028.

Galiano, K., Pleifer, C., Engelhardt, K., Brössner, G., Lackner, P., Huck, C., Lass-Flörl, C., Obwegeser, A., 2008. Silver segregation and bacterial growth of intraventricular catheters impregnated with silver nanoparticles in cerebrospinal fluid drainages. Neurol. Res. 30 (3), 285–287. https://doi.org/10.1179/016164107X229902.

Gao, J., Gu, H., Xu, B., 2009. Multifunctional magnetic nanoparticles: design, synthesis, and biomedical applications. Acc. Chem. Res. 42 (8), 1097–1107. https://doi.org/10.1021/ar9000026.

Gargano, J.M., Ngo, T., Kim, J.Y., Acheson, D.W., Lees, W.J., 2001. Multivalent inhibition of AB_5 toxins. J. Am. Chem. Soc. 123 (51), 12909–12910. https://doi.org/10.1021/ja016305a.

Gearheart, L.A., Ploehn, H.J., Murphy, C.J., 2001. Oligonucleotide adsorption to gold nanoparticles: a surface-enhanced Raman spectroscopy study of intrinsically bent DNA. J. Phys. Chem. B 105 (50), 12609–12615. https://doi.org/10.1021/jp0106606.

Georgieva, J.V., Brinkhuis, R.P., Stojanov, K., Weijers, C.A., Zuilhof, H., Rutjes, F.P., Zuhorn, I.S., 2012. Peptide-mediated blood–brain barrier transport of polymersomes. Angew. Chem. Int. Ed. 51 (33), 8339–8342. https://doi.org/10.1002/anie.201202001.

Ghosh, S., Jiang, W., McClements, J.D., Xing, B., 2011. Colloidal stability of magnetic iron oxide nanoparticles: influence of natural organic matter and synthetic polyelectrolytes. Langmuir 27 (13), 8036–8043. https://doi.org/10.1021/la200772e.

Giljohann, D.A., Seferos, D.S., Patel, P.C., Millstone, J.E., Rosi, N.L., Mirkin, C.A., 2007. Oligonucleotide loading determines cellular uptake of DNA-modified gold nanoparticles. Nano Lett. 7 (12), 3818–3821. https://doi.org/10.1021/nl072471q.

Giljohann, D.A., Seferos, D.S., Prigodich, A.E., Patel, P.C., Mirkin, C.A., 2009. Gene regulation with polyvalent siRNA–nanoparticle conjugates. J. Am. Chem. Soc. 131 (6), 2072–2073. https://doi.org/10.1021/ja808719p.

Goldman, E.R., Medintz, I.L., Whitley, J.L., Hayhurst, A., Clapp, A.R., Uyeda, H.T., Deschamps, J.R., Lassman, M.E., Mattoussi, H., 2005. A hybrid quantum dot – antibody fragment fluorescence resonance energy transfer-based TNT sensor. J. Am. Chem. Soc. 127 (18), 6744–6751. https://doi.org/10.1021/ja043677l.

Gong, L., Zhao, Z., Lv, Y.F., Huan, S.Y., Fu, T., Zhang, X.B., Shen, G.L., Yu, R.Q., 2015. DNAzyme-based biosensors and nanodevices. Chem. Commun. 51 (6), 979–995. https://doi.org/10.1039/C4CC06855F.

Graff, R.A., Swanson, T.M., Strano, M.S., 2008. Synthesis of nickel–nitrilotriacetic acid coupled single-walled carbon nanotubes for directed self-assembly with polyhistidine-tagged proteins. Chem. Mater. 20 (5), 1824–1829. https://doi.org/10.1021/cm702577h.

Green, M.R., Manikhas, G.M., Orlov, S., Afanasyev, B., Makhson, A.M., Bhar, P., Hawkins, M.J., 2006. Abraxane®, a novel Cremophor®-free, albumin-bound particle form of paclitaxel for the treatment of advanced non-small-cell lung cancer. Ann. Oncol. 17 (8), 1263–1268. https://doi.org/10.1093/annonc/mdl104.

Grynkiewicz, G., Poenie, M., Tsien, R.Y., 1985. A new generation of Ca^{2+} indicators with greatly improved fluorescence properties. J. Biol. Chem. 260 (6), 3440–3450. https://www.jbc.org/content/260/6/3440.long.

Guo, S., Wang, E., 2011. Noble metal nanomaterials: controllable synthesis and application in fuel cells and analytical sensors. Nano Today 6 (3), 240–264. https://doi.org/10.1016/j.nantod.2011.04.007.

Gupta, A.K., Gupta, M., 2005. Synthesis and surface engineering of iron oxide nanoparticles for biomedical applications. Biomaterials 26 (18), 3995–4021. https://doi.org/10.1016/j.biomaterials.2004.10.012.

Haase, M., Schäfer, H., 2011. Upconverting nanoparticles. Angew. Chem. Int. Ed. 50 (26), 5808–5829. https://doi.org/10.1002/anie.201005159.

Habibi, N., Kamaly, N., Memic, A., Shafiee, H., 2016. Self-assembled peptide-based nanostructures: smart nanomaterials toward targeted drug delivery. Nano Today 11 (1), 41–60. https://doi.org/10.1016/j.nantod.2016.02.004.

Halo, T.L., McMahon, K.M., Angeloni, N.L., Xu, Y., Wang, W., Chinen, A.B., Malin, D., Strekalova, E., Cryns, V.L., Cheng, C., Mirkin, C.A., 2014. Nanoflares for the detection, isolation, and culture of live tumor cells from human blood. Proc. Natl. Acad. Sci. 111 (48), 17104–17109. https://doi.org/10.1073/pnas.1418637111.

Han, M., Gao, X., Su, J.Z., Nie, S., 2001. Quantum-dot-tagged microbeads for multiplexed optical coding of biomolecules. Nat. Biotechnol. 19 (7), 631–635. https://doi.org/10.1038/90228.

Hao, R., Xing, R., Xu, Z., Hou, Y., Gao, S., Sun, S., 2010. Synthesis, functionalization, and biomedical applications of multifunctional magnetic nanoparticles. Adv. Mater. 22 (25), 2729–2742. https://doi.org/10.1002/adma.201000260.

Harush-Frenkel, O., Altschuler, Y., Benita, S., 2008. Nanoparticle-cell interactions: drug delivery implications. Crit. Rev. Therap. Drug Carrier Syst. 25 (6), 485–544. https://doi.org/10.1615/CritRevTherDrugCarrierSyst.v25.i6.10.

Hassan, M., Walter, M., Moseler, M., 2014. Interactions of polymers with reduced graphene oxide: van der Waals binding energies of benzene on graphene with defects. PCCP 16 (1), 33–37. https://doi.org/10.1039/C3CP53922A.

Heddle, J.G., 2013. Gold nanoparticle-biological molecule interactions and catalysis. Catalysts 3 (3), 683–708. https://doi.org/10.3390/catal3030683.

Ho, Y.P., Leong, K.W., 2010. Quantum dot-based theranostics. Nanoscale 2 (1), 60–68. https://doi.org/10.1039/B9NR00178F.

Holzinger, M., Le Goff, A., Cosnier, S., 2014. Nanomaterials for biosensing applications: a review. Front. Chem. 2, 63. https://doi.org/10.3389/fchem.2014.00063.

Hong, S., Leroueil, P.R., Majoros, I.J., Orr, B.G., Baker Jr, J.R., Holl, M.M.B., 2007. The binding avidity of a nanoparticle-based multivalent targeted drug delivery platform. Chem. Biol. 14 (1), 107–115. https://doi.org/10.1016/j.chembiol.2006.11.015.

Hsueh, P.R., 2010. New Delhi metallo-β-lactamase-1 (NDM-1): an emerging threat among Enterobacteriaceae. J. Formos. Med. Assoc. 109 (10), 685–687. https://doi.org/10.1016/S0929-6646(10)60111-8.

Huang, C.C., Huang, Y.F., Cao, Z., Tan, W., Chang, H.T., 2005. Aptamer-modified gold nanoparticles for colorimetric determination of platelet-derived growth factors and their receptors. Anal. Chem. 77 (17), 5735–5741. https://doi.org/10.1021/ac050957q.

Huang, K.W., Yu, C.J., Tseng, W.L., 2010. Sensitivity enhancement in the colorimetric detection of lead (II) ion using gallic acid-capped gold nanoparticles: improving size distribution and minimizing interparticle repulsion. Biosens. Bioelectron. 25 (5), 984–989. https://doi.org/10.1016/j.bios.2009.09.006.

Huang, S.H., 2007. Gold nanoparticle-based immunochromatographic assay for the detection of Staphylococcus aureus. Sens. Actuators B 127 (2), 335–340. https://doi.org/10.1016/j.snb.2007.04.027.

Huang, Y.F., Lin, Y.W., Lin, Z.H., Chang, H.T., 2009. Aptamer-modified gold nanoparticles for targeting breast cancer cells through light scattering. J. Nanopart. Res. 11 (4), 775–783. https://doi.org/10.1007/s11051-008-9424-x.

Hudhomme, P., Hiroshi, I., Guldi, D., Paolucci, F., Burley, G., Deschenaux, R., Campidelli, S., Martin, N., Da Ros, T., Armaroli, N., de la Cruz, P., 2011. Fullerenes: Principles and Applications. Royal Society of Chemistry, London.

Huh, Y.M., Jun, Y.W., Song, H.T., Kim, S., Choi, J.S., Lee, J.H., Yoon, S., Kim, K.S., Shin, J.S., Suh, J.S., Cheon, J., 2005. In vivo magnetic resonance detection of cancer by using multifunctional magnetic nanocrystals. J. Am. Chem. Soc. 127 (35), 12387–12391. https://doi.org/10.1021/ja052337c.

Hussain, S.M., Hess, K.L., Gearhart, J.M., Geiss, K.T., Schlager, J.J., 2005. In vitro toxicity of nanoparticles in BRL 3A rat liver cells. Toxicol. In Vitro 19 (7), 975–983. https://doi.org/10.1016/j.tiv.2005.06.034.

Jain, P.K., Huang, X., El-Sayed, I.H., El-Sayed, M.A., 2008. Noble metals on the nanoscale: optical and photothermal properties and some applications in imaging, sensing, biology, and medicine. Acc. Chem. Res. 41 (12), 1578–1586. https://doi.org/10.1021/ar7002804.

Jain, P.K., Lee, K.S., El-Sayed, I.H., El-Sayed, M.A., 2006. Calculated absorption and scattering properties of gold nanoparticles of different size, shape, and composition: applications in biological imaging and biomedicine. J. Phys. Chem. B 110 (14), 7238–7248. https://doi.org/10.1021/jp057170o.

Jamdagni, P., Khatri, P., Rana, J.S., 2016. Nanoparticles based DNA conjugates for detection of pathogenic microorganisms. Int. Nano Lett. 6 (3), 139–146. https://doi.org/10.1007/s40089-015-0177-0.

Jamdagni, P., Rana, J.S., Khatri, P., 2017. Rapid optical detection strategy for human pathogens: a brief review. J. Infect. Dis. Diagn 2, 115–121.

Jana, N.R., Peng, X., 2003. Single-phase and gram-scale routes toward nearly monodisperse Au and other noble metal nanocrystals. J. Am. Chem. Soc. 125 (47), 14280–14281. https://doi.org/10.1021/ja038219b.

Jayakumar, M.K.G., Idris, N.M., Zhang, Y., 2012. Remote activation of biomolecules in deep tissues using near-infrared-to-UV upconversion nanotransducers. Proc. Natl. Acad. Sci. 109 (22), 8483–8488. https://doi.org/10.1073/pnas.1114551109.

Jayasena, S.D., 1999. Aptamers: an emerging class of molecules that rival antibodies in diagnostics. Clin. Chem. 45 (9), 1628–1650. https://doi.org/10.1093/clinchem/45.9.1628.

Jazayeri, M.H., Amani, H., Pourfatollah, A.A., Pazoki-Toroudi, H., Sedighimoghaddam, B., 2016. Various methods of gold nanoparticles (GNPs) conjugation to antibodies. Sens. Bio-sens. Res. 9, 17–22. https://doi.org/10.1016/j.sbsr.2016.04.002.

Jeong, U., Teng, X., Wang, Y., Yang, H., Xia, Y., 2007. Superparamagnetic colloids: controlled synthesis and niche applications. Adv. Mater. 19 (1), 33–60. https://doi.org/10.1002/adma.200600674.

Jeong, W.J., Bu, J., Kubiatowicz, L.J., Chen, S.S., Kim, Y., Hong, S., 2018. Peptide–nanoparticle conjugates: a next generation of diagnostic and therapeutic platforms? Nano Converg. 5 (1), 38. https://doi.org/10.1186/s40580-018-0170-1.

Jesus, M., Penadés, S., 2005. Glyco-quantum dots: a new luminescent system with multivalent carbohydrate display. Tetrahedr. Asymm. 16 (2), 387–391. https://doi.org/10.1016/j.tetasy.2004.12.002.

Jesús, M., Penadés, S., 2006. Glyconanoparticles: types, synthesis and applications in glycoscience, biomedicine and material science. Biochim. Biophys. Acta-Gen. Subjects 1760 (4), 636–651. https://doi.org/10.1016/j.bbagen.2005.12.001.

Jiang, S., Win, K.Y., Liu, S., Teng, C.P., Zheng, Y., Han, M.Y., 2013. Surface-functionalized nanoparticles for biosensing and imaging-guided therapeutics. Nanoscale 5 (8), 3127–3148. https://doi.org/10.1039/D0NR90098B.

Jiang, W., Kim, B.Y., Rutka, J.T., Chan, W.C., 2008. Nanoparticle-mediated cellular response is size-dependent. Nat. Nanotechnol. 3 (3), 145–150. https://doi.org/10.1038/nnano.2008.30.

Jiang, X., Bugno, J., Hu, C., Yang, Y., Herold, T., Qi, J., Chen, P., Gurbuxani, S., Arnovitz, S., Strong, J., Ferchen, K., 2016. Eradication of acute myeloid leukemia with FLT3 ligand-targeted miR-150 nanoparticles. Cancer Res. 76 (15), 4470–4480. https://doi.org/10.1158/0008-5472.

Jiang, Y., Zhao, H., Lin, Y., Zhu, N., Ma, Y., Mao, L., 2010. Colorimetric detection of glucose in rat brain using gold nanoparticles. Angew. Chem. Int. Ed. 49 (28), 4800–4804. https://doi.org/10.1002/anie.201001057.

Joshi, S., Sharma, P., Siddiqui, R., Kaushal, K., Sharma, S., Verma, G., Saini, A., 2020. A review on peptide functionalized graphene derivatives as nanotools for biosensing. Microchim. Acta 187 (1), 27. https://doi.org/10.1007/s00604-019-3989-1.

Kajita, M., Hikosaka, K., Iitsuka, M., Kanayama, A., Toshima, N., Miyamoto, Y., 2007. Platinum nanoparticle is a useful scavenger of superoxide anion and hydrogen peroxide. Free Radic. Res. 41 (6), 615–626. https://doi.org/10.1080/10715760601169679.

Kandiel, T.A., Feldhoff, A., Robben, L., Dillert, R., Bahnemann, D.W., 2010. Tailored titanium dioxide nanomaterials: anatase nanoparticles and brookite nanorods as highly active photocatalysts. Chem. Mater. 22 (6), 2050–2060. https://doi.org/10.1021/cm903472p.

Kang, B., Mackey, M.A., El-Sayed, M.A., 2010. Nuclear targeting of gold nanoparticles in cancer cells induces DNA damage, causing cytokinesis arrest and apoptosis. J. Am. Chem. Soc. 132 (5), 1517–1519. https://doi.org/10.1021/ja9102698.

Kar, A., Patra, A., 2012. Impacts of core–shell structures on properties of lanthanide-based nanocrystals: crystal phase, lattice strain, downconversion, upconversion and energy transfer. Nanoscale 4 (12), 3608–3619. https://doi.org/10.1039/C2NR30389B.

Kelly, K.L., Coronado, E., Zhao, L.L., & Schatz, G.C. (2003). The optical properties of metal nanoparticles: the influence of size, shape, and dielectric environment. https://doi.org/10.1021/jp026731y

Khopade, A.J., Caruso, F., 2002. Investigation of the factors influencing the formation of dendrimer/polyanion multilayer films. Langmuir 18 (20), 7669–7676. https://doi.org/10.1021/la020251g.

Kim, B., Han, G., Toley, B.J, 2010. Tuning payload delivery in tumour cylindroids using gold nanoparticles. Nat. Nanotechnol 5, 465. https://doi.org/10.1038/nnano.2010.58.

Kim, D.S., Park, K.S., Jeong, K.C., Lee, B.I., Lee, C.H., Kim, S.Y., 2009. Glucosamine is an effective chemo-sensitizer via transglutaminase 2 inhibition. Cancer Lett. 273 (2), 243–249. https://doi.org/10.1016/j.canlet.2008.08.015.

Kim, J., Takahashi, M., Shimizu, T., Shirasawa, T., Kajita, M., Kanayama, A., Miyamoto, Y., 2008. Effects of a potent antioxidant, platinum nanoparticle, on the lifespan of *Caenorhabditis elegans*. Mech. Ageing Dev. 129 (6), 322–331. https://doi.org/10.1016/j.mad.2008.02.011.

Kitov, P.I., Bundle, D.R., 2003. On the nature of the multivalency effect: a thermodynamic model. J. Am. Chem. Soc. 125 (52), 16271–16284. https://doi.org/10.1021/ja038223n.

Klumpp, C., Lacerda, L., Chaloin, O., Da Ros, T., Kostarelos, K., Prato, M., Bianco, A., 2007. Multifunctionalised cationic fullerene adducts for gene transfer: design, synthesis and DNA complexation. Chem. Commun. (36), 3762–3764. https://doi.org/10.1039/B708435H.

Knobel, M., Nunes, W.C., Socolovsky, L.M., De Biasi, E., Vargas, J.M., Denardin, J.C., 2008. Superparamagnetism and other magnetic features in granular materials: a review on ideal and real systems. J. Nanosci. Nanotechnol. 8 (6), 2836–2857. https://doi.org/10.1166/jnn.2008.15348.

Kose, N., Çaylak, R., Pekşen, C., Kiremitçi, A., Burukoglu, D., Koparal, S., Doğan, A., 2016. Silver ion doped ceramic nano-powder coated nails prevent infection in open fractures: in vivo study. Injury 47 (2), 320–324. https://doi.org/10.1016/j.injury.2015.10.006.

Kumar, C.S., Mohammad, F., 2011. Magnetic nanomaterials for hyperthermia-based therapy and controlled drug delivery. Adv. Drug. Deliv. Rev. 63 (9), 789–808. https://doi.org/10.1016/j.addr.2011.03.008.

Kumar, N., Ray, S.S., 2018. Synthesis and functionalization of nanomaterials. In: Kenig, S. (Ed.), Processing of Polymer-Based Nanocomposites. Springer, Cham, pp. 15–55. https://doi.org/10.1007/978-3-319-97779-9_2.

Kumar, S., Harrison, N., Richards-Kortum, R., Sokolov, K., 2007. Plasmonic nanosensors for imaging intracellular biomarkers in live cells. Nano Lett. 7 (5), 1338–1343. https://doi.org/10.1021/nl070365i.

Lagopati, N., Kitsiou, P.V., Kontos, A.I., Venieratos, P., Kotsopoulou, E., Kontos, A.G., Dionysiou, D.D., Pispas, S., Tsilibary, E.C., Falaras, P., 2010. Photo-induced treatment of breast epithelial cancer cells using nanostructured titanium dioxide solution. J. Photochem. Photobiol. A 214 (2-3), 215–223. https://doi.org/10.1016/j.jphotochem.2010.06.031.

Lansdown, A.B., 2006. Silver in health care: antimicrobial effects and safety in use. Biofunct. Textiles Skin 33, 17–34. https://doi.org/10.1159/000093928.

Larson, T.A., Bankson, J., Aaron, J., Sokolov, K., 2007. Hybrid plasmonic magnetic nanoparticles as molecular specific agents for MRI/optical imaging and photothermal therapy of cancer cells. Nanotechnology 18 (32), 325101. https://doi.org/10.1088/0957-4484/18/32/325101.

Laurent, S., Dutz, S., Häfeli, U.O., Mahmoudi, M., 2011. Magnetic fluid hyperthermia: focus on superparamagnetic iron oxide nanoparticles. Adv. Colloid Interface Sci. 166 (1-2), 8–23. https://doi.org/10.1016/j.cis.2011.04.003.

Laurent, S., Forge, D., Port, M., Roch, A., Robic, C., Vander Elst, L., Muller, R.N., 2008. Magnetic iron oxide nanoparticles: synthesis, stabilization, vectorization, physicochemical characterizations, and biological applications. Chem. Rev. 108 (6), 2064–2110. https://doi.org/10.1021/cr068445e.

Lee, D.K., Kim, S.V., Limansubroto, A.N., Yen, A., Soundia, A., Wang, C.Y., Shi, W., Hong, C., Tetradis, S., Kim, Y., Park, N.H., 2015. Nanodiamond–gutta percha composite biomaterials for root canal therapy. ACS Nano 9 (11), 11490–11501. https://doi.org/10.1021/acsnano.5b05718.

Lee, J.S., Mirkin, C.A., 2008. Chip-based scanometric detection of mercuric ion using DNA-functionalized gold nanoparticles. Anal. Chem. 80 (17), 6805–6808. https://doi.org/10.1021/ac801046a.

Lesch, H.P., Kaikkonen, M.U., Pikkarainen, J.T., Ylä-Herttuala, S., 2010. Avidin-biotin technology in targeted therapy. Expert Opin. Drug Deliv. 7 (5), 551–564. https://doi.org/10.1517/17425241003677749.

Li, C., Fu, R., Yu, C., Li, Z., Guan, H., Hu, D., Zhao, D., Lu, L., 2013. Silver nanoparticle/chitosan oligosaccharide/poly (vinyl alcohol) nanofibers as wound dressings: a preclinical study. Int. J. Nanomed. 8, 4131. https://doi.org/10.2147/IJN.S51679.

Li, C., Lin, J., 2010. Rare earth fluoride nano-/microcrystals: synthesis, surface modification and application. J. Mater. Chem. 20 (33), 6831–6847. https://doi.org/10.1039/C0JM00031K.

Li, D., Zhang, W., Yu, X., Wang, Z., Su, Z., Wei, G., 2016. When biomolecules meet graphene: from molecular level interactions to material design and applications. Nanoscale 8 (47), 19491–19509. https://doi.org/10.1039/C6NR07249F.

Li, H., Rothberg, L., 2004. Colorimetric detection of DNA sequences based on electrostatic interactions with unmodified gold nanoparticles. Proc. Natl. Acad. Sci. 101 (39), 14036–14039. https://doi.org/10.1073/pnas.0406115101.

Li, J., Lu, Y., 2000. A highly sensitive and selective catalytic DNA biosensor for lead ions. J. Am. Chem. Soc. 122 (42), 10466–10467. https://doi.org/10.1021/ja0021316.

Li, X., Kohli, P., 2010. Investigating molecular interactions in biosensors based on fluorescence resonance energy transfer. J. Phys. Chem. C 114 (14), 6255–6264. https://doi.org/10.1021/jp911573g.

Li, Y., Liu, Y., Breaker, R.R., 2000. Capping DNA with DNA. Biochemistry 39 (11), 3106–3114. https://doi.org/10.1021/bi992710r.

Li, Y., Sen, D., 1996. Cleaving DNA with DNA. Nat. Struct. Biol 3, 743–747. https://doi.org/10.1038/nsb0996-743.

Li, Y.C., Lin, Y.S., Tsai, P.J., Chen, C.T., Chen, W.Y., Chen, Y.C., 2007. Nitrilotriacetic acid-coated magnetic nanoparticles as affinity probes for enrichment of histidine-tagged proteins and phosphorylated peptides. Anal. Chem. 79 (19), 7519–7525. https://doi.org/10.1021/ac0711440.

Liandris, E., Gazouli, M., Andreadou, M., Čomor, M., Abazovic, N., Sechi, L.A., Ikonomopoulos, J., 2009. Direct detection of unamplified DNA from pathogenic mycobacteria using DNA-derivatized gold nanoparticles. J. Microbiol. Methods 78 (3), 260–264. https://doi.org/10.1016/j.mimet.2009.06.009.

Link, S., El-Sayed, M.A., 2003. Optical properties and ultrafast dynamics of metallic nanocrystals. Annu. Rev. Phys. Chem. 54 (1), 331–366. https://doi.org/10.1146/annurev.physchem.54.011002.103759.

Link, S., Mohamed, M.B., El-Sayed, M.A., 1999. Simulation of the optical absorption spectra of gold nanorods as a function of their aspect ratio and the effect of the medium dielectric constant. J. Phys. Chem. B 103 (16), 3073–3077. https://doi.org/10.1021/jp990183f.

Liong, M., Lu, J., Kovochich, M., Xia, T., Ruehm, S.G., Nel, A.E., Tamanoi, F., Zink, J.I., 2008. Multifunctional inorganic nanoparticles for imaging, targeting, and drug delivery. ACS Nano 2 (5), 889–896. https://doi.org/10.1021/nn800072t.

Liu, G., Zhao, J., Sun, Q., Zhang, G., 2008. Role of chain interpenetration in layer-by-layer deposition of polyelectrolytes. J. Phys. Chem. B 112 (11), 3333–3338. https://doi.org/10.1021/jp710600f.

Liu, J., Cao, Z., Lu, Y., 2009. Functional nucleic acid sensors. Chem. Rev. 109 (5), 1948–1998. https://doi.org/10.1021/cr030183i.

Liu, J., Lu, Y., 2003. A colorimetric lead biosensor using DNAzyme-directed assembly of gold nanoparticles. J. Am. Chem. Soc. 125 (22), 6642–6643. https://doi.org/10.1021/ja034775u.

Liu, J., Lu, Y., 2004. Adenosine-dependent assembly of aptazyme-functionalized gold nanoparticles and its application as a colorimetric biosensor. Anal. Chem. 76 (6), 1627–1632. https://doi.org/10.1021/ac0351769.

Liu, J., Lu, Y., 2006. Fast colorimetric sensing of adenosine and cocaine based on a general sensor design involving aptamers and nanoparticles. Angew. Chem. Int. Ed. 45 (1), 90–94. https://doi.org/10.1002/anie.200502589.

Liu, J.H., Cao, L., Luo, P.G., Yang, S.T., Lu, F., Wang, H., Meziani, M.J., Haque, S.A., Liu, Y., Lacher, S., Sun, Y.P., 2010. Fullerene-conjugated doxorubicin in cells. ACS Appl. Mater. Interfaces 2 (5), 1384–1389. https://doi.org/10.1021/am100037y.

Liu, Y., Lu, Y., Yang, X., Zheng, X., Wen, S., Wang, F., Vidal, X., Zhao, J., Liu, D., Zhou, Z., Ma, C., 2017. Amplified stimulated emission in upconversion nanoparticles for super-resolution nanoscopy. Nature 543 (7644), 229–233. https://doi.org/10.1038/nature21366.

Lopez, A., Liu, J., 2018. DNA oligonucleotide-functionalized liposomes: bioconjugate chemistry, biointerfaces, and applications. Langmuir 34 (49), 15000–15013. https://doi.org/10.1021/acs.langmuir.8b01368.

Lu, K., Duan, Q.P., Ma, L., Zhao, D.X., 2010. Chemical strategies for the synthesis of peptide–oligonucleotide conjugates. Bioconjug. Chem. 21 (2), 187–202. https://doi.org/10.1021/bc900158s.

Lytton-Jean, A.K., Langer, R., Anderson, D.G., 2011. Five years of siRNA delivery: spotlight on gold nanoparticles. Small 7 (14), 1932–1937. https://doi.org/10.1002/smll.201100761.

Mader, H.S., Kele, P., Saleh, S.M., Wolfbeis, O.S., 2010. Upconverting luminescent nanoparticles for use in bioconjugation and bioimaging. Curr. Opin. Chem. Biol. 14 (5), 582–596. https://doi.org/10.1016/j.cbpa.2010.08.014.

Maeda-Mamiya, R., Noiri, E., Isobe, H., Nakanishi, W., Okamoto, K., Doi, K., Sugaya, T., Izumi, T., Homma, T., Nakamura, E., 2010. In vivo gene delivery by cationic tetraamino fullerene. Proc. Natl. Acad. Sci. 107 (12), 5339–5344. https://doi.org/10.1073/pnas.0909223107.

Mailander, V., Landfester, K., 2009. Interaction of nanoparticles with cells. Biomacromolecules 10 (9), 2379–2400. https://doi.org/10.1021/bm900266r.

Mallakpour, S., Soltanian, S., 2016. Surface functionalization of carbon nanotubes: fabrication and applications. RSC Adv. 6 (111), 109916–109935. https://doi.org/10.1039/C6RA24522F.

Mandal, S., Gole, A., Lala, N., Gonnade, R., Ganvir, V., Sastry, M., 2001. Studies on the reversible aggregation of cysteine-capped colloidal silver particles interconnected via hydrogen bonds. Langmuir 17 (20), 6262–6268. https://doi.org/10.1021/la010536d.

Manivasagan, P., Bharathiraja, S., Moorthy, M.S., Oh, Y.O., Seo, H., Oh, J., 2017. Marine biopolymer-based nanomaterials as a novel platform for theranostic applications. Polym. Rev. 57 (4), 631–667. https://doi.org/10.1080/15583724.2017.1311914.

Marasco, D., Perretta, G., Sabatella, M., Ruvo, M., 2008. Past and future perspectives of synthetic peptide libraries. Curr. Protein Pept. Sci. 9 (5), 447–467. https://doi.org/10.2174/138920308785915209.

Martinez-Veracoechea, F.J., Frenkel, D., 2011. Designing super selectivity in multivalent nano-particle binding. Proc. Natl. Acad. Sci. 108 (27), 10963–10968. https://doi.org/10.1073/pnas.1105351108.

Massich, M.D., Giljohann, D.A., Seferos, D.S., Ludlow, L.E., Horvath, C.M., Mirkin, C.A., 2009. Regulating immune response using polyvalent nucleic acid–gold nanoparticle conjugates. Mol. Pharm. 6 (6), 1934–1940. https://doi.org/10.1021/mp900172m.

Mattoussi, H., Mauro, J.M., Goldman, E.R., Anderson, G.P., Sundar, V.C., Mikulec, F.V., Bawendi, M.G., 2000. Self-assembly of CdSe–ZnS quantum dot bioconjugates using an engineered recombinant protein. J. Am. Chem. Soc. 122 (49), 12142–12150. https://doi.org/10.1021/ja002535y.

McBain, S.C., Yiu, H.H., Dobson, J., 2008. Magnetic nanoparticles for gene and drug delivery. Int. J. Nanomed. 3 (2), 169. https://doi.org/10.2147/ijn.s1608.

Medintz, I.L., Clapp, A.R., Brunel, F.M., Tiefenbrunn, T., Uyeda, H.T., Chang, E.L., Deschamps, J.R., Dawson, P.E., Mattoussi, H., 2006. Proteolytic activity monitored by fluorescence resonance energy transfer through quantum-dot–peptide conjugates. Nat. Mater. 5 (7), 581–589. https://doi.org/10.1038/nmat1676.

Medintz, I.L., Pons, T., Delehanty, J.B., Susumu, K., Brunel, F.M., Dawson, P.E., Mattoussi, H., 2008. Intracellular delivery of quantum dot – protein cargos mediated by cell penetrating peptides. Bioconjug. Chem. 19 (9), 1785–1795. https://doi.org/10.1021/bc800089r.

Medintz, I.L., Pons, T., Susumu, K., Boeneman, K., Dennis, A.M., Farrell, D., Deschamps, J.R., Melinger, J.S., Bao, G., Mattoussi, H., 2009. Resonance energy transfer between luminescent quantum dots and diverse fluorescent protein acceptors. J. Phys. Chem. C 113 (43), 18552–18561. https://doi.org/10.1021/jp9060329.

Medintz, I.L., Uyeda, H.T., Goldman, E.R., Mattoussi, H., 2005. Quantum dot bioconjugates for imaging, labelling and sensing. Nat. Mater. 4 (6), 435–446. https://doi.org/10.1038/nmat1390.

Mei, S.H., Liu, Z., Brennan, J.D., Li, Y., 2003. An efficient RNA-cleaving DNA enzyme that synchronizes catalysis with fluorescence signaling. J. Am. Chem. Soc. 125 (2), 412–420. https://doi.org/10.1021/ja0281232.

Michalet, X., Pinaud, F.F., Bentolila, L.A., Tsay, J.M., Doose, S.J.J.L., Li, J.J., Weiss, S., 2005. Quantum dots for live cells, in vivo imaging, and diagnostics. Science 307 (5709), 538–544. https://doi.org/10.1126/science.1104274.

Miola, M., Fucale, G., Maina, G., Verné, E., 2015. Antibacterial and bioactive composite bone cements containing surface silver-doped glass particles. Biomed. Mater. 10 (5), 055014. https://doi.org/10.1088/1748-6041/10/5/055014.

Mirkin, C.A., Letsinger, R.L., Mucic, R.C., Storhoff, J.J., 1996. A DNA-based method for rationally assembling nanoparticles into macroscopic materials. Nature 382 (6592), 607–609. https://doi.org/10.1038/382607a0.

Modi, D.A., Sunoqrot, S., Bugno, J., Lantvit, D.D., Hong, S., Burdette, J.E., 2014. Targeting of follicle stimulating hormone peptide-conjugated dendrimers to ovarian cancer cells. Nanoscale 6 (5), 2812–2820. https://doi.org/10.1039/C3NR05042D.

Montellano, A., Da Ros, T., Bianco, A., Prato, M., 2011. Fullerene C 60 as a multifunctional system for drug and gene delivery. Nanoscale 3 (10), 4035–4041. https://doi.org/10.1039/C1NR10783F.

Morrison, D., Rothenbroker, M., Li, Y., 2018. DNAzymes: selected for applications. Small Methods 2 (3), 1700319. https://doi.org/10.1002/smtd.201700319.

Mout, R., Moyano, D.F., Rana, S., Rotello, V.M., 2012. Surface functionalization of nanoparticles for nanomedicine. Chem. Soc. Rev. 41 (7), 2539–2544. https://doi.org/10.1039/C2CS15294K.

Moyano, D.F., Rana, S., Bunz, U.H., Rotello, V.M., 2011. Gold nanoparticle-polymer/biopolymer complexes for protein sensing. Faraday Discuss. 152, 33–42. https://doi.org/10.1039/C1FD00024A.

Moyano, D.F., Rotello, V.M., 2011. Nano meets biology: structure and function at the nanoparticle interface. Langmuir 27 (17), 10376–10385. https://doi.org/10.1021/la2004535.

Murphy, C.J., Gole, A.M., Stone, J.W., Sisco, P.N., Alkilany, A.M., Goldsmith, E.C., Baxter, S.C., 2008. Gold nanoparticles in biology: beyond toxicity to cellular imaging. Acc. Chem. Res. 41 (12), 1721–1730. https://doi.org/10.1021/ar800035u.

Murphy, C.J., Sau, T.K., Gole, A.M., Orendorff, C.J., Gao, J., Gou, L., Hunyadi, S.E., & Li, T. (2005). Anisotropic metal nanoparticles: synthesis, assembly, and optical applications. https://doi.org/10.1021/jp0516846

Murray, C.B., Kagan, A.C., Bawendi, M.G., 2000. Synthesis and characterization of monodisperse nanocrystals and close-packed nanocrystal assemblies. Annu. Rev. Mater. Sci. 30 (1), 545–610. https://doi.org/10.1146/annurev.matsci.30.1.545.

Nagasaki, Y., Ishii, T., Sunaga, Y., Watanabe, Y., Otsuka, H., Kataoka, K., 2004. Novel molecular recognition via fluorescent resonance energy transfer using a biotin–PEG/polyamine stabilized CdS quantum dot. Langmuir 20 (15), 6396–6400. https://doi.org/10.1021/la036034c.

Nair, LS., Laurencin, CT., 2007. Biodegradable polymers as biomaterials. Prog. Polym. Sci. 32 (.8-9), 762–798. https://doi.org/10.6092/unibo/amsdottorato/8834.

Nakanishi, T., Iida, H., Osaka, T., 2003. Preparation of iron oxide nanoparticles via successive reduction–oxidation in reverse micelles. Chem. Lett. 32 (12), 1166–1167. https://doi.org/10.1246/cl.2003.1166.

Ndong, C., Toraya-Brown, S., Kekalo, K., Baker, I., Gerngross, T.U., Fiering, S.N., Griswold, K.E., 2015. Antibody-mediated targeting of iron oxide nanoparticles to the folate receptor alpha increases tumor cell association in vitro and in vivo. Int. J. Nanomed. 10, 2595. https://doi.org/10.2147/IJN.S79367.

Nel, A.E., Mädler, L., Velegol, D., Xia, T., Hoek, E.M., Somasundaran, P., Thompson, M., 2009. Understanding biophysicochemical interactions at the nano–bio interface. Nat. Mater. 8 (7), 543–557. https://doi.org/10.1038/nmat2442.

Niidome, T., Yamagata, M., Okamoto, Y., Akiyama, Y., Takahashi, H., Kawano, T., Katayama, Y., Niidome, Y., 2006. PEG-modified gold nanorods with a stealth character for in vivo applications. J. Control. Release 114 (3), 343–347. https://doi.org/10.1016/j.jconrel.2006.06.017.

Nikaido, H., 2009. Multidrug resistance in bacteria. Annu. Rev. Biochem. 78, 119–146. https://doi.org/10.1146/annurev.biochem.78.082907.145923.

Nitta, S.K., Numata, K., 2013. Biopolymer-based nanoparticles for drug/gene delivery and tissue engineering. Int. J. Mol. Sci. 14 (1), 1629–1654. https://doi.org/10.3390/ijms14011629.

Niu, J., Chu, Y., Huang, Y.F., Chong, Y.S., Jiang, Z.H., Mao, Z.W., Peng, L.H., Gao, J.Q., 2017. Transdermal gene delivery by functional peptide-conjugated cationic gold nanoparticle reverses the progression and metastasis of cutaneous melanoma. ACS Appl. Mater. Interfaces 9 (11), 9388–9401. https://doi.org/10.1021/acsami.6b16378.

Oh, E., Hong, M.Y., Lee, D., Nam, S.H., Yoon, H.C., Kim, H.S., 2005. Inhibition assay of biomolecules based on fluorescence resonance energy transfer (FRET) between quantum dots and gold nanoparticles. J. Am. Chem. Soc. 127 (10), 3270–3271. https://doi.org/10.1021/ja0433323.

O'Reilly, R.K., Joralemon, M.J., Wooley, K.L., Hawker, C.J., 2005. Functionalization of micelles and shell cross-linked nanoparticles using click chemistry. Chem. Mater. 17 (24), 5976–5988. https://doi.org/10.1021/cm051047s.

Osman, G., Rodriguez, J., Chan, S.Y., Chisholm, J., Duncan, G., Kim, N., atler, A.L., Shakesheff, K.M., Hanes, J., Suk, J.S., Dixon, J.E., 2018. PEGylated enhanced cell penetrating peptide nanoparticles for lung gene therapy. J. Control. Release 285, 35–45. https://doi.org/10.1016/j.jconrel.2018.07.001.

Pankhurst, Q.A., Connolly, J., Jones, S.K., Dobson, J.J., 2003. Applications of magnetic nanoparticles in biomedicine. J. Phys. D Appl. Phys. 36 (13), R167. https://doi.org/10.1088/0022-3727/36/13/201.

Pankhurst, Q.A., Thanh, N.T.K., Jones, S.K., Dobson, J., 2009. Progress in applications of magnetic nanoparticles in biomedicine. J. Phys. D Appl. Phys. 42 (22), 224001. https://doi.org/10.1088/0022-3727/36/13/201.

Park, J.H., von Maltzahn, G., Xu, M.J., Fogal, V., Kotamraju, V.R., Ruoslahti, E., Bhatia, S.N., Sailor, M.J., 2010. Cooperative nanomaterial system to sensitize, target, and treat tumors. Proc. Natl. Acad. Sci. 107 (3), 981–986. https://doi.org/10.1073/pnas.0909565107.

Park, S.K., Do Kim, K., Kim, H.T, 2002. Preparation of silica nanoparticles: determination of the optimal synthesis conditions for small and uniform particles. Colloids Surf. A 197 (1-3), 7–17. https://doi.org/10.1016/S0927-7757(01)00683-5.

Parker, J.F., Fields-Zinna, C.A., Murray, R.W., 2010. The story of a monodisperse gold nanoparticle: $Au_{25}L_{18}$. Acc. Chem. Res. 43 (9), 1289–1296. https://doi.org/10.1021/ar100048c.

Patel, S.A., Richards, C.I., Hsiang, J.C., Dickson, R.M., 2008. Water-soluble Ag nanoclusters exhibit strong two-photon-induced fluorescence. J. Am. Chem. Soc. 130 (35), 11602–11603. https://doi.org/10.1021/ja804710r.

Patlolla, R.R., Desai, P.R., Belay, K., Singh, M.S., 2010. Translocation of cell penetrating peptide engrafted nanoparticles across skin layers. Biomaterials 31 (21), 5598–5607. https://doi.org/10.1016/j.biomaterials.2010.03.010.

Peer, D., Karp, J.M., Hong, S., Farokhzad, O.C., Margalit, R., Langer, R., 2007. Nanocarriers as an emerging platform for cancer therapy. Nat. Nanotechnol. 2 (12), 751. https://doi.org/10.1038/nnano.2007.387.

Penadés, S., Martín-Lomas, M., Martínez de la Fuente, J., & Rademacher, T.W. (2004). Magnetic nanoparticles. WO 2004/108165 A2.

Peneva, K., Mihov, G., Herrmann, A., Zarrabi, N., Börsch, M., Duncan, T.M., Müllen, K., 2008. Exploiting the nitrilotriacetic acid moiety for biolabeling with ultrastable perylene dyes. J. Am. Chem. Soc. 130 (16), 5398–5399. https://doi.org/10.1021/ja711322g.

Pérez-López, B., Merkoçi, A., 2012. Carbon nanotubes and graphene in analytical sciences. Microchim. Acta 179 (1-2), 1–16. https://doi.org/10.1007/s00604-012-0871-9.

Petros, R.A., DeSimone, J.M., 2010. Strategies in the design of nanoparticles for therapeutic applications. Nat. Rev. Drug Discov. 9 (8), 615–627. https://doi.org/10.1038/nrd2591.

Pinaud, F., King, D., Moore, H.P., Weiss, S., 2004. Bioactivation and cell targeting of semiconductor CdSe/ZnS nanocrystals with phytochelatin-related peptides. J. Am. Chem. Soc. 126 (19), 6115–6123. https://doi.org/10.1021/ja031691c.

Poellmann, M.J., Bu, J., Hong, S., 2018. Would antioxidant-loaded nanoparticles present an effective treatment for ischemic stroke? Nanomedicine 13 (18), 2327–2340. https://doi.org/10.2217/nnm-2018-0084.

Polavarapu, L., Manna, M., Xu, Q.H., 2011. Biocompatible glutathione capped gold clusters as one-and two-photon excitation fluorescence contrast agents for live cells imaging. Nanoscale 3 (2), 429–434. https://doi.org/10.1039/C0NR00458H.

Poole Jr, C.P., Owens, F.J., 2003. Introduction to Nanotechnology. Wiley, Hoboken, New Jersey.

Popov, A.P., Priezzhev, A.V., Lademann, J., Myllylä, R., 2005. TiO_2 nanoparticles as an effective UV-B radiation skin-protective compound in sunscreens. J. Phys. D Appl. Phys. 38 (15), 2564. https://doi.org/10.1088/0022-3727/38/15/006.

Pradeepkumar, P.E., Höbartner, C., Baum, D.A., Silverman, S.K., 2008. DNA-catalyzed formation of nucleopeptide linkages. Angew. Chem. Int. Ed. 47 (9), 1753–1757. https://doi.org/10.1002/anie.200703676.

Putnam, D., Zelikin, A.N., Izumrudov, V.A., Langer, R., 2003. Polyhistidine–PEG: DNA nanocomposites for gene delivery. Biomaterials 24 (24), 4425–4433. https://doi.org/10.1016/s0142-9612(03)00341-7.

Qiao, R., Yang, C., Gao, M., 2009. Superparamagnetic iron oxide nanoparticles: from preparations to in vivo MRI applications. J. Mater. Chem. 19 (35), 6274–6293. https://doi.org/10.1039/B902394A.

Qin, Y., 2005. Silver-containing alginate fibres and dressings. Int. Wound J. 2 (2), 172–176. https://doi.org/10.1111/j.1742-4801.2005.00101.x.

Raj, V., Vijayan, A.N., Joseph, K., 2015. Cysteine capped gold nanoparticles for naked eye detection of E. coli bacteria in UTI patients. Sens. Bio-sens. Res. 5, 33–36. https://doi.org/10.1016/j.sbsr.2015.05.004.

Ramakrishna, G., Varnavski, O., Kim, J., Lee, D., Goodson, T., 2008. Quantum-sized gold clusters as efficient two-photon absorbers. J. Am. Chem. Soc. 130 (15), 5032–5033. https://doi.org/10.1021/ja800341v.

Ramanathan, S., Archunan, G., Sivakumar, M., Selvan, S.T., Fred, A.L., Kumar, S., Gulyás, B., Padmanabhan, P., 2018. Theranostic applications of nanoparticles in neurodegenerative disorders. Int. J. Nanomed. 13, 5561. https://doi.org/10.2147/IJN.S149022.

Rana, S., Bajaj, A., Mout, R., Rotello, V.M., 2012. Monolayer coated gold nanoparticles for delivery applications. Adv. Drug Deliv. Rev. 64 (2), 200–216. https://doi.org/10.1016/j.addr.2011.08.006.

Rao, C.R., Kulkarni, G.U., Thomas, P.J., Edwards, P.P., 2000. Metal nanoparticles and their assemblies. Chem. Soc. Rev. 29 (1), 27–35. https://doi.org/10.1039/A904518J.

Ravindran, A., Chandran, P., Khan, S.S., 2013. Biofunctionalized silver nanoparticles: advances and prospects. Colloids Surf. B 105, 342–352. https://doi.org/10.1016/j.colsurfb.2012.07.036.

Ray, S.C., Saha, A., Jana, N.R., Sarkar, R., 2009. Fluorescent carbon nanoparticles: synthesis, characterization, and bioimaging application. J. Phys. Chem. C 113 (43), 18546–18551. https://doi.org/10.1021/jp905912n.

Redl, F.X., Black, C.T., Papaefthymiou, G.C., Sandstrom, R.L., Yin, M., Zeng, H., Murray, C.B., O'Brien, S.P, 2004. Magnetic, electronic, and structural characterization of nonstoichiometric iron oxides at the nanoscale. J. Am. Chem. Soc. 126 (44), 14583–14599. https://doi.org/10.1021/ja046808r.

Reginster, J.Y., Bruyere, O., Neuprez, A., 2007. Current role of glucosamine in the treatment of osteoarthritis. Rheumatology 46 (5), 731–735. https://doi.org/10.1093/rheumatology/kem026.

Resch-Genger, U., Grabolle, M., Cavaliere-Jaricot, S., Nitschke, R., Nann, T., 2008. Quantum dots versus organic dyes as fluorescent labels. Nat. Methods 5 (9), 763. https://doi.org/10.1038/nmeth.1248.

Reyes, V.C., Opot, S.O., Mahendra, S., 2015. Planktonic and biofilm-grown nitrogen-cycling bacteria exhibit different susceptibilities to copper nanoparticles. Environ. Toxicol. Chem. 34 (4), 887–897. https://doi.org/10.1002/etc.2867.

Reza Mozafari, M., Johnson, C., Hatziantoniou, S., Demetzos, C., 2008. Nanoliposomes and their applications in food nanotechnology. J. Liposome Res. 18 (4), 309–327. https://doi.org/10.1080/08982100802465941.

Riaz, M.K., Riaz, M.A., Zhang, X., Lin, C., Wong, K.H., Chen, X., Zhang, G., Lu, A., Yang, Z., 2018. Surface functionalization and targeting strategies of liposomes in solid tumor therapy: a review. Int. J. Mol. Sci. 19 (1), 195. https://doi.org/10.3390/ijms19010195.

Rizvi, S.A., Saleh, A.M., 2018. Applications of nanoparticle systems in drug delivery technology. Saudi Pharm. J. 26 (1), 64–70. https://doi.org/10.1016/j.jsps.2017.10.012.

Roll, D., Malicka, J., Gryczynski, I., Gryczynski, Z., Lakowicz, J.R., 2003. Metallic colloid wavelength-ratiometric scattering sensors. Anal. Chem. 75 (14), 3440–3445. https://doi.org/10.1021/ac020799s.

Romero, G., Moya, S.E., 2012. Synthesis of organic nanoparticles. Front. Nanosci. 4, 115–141. https://doi.org/10.1016/B978-0-12-415769-9.00004-2.

Roney, C., Kulkarni, P., Arora, V., Antich, P., Bonte, F., Wu, A., Mallikarjuana, N.N., Manohar, S., Liang, H.F., Kulkarni, A.R., Sung, H.W., 2005. Targeted nanoparticles for drug delivery through the blood–brain barrier for Alzheimer's disease. J. Control. Release 108 (2-3), 193–214. https://doi.org/10.1016/j.jconrel.2005.07.024.

Rosenthal, S.J., Chang, J.C., Kovtun, O., McBride, J.R., Tomlinson, I.D., 2011. Biocompatible quantum dots for biological applications. Chem. Biol. 18 (1), 10–24. https://doi.org/10.1016/j.chembiol.2010.11.013.

Rosi, N.L., Giljohann, D.A., Thaxton, C.S., Lytton-Jean, A.K., Han, M.S., Mirkin, C.A., 2006. Oligonucleotide-modified gold nanoparticles for intracellular gene regulation. Science 312 (5776), 1027–1030. https://doi.org/10.1126/science.1125559.

Rosi, N.L., Mirkin, C.A., 2005. Nanostructures in biodiagnostics. Chem. Rev. 105 (4), 1547–1562. https://doi.org/10.1021/cr030067f.

Rouge, J.L., Sita, T.L., Hao, L., Kouri, F.M., Briley, W.E., Stegh, A.H., Mirkin, C.A., 2015. Ribozyme–spherical nucleic acids. J. Am. Chem. Soc. 137 (33), 10528–10531. https://doi.org/10.1021/jacs.5b07104.

Roy, I., Ohulchanskyy, T.Y., Bharali, D.J., Pudavar, H.E., Mistretta, R.A., Kaur, N., Prasad, P.N., 2005. Optical tracking of organically modified silica nanoparticles as DNA carriers: a nonviral, nanomedicine approach for gene delivery. Proc. Natl. Acad. Sci. 102 (2), 279–284. https://doi.org/10.1073/pnas.0408039101.

Ryvkin, A., Ashkenazy, H., Weiss-Ottolenghi, Y., Piller, C., Pupko, T., Gershoni, J.M., 2018. Phage display peptide libraries: deviations from randomness and correctives. Nucleic Acids Res. 46 (9). e52-e52. https://doi.org/10.1093/nar/gky077.

Saallah, S., Lenggoro, I.W., 2018. Nanoparticles carrying biological molecules: recent advances and applications. KONA Powder Particle J. 35, 89–111. https://doi.org/10.14356/kona.2018015.

Sakuma, S., Hayashi, M., Akashi, M., 2001. Design of nanoparticles composed of graft copolymers for oral peptide delivery. Adv. Drug. Deliv. Rev. 47 (1), 21–37. https://doi.org/10.1016/S0169-409X(00)00119-8.

Samanta, D., Sarkar, A., 2011. Immobilization of bio-macromolecules on self-assembled monolayers: methods and sensor applications. Chem. Soc. Rev. 40 (5), 2567–2592. https://doi.org/10.1039/C0CS00056F.

Samuel, U., Guggenbichler, J.P., 2004. Prevention of catheter-related infections: the potential of a new nano-silver impregnated catheter. Int. J. Antimicrob. Agents 23, 75–78. https://doi.org/10.1016/j.ijantimicag.2003.12.004.

Sapsford, K.E., Algar, W.R., Berti, L., Gemmill, K.B., Casey, B.J., Oh, E., Medintz, I.L., 2013. Functionalizing nanoparticles with biological molecules: developing chemistries that facilitate nanotechnology. Chem. Rev. 113 (3), 1904–2074. https://doi.org/10.1021/cr300143v.

Schlosser, K., Li, Y., 2009. Biologically inspired synthetic enzymes made from DNA. Chem. Biol. 16 (3), 311–322. https://doi.org/10.1016/j.chembiol.2009.01.008.

Schlosser, K., Li, Y., 2010. A versatile endoribonuclease mimic made of DNA: characteristics and applications of the 8–17 RNA-cleaving DNAzyme. ChemBioChem 11 (7), 866–879. https://doi.org/10.1002/cbic.200900786.

Schneider, G., Decher, G., 2004. From functional core/shell nanoparticles prepared via layer-by-layer deposition to empty nanospheres. Nano Lett. 4 (10), 1833–1839. https://doi.org/10.1021/nl0490826.

Schoeler, B., Poptoshev, E., Caruso, F., 2003. Growth of multilayer films of fixed and variable charge density polyelectrolytes: effect of mutual charge and secondary interactions. Macromolecules 36 (14), 5258–5264. https://doi.org/10.1021/ma034018g.

Schrooyen, P.M., van der Meer, R., De Kruif, C.G., 2001. Microencapsulation: its application in nutrition. Proc. Nutr. Soc. 60 (4), 475–479. https://doi.org/10.1079/PNS2001112.

Schultz, S., Smith, D.R., Mock, J.J., Schultz, D.A., 2000. Single-target molecule detection with nonbleaching multicolor optical immunolabels. Proc. Natl. Acad. Sci. 97 (3), 996–1001. https://doi.org/10.1073/pnas.97.3.996.

Seferos, D.S., Giljohann, D.A., Hill, H.D., Prigodich, A.E., Mirkin, C.A., 2007. Nano-flares: probes for transfection and mRNA detection in living cells. J. Am. Chem. Soc. 129 (50), 15477–15479. https://doi.org/10.1021/ja0776529.

Selvakannan, P.R., Mandal, S., Phadtare, S., Pasricha, R., Sastry, M., 2003. Capping of gold nanoparticles by the amino acid lysine renders them water-dispersible. Langmuir 19 (8), 3545–3549. https://doi.org/10.1021/la026906v.

Selvan, S.T., Tan, T.T.Y., Yi, D.K., Jana, N.R., 2010. Functional and multifunctional nanoparticles for bioimaging and biosensing. Langmuir 26 (14), 11631–11641. https://doi.org/10.1021/la903512m.

Shahidi, F., Han, X.Q., 1993. Encapsulation of food ingredients. Crit. Rev. Food Sci. Nutr. 33 (6), 501–547. https://doi.org/10.1080/10408399309527645.

Shang, L., Dong, S., Nienhaus, G.U., 2011. Ultra-small fluorescent metal nanoclusters: synthesis and biological applications. Nano Today 6 (4), 401–418. https://doi.org/10.1016/j.nantod.2011.06.004.

Sharma, V., Park, K., Srinivasarao, M., 2009. Colloidal dispersion of gold nanorods: historical background, optical properties, seed-mediated synthesis, shape separation and self-assembly. Mater. Sci. Eng.: R: Rep. 65 (1-3), 1–38. https://doi.org/10.1016/j.mser.2009.02.002.

Sheen, S.H., Kim, J.E., Ryu, H.J., Yang, Y., Choi, K.C., Kang, T.C., 2011. Decrease in dystrophin expression prior to disruption of brain–blood barrier within the rat piriform cortex following status epilepticus. Brain Res. 1369, 173–183. https://doi.org/10.1016/j.brainres.2010.10.080.

Shenton, W., Davis, S.A., Mann, S., 1999. Directed self-assembly of nanoparticles into macroscopic materials using antibody–antigen recognition. Adv. Mater. 11 (6), 449–452. https://doi.org/10.1002/(SICI)1521-4095(199904)11 6≤449::AID-ADMA449≥3.0.CO;2-A.

Shu, J.Y., Panganiban, B., Xu, T., 2013. Peptide-polymer conjugates: from fundamental science to application. Annu. Rev. Phys. Chem. 64, 631–657. https://doi.org/10.1146/annurev-physchem-040412-110108.

Silverman, S.K., 2005. In vitro selection, characterization, and application of deoxyribozymes that cleave RNA. Nucleic Acids Res. 33 (19), 6151–6163. https://doi.org/10.1093/nar/gki930.

Simovic, S., Ghouchi-Eskandar, N., Moom Sinn, A., Losic, D., Prestidge, A.C., 2011. Silica materials in drug delivery applications. Curr. Drug Discov. Technol. 8 (3), 250–268. https://doi.org/10.2174/157016311796799026.

Sitharaman, B., Zakharian, T.Y., Saraf, A., Misra, P., Ashcroft, J., Pan, S., Pham, Q.P., Mikos, A.G., Wilson, L.J., Engler, D.A., 2008. Water-soluble fullerene (C60) derivatives as nonviral gene-delivery vectors. Mol. Pharm. 5 (4), 567–578. https://doi.org/10.1021/mp700106w.

Slocik, J.M., Zabinski Jr, J.S., Phillips, D.M., Naik, R.R., 2008. Colorimetric response of peptide-functionalized gold nanoparticles to metal ions. Small 4 (5), 548–551. https://doi.org/10.1002/smll.200700920.

Song, X., Li, F., Ma, J., Jia, N., Xu, J., Shen, H., 2011. Synthesis of fluorescent silica nanoparticles and their applications as fluorescence probes. J. Fluoresc. 21 (3), 1205–1212. https://doi.org/10.1007/s10895-010-0799-6.

Soppimath, K.S., Kulkarni, A.R., Aminabhavi, T.M., 2001. Chemically modified polyacrylamide-g-guar gum-based crosslinked anionic microgels as pH-sensitive drug delivery systems: preparation and characterization. J. Control. Release 75 (3), 331–345. https://doi.org/10.1016/S0168-3659(01)00404-7.

Sun, Y.P., Zhou, B., Lin, Y., Wang, W., Fernando, K.S., Pathak, P., Meziani, M.J., Harruff, B.A., Wang, X., Wang, H., Luo, P.G., 2006. Quantum-sized carbon dots for bright and colorful photoluminescence. J. Am. Chem. Soc. 128 (24), 7756–7757. https://doi.org/10.1021/ja062677d.

Sundar, S., Kundu, J., Kundu, SC., 2010. Biopolymeric nanoparticles. Sci. Technol. Adv. Mater. 11 (1), 014104. https://doi.org/10.1088/1468-6996/11/1/014104.

Susumu, K., Uyeda, H.T., Medintz, I.L., Pons, T., Delehanty, J.B., Mattoussi, H., 2007. Enhancing the stability and biological functionalities of quantum dots via compact multifunctional ligands. J. Am. Chem. Soc. 129 (45), 13987–13996. https://doi.org/10.1021/ja0749744.

Szwed, A., Milowska, K., Ionov, M., Shcharbin, D., Moreno, S., Gomez-Ramirez, R., Gabryelak, T., 2016. Interaction between dendrimers and regulatory proteins. Comparison of effects of carbosilane and carbosilane–viologen–phosphorus dendrimers. RSC Adv. 6 (100), 97546–97554. https://doi.org/10.1039/C6RA16558C.

Tagmatarchis, N., Shinohara, H., 2001. Fullerenes in medicinal chemistry and their biological applications. Mini Rev. Med. Chem. 1 (4), 339–348. https://doi.org/10.2174/1389557013406684.

Tai, H., Jiang, Y., Xie, G., Yu, J., Chen, X., 2007. Fabrication and gas sensitivity of polyaniline–titanium dioxide nanocomposite thin film. Sens. Actuators B 125 (2), 644–650. https://doi.org/10.1016/j.snb.2007.03.013.

Tan, B., Wu, Y., 2006. Dye-sensitized solar cells based on anatase TiO_2 nanoparticle/nanowire composites. J. Phys. Chem. B 110 (32), 15932–15938. https://doi.org/10.1021/jp063972n.

Tanifum, E.A., Dasgupta, I., Srivastava, M., Bhavane, R.C., Sun, L., Berridge, J., Pourgarzham, H., Kamath, R., Espinosa, G., Cook, S.C., Eriksen, J.L., 2012. Intravenous delivery of targeted liposomes to amyloid-β pathology in APP/PSEN1 transgenic mice. PLoS One 7 (10), e48515. https://doi.org/10.1371/journal.pone.0048515.

Tavares, A.J., Chong, L., Petryayeva, E., Algar, W.R., Krull, U.J., 2011. Quantum dots as contrast agents for in vivo tumor imaging: progress and issues. Anal. Bioanal. Chem. 399 (7), 2331–2342. https://doi.org/10.1007/s00216-010-4010-3.

Thanh, N.T., Green, L.A., 2010. Functionalisation of nanoparticles for biomedical applications. Nano Today 5 (3), 213–230. https://doi.org/10.1016/j.nantod.2010.05.003.

Thanh, N.T.K., Rosenzweig, Z., 2002. Development of an aggregation-based immunoassay for anti-protein A using gold nanoparticles. Anal. Chem. 74 (7), 1624–1628. https://doi.org/10.1021/ac011127p.

Thompson, D.G., Enright, A., Faulds, K., Smith, W.E., Graham, D., 2008. Ultrasensitive DNA detection using oligonucleotide–silver nanoparticle conjugates. Anal. Chem. 80 (8), 2805–2810. https://doi.org/10.1021/ac702403w.

Tiwari, P.M., Vig, K., Dennis, V.A., Singh, S.R., 2011. Functionalized gold nanoparticles and their biomedical applications. Nanomaterials 1 (1), 31–63. https://doi.org/10.3390/nano1010031.

Toghill, K.E., Compton, R.G., 2010. Electrochemical non-enzymatic glucose sensors: a perspective and an evaluation. Int. J. Electrochem. Sci. 5 (9), 1246–1301. http://www.electrochemsci.org/papers/vol5/5091246.pdf.

Tong, L., Wei, Q., Wei, A., Cheng, J.X., 2009. Gold nanorods as contrast agents for biological imaging: optical properties, surface conjugation and photothermal effects. Photochem. Photobiol. 85 (1), 21–32. https://doi.org/10.1111/j.1751-1097.2008.00507.x.

Tram, K., Kanda, P., Li, Y., 2012. Lighting up RNA-cleaving DNAzymes for biosensing. J. Nucleic Acids 2012, 1–8. https://doi.org/10.1155/2012/958683.

Trewyn, B.G., Slowing, I.I., Giri, S., Chen, H.T., Lin, V.S.Y., 2007. Synthesis and functionalization of a mesoporous silica nanoparticle based on the sol–gel process and applications in controlled release. Acc. Chem. Res. 40 (9), 846–853. https://doi.org/10.1021/ar600032u.

Vanhee, P., van der Sloot, A.M., Verschueren, E., Serrano, L., Rousseau, F., Schymkowitz, J., 2011. Computational design of peptide ligands. Trends Biotechnol. 29 (5), 231–239. https://doi.org/10.1016/j.tibtech.2011.01.004.

Vauquelin, G., Charlton, S.J., 2010. Long-lasting target binding and rebinding as mechanisms to prolong in vivo drug action. Br. J. Pharmacol. 161 (3), 488–508. https://doi.org/10.1111/j.1476-5381.2010.00936.x.

Veerapandian, M., Lim, S.K., Nam, H.M., Kuppannan, G., Yun, K.S., 2010. Glucosamine-functionalized silver glyconanoparticles: characterization and antibacterial activity. Anal. Bioanal. Chem. 398 (2), 867–876. https://doi.org/10.1007/s00216-010-3964-5.

Veerapandian, M., Sadhasivam, S., Choi, J., Yun, K., 2012. Glucosamine functionalized copper nanoparticles: preparation, characterization and enhancement of anti-bacterial activity by ultraviolet irradiation. Chem. Eng. J. 209, 558–567. https://doi.org/10.1016/j.cej.2012.08.054.

Verdoodt, N., Basso, C.R., Rossi, B.F., Pedrosa, V.A., 2017. Development of a rapid and sensitive immunosensor for the detection of bacteria. Food Chem. 221, 1792–1796. https://doi.org/10.1016/j.foodchem.2016.10.102.

Vidu, R., Rahman, M., Mahmoudi, M., Enachescu, M., Poteca, T.D., Opris, I., 2014. Nanostructures: a platform for brain repair and augmentation. Front. Syst. Neurosci. 8, 91. https://doi.org/10.3389/fnsys.2014.00091.

Vinogradov, S., Batrakova, E., Kabanov, A., 1999. Poly (ethylene glycol)–polyethyleneimine NanoGel™ particles: novel drug delivery systems for antisense oligonucleotides. Colloids Surf. B 16 (1-4), 291–304. https://doi.org/10.1016/S0927-7765(99)00080-6.

Volicer, L., 2001. Management of severe Alzheimer's disease and end-of-life issues. Clin. Geriatr. Med. 17 (2), 377–391. https://doi.org/10.1016/S0749-0690(05)70074-4.

Walker, D.A., Kowalczyk, B., de La Cruz, M.O., Grzybowski, B.A., 2011. Electrostatics at the nanoscale. Nanoscale 3 (4), 1316–1344. https://doi.org/10.1039/C0NR00698J.

Wang, F., Deng, R.R., Wang, J., Wang, Q.X., Han, Y., Zhu, H.M., 2011. XY UCNPs@ ZnxCd1-xS/TiO$_2$ core-shell nanoparticles.). Nat. Mater 10, 968–973. https://doi.org/10.1038/nmat3149.

Wang, F., Han, Y., Lim, C.S., Lu, Y., Wang, J., Xu, J., Chen, H., Zhang, C., Hong, M., Liu, X., 2010. Simultaneous phase and size control of upconversion nanocrystals through lanthanide doping. Nature 463 (7284), 1061–1065. https://doi.org/10.1038/nature08777.

Wang, F., Liu, X., 2009. Recent advances in the chemistry of lanthanide-doped upconversion nanocrystals. Chem. Soc. Rev. 38 (4), 976–989. https://doi.org/10.1039/B809132N.

Wang, L., Hu, C., Shao, L., 2017. The antimicrobial activity of nanoparticles: present situation and prospects for the future. Int. J. Nanomed. 12, 1227. https://doi.org/10.2147/IJN.S121956.

Wang, L., Zhao, W., Tan, W., 2008. Bioconjugated silica nanoparticles: development and applications. Nano Res. 1 (2), 99–115. https://doi.org/10.1007/s12274-008-8018-3.

Wang, S.H., Yu, J., 2018. Structure-based design for binding peptides in anti-cancer therapy. Biomaterials 156, 1–15. https://doi.org/10.1016/j.biomaterials.2017.11.024.

Wang, Y.Q., Chen, S.G., Tang, X.H., Palchik, O., Zaban, A., Koltypin, Y., Gedanken, A., 2001. Mesoporous titanium dioxide: sonochemical synthesis and application in dye-sensitized solar cells. J. Mater. Chem. 11 (2), 521–526. https://doi.org/10.1039/B006070O.

Wark, A.W., Lee, H.J., Qavi, A.J., Corn, R.M., 2007. Nanoparticle-enhanced diffraction gratings for ultrasensitive surface plasmon biosensing. Anal. Chem. 79 (17), 6697–6701. https://doi.org/10.1021/ac071062b.

Wei, M., Konishi, Y., Zhou, H., Yanagida, M., Sugihara, H., Arakawa, H., 2006. Highly efficient dye-sensitized solar cells composed of mesoporous titanium dioxide. J. Mater. Chem. 16 (13), 1287–1293. https://doi.org/10.1039/B514647J.

Wen, S., Zhou, J., Zheng, K., Bednarkiewicz, A., Liu, X., Jin, D., 2018. Advances in highly doped upconversion nanoparticles. Nat. Commun. 9 (1), 1–12. https://doi.org/10.1038/s41467-018-04813-5.

Wilchek, M., Bayer, E.A., Livnah, O., 2006. Essentials of biorecognition: the (strept) avidin–biotin system as a model for protein–protein and protein–ligand interaction. Immunol. Lett. 103 (1), 27–32. https://doi.org/10.1016/j.imlet.2005.10.022.

Wilhelmsson, L.M., 2010. Fluorescent nucleic acid base analogues. Q. Rev. Biophys. 43 (2), 159–183. https://doi.org/10.1017/S0033583510000090.

Willey, T.M., Vance, A.L., Bostedt, C., van Buuren, T., Meulenberg, R.W., Terminello, L.J., Fadley, C.S., 2004. Surface structure and chemical switching of thioctic acid adsorbed on Au (111) as observed using near-edge X-ray absorption fine structure. Langmuir 20 (12), 4939–4944. https://doi.org/10.1021/la049868j.

Wong, K.K., Cheung, S.O., Huang, L., Niu, J., Tao, C., Ho, C.M., Che, C.M., Tam, P.K., 2009. Further evidence of the anti-inflammatory effects of silver nanoparticles. ChemMedChem: Chem. Enabling Drug Discov. 4 (7), 1129–1135. https://doi.org/10.1002/cmdc.200900049.

Wu, P., Hwang, K., Lan, T., Lu, Y., 2013. A DNAzyme-gold nanoparticle probe for uranyl ion in living cells. J. Am. Chem. Soc. 135 (14), 5254–5257. https://doi.org/10.1021/ja400150v.

Xia, W., Grandfield, K., Hoess, A., Ballo, A., Cai, Y., Engqvist, H., 2012. Mesoporous titanium dioxide coating for metallic implants. J. Biomed. Mater. Res. Part B: Appl. Biomater. 100 (1), 82–93. https://doi.org/10.1002/jbm.b.31925.

Xiao, Y., Reis, L.A., Feric, N., Knee, E.J., Gu, J., Cao, S., Laschinger, C., Londono, C., Antolovich, J., McGuigan, A.P., Radisic, M., 2016. Diabetic wound regeneration using peptide-modified hydrogels to target re-epithelialization. Proc. Natl. Acad. Sci. 113 (40), E5792–E5801. https://doi.org/10.1073/pnas.1612277113.

Xing, Z.C., Chang, Y., Kang, I.K., 2010. Immobilization of biomolecules on the surface of inorganic nanoparticles for biomedical applications. Sci. Technol. Adv. Mater. 11 (1), 014101. https://doi.org/10.1088/1468-6996/11/1/014101.

Xiong, Y., McLellan, J.M., Chen, J., Yin, Y., Li, Z.Y., Xia, Y., 2005. Kinetically controlled synthesis of triangular and hexagonal nanoplates of palladium and their SPR/SERS properties. J. Am. Chem. Soc. 127 (48), 17118–17127. https://doi.org/10.1021/ja056498s.

Xu, H., Suslick, K.S., 2010. Water-soluble fluorescent silver nanoclusters. Adv. Mater. 22 (10), 1078–1082. https://doi.org/10.1002/adma.200904199.

Xu, S.L., Cui, H., 2007. Luminol chemiluminescence catalysed by colloidal platinum nanoparticles. Luminescence 22 (2), 77–87. https://doi.org/10.1002/bio.929.

Yang, H., Fung, S.Y., Sun, W., Mikkelsen, S., Pritzker, M., Chen, P., 2008. Ionic-complementary peptide-modified highly ordered pyrolytic graphite electrode for biosensor application. Biotechnol. Prog. 24 (4), 964–971. https://doi.org/10.1002/btpr.1.

Yang, S.T., Wang, X., Wang, H., Lu, F., Luo, P.G., Cao, L., Meziani, M.J., Liu, J.H., Liu, Y., Chen, M., Huang, Y., 2009. Carbon dots as nontoxic and high-performance fluorescence imaging agents. J. Phys. Chem. C 113 (42), 18110–18114. https://doi.org/10.1021/jp9085969.

Yang, Y., Shao, Q., Deng, R., Wang, C., Teng, X., Cheng, K., Cheng, Z., Huang, L., Liu, Z., Liu, X., Xing, B., 2012. In vitro and in vivo uncaging and bioluminescence imaging by using photocaged upconversion nanoparticles. Angew. Chem. Int. Ed. 51 (13), 3125–3129. https://doi.org/10.1002/anie.201107919.

Yao, H., Wang, K., Wang, Y., Wang, S., Li, J., Lou, J., Ye, L., Yan, X., Lu, W., Huang, R., 2015. Enhanced blood–brain barrier penetration and glioma therapy mediated by a new peptide modified gene delivery system. Biomaterials 37, 345–352. https://doi.org/10.1016/j.biomaterials.2014.10.034.

Yehl, K., Joshi, J.P., Greene, B.L., Dyer, R.B., Nahta, R., Salaita, K., 2012. Catalytic deoxyribozyme-modified nanoparticles for RNAi-independent gene regulation. ACS Nano 6 (10), 9150–9157. https://doi.org/10.1021/nn3034265.

Zempleni, J., Wijeratne, S.S., Hassan, Y.I., 2009. Biotin. Biofactors 35 (1), 36–46. https://doi.org/10.1002/biof.8.

Zeng, S., Yong, K.T., Roy, I., Dinh, X.Q., Yu, X., Luan, F., 2011. A review on functionalized gold nanoparticles for biosensing applications. Plasmonics 6 (3), 491. https://doi.org/10.1007/s11468-011-9228-1.

Zhang, F., Lees, E., Amin, F., Rivera_Gil, P., Yang, F., Mulvaney, P., Parak, W.J., 2011. Polymer-coated nanoparticles: a universal tool for biolabelling experiments. Small 7 (22), 3113–3127. https://doi.org/10.1002/smll.201100608.

Zhang, J., Ge, J., Shultz, M.D., Chung, E., Singh, G., Shu, C., Fatouros, P.P., Henderson, S.C., Corwin, F.D., Geohegan, D.B., Puretzky, A.A., 2010. In vitro and in vivo studies of single-walled carbon nanohorns with encapsulated metallofullerenes and exohedrally functionalized quantum dots. Nano Lett. 10 (8), 2843–2848. https://doi.org/10.1021/nl1008635.

Zhao, W., Karp, J.M., 2009. Nanoantennas heat up. Nat. Mater. 8 (6), 453–454. https://doi.org/10.1038/nmat2463.

Zheng, J., Dickson, R.M., 2002. Individual water-soluble dendrimer-encapsulated silver nanodot fluorescence. J. Am. Chem. Soc. 124 (47), 13982–13983. https://doi.org/10.1021/ja028282l.

Zheng, J., Petty, J.T., Dickson, R.M., 2003. High quantum yield blue emission from water-soluble Au_8 nanodots. J. Am. Chem. Soc. 125 (26), 7780–7781. https://doi.org/10.1021/ja035473v.

Zhou, B., Shi, B., Jin, D., Liu, X., 2015. Controlling upconversion nanocrystals for emerging applications. Nat. Nanotechnol. 10 (11), 924. https://doi.org/10.1038/nnano.2015.251.

Zhou, J., Liu, Z., Li, F., 2012. Upconversion nanophosphors for small-animal imaging. Chem. Soc. Rev. 41 (3), 1323–1349. https://doi.org/10.1039/C1CS15187H.

Zhou, W., Ding, J., Liu, J., 2017. Theranostic dnazymes. Theranostics 7 (4), 1010. https://doi.org/10.7150/thno.17736.

Zhou, Y., Tang, L., Zeng, G., Zhang, C., Zhang, Y., Xie, X., 2016. Current progress in biosensors for heavy metal ions based on DNAzymes/DNA molecules functionalized nanostructures: a review. Sens. Actuators B 223, 280–294. https://doi.org/10.1016/j.snb.2015.09.090.

Zhu, J., Zhou, Z., Yang, C., Kong, D., Wan, Y., Wang, Z., 2011. Folate-conjugated amphiphilic star-shaped block copolymers as targeted nanocarriers. J. Biomed. Mater. Res. Part A 97 (4), 498–508. https://doi.org/10.1002/jbm.a.33071.

Zhu, Y., Shi, J., Zhang, Z., Zhang, C., Zhang, X., 2002. Development of a gas sensor utilizing chemiluminescence on nanosized titanium dioxide. Anal. Chem. 74 (1), 120–124. https://doi.org/10.1021/ac010450p.

Zou, L., Ding, W., Zhang, Y., Cheng, S., Li, F., Ruan, R., Wei, P., Qiu, B., 2018. Peptide-modified vemurafenib-loaded liposomes for targeted inhibition of melanoma via the skin. Biomaterials 182, 1–12. https://doi.org/10.1016/j.biomaterials.2018.08.013.

Zrazhevskiy, P., Sena, M., Gao, X., 2010. Designing multifunctional quantum dots for bioimaging, detection, and drug delivery. Chem. Soc. Rev. 39 (11), 4326–4354. https://doi.org/10.1039/B915139G.

Chapter 20

Nanostructures for the efficient oral delivery of chemotherapeutic agents

Ravindra Satpute[a,#], Nilesh Rarokar[b,#], Sunil Menghani[b], Anjali Ganjare[b], Vivek S. Dave[c], Nishikant A. Raut[b], Pramod B. Khedekar[b]

[a]*Toxicology Laboratory, Defense R & D Establishment, Nagpur, Maharashtra, India*
[b]*Department of Pharmaceutical Sciences, Rashtrasant Tukadoji Maharaj Nagpur University, Nagpur, Maharashtra, India*
[c]*Department of Pharmaceutical Sciences, St. John Fisher College, Wegmans School of Pharmacy, Rochester, NY, USA*
[#]*Equal contribution.*

20.1 Introduction

Drug administration with a suitable method is one of the important aspects for the efficacy of pharmaceuticals and the oral route has been convenient, popular, and widely used for drug administration. The oral route is preferred being the simplest method allowing a variety of dosage forms to be administered. The most important aspect of the oral route is the requirement of the fewest aseptic restrictions during manufacturing and the easiest methods of manufacturing. Oral novel drug delivery systems outperform oral traditional formulations, including tablets, pills, liquids, emulsions, and suspensions. However, due to their peculiar properties and wide range of possible biochemical, optical, and electronic applications, nanoparticles are now the subject of intense science and engineering attention. Drug development is a continuous process indispensable for improving the safety and effectiveness of therapy when given to patients. Along with this, the selection of appropriate dosage forms also plays a vital role for efficient delivery of drugs and therefore several pharmaceutical companies struggle to improve drug properties, delivery systems, and dosage types. One such technique to improve drug efficacy is the use of nanotechnology in pharmaceuticals at different stages of drug development.

The theoretical framework for understanding basic concepts of nanotechnology is essentially needed for the effective development of a nanomedicine into an oral dosage form for delivering a drug and some of the concepts include:

i. Biopharmaceutical properties of the drug molecule
ii. Anatomy and physiology of the gastrointestinal tract and digestive system
iii. Physicochemical characteristics of excipients used for novel dosage form design for the model delivery.

Though it is impractical to change the biopharmaceutical properties of a drug by chemical alterations such as the synthesis of an analog or a prodrug for improving the efficacy of the drug, these methods have been used for several therapeutic agents. Furthermore, it is scientifically impossible to change the anatomy and physiology of the gastrointestinal tract; however, it is imperative to convert drug molecules into such a form that will be a more practical solution to this problem. The design of nanostructured medicine as an oral dosage form by optimizing characteristics of dosage form according to the need of GIT characteristics may provide some opportunities to rationalize systemic drug delivery with maximum therapeutic benefits (Chein, 2005; Wang et al., 2021).

The multifunctional capabilities, flexibility, and alteration in the designing of nanostructures are the key features for therapeutic applications using a variety of ways of interaction of nanostructures with biological systems. These versatile characteristics are essential for treating intricately assorted life-threatening diseases, such as cancer. A combination of varied functionalities for targeting cancerous cells is possible through cancer therapies involving nanostructures. Although, this is not that easy as it appears there are several hurdles and challenges which are required to be taken care of, some of them include absorption from the site of administration and transportation to the targeted tissues due to restrictions at reticuloendothelium and altered membrane permeabilities of tumor cells. Sometimes, the penetration of nanoformulation is less than free drug creating challenges for the formulation scientists; however, the multifunctionality of nanostructures can overcome these limitations (Douglas et al., 2012; Shi et al., 2017; Chen et al., 2017; Anselmo and Mitragotri, 2019; Zhao et al., 2020; Wang et al. 2021). The unique properties of nanostructures incorporate lipophilic and hydrophilic drugs with high carrying capacity and stability, making them suitable for varied routes of administration allowing it for controlled delivery of drugs as per need (Salahuddin and Galal, 2017).

20.1.1 Limitations of conventional chemotherapy

i. A distinguishing characteristic of cancer cells is the rapid and uncontrolled division, which is disrupted by the traditional chemotherapeutic agents. Along with rapidly dividing cancer cells, these chemotherapeutic agents also disrupt healthy cells which divide quickly, and these cells include the cells of the GIT lining, bone marrow, hair follicles, and macrophages.

ii. The biggest disadvantage of traditional chemotherapy includes nonspecific targeting and the inability to target only cancer cells.

iii. Myelosuppression (lower production of white blood cells, which causes immunosuppression), alopecia (hair loss), mucositis (the inflammation of the mucosal lining of the gastrointestinal tract), anemia or thrombocytopenia, and organ dysfunction, are all common adverse effects of most chemotherapeutic agents.

iv. These side effects can lead to dose reductions, treatment delays, or therapy discontinuation.

v. Chemotherapeutic agents become insensitive in solid tumors because the cell division is stopped effectively close to the center, and they often fail to permeate to reach the center of solid tumors failing to destroy cancer cells.

vi. Conventional chemotherapeutic drugs are often flushed out of circulation due to macrophage engulfment. As a result, their period of action and association with cancerous cells are affected, rendering chemotherapy ineffective.

vii. The drugs' low solubility is also a major issue in traditional chemotherapy, preventing them from penetrating biological membranes.

viii. P-glycoprotein is a multidrug-resistant protein that is over-expressed on the surfaces of the cancer cells that acts as an efflux pump preventing the accumulation of drugs within the tumor and mediates the development of anticancer drug resistance. As a result, the medications used are ineffective or do not produce the desired results.

20.1.2 Edges of nanoparticles over the other delivery system

i. The size of nanoparticles and surface properties can be effectively managed to accomplish either active or passive targeting following the administration.

ii. The drug release at the site of absorption and during permeation has been maintained and regulated by the nanoparticles, altering the delivery of drug and subsequent clearance to enhance the effectiveness of the drug, and reduce adverse effects.

iii. The controlled release of drug and degradation characteristics of nanoparticles can be modulated by selecting appropriate matrix constituents with high drug loading without undergoing any chemical reactions; this is a fundamental requirement for maintaining drug activity.

iv. Targeting a drug at a specific site of action is usually achieved by tagging targeting ligand on the surfaces of nanoparticles and/or using magnetic guidance.

v. The nanoparticulate systems are used for a variety of routes of administration that includes intra-ocular, nasal, oral, parenteral, and other routes of administration.

vi. The targetability of nanoparticle delivery systems is restricted mainly to the liver and spleen, and to some extent to the bone marrow.

20.1.3 Components of nanoparticles as a targeting system

The idea of an "ingenious" drug targeting system necessitates the behavior of three components: the nanocarrier, the targeting moiety, and the therapeutic drug, all of which must work together.

Targeting moieties are the first one that identifies and binds the target

Nanocarriers are the second who carries the drugs.

Therapeutic agents, the third one has a therapeutic effect on a particular location.

Targeting moieties - proteins, antibodies, hormones, lipoproteins, polysaccharides, charged molecules, low-molecular-weight ligands.

20.1.4 Characteristics features of ideal targeting moieties

i. Ideal targeting moieties should be capable of crossing blood-brain barriers, and tumor vasculature in the case of chemotherapy of tumor.
ii. It must be selectively and specifically recognized by target cells maintaining the surface ligand specificity.
iii. In plasma, interstitial fluid, and other biofluids, the drug-ligand complex should be stable.
iv. Nontoxic, nonimmunogenic, and biodegradable targeting moieties should be used.
v. Following identification, the drug moiety should be released into the target organs, tissues, or cells by the carrier system.

20.1.5 The potential of nanocarriers as drug delivery systems

i. Exhibit higher intracellular uptake
ii. Can pass through the submucosal membranes, while microcarriers are often found on the epithelial lining.
iii. May be administered into the bloodstream without causing particle accumulation or blocking fine blood capillaries.

20.1.6 Nanoparticle properties

To control the accumulation kinetics of tumors and to prevent the reverse diffusion of drug back to systemic vasculature requires an understanding of tumor size. It has been demonstrated that liposomal formulations with 90 nm liposome size are responsible for the diffusion through tumor vasculature, but it retained into the tumor beyond a week. However, the impact of particle size of nanomaterials on blood circulation is more complicated and hence it does not behave like other small molecules or protein-based chemotherapeutic agents. Particles with a small size distribution were used to investigate the effect of particle size on biodistribution. Particle size has direct relevance with the removal of nanomaterials from the blood circulation as the smaller particles with a diameter in the range of 50 to 300 nm are removed very slowly compared to the bigger particles (Rarokar et al., 2021). To achieve long term circulation of narrow size nanomaterial in blood sustained release drug delivery is warranted. Additionally, nanoparticle stability is ensured in tumors by improving blood circulation and its aggregation. Previous studies revealed drug-loaded polymeric micelles are usually ranging between 100 and 150 nm in size (Rarokar et al., 2019).

Relevance of biological system with the particle size of the nanomaterials (Bae and Park, 2011).

Particle size and anticipated bioactivities in human physiology are illustrated in Fig. 20.1.

20.1.7 Cancer therapy: Selective targeting of tissues by nanotechnology

Cancer is one of the world's most deadly illnesses, claiming the lives of millions of people annually. Since, it can affect to any organ in human beings irrespective of community, region, and social status. For the treatment of cancer, chronic and combined therapies are required as it is uncontrolled cell division without apoptosis. Medical diversity and therapeutic resistance are evident due to the heterogeneity of the genetic and phenotypic stages; having diverse treatments with their own set of limitations and side effects. Surgical intervention, drug treatment including hormonal therapy and radiation are the major

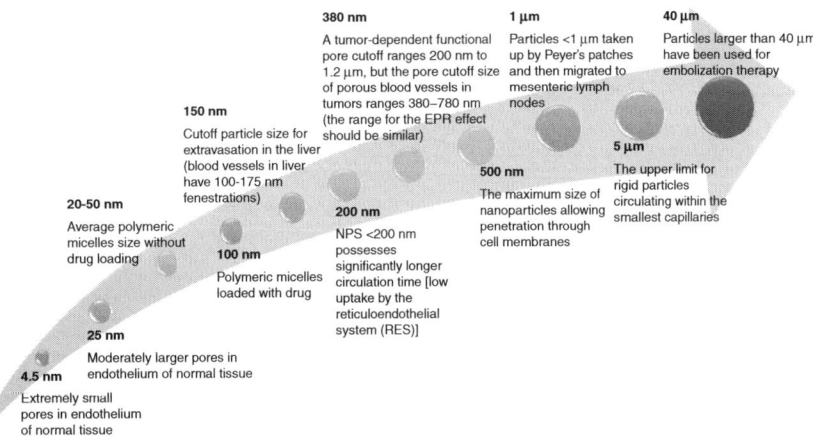

FIG. 20.1 Relevance of biological system with the particle size of the nanomaterial.

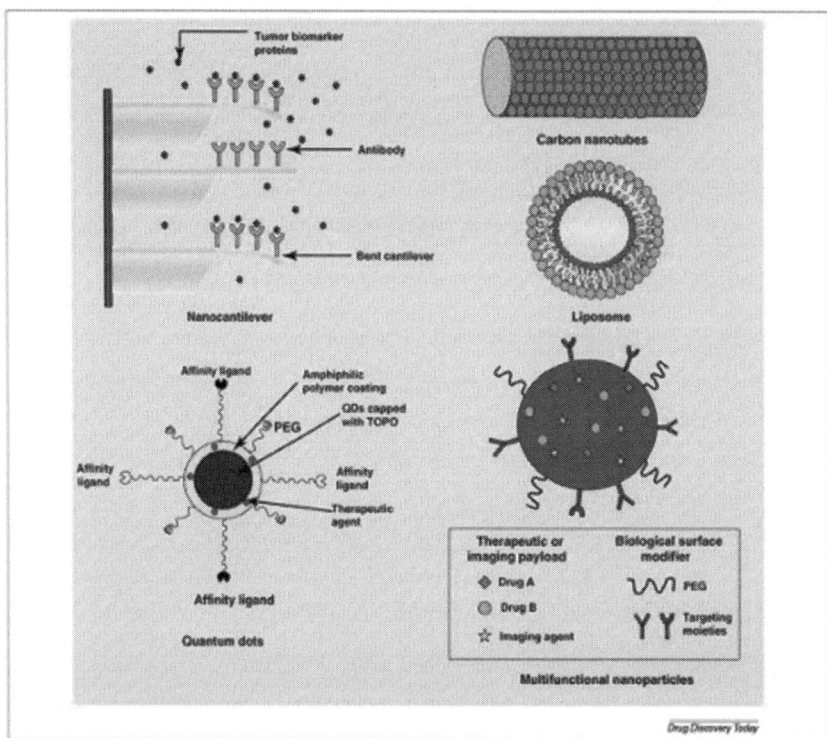

FIG. 20.2 Schematic representation of different nanocarriers like liposomes, carbon nanotubes, multifunctional nanoparticles used in cancer chemotherapy (Misra et al., 2010).

regimen for cancer care. Chemotherapy is a standard procedure that delivers anticancer drugs systemically to patients to stop cancerous cells from proliferating uncontrollably. Unfortunately, many side effects arise because of anticancer agents' nonspecific targeting, and inadequate drug delivery of those agents is unable to achieve the desired outcome usually.

The development of anticancer drugs is a time-consuming process involving advanced electronic engineering and polymer chemistry. For the drug to be efficacious, the most challenging aspect is that the cancer therapy requires distinguishing cancerous cells from healthy body cells. As a result, the primary goal is to engineer the chemotherapeutic agent for recognizing cancer cells and inhibiting the progression and propogation as shown in Fig. 20.2. Selective killing of cancer cells with standard chemotherapy without affecting healthy cells cannot be achieved that leads to severe adverse effects such as organ damage causing medication to be compromised, leading to lower doses and, eventually, lower survival rates.

20.2 Nanodrug carriers

Nanoparticles are colloidal particulate structures with sizes in the sub-micron range that serve as drug carriers comprising negatively charged lipophilic amorphous solid spheres. The scale ranges from 10 nm to 1000 nm, depending on the manufacturing process. The porosity of nanocapsules can range from 3 to 6 nanometers, and the wall thickness can range from 15 to 60 nanometers. The drug is entrapped, dissolved, and/or adsorbed to the macromolecular substance. Parenteral administration is the preferred method. Moreover, it is possible to change the distribution of nanoparticles in the body either may by coating them with components of serum or having antibodies to their surfaces. Magnetic nanoparticles, on the other hand, may be used to improve site-specificity.

20.2.1 Classification of nanoparticles as drug carriers

Nano-drug carriers are classified into two broad categories depending on the composition as lipid nanocarriers and inorganic nanocarriers.

Lipid nanocarriers: The rapidly evolving nanotechnology field majorly comprises lipid nanocarriers with several drug delivery applications for a variety of diseases, especially cancer. Lipid nanoparticles are used to deliver new therapeutic agents efficiently providing an alternate way for colloidal equivalents available for drug delivery. Lipids differ in their structures and design affecting the properties of the nanocarrier and their intended use, and therefore it is imperative to consider their effects while choosing lipids for the carrier system. Depending on the properties, lipid nanocarriers can be further divided into:

Liposomes

Bangham et al. (1965), first described the closed, spherical phospholipid bilayer mimicking cell membrane as liposomes are the earliest nanoscale drug carriers. Water-soluble drugs are encapsulated within hydrophilic cores of liposomes by their lipid bilayer structure, whereas water-insoluble drugs are carried over by lipid bilayers (Ghosh et al., 2008; Haley and Frenkel, 2008). Depending on the type and their lamellar layers liposomes sizes may vary from 500 to 5000 nm (Alexis et al., 2010; Torchilin, 2005). *United States Food and Drug Administration (US-FDA)* approved various biocompatible phospholipids, that includes 1,2-distearoyl-glycero-3-phosphocholine (DSPG) and 1,2-distearoylsn-glycero-3-phosphoethanolamine (DSPE) to use for human (Alexis et al., 2010). Liposomes surfaces are modified using polyethylene glycol (PEGylation) to increase their blood circulation (Arias, 2013; Klibanov et al., 1990), with various targeting moieties like anisamide (ligand to sigma receptors) (Banerjee et al., 2004; Li and Rodriguez et al., 2013) or macromolecules like monoclonal antibodies (immunoliposomes) (Raju et al., 2013; Gao et al., 2010) specifically designed to bind targets on the malignant cells. The anticancer drugs formulated into liposomal drug delivery system are present on the market. One such product is Doxil which is about 100 nm (Liposomes of PEGylated doxorubicin) and are being used against metastatic breast and ovarian cancers, HIV-related Kaposi's sarcoma. Combination of non-PEGylated doxorubicin liposomes (Myocet) with cyclophosphamide is used in metastatic breast cancer whereas, non-PEGylated daunorubicin liposomes (Daunoxome) ranging about 35–65 nm, are used in HIV-related Kaposi's sarcoma.

20.2.2 Micelles

The self-assembled nanoaggregates with a lipophilic core having hydrophilic shell are called micelles and used to solubilize lipophilic therapeutic agents. These are made up of amphiphilic block-copolymers and sometimes lipid-based molecules having self-assembling properties. Several micelles based on poloxamers (Saxena and Hussain, 2012), PCL-PEG (Vasey et al., 1999), or PEGylated polyesters for example, PLA-PEG (Kim et al., 2001) have been reported for delivery of antineoplastics. In the year 2006, PLA-PEG (Genexol-PM) micelles, encapsulated with paclitaxel with particles size ranging around 60 nm, has been approved as a clinical option to treat cancer in humans. A mixed micelle nanostructure is made up of mixed copolymers comprising of block copolymers that have been mixed to give the final micellar structure (Ostacolo, 2010). Hydrophilic polyethylene oxide (PEO) and lipophilic polypropylene oxide (PPO) together forms the poloxamers, wherein the PPO is sandwiched between two molecules of PEO. A mixed micelle comprising Pluronics F127 and P105 formulation has been developed for delivering docetaxel to lung cancer cells resistant to taxol. Because of the high PPO content in P105, a high concentration of docetaxel could be loaded, on the other hand, the short chains of PEO are insufficient for stabilizing the micelles. Pluronic F127's long chains of PEO, when combined with P105, significantly enhanced the stability of the final formulation (Chen et al., 2013).

20.2.3 Solid-lipid nanoparticles (SLNs)

The SLNs are usually spherical, having a diameter of 10 to 1000 nm. They are lipid-based nanocarriers encapsulating lipid-soluble drugs successfully encapsulate lipid-soluble drugs and are solid at room temperature. It has collective advantages of both polymeric nanoparticles and lipid emulsion solving the in vivo stability problems (Mehnert, 2001). It has several benefits over other colloidal carriers, including the incorporation of both lipophilic and hydrophilic drugs, altering the release of the drug, and it is targeting along with ease of large-scale processing (Mehnert, 2001). It can also be used for the solubility enhancement of water-insoluble drugs. Usually, surfactants are used during the formulation of SLNs which help to keep the lipid core stable (emulsifiers). Lipids being used here include monoglycerides, diglycerides, triglycerides, fatty acids, waxes, and steroids. SLNs are prepared using *high-pressure homogenization (HPH),* a is commonly used method for the preparation of SLNs (Mehnert, 2001, Xu et al., 2009), there are some other methods in the literature such as the *emulsion-based method* (Sloat et al., 2011), *ultrasonication* (Mosallaei et al., 2013), and *solvent injection method* (Parveen et al., 2014).

20.2.4 Cubosomes

Cubosomes (cubosome dispersions) have recently been considered as the drug-lipid nanocarrier due to their enormous potential as an alternative drug delivery system relative to liposomes (Rarokar et al., 2016). Cubosomes are made up of binary systems such as monoolein–water that is self-assembled into thermodynamically stable discontinuous cubic liquid crystalline phases (Radiman et al., 1994). Cubosomes have solid-like rheology, viscous, an optically transparent substance with a special structure at the nanometer scale (Jones and Mcleish 1999). Structurally cubosomes are the surfactants organized into bilayers bent into a three-dimensional, tightly packed to form minimal surface and periodic structure, similar to *honeycomb* with discontinuous realms of lipid and water. Cubic hydro-lipid phase is disintegrated to form dispersion of liposomes in a three-phase area for the preparation of Cubosome particles. Its composition differs from that of liposomes in that it can contain lipid-soluble, water-soluble, and also amphiphilic molecules all at the same time. The three structures of cubosomes have been proposed with regard to various nodal surfaces that involves first one as "I$a3d$" called as Gyroid surface or G-surface, second one as "I$m3m$" known as a Primitive surface or P-surface and the third one as "P$n3m$" called as Diamond surface or D-surface. These structures help to maintain the effectiveness and the stability of active ingredients including vitamins and proteins. Cubosome colloidal dispersions can be stabilized by adding polymers (Jones and Mcleish 1999). They also have the capacity for controlled distribution of active pharmaceutical ingredients (API), as it is the tortuous diffusion of the API across the cubic phase's *regular* channels governing the diffusion. To form an aqueous surfactant system, a sufficient molecular orientation at relatively higher amphiphile concentrations is required to differentiate by structural symmetry. Cubosomes can load lipophilic, hydrophilic, and amphiphilic drugs (Spicer, 2004; Rarokar et al., 2016). Due to the 3D nanostructure of hydrophilic and lipophilic realms, the cubic liquid crystalline phases are used in the drug delivery of pharmaceuticals. While lipid constituents possess the properties like bioadhesivity, biocompatibility, and are digestible, the wide interfacial area provides a complicated pathway for the diffusion of the entrapped drug as a sustained release.

20.2.5 Drug-polymer conjugates

Drug-polymer conjugates (DPCs) are composed of an anticancer drug covalently connected to a biocompatible polymeric chain through a bioresponsive linker. DPCs aid in the improvement of the drug's pharmacokinetic profile as well as its stability. It also causes the drugs to accumulate in tumor tissues and malignant cells (Alexis et al., 2010; Sanchis et al., 2010; Pasut and Veronese 2009; Duncan, 2009). The polymers used in DPCs have unique physicochemical properties that they bypass the process of filtration and pass through the kidneys and liver. Newly synthesized polymers are developed to be susceptible to particular enzymes found in diseased tissue. The drugs stay attached to the polymer and aren't activated until they come into close contact with enzymes associated with the diseased tissue. This strategy helps to avoid harmful effects on normal cells and tissues. The copolymer conjugate of doxorubicin with N-(2-hydroxypropyl) methacrylamide (doxorubicin-HPMA, PK1, FCE28068) was prepared to deliver doxorubicin for targeting tumor cells (Vasey et al., 1999). Subsequently, several other conjugates of the drug have been developed including *HPMA-doxorubicin-gemcitabine conjugate* (Lammers et al., 2009), *paclitaxel-polylactide conjugate* (Tong et al., 2010), *camptothecin-polyethylene glycol (Pegamotecan, Camptothecin-PEG)* (Pasut and Veronese 2009), and *paclitaxel-polyglutamic acid conjugate* (Opaxio, PPX, paclitaxel polyglumex) (Sanchis et al., 2010). Though pegamotecan showed promising results in preclinical trials, it was discontinued due to clinical failure when compared to irinotecan treated patients in phase II trials for curing gastroesophageal or gastric adenocarcinoma (Pasut and Veronese 2009). Opaxio (PPX, or Xyotax) is comprised of 37 % w/w conjugated paclitaxel with L-glutamate (Pasut and Veronese 2009; Stirland et al., 2013) could not get through the phase III clinical trials as it failed to show more effects than the existing standard treatment regimen for patients having advanced *nonsmall cell lung cancer* (NSCLC) (Paz-Ares et al., 2008; O'Brien et al., 2008).

20.2.6 Antibody-drug conjugates

Antibody-drug conjugate (ADCs) is also a promising strategy for the targeted delivery of a drug. Monoclonal antibodies (mAbs) are chemically linked to biologically active drugs by chemical linkers with labile bonds (immunoconjugates) to form ADCs. This allows selective differentiation between healthy and diseased tissue by combining the specific targeting of mAbs with the cancer-killing capacity of cytotoxic drugs. The mAbs chosen for this purpose must be unique to target the antigens expressing or overexpressing on cancer cells only excluding healthy tissues nearby (Ricart and Tolcher, 2007). Advances in the binding of antibodies to cytotoxic drugs allow for better control of drug pharmacokinetics and improved delivery to target tissues. This also allows for a much higher cytotoxic agent concentration in tumor tissues, as well as lower systemic toxicity (Wu and Senter, 2005; Junutula et al., 2008; Mosure et al., 1997). ADCs could be the strategy for the extended circulatory half-life of mAbs (Gerber et al., 2013). On the other hand, it must have a stable association between the antibody and the therapeutic agent as shown in Fig. 20.3. Linkers regulate delivery of the cytotoxic agent to the target site using chemical motifs such as disulfides, hydrazones, or peptides (cleavable), or thioethers (noncleavable). In preclinical and clinical trials, cleavable and noncleavable linkers both are effective. The ADC is the combination of

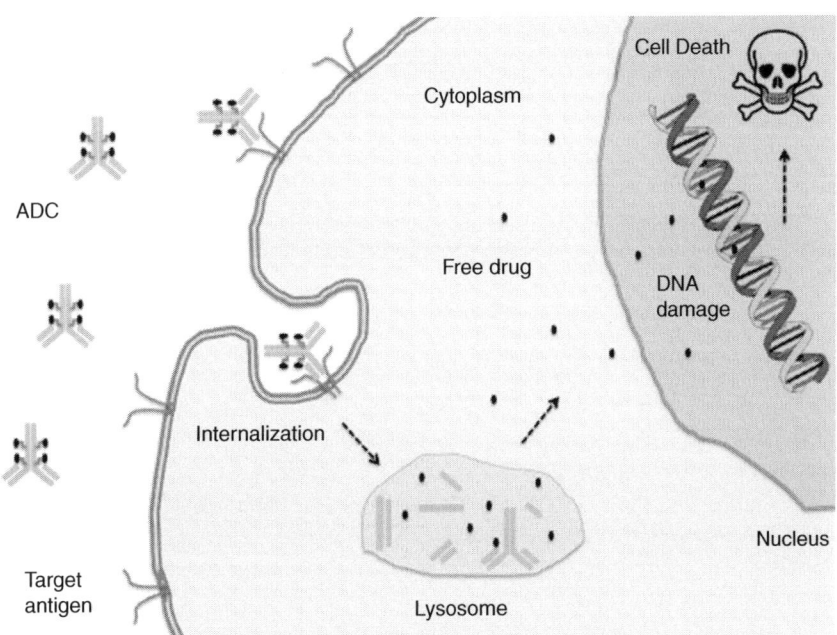

FIG. 20.3 DNA cytotoxics delivery: ADCs' mechanism of action for targeting Hartley (2014).

molecules with a variety of drug-antibody ratios (also known as DARs). Alterations of the linker in between the Ab and the drug molecule have effectively conquered some of the major issues that came across earlier with ADCs. The hydrophobicity of both (drug and linker) and the DAR around 3 to 4 (which is considered to be maximum) and if the ratio is increased further, it causes accumulation with limited time for circulation (Zhao et al., 2011) and therefore, DAR can be improved using the hydrophilic linkers (Kovtun et al., 2010; Zhao et al., 2011; Gerber et al., 2013). Tumor tissues sheds antigen into the bloodstream and the early release of drug resulted in the systemic toxicity, which can lead to ADCs targeting the wrong targets (Carter and Senter, 2008; Chan et al., 2003) and another issue relates to the unexpectedly low ranging accumulation of cytotoxic agent (0.0003–0.08%) as compared to the dose injected per gram of tumor (Carter and Senter, 2008). This resulted in a move away from traditional cytotoxic agents such as doxorubicin and in favor of other agents having better toxicity profiles amid limited clinical use previously (e.g., auristatins, calicheamicins, and maytansinoids) (Ricart and Tolcher, 2007; Lambert, 2005). However, even after mAb conjugation, these agents' systemic toxicity may still be present. Pfizer voluntarily recalled Mylotarg (gemtuzumabozogamicin) from the market in 2010 for the concerns of safety and efficacy, according to the FDA (Kharfan-Dabaja et al., 2013). Furthermore, because of the low tumor accumulation, high doses of the conjugates are required to be administered, which can cause financial issues, particularly for long-term therapy. An auristatin-CD30 conjugate (Brentuximabvedotin) is estimated to cost about $13,500 per dose every 3 weeks (Newland et al., 2013). Despite this, ADCs could be promising alternatives to traditional treatment, with better antiproliferative activity and safety profiles.

20.2.7 Inorganic nanoparticles

These include carbon nanotubes (Liang et al., 2010; Wong et al., 2013), gold nanoparticles (Dreaden et al., 2012), silica nanoparticles (Shen et al., 2013), iron oxide nanoparticles, zirconium nanoplatelets (Diaz et al., 2012), and quantum dots (Fang et al., 2013), some of which are described here.

20.2.8 Carbon nanotubes (CNTs)

CNTs have been a hot topic of cancer research for diagnosis and treatment due to their unusual physicochemical properties. They are thought to be one of the most promising nanomaterials because they can detect cancerous cells and deliver drugs to them. CNTs have been studied in most of the cancer treatment modalities over the last few years, including drug delivery, photodynamic therapy, thermal therapy, lymphatic targeted chemotherapy, and gene therapy. Ji et al. (2010) in-depth explored tumor-targeted drug delivery systems (DDS) based on novel single-walled nanotubes (SWNT). Liu et al. (2008) at Stanford University have done conjugation with paclitaxel (PTX) using cleavable ester bond and branched PEG chains on SWNTs. This results in the formation of water-soluble SWNT-PTX conjugate which also enhanced the permeability and probably responsible for higher uptake of PTX by tumors (10-fold increased) resulting in to suppress the tumor growth efficiently as compared to clinical PTX alone in breast cancer cell lines murine 4T1. The chemotherapeutic molecules are brought to the reticuloendothelium system that is released by the SWNTs, whereas it is excreted through biliary system without harming normal organs by adverse effects. This strengthens the potential of nanotube drug delivery for treating cancer with low doses of the chemotherapeutic agents having high efficacy. Carbon nanotubes have an unusual property in that they absorb near-infrared radiation and become very hot. Once the nanotube is bound to the cancer cells, it is heated with a near-infrared laser beam before the cancer cells are killed. The method is still in its early stages of growth and likely takes two or more years before it is evaluated in human clinical trials.

20.2.9 Gold nanoparticles (GNPs)

GNPs have a rare combination of physical and chemical properties, that includes surface plasmon resonance (SPR) with their capability of binding thiol and amine groups, enabling modification of surface and biomedical applications (Canovi et al., 2012). In pancreatic cancer, 5-nm GNPs are used as a vehicle for delivery of an active targeting agent cetuximab and therapeutic payload gemcitabine (Patra et al., 2010). In an orthotopic pancreatic cancer model, gemcitabine complex at a dose of 2 mg/kg inhibited tumor growth by 80%, compared to 30% inhibition using a combination of nonconjugated agents (Patra et al., 2008; Kullmann et al., 2009). Metal NPs were targeted with lasers that generate nonionizing electromagnetic radiation, causing the tumor's temperature to rise. In an in vivo study, 100 nm gold nanoshells were injected intravenously (IV) and accumulated maximally in human breast tumors (SK-BR-3) after 24 hours. In the nanoshell category, irreparable damage to the tissues was caused by the increase in average temperature of 37°C (in treatment group) and 9°C (in control group) due to the nanoshell resonance with tuned lasers. When opposed to controls, who were euthanized due to uncontrolled tumors, the mice grouped in the nanoshell category, lasted 90 days having no signs of tumor recurrence. However, since a laser can only penetrate a few centimeters into soft tissue, this procedure is limited in its ability to treat deep-seated tumors. Furthermore, according to some reports, at least 5000 nanoshells must be distributed per cell to achieve sufficient heat output for coagulative necrosis to occur (Stern and Cadeddu, 2008).

20.2.10 Porous silicon particles (PSiPs)

PSiPs are a very promising carrier for cancer therapy because of their biodegradability, biocompatibility (Popplewell et al., 1998; Park et al., 2009; Chen et al., 2011), and a readily functionalized surface (Fenollosa et al., 2014) as well as an excellent photothermal property (Jain et al., 2008). PSiPs work as a passive carrier of anticancer cargo and is simply prepared using electrochemical anodization of silicon. It can be activated by a suitable trigger, like acoustic waves or light for particle thermalization (Li et al., 2012; Mackowiak et al., 2013), or singlet oxygen generation in photodynamic therapies (Xiao et al., 2011). PSi must be functionalized with PEG to improve the PSi particles internalization into cells and the binding of PSi particles to antibodies for administration to the cancer patient. The temperature of PSi nanoparticles rises very quickly to about 70°C in around 6 minutes and remains constant for a long time with exposure to an NIR laser of 808 nm at 1.5 W/cm^2. PSi's high photothermal effects are mostly due to its high absorbance and high surface-to-volume ratio, because of multiple micropores. Hong et al. (2011) investigated the efficacy of photothermal therapy using PSi in conjunction with an 808 nm NIR laser discovering the fact that the combination resulted in a significantly higher cell death rate than either technique alone. Tumor CT-26, the murine colon carcinoma is reported to be fully resorbed within five days of start of therapy causing minimal damage to surrounding healthy tissue with NIR laser and PSi. This research shows that photothermotherapy using PSi in conjunction with irradiation using NIR laser can be used to selectively destroy cancer cells while leaving healthy cells unharmed.

20.2.11 Quantum dots (QDs)

The term "quantum dot" (QD) was coined by Mark Reed in 1988 and QDs are the fluorescent semiconductor nanocrystals. One of the major properties and applications of QDs is the optical imaging accompanied with multiplexing, great sensitivity, excellent resolution, and economic costing. The imperceptible nanocarriers can be monitored in real-time only because of their nano size, exceptional optical properties, and flexibility in surface chemistry. As reported by Ye et al. (2012), suitably coated and passivated QDs did not show acute toxicity in experimental animals like rodents and rhesus monkey even though they can release toxic chemicals such as cadmium (Cd) (Derfus et al., 2004) and generate ROS. As QDs can demonstrate fluorescence depending on particle size, they are considered to be the convenient to use for detection and labeling in biological systems (Jin and Ye, 2007; Choi et al., 2007; Choi et al., 2010). If the probes based on QDs are conjugated with biomolecules such as antibodies, peptides, or other small molecules highly specific and sensitive targeting of cancer cells can be achieved.

20.2.12 Iron oxide nanoparticles (IONPs)

IONPs are one of the first nanomaterials used in oncology and bioimaging, like magnetic resonance imaging (MRI) (Na et al., 2009; Rosen et al., 2012). Gilchrist et al. (1957) proposed the lymphatic metastasis that could therapeutically be heated inducing localized IONPs by exposing them to an interchanging magnetic field. Hyperthermia can improve the effectiveness of chemo and radiation therapies in addition to directly destroying cancer cells, and indirectly activating the intrinsic anticancer immune response.

20.2.13 IONPs

i. *Superparamagnetic IONPs (SPIONPs)*:

IONPs are commonly classified based on their size into superparamagnetic IONPs and ferromagnetic IONPs. Small synthetic hematite (-Fe_2O_3), magnetite (Fe_3O_4) or maghemite (-Fe_2O_3) particles having 10 to 20 nm of core diameter of are known as SPIONPs holding a potential for noninvasive detection and treatment of tumors because of possessing capabilities of simultaneous imaging and therapeutic applications. These multifunctional nanoparticles can conjugate with tumor-specific targeting ligands for delivering anticancer drugs to tumors with minimal adverse effects. Targeted IONPs can be used in three different ways to treat tumors. To begin, the specific antibodies can conjugate with IONPs and allow them to bind the related receptors inhibiting tumor development (Huh et al., 2005). Secondly, tumor hyperthermia can be achieved using targeted IONPs (Sonvico et al., 2005) and finally, drugs are loaded onto IONPs for targeted delivery to get therapeutic efficacy.

ii. *Ferromagnetic iron oxide nanoparticles (FIONPs)*:

FIONPs have some intriguing biomedical applications with regulated size varying from a few nanometers to 10 nm, if we place them in such a way to have dimensions equivalent to or lesser than dimensions of a virus (ranging 20–450 nm), a protein (ranging 5–50 nm), a cell (ranging 10–100 nm), or a gene (ranging 5–50 nm; 10–100 nm long, and 2 nm wide) confirming the probability of their penetration in a biological object of interest. The coating of nanomaterials through biomolecules makes them allow a controllable addressing or labeling tool by binding to or communicating with a biological system. Furthermore, external magnetic fields may affect these magnetic NPs. The exciting fields can greatly affect nanoparticles in electromagnetic fields by transferring energy to allow ferromagnetic iron oxide NPs to resonantly react to time-varying magnetic fields. Although SPIONPs are less magnetic, they are often favored for relative instability of FIONPs, which cause aggregation and making them tough to suspend to make biomedical applications more challenging.

iii. *Iron oxide for magnetic hyperthermia*

Magnetic hyperthermia, a well-known complementary cancer treatment, is used alone or by combining with radiation and/or chemotherapy. When exposed to strong magnetic fields, iron oxide NPs generate heat. The energy of alternating magnetic fields is stored in iron oxide NPs magnetic crystal suspensions, which are then released as heat to cause hyperthermic stress in cancer tissues and cells (Laurent et al., 2011). Under an alternating magnetic field, magnetic nanoparticles (MNPs) serve as thermal seeds in cancer treatment using targeted magnetic hyperthermia (Hergt et al., 2006). The radiation effect is improved, and cancer cells are thermoablated when the temperature is raised above 40°C. In one fascinating research, Kossatz et al. studied the in vivo therapeutic effects of hyperthermia induced by SPIONPs in mice on BxPC-3 xenografts. The intratumoral insertion of SPIONPs was done into the pancreatic model and then heated to 43°C cumulatively. When comparing magnetic hyperthermia-treated tumor tissue to controls, histological examination revealed changes in cell viability (apoptotic and necrotic changes) as well as reduced cell proliferation. This emphasizes the use of magnetic hyperthermia for treating pancreatic cancer (Kossatz et al., 2014). To assess the applicability and to optimize MNPs, the main drawback is the incongruities among the estimated heat produced and the actual experimental outcomes (Johannsen et al., 2010).

iv. *IONPs as vehicles for chemotherapy*

The surface modifications of IONPs have been previously done with anticancer agents including doxorubicin (Dox)/Paclitaxel (Jain et al., 2005), and Catcchin–Dextran. Endorem, a dextran-coated IONPs approved by FDA, are Catechin-Dextran conjugated NPs show greater efficiency compared with the free drug. This formulation demonstrated induction of apoptosis in 98% of MIA PaCa-2, the human pancreatic cancer cell lines placed under a magnetic field. The appropriate strategy of conjugation can enhance the anticancer activity of the drug and it could be the novel approach for targeted delivery of anticancer agents to tumor cells and that are magnetic field driven.

References

Alexis, F., Pridgen, E.M., Langer, R., Farokhzad, O.C., 2010. Nanoparticle technologies for cancer therapy. Drug Deliv. 197, 55–86.

Anselmo, A.C., Mitragotri, S., 2019. Nanoparticles in the clinic: an update. Bioeng. Transl. Med. 4 (3), e10143.

Arias, J.L., 2013. Liposomes in drug delivery: a patent review (2007–present). Expert Opin. Ther. Pat. 23 (11), 1399–1414.

Bae, Y.H., Park, K., 2011. Targeted drug delivery to tumors: Myths, reality and possibility. J. Controlled Release 153 (3), 198–205.

Banerjee, R., Tyagi, P., Li, S., Huang, L., 2004. Anisamide-targeted stealth liposomes: a potent carrier for targeting doxorubicin to human prostate cancer cells. Int. J. Cancer 112 (4), 693–700.

Bangham, A.D., Standish, M.M., Watkins, J.C., 1965. Diffusion of univalent ions across the lamellae of swollen phospholipids. J. Mol. Biol. 13 (1) 238-IN27.

Canovi, M., Lucchetti, J., Stravalaci, M., Re, F., Moscatelli, D., Bigini, P., Salmona, M., Gobbi, M., 2012. Applications of surface plasmon resonance (SPR) for the characterization of nanoparticles developed for biomedical purposes. Sensors 12 (12), 16420–16432.

Carter, P.J., Senter, P.D., 2008. Antibody-drug conjugates for cancer therapy. Cancer J. 14 (3), 154–169.

Chan, S.Y., Gordon, A.N., Coleman, R.E., Hall, J.B., Berger, M.S., Sherman, M.L., Eten, C.B., Finkler, N.J., 2003. A phase 2 study of the cytotoxic immunoconjugate CMB-401 (hCTM01-calicheamicin) in patients with platinum-sensitive recurrent epithelial ovarian carcinoma. Cancer Immunol. Immunother. 52 (4), 243–248.

Chein, Y.W., 2005. Novel Drug Delivery Systems, 2nd ed. Marcel Dekker, New York, pp. 157–163.

Chen, H., Zhang, W., Zhu, G., Xie, J., Chen, X., 2017. Rethinking cancer nanotheranostics. Nat. Rev. Mater. 2 (7), 1–18.

Chen, L., Sha, X., Jiang, X., Chen, Y., Ren, Q., Fang, X., 2013. Pluronic P105/F127 mixed micelles for the delivery of docetaxel against Taxol-resistant non-small cell lung cancer: optimization and in vitro, in vivo evaluation. Int. J. Nanomed. 8, 73.

Chen, Y., Wan, Y., Wang, Y., Zhang, H., Jiao, Z., 2011. Anticancer efficacy enhancement and attenuation of side effects of doxorubicin with titanium dioxide nanoparticles. Int. J. Nanomed. 6, 2321.

Choi, H.S., Liu, W., Liu, F., Nasr, K., Misra, P., Bawendi, M.G., Frangioni, J.V., 2010. Design considerations for tumour-targeted nanoparticles. Nat. Nanotechnol. 5 (1), 42–47.

Choi, H.S., Liu, W., Misra, P., Tanaka, E., Zimmer, J.P., Ipe, B.I., Bawendi, M.G., Frangioni, J.V., 2007. Renal clearance of quantum dots. Nat. Biotechnol. 25 (10), 1165–1170.

Derfus, A.M., Chan, W.C., Bhatia, S.N., 2004. Probing the cytotoxicity of semiconductor quantum dots. Nano Lett. 4 (1), 11–18.

Díaz, A., Saxena, V., González, J., David, A., Casañas, B., Carpenter, C., Batteas, J.D., Colón, J.L., Clearfield, A., Hussain, M.D., 2012. Zirconium phosphate nano-platelets: a novel platform for drug delivery in cancer therapy. Chem. Commun. 48 (12), 1754–1756.

Douglas, S.M., Bachelet, I., Church, G.M., 2012. A logic-gated nanorobot for targeted transport of molecular payloads. Science 335 (6070), 831–834.

Dreaden, E.C., Austin, L.A., Mackey, M.A., El-Sayed, M.A., 2012. Size matters: gold nanoparticles in targeted cancer drug delivery. Therapeut. Deliv. 3 (4), 457–478.

Duncan, R., 2009. Development of HPMA copolymer–anticancer conjugates: clinical experience and lessons learnt. Adv. Drug. Deliv. Rev. 61 (13), 1131–1148.

Fang, M., Yuan, J.P., Peng, C.W., Pang, D.W., Li, Y., 2013. Quantum dots-based in situ molecular imaging of dynamic changes of collagen IV during cancer invasion. Biomaterials 34 (34), 8708–8717.

Fenollosa, R., Garcia-Rico, E., Alvarez, S., Alvarez, R., Yu, X., Rodriguez, I., Carregal-Romero, S., Villanueva, C., Garcia-Algar, M., Rivera-Gil, P., de Lera, A.R., 2014. Silicon particles as Trojan horses for potential cancer therapy. J. Nanobiotechnol. 12 (1), 1–10.

Gao, J., Sun, J., Li, H., Liu, W., Zhang, Y., Li, B., Qian, W., Wang, H., Chen, J., Guo, Y., 2010. Lyophilized HER2-specific PEGylated immunoliposomes for active siRNA gene silencing. Biomaterials 31 (9), 2655–2664.

Gerber, H.P., Koehn, F.E., Abraham, R.T., 2013. The antibody-drug conjugate: an enabling modality for natural product-based cancer therapeutics. Nat. Prod. Rep. 30 (5), 625–639.

Ghosh, P., Han, G., De, M., Kim, C.K., Rotello, V.M., 2008. Gold nanoparticles in delivery applications. Adv. Drug. Deliv. Rev. 60 (11), 1307–1315.

Gilchrist, R.K., Medal, R., Shorey, W.D., Hanselman, R.C., Parrott, J.C., Taylor, C.B., 1957. Selective inductive heating of lymph nodes. Ann. Surg. 146 (4), 596.

Haley, B., Frenkel, E., 2008. Nanoparticles for drug delivery in cancer treatmentUrologic Oncology: Seminars and Original Investigations26. Elsevier, pp. 57–64.

Hartley, J.A., 2014. Antibody-drug conjugates delivering DNA cytotoxics. In: Neidle, S. (Ed.), Cancer Drug Design and Discovery. 2nd ed. Academic press, London, UK, pp. 479–490.

Hergt, R., Dutz, S., Müller, R., Zeisberger, M., 2006. Magnetic particle hyperthermia: nanoparticle magnetism and materials development for cancer therapy. J. Phys. Condens. Matter 18 (38), S2919.

Hong, C., Lee, J., Son, M., Hong, S.S., Lee, C., 2011. In-vivo cancer cell destruction using porous silicon nanoparticles. Anticancer Drugs 22 (10), 971–977.

Huh, Y.M., Jun, Y.W., Song, H.T., Kim, S., Choi, J.S., Lee, J.H., Yoon, S., Kim, K.S., Shin, J.S., Suh, J.S., Cheon, J., 2005. In vivo magnetic resonance detection of cancer by using multifunctional magnetic nanocrystals. J. Am. Chem. Soc. 127 (35), 12387–12391.

Jain, P.K., Huang, X., El-Sayed, I.H., El-Sayed, M.A., 2008. Noble metals on the nanoscale: optical and photothermal properties and some applications in imaging, sensing, biology, and medicine. Acc. Chem. Res. 41 (12), 1578–1586.

Jain, T.K., Morales, M.A., Sahoo, S.K., Leslie-Pelecky, D.L., Labhasetwar, V., 2005. Iron oxide nanoparticles for sustained delivery of anticancer agents. Mol. Pharm. 2 (3), 194–205.

Ji, S.R., Liu, C., Zhang, B., Yang, F., Xu, J., Long, J., Jin, C., Fu, D.L., Ni, Q.X., Yu, X.J., 2010. Carbon nanotubes in cancer diagnosis and therapy. Biochimica et Biophysica Acta (BBA)-Rev. Cancer 1806 (1), 29–35.

Jin, S., Ye, K., 2007. Nanoparticle-mediated drug delivery and gene therapy. Biotechnol. Prog. 23 (1), 32–41.

Johannsen, M., Thiesen, B., Wust, P., Jordan, A., 2010. Magnetic nanoparticle hyperthermia for prostate cancer. Int. J. Hyperthermia 26 (8), 790–795.

Jones, J.L., McLeish, T.C.B., 1999. Concentration fluctuations in surfactant cubic phases: theory, rheology, and light scattering. Langmuir 15 (22), 7495–7503.

Junutula, J.R., Raab, H., Clark, S., Bhakta, S., Leipold, D.D., Weir, S., Chen, Y., Simpson, M., Tsai, S.P., Dennis, M.S., Lu, Y., 2008. Site-specific conjugation of a cytotoxic drug to an antibody improves the therapeutic index. Nat. Biotechnol. 26 (8), 925–932.

Kharfan-Dabaja, M.A., Hamadani, M., Reljic, T., Pyngolil, R., Komrokji, R.S., Lancet, J.E., Fernandez, H.F., Djulbegovic, B., Kumar, A., 2013. Gemtuzumab ozogamicin for treatment of newly diagnosed acute myeloid leukaemia: a systematic review and meta-analysis. Br. J. Haematol. 163 (3), 315–325.

Kim, S.C., Kim, D.W., Shim, Y.H., Bang, J.S., Oh, H.S., Kim, S.W., Seo, M.H., 2001. In vivo evaluation of polymeric micellar paclitaxel formulation: toxicity and efficacy. J. Control. Release 72 (1-3), 191–202.

Klibanov, A.L., Maruyama, K., Torchilin, V.P., Huang, L., 1990. Amphipathic polyethyleneglycols effectively prolong the circulation time of liposomes. FEBS Lett. 268 (1), 235–237.

Kossatz, S., Ludwig, R., Dahring, H., Ettelt, V., Rimkus, G., Marciello, M., et al., 2014. High therapeutic efficiency of magnetic hyperthermia in xenograft models achieved with moderate temperature dosages in the tumor area. Pharm. Res. 31 (12), 3274–3288.

Kovtun, Y.V., Audette, C.A., Mayo, M.F., Jones, G.E., Doherty, H., Maloney, E.K., Erickson, H.K., Sun, X., Wilhelm, S., Ab, O., Lai, K.C., 2010. Antibody-maytansinoid conjugates designed to bypass multidrug resistance. Cancer Res. 70 (6), 2528–2537.

Kullmann, F., Hollerbach, S., Dollinger, M.M., Harder, J., Fuchs, M., Messmann, H., Trojan, J., Gäbele, E., Hinke, A., Hollerbach, C., Endlicher, E., 2009. Cetuximab plus gemcitabine/oxaliplatin (GEMOXCET) in first-line metastatic pancreatic cancer: a multicentre phase II study. Br. J. Cancer 100 (7), 1032–1036.

Lambert, J.M., 2005. Drug-conjugated monoclonal antibodies for the treatment of cancer. Curr. Opin. Pharmacol 5, 543–549.

Lammers, T., Subr, V., Ulbrich, K., Peschke, P., Huber, P.E., Hennink, W.E., Storm, G., 2009. Simultaneous delivery of doxorubicin and gemcitabine to tumors in vivo using prototypic polymeric drug carriers. Biomaterials 30 (20), 3466–3475.

Laurent, S., Dutz, S., Häfeli, U.O., Mahmoudi, M., 2011. Magnetic fluid hyperthermia: focus on superparamagnetic iron oxide nanoparticles. Adv. Colloid Interface Sci. 166 (1-2), 8–23.

Li, Z., Barnes, J.C., Bosoy, A., Stoddart, J.F., Zink, J.I., 2012. Mesoporous silica nanoparticles in biomedical applications. Chem. Soc. Rev. 41, 2590–2605.

Liang, X.J., Meng, H., Wang, Y., He, H., Meng, J., Lu, J., Wang, P.C., Zhao, Y., Gao, X., Sun, B., Chen, C., 2010. Metallofullerene nanoparticles circumvent tumor resistance to cisplatin by reactivating endocytosis. Proc. Natl. Acad. Sci. 107 (16), 7449–7454.

Liu, Z., Chen, K., Davis, C., Sherlock, S., Cao, Q., Chen, X., Dai, H., 2008. Drug delivery with carbon nanotubes for in vivo cancer treatment. Cancer Res. 68 (16), 6652–6660.

Mackowiak, S.A., Schmidt, A., Weiss, V., Argyo, C., von Schirnding, C., Bein, T., Bräuchle, C., 2013. Targeted drug delivery in cancer cells with red-light photoactivated mesoporous silica nanoparticles. Nano Lett. 13 (6), 2576–2583.

Mehnert, W., Mader, K., 2001. Solid lipid nanoparticles: production, characterization and applications. Adv. Drug Deliv. Rev. 47, 165–196.

Misra, R., Acharya, S., Sahoo, S.K., 2010. Cancer nanotechnology: application of nanotechnology in cancer therapy. Drug Discov. Today 15 (19-20), 842–850.

Mosallaei, N., Jaafari, M.R., Hanafi-Bojd, M.Y., Golmohammadzadeh, S., Malaekeh-Nikouei, B., 2013. Docetaxel-loaded solid lipid nanoparticles: preparation, characterization, in vitro, and in vivo evaluations. J. Pharm. Sci. 102 (6), 1994–2004.

Mosure, K.W., Henderson, A.J., Klunk, L.J., Knipe, J.O., 1997. Disposition of conjugate-bound and free doxorubicin in tumor-bearing mice following administration of a BR96-doxorubicin immunoconjugate (BMS 182248). Cancer Chemother. Pharmacol. 40 (3), 251–258.

Na, H.B., Song, I.C., Hyeon, T., 2009. Inorganic Nanoparticles for MRI Contrast Agents. Adv. Materials 21, 2133–2148.

Newland, A.M., Li, J.X., Wasco, L.E., Aziz, M.T., Lowe, D.K., 2013. Brentuximab vedotin: A CD 30-directed antibody-cytotoxic drug conjugate. Pharmacotherapy 33 (1), 93–104.

O'Brien, M.E., Socinski, M.A., Popovich, A.Y., Bondarenko, I.N., Tomova, A., Bilynsky, B.T., Hotko, Y.S., Ganul, V.L., Kostinsky, I.Y., Eisenfeld, A.J., Sandalic, L., 2008. Randomized phase III trial comparing single-agent paclitaxel Poliglumex (CT-2103, PPX) with single-agent gemcitabine or vinorelbine for the treatment of PS 2 patients with chemotherapy-naive advanced non-small cell lung cancer. J. Thorac. Oncol. 3 (7), 728–734.

Ostacolo, L., Marra, M., Ungaro, F., Zappavigna, S., Maglio, G., Quaglia, F., Abbruzzese, A., Caraglia, M., 2010. In vitro anticancer activity of docetaxel-loaded micelles based on poly (ethylene oxide)-poly (epsilon-caprolactone) block copolymers: do nanocarrier properties have a role? J. Control. Release 148 (2), 255–263.

Park, J.H., Gu, L., Von Maltzahn, G., Ruoslahti, E., Bhatia, S.N., Sailor, M.J., 2009. Biodegradable luminescent porous silicon nanoparticles for in vivo applications. Nat. Mater. 8 (4), 331–336.

Parveen, R., Ahmad, F.J., Iqbal, Z., Samim, M., Ahmad, S., 2014. Solid lipid nanoparticles of anticancer drug andrographolide: formulation, in vitro and in vivo studies. Drug Dev. Ind. Pharm. 40 (9), 1206–1212.

Pasut, G., Veronese, F.M., 2009. PEG conjugates in clinical development or use as anticancer agents: an overview. Adv. Drug. Deliv. Rev. 61 (13), 1177–1188.

Patra, C.R., Bhattacharya, R., Mukhopadhyay, D., Mukherjee, P., 2010. Fabrication of gold nanoparticles for targeted therapy in pancreatic cancer. Adv. Drug Deliv. Rev. 62, 346–361.

Patra, C.R., Bhattacharya, R., Wang, E., Katarya, A., Lau, J.S., Dutta, S., Muders, M., Wang, S., Buhrow, S.A., Safgren, S.L., Yaszemski, M.J., 2008. Targeted delivery of gemcitabine to pancreatic adenocarcinoma using cetuximab as a targeting agent. Cancer Res. 68 (6), 1970–1978.

Paz-Ares, L., Ross, H., O'brien, M., Riviere, A., Gatzemeier, U., Von Pawel, J., Kaukel, E., Freitag, L., Digel, W., Bischoff, H., Garcia-Campelo, R., 2008. Phase III trial comparing paclitaxel poliglumex vs docetaxel in the second-line treatment of non-small-cell lung cancer. Br. J. Cancer 98 (10), 1608–1613.

Popplewell, J.F., King, S.J., Day, J.P., Ackrill, P., Fifield, L.K., Cresswell, R.G., Di Tada, M.L., Liu, K., 1998. Kinetics of uptake and elimination of silicic acid by a human subject: a novel application of 32Si and accelerator mass spectrometry. J. Inorg. Biochem. 69 (3), 177–180.

Radiman, S., Toprakcioglu, C., McLeish, T., 1994. Rheological study of ternary cubic phases. Langmuir 10 (1), 61–67.

Raju, A., Muthu, M.S., Feng, S.S., 2013. Trastuzumab-conjugated vitamin E TPGS liposomes for sustained and targeted delivery of docetaxel. Expert Opin. Drug Deliv. 10 (6), 747–760.

Rarokar, N., Ravikumar, S., Gurav, S., Khedekar, P., 2021. Meloxicam encapsulated nanostructured colloidal self-assembly for evaluating antitumor and anti-inflammatory efficacy in 3D printed scaffolds. J. Biomed. Mater. Res. A. 109 (8), 1441–1456.

Rarokar, N.R., Khedekar, P.B., Bharne, A.P., Umekar, M.J., 2019. Development of self-assembled nanocarriers to enhance antitumor efficacy of docetaxel trihydrate in MDA-MB-231 cell line. Int. J. Biol. Macromol. 125, 1056–1068.

Rarokar, N.R., Saoji, S.D., Raut, N.A., Taksande, J.B., Khedekar, P.B., Dave, V.S., 2016. Nanostructured cubosomes in a thermoresponsive depot system: an alternative approach for the controlled delivery of docetaxel. AAPS PharmSciTech. 17 (2), 436–445.

Ricart, A.D., Tolcher, A.W., 2007. Technology insight: cytotoxic drug immunoconjugates for cancer therapy. Nat. Clin. Pract. Oncol. 4 (4), 245–255.

Rodriguez, B.L., Blando, J.M., Lansakara-P, D.S., Kiguchi, Y., DiGiovanni, J., Cui, Z., 2013. Antitumor activity of tumor-targeted RNA replicase-based plasmid that expresses interleukin-2 in a murine melanoma model. Mol. Pharm. 10 (6), 2404–2415.

Rosen, J.E., Chan, L., Shieh, D.B., Gu, F.X., 2012. Iron oxide nanoparticles for targeted cancer imaging and diagnostics. Nanomed. Nanotechnol. Biol. Med. 8 (3), 275–290.

Salahuddin, N., Galal, A., 2017. Chapter 4 - Improving chemotherapy drug delivery by nanoprecision tools. In: Ficai, A., Grumezescu, A.M. (Eds.), In Micro and Nano Technologies, Nanostructures for Cancer Therapy. Elsevier, pp. 87–128.

Sanchis, J., Canal, F., Lucas, R., Vicent, M.J., 2010. Polymer–drug conjugates for novel molecular targets. Nanomedicine 5 (6), 915–935.

Saxena, V., Hussain, M.D., 2012. Poloxamer 407/TPGS mixed micelles for delivery of gambogic acid to breast and multidrug-resistant cancer. Int. J. Nanomed. 7, 713.

Shen, J., Song, G., An, M., Li, X., Wu, N., Ruan, K., et al., 2013. The use of hollow mesoporous silica nanospheres to encapsulate bortezomib and improve efficacy for non-small cell lung cancer therapy. Biomaterials 35 (1), 316–326.

Shi, J., Kantoff, P.W., Wooster, R., Farokhzad, O.C., 2017. Cancer nanomedicine: progress, challenges and opportunities. Nat. Rev. Cancer 17 (1), 20–37.

Sloat, B.R., Sandoval, M.A., Li, D., Chung, W.G., Lansakara-P, D.S., Proteau, P.J., Kiguchi, K., DiGiovanni, J., Cui, Z., 2011. In vitro and in vivo antitumor activities of a gemcitabine derivative carried by nanoparticles. Int. J. Pharm. 409 (1-2), 278–288.

Sonvico, F., Mornet, S., Vasseur, S., Dubernet, C., Jaillard, D., Degrouard, J., Hoebeke, J., Duguet, E., Colombo, P., Couvreur, P., 2005. Folate-conjugated iron oxide nanoparticles for solid tumor targeting as potential specific magnetic hyperthermia mediators: synthesis, physicochemical characterization, and in vitro experiments. Bioconjug. Chem. 16 (5), 1181–1188.

Spicer, P.T., 2004. Cubosomes: Bicontinuous Liquid Crystalline Nanoparticles. Dekker Encyclopaedia of Nanoscience and Nanotechnology. Marcel Dekker, New York, pp. 881–892.

Stern, J.M., Cadeddu, J.A., 2008. Emerging use of nanoparticles for the therapeutic ablation of urologic malignanciesUrologic Oncology: Seminars and Original Investigations26. Elsevier, pp. 93–96.

Stirland, D.L., Nichols, J.W., Miura, S., Bae, Y.H., 2013. Mind the gap: a survey of how cancer drug carriers are susceptible to the gap between research and practice. J. Control. Release 172 (3), 1045–1064.

Tong, R., Yala, L., Fan, T.M., Cheng, J., 2010. The formulation of aptamer-coated paclitaxel–polylactide nanoconjugates and their targeting to cancer cells. Biomaterials 31 (11), 3043–3053.

Torchilin, V.P., 2005. Recent advances with liposomes as pharmaceutical carriers. Nat. Rev. Drug Discovery 4 (2), 145–160.

Vasey, P.A., Kaye, S.B., Morrison, R., Twelves, C., Wilson, P., Duncan, R., Thomson, A.H., Murray, L.S., Hilditch, T.E., Murray, T., Burtles, S., 1999. Phase I clinical and pharmacokinetic study of PK1 [N-(2-hydroxypropyl) methacrylamide copolymer doxorubicin]: first member of a new class of chemotherapeutic agents—drug-polymer conjugates. Clin. Cancer Res. 5 (1), 83–94.

Wang, J., Li, Y., Nie, G., 2021. Multifunctional biomolecule nanostructures for cancer therapy. Nat. Rev. Mater., 1–18.

Wong, B.S., Yoong, S.L., Jagusiak, A., Panczyk, T., Ho, H.K., Ang, W.H., Pastorin, G., 2013. Carbon nanotubes for delivery of small molecule drugs. Adv. Drug. Deliv. Rev. 65 (15), 1964–2015.

Wu, A.M., Senter, P.D., 2005. Arming antibodies: prospects and challenges for immunoconjugates. Nat. Biotechnol. 23 (9), 1137–1146.

Xiao, L., Gu, L., Howell, S.B., Sailor, M.J., 2011. Porous silicon nanoparticle photosensitizers for singlet oxygen and their phototoxicity against cancer cells. ACS Nano 5 (5), 3651–3659.

Xu, Z., Chen, L., Gu, W., Gao, Y., Lin, L., Zhang, Z., Xi, Y., Li, Y., 2009. The performance of docetaxel-loaded solid lipid nanoparticles targeted to hepatocellular carcinoma. Biomaterials 30 (2), 226–232.

Ye, L., Yong, K.T., Liu, L., Roy, I., Hu, R., Zhu, J., Cai, H., Law, W.C., Liu, J., Wang, K., Liu, J., 2012. A pilot study in non-human primates shows no adverse response to intravenous injection of quantum dots. Nat. Nanotechnol. 7 (7), 453–458.

Zhao, R.Y., Wilhelm, S.D., Audette, C., Jones, G., Leece, B.A., Lazar, A.C., Goldmacher, V.S., Singh, R., Kovtun, Y., Widdison, W.C., Lambert, J.M., 2011. Synthesis and evaluation of hydrophilic linkers for antibody–maytansinoid conjugates. J. Med. Chem. 54 (10), 3606–3623.

Zhao, Z., Ukidve, A., Kim, J., Mitragotri, S., 2020. Targeting strategies for tissue-specific drug delivery. Cell 181 (1), 151–167.

Chapter 21

Photo-triggered theranostics nanomaterials: Development and challenges in cancer treatment

Neha S. Raut[a], Divya Zambre[b], Milind J. Umekar[b], Sanjay J. Dhoble[c]
[a]Department of Pharmaceutical Chemistry, Smt. Kishoritai Bhoyar College of Pharmacy, Kamptee, India
[b]Department of Pharmaceutics, Smt. Kishoritai Bhoyar College of Pharmacy, Kamptee, India
[c]Department of Physics, Rashtrasant Tukadoji Maharaj Nagpur University, Nagpur, Maharashtra, India

21.1 Introduction of nanomaterials in phototherapeutics

Nanotechnology contributes major applications in the cancer therapy and having advanced approaches than the conventional chemotherapy. Moreover, highly specific treatment is required considering the complexity of the cancer disease and prevalence of diverse cell populations. Nanoconstructs are the type of nanomaterial composed of a core comprising of metal nanoparticle (NP) and a shell made up of ligands for therapeutic, prognostic as well as diagnostic applications (Gao et al., 2014). Phototherapeutics known as the methods for photo-based diagnosis and treatment are gaining more importance as showing better resolution in spatial imaging with a localized treatment and lesser invasive modalities (Menon et al., 2013). The unique architecture of nanoconstructs makes them ideal and preferable for diagnosis and treatment in cancer patients (Giljohann et al., 2010; Dilnawaz and Kumar, 2017).

The clinical safety of chemotherapeutic treatments has been improved greatly by nanoparticle (NP) controlled drug delivery systems; however, their efficacy showed mixed results (Green et al., 2006; Kim et al., 2007). Nanoconstructs have emerged as promising nanotools for cancer diagnostics and treatment due to their capacity to consolidate a wide range of ligands and their adaptability to surface functionalization with diverse chemistries. The theranostic nanomedicines are expected to give real-time data from administration, biodistribution, therapeutic efficacy, drug concentration, doses and drug release kinetics which would be a tool to clinicians for adjusting the treatment regimens accordingly (Fig. 21.1; Moore et al., 2014).

The major issues that chemotherapy encounters are undesirable physical and chemical properties of drugs such as inadequate solubility, improper biodistribution, systemic toxicity, and rapid drug clearance (Allen and Cullis, 2004). Some of these issues have been overcome through nanomedicine. As reported in the literature, the circulation half-life of various chemotherapeutic formulations has been extended through nanomedicine as indicated by the preclinical and clinical pharmacokinetic investigations/phase I data. Nanoparticles offer the advantage of allowing for more medication circulation during treatment.

Protein overexpression or leaky vasculature like physiological anomalies found in tumors makes nanomedicines as a potential drug delivery system. NP increase permeability, retention and act by active targeting which ultimately increase the intratumoral drug concentration (Hrkach et al., 2012; Maeda, 2001). However, as previously mentioned, the clinical performance of NP drug delivery methods has been inconsistent. This is primarily due to intra- and interpatient heterogeneity (Sottoriva et al., 2013; Lammers et al., 2008). As a result, a new paradigm is required to estimate the amount of medicine injected into the tumor noninvasively and adapt subsequent treatment.

In nanoparticles the imaging property is not affected by drug delivery as imaging and drug delivery are separate components. The combined effect such as drug delivery and imaging were observed in theranostics as a determination of nanoparticle drug delivery vehicle concentration and drug biodistribution deposited at the affected site (Chen et al., 2012; Majumdar et al., 2010; Cole et al., 2011).

The fluorescence intensity and the lifespan of drug increase when it is released from the NP. The noninvasive methods available for quantifying drug release involves the combination of MRI contrast agent along with pharmaceuticals considering its optical characteristics. Nanotechnologies provide a platform for noninvasive drug concentration assessment, improved solid tumor imaging and drug delivery, along with improved drug biodistribution and efficacy (Moore et al., 2014). Various nanoconstructs used in cancer therapy include magnetic NPs, polymeric NPs, QDs, lipid gold, and others as shown in Fig. 21.2; Dilnawaz and Kumar, 2017).

21.2 Types of nanomaterials

Different types of nanomaterials include magnetic nanoparticles, gold-based nanoparticles, carbon nanotubes, polymeric nanocarriers for photosensitizer/dye encapsulation, photo-triggered theranostic nanocarriers, radioluminescent nanoparticles, upconversion nanoparticles, fluorescent nanoparticles, and others.

21.2.1 Magnetic nanoparticles

Magnetic NPs can also be coupled with other nanocarriers to improve medication delivery such as nanobubbles, polymeric NPs, gold NPs, carbon nanotubes, quantum dots, and liposomes for imaging and release. Drug delivery, MRI contrast, and in vivo controlled localization are some of the applications of magnetic NPs (Al-Jamal and Kostarelos, 2011).

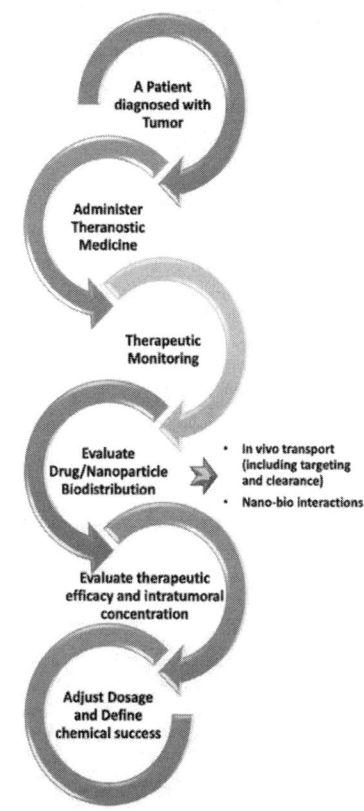

Fig. 21.1 Theranostic nanomedicine: Voyage and clinical success.

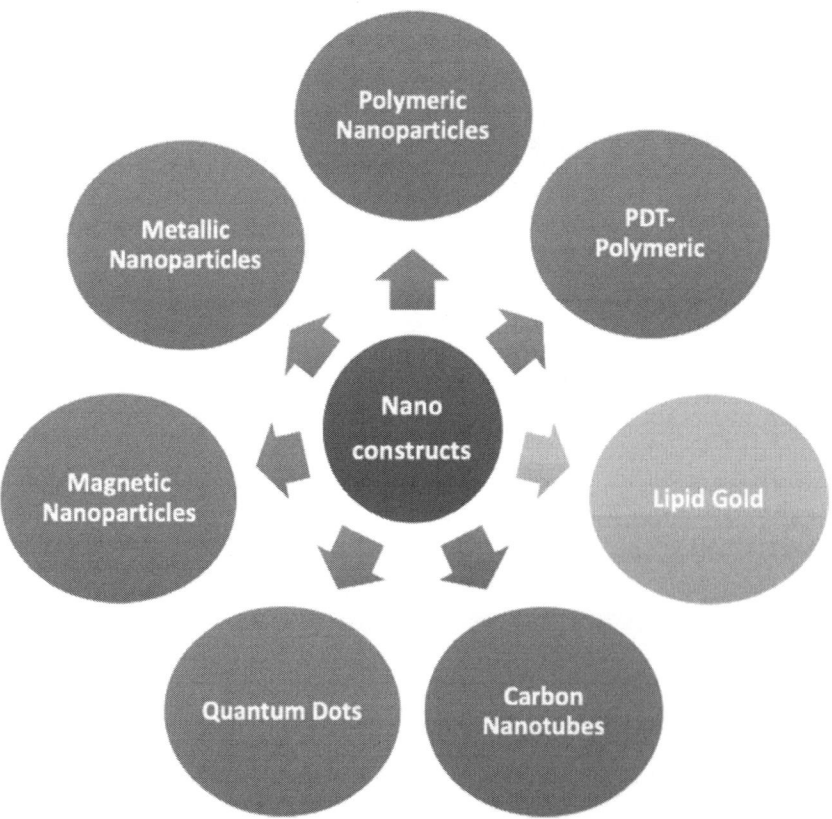

Fig. 21.2 Types of nanoconstructs used in cancer therapy (Dilnawaz and Kumar, 2017).

21.2.2 Properties and materials for preparation of photo-based nanomaterials

For photoimaging and therapeutic applications, nanocarriers such as polymeric nanoparticles delivering fluorescent dyes or photosensitizers, optical imaging, magnetic resonance imaging (MRI), photothermal treatment (PTT), and chemotherapy. Metal-based to polymer-based materials are extensively employed to produce photo-based NPs (Menon et al., 2013).

Noble metals such as silver and gold nanoparticles are well known for excellent and controllable optical features making them ideal for biological imaging, sensing, and therapeutic purposes. Another type of metallic NP employed in the presence of light for hyperthermia is quantum dots (QDs). Biocompatible polymer-based NPs are also used for diagnostic as well as therapeutic purposes which are based on light whereas some of these materials lack imaging capabilities. They can contain both fluorescent dyes and metals for therapy and imaging. The dyes approved by FDA are fluorescent Cyanine7 (Cy7), which absorbed in near infra-red (NIR) region, indocyanine green (ICG), and dialkylcarbocyanine fluorophores (DiI, DiR) are added in the polymeric NPs (Menon et al., 2013). In response to the ever increasing demand for an effective biocompatible treatment, theranostic NPs as a contrast agent have been developed; concurrently delivering molecules used for their therapeutic and diagnostic imaging applications such as fluorescent imaging, MRI, and optical coherence tomography (OCT). In addition, depending on the application and polymer used, treatment is carried out via photo-activated hyperthermia and modifying medication release rates (Menon et al., 2013).

The applications and interest in "smart" materials is still high in biomedical research as it can be regulated by light, temperature as well as pH. Among all the external stimuli, light-based stimuli are quickly gaining prominence for generation of innovative medication delivery systems and biomaterials. The use of visible light in the near-infrared region, great focus and spatial control, least intrusive techniques used, and the materials that may be used and tested are all appealing aspects of light-responsive materials. Furthermore, the biomedical applications such as light-based diagnostics and treatment involves NIR light, noncontact approach for treatment and controlling the rate of release of encapsulated drugs through external source of light are superior to existing modalities (Liu et al., 2012). The light responsiveness of the materials is because of their unique light absorption capabilities. As a result, plasmonic metal compounds with excellent optical properties in the NIR range, at which penetration through tissue is conceivable, account for most light-responsive materials (Katz and Burdick, 2010; Allen and Cullis, 2004).

21.2.3 Gold-based nanoparticles

Gold nanoconstructs with tunable surface plasmon resonance (SPR), which cause coherent electron oscillations in the conduction band, resulting to photothermal conversion (Skrabalak et al., 2009). The two-dimensional gold nanoparticle arrays are embedded in an elastic polydimethylsiloxane (PDMS) membrane for excellent malleability. The interparticle distance in the gold nanoparticle array can be controlled by applying external force, and the change in distance between each particle modifies the interaction and thereby the frequency (Skrabalak et al., 2009). Due to the great transparency of the tissues, gold nanoconstructs phototherapy has emerged as an essential choice for the treatment of cancer, where the LSPR peaks are directed towards the NIR region for better treatment (Yavuz et al., 2009). NIR light photon energy can be converted into heat by thermal ablation in gold nanoparticles (AuNPs) (Bae and Na, 2012).

21.2.4 Carbon nanotubes

Carbon nanotubes (CNTs) a hexagonal network made up of around 1 nm diameter and 1–100 μm of a length of carbon atoms (Dai et al., 2006). CNTs are either single walled or multiwalled, have high thermal and electrical conductivities, and are employed in a variety of applications such as biosensing, photoacoustic imaging, and cancer cell detection, photothermal therapy, and drug delivery (Fernandez-Fernandez et al., 2011; Flores et al., 2019). CNTs can enter cells and even cell nuclei in drug delivery applications, and their inner and outer surfaces can be functionalized with various moieties for targeting and conjugation. As reported by Liu et al., specifically in a murine 4T1 breast cancer model, conjugated chemotherapeutic agent paclitaxel with single-walled CNTs exhibited 10 times greater tumor absorption in comparison with free drug (Saad et al., 2012). Panchapakesan et al. reported a very interesting concept for eliminating cancer cells employing localized nanobombs in the form of CNTs by exposing them to laser (of 800 nm) at 50–200 mW/cm^2 intensity in a PBS solution inducing localized explosions. This nanobomb effect is achieved in a very systematic manner by increasing temperature of adsorbed water molecules on surface of CNTs employing laser heating. When temperature of adsorbed water molecules reaches over 100°C, it results in explosion locally at the surface of CNTs destroying themselves and the host cancer cell (Panchapakesan et al., 2005). Though, targeted cancer cells are destroyed by CNTs while the surrounding cells not having exposure to CNTs survived. Some studies reported significant toxicity to lungs of mice including asbestos like pathology underlying toxicity profile of CNT-based therapeutics indicating major problem in clinical translation (Fernandez-Fernandez et al., 2011).

21.3 Polymeric nanocarriers for photosensitizer/dye encapsulation

Luminescent nanomaterials have recently gotten attention because of their prospective applications in photonics and biophotonics (Armaroli and Balzani, 2007). As an alternative luminescent source the conjugated polymer nanoparticles (PNPs) encapsulated with dye have several advantages such as ease of synthesis, efficient brightness, more photo-stability, color tunability, surface functionalization, and exceptionally high molar extinction coefficient with broad absorption (Bhattacharyya and Patra, 2014;, Hedley et al., 2017). In case of multichromophoric conjugated polymers systems the high absorption cross-section results due to the π-electrons delocalization of each monomeric unit (Collini and Scholes, 2009; Hwang and Scholes, 2011). As a result, the optical characteristics are substantially influenced by the electronic delocalization, degree of conjugation, and inter/intramolecular interactions, with an exciton diffusion length of 5–20 nm (Beljonne et al., 2002; Scholes and Rumbles, 2010). Semiconducting PNPs are made by coiling polymer molecules and changing their extended to collapsed states. The composition of the polymer determines whether the absorption band of PNPs shifts blue or red. The bending and kinking of the polymer backbone cause the blue shifting of absorption (Padmanaban and Ramakrishnan, 2000), whereas the red shifting of absorption is caused by comparatively relaxed and organized conformations with higher inter-chain interactions (Ong et al., 2005; Wang et al., 2005). Encapsulating organic fluorophores inside fluorescent PNPs can also improve the color tunability and brightness of the PNPs. The limited motion of dye molecules and the photophysics of encapsulated dye molecules in PNPs have recently been reported (Martin et al., 2014; Jiang et al., 2010). For practical applications, dye-encapsulated PNPs' photostability and brightness are critical considerations. Furthermore, the excellent energy transfer as well as charge kinetics for a variety of optoelectronic devices of hybrid conjugated PNPs make them more promising (Tang et al., 1989; Kim et al., 2007). As a result, a fundamental understanding of PNP photophysics is critical for their future growth. For the synthesis of fluorophore-doped polymer nanomaterials, mainly miniemulsion and reprecipitation procedures are used.

Dye doped conjugated PNPs are examined and reported in literature for their high photostability, brightness, dye dynamics inside the PNPs, ease of synthesis, nontoxic behavior and employed for in-vitro and in-vivo bioimaging applications. Therefore, they are regarded as alternative next generation luminescent materials. Fluorescent PNPs have also been identified as possible formulations for clinical cancer detection, as well as drug and gene delivery treatments. Furthermore, dye-doped PNPs' promise in cancer treatment is enhanced by Fluorescence Resonance Energy Transfer (FRET) -based photosensitization and single-oxygen production capabilities as well as photodynamic therapy. The most promising application of dye doped PNP's are as a light-emitting device, tunable PL, white light generation in solution and solid state. More study into multiple dye doped PNPs will undoubtedly open new possibilities in other systems such as solar and OLED. For laser emission and other photo-driven devices the arrangement of dye molecules is important inside polymer nanofibers and would be more beneficial. It is critical for building efficient devices to investigate for self-assembled as well as aggregated nanostructures with inter- or intrachain energy transfer using ultrafast spectroscopy and to understand separation of charge in inorganic/organic hybrid systems to create light harvesting devices. Furthermore, organic-inorganic hybrids based on polymers are the most effective materials for increasing quantum efficiency in solar cell-based applications. The challenges in the optoelectronic applications which are not well known in conjugated PNPs are excitation diffusion and diffusion length. To address these critical concerns, ultrafast spectroscopic research is critical. Dye-doped polymers and polymer hybrids with their fundamental photophysical properties are great platforms for future study in photonics, biophotonics, and optoelectronic device development (Jana et al., 2018).

21.4 Nanoconstructs for photodynamic therapy

The possibility for photosensitizer (PS) based photodynamic therapy (PDT) has been examined using nanocarrier platforms such as liposomes, polymeric nanoparticles, and micelles. PS is enclosed in polymers by hydrophobic or chemical conjugation processes (Qian et al., 2011). An aromatic planar tetrapyrrole backbone of porphyrins and phthalocyanines make them ideal for multifunctional imaging, therapeutic and theranostic applications. Endogenous brilliant red porphyrins in the blood could be considered the original old theranostic agent, whereas heme has been utilized in cancer magnetic resonance imaging since the 1980s, and phthalocyanines have been used in positron emission imaging since the 1950s. Porphyrins, when irradiated with light, produce singlet oxygen, which is employed for therapeutic purposes and has made a clinical impression in the form of PDT (Schipper et al., 2009). The potential of nanocarrier platforms such as liposomes, polymeric nanoparticles, and micelles for photosensitizer (PS) based PDT has been investigated. PS is encapsulated by hydrophobic or chemical conjugation processes to polymers (Bae and Na, 2012; Qian et al., 2011; Umeda et al., 2010). As reported in the literature the better polysaccharide based nanogels can be created using phullulan/folate-pheoforbide-a (Pheo-A) Conjugate. In the research conducted using in vivo protocol, the nanogel showed fluorescence activity after 30 minutes, which rose dramatically until 12 hours and was maintained for more than three weeks, whereas after injection the free form of Pheo-A showed fluorescence immediately but could not be sustained for longer. The window of opportunity for PDT with little unfavorable phototoxicity was demonstrated using these newly created nanogels. A cancer cell-specific photosensitizer nanoparticles of pheo-A conjugated glycol chitosan have been created containing disulfide bonds (pheo-A- ss - GC) reducible in nature with switchable photoactivity (Lu, 2010). The nanoparticulate structure dissociates rapidly after uptake by cancer cells due to reductive cleavage and disulfide linker with dequenching action, indicating increased cytotoxicity with light treatment. The photophysical behavior of PLGA nanoparticles encapsulated with Zinc (II) phthalocyanine was better and more stable. These nanoparticles have a three-day sustained release activity and a higher photo cytotoxic activity. Lee et al. created tumor-homing drug carriers utilizing protoporphyrin IX (ppix) GC nanoparticles in another investigation (ppix-GC-NPs). Some nanoparticles demonstrated self-quenching function due to an off state with no fluorescence signal as well as with light exposure phototoxicity (Ren et al., 2011). The compact nanoparticle shape rapidly diminished after cellular uptake, producing a robust fluorescence signal and singlet oxygen production with irradiation, suggesting the efficiency for the synchronized photodynamic therapy as well as dynamics. When comparing free ppix-treated mice to ppix GC NPs treated tumor bearing mice, the ppix-GC-NPs-treated tumor bearing mice showed better blood circulation and therapeutic efficiency. As a bio reducible biarmed mPEG- (ss-PhA) 2 conjugates, a self-quenchable PS carrier system was designed using PEG polymer for cancer photodynamic treatment. Chemically, the PhA molecule was linked to the biarmed linkage at one end of the mPEG via disulfide linkages. Due to their self-quenching capabilities, photoactivity of mPEG-(ss-PhA)2 nanoparticles were reduced under physiological conditions. The bioreducible activation mechanism in cancer cells demonstrated high cytotoxicity during light exposure, resulting in superior photodynamic cancer treatment and fewer adverse effects (Bogart et al., 2012). The recent invention of up conversion nanoparticles (UCNPs) for treatment of inaccessible deep tissues is very interesting and beneficial as it converts near-infrared radiations into UV visible radiations. When PDT-based-UCNPs are excited using NIR radiations, the ROS has been created by the FRET attached to photosensitizers and kills (Boppart et al., 2005). A PDT developed by Cui et al. for deep-tissue treatment using zinc phthalocyanine (ZnPc) (as photosensitizer) loaded UCNPs and coated with FA-modified chitosan (FASOC) for tumor accumulation after NIR irradiation. The NIR fluorescent dye Indocyanin green (ICG-Der-01) was encapsulated to FASOC-UCNPs to test the nanoconstructs capabilities for biodistribution and tumor targeting. The excised organs showed strong fluorescence in essential organs after 24 hours, however spleen and heart exhibited limited uptake. Mice (FR positive) with Bel-7402 tumors were given FASOC-UCNPICG and scanned at various time periods to test the nanoconstruct's tumor selection abilities. The greatest tumor fluorescence was recorded 24 hours after injection and lasted for more than 96 hours. These findings suggested that a multifunctional nanoconstruct could be used to cure tumors (Jiang et al., 2010).

21.5 Photo-triggered theranostic nanocarriers

The combination of the treatment and diagnosis where the nanoparticles can perform disease diagnosis, detection, and treatment simultaneously is termed as theranostic medicine. The objective for the treatment of cancer is to lessen the patient suffering by employing minimally invasive modalities. As shown in Fig. 21.2, For monitoring and treating diseases the light-based approaches are of interest to theranostic NPs. As reported in the literature for tumor monitoring, simultaneous PDT and plasmonic heating therapy, a nanocage core of the gold-silver and a shell made with silica carrying the multifunctional NPs of Yb–2,4-dimethoxyhematoporphyrin (Yb–HP), an NIR photosensitizer designed by Khlebtsov et al. and for the diagnosis, the Yb-HP IR luminescence are recommended. Additionally, Galanzha et al. reported the efficacy of the gold layered nanotubes coupled with folate and specific antibodies CD44 in identification and treatment of circulating stem cells in vivo which can initiate breast cancer (Khlebtsov et al., 2011; Galanzha et al., 2009). A significant (95%) decrease in viability of tumor was observed after NIR laser exposure using direct NPs injection in mice and were diagnosed with nonplasmonic silica NPs containing the Si-naphthalocyanine (heating) dyes and heptamethine cyanine dye which are NIR fluorescent (Galanzha et al., 2009). The multifunctional capabilities of NPs increasing their importance in nanotechnology-based research (Lee et al., 2009). Santra et al. and Huan et al. developed 4-DiR and 4-DiI, the NIR dyes encapsulated in the NPs of the iron oxide having polyacrylic acid coating and GNR-SiO2-FA, gold nanorods modified with silica for individual and dual therapy respectively (Santra et al., 2009; Huang et al., 2011).

The nanogels also have an application for specifically targeting cancer treatment with chemo-photo thermal therapy, temperature sensing by employing optical properties, and fluorescence therapy. For example, targeting ligands based on hyaluronic acid and comprising of a bimetallic core of gold and silver with a shell of thermoresponsive polyethylene glycol (Huang et al., 2011); Temozolomide (TMZ)-loaded NPs (Wu et al., 2010); and a DOX encapsulated PEG-PLGA-gold half-shell nanoparticles (Lee et al., 2010). Chemotherapy or PTT were found to be less effective and induce weight loss or tumor recurrence in the mice during cancer treatment which can be eliminated by a combination of NIR irradiated photothermal

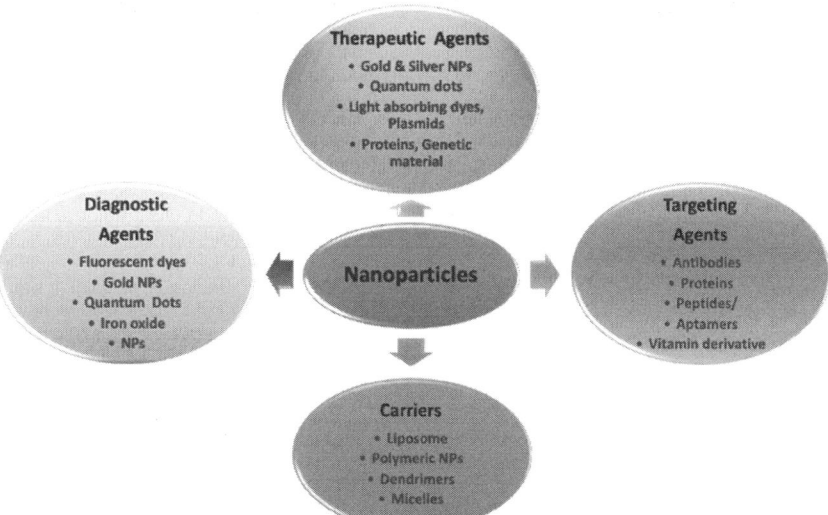

Fig. 21.3 Photo-triggered theranostic NPs as carriers, diagnostic, therapeutic, and targeting agents.

therapy and chemotherapy. Similarly, for radiation therapy or nuclear imaging of malignancies in addition to chemotherapy the NPs encapsulated with radioisotope. For example, folate conjugated cross-linked NPs of 64Cu-radiolabeled shell (SCK) (Rossin R et al., 2005); QD-photosensitizer (Photofrin) conjugate (Yang et al., 2008). Gold NPs in combination with radiation and hyperthermia were studied for lowering tumor volume in radio resistant mouse cell cancer for long-term survival (Hainfeld et al., 2010). The photo-triggered theranostic NPs as carriers, diagnostic, therapeutic and targeting agents are necessary for obtaining efficient diagnostic and therapeutic effects (Fig. 21.3).

21.6 Approaches to measure drug release through theranostic nanomedicine

The optical approaches to measure drug release through theranostic nanomedicine were studied and reported. Theranostic nanotechnologies that can quantitatively quantify medication release in real time would provide vital information about the administered drug dose. The concentration of fluorescent medication in tumors is frequently monitored using noninvasive optical approaches (Al-Jamal and Kostarelos, 2011; Saager et al., 2011; Chen et al., 2013; Reif et al., 2007; Palmer et al., 2010; Diamond et al., 2005).

21.6.1 Silicon photonic crystals with pores

Biocompatible material designed using electrochemical and lithographic techniques, like silicone can be used to produce wide range of drug carriers as they are electrochemically fixed to create nanopores holding pharmaceuticals. Structures with acute reflections are observed which are produced by controlling the width of pores during electrochemical etching and can be compared to photonic pattern seen in beetles and butterfly (Park et al., 2009; Doan and Sailor, 1992).

Inherent photoluminescent features are seen in PSi that are found to be useful for in vivo imaging. A drug release visualizing platform is made consisting of PSi particles. Photoluminescence was used to visualize drug release from PSi NPs but did not proved to be a reliable way to quantify it; spectral reflectance was used to visualize daunorubicin release from PSi microparticles (Park et al., 2009). Conversely, in vivo imaging reflectance to assess drug release is confined to transparent or this layer of tissues and the best example to explain this is none other than intraocular drug delivery (Moore et al., 2014). For various measurements in cell research and histology of thin tissues and to obtain quick, sensitive, submicrometric high resolution pictures the optical microscopes are preferably employed.

For combined photodynamic imaging and treatment protoporphyrin IX (PpIX), a photosensitizer has been designed and delivered employing PEG-co-poly (-amino ester) (PEG-PBA), a pH sensitive nanoparticles. In this system, nanoparticles are stabilized by PEG, whereas PBA loads the hydrophobic drug. At a lower pH protonation at tertiary amine of PBA occurs which destabilizes the polymeric micelle and releases its PpIX cargo. A fluorescent PpIX molecule are also utilized for imaging in addition to being a photoactive medication (Moore et al., 2014). The molecular oxygen quenched the fluorescence of PpIX which results in singlet oxygen and considered as a major effector in cell death.

21.6.2 Fluorescent nanoparticles

An inorganic substance that converts several types of electromagnetic energy (ultraviolet, infrared, or X-ray light) into visible light are called earth-doped nanophosphors. Fluorescent phosphors, upconversion phosphors, and radioluminescent phosphors are examples of nanophosphors that can convert X-ray, infrared or ultraviolet radiations into visible light. Nanophosphors are better alternative to organic dyes because of higher chemical stability and as they do not photobleach or fade (Ukonaho et al., 2007).

Nanophosphors are superior to quantum dots and fluorescent molecules in terms of line width of one nanometer as compared to line width of 20–100 nanometer for their counter parts establishing narrow fine structures having discrete spectrum peaks and low toxicity, attracting favors for their biological uses (Petoral et al., 2009). To evaluate drug loading and release, nanophosphors with excitation wavelengths of 613 nm for visible and 256 nm for shortwave UV light are employed (Wang et al., 2011).

21.6.3 Upconversion nanoparticles

Dispersion and inadequate penetration of UV and visible light having short excitation wavelengths limits the applicability of UV-visible light phosphors for the imaging of deep tissues (Stolik et al., 2000). Upconversion NPs transform NIR with longer wavelength to visible range. Due to this ability of conversion, tissue autofluorescence is less, penetration of light in tissue is profound, and scattering of light is low. Ideal properties of upconversion probes like capability of excitation by longer wavelengths and emission by optical transmission window (650–950 nm) appear favorable for research. Upconversion particles having 980 nm as excitation wavelength and 800 nm as emission wavelength can penetrate as deep as 3.2 cm through hog tissue for imaging. The use of upconversion NPs for tracking release of medication is achieved by determining intensity of luminescence. Coating of organic layers on upconversion NPs is done for increasing water dispersion and loading of lipophilic drug (Tian et al., 2013). The NP's therapeutic efficacy was greatly boosted when Doxorubicin (Dox) and photodynamic therapy were combined. In an in vitro assay, NPs containing Dox (0.082 mmol/g) and Ce6 (0.077 mmol/g) revealed about 15% viability for A-549 cells when treated using laser of wavelength 980 nm indicated better results than Ce6 and Dox nanoparticles that had not been irradiated (~40 percent cell viability) or Ce6 NPs that had been irradiated (~40 percent viability of cell). Whereas these therapies are efficient in vitro, the penetration depth of near-infrared and infrared wavelengths is on the millimeter scale, which limits their application to deep tissue imaging (Stolik et al., 2000). Theranostic modalities are predicted to improve as technology improves its ability to penetrate deep into tissue.

21.6.4 Radioluminescent nanoparticles

Radioluminescence can be defined as conversion of radiation energy into visible light. The focus, here is on X-ray radiation, though X-ray considerably penetrates soft tissue, and utilizing this type of device to measure drug release is advantageous. Radioluminescent nanophosphors are been researched for quantitative drug delivery sensors and in vivo imaging (Chen et al., 2013; Sottoriva et al., 2013). The pH-sensitive drug release was achieved using poly(allyl-amine HCl) (PAH) and poly(styrene sulfonate sodium) (PSS) a layer-by-layer coated with theranostic radioluminescent NPs (Chen et al., 2013).

Nanophosphors (luminescent nanoparticles) are of great potential for a theranostic system that can measure medication delivery quantitatively at the bench. The excitation wavelengths of nanoparticles can be tuned in broader range of wavelengths from X-ray to NIR. It has been demonstrated that using NPs can quantify drug release with luminescence coinciding optical characteristics of therapeutic agent. Specific challenges, such as nanoparticle-associated toxicity and physical restrictions, still limit the utilization of these NPs in therapeutic settings. In tissues, the optical imaging is now confined to millimeter or centimeter depths. As a result, recognizing and addressing current imaging issues will inform research decisions for future.

21.7 Magnetic resonance imaging for monitoring release of drug

Being a theranostic imaging modality, MRI is specifically appealing as it is not restricted to in depth imaging while, quantitative release of drug can be monitored. The behavior of the protons in tissue (mostly hydrogen released through water) in response to electromagnetic field causing magnetization of protons explain how MRI works. When a powerful magnetic field is developed around the patient/person it forces protons in the tissues to align along the magnetic field's direction, and the proton's spin is perturbed by a short electromagnetic pulse. As protons re-align, they release energy, which can be measured in two ways either by returning to longitudinal or transverse magnetization, wherein magnetization in longitudinal way is called as T1 relaxation while by transverse way is termed as T2 decay/relaxation (Skinner et al., 1995).

In theranostic MRI methods, mostly the drugs, that can form a complex with imaging agents or release with the speed with which imaging agents can, have been used. Concurrent triggering of drug and imaging agent has been the result of drug release by external trigger. These studies show how drug release kinetics can be evaluated by calculating the corelease of distinct drugs and imaging agents, or by evaluating the release of a drug in combination with an imaging agent. Furthermore, release could be induced by physiological changes (such as pH variations) or exogenous cues like temperature. Through thick tissue, MRI was useful in imaging drug release with greater than millimeter resolution. Cost, release rate, and differing pharmacodynamics and pharmacokinetic behavior between imaging agent and drug are all limitations of this method (Moore et al., 2014).

21.8 Photo-triggered theranostics nanomaterials: Principle and applications

Photo-triggered release systems are predicted to benefit by providing a noninvasive technique for guaranteeing correct drug localization as well as determining the concentration of the released drug. This technology aids preclinical nanomedicine research by enhancing methods for optimizing kinetics of release, particularly in *in vivo* intricate models. For evaluating rates of accumulation of total drug in tissue, traditional pharmacokinetics methods such as single photon emission tomography (SPECT), positron emission tomography (PET), or radiography are useful. Distinguishing free drug that has released from encapsulated drug or nanoparticles is difficult (Moore et al., 2014).

21.8.1 Applications of photo-triggered theranostics nanomaterials in cancer treatments

Theranostic techniques assessing kinetics of quantitative drug release have a lot of interest since they can help us understanding intratumoral drug concentrations in a better way and it is given that therapeutic drugs remain inactive until nanocarriers release them. Furthermore, quantification of release of drugs from noninvasive theranostic techniques would be very useful to predict the actual amount of drug reaching to the tumor by distinguishing between released drug and encapsulated. (Moore et al., 2014)

21.8.2 Therapeutic applications of photo-based theranostic nanoparticles

Photo-based theranostic NPs are employed for a variety of applications, including detecting and treating circulating tumor cells along with small and late-stage malignancies (Galanzha et al., 2009). For targeting prostate specific membrane antigen (PSMA) -overexpressing cancerous cells of prostate selectively, novel aptamer-QD-Dox conjugates have been developed and imaging is done by the dual FRET (Bi-FRET). The interaction of Dox-aptamer and Dox-QD provide therapeutic effect in cancer cells by release of Dox inside the cell. After endocytic uptake, dissociation of doxorubicin from the nanoconjugates causing QDs to fluoresce distinctly. As a result, nanoparticles can be identified at cellular level using new formulation. An amphiphilic triblock polymer coated semiconductor QDs which are coupled with monoclonal antibodies against PSMA have been developed by Gao et al. for targeting prostate cancer and spectrum imaging of it. In PSMA-overexpressing C4-2 prostate cancer xenografts, ligand conjugated QDs have been accumulated as demonstrated by fluorescence imaging in an in vivo absorption assays carried out in nude mice (Gao et al., 2004). Cheng et al. developed mesoporous silica NPs for three different functions simultaneously and these were consisting of photosensitizer for PDT based on palladium-porphyrin, for imaging ATTO647N, a contrast agent fluoresce by NIR and for specifically targeting v3 integrin, cRGDyK peptides. The NPs have been demonstrated to be compatible with cells wherein, cells overexpressing v3 integrins such as U87-MG human glioblastoma absorbed them more than MCF-7 breast cancer cells lacking integrin proteins (Cheng et al., 2010). For the treatment of cancer guided by imaging performed using concurrently by MRI and NIR fluorescence, multifunctional nanoemulsions comprising nanocrystals of iron oxide and Cy7, a fluorescent dye have been developed by Gianella et al. to target tumor angiogenesis, the nanoemulsions was also containing the RGD peptide and prednisolone acetate valerate (PAV), a hydrophobic glucocorticoid. It could effectively perform in vivo imaging of tumors with significant reduction in size of tumor and found to be biodegradable. Considering the increasing emphasis on establishing an adequate nano-scale diagnostic and therapeutic system for various forms of cancer, substantial studies have also been undertaken for developing theranostic NPs for the management of several other diseases (Gianella et al., 2011). The example includes development of nanoparticles for treating inflammatory atherosclerosis by McCarthy et al. In this work, nanoparticles of crosslinked iron oxide (CLIO) coated with dextran have been developed. For imaging by NIR fluorescence, AlexaFluor 750; the NIR fluorophore encapsulated with CLIO can be employed. In atherosclerotic lesions, inflammation may be caused by macrophages which can be intoxicated using photo-triggered photosensitizers containing chlorin (McCarthy et al., 2010). Macrophages visualized and treated by hollow Au-NPs (gold nanoparticles) coated by dextran have been developed by Lim et al. NPs having ability to scatter NIR radiations for utilizing them for imaging and PTT have been absorbed rapidly by RAW274.7 macrophage cells (Lim et al., 2008).

Visudyne is a liposome activated by light and has been approved by the FDA for the treatment of macular degeneration. It contains a mixture of phospholipids such as dimyristoyl phosphatidyl choline (DMPC) and egg phosphatidyl glycerol (EPG) along with hydromonobenzoporphyrin, a photosensitizer (Rai et al., 2010). The reorganization of components of tissue by the heat generated through irradiation of nanoparticles by laser resulted in repair and closure of wound. SPIO NPs with particle size of 15 nm when exposed to radio waves generated heat and demonstrated to solder albumin onto rabbit aorta. For noninvasive bonding of the tissues including skin lesions to explanted lens capsular tissues, heat is generated through photothermal transducers which are nothing but the small gold NPs (Reinert et al., 2008). The NPs based on light radiations can also be used in the infections caused by bacteria and example includes anti-protein A antibodies coated gold nanoparticles, which have been demonstrated for targeting and killing Gram-positive *S. aureus*. In this case, after absorption of NPs by bacteria they are exposed to the concentrated pulses of laser in the wavelength range of 420 to 520 nm causing generation of heat resulting in irreversible destruction of bacteria (Zharov et al., 2006). To treat bacterial infection, surface adsorbed vancomycin on nanoeggs made up of Fe_3O_4-gold, which are multifunctional in nature and have been developed by Huang et al. They also proved that using an external magnet helps with bacterial aggregation and PTT efficiency (Huang et al., 2009; Menon et al., 2013).

21.9 Opportunities and limitations of nanomaterials

Real time assessment of kinetics of drug release is the major limitation of nanomaterial system though the imaging has been integrated with drug delivery using novel technologies. Imaging through thick tissue is very challenging since the intensity of light decreases as it passes through the tissues and analyte signals are limited by autofluorescence backgrounds and quality of image is compromised by diffuse scattering. X-rays have excellent penetration power to pass through soft tissues, emitted visible light lacks the infiltration from depth of the tissue after optical excitation by X-ray for luminescence imaging restricting the radio-luminescent system. Measurement of noninvasive drug delivery through theranostic nanomedicines is a challenging process as several imaging particles have proven efficacy in vitro for cellular labeling, but their application to more complicated biological systems is limited due to toxicity (Moore et al., 2014).

21.10 Preclinical challenges

For developing effective theranostic technologies, a fundamental understanding of nanomaterials is required. This involves a grasp of material selection, physicochemical property constraints and advantages, and potential nanomaterial-related toxicities. The composition, in vivo stability, shape and size of nanoparticles are all factors that contribute to their toxicity. The toxicity of nanomaterials, as well as the physical constraints of optical technologies and the difficulty of deep tissue well resolved and sensitive images are the major challenges which need to be addressed.

21.11 Future aspects of nanomaterials in the therapeutics

The consistent expansion in the utility domain of nanotechnology with persistent advancement in nanomaterials and methods of preparation leads to its greater scope in diagnostics and therapeutics. The integrated research endeavor of photodynamic and nanostructures development successfully resulted into the bifunctional system which can be traced upon administration that can be targeted in cells for diagnosis and treatment. However, more precision must be aimed and worked upon to translate it into effective clinical practice.

A thorough investigation on biocompatibility of the nanoparticulate system and interactions with cells and tissues is also required before it can be used in humans. The NPs' components would have to be thoroughly examined for in vivo immunological responses and toxicity. Even though theranostic NPs are in the early stage of clinical trials, they have high potential to overcome the pharmacokinetic limitations, assure cellular safety and precision in real time diagnosis and treatment. The photo-based theranostic NPs system is constantly getting evolved with development of knowledge and technology in the field of basic and advanced sciences such as biology, chemistry, physics, and medical physics. Importantly, several modalities and nanoparticle components must be evaluated in order to optimize the depth to which photosensitizer- and metal-based imaging agents can have access. Imaging techniques must also be assessed for their sensitivity with different nanomaterials for accurate detection of signals. The use of photo-based theranostic NPs in phototherapies like PTT and PDT should cause negligible damage to surrounding healthy tissue. It has been observed that some of the encumbrances sensitive to light, such as QDs, are hazardous and critical to remove NPs before the complete degradation of the polymeric shell and the encapsulated agents are released.

Minimizing the hurdles of penetration and sensitivity will clearly uphold theranostic NPs as a system with good spatial resolution. It will also provide more controllable drug release along with specific targeting encapsulated agent (Menon et al., 2013). Over the last decade, consistent and steady advances in nanocarrier creation have resulted in a paradigm change in cancer treatment, diagnostics, and imaging. According to studies, the foundation of nanocarriers' performance as a delivery platform is the optimization of shape, size, composition, drug payloads, biocompatibility, biodegradability, cellular interaction, and penetration of the system into the affected tumor tissue. Nanoconstructs such as nanostars, nanoshells, nanoflares, magnetic, polymeric, and/or lipidic nanoconstructs are said to have surpassed previous nanomedicine systems by presenting "an all-encompassing" cancer therapy approach. These systems are explored in depth in this review, with a focus on their ability to cater to various areas of cancer treatment. The strategies range from interacting with intracellular organelles to delivering drugs and imaging agents to specific areas and collecting their effects utilizing advanced instruments for converting irradiation light onto nanoconstructs to release heat, causing tumor shrinkage. However, it is recognized that practical translation of such technologies can only be achieved by fullproof optimization, safety assessments, and appropriate animal development, all of which can instill confidence in researchers, practitioners, and patients.

References

Al-Jamal, W.T., Kostarelos, K., 2011. Liposomes: from a clinically established drug delivery system to a nanoparticle platform for theranostic nanomedicine. Acc. Chem. Res. 44 (10), 1094–1104. Oct. http://doi.org/10.1021/ar200105p.

Allen, T.M., Cullis, P.R., 2004. Drug delivery systems: entering the mainstream. Science 303 (5665), 1818–1822. Mar. http://doi.org/10.1126/science.1095833.

Armaroli, N., Balzani, V., 2007. The future of energy supply: challenges and opportunities. Angew. Chem. Int. Ed. 46 (1–2), 52–66. Jan. http://doi.org/10.1002/anie.200602373.

Bae, B., Na, K., 2012. Development of polymeric cargo for delivery of photosensitizer in photodynamic therapy. Int. J. Photoenergy 2012, 1–14. http://doi.org/10.1155/2012/431975.

Beljonne, D., et al., 2002. Interchain vs. intrachain energy transfer in acceptor-capped conjugated polymers. Proc. Natl. Acad. Sci. 99 (17), 10982–10987. Aug. http://doi.org/10.1073/pnas.172390999.

Bhattacharyya, S., Patra, A., 2014. Interactions of π-conjugated polymers with inorganic nanocrystals. J. Photochem. Photobiol. C 20 (1), 51–70. Sep. http://doi.org/10.1016/J.JPHOTOCHEMREV.2014.05.001.

Bogart, L.K., Taylor, A., Cesbron, Y., Murray, P., Lévy, R., 2012. Photothermal microscopy of the core of dextran-coated iron oxide nanoparticles during cell uptake. ACS Nano 6 (7), 5961–5971. Jul. http://doi.org/10.1021/nn300868z.

Boppart, S.A., Oldenburg, A.L., Xu, C., Marks, D.L., 2005. Optical probes and techniques for molecular contrast enhancement in coherence imaging. J. Biomed. Opt. 10 (4), 041208. http://doi.org/10.1117/1.2008974.

Chen, H., et al., 2012. Magnetic and optical properties of multifunctional core–shell radioluminescence nanoparticles. J. Mater. Chem. 22 (25), 12802. http://doi.org/10.1039/c2jm15444g.

Chen, H., et al., 2013. Monitoring pH-triggered drug release from radioluminescent nanocapsules with X-ray excited optical luminescence. ACS Nano 7 (2), 1178–1187. Feb. http://doi.org/10.1021/nn304369m.

Cheng, S.-H., et al., 2010. Tri-functionalization of mesoporous silica nanoparticles for comprehensive cancer theranostics—the trio of imaging, targeting and therapy. J. Mater. Chem. 20 (29), 6149. http://doi.org/10.1039/c0jm00645a.

Cole, A.J., Yang, V.C., David, A.E., 2011. Cancer theranostics: the rise of targeted magnetic nanoparticles. Trends Biotechnol. 29 (7), 323–332. Jul. http://doi.org/10.1016/J.TIBTECH.2011.03.001.

Collini, E., Scholes, G.D., 2009. Coherent intrachain energy migration in a conjugated polymer at room temperature. Science 323 (5912), 369–373. Jan. http://doi.org/10.1126/science.1164016.

Dai, H., et al., 2006. Chitosan-DNA nanoparticles delivered by intrabiliary infusion enhance liver-targeted gene delivery. Int. J. Nanomed. 1 (4), 507–522. Dec. http://doi.org/10.2147/nano.2006.1.4.507.

Diamond, K.R., Malysz, P.P., Hayward, J.E., Patterson, M.S., 2005. Quantification of fluorophore concentration in vivo using two simple fluorescence-based measurement techniques. J. Biomed. Opt. 10 (2), 024007. http://doi.org/10.1117/1.1887932.

Dilnawaz, F. and Kumar, D., "Nanoconstructs for Cancer Therapy and Imaging," 2017. [Online]. Available: http://elynsgroup.com

Doan, V.v., Sailor, M.J., 1992. Luminescent color image generation on porous silicon. Science 256 (5065), 1791–1792. Jun. http://doi.org/10.1126/science.256.5065.1791.

Fernandez-Fernandez, A., Manchanda, R., McGoron, A.J., 2011. Theranostic applications of nanomaterials in cancer: drug delivery, image-guided therapy, and multifunctional platforms. Appl. Biochem. Biotechnol. 165 (7–8), 1628–1651. Dec. http://doi.org/10.1007/s12010-011-9383-z.

Flores, A.M., Ye, J., Jarr, K.-U., Hosseini-Nassab, N., Smith, B.R., Leeper, N.J., 2019. Nanoparticle therapy for vascular diseases. Arterioscler. Thromb. Vasc. Biol. 39 (4), 635–646. Apr. http://doi.org/10.1161/ATVBAHA.118.311569.

Galanzha, E.I., Kim, J.-W., Zharov, V.P., 2009. Nanotechnology-based molecular photoacoustic and photothermal flow cytometry platform for *in-vivo* detection and killing of circulating cancer stem cells. J. Biophotonics 2 (12), 725–735. Dec. http://doi.org/10.1002/jbio.200910078.

Gao, X., Cui, Y., Levenson, R.M., Chung, L.W.K., Nie, S., 2004. In vivo cancer targeting and imaging with semiconductor quantum dots. Nat. Biotechnol. 22 (8), 969–976. Aug. http://doi.org/10.1038/nbt994.

Gao, Y., et al., 2014. Nanotechnology-based intelligent drug design for cancer metastasis treatment. Biotechnol. Adv. 32 (4), 761–777. Elsevier Inc. http://doi.org/10.1016/j.biotechadv.2013.10.013.

Gianella, A., et al., 2011. Multifunctional nanoemulsion platform for imaging guided therapy evaluated in experimental cancer. ACS Nano 5 (6), 4422–4433. Jun. http://doi.org/10.1021/nn103336a.

Giljohann, D.A., Seferos, D.S., Daniel, W.L., Massich, M.D., Patel, P.C., Mirkin, C.A., 2010. Gold Nanoparticles for Biology and Medicine. Angew. Chem. Int. Ed. 49 (19), 3280–3294. Apr. http://doi.org/10.1002/anie.200904359.

Green, M.R., et al., 2006. Abraxane®, a novel Cremophor®-free, albumin-bound particle form of paclitaxel for the treatment of advanced non-small-cell lung cancer. Ann. Oncol. 17 (8), 1263–1268. Aug. http://doi.org/10.1093/ANNONC/MDL104.

Hainfeld, J.F., Dilmanian, F.A., Zhong, Z., Slatkin, D.N., Kalef-Ezra, J.A., Smilowitz, H.M., 2010. Gold nanoparticles enhance the radiation therapy of a murine squamous cell carcinoma. Phys. Med. Biol. 55 (11), 3045–3059. Jun. http://doi.org/10.1088/0031-9155/55/11/004.

Hedley, G.J., Ruseckas, A., Samuel, I.D.W., 2017. Light harvesting for organic photovoltaics. Chem. Rev. 117 (2), 796–837. Jan. http://doi.org/10.1021/acs.chemrev.6b00215.

Hrkach, J., et al., 2012. Preclinical development and clinical translation of a PSMA-targeted docetaxel nanoparticle with a differentiated pharmacological profile. Sci. Transl. Med. 4 (128), 128–139. Apr. http://doi.org/10.1126/scitranslmed.3003651.

Huang, P., et al., 2011. Folic acid-conjugated silica-modified gold nanorods for X-ray/CT imaging-guided dual-mode radiation and photo-thermal therapy. Biomaterials 32 (36), 9796–9809. Dec. http://doi.org/10.1016/j.biomaterials.2011.08.086.

Huang, W.-C., Tsai, P.-J., Chen, Y.-C., 2009. Multifunctional Fe_3O_4@Au nanoeggs as photothermal agents for selective killing of nosocomial and antibiotic-resistant bacteria. Small 5 (1), 51–56. Jan. http://doi.org/10.1002/smll.200801042.

Hwang, I., Scholes, G.D., 2011. Electronic energy transfer and quantum-coherence in π-conjugated polymers. Chem. Mater. 23 (3), 610–620. Feb. http://doi.org/10.1021/cm102360x.

Jana, B., Bhattacharyya, S., Patra, A., 2018. Perspective of dye-encapsulated conjugated polymer nanoparticles for potential applications. Bull. Mater. Sci. 41 (5), 122. Oct. http://doi.org/10.1007/s12034-018-1643-x.

Jiang, S., Gnanasammandhan, M.K., Zhang, Y., 2010. Optical imaging-guided cancer therapy with fluorescent nanoparticles. J. R. Soc., Interface 7 (42), 3–18. Jan. http://doi.org/10.1098/rsif.2009.0243.

Katz, J.S., Burdick, J.A., 2010. Light-responsive biomaterials: development and applications. Macromol. Biosci. 10 (4), 339–348. Apr. http://doi.org/10.1002/mabi.200900297.

Khlebtsov, B., et al., 2011. Nanocomposites containing silica-coated gold–silver nanocages and Yb–2,4-dimethoxyhematoporphyrin: multifunctional capability of IR-luminescence detection, photosensitization, and photothermolysis. ACS Nano 5 (9), 7077–7089. Sep. http://doi.org/10.1021/nn2017974.

Kim, D.W., et al., 2007. Multicenter phase II trial of Genexol-PM, a novel Cremophor-free, polymeric micelle formulation of paclitaxel, with cisplatin in patients with advanced non-small-cell lung cancer. Ann. Oncol. 18 (12), 2009–2014. Dec. http://doi.org/10.1093/ANNONC/MDM374.

Lammers, T., Hennink, W.E., Storm, G., 2008. Tumour-targeted nanomedicines: principles and practice. Br. J. Cancer 99 (3), 392–397. Aug. http://doi.org/10.1038/sj.bjc.6604483.

Lee, C.-M., et al., 2009. Superparamagnetic iron oxide nanoparticles as a dual imaging probe for targeting hepatocytes in vivo. Magn. Reson. Med. 62 (6), 1440–1446. Dec. http://doi.org/10.1002/mrm.22123.

Lee, S.-M., Park, H., Yoo, K.-H., 2010. Synergistic cancer therapeutic effects of locally delivered drug and heat using multifunctional nanoparticles. Adv. Mater. 22 (36), 4049–4053. Sep. http://doi.org/10.1002/adma.201001040.

Lim, Y.T., Cho, M.Y., Choi, B.S., Noh, Y.-W., Chung, B.H., 2008. Diagnosis and therapy of macrophage cells using dextran-coated near-infrared responsive hollow-type gold nanoparticles. Nanotechnology 19 (37), 375105. Sep. http://doi.org/10.1088/0957-4484/19/37/375105.

Liu, G.-Y., Chen, C.-J., Li, D.-D., Wang, S.-S., Ji, J., 2012. Near-infrared light-sensitive micelles for enhanced intracellular drug delivery. J. Mater. Chem. 22 (33), 16865. http://doi.org/10.1039/c2jm00045h.

Lu, Z.R., 2010. Molecular imaging of HPMA copolymers: Visualizing drug delivery in cell, mouse and man. Adv. Drug. Deliv. Rev. 62 (2), 246–257. Feb. http://doi.org/10.1016/J.ADDR.2009.12.007.

Maeda, H., 2001. The enhanced permeability and retention (EPR) effect in tumor vasculature: the key role of tumor-selective macromolecular drug targeting. Adv. Enzyme. Regul. 41 (1), 189–207. May. http://doi.org/10.1016/S0065-2571(00)00013-3.

Majumdar, D., Peng, X.-H., Shin, D.M., 2010. The medicinal chemistry of theragnostics, multimodality imaging and applications of nanotechnology in cancer. Curr. Top. Med. Chem. 10 (12), 1211–1226. Aug. http://doi.org/10.2174/156802610791384171.

Martin, C., Bhattacharyya, S., Patra, A., Douhal, A., 2014. Single and multistep energy transfer processes within doped polymer nanoparticles. Photochem. Photobiol. Sci. 13 (9), 1241–1252. http://doi.org/10.1039/C4PP00086B.

McCarthy, J.R., Korngold, E., Weissleder, R., Jaffer, F.A., 2010. A light-activated theranostic nanoagent for targeted macrophage ablation in inflammatory atherosclerosis. Small 6 (18), 2041–2049. Sep. http://doi.org/10.1002/smll.201000596.

Menon, J.U., Jadeja, P., Tambe, P., Vu, K., Yuan, B., Nguyen, K.T., 2013. Nanomaterials for photo-based diagnostic and therapeutic applications. Theranostics 3 (3), 152–166. http://doi.org/10.7150/thno.5327.

Moore, T., Chen, H., Morrison, R., Wang, F., Anker, J.N., Alexis, F., 2014. Nanotechnologies for noninvasive measurement of drug release. Mol. Pharmaceutics 11 (1), 24–39. Jan. 06,. http://doi.org/10.1021/mp400419k.

Ong, B.S., Wu, Y., Liu, P., Gardner, S., 2005. Structurally ordered polythiophene nanoparticles for high-performance organic thin-film transistors. Adv. Mater. 17 (9), 1141–1144. May. http://doi.org/10.1002/adma.200401660.

Padmanaban, G., Ramakrishnan, S., 2000. Conjugation length control in soluble poly[2-methoxy-5-((2'-ethylhexyl)oxy)-1,4-phenylenevinylene] (MEHP-PV): synthesis, optical properties, and energy transfer. J. Am. Chem. Soc. 122 (10), 2244–2251. Mar. http://doi.org/10.1021/ja9932481.

Palmer, G.M., Boruta, R.J., Viglianti, B.L., Lan, L., Spasojevic, I., Dewhirst, M.W., 2010. Non-invasive monitoring of intra-tumor drug concentration and therapeutic response using optical spectroscopy. J. Controlled Release 142 (3), 457–464. Mar. http://doi.org/10.1016/J.JCONREL.2009.10.034.

Panchapakesan, B., Lu, S., Sivakumar, K., Teker, K., Cesarone, G., Wickstrom, E., 2005. Single-wall carbon nanotube nanobomb agents for killing breast cancer cells. NanoBiotechnology 1 (2), 133–140 http://doi.org/10shrink2:133.

Park, J.-H., Gu, L., von Maltzahn, G., Ruoslahti, E., Bhatia, S.N., Sailor, M.J., 2009. Biodegradable luminescent porous silicon nanoparticles for in vivo applications. Nat. Mater. 8 (4), 331–336. Apr. http://doi.org/10.1038/nmat2398.

Petoral, R.M., et al., 2009. Synthesis and characterization of Tb^{3+}-doped Gd_2O_3 nanocrystals: a bifunctional material with combined fluorescent labeling and MRI contrast agent properties. J. Phys. Chem. C 113 (17), 6913–6920. Apr. http://doi.org/10.1021/jp808708m.

Qian, W., Murakami, M., Ichikawa, Y., Che, Y., 2011. Highly efficient and controllable PEGylation of gold nanoparticles prepared by femtosecond laser ablation in water. J. Phys. Chem. C 115 (47), 23293–23298. Dec. http://doi.org/10.1021/jp2079567.

Rai, P., et al., 2010. Development and applications of photo-triggered theranostic agents. Adv. Drug. Deliv. Rev. 62 (11), 1094–1124. Aug. http://doi.org/10.1016/j.addr.2010.09.002.

Reif, R., Wang, M., Joshi, S., A'Amar, O., Bigio, I.J., 2007. Optical method for real-time monitoring of drug concentrations facilitates the development of novel methods for drug delivery to brain tissue. J. Biomed. Opt. 12 (3), 034036. http://doi.org/10.1117/1.2744025.

Reinert, M., et al., 2008. Electromagnetic tissue fusion using superparamagnetic iron oxide nanoparticles: first experience with rabbit aorta. Open Surg. J. 2 (1), 3–9. Oct. http://doi.org/10.2174/1874300500802010003.

Ren, K., et al., 2011. Early detection and treatment of wear particle-induced inflammation and bone loss in a mouse calvarial osteolysis model using HPMA copolymer conjugates. Mol. Pharmaceutics 8 (4), 1043–1051. Aug. http://doi.org/10.1021/mp2000555.

Rossin, R., Pan, D., Qi, K., et al., 2005. 64Cu-labeled folate-conjugated shell cross-linked nanoparticles for tumor imaging and radiotherapy: synthesis, radiolabeling, and biologic evaluation. J. Nucl. Med. 46 (7), 1210–1218. PMID: 16000291.

Saad, Z.H., Jahan, R., Bagul, U., 2012. Nanopharmaceuticals: a new perspective of drug delivery system. Asian J. Biomed. Pharmaceut. Sci. 14 (2), 11–20.

Saager, R.B., Cuccia, D.J., Saggese, S., Kelly, K.M., Durkin, A.J., 2011. Quantitative fluorescence imaging of protoporphyrin IX through determination of tissue optical properties in the spatial frequency domain. J. Biomed. Opt. 16 (12), 126013. http://doi.org/10.1117/1.3665440.

Santra, S., Kaittanis, C., Grimm, J., Perez, J.M., 2009. Drug/dye-loaded, multifunctional iron oxide nanoparticles for combined targeted cancer therapy and dual optical/magnetic resonance imaging. Small 5 (16), 1862–1868. Aug. http://doi.org/10.1002/smll.200900389.

Schipper, M.L., et al., 2009. Particle size, surface coating, and PEGylation influence the biodistribution of quantum dots in living mice. Small 5 (1), 126–134. Jan. http://doi.org/10.1002/smll.200800003.

Scholes, G.D., Rumbles, G., 2010. Excitons in nanoscale systems Materials for Sustainable Energy. Co-Published with Macmillan Publishers Ltd, UK, pp. 12–25. http://doi.org/10.1142/9789814317665_0002.

Skinner, M.P., et al., 1995. Serial magnetic resonance imaging of experimental atherosclerosis detects lesion fine structure, progression and complications in vivo. Nat. Med. 1 (1), 69–73. Jan. http://doi.org/10.1038/nm0195-69.

Skrabalak, S.E., et al., 2009. ChemInform abstract: gold nanocages: synthesis, properties, and applications. ChemInform 40 (14). Apr. http://doi.org/10.1002/chin.200914224.

Sottoriva, A., et al., 2013. Intratumor heterogeneity in human glioblastoma reflects cancer evolutionary dynamics. Proc. Natl. Acad. Sci. 110 (10), 4009–4014. Mar. http://doi.org/10.1073/pnas.1219747110.

Stolik, S., Delgado, J.A., Pérez, A., Anasagasti, L., 2000. Measurement of the penetration depths of red and near infrared light in human 'ex vivo' tissues. J. Photochem. Photobiol. B 57 (2–3), 90–93. Sep. http://doi.org/10.1016/S1011-1344(00)00082-8.

Tang, C.W., VanSlyke, S.A., Chen, C.H., 1989. Electroluminescence of doped organic thin films. J. Appl. Phys. 65 (9), 3610–3616. May. http://doi.org/10.1063/1.343409.

Tian, G., et al., 2013. Red-emitting upconverting nanoparticles for photodynamic therapy in cancer cells under near-infrared excitation. Small 9 (11), 1929–1938. Jun. http://doi.org/10.1002/smll.201201437.

Ukonaho, T., et al., 2007. Comparison of infrared-excited up-converting phosphors and europium nanoparticles as labels in a two-site immunoassay. Anal. Chim. Acta 596 (1), 106–115. Jul. http://doi.org/10.1016/J.ACA.2007.05.060.

Umeda, Y., Kojima, C., Harada, A., Horinaka, H., Kono, K., 2010. PEG-attached PAMAM dendrimers encapsulating gold nanoparticles: growing gold nanoparticles in the dendrimers for improvement of their photothermal properties. Bioconjugate Chem. 21 (8), 1559–1564. Aug. http://doi.org/10.1021/bc1001399.

Wang, F., et al., 2011. Synthesis of magnetic, fluorescent and mesoporous core-shell-structured nanoparticles for imaging, targeting and photodynamic therapy. J. Mater. Chem. 21 (30), 11244. http://doi.org/10.1039/c1jm10329f.

Wang, F., Han, M.-Y., Mya, K.Y., Wang, Y., Lai, Y.-H., 2005. Aggregation-driven growth of size-tunable organic nanoparticles using electronically altered conjugated polymers. J. Am. Chem. Soc. 127 (29), 10350–10355. Jul. http://doi.org/10.1021/ja0521730.

Wu, W., Shen, J., Banerjee, P., Zhou, S., 2010. Core–shell hybrid nanogels for integration of optical temperature-sensing, targeted tumor cell imaging, and combined chemo-photothermal treatment. Biomaterials 31 (29), 7555–7566. Oct. http://doi.org/10.1016/j.biomaterials.2010.06.030.

Yang, W., et al., 2008. Semiconductor nanoparticles as energy mediators for photosensitizer-enhanced radiotherapy. Int. J. Radiat. Oncol. Biol. Phys. 72 (3), 633–635. Nov. http://doi.org/10.1016/J.IJROBP.2008.06.1916.

Yavuz, M.S., et al., 2009. Gold nanocages covered by smart polymers for controlled release with near-infrared light. Nat. Mater. 8 (12), 935–939. Dec. http://doi.org/10.1038/nmat2564.

Zharov, V.P., Mercer, K.E., Galitovskaya, E.N., Smeltzer, M.S., 2006. Photothermal nanotherapeutics and nanodiagnostics for selective killing of bacteria targeted with gold nanoparticles. Biophys. J. 90 (2), 619–627. Jan. http://doi.org/10.1529/BIOPHYSJ.105.061895.

Chapter 22

Nanocrystals in the drug delivery system

Raju Ramesh Thenge[a], Amar Patel[b], Gautam Mehetre[a]

[a]Dr. Rajendra Gode College of Pharmacy, Malkapur, Buldana, MS, India
[b]Bristol Myers Squibb, New Jersey, USA

22.1 Introduction to nanocrystals and nanosuspension

Nanotechnology is the emerging field since last two decades. The Nanomedicine was derived when nanotechnology was applied to the discipline of pharmaceutical science (Lu,2016; Zhou et al., 2014, 2016). The ability of nanotechnology in improving the solubility of insoluble drug substance has become increasingly apparent (Gauniya et al., 2016). Nanotechnology used in pharmaceuticals includes nanocarrier drugs (e.g., liposomes, polymer micelles, nanoparticles, etc.) and nanocrystal drugs (e.g., nanosuspensions and nanosolid preparations, etc.) (Hollis et al., 2013; Chen et al., 2015; Yang et al., 2013; Yu et al., 2016). The former (nanocarrier drugs) belongs to the matrix skeleton type or vesicular type nanodrug product; the preparation process is complex and the loading of drug substances is relatively low (Mishra B and Kale 2016).

Drug nanocrystals are a versatile option for drug delivery purposes. They have mostly been utilized for improving poor solubility properties of drugs, but there are also controlled release applications, for example, based on nanocrystal technologies. In order to reach the desired formulation properties with drug nanocrystals, thorough physicochemical analyses both in vitro and in vivo are needed. In this chapter, drug nanocrystals and their benefits for drug delivery purposes are described. Then the techniques to produce drug nanocrystals are reviewed and the most important applications for drug delivery presented (Peltonen and Hirvonen, 2018).

Drug Nanocrystals are essentially nanoscopic crystals of the parent compound. By definition, the dimensions of nanocrystals are <1 μm but for practical purposes they are often ~500 nm in dimensions.

Nanosuspensions are colloidal dispersions of nanosized drug particles stabilized by surfactants. They can also be defined as a biphasic system consisting of pure drug particles dispersed in an aqueous vehicle in which the diameter of the suspended particle is less than 1μm in size. The Nanosuspensions can also be lyophilized or spray dried and the nanoparticles of nanosuspension can also be incorporated in a solid matrix (Martena et al. 2012).

22.1.1 Properties of nanocrystals (Colombo, 2017; Mitri et al., 2011)

1. The size reduction leads to an increased surface area and thus an increased dissolution velocity. Therefore, micronization is a suitable way to successfully enhance the bioavailability of drugs where the dissolution velocity is the rate limiting step. By moving from micronization further down to nanonization, the particle surface is further increased and thus the dissolution velocity increases too.
2. It increases with decreasing particle size below 1000 nm. Therefore, drug nanocrystals possess increased saturation solubility.
3. The nanocrystals of drug having size below 1 μm
4. The nanocrystals consist of 100% drug, no carrier
5. The nanocrystals generally needed to be stabilized using stabilizer
6. Nanocrystals may be in the form of crystalline or amorphous structure
7. Nanocrystals should generally increase of dissolution velocity
8. Nanocrystals should increase in saturation solubility
9. Nanocrystals in the amorphous particle state offers advantages

22.1.2 Nanocrystals and bioavailability

The bioavailability of a drug depends on its ability to dissolve in biological fluids, to cross membranes, and to efficiently reach its pharmacological target. In the biopharmaceutical classification of drugs (Dressman et al., 2001), drugs of Class II group are characterized by poor solubility, but have a good ability to cross membranes. Thus, to improve the bioavailability of a Class II drug, it is necessary to increase drug solubility and/or drug dissolution rate.

In particular, for nanocrystals, it is possible to consider the following scenarios:

1. A decrease in particle size leads to an increase in surface area available for the interaction with the dissolution media and thus an increase in particle dissolution rate, in accordance with the Noyes-Whitney law (Mauludin et al., 2009; Noyes and Whitney, 1897).
2. An increase in the particle curvature (particularly pronounced for colloidal particles) leads to an increase in dissolution pressure, according to the Kelvin's equation (Muller, 2004).
3. When kinetic saturation solubility is greater than thermodynamic equilibration solubility, this leads to an increased concentration gradient at membranes and thus subsequently to higher penetration or permeation through membranes (Zhai et al., 2014).
4. High penetration through membranes is also favored by high adhesion to biological
    1. Membranes of nanocrystals (Gao et al., 2008; Junyaprasert and Morakul, 2015; Moschwitzer and Muller, 2007; Ponchel et al., 1997).

22.1.3 Various methods of characterization of nanocrystals formulations (doi:10.3390/molecules201219851)

Sr. No.	Category	Characterization method	Principle	Information	References
1.	Size and morphology	Dynamic light scattering (photon correlation spectroscopy)	Fluctuation of Rayleigh scattering of light associated with Brownian motion of nanoparticles	Particle size, particle size distribution	(Lu et al., 2014)
		Scanning electron microscopy (SEM)	Backscattering of electrons	Topographical information about particles	(Sarnes et al., 2014)
		Transmission Electron microscopy	Transmission of electrons	Density information	(Laaksonen et al., 2011)
2.	Solid state forms	X-ray powder diffraction (XRPD)	Diffraction of x-rays from lattice planes	Polymorphic form (unique diffraction peaks), amorphous form (no peaks)	(Valo et al., 2010)
		Differential scanning calorimetry (DSC)	Change in heat flow due to sample changes during heat/cooling	Polymorphic form (melting temperature, Crystallization temperature) Amorphous form (glass transition temperature), crystallinity (enthalpy of fusion, enthalpy of crystallisation)	(Valo et al., 2011)
		Infrared (IR) spectroscopy (mid-IR spectroscopy)	Change in dipole moment during molecular vibrations	Polymorphic form (peak shifts and relative intensities), crystallinity (broadening of bands, peak shifts and relative intensities)	(Valo et al., 2013)
		Raman spectroscopy	Change in polarizability during molecular vibrations	Polymorphic form, crystallinity	(Heinz et al., 2007)
3.	Surface properties	Zeta-potential	Dynamic electrophoretic mobility under electric field	Surface charge (zeta potential)	(Van Eerdenbrugh et al., 2009)
		Surface plasmon resonance (SPR)	Changes in refractive index in the vicinity of a planar sensor surface	Surface adsorption	(Liu et al., 2015)

Sr. No.	Category	Characterization method	Principle	Information	References
4.	Drug delivery	Dissolution testing	Dissolved drug analysed over time, usually using UV spectroscopy or HPLC	Dissolution profile	(Liu et al., 2013)
		Fluorescence microscopy	Fluorescence by endogenous or added fluorophores	Localization of nanocrystals in relation to cells and tissues	(Valo et al., 2010)
		Nonlinear Raman microscopy	Change in polarizability during molecular vibrations.	Label free localisation of particles	(Darville et al., 2014)

22.2 Production methods and technology of nanocrystals

Methods for the production of nanosuspensions can be categorized as top-down and bottom-up methods, depending on the starting material. In top-down methods, such as wet milling, high-pressure homogenization, and microfluidization, the starting material comprises larger solid particles than the resulting NPs and mechanical processes are the fundamental mechanism causing particle size reduction. In bottom-up methods, particles are formed from the molecular level. Such methods are further sub-divided into solvent evaporation (e.g., spray drying, electrospraying, cryogenic solvent evaporation, etc.) and antisolvent methods (e.g., liquid antisolvent, supercritical antisolvent, etc.)

22.2.1 Top down technology

"Top-down" technology employs various types of grinding and homogenisation techniques for dispersion methods. "Top-down" technology is more common than technology called "Bottom-up." This is a mechanism that splits large crystal particles into tiny pieces, in other words. Both approaches are used together in "top-down and bottom-up" research. Homogenization or framing can apply top-down technology.

"Top down" technology can be applied by either homogenization or milling.

22.2.1.1 Homogenization

High-pressure homogenization is a technology that has been applied for many years in various areas for the production of emulsions and suspensions. A distinct advantage of this technology is its ease for scaling up, even to very large volumes. High-pressure homogenization is currently used in the food industry, e.g., homogenization of milk. In the pharmaceutical industry parenteral emulsions are produced by this technology. Commercial products such as Intralipid and Lipofundin possess a mean droplet diameter in the range of 200 to 400 nm (photon correlation spectroscopy data) (Muller and Heinemann, 1989). In the mid-1990s of the last century drug nanosuspensions produced with high-pressure homogenization were developed (Muller and Peters, 1998; Muller et al., 2001; Jacobs and Muller et al., 2002; Jacobs et al., 2000; Peter and Muller, 1996). Typical pressures for the production of drug nanosuspensions are 1000 to 1500 bar (corresponding to 100–150 Mpa, 14504–21756 psi); the number of required homogenization cycles varies from 10 to 20 depending on the properties of the drug. Most of the homogenizers used are based on the piston-gap principle; an alternative is the jet-stream technology (Fig. 22.1).

The Microfluidizer (Microfluidics TM Inc., USA) is based on the jet-stream principle. Two streams of liquid collide, diminution of droplets or crystals is achieved mainly by particle collision, but occurrence of cavitation is also considered. The Micro fluidizer has also been described for the production of drug nanosuspensions; however, according to the patent 10 to 50 cycle passes were required (Dearn, 2000). Such a high cycle number is not convenient for the production scale. The microfluidizer can be used for the production of drug nanosuspensions in the case of soft drugs. In the case of harder drugs, a larger fraction of particles in the micrometer range remain, which do not exhibit the increase in saturation solubility because of their too large size.

In the piston-gap homogenizer the liquid is forced through a tiny homogenization gap, typically in the size range of 5 to 20 mm (depending on the pressure applied and the viscosity of the dispersion medium). Using a Micron Lab 40 the suspension is supplied from a metal cylinder by a piston, the cylinder diameter is approximately 3 cm. The suspension is moved by the piston having an applied pressure between 100 and 1500 bar. In principle the pistongap homogenizer corresponds to a tube system in which the tube diameter narrows from 3 to 5-20 mm. According to the Bernoulli equation, the streaming velocity and dynamic pressure increase extremely, the static pressure in the gap falls below the vapor pressure of water at room temperature. A liquid boils when its vapor pressure is equal to the static pressure, which means water starts boiling

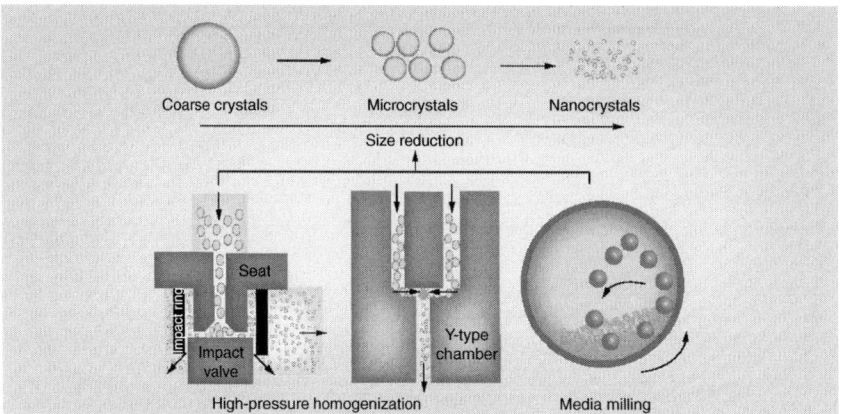

FIG. 22.1 High pressure homogenization (https://doi.org/10.2217/nnm.15.73).

in the gap at room temperature leading to the formation of gas bubbles. The formation of gas bubbles leads to pressure waves disrupting oil droplets or disintegrating crystals. When leaving the homogenization gap, the static pressure increases to normal air pressure, which means the water does not boil anymore and the gas bubbles collapse. Collapsing of the gas bubbles (implosion) leads again to shock waves contributing to diminution. There are different definitions of cavitation in the literature, describing cavitation either as the formation of gas bubbles in high streaming liquids or as the formation and subsequent implosion of these gas bubbles.

22.2.2 Bottom up technology

The theory for the process is that the active material is dissolved in an organic solvent that is then added in a nonsolvent (miscible with an organic solvent). The nanocrystals then precipitate in the presence of stabilizers. This is quick and has a low cost as a primary benefit of the precipitate technique. In this process, scale-up is also simple. We must note that several parameters should be tested to obtain homogenous nanocrystals, such as stirring rate, temperature, solvent/nonsolvent rate, medicine concentration, viscosity, a form of solvent and stabilizer.

The crystalline state of the particles obtained by the precipitation process can be controlled. Depending on the employed method, amorphous drug nanoparticles can also be generated (Auweter et al., 2000). Beta-carotene is dissolved in a water-miscible organic solvent together with digestible oil. This solution is admixed to an aqueous solution of a protective colloid (gelatin) causing a precipitation of amorphous nanoparticulate beta-carotene. After an annealing step and spray drying, a stable amorphous product can be obtained. This NanoMorph technology, invented by Auweter et al., is used by the company Soliqs. Another approach to preserve the size of the precipitated nanocrystals is the use of polymeric growth inhibitors, which are preferably soluble in the aqueous phase. The increased viscosity of the aqueous phase can reduce particle growing. The resulting suspension is subsequently spray-dried to obtain a dry powder with a relatively high drug loading (Mueller and Rasenack, 2003). Using this technique, a tremendous increase in dissolution rate (from 4% to 93% within 20 minutes) was shown for a poor water-soluble drug ECU-01 (Rasenack et al., 2003). Although the feasibility of preparing drug nanocrystals by precipitation has been shown by many groups, no commercial drug product using this technology has entered the market. To use the prementioned methods, it is required that the drug is soluble in at least one water-miscible solvent. This is often not the case for NCEs. Many drugs are simultaneously poorly soluble in aqueous and nonaqueous media.

Even if there is a suitable solvent available, it is difficult to remove this solvent completely. Solvent residues can be potential risk factors for drug alteration and toxic side effects. In addition, in cases of amorphous drug nanoparticles, it is seen as very critical to preserve the amorphous character throughout the shelf life of a product. Recrystallization would impair the oral bioavailability. This effect is less critical in food products because of less strict regulatory requirements that allow more tolerance.

22.2.3 Top down and bottom up technology

In "top down and bottom up" technology, both methods are used together. NanoEdge is a product obtained by such a combination technology. As can be inferred, precipitation is followed by high pressure homogenization in this technology (Keck and Müller, 2006) (Fig. 22.2).

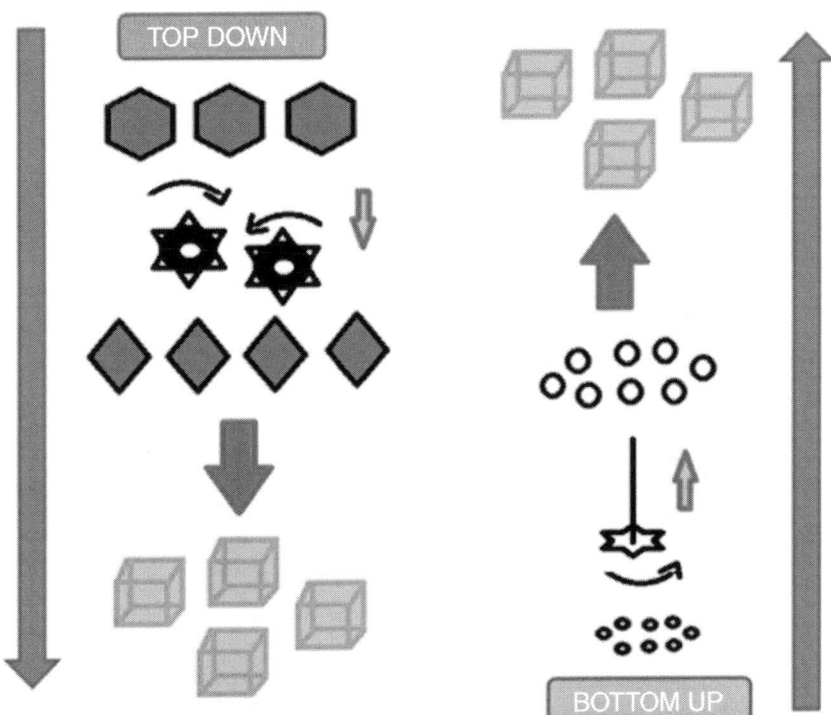

FIG. 22.2 Top down method (http://dx.doi.org/10.1016/j.jsps.2014.04.007) (Yi, 2015).

Nanoedge technology

Baxter's Nanoedge process relies on the combination of a micro precipitation technique with a subsequent annealing step by applying high shear and/or thermal energy. A fine suspension is formed by adding an organic solution of the water-insoluble drug to an antisolvent, for example, aqueous surfactant solution. Depending on the precipitation conditions, either small amorphous or crystalline drug particles in the nanometer range or friable needle-like crystals in the micrometer range are formed. Consequently, the following high-energy input can have two effects on the preformed particles. Small amorphous or crystalline drug particles will be preserved in size by an annealing step without changing the mean diameter. It could be shown that the tendency to crystal growth can be reduced by energy input after the precipitation step. In case long friable needle-like crystals are obtained, they will be reduced in size by the high-energy input using high-pressure homogenizers. According to the patent (Kipp et al., 2001), particle sizes in the range of 400 to 2000 nm can be obtained. The organic solvent utilized has to be carefully removed from the final nanosuspension without changing the particle size of the drug nanocrystals. Otherwise, crystal growth can be promoted by an increased solubility of the drug. Any content of the organic solvent dissolved in the aqueous phase can act as a "cosolvent," leading to an increased tendency to Ostwald ripening. Also, toxic effects can be caused by potential solvent residues, especially if the nanosuspension is the final product. For these reasons, the NANOEDGE process is particularly suitable for drugs that are soluble in nonaqueous media possessing low toxicity, such as N-methyl-2-pyrrolidinone.

22.2.4 Spray drying

For the production of tablets, an aqueous nanosuspension can be used as granulation fluid or a dry form of the nanosuspension, powder, or granulate can be employed. Starting from an aqueous macrosuspension containing the original coarse drug powder, surfactant, and water-soluble excipient, the homogenization process can be performed in an easy one step yielding a fine aqueous nanosuspension. In a subsequent step the water has to be removed from the suspension to obtain a dry powder. One method of removing the water from the formulation is freeze drying, but it is complex and cost-intensive leading to a highly sensitive product (Peters, 1999; Freitas and Muller, 1998). Another simple and most suitable method for the industrial production is spray drying. The drug nanosuspension can directly be produced by high-pressure homogenization in aqueous solutions of water-soluble matrix materials, for example, polymers (PVP, poly-vinylalcohol or long chained PEG, sugars [saccharose, lactose], or sugar alcohols [mannitol, sorbitol]). Afterward the aqueous drug nanosuspension can be spray dried under adequate conditions; the resulting dry powder is composed of drug nanocrystals embedded in a water-soluble matrix (Bushrab and Muller, 2004).

22.3 Advantages and Disadvantages of nanocrystals

22.3.1 Potential advantages and disadvantages of nanocrystals

1. Nanocrystals increased rate of absorption and thus increased oral bioavailability,
2. Improved dose proportionality and reduction in required dose.
3. Reduction in fed/fasted variability.
4. Rapid, simple and cheap formulation development
5. Possibility of high amounts (30–40%) of drug loading.
6. Increased reliability. Usually side effects are proportional to drug concentration, so decreasing the concentration of active drug substances leads to an increased reliability for patients.
7. Applicability to all poorly soluble drugs because all these drugs could be directly disintegrated into nanometer-sized particles.
8. Sustained crystal structure. Nanocrystal technology leads to an increase in dissolution rate depending on the increase in surface area obtained by reduction of the particle size of the active drug substance down to the nano size range preserving the crystal morphology of the drug.
9. Improved stability. They are stable systems because of the use of a stabilizer that prevents reaggregation of active drug substances during preparation. Suspension of drug nanocrystals in liquid can be stabilized by adding surface active substances or polymers.
10. Applicability to all routes of administration in any dosage form. Contrary to micronized drugs, nanocrystals can be administered via several routes. Oral administration is possible in the form of tablets, capsules, sachets or powder; preferably in the form of a tablet.
11. Nanosuspensions can also be administered via the intravenous route due to very small particle size, and in this way, bioavailability can reach 100%.

22.3.2 Disadvantages of nanocrystals

1. Physicochemical related stability problems
2. Bulking sufficient care must be taken during handling and transport
3. Uniform and accurate dose cannot be achieved

22.4 Pharmaceutical Nanocrystals of API

22.4.1 Case studies of drug loaded in the nanocrystals

Girdhar et al. synthesized two nanocrystal-based formulations, that is, nanosuspension and NanoCrySP-generated nanocrystalline solid dispersion, were compared for their permeability behavior, cellular internalization, and mucoadhesion. The interplay of concentration-gradient created by the solubility values and mucoadhesion imparted their permeability behavior. Everted gut sac experiments enabled an understanding of the role of mucoadhesion in the overall permeability behavior. Though the particle size obtained in the NanoCrySP-based formulation was higher than the nanosuspension, the permeability of the former was comparable to the latter (Girdhar et al., 2018).

Ige et al. were prepared a nanosuspension of Fenofibrate, a lipophilic drug used in hypercholesterolemia and hypertriglyceridemia, and which is practically insoluble in water through processing in a probe sonicator and subsequent freeze-drying to transform it into a dry powder. A decrease in particle size significantly increased its saturation solubility. Pharmacokinetic studies conducted in white rabbits confirmed a 4.73-fold increase in relative bioavailability compared to the pure drug form (Ige et al., 2013).

Chen et al. prepared nanocrystals of bexarotene, a potent antitumor drug of poor solubility and bioavailability, were obtained under a method combining precipitation and microfluidization. The decreased particle size thus achieved afforded a significant increase in the dissolution rate, with improved in vivo results in rats. The higher AUC and lower Cmax indicated that oral bexarotene nanocrystals significantly increased the bioavailability of this important drug and decreased its side effects. Nanocrystals administered through intravenous injection showed higher bioavailability because of the absence of first-pass effect and enterohepatic circulation (Chen et al., 2014).

Liu et al. formulated paclitaxel nanocrystals in the presence of D-R-tocopheryl polyethylene glycol 1000 succinate as a surfactant to stabilize the nanocrystals. Those nanocrystals exhibited a variety of benefits, including a high drug-loading capacity, high stability, and sustained release. Most important, the surfactant was responsible for successfully reversing the multidrug resistance generally observed in the presence of paclitaxel formulations, in both in vitro and in vivo experiments. This effect can probably be explained by the surfactant over a certain percentage coating the paclitaxel, minimizing the interaction with those biological substrates that are responsible for multidrug resistance. Nonetheless, the paclitaxel nanocrystals exhibited improved antitumor effect compared to Taxol. (Liu et al., 2010)

Fu et al. were prepared nimodipine nanocrystals of different sizes (159.0 nm, 503.0 nm, and 833.3 nm) by a combination of microprecipitation and high-pressure homogenization. The in vitro and in vivo behaviors were compared to Nimotop, which is a commercially available formulation of nimodipine. Even if Nimotop exhibited a higher dissolution rate than the three different nanocrystal batches, the bioavailability measured by the plasma concentration-time curves determined in beagle dogs was significantly higher for optimized nanocrystals (159.0 nm and 833.3 nm) than for Nimotop (Fu et al., 2013).

Lai et al. formulated piroxicam nanocrystals with poloxamer 188 as a stabilizer by high pressure homogenization. While the raw material was form I, the resulting nanocrystals were a mixture of monohydrate and form III. The solubility of form I is 14.3 mg/L, while that of form III is 17.0 mg/L. In this case the solubility was increased not only due to the smaller particle size, but also due to the formation of the higher energy solid-state forms (Lai et al., 2014).

Ali et al. prepared hydrocortisone nanosuspensions by both wet-milling and microfluidic nanoprecipitation. With both methods, the particle sizes were approximately 300 nm, yet with milling the product was crystalline, while precipitation resulted in a predominantly amorphous product. In in vivo tests with rabbits, the bioavailability during ocular delivery was comparable with both the formulations and when compared to drug solution almost doubled. Differences were clear in stability tests: the crystalline wet-milled nanosuspension was stable for two months (unaltered particle size), but the particle size of the amorphous precipitated nanosuspension had increased to 440 nm. (Ali et al., 2011)

Nanocrystals product in the clinical trial phase (Shegokar and Müller, 2010; Malamatari et al., 2016).

Trade name	Drug	Indication	Applied technology	Company	Status	Route
Paxceed	Paclitaxel	Anti-inflammatory	Unknown	Angiotech	Phase III	Intravenous
Theralux	Thymectacin	Anticancer	Nanocrystal	Celmed biosciences	Phase II	Intravenous
NPI 32101	Silver	Antibacterial	Magnetron sputtering	Nucryst Pharmaceuticals	Phase II	Topical
Semapimod	Guanylhydrazone	TNF-α inhibitor	Self-developed	Cytokine Pharmasciences	Phase II	Oral
	Insulin	Diabetes	Self-developed	BioSante pharmaceuticals	Phase II	Oral
Panzem NCD	2-methoxy estradiol	Ovarian cancer	Nanocrystal	EntreMed	Phase II	Oral
	Undisclosed	Anti-infective	Nanoedge	Undisclosed	Phase II	
Zolip	Fenofibrate/Simvastatin	Antihyperlipidemic	Nanocrystal	Abbott	Phase III	Oral
Budesonide	Budenoside	Antiasthmatic	Nanocrystal	Sheffield Pharmaceuticals	Phase III	Pulmonary

22.4.2 Application of nanocrystals-loaded carrier

22.4.2.1 Nanocrystals in oral delivery system

The oral route is the most important and preferred route of administration. The formulation of drug nanocrystals can impressively improve the bioavailability of perorally administered poorly soluble drugs. In 1995, Liversidge and Cundy reported an increase in bioavailability for the drug Danazol from 5.1 ± 1.9% for the conventional suspension to 82.3 ± 10.1% for the nanosuspension (Liversidge et al., 1995). The increased dissolution velocity and saturation solubility lead to fast and complete drug dissolution, an important prerequisite for drug absorption. Whenever a rapid onset of a poorly soluble drug is desired, the formulation of drug nanocrystals can be beneficial, for example, in case of analgesics. The analgesic naproxen, formulated as a nanosuspension, has shown a reduced tmax but simultaneously approximately threefold increased AUC in comparison to a normal suspension (Naprosyn) (Merisko-Liversidge et al., 2003). Besides the faster onset of action, the naproxen nanosuspension has also shown a reduced gastric irritancy (Liversidge and Conzentino, 1995; Eickhoff, 1996). If absorption windows limit the drug absorption or by food effects, drug nanocrystals have advantages in comparison to conventional suspensions. Wu et al. have reported reduced fed-fasted ratio and an improved bioavailability for nanocrystalline aprepitant (MK-0869), the active ingredient in Emend, in beagle dogs. Another important advantage of drug nanocrystals is their adhesiveness and the increased residence time, which can positively influence the bioavailability. The mucoadhesiveness can be raised by the use of mucoadhesive polymers in the dispersion medium (Jacobs et al., 2001; Müller and Jacobs, 2002). Additionally the utilized mucoadhesive polymers can prevent the drug from degradation. The reduced particle size can be also exploited for improved drug targeting, as reported for inflammatory tissues (Lamprecht et al. 2001) or the lymphatic drug uptake (Hussain et al., 2001). Muller et al. the use of Mucoadhesive nanosuspensions as layering dispersions for preparation of multiparticulate drug delivery systems was investigated. Nanosuspensions on the other hand, enable incorporation of all hydrophobic drugs in well-established sustained release technologies. However, whole doing so, the effect and interaction of dosage form excipients with the nanocrystalline drug must be critically investigated. Drug nanosuspensions can be incorporated into dosage forms, such as tablets, capsules, and fast melts by means of standard manufacturing technologies. A ketoprofen nanosuspension has been successfully incorporated into pellets to release drug over a period of 24 hours (Vergote, 2001). O. Kayser prepared bupravaquone mucoadhesive nanosuspensions; a potential drug delivery system for poorly soluble drugs has been investigated to overcome bioavailability problems caused by the pathophysiological diarrheic situation in patients suffering from cryptosporidiosis. Cryptosporidium parvum identified as the main pathogen causing, severe diarrhea in immune suppressant HIV patients has attracted much interest. Adapting drug delivery systems to the situation of Cryptosporidium parvum infections in man allows increased retention times with a prolonged action at reduced elimination in the gastrointestinal tract. In this communication, in vivo data are presented to document the efficiency of bupravaquone formulated as mucoadhesive polymers to improve its activity against C. Parvum (Jacobs et al., 2001).

22.4.2.2 Parenteral administration of drug nanocrystals

The parenteral application of poorly soluble drugs, particularly intravenous (IV) administration of practically insoluble compounds, using cosolvents, surfactants, liposomes, or cylcodextrines, is often associated with large injection volumes or toxic side effects. Carrier-free nanosuspensions enable potential higher loading capacity compared to other parenteral application systems. Using nanosuspensions, the application volume can be distinctly reduced compared to solutions (Moschwitzer et al., 2004). To fulfill the distinctly higher regulatory hurdles, the drug nanocrystals need to be produced in an aseptic process. Alternatively, nanosuspensions can be sterilized by autoclaving or alternatively by gamma irradiation as well as sterile filtration (Muller et al., 2000). When a drug is administered as a nanosuspension, the rapid dissolution of the nanocrystals will mimic the plasma concentration profile of a solution. Drug nanosuspensions can be formulated with accepted surfactants and polymeric stabilizers for IV injection. In contrast, solutions of poorly soluble drugs require the use of cosolvents and/or high surfactant contents (e.g., Chremophor EL in Taxol), which can cause undesired side effects (Singla, 2002).

Comparing a clofazimine nanosuspension with a liposomal formulation, both are similarly effective in the treatment of artificially induced Mycobacterium avium infections. The targeting to the reticuloendothelial system, the lung, liver, and spleen was comparable to the liposomal formulation (Peters K et al.). Furthermore, a special targeting can be achieved by a surface modification using the concept of .differential protein adsorption. A surface modification of drug nanocrystals with the surfactant Tween 80 leads to a preferential adsorption of apolipoprotein E. This protein adsorption enables a targeted delivery of drug nanocrystals to efficacy in the treatment of Toxoplasmosis (Scholer et al., 2001). Administration of nanosuspsensions into body cavities is also of great interest, for example, to increase the tolerability of the drug, to achieve a local treatment or to have a depot with slow release (e.g., into the blood). It could be shown that intraperitonal administration of a nanosuspension was well tolerated, whereas administration of a macrosuspension leads to irritancy Intraperitonal administration can be used for local treatment or to obtain a depot with prolonged release into the blood. Interesting therapeutic targets include local inflammations, e.g. in joints. For instance, arthritic joint inflammations are caused by secretion products of activated macrophages. An interesting approach is therefore the administration of a corticoid nanosuspension directly into the joint capsule. The drug particles will be phagocytosed, the drug dissolves and reduces the hyperactivity of the macrophages. This concept is not new, being adopted by the company Boots in the 80s in an attempt to incorporate the corticoid prednisolone into polymeric Nanoparticles made from PLA-GA-copolymer (Smith and Hunneyball, 1986). Moschwitzer et al. developed intravenously injectable and chemically stable aqueous omeprazole formulations using nanosuspension technology. The researchers stated that even after 1 month of production, no discoloration or recognizable drug loss was observed when nanosuspensions were formulated at 0°c. As a result, it can be proven that the production of nanosuspension by high pressure homogenization is suitable for preventing degradation of labile drugs (Jan, 2004).

22.4.2.3 Drug nanocrystals for pulmonary drug delivery

Delivery of water-insoluble drugs to the respiratory tract is very important for the local or systemic treatment of diseases. Many important drugs for pulmonary delivery show poor solubility simultaneously in water and nonaqueous media, for example, important corticosteroids such as budesonide or beclomethasone dipropionate. Poorly soluble drugs could be inhaled as drug nanosuspension. The drug nanosuspension can be nebulized using commercially available nebulizers. Disposition in the lungs can be controlled via the size distribution of the generated aerosol droplets. Compared with microcrystals, the drug is more evenly distributed in the droplets when using a nanosuspension. The number of crystals are higher, consequently, the possibility that one or more drug crystals are present in each droplet is higher. Besides this, drug nanocrystals show an increased mucoadhesiveness, leading to a prolonged residence time at the mucosal surface of the lung (Hernandez-Trejo et al., 2004). Jacobs et al. prepared budenoside nanosuspension by high-pressure homogenization. It was possible to obtain a long-term stable budesonide nanosuspension. Mean particle size of this nanosuspension was about 500 to 600 nm, analyzed by photon correlation spectroscopy. Analysis by laser diffraction showed that the diameters 95% and 99% were below 3 m. Budesonide nanosuspension showed a long-term stability; no aggregates and particle growth occurred over the examined period of 1 year (Claudia, 2002). Hernandez-Trejo et al. stated that the poorly soluble drug bupravaquone is proposed for an alternative treatment of lung infection(pneumonia), which is caused by Pnemocystis carinii. Physically stable nanosuspensions were formulated to deliver bupravaquone at the site of lung infection using nebulization (Hernandzo).

22.4.2.4 Drug nanocrystals for ophthalmic drug delivery

It could be shown that nanoparticles possess a prolonged retention time in the eye, most likely due to their adhesive properties. From this, poorly soluble drugs could be administered as a nanosuspension. The development of such colloidal delivery systems for ophthalmic use aims at dropable dosage forms with a high drug loading and a long-lasting drug action. The nanosuspensions were prepared by a modification of the quasi-emulsion solvent diffusion technique using variable formulation parameters (drug-to-polymer ratio, total drug and polymer amount, stirring speed). Nanosuspensions had mean sizes around 100 nm and a positive charge (zeta-potential of +40/+60 mV), this makes them suitable for ophthalmic applications. Stability tests (up to 24 months storage at 4°C or at room temperature) or freeze-drying were carried out to optimize a suitable pharmaceutical preparation. In vitro dissolution tests indicated a controlled release profile of IBU from nanoparticles. In vivo efficacy was assessed on the rabbit eye after induction of an ocular trauma (paracentesis). An inhibition of the miotic response to the surgical trauma was achieved, comparable to a control aqueous eye-drop formulation, even though a lower concentration of free drug in the conjunctival sac was reached from the nanoparticle system. Drug levels in the aqueous humor were also higher after application of the nanosuspensions; moreover, IBU-loaded nanosuspensions did not show toxicity on ocular tissues (Pignatello, 2002).

22.4.2.4a Drug nanocrystals for dermal drug delivery

Dermal nanosuspensions are mainly of interest if conventional formulation approaches fail. The use of drug nanocrystals leads to an increased concentration gradient between the formulation and the skin. The increased saturation solubility leads to supersaturated formulations, enhancing the drug absorption through the skin. This effect can further be enhanced by the use of positively charged polymers as stabilizers for the drug nanocrystals. The opposite charge leads to an increased affinity of the drug nanocrystals to the negatively charged stratum corneum (Kayser, 2000).

22.4.2.5 Drug nanocrystals for targeted drug delivery

Nanosuspension can be used for targeted deliver as their surface properties and changing of the stabilizer can easily alter in vivo behavior. Their versatility and ease of scale up and commercial production enables the development of commercially viable nanosuspensions for targeted drug delivery. The natural targeting process could pose obstacles when macrophages are not the desired targets. Hence, in order to bypass the phagocytic uptake of drugs, its surface potential needs to be altered. Kayser developed the formulation of aphidicolin as a nanosuspension to improve the drug targeting effect against Leishmania-infected macrophages. He stated that aphidicolin was highly active at a concentration in the microgram range. Nanosuspensions afford a means of administrating poorly soluble drugs to brain with decreased side effects. Significant efficiency has been associated with microparticulate busulfan in mice administered intrathecally. Another example is successful targeting of the peptide Dalargin to the brain by employing surface modified polyisobutyl cyanoacrylate nanoparticles.

22.5 Conclusion

Poor aqueous solubility is rapidly becoming the leading hurdle for formulation scientists working on oral delivery of drug compounds and leads to employment of novel formulation technologies. The use of drug nanocrystals is a universal formulation approach to increase the therapeutic performance of these drugs in any route of administration. Almost any drug can be reduced in size to the nanometer range. Owing to their great formulation versatility drug nanocrystals are no longer only the last chance rescue for a few drugs. Many insoluble drug candidates are in clinical trials formulated as drug nanocrystals. Currently, attention is turned to improving the diminution performance to produce drug nanocrystals well below 100 nm, also in cases of very hard drugs. First approaches were already successful. New technologies are in development to produce final dosage forms with higher drug loadings, better redispersability at their site of action, and an improved drug targeting.

References

Ali, H.S.M., York, P., Ali, A.M.A., Blagden, N., 2011. Hydrocortisone nanosuspensions for ophthalmic delivery: a comparative study between microfluidic nanoprecipitation and wet milling. J. Control. Release 149, 175–181.

Auweter, H., Bohn, H., Luddecke, E. Stable, aqueous dispersions and stable, waterdispersible dry powders of xanthophylls, their production and their use, Application. WO 2000-EP3467, 2000066665, 2000.

Bushrab, F.N., Muller, RH., 2004. Drug Nanocrystals for Oral Delivery—Compounds by Spray Drying. AAPS, Philadelphia.

Chen, L., Wang, Y., Zhang, J., Hao, L., Guo, H., Lou, H., Zhang, D., 2014. Bexarotene nanocrystals Oral and parenteral formulation development, characterization and pharmacokinetic evaluation. Eur. J. Pharm. Biopharm. 87, 160–169.

Colombo, M., Staufenbiel, S., Rühl, E., Bodmeier, R., 2017. In situ determination of the saturation solubility of nanocrystals of poorly soluble drugs for dermal application. Int. J. Pharm. 521, 156–166.

Chen, A., Shi, Y., Yan, Z., et al., 2015. Dosage form developments of nanosuspension drug delivery system for oral administration route. Curr. Pharm. Des. 21 (29), 4355–4365.

Claudia, J., 2002. Production and characterization of a budesonide nanosuspension for pulmonary administration. Pharma Res 19, 189–194.

Darville, N., van Heerden, M., Vynckier, A., de Meulder, M., Sterkens, P., Annaert, P., van den Mooter, G., 2014. Intramuscular administration of paliperidone palmitate extended-release injectable microsuspension induces a subclinical inflammatory reaction modulating the pharmacokinetics in rats. J. Pharm. Sci. 103, 2072–2087.

Dearn, A.R., inventor Glaxo Wellcome Inc., assignee. Atovaquone pharmaceutical compositions. U.S. Patent 6,018,080, 2000.

Dressman, J., et al., 2001. The BCS: Where do we go from here? Pharm. Technol. 25 (7), 68–77.

Eickhoff, W.M., Engers, D.A., Mueller, KR. Nanoparticulate NSAID compositions, application. US 95-385614, 5518738, 1996.

Fu, Q., Sun, J., Zhang, D., Li, M., Wang, Y., Ling, G., Liu, X., Sun, Y., Sui, X., Luo, C., 2013. Nimodipine nanocrystals for oral bioavailability improvement: preparation, characterization and pharmacokinetic studies. Colloids Surf. B Biointerfaces 109, 161–166.

Freitas, C., Muller, RH., 1998. Spray-drying of solid lipid nanoparticles (SLNTM). Eur. J. Pharm. Biopharm. 46 (2), 145 –15.

Gao, L., Zhang, D., Chen, M., 2008. Drug nanocrystals for the formulation of poorly soluble drugs and its application as a potential drug delivery system. J. Nanopart. Res. 10 (5), 845–862.

Gauniya, A., Mazumder, R., Pathak, K., 2016. Enhancement of dissolution profile of poorly water soluble drugs by using techniques of nanocrystallization. Curr. Nanomed 6 (3), 240–250.

Girdhar, P.S.T, Sheokand, S., Bansal, A.K., 2018. Permeability behavior of nanocrystalline solid dispersion of dipyridamole generated using NanoCrySP technology. Pharmaceutics 10, 160.

Heinz, A., Strachan, C.J., Gordon, K.C., Rades, T., 2007. Analysis of solid-state transformations of pharmaceutical compounds using vibrational spectroscopy. J. Pharm. Pharmacol. 61, 971–988.

Hernandez-Trejo, N., Kayser, O., Miiller, R.H., Steckel, H., 2004. Physical stability of buparvaquone nanosuspensions following nebulization with jet and ultrasonic nebulizers, Proceedings of the International Meeting on Pharmaceutics, Biopharmaceutics and Pharmaceutical Technology. Nuremberg,Germany.

Hollis, C.P., Zhao, R., Li, T., 2013. 9 - Hybrid nanocrystal as a versatile platform for cancer theranostics. Biomaterials for Cancer Therapeutics. Woodhead Publishing, Cambridge UK, pp. 188–207e.

Hussain, N., Jaitley, V., Florence, AT., 2001. Recent advances in the understanding of uptake of microparticulates across the gastrointestinal lymphatics. Adv. Drug. Deliv. Rev. 50 (12), 107–142.

Ige, P.P., Baria, R.K., Gattani, S.G., 2013. Fabrication of fenofibrate nanocrystals by probe sonication method for enhancement of dissolution rate and oral bioavailability. Colloids Surf. B Biointerfaces 108, 366–373.

Jacobs, C., Muller, RH., 2002. Production and characterization of a budesonide nanosuspension for pulmonary administration. Pharm. Res. 19 (2), 189–194.

Jacobs, C., Kayser, O., Muller, RH., 2000. Nanosuspensions as a new approach for the formulation for the poorly soluble drug tarazepide. Int. J. Pharm. 196 (2), 161–164.

Jacobs, C., Kayser, O., Müller, RH., 2001. Production and characterisation of mucoadhesive nanosuspensions for the formulation of bupravaquone. Int. J. Pharm. 214 (1–2), 3–7.

Jan, M., 2004. Development of an intravenously injectable chemically stable aqueous omeprazole formulation using nanosuspension technology. Eur. J. Pharm. Sci. 58, 615–619.

Junyaprasert, V.B., Morakul, B., 2015. Nanocrystals for enhancement of oral bioavailability of poorly water-soluble drugs. Asian J. Pharm. Sci. 10 (1), 13–23.

Kayser, O., 2000. Nanosuspensions for the formulation of aphidicolin to improve drug targeting effects against Leishmania infected macrophages. Int. J. Pharm. 196 (2), 253–256.

Keck, C.M., Müller, RH., 2006. Drug nanocrystals of poorly soluble drugs produced by high pressure homogenization. Eur. J. Pharm. Biopharm. 62, 3–16.

Kipp, J.E.W., Joseph Chung, T., Doty, M.J., Rebbeck, CL. Microprecipitation method for preparing submicron suspensions. US 6,869, 617, 2001.

Laaksonen, T., Liu, P., Rahikkala, A., Peltonen, L., Kauppinen, E.I., Hirvonen, J., Järvinen, K., Raula, J., 2011. Intact nanoparticulate indomethacin in fast-dissolving carrier particles by combined wet milling and aerosol flow reactor methods. Pharm. Res. 28, 2403–2411.

Lai, F., Pini, E., Corrias, F., Perricci, J., Manconi, M., Fadda, A.M., Sinico, C., 2014. Formulation strategy and evaluation of nanocrystal piroxicam orally disintegrating tablets manufacturing by freeze-drying. Int. J. Pharm. 467, 27–33.

Lamprecht, A., Ubrich, N., Yamamoto, H., et al., 2001. Biodegradable nanoparticles for targeted drug delivery in treatment of inflammatory bowel disease. J. Pharmacol. Exp. Ther. 299 (2), 775–781.

Liu, P., Viitala, T., Kartal-Hodzig, A., Liang, H., Laaksonen, T., Hirvonen, J., Peltonen, L., 2015. Interaction studies between indomethacin nanocrystals and PEO/PPO copolymer stabilizers. Pharm. Res. 32, 628–639.

Liversidge, G.G., Cundy, K.C., 1995. Particle size reduction for improvement of oral bioavailability of hydrophobic drugs: I. Absolute oral bioavailability of nanocrystalline danazol in beagle dogs. Int. J. Pharm. 125, 91–7.

Liversidge, G.G., Conzentino, P., 1995. Drug particle size reduction for decreasing gastric irritancy and enhancing absorption of naproxen in rats. Int. J. Pharm. 125, 309–313.

Liu, Y., Huang, L., Liu, F., 2010. Paclitaxel nanocrystals for overcoming multidrug resistance in cancer. Mol. Pharm. 7, 863–869.

Lu, H., Wang, J., Wang, T., Zhong, J., Bao, Y., Hao, H., 2016. Recent progress on nanostructures for drug delivery applications. J. Nanomater, 20.

Lu, Y., Wang, Z.-H., Li, T., McNally, H., Park, K., Sturek, M., 2014. Development and evaluation of transferrin-stabilized paclitaxel nanocrystal formulation. J. Control Release 176, 76–85.

Malamatari, M., Somavarapu, S., Taylor, K.M., Buckton, G., 2016. Solidification of nanosuspensions for the production of solid oral dosage forms and inhalable dry powders. Expert Opin. Drug Deliv. 13 (3), 435–450.

Martena, V., Censi, R., Hoti, E., Malaj, L., Di Martino, P., 2012. A new nanospray drying method for the preparation of nicergoline pure nanoparticles. J. Nanopart. Res. 14, 934.

Mauludin, R., Müller, R.H., Keck, C.M., 2009. Development of an oral rutin nanocrystal formulation. Int. J. Pharm. 370 (1-2), 202–209.

Merisko-Liversidge, E., Liversidge, G.G., Cooper, ER., 2003. Nanosizing: a formulation approach for poorly watersoluble compounds. Eur. J. Pharm. Sci. 18 (2), 113–120.

Mitri, K., Shegokar, R., Gohla, S., Anselmi, C., Müller, R.H., 2011. Lutein nanocrystals as antioxidant formulation for oral and dermal delivery. Int. J. Pharm. 420, 141–146.

Mishra, B., Kale, M.R.S., 2016. Drug nanocrystals: A way toward scale-up. Saudi Pharm. J. 24 (4), 386–404.

Moschwitzer, J., Achleitner, G., Pomper, H., Muller, RH., 2004. Development of an intravenously injectable chemically stable aqueous omeprazole formulation using nanosuspension technology. Eur. J. Pharm. Biopharm. 58 (3), 615–619.

Moschwitzer, J., Muller, R., 2007. Drug nanocrystals-the universal formulation approach for poorly soluble drugs. Drugs Pharmaceut. Sci. 166, 71.

Muller, R.H., Heinemann, S., 1989. Surface modelling of microparticles as parenteral systems with high tissue affinity. In: Gurny RaJ, H.E. (Ed.), Bioadhesion Possibilities and Future Trends. Wissenschaftliche Verlagsgesellschaft, Stuttgart, pp. 202–214.

Muller, R.H., Peters, K., 1998. Nanosuspensions for the formulation of poorly soluble drugs: I. Preparation by a size-reduction technique. Int. J. Pharm. 160 (2), 229–237.

Muller, R.H., Jacobs, C., Kayser, O., 2001. Nanosuspensions as particulate drug formulations in therapy: rationale for development and what we can expect for the future. Adv. Drug Deliv. Rev. 47 (1), 3–19.

Mueller, B.W., Rasenack, N. Method for the production and the use of microparticles and nano-particles by constructive micronisation. WO 002003080034A3, 2003.

Muller, R., Drug nanocrystals of poorly soluble drugs. Encyclopedia of nanoscience and nanotechnology, 627–638, 2004.

Müller, R.H., Jacobs, C., 2002. Buparvaquone Mucoadhesive nanosuspension: preparation, optimisation and long-term stability. Int. J. Pharm. 237 (1–2), 151–161.

Müller, R.H., Jacobs, C., Kayser, O., 2000. Nanosuspensions for the formulation of poorly soluble drugs. In: Nielloud, F, Marti-Mestres, G (Eds.), Pharmaceutical Emulsions and Suspensions. Marcel Dekker, New York, pp. 383–407.

Noyes, A.A., Whitney, W.R., 1897. The rate of solution of solid substances in their own solutions. J. Am. Chem. Soc. 19 (12), 930–934.

Pignatello, R., 2002. Eudragit RS100 nanosuspensions for the ophthalmic controlled delivery of ibuprofen. Eur. J. Pharm. Sci. 16 (1–2), 53–61.

Ponchel, G., et al., 1997. Mucoadhesion of colloidal particulate systems in the gastro-intestinal tract. Eur. J. Pharm. Biopharm. 44 (1), 25–31.

Peltonen, L., Hirvonen, J., 2018. Drug nanocrystals-versatile option for formulation of poorly soluble materials. Int. J. Pharm. 537, 73–83.

Peters, K., Muller, RH., 1996. Nanosuspensions for the oral application of poorly soluble drugs, Proceeding European Symposium on Formulation of Poorly- Available Drugs for Oral Administration. Paris. APGI.

Peters, K., 1999. Nanosuspension—ein neues Formulierungsprinzip fur schwerslosliche Arzneistoffe. Freie Universitat Berlin, Berlin.

Rasenack, N., Hartenhauer, H., Müller, BW., 2003. Microcrystals for dissolution rate enhancement of poorly water-soluble drugs. Int. J. Pharm. 254 (2), 137–145.

Sarnes, A., Kovalainen, M., Hakkinen, M.R., Laaksonen, T., Laru, J., Kiesvaara, J., Ilkka, J., Oksala, O., Ronkko, S., Järvinen, K., 2014. Nanocrystal-based per-oral itraconazole delivery: Superior in vitro dissolution enhancement versus Sporanoxr is not realized in in vivo drug absorption. J. Control. Release. 180, 109–116.

Shegokar, R., Müller, RH., 2010. Nanocrystals: industrially feasible multifunctional formulation technology for poorly soluble actives. Int. J. Pharm. 399 (1), 129–139.

Scholer, N., et al., 2001. Atovaquone nanosuspensions show excellent therapeutic effect in a new murine model of reactivated toxoplasmosis. Antimicrob. Agents Chemother. 45 (6), 1771–1779.

Smith, A., Hunneyball, L.M., 1986. Evaluation of poly(lactic acid) as a biodegradable drug delivery system for parenteral administration. Int. J. Pharm. 30 (2-3), 215–220.

Valo, H.K., Laaksonen, P.H., Peltonen, L.J., Linder, M.B., Hirvonen, J.T., Laaksonen, T.J., 2010. Multifunctional hydrophobin: toward functional coatings for drug nanoparticles. ACS Nano 4, 1750–1758.

Valo, H., Kovalainen, M., Laaksonen, P., Häkkinen, M., Auriola, S., Peltonen, L., Linder, M., Järvinen, K., Hirvonen, J., Laaksonen, T., 2011. Immobilization of protein-coated drug nanoparticles in nanofibrillar cellulose matrices—enhanced stability and release. J. Control. Release 156, 390–397.

Valo, H., Arola, S., Laaksonen, P., Torkkeli, M., Peltonen, L., Linder, M.B., Serimaa, R., Kuga, S., Hirvonen, J., Laaksonen, T., 2013. Drug release from nanoparticles embedded in four different nanofibrillar cellulose aerogels. Eur. J. Pharm. Sci. 50, 69–77.

Van Eerdenbrugh, B., Vermant, J., Martens, J.A., Froyen, L., Van Humbeeck, J., Augustijns, P., Van den Mooter, G., 2009. A screening study of surface stabilization during the production of drug nanocrystals. J. Pharm. Sci. 98, 2091–2103.

Vergote, G.J., 2001. An oral controlled release matrix pellet formulation containing nanocrystalline ketoprofen. Int. J. Pharm. 219 (1-2), 81–87.

Yang, Y., Yan, Z., Wei, D., et al., 2013. Tumor-penetrating peptide functionalization enhances the anti-glioblastoma effect of doxorubicin liposomes. Nanotechnology 24 (40), 405101.

Yi, L., Yan, C., Richard, A.G., Wei, Wu., Tonglei, L., 2015. Developing nanocrystals for Cancer treatment. Nanomedoicine. Nanomedicine 10 (16), 2537–2552.

Yu, L., Sheng, J., Zhou, J., et al., 2016. Fluorescent magnetic nanoparticles based on supramolecular interactions as a prospective imaging biomaterial. Sci. Adv. Mater. 8 (10), 1893–1900.

Zhou, J., Zhang, B., Shi, L., et al., 2014. Regenerated silk fibroin films with controllable nanostructure size and secondary structure for drug delivery. ACS Appl. Mater. Interfaces 6 (24), 21813–21821.

Zhou, J., Zhang, B., Liu, X., et al., 2016. Facile method to prepare silk fibroin/hyaluronic acid films for vascular endothelial growth factor release. Carbohydr. Polym. 143, 301–309.

Zhai, X., et al., 2014. Dermal nanocrystals from medium soluble actives–Physical stability and stability affecting parameters. Eur. J. Pharm. Biopharm. 88 (1), 85–91.

Index

Page numbers followed by "*f*" and "*t*" indicate, figures and tables respectively.

A

Absorptive-mediated transcytosis (AMT), 223
Acidosis, 315
Acne vulgaris, 25
Aerosol therapy, 281
Alkaloid, 101
Alzheimer's disease (AD), 357–358
Aminolevulinic acid, 99
Angiogenesis, 39
Angiogenesis process, 316
Anthraquinones, 101
Antiangiogenesis, 318
Antibody-drug conjugate (ADCs), 424–425
Antimicrobial peptides (AMPs), 333
Antiretroviral drugs delivery, 295
Antiretroviral therapy, 293–295, 299*f*
Apoptosis, 65
    death receptor-mediated, 65
    mitochondria-mediated, 65
    in photodynamic therapy, 65
Aptamers, 332
Arthritis-specific antigens, 321
Atom, 163
Atomic number, 164
Atopic dermatitis (AD), 19, 24
    heliotherapy, 3
Attenuation coefficient, 168*t*
Autophagy, 65 *See also* Photodynamic therapy (PDT)

B

Bacterial infections, 402
Bacterial skin diseases, 19
    cellulitis, 19–20
    dermatitis, 19
    folliculitis, 20
    furuncle, 20
    impetigo, 19
    pyoderma, 19
    rashes, 20
    scabies, 19
Bacteriochlorins, 57–58
Bacteriophages, 332
Benzoporphyrin derivative monoacid (BPDMA), 42
Betatron, 176
Biocompatible material, 436
Biological probes, 403
Bioluminescence resonance energy transfer (BRET), 40
Biosensor, 330

Bladder cancer, 104
Blood Brain Barrier (BBB), 217
Blood Brain -Tumor Barrier (BBTB), 217
Bottom up technology, 446
Brachytherapy, 143, 184, 185
    high-dose rate implants, 143
    low-dose rate implants, 143
    permanent implants, 143
    types, 143
Brain cancer treatment
    artificial intelligence, 229
    barriers and challenges, 219
    blood brain barrier, 219
    cancer stem cells, 220
    carbon nanotubes and nanodots, 226
    chemoresistance and efflux, 220
    CRISPR, 232
    dendrimers, 229
    drug delivery, 221
    future prospects, 232
    gene-based nanotherapy, 231
    gold nanoparticles, 226
    inorganic (metallic) nanoparticles, 225
    lipid-based and polymeric nanoparticles, 228
    liposomes, 228
    mesoporous silica nanoparticles, 227
    mimicking models, 221
    nanoliposomes, 229
    nanotherapeutic platforms, 225
    nanotherapeutic systems, 232
    nose to brain drug delivery, 232
    novel therapies, 229
    polymeric micelles, 228
    quantum dots, 227
    superparamagnetic iron oxide nanoparticles, 227
    tumor microenvironment, 220
    zinc oxide NP, 227
Brain tumors, 218
    absorptive-mediated transcytosis, 223
    active targeting, 222
    benign, 218
    cancer stem cells, 224
    chemotherapeutic drug delivery, 222
    cytokine-targeted nanocarriers, 224
    dual-targeted/multifunctional nanocarriers, 224
    malignant, 218
    oligonucleotide (aptamer) mediated, 223
    passive targeting, 222
    peptide conjugated, 223
    photodynamic therapy, 104

    receptor-mediated endocytosis, 223
    small molecule ligand mediated, 223
    stimuli responsive nanocarriers systems, 224
    transporter- or carrier-mediated transcytosis, 223
Breast cancer, 245

C

Calcium-based phosphors, 166
Cancer
    active targeting, 242
    nanotechnology-based drug targeting strategies, 240, 241
    passive targeting, 241
    phototherapy, 8
    physical targeting, 245
    therapy, 421–422
    treatment, nanotechnology, 247
    tumoral endothelium targeting, 244
    tumor cell targeting, 243
    and tumors, 21
Cancer treatment
    bladder cancer, 104
    brain tumors, 104
    calcium, 37
    cellular adhesion, 39
    Cherenkov radiation energy transfer, 41
    cytokines, 39
    digestive system tumors, 103
    electroporation, 43
    external beam radiation treatment, 141
    head and neck tumors, 103
    hypoxia and angiogenesis, 39
    immune system mediated, 92
    intermolecular chemically induced electronic excitation, 40
    light source, 35
    lipid metabolism, 38
    liposomes and lipoproteins, 42
    nanotechnology, 42
    nonsmall cell lung cancer and mesothelioma, 105
    nucleic acids, photosensitized modification, 34
    photodynamic reaction, 32, 33
    photodynamic targets, 33
    photodynamic therapy, 37, 51, 92, 102
    photodynamic therapy-induced lipid peroxidation, 34
    photophysics and photochemistry, 32
    photosensitizing agents, 40
    prostate cancer, 103

455

Cancer treatment (Continued)
  proteins, 33
  radiotherapy, 140, 144
  resonance energy transfer excitation, 40
  skin tumor, 103
  stress response, 39
  transcription factors, 39
  tumor cell destruction, 92
  two-stage photosensitizer excitation, 41
  types, 144
  tyrosine kinases, 38
  urinary system tumors, 103
  vascular events, 92
Carbon-based nanoparticles, 335
Carbon ion beam therapy, 169–170
Carbon nanomaterials, 385
Carbon nanotube-based theranostic agents, 367
Carbon nanotubes, 226, 427, 433
Cell surface carbohydrates, 332
Cellular adhesion, 39
Cellulitis, 19–20
Chemical exchange saturation transfer (CEST), 362
Chemically Initiated Electron Exchange Luminescence (CIEEL), 40
Chemokine receptors, 312
Chemotherapy, 270–271
Cherenkov radiation energy transfer, 40, 41
Chlorins, 57
Citicoline, 362
Coherent scattering, 169
Collagen, 15–17
Colorectal cancers (CRC), 251
  anatomy, 251
  chemotherapy, 256
  conventional treatment strategies, 256
  current therapies, 256
  diagnosis, 254
  endoscopy, 255
  imaging, 255
  immunotherapy, 259, 260
  laboratory, 255
  nanodrug delivery, 261
  pathogenesis and molecular pathways, 252
  pathology, 255
  polymers, 261
  polypectomy and surgery, 256
  radiation therapy, 256
  risk factors, 253
  signs and symptoms, 254
  stages, 254
  targeted therapies, 259, 260
Colorimetric approaches, 337–338
Complement system, 321
Compton effects, 163, 168, 170
Compton scattering, 126
Contact therapy, 147
  simulator, 179–180
Contemporary treatment methods, 350
Conventional simulator, 179–180
Cubosomes, 423
Curcumin, 101
Cutaneous T cell lymphoma (CTCL), 25
Cyclotron, 178
Cytokines, 39

D

Deep therapy machines, 175–176
Dendrimers, 229, 299, 319
  based approaches, 277–278, 278t
Dendrimers-based drug delivery, 277–278
Dermal papillae, 15–17
Dermatitis, 19
Dermis, 15
  appendages, 17f
  papillary region, 15–17
  reticular layer, 15–17
Diagnostic techniques, light, 116
  target material, 120
  voltage applied, 120
  X-ray beam intensity, 119
  X-ray production, 118
  X-ray tube current, 120
Digestive system tumors, photodynamic therapy, 103
Digital radiography, 124
Digital subtraction angiography, 123
DNA sensing systems,, 400
DNAzymes, 394
Double strands break (DSB), 173
Drug
  delivery systems, 272
  molecules, 363
  nanocrystals, 450
  nanoparticle, 296
  particle, 273f
  polymer conjugates, 424
  resistance, 275, 276
Drug administration, 419
Dual energy computed tomography, 124
Dual energy X-ray absorptiometry, 124

E

Eczema, 19
Efavirenz (EFV), 301–302
Electrochemical impedance spectroscopy (EIS), 334
Electrochemotherapy (ECT), 43
Electromagnetic radiation, 164, 165
  classification, 165f
Electrons, 168, 172
  interaction, 172
Electroporation, 43
Electrostatic attachment, 390
Electrostatic coulombic interaction, 174
Energy transfer, 8–9
Enhanced permeability and retention (EPR), 315
Enrichment media, 328
Enzyme-linked immunosorbent assay, 328, 329t
Epidermal growth factor receptor (EGFR), 38
Epidermis, 15
  cross-section, 16f
  dermis, 15
Excimer laser, 10 *See also* Phototherapy
Exposure-dose-response relationships, 284f
External beam radiation treatment (EBRT), 141
  brachytherapy, 143
  for cancer, 141
  image-guided radiation therapy, 142
  intensity-modulated radiation therapy, 142
  stereotactic body radiation therapy, 142
  stereotactic radiosurgery, 142
  three dimensional conformal radiation therapy, 141
  tomotherapy, 142
  types, 141

F

Fiberoptic blankets, 7
Fiber-optic phototherapy systems, 5–6
Film dosimetry, 189
Filtered sunlight, 7
First-line antiretroviral drug, 298
Fluorescence-based detection systems, 338
Fluorescent nanoparticles, 437
Fluorescent tubes, 7
Fluoroscopy, 123
Folate, 279
Folate receptor, 317, 318
Folliculitis, 20, 25
Forster resonance energy transfer (FRET), 40
Fungal skin diseases, 20, 25
  sporotrichosis, 20
  tinea capitis, 20
  tinea imbricata, 20
  tinea versicolor, 20
Furanocoumarins, 101
Furuncle, 20

G

Gallium nitride, 7
Gamma rays, 166
  therapy, 163
Gardener's disease, 20
Gene therapy, 305, 364
Global Cancer Observatory (GCO), 217
Glucosamine, 395
Gold
  nanoconstructs, 433
  nanoparticle-based theranostic agents, 366
  nanoparticles, 226, 336, 366, 381, 426
Graphene oxide (GO), 335
Graphitic derivative, 335
Grenz-ray therapy, 147

H

Halogen spotlights, 7
Handheld phototherapy, 10
Head and neck malignancies (HNC), 37
Heavy charged particle, 172
Heliotherapy, 3, 21
Hematoporphyrin derivative, 42, 51–52, 97, 102
High-intensity phototherapy, 6
High-pressure homogenization technology, 445
Home phototherapy, 6, 7
  limitations, 9
Homogenization, 446f
Human immunodeficiency virus (HIV), 291
  life cycle and pathogenesis, 291
Hydrogels, 281
  based drug delivery, 281
Hypericin, 61
Hypodermis, 15, 17
Hypoxia, 39, 315
  inducible factor, 39

I

Image-guided radiation therapy (IGRT), 142, 184–185
Image-guided radiotherapy treatment, 151
    intertreatment motion, 152
    intratreatment motion, 152
Immunoassays, 328
Immunotherapy, 304
Impetigo, 19, 25
Incident beam, 167
Inherent photoluminescent, 436
Inorganic nanoparticles, 425
Intense pulse light (IPL), 3
Intensity-modulated radiation therapy (IMRT), 142, 150, 182, 184–185
    dynamic, 151
    segmental, 151
Internal light, 37
Interstitial light, 37
Intracellular cell adhesion molecule-1 (ICAM-1), 39
Ion beam therapy, 164
Ionization chamber dosimetry, 189
Ionizing radiations, 115, 174, 175
    direct and indirect action of, 173
Iron oxide nanoparticles (IONPs), 384, 427
    based theranostic agents, 363

K

Keratinocytes, 15
Kerma, 188
Kilovoltage units, 147

L

Langerhans cells, 15
Layered double hydroxides (LDH), 35–36
Light-emitting diode (LED), 5–6, 7
Light source, phototherapy, 35
    internal light, 37
    interstitial light, 37
    near infrared light, 35
    X-ray, 36
Linear accelerator, 177
Linear energy transfer, 173, 174
Lipid-based nanoparticles, 319
Lipid metabolism, 38
Lipoproteins, 42
Liposomes, 228, 302–303, 318
    application, 42
Lithium fluoride, 164
Low-density lipoproteins (LDL), 42
Low-level light/laser treatments (LLLT), 3
Luminescence dosimetry, 190
Lung cancer (LC), 246, 269
    diagnosis and treatment, 273
    diagnosis of, 274
    staging system, 269–270
    treatment, 283t
    treatment strategies, 270
Lutex, 99–100
Lymphoma, 25

M

Macrophages, 313f
Magnetic
    hyperthermia, 427
    nanoparticles, 336, 432
    resonance imaging, 350, 437
    targeted chemo-photothermal nanotherapy, 367
Mass number, 164
Mass stopping power, 172
Materials, 209
    electrical properties, 211
    monolithic and hybrid classification, 209f
    nanomaterial properties, 210
    optical properties, 210
Matter, 163
Megavoltage therapy, 148
Melanocytes, 15
Melanogenesis, 6
Melanoma, 21
Membrane-coated spherical particle, 293
Merkel cells, 15
Mesoporous silica nanoparticles (MSN), 227
Metallic nanoparticles, 379
Metal-organic framework (MOF), 35, 74
Metal oxide nanoparticles, 384
Meta-tetrahydroxyphenylchlorin, 57
Metronomic photodynamic therapy, 93
Microscopic techniques, 338
Microtron, 176, 177
Minimal erythema dose (MED), 3
Mitochondria-mediated apoptosis, 65
Mitogen-activated protein kinase (MAPK) signaling pathway, 38
Monoclonal antibodies, 322
Monocytes, 313f
Morphea, 8
Mycosis fungoides (MF), 25

N

Nanocomposites, 366
Nanocrystalline materials, 209
Nanodrug carriers, 422
Nanoemulsions, 276
Nanogels, 435–436
Nanoliposomes, 229
Nanomaterial physics, 208
Nanomaterials, 379, 432
Nanoparticles, 400, 422
    based immunoengineering approach, 304f
    based theranostics, 363
    properties, 421
    systems, 320
Nanoparticulate
    drug delivery, 297
    systems, 280t
Nanosuspension, 451
Nanotechnology, 306, 431, 443
    photodynamic therapy, 93
Nanotheranostics, 349
Nanotheranostics and neurological disorders, 350
Nanotheranostics approach, 363
Naturopathy, 22
Near infrared (NIR) light, 35
Necrosis, 65 See also Photodynamic therapy (PDT)
Neonatal jaundice, phototherapy, 8
Neonatal phototherapy, 5
Neurovascular diseases, 360
Noncovalent attachment, 390
Non-fluorescent halogen lamps, 5
Nonsmall cell lung cancer (NSCLC), 105
Nonsteroidal anti-inflammatory drugs (NSAIDs), 363
Nuclear stability, 164
Nucleus, 164

O

Optical window, 35 See also Near infrared (NIR) light
Organic nanoparticles, 385
Orthopantomography, 124
Orthovoltage therapy, 147

P

Pair production phenomenon, 171
Parasitic infections, 21
Parkinson's disease (PD), 358
Particle therapy, 152
Particulate radiation, 166
Pathogenic microorganisms, 327
Peptide-based detection of microorganisms, 335
Peptides, 396
Permeability Transition Pore Complex (PTPC), 65
Phenothiazinium salts, 61–62
    methyelene blue, 61–62
    toluidine blue, 62
Pheophorbides, 59, 102
Phosphate materials, 169
Photo diagnostic techniques, 120
    atomic number, 128
    attenuation, 125
    Compton scattering, 126
    computed tomography, 121
    cone beam computed tomography, 122
    differential absorption, 127
    digital radiography, 124
    digital subtraction angiography, 123
    dual energy computed tomography, 124, 133
    dual energy X-ray absorptiometry, 124
    fluoroscopy, 123, 133
    mass density, 128
    opportunities, challenges and limitations, 134
    orthopantomography, 124, 133
    pair production, 127
    perfusion imaging, 122
    photoelectric effect, 127
    photon energy, 128
    physics, 125
    picture archiving and communication system, 130
    plain radiography and digital radiography, 120
    positron emission tomography, 122
    prospective techniques, 134
    radiation with matter, 125
    Rayleigh or coherent scattering, 125
    retrospective techniques, 134

Photodisintegration
  process, 171f
  reaction, 171
Photodynamic reaction
  mechanism, 33f
  type II mechanism, 33
  type I mechanism, 32
Photodynamic targets, 33
  nucleic acids, photosensitized modification, 34
  photodynamic therapy-induced lipid peroxidation, 34
  proteins, 33
Photodynamic therapy (PDT), 8–9, 31, 37, 51
  advantages and disadvantages, 65, 71, 72
  apoptosis, 65
  autophagy, 65
  background, 90
  bacteriochlorins, 57–58
  biological effects, 69
  calcium, 37
  in cancer treatment, 92, 102
  cell death mechanism, 64
  cellular adhesion, 39
  cellular effects, 70
  chlorins, 57
  components, 91f
  concept, 52
  cytokines, 39
  death receptor-mediated apoptosis, 65
  DNA damage, 71
  electroporation, 43
  essential wavelength region, 72
  first generation photosensitizers, 53
  future scope, and challenges, 105
  future scopes, 77
  history and development, 94
  hypericin, 61
  hypoxia and angiogenesis, 39
  hypoxia-controlled nanomedicine, 76
  immunological effects, 67
  immunological responses, 32
  Jablonski diagram, 91f
  lipid metabolism, 38
  liposomes and lipoproteins, 42
  mechanism, 90
  merocyanine 540, 62
  metal-organic frameworks, 74
  methylene blue, 61–62
  metronomic, 93
  mitochondria-mediated apoptosis, 65
  modifications, 42
  molecular beacons, 93
  molecular mechanism, 32
  nanoconstructs for, 434
  nanotechnology, 42, 93
  necrosis, 65
  nonporphyrin-based photosensitizers, 60
  novel strategies, 93
  origin, 90
  outcomes, 31–32
  phenothiazinium salts, 61–62
  pheophorbides, 59
  photosensitizers, 52
  photosensitizing agents, 94
  phototherapy in, 31–32
  porphyrin-based photosensitizers, 53
  protoporphyrins, 55–56
  pthalocyanines, 58
  purpurins, 58–59
  recent developments, 74
  second generation photosensitizers, 54
  stress response, 39
  texaphyrins, 59
  third generation photosensitizers, 60
  toluidine blue, 62
  transcription factors, 39
  tyrosine kinases, 38
  UNCP in, 74
  vascular and inflammatory effects, 69–70
  wavelength response, 75
  working mechanism, 62
  working principle, 90
Photoelectric
  effect process, 127, 169
  interaction, 169
Photons, 168
  energy, 170
Photonuclear reactions, 171
Photosensitizers (PS), 8, 9, 52
  first generation, 53, 54f
  generation, 95
  nonporphyrin-based, 60, 61f
  porphyrin-based, 53
  second generation, 54, 55f
  third generation, 60
Photosensitizing agents (PA), 31, 40
  Cherenkov radiation energy transfer, 41
  intermolecular chemically induced electronic excitation, 40
  resonance energy transfer excitation, 40
  two-stage photosensitizer excitation, 41
Phototherapy, 3
  acne vulgaris, 25
  applications, 8
  atopic dermatitis, 24
  and biologic agents, 11
  blankets, 7
  cancer, 8
  clinical applications, 3
  defined, 3
  diseases and treatment, 24
  excimer laser, 10
  features, 22
  fiber optic, 5, 6
  filtered sunlight, 7
  fluorescent tubes, 7
  folliculitis, 25
  fungal skin diseases, 25
  future scope, 11
  halogen spotlights, 7
  handheld, 10
  high-intensity, 6
  historical perspective, 4
  immunoregulatory effects, 9
  impetigo, 25
  light-emitting diodes, 7
  light sources, 7, 35
  limitations, 8, 25
  low intensity, 6
  lymphoma, 25
  methods, 7, 23
  morphea and scleroderma, 8
  neonatal jaundice, 8
  origin, 3
  photodynamic reaction, 32, 33
  photodynamic targets, 33
  photophysics and photochemistry, 32
  principle, 5f
  psoriasis, 24
  recent development and future scope, 9, 26
  scleroderma, 25
  selective UV, 3
  side effects, 3, 26
  skin diseases, 22
  sunlight, 9
  treatments, 3
  types, 5
  various diseases, 24
  vitiligo, 24
Photothermal therapy (PTT), 9, 31
Photo-triggered theranostic nanocarriers, 435
Phthalocyanines, 101–102
Picture archiving and communication system (PACS), 124, 130
Pigmentation disorders, 21
Polyacetylene, 101
Polyhistidine-nitrilotriacetic acid chelation, 398
Polymerase chain reactions (PCR), 327
Polymer encapsulation, 393
Polymeric micelles, 228, 301
Polymeric nanocarriers, 434
Polymeric nanoparticles (PNP), 262, 276, 297
  enteric-coated nanoparticles, 264
  gold nanoparticles, 263
  lipid-based nanoparticles, 263
  superparamagnetic iron oxide nanoparticles, 263
Porous silicon particles (PSiPs), 426
Porphine macrocycle, 53f
Positron emission tomography (PET), 122, 438
Prostate cancer, 103
Proteins, 401
  and antibody functionalized nanoparticles, 395
  overexpression, 431
Proton and light ion therapy, 178
Proton therapy, 163–164
Protoporphyrins, 55–56
Psoriasis, 24
  heliotherapy, 3
Pthalocyanines, 58
Pulmonary
  gene delivery, 276
  physiology and drug absorption, 273
Pure culture-based protocols, 327
Purpurins, 58–59
Pyoderma, 19

Q

Quantum dots (QDs), 227, 279, 337, 365, 426–427

R

Radiant energy, 188
Radiation
  biological effect, 173
  therapy, 165, 271
    external beam, 181
Radiography
  conventional, 120–121
  digital, 120
  plain, 120
  projectional, 120–121
Radioluminescent nanoparticles, 437
Radiosurgery, 153
  history, 153
  systems overview, 154
  treatment, 154
Radiotherapy, 139, 163, 186
  adaptive, 152
  cancer control, 144
  cancer treatment, 140
  cancer types, 144
  contact therapy, 147
  development, 145
  dose planning, 148
  dynamic IMRT, 151
  external, 146
  external beam radiation therapy, 141
  facility and quality assurance, 180
  general indications, 143
  Grenz-ray therapy, 147
  history, 145
  instigation, 149
  intensity modulated radiotherapy, 150
  intent, 143
  kilovoltage units, 147
  megavoltage therapy, 148
  orthovoltage therapy, 147
  particle therapy, 152
  principle and mechanism, 149
  principle of, 174
  recent advancement, 149
  segmental IMRT, 151
  stereotactic radiosurgery, 152
  superficial therapy, 147
  supervoltage therapy, 148
  technique, 180
  technology development, 150
  three-dimensional conformal radiotherapy, 150
  traditional facility, 175
  types, 141
  works, 140
Rashes, 20
Rayleigh coherent scattering, 169
Rayleigh scattering phenomenon, 169
Reactive oxygen species (ROS), 8–9
Receptor-mediated endocytosis (RME), 223
Recognition components, 331
Relative biological effectiveness (RBE), 164, 174
Resonance energy transfer excitation, 40
Retention effect, 320
Reverse transcription, 292
Rheumatoid arthritis (RA), 311, 362
  management of, 314, 315t
    nanotheranostic approach, 363

  pathology, 312
  symptoms, 312

S

Scabies, 19
Scleroderma, 25
  phototherapy, 8
Semiconductor diode detector, 191
Semiconductor quantum dots, 383
Silica nanoparticle-based theranostic agents, 368
Silicon dioxide, 384
Single photon emission tomography (SPECT), 350, 438
Single strand break (SSB), 173
Skin
  cross-section, 16f
  functions, 17
  metabolism, 17
  protection, 18
  sensation, 18
  test, 21
  thermoregulation, 17
  tumor, photodynamic therapy, 103
  vitamin D, synthesis, 18
Skin disease
  acne vulgaris, 25
  atopic dermatitis, 24
  etiology, 18
  folliculitis, 25
  fungal skin diseases, 25
  HIV related, 20
  impetigo, 25
  lymphoma, 25
  phototherapy, 22
  primary diseases, 18
  psoriasis, 24
  recent development and future scope, 26
  scleroderma, 25
  secondary diseases, 18
  side effects, 26
  treatment, 18
  vitiligo, 24
Solid-lipid nanocarriers, 275
Solid lipid nanoparticles (SLNs), 303, 423
Solid state reaction method, 167
Solution combustion method, 164
Spherical fullerenes, 387f
Spray drying, 447–448
Squamous cell carcinoma, 21
Stereotactic radiosurgery (SRS), 184
Stereotactic radiotherapy, 184
Stimuli responsive nanocarriers systems, 224
  photosensitive (physical) drug delivery systems, 224
  redox-sensitive nanocarriers, 225
Stratum basale, 15
Stratum corneum, 15
Stratum granulosum, 15
Stratum lucidum, 15
Stratum spinosum, 15
Superficial therapy, 147
Superparamagnetic iron oxide nanoparticles (SPION), 227
Supervoltage therapy machines, 148, 176

Surface-enhanced Raman spectroscopy (SERS), 338
Synchrotrons, 179

T

Tanning beds, 6–7
Target-mediated targeted therapy, 279
Teletherapy machines, 176
Texaphyrins, 59
Theranostics, 403
  nanoagents, 349
  nanoparticles, 351
Thermoluminescence, 190
Three-dimensional conformal radiation therapy, 181
Tomotherapy, 178
Top down technology, 445
Transcription factors, 39
Transrectal ultrasound and photoacoustic (TRUSPA) device, 241
Trauma, 21
Treatment verification process, 187
Tropical ulcers, 20
Tumor control probability (TCP), 183–184
Tumor microenvironment, 239
  angiogenesis and endothelial permeability, 240
  microenvironment pH, 240
  microenvironment temperature, 240
Tumors
  cell destruction, 92
  staging system, 270
Tumors and cancers, 21
Tyrosine kinases, 38

U

Ultraviolet A (UVA) therapy, 6
Ultraviolet B (UVB) therapy, 3, 6
Upconversion nanoparticles, 437
Upconversion nanoparticles (UCNP), 35, 66–67
Urinary system tumors, photodynamic therapy, 103
  bladder cancer, 104
  prostate cancer, 103

V

Vaccines, 305
Vascular cell adhesion molecule-1 (VCAM-1), 39
Vasoactive intestinal peptide (VIP), 318
Viral skin diseases, 20
Virus detection, 340
Visudyne, 100
Vitiligo, 24
Volumetric modulated arc therapy, 184

W

Water-insoluble drugs, 297, 451

X

Xanthenes, 102
X-rays, 36, 165, 166
  emission, 165
  machines, 176
  mediated PDT, 36

CPI Antony Rowe
Eastbourne, UK
May 13, 2022